MONOGRAPHS IN CONTACT ALLERGY
Anton C. de Groot

Volume 1: Non-Fragrance Allergens in Cosmetics
Volume 2: Fragrances and Essential Oils
Volume 3: Topical Drugs

MONOGRAPHS IN CONTACT ALLERGY
VOLUME 3

TOPICAL DRUGS

MONOGRAPHS IN CONTACT ALLERGY
VOLUME 3

TOPICAL DRUGS

Anton C. de Groot

With the help of Heleen de Jong in drawing the structural formulas

CRC Press
Taylor & Francis Group
Boca Raton London New York

CRC Press is an imprint of the
Taylor & Francis Group, an **informa** business

CRC Press
Taylor & Francis Group
6000 Broken Sound Parkway NW, Suite 300
Boca Raton, FL 33487-2742

© 2021 by Taylor & Francis Group, LLC
CRC Press is an imprint of Taylor & Francis Group, an Informa business

No claim to original U.S. Government works

Printed on acid-free paper

International Standard Book Number-13: 978-0-367-23693-9 (Hardback)
International Standard Book Number-13: 978-0-367-74761-9 (Paperback)

Visit the Taylor & Francis Web site at
http://www.taylorandfrancis.com

and the CRC Press Web site at
http://www.crcpress.com

Contents

PREFACE

Long before I finished writing the second book in the *Monographs in Contact Allergy* series on Fragrances and Essential oils in December 2018, I had decided that, in the third volume, topical and systemic drugs were to be presented. That made perfect sense to me. The drugs in topical pharmaceuticals are important contact allergens: in up to 15% of all patients with allergic contact dermatitis, topical drugs are the cause of the dermatitis or contribute to it (iatrogenic contact dermatitis). Indeed, some drugs are important enough to be included in the baseline series for routine testing, including neomycin, the caine mix, budesonide and tixocortol pivalate in Europe, and neomycin, bacitracin, tocopherol, benzocaine, lidocaine, and the corticosteroids budesonide, tixocortol pivalate, hydrocortisone butyrate, clobetasol propionate and desoximetasone in the USA. As to the systemic drugs, much research has been performed in the last 25 years investigating the role of 'contact allergy', better termed 'delayed-type hypersensitivity' or 'type IV hypersensitivity' in this context, in a variety of drug eruptions. Excellent material for Volume 3 of the *Monographs in Contact Allergy*.

After having worked my way through all issues of the journals *Contact Dermatitis*, the *American Journal of Contact Dermatitis* and *Dermatitis*, recent textbooks, the literature lists of all relevant articles found there and 'Similar articles', kindly brought to my attention by PubMed, I estimated that the total number of topical and systemic drugs to be discussed would reach 750. An average of 2 pages per monographs plus additional chapters would result in a book too heavy to lift for most dermatologists and other scientists. As a consequence, the subject is now divided over 2 books: volume 3 (Topical drugs) and volume 4 (Systemic drugs). This too seems the logical thing to do, as the clinical manifestations of topical and systemic drugs are quite different: allergic contact dermatitis (topical) versus (a variety of) drug eruptions (systemic). Occupational allergic contact dermatitis to systemic drugs, well known in pharmaceutical and health care workers, will also be discussed in volume 4.

Is there enough material in contact allergy to topical drugs, which include all pharmaceuticals applied to the skin and the mucous membranes of the eyes, nose, mouth, lower airways (by inhalation) and the anogenital area, to write a book of proper weight and size? Most certainly. I have found over 380 topical drugs that have caused contact allergy/allergic contact dermatitis. Currently, the most important groups of sensitizers in topical drugs are – depending on the countries – antibiotics, corticosteroids (budesonide and tixocortol pivalate, albeit more often as indicators of corticosteroid allergy from cross-reactivity than from primary sensitization), topical anesthetics (dibucaine, benzocaine), and non-steroidal anti-inflammatory drugs (also photosensitizers), notably ketoprofen.

The NSAID bufexamac has caused so many cases of sensitization in some European countries, that the drug was withdrawn by the European Medicines Agency (EMA). Contrary to what is often believed, antihistamines are currently infrequent causes of contact allergy with the exception of doxepin. Promethazine formerly caused many cases of sensitization and photosensitization, but its use has been abandoned in most countries. Indeed, there are quite a few examples of topical drugs that gave so many problems with (photo)sensitization, including penicillins and sulfonamides, that their use was abandoned, making them historical allergens. Others died of natural causes, being replaced by newer, more effective and safer drugs, e.g. the antimycotics pyrrolnitrin and haloprogin replaced by the imidazoles. Most topical drugs, fortunately, only occasionally, infrequently or rarely cause allergic contact dermatitis.

All 384 topical drugs that have been reported as sensitizers are discussed in this book in 369 monographs (in one, 16 'historical' allergens are briefly presented). Next to the monographs, there are additional chapters presenting a rationale for this book and the data presented therein (chapter 1); an overview of allergic contact dermatitis discussing the prevalence of sensitization, predisposing factors, clinical manifestations, the allergens and diagnostic procedures (chapter 2); allergic contact dermatitis to ingredients other than the active drug such as vehicle ingredients (lanolin alcohol, propylene glycol, cetearyl alcohol), preservatives and antioxidants (thimerosal, benzalkonium chloride, parabens, sodium metabisulfite) and fragrances (chapter 4); and immediate contact reactions to topical drug (chapter 5). In addition, chapter 6 gives a preview of the systemic drugs found thus far by the author to have caused allergic cutaneous adverse drug reactions, occupational allergic contact dermatitis or both, and which will be the subject of Volume 4.

It has been 2 years since the previous volume of *Monographs in Contact Allergy* (2, Fragrances and Essential Oils) was published. As much work on it has already been done, the release of the next volume (4, Systemic Drugs) may be realized in a far shorter period of time, unexpected events aside, of course.

Anton de Groot, MD, PhD
Wapserveen, The Netherlands, October 2020

ACKNOWLEDGMENTS

The author wishes to express his gratitude to Heleen de Jong BSc, for drawing a number of the structural formulas shown in this book.

I am also very grateful to Prof. dr. An Goossens, Department of Dermatology, Contact Allergy Unit, University Hospitals Leuven, Leuven, Belgium, for providing information on corticosteroids and allowing me to use her previously published data. An and her team have for over 40 years greatly contributed to the understanding and knowledge of contact allergy and allergic contact dermatitis, especially in the field of drug and cosmetic allergy. Numerous allergens have been discovered and first described by her and much of what we know about contact allergy to corticosteroids stems from her excellent work, which I greatly admire.

Heleen de Jong BSc

Prof. dr. An Goossens

ABOUT THE AUTHOR

Anton C. de Groot, MD, PhD (1951) received his medical and specialist training at the University of Groningen, The Netherlands. In 1980, he started his career as dermatologist in private practice in 's-Hertogenbosch. At that time, he had already become interested in contact allergy and in side effects of drugs by writing the chapter *Drugs used on the skin* with his mentor prof. Johan Nater, for the famous *Meyler's Side Effects of Drugs* series. Soon, the subject of this chapter in new Editions and the yearly *Side Effects of Drugs Annuals* would be expanded to include cosmetics and oral drugs used in dermatology (1980-2000). Contact allergy to cosmetics would become de Groot's main area of interest and expertise and in 1988, he received his PhD degree on his Thesis entitled *Adverse Reactions to Cosmetics*, supervised by prof. Nater.

Frustrated by the lack of easily accessible information on the ingredients of cosmetic products, and convinced that compulsory ingredient labeling of cosmetics (which at that time was already implemented in the USA) would benefit both consumers and allergic patients and would lead to only slight and temporary disadvantages to the cosmetics industry, De Groot approached the newly founded European Society of Contact Dermatitis and became Chairman of the Working Party European Community Affairs. The European Commission and its committees, elected legislators, national trade, health departments and the cosmetics industries were extensively lobbied. This resulted in new legislation by the Commission of the European Communities in 1991, making ingredient labeling mandatory for all cosmetic products sold in EC Member States by December 31, 1997.

Anton has been the chairman of the 'Contact Dermatitis Group' of the Dutch Society for Dermatology and Venereology from 1984 to 1998. In 1990, he was one of the founders of the *Nederlands Tijdschrift voor Dermatologie en Venereologie* (Dutch Journal of Dermatology and Venereology) and was Editor of this scientific journal for 20 years, of which he served 10 years as Editor-in-chief.

De Groot has authored eighteen book titles, eleven of which – all co-authored by Johan Toonstra MD, PhD – are general dermatology books in Dutch for medical students, general practitioners, 'skin therapists' (huidtherapeuten, a paramedical profession largely restricted to The Netherlands) and pedicures and podotherapists. He also co-authored a Dutch booklet for (parents of) patients with atopic dermatitis.

Anton has written six (including the current one) international books, of which one has had three Editions: *Unwanted Effects of Cosmetics and Drugs used in Dermatology* (first Edition 1983, second 1985, third 1994). Of his best known book *Patch Testing* 4 editions have been published (first Edition 1986, second 1994, third 2008, fourth 2018) (www.patchtesting.info).

After writing a book entitled *Essential Oils: Contact Allergy and Chemical Composition* with Erich Schmidt, which appeared in 2016, he started working on a series of *Monographs in Contact Allergy*. The first Volume discussing 'Non-fragrance allergens in cosmetics' was released in 2018, followed by 'Fragrances and essential oils' in 2019. The current book presenting contact allergy to topical drugs is Volume 3. A fourth one on 'Systemic Drugs' is planned to be finished in 2021.

In addition to these books, Anton has written over 70 book chapters (mostly in international books), 135 articles in international journals and some 235 articles in Dutch medical and paramedical journals. He served as board member of several journals including *Dermatosen* and is currently member of the Editorial Advisory Board of the Journal *Dermatitis*.

In 2019, the author received the American Contact Dermatitis Society Honorary Membership status for his 'vast contributions to contact dermatitis'.

Anton de Groot has retired from dermatology practice, but since 2008 regularly teaches general dermatology to junior medical doctors at the University of Groningen, which unfortunately is temporarily interrupted by the covid-19 pandemic. He and his wife Janny have two daughters, both lawyers.

CHAPTER 1   INTRODUCTION

1.1 WHY A BOOK WITH MONOGRAPHS ON TOPICAL DRUGS?

After the first volume in the series *Monographs in Contact Allergy* on the 'non-fragrance allergens in cosmetics' and the second on 'fragrances and essential oils', the choice for the subject 'topical drugs' (pharmaceuticals) seemed a logical one. Indeed, topical drugs are an important and frequent cause of allergic contact dermatitis. In an estimated 10-15% of all patients with allergic contact dermatitis seen by dermatologists, topical drugs are the cause (iatrogenic allergic contact dermatitis) (1,2). There is a reason for this. Topical drugs are not strong or even moderate sensitizers, but they are virtually always used on diseased (often inflamed) or non-intact (wounds, burns, leg ulcers) skin or mucous membranes, often with a damaged or insufficiently functioning skin barrier. This facilitates penetration of the drug and access to the antigen-presenting cells, thus favoring sensitization. The most notable risk factor for sensitization is treatment of leg ulcers and/or stasis dermatitis. In epidemiological studies, these patients often have very high prevalences of positive patch tests with multi-sensitization to active drugs, excipients of topical drugs (lanolin alcohol, propylene glycol, fragrances) and unrelated allergens (3-5).

To screen for the most important drug allergies, the European baseline series contains several topical drugs: neomycin, the caine mix III (benzocaine, dibucaine, tetracaine), budesonide, and tixocortol pivalate. In addition, it screens for some non-drug constituents that may be the allergenic culprit in topical medicaments such as parabens, lanolin alcohol, and fragrances (6). The North American Contact Dermatitis Group (NACDG) screening series includes as topical drugs neomycin, bacitracin, tocopherol (vitamin E, not 'officially' a topical drug, but frequently used as such, also by physicians), benzocaine, lidocaine, and the corticosteroids budesonide, tixocortol pivalate, hydrocortisone butyrate, clobetasol propionate and desoximetasone; non-drug ingredients include lanolin alcohol, parabens, ethylenediamine and propylene glycol (7).

However, whereas these drugs are well-known causes of allergic contact dermatitis and 'covered for' by routine testing, there are many other topical drugs that also cause allergic contact dermatitis, either frequently, sometimes, or rarely. In fact, in preparing this book, the author has found over 380 such drug allergens. Literature on these are sometimes easy to find in digital databases, others more difficult. In sensitized patients, the reactions to pharmaceuticals often present with classic allergic contact dermatitis at the site of application, which can easily be recognized and ascribed to the topical drug (followed by confirmation through patch testing, of course). Others are less easy to identify, e.g. in the case of corticosteroids, where contact allergy may present merely as failure to improve the treated dermatosis. Atypical manifestations, e.g. erythema multiforme-like allergic contact dermatitis, are not rare and absorption of the drug through the skin or mucous membranes may lead to 'systemic contact dermatitis' ('systemic allergic dermatitis') with a variety of possible clinical manifestations including maculopapular exanthema, urticaria, extensive eczematous reactions, worsening of existing dermatitis, recurrence of previous dermatitis or positive patch tests, or the 'baboon syndrome'. Diagnosis with the aid of patch testing may sometimes be easy but, especially in the case of ophthalmic drugs, very difficult and the allergic contact dermatitis may frequently be missed when the dermatologist is not aware of this and does not know how to overcome the problem.

Therefore, it is important that dermatologists, allergists and other professionals performing patch tests should have a thorough knowledge of the subject and of individual topical drugs, or at least have access to comprehensive relevant information. In recent textbooks on contact dermatitis, allergy to topical drugs is discussed in separate chapters (2,8,9). Also, in scientific journals, in a number of articles, the subject of sensitization to topical drugs (10,11) or a specific group of drugs (12-17), notably the corticosteroids (18-22) and ophthalmic drugs (23-25), has been reviewed. Although these are very useful, they are unavoidably somewhat superficial and are always incomplete. Not a single individual topical drug (of any importance) has been thoroughly and completely reviewed recently and relevant information, if searched for, has to be obtained from many different sources. For some important sensitizers such as budesonide, neomycin and bacitracin, there is so much literature available that it would take an interested investigator wanting to publish a huge amount of time to review the subject thoroughly.

Therefore, the author, who has been interested in contact allergy to topical drugs since the early 1980s (26-30), decided to make this third volume in the *Monographs in Contact Allergy* series an all-encompassing reference work covering the full spectrum of topical drugs that have caused contact allergy.

The author is aware that the terms 'allergen' and 'allergens' are often *sensu stricto* not correct and should read 'hapten' and 'haptens'. However, in scientific journals dedicated to contact dermatitis and contact allergy such as *Contact Dermatitis* and *Dermatitis* the term allergen is still often used and is also familiar to physicians, who are not experts in the field, and to non-physicians. Therefore, the term allergen is used throughout this book.

1.2 DATA PROVIDED IN THIS BOOK

SCOPE AND DATA COLLECTING

This book provides monographs of all drugs in topical pharmaceuticals that have caused contact allergy/allergic contact dermatitis. Topical pharmaceuticals include products applied to the skin or mucous membranes of the eyes, nose, mouth, lower airways (by inhalation), genitals, or anal mucosa in the form of cream, ointment, lotion, spray, eye drops, nose drops, ear drops, inhalation preparation, transdermal therapeutic system, or any other form of drug delivery to the skin or mucosae affected by disease, the presence of signs such as itch or pain or as preventive measure. Vaginal tablets and suppositories and rectal suppositories are considered as topical drugs. Excluded are drugs administered by mouth or by injection (intralesional, subcutaneous, intra-articular, intravenous, intramuscular, intrathecal). These will be discussed in the next volume of the *Monographs in Contact Allergy* series on Systemic drugs.

Allergic contact dermatitis from topical drugs may be caused by the active drug, but also by non-drug ingredients such as vehicle components, preservatives, antioxidants or fragrances. This subject is discussed in Chapter 4 Allergic contact dermatitis to ingredients other than the active drug.

The main sources of information are:
- the journals *Contact Dermatitis* and *Dermatitis*, that were fully screened from their start in 1975 resp. 1990 through August (*Dermatitis*) resp. September 2020 (*Contact Dermatitis*);
- the author's book *Unwanted Effects of Cosmetics and Drugs used in Dermatology* (26) and other reference works: several editions of *Fisher's Contact Dermatitis* (31), Etain Cronin's 1980 book *Contact Dermatitis* (32) and the most recent editions of the – digital - books *Contact Dermatitis,* 6th Edition, 2018 (33) and *Kanerva's Occupational Dermatology*, 3rd Edition, 2018 (34);
- relevant articles found in literature lists of journal publications used for the book; most of these could be accessed on-line through the Medical Library of the University of Groningen; a limited number of important articles that were not accessible on-line were requested from the library;
- all journals available on-line through the Medical Library of the University of Groningen; before finishing and closing a monograph (in the period August 2020-September 2020), the drug name was searched for in PubMed and any additional relevant articles and data were added.

A number of articles could neither be accessed on-line nor were they requested. If some relevant information was available, e.g. from the Abstract or from being cited in other sources, it was included, in the latter situation mentioning that the information was cited in another article (often in non-English literature including Japanese, French or German). If data from articles are missing, this is indicated by 'data unavailable to the author', 'no details available' or similar text in the Monographs.

CRITERIA FOR INCLUSION

To be included in this book, topical drugs (which include vitamins and veterinary medications) had to meet one or more of the following criteria:
- they were presented in a case report as cause of allergic contact dermatitis;
- the chemical/substance was mentioned in a list of chemicals that had caused contact allergy/allergic contact dermatitis in groups of patients
- when a positive patch test to a topical drug was specifically stated or implied to be the cause of allergic contact dermatitis by the author(s);
- the substance caused one or more positive patch test reactions in groups of patients tested with it, either consecutive patients suspected of contact dermatitis (routine testing) or selected patient groups, also when no comments were made on their relevance.

Excluded from discussion in this book are modern wound dressings such as hydrocolloids, topical mercurials, herbal medicines, 'alternative' medicines, vaccines, DMSO, sensitizers used in the treatment of alopecia areata and warts (diphencyprone, squaric acid dibutyl ester, DNCB [1-chloro-2,4-dinitrobenzene]), medicated bandages and disinfectants used on inanimate subjects only, not on human skin (contrary to 'anti-infectives'). A number of drugs and chemicals also used as drugs have been discussed – as 'non-fragrance allergen in cosmetics' or as 'fragrance' (benzyl benzoate) - in the 2 previous volumes of this series (35,36). Most are not again presented in this book (table 1.1)

Table 1.1 Chemicals used as drugs that have been discussed in previous volumes of Monographs in Contact Allergy (35,36)

Ammoniated mercury	Hexylresorcinol
Benzyl benzoate	Kojic acid
Bithionol	Panthenol
5-Bromo-4'-chlorosalicylanilide (Multifungin)	Phenyl salicylate
Cetrimonium bromide (cetrimide)	Phytonadione
Fenticlor (fentichlor)	Topical mercurials used as anti-infectives
Hexachlorophene	Quinine

DATA PROVIDED IN MONOGRAPHS

The side effects of topical drugs discussed in this book are contact allergy/allergic contact dermatitis, photocontact allergy/photoallergic contact dermatitis and contact urticaria. The literature search resulted in the identification of 384 topical drugs (Chapter 3, 368 Monographs for individual drugs, 16 presented in Chapter 3.228 Miscellaneous topical drugs). For each chemical included, the data shown in table 1.2 – when available and applicable - were searched for. For most drugs, is has been attempted to provide a comprehensive or (nearly) full literature review. However, for a number of drugs, especially historical allergens (drugs that were formerly in use, but have been discontinued), drugs with a large number of articles in (very) early literature and for topical pharmaceuticals with a vast number of available data (e.g. neomycin, bacitracin), a full review was not attempted.

Table 1.2 Information provided (when available and applicable) in the monographs

IDENTIFICATION	CONTACT ALLERGY
Description/definition	General
Pharmacological classes	Patch testing in the general population
IUPAC name	Patch testing in groups of patients
Other names	Case reports and case series
CAS registry number	Cross-reactions, pseudo-cross-reactions and co-reactions
EC number	
Merck Index monograph	Patch test sensitization
Patch testing	**PHOTOSENSITIVITY**
Molecular formula	Immediate contact reactions (contact urticaria)
Structural formula	Cutaneous adverse drug reactions from systemic administration
GENERAL	

IDENTIFICATION

For the monographs' titles, the names shown in the ChemIDplus database of the NIH U.S. National Library of Medicine were used (https://chem.nlm.nih.gov/chemidplus/). In the section IDENTIFICATION the following data are provided: Pharmacological classes, IUPAC (International Union of Pure and Applied Chemistry) name, other names (synonyms), CAS (Chemical Abstract Service) registry number (www.cas.org), EC/EINECS number (ECHA European Chemicals Agency, EC Inventory: https://echa.europa.eu/information-on-chemicals/ec-inventory), Merck Index monographs number (The Merck Index online: https://www.rsc.org/merck-index) and their molecular and structural formulas. Also, a general description/definition of the compounds is provided. Most of these data were readily available from ChemIDPlus and from PubChem, the NIH National Library of Medicine, National Center for Biotechnology Information online database (https://pubchem.ncbi.nlm.nih.gov/).

Finally, advice on how to patch test the drug is provided based on the publications presented, the author's book *Patch Testing*, 4th Edition (37), and commercial availability (Chemotechnique Diagnostics, www.chemotechnique.se; SmartPractice Europe, www.smartpracticeeurope.com; and SmartPractice Canada, www.smartpracticecanada.com).

GENERAL

A general description of the index drug with indications, mechanism of action and sometimes additional relevant data are provided here. Important sources were ChemIDPlus, PubChem, DrugBank (https://go.drugbank.com/), Drug Central (https://drugcentral.org/), Drugs.com (https://www.drugs.com/), and Wikipedia (https://www.wikipedia.org/).

CONTACT ALLERGY

General
In a limited number of the larger monographs, a summary of the data provided in that chapter is given for a quick overview.

Patch testing in the general population
A few studies have investigated the frequency of contact allergy to topical drugs in the general population or subgroups, including neomycin, parabens, tixocortol pivalate, hydrocortisone butyrate, budesonide, benzocaine, the quinoline mix and the caine mix III. Their results are presented in this section.

Patch testing in groups of patients
This section shows the results of studies performing patch testing in groups of consecutive patients suspected of contact dermatitis (routine testing) and in groups of selected patients, e.g. patients suspected of iatrogenic contact dermatitis, patient with periorbital eczema, individuals with leg ulcers/stasis dermatitis or patients with anogenital dermatitis/dermatoses/symptoms. Data provided include country/countries of the study, period of study, patch test concentration and vehicle used, number of patients tested, number of patients and percentage with positive patch tests, mode of selection, relevance (percentage), other pertinent information and literature references.

Case reports and case series
Case reports are mostly clinical descriptions of one or two patients with allergic contact dermatitis from the index topical drug. Where a larger number of case reports from different publications are available, some are presented with clinical descriptions and the rest in tabular format summarizing the relevant data. Case series are mostly 3 or more patients with clinical data, but far more often the number of patients allergic to that particular drug in groups of individuals with iatrogenic allergic contact dermatitis (from any medicament).

Cross-reactions, pseudo-cross-reactions and co-reactions
It may be difficult to determine whether co-reactions (concomitant positive patch tests) to other drugs in addition to the index chemical are the result of cross-reactivity (contact allergic reaction to a structurally similar chemical to which the individual has not yet been exposed), of pseudo-cross-reactions (the index chemical and the co-reacting chemical have the same allergenic constituent, contaminant, or metabolite [frequent with fragrances and essential oils, but rare with drugs]) or co-reactions, which are independent of each other. Co-reactivity from independent sensitization is frequent with neomycin and bacitracin from their presence in the same topical pharmaceutical. Sensitization to corticosteroids often results in a large number of cross-reactions to other steroids, but at the same time in co-reactions to non-related chemicals such as neomycin, an antifungal, base ingredients of creams or ointments (lanolin alcohol, propylene glycol, cetearyl alcohol), antioxidants (sodium metabisulfite) or preservatives (thimerosal, benzalkonium chloride, parabens).

    In this monograph section, established cross-reactions are mentioned, as are co-reactions to drugs other than the index drug, that are *likely* or *possibly* the result of cross-reactivity.

Patch test sensitization
Patch test sensitization to topical drugs appears to be very rare. Formerly, positive patch test reactions first appearing after D7 were usually considered to be the result of active sensitization from the patch test. However, it is now well known that late reactions to neomycin and related aminoglycosides are very common. The same applies to corticosteroids, where reactions starting at D7 or even later are by no means rare and can be explained by their anti-inflammatory action, which in the earlier stages suppresses the inflammatory positive patch test from evolving.

PHOTOSENSITIVITY
Photosensitivity reactions include photocontact allergy (causing photoallergic contact dermatitis), phototoxicity (causing phototoxic dermatitis) and combined photocontact allergy and contact allergy, causing photoaggravated (synonym: photoaugmented) allergic contact dermatitis. This section shows the results of studies performing photopatch testing in groups of selected patients, usually individuals suspected of photosensitivity or photoallergic reactions to topical drugs. Data provided include country/countries of the study, period of study, photopatch test concentration and vehicle used, number of patients tested, number of patients and percentage with positive patch tests, mode of selection, relevance (percentage) and literature references. Monographed chemicals that have caused photoallergic contact dermatitis, most of which are NSAIDs (non-steroidal anti-inflammatory drugs) are summarized in table 2.6.

Immediate contact reactions

Many drugs have caused immediate-type reactions (either type I IgE-mediated reactions or from other mechanisms), usually from systemic administration (injection), but some from application to the skin or mucous membranes. Possible symptoms (single or in combination) of such reactions include localized erythema, itching or tingling, localized urticaria, angioedema, generalized urticaria, respiratory symptoms (wheezing, dyspnea, asthma, rhinitis, nasal discharge), cardiac problems (hypotension, bradycardia, ventricular fibrillation or cardiac arrest), gastrointestinal symptoms (abdominal pain, diarrhea, nausea, vomiting) or even anaphylactic shock.

Because of the focus of this book on contact allergy, it was decided to only briefly discuss this issue (Chapter 5 Immediate contact reactions [contact urticaria]) and limit it to immediate contact reactions (contact urticaria) from application of topical drugs to intact skin or mucosa. Reactions to topical medicaments applied to non-intact skin such as wounds or burns that cause immediate reactions (well know are cases of anaphylaxis from bacitracin) were excluded.

Cutaneous adverse drug reactions from systemic administration caused by type IV (delayed-type) hypersensitivity

Approximately 50 drugs are used both in topical and in systemic applications. Systemic drugs causing delayed-type (type IV) hypersensitivity are planned to be discussed in volume 4 of the *Monographs in Contact Allergy* series. During the preparation of this book, the author has already collected many cases of allergic cutaneous adverse drug reactions to systemic drugs and occupational allergic contact dermatitis (which will also be presented in volume 4). These are, as a 'preview', summarized in this section with the type of allergic reaction to the systemic drug and with references. A list of pertinent drugs found thus far is shown in Chapter 6 Systemic drugs and delayed-type hypersensitivity: A preview. Of course, additional cases will be found during the writing of the next book.

LITERATURE

1 Gilissen L, Goossens A. Frequency and trends of contact allergy to and iatrogenic contact dermatitis caused by topical drugs over a 25-year period. Contact Dermatitis 2016;75:290-302

2 Goossens A, Gonçalo M. Topical drugs. In: Johansen J, Mahler V, Lepoittevin JP, Frosch P, Eds. Contact Dermatitis, 6th Edition. Springer: Cham, 2020

3 Erfurt-Berge C, Geier J, Mahler V. The current spectrum of contact sensitization in patients with chronic leg ulcers or stasis dermatitis – new data from the Information Network of Departments of Dermatology (IVDK). Contact Dermatitis 2017;77:151-158

4 Erfurt-Berge C, Mahler V. Contact Sensitization in patients with lower leg dermatitis, chronic venous insufficiency, and/or chronic leg ulcer: Assessment of the clinical relevance of contact allergens. J Investig Allergol Clin Immunol 2017;27:378-380

5 D'Erme A, Iannone M, Dini V, Romanelli M. Contact dermatitis in patients with chronic leg ulcers: a common and neglected problem: a review 2000-2015. J Wound Care 2016;25(S9):S23-S29

6 Wilkinson M, Gonçalo M, Aerts O, Badulici S, Bennike NH, Bruynzeel D, et al. The European baseline series and recommended additions: 2019. Contact Dermatitis 2019;80:1-4

7 DeKoven JG, Warshaw EM, Zug KA, Maibach HI, Belsito DV, Sasseville D, et al. North American Contact Dermatitis Group patch test results: 2015-2016. Dermatitis 2018;29:297-309

8 Kraft M, Soost S, Worm M. Topical and systemic corticosteroids. In: John S, Johansen J, Rustemeyer T, Elsner P, Maibach H, Eds. Kanerva's Occupational Dermatology, 3rd Ed. Cham: Springer, 2018

9 Rietschel RL, Fowler JF Jr, Eds. Fisher's Contact Dermatitis, 6th Edition. Hamilton, USA: BC Decker Inc., 2008

10 Nguyen H, Yiannias J. Contact dermatitis to medications and skin products. Clin Rev Allergy Immunol 2019;56:41-59

11 Davis MDP. Unusual patterns in contact dermatitis: medicaments. Dermatol Clin 2009;27:289-297

12 Bershow A, Warshaw E. Cutaneous reactions to transdermal therapeutic systems. Dermatitis 2011;22:193-203

13 Musel AL, Warshaw EM. Cutaneous reactions to transdermal therapeutic systems. Dermatitis 2006;17:109-122

14 Murphy M, Carmichael AJ. Transdermal drug delivery systems and skin sensitivity reactions. Incidence and management. Am J Clin Dermatol 2000;1:361-368

15 Veraldi S, Brena M, Barbareschi M. Allergic contact dermatitis caused by topical antiacne drugs. Expert Rev Clin Pharmacol 2015;8:377-381

16 Foti C, Romita P, Borghi A, Angelini G, Bonamonte D, Corazza M. Contact dermatitis to topical acne drugs: a review of the literature. Dermatol Ther 2015;28:323-329

17 Gehrig KA, Warshaw EM. Allergic contact dermatitis to topical antibiotics: Epidemiology, responsible allergens, and management. J Am Acad Dermatol 2008;58:1-21

18 Vatti RR, Ali F, Teuber S, Chang C, Gershwin ME. Hypersensitivity reactions to corticosteroids. Clin Rev Allergy Immunol 2014;47:26-37

19   Otani IM, Banerji A. Immediate and delayed hypersensitivity reactions to corticosteroids: Evaluation and management. Curr Allergy Asthma Rep 2016 Mar;16(3):18

20   Isaksson M. Skin reactions to inhaled corticosteroids. Drug Saf 2001;24:369-373

21   Baeck M, Marot L, Nicolas J-F, Pilette C, Tennstedt D, Goossens A. Allergic hypersensitivity to topical and systemic corticosteroids: a review. Allergy 2009;64:978-994

22   Browne F, Wilkinson S. Effective prescribing in steroid allergy: controversies and cross-reactions. Clin Dermatol 2011;29:287-294

23   Zheng Y, Chaudhari PR, Maibach HI. Allergic contact dermatitis from ophthalmics. Cutan Ocul Toxicol 2017;36:317-320

24   Grey KR, Warshaw EM. Allergic contact dermatitis to ophthalmic medications: Relevant allergens and alternative testing methods. Dermatitis 2016;27:333-347

25   Erdinest N, Nche E, London N, Solomon A. Ocular allergic contact dermatitis from topical drugs. Curr Opin Allergy Clin Immunol 2020;20:528-538

26   De Groot AC, Nater JP, Weijland JW. Unwanted effects of cosmetics and drugs used in dermatology, 3rd Edition. Amsterdam: Elsevier Science, 1994 (ISBN 0444897755)

27   Nater JP, de Groot AC. Drugs used on the skin. In: MNG Dukes, red. Side effects of drugs Annual 4. Amsterdam: Excerpta Medica, 1980:104-110

28   De Groot AC, Nater JP. Contact allergy to dithranol. Contact Dermatitis 1981;7:5-8

29   De Groot AC. Contact allergy to clindamycin. Contact Dermatitis 1982;8:428

30   De Groot AC. Contact allergy to sodium fusidate. Contact Dermatitis 1982;8:429

31   Rietschel RL, Fowler JF Jr, Eds. Fisher's Contact Dermatitis, 6th Edition. Hamilton, USA: BC Decker Inc., 2008

32   Cronin E. Contact Dermatitis. Edinburgh London New York: Churchill Livingstone, 1980

33   Johansen JD, Mahler V, Lepoittevin J-P, Frosch PJ, Eds. Contact Dermatitis, 6th Edition. Berlin Heidelberg: Springer-Verlag, 2018

34   John S, Johansen J, Rustemeyer T, Elsner P, Maibach H, Eds. Kanerva's Occupational Dermatology, 3rd Ed. Cham: Springer, 2018

35   De Groot AC. Monographs in Contact Allergy Volume I. Non-Fragrance Allergens in Cosmetics (Part I and Part 2). Boca Raton, Fl, USA: CRC Press Taylor and Francis Group, 2018 (ISBN 978-1-138-57325-3 and 9781138573383)

36   De Groot AC. Monographs in Contact Allergy, Volume II. Fragrances and Essential Oils. Boca Raton, Fl, USA: CRC Pess Taylor and Francis Group, 2019 (ISBN 9780367149802)

37   De Groot AC. Patch Testing, 4th Edition. Wapserveen, The Netherlands: acdegroot publishing, 2018 (ISBN 978-90-813233-1-4) (www.patchtesting.info)

Chapter 2   AN OVERVIEW OF ALLERGIC CONTACT DERMATITIS FROM TOPICAL DRUGS

2.1 SIDE EFFECTS OF TOPICAL DRUGS

Topical pharmaceuticals are applied in various forms to the skin (e.g. cream, lotion, ointment, fatty ointment, solution, foam, ear drops, suspension) or to the mucous membranes of the eyes, nose, mouth, vulva, vagina, penis, and anus (e.g. eye drops, nose drops, ointment and creams, solutions, gargles, suppositories, vaginal suppositories) to treat a variety of skin and mucous membrane diseases. Frequently used topical drugs are corticosteroids, antifungal agents and antibiotics, often in combination preparations. The use of topical pharmaceuticals can lead to a variety of local side effects including itching, burning, irritant dermatitis (either from the entire formulation or from the active principle, e.g. dithranol and retinoids), sensitization and allergic contact dermatitis and – far less frequently – phototoxicity, photoallergic contact dermatitis, immediate-type reactions such as contact urticaria or anaphylaxis (Chapter 5 Immediate contact reactions), acneiform and rosacea-like eruptions and striae (from topical corticosteroids). The fact that topical drugs are by definition applied to non-intact or non-healthy skin with a compromised barrier favors possible adverse reactions.

Whereas most side effects of topical drugs affect the skin or mucous membrane, systemic side effects from transcutaneous or transmucosal absorption may also occur. Such effects from corticosteroids have been well known since soon after their introduction in the beginning of the 1950s. Absorption was facilitated by long-term application to large areas of inflamed skin such as in atopic dermatitis in children. Treatment with salicylic acid-containing topical drugs led to intoxication ('salicylism') in many children (156). Also, drugs applied in the eyes, the nose or the deeper airways by inhalation, are very rapidly absorbed into the bloodstream and may reach levels that are able to induce serious systemic side effects.

Contact allergy and allergic contact dermatitis is the side effect of topical drugs most frequently observed by dermatologists and reported in literature and is the main subject of this chapter and this book. Further contents are limited to photocontact allergy and – in tabular format – contact urticaria (immediate contact reactions).

2.2 ALLERGIC CONTACT DERMATITIS FROM TOPICAL DRUGS

In this section, various aspects of allergic contact dermatitis to topical drugs are discussed. Because of their qualitative and quantitative importance, all aspects of (allergy to) topical corticosteroids are discussed in a separate part in this chapter (see the section 'Corticosteroids' below), but some overlap does occur.

2.2.1 Prevalence

There is little information on the prevalence of sensitization to (individual) topical drugs in the general population. In a 2008-2011 cross-sectional study, in which a random sample of 3119 individuals from the general population aged 18-74 years from five different European countries (Sweden, the Netherlands, Germany, Italy and Portugal) were patch tested, the prevalence of sensitization to 6 (mixtures of) topical drugs (caine mix, neomycin, quinoline mix, budesonide, hydrocortisone butyrate and tixocortol pivalate) ranged from 0.2% to 0.4%. The frequencies in men ranged from 0.1% (quinoline mix) to 0.6% (budesonide and hydrocortisone butyrate) and in women from 0.2% (quinoline mix) to 0.6% (caine mix) (145). It should be realized that the results were read once, at D3. As reactions to neomycin and the corticosteroids often start after D3, it is quite certain that a number of sensitizations have gone unnoticed.

The prevalence of allergic contact dermatitis to topical drugs in *patients patch tested by dermatologists* for suspected contact dermatitis is estimated at about 15% (60). Just as with the nature of the allergens involved (paragraph 2.2.4), the prevalence in individual countries, regions or hospitals is influenced by local and regional prescribing and self-medication habits (over-the-counter products) and availability of products that may be prone to cause (photo)allergic contact dermatitis such as neomycin, bacitracin, benzocaine, dibucaine, bufexamac, or ketoprofen. Other crucial factors are the interest of the investigator in this specific topic, whether the patients' contact materials are patch tested, routine testing of an extended baseline series or targeted additional testing (medicament series), patch testing techniques (if reactions are not read at D7, many cases of neomycin and corticosteroid allergy may be missed), the profile of the hospital (general versus academic), patient population (elderly people, leg ulcer patients) and the selection of patients for patch testing (liberal versus strongly selected).

2.2.2 Predisposing factors

Topical drugs are virtually always used on diseased (often inflamed) or non-intact (wounds, burns, leg ulcers) skin or mucous membranes, often with a damaged or insufficiently functioning skin barrier. This facilitates penetration and access to the antigen-presenting cells (Langerhans cells, dendritic cells). The presence of inflammation in the skin treated with the topical pharmaceutical (stasis dermatitis, leg ulcer, dermatitis, skin infections) activates the innate immune system which may favor the presentation of the chemical to the immune system, and subsequently the sensitization of T-cells (109). Predisposing factors for sensitization to topical drugs are summarized in table 2.1. The most notable of these is treatment of leg ulcers and/or stasis dermatitis. In epidemiological studies, these patients often have very high prevalences of positive patch tests with multi-sensitization to active drugs, excipients of topical drugs (lanolin alcohol, propylene glycol, fragrances) and unrelated allergens (111,112,113).

Table 2.1 Factors predisposing to sensitization, or often involved in medicament allergy

Strong intrinsic sensitizing potential of applied drugs
Application to damaged skin
Application to inflamed skin (notably stasis dermatitis, leg ulcers [111-113])
Penetration enhancing factors:
- irritant materials in the topical preparation (e.g. propylene glycol)
- use of occlusive dressing
- application to large body folds
- transdermal therapeutic systems

Application to specific anatomic locations (penis, vulva [138], anus/perianal region, conjunctivae, ear canal)
Long term treatment (chronic eczema)
Advanced age (110,131)

2.2.3 Clinical manifestations

Reported clinical manifestations of allergic contact dermatitis from topical drugs are summarized in table 2.2. In its classic form, it presents as acute itchy dermatitis with erythema, edema, papules and sometimes with vesicles and exudation. Subacute forms with papules and scaling also occur. Often, the dermatitis stays limited to the site of application. However, spreading of the dermatitis to other locations and sometimes generalization is far from rare, especially with strong sensitizers (e.g., bufexamac) or in the case of sensitization to topical pharmaceuticals applied to leg ulcers or stasis dermatitis. Spreading may occur from accidental contact, transmission through clothing or hands and as a manifestation of systemic contact dermatitis, caused by absorption of the culprit drug (see below).

Table 2.2 The spectrum of clinical manifestations of allergic contact dermatitis from topical drugs

Acute or subacute allergic contact dermatitis limited to the site(s) of application
Dermatitis spreading to other (non-application) sites or generalization
Therapy-resistance of treated dermatosis (corticosteroids, allergy to other drugs in combination preparations with
   corticosteroids)
Worsening of the treated dermatosis (eczema, skin infection)

Special and atypical forms of contact dermatitis
Connubial contact dermatitis (consort contact dermatitis, dermatitis 'by proxy')
Erythema multiforme-like allergic contact dermatitis/urticarial papular and plaque eruption
Ectopic dermatitis (114)
Pustular contact dermatitis
Acute generalized exanthematous pustulosis-like allergic contact dermatitis (AGEP)
Other atypical forms of allergic contact dermatitis (table 2.4)

Allergic reactions to the mucous membranes of the eyes, nose, mouth and airways (inhalation)
Allergic reactions to eye drops: itching, burning, tearing, red eyes, conjunctivitis, conjunctival injection, erythema and
   edema of the eyelids, periorbital dermatitis, systemic contact dermatitis
Allergic reactions to nose drops or nasal spray: nasal itching, dryness, burning, nasal congestion, worsening of
   rhinitis, dysphagia, edema of the tongue, lips and face, and dermatitis on the nose, upper lip and face, systemic
   contact dermatitis
Allergic reactions from inhalation: stomatitis, perioral eczema, oropharyngeal pruritus, dryness, mouth erythema and
   edema, dry cough, dysphonia, dysphagia and odynophagia, dyspnea, and wheezing (152), systemic contact
   dermatitis

Other manifestations of allergic contact dermatitis
Photoallergic contact dermatitis: eczema limited to light-exposed areas (photocontact allergy) or secondary spread
   to light-exposed areas (photoaugmented contact allergy)
Occupational allergic contact dermatitis (rare with topical drugs, frequent with systemic drugs)
Systemic contact dermatitis (135,141,142,149)

ACD: allergic contact dermatitis

Allergy to corticosteroids is different from other drugs because of their anti-inflammatory action. It should be suspected when skin diseases such as eczema and psoriasis respond poorly or not at all to treatment (32,81), become worse during therapy, or reoccur rapidly after withdrawal of the corticosteroid. Contact allergy to an antibiotic or antifungal drug in a combination preparation with a corticosteroid may, for the same reason, be suppressed and go unnoticed. Conversely, symptoms of allergic contact dermatitis may be interpreted by both patient and physician as spontaneous exacerbation of the treated dermatosis, by which the allergic reaction goes unrecognized.

Special and atypical forms of allergic contact dermatitis

There are various special and atypical forms and manifestations of allergic contact dermatitis. Well-known are erythema multiforme-like eruptions. These are not well-defined and are usually based on the morphological interpretation of the investigator. However, besides the occasional target-like lesions, the morphology (with urticarial papules and plaques), clinical course and history of erythema multiforme-like eruptions of contact allergy are not characteristic of classical erythema multiforme. They also lack the histopathology of 'real' erythema multiforme. Some authors therefore prefer the term 'urticarial papular and plaque eruptions'. The exact mechanism is unknown, but allergic immune complexes are thought to play a role (153). Topical drugs that have caused EM-like allergic contact dermatitis are shown in table 2.3.

In 'connubial' allergic contact dermatitis, also termed 'consort' contact dermatitis and more recently allergic contact dermatitis 'by proxy', the culprit drug causing allergic contact dermatitis is not used by the patient him- or herself, but by a partner, child, parent, sibling or someone else in the patient's close personal environment (table 2.3). Ectopic dermatitis is dermatitis located only at a part of the body distant from the application site, which has been observed with tioconazole in nail lacquer (114) and with corticosteroids.

Some eruptions of allergic contact dermatitis show pustules or the clinical picture is even dominated by them (pustular contact dermatitis). Rarely, the allergic contact dermatitis resembles an eruption which is usually caused by systemic drugs: acute generalized exanthematous pustulosis (AGEP). Fever, however, a frequent sign of AGEP, is absent. The culprit products for these and other atypical forms of allergic contact dermatitis are shown in tables 2.3 and 2.4. See also table 2.24 for atypical manifestations of corticosteroids.

Table 2.3 Atypical presentations of allergic contact dermatitis

Erythema multiforme-like ACD	Sulfanilamide
Amlexanox	Sulfonamide (153)
Bufexamac	Tocopherol
Clioquinol	Tocopheryl acetate
Econazole (153)	
Fluorouracil	
Furazolidone (153)	**Connubial allergic contact dermatitis**
Glycol salicylate	Benzoyl peroxide
Idoxuridine	Bufexamac
Ketoprofen	Clotrimazole
Lincomycin	Diphenhydramine
Mafenide	Ketoprofen (also photoallergic)
Mechlorethamine (153)	Miconazole
Mephenesin	Nifuratel
Neomycin	Procaine
Nifuroxime (153)	Resorcinol
Nitroglycerin	
Permethrin (153)	**Pustular contact dermatitis (PCD) and acute**
Phenylbutazone	**generalized exanthematous pustulosis (AGEP)**
Povidone-iodine	Benzocaine (AGEP)
Proflavine	Bufexamac (AGEP)
Promethazine	Dibucaine (AGEP)
Pyrrolnitrin	Fluorouracil (PCD)
Scopolamine (153)	Minoxidil (PCD)
	Nitrofurazone (AGEP, PCD)

<u>Allergic reactions to the mucous membranes of the eyes, nose, mouth and airways (inhalation)</u>
The manifestations of allergic reactions to eye drops, nose drops and inhaled medications (mostly budesonide) are summarized in table 2.2. Contact allergic reactions of the oral cavity from topical drugs appear to be very rare. Drugs that have caused allergic contact dermatitis from their use in eye drops or other ophthalmic pharmaceuticals are shown in table 2.5.

<u>Other manifestations of allergic contact dermatitis</u>
Photoallergic contact dermatitis to topical drugs, formerly frequent with topical phenothiazines, is currently rare with the notable exception of some non-steroidal anti-inflammatory drugs such as ketoprofen, piroxicam, diclofenac and benzydamine. Photoallergic contact dermatitis mainly presents as acute eczema, or exaggerated sunburn in areas of drug application that are also exposed to sunlight or artificial UV-light (sunbed, solarium), sparing naturally shaded areas such as the body folds and skin covered by dense hair, and is absent on skin protected by cloths or other personal objects including jewelry or a wrist watch (60). Topical drugs that have caused photocontact allergy are shown in table 2.6.

Occupational allergic contact dermatitis, which is well-known in health care professionals and workers in the pharmaceutical industry with systemic drugs (Chapter 6), is rare with topical drugs, although seen occasionally with corticosteroids. Topical drugs that have caused occupational sensitization or allergic contact dermatitis are shown in table 2.6.

Systemic contact dermatitis, its manifestations and topical drugs that have caused it are discussed below.

Table 2.4 Miscellaneous forms of atypical allergic contact dermatitis

3-(Aminomethyl)pyridyl salicylate	airborne allergic contact dermatitis
Benzocaine	orodynia
Benzoyl peroxide	purpuric contact dermatitis (probably irritant)
Benzydamine	lymphomatoid photocontact dermatitis; airborne allergic contact dermatitis
Bufexamac	generalized rash resembling the baboon-syndrome; pigmented purpuric dermatosis
Clindamycin	rosacea-like rash; allergic contact dermatitis mimicking a 'retinoid flare'
Ketoprofen	leukomelanoderma following photoallergic contact dermatitis
Lidocaine	fixed drug eruption (from lidocaine suppository)
Mephenesin	purpuric contact dermatitis
Minoxidil	psoriasiform dermatitis; pigmented contact dermatitis; persisting pseudolymphoma-like patch test reactions
Neomycin	lichenoid allergic contact dermatitis
Phenylbutazone	purpuric contact dermatitis
Proflavine	pigmented contact dermatitis
Sulfiram	toxic epidermal necrolysis-like allergic contact dermatitis

Table 2.5 Drugs that have caused allergic contact dermatitis in eye drops or other ophthalmic pharmaceuticals

Aceclidine	Brimonidine	Framycetin
Acetylcysteine	Carteolol	Gentamicin
Acyclovir	Chloramphenicol	Homatropine
Alcaftadine	Chlorpheniramine	Hydrocortisone acetate
Aminocaproic acid	Colistimethate	Hydrocortisone sodium
Amlexanox	Cromoglicic acid	phosphate (see Hydrocortisone)
Antazoline	Cyclopentolate	Idoxuridine
Apraclonidine	Dexamethasone sodium	Interferon beta
Atropine	phosphate	Ketorolac tromethamine
Azidamfenicol	Dibekacin	Ketotifen fumarate
Azithromycin	Diclofenac	Latanoprost
Befunolol	Dipivefrin	Levobunolol
Betamethasone valerate	Dorzolamide	Levocabastine
Betaxolol	Echothiophate	Methylprednisolone acetate
Bibrocathol	Ephedrine	Metipranolol
Bimatoprost	Epinephrine	Metoprolol

Table 2.5 Drugs that have caused allergic contact dermatitis in eye drops or other ophthalmic pharmaceuticals (continued)

Mitomycin C	Pilocarpine	Rifamycin
Neomycin	Pirenoxine	Ripasudil
Nitrofurazone	Polymyxin B	Rubidium iodide
Oxybuprocaine	Povidone-iodine	Scopolamine
Oxytetracycline	Prednisolone acetate	Spaglumic acid
Penicillamine	Prednisolone pivalate	Sulfacetamide
Pheniramine	Proparacaine	Tetracaine
Phenylephrine	Resorcinol	Tetrahydrozoline
Thioctic acid	Travoprost	Tromethamine
Timolol	Tretinoin	Tropicamide
Tobramycin	Trifluridine	Vancomycin
Tolazoline	Triamcinolone acetonide	

Table 2.6 Topical drugs that have caused photoallergic contact dermatitis or occupational contact dermatitis

Photoallergic contact dermatitis	Occupational allergic contact dermatitis
Aceclofenac	Benzydamine
Benzocaine	Budesonide
Benzydamine	Chloramphenicol
Coal tar	Coal tar
Chlorproethazine	Diclofenac
Dexketoprofen	Enilconazole
Dibucaine	Ethacridine lactate
Diclofenac	Gentamicin (in bone cement)
Dioxopromethazine	Homatropine
Etofenamate	Hydrocortisone
Fepradinol	Ichthammol
Flufenamic acid	Methyl aminolevulinate
Hexamidine diisethionate	Minoxidil
Indomethacin	Neomycin
Isothipendyl	Phenylephrine
Ketoprofen (and persistent light reaction)	Piroxicam
Methoxsalen	Povidone-iodine
Methyl aminolevulinate	Proflavine
Minoxidil	Proparacaine
Piketoprofen	Tetracaine
Pilocarpine	Thiocolchicoside
Piroxicam	Tocopherol
Promethazine (and persistent light reaction)	Tropicamide
Pyrithione zinc	Virginiamycin
Sulfanilamide	Thiocolchicoside
Suprofen	
Thiabendazole	

Systemic contact dermatitis

Systemic contact dermatitis (also termed systemic allergic contact dermatitis, systemic allergic dermatitis) is a condition that occurs when an individual sensitized to a contact allergen is exposed to that same allergen or a cross-reacting molecule through a systemic route. Systemic exposure to allergens can include transcutaneous, transmucosal, oral, intravenous, intramuscular, and inhalational routes, as well as implants (134,135,137,141,142, 149). Possible manifestations are shown in table 2.7 and include reactivation of previous eczema and positive patch tests, acrovesicular dermatitis, systemic symptoms and various drug exanthemas including maculopapular rashes, urticaria, erythema multiforme and vasculitis. Such reactions are most frequently caused by drugs given orally or parenterally. However, topical drugs absorbed through the skin or the mucosae may also induce such reactions. Well-known causes are the vasodilator diltiazem used for anal fissure (Chapter 3.108 Diltiazem) and the topical anesthetic dibucaine (Chapter 3.101 Dibucaine), which are absorbed through the anal mucosa and the inflamed perianal skin, and budesonide in inhalation preparations (Chapter 3.49 Budesonide).

The most characteristic manifestation of systemic contact dermatitis is the so-called baboon syndrome (136). It presents as diffuse pink or dark violet erythema of the buttocks and inner thighs, like an inverted triangle or V-shaped – resembling the red bottom of a baboon – often accompanied by dermatitis in the axillae and sometimes other body folds (142). A similar eruption can be caused by systemic drugs, notably antibiotics, without previous sensitization (61). For this variant of the baboon syndrome, in 2004 the alternative term 'symmetric drug-related intertriginous and flexural exanthema' (SDRIFE) was coined (95). One of the considerations of the authors was that the term baboon syndrome was ethically incorrect and possibly offensive. Unfortunately, previous sensitization was excluded by the authors introducing this term (95), so the 'baboon syndrome' is mostly treated as a separate entity for cases with prior sensitization. Therefore, the term remains in general use (61), although some authors treat these names as synonyms (60).

Topical drugs (including in rectal and vaginal suppositories) that have caused systemic allergic contact dermatitis from resorption through the mucous membranes and skin are shown in table 2.8.

Table 2.7 Symptoms and signs of systemic contact dermatitis from topical drugs

Reactivation of previous allergic contact dermatitis
Reactivation of previous positive patch test
Worsening of existing eczema
Vesicular dermatitis of the palms of the hands, sides of the fingers and soles of the feet, with or without erythema
Drug eruptions
- generalized eczema
- maculopapular exanthema
- urticaria
- baboon syndrome
- erythema multiforme
- purpura
- vasculitis
- acute generalized exanthematous pustulosis (AGEP)

Systemic symptoms: fever, malaise, nausea, vomiting, diarrhea, headache, arthralgia, (rarely) syncope

Table 2.8 Topical drugs that have caused systemic contact dermatitis

Acetarsone	Diltiazem	Nifuroxime
Amlexanox	Dimethindene	Nylidrin
Bacitracin	Dorzolamide	Nystatin
Benzydamine (systemic	Ephedrine	Oxyphenbutazone
*photo*contact dermatitis)	Estradiol	Phenylbutazone
Budesonide	Eucaine	Phenylephrine
Bufexamac	Framycetin	Piperazine
Buprenorphine	Gentamicin	Prednisolone acetate
Carbarsone	Hydrocortisone aceponate	Promestriene
Chloral hydrate	Iodine	Pyrazinobutazone
Chloramphenicol	Iodoquinol	Sisomicin
Chlorquinaldol	Lidocaine	Stannous fluoride
Clioquinol	Methyl aminolevulinate	Testosterone
Clonidine	Methylphenidate	Tetracaine
Dibucaine (contact and	Neomycin	Triamcinolone acetonide
*photo*contact dermatitis)	Nicotine	Trimebutine

2.2.4 The allergens

The nature of the allergens depends *inter alia* on and varies between countries/areas of countries and in time. Formerly, the topical use of penicillin, sulfonamides and promethazine caused a large number of sensitizations, but they are unimportant now. Neomycin and bacitracin were very common sensitizers in the 1950s and 1960s in Finland from their combined presence in a very popular over-the-counter antibacterial product (154,155), but since stricter regulations were enforced, their prevalence of sensitization dropped considerably. In fact, in the European Union, thanks to legislation, bacitracin, and to a lesser degree neomycin, are not important sensitizers anymore. However, in the USA, where such OTC pharmaceuticals still are readily available, these antibiotics both belong to the

Top-6 of common sensitizers after nickel, methyl(chloro)isothiazolonone, fragrances (fragrance mix I and Myroxylon pereirae resin) and formaldehyde with currently some 7% positive reactions in patients routinely tested by the members of the North American Contact Dermatitis Group (108).

Since the mid-1980s, non-steroidal anti-inflammatory drugs (NSAIDs) such as ketoprofen have caused many cases of photoallergic contact dermatitis (and in lesser numbers contact allergy) in some countries, notably Italy, France, Japan, and Belgium, where these products are widely used, but also in Sweden, Croatia and Spain (Chapter 3.193 Ketoprofen). Bufexamac caused so many cases of contact allergic reactions in Germany and Austria, that thedrug was withdrawn by the European Medicines Agency (EMA) in April 2010 (Chapter 3.50 Bufexamac). Several drug allergens have been reported mainly or exclusively from Japan (e.g. prednisolone valerate acetate), presumably because they are or were used in that country only, or on a large scale.

Currently, the most important groups of sensitizers in topical drugs are – depending on the countries – antibiotics, corticosteroids, topical anesthetics, and non-steroidal anti-inflammatory drugs (both sensitizers and photosensitizers). It should be realized that, with the exception of the NSAIDs, their most important allergenic representatives (neomycin, bacitracin, benzocaine), or markers of sensitization (caine mix III, budesonide, tixocortol pivalate) are present in some or most routine series, thus identifying sensitization to them, whereas other (important) drug allergens may be missed when suspected pharmaceutical products or appropriate additional series are not patch tested.

In most cases, the culprit in a topical medicament causing allergic contact dermatitis is the active principle, but the reactions may also be caused by excipients (133) such as ethylenediamine, lanolin alcohol, propylene glycol, fragrances or essential oils, or parabens and other preservatives (thimerosal and benzalkonium chloride in eye drops) (Chapter 4).

Topical drugs that have caused contact allergy/allergic contact dermatitis are shown in tables 2.9 – 2.20 and include antibiotics (table 2.9), anti-infective agents (table 2.10), antifungal, antiviral and antiparasitic drugs (table 2.11), antihistamines (table 2.12), local (topical) anesthetics (table 2.13), non-steroidal anti-inflammatory drugs (table 2.14), drugs used in the treatment of eczema, acne and psoriasis and anti-neoplastic and immunosuppressive drugs (table 2.15), antiglaucoma drugs and ophthalmological drugs other than for glaucoma (table 2.16), sex hormones, vitamins and muscle relaxants (table 2.17), therapeutic agents for the nose, mouth and airways (table 2.18), drugs used in transdermal therapeutic systems (table 2.19) and miscellaneous therapeutic agents (table 2.20). Corticosteroids that have caused contact allergy/allergic contact dermatitis are shown in table 2.25. Topical drugs that have caused photocontact allergy/photoallergic contact dermatitis can be found in table 2.6. Topical drugs that have been associated with contact urticaria are presented in Chapter 5.

Table 2.9 Antibiotics

Arbekacin	Fusidic acid	Polymyxin B
Azidamfenicol	Gentamicin	Retapamulin
Azithromycin	Gramicidin	Rifamycin
Bacitracin	Kanamycin	Sisomicin
Chloramphenicol	Lincomycin	Sulfadiazine silver
Chlortetracycline	Mafenide	Sulfanilamide
Clindamycin	Metronidazole	Sulfathiourea
Colistimethate sodium	Mupirocin	Tetracycline
Colistin	Neomycin	Tobramycin
Dibekacin	Nitrofurazone	Tyrothricin
Erythromycin	Oxytetracycline	Vancomycin
Framycetin	Paromomycin	Virginiamycin
Furaltadone	Penicillins, unspecified	Xantocillin

Table 2.10 Anti-infective agents

Bibrocathol	Ethacridine	Iothion
Bismuth tribromophenate	Halquinol	Oxyquinoline
Chlorquinaldol	Hexamidine	Povidone-iodine
Clioquinol	Hexamidine diisethionate	Proflavine
Cloxyquin	Hexetidine	Silver nitrate
Dibrompropamidine	Iodoform	

Table 2.11 Antifungal, antiviral and antiparasitic drugs

Antifungal drugs	Lanoconazole	
Amorolfine	Luliconazole	**Anti-parasitic drugs**
Bifonazole	Miconazole	Acetarsone
Buclosamide	Mycanodin	Carbarsone
Castellani's solution	Naftifine	Mesulfen
Chlordantoin	Neticonazole	Nifuratel
Ciclopirox olamine	Nystatin	Nifuroxime
Clofenoxyde	Oxiconazole	Nifurprazine
Clotrimazole	Pecilocin	Piperazine
Croconazole	Pyrrolnitrin	Sulfiram
Econazole	Sertaconazole	Thiabendazole
Efinaconazole	Sulbentine	
Enilconazole	Sulconazole	**Antiviral drugs**
Etisazole	Terconazole	Acyclovir
Fenticonazole	Tioconazole	Ibacitabine i
Haloprogin	Tolciclate	Idoxuridine
Isoconazole	Tolnaftate	Trifluridine
Ketoconazole	Undecylenic acid	Tromantadine

Table 2.12 Antihistamines

Alcaftadine	Dimethindene	Methapyrilene
Antazoline	Dioxopromethazine	Phenindamine tartrate
Chlorcyclizine	Diphenhydramine	Pheniramine
Chlorpheniramine	Doxepin	Promethazine
Clemizole	Isothipendyl	Pyrilamine
Cyproheptadine	Ketotifen	Spaglumic acid
Dexchlorpheniramine	Levocabastine	Tripelennamine

Table 2.13 Local (topical) anesthetics

Amylocaine	Dyclonine	Prilocaine
Benzocaine	Eucaine	Procaine
Butacaine	Lidocaine	Propanocaine
Butamben	Meprylcaine	Proparacaine
Butethamine	Orthocaine	Propipocaine
Dibucaine	Oxybuprocaine	Tetracaine
Dimethisoquin	Polidocanol	
Diperodon	Pramoxine	

Table 2.14 Non-steroidal anti-inflammatory drugs

Aceclofenac	Felbinac	Methyl salicylate 2-ethylbutyrate
Acexamic acid	Fepradinol	Oxyphenbutazone
3-(Aminomethyl)pyridyl	Feprazone	Phenylbutazone
salicylate	Flufenamic acid	Piketoprofen
Bendazac	Glycol salicylate	Piroxicam
Benzydamine	Ibuproxam	Piroxicam cinnamate
Bufexamac	Indomethacin	Pyrazinobutanone
Dexketoprofen	Ketoprofen	Salicylamide
Diclofenac	Ketorolac tromethamine	Suprofen
Enoxolone	Mabuprofen	Tribenoside
Etofenamate	Methyl salicylate	

Table 2.15 Drugs used in the treatment of eczema, acne and psoriasis and anti-neoplastic and immunosuppressive drugs

Drugs used in the treatment of eczema, acne and psoriasis		Anti-neoplastic and immunosuppressive drugs
Adapalene	Salicylic acid	Carmustine
Anthrarobin	Selenium sulfide	Fluorouracil
Benzoyl peroxide	Tacalcitol	Interferons
Calcipotriol	Tioxolone	Mechlorethamine
Coal tar	Tretinoin	Methyl aminolevulinate
Dithranol		Mitomycin
Ethyl lactate		Pimecrolimus
Ichthammol		Retinyl palmitate
Methoxsalen		Stepronin
Pyrithione zinc		Tacrolimus
Resorcinol		Triaziquone

Table 2.16 Antiglaucoma drugs and ophthalmological drugs other than for glaucoma

Anti-glaucoma drugs

Aceclidine	Epinephrine
Befunolol	Latanoprost
Betaxolol	Levobunolol
Bimatoprost	Metipranolol
Brimonidine	Metoprolol
Carteolol	Pilocarpine
Dipivefrin	Ripasudil
Dorzolamide	Timolol
Echothiophate	Travoprost

Ophthalmological drugs other than for glaucoma

Acetazolamide	carbonic anhydrase inhibitor; prevention or treatment of high intraocular pressure after cataract surgery
Aminocaproic acid	antifibrinolytic agent; acts as buffer and anti-inflammatory drug in eye drops
Atropine	mydriatic and cycloplegic drug for diagnostic purposes, for prevention of synechiae in inflammatory processes, for mild amblyopia and serious myopia
Cyclopentolate	for mydriasis and cycloplegia during ophthalmic diagnostic procedures
Homatropine	parasympatholytic for mydriasis during ophthalmic diagnostic procedures
Penicillamine	chelating agent; prevention of corneal fibrosis after chemical trauma
Phenylephrine	mydriatic; dilation of the pupil prior to intraocular surgery and diagnostic examination
Pirenoxine	antioxidant; inhibits the development of cataracts in patients at risk
Rubidium iodide	iodine source; used to retard or prevent the formation of cataracts
Thioctic acid	antioxidant; treatment of cataract, diabetic retinopathy, macular degeneration
Tropicamide	muscarinic antagonist; mydriatic; for mydriasis and cycloplegia during diagnostic procedures

Table 2.17 Sex hormones, vitamins and muscle relaxants

Sex hormones	Vitamins	Muscle relaxants
Diethylstilbestrol	Pyridoxine	Belladonna extract
Estradiol	Retinol	Chlorphenesin
Estradiol benzoate	Thiamine	Chlorproethazine
Estrogens	Tocopherol	Cyclobenzaprine
Estrogens, conjugated	Tocopheryl acetate	Mephenesin
Norethisterone		Methyldiphenhydramine
Testosterone		Thiocolchicoside

Table 2.18 Therapeutic agents for the nose, mouth and airways

Ambroxol	expectorant
Amlexanox	anti-allergic agent; treatment of allergic rhinitis and aphthous ulcers
Cromoglicic acid	anti-asthmatic agent
4,5-Dihydro-1*H*-imida-zole monohydrochloride	sympathomimetic agent; nasal decongestant
Ephedrine	vasoconstrictor agent; nasal decongestant
Fenoterol	bronchodilator agent
Naphazoline	nasal decongestant
Olaflur	caries prophylactic agent
Sodium fluoride	cariostatic agent
Stannous fluoride	cariostatic agent
Tetrahydrozoline	nasal decongestant
Tiopronin	mucolytic agent
Tolazoline	vasodilator agent; treatment of circulatory changes of the retina, chorioid and optic tract, caustic trauma and degenerative changes of the cornea

Table 2.19 Drugs used in transdermal therapeutic systems

Buprenorphine	opioid analgesic; treatment of pain, peri-operative analgesia, and opioid dependence
Clonidine	antihypertensive agent
Methylphenidate	central nervous system stimulant; for attention deficit hyperactivity disorder (ADHD) and for the treatment of narcolepsy
Nicotine	ganglionic stimulant; nicotinic agonist; to help smoking cessation
Nitroglycerin	vasodilator agent; for prevention and treatment of angina pectoris
Rotigotine	dopamine agonist; for treatment of Parkinson's disease and primary restless legs syndrome
Scopolamine	antiemetics; cholinergic antagonist; prevention of motion sickness

Table 2.20 Miscellaneous therapeutic agents

Aluminum acetate	adstringent
Benzarone	vasoprotective agent; treatment of capillary fragility and capillary bleeding
Bismuth subnitrate	adstringent and antiseptic
Bromelains	enzymes in pineapples used as anti-inflammatory drug for musculoskeletal disorders
Canrenone	anti-androgenic properties; treatment of acne, hirsutism and alopecia androgenetica
Chloral hydrate	hypnotic; formerly used for insomnia and topically as counter-irritant
Collagenase clostridium histolyticum	enzyme; treatment of leg ulcers
Crotamiton	anti-itch; pesticide (scabies)
Diacetazotol	stimulant of wound epithelialization (misc.)
Diltiazem	vasodilator; treatment of fissura ani
Hirudoid ® cream	anticoagulant; used for the treatment of phlebitis
Hydroquinone	depigmenting agent
Lactic acid	treatment of warts; also gynecological anti-infective
Methyl nicotinate	rubefacient
Minoxidil	vasodilator agent; treatment of alopecia androgenetica
Monobenzone	depigmenting agent
Nicoboxil	rubefacient
Nonoxynol-9	spermatocidal agent
Nylidrin	vasodilator agent
Propantheline	muscarinic antagonist; anticholinergic agent; formerly used to treat hyperhidrosis
Propranolol	treatment of hemangiomas
Sirolimus	antiangiogenic properties; adjuvant for the laser treatment of port wine stains
Spironolactone	mineralocorticoid receptor antagonist; because of antiandrogenic effects used to treat acne vulgaris, idiopathic hirsutism and androgenic alopecia
Tetrachlorodecaoxide	radiation-protective agent; wound-healing agent
Triethanolamine polypeptide oleate condensate	cerumenolytic

Table 2.20 Miscellaneous therapeutic agents (continued)

Trimebutine	parasympatholytic; used rectally and topically for anal fissures and hemorrhoids
Tromethamine	buffer; excipient in eye gel
Troxerutin	anticoagulant, vasoprotective agent; treatment of chronic venous insufficiency, topically as analgesic for pain, swelling and bruising

2.3 DIAGNOSTIC PROCEDURES

Diagnostic procedures in cases of suspected corticosteroid allergy are discussed in Paragraph 2.7. When allergic contact dermatitis to one or more topical drugs is suspected (see Paragraph 2.2.3), the first step is to patch test the routine series and all pharmaceutical products (previously) used by the patient. Most of these can be tested 'as is', although some may produce irritant reactions, e.g. povidone-iodine solution, permethrin cream, and pharmaceuticals containing calcipotriol, tretinoin, or 5-fluorouracil. The European baseline series (28) contains several topical drugs: neomycin, the caine mix III (benzocaine, dibucaine, tetracaine), budesonide, and tixocortol pivalate. In addition, it screens for some excipients that may be the allergenic culprit in topical medicaments (parabens, lanolin alcohol, fragrances). The NACDG screening series includes as topical drugs neomycin, bacitracin, tocopherol (vitamin E, not 'officially' a topical drug, but frequently used as such, also by physicians), benzocaine, lidocaine, and the corticosteroids budesonide, tixocortol pivalate, hydrocortisone butyrate, clobetasol propionate and desoximetasone; base ingredients include lanolin alcohol, parabens, ethylenediamine and propylene glycol (108). In a number of cases, the final diagnosis can be made on the basis of the tests results of the routine series plus positive reactions to the patient's own product(s) and knowledge of their contents. Additional testing with a 'medicament series' or a 'corticosteroid series' may be advisable to detect additional unrelated drug allergens (e.g. used by the patient in the past), cross-reacting chemicals (local anesthetics, aminoglycoside antibiotics, corticosteroids), but also to find suitable alternatives that are negative on patch testing. Approximately 125 drugs are available for patch testing from the major suppliers of patch test materials, of which nearly 95 are (also) used as topical drugs, including corticosteroids (tables 2.21 and 2.26).

When a topical drug gives a positive patch test reaction, but the allergenic culprit is unknown, all ingredients should be patch tested separately, preferably after having confirmed sensitization by a ROAT or reusing the drug and exclusion of irritancy by adequate control testing. The best approach is to contact the supplier or manufacturer and request samples of all ingredients. It is advisable to include a list of all known ingredients with the concentration and vehicle suitable for the patch tests, e.g. based on the information in this book and *Patch Testing*, 4th edition (2018) (www.patchtesting.info). If no (gentle) directions are given to the manufacturer, most likely the materials will be provided in the concentrations as present in the topical formulation, which entails a great risk of false-negative reactions. An easier alternative is to purchase the active ingredient, if available, from one of the providers of patch test materials (table 2.21), but, in the case of a positive result, additional sensitizations to excipients will go unnoticed Indeed, patients with drugs allergies, especially those with leg ulcers, stasis dermatitis and corticosteroid allergy, often have polysensitization to drugs and base materials.

False-negative patch test reactions

Patch testing with topical drugs may be problematic and result in false-negative results, for example in the case of contact allergy to an antibiotic or antifungal in a combination preparation with a corticosteroid, where the anti-inflammatory action of the latter suppresses the inflammatory patch test reaction. Patch tests with commercial eye drop preparations containing beta-blockers may also – not infrequently -- be false-negative (115,116,119-123,125, 128). The likely explanation is that, whereas the anatomic and physiologic properties of eyelid skin and conjunctivae are associated with a lower threshold to development of sensitization and allergic contact dermatitis, the low concentration of the allergens (usually 0.25% to 0.5%) in commercial products is insufficient to elicit an allergic reaction in the far thicker skin of the back. When contact allergy is strongly suspected, but patch tests are negative, the following alternative diagnostic methods have been suggested (139,140): patch test on adhesive tape-stripped (117,122), pricked (10x with a prick-test lancet) (124) or scarified skin (122); patch test after pretreatment with 0.5% aqueous sodium lauryl sulfate solution for 24 hours (127); enlarge the patch test area (121); or perform intradermal testing (118). By far the best method, however, is to patch test the active principle itself at a concentration higher than the one present in the commercial preparation (115,116,119,125), up to 10% active drug (116,129). ROATs are usually (ergo not always [130]) negative (115,125,126,132), but provocative use testing may be positive and aid in diagnosis (120,126). The same problem of false-negative reactions may be encountered with other ophthalmic drugs.

A new discovery!

When a 'new' or rare allergen is found, either the active drug or an excipient, contact allergy should be confirmed by (a combination of) retesting, testing with a dilution series, positive ROAT and exclusion of irritancy by testing the material in at least 20 unexposed controls. Patients with drug allergies, especially those with leg ulcers, stasis dermatitis or corticosteroid allergy, often have multiple positive patch test reactions. In all such cases, retesting with a few of the haptens at a time is strongly recommended to exclude false-positive reactions due to the excited skin syndrome. Patch tests should always be read at D7-D8 to avoid missing late-developing positive reactions, notably with neomycin, other aminoglycosides and corticosteroids.

Compound allergy
Sometimes, patch testing with a topical drug is positive but with the ingredients negative. This is often reported as 'compound allergy', i.e. that a new chemical is formed by the interaction of 2 or more ingredients. This is probably a very exceptional situation and has rarely been well documented (Chapter 3.167 Hirudoid cream). It is far more likely that the positive patch test to the pharmaceutical was false-positive or testing with the ingredients false-negative (concentration too low, inadequate vehicle, not all ingredients tested, batch variations).

Photopatch testing
When photoallergic contact dermatitis is suspected on the basis of patient history and clinical manifestations (allergic contact dermatitis at sites exposed to sunlight or artificial UV-radiation), photopatch tests should be performed (143,144). As photoallergic contact dermatitis is relatively infrequent, photopatch testing is fairly complicated, time-consuming and laborious, and the yield of positive reactions is usually <20% (148), not many dermatologists and university centers perform this diagnostic procedure. In photopatch testing, two sets of allergens are applied for 1 or 2 days, after which one set is irradiated with 5J/cm^2 of UVA. Readings are performed immediately after irradiation and 2 or more days thereafter (144). Patients may be routinely tested with the European photopatch test baseline series (n=20) and – either routinely or on indication – with the extended series of photoallergens (n=15), consisting mostly of UV filters and (topical) drugs, particularly NSAIDs (table 2.22) (143).

   Chemicals that cause photoallergic contact dermatitis have varied and vary in time and region. The antihistamines promethazine and chlorpromazine were formerly widely used in topical pharmaceuticals and caused many contact and photocontact allergic reactions. The latter is not used as such anymore, but promethazine is still on the market in some South-European countries like Portugal and Greece. Formerly, the halogenated salicylanilides were important sensitizers, but these have largely been abandoned, which also applies to fragrance photosensitizers such as musk ambrette and 6-methylcoumarin. Currently, topical NSAIDs are the most important causes of photoallergic contact dermatitis, notably ketoprofen. Most such reactions have been reported from southern European countries including Spain and Italy, but ketoprofen has also caused many photoallergic reactions in France, Belgium and Sweden (Chapter 3.193 Ketoprofen). In most other countries, these preparations are not or little used and pose no health threat. Obviously, as only few centers perform photopatch tests, (many) cases of photoallergic contact dermatitis may go unnoticed. Apart from the baseline and the extended baseline series, patients' own products or other chemicals to which he or she was exposed to should also be considered for photopatch testing (144).

Table 2.21 Topical drugs commercially available for patch testing [a,c]

Patch test allergen (hapten)	Chemotech	SPCanada	SPEurope
Acyclovir	10.0%		
Amylocaine hydrochloride	5.0%		
Atropine sulfate		1% water	1% water
Bacitracin	5.0%; 20%	20%	20%
Benzocaine	5.0%	5%	5%
Benzoyl peroxide	1.0%	1%	1%
Benzydamine hydrochloride	2.0%	1%; 2%	
Bufexamac	5.0%	5%	5%
Chloramphenicol	5.0%	2% alc.; 5%	5%
Chlorpheniramine maleate		5%	
Chlorquinaldol	5.0%	5%	5%
Chlortetracycline hydrochloride		1%	
Clindamycin phosphate	10.0%		
Clioquinol	5.0%	5%	5%
Clotrimazole		1%	1%
Coal tar	5.0%	5%	5%
Dexpanthenol		5%	5%
Dibucaine (cinchocaine) hydrochloride	2.5%; 5.0%	2.5%; 5%	5%
Diclofenac sodium salt	1.0%; 5.0%	2.5%; 5%	2.5%

Table 2.21 Topical drugs commercially available for patch testing (continued) [a,c]

Patch test allergen (hapten)	Chemotech	SPCanada	SPEurope
Diltiazem hydrochloride	10.0%		
Diphenhydramine hydrochloride	1.0%		
Econazole nitrate	1.0% alc.		
Erythromycin base	10.0%	1%; 2%	2%
Etofenamate	2.0%	2%	
Framycetin sulfate	20.0%	10%	10%
Fusidic acid sodium salt	2.0%	2%	2%
Gentamicin sulfate	20.0%	20%	20%
Hexylresorcinol		0.25%	0.25%
Hydroquinone	1.0%	1%	1%
Ibuprofen	5.0%; 10.0%	5%	
Idoxuridine		1%	
Indomethacin		1%	1%
Iodoform		5%	
Kanamycin sulfate	10.0%	10%	10%
Ketoprofen	1.0%	1%; 2.5%	
Lidocaine hydrochloride	5.0%; 15.0%	15%	15%
Mafenide		10%	10%
Metronidazole		1%	1%
Miconazole	1.0% alc		
Monobenzone	1.0%	1%	1%
Neomycin sulfate	20.0%	20%	20%
Nitrofurazone	1.0%	1%	1%
Nystatin		2%	2%
Oxytetracycline		3%	3%
Panthenol	5.0%		
Penicillamine		1%	
Phenylbutazone	10.0%	10%	10%
Phenylephrine hydrochloride		10% water	10% water
Piketoprofen		1%	
Pilocarpine hydrochloride		1% water	1%
Piperazine		1%	1%
Piroxicam	1.0%	1%	
Polidocanol		3%	3%
Polymyxin B sulfate	5.0%	3%	3%
Povidone-iodine		10% water	10%
Pramocaine hydrochloride	2.0%		
Prilocaine hydrochloride	5.0%		
Procaine hydrochloride	1.0%	1%; 2%	1%; 2%
Promethazine hydrochloride	0.1%	2%	
Propranolol hydrochloride		2%	
Resorcinol	1.0%	1%; 2%	2%
Salicylamide		2%	2%
Salicylic acid		5%	
Silver nitrate	1.0% water	1% water	1%
Spiramycin base	10.0%		
Streptomycin sulfate		5%	
Sulfanilamide	5.0%	5%	5%
Sulfur, precipitated		10%	
Tetracaine hydrochloride	5.0%	1%	1%
Tetracycline hydrochloride		2%	2%
Tioconazole	1.0%		
Tobramycin	20.0%	20%	
Tocopherol (vitamin E)	100%	100%	
Tocopheryl acetate	10.0%		
Vancomycin hydrochloride	10.0% water		

Table 2.21 Topical drugs commercially available for patch testing (continued) [a,c]

Patch test allergen (hapten)		Chemotech	SPCanada	SPEurope
Zinc pyrithione (pyrithione zinc)		1.0%	0.1%	
Mixes of haptens [b]				
Caine mix II		10.0%		
- lidocaine	5.0%			
- dibucaine hydrochloride	2.5%			
- tetracaine hydrochloride	2.5%			
Caine mix III		10.0%		
- benzocaine	5.0%			
- dibucaine hydrochloride	2.5%			
- tetracaine hydrochloride	2.5%			
Caine mix IV		10.0%		
- lidocaine	5.0%			
- amylocaine hydrochloride	2.5%			
- prilocaine hydrochloride	2.5%			
Caine mix V		7.0%		
- benzocaine	5.0%			
- dibucaine hydrochloride	1.0%			
- tetracaine hydrochloride	1.0%			
Caine mix A			7%	
- benzocaine	5%			
- dibucaine hydrochloride	1%			
- tetracaine hydrochloride	1%			
Caine mix C			7%	
- benzocaine	5%			
- dibucaine hydrochloride	1%			
- procaine hydrochloride	1%			
Oxyquinoline mix			6%	
- clioquinol	3%			
- chlorquinaldol	3%			
Quinoline mix		6.0%		
- clioquinol	3.0%			
- chlorquinaldol	3.0%			

[a] The vehicle for all haptens is petrolatum, unless otherwise indicated
[b] Not a drug hapten *per se*, but a mixture of drugs for diagnostic patch testing
[c] For the commercially available corticosteroids see Paragraph …. Corticosteroids
alc.: alcohol; HCl: hydrochloride; pet.: petrolatum
Chemotech: Chemotechnique Diagnostics (www.chemotechnique.se)
SPCanada: SmartPractice Canada (www.smartpracticecanada.com)
SPEurope: SmartPractice Europe (www.smartpracticeeurope.com)

Table 2.22 Topical drugs present in the European photopatch test baseline and extended baseline series (143)

Baseline series		Drug type	Extended series		Drug type
Benzydamine	2% pet.	NSAID	Dexketoprofen	1% pet.	NSAID
Etofenamate	2% pet.	NSAID	Diclofenac	5% pet.	NSAID, treatment of AK
Ketoprofen	1% pet.	NSAID	Ibuprofen	5% pct.	NSAID
Piroxicam	1% pet.	NSAID	Piketoprofen	1% pet.	NSAID
Promethazine	0.1% pet.	Antihistamine			

AK: actinic keratosis; NSAID: non-steroidal anti-inflammatory drug

CORTICOSTEROIDS

2.4 USES OF CORTICOSTEROIDS

Glucocorticoids (usually termed 'corticosteroids') are synthetic derivatives of the endogenous adrenal cortex hormone cortisone. They have anti-inflammatory, vasoconstrictive, antiproliferative and antipruritic activities. Topical corticosteroids may be applied to the skin (dermatocorticosteroids) or to the mucosae of the nose, eyes, airways (inhalation), genitals, anus or mouth (table 2.23). The most important indications for topical use to the skin are inflammatory skin diseases including (various forms of) eczema (dermatitis) and psoriasis. Other dermatoses that can be treated with − the more potent − corticosteroids include lichen planus, lichen sclerosus, pustulosis palmoplantaris and lupus erythematosus. Application forms for the skin include cream, fatty cream, ointment, emulsion, lotion, scalp lotion, solution for cutaneous application, shampoo, and foam. Dermatocorticosteroids may be combined with other drugs including salicylic acid, calcipotriol, antibiotics, antifungals or local anesthetics. Preparations to treat (secondarily infected) otitis externa (and otitis media) are (virtually) always combined with antibiotics including chloramphenicol, framycetin/neomycin, gramicidin, oxytetracycline, polymyxin B or other antimicrobials such as clioquinol or acetic acid. These ear drops and ointments clear or prevent infections, and reduce the swelling and itching of the ear canal caused by inflammation of the auricular skin. They also prevent epidermal hyperplasia, granulation tissue, the development of polyps and inflammation of the middle ear.

Applied to the mucosae of the nose, eyes, airways or mouth, corticosteroids suppress and diminish (allergic and non-allergic) inflammatory processes. The main indications for corticosteroids applied in the nasal cavity (nose drops, nasal spray, nasal inhalation powder, suspension) are allergic and non-allergic rhinitis. They improve congestion, sneezing, itch and rhinorrhea. Corticosteroids in eye drops, gels, and suspensions are indicated for the treatment of allergic conjunctivitis, keratitis, uveitis anterior, episcleritis, scleritis, iridocyclitis and inflammation of the cornea. They may be combined with antibiotics such as chloramphenicol, polymyxin B, gentamicin, tobramycin, framycetin, gramicidin, neomycin or oxytetracycline. Asthma and chronic obstructive pulmonary disease (COPD) may be treated with inhalation of corticosteroid-containing aerosols and powders, sometimes combined with β2-sympathomimetics (e.g. formoterol, salmeterol, vilanterol). Mouth paste and other corticosteroid applications may be used for the treatment of aphthae or lichen oris. Rectal applications (enema, foam) are sometimes indicated for colitis ulcerosa.

Examples of dermatocorticosteroids used on the skin and mucosae are shown in table 2.23. Some corticosteroids (e.g. betamethasone, cortisone, deflazacort, dexamethasone, fludrocortisone, hydrocortisone, methylprednisolone, prednisolone, prednisone, triamcinolone acetonide, triamcinolone hexacetonide) are also used for oral or parenteral administration (8).

Table 2.23 Examples of corticosteroids for application to the skin and mucous membranes (8)

Application to:	Corticosteroids
Skin	betamethasone valerate, betamethasone dipropionate, clobetasol propionate, clobetasone butyrate, desoximetasone, flumethasone pivalate, fluticasone propionate, hydrocortisone acetate, hydrocortisone butyrate, mometasone furoate, triamcinolone acetonide
Auricular skin (ear)	dexamethasone sodium phosphate, dexamethasone metasulfobenzoate sodium, fludrocortisone acetate, hydrocortisone acetate
Nasal mucosa	budesonide, beclomethasone, fluticasone furoate, fluticasone propionate, mometasone furoate, triamcinolone acetonide
Conjunctivae (eyes)	dexamethasone sodium phosphate, fluorometholone, hydrocortisone sodium phosphate, prednisolone acetate, prednisolone sodium phosphate, prednisolone pivalate
Airways (inhalation)	beclomethasone dipropionate, budesonide, ciclesonide, desonide, fluticasone propionate, mometasone furoate
Oral mucosa	triamcinolone acetonide
Rectal mucosa	beclomethasone dipropionate, betamethasone sodium phosphate, budesonide

2.5 ALLERGIC CONTACT DERMATITIS FROM CORTICOSTEROIDS

Corticosteroids (CSs) can cause allergic reactions, mostly type IV hypersensitivities; the first cases were reported in 1959 (79,80). The prevalence of contact allergy to corticosteroids is very variable: whereas low rates are found in Spain (30,57,97) and some other countries like Portugal (97), frequencies of 4-5% of patients attending patch test clinics have been reported from the United Kingdom (32,34,38), the USA (33), Belgium (4.4% [101], Finland (67) and Hungary (94). Even higher frequencies have been found in the U.K. (6% [99], Belgium (6.4% [97]) and in one − very small − group of Polish consecutive patients tested (7.4% [78]). These differences can be explained by patient selection (82), local, regional or national prescribing habits (82,91), physicians' awareness of corticosteroid allergy, availability of prescription and non-prescription over-the-counter drugs, quantitative use of corticosteroids in the population, which corticosteroids are patch tested, number of CSs tested, test concentration and vehicle, test methods (patch testing, intradermal), interpretation of patch test results and whether or not late readings are performed, as up to 30% of sensitizations can be missed when reactions are not read at D6-10 (1,2,10,31,37). In some studies with high frequencies and using alcohol as vehicle, some reactions may have been false-positive, as alcohol can induce such irritant reactions and was not tested as control (99).

Over 80% of sensitized patients have multiple steroid allergies (either from cross-reactions or from independent sensitization) (32,58); in addition, co-sensitization to multiple allergens such as preservatives, excipients and antibiotics are often found during patch testing (1,2,58,104,106).

By far, most cases of sensitization are caused by the cutaneous use of CSs (58). Patients who suffer from a long-term disease are at a higher risk of sensitization (64). These include chronic eczema (notably atopic dermatitis from its skin barrier defects and pro-inflammatory environment), stasis dermatitis, chronic ulceration, chronic actinic dermatitis, facial, anogenital, and hand and foot dermatitis (1,2,71,104).

Clinical picture of allergic contact dermatitis to corticosteroids

The clinical presentation of allergic contact dermatitis to corticosteroids is often specific nor spectacular, which is due to the anti-inflammatory properties of corticosteroids, which suppress the eczematous reactions at the same time as inducing them (100,103). Acute weeping dermatitis is not to be expected (103). Allergy to CSs should be suspected when skin diseases such as eczema and psoriasis respond poorly or not at all to CS treatment (32,81), become worse during therapy, or reoccur rapidly after withdrawal of the corticosteroid. Allergic contact dermatitis caused by CSs may present as chronic eczema, which is often more pronounced at the periphery of the treated zone, which is called the 'edge effect' (32). As there is frequently a long delay before the diagnosis of contact allergy is suspected, prolonged use of the corticosteroids has often occurred which can lead to 'classic' side effects, such as cutaneous atrophy, rosacea, and perioral or perinasal dermatitis dominating the clinical presentation (1,2).

Another route of sensitization is through the mucosae, notably from corticosteroids used as spray in the nasal cavity for allergic or non-allergic rhinitis (especially budesonide and tixocortol pivalate), as inhalation therapy in the pulmonary tract for the treatment of asthma (budesonide) and in eye drops for allergic conjunctivitis (93) and in the mouth for aphthous stomatitis or lichen oris (96). Although these treatments are used extensively, sensitization is relatively infrequent (58,84,85,92). Application in the nose may cause nasal itching, dryness, burning, nasal congestion, worsening of rhinitis, dysphagia, edema of the tongue, lips and face, and dermatitis on the nose, upper lip and face (16,23,25,35,36). Local allergic side effects of inhaled corticosteroids include stomatitis, perioral eczema, oropharyngeal pruritus, dryness, mouth erythema and edema, dry cough, dysphonia, dysphagia and odynophagia (pain when swallowing) (17,18,19,61,146). Symptoms often start early in the treatment (after several days) and may develop hours after inhalation (19). Systemic contact dermatitis from resorption through the nasal and pulmonary mucosae has been observed with swelling of the eyelids, urticaria, general pruritus, maculopapular exanthemas, flare-up of previous dermatitis and flare-up of previous positive patch tests (15,19,20,21,22,24,26,27,59,61). See also tables 2.7 and 2.8 and the section 'Systemic contact dermatitis' above.

Allergic contact dermatitis from corticosteroids in sensitized patients may present atypical manifestations; examples are given in table 2.24. See also the section 'Special and atypical forms of contact dermatitis' above.

Table 2.24 Allergic contact dermatitis from corticosteroids: examples of atypical manifestations [a]

Acute dermatitis (atypical for corticosteroid ACD) (41)
Acute generalized exanthematous pustulosis-like ACD (45,46)
Allergic contact dermatitis 'by proxy' (connubial contact dermatitis, consort contact dermatitis) (51-55,58)
Allergic contact dermatitis mimicking papular rosacea (50)
Allergic contact dermatitis presenting as lupus erythematosus (47)
Eczema only at a site distant from the application site, not at the application site itself or around it (87)
Edema of the lips and face mimicking type I allergy (14,42)
Erythema multiforme-like eruptions (44,48,49)
Occupational allergic contact dermatitis (51,56,65)
Systemic contact dermatitis from topical application (15,19,20,21,22,24,26,27,59,61,84)
Urticaria as sign of systemic contact dermatitis (88)

[a] with examples of references, not full listing; ACD: allergic contact dermatitis

Other loco-regional routes, such as administration in the digestive tract (treatment of Crohn's disease or colitis ulcerosa) (62) and intra-articular administration have been less frequently implicated (58). Sensitization related to *systemic* (oral, parenteral) administration of a CS, of which the intravenous route is the most common, does occur, but is not frequent (58,146). In many such cases, the patients have previously become sensitized to the corticosteroid or a cross-reacting CS from topical administration, and systemic administration subsequently leads to systemic contact dermatitis (60). However, some patients, who have never used topical steroids before, develop a cutaneous adverse drug reaction such as maculopapular exanthema and prove to be allergic to it when patch tested (146). Such reactions to non-topical administration of CSs (oral, intravenous, intramuscular, intra-articular, subcutaneous, intrathecal) CSs are planned to be discussed in Volume 4 of the *Monographs in Contact Allergy* series on Systemic drugs.

2.6 THE CORTICOSTEROID SENSITIZERS

Corticosteroids that have caused contact allergy/allergic contact dermatitis are shown in table 2.25. The corticosteroids that most frequently cause allergic reactions belong to the hydrocortisone – prednisolone – tixocortol pivalate group (table 2.27, Group 1). These are (mostly) not halogenated and do not have a methyl group at the C_{16} position of the corticosteroid molecule. Far less allergenic are corticosteroids that are halogenated and have a methyl group at the C_{16} position. To this group (table 2.27, Group 3) belong *inter alia* betamethasone (esters), dexamethasone (esters), clobetasol propionate, clobetasone butyrate, desoximetasone, flumethasone pivalate, fluocortolone esters and mometasone furoate (1,2).

Table 2.25 Corticosteroids that have caused contact allergy/allergic contact dermatitis

Alclometasone dipropionate	Fluclorolone acetonide	Isoflupredone acetate
Amcinonide	Fludrocortisone acetate	Mazipredone
Beclomethasone dipropionate	Flumethasone pivalate	Medrysone
Betamethasone	Flunisolide	Methylprednisolone
Betamethasone dipropionate	Fluocinolone acetonide	Methylprednisolone aceponate
Betamethasone sodium	Fluocinonide	Methylprednisolone acetate
phosphate	Fluocortolone	Methylprednisolone
Betamethasone valerate	Fluocortolone caproate	hemisuccinate
Budesonide	Fluocortolone pivalate	Mometasone furoate
Ciclesonide	Fluorometholone	Prednicarbate
Clobetasol propionate	6α-Fluoroprednisolone-21-	Prednisolone
Clobetasone butyrate	acetate	Prednisolone acetate
Clocortolone pivalate	Fluprednidene acetate	Prednisolone caproate
Cloprednol	Fluprednisolone	Prednisolone hemisuccinate
Cortisone acetate	Flurandrenolide	Prednisolone pivalate
Desonide	Fluticasone propionate	Prednisolone sodium metazoate
Desoximetasone	Halcinonide	Prednisolone valerate acetate
Dexamethasone	Halometasone	Prednisone
Dexamethasone acetate	Hydrocortisone	Procinonide
Dexamethasone phosphate	Hydrocortisone aceponate	Tixocortol pivalate
Dexamethasone sodium	Hydrocortisone acetate	Triamcinolone
phosphate	Hydrocortisone butyrate	Triamcinolone acetonide
Diflorasone diacetate	Hydrocortisone hemisuccinate	Triamcinolone diacetate
Diflucortolone valerate	Hydrocortisone probutate	Triamcinolone hexacetonide
Difluprednate	Hydrocortisone valerate	

2.7 DIAGNOSTIC PROCEDURES IN SUSPECTED CORTICOSTEROID ALLERGY

Allergic contact dermatitis from corticosteroids (CSs) should be suspected when a chronic dermatitis is exacerbated during corticosteroid treatment, or when it does not respond to therapy. Most screening series including the European baseline series (28) contain tixocortol pivalate and budesonide as 'markers' for corticosteroid allergy, sometimes supplemented with hydrocortisone butyrate and other corticosteroids. These will probably detect a large portion of all corticosteroid allergies (29), but also fail to detect many (30).

All commercial preparations used by the patients should also be patch tested, and preferably the corticosteroids themselves as well, as many reactions are not due to the corticosteroid, but to another active ingredient (e.g. neomycin) or an excipient (e.g. propylene glycol, lanolin alcohol, cetearyl alcohol) (105). A large number of corticosteroids is commercially available for patch testing (table 2.26). These (or a selection) can be tested in a second patch test session, when a patient has reacted to one of the markers for corticosteroid allergy and/or to CSs used by him or her, both to identify other allergies and to find corticosteroids which can safely be used. Most of the commercial test preparations have petrolatum as vehicle. This works well for budesonide (Chapter 3.49) and tixocortol pivalate (Chapter 3.342), but most other corticosteroids can better be tested in alcohol 70% in a concentration of 0.1-1% (1,2). The lower concentration of 0.1% (and sometimes even lower) may often perform better, as the anti-inflammatory action of the molecule, which suppresses the allergic reaction, is less pronounced. It is very important to test alcohol 70% as a control, as it not infrequently causes false-positive, irritant, reactions (86) and occasionally allergic reactions.

Intradermal testing may reveal additional sensitizations not picked up by patch tests (not in all studies [75]), especially in the case of hydrocortisone (Chapter 3.169). However, especially with the stronger corticosteroids, there is a risk of skin atrophy (1,2,10) and, in the case of hydrocortisone, these allergies are almost always picked up by the tixocortol pivalate marker in the baseline series. Many patch tests become positive only after D3-D4 (due to the anti-inflammatory effect of the molecule) and later readings (D6-D10) are imperative to avoid missing late positive patch tests (1,2,10,31,37,76,77). Unique for patch testing with corticosteroids is the 'edge effect': at the first reading, there is often infiltration and erythema only on and around the edge of the patch-test chamber with a completely clear center. Later, this center also becomes eczematous as well (5,32,40,76,103). At the same time, the anti-inflammatory action of corticosteroids may also occasionally result in false-negative patch test reactions.

Table 2.26 Corticosteroids commercially available for patch testing [a]

Patch test allergen (hapten)		Chemotech	SPCanada	SPEurope
Alclometasone dipropionate		1.0%		
Amcinonide			0.1%	0.1%
Betamethasone dipropionate		1.0%	0.5%; 0.1% alc.	
Betamethasone valerate		1.0%	0.12%; 1%; 0.1% alc.	0.12%
Budesonide		0.01%; 0.1%	0.01%; 0.1%	0.1%
Clobetasol propionate		1.0%	0.25%; 1%; 0.1% alc.	0.25%; 1%
Desoximetasone		1.0%	1%	
Dexamethasone			0.5%	
Dexamethasone phosphate			1%	
Dexamethasone sodium phosphate		1.0%		1%
Hydrocortisone			1%	1%
Hydrocortisone acetate		1.0%	1%	
Hydrocortisone butyrate		1.0% alc.; 1.0%	0.1%; 1%: 1% alc.	0.1%
Methylprednisolone aceponate		1.0%	0.1% alc.	
Mometasone furoate			0.1% alc.	
Prednicarbate			1% alc.	
Prednisolone			1%	1%
Tixocortol pivalate		0.1%; 1.0%	0.1%; 1%	1%
Triamcinolone acetonide		1.0%	0.1%; 1%	1%
Corticosteroid mix		2.1%		
- hydrocortisone butyrate	1.0%			
- tixocortol pivalate	1.0%			
- budesonide	0.1%			
Corticosteroid mix			2.01%	
- hydrocortisone acetate	1%			
- hydrocortisone butyrate	1%			
- budesonide	0.01%			

[a] The vehicle for all haptens is petrolatum, unless otherwise indicated; alc.: alcohol; Chemotech: Chemotechnique Diagnostics (www.chemotechnique.se); SPCanada: SmartPractice Canada (www.smartpracticecanada.com); SPEurope: SmartPractice Europe (www.smartpracticeeurope.com);

2.8 CROSS-REACTIVITY BETWEEN CORTICOSTEROIDS

Multiple reactions to corticosteroids are frequent, mostly from cross-reactions. The allergens are probably not the corticosteroids themselves, but a byproduct from their skin metabolism. The principal metabolites, steroid glyoxals or 21-dehydrocorticosteroids (aldehydes), are the most probably haptens (1,2,9). Based on results of patch testing, molecular modelling and previous work (3-7), corticosteroids have been divided into three groups (table 2.27). Group 1 consists of (mostly) non-methylated, non-halogenated molecules. Group 2 are (mostly) halogenated molecules with a C_{16}/C_{17} cis-ketal/diol structure. Group 3 consists of halogenated and C_{16}-methylated molecules.

C_{16}-methyl substitution and halogenation seem to reduce the allergenicity of corticosteroid molecules (9). Indeed, by far most allergic reactions are caused by the corticosteroids in group 1, whereas the molecules in group 3 rarely sensitize. As to the cross-reaction pattern: patients sensitized from one or more corticosteroids in group 1 often cross-react with other chemicals in group 1. However, another cross-reactivity profile is that patients may be sensitized to steroids in group 1 with cross-reactions not only in group 1, but also to group 2, group 3 or both. The classification presented in table 2.27 cannot explain all observed cross-reactions and often does not accurately predict cross-reactivity (11,12,13). Therefore, patients with positive patch test reactions to tixocortol pivalate and/or budesonide in the baseline series and/or to corticosteroids used by them, should also be tested with other corticosteroids to determine the cross-reactivity pattern and to establish which CSs can safely be used for continued treatment. Cross-reactivity can sometimes also be observed to endogenous steroidal sex hormones and derivatives including progesterone, 17-α-hydroxyprogesterone, and testosterone (1,2,89,90,150,151).

Table 2.27 Corticosteroid classification based on cross-reaction pattern (adapted from refs. 1 and 2)

GROUP 1	GROUP 2	GROUP 3
No C_{16}-methyl substitution	C_{16}/C_{17} cis-ketal or diol structure	C_{16}-methyl substitution
No halogen substitution in most cases	Halogen substitution	Halogen substitution
Budesonide (S-isomer)	Amcinonide	Alclomethasone dipropionate [c]
Cloprednol	(Budesonide, R isomer) [a]	Beclomethasone dipropionate
Cortisone acetate	Ciclesonide	Betamethasone
Dichlorisone acetate	Desonide [b]	Betamethasone 17-valerate
Difluprednate	Fluclorolone acetonide	Betamethasone dipropionate
Fludrocortisone acetate	Flumoxonide [d]	Betamethasone sodium phosphate
Fluorometholone	Flunisolide	Clobetasol propionate
Fluprednisolone acetate	Fluocinonide acetonide	Clobetasone butyrate
Hydrocortisone	Fluocinonide	Cortivazol
Hydrocortisone aceponate	Halcinonide [b]	Desoximetasone
Hydrocortisone acetate	Triamcinolone acetonide	Dexamethasone
Hydrocortisone-17-butyrate	Triamcinolone benetonide [d]	Dexamethasone acetate
Hydrocortisone-21-butyrate	Triamcinolone diacetate	Dexamethasone sodium phosphate
Hydrocortisone hemisuccinate	Triamcinolone hexacetonide	Diflucortolone valerate
Isoflupredone acetate		Diflorasone diacetate
Mazipredone		Flumethasone pivalate
Medrysone		Fluocortin butyl
Methylprednisolone aceponate		Fluocortolone
Methylprednisolone acetate		Fluocortolone caproate
Methylprednisolone hemisuccinate		Fluocortolone pivalate
Prednicarbate		Fluprednidene acetate
Prednisolone		Fluticasone propionate
Prednisolone caproate		Halometasone
Prednisolone hemisuccinate		Meprednisone
Prednisolone pivalate		Mometasone furoate
Prednisolone sodium metazoate		
Prednisone		
Tixocortol pivalate		
Triamcinolone		

[a] also included in group 2, as it may - in exceptional cases - cross-react with the acetonides; [b] No halogen substitution; [c] Unexpectedly in group 3, as alclomethasone dipropionate often co-reacts with group 1; [d] Not used in pharmaceuticals

Review articles
Several review articles published after 2000 have discussed delayed-type allergy to corticosteroids (5,11,12,13,18, 29,32,63,73,83,147). Earlier literature reviews on the subject are provided in refs. 36,39,43,66,67,68,69,70,74,98, 102, and 107.

LITERATURE

1   Baeck M, Goossens A. Immediate and delayed allergic hypersensitivity to corticosteroids: practical guidelines. Contact Dermatitis 2012;66:38-45

2   Goossens A, Gonçalo M. Topical drugs. In: Johansen J, Mahler V, Lepoittevin JP, Frosch P, Eds. Contact Dermatitis, 6th Edition. Springer: Cham, 2020

3   Coopman S, Degreef H, Dooms-Goossens A. Identification of cross-reaction patterns in allergic contact dermatitis from topical corticosteroids. Br J Dermatol 1989;121:27-34

4   Lepoittevin JP, Drieghe J, Dooms-Goossens A. Studies in patients with corticosteroid contact allergy: understanding cross-reactivity among different steroids. Arch Dermatol 1995;131:31-37

5   Matura M, Goossens A. Contact allergy to corticosteroids. Allergy 2000;55:698-704

6   Wilkinson SM. Corticosteroid cross-reactions: an alternative view. Contact Dermatitis 2000;42:59-63

7   Baeck M, Chemelle JA, Goossens A, Nicolas JF, Terreux R. Corticosteroid cross-reactivity: clinical and molecular modelling tools. Allergy 2011;66:1367-1374

8   Zorginstituut Nederland. Farmacotherapeutisch Kompas. www.farmacotherapeutischkompas.nl

9   Baeck M, Chemelle JA, Rasse C, Terreux R, Goossens A. C(16)-methyl corticosteroids are far less allergenic than the non-methylated molecules. Contact Dermatitis 2011;64:305-312

10   Soria A, Baeck M, Goossens A, Marot L, Duveille V, Derouaux AS, et al. Patch, prick or intradermal tests to detect delayed hypersensitivity to corticosteroids? Contact Dermatitis 2011;64:313-324

11   Vatti RR, Ali F, Teuber S, Chang C, Gershwin ME. Hypersensitivity reactions to corticosteroids. Clin Rev Allergy Immunol 2014;47:26-37

12   Otani IM, Banerji A. Immediate and delayed hypersensitivity reactions to corticosteroids: Evaluation and management. Curr Allergy Asthma Rep 2016 Mar;16(3):18

13   Nguyen H, Yiannias J. Contact dermatitis to medications and skin products. Clin Rev Allergy Immunol 2019;56:41-59

14   Opstrup MS, Garvey LH, Johansen JD, Bregnbak DK, Thyssen JP. A contact allergic reaction to budesonide mimicking immediate-type allergy. Contact Dermatitis 2017;77:62-63

15   Salava A, Alanko K, Hyry H. A case of systemic allergic dermatitis caused by inhaled budesonide: cross-reactivity in patch tests with the novel inhaled corticosteroid ciclesonide. Contact Dermatitis 2012;67:244-246

16   Pitsios C, Stefanaki EC, Helbling A. Type IV delayed-type hypersensitivity of the respiratory tract due to budesonide use: report of two cases and a literature review. Prim Care Respir J 2010;19:185-188

17   García AP, Tovar V, de Barrio M, Villanueva A, Tornero P. Contact allergy to inhaled budesonide. Contact Dermatitis 2008;59:60-61

18   Isaksson M. Skin reactions to inhaled corticosteroids. Drug Saf 2001;24:369-373

19   Pirker C, Misić A, Frosch PJ. Angioedema and dysphagia caused by contact allergy to inhaled budesonide. Contact Dermatitis 2003;49:77-79

20   Isaksson M, Bruze M. Allergic contact dermatitis in response to budesonide reactivated by inhalation of the allergen. J Am Acad Dermatol 2002;46:880-885

21   Isaksson M, Bruze M. Allergic contact dermatitis to budesonide reactivated by inhalation of the allergen. Am J Contact Dermat 2001;12:130 (Abstract)

22   Goossens A, Huygens S, Matura M, Degreef H. Fluticasone propionate: a rare contact sensitizer. Eur J Dermatol 2001;11:29-34

23   Isaksson M, Bruze M, Wihl JA. Contact allergy to budesonide and perforation of the nasal septum. Contact Dermatitis 1997;37:133

24   Iglesias-Cadarso A, Diaz C, Laguna JJ, Hernandez-Weigand P. Allergic contact dermatomucositis to budesonide. J Allergy Clin Immunol 1994;94(3Pt.1):559-560

25   Peris-Tortajada A, Giner A, Perez C, Hernandez D, Basomba A. Contact allergy to topical budesonide. J Allergy Clin Immunol 1991;87:597-598

26   Jerez J, Rodríguez F, Garcés M, Martín-Gil D, Jiménez I, Antón E, Duque S. Allergic contact dermatitis from budesonide. Contact Dermatitis 1990;22:231-232

27   Al-Shaikhly T, Rosenthal JA, Chau AS, Ayars AG, Rampur L. Systemic contact dermatitis to inhaled and intranasal corticosteroids. Ann Allergy Asthma Immunol 2020;125:103-105

28   Wilkinson M, Gonçalo M, Aerts O, Badulici S, Bennike NH, Bruynzeel D, et al. The European baseline series and recommended additions: 2019. Contact Dermatitis 2019;80:1-4

29   Baeck M, Marot L, Nicolas J-F, Pilette C, Tennstedt D, Goossens A. Allergic hypersensitivity to topical and systemic corticosteroids: a review. Allergy 2009;64:978-994

30   Mercader-García P, Pastor-Nieto MA, García-Doval I, Giménez-Arnau A, González-Pérez R, Fernández-Redondo V, et al. Are the Spanish baseline series markers sufficient to detect contact allergy to corticosteroids in Spain ? A GEIDAC prospective study. Contact Dermatitis 2017;78:76-82

31   Isaksson M. Corticosteroid contact allergy – the importance of late readings and testing with corticosteroids used by the patients. Contact Dermatitis 2007;56:56-57

32   Browne F, Wilkinson S. Effective prescribing in steroid allergy: controversies and cross-reactions. Clin Dermatol 2011;29:287-294

33   Pratt MD, Mufti A, Lipson J, Warshaw EM, Maibach HI, Taylor JS, et al. Patch test reactions to corticosteroids: Retrospective analysis from the North American Contact Dermatitis Group 2007-2014. Dermatitis 2017;28:58-63

34   Burden AD, Beck MH. Contact hypersensitivity to topical corticosteroids. Br J Dermatol 1992;127:497-500

35   Bircher AJ. Short induction phase of contact allergy to tixocortol pivalate in a nasal spray. Contact Dermatitis 1990;22:237-238

36   Boujnah-Khouadja A, Brändle I, Reuter G, Foussereau J. Allergy to 2 new corticoid molecules. Contact Dermatitis 1984;11:83-87

37 Isaksson M, Andersen K E, Brandão FM, Goossens A. Patch testing with corticosteroid mixes in Europe. Contact Dermatitis 2000;42:27-35

38 Wilkinson SM, Cartwright PH, English JSC. Hydrocortisone: an important cutaneous allergen. Lancet 1991;337:761-762

39 Dooms-Goossens A, Vanhee J, Vanderheyden D, Gevers D, Willems L, Degreef H. Allergic contact dermatitis to topical corticosteroids: clobetasol propionate and clobetasone butyrate. Contact Dermatitis 1983;9:470-478

40 Bjarnason B, Flosadóttir E, Fischer T. Reactivity at edges of corticosteroid patch tests may be an indicator of a strong positive test response. Dermatology 1999;199:130-134

41 Rodríguez-Serna M, Silvestre JF, Quecedo E, Martinez A, Miguel FJ, Gauchía R. Corticosteroid allergy: report of 3 unusually acute cases. Contact Dermatitis 1996;35:361-362

42 Miranda-Romero A, Sanchez-Sambucety P, Bajo C, Martinez M, Garcia-Munoz M. Genital oedema from contact allergy to prednicarbate. Contact Dermatitis 1998;38:228-229

43 Whitmore SE. Delayed systemic allergic reactions to corticosteroids. Contact Dermatitis 1995;32:193-198

44 Stingeni L, Hansel K, Lisi P. Morbilliform erythema-multiforme-like eruption from desoxymethasone. Contact Dermatitis 1996;35:363-364

45 Chavarria Mur E, Gonzalez-Carrascosa Ballesteros M, Suarez Fernandez R, Bueno Marco C. Generalized exanthematous reaction with pustulosis induced by topical corticosteroids. Contact Dermatitis 2005;52:114-115

46 Broesby-Olsen S, Clemmensen O, Andersen KE. Allergic contact dermatitis from a topical corticosteroid mimicking acute generalized exanthematous pustulosis. Acta Derm Venereol 2005;85:444-445

47 Sánchez-Pérez J, Gala SP, Jiménez YD, Fraga J, Diez AG. Allergic contact dermatitis to prednicarbate presenting as lupus erythematosus. Contact Dermatitis 2006;55:247-249

48 Smart DR, Powell DL. Erythema multiforme-like allergic contact reaction to topical triamcinolone. Dermatitis 2014;25:89-90

49 Valsecchi R, Reseghetti A, Leghissa P, Cologni L, Cortinovis R. Erythema-multiforme-like lesions from triamcinolone acetonide. Contact Dermatitis 1998;38:362-363

50 D'Erme AM, Gola M. Allergic contact dermatitis induced by topical hydrocortisone-17-butyrate mimicking papular rosacea. Dermatitis 2012;23:95-96

51 Baeck M, Goossens A. Patients with airborne sensitization/contact dermatitis from budesonide-containing aerosols 'by proxy'. Contact Dermatitis 2009;61:1-8

52 Teixeira V, Coutinho I, Gonçalo M. Budesonide allergic contact dermatitis "by proxy"? Dermatitis 2013;24:144-146

53 Corazza M, Baldo F, Osti F, Virgili A. Airborne allergic contact dermatitis due to budesonide from professional exposure. Contact Dermatitis 2008;59:318-319

54 O'Hagan AH, Corbett JR. Contact allergy to budesonide in a breath-actuated inhaler. Contact Dermatitis 1999;41:53

55 Raison-Peyron N, Co Minh HB, Vidal-Mazuy A, Guilhou JJ, Guillot B. Connubial contact dermatitis to an inhaled corticosteroid. Ann Dermatol Venereol 2005;132:143-146 (Article in French)

56 Pontén A. Airborne occupational contact dermatitis caused by extremely low concentrations of budesonide. Contact Dermatitis 2006;55:121-124

57 Berbegal L, DeLeon FJ, Silvestre JF. Corticosteroid hypersensitivity studies in a skin allergy clinic. Actas Dermosifiliogr 2015;106:816-822

58 Baeck M, Chemelle JA, Terreux R, Drieghe J, Goossens A. Delayed hypersensitivity to corticosteroids in a series of 315 patients: clinical data and patch test results. Contact Dermatitis 2009;61:163-175

59 Faber MA, Sabato V, Ebo DGD, Verheyden M, Lambert J, Aerts O. Systemic allergic dermatitis caused by prednisone derivatives in nose and eardrops. Contact Dermatitis 2015;73:317-320.

60 Goossens A, Gonçalo M. Topical drugs. In: Johansen J, Mahler V, Lepoittevin JP, Frosch P, Eds. Contact Dermatitis, 6th Edition. Springer: Cham, 2020

61 Winnicki M, Shear NH. A systematic approach to systemic contact dermatitis and symmetric drug-related intertriginous and flexural exanthema (SDRIFE). Am J Clin Dermatol 2011;12:171-180

62 Malik M, Tobin AM, Shanahan F, O'Morain C, Kirby B, Bourke J. Steroid allergy in patients with inflammatory bowel disease. Br J Dermatol 2007;157:967-969

63 Isaksson M. Corticosteroids. Dermatol Ther 2004;17:314-320

64 Corazza M, Mantovani L, Maranini C, Bacilieri S, Virgili A. Contact sensitization to corticosteroids: increased risk in long term dermatoses. Eur J Dermatol 2000;10:533-535

65 Lauerma AI. Occupational contact sensitization to corticosteroids. Contact Dermatitis 1998;39:328-329

66 Wilkinson SM. Hypersensitivity from topical corticosteroids. Clin Exp Dermatol 1994;19:1-11

67 Lauerma AI. Contact hypersensitivity to glucocorticosteroids. Am J Contact Dermat 1992;3:112-132

68 Preuss L. Allergic reactions to systemic glucocorticoids: a review. Ann Allergy 1985;55:772-775

69   Guin JD. Contact sensitivity to topical corticosteroids. J Am Acad Dermatol 1984;10(5Pt.1):773-782

70   Tegner E. Contact allergy to corticosteroids. Int J Dermat 1976;15:520-523

71   Vind-Kezunovic D, Johansen JD, Carlsen BC. Prevalence of and factors influencing sensitization to corticosteroids in a Danish patch test population. Contact Dermatitis 2011;64:325-329

72   Baeck M, Pilette C, Drieghe J, Goossens A. Allergic contact dermatitis to inhalation corticosteroids. Eur J Dermatol 2010;20:102-108

73   Foti C, Calogiuri G, Cassano N, Buquicchio R, Vena GA. Contact allergy to topical corticosteroids: update and review on cross-sensitization. Recent Pat Inflamm Allergy Drug Discov 2009;3:33-39

74   Uter W. Allergische reaktionen auf glukokortikoide. Derm Beruf Umwelt 1990;38:75-90

75   Mimesh S, Pratt M. Allergic contact dermatitis from corticosteroids: reproducibility of patch testing and correlation with intradermal testing. Dermatitis 2006;17:137-142

76   Isaksson M, Bruze M. Corticosteroids. Dermatitis 2005;16:3-5

77   Subramanian S, Jerajani HR, Chowkekar S, Karkhanis A. Delayed patch positivity to corticosteroids in chronic eczema. Contact Dermatitis 2006;54:227-228

78   Reduta T, Laudanska H. Contact hypersensitivity to topical corticosteroids – frequency of positive reactions in patch-tested patients with allergic contact dermatitis. Contact Dermatitis 2005;52:109-110

79   Burckhardt W. Kontaktekzem durch Hydrocortison. Hautarzt 1959;10:42-43 (Article in German)

80   Kooij R. Hypersensitivity to hydrocortisone. Br J Dermatol 1959;71:392-394

81   Gönül M, Gül U. Detection of contact hypersensitivity to corticosteroids in allergic contact dermatitis patients who do not respond to topical corticosteroids. Contact Dermatitis 2005;53:67-70

82   Keegel T, Saunders H, Milne R, Sajjachareonpong P, Fletcher A, Nixon R. Topical corticosteroid allergy in an urban Australian centre. Contact Dermatitis 2004;50:6-14

83   Scheuer E, Warshaw E. Allergy to corticosteroids: Update and review of epidemiology, clinical characteristics, and structural cross-reactivity. Am J Contact Dermat 2003;14:179-187

84   Kilpio K, Hannuksela M. Corticosteroid allergy in asthma. Allergy 2003;58:1131-1135

85   Wilkinson SM, Higgins B, Owen S, Mattey D, Woodcock A. An assessment of steroid hypersensitivity in asthma. Respir Med 1997;91:231-233

86   Devos SA, Van der Valk PG. Relevance and reproducibility of patch-test reactions to corticosteroids. Contact Dermatitis 2001;44:362-365

87   Weber F, Barbaud A, Reichert-Penetrat S, Danchin A, Schmutz JL. Unusual clinical presentation in a case of contact dermatitis due to corticosteroids diagnosed by ROAT. Contact Dermatitis 2001;44:105-106

88   Fuchs T, Uter W, Sprotte U. Generalisierte Urtikaria durch Budesonid -verzögerte, IgE-vermittelte Sofortreaktion? Allergologie 1991;14:234-238 (Article in German)

89   Wilkinson SM, Beck MH. The significance of positive patch tests to 17-hydroxyprogesterone. Contact Dermatitis 1994;30:302-303

90   Schoenmakers A, Vermorken A, DeGreef H, Dooms-Goossens A. Corticosteroid or steroid allergy? Contact Dermatitis 1992;26:159-162

91   Thomson KF, Wilkinson SM, Powell S, Beck MH. The prevalence of corticosteroid allergy in two U.K. centres: prescribing implications. Br J Dermatol 1999;141:863-866

92   Isaksson M, Bruze M, Hörnblad Y, Svenonius E, Wihl JA. Contact allergy to corticosteroids in asthma/rhinitis patients. Contact Dermatitis 1999;40:327-328

93   Lyon CC, Beck MH. Allergic contact dermatitis reactions to corticosteroids in periorbital inflammation and conjunctivitis. Eye (Lond) 1998;12(Pt.1):148-149

94   Matura M. Corticosteroid contact allergy in Hungary. Contact Dermatitis 1998;38:225-226

95   Häusermann P, Harr T, Bircher AJ. Baboon syndrome resulting from systemic drugs: Is there strife between SDRIFE and allergic contact dermatitis syndrome? Contact Dermatitis 2004;51:297-310

96   Bircher AJ, Pelloni F, Langauer Messmer S, Müller D. Delayed hypersensitivity reactions to corticosteroids applied to mucous membranes. Br J Dermatol 1996;135:310-313

97   Dooms-Goossens A, Andersen KE, Brandão FM, Bruynzeel D, Burrows D, Camarasa J, et al. Corticosteroid contact allergy: an EECDRG multicentre study. Contact Dermatitis 1996;35:40-44

98   Dooms-Goossens A. Allergy to inhaled corticosteroids: A review. Am J Contact Dermat 1995;6:1-3

99   Boffa MJ, Wilkinson SM, Beck MH. Screening for corticosteroid contact hypersensitivity. Contact Dermatitis 1995;33:149-151

100 Dooms- Goossens A, Degreef H. Clinical aspects of contact allergy to corticosteroids. Dermatology 1994;189:54-55

101 Dooms-Goossens A, Meinardi MM, Bos JD, Degreef H. Contact allergy to corticosteroids: the results of a two-centre study. Br J Dermatol 1994;130:42-47

102 Lauerma AI, Reitamo S. Contact allergy to corticosteroids. J Am Acad Dermatol 1993;28:618-622

103 Dooms-Goossens A. Corticosteroid contact allergy: A challenge to patch testing. Am J Contact Dermat 1993;4:120-122

104 Dooms-Goossens A, Morren M. Results of routine patch testing with corticosteroid series in 2073 patients. Contact Dermatitis 1992;26:182-191

105 Sasaki E. Corticosteroid sensitivity and cross-sensitivity. A review of 18 cases 1967-1988. Contact Dermatitis 1990;23:306-315

106 Dooms-Goossens AE, Degreef HJ, Marien KJC, Coopman SA. Contact allergy to corticosteroids: a frequently missed diagnosis? J Am Acad Dermatol 1989;21:538-543

107 Rivara G, Tomb RR, Foussereau J. Allergic contact dermatitis from topical corticosteroids. Contact Dermatitis 1989;21:83-91

108 DeKoven JG, Warshaw EM, Zug KA, Maibach HI, Belsito DV, Sasseville D, et al. North American Contact Dermatitis Group patch test results: 2015-2016. Dermatitis 2018;29:297-309

109 Martin S, Rustemeyer T, Thyssen J. Recent advances in understanding and managing contact dermatitis. F1000Res 2018 Jun 20;7:F1000 Faculty Rev-810. doi: 10.12688/f1000research.13499.1

110 Green C, Holden C, Gawkrodger D. Contact allergy to topical medicaments becomes more common with advancing age: an age-stratified study. Contact Dermatitis 2007;56:229-231

111 Erfurt-Berge C, Geier J, Mahler V. The current spectrum of contact sensitization in patients with chronic leg ulcers or stasis dermatitis – new data from the Information Network of Departments of Dermatology (IVDK). Contact Dermatitis 2017;77:151-158

112 Erfurt-Berge C, Mahler V. Contact Sensitization in patients with lower leg dermatitis, chronic venous insufficiency, and/or chronic leg ulcer: Assessment of the clinical relevance of contact allergens. J Investig Allergol Clin Immunol 2017;27:378-380

113 D'Erme A, Iannone M, Dini V, Romanelli M. Contact dermatitis in patients with chronic leg ulcers: a common and neglected problem: a review 2000-2015. J Wound Care 2016;25(S9):S23-S29

114 Faria A, Gonçalo S, Gonçalo M, Freitas C, Baptista PP. Allergic contact dermatitis from tioconazole. Contact Dermatitis 1996;35:250-252

115 Kalavala M, Statham BN. Allergic contact dermatitis from timolol and dorzolamide eye drops. Contact Dermatitis 2006;54:345

116 De Groot AC, van Ginkel CJ, Bruynzeel DP, Smeenk G, Conemans JM. Contact allergy to eyedrops containing beta-blockers. Ned Tijdschr Geneeskd 1998;142:1034-1036 (Article in Dutch)

117 Koch P. Allergic contact dermatitis due to timolol and levobunolol in eyedrops, with no cross-sensitivity to other ophthalmic beta-blockers. Contact Dermatitis 1995;33:140-141

118 O'Donnell BF, Foulds IS. Contact allergy to beta-blocking agents in ophthalmic preparations. Contact Dermatitis 1993;28:121-122

119 De Groot AC, Conemans J. Contact allergy to metipranolol. Contact Dermatitis 1988;18:107-108

120 Corazza M, Virgili A, Mantovani L, Masieri LT. Allergic contact dermatitis from cross-reacting beta-blocking agents. Contact Dermatitis 1993;28:188-189

121 Gailhofer G, Ludvan M. 'Beta-blockers': sensitizers in periorbital allergic contact dermatitis. Contact Dermatitis 1990;23:262

122 Frosch PJ, Weickel R, Schmitt T, Krastel H. Side effects of external ophthalmologic drugs. Z Hautkr 1988;63:126, 129-132, 135-136 (Article in German)

123 Gonzalo-Garijo MA, Zambonino MA, Pérez-Calderón R, Pérez-Rangel I, Sánchez-Vega S. Allergic contact dermatitis due to carteolol, with good tolerance to betaxolol. Dermatitis 2011;22:232-233

124 Wilkinson SM. False-negative patch test with levobunolol. Contact Dermatitis 2001;44:264

125 Statham BN. Failure of patch testing with levobunolol eyedrops to detect contact allergy. Contact Dermatitis 2000;43:365-366

126 Sánchez-Pérez J, Jesús Del Río M, Fernández-Villalta MJ, García-Díez A. Positive use test in contact dermatitis from betaxolol hydrochloride. Contact Dermatitis 2002;46:313-314

127 Corazza M, Virgili A. Allergic contact dermatitis from ophthalmic products: can pre-treatment with sodium lauryl sulfate increase patch test sensitivity? Contact Dermatitis 2005;52:239-241

128 Corazza M, Levratti A, Zampino MR, Virgili A. Conventional patch tests are poor detectors of contact allergy from ophthalmic products. Contact Dermatitis 2002;46:298-299

129 Hashimoto Y, Aragane Y, Kawada A. Allergic contact dermatitis due to levobunolol in an ophthalmic preparation. J Dermatol 2006;33:507-509

130 Vandebuerie L, Kerre S. Allergic contact dermatitis due to betablocker agents in eye drops: relevance of patch testing versus open use testing (ROAT). Ned Tijdschr Derm Venereol 2011;21:328-330 (Article in Dutch)

131 Uter W, Geier J, Pfahlberg A, Effendy I. The spectrum of contact allergy in elderly patients with and without lower leg dermatitis. Dermatology 2002;204:266-272

132 Carrière M, Giordano-Labadie F, Schwarze HP, Loche F, Bazex J. Difficulties in the interpretation of patch test reactions to ophthalmic beta-blockers. Contact Dermatitis 1998;39:319-320.

133 Goossens A. Allergic contact dermatitis from the vehicle components of topical pharmaceutical products. Immunol Allergy Clin North Am 2014:34: 663-670

134 Wolf R, Tüzün Y. Baboon syndrome and toxic erythema of chemotherapy: Fold (intertriginous) dermatoses. Clin Dermatol 2015;33:462-465

135 Aquino M, Rosner G. Systemic contact dermatitis. Clin Rev Allergy Immunol 2019;56:9-18

136 Andersen KE, Hjorth N, Menné T. The baboon syndrome: systemically-induced allergic contact dermatitis. Contact Dermatitis 1984;10:97-100

137 Menné T, Veien N, Sjølin K-E, Maibach HI. Systemic contact dermatitis. Am J Contact Dermat 1994;5:1-12

138 Farage MA. Vulvar susceptibility to contact irritants and allergens: a review. Arch Gynecol Obstet 2005;272:167-172

139 Grey KR, Warshaw EM. Allergic contact dermatitis to ophthalmic medications: Relevant allergens and alternative testing methods. Dermatitis 2016;27:333-347

140 Mughal AA, Kalavala M. Contact dermatitis to ophthalmic solutions. Clin Exp Dermatol 2012;37:593-597

141 Thyssen JP, Maibach HI. Drug-elicited systemic allergic (contact) dermatitis--update and possible pathomechanisms. Contact Dermatitis 2008;59:195-202

142 Kulberg A, Schliemann S, Elsner P. Contact dermatitis as a systemic disease. Clin Dermatol 2014;32:414-419

143 Gonçalo M, Ferguson J, Bonevalle A, Bruynzeel DP, Giménez-Arnau A, Goossens A, et al. Photopatch testing: Recommendations for a European photopatch test baseline series. Contact Dermatitis 2013;68:239-243

144 Gonçalo M. Photopatch testing. In: Johansen J, Mahler V, Lepoittevin JP, Frosch P, Eds. Contact Dermatitis, 6th Edition. Springer: Cham, 2020

145 Diepgen TL, Ofenloch RF, Bruze M, Bertuccio P, Cazzaniga S, Coenraads P-J, et al. Prevalence of contact allergy in the general population in different European regions. Br J Dermatol 2016;174:319-329

146 Barbaud A, Waton J. Systemic allergy to corticosteroids: Clinical features and cross reactivity. Curr Pharm Des 2016;22:6825-6831

147 Nettis E, Colanardi MC, Calogiuri GF, Ferrannini A, Vacca A, Tursi A. Allergic reactions to inhalant glucocorticosteroids: a hot topic for pneumologists and allergologists. Immunopharmacol Immunotoxicol 2006;28:511-534

148 The European Multicentre Photopatch Test Study (EMCPPTS) Taskforce. A European multicentre photopatch test study. Br J Dermatol 2012;166:1002-1009

149 Veien N. Systemic contact dermatitis. In: Johansen J, Mahler V, Lepoittevin JP, Frosch P, Eds. Contact Dermatitis, 6th Edition. Springer: Cham, 2020

150 Ingber A, Trattner A, David M. Hypersensitivity to an oestrogen-progesterone preparation and possible relationship to autoimmune progesterone dermatitis and corticosteroid hypersensitivity. J Derm Treat 1999;10:139-140

151 Lamb SR, Wilkinson SM. Contact allergy to progesterone and estradiol in a patient with multiple corticosteroid allergies. Dermatitis 2004;15:78-81

152 Romano A, Di Fonso M, Mormile F, Quaratino D, Giuffreda F, Venuti A. Accelerated cell-mediated broncho-obstructive reaction to inhaled stepronin: a case report. Allergy 1996;51:269-271

153 Goon A, Goh C-L. Non-eczematous contact reactions. In: Johansen J, Mahler V, Lepoittevin JP, Frosch P, Eds. Contact Dermatitis, 6th Edition. Springer: Cham, 2020

154 Pirilä V, Rouhunkoski S. On sensitivity to neomycin and bacitracin. Acta Dermato-Venereologica 1959;39:470-476

155 Pirilä V, Förström L, Rouhunkosky S. Twelve years of sensitization to neomycin in Finland. Report of 1760 cases of sensitivity to neomycin and-or bacitracin. Acta Derm Venereol 1967;47:419-425

156 Madan RK, Levitt J. A review of toxicity from topical salicylic acid preparations. J Am Acad Dermatol 2014;70:788-792

Chapter 3 MONOGRAPHS OF TOPICAL DRUGS THAT HAVE CAUSED CONTACT ALLERGY/ALLERGIC CONTACT DERMATITIS

3.1 INTRODUCTION

In this chapter, monographs of 368 topical drugs and a monograph 'Miscellaneous' with 16 topical pharmaceuticals that have caused contact allergy/allergic contact dermatitis are presented. They have a standardized format, which is explained and detailed in Chapter 1.2. The topical drugs discussed here are shown in table 3.1.1. An overview of various aspects of allergic contact dermatitis from topical pharmaceuticals is provided in Chapter 2, discussing the spectrum of topical drugs, the prevalence of contact allergic reactions, predisposing factors, clinical picture of allergic contact dermatitis (classic and atypical manifestations), systemic allergic reactions from absorption of the drug through the skin and mucous membranes, pharmaceutical classes, the main sensitizers, diagnostic procedures and the commercial availability of topical drugs for patch testing. Corticosteroids are discussed in a separate section because of their quantitative importance and special and specific features.

Allergic reactions to topical pharmaceutical products can be caused by the active drug, but also by another constituent. Examples include vehicle materials (lanolin alcohol, cetearyl alcohol, propylene glycol), preservatives (thimerosal, parabens, benzalkonium chloride), antioxidants (sodium metabisulfite), fragrances and essential oils. This topic is discussed in Chapter 4. Contact urticaria (immediate contact reactions) from topical drugs is shortly summarized in Chapter 5. A preview of allergic reactions to *systemic* drugs, which will be the subject of Volume 4 of the *Monographs in Contact Allergy* series, can be found in Chapter 6.

Table 3.1.1 Topical drugs presented in monographs in this chapter

Aceclidine	Betamethasone sodium phosphate
Aceclofenac	Betamethasone valerate
Acetarsone	Betaxolol
Acetazolamide	Bibrocathol
Acexamic acid	Bifonazole
Acyclovir	Bimatoprost
Adapalene	Bismuth subnitrate
Alcaftadine	Bismuth tribromophenate
Alclometasone dipropionate	Brimonidine
Aluminum acetate	Bromelains
Ambroxol	Budesonide
Amcinonide	Bufexamac
Aminocaproic acid	Buprenorphine
Aminolevulinic acid	Butamben
3-(Aminomethyl)pyridyl salicylate	Caine mix III
Amlexanox	Calcipotriol
Amorolfine	Canrenone
Amylocaine	Carbarsone
Antazoline	Carmustine
Anthrarobin	Carteolol
Apraclonidine	Chloral hydrate
Arbekacin	Chloramphenicol
Atropine	Chlordantoin
Azidamfenicol	Chlorphenesin
Azithromycin	Chlorpheniramine
Bacitracin	Chlorphenoxamine
Beclomethasone dipropionate	Chlorproethazine
Befunolol	Chlorquinaldol
Belladonna extract	Chlortetracycline
Bendazac	Ciclesonide
Benzarone	Ciclopirox olamine
Benzocaine	Clemizole
Benzoyl peroxide	Clindamycin
Benzydamine	Clioquinol
Betamethasone	Clobetasol propionate
Betamethasone dipropionate	Clobetasone butyrate

Table 3.1.1 Topical drugs presented in monographs in this chapter (continued)

Clocortolone pivalate	Eucaine
Clonidine	Felbinac
Cloprednol	Fenoterol
Clotrimazole	Fenticonazole
Cloxyquin	Fepradinol
Coal tar	Feprazone
Colistimethate sodium	Fluclorolone acetonide
Colistin	Fludrocortisone acetate
Collagenase clostridium histolyticum	Flufenamic acid
Cortisone acetate	Flumethasone pivalate
Croconazole	Flunisolide
Cromoglicic acid	Fluocinolone acetonide
Crotamiton	Fluocinonide
Cyclobenzaprine	Fluocortolone
Cyclopentolate	Fluocortolone caproate
Cyproheptadine	Fluocortolone pivalate
Desonide	Fluorometholone
Desoximetasone	Fluorouracil
Dexamethasone	Fluprednidene acetate
Dexamethasone acetate	Fluprednisolone
Dexamethasone phosphate	Flurandrenolide
Dexamethasone sodium phosphate	Flurbiprofen
Dexchlorpheniramine	Fluticasone propionate
Dexketoprofen	Framycetin
Dibekacin	Furaltadone
Dibrompropamidine	Fusidic acid
Dibucaine	Gentamicin
Diclofenac	Glycol salicylate
Diethylstilbestrol	Gramicidin
Diflorasone diacetate	Halcinonide
Diflucortolone valerate	Halometasone
Difluprednate	Haloprogin
4,5-Dihydro-1*H*-imidazole monohydrochloride	Halquinol
Diltiazem	Hexamidine
Dimethindene	Hexamidine diisethionate
Dioxopromethazine	Hirudoid ® cream
Diperodon	Homatropine
Diphenhydramine	Hydrocortisone
Dipivefrin	Hydrocortisone aceponate
Dithranol	Hydrocortisone acetate
Dorzolamide	Hydrocortisone butyrate
Doxepin	Hydrocortisone hemisuccinate
Dyclonine	Hydrocortisone probutate
Echothiophate	Hydrocortisone valerate
Econazole	Hydroquinone
Efinaconazole	Ibacitabine
Enilconazole	Ibuprofen piconol
Enoxolone	Ibuproxam
Ephedrine	Ichthammol
Epinephrine	Idoxuridine
Erythromycin	Indomethacin
Estradiol	Interferons
Estradiol benzoate	Iodine
Estrogens, conjugated	Iodoform
Ethacridine	Iodoquinol
Etisazole	Iothion
Etofenamate	Isoconazole

Table 3.1.1 Topical drugs presented in monographs in this chapter (continued)

Isoflupredone acetate	Isothipendyl
Kanamycin	Olaflur
Ketoconazole	Oxiconazole
Ketoprofen	Oxybuprocaine
Ketorolac tromethamine	Oxyphenbutazone
Ketotifen	Oxyquinoline
Lactic acid	Oxytetracycline
Lanoconazole	Paromomycin
Latanoprost	Pecilocin
Levobunolol	Penicillamine
Levocabastine	Penicillins, unspecified
Lidocaine	Pheniramine
Lincomycin	Phenylbutazone
Luliconazole	Phenylephrine
Mabuprofen	Piketoprofen
Mafenide	Pilocarpine
Mazipredone	Pimecrolimus
Mechlorethamine	Piperazine
Medrysone	Pirenoxine
Mephenesin	Piroxicam
Meprylcaine	Piroxicam cinnamate
Mesulfen	Polidocanol
Methoxsalen	Polymyxin B
Methyl aminolevulinate	Povidone-iodine
Methyldiphenhydramine	Pramoxine
Methyl nicotinate	Prednicarbate
Methylphenidate	Prednisolone
Methylprednisolone	Prednisolone acetate
Methylprednisolone aceponate	Prednisolone caproate
Methylprednisolone acetate	Prednisolone hemisuccinate
Methylprednisolone hemisuccinate	Prednisolone pivalate
Methyl salicylate	Prednisolone sodium metazoate
Methyl salicylate 2-ethylbutyrate	Prednisolone valerate acetate
Metipranolol	Prednisone
Metoprolol	Prilocaine
Metronidazole	Procaine
Miconazole	Procinonide
Minoxidil	Proflavine
Miscellaneous topical drugs	Promestriene
Mitomycin	Promethazine
Mometasone furoate	Propanocaine
Monobenzone	Propantheline
Mupirocin	Proparacaine
Naftifine	Propipocaine
Naphazoline	Propranolol
Neomycin	Pyrazinobutanone
Neticonazole	Pyridoxine
Nicoboxil	Pyrilamine
Nicotine	Pyrithione zinc
Nifuratel	Pyrrolnitrin
Nifuroxime	Quinoline mix
Nifurprazine	Resorcinol
Nitrofurazone	Retapamulin
Nitroglycerin	Retinol
Nonoxynol-9	Retinyl palmitate
Norethisterone	Rifamycin
Nylidrin	Ripasudil
Nystatin	Rotigotine

Table 3.1.1 Topical drugs presented in monographs in this chapter (continued)

Rubidium iodide	Salicylic acid
Salicylamide	
Scopolamine	Tioconazole
Selenium sulfide	Tiopronin
Sertaconazole	Tioxolone
Sirolimus	Tixocortol pivalate
Silver nitrate	Tobramycin
Sisomicin	Tocopherol
Sodium fluoride	Tocopheryl acetate
Spaglumic acid	Tolazoline
Spironolactone	Tolciclate
Stannous fluoride	Tolnaftate
Stepronin	Travoprost
Sulbentine	Tretinoin
Sulconazole	Triafur
Sulfacetamide	Triamcinolone
Sulfadiazine silver	Triamcinolone acetonide
Sulfanilamide	Triamcinolone diacetate
Sulfiram	Triamcinolone hexacetonide
Suprofen	Tribenoside
Tacalcitol	Triethanolamine polypeptide oleate condensate
Tacrolimus	Trifluridine
Terconazole	Trimebutine
Testosterone	Tripelennamine
Tetracaine	Tromantadine
Tetrachlorodecaoxide	Tromethamine
Tetracycline	Tropicamide
Tetrahydrozoline	Troxerutin
Thiabendazole	Tyrothricin
Thiamine	Undecylenic acid
Thiocolchicoside	Vancomycin
Thioctic acid	Virginiamycin
Timolol	Xantocillin

Chapter 3.2   ACECLIDINE

IDENTIFICATION

Description/definition   : Aceclidine is a member of the quinuclidines (compounds containing a 1-azabicy-
                              clo[2.2.2]octane moiety) that conforms to the structural formula shown below
Pharmacological classes   : Miotics; parasympathomimetics
IUPAC name   : 1-Azabicyclo[2.2.2]octan-3-yl acetate
Other names   : 3-Acetoxyquinuclidine
CAS registry number   : 827-61-2
EC number   : 212-574-1
Patch testing   : No data available; generally speaking, the test concentration of ophthalmic medications
                              should higher than their concentration in the eye drops
Molecular formula   : $C_9H_{15}NO_2$

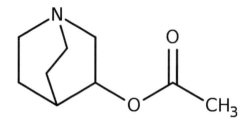

GENERAL

Aceclidine is a parasympathomimetic agent that is used in the treatment of open-angle glaucoma. In pharmaceutical products, aceclidine is employed as aceclidine hydrochloride (CAS number 6109-70-2, EC number 228-071-5, molecular formula $C_9H_{16}ClNO_2$) (1).

CONTACT ALLERGY

Case report

A 69-year old woman had bilateral periorbital redness, edema and severe pruritus for a few months. For glaucoma she used eye drops containing aceclidine hydrochloride. Earlier, she had been treated with latanoprost and dorzolamide containing topical ophthalmic medications. Patch testing gave a positive (D2 +++, D3 +++) reaction to the aceclidine eye drops after tape stripping. These eye drops are prepared from a dry substance and a solvent. The solvent contains benzalkonium chloride, boric acid and sodium tetraborate. The dry substance also contains boric acid and sodium tetraborate in addition to the active principle aceclidine hydrochloride. Both substances were patch tested (probably pure, no concentrations or vehicles mentioned) and there was a positive reaction to the dry substance only. It was concluded *per exclusionem* that the patient was allergic to aceclidine hydrochloride. Control tests were not performed. After the use of the eye drops was discontinued, the dermatitis quickly subsided (2).

LITERATURE

1   The data in the section 'General' may have been obtained from literature discussed in this chapter, but mostly also or exclusively from one or more of the following online sources: ChemIDPlus Advanced, PubChem, DrugBank, RxList, Drug Central, Drugs.com, and Wikipedia
2   Mayer K , Linse R, Harth W. Periorbitales Kontaktekzem durch das Cholinergicum Aceclidin. Akt Dermatol 2001;27:420-422 (Article in German)

Chapter 3.3   ACECLOFENAC

IDENTIFICATION
Description/definition   : Aceclofenac is the monocarboxylic acid that conforms to the structural formula shown
                             below
Pharmacological classes : Anti-inflammatory agents, non-steroidal
IUPAC name   : 2-[2-[2-(2,6-Dichloroanilino)phenyl]acetyl]oxyacetic acid
Other names   : 2-((2,6-Dichlorophenyl)amino)benzeneacetic acid carboxymethyl ester
CAS registry number   : 89796-99-6
EC number   : Not available
Merck Index monograph : 1293
Patch testing   : 10% pet.
Molecular formula   : $C_{16}H_{13}Cl_2NO_4$

GENERAL
Aceclofenac, the carboxymethyl ester of diclofenac, is a nonsteroidal anti-inflammatory drug (NSAID) with marked anti-inflammatory and analgesic properties. It is orally administered for the relief of pain and inflammation in osteoarthritis, rheumatoid arthritis and ankylosing spondylitis. Aceclofenac is also reported to be effective in other painful conditions such as dental and gynecological conditions (1). Aceclofenac is also widely used in topical form for acute soft trauma and inflammatory or degenerative musculoskeletal disorders (3). Aceclofenac is metabolized into diclofenac after systemic administration; possibly, hydrolysis also occurs in the skin (4).

CONTACT ALLERGY

Case reports and case series
In the period 1996-2001, in 2 hospitals in Spain, one patient was diagnosed with contact allergy and zero with photocontact allergy to aceclofenac. The accumulated incidence per million inhabitants (catchment population of the hospitals) of both side effects together was 1.2 (7).

A 71-year-old-man developed an acute dermatitis after local application of a cream containing 1.5% aceclofenac for 8 days. He had never used topical non-steroidal anti-inflammatory drugs and had not exposed himself to sunlight. Patch tests were positive to the cream and to aceclofenac 5% pet. (D2 ++, D4 ++) and to diclofenac 5% pet. (D2 and D4 +). Controls tests in 24 'healthy volunteers' were negative to aceclofenac 5% pet. (5).

Cutaneous adverse drug reactions from systemic administration caused by type IV (delayed-type) hypersensitivity
Cutaneous adverse drug reactions from systemic administration of aceclofenac caused by type IV (delayed-type) hypersensitivity, including fixed drug eruption (2,4), are planned to be discussed in Volume IV of the *Monographs in Contact Allergy* series on Systemic drugs.

Cross-reactions, pseudo-cross-reactions and co-reactions
Cross-allergy to diclofenac in patients with aceclofenac allergy (4,5). Two patients with photocontact allergy to diclofenac had photocross-reactions to aceclofenac (3).

PHOTOSENSITIVITY

Case reports

A 61-year-old woman presented with an eczematous rash on her legs which subsequently spread to her face and neck. For several days, she had been applying diclofenac gel on her knee because of osteoarthritic pain. The skin reaction appeared several hours after sun exposure. Topical application was stopped, sun exposure avoided, and the eczema was controlled by topical corticosteroids. A year later, a similar eczematous eruption developed on her right knee after the application of a cream containing 15 mg/g aceclofenac. Patch and photopatch tests showed positive photopatch tests to diclofenac (1%, 5% and 10% pet.) and aceclofenac (10% pet. D1 +, D3 ++). Because the patient had never used aceclofenac before and the dermatitis started soon after the first use of aceclofenac cream, it was concluded that the reaction to aceclofenac was a photocross-reaction to diclofenac (3).

An 81-year-old-woman had been treated for osteoarthrosis with various oral non-steroidal anti-inflammatory drugs (NSAIDs), including aceclofenac, with good tolerance. However, after local application of a cream containing 1.5% aceclofenac twice daily, acute eczema appeared on sun-exposed areas of her legs. Cream application was stopped, sun exposure avoided and topical corticosteroid treatment started, with rapid resolution. She was patch and photopatch tested and had positive photopatch tests to aceclofenac 10% pet. (D1 +, D3 ++) but negative to 5% and 1% pet. There were no reactions to any of 13 other NSAIDs including diclofenac (6).

LITERATURE

1   The data in the section 'General' may have been obtained from literature discussed in this chapter, but mostly also or exclusively from one or more of the following online sources: ChemIDPlus Advanced, PubChem, DrugBank, RxList, Drug Central, Drugs.com, and Wikipedia

2   Prabha N, Chhabra N. Non-pigmenting fixed drug eruption: An unusual adverse reaction to aceclofenac. Curr Drug Saf 2018;13:224-225

3   Fernández-Jorge B, Goday-Buján JJ, Murga M, Molina FP, Pérez-Varela L, Fonseca E. Photoallergic contact dermatitis due to diclofenac with cross-reaction to aceclofenac: two case reports. Contact Dermatitis 2009;61:236-237

4   Linares T, Marcos C, Gavilan MJ, Arenas L. Fixed drug eruption due to aceclofenac. Contact Dermatitis 2007;56:291-292

5   Pitarch Bort G, de la Cuadra Oyanguren J, Torrijos Aguilar A, García-Melgares Linares ML. Allergic contact dermatitis due to aceclofenac. Contact Dermatitis 2006;55:365-366

6   Goday Buján JJ, García Alvarez-Eire GM, Martinez W, del Pozo J, Fonseca E. Photoallergic contact dermatitis from aceclofenac. Contact Dermatitis 2001;45:170

7   Diaz RL, Gardeazabal J, Manrique P, Ratón JA, Urrutia I, Rodríguez-Sasiain JM, Aguirre C. Greater allergenicity of topical ketoprofen in contact dermatitis confirmed by use. Contact Dermatitis 2006;54:239-243

Chapter 3.4   ACETARSONE

IDENTIFICATION

Description/definition   : Acetarsone is the pentavalent arsenical and organometallic compound that conforms to the structural formula shown below
Pharmaceutical classes   : Anti-infective agents
IUPAC name   : (3-Acetamido-4-hydroxyphenyl)arsonic acid
Other names   : Acetarsol; 3-Acetamido-4-hydroxyphenylarsonic acid
CAS registry number   : 97-44-9
EC number   : 202-582-3
Merck Index monograph   : 1321
Patch testing   : 1% and 5% pet.
Molecular formula   : $C_8H_{10}AsNO_5$

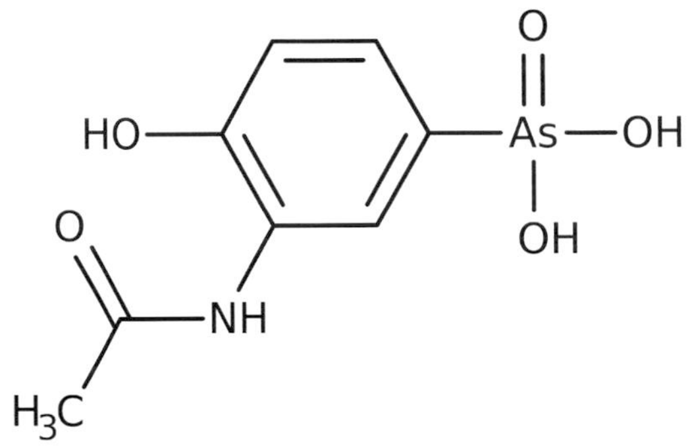

GENERAL

Acetarsone is an organic pentavalent arsenical compound. In the form of pessaries, powder, or vaginal tablets, it used to be commonly employed in the treatment of vaginitis due to *Trichomonas vaginalis* and *Candida albicans* infections. Acetarsone has also been used for the treatment of various other diseases such as syphilis, amoebiasis, yaws, trypanosomiasis, and malaria. It was withdrawn from the market in 1997 (11).

CONTACT ALLERGY

General

Acetarsone present in medications to treat vaginal infections with *Trichomonas vaginalis* or *Candida albicans* has formerly caused many cases of allergic contact dermatitis and systemic contact dermatitis from absorption through the vaginal mucosa. It is now a historical allergen.

Case reports

A woman was treated with acetarsone pessaries for vaginal infection with *Trichomonas vaginalis*. After 2 months, she presented complaining of vulval irritation and of a rash on the hands, legs, and trunk. Examination showed a symmetrical erythematous and vesicular eruption involving the back of the hands, both axillae, cubital fossae, dorsal aspects of the feet, and the vulva. The pessaries were discontinued. Nine days later she returned stating that 4 days previously her eyes had become swollen up, and almost completely closed. Now there was edema of the eyelids and a generalized exfoliative dermatitis especially marked on the face, scalp, dorsal aspects of hands and feet, cubital fossae, and axillae. After healing, a patch test using an acetarsone pessary dissolved in distilled water gave a strongly positive reaction within 24 hours. This was a case of systemic contact dermatitis (10).

A 38-year-old woman also had allergic contact dermatitis and systemic contact dermatitis from acetarsone vaginal tablets used to treat trichomoniasis vaginalis. She had positive patch tests to the vaginal tablets and its ingredient acetarsone 1% and 5% pet. (2). Another woman, known to be allergic to neoarsphenamine (Neosalvarsan, a trivalent arsenical compound), was treated with a pessary containing acetarsone for leucorrhea. Some days later, she developed a generalized erythema with edema of the eyelids. Later, a patch test to the pessary was positive (3).

A female patient aged 47 years had a history of allergy to toothpastes containing acetarsone. She presented with acute dermatitis of the inner thighs from vaginal tablets containing acetarsone. Patch tests were strongly positive to the vaginal tablets, to the toothpaste containing acetarsone and to acetarsone itself (no details provided). At that time (1978), acetarsone was still present in France in various vaginal products, toothpastes, mouth washes and nasal drops (11).

In early literature, there have been several other reports of female patients with local skin and mucous membrane reactions and often generalized dermatitis after the use of acetarsone in the form of vaginal powder, tablets or pessaries (4-9,12). However, in these cases, either no patch tests were performed (4-9), they were negative (one patient [4]) or details are unknown (12). Interestingly, in some patients, dermatitis appeared in the antecubital fossae and/or or the axillae and/or hands, suggesting systemic allergic contact dermatitis (4,6,10). In a number of cases, the patients had previously had generalized rashes from treatment of syphilis with the trivalent arsenic compound neoarsphenamine (3,5), again suggesting contact allergy to acetarsone and systemic allergic contact dermatitis from absorption of arsenic from the vagina mucosa.

LITERATURE

1   The data in the section 'General' may have been obtained from literature discussed in this chapter, but mostly also or exclusively from one or more of the following online sources: ChemIDPlus Advanced, PubChem, DrugBank, RxList, Drug Central, Drugs.com, and Wikipedia
2   Sasseville D, Carey WD, Singer MI. Generalized contact dermatitis from acetarsone. Contact Dermatitis 1995;33:431-432
3   Orchard WE. Dermatitis after use of pentavalent arsenicals per vaginam. Br Med J 1951;2(4745):1444
4   Kesten BM. Arsenical dermatitis produced in treatment of trichomonas vaginitis. Report of five cases. Arch Derm Syphilol 1938;38:198-199
5   Campbell CHG. Arsenical intolerance and the treatment of Trichomonas vaginalis infection. Lancet 1937;2(5951):688-689
6   Peck BJ. Exfoliative dermatitis after acetarsol vaginal pessaries. Brit Med J 1954;2:850-851
7   White A. Acute systemic reaction to acetarsol. Br Med J 1956;29:2(5008):1528-1529
8   Crosswell HD. Jaundice following acetarsol pessaries. Br J Vener Dis 1951;27:150-153
9   Long MCW. Arsenical intolerance and the treatment of *Trichomonas vaginalis* infection (Letter). Lancet 1937: 2(5953):828.
10 Doyle JO. Acetarsol pessary dermatitis. Brit J Vener Dis 1952;28:210-212
11 Robin J. Contact dermatitis to acetarsol. Contact Dermatitis 1978;4:309-310
12 Thompson TA, Marshall RJ. Hypersensitivity reaction from acetarsol pessaries. J Obstet Gynaecol Br Emp 1958;65:475-478

Chapter 3.5   ACETAZOLAMIDE

IDENTIFICATION

Description/definition   : Acetazolamide is the thiadiazole sulfonamide that conforms to the structural formula
                                           shown below
Pharmacological classes   : Carbonic anhydrase inhibitors; anticonvulsants; diuretics
IUPAC name   : N-(5-Sulfamoyl-1,3,4-thiadiazol-2-yl)acetamide
Other names   : 2-Acetylamino-1,3,4-thiadiazole-5-sulfonamide
CAS registry number   : 59-66-5
EC number   : 200-440-5
Merck Index monograph : 1322
Patch testing   : 5% pet.; if pure material is not available, crushed tablet or content of capsule in pet. with
                                           10% active drug
Molecular formula   : $C_4H_6N_4O_3S_2$

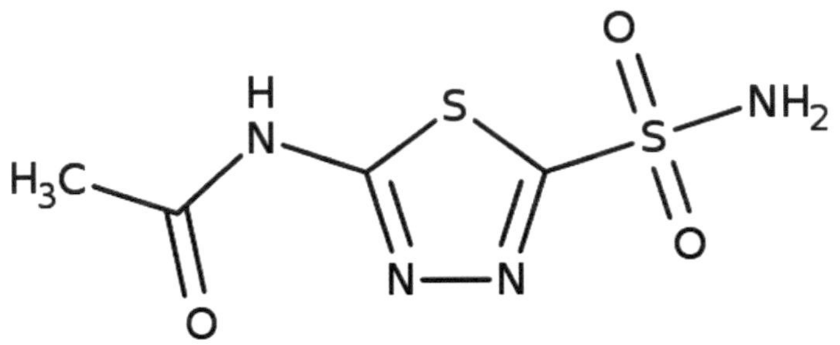

GENERAL

Acetazolamide is a sulfonamide derivative and non-competitive inhibitor of carbonic anhydrase with diuretic, antiglaucoma, and anticonvulsant properties. Acetazolamide is Indicated for the treatment of hypercapnia due to chronic obstructive pulmonary disease, idiopathic intracranial hypertension, prevention or treatment of postoperative intraocular pressure after cataract surgery, absence seizures and prophylaxis of acute mountain sickness (1,2). In pharmaceutical products, acetazolamide is employed as acetazolamide sodium (CAS number 1424-27-7, EC number not available, molecular formula $C_4H_5N_4NaO_3S_2$) (1).

CONTACT ALLERGY

Case reports and case series

One day after application of 10% acetazolamide cream under a compression panty following liposuction of the thighs, a 32-year-old woman developed a skin eruption at the application site, which spread to the face, arms and neck, but not to the trunk. After stopping the use of the cream and with local corticosteroid therapy, the rash had nearly resolved after 13 days. Patch tests were positive to the cream at D2 (no second reading mentioned). The ingredients of the cream were not tested separately (4). A second patient with a similar history, a woman of 39, was also presented. A patch test with the commercial cream was positive (no details provided). The application of the cream under occlusion, which is contra-indicated according to the prescription of the pharmaceutical, may have facilitated spreading of the dermatitis (4).

Other cases of contact dermatitis from this drug have apparently not been published, but French adverse drug reaction reporting data up to 2007 include 10 other cases of eczema or rash from the cream at the application site. In one of these, a positive reaction was observed on re-administration, and in 2 cases allergy skin tests (not specified) were positive (4). It should be mentioned that, where the authors claim these to be 2 cases of contact allergy to acetazolamide, this is in fact uncertain, as the other – unspecified – ingredients of the cream have not been tested.

Cutaneous adverse drug reactions from systemic administration caused by type IV (delayed-type) hypersensitivity
Cutaneous adverse drug reactions from systemic administration of acetazolamide caused by type IV (delayed-type) hypersensitivity, including maculopapular exanthema (2,3) and acute generalized exanthematous pustulosis (AGEP) (3), are planned to be discussed in Volume IV of the *Monographs in Contact Allergy* series on Systemic drugs.

LITERATURE

1   The data in the section 'General' may have been obtained from literature discussed in this chapter, but mostly also or exclusively from one or more of the following online sources: ChemIDPlus Advanced, PubChem, DrugBank, RxList, Drug Central, Drugs.com, and Wikipedia
2   Dequidt L, Milpied B, Chauvel A, Seneschal J, Taieb A, Darrigade AS. A case of lichenoid and pigmented drug eruption to acetazolamide confirmed by a lichenoid patch test. J Allergy Clin Immunol Pract 2018;6:283-285
3   Jachiet M, Bellon N, Assier H, Amsler E, Gaouar H, Pecquet C, et al. Cutaneous adverse drug reaction to oral acetazolamide and skin tests. Dermatology 2013;226:347-352
4   Daveluy A, Vial T, Marty L, Miremont-Salamé G, Moore N, Haramburu F. Contact dermatitis caused by acetazolamide under occlusion. Presse Med 2007;36(12 Pt.1):1756-1758 (Article in French)

Chapter 3.6   ACETYLCYSTEINE

IDENTIFICATION

Description/definition   : Acetylcysteine is the synthetic *N*-acetyl derivative of the endogenous amino acid
                                        L-cysteine; it conforms to the structural formula shown below
Pharmacological classes : Antiviral agents; expectorants; free radical scavengers
IUPAC name   : (2*R*)-2-Acetamido-3-sulfanylpropanoic acid
Other names   : L-α-Acetamido-β-mercaptopropionic acid; *N*-acetyl-L-cysteine; mercapturic acid
CAS registry number   : 616-91-1
EC number   : 210-498-3
Merck Index monograph : 1353
Patch testing   : 10% water or pet.
Molecular formula   : $C_5H_9NO_3S$

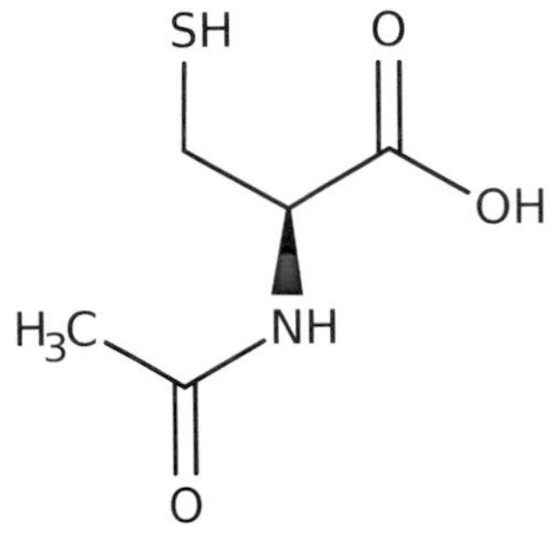

GENERAL

Acetylcysteine is a synthetic *N*-acetyl derivative of the endogenous amino acid L-cysteine, a precursor of the antioxidant enzyme glutathione. Acetylcysteine is used mainly as a mucolytic drug to reduce the viscosity of mucous secretions, in the management of paracetamol (acetaminophen) overdose, and as a protective agent for renal function in contrast medium-induced nephropathy. It has also been shown to have antiviral effects in patients with HIV due to inhibition of viral stimulation by reactive oxygen intermediates. Acetylcysteine is essentially a prodrug that is converted to cysteine in the intestine and absorbed there into the blood stream. In combination with hypromellose eye drops, is commonly used to alleviate the chronic soreness associated with dry eyes (1,2).

CONTACT ALLERGY

Case reports

A 30-year-old woman with lifelong atopic eczema had a 10-year history of peri-orbital eczema, associated with severe atopic keratoconjunctivitis and keratoconus. Her left eye was blind from previous corticosteroid glaucoma, and her right eye had moderately advanced glaucoma, which had been controlled surgically. In addition to glaucoma, topical corticosteroids had resulted in early cataracts. Her treatment at presentation consisted of fluorometholone and sodium cromoglycate eye drops, in addition to hypromellose and *N*-acetylcysteine drops as tear replacement. Patch tests were positive to *N*-acetylcysteine eye drops (10% water, preservative free); 14 controls were negative. A marked reduction in redness and itching of the eyes was noted by the patient within a week of discontinuation, and this clinical improvement has been maintained at 4 months (2).

A 32-year-old nurse presented with itchy eczematous skin lesions on exposed areas of the hands, arms and face that had been present for several weeks. The symptoms had developed since she started working in a general ward, where she mixed *N*-acetylcysteine (NAC) with normal saline, and improved during the weekends. Patch tests were positive to NAC 10% pet., as were intradermal tests with NAC at 1 and 10 mg/ml in 0.9% saline, read at D2 (3). This was a case of occupational airborne allergic contact dermatitis.

LITERATURE

1   The data in the section 'General' may have been obtained from literature discussed in this chapter, but mostly also or exclusively from one or more of the following online sources: ChemIDPlus Advanced, PubChem, DrugBank, RxList, Drug Central, Drugs.com, and Wikipedia
2   Davison SC, Wakelin SH. Allergic contact dermatitis from *N*-acetylcysteine eye drops. Contact Dermatitis 2002;47:238
3   Kim JH, Kim SH, Yoon MG, Jung HM, Park HS, Shin YS. A case of occupational contact dermatitis caused by *N*-acetylcysteine. Contact Dermatitis 2016;74:373-374

Chapter 3.7 ACEXAMIC ACID

IDENTIFICATION

Description/definition : Acexamic acid is the medium chain fatty acid that conforms to the structural formula shown below

Pharmacological classes : Anti-inflammatory agents, non-steroidal

IUPAC name : 6-Acetamidohexanoic acid

Other names : 6-Acetylaminocaproic acid; ε-acetamidocaproic acid; 6-(acetylamino)hexanoic acid

CAS registry number : 57-08-9

EC number : 200-310-8

Merck Index monograph : 1316

Patch testing : Acexamic acid 5% and 10% pet.; zinc acexamate 5% water

Molecular formula : $C_8H_{15}NO_3$

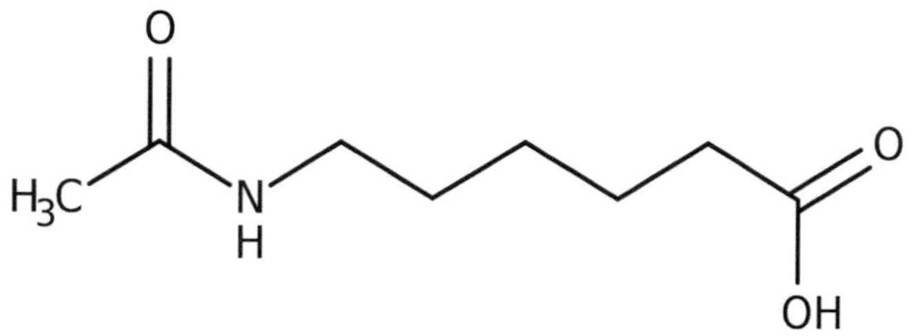

GENERAL

Acexamic acid is a drug used as a cicatrization helper. Its salt zinc acexamate has been employed in the treatment of gastric and duodenal ulcers and in the prevention of gastric ulcer induced by nonsteroidal anti-inflammatory drugs in topical formulations, it has been used as a 'cicatrization helper' (2).

In pharmaceutical products, acexamic acid may be used as sodium acexamate (CAS number 7234-48-2, EC number 230-635-0, molecular formula $C_8H_{14}NNaO_3$) or zinc acexamate (CAS number 70020-71-2, EC number not available, molecular formula $C_{16}H_{28}N_2O_6Zn$) (1).

CONTACT ALLERGY

Case report

A 66-year-old woman had been applying an ointment containing acexamic acid and neomycin for several months to a leg ulcer until complete healing. Meanwhile, she developed erythematous, itching papules, that spread to the trunk and arms, and edema of the legs. Her condition only improved with systemic corticosteroid therapy. Some months later, after a trauma on the leg, another small ulcer appeared. She applied the same ointment and dermatitis reappeared. Patch tests were strongly positive to the ointment and to its ingredient acexamic acid 5% and 10% pet. (D2 ++, D4 +++) but negative to neomycin. Ten controls were negative (2).

Cutaneous adverse drug reactions from systemic administration caused by type IV (delayed-type) hypersensitivity

Cutaneous adverse drug reactions from systemic administration of zinc acexamate caused by type IV (delayed-type) hypersensitivity, including a generalized maculopapular eruption with urticarial characteristics (3), are planned to be discussed in Volume IV of the *Monographs in Contact Allergy* series on Systemic drugs.

LITERATURE

1 The data in the section 'General' may have been obtained from literature discussed in this chapter, but mostly also or exclusively from one or more of the following online sources: ChemIDPlus Advanced, PubChem, DrugBank, RxList, Drug Central, Drugs.com, and Wikipedia

2 Reis AM, Silva R, Pignatelli J. Allergic contact dermatitis to acexamic acid. Contact Dermatitis 2008;58:241-242

3 Galindo PA, Garrido JA, Gómez E, Borja J, Feo F, Encinas C, et al. Zinc acexamate allergy. Contact Dermatitis 1998;38:301-302

Chapter 3.8   ACYCLOVIR

IDENTIFICATION

Description/definition   : Acyclovir is the synthetic analog of the purine nucleoside, guanosine, that conforms to the
structural formula shown below
Pharmacological classes : Antiviral agents
IUPAC name   : 2-Amino-9-(2-hydroxyethoxymethyl)-3*H*-purin-6-one
Other names   : Acycloguanosine
CAS registry number   : 59277-89-3
EC number   : 261-685-1
Merck Index monograph   : 1404
Patch testing   : 10.0% pet. (Chemotechnique); scratch-patch tests may be performed in case of a ?+
reaction or a negative reaction to acyclovir with highly suspect history for allergy (13)
Molecular formula   : $C_8H_{11}N_5O_3$

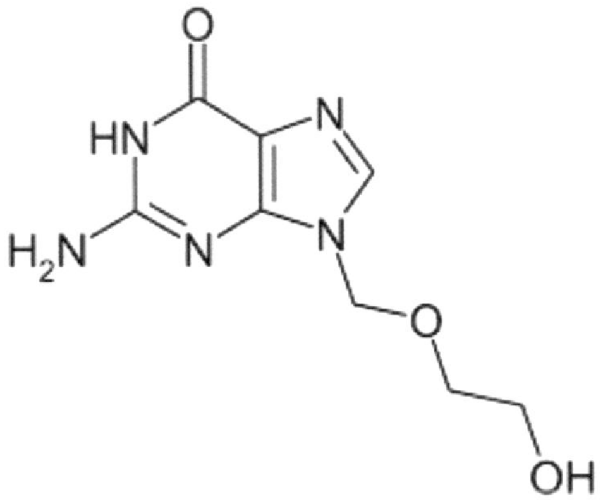

GENERAL

Acyclovir is a synthetic analog of the purine nucleoside, guanosine, with potent antiviral activity against *Herpes simplex* viruses type 1 and 2, *Varicella zoster* virus and other viruses of the herpesvirus family. After conversion *in vivo* to the active metabolite acyclovir triphosphate by viral thymidine kinase, acyclovir competitively inhibits viral DNA-polymerase by incorporating into the growing viral DNA chain and terminating further polymerization. Acyclovir is used for the treatment of herpes simplex virus infections, varicella and herpes zoster. In topical pharmaceutical products, acyclovir base is used; in powder for injection fluids, acyclovir sodium is employed (CAS number 69657-51-8, EC number 614-996-5, molecular formula $C_8H_{10}N_5NaO_3$) (1).

CONTACT ALLERGY

General

Contact allergy to and allergic contact dermatitis from acyclovir, although well-known, are not frequent, related to its widespread use. In Zovirax cream, an excipient may also cause contact allergy, notably propylene glycol. Patch tests with the cream itself may be irritant due to its high concentration of propylene glycol. Oral or parenteral administration of acyclovir, valaciclovir or famciclovir can cause systemic contact dermatitis in individuals previously sensitized to acyclovir.

Case series

In Leuven, Belgium, in the period 1990-2014, iatrogenic contact dermatitis was diagnosed in 2600 individuals (17% of the total patch test population). 96% of all positive patch test reactions to topical drugs and antiseptics were considered to be relevant. Acyclovir (10% pet.) was tested in 98 patients and there were 5 positive reactions to it (2). A small series of 3 patients sensitized to acyclovir was reported from Italy in 1992 (26). Fourteen patients suspected to be allergic to acyclovir cream were reported from Spain (9). Some were allergic to the active principle, others to excipients (10), but details are unknown.

Case reports

Several single case reports (13,25,27,28,29) have described contact allergy to and allergic contact dermatitis from acyclovir, most often from the use of a cream containing 5% acyclovir (Zovirax) on herpes simplex labialis (cold sores). Patch tests have been positive to acyclovir 1% and 2% water (25), 1% and 5% pet. (13), 1%, 3% and 5% pet. (26,28), and 5% water (27,29). Some patients both reacted to acyclovir and an excipient: propylene glycol (20,21,22), sodium lauryl sulfate (25) or cetearyl alcohol (23,27).

In one report, a scratch-patch test resulted in far stronger reactions to 1% and 5% acyclovir in petrolatum than conventional patch tests (13). In another case, the diagnosis of acyclovir allergy was – possibly erroneously - made *per exclusionem*: positive patch test to the cream, negative to the excipients, acyclovir itself not tested (8).

Systemic contact dermatitis

A 23-year-old woman developed a severe itchy erythematous vesicular dermatitis of the upper lip and right cheek while treating recurrent herpes simplex of the lips with acyclovir 5% cream. A few months after the resolution of the dermatitis, 6 hours after beginning oral acyclovir, the patient developed urticaria with itchy, erythematous, edematous lesions on the trunk and extremities. Patch tests were positive to commercial acyclovir cream (50% pet.), acyclovir ophthalmic ointment ('as is'), and acyclovir tablet, probably 100% and crushed. The patient also reacted to commercial valaciclovir (tablet 30% pet., water, alcohol) and to propylene glycol, which is present in the cream. She had a negative patch test to famciclovir, but when oral administration of famciclovir, with progressively increased dosages, was carried out, an itchy erythematous dermatitis occurred on the trunk 15 hr after administration of the last dose of famciclovir 500 mg (20). This was a case of allergic contact dermatitis and systemic dermatitis from acyclovir with cross-reactivity to valaciclovir and famciclovir.

Several other case reports of systemic contact dermatitis from oral or parenteral administration of acyclovir after previous sensitization from topical medicaments containing acyclovir have been reported (21,22,23). A 19-year-old woman had a lip dermatitis after application of 5% acyclovir cream to herpes labialis. Clinical suspicion that the herpetic infection initially diagnosed had worsened, therapy was given with 2 intravenous infusions of acyclovir 250 mg at an interval of 8 hours. These were both followed, some 15 minutes later, by diffuse urticaria. Patch tests were positive to acyclovir cream and ophthalmic ointment, to acyclovir 1%, 3% and 5% pet. and to propylene glycol 10% water (21).

A 29-year-old woman presented with labial eczema after application of 5% acyclovir cream, then with edematous eczema of the upper eyelid following the application of acyclovir ophthalmic ointment. A few months later, 24 hours after the intake of acyclovir tablets, she developed a pruriginous maculopapular rash, with secondary eczematization on the arms, trunk, and inner thigh. Patch tests were positive to propylene glycol 5% pet. and to acyclovir 10% water. There was a cross-reaction to valaciclovir but not to famciclovir (22). The authors also presented a 28-year-old woman who was sensitized to topical acyclovir and developed a widespread eczema that had started 6 hours after taking one acyclovir tablet. Patch tests showed the same pattern of sensitization as in the first patient: ++ to propylene glycol, acyclovir and valaciclovir and negative to famciclovir. A provocation test with famciclovir, however, elicited a pruriginous rash 12 hours after its first intake (22).

A 44-year-old woman used acyclovir cream for 2 weeks on her first attack of genital herpes without any improvement. Oral valaciclovir (500 mg 2x daily) was started. After the first 2 tablets, an itchy, symmetrical exanthem appeared on the face, trunk and extremities. One month later, the patient developed a labial herpes infection. She used acyclovir cream and vesicobullous lesions with erythema appeared in the labial and perioral skin, with an exanthem on the upper trunk and extremities. Patch tests were positive to cetearyl alcohol (present in the cream), acyclovir and valaciclovir as is, 20%, 10% and 1% water and pet., and ganciclovir 'as is' and 20% pet. (no lower concentrations tested). There was no reaction to famciclovir, but when orally challenged with famciclovir, itchy erythematous patches appeared on the upper lip and upper trunk after 325 mg famciclovir given in 3 doses over 2 days (23).

Contact allergy to excipients in acyclovir cream

Some patients who had become sensitized to topical pharmaceuticals with acyclovir both reacted to acyclovir and an excipient: propylene glycol (20,21,22), sodium lauryl sulfate (25) or cetearyl alcohol (23,27). Propylene glycol has also been the allergenic excipient in Zovirax cream several times, when acyclovir was negative (14-19).

Positive patch test reactions to acyclovir cream, negative to all ingredients

There have been several case reports of patients who developed acute dermatitis of the lips from the use of a cream containing acyclovir (Zovirax), where the cream itself gave a positive patch test but all ingredients including acyclovir were negative (3-7,26). Possible explanations given were compound allergy (5,7,26), insufficient penetration of acyclovir through the epidermis (4,6) and an irritant contact dermatitis and patch test reaction to the cream, possibly because of its high content (40% in 1995) of propylene glycol (6). In favor of the latter explanation is that of 10 controls, 5 had a weak positive (+) reaction to acyclovir cream (6) and that all 'positive' patch test reactions to the cream in the studies mentioned here have only been weakly positive (+).

Cross-reactions, pseudo-cross-reactions and co-reactions

Cross-reactions between acyclovir and valaciclovir have been documented (20,22,23). One patient sensitized to acyclovir cross-reacted to ganciclovir (23). In acyclovir-allergic individuals who do *not* cross-react when patch tested with famciclovir (i.e., no positive patch test to famciclovir), oral provocation tests may yet be positive (20,22,23). Valaciclovir is the L-valine ester of acyclovir and is almost completely metabolized to acyclovir after oral administration. The chemical structure common to acyclovir, valaciclovir, ganciclovir and famciclovir is the 2-aminopurine nucleus, and this is probably the allergenic determinant of the molecules (20).

Cutaneous adverse drug reactions from systemic administration caused by type IV (delayed-type) hypersensitivity

Cutaneous adverse drug reactions from systemic administration of acyclovir caused by type IV (delayed-type) hypersensitivity, including acute generalized exanthematous pustulosis (AGEP) (12) and fixed drug eruption (24), are planned to be discussed in Volume IV of the *Monographs in Contact Allergy* series on Systemic drugs.

PHOTOSENSITIVITY

A patient with suspected photosensitivity to acyclovir cream had a positive photopatch test to the cream, but not to any of its ingredients. A diagnosis of compound photoallergy was made (10). Possibly, a similar case has been observed in France (11).

LITERATURE

1   The data in the section 'General' may have been obtained from literature discussed in this chapter, but mostly also or exclusively from one or more of the following online sources: ChemIDPlus Advanced, PubChem, DrugBank, RxList, Drug Central, Drugs.com, and Wikipedia

2   Gilissen L, Goossens A. Frequency and trends of contact allergy to and iatrogenic contact dermatitis caused by topical drugs over a 25-year period. Contact Dermatitis 2016;75:290-302

3   Schenkelberger V, Wapenhans A, Schindera I. Kontaktallergie auf Zovirax Creme. Eine Compound-Allergie gegenüber Acyclovir? Z Hautkr 1994;69:771-773 (Article in German)

4   Baes H, van Hecke E. Contact dermatitis from Zovirax cream. Contact Dermatitis 1990;23:200-201

5   Goh CL. Compound allergy to Spectraban® 15 lotion and Zovirax® cream. Contact Dermatitis 1990;22:61-62

6   Koch P. No evidence of contact sensitization to acyclovir in acute dermatitis of the lips following local application of Zovirax cream. Contact Dermatitis 1995;33:255-257

7   Serpentier-Daude A, Collet E, Didier AF, Touraud JP, Sgro C, Lambert D. Contact dermatitis to topical antiviral drugs. Ann Dermatol Venereol 2000;127:191-193 (Article in French)

8   Chiriac A, Chiriac AE, Pinteala T, Moldovan C, Stolnicu S. Allergic contact dermatitis from topical acyclovir: Case series. J Emerg Med 2017;52:e37-e39

9   Gonzalez-Perez R, Aguirre A, Gonzalez-Guemes M, Soleta R, Diaz Perez JL. Eczema de contacto por cremas de aciclovir. Estudio de 14 casos. Actas Sifiliográficas 1994;8:541-544 (Article in Spanish)

10 Rodriguez-Serna M, Velasco M, Miquel J, de la Cuadra J, Aliaga A. Photoallergic contact dermatitis from Zovirax cream. Contact Dermatitis 1999;41:54-55

11 No authors listed. Photoallergy to Zovirax cream. Ann Dermatol Venereol 2001;128:184 (Article in French)

12 Serra D, Ramos L, Brinca A, Gonçalo M. Acute generalized exanthematous pustulosis associated with acyclovir, confirmed by patch testing. Dermatitis 2012;23:99-100

13 Nino M, Balato N, Di Costanzo L, Gaudiello F. Scratch-patch test for the diagnosis of allergic contact dermatitis to aciclovir. Contact Dermatitis 2009;60:56-57

14 Ozkaya E, Topkarci Z, Ozarmağan G. Allergic contact cheilitis from a lipstick misdiagnosed as herpes labialis: Subsequent worsening due to Zovirax contact allergy. Australas J Dermatol 2007;48:190-192

15 Kim YJ, Kim JH. Allergic contact dermatitis from propylene glycol in Zovirax cream. Contact Dermatitis 1994;30:119-120

16 Bourezane Y, Girardin P, Aubin F, Vigan M, Adessi B, Humbert P, et al. Allergic contact dermatitis to Zovirax cream. Allergy 1996;51:755-756

17 Claverie F, Giordano-Labadie F, Bazex J. Contact eczema induced by propylene glycol. Concentration and vehicle adapted for patch tests. Ann Dermatol Venereol 1997;124:315-317 (Article in French)

18 Corazza M, Virgili A, Mantovani L, La Malfa W. Propylene glycol allergy from acyclovir cream with cross-reactivity to hydroxypropyl cellulose in a transdermal estradiol system? Contact Dermatitis 1993;29:283-284

19 Hernández N, Hernández Z, Liuti F, Borrego L. Intolerance to cosmetics as key to the diagnosis in a patient with allergic contact dermatitis caused by propylene glycol contained in a topical medication. Contact Dermatitis 2017;76:246-247

20 Vernassiere C, Barbaud A, Trechot PH, Weber-Muller F, Schmutz JL. Systemic acyclovir reaction subsequent to acyclovir contact allergy: which systemic antiviral drug should then be used? Contact Dermatitis 2003;49:155-157

21 Gola M, Francalanci S, Brusi C, Lombardi P, Sertoli A. Contact sensitization to acyclovir. Contact Dermatitis 1989;20:394-395

22 Bayrou O, Gaouar H, Leynadier F. Famciclovir as a possible alternative treatment in some cases of allergy to acyclovir. Contact Dermatitis 2000;42:42

23 Lammintausta K, Mäkelä L, Kalimo K. Rapid systemic valaciclovir reaction subsequent to aciclovir contact allergy. Contact Dermatitis 2001;45:181

24 Montoro J, Basomba A. Fixed drug eruption due to acyclovir. Contact Dermatitis 1997;36:225

25 Aguirre A, Manzano D, Izu R, Gardeazabal J, Díaz Pérez JL. Allergic contact cheilitis from mandelic acid. Contact Dermatitis 1994;31:133-134

26 Vincenzi C, Peluso AM, Cameli N, Tosti A. Allergic contact dermatitis caused by acyclovir. Am J Contact Dermat 1992;3:105-107

27 Goday J, Aguirre A, Ibarra NG, Eizaguirre X. Allergic contact dermatitis from acyclovir. Contact Dermatitis 1991;24:380-381

28 Valsecchi R, Imberti G, Cainelli T. Contact allergy to acyclovir. Contact Dermatitis 1990;23:372-373

29 Camarasa JG, Serra-Baldrich E. Allergic contact dermatitis from acyclovir. Contact Dermatitis 1988;19:235-236

Chapter 3.9   ADAPALENE

IDENTIFICATION

Description/definition   : Adapalene is the naphthalene derivative and retinoid that conforms to the structural
                            formula shown below
Pharmacological classes   : Anti-inflammatory agents, non-steroidal; dermatological agents
IUPAC name   : 6-[3-(1-Adamantyl)-4-methoxyphenyl]naphthalene-2-carboxylic acid
Other names   : 6-(3-(Adamantan-1-yl)-4-methoxyphenyl)-2-naphthoic acid
CAS registry number   : 106685-40-9
EC number   : Not available
Merck Index monograph   : 1408
Patch testing   : 0.1% and 0.01% pet.
Molecular formula   : $C_{28}H_{28}O_3$

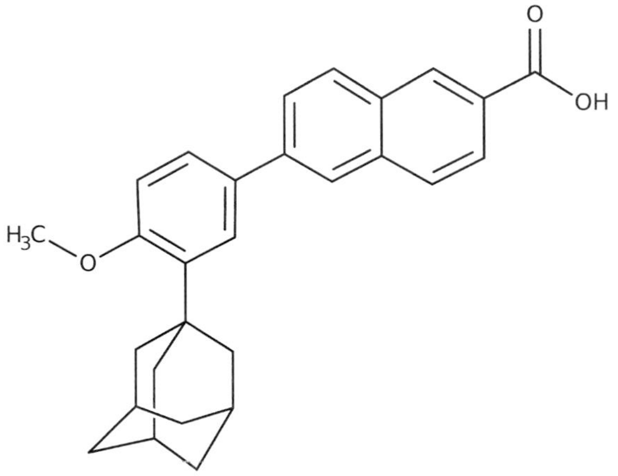

GENERAL

Adapalene is a member of the retinoid drug class that is typically found in topical formulations used for the treatment of acne. It is a modulator of cellular differentiation, keratinization, and inflammatory processes, all of which represent important features in the pathology of acne vulgaris. Adapalene may also be used off-label to treat keratosis pilaris and certain other skin conditions (1).

CONTACT ALLERGY

Case report

A 22-year-old woman presented with a 4-month history of facial erythema and papules with scaling. She had started using adapalene gel for the treatment of acne vulgaris before the appearance of the eruptions, which remitted after discontinuation of treatment, only to recur 2 months later when she resumed the use of the gel. Patch tests were positive to adapalene gel (D2 +, D3 +, D7 +). Ingredient patch testing gave positive results for adapalene 0.1% pet. (D2 +, D3 + D7 +), but not for the other five ingredients. Subsequent tests were positive to adapalene 0.01% pet., 0.001% pet. and 0.0001% pet. Eight healthy controls showed negative reactions to the substance (2).

LITERATURE

1   The data in the section 'General' may have been obtained from literature discussed in this chapter, but mostly
     also or exclusively from one or more of the following online sources: ChemIDPlus Advanced, PubChem,
     DrugBank, RxList, Drug Central, Drugs.com, and Wikipedia
2   Numata T, Jo R, Kobayashi Y, Tsuboi R, Okubo Y. Allergic contact dermatitis caused by adapalene. Contact
     Dermatitis 2015;73:187-188

Chapter 3.10   ALCAFTADINE

IDENTIFICATION

Description/definition : Alcaftadine is the imidazobenzazepine that conforms to the structural formula shown below

Pharmacological classes : Histamine H1 antagonists

IUPAC name : 11-(1-Methylpiperidin-4-ylidene)-5,6-dihydroimidazo[2,1-b][3]benzazepine-3-carbaldehyde

CAS registry number : 147084-10-4

EC number : Not available

Patch testing : No data available; generally speaking, the test concentration of ophthalmic medications should higher than their concentration in the eye drops

Molecular formula : $C_{19}H_{21}N_3O$

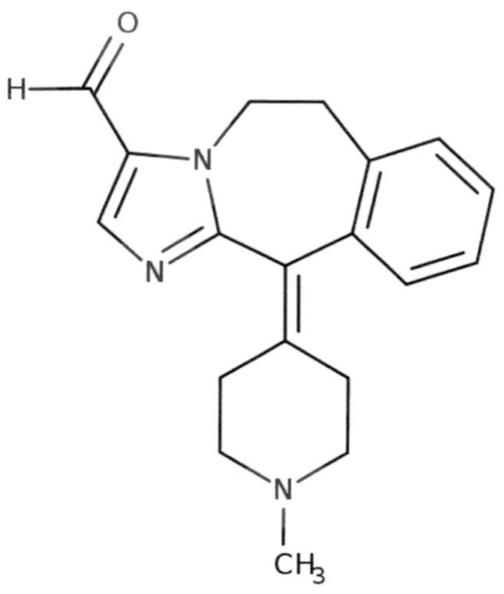

GENERAL

Alcaftadine is a H1 histamine receptor antagonist indicated for the prevention of itching associated with allergic conjunctivitis (1).

CONTACT ALLERGY

Case report

A 51-year-old woman had been prescribed alcaftadine 0.25% for allergic conjunctivitis. After the second application, both eyelids became swollen, and erythematous changes were evident. On slit-lamp examination, conjunctival injection was noted in the absence of conjunctival swelling or any other findings. Fundus examination was unremarkable. A – rather amateurish – patch test with the commercial alcaftadine 0.25% was positive at D2 (no other readings mentioned). The other possibly allergenic ingredients benzalkonium chloride and edetate disodium were not tested, but these were also present in 2 other ophthalmic preparations that were patch tested and found to be negative (2). Thus, although contact allergy to alcaftadine in this case is possible, the evidence for it is rather circumstantial and weak.

LITERATURE

1   The data in the section 'General' may have been obtained from literature discussed in this chapter, but mostly also or exclusively from one or more of the following online sources: ChemIDPlus Advanced, PubChem, DrugBank, RxList, Drug Central, Drugs.com, and Wikipedia

2   Kim JH, Kim HJ, Kim SW. Allergic contact dermatitis of both eyes caused by alcaftadine 0.25%: a case report. BMC Ophthalmol 2019;19:158

Chapter 3.11   ALCLOMETASONE DIPROPIONATE

IDENTIFICATION

Description/definition	: Alclometasone dipropionate is the 17,21- dipropionate ester of the synthetic glucocorticoid alclometasone that conforms to the structural formula shown below
Pharmacological classes	: Anti-inflammatory agents; glucocorticoids
IUPAC name	: [2-[(7R,8S,9S,10R,11S,13S,14S,16R,17R)-7-Chloro-11-hydroxy-10,13,16-trimethyl-3-oxo-17-propanoyloxy-7,8,9,11,12,14,15,16-octahydro-6H-cyclopenta[a]phenanthren-17-yl]-2-oxoethyl] propanoate
Other names	: Alclometasone 17,21-dipropionate; 7α-chloro-11β,17,21-trihydroxy-16α-methylpregna-1,4-diene-3,20-dione 17,21-di(propionate)
CAS registry number	: 66734-13-2
EC number	: 266-464-3
Merck Index monograph	: 1483 (Alclometasone)
Patch testing	: 1.0% pet. (Chemotechnique)
Molecular formula	: $C_{28}H_{37}ClO_7$

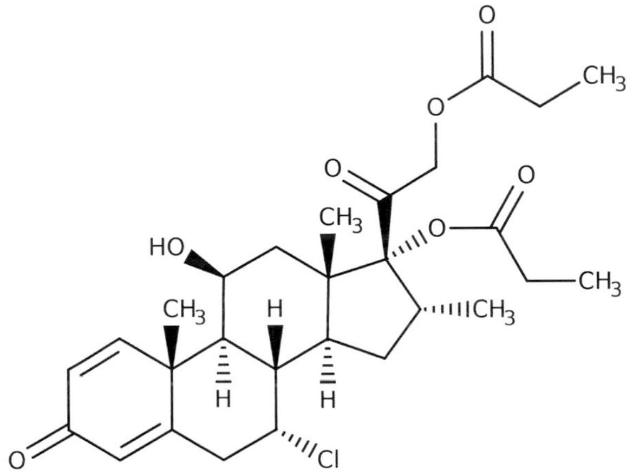

GENERAL

General aspects of corticosteroids used on the skin and mucous membranes are discussed in Chapter 2.4. A practical guideline for diagnosing allergic reactions to corticosteroids is presented in ref. 1.

CONTACT ALLERGY

Patch testing in groups of patients

Results of patch testing alclometasone dipropionate in consecutive patients suspected of contact dermatitis (routine testing) are shown in table 3.11.1. In a joint study performed in Belgium and The Netherlands in 1991 (19), a prevalence of sensitization of 0.4% was found in The Netherlands with both test materials used (alclometasone dipropionate 0.5% alc. and 1% pet.). In Belgium, 0.5% alcohol scored 1% positive reactions and alclometasone dipropionate 1% pet. 1.8% reactions. The relevance of the positive patch tests was not mentioned, so it is unclear how many (possibly all) were cross-reactions to other corticosteroids (19).

Results of testing alclometasone dipropionate in groups of selected patients (individuals (strongly) suspected of corticosteroid allergy) are also shown in table 3.11.1. In a USA study, only 11 positive reactions were observed in a group of 1187 patients (0.9%), but they were all considered to be relevant. In Finland, in a more strictly selected population (and tested with the corticosteroid in a vehicle of alcohol, which is far more reliable than petrolatum), 15% of 34 individuals had positive patch test reactions to alclometasone dipropionate 0.5% and/or 0.05% alc. Whether these were relevant was not mentioned (20).

Table 3.11.1 Patch testing in groups of patients

Years and Country	Test conc. & vehicle	Number of patients		Selection of patients (S); Relevance (R); Comments (C)	Ref.
		tested	positive (%)		
Routine testing					
1991 Belgium	0.5% alc.	610	6 (1.0%)	R: not stated	19
	1% pet.	610	11 (1.8%)	R: not stated	
1991 The Netherlands	0.5% alc.	533	2 (0.4%)	R: not stated	19
	1% pet.	533	2 (0.4%)	R: not stated	
Testing in groups of selected patients					
2000-2005 USA	1% pet.	1187	11 (0.90%)	S: patients suspected of corticosteroid allergy; R: 100%	17
1985-1990 Finland	0.5% and 0.05% alc.	34	5 (15%)	S: patients very likely to be corticosteroid-allergic; R: not stated	20

Case series

From January 1990 to June 2008, in Leuven, Belgium, 315 patients were diagnosed with contact allergy to / allergic contact dermatitis from corticosteroids (CSs) from routine patch testing with a baseline series including tixocortol pivalate, budesonide, hydrocortisone butyrate and prednisone caproate, patch testing with patients' own CS preparations, and testing those with proven contact allergy to a corticosteroid or strongly suspected of CS allergy later with a series of 66 CSs, including two sex hormones (progesterone and testosterone). 71% of the patients had relevant reactions, but these were not specified. In this group of 315 CS allergic patients, 57 had positive patch tests to alclometasone dipropionate 0.5% alc. (16). It is unknown how many of these reactions were caused by the use of a pharmaceutical product containing alclometasone dipropionate and how many were cross-reactions to other corticosteroids.

In Japan, alclometasone dipropionate ointment was introduced in 1988. In that country, between 1991 and 1997, 10 cases of allergic contact dermatitis and one of photoallergic contact dermatitis (9) have been reported in Japanese literature (6-13), including small case series of 2 (10) and 3 patients (6). Eight were female; ages ranged from 20 to 71 years. In nine individuals, the face was affected by dermatitis. The ointment had been used from 2 days to 2 years before allergic contact dermatitis occurred, median 3 months (6-13, data cited in ref. 4).

In Sydney, Australia, from 1988 to 1994, 19 cases of corticosteroid allergy were diagnosed, including 7 to alclometasone dipropionate (1% alc. and pet.) (18).

In Japan, before 1994, a group of 74 patients with severe refractory atopic dermatitis involving the face were patch tested. There were 4 relevant reactions to alclometasone. It is uncertain whether all patients were tested with this hapten and the test concentrations and vehicles were not mentioned (3, possibly overlap with ref. 11).

Case reports

A 28-year-old man with atopic dermatitis presented with a 1-month history of edematous erythema of the face and trunk. He had been treated with two alclometasone dipropionate ointments on the face and trunk and bendazac ointment on the face for the last 6 months. Patch testing was positive to the 3 topical preparations, to alclometasone dipropionate 0.1%, 0.05% and 0.01% pet. and to bendazac 30%, 1.5% and 0.3% pet. (4).

A 25-year-old woman presented with dermatitis lasting for 1 year localized to the face, earlobes, and neck, characterized by severe erythema and edema as well as vesicles, oozing and itching. She had previously used alclomethasone dipropionate 0.05% cream and hydrocortisone butyrate 0.1% cream. Patch tests were positive to budesonide (indicator for corticosteroid allergy), alclometasone 1% pet. and hydrocortisone butyrate 1% alcohol. A ROAT with the cream containing alclomethasone was positive after 2 days (14).

A 22-year-old woman noticed worsening of atopic dermatitis of the face when she started using alclometasone dipropionate 0.1% ointment. Patch tests were positive to the ointment (+) and to the corticosteroid 0.1% pet. (+ only at D4) but negative to the ointment base (5).

Cross-reactions, pseudo-cross-reactions and co-reactions

Cross-reactions between corticosteroids are discussed in Chapter 2.8.

PHOTOSENSITIVITY

One case of photoallergic contact dermatitis from alclometasone dipropionate has been reported from Japan (9). The patient was a 71-year-old woman who had used alclometasone ointment for 1.5 months when she developed eczema in the face. Clinical and (photo)patch testing details are not available (article in Japanese, data cited in ref. 4).

LITERATURE

1 Baeck M, Goossens A. Immediate and delayed allergic hypersensitivity to corticosteroids: practical guidelines. Contact Dermatitis 2012;66:38-45

2 Van Ginkel CJ, Bruintjes TD, Huizing EH. Allergy due to topical medications in chronic otitis externa and chronic otitis media. Clin Otolaryngol Allied Sci 1995;20:326-328

3 Tada J, Toi Y, Arata J. Atopic dermatitis with severe facial lesions exacerbated by contact dermatitis from topical medicaments. Contact Dermatitis 1994;31:261-263

4 Iwakiri K, Hata M, Miura Y, Numano K, Yuge M, Sasaki E. Allergic contact dermatitis due to bendazac and alclometasone dipropionate. Contact Dermatitis 1999;41:218-219

5 Kabasawa Y, Kanzaki T. Allergic contact dermatitis from alclometasone dipropionate. Contact Dermatitis 1990;23:374-375

6 Yoshioka T, Yasutomi H, Miyashita M, Hirano N, Kuyama M, Kondo A, et al. Three cases of contact dermatitis due to Almeta ® ointment. Hifu 1991;33(Suppl.10):118-119 (Article in Japanese, data cited in ref. 4)

7 Mita T, Sugai T, Shoji A, Mochida K. A case of contact dermatitis from alclometasone 17,21-dipropionate and 4 cases of positive reaction to budesonide in patch tests. Hifu 1991;33(Suppl.11):281-289 (Article in Japanese, data cited in ref. 4)

8 Nishioka K. A few asides in the practice of contact dermatitis. Hifubyoh-Shinryoh 1994;16:107-111 (Article in Japanese, data cited in ref. 4)

9 Tamura T, Iizuka H, Kishiyama K. A case of photo-contact dermatitis due to alclometasone dipropionate. Rinsho Hifuka 1992;46:629-633 (Article in Japanese, data cited in ref. 4)

10 Tsuchida Y, Niimi Y, Sasaki E, Hata M, Yajima J, Hattori S, et al. Allergic contact dermatitis due to alclometasone dipropionate. Rinsho Derma 1993;35:1715-1718 (Article in Japanese, data cited in ref. 4)

11 Yoshioka T, Tada J, Arata J. 4 cases of allergic contact dermatitis to alclometasone dipropionate. Nishinihon J Dermatol 1993;55:459-463 (Article in Japanese, data cited in ref. 4)

12 Endo Y, Egawa Y, Yokoyama I, Ito F, Moriyama M. A case of allergic contact dermatitis due to alclometasone dipropionate. Rinsho Hifuka 1995;49:703-705 (Article in Japanese, data cited in ref. 4)

13 Sasao Y, Nakamori M, Iizuka M, Urano K, Ozawa A, Okido M. A case of allergic contact dermatitis due to alclometasone dipropionate. The 731st Local Society of the Japanese Dermatological Association in Tokyo, 1997 (Article in Japanese, data cited in ref. 4)

14 Blancas R, Ancona A, Arévalo A. Allergic contact dermatitis from alclomethasone dipropionate. Am J Contact Dermat 1995;6:115-116

15 Reitamo S, Lauerma AI, Förström L. Alclometasone dipropionate allergy. Contact Dermatitis 1991;25:78

16 Baeck M, Chemelle JA, Terreux R, Drieghe J, Goossens A. Delayed hypersensitivity to corticosteroids in a series of 315 patients: clinical data and patch test results. Contact Dermatitis 2009;61:163-175

17 Davis MD, El-Azhary RA, Farmer SA. Results of patch testing to a corticosteroid series: a retrospective review of 1188 patients during 6 years at Mayo Clinic. J Am Acad Dermatol 2007;56:921-927

18 Freeman S. Corticosteroid allergy. Contact Dermatitis 1995;33:240-242

19 Dooms-Goossens A, Meinardi MM, Bos JD, Degreef H. Contact allergy to corticosteroids: the results of a two-centre study. Br J Dermatol 1994;130:42-47

20 Lauerma AI. Contact hypersensitivity to glucocorticosteroids. Am J Contact Dermat 1992;3:112-132

Chapter 3.12   ALUMINUM ACETATE

IDENTIFICATION

Description/definition	: Aluminum acetate is the aluminum salt of acetic acid that conforms to the structural formula shown below
Pharmacological classes	: Dermatological agents; adstringents
IUPAC name	: Aluminum;triacetate
Other names	: Aluminum triacetate
CAS registry number	: 139-12-8
EC number	: Not available
Merck Index monograph	: 1590
Patch testing	: 1% and 5% water
Molecular formula	: $C_6H_9AlO_6$

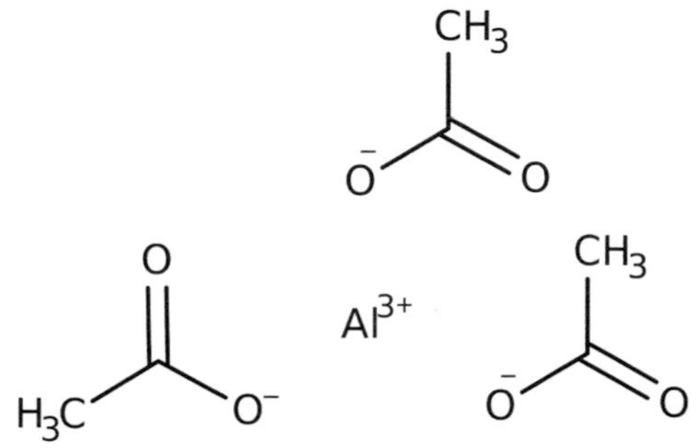

GENERAL

Aluminum acetate is a topical astringent that is used as an antiseptic agent. It reduces inflammation, itching, and stinging of the infected skin and is said to promote healing (1).

CONTACT ALLERGY

Case report

A 9-year-old boy had a 2-year history of eczema of the external auditory canals. He was congenitally deaf and had worn hearing aids since infancy. For the last 3 years, the patient had had chronic bilateral discharging otitis media. 5% aluminum acetate drops were effective in drying up the secretions, but otitis externa developed and persisted. He was patch tested to 126 haptens. At D2, all tests were negative, but at D4 all tests were positive, in each case with an eczematous ring corresponding to the rim of the Finn Chamber ®. Subsequent patch testing to aluminum acetate 1% and 5% water in plastic chambers showed ++ reactions at 2 and 4 days. The use of aluminum acetate ear drops was discontinued and the otitis externa has resolved, though chronic otitis media continues with variable severity. The boy's mother recollected that when aged 4 year, he had suffered a severe reaction with localized erythema and swelling accompanied by systemic upset to a diphtheria/tetanus/pertussis triple booster vaccination. This vaccine contains aluminum hydroxide and was considered to be the primary source of sensitization (2).

LITERATURE

1   The data in the section 'General' may have been obtained from literature discussed in this chapter, but mostly also or exclusively from one or more of the following online sources: ChemIDPlus Advanced, PubChem, DrugBank, RxList, Drug Central, Drugs.com, and Wikipedia
2   O'Driscoll JB, Beck MB, Kesseler ME, Ford G. Contact sensitivity to aluminium acetate eardrops. Contact Dermatitis 1991;24:156-157

Chapter 3.13   AMBROXOL

IDENTIFICATION

Description/definition   : Ambroxol is the aromatic amine and metabolite of bromhexine that conforms to the
                             structural formula shown below
Pharmacological classes   : Expectorants
IUPAC name   : 4-[(2-Amino-3,5-dibromophenyl)methylamino]cyclohexan-1-ol
Other names   : N-(2-Amino-3,4-dibromocyclohexyl)-trans-4-aminocyclohexanol
CAS registry number   : 18683-91-5
EC number   : 242-500-3
Merck Index monograph   : 1650
Patch testing   : 0.75% water when ambroxol is administered topically; 10% pet. when ambroxol is
                             given orally
Molecular formula   : $C_{13}H_{18}Br_2N_2O$

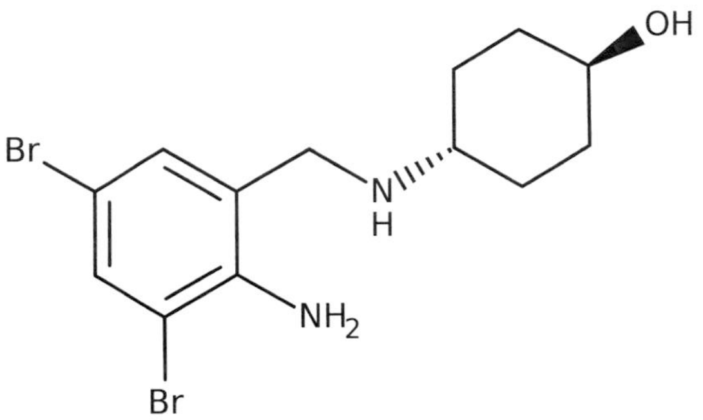

GENERAL

Ambroxol is an aromatic amine and a metabolite of bromhexine that stimulates mucociliary action and clears the air passages in the respiratory tract. As a secretolytic agent it is indicated for treatment of bronchopulmonary diseases with abnormal or excessive mucus secretion and transport. It allows the mucus to be more easily cleared and eases a patient's breathing. In pharmaceutical products, ambroxol is employed as ambroxol hydrochloride (CAS number 23828-92-4, EC number 245-899-2, molecular formula $C_{13}H_{19}Br_2ClN_2O$) (1).

CONTACT ALLERGY

Case report

A 60-year-old woman developed itchy erythema and vesicles around the nose, upper lip and cheeks. She also had pharyngeal soreness and spasm. 10 days earlier, she had started aerosol therapy for chronic sinusitis with ambroxol 0.75%, which she had used before. She was patch tested with this solution and freshly prepared aqueous solutions of ambroxol (0.75%, 0.25%, 0.125%, 0.0125%), the related bromhexine (0.75%, 0.25%) and the other ingredients of the commercial ambroxol fluid. Positive reactions were seen to the commercial ambroxol solution and ambroxol at all concentrations, but not to the related bromhexine. Concurrently with the positive patch tests to ambroxol, the patient experienced a focal flare of perinasal eczema. Twenty control subjects did not react to ambroxol 0.75% (3).

Cutaneous adverse drug reactions from systemic administration caused by type IV (delayed-type) hypersensitivity
Cutaneous adverse drug reactions from systemic administration of ambroxol caused by type IV (delayed-type) hypersensitivity, including maculopapular exanthema (4) and photoallergy (2), are planned to be discussed in Volume IV of the Monographs in Contact Allergy series on Systemic drugs.

LITERATURE

1   The data in the section 'General' may have been obtained from literature discussed in this chapter, but mostly
    also or exclusively from one or more of the following online sources: ChemIDPlus Advanced, PubChem,
    DrugBank, RxList, Drug Central, Drugs.com, and Wikipedia
2   Fujimoto N, Danno K, Wakabayashi M, Uenishi T, Tanaka T. Photosensitivity with eosinophilia due to ambroxol
    and UVB. Contact Dermatitis 2009;60:110-113
3   Mancuso G, Berdondini RM. Contact allergy to ambroxol. Contact Dermatitis 1989;20:154
4   Monzón S, Del Mar Garcés M, Lezaun A, Fraj J, Asunción Dominguez M, Colás C. Ambroxol-induced systemic
    contact dermatitis confirmed by positive patch test. Allergol Immunopathol (Madr) 2009;37:167-168

Chapter 3.14   AMCINONIDE

IDENTIFICATION

Description/definition : Amcinonide is the synthetic glucocorticoid that conforms to the structural formula shown below

Pharmacological classes : Glucocorticoids

IUPAC name : [2-[(1S,2S,4R,8S,9S,11S,12R,13S)-12-Fluoro-11-hydroxy-9,13-dimethyl-16-oxospiro[5,7-dioxapentacyclo[10.8.0.0^{2,9}.0^{4,8}.0^{13,18}]icosa-14,17-diene-6,1'-cyclopentane]-8-yl]-2-oxo-ethyl] acetate

Other names : 9-Fluoro-11β,16α,17,21-tetrahydroxypregna-1,4-diene-3,20-dione cyclic 16,17-acetal with cyclopentanone, 21-acetate; triamcinolone 16,17-cyclopentylidenedioxy-21-acetate

CAS registry number : 51022-69-6

EC number : 256-915-2

Merck Index monograph : 1652

Patch testing : 0.1% pet. (SmartPracticeCanada, SmartPracticeEurope)

Molecular formula : C$_{28}$H$_{35}$FO$_7$

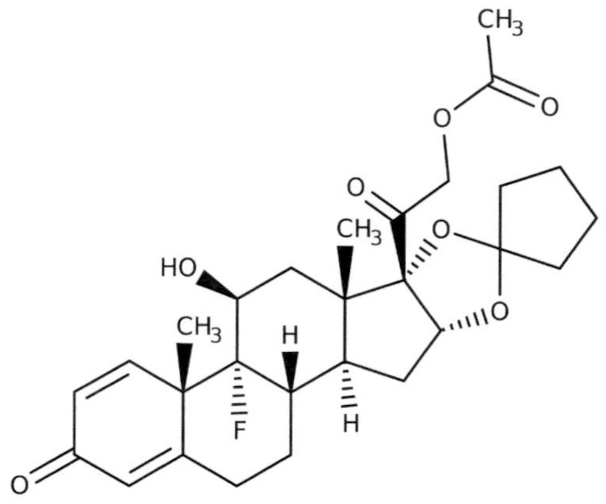

GENERAL

General aspects of corticosteroids used on the skin and mucous membranes are discussed in Chapter 2.4. A practical guideline for diagnosing allergic reactions to corticosteroids is presented in ref. 3.

CONTACT ALLERGY

Contact allergy in the general population

With the CE-DUR approach, the incidence of sensitization to amcinonide in the German population was estimated to range from 1 to 3 cases/100,000/year in the period 1995-1999 and from 1 to 3 cases/100,000/year in the period 2000-2004 (1). Also in Germany, for the period 1995-2004, the population-based relative incidence (RI) of contact sensitization to amcinonide (cases/100,000 defined daily doses (DDDs) per year) was estimated to be 23.6. In the group of corticosteroids, the RI ranged from 0.3 (dexamethasone sodium phosphate) to 43.2 (budesonide) (2).

Patch testing in consecutive patients suspected of contact dermatitis: Routine testing

In Poland, before 2005, 257 consecutive patients were patch tested with amcinonide 0.1% pet. and there were 7 positive reactions, one of which was considered to be relevant (20). In 1988-1990, in Belgium, 1960 consecutive patients were tested with amcinonide 1% alc. and there were 10 (0.5%) positive reactions; no relevance data were provided (25).

Patch testing in groups of selected patients

The results of patch testing amcinonide in groups of selected patients (patients with anogenital dermatitis, individuals suspected of corticosteroid allergy) are shown in table 3.14.1. In 6 such investigations (5 performed by the IVDK: Germany, Austria, Switzerland) the prevalences of sensitization ranged from 1.3% to 3.0%. In 5 of the studies, no relevance data were provided, but in a USA study, 23 of 24 positive reactions (96%) were considered to be relevant (18).

Table 3.14.1 Patch testing in groups of patients: Selected patient groups

Years and Country	Test conc. & vehicle	Number of patients tested	positive (%)	Selection of patients (S); Relevance (R); Comments (C)	Ref.
2004-2008 IVDK			(1.3%)	S: patients with anogenital dermatoses tested with a medicament series; R: not stated; C: number of patients tested unknown	5
2000-2005 USA	1% alc. 70%	1184	24 (2.0%)	S: patients suspected of corticosteroid allergy; R: 96%	18
1995-2004 IVDK	0.1% pet.	6130	(1.6%)	S: not stated; R: not stated	1
1995-2004 IVDK	0.1% pet.	5925	94 (1.6%)	S: patients tested with a corticosteroid series; R: not stated	19
1990-2003 IVDK		193	3 (1.6%)	S: patients with perianal dermatoses; R: not stated; C: the frequency was not higher than in a control group	4
1996-1997 IVDK	0.1% pet.	608	18 (3.0%)	S: patients tested with a corticosteroid series; R: not stated	22

IVDK: Information Network of Departments of Dermatology (Germany, Austria, Switzerland)

Case series
From January 1990 to June 2008, in Leuven, Belgium, 315 patients were diagnosed with contact allergy to/allergic contact dermatitis from corticosteroids (CSs) from routine patch testing with a baseline series including tixocortol pivalate, budesonide, hydrocortisone butyrate and prednisone caproate, patch testing with patients' own CS preparations, and testing those with proven contact allergy to a corticosteroid or strongly suspected of CS allergy later with a series of 66 CSs, including two sex hormones (progesterone and testosterone). 71% of the patients had relevant reactions, but these were not specified. In this group of 315 CS allergic patients, 91 had positive patch tests to amcinonide 0.1% alc. (17). It is unknown how many of these reactions were caused by the use of a pharmaceutical product containing amcinonide and how many were cross-reactions to other corticosteroids.

In the period 1982-1994, in a center in Tokyo, 11 patients were diagnosed with allergic contact dermatitis from topical corticosteroids, including two cases caused by amcinonide (29). In reviewing the Japanese literature up to 1994, 43 patients with allergic contact dermatitis from topical corticosteroid were identified, including 8 caused by amcinonide (29).

In a hospital in the USA, between 1981 and 1991 (before the introduction of screening markers), 19 patients were found to have positive patch test reactions to commercial topical steroid preparations. Eleven were allergic to the corticosteroid ingredients: 3 to amcinonide, 5 to hydrocortisone, 1 to hydrocortisone butyrate and 2 to multiple CSs. Ten co-reacted or reacted alone to another active ingredient (neomycin) or one or more excipients (23).

Case reports
A 20-year-old atopic woman was treated for eczema on the backs of the hands with amcinonide 0.05% cream. Within 24 hours, her lesions acutely worsened, with massive edema and copious exudation. Edema and erythema also appeared on the eyelids and lips, sites never previously involved. Patch tests revealed contact allergy to 3 formulations of amcinonide (cream, ointment and lotion), budesonide, and triamcinolone acetonide. Simultaneously, the patient experienced exacerbation of her hand and face dermatitis, accompanied by distant spread to the legs (6). Short summaries of other case reports of allergic contact dermatitis from amcinonide are shown in table 3.14.2. Other case reports of contact allergy to/allergic contact dermatitis from amcinonide, adequate data of which are not available to the author, can be found in refs. 10,11,12, 30 and 31. One Japanese author in 1994 reviewed the Japanese literature and found 10 reported cases of allergic contact dermatitis to amcinonide in 8 articles published in the period 1984-1992), probably including refs. 10-12) (21).

Table 3.14.2 Short summaries of case reports of allergic contact dermatitis from amcinonide

Year and country	Sex	Age	Positive patch tests	Clinical data and comments	Ref.
1993 Japan	F	34	AM cream and ointment 'as is'; AM 5%, 1% and 0.1% pet.	worsening of allergic contact dermatitis of the earlobes from golden earrings	24
1991 Germany	F	68	AM comm prep 'as is'; amcinonide 2% pet.	contact dermatitis around leg ulcers	26
	F	36	AM comm prep 'as is'; amcinonide 2% pet.	contact dermatitis on psoriasis	
1990 Japan	F	24	AM cream and ointment 'as is'; AM 1% pet.	eczema of the face worsened by AM ointment	7
1990 Germany	F	36	AM ointment 'as is'; AM powder 2% pet.	acute dermatitis after 3 applications of AM ointment to psoriasis on the elbows, spreading to the entire body except the back	8

Table 3.14.2 Short summaries of case reports of allergic contact dermatitis from amcinonide (continued)

Year and country	Sex	Age	Positive patch tests	Clinical data and comments	Ref.
1989 France	M	26	AM 0.1% pet.	generalized atopic dermatitis aggravated by AM cream; the patch test was still positive after 6 months	27
	F	71	AM 0.1% pet.	dermatitis around leg ulcer after AM cream	
1988 Japan	M	39	AM ointment and cream 'as is'; ointment diluted 2x and 10x	erythematous eruption after one application of amcinonide; never used before; many other CS patch test reactions; negative to the AM cream and ointment bases	9
1988 Japan	M	35	AM cream 'as is'; amcinonide 0.1% pet.	worsening of pre-existing non-specific eczema	28
1987 Japan	F	46	AM ointment and cream 'as is'; AM 0.1% pet.	worsening of red papules on the face one day after using AM ointment; negative to the AM cream/ointment bases	13
1986 Japan	M	12	AM ointment 'as is'; AM 0.01%, 0.05%, 0.1% and 1% pet.	swelling, erythema and papules on the face after treating dermatitis perioralis with AM ointment for a month	14
1985 Japan	F	35	AM ointment 'as is'; AM 1% and 0.1% pet.	worsening of eczema on the back of the hands after 2 days of AM cream use	15
1984 France	F	36	AM cream 'as is'; AM 0.1% pet.	exacerbation of hand dermatitis spreading to the face 1 day after first use of AM cream	16
	F	28	AM cream 'as is'; AM 0.1% pet.	exacerbation of dyshidrotic eczema of the hands within one day after using AM cream	
	F	48	AM cream 'as is'; AM 0.1% pet.	acute dermatitis of the face and left hand after using AM cream for 5-6 days; also allergic to neomycin in the combination preparation AM-neomycin cream	

AM: amcinonide; CS corticosteroid(s); comm prep: commercial preparation

Cross-reactions, pseudo-cross-reactions and co-reactions
Cross-reactions between corticosteroids are discussed in Chapter 2.8.

LITERATURE

1   Menezes de Padua CA, Uter W, Schnuch A. Contact allergy to topical drugs: prevalence in a clinical setting and estimation of frequency at the population level. Pharmacoepidemiol Drug Saf 2007;16:377-384

2   Menezes de Padua CA, Schnuch A, Nink K, Pfahlberg A, Uter W. Allergic contact dermatitis to topical drugs – Epidemiological risk assessment. Pharmacoepidemiol Drug Saf 2008;17:813-821

3   Baeck M, Goossens A. Immediate and delayed allergic hypersensitivity to corticosteroids: practical guidelines. Contact Dermatitis 2012;66:38-45

4   Kügler K, Brinkmeier T, Frosch PJ, Uter W. Anogenital dermatoses—allergic and irritative causative factors. Analysis of IVDK data and review of the literature. J Dtsch Dermatol Ges 2005;3:979-986

5   Bauer A. Contact sensitization in the anal and genital area. Curr Probl Dermatol 2011;40:133-141

6   Sasseville D. Exacerbation of allergic contact dermatitis from amcinonide triggered by patch testing. Contact Dermatitis 2001;45:232-233

7   Hayakawa R, Matsunaga K, Suzuki M, Ogino Y, Arisu K, Arima Y, et al. Allergic contact dermatitis due to amcinonide. Contact Dermatitis 1990;23:49-50

8   Dunkel FG, Elsner P, Pevny I, Röger J, Burg G. Amcinonide-induced contact dermatitis—Possibly due to the ketal structure. Am J Contact Dermat 1990;1:246-249

9   Sasaki E, Shinya M, Fujioka M, Sato M, Hata M, Yajima J, et al. Contact dermatitis due to amcinonide. Contact Dermatitis 1988;18:61-62

10 Oh Y, Yoshii T, Tamoku K. A case of contact dermatitis due to amcinonide cream: conceivable cross reactions between some topical steroid preparations. Skin Research 1984;26:648-653 (Article in Japanese, data cited in ref. 9)

11 Hosokawa K, Uke C, Matsunaga K, Hayakawa R. Amcinonide contact dermatitis. Skin Research 1985;27:564-570 (Article in Japanese, data cited in ref. 9)

12 Kubo Y. A case of contact dermatitis from amcinonide. Nishinihon J Dermatol 1986;48:1075-1078 (Article in Japanese, data cited in ref. 9)

13 Hayakawa R, Matsunaga K, Suzuki M, Hosokawa K. Contact dermatitis from amcinonide. Contact Dermatitis 1987;16:48-49

14 Kubo Y, Nonaka S, Yoshida H. Contact allergy to amcinonide. Contact Dermatitis 1986;15:109-111

15 Hayakawa R, Matsunaga K, Ukei C, Hosokawa K. Allergic contact dermatitis from amcinonide. Contact Dermatitis 1985;12:213-214

16 Boujnah-Khouadja A, Brändle I, Reuter G, Foussereau J. Allergy to 2 new corticoid molecules. Contact Dermatitis 1984;11:83-87

17 Baeck M, Chemelle JA, Terreux R, Drieghe J, Goossens A. Delayed hypersensitivity to corticosteroids in a series of 315 patients: clinical data and patch test results. Contact Dermatitis 2009;61:163-175

18 Davis MD, El-Azhary RA, Farmer SA. Results of patch testing to a corticosteroid series: a retrospective review of 1188 patients during 6 years at Mayo Clinic. J Am Acad Dermatol 2007;56:921-927

19 Uter W, de Pádua CM, Pfahlberg A, Nink K, Schnuch A, Lessmann H. Contact allergy to topical corticosteroids – results from the IVDK and epidemiological risk assessment. J Dtsch Dermatol Ges 2009;7:34-41

20 Reduta T, Laudanska H. Contact hypersensitivity to topical corticosteroids – frequency of positive reactions in patch-tested patients with allergic contact dermatitis. Contact Dermatitis 2005;52:109-110

21 Okano M. Contact dermatitis due to budesonide: report of five cases and review of the Japanese literature. Int J Dermatol 1994;33:709-715

22 Uter W, Geier J, Richter G, Schnuch A; IVDK Study Group, German Contact Dermatitis Research Group. Patch test results with tixocortol pivalate and budesonide in Germany and Austria. Contact Dermatitis 2001;44:313-314

23 Jagodzinski LJ, Taylor JS, Oriba H. Allergic contact dermatitis from topical corticosteroid preparations. Am J Contact Dermat 1995;6:67-74

24 Hisa T, Katoh J, Yoshioka K, Taniguchi S, Mochida K, Nishimura T, et al. Contact allergies to topical corticosteroids. Contact Dermatitis 1993;28:174-179

25 Dooms-Goossens A, Morren M. Results of routine patch testing with corticosteroid series in 2073 patients. Contact Dermatitis 1992;26:182-191

26 Dunkel FG, Elsner P, Burg G. Contact allergies to topical corticosteroids: 10 cases of contact dermatitis. Contact Dermatitis 1991;25:97-103

27 Rivara G, Tomb RR, Foussereau J. Allergic contact dermatitis from topical corticosteroids. Contact Dermatitis 1989;21:83-91

28 Feldman SB, Sexton FM, Buzas J, Marks JG Jr. Allergic contact dermatitis from topical steroids. Contact Dermatitis 1988;19:226-228

29 Oh-i T. Contact dermatitis due to topical steroids with conceivable cross reactions between topical steroid preparations. J Dermatol 1996;23:200-208

30 Senff H, Kunze J, Kreyes E, Hausen BM. Allergisches Kontaktekzem gegen Bufexamac und Amcinonid mit Kreuzreaktionen zu chemisch verwandten Glukokortikosteroiden. Dermatosen 1992;40:62-65 (Article in German)

31 Fedler R, Pilz B, Frosch PJ. Contact allergy to topical glucocorticoids. Hautarzt 1993;44:91-95 (Article in German)

Chapter 3.15   AMINOCAPROIC ACID

IDENTIFICATION

Description/definition   : Aminocaproic acid is the synthetic lysine derivative that conforms to the structural
                            formula shown below
Pharmacological classes : Antifibrinolytic agents
IUPAC name   : 6-Aminohexanoic acid
Other names   : epsilon-Aminocaproic acid; ε-aminocaproic acid
CAS registry number   : 60-32-2
EC number   : 200-469-3
Merck Index monograph   : 1697
Patch testing   : 5% pet.; 10% water
Molecular formula   : $C_6H_{13}NO_2$

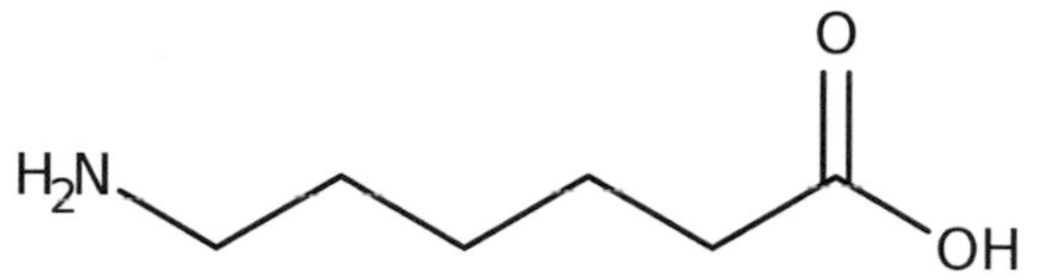

GENERAL

Aminocaproic acid (ACA) is an antifibrinolytic agent that acts by inhibiting plasminogen activators which have fibrinolytic properties. This drug is used in the treatment and prophylaxis of hemorrhage, including hyperfibrinolysis-induced hemorrhage and excessive postoperative bleeding (1). ε-Aminocaproic acid was discovered in Japan in 1953. In addition to its antifibrinolytic properties, it acts as buffer and has an anti-inflammatory effect. Therefore, ACA is incorporated in various products in Japan, such as eye drops, eye rinses, detergents, toothpastes, soaps, creams, and cosmetics (2).

CONTACT ALLERGY

General

There have been some 15 case reports of allergic contact dermatitis from aminocaproic acid (ACA). Most cases have been observed in Japan. Eye drops were the most common cause, allergy resulting in – mostly mild – erythematous dermatitis around the eyes and on the cheeks. Sensitization usually occurred after a prolonged period of contact with the incriminated eye drops (3).

Case reports

A 78-year-old woman presented with bilateral periocular pruritic erythema and edema. She had used purified sodium hyaluronate ophthalmic solution 0.1% and some other pharmaceuticals for the treatment of 'dry eyes' and keratoconjunctivitis sicca for 2 years. Patch tests were positive tot the sodium hyaluronate solution only. When tested later with its ingredients, the patient showed positive patch tests to aminocaproic acid 1% and 2% water, which functions as a buffer in the eye solution (2).

  Short summaries of other case reports of allergic contact dermatitis from ε-aminocaproic acid are shown in table 3.15.1.

Table 3.15.1 Short summaries of case reports of allergic contact dermatitis from aminocaproic acid

Year	Sex	Age	Clinical picture	Positive patch tests	Comments	Ref.
2014	F	67	recurrent blepharitis	aminocaproic acid, test concentration and vehicle unknown	article not read; ACA was present in sodium hyaluronate eye drops as buffer	12
2013	F	78	pruritic erythema and licheni-fication around the eyes	sodium hyaluronate eye drops, aminocaproic acid 0.2% water	eye drops only positive in scratch-patch test	3
2000	F	63	itchy erythema in the peri-orbital areas and on cheeks	ACA 1% eye drops, ACA 1% pet.		5
1993	F	53	itchy erythematous lesions around the eyes and on the cheeks	ACA 3% eye drops, ACA 3% and 6% petrolatum		9
1989	F	60	pruritic erythematous ede-matous periocular dermatitis	ACA 1% eye drops, ACA 0.1%, 1% and 10% water	20 controls were negative to ACA 10% water	10

ACA: aminocaproic acid

Two more case report (2 women, 59 and 60 years) of contact allergy to aminocaproic acid from its presence in eye drops, adequate data of which are not available to the author, can be found in ref. 11. Japanese authors (5) mention in the table 'Previous reports of allergic contact dermatitis from epsilon-aminocaproic acid' in their publication 7 patients sensitized to aminocaproic acid from its presence in one particular brand of eye drops, which is widely used in Japan for cataracts. These were apparently published by various Japanese authors between 1992 and 1997, but bibliographical references were not provided (5).

Polymerized aminocaproic acid is used in nylon, which has caused a case of non-pharmaceutical sensitization to ACA. The patient, who developed scaly erythema on her trunk from wearing nylon body stocking, reacted to 3% and 6% aminocaproic acid pet. (8). Another female patient sensitized to aminocaproic acid from eye drops experienced occasional dermatitis on her legs after wearing nylon stockings. It was hypothesized that this was caused by aminocaproic acid monomer released into sweat from the stockings (11).

Cutaneous adverse drug reactions from systemic administration caused by type IV (delayed-type) hypersensitivity
Cutaneous adverse drug reactions from systemic administration of aminocaproic acid caused by type IV (delayed-type) hypersensitivity, including symmetrical drug-related intertriginous and flexural exanthema (SDRIFE) (4), generalized micropapular eruption (6), and maculopapular eruptions (7), are planned to be discussed in Volume IV of the *Monographs in Contact Allergy* series on Systemic drugs.

LITERATURE

1   The data in the section 'General' may have been obtained from literature discussed in this chapter, but mostly also or exclusively from one or more of the following online sources: ChemIDPlus Advanced, PubChem, DrugBank, RxList, Drug Central, Drugs.com, and Wikipedia
2   Mitsuyama S, Abe F, Kimura M, Higuchi T. Allergic contact dermatitis caused by ∈-aminocaproic acid in a purified sodium hyaluronate ophthalmic solution. Contact Dermatitis 2017;77:191-192
3   Yamamoto Y, Wada M, Nakai N, Katoh N. Allergic contact dermatitis due to epsilon-aminocaproic acid: a case report and mini-review of the published work. J Dermatol 2013;40:301-303
4   Cunha D, Carvalho R, Santos R, Cardoso J. Systemic allergic dermatitis to epsilon-aminocaproic acid. Contact Dermatitis 2009;61:303-304
5   Miyamoto H, Okajima M. Allergic contact dermatitis from epsilon-aminocaproic acid. Contact Dermatitis 2000;42:50
6   Villarreal O. Systemic dermatitis with eosinophilia due to epsilon-aminocaproic acid. Contact Dermatitis 1999;40:114
7   Gutiérrez M, López M, Ruiz M. Positivity of patch tests in cutaneous reaction to aminocaproic acid: two case reports. Allergy 1995;50:745-746
8   Tanaka M, Kobayashi S, Miyakawa S. Contact dermatitis from nylon 6 in Japan. Contact Dermatitis 1993;28:250
9   Tanaka M, Niizeki H, Miyakawa S. Contact dermatitis from epsilon-aminocaproic acid. Contact Dermatitis 1993;28:124
10 Shono M. Allergic contact dermatitis from epsilon-aminocaproic acid. Contact Dermatitis 1989;21:106-107
11 Sugai T. Allergic contact dermatitis to medicaments. In: Atlas of contact dermatitis. Tokyo: Kanehara Shuppan, 1986:254-256
12 Sukegawa T. A case of allergic contact dermatitis due to unit dose type purified sodium hyaluronate ophthalmic solution. Nippon Ganka Gakkai Zasshi 2014;118:111-115 (Article in Japanese)

Chapter 3.16   AMINOLEVULINIC ACID

IDENTIFICATION

Description/definition   : Aminolevulinic acid is the delta-amino acid and a 4-oxo monocarboxylic acid that conforms to the structural formula shown below
Pharmacological classes : Photosensitizing agents
IUPAC name   : 5-Amino-4-oxopentanoic acid
Other names   : δ-Aminolevulinic acid
CAS registry number   : 106-60-5
EC number   : 203-414-1
Merck Index monograph : 1713
Patch testing   : 20% water; a concentration of 10% in petrolatum causes irritant reactions (3)
Molecular formula   : $C_5H_9NO_3$

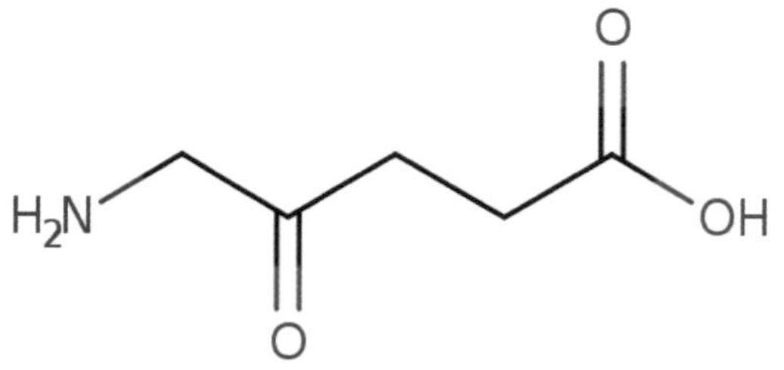

GENERAL

Aminolevulinic acid is a topically administered metabolic precursor of protoporphyrin IX. After topical administration, aminolevulinic acid is converted to protoporphyrin IX, which is a photosensitizer. When the proper wavelength of light activates protoporphyrin IX, singlet oxygen is produced, resulting in a local cytotoxic effect. Aminolevulinic acid is indicated for use in photodynamic therapy to treat actinic keratoses and superficial basal cell carcinomas. In pharmaceutical products, aminolevulinic acid is employed as aminolevulinic acid hydrochloride (CAS number 5451-09-2, EC number 226-679-5, molecular formula $C_5H_{10}ClNO_3$) (1).

See also Chapter 3.213 Methyl aminolevulinate.

CONTACT ALLERGY

Case report

A 54-year-old woman was treated for the first time with photodynamic therapy using 20% aminolevulinic acid (ALA) gel for Bowen's disease of the vulva. By the following day, an intensely itchy, erythematous and papular dermatitis had developed in the genital region, spreading to the groins, abdomen, buttocks, and medial thighs. Patch testing showed positive reactions to the photosensitizing gel 'as is' and to a 20% aqueous solution of ALA, but negative reactions to the other ingredients. Ten controls were negative to 20% ALA in water. How the patient had previously become sensitized to ALA was unknown (2).

Cross-reactions, pseudo-cross-reactions and co-reactions

Most patients sensitized to methyl aminolevulinate probably do not cross-react to aminolevulinic acid, but data are conflicting. This topic is discussed in Chapter 3.213 Methyl aminolevulinate.

PHOTOSENSITIVITY

While driving to work on a sunny morning, a 30-year-old dermatology resident experienced intense burning on his forehead that abated abruptly upon arrival at clinic. A sickle-shaped erythematous patch was noted. A diagnosis of phytophotodermatitis was proposed; however, the resident did not recall recent contact with plants. However, the diagnosis of photosensitivity dermatitis was confirmed when the resident remembered being inadvertently squirted with 5-aminolevulinic acid while in clinic 24 hours prior and neglecting to wash it off. The lesion resolved spontaneously within a week (4).

Immediate contact reactions

Immediate contact reactions (photocontact urticaria) to aminolevulinic acid are presented in Chapter 5.

LITERATURE

1   The data in the section 'General' may have been obtained from literature discussed in this chapter, but mostly also or exclusively from one or more of the following online sources: ChemIDPlus Advanced, PubChem, DrugBank, RxList, Drug Central, Drugs.com, and Wikipedia

2   Gniazdowska B, Ruëff F, Hillemanns P, Przybilla B. Allergic contact dermatitis from delta-aminolevulinic acid used for photodynamic therapy. Contact Dermatitis 1998;38:348-349

3   Jungersted JM, Dam TN, Bryld LE, Agner T. Allergic reactions to Metvix (ALA-ME). Contact Dermatitis 2008;58:184-186

4   Zeltser R, Gilchrest BA. Photosensitivity dermatitis from inadvertent exposure to aminolevulinic acid. Cutis 2007;80:124

Chapter 3.17 3-(AMINOMETHYL)PYRIDYL SALICYLATE

IDENTIFICATION

Description/definition : 3-(Aminomethyl)pyridyl salicylate is salicylic acid, compounded with pyridine-3-methylamine; it conforms to the structural formula shown below

Pharmacological classes : Dermatological agents; topical analgesics; rubefacients

IUPAC name : 2-Hydroxybenzoic acid;pyridin-3-ylmethanamine

Other names : Salicylic acid, compound with pyridine-3-methylamine (1:1); 2-hydroxybenzoic acid-1-(pyridin-3-yl)methanamine(1:1); 3-pyridylmethylamine salicylate; picolamine salicylate

Synonyms : Synonyms for pyridine-3-methylamine: picolamine; 3-(aminomethyl)pyridine; 3-pyridinemethylamine

CAS registry number : 34148-38-4

EC number : 251-852-7

Merck Index monograph : 9739 (Salicylic acid)

Patch testing : 3-(Aminomethyl)pyridyl salicylate 1% water; it is preferable to test 3-(aminomethyl)-pyridine 1% water, as this is the allergenic part of the compounded formula (4)

Molecular formula : $C_{13}H_{14}N_2O_3$

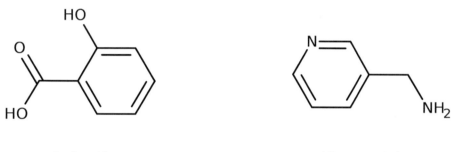

salicylic acid   pyridine-3-methylamine

GENERAL

3-(Aminomethyl)pyridyl salicylate is a salicylic acid derivative which is used in certain countries in topical pharmaceuticals as topical analgesic and anti-inflammatory agent.

CONTACT ALLERGY

Case series

In 1987, in a report from Spain, 11 patients were described who were allergic to 3-(aminomethyl)pyridyl salicylate present in 2 proprietary analgesic sprays available in pharmacies there. The group consisted of 7 women and 4 men, with an age range of 26-68. The sprays had been applied to various body parts, e.g. for arthrosis, trauma or contusion. In a number of persons the dermatitis spread extensively and sometimes lasted for 4-6 weeks despite treatment. In 3, the spray had been applied to others, which led to airborne contact dermatitis of the face. In some patients, the pharmaceuticals had never been used before. Patch testing with the sprays in 4 patients gave such strong reactions that testing the pharmaceuticals themselves was abandoned. It was found that that the sensitizer in the sprays was the 3-(aminomethyl)pyridine moiety (tested at 1% water), not the salicylate part (4).

Case reports

A 35-year-old male with a painful shoulder was treated with a topical medicament containing picolinamine (should be 'picolamine') salicylate 1% (3-(aminomethyl)pyridyl salicylate). On the 1st day of treatment, he developed erythema, pruritus, and vesicular skin lesions over the treated area, which cleared within 7 days with a topical corticosteroid. Patch tests were positive to picolinamine salicylate 2% oil (D2 ++, D3 +++) and (possibly) the cream. The excipients were negative. Four controls were negative to the cream and the salicylate. The primary source of sensitization was not mentioned (1).

A 49-year-old woman had suffered acute severe itchy bullous dermatitis on the thorax after a few applications of an analgesic spray for neuralgia, containing 3-(aminomethyl)pyridyl salicylate (3-APS). Patch tests with all the components of the spray gave a positive reaction (D4 ++) to 3-APS 1% water. A lymphocyte transformation test (LTT) with 1 x 10^{-3} aq. solution of 3-APS was strongly positive, with a response index (RI) of 8.69 (normal RI < 2). 35 controls were negative to 3-APS 1% water (2,3,5).

The same author presented a 39-year-old housewife known with idiopathic angioedema, who had suffered two acute attacks of skin allergy. First, a severe itchy vesicular dermatitis had appeared on the abdomen after connubial contact. Her husband had applied an analgesic spray to treat neuralgia, containing 3-(aminomethyl)pyridyl salicylate, a few days before. Sometime after this episode, an acute relapse of her angioedema, followed by respiratory symptoms, occurred immediately after her son used the same spray to treat a painful traumatic lesion. She had not touched the spray, nor the skin of her son, suggesting that the skin and respiratory symptoms were mediated by airborne contact. Patch tests were positive to 3-APS 1% water (D4 ++). A LTT with 2 x 10^{-3} aq. solution of 3-APS showed a strongly positive reaction, with an RI of 7.75 (2,3,5).

Cross-reactions, pseudo-cross-reactions and co-reactions
A cross-reaction has been suggested with *p*-phenylenediamine (PPD) and other para-compounds (2,4). Two of two patients (2), one of one (1) and 9 of 11 (4) were allergic to both 3-(aminomethyl)pyridyl salicylate and PPD, some of who also to other para-substances such as benzocaine and aniline (2,4). No cross-reaction to ethyl salicylate or methyl salicylate (4).

LITERATURE

1   Gamboa P, Jáuregui I, Fernández JC, Antépara I, Urrutia I. Contact sensitization to topical salicylate. Contact Dermatitis 1995;33:52
2   Camarasa JG, Lluch M, Serra-Baldrich E, Zamorano M, Malet A, García-Calderón PA. Allergic contact dermatitis from 3-(aminomethyl)-pyridyl salicylate. Contact Dermatitis 1989;20:347-351
3   Camarasa JG. Allergic contact dermatitis to 3-(amino methyl)-pyridyl salicylate. Am J Cont Dermatit 1990;1(4):254
4   Schmidt RJ, Fernández de Corres L. Allergic contact dermatitis from proprietary topical analgesic sprays containing 3-(aminomethyl)-pyridyl salicylate. Dermatologica 1987;174:272-279
5   Camarasa JG. Analgesic spray contact dermatitis. Dermatologic Clinics 1990;8:137-138

Chapter 3.18 AMLEXANOX

IDENTIFICATION

Description/definition	: Amlexanox is the carboxylic acid derivative that conforms to the structural formula shown below
Pharmacological classes	: Anti-allergic agents
IUPAC name	: 2-Amino-5-oxo-7-propan-2-ylchromeno[2,3-b]pyridine-3-carboxylic acid
Other names	: 2-Amino-7-isopropyl-5-oxo-5H-(1)benzopyrano(2,3-b)pyridine-3-carboxylic acid
CAS registry number	: 68302-57-8
EC number	: Not available
Merck Index monograph	: 1756
Patch testing	: 1% water; petrolatum is less suitable as patch testing vehicle (13)
Molecular formula	: $C_{16}H_{14}N_2O_4$

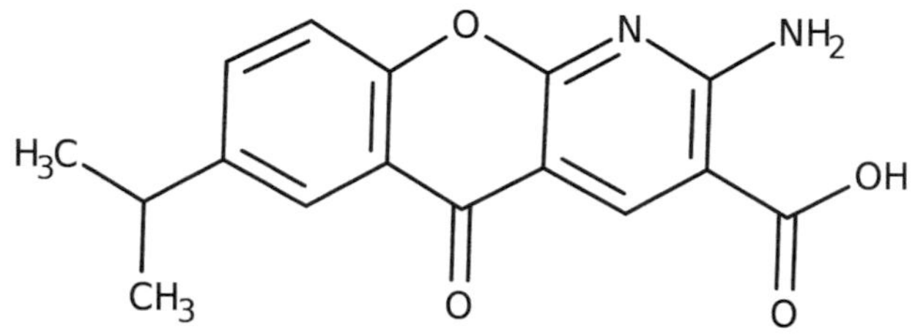

GENERAL

Amlexanox is an anti-allergic and anti-inflammatory drug. It is a strong inhibitor of histamine release and a competitive inhibitor of leukotriene. In a mucoadhesive oral paste, this agent has been clinically proven to abort the onset, accelerate healing and resolve the pain of aphthous ulcers (canker sores). Tablets have been on the market since 1987 in Japan and are used for the treatment of asthma and allergic rhinitis. Amlexanox is (or was) also available as ophthalmic solution for allergic conjunctivitis and as nasal douche for treatment of allergic rhinitis (1,2). Probably, the ophthalmic solution is not used anymore, but the oral paste and tablets appear to be available in some countries (drugs.com).

CONTACT ALLERGY

General

The ophthalmic solution became available in Japan in April 1989. Within 18 months, 24 cases of allergic contact dermatitis from amlexanox were reported from 11 Japanese hospitals (16). In 1994, animal experiments showed the sensitizing capacity of amlexanox to be comparable to that of p-phenylenediamine and DNCB, which are very strong contact sensitizers in man as well as in experimental animals. Oral or intraperitoneal administration in sensitized animals led to systemic contact dermatitis with generalized rash, usually accompanied by a flare-up reaction at the previous patch test sites (17).

All but one (nasal douche, ref. 18) cases of allergic contact dermatitis to amlexanox have been caused by one brand of ophthalmic solution containing 0.25% amlexanox. All cases were reported from Japan. After 1994, no new cases of allergy to amlexanox appear to have been published.

Case reports

A 23-year-old woman with allergic conjunctivitis had been treated with 0.25% amlexanox ophthalmic solution for 1½ year, when she woke up with itchy redness around her eyes. Patch tests showed a positive reaction to the solution. Later, the ingredients of the product were tested separately, and the patient reacted to amlexanox 1%, 0.25%, 0.1% and 0.025% water and to 1% and 0.25% pet. Her eyelid dermatitis cleared with discontinuance of the ophthalmic solution and topical corticosteroids. Four months later, an otologist prescribed tablets containing 50 mg amlexanox. The first tablet resulted in itching after an hour and the second, a day later, was followed by an erythema multiforme-like eruption on the patient's ears, neck, breasts and trunk. A diagnosis of systemic contact dermatitis was made (6).

The authors of ref. 6 mention that 'many cases of contact dermatitis from amlexanox 0.25% ophthalmic solution have been reported' and cited 6 Japanese literature references from 1990 and 1991 describing at least 13 sensitized patients, including a series of three and one of four. As all reports are in Japanese, details are unknown (7-12). The authors of ref. 13 refer to another study with 4 patients allergic to the same ophthalmic solution (14) and an additional literature reference was found online (15).

Three patients, 2 men and a woman, were investigated for pruritic erythema and in one edema of the eyelids while using amlexanox 0.25% ophthalmic solution for allergic conjunctivitis. Patch testing with the product was positive in all. On discontinuing the eye drops, all patients' eruptions disappeared with topical corticosteroid therapy in a few days. Ingredient patch testing showed the following results: patient 1 reacted strongly (++) to amlexanox 1% and 0.25% water and had weaker (+) positive patch tests to amlexanox 1%, 0.25% and 0.1% pet. Patient 2 was tested with amlexanox in pet. only and had + reactions to 1% and 0.25% pet. The third subject, who only had a weak (+) reaction to the ophthalmic solution, had negative reactions to amlexanox in pet. and dubious-positive (?+) reactions to amlexanox 1% and 0.25% water. Combined with methylparaben (0.026%) or propylparaben (0.014%) in water, however, the patch tests were clearly, albeit weakly, positive (+). The other ingredients, including the parabens, were also tested and gave negative results (13).

Two more Japanese patients with dermatitis of the eyelids while on amlexanox ophthalmic solution were reported in 1991 in *Contact Dermatitis*. When patch tested, there were positive reactions (+) to the solution, but only ?+ reactions to amlexanox 1% and 0.25% pet. (16). Water appears to be a better vehicle for patch testing amlexanox (13).

In 1994, when the strong sensitizing capacity of amlexanox had been demonstrated in animal assays, the number of sensitizations from the ophthalmic solution had risen to 41 (17; all references can be found there). The amlexanox nasal douche apparently has caused only 1 case of sensitization (18).

Cutaneous adverse drug reactions from systemic administration caused by type IV (delayed-type) hypersensitivity
Cutaneous adverse drug reactions from systemic administration of amlexanox caused by type IV (delayed-type) hypersensitivity, including fixed drug eruption (2), eczematous drug eruption (3,4) and pityriasis rosea-like drug eruption (5), are planned to be discussed in Volume IV of the *Monographs in Contact Allergy* series on Systemic drugs.

LITERATURE

1   The data in the section 'General' may have been obtained from literature discussed in this chapter, but mostly also or exclusively from one or more of the following online sources: ChemIDPlus Advanced, PubChem, DrugBank, RxList, Drug Central, Drugs.com, and Wikipedia
2   Sugiura M, Hayakawa R, Osada T. Fixed drug eruption due to amlexanox. Contact Dermatitis 1998;38:65-67
3   Ishiguro N, Kato T, Nogita T, Kawashima M, Hidano S. A case of drug eruption from amlexanox (SOLFA). Rinsho Derma 1991;33:1602-1603 (Article in Japanese, data cited in ref. 2)
4   Taniguchi H, Otaki R, Takino C. A case of drug eruption due to amlexanox with drug induced hepatopathy. Rinsho Derma 1992;34:1745-1749 (Article in Japanese, data cited in ref. 2)
5   Inoue N, Makino H, Kamide R. 2 cases of drug eruption due to amlexanox. Rinsho Hifuka 1993;47:873-876 (Article in Japanese, data cited in ref. 2)
6   Hayakawa R, Ogino Y, Aris K, Matsunaga K. Systemic contact dermatitis due to amlexanox. Contact Dermatitis 1992;27:122-123
7   Kawai K, Kawai K, Nakagawa M. Contact dermatitis due to amlexanox. Hifukakiyo 1990;85:433-436 (Article in Japanese)
8   Nagano T. Three cases of contact dermatitis due to Elics® ophthalmic solution. Skin Research 1991;33(suppl.10):15-18 (Article in Japanese)
9   Oka K, Saito F. Allergic contact dermatitis due to amlexanox ophthalmic solution. Skin Research 1991;33(suppl.10): 19-23 (Article in Japanese)
10 Iwasa M, Asakawa Y, Okumura H, Yoshikawa K. Two cases of contact dermatitis due to amlexanox in Elics® ophthalmic solution. Skin Research 1991;33(suppl.10):24-29 (Article in Japanese)
11 Mita T, Sugai T. Two cases of contact dermatitis from amlexanox in eyedrops. Skin Research 1991;33(suppl.10):30-34 (Article in Japanese)
12 Miyazaki M, Takeda A, Ogino A. Four cases of contact dermatitis due to Elics® (amlexanox) ophthalmic solution. Hifukakiyo 1991;86:13-16 (Article in Japanese)

13 Yamashita H, Kawashima M. Contact dermatitis from amlexanox eyedrops. Contact Dermatitis 1991;25:255-256

14 Ooe M, Ichikawa E, Asano S, Okabe S. 4 cases of contact dermatitis from amlexanox (Elics® eyedrops). Hifubyoh Shinryoh 1990;12:1119-1122 (Article in Japanese)

15 Yano E, Higuchi M, Tsuda S, Tanigawa E. Contact dermatitis caused by amlexanox (ELICS) ophthalmic solution. Nishi Nihon Hifuka 1992;54:686-690 (Article in Japanese)

16 Kabasawa Y, Kanzaki T. Allergic contact dermatitis from amlexanox (Elics) ophthalmic solution. Contact Dermatitis 1991;24:148

17 Hariya T, Ikezawa Z, Aihara M, Kitamura K, Osawa J, Nakajima H. Allergenicity and tolerogenicity of amlexanox in the guinea pig. Contact Dermatitis 1994;31:31-36

18 Kitamura K, Oosawa J, Aihara M, Ikezawa Z. One case of contact dermatitis due to amlexanox nasal douche (Solfa ®). The Allergy in Practice 1991;11:48-49 (Article in Japanese)

Chapter 3.19   AMOROLFINE

IDENTIFICATION

Description/definition   : Amorolfine is the morpholine that conforms to the structural formula shown below
Pharmacological classes   : Antifungal agents
IUPAC name   : (2R,6S)-2,6-Dimethyl-4-[2-methyl-3-[4-(2-methylbutan-2-yl)phenyl]propyl]morpholine
Other names   : 2,6-Dimethyl-4-(2-methyl-3-(p-tert-pentylphenyl)propyl)morpholine
CAS registry number   : 78613-35-1
EC number   : Not available
Merck Index monograph   : 1840
Patch testing   : 1% pet.
Molecular formula   : $C_{21}H_{35}NO$

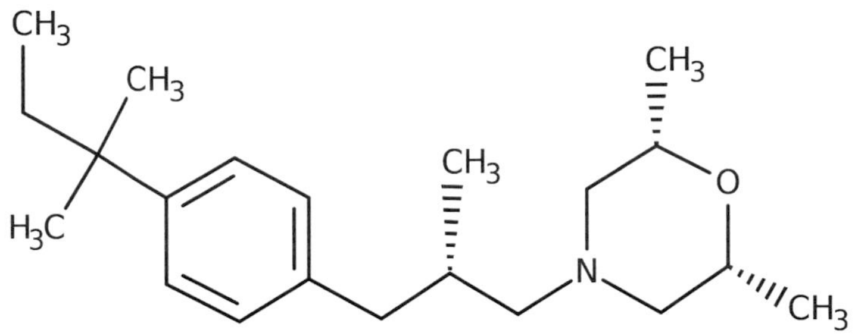

GENERAL

Amorolfine is a morpholine antifungal agent for topical use. It is commonly available in the form of a nail lacquer, containing 5% amorolfine as the active ingredient to treat onychomycosis. In these pharmaceutical products, amorolfine is employed as amorolfine hydrochloride (CAS number 78613-38-4, EC number not available, molecular formula $C_{21}H_{36}ClNO$) (1).

CONTACT ALLERGY

General

Amorolfine appears to have a low allergenic potential. However, amorolfine nail lacquers could have higher allergenicity as they contain excipients such as butyl acetate, ethyl acetate and ethanol that evaporate after the application of the product, thereby increasing the concentration of amorolfine in the nail plate (which is no problem) and in the perionychium (which may be a problem). Also, topical antifungal lacquers are commonly provided with nail files to rasp the nail plate before drug application. This procedure may accidentally cause disruption of the cutaneous barrier of the perionychium (2).

Case reports

A 45-year-old woman presented with itchy vesicles and blisters containing partly purulent fluid on her right first toe. She reported that she had been applying a nail lacquer containing amorolfine 5.0% to treat a suspected fungal nail infection for a couple of weeks. Patch tests were positive to the nail lacquer 'as is' and – later – to amorolfine 1.0% pet. Seven controls were negative (2).

A 29-year-old woman had treated a rash on the right leg with amorolfine 0.5% cream because of suspicion of candidiasis. After 5 weeks of use, she had developed pruritic erythema and papules at the application sites. A patch test with the cream was positive. Additional patch tests were carried out with amorolfine and the other ingredients of the commercial cream and showed positive reactions to amorolfine 0.25% and 0.5% pet. Thirty controls were negative to amorolfine 0.5% pet. (3).

A 45-year-old man had developed contact dermatitis of the left leg and the nail folds. He had used amorolfine cream for tinea corporis on his leg and amorolfine nail lacquer for onychomycosis for 3 months. Patch tests showed positive reactions to the pharmaceutical cream and – later – to amorolfine 0.3%, 0.5% and 1% pet. and 1% water. There were no reactions to the other ingredients of amorolfine cream (4).

Another case of allergic contact dermatitis from amorolfine nail lacquer was apparently published in *Dermatitis* in 2004, but at the journal site, the relevant issue of the journal is missing (5). A 36-year-old woman from Spain developed allergic contact dermatitis from amorolfine in a nail lacquer (6). Details are not available to the author.

LITERATURE

1   The data in the section 'General' may have been obtained from literature discussed in this chapter, but mostly also or exclusively from one or more of the following online sources: ChemIDPlus Advanced, PubChem, DrugBank, RxList, Drug Central, Drugs.com, and Wikipedia

2   Romita P, Stingeni L, Hansel K, De Prezzo S, Ambrogio F, Bonamonte D, et al. Allergic contact dermatitis caused by amorolfine in a nail lacquer. Contact Dermatitis 2019;81:407-408

3   Kaneko K, Aoki N, Hata M, Yajima J, Kawana S, Hattori S. Allergic contact dermatitis from amorolfine cream. Contact Dermatitis 1997;37:307

4   Kramer K, Paul E. Contact dermatitis from amorolfine-containing cream and nail lacquer. Contact Dermatitis 1996;34:145

5   Fidalgo A, Lobo L. Allergic contact dermatitis due to amorolfine nail lacquer. Dermatitis 2004;15:54

6   Pérez-Varela L, Goday-Buján J, Piñeyro-Molina F, Costa-Domínguez C. Allergic contact dermatitis due to amorolfine in nail lacquer. Actas Dermosifiliogr 2010;101:281-283 (Article in Spanish)

Chapter 3.20   AMYLOCAINE

IDENTIFICATION

Description/definition   : Amylocaine is the benzoic acid ester that conforms to the structural formula shown below
Pharmacological classes   : Local anesthetics
IUPAC name   : [1-(Dimethylamino)-2-methylbutan-2-yl] benzoate
CAS registry number   : 644-26-8
EC number   : 211-411-1
Merck Index monograph : 1877
Patch testing   : Hydrochloride, 5.0% pet. (Chemotechnique)
Molecular formula   : $C_{14}H_{21}NO_2$

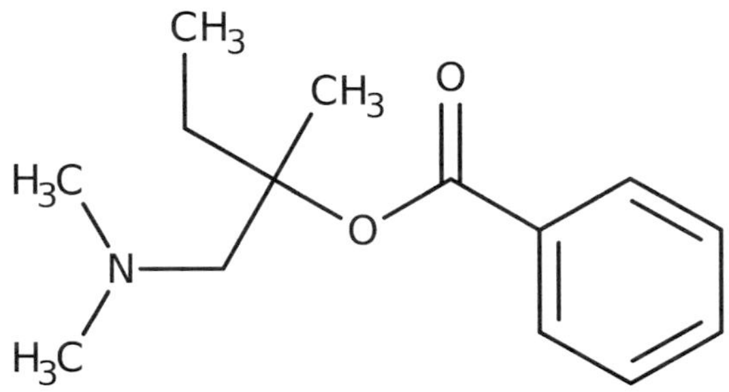

GENERAL

In 1903, amylocaine, the world's first synthetic and non-addictive local anesthetic, was synthesized and patented under the name Forneaucaine by Ernest Fourneau at the Pasteur Institute. Elsewhere, in English speaking countries, it was referred to as Stovaine, given the meaning of the French word 'fourneau' as 'stove' in English. Although amylocaine could be administered topically or injected, it was most widely used for spinal anesthesia. The development and clinical use of newer, more effective, and safer local anesthetics like lidocaine, bupivacaine, and prilocaine in the 1940s and 1950s superseded and made the use of amylocaine obsolete. However, it is still used in a number of countries in topical preparations, usually in combination with neomycin and bacitracin (www.drugs.com). In such pharmaceutical products, amylocaine is employed as amylocaine hydrochloride (CAS number 532-59-2, EC number 208-541-6, molecular formula $C_{14}H_{22}ClNO_2$) (1).

CONTACT ALLERGY

Case reports and case series
Two of 20 patients sensitized by ointments containing local anesthetics were allergic to amylocaine hydrochloride, tested 1% pet. (2). Details are not available (data cited in ref. 3).

LITERATURE
1   The data in the section 'General' may have been obtained from literature discussed in this chapter, but mostly also or exclusively from one or more of the following online sources: ChemIDPlus Advanced, PubChem, DrugBank, RxList, Drug Central, Drugs.com, and Wikipedia
2   Wilson HTH. Dermatitis from anaesthetic ointments. Practitioner 1966;197:673 (cited in ref. 3)
3   Cronin E. Contact Dermatitis. Edinburgh: Churchill Livingstone, 1980:198

Chapter 3.21   ANTAZOLINE

IDENTIFICATION

Description/definition   : Antazoline is the phenylbenzamine that conforms to the structural formula shown below
Pharmacological classes   : Anti-allergic agents; histamine H1 antagonists
IUPAC name   : N-Benzyl-N-(4,5-dihydro-1H-imidazol-2-ylmethyl)aniline
Other names   : Phenazoline; Antistine ®
CAS registry number   : 91-75-8
EC number   : 202-094-0
Merck Index monograph   : 1942
Patch testing   : 1% pet.
Molecular formula   : $C_{17}H_{19}N_3$

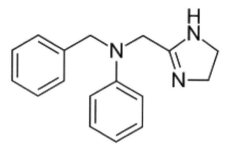

GENERAL

Antazoline is an ethylenediamine derivative with histamine H1 antagonistic, anticholinergic and sedative properties. It is used to relieve nasal congestion. It is also formulated as eye drops with naphazoline to relieve allergic conjunctivitis. In pharmaceutical products, antazoline is employed as antazoline phosphate (CAS number 154-68-7, EC number 205-831-4, molecular formula $C_{17}H_{22}N_3O_4P$) or as antazoline sulfate (CAS number 24359-81-7, EC number not available, molecular formula $C_{17}H_{21}N_3O_4S$) (1).

CONTACT ALLERGY

Case reports

A 53-year-old woman presented with photophobia and diplopia with erythema, edema, and itching of the eyelids. Previously, she had used antazoline 5% eye drops (Antistine ®) without problems, but now they caused a marked exacerbation of all symptoms and an extension of the dermatitis and conjunctivitis. Physical examination showed a rather sharply demarcated dermatitis involving the lower forehead, both upper and lower lids, sides of the cheeks and the upper lip, with erythema, edema, vesiculation, crusting and oozing. Patch tests with the eye drops were strongly positive on two occasions, both times accompanied by a flareup of the dermatitis around the eyes. A patch test with antazoline crystals was strongly positive, as was an intradermal test read after 2 days (2).

A man previously sensitized to 2% antazoline ointment had a severe exacerbation when he used antazoline-containing eye drops. Patch tests were positive to the ointment and several other antihistamines (3; details unknown).

A 54-year-old woman complained of swelling and redness of the eyelids. Antazoline eye drops were prescribed and the inflammation subsided in 4 days. She remained well for one month, then redness of the eyes recurred. The patient again used the eye drops and was prescribed oral tripelennamine (pyribenzamine). Within 24 hours she presented with marked edema of the eyelids with an area of red, oozing, vesicular dermatitis covering the lids and upper parts of the cheeks. The skin of the back and chest showed a diffuse maculopapular rash most marked below the axillae. Patch tests were positive to antazoline, pyribenzamine and various other antihistamines (test concentrations/vehicles not mentioned). The maculopapular rash was probably systemic contact dermatitis caused by oral tripelennamine (4).

LITERATURE

1   The data in the section 'General' may have been obtained from literature discussed in this chapter, but mostly also or exclusively from one or more of the following online sources: ChemIDPlus Advanced, PubChem, DrugBank, RxList, Drug Central, Drugs.com, and Wikipedia
2   Mosko MM, Peterson WL. Sensitization to antistine. J Invest Dermatol 1950;14:1. doi: 10.1038/jid.1950.1
3   Rajka G, Pallin 0. Sensitization to locally applied Antistine. Acta Derm Venereol (Stockh) 1964;44:255-256
4   Sherman WB, Cooke RA. Dermatitis following the use of pyribenzamine and antistine. J Allergy 1950;21:63-67

Chapter 3.22   ANTHRAROBIN

IDENTIFICATION

Description/definition   : Anthrarobin is the tricyclic compound that conforms to the structural formula shown
                              below
Pharmacological classes : Dermatological agents; antipsoriatic agents
IUPAC name   : Anthracene-1,2,10-triol
Other names   : 1,2,10-Trihydroxyanthracene; desoxyalizarin
CAS registry number   : 577-33-3
EC number   : 209-410-6
Merck Index monograph : 1011
Patch testing   : 1% pet.; perform control testing
Molecular formula   : $C_{14}H_{10}O_3$

GENERAL

Anthrarobin is a mixture of isomeric anthratriols. Formerly is was commonly used as an antiparasitic drug and was a component of Arning's tincture, a topical medication for the treatment of eczema and other skin diseases (1).

CONTACT ALLERGY

Case report

A 52-year-old man, with a known history of palmoplantar eczema for several years and contact sensitivities to Myroxylon pereirae resin and the fragrance mix, developed eczema on his trunk, face and lower extremities. For 6 months, he had applied Arning's solution NRF 11.13 and topical corticosteroids. Having discontinued Arning's solution, the clinical findings improved. Four weeks later, after complete healing, patch testing was performed with the solution and all its components (anthrarobin, ammonium tumenol [unknown drug], propylene glycol, isopropyl alcohol, ether). A positive reaction was observed to Arning's solution and its ingredient anthrarobin 1% and 10% pet., both ++/++ at D2 and D3. The patient also reacted to tincture of benzoin, which was an ingredient of the *original* Arning's tincture, but not of the solution the patient had used (1). Possibly, he had used the original tincture before, but the positive patch test may also have been a cross-reaction to Myroxylon pereirae resin, which frequently occurs (2).

LITERATURE

1   Keller-Melchior R, Bräuninger W. Allergic contact dermatitis from anthrarobin. Contact Dermatitis 1995;33:361
2   Hjorth N. Eczematous allergy to balsams. Acta Dermato-Venereologica 1961:46(suppl.):1-216

Chapter 3.23   APRACLONIDINE

IDENTIFICATION

Description/definition : Apraclonidine is the clonidine derivative that conforms to the structural formula shown below
Pharmacological classes : α_2-Adrenergic receptor agonists
IUPAC name : 2,6-Dichloro-1-N-(4,5-dihydro-1H-imidazol-2-yl)benzene-1,4-diamine
Other names : 4-Aminoclonidine; 1,4-benzenediamine, 2,6-dichloro-N'-(4,5-dihydro-1H-imidazol-2-yl)-
CAS registry number : 66711-21-5
EC number : Not available
Merck Index monograph : 2008
Patch testing : 10% water
Molecular formula : $C_9H_{10}Cl_2N_4$

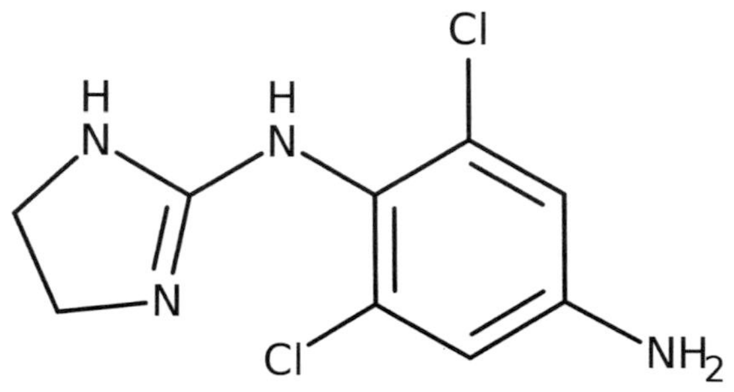

GENERAL

Apraclonidine is a clonidine derivative with selective α_2-adrenergic agonistic activity. Upon ocular administration, apraclonidine enhances aqueous humor uveoscleral outflow and decreases aqueous production by vasoconstriction, thereby decreasing intraocular pressure (IOP). Apraclonidine is used for prevention or reduction of intraoperative and postoperative increases in intraocular pressure before and after ocular laser surgery. It has also application as a short-term adjunctive therapy in patients with open-angle glaucoma who are on maximally tolerated medical therapy requiring additional IOP reduction. In pharmaceutical products, apraclonidine is employed as apraclonidine hydrochloride (CAS number 73218-79-8, EC number not available, molecular formula $C_9H_{11}Cl_3N_4$) (1).

CONTACT ALLERGY

General

In ophthalmological literature, high percentages of 'allergy' to apraclonidine have been reported, up to 48% with apraclonidine 1% eye drops. The types of allergy reported are usually follicular conjunctivitis or allergic contact dermatoconjunctivitis, and may include itching, foreign-body sensation, tearing or discharge, chemosis, hyperemia, and eyelid edema and erythema. However, patch tests with apraclonidine or the commercial eye drops were never performed (6,7).

Case series

In Ferrara, Italy, over a 65-month period before 2005, 50 patients affected by periorbital dermatitis while using topical ocular products were patch tested, including with their own ophthalmic medications (n=210). Only 15 positive reactions were detected in 12 subjects, including 14 reactions to commercial eye drops. There was one reaction to apraclonidine hydrochloride. The active ingredient was not tested separately, but contact allergy to the excipients and preservatives was excluded by patch testing. The authors concluded that patch testing with commercial eye drops has doubtful value (2).

Case reports

A 59-year-old man presented with right sided conjunctivitis and lichenified dry eczema on the right lower eyelid and cheek present for 2 years. Past medical history included glaucoma of the right eye on treatment with timolol eyedrops (applied to both eyes) and apraclonidine eye drops (applied only to the right eye) for 7 years. Patch testing revealed positive reactions to the 0.5% apraclonidine eye drops and – in a second session – to apraclonidine 10% water (negative to 1%, 0.25% and 0.1% water). Twenty controls were negative to the 10% test material (3). A similar patient has been reported in Spanish literature. A 63-year-old woman presented with eczematous lesions on the upper and lower left eyelids while on treatment with apraclonidine for glaucoma. She had a positive patch test to 0.5% apraclonidine eye drops, but the active principle itself was not tested (5).

An 83-year-old woman, with glaucoma and awaiting surgery, was treated with 0.5% apraclonidine and timolol eye drops. 15 days after beginning this treatment, she presented with ocular pruritus, conjunctival and palpebral edema, erythema and lacrimation, which disappeared 7 days after stopping these drugs. Patch tests were positive to the 0.5% apraclonidine eye drops. The positive reaction consisted of erythema and papules which persisted for 7 days. 25 subjects were tested as controls with negative results. Apraclonidine itself nor the ingredients of the commercial product were tested separately (4).

LITERATURE

1   The data in the section 'General' may have been obtained from literature discussed in this chapter, but mostly also or exclusively from one or more of the following online sources: ChemIDPlus Advanced, PubChem, DrugBank, RxList, Drug Central, Drugs.com, and Wikipedia
2   Corazza M, Massieri LT, Virgili A. Doubtful value of patch testing for suspected contact allergy to ophthalmic products. Acta Derm Venereol 2005;85:70-71
3   Silvestre JF, Carnero L, Ramón R, Albares MP, Botella R. Allergic contact dermatitis from apraclonidine in eyedrops. Contact Dermatitis 2001;45:251
4   Armisen M, Vidal C, Quintans R, Suarez A, Castroviejo M. Allergic contact dermatitis from apraclonidine. Contact Dermatitis 1998;39:193
5   Bordel-Gómez MT, Sánchez-Estella J, Santos-Durán JC. Unilateral allergic contact dermatitis of the eyelid caused by Iopimax. Actas Dermosifiliogr 2009;100:917-918 (Article in Spanish)
6   Williams GC, Orengo-Nania S, Gross RL. Incidence of brimonidine allergy in patients previously allergic to apraclonidine. J Glaucoma 2000;9:235-238
7   Butler P, Mannschreck M, Lin S, Hwang I, Alvarago J. Clinical experience with the long-term use of 1% apraclonidine. Arch Ophthalmol 1995;113:293-296

Chapter 3.24 ARBEKACIN

IDENTIFICATION

Description/definition : Arbekacin is the 4,6-disubstituted 2-deoxystreptamine that conforms to the structural
formula shown below

Pharmacological classes : Anti-infective agents

IUPAC name : (2S)-4-Amino-N-[(1R,2S,3S,4R,5S)-5-amino-4-[(2R,3R,6S)-3-amino-6-(aminomethyl)oxan-2-
yl]oxy-2-[(2S,3R,4S,5S,6R)-4-amino-3,5-dihydroxy-6-(hydroxymethyl)oxan-2-yl]oxy-3-
hydroxycyclohexyl]-2-hydroxybutanamide

Other name(s) : 1-N-((S)-4-Amino-2-hydroxybutyryl)dibekacin

CAS registry number : 51025-85-5

EC number : Not available

Merck Index monograph : 2030

Patch testing : No data available; suggested: 5%, 10% and 20% pet.

Molecular formula : $C_{22}H_{44}N_6O_{10}$

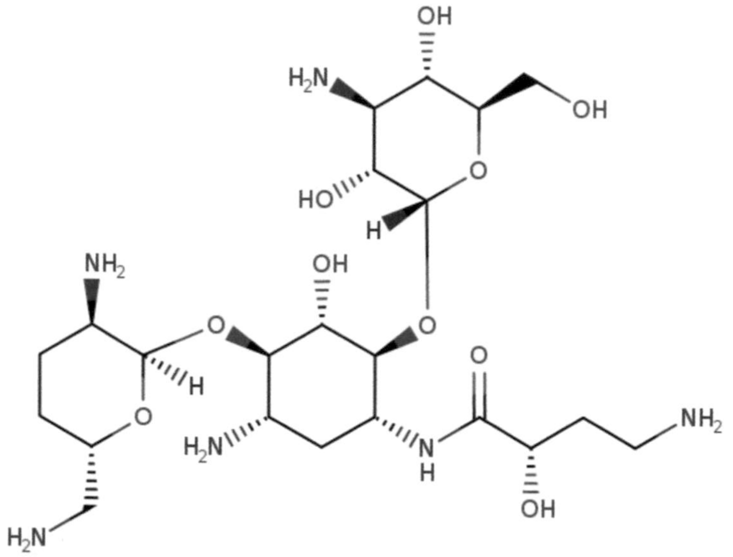

GENERAL

Arbekacin is a semisynthetic aminoglycoside antibiotic derived from dibekacin with activity against both gram-positive and gram-negative bacteria. Arbekacin is most commonly used in the treatment of multi-resistant bacterial infections such as methicillin-resistant *Staphylococcus aureus* (MRSA). In pharmaceutical products, arbekacin is employed as arbekacin sulfate (CAS number 104931-87-5, EC number not available, molecular formula $C_{22}H_{46}N_6O_{14}S$) (1).

CONTACT ALLERGY

Case report

A 43-year-old woman suffering from CREST syndrome had a refractory cutaneous ulcer, 1 × 2 cm in size, on the left leg, colonized with MRSA. Treatment was initiated with an ointment containing 0.2% arbekacin sulfate once daily. On the 3rd day, a well-defined erythematous eruption with infiltration appeared at the site where the ointment had been applied. Patch testing showed a ?+ reaction at D2 and a + reaction at D3 to the arbekacin ointment. Later, patch tests were performed with the ingredients of the ointment (arbekacin solution [concentration not specified], hydrophilic petrolatum and white petrolatum), and two other aminoglycosides: gentamicin sulfate 20% pet. and 'fradiomycin sulfate' (neomycin sulfate) 20% pet. The patient had positive reactions (D2 ?+, D3 +) to the arbekacin ointment and solution, and also reacted to neomycin sulfate, but not to gentamicin sulfate. She had previously received intravenous arbekacin without any untoward effects after an operation (2).

LITERATURE

1   The data in the section 'General' may have been obtained from literature discussed in this chapter, but mostly also or exclusively from one or more of the following online sources: ChemIDPlus Advanced, PubChem, DrugBank, RxList, Drug Central, Drugs.com, and Wikipedia

2   Akaki T, Dekio S. Contact dermatitis from arbekacin sulfate: report of a case. J Dermatol 2002;29:674-675

Chapter 3.25   ATROPINE

IDENTIFICATION

Description/definition   : Atropine is the naturally-occurring alkaloid isolated from the plant *Atropa belladonna* that
                             conforms to the structural formula shown below
Pharmacological classes : Muscarinic antagonists; parasympatholytics; bronchodilator agents; adjuvants,
                             anesthesia; mydriatics
IUPAC name   : [(1*S*,5*R*)-8-Methyl-8-azabicyclo[3.2.1]octan-3-yl] 3-hydroxy-2-phenylpropanoate
CAS registry number   : 51-55-8
EC number   : 200-104-8
Merck Index monograph   : 2136
Patch testing   : Sulfate, 1% water (SmartPracticeCanada, SmartPracticeEurope)
Molecular formula   : $C_{17}H_{23}NO_3$

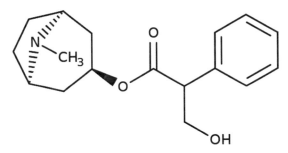

GENERAL

Atropine is an alkaloid, originally derived from the plant *Atropa belladonna*, but the compound is also found in other plants, mainly of the *Solanaceae* family. Atropine functions as a sympathetic, competitive antagonist of muscarinic cholinergic receptors, thereby abolishing the effects of parasympathetic stimulation. This agent may induce tachycardia, inhibit secretions, and relax smooth muscles. It is indicated for the treatment of poisoning by susceptible organophosphorous nerve agents having anti-cholinesterase activity (cholinesterase inhibitors) as well as organophosphorous or carbamate insecticides. In ocular pharmaceuticals, it is used as mydriatic and cycloplegic drug for diagnostic purposes, for the prevention of synechiae in inflammatory processes (e.g. uveitis), for mild amblyopia and serious myopia. Systemic administration is sometimes done in the case of sinus bradycardia or atrioventricular block and as premedication for anesthesia. In pharmaceutical products, atropine is usually employed as atropine sulfate (CAS number 5908-99-6, EC number 200-235-0, molecular formula $C_{34}H_{50}N_2O_{11}S$), occasionally as atropine hydrochloride (1).

CONTACT ALLERGY

Contact allergy in the general population

In Germany, for the period 1995-2004, the population-based relative incidence (RI) of contact sensitization to atropine (cases/100,000 defined daily doses (DDDs) per year) was estimated to be 2.7. In the group of ophthalmic drugs, the RI ranged from 0.3 (pilocarpine) to 21.6 (phenylephrine) (5).

Patch testing in groups of selected patients

In three groups of selected patients investigated by the IVDK, low frequencies of sensitization to atropine ranging from 0.3% to 0.5% were found (table 3.25.1).

Table 3.25.1 Patch testing in groups of patients: Selected patient groups

Years and Country	Test conc. & vehicle	Number of patients tested	positive (%)	Selection of patients (S); Relevance (R); Comments (C)	Ref.
2001-2010 IVDK	1% water	1144	6 (0.5%)	S: patients with periorbital dermatitis tested with an ophthalmic tray; R: not stated	7
1995-2004 IVDK	1% water	3873	(0.3%)	S: patients patch tested for suspected contact allergy to ophthalmological drugs; R: not stated; the estimated number of sensitizations per 100,000 prescriptions was 2.7	3
1995-2004 IVDK	1% water	4280	(0.3%)	S: not stated; R: not stated	4

IVDK: Information Network of Departments of Dermatology (Germany, Austria, Switzerland)

Case series

In Leuven, Belgium, between 1990 and 2017, 16,065 patients were investigated for contact allergy and 118 (0.7%) showed positive patch test reactions to topical ophthalmic medications and/or to their ingredients. Eighty-four individuals (71%) reacted to an active principle. Atropine was tested in 39 patients and was the allergen in eye medications in five. There were no reactions to atropine in other types of medications (2; overlap with data in ref. 6).

In Leuven, Belgium, in the period 1990-2014, iatrogenic contact dermatitis was diagnosed in 2600 individuals (17% of the total population). 96% of all positive patch test reactions to topical drugs and antiseptics were considered to be relevant. Atropine (1% water) was tested in 43 patients and there were 6 positive reactions to it (6; overlap with data in ref. 2).

In Pamplona, Spain, in one year's time (1992), 13 patients were diagnosed with contact allergy to ophthalmic medications, nearly 3% of all patients investigated for suspected contact dermatitis. There were 2 reactions to atropine 2% water, both in individuals who had previously used atropine eye drops (9).

In an early study from France, 3% of hospitalized patients had contact allergy to atropine in eye drops. Details are not available (article not read) (18). Before 1982, in Germany, 11 of 25 patients with periocular dermatitis due to ophthalmological treatment had positive patch tests to atropine eye drops, but the ingredients including atropine itself were not tested (17, data cited in refs. 15 and 16).

Case reports

A 23-year-old man presented with severe dermatitis one day after vitrectomy of his right eye. First, the eye had become red and swollen, and then an id-like dermatitis gradually developed over the entire body. Several topical drugs had been used during the surgical procedure, including eye drops containing atropine. Patch tests with the ophthalmologic preparations he had received showed a positive patch test reaction to atropine sulfate 1% water. Additional testing with compounds chemically related to atropine resulted in positive reactions to scopolamine 0.2% water, homatropine hydrobromide 1% water and belladonna extract 5% pet. (14).

A 12-year-old boy was treated with atropine sulfate 1% drops for retardation of progressive myopia when he developed redness, swelling, superficial erosions, and hemorrhagic crusting of both eyes. A non-conventional patch test with the medication was weakly positive, with 0.01% atropine negative (10). A 65-year-old woman presented with a 3-year history of itchy, bilateral, eyelid dermatitis. Patch tests showed positive reactions to the fragrance mix and 3 scented cosmetics. All cosmetics were withdrawn but after 4 weeks the skin lesions were unchanged. The patient was at that time being treated with diclofenac-containing eye drops and atropine-containing eye drops for iridocyclitis and glaucoma. Patch tests with both preparations were performed and a positive reaction to the 1% atropine preparation, and later to atropine sulfate 1% water, was elicited (11).

A 72-year-old woman had redness and swelling of the eyelids of both eyes. When patch tested, she strongly reacted to her atropine eye drops 1%, but atropine itself was not tested (15). The same authors also described a 69-year-old man with swelling, erythema and crusting of the eyelids of the right eye of one month duration. Twelve years previously, a new lens had been implanted and he had used 1% atropine eye drops for 6 months. Patch tests were positive to the eye drops, atropine powder pure and atropine 2% and 1%, both in water and petrolatum, but negative to 0.5% in both vehicles (15).

A 48-year-old man developed severe dermatitis 2 days after photocoagulation therapy of his retina. Topical drugs used at that time were atropine and 2 antibacterial agents. He had previously used eye drops containing tropicamide and phenylephrine for an eye examination. Discontinuation of all eye drops and application of a steroid ointment resulted in resolution within a few days. Patch tests showed positive reactions to atropine (++ reaction to 0.0006%), homatropine 1%, tropicamide 1% and phenylephrine 1%. Compounds structurally similar to atropine such as tropine, tropic acid, and homatropine were positive to homatropine (1%) only (16).

A patient with allergic contact dermatitis caused by idoxuridine eye drops, used for treating herpetic keratitis, had been investigated 1 year previously for dermatitis of the right orbit suspected to have been induced by atropine sulfate 0.25%. The suspected diagnosis had been confirmed by a strongly positive patch test to atropine sulfate. Clinical details were not provided (19).

A 65-year-old painter accidentally splashed lime in his face. Immediately, he rinsed the eyes with water. The ophthalmologist prescribed 4 different medications including atropine and penicillamine eye drops. Six weeks later, the patient was referred to the dermatology department because he had developed erythema, edema and vesicles of the eyelids and surrounding skin. Patch tests showed a weakly positive reaction to atropine 1% saline and extremely strong positive reactions to penicillamine 0.15 M water (D3 +++, D4 ++++). All medications had been stopped, hydrocortisone cream was prescribed and the skin healed within a few days (20). It should be appreciated that, because of the extremely strong reaction to penicillamine, it cannot be excluded that the + reaction to atropine was in fact false-positive.

A 43-year-old man accidentally splashed caustic soda in his right eye. The eye was immediately rinsed with water, and subsequently treated with atropine eye drops, corticosteroid eye drops and antibiotic ointment. Later, because of corneal vascularization and ulceration, a cornea transplantation and lens extraction, with implantation of an artificial lens, was performed. Treatment with indomethacin, atropine and chloramphenicol eye drops seemed to cause an allergic reaction. Patch testing showed positive reactions to atropine sulfate 1% water and d-penicillamine (8).

Single case reports of contact allergy to atropine have apparently also been described in refs. 12 and 13 (data cited in ref. 11).

Cross-reactions, pseudo-cross-reactions and co-reactions
Possible cross-reactivity to scopolamine, homatropine and belladonna in a patient with allergic contact dermatitis from atropine (14). Cross-reaction to homatropine, not to tropine and tropic acid (16).

Immediate contact reactions
Immediate contact reactions (contact urticaria) to atropine are presented in Chapter 5.

LITERATURE

1 The data in the section 'General' may have been obtained from literature discussed in this chapter, but mostly also or exclusively from one or more of the following online sources: ChemIDPlus Advanced, PubChem, DrugBank, RxList, Drug Central, Drugs.com, and Wikipedia
2 Gilissen L, De Decker L, Hulshagen T, Goossens A. Allergic contact dermatitis caused by topical ophthalmic medications: Keep an eye on it! Contact Dermatitis 2019;80:291-297
3 Uter W, Menezes de Pádua C, Pfahlberg A, Nink K, Schnuch A, Behrens-Baumann W. Contact allergy to topical ophthalmological drugs - epidemiological risk assessment. Klin Monbl Augenheilkd 2009;226:48-53 (article in German)
4 Menezes de Padua CA, Uter W, Schnuch A. Contact allergy to topical drugs: prevalence in a clinical setting and estimation of frequency at the population level. Pharmacoepidemiol Drug Saf 2007;16:377-384
5 Menezes de Padua CA, Schnuch A, Nink K, Pfahlberg A, Uter W. Allergic contact dermatitis to topical drugs – Epidemiological risk assessment. Pharmacoepidemiol Drug Saf 2008;17:813-821
6 Gilissen L, Goossens A. Frequency and trends of contact allergy to and iatrogenic contact dermatitis caused by topical drugs over a 25-year period. Contact Dermatitis 2016;75:290-302
7 Landeck L, John SM, Geier J. Topical ophthalmic agents as allergens in periorbital dermatitis. Br J Ophthalmol 2014;98:259-262
8 Coenraads PJ, Woest TE, Blanksma LJ, Houtman WA. Contact allergy to d-penicillamine. Contact Dermatitis 1990;23:371-372
9 Tabar AI, García BE, Rodríguez A, Quirce S, Olaguibel JM. Etiologic agents in allergic contact dermatitis caused by eyedrops. Contact Dermatitis 1993;29:50-51
10 Kothari M, Jain R, Khadse N, Rathod V, Mutha S. Allergic reactions to atropine eye drops for retardation of progressive myopia in children. Indian J Ophthalmol 2018;66:1446-1450
11 De Misa RF, Suárez J, Feliciano L, López B. Allergic periocular contact dermatitis due to atropine. Clin Exp Dermatol 2003;28:97-98
12 Haetinen A, Teraesvirta M, Fraeki JE. Contact allergy to components in topical ophthalmologic preparations. Acta Derm Venereol 1955;63:424-426
13 Gutierrez Ortega MC, Hasson Nisis A, Zamora Martinez E, Martín Moreno L, de Castro Torres A. Dermatitis alérgica de contacto a la atropina (Contact allergic dermatitis caused by atropine). Med Cutan Ibero Lat Am 1988;16:430-431 (article in Spanish)
14 Decraene T, Goossens A. Contact allergy to atropine and other mydriatic agents in eye drops. Contact Dermatitis 2001;45:309-310
15 Van der Willigen AH, de Graaf YP, van Joost T. Periocular dermatitis from atropine. Contact Dermatitis 1987;17:56-57
16 Yoshikawa K, Kawahara S. Contact allergy to atropine and other mydriatic agents. Contact Dermatitis 1985;12:56-57
17 Maucher OM. Periorbitalekzem als iatrogene Erkrankung. Klin Mbl Augenheilk 1982;164:350-356 (article in German)
18 Vadot E, Piasentin D. Incidence of allergy to eyedrops. Results of a prospective survey in a hospital milieu. J Fr Ophtalmol 1986;9:41-43 (article in French)
19 Van Ketel WG. Allergy to idoxuridine eyedrops. Contact Dermatitis 1977;3:106-107
20 De Moor A, Van Hecke E, Kestelyn P. Contact allergy to penicillamine in eyedrops. Contact Dermatitis 1993;29:155-156

Chapter 3.26   AZIDAMFENICOL

IDENTIFICATION

Description/definition : Azidamfenicol is the amphenicol antibiotic that conforms to the structural formula shown below

Pharmacological classes : Anti-bacterial agents

IUPAC name : 2-Azido-N-[(1R,2R)-1,3-dihydroxy-1-(4-nitrophenyl)propan-2-yl]acetamide

Other names : Azidamphenicol; threo-2-azido-N-(β-hydroxy-α-(hydroxymethyl)-p-nitrophenethyl)acet-amide

CAS registry number : 13838-08-9

EC number : 237-552-9

Merck Index monograph : 1018

Patch testing : 2% and 5% pet.

Molecular formula : $C_{11}H_{13}N_5O_5$

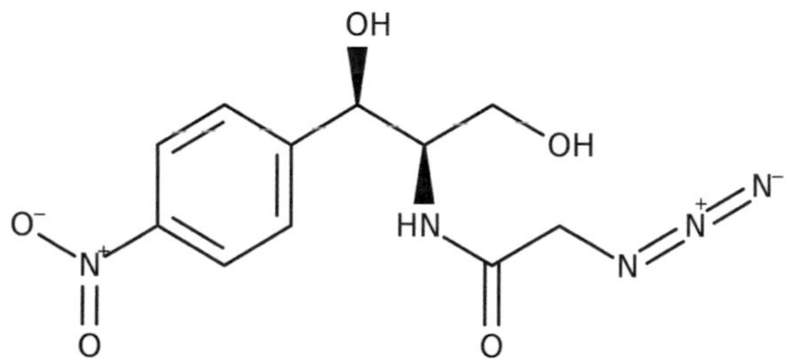

GENERAL

Azidamfenicol is an amphenicol antibiotic, which has a profile similar to chloramphenicol. It is used only topically in eye drops and ointment for treatment of bacterial infections, though probably rarely (1).

CONTACT ALLERGY

Case reports and case series

In 1967, in Norway, a steroid-antibiotic combination cream was put on the market containing dexamethasone 0.01% and azidamphenicol 1.5%. In a hospital in Oslo, from 1968 to 1975, in 6 patients, contact allergy to the topical pharmaceutical demonstrated. The patients were 4 women and 2 men, in age ranging from 20 to 57 years. Five had lower leg dermatitis spreading to other parts of the body and one had hand and arm dermatitis. All had positive patch tests to the cream and a negative reaction to the cream base. Two patients had positive patch tests to azidamphenicol 2% (vehicle not mentioned, probably petrolatum). Of four tested with the antibiotic 1%, one was positive and three negative. However, one of the negatives was later retested and now had positive patch tests to azidamphenicol 2% (+) and 5% (++). (2). Many cases of azidamphenicol contact allergy have passed undiagnosed due to a false-negative reactions to the cream from the presence of dexamethasone as ingredient (2).

An 80-year-old man developed periorbital contact dermatitis after application of azidamphenicol-containing ointment. Patch tests were positive to azidamphenicol 5% pet. The patient co-reacted to chloramphenicol but not to thiamphenicol 5% pet. Lymphocyte stimulation tests were positive to both azidamphenicol and chloramphenicol (4).

One patient was sensitized to azidamphenicol and clotrimazole, both present in an antifungal preparation; details are unknown (3).

Cross-reactions, pseudo-cross-reactions and co-reactions

Most patients with contact allergy to azidamphenicol will cross-react to chloramphenicol (2,4). One patient sensitized to chloramphenicol cross-reacted to azidamphenicol (4).

LITERATURE

1   The data in the section 'General' may have been obtained from literature discussed in this chapter, but mostly also or exclusively from one or more of the following online sources: ChemIDPlus Advanced, PubChem, DrugBank, RxList, Drug Central, Drugs.com, and Wikipedia

2   Wereide K. Sensitivity to azidamphenicol. Contact Dermatitis 1975;1:271-272

3   Raulin C, Frosch PJ. Kontaktallergie auf Clotrimazol und Azidamfenicol. Dermatosen in Beruf und Umwelt 1987;35:64-66 (Article in German)

4   Sachs B, Erdmann S, al Masaoudi T, Merk HF. Molecular features determining lymphocyte reactivity in allergic contact dermatitis to chloramphenicol and azidamphenicol. Allergy 2001;56:69-72

Chapter 3.27 AZITHROMYCIN

IDENTIFICATION

Description/definition : Azithromycin is the semisynthetic macrolide antibiotic that conforms to the structural formula shown below
Pharmacological classes : Anti-bacterial agents
IUPAC name : (2R,3S,4R,5R,8R,10R,11R,12S,13S,14R)-11-[(2S,3R,4S,6R)-4-(Dimethylamino)-3-hydroxy-6-methyloxan-2-yl]oxy-2-ethyl-3,4,10-trihydroxy-13-[(2R,4R,5S,6S)-5-hydroxy-4-methoxy-4,6-dimethyloxan-2-yl]oxy-3,5,6,8,10,12,14-heptamethyl-1-oxa-6-azacyclopentadecan-15-one
CAS registry number : 83905-01-5
EC number : 617-500-5
Merck Index monograph : 2177
Patch testing : 10% pet.; this may occasionally lead to a false-negative reaction (5); 20% pet. is probably not irritant when used for patch testing (5)
Molecular formula : $C_{38}H_{72}N_2O_{12}$

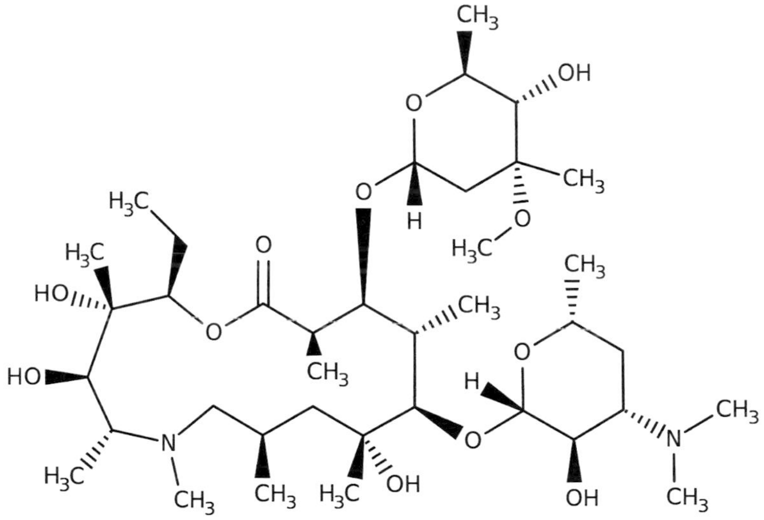

GENERAL

Azithromycin is a semisynthetic macrolide antibiotic structurally related to erythromycin. It has been used in the treatment of *Mycobacterium avium intracellulare* infections, toxoplasmosis, and cryptosporidiosis. Indications for its use include acute bacterial exacerbations of chronic obstructive pulmonary disease, acute bacterial sinusitis and community-acquired pneumonia, pharyngitis/tonsillitis, uncomplicated skin and skin structure infections (all due to specific species of bacteria), urethritis and cervicitis due to *Chlamydia trachomatis* or *Neisseria gonorrhoeae* and genital ulcer disease in men due to *Haemophilus ducreyi* (chancroid). In pharmaceutical products, both azithromycin and azithromycin dihydrate (CAS number 117772-70-0, EC number not available, molecular formula $C_{38}H_{76}N_2O_{14}$) may be employed (1).

CONTACT ALLERGY

Case series

Three women aged 73 to 77 years developed acute eczema of the upper and lower eyelids 24 hours after intravitreal injections containing aflibercept or ranibizumab, where azithromycin eye drops were used to prevent infection. When patch tested, patients 1 and 2 reacted to the eye drops 'as is' and patient 2 had also a positive ROAT. Azithromycin 30% pet. gave positive reactions in the 2 patients tested with it (patients 2 and 3) (2).

Case reports

An 85-year-old woman with a past history of bilateral phacoemulsification (cataracts) and chronic dacryocystitis had bilateral eyelid eczema for 3 years. The patient had applied eye drops with azithromycin dehydrate 15 mg/g recently for the second time, denying any worsening after the first application a year before. Patch tests were positive to the eye drops, and to azithromycin 1%, 5% and 10% pet. Ten controls were negative to the antibiotic 10% pet. (3).

A 76-year-old woman presented with acute conjunctivitis and acute eczema affecting the eyelids and cheeks. Three weeks earlier, she had undergone pseudophakia surgery, and she was then prescribed two ophthalmic preparations, including eye drops containing azithromycin dihydrate 15 mg/g. Patch tests were positive to the eye drops and to azithromycin 1%, 5% and 10% pet., but negative to the other ophthalmic, to erythromycin and clarithromycin (4).

Cutaneous adverse drug reactions from systemic administration caused by type IV (delayed-type) hypersensitivity
Cutaneous adverse drug reactions from systemic administration of azithromycin caused by type IV (delayed-type) hypersensitivity and occupational allergic contact dermatitis (5-8) are planned to be discussed in Volume IV of the *Monographs in Contact Allergy* series on Systemic drugs.

Cross-reactions, pseudo-cross-reactions and co-reactions
Not to erythromycin (2,3,4,6) and clarithromycin (2,3,4,5). Two patients who had occupational allergic contact dermatitis from azithromycin also reacted to one or more intermediates in azithromycin synthesis (hydroxylamine hydrochloride, erythromycin A oxime hydrochloride, erythromycin A iminoether and azaerythromycin A, all tested 1% and 5% pet.) but not to erythromycin (6).

LITERATURE

1   The data in the section 'General' may have been obtained from literature discussed in this chapter, but mostly also or exclusively from one or more of the following online sources: ChemIDPlus Advanced, PubChem, DrugBank, RxList, Drug Central, Drugs.com, and Wikipedia
2   de Risi-Pugliese T, Amsler E, Collet E, Francès C, Barbaud A, Pecquet C, et al. Eyelid allergic contact dermatitis after intravitreal injections of anti-vascular endothelial growth factor: What is the culprit? A report of 3 cases. Contact Dermatitis 2018;79:103-104
3   Mendes-Bastos P, Brás S, Amaro C, Cardoso J. Non-occupational allergic contact dermatitis caused by azithromycin in an eye solution. J Dtsch Dermatol Ges 2014;12:729-730
4   Flavia Monteagudo Paz A, Francisco Silvestre Salvador J, Latorre Martínez N, Cuesta Montero L, Toledo Alberola F. Allergic contact dermatitis caused by azithromycin in an eye drop. Contact Dermatitis 2011;64:300-301
5   López-Lerma I, Romaguera C, Vilaplana J. Occupational airborne contact dermatitis from azithromycin. Clin Exp Dermatol 2009;34:e358-e359
6   Milković-Kraus S, Macan J, Kanceljak-Macan B. Occupational allergic contact dermatitis from azithromycin in pharmaceutical workers: a case series. Contact Dermatitis 2007;56:99-102
7   Milkovic-Kraus S, Kanceljak-Macan B. Occupational airborne allergic contact dermatitis from azithromycin. Contact Dermatitis 2001;45:184
8   Mimesh S, Pratt M. Occupational airborne allergic contact dermatitis from azithromycin. Contact Dermatitis 2004;51:151

Chapter 3.28 BACITRACIN

Bacitracin is a mixture of related cyclic polypeptides produced by organisms of the licheniformis group of *Bacillus subtilis* var. *Tracy*. Its unique name derives from the fact that the bacillus producing it was first isolated in 1943 from a knee scrape from a girl named Margaret Tracy (PubChem). The data in the identification section relate to Bacitracin A, the main ingredient of bacitracin.

IDENTIFICATION

Description/definition : Bacitracin A is a homodetic cyclic peptide and the main component of bacitracin, which is a mixture of related cyclic polypeptides produced by organisms of the licheniformis group of *Bacillus subtilis* var. *Tracy*

Pharmacological classes : Anti-bacterial agents; anti-infective agents, local

IUPAC name : 4-[[2-[[2-(1-Amino-2-methylbutyl)-4,5-dihydro-1,3-thiazole-4-carbonyl]amino]-4-methyl-pentanoyl]amino]-5-[[1-[[3-(2-amino-2-oxoethyl)-18-(3-aminopropyl)-12-benzyl-15-butan-2-yl-6-(carboxymethyl)-9-(1*H*-imidazol-5-ylmethyl)-2,5,8,11,14,17,20-heptaoxo-1,4,7,10,13,16,19-heptazacyclopentacos-21-yl]amino]-3-methyl-1-oxopentan-2-yl]amino]-5-oxo-pentanoic acid

CAS registry number : 1405-87-4; 22601-59-8 (Bacitracin A)

EC number : 215-786-2; 245-115-9 (Bacitracin A)

Merck Index monograph : 2197

Patch testing : 5.0% pet. (Chemotechnique); 20% pet. (Chemotechnique, SmartPracticeCanada, SmartPracticeEurope); patch tests may become positive only at D4 (51)

Molecular formula : $C_{66}H_{103}N_{17}O_{16}S$

GENERAL

Bacitracin is a mixture of at least nine related cyclic polypeptides (of which bacitracin A is the major constituent) produced by organisms of the licheniformis group of *Bacillus subtilis* var. *Tracy*. As a toxic and difficult-to-use antibiotic, bacitracin doesn't work well orally. It is mainly used in ointment form for topical treatment of a variety of localized skin and eye infections caused by gram-positive bacteria, as well as for the prevention of wound infections In pharmaceutical products, both bacitracin and bacitracin zinc (CAS number 1405-89-6, EC number 215-787-8, molecular formula $C_{122}H_{201}N_{33}O_{33}S_2Zn$) may be employed (1).

Because of the abundance of data on bacitracin allergy, it has not been attempted to provide a full literature review.

CONTACT ALLERGY

GENERAL

Bacitracin is a frequent sensitizer in North America, where it is widely used as an over-the-counter topical antibiotic preparation, often in combination with neomycin, for minor cuts and wounds; it is also often used for surgical wound dressings (for the allergenic risk of which dermatologists have repeatedly warned [58,59,60]), for irrigation during surgical procedures and in ophthalmological preparations. In most other countries, bacitracin is used far less often and consequently is not an important allergen. However, in the 1950s and 1960s, combination products of bacitracin and neomycin were available without prescription in Finland, which led to many cases of sensitization (33). Indeed, in the 1960s, a sensitization rate of 7.8% to bacitracin was found in routine testing of 17,500 patients in a period of 12 years. An increase in frequency of sensitivity to bacitracin closely paralleled the increase in consumption of neomycin-bacitracin ointment (47). Because of the combined presence of neomycin and bacitracin in such products, many patients became sensitized to both antibiotics. Thus, of patients with positive patch tests to neomycin, 66% (45) and 88% (44) co-reacted to bacitracin. Conversely, in one study, all 99 patients with positive bacitracin patch tests also reacted to neomycin (33). This is not the result of cross-sensitization (dissimilar chemical structures), but of concomitant sensitization to neomycin and bacitracin in the same product.

Bacitracin has also caused many cases of immediate-type reactions such as contact urticaria and anaphylaxis (Chapter 5). A review article on both delayed-type and immediate-type reactions to bacitracin was published in 2004 (49). A year earlier, bacitracin had been named 'Allergen of the year' by the American Contact Dermatitis Society (50). Unfortunately, no data are available, as the issue in question is cannot be found on the website of the publisher. A small review was also released in 2005 (55).

Contact allergy in the general population

With the CE-DUR approach, the incidence of sensitization to bacitracin in the German population was estimated to range from 1 to 5 cases/100,000/year in the period 1995-1999 and from 1 to 4 cases/100,000/year in the period 2000-2004 (26). Also in Germany, for the period 1995-2004, the population-based relative incidence (RI) of contact sensitization to bacitracin (cases/100,000 defined daily doses (DDDs) per year) was estimated to be 16.7. In the group of antibiotics, the RI ranged from 1.6 (oxytetracycline) to 86.2 (framycetin) (27).

Patch testing in groups of patients

Results of patch testing bacitracin in consecutive patients suspected of contact dermatitis (routine testing) back to 2000 are shown in table 3,28.1. Results of testing in groups of *selected* patients (e.g., patients with leg ulcers/stasis dermatitis, patients with perianal, genital or anogenital dermatitis, individuals with chronic otitis externa) back to 1990 are shown in table 3.28.2.

Patch testing in consecutive patients suspected of contact dermatitis: routine testing

Results of patch testing bacitracin in consecutive patients suspected of contact dermatitis (routine testing) back to 2000 are shown in table 3.28.1.Bacitracin is tested in a routine series only in North America (NACDG screening series), hence all available data are from the USA and Canada. Prevalences of sensitization in all studies since 2000 have been high, ranging from 6.4% to 9.2%. Since 2012, the frequencies of sensitization are declining somewhat. Relatively few reactions were considered to be relevant, 'definite + probable' relevance in the NACDG studies having been mostly between 20% and 30%. Relevance figures from non-NACDG studies were higher, ranging from 45% to 81% (7,10,61). Details on the culprit products were never provided.

Table 3.28.1 Patch testing in groups of patients: Routine testing

Years and Country	Test conc. & vehicle	Number of patients tested | positive (%)		Selection of patients (S); Relevance ©; Comments ©	Ref.
2015-2017 NACDG	20% pet.	5589	386 (6.9%)	R: definite + probable relevance: 29%	2
2010-2016 NACDG		585	35 (6.5%)	R: not stated	54
2007-2016 USA	20% pet.	2312	(7.7%)	R: not stated	38
2011-2015 USA	20% pet.	2572	165 (6.4%)	R: not stated	3
2013-2014 NACDG	20% pet.	4858	360 (7.4%)	R: definite + probable relevance: 27%	4
2011-2012 NACDG	20% pet.	4234	330 (7.8%)	R: definite + probable relevance: 30%	5
2009-2010 NACDG	20% pet.	4305	(8.3%)	R: definite + probable relevance: 34%	6
2006-2010 USA	20% pet.	3084	(7.3%)	R: 45%	7
2007-2008 NACDG	20% pet.	5077	(7.9%)	R: definite + probable relevance: 30%	8
2005-2006 NACDG	20% pet.	4437	(9.2%)	R: definite + probable relevance: 20%	9
2001-2005 USA	20% pet.	3844	(8.0%)	R: 52%	10
2003-2004 NACDG	20% pet.	5143	407 (7.9%)	R: not stated	11
2001-2002 NACDG	20% pet.	4909	(7.9%)	R: definite + probable relevance: 26%	12
1995-2001 USA	20% pet.	898	92 (9.2%)	R: 81%	61
1998-2000 USA	20% pet.	5812	(9.2%)	R: definite + probable relevance: 24%	13
1998-2000 USA	20% pet.	1321	(8.7%)	R: not stated	14

NACDG: North American Contact Dermatitis Group (USA, Canada)

Patch testing in groups of selected patients

Results of testing in groups of *selected* patients (e.g., patients with leg ulcers/stasis dermatitis, patients with perianal, genital or anogenital dermatitis, individuals with chronic otitis externa) back to 1987 are shown in table 3.28.2. High frequencies of sensitization have been observed in patients with leg ulcers with or without stasis dermatitis in the USA and Canada (24%, ref. 16) and the U.K. (22%, ref. 18), but not in other countries, which may reflect differences in exposure. In all other patient groups low frequencies have been observed with the exception of a 2000-2010 Canadian study, where 44% of the patients were sensitized to bacitracin. However, the patients had – rather unusual – been selected on the basis of a chart review and on at least one positive reaction to a topical drug demonstrated in previous patch testing (29).

Table 3.28.2 Patch testing in groups of patients: Selected patient groups

Years and Country	Test conc. & vehicle	Number of patients tested | positive (%)			Selection of patients (S); Relevance ©; Comments ©	Ref.
Patients with leg ulcers/stasis dermatitis						
<2017 India	5% pet.	172	4	(2.3%)	S: patients with venous leg ulcers of over 6 weeks' duration; R: 'the majority of reactions were relevant'	19
2003-2014 IVDK	20% pet.	2029		(1.3%)	S: patients with stasis dermatitis/chronic leg ulcers; R: not stated	15
2006-2007 Canada		100	8	(8%)	S: patients with leg ulcers or venous disease; R: not stated	17
<2004 USA, Canada	20% pet.	54	13	(24%)	S: patients with past or present leg ulcers with or without dermatitis; R: definite + probable 31%	16
<1994 U.K.	20% pet.	85		(22%)	S: patients with longstanding venous ulceration or eczema complicating leg ulcers; R: not stated	18
1984-1987 Germany	5% pet.	317	25	(7.9%)	S: patients with leg ulcers; in many cases, the patients had used bacitracin ointment	42
Patients with perianal, genital or anogenital dermatitis						
2005-2016 NACDG	20% pet.	449	6	(1.3%)	S: patients with only anogenital dermatitis; R: all positives represent relevant reactions; C: the frequency was significantly lower than in a control group	23
2013-2015 Ireland	5% pet.	99	1	(1%)	S: patients patch tested for perianal and/or genital symptoms; R: all reactions to medicaments were relevant	24
2003-2010 USA	20% pet.	90	1	(1%)	S: women with (predominantly) vulvar symptoms; R: 0%	22
1994-2004 NACDG	20% pet.	344	9	(2.6%)	S: patients with anogenital signs or symptoms; R: only clinically relevant reactions were given; C: the frequency in a control group was 3.8%	21
Other patient groups						
1990-2014 Belgium	20% pet.	492	12	(2.4%)	S: patients suspected of iatrogenic contact dermatitis and tested with a pharmaceutical series and their own products; R: 96% of the positive patch test reactions to all topical drugs and antiseptics were considered to be relevant	28
2000-2010 Canada		100	44	(44%)	S: charts reviewed and included in the study when there was at least one positive reaction to a topical drug; R: not stated; C: the high percentages to all drugs are obviously the result of the – rather unusual – selection procedure targeting at previously diagnosed topical drug allergy	29
1995-2004 IVDK	20% pet.	10,652		(2.0%)	S: not stated; R: not stated	26
1997-2001 USA	20% pet.	203	13	(6%)	S: patients with eyelid dermatitis; R: not stated	20
1996-2001 USA		70	7	(10%)	S: patients with foot dermatitis suspected of contact allergy	53
1993-4 Netherlands	5% pet.	34	2	(6%)	S: patients with chronic otitis externa or media; R: not stated	25
1986-1988 Italy	15% pet.	204	1	(0.5%)	S: animal feed mill workers; R: the reaction was relevant; C: there were 36 individuals with clinical complaints, of who this individual was one; occupational contact dermatitis	31

IVDK: Information Network of Departments of Dermatology, Germany, Austria, Switzerland; NACDG: North American Contact Dermatitis Group (USA, Canada)

Case series

In Finland, in 1959, data were presented on 99 patients with contact allergy to bacitracin; all of them co-reacted to neomycin (33). Nine patients with contact allergy from bacitracin were reported from the USA in 1987 (51). The group consisted of 8 women and one man, ages ranging from 14 to 89 years. They had been diagnosed in 2 centers in New York in a period of one year (1986). All had previously used bacitracin ointment, of whom 7 only briefly for surgical wounds (curettage and desiccation, excision, electrocoagulation). Only one co-reacted to neomycin (51).

Of 215 patients who had undergone a surgical procedure and who had applied an antibiotic ointment to the wound, 9 (4.2%) had signs of allergic contact dermatitis at the follow-up visit. Eight of these had been treated with an ointment containing bacitracin, neomycin and polymyxin B, the ninth had used bacitracin. Seven were patch tested of who 5 proved to be allergic to neomycin and 4 to bacitracin. All had a prior history of topical antibiotic use (52). Additional small case series can be found in refs. 39 (n=2) and 40 (n=3).

Case reports

A 52-year-old male to female transgender patient presented with genital dermatitis. She complained of a red, papular, itchy rash around the newly constructed vulva since undergoing gender affirmation surgery in Thailand. She was prescribed a topical antibacterial ointment with neomycin and bacitracin immediately after surgery. Later, she started using a cream with the anesthetic pramoxine. Physical examination showed excellent anatomical reconstruction with diffuse erythema of the introitus, the labia majora and inguinal skin with scattered 2-3mm erosions. Day 5 patch testing results showed positive reactions to the antibacterial ointment, the anesthetic cream, bacitracin, neomycin and pramoxine (34).

A 27-year-old woman had both ears pierced followed by prophylactic application of an ointment containing bacitracin 250 U, colistin sulfate 5×10^4 U, liquid paraffin and white petrolatum to 1 gram for 1 week. Three years later, both ears were pierced again followed by application of the same ointment, which resulted in pruritic exudative erythema on both auricles. Patch testing with the ointment and its ingredients resulted in positive reactions to the ointment 'as is', bacitracin 5% and 0.5% pet. (negative to 0.05%) and colistin sulfate 5% and 0.5% pet. (negative to 0.05%) (30).

A 46-year old woman was sensitized to bacitracin from an ointment applied to a stasis ulcer (32). Another female patient had applied bacitracin to her lips because of chronic dryness, scaling and fissuring. This resulted in severe erythema and crusting of the lips with some erythema and scaling on the perioral skin. A patch test with bacitracin ointment was positive, but the antibiotic itself was not tested (32). Two patients, who were patch tested because of suspected allergy to bandages, had positive reactions to bacitracin and neomycin, both of which (probably in a combination product) had been applied under adhesive bandages following biopsies and had resulted in pruritic plaques extending beyond the bandage (35).

In a 39-year-old man who was previously shown to be allergic to bacitracin and neomycin, systemic contact dermatitis developed from a root canal paste filling containing both antibiotics. Symptoms included generalized itching, erythema and edema of the face, inflammation of the oral mucosa with difficulty in swallowing and breathing, and aggravation of previously existing eczema on the amputation stump of a leg (where he had become sensitized to neomycin and bacitracin). Oral treatment with bacitracin had considerably aggravated the dermatitis and oral provocation with neomycin resulted in itching of the skin in the popliteal folds, follicular eczema at those locations and edema of the face (43). This case of systemic contact dermatitis was caused by resorption of bacitracin and neomycin from the dental root canal paste.

A female patient aged 28 years was treated for a tinea pedis with an ointment containing bacitracin (500 U/gr) and polymyxin B sulfate (10.000 U/gr), which resulted in contact dermatitis. Patch tests were positive to the ointment, to bacitracin and to polymyxin B sulfate, both tested at 1%, 5% and 30% pet. (36). A 27-year-old woman had recurrent lip dermatitis, to which she had applied a large number of cosmetic lip balms and topical pharmaceuticals. When patch tested, she reacted to bacitracin, an ointment containing bacitracin, and to lanolin alcohol, neomycin and tixocortol pivalate, all of which were also relevant (56).

A 39-year-old woman had developed an acute, erythematous, vesicular dermatitis on the right arm, perioral region, and right side of the forehead, while being treated with an ointment containing bacitracin, polymyxin B, and neomycin for impetigo. Patch tests were positive to bacitracin 1%, polymyxin B sulfate 5% pet., and neomycin 20% (57). Another patient of these authors, a 53-year-old woman had a subacute, scaly, edematous dermatitis of the upper lip and nares of several months' duration. It began after she had been given an antibiotic ointment for skin irritation that resulted from nasal oxygen administration. Patch tests were positive to bacitracin, neomycin and polymyxin B sulfate. It was not mentioned whether the antibiotic ointment contained all 3 antibiotics (57).

A patient with combined type I (anaphylaxis) and type IV allergy was presented in 1981 (41).

Cross-reactions, pseudo-cross-reactions and co-reactions

Patients sensitized to bacitracin very often co-react to neomycin and vice versa. In an older publication, for example, all 99 patients with positive bacitracin patch tests also reacted to neomycin (33). Conversely, 66-88% of patients with positive reactions to neomycin co-react to bacitracin (44,45,46,48). This is not the result of cross-sensitization (dissimilar chemical structures), but is caused by the fact that neomycin and bacitracin are often contained in one product, leading to concomitant sensitization.

In an American study, of 16 patients with positive patch tests to lidocaine, 10 co-reacted to neomycin and 9 to bacitracin. In the USA, topical pharmaceuticals containing lidocaine as well as neomycin and bacitracin are available (48).

Immediate contact reactions

Immediate contact reactions (contact urticaria) to bacitracin are presented in Chapter 5.

LITERATURE

1   The data in the section 'General' may have been obtained from literature discussed in this chapter, but mostly also or exclusively from one or more of the following online sources: ChemIDPlus Advanced, PubChem, DrugBank, RxList, Drug Central, Drugs.com, and Wikipedia

2   DeKoven JG, Warshaw EM, Zug KA, Maibach HI, Belsito DV, Sasseville D, et al. North American Contact Dermatitis Group patch test results: 2015-2016. Dermatitis 2018;29:297-309

3   Veverka KK, Hall MR, Yiannias JA, Drage LA, El-Azhary RA, Killian JM, et al. Trends in patch testing with the Mayo Clinic standard series, 2011-2015. Dermatitis 2018;29:310-315

4   DeKoven JG, Warshaw EM, Belsito DV, Sasseville D, Maibach HI, Taylor JS, et al. North American Contact Dermatitis Group Patch Test Results: 2013-2014. Dermatitis 2017;28:33-46

5   Warshaw EM, Maibach HI, Taylor JS, Sasseville D, DeKoven JG, Zirwas MJ, et al. North American Contact Dermatitis Group patch test results: 2011-2012. Dermatitis 2015;26:49-59

6   Warshaw EM, Belsito DV, Taylor JS, Sasseville D, DeKoven JG, Zirwas MJ, et al. North American Contact Dermatitis Group patch test results: 2009 to 2010. Dermatitis 2013;24:50-59

7   Wentworth AB, Yiannias JA, Keeling JH, Hall MR, Camilleri MJ, Drage LA, et al. Trends in patch-test results and allergen changes in the standard series: a Mayo Clinic 5-year retrospective review (January 1, 2006, to December 31, 2010). J Am Acad Dermatol 2014;70:269-275

8   Fransway AF, Zug KA, Belsito DV, Deleo VA, Fowler JF Jr, Maibach HI, et al. North American Contact Dermatitis Group patch test results for 2007-2008. Dermatitis 2013;24:10-21

9   Zug KA, Warshaw EM, Fowler JF Jr, Maibach HI, Belsito DL, Pratt MD, et al. Patch-test results of the North American Contact Dermatitis Group 2005-2006. Dermatitis 2009 ;20 :149-160

10 Davis MD, Scalf LA, Yiannias JA, Cheng JF, El-Azhary RA, Rohlinger AL, et al. Changing trends and allergens in the patch test standard series. Arch Dermatol 2008;144:67-72

11 Warshaw EM, Belsito DV, DeLeo VA, Fowler JF Jr, Maibach HI, Marks JG, et al. North American Contact Dermatitis Group patch-test results, 2003-2004 study period. Dermatitis 2008;19:129-136

12 Pratt MD, Belsito DV, DeLeo VA, Fowler JF Jr, Fransway AF, Maibach HI, et al. North American Contact Dermatitis Group patch-test results, 2001-2002 study period. Dermatitis 2004;15:176-183

13 Marks JG Jr, Belsito DV, DeLeo VA, Fowler JF Jr, Fransway AF, Maibach HI, et al. North American Contact Dermatitis Group patch-test results, 1998–2000. Am J Contact Dermat 2003;14:59-62

14 Wetter DA, Davis MDP, Yiannias JA, Cheng JF, Connolly SM, el-Azhary RA, et al. Patch test results from the Mayo Contact Dermatitis Group, 1998–2000. J Am Acad Dermatol 2005;53:416-421

15 Erfurt-Berge C, Geier J, Mahler V. The current spectrum of contact sensitization in patients with chronic leg ulcers or stasis dermatitis - new data from the Information Network of Departments of Dermatology (IVDK). Contact Dermatitis 2017;77:151-158

16 Saap L, Fahim S, Arsenault E, Pratt M, Pierscianowski T, Falanga V, Pedvis-Leftick A. Contact sensitivity in patients with leg ulcerations: a North American study. Arch Dermatol 2004;140:1241-1246

17 Smart V, Alavi A, Coutts P, Fierheller M, Coelho S, Holness LD, et al. Contact allergens in persons with leg ulcers: a Canadian study in contact sensitization. Int J Low Extrem Wounds 2008;7:120-125

18 Zaki I, Shall L, Dalziel KL. Bacitracin: a significant sensitizer in leg ulcer patients? Contact Dermatitis1994;31:92-94

19 Rai R, Shenoy MM, Viswanath V, Sarma N, Majid I, Dogra S. Contact sensitivity in patients with venous leg ulcer: A multi-centric Indian study. Int Wound J 2018;15:618-622

20 Guin JD. Eyelid dermatitis: experience in 203 cases. J Am Acad Dermatol 2002;47:755-765

21 Warshaw EM, Furda LM, Maibach HI, Rietschel RL, Fowler JF Jr, Belsito DV, et al. Anogenital dermatitis in patients referred for patch testing: retrospective analysis of cross-sectional data from the North American Contact Dermatitis Group, 1994-2004. Arch Dermatol 2008;144:749-755

22 O'Gorman SM, Torgerson RR. Allergic contact dermatitis of the vulva. Dermatitis 2013;24:64-72

23 Warshaw EM, Kimyon RS, Silverberg JI, Belsito DV, DeKoven JG, Maibach HI, et al. Evaluation of patch test findings in patients with anogenital dermatitis. JAMA Dermatol 2019;156:85-91

24 Foley CC, White S, Merry S, Nolan U, Moriarty B, et al. Understanding the role of cutaneous allergy testing in anogenital dermatoses: a retrospective evaluation of contact sensitization in anogenital dermatoses. Int J Dermatol 2019;58:806-810

25 Van Ginkel CJ, Bruintjes TD, Huizing EH. Allergy due to topical medications in chronic otitis externa and chronic otitis media. Clin Otolaryngol Allied Sci 1995;20:326-328

26 Menezes de Padua CA, Uter W, Schnuch A. Contact allergy to topical drugs: prevalence in a clinical setting and estimation of frequency at the population level. Pharmacoepidemiol Drug Saf 2007;16:377-384

27 Menezes de Padua CA, Schnuch A, Nink K, Pfahlberg A, Uter W. Allergic contact dermatitis to topical drugs – Epidemiological risk assessment. Pharmacoepidemiol Drug Saf 2008;17:813-821

28 Gilissen L, Goossens A. Frequency and trends of contact allergy to and iatrogenic contact dermatitis caused by topical drugs over a 25-year period. Contact Dermatitis 2016;75:290-302

29 Spring S, Pratt M, Chaplin A. Contact dermatitis to topical medicaments: a retrospective chart review from the Ottawa Hospital Patch Test Clinic. Dermatitis 2012;23:210-213

30 Sowa J, Tsuruta D, Kobayashi H, Ishii M. Allergic contact dermatitis caused by colistin sulfate & bacitracin. Contact Dermatitis 2005;53:175-176

31 Mancuso G, Staffa M, Errani A, Berdondini RM, Fabbri P. Occupational dermatitis in animal feed mill workers. Contact Dermatitis 1990;22:37-41

32 Binnick AN, Clendenning WE. Bacitracin contact dermatitis. Contact Dermatitis 1978;4:180-181

33 Pirilä V, Rouhunkoski S. On sensitivity to neomycin and bacitracin. Acta Dermato-Venereologica 1959;39:470-476

34 Schlarbaum JP, Kimyon RS, Liou YL, Becker O'Neill L, Warshaw EM. Genital dermatitis in a transgender patient returning from Thailand: A diagnostic challenge. Travel Med Infect Dis 2019;27:134-135

35 Widman TJ, Oostman H, Storrs FJ. Allergic contact dermatitis from medical adhesive bandages in patients who report having a reaction to medical bandages. Dermatitis 2008;19:32-37

36 Van Ketel WG. Polymixine B-sulfate and bacitracin. Contact Dermatitis Newsletter 1974;15:445

37 Amado A, Sood A, Taylor JS. Contact allergy to lidocaine: a report of sixteen cases. Dermatitis 2007;18:215-220

38 Tam I, Schalock PC, González E, Yu J. Patch testing results from the Massachusetts General Hospital Contact Dermatitis Clinic, 2007-2016. Dermatitis 2020;31:202-208

39 Held JL, Kalb RE, Ruszkowski AM, DeLeo V. Allergic contact dermatitis from bacitracin. J Am Acad Dermatol 1987;17:592-594

40 Björkner B, Möller H. Bacitracin: a cutaneous allergen and histamine liberator. Acta Derm Venereol (Stockh) 1973;53:487-492

41 Schecter JF, Wilkinson RD, Del Carpio J. Anaphylaxis following the use of bacitracin ointment. Arch Dermatol 1984;120:909-911

42 Kleinhans D. Bacitracin and polymyxin B: Important contact allergens in patients with leg ulcers. In: Frosch PJ, Dooms-Goossens A, Lachapelle JM, Rycroft RJG, Scheper RJ (eds). Current Topics in Contact Dermatitis. Berlin: Springer-Verlag, 1989: 258-260

43 Pirilä V, Rantanen AV. Root canal treatment with bacitracin-neomycin as a cause of flare-up of allergic eczema. Oral Surg 1960;13:589-593

44 Pirilä V, Rouhunkoski S. On cross-sensitization between neomycin, bacitracin, kanamycin and framycetin. Dermatologica 1960;121:335-342

45 Förström L, Pirilä V. Cross-sensitivity within the neomycin group of antibiotics. Contact Dermatitis 1978;4:312

46 Reitamo S, Lauerma AI, Stubb S, Käyhkö K, Visa K, Förström L. Delayed hypersensitivity to topical corticosteroids. J Am Acad Dermatol 1986;14:582-589

47 Pirilä V, Förström L, Rouhunkosky S. Twelve years of sensitization to neomycin in Finland. Report of 1760 cases of sensitivity to neomycin and-or bacitracin Acta Derm Venereol 1967;47:419-425

48 Amado A, Sood A, Taylor JS. Contact allergy to lidocaine: a report of sixteen cases. Dermatitis 2007;18:215-220

49 Jacob SE, James WD. From road rash to top allergen in a flash: bacitracin. Dermatol Surg 2004;30:521-524

50 Sood A, Taylor JS. Bacitracin: allergen of the year. Am J Contact Dermat 2003;14:3-4

51 Katz BE, Fisher AA. Bacitracin: a unique topical antibiotic sensitizer. J Am Acad Dermatol 1987;17:1016-1024

52 Gette MT, Marks JG, Maloney ME. Frequency of postoperative allergic contact dermatitis to topical antibiotics. Arch Dermatol 1992;128:365-367

53 Shackelford KE, Belsito DV. The etiology of allergic-appearing foot dermatitis: a 5-year retrospective study. J Am Acad Dermatol 2002;47:715-721

54 Sundquist BK, Yang B, Pasha MA. Experience in patch testing: A 6-year retrospective review from a single academic allergy practice. Ann Allergy Asthma Immunol 2019;122:502-507

55 Schalock PC, Zug KA. Bacitracin. Cutis 2005;76:105-107

56 Fraser K, Pratt M. Polysensitization in recurrent lip dermatitis. J Cutan Med Surg 2015;19:77-80

57 Grandinetti PJ, Fowler JF Jr. Simultaneous contact allergy to neomycin, bacitracin, and polymyxin. J Am Acad Dermatol 1990;23:646-647

58 Aberer W. Bacitracin for lubrication: an allergen for more convenience? J Am Acad Dermatol 2005;52:1114-1115

59 Jacob SE, James WD. Bacitracin after clean surgical procedures may be risky. J Am Acad Dermatol 2004;51:1036

60 Kim B, James W. Postoperative use of topical antimicrobials. Dermatitis 2009;20:174

61 Saripalli YU, Achen F, Belsito DV. The detection of clinically relevant contact allergens using a standard screening tray of twenty-three allergens. J Am Acad Dermatol 2003;49:65-69

Chapter 3.29   BECLOMETHASONE DIPROPIONATE

IDENTIFICATION

Description/definition   : Beclomethasone dipropionate is the dipropionate ester of the synthetic glucocorticoid
                             beclomethasone that conforms to the structural formula shown below
Pharmacological classes : Anti-inflammatory agents; glucocorticoids; anti-asthmatic agents
IUPAC name   : [2-[(8S,9R,10S,11S,13S,14S,16S,17R)-9-Chloro-11-hydroxy-10,13,16-trimethyl-3-oxo-17-
                             propanoyloxy-6,7,8,11,12,14,15,16-octahydrocyclopenta[a]phenanthren-17-yl]-2-
                             oxoethyl] propanoate
Other names   : 9-Chloro-11β-hydroxy-16β-methylpregna-1,4-diene-3,20-dione 17,21-dipropionate;
                             beclometasone dipropionate
CAS registry number   : 5534-09-8
EC number   : 226-886-0
Merck Index monograph : 2287 (Beclomethasone)
Patch testing   : In general, corticosteroids may be tested at 0.1% and 1% in alcohol; late readings (6-10
                             days) are strongly recommended
Molecular formula   : $C_{28}H_{37}ClO_7$

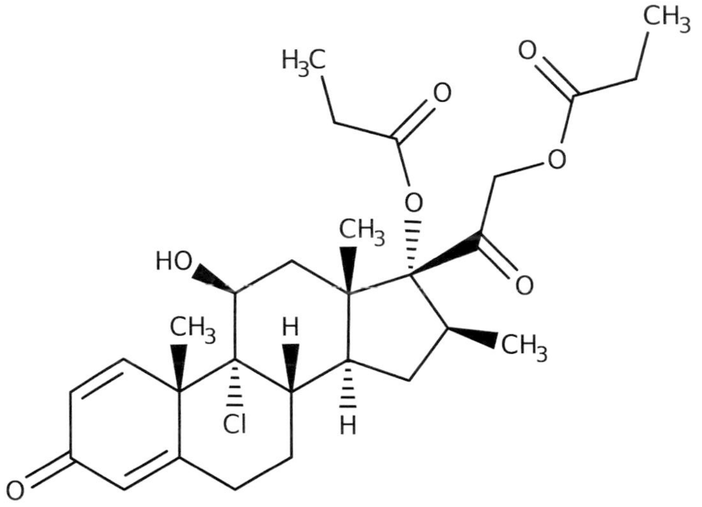

GENERAL

General aspects of corticosteroids used on the skin and mucous membranes are discussed in Chapter 2.4. A practical guideline for diagnosing allergic reactions to corticosteroids is presented in ref. 1.

CONTACT ALLERGY

Case series

From January 1990 to June 2008, in Leuven, Belgium, 315 patients were diagnosed with contact allergy to/allergic contact dermatitis from corticosteroids (CSs) from routine patch testing with a baseline series including tixocortol pivalate, budesonide, hydrocortisone butyrate and prednisone caproate, patch testing with patients' own CS preparations, and testing those with proven contact allergy to a corticosteroid or strongly suspected of CS allergy later with a series of 66 CSs, including two sex hormones (progesterone and testosterone). 71% of the patients had relevant reactions, but these were not specified. In this group of 315 CS allergic patients, 22 had positive patch tests to beclomethasone dipropionate 0.1% alc. (5). It is unknown how many of these reactions were caused by the use of a pharmaceutical product containing beclomethasone dipropionate and how many were cross-reactions to other corticosteroids.

Case reports

A 61-year-old man was referred for a pruritic eruption around his tracheostoma. Since one year, the patient used inhaled salbutamol sulfate and beclomethasone dipropionate aerosols for allergic asthma. After that, he noticed a worsening itch around the tracheostoma. On examination, there was lichenified, exudative erythema with pigmentation. Patch tests were positive to both commercial inhalers tested 1% water and later to salbutamol sulfate 1% pet. Beclomethasone dipropionate and the other ingredients of this inhaler could not be tested, so contact allergy to the corticosteroid was not ascertained, although likely (the other ingredients are very unlikely haptens). Direct insertion of the inhalation device into the trachea prevented the cutaneous symptoms (2).

A 51-year-old woman, using an inhaler containing beclomethasone dipropionate for asthma, noted scales and fissures on the lips and erythema and pigmentation on the perioral area and jaw. Patch tests were positive to the inhaler, a cream with beclomethasone dipropionate and the corticosteroid itself tested at 20% water. The cheilitis and dermatitis cleared after using another inhalant (3).

A 57-year-old atopic man had a 25-year history of hand and foot dermatitis, recently spreading to the backs of the hands and the face in an exposed-site distribution. His asthma was treated with a beclomethasone dipropionate inhaler. He had used a variety of topical corticosteroids. Patch testing with the European baseline series was positive to tixocortol pivalate. Further testing to 18 individual corticosteroids revealed positive reactions to beclomethasone dipropionate 2% alcohol and 3 other glucocorticoids. Photopatch testing was negative but phototesting to UV light was positive at 330, 360 and 400 nm. In view of his multiple contact sensitivities and photosensitivities, a diagnosis of chronic actinic dermatitis was made (4).

A 61-year-old woman found that topical steroids aggravated her current hand dermatitis. She was patch tested and reacted to an ointment she used containing 0.025% beclomethasone dipropionate and to the steroid itself at 20% pet. She was also allergic to prednisolone; oral prednisolone 20 mg daily after 2 days resulted in systemic contact dermatitis and worsening of existing dermatitis (6).

Cross-reactions, pseudo-cross-reactions and co-reactions
Cross-reactions between corticosteroids are discussed in Chapter 2.8.

LITERATURE

1   Baeck M, Goossens A. Immediate and delayed allergic hypersensitivity to corticosteroids: practical guidelines. Contact Dermatitis 2012;66:38-45
2   Tsuruta D, Sowa J, Kobayashi H, Ishii M. Contact dermatitis around a tracheostoma due to salbutamol sulfate and Aldecin. Contact Dermatitis 2006;54:121-122
3   Tani A, Miyoshi H, Kanzaki T. Allergic contact dermatitis due to beclometasone dipropionate in an inhalant for ` asthma. Contact Dermatitis 2000;43:363
4   Cooper SM, Shaw S. Contact allergy to beclomethasone dipropionate. Contact Dermatitis 1998;39:271
5   Baeck M, Chemelle JA, Terreux R, Drieghe J, Goossens A. Delayed hypersensitivity to corticosteroids in a series of 315 patients: clinical data and patch test results. Contact Dermatitis 2009;61:163-175
6   English JS, Ford G, Beck MH, Rycroft RJ. Allergic contact dermatitis from topical and systemic steroids. Contact Dermatitis 1990;23:196-197

Chapter 3.30   BEFUNOLOL

IDENTIFICATION

Description/definition	: Befunolol is the benzofuran that conforms to the structural formula shown below
Pharmacological classes	: β-Adrenergic antagonists
IUPAC name	: 1-[7-[2-Hydroxy-3-(propan-2-ylamino)propoxy]-1-benzofuran-2-yl]ethanone
Other names	: 1-(7-(2-Hydroxy-3-((1-methylethyl)amino)propoxy)-2-benzofuranyl)ethanone
CAS registry number	: 39552-01-7
EC number	: Not available
Merck Index monograph	: 2291
Patch testing	: 1% water; when patch tests are negative but there is suspicion of contact allergy, use 5%
Molecular formula	: $C_{16}H_{21}NO_4$

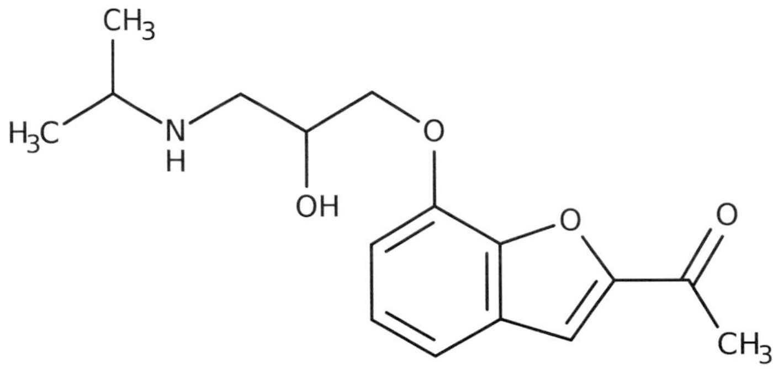

GENERAL

Befunolol is a β-blocker used in the management of open-angle glaucoma. In pharmaceutical products, befunolol is employed as befunolol hydrochloride (CAS number 39543-79-8, EC number not available, molecular formula $C_{16}H_{22}ClNO_4$) (1).

CONTACT ALLERGY

General

For information on patch testing with commercial eye drops containing beta-blockers see Chapter 3.338 Timolol.

Case series

In Ferrara, Italy, over a 65-month period before 2005, 50 patients affected by periorbital dermatitis while using topical ocular products were patch tested, including with their own ophthalmic medications (n=210). Only 15 positive reactions were detected in 12 subjects, including 14 reactions to commercial eye drops. There were 3 reactions to befunolol eye drops. The active ingredients were not tested separately, but contact allergy to the excipients and preservatives was excluded by patch testing. The authors concluded that patch testing with commercial eye drops has doubtful value (14).

In Graz, Austria, before 1990, 7 patients with periorbital dermatitis had allergic contact dermatitis from topical beta-blockers used for treatment of glaucoma. Treatment preceded the appearance of the skin lesions for one month up to one year. The beta-blockers were patch tested as hydrous solutions with the same concentrations as present in therapeutic preparations. In 3 patients, befunolol HCl could be identified as the causative allergen. There were no cross-reactions between befunolol, metipranolol, and timolol (13).

In Tokyo, Japan, out of 3903 patients who were patch tested in the period January 1987 to December 1995, 141 (3.6%) were patch tested with eye drops and 49 individuals (35%) reacted positively and were diagnosed with allergic contact dermatitis. In 36 cases ingredient patch testing was performed and there were three reactions to befunolol HCl (2).

In Nagoya, Japan, 6 patients were diagnosed in a 2-year-period with allergic contact dermatitis from using levobunolol eye drops, developing after 1-9 months use of this medication. All patients reacted to befunolol 1% water (20 controls were negative) but not to timolol or carteolol. Clinical data were not provided except that one patient had erythema and edema of 'both eyelids' (12).

From Bologna, Italy, in 1994 three patients were reported with contact allergy to befunolol in eye drops used for the treatment of glaucoma. A 67-year old woman (patient 1) had acute periorbital and facial contact dermatitis, a 68-year-old man (patient 2) had a 4-month history of periorbital dermatitis and patient 3, a 61-year old man, had eczematous lesions on his face, palms and forearms. They all had positive patch tests to their eye drops containing 0.5% befunolol and to befunolol 0.25%, 0.5% and 1% water. Patient 3 also reacted to timolol, which he had also used (3).

Case reports

A 65-year-old man developed eczema localized on the upper and lower eyelids. Patch testing showed contact allergy to befunolol and timolol, both of which had been used in eye drops for the treatment of glaucoma. Later prescription of eye drops containing carteolol led to recurrence of the eczema, which was considered to be a cross-reaction (4).

A 45-year-old woman, affected by a 6-year primary open-angle glaucoma and treated for years with levobunolol eye drops, presented with bilateral itching erythematosquamous edematous plaques on the upper and lower eyelids. Patch testing with the constituents of the eye drops gave a positive reaction to levobunolol 1% water. Levobunolol was discontinued and the clinical manifestations completely resolved within 4 weeks. Afterwards, the patient used an ophthalmic solution containing befunolol, which had not been used previously, and she developed similar lesions 7 days later. Patch testing with a series of β-blockers showed a positive reaction to befunolol 1% water, which was considered to be a cross-reaction to levobunolol (5).

A 68-year-old woman had long been treated for open-angle glaucoma with eye drops containing the β-blockers befunolol and timolol, when she developed acute eyelid dermatitis, associated with hyperemia and conjunctival chemosis. Her condition rapidly improved after suspending befunolol and following therapy with topical corticosteroids. Patch testing showed positive reactions to befunolol eye drops, befunolol 1% water, carteolol eye drops and carteolol 1% water, but not to timolol. As the patient had not previously used carteolol, these positive reactions were interpreted as resulting from cross-sensitivity (7).

A 29-year-old man had been using levobunolol eye drops for the treatment of open-angle glaucoma for 5.5 years, when he developed blepharoconjunctivitis. He was then treated with various other eye drops including befunolol eye drops, but intolerance persisted. Patch tests showed positive reactions to levobunolol and to befunolol hydrochloride 0.25%, 0.5%, 1%, 2.5% and 5% water (8).

A 70-year old man with glaucoma developed symmetrical eyelid dermatitis and conjunctival hyperemia, while using levobunolol eye drops. Patch testing showed contact allergy to this medication. The patient was then placed on befunolol eye drops with initial good tolerance. However, after 4 months, the symptoms relapsed. Patch tests were positive to the eye drops and befunolol 0.25%, 0.5% and 1% water, indicating that the patient had become allergic to this second beta-blocker (8).

A 54-year-old woman treated for chronic open-angle glaucoma with the beta-blockers befunolol and timolol developed relapsing eyelid itching, swelling and eczema, hyperemia and conjunctival chemosis. She showed positive patch tests to befunolol 0.25% eye drops and befunolol 1% in water and a ?+ reaction to timolol eye drops. Searching for a safe alternative, she was patch tested with 3 beta-blockers she had never used before, but showed a positive reaction (cross-reaction) to levobunolol 1% water (10).

A man aged 53 developed dermatitis of the eyelids and the periorbital area while using befunolol eye drops 0.5%. Patch tests showed positive reactions to the eye drops and to befunolol 0.5%, 0.05% and 0.005% water (11). Another case of allergic contact dermatitis from levobunolol eye drops was reported in Japanese literature (15).

Cross-reactions, pseudo-cross-reactions and co-reactions

Generally speaking, cross-reactions between beta-blockers appear to be infrequent (6). A patient sensitized to befunolol and timolol may have cross-reacted to carteolol (4). A woman sensitized to levobunolol cross-reacted to befunolol (5). A woman sensitized to befunolol had co-reactivity to carteolol, which she had never used before (7). One patient who had developed allergic contact dermatitis from befunolol cross-reacted to levobunolol (10).

LITERATURE

1   The data in the section 'General' may have been obtained from literature discussed in this chapter, but mostly also or exclusively from one or more of the following online sources: ChemIDPlus Advanced, PubChem, DrugBank, RxList, Drug Central, Drugs.com, and Wikipedia
2   Aoki J. [Allergic contact dermatitis due to eye drops. Their clinical features and the patch test results]. Nihon Ika Daigaku Zasshi 1997;64:232-237 (article in Japanese)
3   Vincenzi C, Ricci C, Peluso AM, Tosti A. Allergic contact dermatitis caused by β-blockers in eyedrops. Am J Contact Derm 1994;5:102-103
4   Giordano-Labadie F, Lepoittevin JP, Calix I, Bazex J. Contact allergy to beta blockaders in eye drops: cross allergy?. Ann Dermatol Venereol 1997;124:322-324 (Article in French)

5   Nino M, Balato A, Ayala F, Balato N. Allergic contact dermatitis due to levobunolol with cross-sensitivity to befunolol. Contact Dermatitis 2007;56:53-54

6   Jappe U, Uter W, Menezes de Pádua CA, Herbst RA, Schnuch A. Allergic contact dermatitis due to beta-blockers in eye drops: a retrospective analysis of multicentre surveillance data 1993-2004. Acta Derm Venereol 2006;86:509-514

7   Nino M, Suppa F, Ayala F, Balato N. Allergic contact dermatitis due to the beta-blocker befunolol in eyedrops, with cross-sensitivity to carteolol. Contact Dermatitis 2001;44:369

8   Zucchelli V, Silvani S, Vezzani C, Lorenzi S, Tosti A. Contact dermatitis from levobunolol and befunolol. Contact Dermatitis 1995;33:66-67

9   Morelli R, Arcangeli F, Brunelli D, Vincenzi C, Landi G. Contact allergy to β-blocking agents in eyedrops. Dermatitis 1995;6:172-173

10   Corazza M, Virgili A, Mantovani L, Masieri LT. Allergic contact dermatitis from cross-reacting beta-blocking agents. Contact Dermatitis 1993;28:188-189

11   Mancuso G. Allergic contact dermatitis due to befunolol in eyedrops. Contact Dermatitis 1992;27:198

12   Kanzaki T, Kato N, Kabasawa Y, Mizuno N, Yuguchi M, Majima A. Contact dermatitis due to the beta-blocker befunolol [corrected] in eyedrops. Contact Dermatitis 1988;19:388
       Erratum in: Contact Dermatitis 1989;20:320 (the original title stated timolol instead of befunolol)

13   Gailhofer G, Ludvan M. 'Beta-blockers': sensitizers in periorbital allergic contact dermatitis. Contact Dermatitis 1990;23:262

14   Corazza M, Massieri LT, Virgili A. Doubtful value of patch testing for suspected contact allergy to ophthalmic products. Acta Derm Venereol 2005;85:70-71

15   Katoh N, Kanzaki T. Contact dermatitis due to befunolol hydrochloride eyedrops. Nihon Hifuka Gakkai Zasshi 1987;97:1113-11116 (Article in Japanese)

Chapter 3.31   BELLADONNA EXTRACT

IDENTIFICATION

Description/definition   : Belladonna extract is an extract of the plant species *Atropa belladonna*
Pharmacological classes : Antimuscarinics; antispasmodics
CAS registry number   : 8007-93-0
EC number   : 232-365-9
Patch testing   : Atropine 1% water, scopolamine 1% pet.

GENERAL

Belladonna extract is the extract of the belladonna, *Atropa belladonna*. The belladonna, also known as deadly nightshade, is a perennial herbaceous plant in the nightshade family Solanaceae. Its roots, leaves and fruits contain hyoscyamine, scopolamine, and mostly, atropine, all alkaloids and muscarinic antagonists. The name 'belladonna' originates from the Italian words 'beautiful woman' and the historical use of herb eye-drops by women to dilate the pupils of the eyes for esthetic purposes. Belladonna is a poisonous plant and belladonna intoxication from accidental ingestion may result in a severe anticholinergic syndrome, which is associated with both central and peripheral manifestations. The extract consists mostly of atropine and hyoscine (scopolamine) (1).

According to DrugBank, there are no therapeutic indications. However, belladonna plaster is as over-the-counter medicine available online. Indications mentioned on the packaging are stiff neck, aching shoulders, muscular tension and strain, swelling, pain, rheumatism, backache, lumbago, sciatica, mumps and boils.

CONTACT ALLERGY

Case report

A 61-year-old man presented with an area of dermatitis on his lower back. He had used a belladonna plaster for backache at this site for 3 years with some initial benefit. Over the last year, he had developed headaches. Examination revealed a large rectangular patch of excoriated lichenified eczema on the lower back, corresponding perfectly to the shape of the belladonna plaster. He was patch tested with the plaster and a strong allergic response, consisting of erythema, infiltration and vesicles, developed at D2. Further patch testing to the plaster constituents (scopolamine [hyoscine], atropine and the adhesive base) showed a positive reaction at D2 to atropine 1% pet. only. The eczema cleared on stopping the plaster, although extensive post-inflammatory pigmentation and scarring from excoriations have persisted. His headaches have since also improved (3). It is uncertain whether this plaster actually contained a belladonna extract or that atropine and scopolamine were added individually.

Cross-reactions, pseudo-cross-reactions and co-reactions

A patient sensitized to atropine may have cross-reacted to Belladonna extract (2).

LITERATURE

1   The data in the section 'General' may have been obtained from literature discussed in this chapter, but mostly also or exclusively from one or more of the following online sources: ChemIDPlus Advanced, PubChem, DrugBank, RxList, Drug Central, Drugs.com, and Wikipedia
2   Decraene T, Goossens A. Contact allergy to atropine and other mydriatic agents in eye drops. Contact Dermatitis 2001;45:309-310
3   Williams HC, du Vivier A. Belladonna plaster – not as *bella* as it seems. Contact Dermatitis 1990;23:119-120

Chapter 3.32   BENDAZAC

IDENTIFICATION

Description/definition	: Bendazac is the monocarboxylic acid that conforms to the structural formula shown below
Pharmacological classes	: Anti-inflammatory agents, non-steroidal
IUPAC name	: 2-(1-Benzylindazol-3-yl)oxyacetic acid
Other names	: ((1-(Phenylmethyl)-1H-indazol-3-yl)oxy)acetic acid
CAS registry number	: 20187-55-7
EC number	: 243-569-2
Merck Index monograph	: 2307
Patch testing	: 30% pet., 1.5% pet. and 0.3% pet.; perform controls
Molecular formula	: $C_{16}H_{14}N_2O_3$

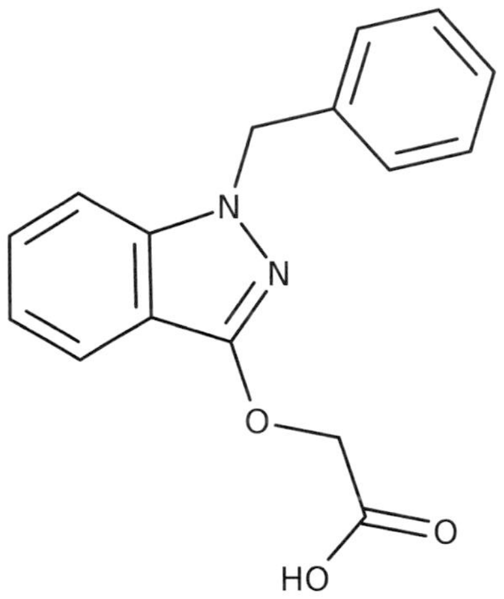

GENERAL

Bendazac is a monocarboxylic acid possessing anti-inflammatory, anti-necrotic, choleretic, and anti-lipidemic characteristics. However, most research has revolved around studying and demonstrating the agent's principal action in inhibiting the denaturation of proteins - an effect that has primarily proven useful in managing and delaying the progression of ocular cataracts and delaying the need for surgical intervention. Bendazac, however, has been withdrawn or discontinued in various international regions due to the risk for causing hepatotoxicity. Nevertheless, a small number of regions may continue to have the medication available for non-prescription use either as a topical anti-inflammatory and analgesic cream or eye drop formulation (in eye drops as bendazac lysine) (1).

CONTACT ALLERGY

Case report

A 28-year-old man with atopic dermatitis presented with a 1-month history of edematous erythema of the face and trunk. He had been treated with two alclometasone dipropionate ointments on the face and trunk and bendazac ointment on the face for the last 6 months. Patch testing was positive to the 3 topical preparations, bendazac 30%, 1.5% and 0.3% pet. and to alclometasone dipropionate 0.1%, 0.05% and 0.01% pet. (2).

LITERATURE

1   The data in the section 'General' may have been obtained from literature discussed in this chapter, but mostly also or exclusively from one or more of the following online sources: ChemIDPlus Advanced, PubChem, DrugBank, RxList, Drug Central, Drugs.com, and Wikipedia
2   Iwakiri K, Hata M, Miura Y, Numano K, Yuge M, Sasaki E. Allergic contact dermatitis due to bendazac and alclometasone dipropionate. Contact Dermatitis 1999;41:218-219

Chapter 3.33   BENZARONE

IDENTIFICATION

Description/definition   : Benzarone is the member of the 1-benzofurans that conforms to the structural formula
                              shown below
Pharmacological classes : Fibrinolytic agents (PubChem; doubtful whether this is correct; see 'General')
IUPAC name   : (2-Ethyl-1-benzofuran-3-yl)-(4-hydroxyphenyl)methanone
Other names   : 2-Ethyl-3-(4-hydroxybenzoyl)benzofuran
CAS registry number   : 1477-19-6
EC number   : 216-026-2
Merck Index monograph   : 580
Patch testing   : 2% pet. and pure (2); perform controls
Molecular formula   : $C_{17}H_{14}O_3$

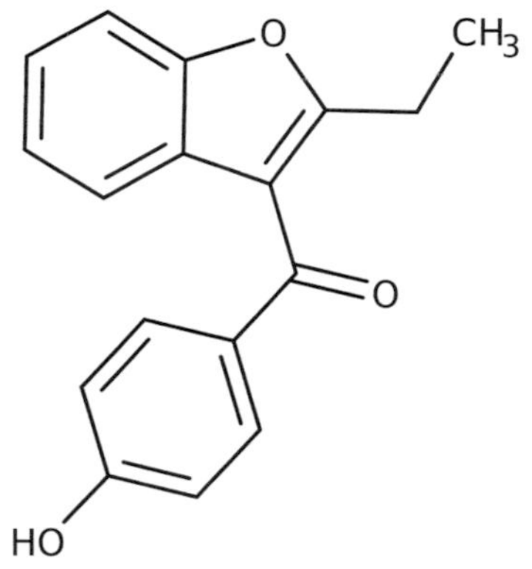

GENERAL

Benzarone is a member of the 1-benzofurans. It has been described as antiarrhythmic, agent for anti-varicose therapy, antihemorrhagic drug and vasoprotective agent. In Taiwan, it may be used for the treatment of capillary fragility and capillary bleeding (1).

CONTACT ALLERGY

Case report

A 59-year-old woman was treated for varicose veins and venous stasis with a cream containing benzarone, when she developed a papulovesicular eruption on the legs, spreading to the hands, arms, face and abdomen. A patch test with the cream was strongly positive (+++/+++). In patch tests with the constituents, benzarone as is and 2% in the ointment base were also strongly positive, but the ointment base alone was negative. Controls were not mentioned (2).

LITERATURE

1   The data in the section 'General' may have been obtained from literature discussed in this chapter, but mostly also or exclusively from one or more of the following online sources: ChemIDPlus Advanced, PubChem, DrugBank, RxList, Drug Central, Drugs.com, and Wikipedia
2   Pevny I. Benzaron allergy. Contact Dermatitis 1984;11:122

Chapter 3.34   BENZOCAINE

IDENTIFICATION

Description/definition	: Benzocaine is the ethyl ester of *p*-aminobenzoic acid, that conforms to the structural formula shown below
Pharmacological classes	: Local anesthetics
IUPAC name	: Ethyl 4-aminobenzoate
Other names	: Ethyl *p*-aminobenzoate
CAS registry number	: 94-09-7
EC number	: 202-303-5
Merck Index monograph	: 2358
Patch testing	: 5.0% pet. (Chemotechnique, SmartPracticeCanada, SmartPracticeEurope)
Molecular formula	: $C_9H_{11}NO_2$

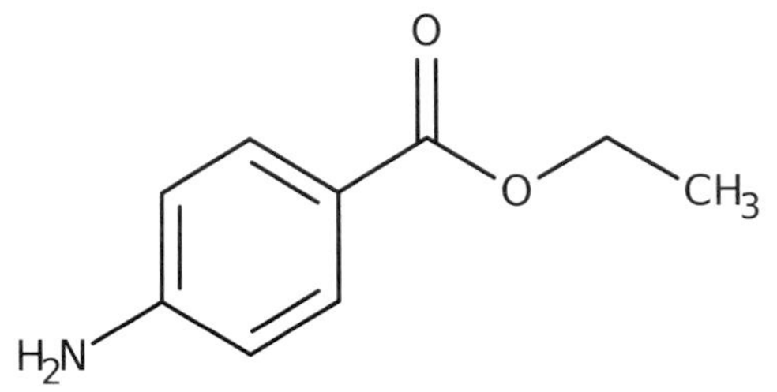

GENERAL

Benzocaine is a local anesthetic and the prototype of the PABA esters commonly used as a topical pain and itch reliever. It is the active ingredient in many over-the-counter analgesic ointments and anti-itch preparations, e.g. for insect bites, hemorrhoids and anal pruritus. It is also indicated for general use as a lubricant and topical anesthetic on intratracheal catheters and pharyngeal and nasal airways to obtund the pharyngeal and tracheal reflexes, on nasogastric and endoscopic tubes, urinary catheters, laryngoscopes, proctoscopes, sigmoidoscopes and vaginal specula (1). The high risk of sensitization to and allergic contact dermatitis from benzocaine has been well recognized for a long time and therefore, in recent decades, benzocaine has been largely replaced in many countries with other less sensitizing local anesthetics such as lidocaine.

CONTACT ALLERGY

General

Benzocaine, the ethyl ester of *p*-aminobenzoic acid, was the first synthetic local anesthetic, introduced into clinical practice as Anaesthesin ® in 1902. Because of poor water solubility its use was restricted to topical application in creams or powders (103). Unfortunately, benzocaine has become a very well-known cause of contact sensitization. The chemical is often stated to be a strong sensitizer, but this does not show from animal experiments (82). Nevertheless, benzocaine must have caused huge numbers of cases of allergic contact dermatitis from its use in topical pharmaceuticals. In a University hospital in Munich, Germany, for example, in the years 1960-1965, benzocaine was by far the most frequent cause of contact allergy to drugs with 765 positive patch test reactions, followed by mafenide (n=437) (99). Many cases of sensitization probably resulted from its use for perianal pruritus.

The high risk of sensitization to and allergic contact dermatitis from benzocaine has been well recognized for a long time and therefore, in recent decades, benzocaine has been gradually replaced with other less sensitizing local anesthetics such as lidocaine. As a consequence, in most European countries, benzocaine is little used anymore and frequencies of sensitization are low (5,33,62,78,80). However, in certain other countries the situation may be quite different. In India, for example, numerous formulations containing benzocaine are still easily available as over-the-counter preparations: oral preparations such as antitussives, astringents and antibacterial mouthwashes, anesthetic gels for toothache, cold sores and denture irritation, as well as skin creams and gels for abrasions, burns, insect bites and leg ulcers, antihemorrhoidal creams, soothing creams for anal and vulvar pruritus, anesthetic gels in condoms and many other applications, and this practice still causes many cases of sensitization there (102).

Benzocaine has long been included in most routine patch test series in a concentration of 5% in petrolatum (78). In 2019 it was replaced in the European baseline series with the caine mix III 10% pet. (containing 5% benzocaine, 2.5% dibucaine [cinchocaine] and 2.5% tetracaine) (77), as it had been shown – already 30 years ago (78,80,81) – and later confirmed (79), that a mix of these 3 local anesthetics detects many cases of sensitization to local anesthetics not identified by benzocaine alone.

Case reports and case series of allergy to benzocaine published in early literature can be found in ref. 98. Because of the abundance of relevant literature, it has not been attempted to provide a full literature review.

Patch testing in the general population and in subgroups

In Australia, before 1999, 219 self-selected adult healthy volunteers, ages ranging from 18 to 82 years, were patch tested with benzocaine and 0.9% had a positive reaction, 1.2% of the men and 0.7% of the women (84). In Germany, in 1997 and 1998, 1141 adults ageing 28-78 years, with a large percentage (>50%) of atopic individuals, were tested with benzocaine and 1.0% (women 1.0%, men 1.0%) had a positive reaction (50). In neither study was mentioned how many of the positive patch tests were of current or past relevance.

Estimates of the 10-year prevalence (1997-2006) of contact allergy to benzocaine in the general population of Denmark based on the CE-DUR method ranged from 0.08 to 0.11% (51). In a similar study from Germany, the estimated prevalence in the general population in the period 1992-2000 ranged from 0.2% to 0.6% (52). Also in Germany, for the period 1995-2004, the population-based relative incidence (RI) of contact sensitization to benzocaine (cases/100,000 defined daily doses (DDDs) per year) was estimated to be 413.9. In the group of local anesthetics, the RI ranged from 1.5 (lidocaine) to 413.9 (benzocaine) (57), indicating that benzocaine at that moment was the most frequent local anesthetic sensitizer.

Patch testing in groups of patients

Results of patch testing benzocaine in consecutive patients suspected of contact dermatitis (routine testing) back to 1999 are shown in table 3.34.1. Results of testing in groups of *selected* patients (patients with leg ulcers/stasis dermatitis, individuals with perianal, genital or anogenital dermatitis, patients with periorbital dermatitis or suspected of medicament allergy) are shown in table 3.34.2.

Patch testing in consecutive patients suspected of contact dermatitis: routine testing

As benzocaine is (or until recently was) present in most baseline/routine/screening/standard series tested worldwide, data on testing this local anesthetic in consecutive patients (routine testing) is abundant. The results of over 35 such published investigations back to 1999 are shown in table 3.34.1. Prevalences of positive patch tests have ranged from 0.5% to 12.7%. The latter percentage is baffling. In that study, all other haptens also had extremely high frequencies of sensitization, indicating that either the methods used were invalid or that the patients had undergone an extremely severe selection procedure before being admitted to routine testing (15).

In European countries, Turkey and Israel, frequencies of positive patch test reactions to benzocaine have ranged from 0.5% to 1.8%. Generally speaking, after 2008 there was a tendency to lower rates. In Belgium, in the period 1990-2014, the frequency of sensitization was 1.8% (58). However, the university hospital where the study was performed is specialized in contact allergy to drugs and cosmetics and, in addition, there was a significant decrease in the frequency of positive reactions from 2.3% in 1990-1994 to 1.3% in 2010-2014 (58).

In the USA + Canada, however, prevalences from the NACDG and some other USA centers were rather constant at a higher level, ranging from 1.4% to 2.1%, mostly in the 1.4-1.7% region, with no apparent tendency to lower rates in recent periods. Relevance data are largely lacking: in most studies, they are not mentioned at all. In the Belgian study, relevance was extremely high (58). In the NACDG studies, 'definite' + 'probable' relevance nearly always was lower than 20%, which may indicate that a number of reactions were due to cross-reactivity, for example to *p*-phenylenediamine. Details on the incriminated products were never provided (table 3.34.1).

Table 3.34.1 Patch testing in groups of patients: Routine testing

Years and Country	Test conc. & vehicle	Number of patients tested | positive (%)		Selection of patients (S); Relevance (R); Comments (C)	Ref.
2013-2019 Turkey	10% pet.	1309	10 (0.8%)	R: not stated	101
2015-2017 NACDG	5% pet.	5583	81 (1.5%)	R: definite + probable relevance: 20%	2
2014-2016 Greece	5% pet.	1978	22 (1.1%)	R: not stated	54
2007-2016 USA	5% pet.	2312	(1.4%)	R: not stated	95
2011-2015 USA	5% pet.	2569	35 (1.4%)	R: not stated	3
2013-2014 NACDG	5% pet.	4851	79 (1.6%)	R: definite + probable relevance: 19%	4
2013-2014 12 European countries, 46 departments [b]	5% pet.	13,085	(0.6%)	R: not stated; C: range of positive reactions: 0% - 1.8%	5
1990-2014 Belgium	5% pet.	14,503	267 (1.8%)	R: 96% of the positive patch test reactions to all topical drugs and antiseptics were considered to be relevant; C: there was a significant decrease in the frequency of positive reactions from 2.3% in 1990-1994 to 1.3% in 2010-2014	58

Table 3.34.1 Patch testing in groups of patients: Routine testing (continued)

Years and Country	Test conc. & vehicle	Number of patients tested	positive (%)	Selection of patients (S); Relevance (R); Comments (C)	Ref.
2011-2012 NACDG	5% pet.	4236	60 (1.4%)	R: definite + probable relevance: 20%	6
2009-12, seven European countries [a]	5% pet.	24,670	(0.7%)	R: not stated; C: range per country: 0.4% - 2.4%	33
2009-1020 NACDG	5% pet.	4306	(1.8%)	R: definite + probable relevance: 18%	7
2006-2010 USA	5% pet.	3086	(2.1%)	R: 30%	8
2001-2010 Australia	5% pet.	5124	80 (1.6%)	R: 13%	9
1985-2010 Denmark	5% pet.	19,347	103 (0.5%)	R: current 10%, past 12%	62
2004-2009 China	5% pet.	2758	20 (0.7%)	R: 41% relevance for all positive patch test reactions together	10
2007-2008 9 European countries [b]	5% pet.	13,197	76 (0.6%) [a]	R: not stated; C: prevalences ranged from 0% to 2.5%	11
2007-2008 NACDG	5% pet.	5082	(1.3%)	R: definite + probable relevance: 18%	12
2005-2006 NACDG	5% pet.	4446	(1.9%)	R: definite + probable relevance: 18%	13
2005-2006 9 European countries [b]	5% pet.	4404	51 (1.2%)	R: not stated; C: prevalences were 1.3% in Central Europe, 1.3% in Northeast and 1.0% in South Europe; in Western Europe (UK), benzocaine was not tested	14
2001-2006 China	5% pet.	1354	(12.7%)	R: not stated; C: all other tested haptens also had very high prevalence scores, suggesting that the patients were highly selected for (routine) patch testing	15
2001-2005 USA	5% pet.	2153	(1.5%)	R: 42%	16
1989-2005 Canada		4368	37 (0.8%)	R: not stated	66
1985-2005 Denmark	5% pet.	14,995	(0.5%)	R: not stated	17
2004, 11 European countries [b]	5% pet.	4564	51 (1.1%) [a]	R: not stated; C: range positives per center: 0.4% - 2.8%	18
2003-2004 NACDG	5% pet.	5141	92 (1.6%)	R: not stated	19
2001-2004 IVDK	5% pet.	31,025	465 (1.7%)	R: not stated	20
1998 2004 Israel	5% pet.	2156	19 (0.9%)	R: not stated	21
1995-2004 IVDK	5% pet.	80,900	(1.5%)	R: not stated	56
1992-2004 Turkey	5% pet.	1038	8 (0.7%)	R: not stated	22
2002-2003 Europe [b]	5% pet.	7224	(0.8%)	R: not stated	23
2001-2002 NACDG	5% pet.	4908	(1.7%)	R: definite + probable relevance: 33%	24
1997-2001 Czech Rep.	5% pet.	12,058	57 (0.5%)	R: not stated	25
1998-2000 USA	5% pet.	5833	(1.7%)	R: definite + probable relevance: 13%	26
1998-2000 USA	5% pet.	1323	(1.6%)	R: not stated	27
1996-2000 Europe	5% pet.	26,210	(1.6%)	R: not stated; C: ten centers, seven countries, EECDRG study; an unspecified number of these patients had been tested with and reacted to the caine mix	28
1996-1999 IVDK	5% pet.	34,859	(1.5%)	R: not stated; C: the frequency of sensitization rose from 1.1% in patients of 60 years or younger to 2.5% in the age group of 61-75 years and to 4.5% in patients over 75 years old; the increase was both dependent and independent of chronic leg ulcers	29

[a] age-standardized and sex-standardized proportions; [b] study of the ESSCA (European Surveillance System on Contact Allergies); EECDRG: European Environmental and Contact Dermatitis Research Group; IVDK: Information Network of Departments of Dermatology, Germany, Austria, Switzerland; NACDG: North American Contact Dermatitis Group (USA, Canada)

Patch testing in groups of selected patients

The results of studies in which groups of selected patients were tested with benzocaine are shown in table 3.34.2. In patients with leg ulcers and/or stasis dermatitis, the frequency of sensitization ranged from 1.2% to 20%, probably reflecting the mode of selection and local, regional or national prescription habits. Testing patients with perianal, genital or anogenital dermatitis yielded mostly rates between 2% and 3%, with a high frequency of 12% in a small study from the USA in women with (predominantly) vulvar symptoms (44). In this patient group also, the higher frequencies were reported from the USA. The relevance rates varied from 0% to >70%. In other patient groups, low rates were found with the exception of a Canadian study (10%) in which all patient charts were screened for previously established contact allergy to a pharmaceutical until they had 100 cases, which would appear to be a rather unorthodox study method (60). Culprit products were never mentioned.

Table 3.34.2 Patch testing in groups of patients: Selected patient groups

Years and Country	Test conc. & vehicle	Number of patients tested	positive (%)	Selection of patients (S); Relevance (R); Comments (C)	Ref.
Patients with leg ulcers/stasis dermatitis					
2003-2014 IVDK	5% pet.	5202	(3.0%)	S: patients with stasis dermatitis/chronic leg ulcers; R: not stated; C: percentage of reactions significantly higher than in a control group of routine testing (1.4%)	30
<2008 Poland	5% pet.	50	2 (4%)	S: patients with venous leg ulcers; R: not stated	32
2005-2008 France	5% pet.	423	5 (1.2%)	S: patients with leg ulcers; R: not stated	31
<2007 Croatia	5% pet.	60	1 (2%)	S: patients with venous leg ulcers with or without (allergic contact) dermatitis; R: not stated	36
2001-2002 France		106	3 (2.8%)	S: patients hospitalized for leg ulcers; R: not mentioned; C: the frequency of sensitization in a dermatitis control group was 1.1%	34
<1999 Croatia	5% pet.	100	20 (20%)	S: patients with leg ulcers; R: not stated	35
1992-1997 IVDK	5% pet.	3227	157 (4.9%)	S: patients with leg ulcers and/or leg eczema; R: not stated	37
1992-1997 Germany	5% pet.	131	11 (8.4%)	S: patients with leg ulcers and/or leg eczema; R: not stated	104
Patients with perianal, genital or anogenital dermatitis					
2005-2016 NACDG	5% pet.	449	15 (3.3%)	S: patients with only anogenital dermatitis; R: all positives represent relevant reactions; C: the frequency was significantly higher than in a control group	49
2004-2016 Spain		124	1 (0.8%)	S: patients with perianal dermatitis lasting >4 weeks; R: 0%	43
2003-2010 USA	5% pet.	59	7 (12%)	S: women with (predominantly) vulvar symptoms; R: 71%	44
2003-2010 USA	5% pet.	55	1 (2%)	S: women with (ichty) vulvar dermatoses; R: not stated	45
2004-2008 IVDK			(2.8%)	S: patients with anogenital dermatoses tested with a medicament series; R: not stated; C: number of patients tested unknown	46
1994-2004 NACDG	5% pet.	342	13 (3.8%)	S: patients with anogenital signs or symptoms; R: 13%; C: the frequency was significantly higher than in a control group	42
1990-2003 IVDK		1168	(2.4%)	S: patients with perianal dermatoses; R: not stated; C: the frequency was significantly higher than in a control group	47
1990-2003 Belgium	5% pet.	92	3 (3%)	S: women suffering from vulval complaints referred for patch testing; R: 0%	53
1992-1997 IVDK		1008	(2.7%)	S: patients evaluated for allergic anogenital contact dermatitis; R: unknown; C: the frequency was higher than in a control group	48
1991-1995 U.K.		121	3 (2.5%)	S: women with pruritus vulvae and primary vulval dermatoses suspected of secondary allergic contact dermatitis; R: 49% for all positive reactions together	40
1992-1994 U.K.	5% pet.	69	1 (1%)	S: women with 'vulval problems'; R: 58% for all allergens together	41
Other patient groups					
2000-2010 Canada		100	10 (10%)	S: charts reviewed and included in the study when there was at least one positive reaction to a topical drug; R: not stated; C: the high percentages to all drugs are obviously the result of the – rather unusual – selection procedure targeting at previously demonstrated topical drug allergy	60
1996-2005 Hungary	5% pet.	401	3 (0.7%)	S: patients with periorbital dermatitis; R: not stated	39
1978-2005 Taiwan		603	11 (1.8%)	S: patients suspected of contact allergy to medicaments; R: 65% of the reactions to all medicaments were considered to be relevant	59
1995-1999 IVDK	5% pet.	970	(1.7%)	S: patients with *allergic* periorbital contact dermatitis; the frequency was not significantly higher than in a control group	38

IVDK: Information Network of Departments of Dermatology, Germany, Austria, Switzerland
NACDG: North American Contact Dermatitis Group (USA, Canada)

Case series
In a study in the United kingdom in the period 1988-1998, 63 patients reacting to a mix of benzocaine, tetracaine and dibucaine were tested with the three active ingredients and there were 22 reactions to benzocaine, 23 to tetracaine and 28 to dibucaine. Of the entire group, 55% were interpreted as either of current or of past relevance (80).

In another study in the United Kingdom, before 1988, 40 patients reacting to a 'caine mix' containing 5% benzocaine, 1% dibucaine (cinchocaine) and 1% tetracaine (amethocaine) were tested with its ingredients and there were 19 reactions to benzocaine, 12 to dibucaine and 16 to tetracaine. The most involved primary sites were the legs (29%) and the anogenital region (27%). Relevance for the entire group was a little over 50%, past relevance 19% (78).

Again in the United Kingdom, in the 1970s, relatively few preparations containing benzocaine were available and consequently it was not a common allergen. In the period 1971-1976 , at St. John's Hospital, London, only 9 patients were diagnosed with contact allergy to benzocaine; 8 had used a cream containing the anesthetic for various purposes (97).

Case reports

A 22-year-old man presented with recurrent erythematous and edematous dermatitis of the shaft of the penis, and an associated balanoposthitis. His female partner had noticed the onset of his symptoms a few hours after every sexual intercourse. She never used spermicidal contraceptives but he always used a condom. Patch tests were positive to the thiuram mix and benzocaine 5% pet. The condom contained a 'retarding' gel with 5% benzocaine to improve the patient's performance (63). Two similar case reports, also from Italy, had been published before (64,65). The first patient had erythematous desquamative lesions on the penis and scrotum and also erythematous desquamative lesions in the periorbital region and on the frontal area of the hairline, which was ascribed to transfer of benzocaine by the hands. Patch tests with benzocaine and the condom containing the gel were positive, but negative to the same condom washed clean of lubricant (64). In the other patient, perianal eczema from benzocaine in antihemorrhoidal ointments had previously occurred. Patch tests were positive to the retarding gel and its constituents benzocaine and parabens (65).

Recently, two more such cases were presented (90,91). In one case, allergic contact dermatitis from benzocaine in the 'extended-pleasure condom' was followed by penile skin gangrene, probably from superinfection with group A beta-hemolytic *Streptococcus* (*S. pyogenes*) (90).

A 43-year-old man developed a perianal eruption 2 hours after applying an ointment containing benzocaine 0.25%, enoxolone 0.7%, procaine HCl 1%, allantoin 0.8%, menthol 0.2% and zinc oxide 20%, which he had previously used for hemorrhoids for 2 years without problems. Patch tests were positive to the ointment 'as is' and 10%, and to benzocaine 1% pet., enoxolone 10% pet., and procaine HCl 1% pet. The other ingredients were all negative (61).

Two women aged 52 and 58 had extreme burning and stinging in the oral cavity (orodynia), in spite of regular application of benzocaine 20% gel and triamcinolone acetonide 0.1% paste for oral lichen planus. One developed a highly pruritic erythematous rash on her lips and perioral area, with predominant involvement of the angles of the mouth, and she had similar lesions on her right index finger and the lateral aspect of her right middle finger with which she would apply the gel and the paste to the mouth. In both patients, the oral mucosa was relatively normal with no clinical signs of inflammation except subsiding lesions of oral lichen planus. Patch tests were positive in both to benzocaine 5% pet. and negative to triamcinolone acetonide. After stopping the use of the benzocaine gel, the orodynia gradually disappeared in both women as did the cutaneous eruption around the mouth and on the fingers of the first patient. It was concluded that contact allergy to benzocaine was the cause of the extreme burning and stinging (orodynia) (102).

Benzocaine has caused many cases of allergic contact dermatitis from its presence in creams and ointments to treat pain and itching caused by various skin disorders such as herpes zoster (67), hemorrhoids and pruritus ani (83,87), insect bites (84), burns (85), and irritated skin (86, the ointment probably contained benzocaine). Contact allergy to benzocaine in a tablet may have contributed in a female patient to severe swelling of the lips and the left side of the face (the side on which the tablets had been left to dissolve), accompanied by inflammation of the oral mucosa. There were no systemic effects (88).

Acute generalized exanthematous pustulosis (AGEP)

A 67-year-old man presented with a severe, widespread, pustular eruption with associated fevers, rigors, watery diarrhea, and malaise. Twenty-four hours before the drug eruption, the patient had dental extraction and received a benzocaine spray. Examination demonstrated confluent erythema studded with non-follicular pustules distributed on the face, trunk, arms and legs. Biopsies demonstrated histologic features consistent with acute generalized exanthematous pustulosis (AGEP). The patient was treated with oral prednisone and topical corticosteroids. After a 3-day hospitalization, he clinically improved and was discharged home. A patch tests was positive to benzocaine 5% at D4 with pustules; a biopsy was consistent with allergic contact dermatitis (92,93).

Benzocaine as impurity in sunscreens

Formerly, many patients with allergic contact dermatitis from the UV-filter glyceryl PABA showed strong co-reactions to benzocaine (68-75), suggesting that the sensitization may be due to the presence of impurities, notably benzocaine, in the glyceryl PABA product (73). Indeed, benzocaine was found in many commercial sources of glyceryl PABA and sunscreens containing glyceryl PABA in concentrations ranging from 1 to 18% (75).

Cross-reactions, pseudo-cross-reactions and co-reactions

Benzocaine is the ethyl ester of *p*-aminobenzoic acid, a so-called 'para-compound'. A such, it can cross-react with related para-compounds, especially if they have an amino group in the para-position of the benzene ring. Cross-reactivity may be observed with other PABA ester-type local anesthetics (tetracaine, procaine, butamben, proparacaine [proxymetacaine], oxybuprocaine [78,80,81,94]), *p*-phenylenediamine and related chemicals, other (hair) dyes (*p*-toluenediamine, *p*-aminophenol, *p*-aminoazobenzene), azo dyes (disperse blue 106, disperse orange 1, disperse orange 3, disperse red 1, disperse yellow 3), certain sulfonamides, and probably to many other chemicals which may also cross-react to *p*-phenylenediamine (105).

Small number of cross-reactions to parabens (66). No cross-reactions to amide-type local anesthetics such as lidocaine, prilocaine, bupivacaine, mepivacaine, articaine or dibucaine (cinchocaine).

PHOTOSENSITIVITY

Formerly, sunscreen products with glyceryl PABA contained high concentrations of benzocaine as impurity (75). One patient had photocontact allergy to benzocaine in the sunscreen (76), another combined contact allergy and photocontact allergy (72). An 81-year-old woman had photoallergic contact dermatitis from benzocaine in an antiseptic cream (89).

Immediate contact reactions

Immediate contact reactions (contact urticaria) to benzocaine are presented in Chapter 5.

LITERATURE

1 The data in the section 'General' may have been obtained from literature discussed in this chapter, but mostly also or exclusively from one or more of the following online sources: ChemIDPlus Advanced, PubChem, DrugBank, RxList, Drug Central, Drugs.com, and Wikipedia

2 DeKoven JG, Warshaw EM, Zug KA, Maibach HI, Belsito DV, Sasseville D, et al. North American Contact Dermatitis Group patch test results: 2015-2016. Dermatitis 2018;29:297-309

3 Veverka KK, Hall MR, Yiannias JA, Drage LA, El-Azhary RA, Killian JM, et al. Trends in patch testing with the Mayo Clinic standard series, 2011-2015. Dermatitis 2018;29:310-315

4 DeKoven JG, Warshaw EM, Belsito DV, Sasseville D, Maibach HI, Taylor JS, et al. North American Contact Dermatitis Group Patch Test Results: 2013-2014. Dermatitis 2017;28:33-46

5 Uter W, Amario-Hita JC, Balato A, Ballmer-Weber B, Bauer A, Belloni Fortina A, et al. European Surveillance System on Contact Allergies (ESSCA): results with the European baseline series, 2013/14. J Eur Acad Dermatol Venereol 2017;31:1516-1525

6 Warshaw EM, Maibach HI, Taylor JS, Sasseville D, DeKoven JG, Zirwas MJ, et al. North American Contact Dermatitis Group patch test results: 2011-2012. Dermatitis 2015;26:49-59

7 Warshaw EM, Belsito DV, Taylor JS, Sasseville D, DeKoven JG, Zirwas MJ, et al. North American Contact Dermatitis Group patch test results: 2009 to 2010. Dermatitis 2013;24:50-59

8 Wentworth AB, Yiannias JA, Keeling JH, Hall MR, Camilleri MJ, Drage LA, et al. Trends in patch-test results and allergen changes in the standard series: a Mayo Clinic 5-year retrospective review (January 1, 2006, to December 31, 2010). J Am Acad Dermatol 2014;70:269-275

9 Toholka R, Wang Y-S, Tate B, Tam M, Cahill J, Palmer A, Nixon R. The first Australian Baseline Series: Recommendations for patch testing in suspected contact dermatitis. Australas J Dermatol 2015;56:107-115

10 Yin R, Huang XY, Zhou XF, Hao F. A retrospective study of patch tests in Chongqing, China from 2004 to 2009. Contact Dermatitis 2011;65:28-33

11 Uter W, Aberer W, Armario-Hita JC, , Fernandez-Vozmediano JM, Ayala F, Balato A, et al. Current patch test results with the European baseline series and extensions to it from the 'European Surveillance System on Contact Allergy' network, 2007-2008. Contact Dermatitis 2012;67:9-19

12 Fransway AF, Zug KA, Belsito DV, Deleo VA, Fowler JF Jr, Maibach HI, et al. North American Contact Dermatitis Group patch test results for 2007-2008. Dermatitis 2013;24:10-21

13 Zug KA, Warshaw EM, Fowler JF Jr, Maibach HI, Belsito DL, Pratt MD, et al. Patch-test results of the North American Contact Dermatitis Group 2005-2006. Dermatitis 2009;20:149-160

14 Uter W, Rämsch C, Aberer W, Ayala F, Balato A, Beliauskiene A, et al. The European baseline series in 10 European Countries, 2005/2006 – Results of the European Surveillance System on Contact Allergies (ESSCA). Contact Dermatitis 2009;61:31-38

15 Cheng S, Cao M, Zhang Y, Peng S, Dong J, Zhang D, et al. Time trends of contact allergy to a modified European baseline series in Beijing between 2001 and 2006. Contact Dermatitis 2011;65:22-27

16 Davis MD, Scalf LA, Yiannias JA, Cheng JF, El-Azhary RA, Rohlinger AL, et al. Changing trends and allergens in the patch test standard series. Arch Dermatol 2008;144:67-72

17   Carlsen BC, Menné T, Johansen JD. 20 Years of standard patch testing in an eczema population with focus on patients with multiple contact allergies. Contact Dermatitis 2007;57:76-83

18   ESSCA Writing Group. The European Surveillance System of Contact Allergies (ESSCA): results of patch testing the standard series, 2004. J Eur Acad Dermatol Venereol 2008;22:174-181

19   Warshaw EM, Belsito DV, DeLeo VA, Fowler JF Jr, Maibach HI, Marks JG, et al. North American Contact Dermatitis Group patch-test results, 2003-2004 study period. Dermatitis 2008;19:129-136

20   Worm M, Brasch J, Geier J, Uter W, Schnuch A. Epikutantestung mit der DKG-Standardreihe 2001-2004. Hautarzt 2005;56:1114-1124 (Article in German)

21   Lazarov A. European Standard Series patch test results from a contact dermatitis clinic in Israel during the 7-year period from 1998 to 2004. Contact Dermatitis 2006;55:73-76

22   Akyol A, Boyvat A, Peksari Y, Gurgey E. Contact sensitivity to standard series allergens in 1038 patients with contact dermatitis in Turkey. Contact Dermatitis 2005;52:333-337

23   Uter W, Hegewald J, Aberer W et al. The European standard series in 9 European countries, 2002/2003 – First results of the European Surveillance System on Contact Allergies. Contact Dermatitis 2005;53:136-145

24   Pratt MD, Belsito DV, DeLeo VA, Fowler JF Jr, Fransway AF, Maibach HI, et al. North American Contact Dermatitis Group patch-test results, 2001-2002 study period. Dermatitis 2004;15:176-183

25   Machovcova A, Dastychova E, Kostalova D, et al. Common contact sensitizers in the Czech Republic. Patch test results in 12,058 patients with suspected contact dermatitis. Contact Dermatitis 2005;53:162-166

26   Marks JG Jr, Belsito DV, DeLeo VA, Fowler JF Jr, Fransway AF, Maibach HI, et al. North American Contact Dermatitis Group patch-test results, 1998–2000. Am J Contact Dermat 2003;14:59-62

27   Wetter DA, Davis MDP, Yiannias JA, Cheng JF, Connolly SM, el-Azhary RA, et al. Patch test results from the Mayo Contact Dermatitis Group, 1998–2000. J Am Acad Dermatol 2005;53:416-421

28   Bruynzeel DP, Diepgen TL, Andersen KE, Brandão FM, Bruze M, Frosch PJ, et al (EECDRG). Monitoring the European Standard Series in 10 centres 1996–2000. Contact Dermatitis 2005;53:146-152

29   Uter W, Geier J, Pfahlberg A, Effendy I. The spectrum of contact allergy in elderly patients with and without lower leg dermatitis. Dermatology 2002;204:266-272

30   Erfurt-Berge C, Geier J, Mahler V. The current spectrum of contact sensitization in patients with chronic leg ulcers or stasis dermatitis - new data from the Information Network of Departments of Dermatology (IVDK). Contact Dermatitis 2017;77:151-158

31   Barbaud A, Collet E, Le Coz CJ, Meaume S, Gillois P. Contact allergy in chronic leg ulcers: results of a multicentre study carried out in 423 patients and proposal for an updated series of patch tests. Contact Dermatitis 2009;60:279-287

32   Zmudzinska M, Czarnecka-Operacz M, SilnyW. Contact allergy to glucocorticosteroids in patients with chronic venous leg ulcers, atopic dermatitis and contact allergy. Acta Dermatovenerol Croat 2008;16:72-78

33   Uter W, Spiewak R, Cooper SM, Wilkinson M, Sánchez Pérez J, Schnuch A, et al. Contact allergy to ingredients of topical medications: results of the European Surveillance System on Contact Allergies (ESSCA), 2009-2012. Pharmacoepidemiol Drug Saf 2016;25:1305-1312

34   Machet L, Couhe C, Perrinaud A, Hoarau C, Lorette G, Vaillant L. A high prevalence of sensitization still persists in leg ulcer patients: a retrospective series of 106 patients tested between 2001 and 2002 and a meta-analysis of 1975-2003. Br J Dermatol 2004;150:929-935

35   Marasovic D, Vuksic I. Allergic contact dermatitis in patients with leg ulcers. Contact Dermatitis1999;41:107-109

36   Tomljanović-Veselski M, Lipozencić J, Lugović L. Contact allergy to special and standard allergens in patients with venous ulcers. Coll Antropol 2007;31:751-756

37   Renner R, Wollina U. Contact sensitization in patients with leg ulcers and/or leg eczema: comparison between centers. Int J Low Extrem Wounds 2002;1:251-255

38   Herbst RA, Uter W, Pirker C, Geier J, Frosch PJ. Allergic and non-allergic periorbital dermatitis: patch test results of the Information Network of the Departments of Dermatology during a 5-year period. Contact Dermatitis 2004;51:13-19

39   Temesvári E, Pónyai G, Németh I, Hidvégi B, Sas A, Kárpáti S. Periocular dermatitis: a report of 401 patients. J Eur Acad Dermatol Venereol 2009;23:124-128

40   Lewis FM, Shah M, Gawkrodger DJ. Contact sensitivity in pruritus vulvae: patch test results and clinical outcome. Dermatitis 1997;8:137-140

41   Lewis FM, Harrington CI, Gawkrodger DJ. Contact sensitivity in pruritus vulvae: a common and manageable problem. Contact Dermatitis 1994;31:264-265

42   Warshaw EM, Furda LM, Maibach HI, Rietschel RL, Fowler JF Jr, Belsito DV, et al. Anogenital dermatitis in patients referred for patch testing: retrospective analysis of cross-sectional data from the North American Contact Dermatitis Group, 1994-2004. Arch Dermatol 2008;144:749-755

43   Agulló-Pérez AD, Hervella-Garcés M, Oscoz-Jaime S, Azcona-Rodríguez M, Larrea-García M, Yanguas-Bayona JI. Perianal dermatitis. Dermatitis 2017;28:270-275

44 O'Gorman SM, Torgerson RR. Allergic contact dermatitis of the vulva. Dermatitis 2013;24:64-72

45 Lucke TW, Fleming CJ, McHenry P, Lever R. Patch testing in vulval dermatoses: how relevant is nickel? Contact Dermatitis 1998;38:111-112

46 Bauer A. Contact sensitization in the anal and genital area. Curr Probl Dermatol 2011;40:133-141

47 Kügler K, Brinkmeier T, Frosch PJ, Uter W. Anogenital dermatoses—allergic and irritative causative factors. Analysis of IVDK data and review of the literature. J Dtsch Dermatol Ges 2005;3:979-986

48 Bauer A, Geier J, Elsner P. Allergic contact dermatitis in patients with anogenital complaints. J Reprod Med 2000;45:649-654

49 Warshaw EM, Kimyon RS, Silverberg JI, Belsito DV, DeKoven JG, Maibach HI, et al. Evaluation of patch test findings in patients with anogenital dermatitis. JAMA Dermatol 2019;156:85-91

50 Schäfer T, Böhler E, Ruhdorfer S, Weigl L, Wessner D, Filipiak B, et al. Epidemiology of contact allergy in adults. Allergy 2001;56:1192-1196

51 Thyssen JP, Uter W, Schnuch A, Linneberg A, Johansen JD. 10-year prevalence of contact allergy in the general population in Denmark estimated through the CE-DUR method. Contact Dermatitis 2007;57:265-272

52 Schnuch A, Uter W, Geier J, Gefeller O (for the IVDK study group). Epidemiology of contact allergy: an estimation of morbidity employing the clinical epidemiology and drug-utilization research (CE-DUR) approach. Contact Dermatitis 2002;47:32-39

53 Nardelli A, Degreef H, Goossens A. Contact allergic reactions of the vulva: a 14-year review. Dermatitis 2004;15:131-136

54 Tagka A, Stratigos A, Lambrou GI, Nicolaidou E, Katsarou A, Chatziioannou A. Prevalence of contact dermatitis in the Greek population: A retrospective observational study. Contact Dermatitis 2019;81:460-462

55 Greig JE, Carson CF, Stuckey MS, Riley TV. Prevalence of delayed hypersensitivity to the European standard series in a self-selected population. Australas J Dermatol 2000;41:86-89

56 Menezes de Padua CA, Uter W, Schnuch A. Contact allergy to topical drugs: prevalence in a clinical setting and estimation of frequency at the population level. Pharmacoepidemiol Drug Saf 2007;16:377-384

57 Menezes de Padua CA, Schnuch A, Nink K, Pfahlberg A, Uter W. Allergic contact dermatitis to topical drugs – Epidemiological risk assessment. Pharmacoepidemiol Drug Saf 2008;17:813-821

58 Gilissen L, Goossens A. Frequency and trends of contact allergy to and iatrogenic contact dermatitis caused by topical drugs over a 25-year period. Contact Dermatitis 2016;75:290-302

59 Shih Y-H, Sun C-C, Tseng Y-H, Chu C-Y. Contact dermatitis to topical medicaments: a retrospective study from a medical center in Taiwan. Dermatol Sinica 2015;33:181-186

60 Spring S, Pratt M, Chaplin A. Contact dermatitis to topical medicaments: a retrospective chart review from the Ottawa Hospital Patch Test Clinic. Dermatitis 2012;23:210-213

61 Tanaka S, Otsuki T, Matsumoto Y, Hayakawa R, Sugiura M. Allergic contact dermatitis from enoxolone. Contact Dermatitis 2001;44:192

62 Thyssen JP, Engkilde K, Menné T, Johansen JD. Prevalence of benzocaine and lidocaine patch test sensitivity in Denmark: temporal trends and relevance. Contact Dermatitis 2011;65:76-80

63 Muratore L, Calogiuri G, Foti C, Nettis E, Di Leo E, Vacca A. Contact allergy to benzocaine in a condom. Contact Dermatitis 2008;59:173-174

64 Placucci F, Lorenzi S, La Placa M, Vincenzi C. Sensitization to benzocaine on a condom. Contact Dermatitis 1996;34:293

65 Foti C, Bonamonte D, Antelmi A, Conserva A, Angelini G. Allergic contact dermatitis to condoms: description of a clinical case and analytical review of current literature. Immunopharm Immunotoxol 2004;26:479-483

66 Turchin I, Moreau L, Warshaw E, Sasseville D. Cross-reactions among parabens, para-phenylenediamine, and benzocaine: a retrospective analysis of patch testing. Dermatitis 2006;17:192-195

67 Roos TC, Merk HF. Allergic contact dermatitis from benzocaine ointment during treatment of herpes zoster. Contact Dermatitis 2001;44:104

68 Meltzer L, Baer RL. Sensitization to monoglycerol para-aminobenzoate. J Invest Dermatol 1949;12:31-39

69 Baer RL, Meltzer L. Sensitization to monoglyceryl para-aminobenzoate. J Invest Dermatol 1948;11:5

70 Curtis GH, Crawford PF. Cutaneous sensitivity to monoglycerol paraminobenzoate. Cleveland Clinical Quarterly 1951;18:35-41

71 Fisher AA. Sunscreen dermatitis due to glyceryl PABA: significance of cross-reactions to this PABA ester. Cutis 1976;18:495-500

72 Caro I. Contact allergy/photoallergy to glyceryl PABA and benzocaine. Contact Dermatitis 1978;4:381-382

73 Fisher AA. Dermatitis due to benzocaine present in sunscreens containing glyceryl PABA (Escalol 106). Contact Dermatitis 1977;3:170-171

74 Hjorth N, Wilkinson D, Magnusson B, Bandmann HJ, Maibach H. Glyceryl-p-aminobenzoate patch testing in benzocaine-sensitive subjects. Contact Dermatitis 1978;4:46-48

75 Fisher AA. The presence of benzocaine in sunscreens containing glyceryl PABA (Escalol 106). Arch Dermatol

1977;113:1299-1300

76 Kaidbey KH, Allen H. Photocontact dermatitis to benzocaine. Arch Dermatol 1981;117:77-79

77 Wilkinson M, Gonçalo M, Aerts O, Badulici S, Bennike NH, Bruynzeel D, et al. The European baseline series and recommended additions: 2019. Contact Dermatitis 2019;80:1-4

78 Beck MH, Holden A. Benzocaine—an unsatisfactory indicator of topical local anaesthetic sensitization for the U.K. Br J Dermatol 1988;118:91-94

79 Brinca A, Cabral R, Gonçalo M. Contact allergy to local anaesthetics-value of patch testing with a caine mix in the baseline series. Contact Dermatitis 2013;68:156-162

80 Sidhu SK, Shaw S, Wilkinson JD. A 10-year retrospective study on benzocaine allergy in the United Kingdom. Am J Cont Dermat 1999;10:57-61

81 Wilkinson JD, Andersen KE, Lahti A, Rycroft RJ, Shaw S, White IR. Preliminary patch testing with 25% and 15% 'caine'-mixes. The EECDRG. Contact Dermatitis 1990;22:244-245

82 Basketter DA, Scholes EW, Wahlkvist H, Montelius J. An evaluation of the suitability of benzocaine as a positive control skin sensitizer. Contact Dermatitis 1995;33:28-32

83 Viraben R, Aquilina C, Cambon L, Bazex J. Allergic contact dermatitis in HIV-positive patients. Contact Dermatitis 1994;31:326-327

84 Van der Schroeff JG, Van Driel LM. Allergy to Belgian but not Dutch Nestosyl ointment®. Contact Dermatitis 1981;7:120-121

85 Van Hecke E. Allergy to Nestosyl ointments. Contact Dermatitis 1981;7:361-362

86 Van Ketel WG. Allergy to Nestosyl ointment. Contact Dermatitis 1979;5:193

87 De Groot AC, Duchateau AMJA. Contactallergie voor butylcaine en benzocaine in een aambeienzalf. Nieuwsbrief Contactdermatologie 1984;20:321-324 (Article in Dutch)

88 Villas-Martinez F, Joral Badas A, Garmendia Goitia JF, Aguirre I. Sensitization to oral enoxolone. Contact Dermatitis 1994;30:124

89 Epstein S. Photocontact dermatitis from benzocaine. Arch Dermatol 1965;92:591

90 Sharma A, Agarwal S, Garg G, Pandey S. Desire for lasting long in bed led to contact allergic dermatitis and subsequent superficial penile gangrene: a dreadful complication of benzocaine-containing extended-pleasure condom. BMJ Case Rep 2018 Sep 27;2018. pii: bcr-2018-227351.

91 Ljubojević Hadžavdić S, Gojčeta Burnić A, Hadžavdić A, Marinović Kulišić S, Jurakić Tončić R. Erythema of the penis after use of a latex condom—latex allergy or something else? Contact Dermatitis 2018;78:168-169

92 O'Toole A, Lacroix J, Pratt M, Beecker J. Acute generalized exanthematous pustulosis associated with 2 common medications: hydroxyzine and benzocaine. J Am Acad Dermatol 2014;71(4):e147-e149

93 O'Toole AC, LaCroix J, Pratt M. Acute generalized exanthematous pustulosis (AGEP) caused: A case series and review of the guidelines for patch testing in cutaneous drug eruptions. Dermatitis 2017;27(5):e4

94 Van Ketel WG, Bruynzeel DP. A 'forgotten' topical anaesthetic sensitizer: butyl aminobenzoate. Contact Dermatitis 1991;25:131-132

95 Tam I, Schalock PC, González E, Yu J. Patch testing results from the Massachusetts General Hospital Contact Dermatitis Clinic, 2007-2016. Dermatitis 2020;31:202-208

96 Warshaw EM, Schram SE, Belsito DV, DeLeo VA, Fowler JF Jr, Maibach HI, et al. Patch-test reactions to topical anesthetics: retrospective analysis of cross-sectional data, 2001 to 2004. Dermatitis 2008;19:81-85

97 Cronin E. Contact Dermatitis. Edinburgh: Churchill Livingstone, 1980:196

98 Lane CG, Luikart R. Dermatitis from local anesthetics, with a review of one hundred and seven cases from the literature. J Am Med Assoc 1951;146:717-720

99 Bandmann H-J. Die Kontaktallergie durch Arzneimitteln. Pharmazeutische Zeitung, 1966;iii:1470 (data cited in ref. 100).

100 Bandmann H-J, Breit R. The mafenide story. Br J Dermatol 1973;89:219-221

101 Boyvat A, Kalay Yildizhan I. Patch test results of the European baseline series among 1309 patients in Turkey between 2013 and 2019. Contact Dermatitis 2020 Jul 3. doi: 10.1111/cod.13653. Online ahead of print.

102 Arshdeep DD, Handa S. Does contact allergy to benzocaine cause orodynia? Ind J Dermatol Venereol Leprol 2015;81:84-86

103 Ruzicka T, Gerstmeier M, Przybilla B, Ring J. Allergy to local anaesthetics: comparison of patch test with prick and intradermal test results. J Am Acad Dermatol 1987;16:1202-1208

104 Reference lost during writing

105 De Groot AC. Monographs in Contact Allergy Volume I. Non-Fragrance Allergens in Cosmetics (Part I and Part 2). Boca Raton, Fl, USA: CRC Press Taylor and Francis Group, 2018:996-997

Chapter 3.35   BENZOYL PEROXIDE

IDENTIFICATION

Description/definition   : Benzoyl peroxide is the peroxide derivative that conforms to the structural formula
                            shown below
Pharmacological classes   : Dermatological agents
IUPAC name   : Benzoyl benzenecarboperoxoate
Other names   : Dibenzoyl peroxide
CAS registry number   : 94-36-0
EC number   : 202-327-6
Merck Index monograph   : 2389
Patch testing   : 1.0% pet. (Chemotechnique, SmartPracticeEurope, SmartPracticeCanada); great risk of
                            irritant reactions (2,28,58)
Molecular formula   : $C_{14}H_{10}O_4$

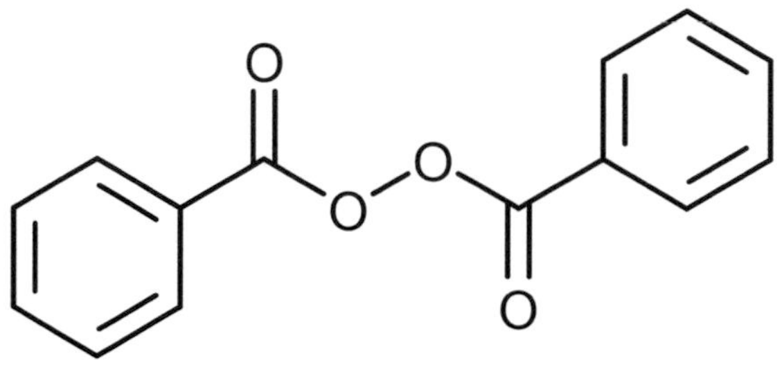

GENERAL

Benzoyl peroxide is an organic compound in the peroxide family and one of the most important in terms of applications. In medicine, benzoyl peroxide is used as a topical treatment for acne, either in combination with antibiotics or as a single agent; it has also been used in burns and for the treatment of leg ulcers. Like most peroxides, benzoyl peroxide is a powerful bleaching agent. Contact with fabrics or hair can cause permanent color dampening almost immediately; even secondary contact can cause bleaching. Because of this quality, benzoyl peroxide is also used for bleaching flour and cheese in the food industry, bleaching of hair and for teeth whitening. Other applications include the cross-linking of polyester resins, an initiator for polymerization of (meth)acrylates in artificial nail material (1), in fillers, putties, plasters, modelling clay, adhesives and sealants, biocides (e.g. disinfect-tants, pest control products), coating products, finger paints and inks and toners. Benzoyl peroxide is found naturally in cereals and cereal products (29).

CONTACT ALLERGY

General

Contact allergy to benzoyl peroxide was common in the 1970s and 1980s, when 20% benzoyl peroxide was used to treat leg ulcers in chronic venous insufficiency. In these patients, (very) high sensitization rates (up to 75%) have been described (3,4,19,20,21), although in some studies, irritant concentrations of benzoyl peroxide were used for patch testing (2-10%) and there were often no clinical signs of allergic contact dermatitis (4). Still earlier, many bakers had developed occupational allergic contact dermatitis from benzoyl peroxide used for flour improvement (26). However, this compound is often cited to have been banned as a flour bleaching agent more than 60 years ago, and nowadays, benzoyl peroxide is no longer considered an occupational allergen in bakers (5). However, in 2005, the Joint Food and Agriculture Organization/World Health Organization Committee on Food Additives has evaluated the use of benzoyl peroxide as a bleaching agent in flour and concluded that 40 mg/kg of flour is acceptable, contradicting a total ban (41). Indeed, in 2019 a new case of allergic contact dermatitis from benzoyl peroxide in flour in a baker was published (40).

There are several case reports of allergic contact dermatitis due to benzoyl peroxide following topical treatment in acne patients (e.g. 12,13,23,50,52,59,60). Larger studies on the frequency of contact sensitization to benzoyl peroxide in acne patients having undergone treatment with it revealed few positive reactions to benzoyl peroxide (14,15,17,18,37), although in a clinical trial, 2 of 44 volunteers (4.5%) treated with BP 5% became – convincingly - sensitized (58). Thus, there seems to be consensus that the widely used acne treatments with benzoyl peroxide rarely sensitize (2,22). In addition, it should be appreciated that patch testing with benzoyl peroxide 1% pet. (the commonly used and commercially available product) frequently leads to weak positive reactions (?+ but also +), especially in patients with atopic dermatitis but certainly also in patients with acne who have used BP (58). Mostly, their clinical relevance remains uncertain, patients with such reactions often have no clinical signs of allergic contact dermatitis and the majority may be irritant patch test responses (2,58). Patch testing with commercial products is notoriously unreliable (high concentrations of 5% or 10% will irritate), and the results do not correspond with the results of patch testing BP 1% pet., a history of intolerance to the products or use tests (15).

Despite all of this, it should be mentioned that a convincing case series of 7 patients with contact allergy from BP used for treating their acne has been reported (51). It is conceivable that sensitization and allergic contact dermatitis occurs occasionally and more often than is generally assumed, but is interpreted as irritant dermatitis, hence no patch tests are performed and the contact allergy goes unrecognized.

Some patients had connubial allergic contact dermatitis (contact dermatitis 'by proxy', caused by the use of BP by a partner, child or someone else nearby) (51,53,54,55).

Patch testing in groups of selected patients

Results of studies in which groups of selected patients were patch tested with benzoyl peroxide, mostly at 1% pet. (which causes many false-positive, irritant reactions), are shown in table 3.35.1. In 4 of 5 investigations in patients with leg ulcers, high frequencies of sensitization ranging from 7%-15% have been observed. However, the issue of relevance was never addressed. Nevertheless, rates of sensitization in most other patient groups were far lower. Percentages of 12% for patients with allergic contact cheilitis (16) and 15.4% in a group of individuals with periorbital eczema (34) must have been artifacts and caused by a large number of irritant patch test reactions. Indeed, in the latter study, 8.5% of controls also had 'positive' patch tests to benzoyl peroxide and in some studies, a great number of irritant reactions was noted (2,37).

Table 3.35.1 Patch testing in groups of patients: Selected patient groups

Years and Country	Test conc. & vehicle	Number of patients tested | positive (%)			Selection of patients (S); Relevance (R); Comments (C)	Ref.
Patients with venous leg ulcers/stasis dermatitis						
<2017 India	1% pet.	172	2	(1.2%)	S: patients with venous leg ulcers of over 6 weeks duration; R: 'the majority of reactions were relevant'	32
2006-2007 Canada		100	7	(7%)	S: patients with leg ulcers or venous disease; R: not stated	33
<1999 Croatia	1% pet.	100	10	(10%)	S: patients with leg ulcers; R: not stated	30
<1994 U.K.		85	13	(15%)	S: patients with longstanding venous ulceration or eczema complicating leg ulcers; R: not stated	31
1982 Italy	1% pet.	120	12	(10%)	S: patients with chronic leg ulcers complicated by contact dermatitis; R: not stated, but assumedly all were considered to be relevant	20
Patients with acne or suspected of medicament allergy						
1990-2014 Belgium	1% pet.	1968	72	(3.7%)	S: patients suspected of iatrogenic contact dermatitis and tested with a pharmaceutical series and their own products; R: 96% of the positive patch test reactions to all topical drugs and antiseptics were considered to be relevant	35
1978-2005 Taiwan		603	11	(1.8%)	S: patients suspected of contact allergy to medicaments; R: 65% of the reactions to all medicaments were considered to be relevant	36
<1996 Italy	5% gel	204	2	(1.0%)	S: patients with acne; R: all reactions were relevant; C: large number of irritant reactions	37
	2% pet.	204	2	(1.0%)		
Other groups						
2001-2011 USA	1% pet.	41	5	(12%)	S: patients with allergic contact cheilitis; R: 60%	16
2004-2009 USA	1% pet.	?	85	(?)	S: not stated; R: not stated	48
1992-2007 IVDK	1% pet.	29,758	2316	(7.8%)	S: not specified; R: not stated; C: there were 9.7% doubtful or irritant reactions and 1.3% ++ or +++ patch tests; higher percentage of patients with leg dermatitis in the group with strong reactions	2
1990-1994 IVDK		136	21	(15.4%)	S: patients with periorbital eczema; R: not stated; C: frequency of sensitization in dermatitis controls: 8.5%; many reactions must have been irritant	34
1968-1977 Italy	1% pet.	497	26	(5.2%)	S: not stated; R: not stated	38

IVDK: Information Network of Departments of Dermatology (Germany, Austria, Switzerland)

Case series: Leg ulcers

In Germany, before 1984, 41 patients with leg ulcers, who had negative patch tests to BP 1%, were treated with 20% benzoyl peroxide lotion for 4 weeks. It was used only on the ulcer, the surrounding skin being covered very carefully with zinc oxide. In 4 cases the treatment had to be stopped because of severe surrounding dermatitis. After treatment, all patients were retested with benzoyl peroxide. Eleven were tested with 1% only and 9 were positive. The other 30 patients were retested with BP 1%, 0.5% and 0.25%. In 16 cases, all 3 concentrations produced positive reactions and in 6, one or 2 concentrations gave positive patch tests. Thus, it appeared that 31 of the 41 patients (76%) treated, who had been negative before treatment, showed positive reactions afterwards and were probably sensitized by the ulcer treatment (21).

Sixteen patients with leg ulcers were treated with 10% benzoyl peroxide (BP) gel for 6 weeks, after which they were patch tested with BP 2% pet. and the 10% BP gel. Nine (56%) had positive patch tests, of whom 8 to both preparations. However, these concentrations are certainly irritant and only one patient had signs of allergic contact dermatitis around the ulcers (4).

Case series: Acne

Between January 2009 and December 2010, in two university hospitals in Belgium, six women suffering from mild facial and/or chest acne presented with a similar history of an acute erythematovesicular eruption on the areas treated with a gel containing 0.1% adapalene and 2.5% benzoyl peroxide. One additional woman, who presented with a recurrent eczematous eruption of the cheeks, was not herself treated with the gel but was living with a partner who had used it for facial acne vulgaris. All patients were patch tested with the European baseline series, with BP 1% pet., adapalene 0.1% pet. and with the gel, tested 'open', 'as is', and diluted 30, 10 and 1% in petrolatum. All 7 women reacted to BP 1% pet., the gel 'as is' and 30% pet., and 5 of the 7 also to 10% pet. The open tests were also positive in all women. Adapalene produced no positive patch test reactions (51).

Five women with contact dermatitis on the face secondary to prolonged use of topical 5% BP preparations for acne vulgaris were patch tested with benzoyl peroxide 1% pet. and 3 (60%) had a positive reaction; the other 2 had irritant reactions (20).

Case reports: Acne

A 20-year-old female patient presented with progressive facial erythema and edema after 2 weeks previously having commenced topical treatment with 5% benzoyl peroxide (BP) and 1% clindamycin phosphate gel for facial acne. Patch testing showed positive reactions to BP 1% pet. (++ at D4) and to the gel 10% water (+++). Five controls were negative to the undiluted gel (23). The authors also described a 42-year-old woman who developed intense irritation, erythema, and scaling of the face while using the same gel and 10% BP cream topically for facial acne. An open patch test with the cream performed by the patient on the forearm provoked similar redness and scaling. Patch testing revealed positive reactions at D4 to BP 1% pet. (++), to the 5% gel diluted 10x with water (++) and to the 10% cream (++) (23).

A 23-year-old woman with cystic acne was treated with a 5% topical BP solution. After 7 days, she noted intolerance with redness, scaling and severe itching. When the medication was discontinued, the reaction cleared in 24-48 hours, but relapses appeared when therapy was restarted. A patch test with BP 1% was negative. She did react, however, to serial dilutions of the 5% BP solution (use concentrations not specified). She had also a positive patch test to another commercial preparation containing 1% BP, but was negative to its base (12).

A woman presented with microvesicular lesions on an erythematous and edematous base with marked itching affecting the face, ears, neck and hair line on the forehead, which had started 4 days earlier. About 20 days previously, she had started treatment for facial acne with alternating applications of a preparation containing retinoic acid and one containing 5% benzoyl peroxide. Patch tests were positive to benzoyl peroxide 1% in olive oil. A second patch test with BP 1% pet. was also positive and she additionally reacted to tretinoin (retinoic acid) (13).

A 26-year-old woman developed an itchy erythematous reaction and strong edema localized to the face 2 weeks after having used BP 10% gel for acne. Patch tests were positive to BP 1% and to the gel (50). As the reactions were read at D2 only and the authors called the swelling 'angioedema', this report is unreliable. One patient had 'an incipient edematous reaction to topical BP used for acne therapy', apparently caused by delayed-type hypersensitivity; details are not available to the author (52).

In some cases of allergic contact dermatitis of the face from BP, the clinical picture resembled (59) or was mistaken for (60) impetigo vulgaris. A case of purpuric contact dermatitis to a topical pharmaceutical containing 5% BP has been reported (39). Patch tests to the product and to BP 5% pet. were positive, but the patient did not react to BP 1%. Based on these findings, the clinical picture and histopathology of a positive patch test reaction, an irritant rather than an allergic reaction seems likely. Quite surprisingly, this issue was hardly addressed in the discussion by the authors (39).

Case reports: Other skin diseases

A 32-year-old man developed an erythematous and papular dermatitis in the groins and on the scrotum from self-treatment with 5% benzoyl peroxide gel after concluding from a Google search on 'burning and redness' that he had an infection. Patch tests were strongly positive to the gel (combined irritant/allergic patch test) and BP 2% pet. (42).

A 40-year-old woman had a dermatitis of the face, arms and backs of the hands. She had been using topical benzoyl peroxide 10% gel for rosacea for 3 months. Patch tests were positive (++) to the gel and to BP 1% pet. at D2 and D3 (57). A 9-year-old boy had a dermatitis of his left hand. He was patch tested with several topical medicaments applied for chronic paronychia and reacted to BP gel 10% pet. and its ingredient BP 1% pet at D2 (++) and D3 (++) (57).

Some patients allergic to benzoyl peroxide developed allergic contact dermatitis not from products used by themselves, but by a child (53,55) or partner (51,54), which is termed 'connubial' or 'consort' contact dermatitis (56), or more recently 'contact dermatitis 'by proxy''.

Allergic contact dermatitis from benzoyl peroxide in non-pharmaceutical products

Patients with complications from cemented arthroplasty and patch tested with various bone cement components not infrequently have positive reactions to benzoyl peroxide 1% pet. Whether these are relevant for the complications of the arthroplasty is debated (24,43,49). Allergic reactions from BP have also been observed in patients wearing dental prostheses (25), from its presence in an adhesive tape (45), in the rubber or glue of swimming goggles (presence not confirmed, but assumed!) (46), and in an arm prosthesis (47).

Occupational allergic contact dermatitis

Allergic contact dermatitis due to benzoyl peroxide may develop as a consequence of occupational exposure, affecting mainly workers in the electronics and plastics industries, electricians, dentists and dental technicians (2,11), bakers (formerly), and laboratory technicians (8). A sacristan in a cathedral developed airborne allergic contact dermatitis from benzoyl peroxide used as bleaching agent in candle wax (6). Two podiatrists were sensitized to BP in thermoplastic resins and developed (airborne) allergic contact dermatitis on the face, neck, hands and arms (7). A 25-year-old technician developed airborne allergic contact dermatitis from BP in insulation plastics, cables and wires that he had to saw and scrape (8). BP in a glue sensitized a marbler and caused typical airborne allergic contact dermatitis involving the skin not protected by clothes: face (notably the eyelids), hands and forearms, but also the trunk (9). BP in a hardening gel sensitized an orthopedic technician (10) and an artificial nail technician developed allergic contact dermatitis from BP present as an initiator in the materials she worked with (1). Occupational sensitization has also occurred from its presence in adhesive tape (44).

LITERATURE

1   Andersen SL, Rastogi SC, Andersen KE. Occupational allergic contact dermatitis to hydroxyethyl methacrylate (2-HEMA) in a manicurist. Contact Dermatitis 2009;61:48-50

2   Ockenfels H-M, Uter W, Lessmann H, Schnuch A, Geier J. Patch testing with benzoyl peroxide: reaction profile and interpretation of positive patch test reactions. Contact Dermatitis 2009;61:209-216

3   Bandmann HJ, Agathos M. Die posttherapeutische Benzoylperoxidkontaktallergie bei Ulcus-cruris-Patienten. Hautarzt 1985;36:670-674 (Article in German)

4   Jensen O, Petersen SH, Vesterager L. Contact sensitization to benzoyl peroxide following topical treatment of chronic leg ulcers. Contact Dermatitis 1980;6:179-182

5   Bauer A, Geier J, Elsner P. Type IV allergy in the food processing industry: sensitization profiles in bakers, cooks, and butchers. Contact Dermatitis 2002;46:228-235

6   Bonnekoh B, Merk HF. Airborne allergic contact dermatitis from benzoyl peroxide as a bleaching agent of candle wax. Contact Dermatitis 1991;24:367-368

7   Dejobert Y, Martin P, Piette F, Thomas P, Bergoend H. Contact dermatitis caused by benzoyl peroxide in podiatrists. Contact Dermatitis 1999;40:163

8   Quirce S, Olaguibel JM, Garcia BE, Tabar AI. Occupational airborne contact dermatitis due to benzoyl peroxide. Contact Dermatitis 1993;29:165-166

9   Tsovilis E, Crepy MN, Jonathan AM, Ameille J. Occupational contact dermatitis due to a marbler's exposure to benzoyl peroxide. Contact Dermatitis 2005;52:117-118

10   Forschner K, Zuberbier T, Worm M. Benzoyl peroxide as a cause of airborne contact dermatitis in an orthopaedic technician. Contact Dermatitis 2002;47:241

11   Rustemeyer T, Frosch PJ. Occupational skin disease in dental laboratory technicians. (I). Clinical picture and causative factors. Contact Dermatitis 1996;34:125-134

12 Mora Morillas I, Aguilar Martinez A, Sanchez Lozano JL, Garcia Perez A. Is benzoyl peroxide an irritant or sensitizer? Contact Dermatitis 1987;16:232-233

13 Romaguera C, Grimalt F. Sensitization to benzoyl peroxide, retinoic acid and carbon tetrachloride. Contact Dermatitis 1980;6:442

14 Cunliffe WJ, Burke B. Benzoyl peroxide: lack of sensitization. Acta Derm Venereol 1982;62:458-459

15 Balato N, Lembo G, Nappa P, Ayala F. Benzoyl peroxide reactions in acne patients. Contact Dermatitis 1984;10:255

16 O'Gorman SM, Torgerson RR. Contact allergy in cheilitis. Int J Dermatol 2016;55:e386-e391

17 Haustein UF, Tegetmeyer L, Ziegler V. Allergic and irritant potential of benzoyl peroxide. Contact Dermatitis 1985;13:252-257

18 Lindemayr H, Dobril M. Contact sensitization to benzoyl peroxide. Contact Dermatitis 1981;7:137-140

19 Bahmer FA, Schulze-Dirks A, Zaun H. Sensibilisierende Wirkung einer für die Behandlung des Ulcus cruris verwendeten 20%igen Benzoylperoxid-Zubereitung. Derm Beruf Umwelt 1984;32:21-24 (Article in German)

20 Vena GA, Angelini G, Meneghini CL. Contact dermatitis to benzoyl peroxide. Contact Dermatitis 1982;8:338

21 Agathos M, Bandmann H-J. Benzoyl peroxide contact allergy in leg ulcer patients. Contact Dermatitis 1984;11:316-317

22 Veraldi S, Brena M, Barbareschi M. Allergic contact dermatitis caused by topical antiacne drugs. Expert Rev Clin Pharmacol. 2015;8:377-381

23 Felton SJ, Orton D, Williams JD. Benzoyl peroxide in topical acne preparations: an underreported contact allergen? Dermatitis 2013;24:146-147

24 Bircher A, Friederich NF, Seelig W, Scherer K. Allergic complications from orthopaedic joint implants: the role of delayed hypersensitivity to benzoyl peroxide in bone cement. Contact Dermatitis 2012;66:20-26

25 Dejobert Y, Piette F, Thomas P. Contact dermatitis from benzoyl peroxide in dental prostheses. Contact Dermatitis 2002;46:177-178

26 Fisher AA. Hand dermatitis- A 'baker' dozen'. Cutis 1982;29:214-221; data cited in ref. 27

27 Scheman A, Cha C, Jacob SE, Nedorost S. Food avoidance diets for systemic, lip, and oral contact allergy: an American Contact Alternatives Group article. Dermatitis 2012;23:248-257

28 Kanerva L, Jolanki R, Alanko K, Estlander T. Patch-test reactions to plastic and glue allergens. Acta Derm Venereol 1999;79:296-300

29 The data in the section 'General' may have been obtained from literature discussed in this chapter, but mostly also or exclusively from one or more of the following online sources: ChemIDPlus Advanced, PubChem, DrugBank, RxList, Drug Central, Drugs.com, and Wikipedia

30 Marasovic D, Vuksic I. Allergic contact dermatitis in patients with leg ulcers. Contact Dermatitis1999;41:107-109

31 Zaki I, Shall L, Dalziel KL. Bacitracin: a significant sensitizer in leg ulcer patients? Contact Dermatitis1994;31:92-94

32 Rai R, Shenoy MM, Viswanath V, Sarma N, Majid I, Dogra S. Contact sensitivity in patients with venous leg ulcer: A multi-centric Indian study. Int Wound J 2018;15:618-622

33 Smart V, Alavi A, Coutts P, Fierheller M, Coelho S, Holness LD, et al. Contact allergens in persons with leg ulcers: a Canadian study in contact sensitization. Int J Low Extrem Wounds 2008;7:120-125

34 Ockenfels H, Seemann U, Goos M. Contact allergy in patients with periorbital eczema: an analysis of allergens. Dermatology 1997;195:119-124

35 Gilissen L, Goossens A. Frequency and trends of contact allergy to and iatrogenic contact dermatitis caused by topical drugs over a 25-year period. Contact Dermatitis 2016;75:290-302

36 Shih Y-H, Sun C-C, Tseng Y-H, Chu C-Y. Contact dermatitis to topical medicaments: a retrospective study from a medical center in Taiwan. Dermatol Sinica 2015;33:181-186

37 Balato N, Lembo G, Cuccurullo FM, Patruno C, Nappa P, Ayala F. Acne and allergic contact dermatitis. Contact Dermatitis 1996;34:68-69

38 Angelini G, Vena GA, Meneghini CL. Allergic contact dermatitis to some medicaments. Contact Dermatitis 1985;12:263-269

39 Van Joost T, van Ulsen J, Vuzevski VD, Naafs B, Tank B. Purpuric contact dermatitis to benzoyl peroxide. J Am Acad Dermatol 1990;22:359-361

40 Adelman M, Mohammad T, Kerr H. Allergic contact dermatitis due to benzoyl peroxide from an unlikely source. Dermatitis 2019;30:230-231

41 World Health Organization. Evaluation of certain food additives and contaminants: Sixty-third report of the Joint FAO/WHO Expert committee on food additives. Geneva: World Health Organization, 2005

42 Corazza M, Amendolagine G, Musmeci D, Forconi R, Borghi A. Sometimes even Dr Google is wrong: An unusual contact dermatitis caused by benzoyl peroxide. Contact Dermatitis 2018;79:380-381

43   Thomas B, Kulichova D, Wolf R, Summer B, Mahler V, Thomas P. High frequency of contact allergy to implant and bone cement components, in particular gentamicin, in cemented arthroplasty with complications: usefulness of late patch test reading. Contact Dermatitis 2015;73:343-349

44   Elangasinghe V, Johnston G. Occupational hand dermatitis caused by benzoyl peroxide in transformer coil binding tape. Contact Dermatitis 2012;66:293

45   Greiner D, Weber J, Kaufmann R, Boehncke WH. Benzoyl peroxide as a contact allergen in adhesive tape. Contact Dermatitis 1999;41:233

46   Azurdia RM, King CM. Allergic contact dermatitis due to phenol-formaldehyde resin and benzoyl peroxide in swimming goggles. Contact Dermatitis 1998;38:234-235

47   Vincenzi C, Cameli N, Vassilopoulou A, Tosti A. Allergic contact dermatitis due to benzoyl peroxide in an arm prosthesis. Contact Dermatitis 1991;24:66-67

48   De Souza A, Cohen N, Franklin B, Cohen DE, Perelman RO. Allergic contact dermatitis to benzoyl peroxide: A retrospective review of cases. Dermatitis 2011;22:184-185

49   Treudler R, Simon JC. Benzoyl peroxide: is it a relevant bone cement allergen in patients with orthopaedic implants? Contact Dermatitis 2007;57:177-180

50   Minciullo PL, Patafi M, Giannetto L, Ferlazzo B, Trombetta D, Saija A, et al. Allergic contact angioedema to benzoyl peroxide. J Clin Pharm Ther 2006;31:385-387

51   Bulinckx A, Dachelet C, Leroy A, Goossens A, Tennstedt D, Baeck M. Contact dermatitis to the combination gel of adapalene 0.1% and benzoyl peroxide (PBO) 2.5%. Eur J Dermatol 2012;22:139-140

52   Shwereb C, Lowenstein EJ. Delayed type hypersensitivity to benzoyl peroxide. J Drugs Dermatol 2004;3:197-199

53   Hernández-Núñez A, Sánchez-Pérez J, Pascual-López M, Aragüés M, García-Díez A. Allergic contact dermatitis from benzoyl peroxide transferred by a loving son. Contact Dermatitis 2002;46:302

54   Caro I. Connubial contact dermatitis to benzoyl peroxide. Contact Dermatitis 1976;2:362

55   Mann RJ, Peachey RDG. Allergen transfer between individuals as a cause of contact dermatitis. Contact Dermatitis 1981;7:164-165

56   Fischer AA. Consort contact dermatitis. Cutis 1979;24:595-596,668

57   Morelli R, Lanzarini M, Vincenzi C, Reggiani M. Contact dermatitis due to benzoyl peroxide. Contact Dermatitis 1989;20:238-239

58   Rietschel RL, Duncan SH. Benzoyl peroxide reactions in an acne study group. Contact Dermatitis 1982;8:323-326

59   Sandre M, Skotnicki-Grant S. A case of a paediatric patient with allergic contact dermatitis to benzoyl peroxide. J Cutan Med Surg 2018;22:226-228

60   Kim C, Craiglow BG, Watsky KL, Antaya RJ. Allergic contact dermatitis to benzoyl peroxide resembling impetigo. Pediatr Dermatol 2015;32:e161-e162.

Chapter 3.36   BENZYDAMINE

IDENTIFICATION

Description/definition	: Benzydamine is the benzyl-indazole that conforms to the structural formula shown below
Pharmacological classes	: Anti-inflammatory agents
IUPAC name	: 3-(1-Benzylindazol-3-yl)oxy-*N,N*-dimethylpropan-1-amine
CAS registry number	: 642-72-8
EC number	: 211-388-8
Merck Index monograph	: 2395
Patch testing	: Hydrochloride, 2.0% pet.(Chemotechnique, SmartPracticeCanada); 1% pet. (SmartPracticeCanada)
Molecular formula	: $C_{19}H_{23}N_3O$

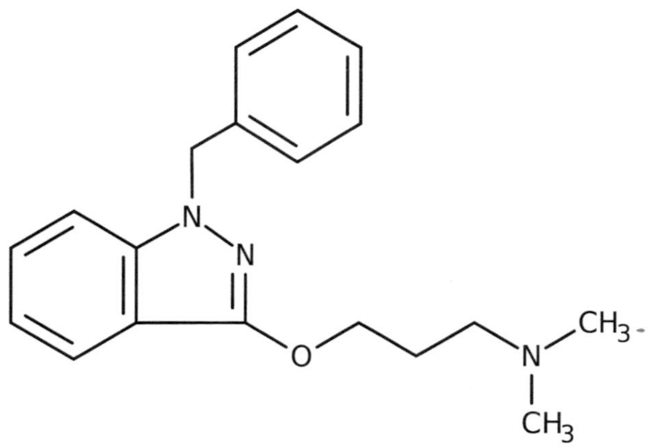

GENERAL

Benzydamine is an indazole non-steroidal anti-inflammatory drug with analgesic, antipyretic, and anti-edema properties. Available as a liquid mouthwash, spray for mouth and throat, topical cream, and vaginal irrigation (formerly also available in tablets, suppositories and intramuscular injections), benzydamine is most frequently employed for the relief of painful inflammatory conditions of the mouth and the musculoskeletal system, respectively. It is also said to promote healing. In pharmaceutical products, benzydamine is employed as benzydamine hydrochloride (CAS number 132-69-4, EC number 205-076-0, molecular formula $C_{19}H_{24}ClN_3O$) (1).

CONTACT ALLERGY

GENERAL

Few cases of allergic contact dermatitis to benzydamine have been reported. However, there are many case reports and case series of *photo*allergic reactions to this NSAID, mostly from topical administration, especially from southern European countries such as Spain, Portugal and Italy, where these products are frequently used and sunlight is abundant.

Case reports and case series

In the period 1996-2001, in 2 hospitals in Spain, one patient was diagnosed with contact allergy and one with photocontact allergy to benzydamine. The accumulated incidence per million inhabitants (catchment population of the hospitals) of both side effects together was 2.4 (26).

A 58-year-old man developed eczema on the lower left leg after the application of a cream containing 5% benzydamine hydrochloride, without preceding sun exposure. Patch testing with the cream, the active ingredient and related chemicals showed positive reactions to the cream, benzydamine HCl 5% (+++), 1% (+++) and 0.1% (+) both in water and in petrolatum, to benzydamine salicylate 5% pet. and to indomethacin 1% pet. and an excipient, an 'essential oil' (which it probably was not). The reaction to indomethacin was considered to be a cross-reaction (4).

A 30-year-old nurse presented with severe exudative dermatitis of the right hand, arms, face and eyelids. She had been using a powder containing benzydamine HCl for over a month, in 0.1% solutions, for performing vaginal irrigations on her patients. Patch tests were positive to the powder and benzydamine HCl at 0.5% and 0.1% water and pet., with a flare-up of the dermatitis on the face and eyelids. Fifteen controls were negative (5). Although not stated as such by the authors, this occupational allergic contact dermatitis was *partly airborne*.

A 74-year old man with dermatitis of the face and back of the hands, using benzydamine 5% cream for musculo-skeletal pain in his finger, knees and elbows, had photoaggravated contact allergy to benzydamine, with a slightly stronger reaction to one photopatch test with benzydamine 5% water (6). A 79-year old man developed an acute dermatitis starting on the back of the hands, with secondary spread to the face with periorbital edema and to the arms, shoulder, flanks and legs, after having applied 5% benzydamine cream for arthritis on the finger joints for years. Patch tests were positive to the cream and benzydamine HCl 5% (++), 1% (++) and 0.1% (+) in pet., resulting in a generalized flare-up of dermatitis (7). A 64-year-old man had been treated with 5% benzydamine HCl gel for edema of the lower leg when he developed eczematous dermatitis on the legs, arms, trunk and face. He had positive patch tests to the gel and benzydamine HCl 5% pet. (8). A 64-year-old woman had developed dermatitis after applying benzydamine 3% cream. Patch tests were positive to the cream and benzydamine HCl 5% water, but only after 5 days (9).

In Italy, before 1993, the members of the GIRDCA Multicentre Study Group diagnosed 102 patients (49 men, 53 women), aged 16 to 66 years (mean 37 years), with (photo)dermatitis induced by systemic or topical NSAIDs. Benzydamine hydrochloride caused one contact allergic and zero photocontact allergic reactions (2).

PHOTOSENSITIVITY
Benzydamine has been shown to have phototoxic properties *in vitro* and to cause both phototoxicity and photoallergy (11).

Photopatch testing in groups of selected patients

In Portugal, in the period 2006-2008, 74 patients with photodermatoses or facial dermatitis with aggravation following sun exposure were patch and photopatch tested with benzydamine 1% and/or 5% pet. and, in suspicious cases, with drugs containing it 'as is'. Ten patients (six women, 4 men), aged 21-84 years (mean 65) had a positive photopatch test to benzydamine and to drugs that contain it (benzydamine oral solution and gel) and were diagnosed with photoallergic contact dermatitis. In nine patients there was a subacute, pruriginous cheilitis, with erythema and scaling on the lower lip in all and superficial erosions in four. Eczema extended to the chin in seven patients, to the malar regions in three and was associated with erythema or eczema on the dorsum of the hands and forearms in two individuals. They all had a long-term history of regular gargling with an oral antiseptic solution containing 1.5% benzydamine. The tenth patient with the most extensive lesions had regularly applied benzydamine 3% gel to his wife's back (10).

In a 2004-2005 study from Spain, 224 patients were photopatch tested and 88 had one or more positive reactions. Seven of these reacted to benzydamine, this drug being the second most frequent photoallergen after ketoprofen (20).

Case series

Two cases of photoallergic contact dermatitis to benzydamine hydrochloride in a mouthwash were reported from Italy in 2020 (27). A 56-year-old man presented with eczematous lesions of his lips, which had appeared one month after using a mouthwash containing 0.15% benzydamine HCl for daily oral hygiene. A 69-year-old woman presented with erythematous lip dermatitis. The patient was wearing an oral prosthesis and reported long-term regular use of the same mouthwash. Patch tests were negative; photopatch tests were positive to benzydamine HCl 2% pet., but negative to the mouthwash 'as is' (27).

In Coimbra, Portugal, in the period 2003-2007, photoallergic contact dermatitis from benzydamine occurred in eight patients. One had facial, forearm, and hand dermatitis from regular application of a topical gel containing benzydamine on his wife. The other seven patients had presented with facial dermatitis involving mainly the chin and lower lip from regular gargling with an antiseptic and anti-inflammatory benzydamine-containing oral solution (25).

In the period 1996-2001, in 2 hospitals in Spain, one patient was diagnosed with contact photoallergy and one with contact allergy to benzydamine. The accumulated incidence per million inhabitants (catchment population of the hospitals) of both side effects together was 2.4 (26).

Three women with photoallergic dermatitis precipitated by benzydamine were reported from Germany. In one patient the cause of the photosensitization was a cream containing benzydamine, whereas in the other two the reaction occurred after oral administration. In all three cases photopatch tests were positive to benzydamine (21).

Case reports

A 73-year-old woman presented with an eczematous eruption involving both cheeks, the anterior and lateral aspects of the neck, upper chest, and back, that had started 3 days after applying a cream containing 3% benzydamine hydrochloride twice daily for treating neck pain. Patch tests and photopatch tests showed a doubtful positive patch test to the cream and strong (++) photopatch tests to the same cream and its ingredients benzydamine HCl 10% pet. (negative at 1% and 0.1%), polysorbate 60 (Tween® 60) 20% pet. and sorbitan monostearate (Span® 60) 50% and 10% pet. (negative at 1% pet.). Twenty controls were negative (3).

An 81-year-old woman presented with pruritic, erythematous, scaly and enlarging plaques over the dorsum of her hands and forearms, on her chin and the centre of the forehead. Histopathology showed a dense lymphoid infiltrate in the superficial papillary dermis with focal interstitial and perivascular extension. The infiltrate was largely composed of lymphocytes. Immunophenotyping revealed a T-lymphocyte CD4/CD8 ratio of 5:6. The direct immunofluorescence study was negative. A T-cell receptor gamma gene rearrangement showed a polyclonal pattern. Patch and photopatch testing revealed a positive photopatch to benzydamine 1% pet. The patient now told that for several years she had used a gynecological solution containing 0.1% benzydamine HCl for vaginal hygiene. After ceasing its use, complete remission was achieved without recurrence in the following year. The patient was diagnosed with lymphomatoid photocontact dermatitis from benzydamine (12).

Previously, 3 case reports had already described the occurrence of photoallergic contact dermatitis of the hands from the use of the same gynecological solution containing 0.1% benzydamine HCl (13,14,15) and a group of Spanish investigators had documented 6 similar cases (16,17).

A 67-year-old woman gargled with a solution containing 0.15% benzydamine HCl for pharyngitis. An erythematous rash on sun-exposed skin developed within the third week of treatment. On examination, there were mainly well-demarcated areas of eczema on the face, neck, neckline, forearms and lower legs. Photopatch tests were positive to the solution 'as is' and 10% water. Additional testing of benzydamine itself was not performed, because 'there are no reported cases of photoallergic reactions produced by any other components of the lotion' (18).

A 23-year-old woman had developed dermatitis of the light-exposed areas (face, dorsa of hands, forearms and legs), 24 hours after taking a preparation made up of 250 mg of ampicillin and 50 mg of benzydamine for pharyngitis. An oral provocation with benzydamine was positive, with ampicillin negative. Years later, dermatitis involving the same areas appeared 48 hours after the administration of benzydamine inhalation spray. A photopatch test with a solution containing 1.5 mg/ml benzydamine (probably the same lotion as in the previous case) was positive (19). The latter episode was a case of systemic photocontact dermatitis. A second patient presented by the authors had photoallergic contact dermatitis from an oral benzydamine preparation with blistering dermatitis on the face (19).

Additional cases of benzydamine photosensitization from topical products can be found in refs. 22 and 23 and from both oral and topical administration in ref. 24. A case of unilateral photoallergic contact dermatitis of the dorsum of one hand was described in 2014 (28).

Cross-reactions, pseudo-cross-reactions and co-reactions
A patient sensitized to benzydamine HCl co-reacted to indomethacin, which was considered to be a cross-reaction (4). Cross-reactivity to benzydamine salicylate (4).

LITERATURE
1   The data in the section 'General' may have been obtained from literature discussed in this chapter, but mostly also or exclusively from one or more of the following online sources: ChemIDPlus Advanced, PubChem, DrugBank, RxList, Drug Central, Drugs.com, and Wikipedia
2   Pigatto PD, Mozzanica N, Bigardi AS, Legori A, Valsecchi R, Cusano F, et al. Topical NSAID allergic contact dermatitis. Italian experience. Contact Dermatitis 1993;29:39-41
3   Giménez-Arnau A, Gilaberte M, Conde D, Espona M, Pujol RM. Combined photocontact dermatitis to benzydamine hydrochloride and the emulsifiers, Span 60 and Tween 60 contained in Tantum cream. Contact Dermatitis 2007;57:61-62
4   Goday Buján JJ, Ilardia Lorentzen R, Soloeta Arechavala R. Allergic contact dermatitis from benzydamine with probable cross-reaction to indomethacin. Contact Dermatitis 1993;28:111-112
5   Foti C, Vena GA, Angelini G. Occupational contact allergy to benzydamine hydrochloride. Contact Dermatitis 1992;27:328-329
6   Vincenzi C, Cameli N, Tardio M, Piraccini BM. Contact and photocontact dermatitis due to benzydamine hydrochloride. Contact Dermatitis 1990;23:125-126
7   Christophersen J. Allergic contact dermatitis to benzydamine. Contact Dermatitis 1987;16:106-107
8   Balato N, Lembo G, Patruno C, Bordone F, Ayala F. Contact dermatitis from benzydamine hydrochloride. Contact Dermatitis 1986;15:105
9   Bruynzeel DP. Contact allergy to benzydamine. Contact Dermatitis 1986;14:313-314
10 Canelas MM, Cardoso JC, Gonçalo M, Figueiredo A. Photoallergic contact dermatitis from benzydamine presenting mainly as lip dermatitis. Contact Dermatitis 2010;63:85-88

11 Moore DE, Wang J. Electron-transfer mechanisms in photosensitization by the anti-inflammatory drug
    benzydamine. J Photochem Photobiol B 1998;43:175-180

12 Alvarez-Garrido H, Sanz-Muñoz C, Martínez-García G, Miranda-Romero A. Lymphomatoid photocontact
    dermatitis to benzydamine hydrochloride. Contact Dermatitis 2010;62:117-119

13 Lasa Elgezua O, Gorrotxategi PE, Gardeazabal García J, Ratón Nieto JA, Pérez JL. Photoallergic hand eczema due
    to benzydamine. Eur J Dermatol 2004;14:69-70

14 Conde-Salazar L, Guimaraens D, Gonzalez M, et al. Fotodermatitis alérgica de contacto por bencidamina. Actas
    Dermo-Sifiliográficas 1996;87:310-314 (Article in Spanish, data cited in ref. 13)

15 Navarro LA, Jorro J, Morales C, et al. Fotodermatitis alérgica de contacto por clorhidrato de bencidamima.
    Revista Española de Alergología e Inmumología Clínica 1996;11:5:239-242 (Article in Spanish, data cited in ref.
    13)

16 De la Cuadra J, Rodríguez M, Pérez A. Fotoalergia a un jabón vaginal con bencidamina. Boletín del GEF: VII
    Reunión del Grupo Español de Fotobiología, Madrid 1993 (Article in Spanish, data cited in ref. 13)

17 De la Cuadra J, Pérez A, Escutia B, et al. Fotoalergia por jabón vaginal con bencidamina (Rosalgin®): una década
    más tarde. Boletín del GEF: XVI Reunión del Grupo Español de Fotobiología, Madrid 2001 (Article in Spanish, data
    cited in ref. 13)

18 Henschel R, Agathos M, Breit R. Photocontact dermatitis after gargling with a solution containing benzydamine.
    Contact Dermatitis 2002;47:53

19 Fernandez de Corres L. Photodermatitis from benzydamine. Contact Dermatitis 1980;6:285

20 Cuadra-Oyanguren J, Pérez-Ferriols A, Lecha-Carrelero M, Giménez-Arnau AM, Fernández-Redondo V, Ortiz de
    Frutos FJ, et al. Results and assessment of photopatch testing in Spain: towards a new standard set of
    photoallergens. Actas Dermosifilogr 2006;98:96-101

21 Frosch PJ, Weickel R. Photocontact allergy caused by benzydamine (Tantum). Hautarzt 1989;40:771-773 (Article
    in German)

22 Motley RJ, Reynolds AJ. Photodermatitis from benzydamine cream. Contact Dermatitis 1988;19:66

23 Goncalo S, Sousa I, Freitas JDG, et al. Photoallergie à la benzydamine. Actualité en Dermo-Allergie 1982;5:61-64
    (Article in French, data cited in ref. 13).

24 Ikemura I. Contact and photocontact dermatitis due to benzydamine hydrochloride. Jpn J Clin Dermatol
    1971;25:129 (Article in Japanese, data cited in ref. 13)

25 Cardoso J, Canelas MM, Gonçalo M, Figueiredo A. Photopatch testing with an extended series of photoallergens:
    a 5-year study. Contact Dermatitis 2009;60:325-329

26 Diaz RL, Gardeazabal J, Manrique P, Ratón JA, Urrutia I, Rodríguez-Sasiain JM, Aguirre C. Greater allergenicity of
    topical ketoprofen in contact dermatitis confirmed by use. Contact Dermatitis 2006;54:239-243

27 Romita P, Barlusconi C, Mercurio CS, Hansel K, Stingeni L, Foti C. Photoallergic contact cheilitis from
    benzydamine hydrochloride contained in a mouthwash. Contact Dermatitis 2020;83:130-132

28 Utrera-Busquets M, Córdoba S, Borbujo-Martínez J. Unilateral eczema on the dorsum of the hand. Actas
    Dermosifiliogr 2014;105:884-885 (Article in English and Spanish)

Chapter 3.37   BETAMETHASONE

IDENTIFICATION

Description/definition   : Betamethasone is the synthetic glucocorticoid that conforms to the structural formula shown below

Pharmacological classes   : Anti-asthmatic agents; glucocorticoids; anti-inflammatory agents

IUPAC name   : (8S,9R,10S,11S,13S,14S,16S,17R)-9-Fluoro-11,17-dihydroxy-17-(2-hydroxyacetyl)-10,13,16-trimethyl-6,7,8,11,12,14,15,16-octahydrocyclopenta[a]phenanthren-3-one

Other names   : 9-Fluoro-11β,17,21-trihydroxy-16β-methylpregna-1,4-diene-3,20-dione

CAS registry number   : 378-44-9

EC number   : 206-825-4

Merck Index monograph   : 2452

Patch testing   : In general, corticosteroids may be tested at 0.1% and 1% in alcohol; late readings (6-10 days) are strongly recommended

Molecular formula   : $C_{22}H_{29}FO_5$

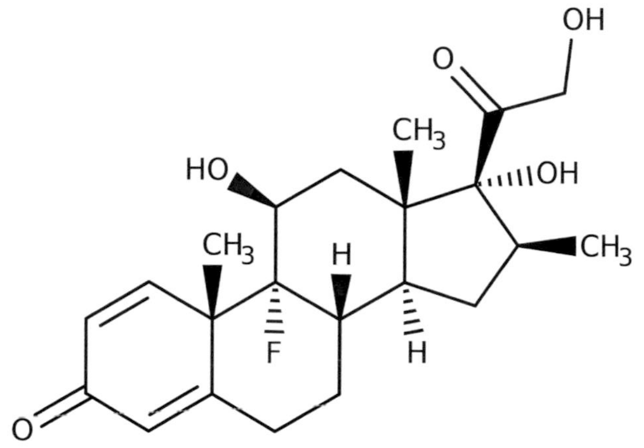

GENERAL

General aspects of corticosteroids used on the skin and mucous membranes are discussed in chapter 3.4. A practical guideline for diagnosing allergic reactions to corticosteroids is presented in ref. 2. For injectable suspensions, betamethasone sodium phosphate (Chapter 3.39), acetate or valerate may be used. Topical preparations mostly either contain betamethasone valerate (Chapter 3.40) or betamethasone dipropionate (Chapter 3.38). Betamethasone *base* is used as tablet only, which implies that by far most allergic reactions to 'betamethasone' have in fact been the result of sensitization to an ester of betamethasone or of cross-reactivity to another corticosteroid. It is also likely that there has been confusion in some publications on the correct forms of the drugs used, e.g. that betamethasone was mentioned where an ester form should have been mentioned.

CONTACT ALLERGY

Patch testing in groups of patients

In the United Kingdom, in the period 1997-2001, 200 patients with venous or mixed venous/arterial leg ulcers were patch tested with various series including a medicament series and 2 (1%) reacted to 'betamethasone'. All reactions to topical drugs were considered to be of probable, past or current relevance (1). As betamethasone base is used in oral preparations only, these patients must most likely have previously become sensitized from either a betamethasone ester or have cross-reacted to other corticosteroids.

Case reports and case series

From January 1990 to June 2008, in Leuven, Belgium, 315 patients were diagnosed with contact allergy to/allergic contact dermatitis from corticosteroids (CSs) from routine patch testing with a baseline series including tixocortol pivalate, budesonide, hydrocortisone butyrate and prednisone caproate, patch testing with patients' own CS preparations, and testing those with proven contact allergy to a corticosteroid or strongly suspected of CS allergy later with a series of 66 CSs, including two sex hormones (progesterone and testosterone). 71% of the patients had

relevant reactions, but these were not specified. In this group of 315 CS allergic patients, 36 had positive patch tests to betamethasone 0.1% alc., which most likely represent cross-reactions (4).

Cross-reactions, pseudo-cross-reactions and co-reactions
Cross-reactions between corticosteroids are discussed in Chapter 2.8.

Cutaneous adverse drug reactions from systemic administration caused by type IV (delayed-type) hypersensitivity
Cutaneous adverse drug reactions from systemic administration of betamethasone caused by type IV (delayed-type) hypersensitivity, including maculopapular exanthema (3), delayed urticaria or exanthema (7) and systemic contact dermatitis (5,6 [baboon syndrome]), are planned to be discussed in Volume IV of the *Monographs in Contact Allergy* series on Systemic drugs.

LITERATURE

1   Tavadia S, Bianchi J, Dawe RS, McEvoy M, Wiggins E, Hamill E, et al. Allergic contact dermatitis in venous leg ulcer patients. Contact Dermatitis 2003;48:261-265
2   Baeck M, Goossens A. Immediate and delayed allergic hypersensitivity to corticosteroids: practical guidelines. Contact Dermatitis 2012;66:38-45
3   Nucera E, Buonomo A, Pollastrini E, De Pasquale T, Del Ninno M, Roncallo C, et al. A case of cutaneous delayed-type allergy to oral dexamethasone and to betamethasone. Dermatology 2002;204:248-250
4   Baeck M, Chemelle JA, Terreux R, Drieghe J, Goossens A. Delayed hypersensitivity to corticosteroids in a series of 315 patients: clinical data and patch test results. Contact Dermatitis 2009;61:163-175
5   Isaksson M. Systemic contact allergy to corticosteroids revisited. Contact Dermatitis 2007;57:386-388
6   Armingaud P, Martin L, Wierzbicka E, Esteve E. Baboon syndrome due to a polysensitization with corticosteroids. Ann Dermatol Venereol 2005;132:675-677 (Article in French)
7   Padial A, Posadas S, Alvarez J, Torres M-J, Alvarez JA, Mayorga C, Blanca M. Nonimmediate reactions to systemic corticosteroids suggest an immunological mechanism. Allergy 2005;60:665-670

Chapter 3.38   BETAMETHASONE DIPROPIONATE

IDENTIFICATION

Description/definition   : Betamethasone dipropionate is the 17,21-dipropionate ester of the synthetic
                               glucocorticoid betamethasone that conforms to the structural formula shown below
Pharmacological classes   : Anti-inflammatory agents
IUPAC name   : [2-[(8S,9R,10S,11S,13S,14S,16S,17R)-9-Fluoro-11-hydroxy-10,13,16-trimethyl-3-oxo-17-
                               propanoyloxy-6,7,8,11,12,14,15,16-octahydrocyclopenta[a]phenanthren-17-yl]-2-
                               oxoethyl] propanoate
Other names   : 9-Fluoro-11β,17,21-trihydroxy-16β-methylpregna-1,4-diene-3,20-dione 17,21-
                               di(propionate; betamethasone-17,21-dipropionate
CAS registry number   : 5593-20-4
EC number   : 227-005-2
Merck Index monograph   : 2452 (Betamethasone)
Patch testing   : 1.0% pet. (Chemotechnique); 0.5% pet., 0.1% alc. (SmartPracticeCanada)
Molecular formula   : $C_{28}H_{37}FO_7$

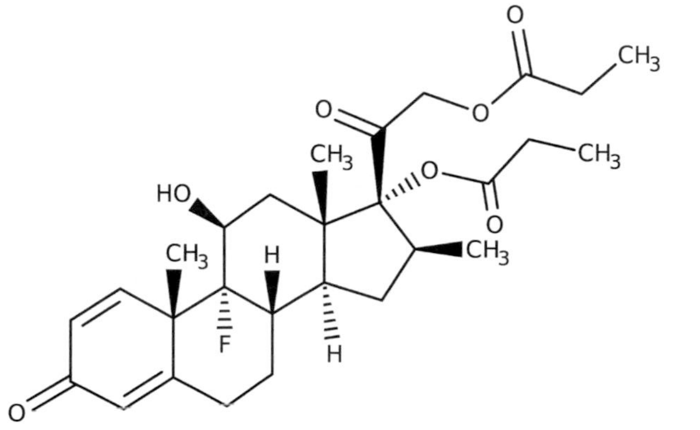

GENERAL

General aspects of corticosteroids used on the skin and mucous membranes are discussed in Chapter 2.4. A practical guideline for diagnosing allergic reactions to corticosteroids is presented in ref. 2. See also betamethasone (Chapter 3.37), betamethasone sodium phosphate (Chapter 3.39) and betamethasone valerate (Chapter 3.40).

CONTACT ALLERGY

Patch testing in groups of patients

The results of patch testing with betamethasone dipropionate in consecutive patients suspected of contact dermatitis (routine testing) and of testing in groups of *selected* patients are shown in table 3.38.1. In a study from Spain, 12 of 3699 consecutive patients (0.3%) reacted to betamethasone dipropionate 0.1% alc.; the relevance of the positive patch tests was not mentioned (7). Prevalences in groups of selected patients ranged from 0.4% to 3%, the latter in a highly selected patient group (12). The relevance rates of the positive patch tests were either not mentioned (11,12) or very high (6,9).

Table 3.38.1 Patch testing in groups of patients

Years and Country	Test conc. & vehicle	Number of patients tested	positive (%)	Selection of patients (S); Relevance (R); Comments (C)	Ref.
Routine testing					
2015-2016 Spain	0.1% alc.	3699	12 (0.3%)	R: not stated	7
Testing in groups of selected patients					
2000-2005 USA	0.5 alc. 70%	1185	16 (1.4%)	S: patients suspected of corticosteroid allergy; R: 94%	6
1995-9 Netherlands	1% alc. 97%	220	2 (0.9%)	S: not stated; R: definite 100%	9
1988-1991 U.K.	20% pet.	528	2 (0.4%)	S: patients with a positive patch test to tixocortol pivalate or suspected of corticosteroid allergy; R: not stated	11
1985-1990 Finland	0.5% and 0.05% alc.	66	2 (3%)	S: patients very likely to be corticosteroid-allergic; R: not stated	12

Case series
From January 1990 to June 2008, in Leuven, Belgium, 315 patients were diagnosed with contact allergy to/allergic contact dermatitis from corticosteroids (CSs) from routine patch testing with a baseline series including tixocortol pivalate, budesonide, hydrocortisone butyrate and prednisone caproate, patch testing with patients' own CS preparations, and testing those with proven contact allergy to a corticosteroid or strongly suspected of CS allergy later with a series of 66 CSs, including two sex hormones (progesterone and testosterone). 71% of the patients had relevant reactions, but these were not specified. In this group of 315 CS allergic patients, 19 had positive patch tests to betamethasone dipropionate 0.1% alc. (5). It is unknown how many of these reactions were caused by the use of a pharmaceutical product containing betamethasone dipropionate and how many were cross-reactions to other corticosteroids. In Sydney, Australia, from 1988 to 1994, 19 cases of corticosteroid allergy were diagnosed, including 2 to betamethasone dipropionate (1% alc.) (10).

Case reports
A 60-year-old woman presented with a 3-year history of recalcitrant vulval dermatitis treated with various topical corticosteroids and antifungals with very little improvement. She had used different combinations of hydrocortisone 1% cream, betamethasone dipropionate 0.05% cream, mometasone furoate 0.1% cream, and betamethasone valerate 0.1% cream/ointment. One month prior to presentation, her treatment was changed to strictly ichthammol in zinc cream and hydrocortisone ointment only and this dramatically cleared her vulval dermatitis. Patch tests were negative to hydrocortisone and the antifungals, but positive to budesonide and the commercial betamethasone dipropionate and mometasone furoate preparations. ROATs were positive after 8 days (4).

A patient with flexural eczema was treated in a dermatological department with 3 (!) topical corticosteroid preparations, which resulted in a maculopapular rash, first in the flexural areas, then with spread to the legs, arms, abdomen and eyelids after 2 days. There were no pustules and the patient had no fever. In a lesional skin biopsy, under the stratum corneum and in the stratum spinosum some pools of neutrophils were found, which formed variable sized pustules. Patch tests were positive to the 3 steroid preparations and their ingredients betamethasone valerate and betamethasone dipropionate. The authors diagnosed 'generalized exanthematous reaction with pustulosis induced by topical corticosteroids' (8). That no pustules were seen clinically seems somewhat contradictory to the 'with pustulosis' part.

Cutaneous adverse drug reactions from systemic administration caused by type IV (delayed-type) hypersensitivity
A 56-year-old man had developed generalized skin eruptions after a betamethasone dipropionate injection and after administration of a budesonide-containing inhalation product. Patch tests were positive to betamethasone dipropionate 1% alcohol and to budesonide 0.1% pet. (3).

Cross-reactions, pseudo-cross-reactions and co-reactions
Cross-reactions between corticosteroids are discussed in Chapter 2.8.

LITERATURE

1   Tomljanović-Veselski M, Lipozencić J, Lugović L. Contact allergy to special and standard allergens in patients with venous ulcers. Coll Antropol 2007;31:751-756
2   Baeck M, Goossens A. Immediate and delayed allergic hypersensitivity to corticosteroids: practical guidelines. Contact Dermatitis 2012;66:38-45
3   Goossens A, Huygens S, Matura M, Degreef H. Fluticasone propionate: a rare contact sensitizer. Eur J Dermatol 2001;11:29-34
4   Chow ET. Multiple corticosteroid allergies. Australas J Dermatol 2001;42:62-63
5   Baeck M, Chemelle JA, Terreux R, Drieghe J, Goossens A. Delayed hypersensitivity to corticosteroids in a series of 315 patients: clinical data and patch test results. Contact Dermatitis 2009;61:163-175
6   Davis MD, El-Azhary RA, Farmer SA. Results of patch testing to a corticosteroid series: a retrospective review of 1188 patients during 6 years at Mayo Clinic. J Am Acad Dermatol 2007;56:921-927
7   Mercader-García P, Pastor-Nieto MA, García-Doval I, Giménez-Arnau A, González-Pérez R, Fernández-Redondo V, et al. GEIDAC. Are the Spanish baseline series markers sufficient to detect contact allergy to corticosteroids in Spain? A GEIDAC prospective study. Contact Dermatitis 2018;78:76-82
8   Chavarría Mur E, González-Carrascosa Ballesteros M, Suárez Fernández R, Bueno Marco C. Generalized exanthematous reaction with pustulosis induced by topical corticosteroids. Contact Dermatitis 2005;52:114-115
9   Devos SA, Van der Valk PG. Relevance and reproducibility of patch-test reactions to corticosteroids. Contact Dermatitis 2001;44:362-365
10   Freeman S. Corticosteroid allergy. Contact Dermatitis 1995;33:240-242
11   Burden AD, Beck MH. Contact hypersensitivity to topical corticosteroids. Br J Dermatol 1992;127:497-501
12   Lauerma AI. Contact hypersensitivity to glucocorticosteroids. Am J Contact Dermat 1992;3:112-132

Chapter 3.39 BETAMETHASONE SODIUM PHOSPHATE

IDENTIFICATION

Description/definition	: Betamethasone sodium phosphate is the disodium salt of the 21-phosphate ester of the synthetic glucocorticoid betamethasone, that conforms to the structural formula shown below
Pharmacological classes	: Glucocorticoids
IUPAC name	: Disodium;[2-[(8S,9R,10S,11S,13S,14S,16S,17R)-9-fluoro-11,17-dihydroxy-10,13,16-trimethyl-3-oxo-6,7,8,11,12,14,15,16-octahydrocyclopenta[a]phenanthren-17-yl]-2-oxoethyl] phosphate
Other names	: 9-Fluoro-11β,17,21-trihydroxy-16β-methylpregna-1,4-diene-3,20-dione 21-(disodium phosphate)
CAS registry number	: 151-73-5
EC number	: 205-797-0
Merck Index monograph	: 2452 (Betamethasone)
Patch testing	: In general, corticosteroids may be tested at 0.1% and 1% in alcohol; late readings (6-10 days) are strongly recommended
Molecular formula	: $C_{22}H_{28}FNa_2O_8P$

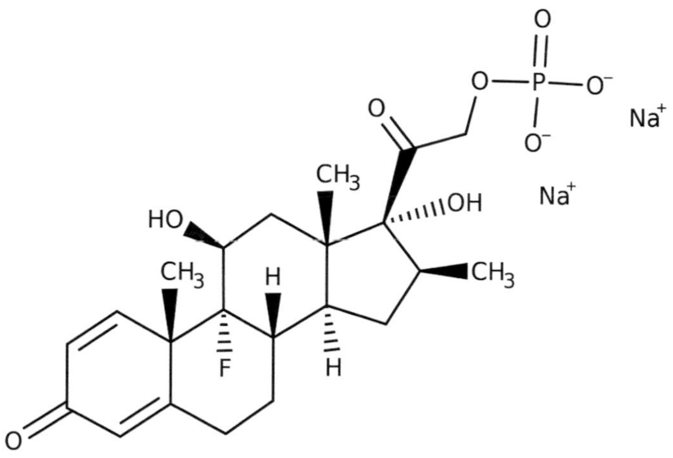

GENERAL

General aspects of corticosteroids used on the skin and mucous membranes are discussed in Chapter 2.4. A practical guideline for diagnosing allergic reactions to corticosteroids is presented in ref. 1. Betamethasone sodium phosphate is mostly used in injection fluids. See also betamethasone (Chapter 3.37), betamethasone valerate (Chapter 3.49) and betamethasone dipropionate (Chapter 3.38).

CONTACT ALLERGY

Case reports and case series

From January 1990 to June 2008, in Leuven, Belgium, 315 patients were diagnosed with contact allergy to/allergic contact dermatitis from corticosteroids (CSs) from routine patch testing with a baseline series including tixocortol pivalate, budesonide, hydrocortisone butyrate and prednisone caproate, patch testing with patients' own CS preparations, and testing those with proven contact allergy to a corticosteroid or strongly suspected of CS allergy later with a series of 66 CSs, including two sex hormones (progesterone and testosterone). 71% of the patients had relevant reactions, but these were not specified. In this group of 315 CS allergic patients, 9 had positive patch tests to betamethasone sodium phosphate 0.1% saline (2). It is unknown how many of these reactions were caused by the use of a pharmaceutical product containing betamethasone sodium phosphate and how many were cross-reactions to other corticosteroids.

In reviewing the Japanese literature up to 1994, 43 patients with allergic contact dermatitis from topical corticosteroid were identified, including one caused by betamethasone sodium phosphate (3).

Cross-reactions, pseudo-cross-reactions and co-reactions
Cross-reactions between corticosteroids are discussed in Chapter 2.8.

Cutaneous adverse drug reactions from systemic administration caused by type IV (delayed-type) hypersensitivity
Cutaneous adverse drug reactions from systemic administration of betamethasone sodium phosphate caused by type IV (delayed-type) hypersensitivity, including systemic contact dermatitis (4), are planned to be discussed in Volume IV of the *Monographs in Contact Allergy* series on Systemic drugs.

LITERATURE

1   Baeck M, Goossens A. Immediate and delayed allergic hypersensitivity to corticosteroids: practical guidelines. Contact Dermatitis 2012;66:38-45
2   Baeck M, Chemelle JA, Terreux R, Drieghe J, Goossens A. Delayed hypersensitivity to corticosteroids in a series of 315 patients: clinical data and patch test results. Contact Dermatitis 2009;61:163-175
3   Oh-i T. Contact dermatitis due to topical steroids with conceivable cross reactions between topical steroid preparations. J Dermatol 1996;23:200-208
4   Gambini D, Sena P, Raponi F, Bianchi L, Hansel K, Tramontana M, et al. Systemic allergic dermatitis presenting as acute generalized exanthematous pustulosis due to betamethasone sodium phosphate. Contact Dermatitis 2020;82:250-252

Chapter 3.40   BETAMETHASONE VALERATE

IDENTIFICATION

Description/definition   : Betamethasone valerate is the 17-valerate ester of the synthetic glucocorticoid betamethasone that conforms to the structural formula shown below
Pharmacological classes : Glucocorticoids; anti-inflammatory agents
IUPAC name   : [(8S,9R,10S,11S,13S,14S,16S,17R)-9-Fluoro-11-hydroxy-17-(2-hydroxyacetyl)-10,13,16-trimethyl-3-oxo-6,7,8,11,12,14,15,16-octahydrocyclopenta[a]phenanthren-17-yl] pentanoate
Other names   : 9-Fluoro-11β,17,21-trihydroxy-16β-methylpregna-1,4-diene-3,20-dione 17-valerate; betamethasone 17-valerate
CAS registry number   : 2152-44-5
EC number   : 218-439-3
Merck Index monograph   : 2452 (Betamethasone)
Patch testing   : 1.0% pet. (Chemotechnique, SmartPracticeCanada); 0.12% pet. (SmartPracticeCanada, SmartPracticeEurope); 0.1% alc. (SmartPracticeCanada)
Molecular formula   : $C_{27}H_{37}FO_6$

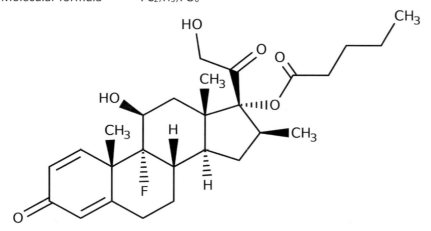

GENERAL

General aspects of corticosteroids used on the skin and mucous membranes are discussed in Chapter 2.4. A practical guideline for diagnosing allergic reactions to corticosteroids is presented in ref. 5. Intradermal testing may detect cases of sensitization to betamethasone valerate not identified by patch testing with the corticosteroid 1% alc. (37).See also betamethasone (Chapter 3.37), betamethasone sodium phosphate (Chapter 3.39) and betamethasone dipropionate (Chapter 3.38).

CONTACT ALLERGY

Contact allergy in the general population

In Germany, for the period 1995-2004, the population-based relative incidence (RI) of contact sensitization to betamethasone valerate (cases/100,000 defined daily doses (DDDs) per year) was estimated to be 0.6. In the group of corticosteroids, the RI ranged from 0.3 (dexamethasone sodium phosphate) to 43.2 (budesonide) (4).

Patch testing in groups of patients

Results of patch testing betamethasone valerate in consecutive patients suspected of contact dermatitis (routine testing) and of testing in groups of *selected* patients (individuals with leg ulcers, patients with anogenital dermatoses or suspected of corticosteroid allergy) are shown in table 3.40.1. In routine testing, 7 of 9 studies had low prevalences of sensitization to betamethasone valerate ranging from 0.1% to 0.7%. Higher rates were seen in Australia (2 [1.2%]) and – in a very small sample – in Poland (24 [1.9%]). Relevance rates, where mentioned, ranged from 0% to 100%, but the numbers of positive reactions were small.

In groups of selected patients, most rates of positive patch tests were <1% (table 3.40.1). Higher percentages (4%) were found in patients with venous leg ulcers in 2 studies from Poland, but the sample sizes were small (n=50) and the relevance of the positive patch tests was not mentioned (6,7). A very high prevalence of 20% was found in a Finnish study, which is probably the result of selecting patients based on a very strong suspicion of contact allergy to corticosteroids (36). Relevance was hardly ever addressed in studies in selected patients.

Table 3.40.1 Patch testing in groups of patients

Years and Country	Test conc. & vehicle	Number of patients tested | positive (%)		Selection of patients (S); Relevance (R); Comments (C)	Ref.
Routine testing					
2015-2016 Spain	0.1% alc.	3699	9 (0.2%)	R: not stated	22
2009-2013 Singapore		2598	(0.1%)	R: present + past relevance: 0%; C: range of positive reactions per year 0 - 0.2%	1
2001-2010 Australia	1% pet.	1322	16 (1.2%)	R: 38%	2
<2005 Poland	0.12% pet.	257	5 (1.9%)	R: 60%	24
<2002 U.K	various	1562	10 (0.7%)	R: not stated; 1% pet. was found to be the best test material (Abstract in ref. 12)	26
1995-1998 Israel		660	4 (0.6%)	R: not stated	28
1994-1996 Singapore	1% pet.	3603	3 (0.1%)	R: all three were relevant	30
1993-1994 Europe	1% pet.	7238	8 (0.1%)	R: 38%	31
1992-1993 U.K	1% alc.	2123	11 (0.5%)	R: not stated	32
Testing in groups of selected patients					
<2008 Poland	1% pet.	50	2 (4%)	S: patients with venous leg ulcers; R: not stated	7
2004-2008 IVDK			(0.4%)	S: patients with anogenital dermatoses tested with a medicament series; R: not stated; C: number of patients tested unknown	8
2000-2005 USA	1% pet.	1185	17 (1.4%)	S: patients suspected of corticosteroid allergy; R: 94%	20
<2005 Poland	1% pet.	50	2 (4%)	S: patients with chronic venous leg ulcers; R: not stated	6
1995-2004 IVDK	0.12% pet.	6143	(0.5%)	S: not stated; R: not stated	3
1995-2004 IVDK	0.12% pet.	5938	28 (0.5%)	S: patients tested with a corticosteroid series; R: not stated	23
1996-1997 IVDK	0.12% pet.	608	4 (0.7%)	S: patients tested with a corticosteroid series; R: not stated	27
1988-1991 U.K.	20% pet.	528	4 (0.8%)	S: patients with a positive patch test to tixocortol pivalate or suspected of corticosteroid allergy; R: not stated	35
1985-1990 Finland	0.1% and 1% alc. 94%	66	13 (20%)	S: patients very likely to be corticosteroid-allergic; R: not stated	36

IVDK: Information Network of Departments of Dermatology (Germany, Austria, Switzerland)

Case series

From January 1990 to June 2008, in Leuven, Belgium, 315 patients were diagnosed with contact allergy to/allergic contact dermatitis from corticosteroids (CSs) from routine patch testing with a baseline series including tixocortol pivalate, budesonide, hydrocortisone butyrate and prednisone caproate, patch testing with patients' own CS preparations, and testing those with proven contact allergy to a corticosteroid or strongly suspected of CS allergy later with a series of 66 CSs, including two sex hormones (progesterone and testosterone). 71% of the patients had relevant reactions, but these were not specified. In this group of 315 CS allergic patients, 19 had positive patch tests to betamethasone valerate 0.1% alc. (19). It is unknown how many of these reactions were caused by the use of a pharmaceutical product containing betamethasone valerate and how many were cross-reactions to other corticosteroids.

Twenty-five cases of allergic contact dermatitis to betamethasone valerate have been reported from Japan up to 2001 (10,11,13).

In Sydney, Australia, from 1988 to 1994, 19 cases of corticosteroid allergy were diagnosed: 4 to betamethasone valerate (1% alc.), 11 to hydrocortisone (1% pet.), 7 to alclometasone dipropionate (1% alc. and pet.), 2 to betamethasone dipropionate (1% alc.) and 2 to triamcinolone acetonide (cream and ointment). 15 patients showed positive patch tests to the corticosteroid in the cream base, but were negative to the same corticosteroid in the ointment base. Of 8 patients performing a ROAT with the commercial cream, 7 were strongly positive (33).

In the period 1982-1994, in a center in Tokyo, 11 patients were diagnosed with allergic contact dermatitis from topical corticosteroids, including two cases caused by betamethasone valerate (43). In reviewing the Japanese literature up to 1994, 43 patients with allergic contact dermatitis from topical corticosteroid were identified, including 9 caused by betamethasone valerate (43).

In Tokyo, Japan, between 1967 and 1988, 69 patients had positive patch tests to topical corticosteroid preparations used by them. 44 were sensitive to vehicle components or other non-steroid active ingredients and 18 were allergic to the corticosteroid itself. In 7 cases, the allergen could not be detected. The most frequently implicated CSs were prednisolone (n=7) and betamethasone valerate (n=5) (39).

Case reports

A 62-year-old man presented with a traumatic ulcer 2 cm in diameter on his right ankle and diffuse erythema with miliary-sized red papules at the periphery on his right leg. He was treated with betamethasone valerate (BV) ointment, but the erythema enlarged further, and blisters and swelling were also noted. Patch testing revealed positive reactions to the ointment and to its active ingredient betamethasone valerate 0.012%, 0.12% and 1.2% pet. (9).

In a trial of the anti-inflammatory activity of some cream preparations containing steroids in volunteers, a 29-year-old woman developed some vesicles and redness on the 3rd day, only on the area of the forearm treated with 0.1% betamethasone valerate cream. Patch tests were positive to the cream and to betamethasone valerate 5% pet. (14). A 35-year-old man developed worsening of itching of the scrotum and erythema after applying betamethasone valerate (BV) ointment. Patch tests were positive to the ointment, BV cream, BV 0.05% and 0.12% propylene glycol (negative to 0.02% and 0.01%), but also (weaker) reactions to the cream and ointment bases (not further investigated) (15).

A 36-year-old woman applied betamethasone valerate cream and pyrrolnitrin solution to herpes zoster on the right side of the trunk. After 5 days an itching, erythematous, vesicular dermatitis appeared. Patch tests were positive to betamethasone valerate 0.1% alcohol and pyrrolnitrin 1% alcohol (16). One individual noticed an exacerbation of dermatitis of the ear after using BV lotion. Patch tests were positive to betamethasone valerate 0.1% pet. (17). A 22-year-old man had dermatitis treated with various ointments. When patch tested, he reacted to a commercial ointment containing betamethasone valerate. Later patch tests were positive to BV cream, BV 1% water, BV powder (probably undiluted) and hydrogenated lanolin, that was presumably present in the ointment (18).

A patient with flexural eczema was treated in a dermatological department with 3 (!) topical corticosteroid preparations, which resulted in a maculopapular rash, first in the flexural areas, then with spread to the legs, arms, abdomen and eyelids after 2 days. There were no pustules and the patient had no fever. In a lesional skin biopsy, under the stratum corneum and in the stratum spinosum some pools of neutrophils were found, which formed variable sized pustules. Patch tests were positive to the 3 steroid preparations and their ingredients betamethasone valerate and betamethasone dipropionate. The authors diagnosed 'generalized exanthematous reaction with pustulosis induced by topical corticosteroids' (25). That no pustules were seen clinically seems somewhat contradictory to the 'with pustulosis' part.

A 52-year-old man with conjunctivitis developed marked periorbital edema and inflammation from prednisolone plus neomycin eyedrops. When betamethasone valerate 0.1% eye drops were ineffective, patch tests were performed yielding positive reactions to neomycin and betamethasone valerate (29). More case reports of allergic contact dermatitis from betamethasone valerate can be found in refs. 34,38,40,41, and 42. A Japanese author in 1994 reviewed the Japanese literature and found 11 reported cases of allergic contact dermatitis to betamethasone valerate in 10 articles published in the period 1986-1992 (21). Additional Japanese literature can be found in ref. 39.

Cross-reactions, pseudo-cross-reactions and co-reactions
Cross-reactions between corticosteroids are discussed in Chapter 2.8.

LITERATURE

1   Ochi H, Cheng SWN, Leow YH, Goon ATJ. Contact allergy trends in Singapore – a retrospective study of patch test data from 2009 to 2013. Contact Dermatitis 2017;76:49-50
2   Toholka R, Wang Y-S, Tate B, Tam M, Cahill J, Palmer A, Nixon R. The first Australian Baseline Series: Recommendations for patch testing in suspected contact dermatitis. Australas J Dermatol 2015;56:107-115
3   Menezes de Padua CA, Uter W, Schnuch A. Contact allergy to topical drugs: prevalence in a clinical setting and estimation of frequency at the population level. Pharmacoepidemiol Drug Saf 2007;16:377-384
4   Menezes de Padua CA, Schnuch A, Nink K, Pfahlberg A, Uter W. Allergic contact dermatitis to topical drugs – Epidemiological risk assessment. Pharmacoepidemiol Drug Saf 2008;17:813-821
5   Baeck M, Goossens A. Immediate and delayed allergic hypersensitivity to corticosteroids: practical guidelines. Contact Dermatitis 2012;66:38-45
6   Zmudzinska M, Czarnecka-Operacz M, Silny W, Kramer L. Contact allergy in patients with chronic venous leg ulcers – possible role of chronic venous insufficiency. Contact Dermatitis 2006;54:100-105
7   Zmudzinska M, Czarnecka-Operacz M, SilnyW. Contact allergy to glucocorticosteroids in patients with chronic venous leg ulcers, atopic dermatitis and contact allergy. Acta Dermatovenerol Croat 2008;16:72-78
8   Bauer A. Contact sensitization in the anal and genital area. Curr Probl Dermatol 2011;40:133-141
9   Tajima M, Murata T, Suzuki Y, Tanikawa A, Amagai M, Tanaka M. Allergic contact dermatitis from betamethasone valerate in Japan. Clin Exp Dermatol 2001;26:220-221
10 Naruse T, Kanai T, Oh IT. A case of contact dermatitis due to betamethasone valerate. Rinsho Derma (Tokyo) 1994;36:737-740 (Article in Japanese, data cited in ref. 9)
11 Egawa Y, Ito Y, Sasaki E, et al. Allergic contact dermatitis due to betamethasone-17-valerate. Hifu 1989;6(Suppl.):212-217 (Article in Japanese, data cited in ref. 9)
12 Sommer S, Wilkinson SM, English J, Gawkrodger DJ, Green C, king CM. Type-IV hypersensitivity to betamethasone valerate and clobetasol propionate: results of a multicentre study. Am J Contact Dermat 2001;12:125

13 Nakajima H, Mohri S, Nagai R. A case of contact dermatitis due to betamethasone valerate. J Dermatol 1986;13:467-470 (Article in Japanese, data cited in ref. 9)

14 Koch EM. Contact allergic reaction to valerate esters of betamethasone and hydrocortisone. Contact Dermatitis 1985;12:58

15 Pasricha JS, Gupta R. Contact sensitivity to betamethasone 17-valerate and fluocinolone acetonide. Contact Dermatitis 1983;9:330-331

16 Romaguera C, Grimalt F. Five cases of contact dermatitis from pyrrolnitrine. Contact Dermatitis 1980;6:352-353

17 Malten KE. Betnelan V lotion contact sensitivity. Contact Dermatitis Newsletter 1973;13:360

18 Bunney MH. Contact dermatitis due to Betamethasone 17-valerate (Betnovate). Contact Dermatitis Newsletter 1972;12:318

19 Baeck M, Chemelle JA, Terreux R, Drieghe J, Goossens A. Delayed hypersensitivity to corticosteroids in a series of 315 patients: clinical data and patch test results. Contact Dermatitis 2009;61:163-175

20 Davis MD, El-Azhary RA, Farmer SA. Results of patch testing to a corticosteroid series: a retrospective review of 1188 patients during 6 years at Mayo Clinic. J Am Acad Dermatol 2007;56:921-927

21 Okano M. Contact dermatitis due to budesonide: report of five cases and review of the Japanese literature. Int J Dermatol 1994,33.709-715

22 Mercader-García P, Pastor-Nieto MA, García-Doval I, Giménez-Arnau A, González-Pérez R, Fernández-Redondo V, et al. GEIDAC. Are the Spanish baseline series markers sufficient to detect contact allergy to corticosteroids in Spain? A GEIDAC prospective study. Contact Dermatitis 2018;78:76-82

23 Uter W, de Pádua CM, Pfahlberg A, Nink K, Schnuch A, Lessmann H. Contact allergy to topical corticosteroids – results from the IVDK and epidemiological risk assessment. J Dtsch Dermatol Ges 2009;7:34-41

24 Reduta T, Laudanska H. Contact hypersensitivity to topical corticosteroids – frequency of positive reactions in patch-tested patients with allergic contact dermatitis. Contact Dermatitis 2005;52:109-110

25 Chavarría Mur E, González-Carrascosa Ballesteros M, Suárez Fernández R, Bueno Marco C. Generalized exanthematous reaction with pustulosis induced by topical corticosteroids. Contact Dermatitis 2005;52:114-115

26 Sommer S, Wilkinson S M, English JS, et al. Type-IV hypersensitivity to betamethasone valerate and clobetasol propionate: results of a multicentre study. Br J Dermatol 2002;147:266-269

27 Uter W, Geier J, Richter G, Schnuch A; IVDK Study Group, German Contact Dermatitis Research Group. Patch test results with tixocortol pivalate and budesonide in Germany and Austria. Contact Dermatitis 2001;44:313-314

28 Weltfriend S, Marcus-Farber B, Friedman-Birnbaum R. Contact allergy to corticosteroids in Israeli patients. Contact Dermatitis 2000;42:47

29 Lyon CC, Beck MH. Allergic contact dermatitis reactions to corticosteroids in periorbital inflammation and conjunctivitis. Eye (Lond) 1998;12(Pt.1):148-149

30 Khoo B-P, Leow Y-H, Ng S-K, Goh C-L. Corticosteroid contact hypersensitivity screening in Singapore. Am J Contact Dermat 1998;9:87-91

31 Dooms-Goossens A, Andersen KE, Brandão FM, Bruynzeel D, Burrows D, Camarasa J, et al. Corticosteroid contact allergy: an EECDRG multicentre study. Contact Dermatitis 1996;35:40-44

32 Boffa MJ, Wilkinson SM, Beck MH. Screening for corticosteroid contact hypersensitivity. Contact Dermatitis 1995;33:149-151

33 Freeman S. Corticosteroid allergy. Contact Dermatitis 1995;33:240-242

34 Hisa T, Katoh J, Yoshioka K, Taniguchi S, Mochida K, Nishimura T, et al. Contact allergies to topical corticosteroids. Contact Dermatitis 1993;28:174-179

35 Burden AD, Beck MH. Contact hypersensitivity to topical corticosteroids. Br J Dermatol 1992;127:497-501

36 Lauerma AI. Contact hypersensitivity to glucocorticosteroids. Am J Contact Dermat 1992;3:112-132

37 Wilkinson SM, English JS. Patch tests are poor detectors of corticosteroid allergy. Contact Dermatitis 1992;26:67-68

38 Dunkel FG, Elsner P, Burg G. Contact allergies to topical corticosteroids: 10 cases of contact dermatitis. Contact Dermatitis 1991;25:97-103

39 Sasaki E. Corticosteroid sensitivity and cross-sensitivity. A review of 18 cases 1967-1988. Contact Dermatitis 1990;23:306-315

40 English JS, Ford G, Beck MH, Rycroft RJ. Allergic contact dermatitis from topical and systemic steroids. Contact Dermatitis 1990;23:196-197

41 Goh CL. Cross-sensitivity to multiple topical corticosteroids. Contact Dermatitis 1989;20:65-67

42 Alani SD, Alani MD. Allergic contact dermatitis and conjunctivitis to corticosteroids. Contact Dermatitis 1976;2:301-304

43 Oh-i T. Contact dermatitis due to topical steroids with conceivable cross reactions between topical steroid preparations. J Dermatol 1996;23:200-208

Chapter 3.41   BETAXOLOL

IDENTIFICATION

Description/definition   : Betaxolol is the cardioselective β_1-adrenergic receptor antagonist that conforms to the structural formula shown below

Pharmacological classes : β_1-Adrenergic receptor antagonists; sympatholytics; antihypertensive agents

IUPAC name   : 1-[4-[2-(Cyclopropylmethoxy)ethyl]phenoxy]-3-(propan-2-ylamino)propan-2-ol

Other names   : 1-(4-(2-(Cyclopropylmethoxy)ethyl)phenoxy)-3-((1-methylethyl)amino)-2-propanol

CAS registry number   : 63659-18-7

EC number   : 613-310-1

Merck Index monograph   : 2454

Patch testing   : 5% water

Molecular formula   : $C_{18}H_{29}NO_3$

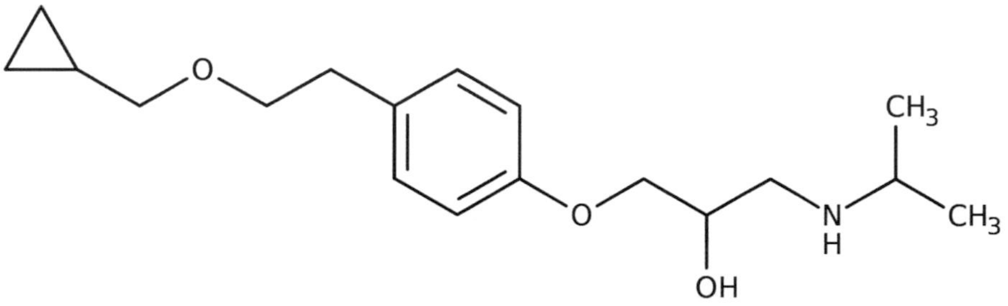

GENERAL

Betaxolol is a racemic mixture and selective β_1-adrenergic receptor antagonist with antihypertensive and anti-glaucoma activities, which is devoid of intrinsic sympathomimetic activity. The drug is used to treat hypertension, arrhythmias, coronary heart disease, glaucoma, and is also used to reduce non-fatal cardiac events in patients with heart failure. In pharmaceutical products, betaxolol is employed as betaxolol hydrochloride (CAS number 63659-19-8, EC number 264-384-3, molecular formula $C_{18}H_{30}ClNO_3$) (1).

CONTACT ALLERGY

General

For information on patch testing with commercial eye drops containing beta-blockers see Chapter 3.338 Timolol.

Case series

In Leuven, Belgium, in the period 1990-2014, iatrogenic contact dermatitis was diagnosed in 2600 individuals (17% of the total patch test population). 96% of all positive patch test reactions to topical drugs and antiseptics were considered to be relevant. Betaxolol (5% water) was tested in 3 patients and there was one positive reaction to it (2).

In Ferrara, Italy, over a 65-month period before 2005, 50 patients affected by periorbital dermatitis while using topical ocular products were patch tested, including with their own ophthalmic medications (n=210). Only 15 positive reactions were detected in 12 subjects, including 14 reactions to commercial eye drops. There was one reaction to betaxolol eye drops. The active ingredient was not tested separately, but contact allergy to the excipients and preservatives was excluded by patch testing. The authors concluded that patch testing with commercial eye drops has doubtful value (4).

Case reports

A 70-year-old man was referred for recurrent bilateral upper and lower eyelid eczema. He had had glaucoma for many years and had been treated with timolol eye drops, which seemed to aggravate the complaints. These were replaced with betaxolol hydrochloride eye drops, but some months later, the dermatitis relapsed. Contact allergy to timolol was established by a positive intracutaneous test at D2 and D4 (patch test negative). The patient also reacted to a patch test with the betaxolol eye drops and – in a second test session – to betaxolol hydrochloride 1%, 2% and 5% water. He was then treated with metipranolol eye drops but experienced a further relapse; contact allergy to metipranolol was shown by patch testing. Thus, it appears, that the patient developed successive independent sensitizations to three beta-blockers. There were no cross-reactions to a battery of beta-blockers to which he had not been exposed (3).

A 59-year-old man had long been treated for open-angle glaucoma with eye drops containing the β-blocker betaxolol, when he developed acute eyelid dermatitis associated with conjunctival hyperaemia. His condition rapidly improved after withdrawing betaxolol and using topical corticosteroids. Patch testing showed a positive reaction to his eye drops and, in a second patch test session, to betaxolol 1% water. There was also a positive reaction to timolol eye drops and timolol 1% water, which was considered to be cross-reactivity, as the patient had never used timolol before (6).

An 87-year-old woman with a 17-year primary open-angle glaucoma presented with severe dermatitis of the eyelids and periorbital areas. Patch testing showed a doubtful positive (?+) reaction at D2 and D4 to the betaxolol eye drops she had used, but its individual components were all negative (uncertain whether betaxolol itself was also tested). The eye drops were discontinued and the skin lesions completely resolved within 2 weeks. A ROAT with the eye drops was negative, but a use test resulted in erythemato-edematous plaques on the upper and lower eyelids within hours after receiving a drop in each eye of this medication. Subsequently, a sterile ophthalmic solution containing the excipients at the same concentrations as found in the commercial ophthalmic preparation was tolerated well and it was decided that the patient was contact allergic to betaxolol (5).

Cross-reactions, pseudo cross-reactions and co-reactions
Cross-reactions between beta-blockers appear to be infrequent (7). An individual sensitized to betaxolol may have cross-reacted to timolol (6).

LITERATURE
1   The data in the section 'General' may have been obtained from literature discussed in this chapter, but mostly also or exclusively from one or more of the following online sources: ChemIDPlus Advanced, PubChem, DrugBank, RxList, Drug Central, Drugs.com, and Wikipedia
2   Gilissen L, Goossens A. Frequency and trends of contact allergy to and iatrogenic contact dermatitis caused by topical drugs over a 25-year period. Contact Dermatitis 2016;75:290-302
3   O'Donnell BF, Foulds IS. Contact allergy to beta-blocking agents in ophthalmic preparations. Contact Dermatitis 1993;28:121-122
4   Corazza M, Massieri LT, Virgili A. Doubtful value of patch testing for suspected contact allergy to ophthalmic products. Acta Derm Venereol 2005;85:70-71
5   Sánchez-Pérez J, Jesús Del Río M, Fernández-Villalta MJ, García-Díez A. Positive use test in contact dermatitis from betaxolol hydrochloride. Contact Dermatitis 2002;46:313-314
6   Nino M, Napolitano M, Scalvenzi M. Allergic contact dermatitis due to the beta-blocker betaxolol in eyedrops, with cross-sensitivity to timolol. Contact Dermatitis 2010;62:319-320
7   Jappe U, Uter W, Menezes de Pádua CA, Herbst RA, Schnuch A. Allergic contact dermatitis due to beta-blockers in eye drops: a retrospective analysis of multicentre surveillance data 1993-2004. Acta Derm Venereol 2006;86:509-514

Chapter 3.42   BIBROCATHOL

IDENTIFICATION

Description/definition : Bibrocathol is the organobromine compound that conforms to the structural formula shown below

Pharmacological classes : Anti-infectives; ophthalmological agents

IUPAC name : 4,5,6,7-Tetrabromo-1,3,2λ^2-benzodioxabismole;hydrate

Other names : Bismuth derivative of tetrabromopyrocatechol

CAS registry number : 6915-57-7

EC number : 230-023-3

Merck Index monograph : 2473

Patch testing : Ointment as is; bismuth oxide 5% pet.

Molecular formula : $C_6H_2BiBr_4O_3$

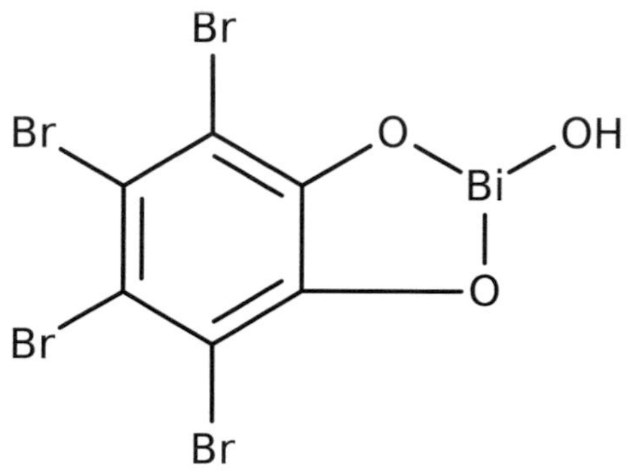

GENERAL

Bibrocathol is a bismuth derivative of tetrabromopyrocatechol that is still used in some countries as a topical ointment for the treatment of blepharitis (1).

CONTACT ALLERGY

Case report

A 33-year-old woman presented with a history of atopic hand eczema and allergic rhinitis. Because of periorbital dermatitis, she was treated with a 5% bibrocathol-containing eye ointment. She noticed an exacerbation of her dermatitis after such treatment. A patch test was positive to the ointment at D4 and D7 (++). The molecules which form bibrocathol, bismuth oxide and tetrabromocatechol (correct name: tetrabromopyrocatechol), were tested together with the ointment in a 2nd series of patch tests. On D4 and D7, the tests were positive to bismuth oxide 5% pet. (+++), 2% pet. (++) and 0.5% (+) and to the commercial ointment 'as is' and 30% pet. No reaction was found to tetrabromocatechol. Ten controls were negative. It was concluded that the inorganic bismuth oxide was the main hapten rather than the complex bibrocathol (2).

LITERATURE

1   The data in the section 'General' may have been obtained from literature discussed in this chapter, but mostly also or exclusively from one or more of the following online sources: ChemIDPlus Advanced, PubChem, DrugBank, RxList, Drug Central, Drugs.com, and Wikipedia

2   Wictorin A, Hansson C. Allergic contact dermatitis from a bismuth compound in an eye ointment. Contact Dermatitis 2001;45:318

Chapter 3.43   BIFONAZOLE

IDENTIFICATION

Description/definition   : Bifonazole is the phenylmethyl imidazole derivative that conforms to the structural
                               formula shown below
Pharmacological classes   : Antifungal agents
IUPAC name   : 1-[Phenyl-(4-phenylphenyl)methyl]imidazole
Other names   : 1-((4-Biphenylyl)phenylmethyl)-1*H*-imidazole
CAS registry number   : 60628-96-8
EC number   : 262-336-6
Merck Index monograph : 2486
Patch testing   : 1% alcohol or MEK (methyl ethyl ketone)
Molecular formula   : $C_{22}H_{18}N_2$

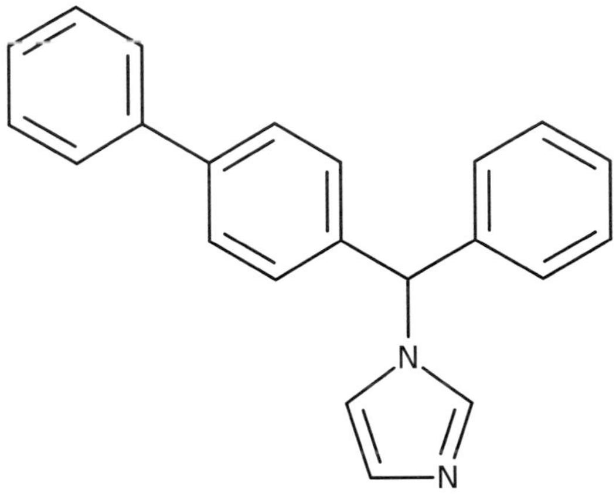

GENERAL
Bifonazole is an imidazole-type antifungal drug that kills fungi and yeasts by interfering with their cell membranes. It
is used for the topical treatment of various superficial fungal infections, including tinea pedis (athlete's foot) (1).

CONTACT ALLERGY

Case series
In a review of imidazoles used as antimycotic agents up to 1994, one positive patch test to bifonazole is mentioned,
but the literature source was not specified (3).

In Japan, in the period 1984 to 1994, 3049 outpatients were patch tested for suspected contact dermatitis and
218 of these with topical antifungal preparations. Thirty-five were allergic to imidazoles, including bifonazole in one
individual. In 60% of the cases, there were cross-reactions between imidazoles (2).

Cross-reactions, pseudo-cross-reactions and co-reactions
A patient sensitized to tioconazole had a cross-reaction to bifonazole 2% pet. (4). Cross-reactions are very unlikely
between bifonazole and other antifungals (3).

LITERATURE
1   The data in the section 'General' may have been obtained from literature discussed in this chapter, but mostly
    also or exclusively from one or more of the following online sources: ChemIDPlus Advanced, PubChem,
    DrugBank, RxList, Drug Central, Drugs.com, and Wikipedia
2   Yoneyama E. Allergic contact dermatitis due to topical imidazole antimycotics. The sensitizing ability of active
    ingredients and cross-sensitivity. Nippon Ika Daigaku Zasshi 1996;63:356-364 (Article in Japanese)
3   Dooms-Goossens A, Matura M, Drieghe J, Degreef H. Contact allergy to imidazoles used as antimycotic agents.
    Contact Dermatitis 1995;33:73-77
4   Izu R, Aguirre A, González M, Díaz-Pérez JL. Contact dermatitis from tioconazole with cross-sensitivity to other
    imidazoles. Contact Dermatitis 1992;26:130-131

Chapter 3.44   BIMATOPROST

IDENTIFICATION

Description/definition : Bimatoprost is a synthetic prostamide and structural prostaglandin analog that conforms to the structural formula shown below

Pharmacological classes : Antihypertensive agents

IUPAC name : (Z)-7-((1R,2R,3R,5S)-3,5-Dihydroxy-2-((1E,3S)-3-hydroxy-5-phenyl-1-pentenyl)cyclopentyl)-N-ethyl-5-heptenamide

CAS registry number : 155206-00-1

EC number : Not available

Patch testing : No data available

Molecular formula : $C_{25}H_{37}NO_4$

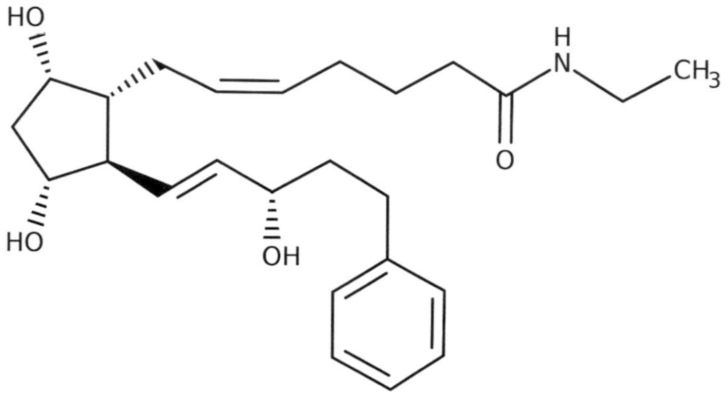

GENERAL

Bimatoprost is a synthetic prostamide and structural prostaglandin analog with ocular hypotensive activity. It mimics the effects of the endogenous prostamides and reduces intraocular pressure by increasing outflow of aqueous humor through both the pressure-sensitive outflow pathway (the trabecular meshwork), and the pressure-insensitive outflow pathway (the uveoscleral routes). Bimatoprost ophthalmic solution is used as an antihypertensive agent for controlling the progression of open-angle glaucoma or ocular hypertension (1).

CONTACT ALLERGY

Case report

A 52-year-old woman, while using 0.03% bimatoprost for open angle glaucoma, developed conjunctival hyperemia, foreign body sensation, blepharitis and ocular pruritis, as well as increased pigmentation of the periocular skin. A patch test with bimatoprost at a concentration of 0.003% was 'positive' at D2, causing pigmentation and local itching. Patch tests with the other ingredients of the commercial eye drops (including benzalkonium chloride, tested at a too low concentration) were negative. Seven controls were negative (2). Obviously, contact allergy to bimatoprost was not proven in this case and the article should never have been accepted for publication.

LITERATURE

1   The data in the section 'General' may have been obtained from literature discussed in this chapter, but mostly also or exclusively from one or more of the following online sources: ChemIDPlus Advanced, PubChem, DrugBank, RxList, Drug Central, Drugs.com, and Wikipedia

2   Sodhi PK, Verma L, Ratan SK. Contact dermatitis from topical bimatoprost. Contact Dermatitis 2004;50:50

Chapter 3.45   BISMUTH SUBNITRATE

IDENTIFICATION

Description/definition   : Bismuth subnitrate is the bismuth-containing compound that conforms to the structural formula shown below
Pharmacological classes : Antacids; adstringents and antiseptics
IUPAC name   : Bismuth(3+) nitrate [hydroxy(oxido)-λ^5-azanyl]oxidanide (1:2:1) (Chemspider); pentabismuth(3+) ion nonahydroxide tetranitrate oxidandiide (DrugBank)
Other names   : Bismuth hydroxide nitrate oxide ($Bi_5(OH)_9(NO_3)_4O$) (ChemIDPlus); CI 77169; bismuth oxynitrate
CAS registry number   : 1304-85-4
EC number   : 215-136-8
Merck Index monograph   : 2554
Patch testing   : Pure; perform controls
Molecular formula   : $Bi_5H_9N_4O_{22}$ (ChemIDPlus)

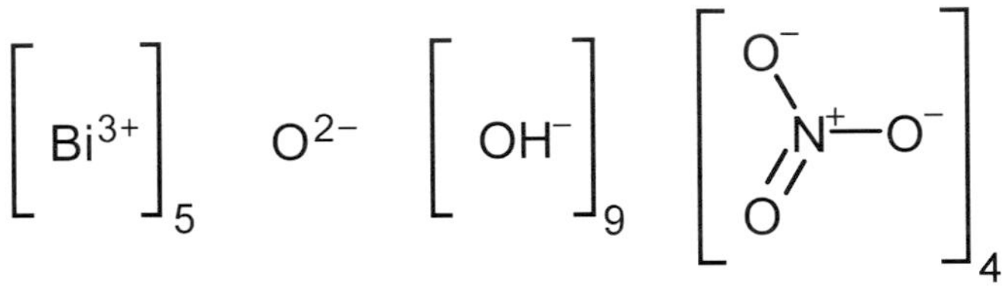

$$\left[Bi^{3+} \right]_5 \quad O^{2-} \quad \left[OH^- \right]_9 \quad \left[\begin{matrix} O^- \\ N^+ - O^- \\ O \end{matrix} \right]_4$$

GENERAL

Bismuth subnitrate is a highly water-soluble crystalline compound that has been used as a treatment for duodenal ulcers and anti-diarrheic agent. It is indicated for over-the-counter use as an antacid (1). In otolaryngology, it is used in bismuth iodoform paraffin paste for its adstringent and antiseptic properties (3).

CONTACT ALLERGY

Case report

A 50-year-old man had a radical mastoidectomy right for chronic suppurative otitis media. Post-operatively, a bismuth and iodoform paraffin paste-impregnated gauze was packed into the external auditory canal, left *in situ* for 2 weeks and changed biweekly. Four weeks after the operation, the patient developed acute dermatitis on his external auditory canal, right pinna and right side of the face, where the bismuth iodoform paraffin pack was applied. Patch testing showed positive reactions to the bismuth iodoform paraffin pack, bismuth subnitrate (as is), iodoform 5%, 10% and 25% pet. and to potassium iodide 25% pet. Twenty controls were negative to bismuth subnitrate (pure) and iodoform 10% and 25% pet. A repeat patch test with bismuth subnitrate (undiluted) on the patient 4 weeks after the first test produced a +++ reaction (2).

LITERATURE

1   The data in the section 'General' may have been obtained from literature discussed in this chapter, but mostly also or exclusively from one or more of the following online sources: ChemIDPlus Advanced, PubChem, DrugBank, RxList, Drug Central, Drugs.com, and Wikipedia
2   Goh CL, Ng SK. Contact allergy to iodoform and bismuth subnitrate. Contact Dermatitis 1987;16:109-110
3   Crossland GJ, Bath AP. Bismuth iodoform paraffin paste: a review. J Laryngol Otol 2011;125:891-895

Chapter 3.46   BISMUTH TRIBROMOPHENATE

IDENTIFICATION

Description/definition : Bismuth tribromophenate is the bismuth salt that conforms to the structural formula shown below

Pharmacological classes : Anti-infective agents

IUPAC name : Bismuth(3+) tris(2,4,6-tribromophenolate)

Other names : Bismuth tris(2,4,6-tribromophenoxide); bismuth tribromophenol; tribromophenol-bismuth; 2,4,6-tribromophenol, bismuth (3+) salt; Xeroform®

CAS registry number : 5175-83-7

EC number : 225-958-9

Merck Index monograph : 2561

Patch testing : Commercial powder (Xeroform ®, containing 5% bismuth tribromophenate) pure and 5% pet.

Molecular formula : $C_{18}H_6BiBr_9O_3$

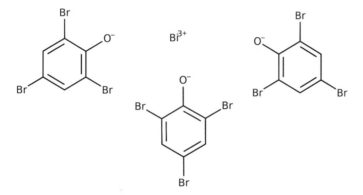

GENERAL

Bismuth tribromophenate is the bismuth salt of tribromophenol that has antimicrobial activity. It is used in topical anti-infective agents, available as powder and petrolatum-based fine mesh gauze (1).

CONTACT ALLERGY

Case reports

A 54-year-old woman for the previous 2 years had suffered from a therapy-resistant post-thrombotic ulcer above the left lateral malleolus, surrounded by a marked oozing eczematous dermatitis. It had been treated with a variety of remedies. Treatment with bismuth tribromophenate powder produced no improvement during 8 weeks. After the applications of the powder were discontinued and topical treatment was changed to an antibiotic powder, the ulcer healed within a few weeks. Patch tests were positive to bismuth tribromophenate powder 100% and 5% pet. and to several other medicinal and vehicle compounds (3).

A 71-year-old woman presented with a relapsing ulcer on the medial aspect of the left lower leg for the past 10 years. The current ulceration had been present for 3½ months and was 2x2.2 centimeter in size, surrounded by a scaly vesicular dermatitis. She had been using 0.1% cetylpyridinium chloride solution for daily cleansing of the ulcer area. Following clearance of the dermatitis, her ulcer was treated with bismuth tribromophenate powder. The ulcer size diminished somewhat during 5 weeks, but then healing stopped and a persistent dermatitis recurred on the surrounding skin. When the bismuth tribromophenate powder was replaced with an antibiotic, the dermatitis disappeared and the ulcer healed in a couple of weeks. Patch tests showed allergy to cetylpyridinium chloride 0.05% water and the commercial powder 100% and 5% pet. (3).

LITERATURE

1   The data in the section 'General' may have been obtained from literature discussed in this chapter, but mostly also or exclusively from one or more of the following online sources: ChemIDPlus Advanced, PubChem, DrugBank, RxList, Drug Central, Drugs.com, and Wikipedia

2   Fräki JE, Peltonen L, Hopsu-Havu VK. Allergy to various components of topical preparations in stasis dermatitis and leg ulcer. Contact Dermatitis 1979;5:97-100

3   Wereide K, Thune P, Hanstad I. Contact allergy to xeroform in leg ulcer patients. Contact Dermatitis 1983;9:525-526

Chapter 3.47   BRIMONIDINE

IDENTIFICATION

Description/definition   : Brimonidine is the quinoxaline that conforms to the structural formula shown below
Pharmacological classes   : α_2-Adrenergic receptor agonists; antihypertensive agents
IUPAC name   : 5-Bromo-N-(4,5-dihydro-1H-imidazol-2-yl)quinoxalin-6-amine
Other names   : 5-Bromo-6-(2-imidazolin-2-ylamino)quinoxaline
CAS registry number   : 59803-98-4
EC number   : Not available
Merck Index monograph   : 2651
Patch testing   : 1%, 5% and 10% pet.
Molecular formula   : $C_{11}H_{10}BrN_5$

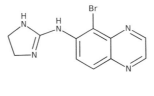

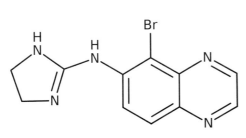

Brimonidine   Brimonidine tartrate

GENERAL

Brimonidine is a quinoxaline derivative and a selective α_2-adrenergic receptor agonist. Upon ocular administration, brimonidine acts on the blood vessels causing them to constrict which leads to a decrease in the production of aqueous humor; it also enhances its outflow. Brimonidine ophthalmic solution is indicated for patients with open-angle glaucoma or ocular hypertension to lower intraocular pressure. A topical gel is indicated for the treatment of persistent facial erythema of rosacea in adults, where it reduces erythema through direct vasoconstriction. In pharmaceutical products, brimonidine is employed as brimonidine tartrate (CAS number 70359-46-5, EC number not available, molecular formula $C_{15}H_{16}BrN_5O_6$) (1).

CONTACT ALLERGY

General

In ophthalmological literature, high percentages of 'allergy' to brimonidine have been reported. The symptoms are usually follicular conjunctivitis or allergic contact dermatoconjunctivitis, but may include itching, foreign-body sensation, tearing or discharge, chemosis, hyperemia, and eyelid edema and erythema. However, patch tests with brimonidine or the commercial eye drops were never performed (2,3). Cases of proven (sometimes possible or likely) contact allergy and allergic contact dermatitis started to emerge, when a gel containing 0.3% brimonidine tartrate had become available for the treatment of persistent facial erythema of rosacea (4,5,6,7,10,11)

Case series

In phase 3 clinical trials of brimonidine gel, 1% of patients experienced contact dermatitis. One patient was diagnosed with contact allergy to the active ingredient brimonidine tartrate, and another patient had a positive patch test reaction to phenoxyethanol, the preservative in the vehicle gel. Further to these findings, a larger, open label study of 345 patients who applied the commercial gel for a 12-month period reported a rate of contact allergy of 2.2% (9).

Case reports

Two men, aged 50 and 60 years, respectively, with long-standing rosacea were treated with brimonidine gel with an excellent effect on facial erythema. After 3 months, the patients suddenly developed facial dermatitis in the treated area. Both were patch tested with the gel 'as is', the active ingredient brimonidine tartrate 0.5% aq., and all excipients. The results were as follows: for patient 1, a + positive test reaction to brimonidine gel on D2 and negative results on D3 and D7; and for patient 2, a ++ positive test reaction to the gel on D3, and a negative result on D7. Both had negative results for all individual ingredients. A repeated open application test (ROAT) with brimonidine gel on the upper inner arm of the patients twice daily yielded a positive reaction within one week (5). Contact allergy to brimonidine was not proven in these cases.

A 24-year-old woman presented with facial dermatitis 6 months after starting to use brimonidine tartrate 0.33% gel once daily for the treatment of her rosaceal erythema. Examination showed a florid exudative dermatitis affecting the malar region, the bridge of the nose, and the forehead. Patch tests gave a positive (+) reaction to the gel on day 2 (D3, D4 or D7 not mentioned). Later, the test was repeated plus additional testing with relevant ingredients of the vehicle gel. On D4, the only positive reaction was to the brimonidine gel itself. A ROAT resulted in an eczematous reaction after 4 days of application. Brimonidine itself was not tested (6).

A 75-year-old woman had a 3-month history of pruritic dermatitis on her malar cheeks and nose, which had started within a few days of initiating treatment of rosacea with brimonidine tartrate gel 0.33%. She had a history of intolerance to eye drops used to treat her glaucoma. Patch testing showed many positive reactions, but not to the brimonidine gel. Because of the high suspicion for brimonidine allergy, a ROAT was performed with brimonidine tartrate gel as well as brimonidine ophthalmic solution, which the patient was suspicious caused previous issues. On day 3 of the test, a red, raised, vesicular dermatitis developed in response to both the gel and the eye drops (7). Brimonidine itself was not tested, but contact allergy to it is highly likely.

A bullous, hyperemic, pigmented, itching lesion discharging watery fluid at the lateral canthus of the left eye in a 67-year-old woman and periorbital pigmentation in a 25-year old man have – rather unconvincingly – been ascribed to contact allergy to brimonidine tartrate 0.2% eye drops on the basis of a positive patch test to the commercial product. Upon re-introduction of the eye drops, the side effects reappeared within an average period of two months (8). Obviously, this prolonged time frame would be highly atypical for a contact allergic reaction (8).

A 61-year-old man presented with 4 days of burning sensation, swelling and erythema with scaling and mild crusting of his face, especially his eyelids and cheeks. He had treated facial flushing associated with rosacea with a formulation of brimonidine eye drops in aqueous cream. Patch tests were positive to the eye drops and the aqueous cream formulation, but brimonidine itself was not tested (10).

A 59-year-old woman with rosacea presented with bouts of very severe facial dermatitis while using brimonidine tartrate gel 0.33%. She developed an eczematous dermatitis in the treated area 24 hours after use, which worsened with each application and possibly with sunlight exposure. Patch tests gave positive reactions to the brimonidine gel, and to brimonidine 1%, 5% and 10% pet.; lower concentrations (0.33% and 0.5%) were ?+ at D4. However, photopatch testing enhanced the reactions to these lower concentrations from ?+ to ++, which was ascribed to phototoxicity. The photopatch test reactions to the gel and to brimonidine 1%, 5% and 10% pet. were the same as in the conventional patch tests. Six controls were negative (11).

Another patient with allergic contact dermatitis from brimonidine tartrate was reported from France in 2017 (4, no details available).

LITERATURE

1   The data in the section 'General' may have been obtained from literature discussed in this chapter, but mostly also or exclusively from one or more of the following online sources: ChemIDPlus Advanced, PubChem, DrugBank, RxList, Drug Central, Drugs.com, and Wikipedia

2   Blondeau P, Rosseau JA. Allergic reactions to brimonidine in patients treated for glaucoma. Can J Ophthalmol 2002; 37:21-26

3   Manni G, Centofanti M, Sacchetti M, Oddone F, Bonini S, Parravano M, et al. Demographic and clinical factors associated with development of brimonidine tartrate 0.2%-induced ocular allergy. J Glaucoma 2004;13:163-167

4   Schmutz JL. Allergic contact dermatitis to Mirvaso® (brimonidine tartrate). Ann Dermatol Venereol 2017;144:86-87 (Article in French)

5   Bangsgaard N, Fischer LA, Zachariae C. Sensitization to and allergic contact dermatitis caused by Mirvaso® (brimonidine tartrate) for treatment of rosacea - 2 cases. Contact Dermatitis 2016;74:378-379

6   Cookson H, McFadden J, White J, White IR. Allergic contact dermatitis caused by Mirvaso®, brimonidine tartrate gel 0.33%, a new topical treatment for rosaceal erythema. Contact Dermatitis 2015;73:366-367

7   Swanson LA, Warshaw EM. Allergic contact dermatitis to topical brimonidine tartrate gel 0.33% for treatment of rosacea. J Am Acad Dermatol 2014;71:832-833

8   Sodhi PK, Verma L, Ratan J. Dermatological side effects of brimonidine: a report of three cases. J Dermatol 2003;30:697-700

9   Moore A, Kempers S, Murakawa G, Weiss J, Tauscher A, Swinyer L, et al. Long-term safety and efficacy of once-daily topical brimonidine tartrate gel 0.5% for the treatment of moderate to severe facial erythema of rosacea: results of a 1-year open-label study. J Drugs Dermatol 2014;13:56-61

10 Rajagopalan A, Rajagopalan B. Allergic contact dermatitis to topical brimonidine. Australas J Dermatol 2015;56:235

11 Ringuet J, Houle MC. Case report: Allergic contact dermatitis to topical brimonidine demonstrated with patch testing: insights on evaluation of brimonidine sensitization. J Cutan Med Surg 2018;22:636-638

Chapter 3.48   BROMELAINS

IDENTIFICATION

Description/definition	: Bromelains are protein-digesting and milk-clotting enzymes found in pineapple fruit juice and stem tissue
Pharmacological classes	: Enzymes
Other names	: Ananase; pineapple protease
CAS registry number	: 9001-00-7
EC number	: 232-572-4
Merck Index monograph	: 2665
Patch testing	: 2.5% pet.
Molecular formula	: Not specified

GENERAL

Bromelain is a protease enzyme derived from the stems of pineapples that is composed of a mixture of different thiol endopeptidases and other components like phosphatase, glucosidase, peroxidase, cellulase, escharase, and several protease inhibitors. It works by selectively inhibiting the biosynthesis of pro-inflammatory prostaglandins and also has analgesic properties, as well as possible anti-cancerous and pro-apoptotic effects. Bromelain has potential therapeutic value as a treatment for angina pectoris, bronchitis, sinusitis, surgical trauma, and osteoarthritis. It is available as an anti-inflammatory drug in France for musculoskeletal disorders and is also employed as food supplement (1).

CONTACT ALLERGY

Case report

A 56-year-old woman had cheilitis associated with perlèches for 2 months. She denied the use of cosmetics and had always the same toothpaste. Since 3 months, the patient every day used a mouthwash with a powder diluted in water. Patch tests were positive (++) at D2 and D3 to the powder diluted with water. The powder contained sodium bicarbonate and bromelain; patch tests with these ingredients were positive to bromelain 2.5% pet. (D3 ++). Ten controls were negative. The patient was advised to avoid products containing bromelain and she has since been symptom-free (2).

LITERATURE

1   The data in the section 'General' may have been obtained from literature discussed in this chapter, but mostly also or exclusively from one or more of the following online sources: ChemIDPlus Advanced, PubChem, DrugBank, RxList, Drug Central, Drugs.com, and Wikipedia

2   Raison-Peyron N, Roulet A, Guillot B, Guilhou JJ. Bromelain: an unusual cause of allergic contact cheilitis. Contact Dermatitis 2003;49:218-219

Chapter 3.49   BUDESONIDE

IDENTIFICATION

Description/definition : Budesonide is the synthetic non-halogenated corticosteroid and the butyraldehyde of prednisone acetonide that conforms to the structural formula shown below

Pharmacological classes : Glucocorticoids; bronchodilator agents; anti-inflammatory agents

IUPAC name : (1S,2S,4R,8S,9S,11S,12S,13R)-11-Hydroxy-8-(2-hydroxyacetyl)-9,13-dimethyl-6-propyl-5,7-dioxapentacyclo[10.8.0.0^{2,9}.0^{4,8}.0^{13,18}]icosa-14,17-dien-16-one

Other names : (11β,16α)-16,17-(Butylidenebis(oxy))-11,21-dihydroxypregna-1,4-diene-3,20-dione

CAS registry number : 51333-22-3

EC number : 257-139-7

Merck Index monograph : 2746

Patch testing : 0.01% pet. (Chemotechnique, SmartPracticeCanada); 0.1% pet. (Chemotechnique; SmartPracticeEurope); 0.2% pet. (SmartPracticeCanada); ADVICE: 0.01% pet., always perform late readings

Molecular formula : $C_{25}H_{34}O_6$

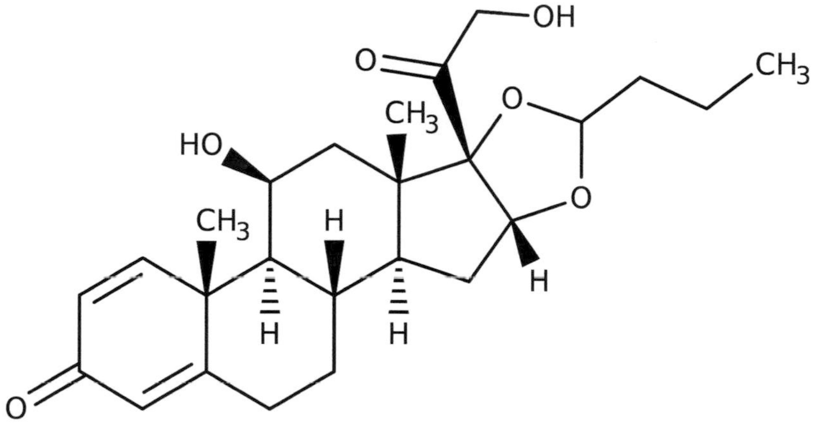

GENERAL

General aspects of corticosteroids used on the skin and mucous membranes are discussed in Chapter 2.4. A practical guideline for diagnosing allergic reactions to corticosteroids is presented in ref. 52. Budesonide 0.01% pet. is included in the European baseline series (51) and at a concentration of 0.1% pet. in the American core allergen series (www.smartpracticecanada.com) as screening agent for corticosteroid hypersensitivity.

Many studies have been performed to find the optimal concentration and vehicle for patch testing (100-109). Contrary to many other corticosteroids, petrolatum performs equally well as alcohol with budesonide (100,101,102). Because petrolatum has the advantage of stability and ease of handling, and also because alcohol may itself give irritant and occasionally allergic reactions and must therefore be tested as a control (102), petrolatum is the preferred vehicle. A concentration of 1% (often used for corticosteroids) performed less well than 0.1% and 0.01% in a multicenter study performed by the EECDRG (104); most studies have compared the latter 2 concentrations by simultaneous testing (4,8,10,12,13,29,103,104). In two studies (4,29) 0.01% detected more patients sensitized to budesonide than the 0.1% test material. In most investigations, however, the results were identical (8,12) or virtually equal (10,13,104). In only one study did the higher concentration score more positive results (0.62% versus 0.43%), but the difference was not significant (103).

To minimize the risk of patch test sensitization, the EECDRG chose to include budesonide in the lower concentration of 0.01% in petrolatum in the European baseline series (104,110). However, some sensitized individuals are traced only (or better) with very low concentrations of budesonide (107) and therefore, when budesonide allergy is strongly suspected, but the routine test concentration (0.01%) is negative, testing also with a lower concentration (0.001%) is advocated (110). Lower concentrations may detect budesonide better at early readings (D3,D4) and may show a positive reaction instead of an 'edge reaction' seen with a higher concentration. This may be explained by lesser immunosuppression of the patch test reaction by lower doses of the corticosteroid (29). Alternatively, an intradermal test with budesonide may be performed (112). A considerable number of positive patch test reactions to budesonide can be missed when readings are not performed at day 7 (43,44).

CONTACT ALLERGY

General

Budesonide is a 1:1 mixture of 2 diastereomers, the *R* and *S* diastereomers, which can both induce positive patch test reactions by themselves (99). Budesonide is available as inhalation powder and aerosol for the treatment of asthma, as nasal spray for allergic rhinitis and may also be administered orally or rectally for treating Crohn's disease and colitis ulcerosa. It is the most commonly used corticosteroid molecule in liquid aerosols, but used in only a few countries for skin diseases in cream and ointment (e.g. Japan, Spain, Italy). Many case reports of contact allergic reactions to budesonide have been reported, mostly from nasal sprays (table 3.49.4) and from creams and ointments (table 3.49.6), and a lower number from inhalation sprays (table 3.49.5). Local allergic side effects of inhaled corticosteroids include stomatitis, perioral eczema, oropharyngeal pruritus, dryness, mouth erythema and edema, dry cough, dysphonia, dysphagia and odynophagia (pain when swallowing) (62,63,65). Symptoms often start early in the treatment (after several days) and may develop hours after inhalation (65).

Application in the nose may cause nasal itching, dryness, burning, nasal congestion, worsening of rhinitis, dysphagia, edema of the tongue, lips and face, and dermatitis on the nose, upper lip and face (59,73,75). Systemic contact dermatitis from resorption through the nasal and pulmonary mucosae has been observed with swelling of the eyelids, urticaria, general pruritus, maculopapular exanthemas, flare-up of previous dermatitis and flare-up of previous positive patch tests (55,58,65,68,69,74,88). In one study, in 4 of 7 patients who had previously had a positive patch test to budesonide, inhalation of budesonide resulted in reactivation of these patch tests within 24 hours, proving systemic contact dermatitis from mucosal absorption of the corticosteroid (67).

A peculiar feature of budesonide allergy are cases of allergic contact dermatitis 'by proxy', which means that the budesonide was not used by the patient himself, but by a partner ('consort dermatitis', 'connubial dermatitis'), a child or someone else near-by (49,56,60,61,71,91; table 3.49.7). Occupational allergic contact dermatitis has also been reported several times (60,61,64).

Patch testing in the general population and in subgroups

The results of patch testing budesonide in (subgroups of) the general population are shown in table 3.49.1. In adolescents from Sweden, the prevalence of positive patch tests was <0.1% (42) and in 8th grade school children from Denmark, not a single positive reaction was observed (39). Of adults from 5 European countries, 0.4% had positive reactions to budesonide, 0.2% of the women and 0.6% of the men (38). It should be appreciated that the reactions were read at D3 or even once at D2. It is well known that many patch test reactions to corticosteroids first develop after D3-4, which means that a (possibly considerable) number of positive patch test reactions may have gone unnoticed.

Table 3.49.1 Contact allergy in the general population and in subgroups

Year and country	Selection and number tested	Prevalence of contact allergy			Comments	Ref.
		Total	Women	Men		
General population						
2008-11 five Euro-pean countries	general population, random sample, 18-74 years, n=3119	0.4%	0.2%	0.6%	TRUE test	38
Subgroups						
2011-2013 Sweden	adolescents from a birth cohort, 15.8-18.9 years, n=2285	0.09%	0.17%	0.0%	TRUE test; patch tests were read at day 2 only	42
2010 Denmark	unselected population of 8th grade schoolchildren in Den-mark, 15 years later; n=442	0%	0%	0%	follow-up study; TRUE test	39

In Germany, for the period 1995-2004, the population-based relative incidence (RI) of contact sensitization to budesonide (cases/100,000 defined daily doses (DDDs) per year) was estimated to be 43.2. In the group of corticosteroids, the RI ranged from 0.3 (dexamethasone sodium phosphate) to 43.2 (budesonide) (47).

Patch testing in groups of patients

Results of patch testing budesonide in consecutive patients suspected of contact dermatitis (routine testing) back to 2000 are shown in table 3.49.2. Results of testing in groups of *selected* patients (individuals with perianal dermatitis or leg ulcers, patient suspected of corticosteroid allergy) are shown in table 3.49.3.

Table 3.49.2 Patch testing in groups of patients: Routine testing

Years and Country	Test conc. & vehicle	Number of patients tested	positive (%)		Selection of patients (S); Relevance (R); Comments (C)	Ref.
Europe						
2008-18 Netherlands	0.1% pet.	3276	26	(0.8%)	R: not stated; C: >40% of the reactions may be missed when patch are not read at D7	43
2015-2016 Spain		3699	23	(0.6%)	R: not stated; TRUE test or 0.01% pet. tested	118
2014-2016 Greece	0.01% pet.	1978	55	(2.8%)	R: not stated	41
2013-2014 12 European countries [b]	0.01% pet.	12,863		(0.8%)	R: not stated; C: range of positive reactions: 0% - 2.4%	4
	0.1% pet.	9009		(0.4%)	R: not stated; C: range of positive reactions: 0% - 1.0%	
1990-2014 Belgium	various	14,192	274	(1.9%)	R: 96% of the positive patch test reactions to all topical drugs and antiseptics were considered to be relevant; C: there was a significant decrease in the frequency of positive reactions from 3.3% in 1990-1994 to 1.3% in 2010-2014	48
2009-12, six European countries [b]	0.1% pet.	21,455		(0.4%)	R: not stated; C: range per country: 0% - 0.9%	29
	0.01% pet.	7794		(1.5%)	R: not stated; C: range per country: 0% - 2.6%	
2007-2008 11 European countries [b]	0.1% pet.	20,985	61	(0.3%) [a]	R: not stated; C: prevalences ranged from 0% to 1.8%	10
	0.01% pet.	21,575	43	(0.2%) [a]	R: not stated; C: prevalences ranged from 0% to 1.5%	
2005-2008 Denmark	0.01% pet.	3594	34	(1.0%)	R: not stated	120
2005-2006 10 European countries [b]	0.1% pet.	8547	56	(0.6%)	R: not stated; C: prevalences were 0.7% in Central Europe, 0.6% in West (tested 0.1% pet.), 0.6% in Northeast and 0.7% in South Europe	13
	0.01% pet.	4190	28	(0.7%)		
<2005 Poland	0.1% pet.	257	10	(3.9%)	R: 10%	123
2004, 11 European countries [b]	0.01% pet.	1755	26	(1.5%) [a]	R: not stated; C: range positives per center: 0.6%-2.4%	15
2002-2003 Europe [b]	0.01% pet.	1328		(1.1%)	R: not stated	17
1999-2001 Sweden	0.01% pet.	3790		(1.3%)	R: not stated; C: prevalence in women 2.3%, in men 1.7% (standardized prevalences)	19
2000 United Kingdom	0.1% pet.	3063		(0.6%)	R: 90% (current and past relevance in one center); C: range of positive reactions per center 0.2% - 3.2%	20
1996-2000 Europe	0.01% pet.	26,210		(1.6%)	R: not stated; C: ten centers, seven countries, EECDRG study	23
USA, Canada						
2015-2017 NACDG	0.1% pet.	5593	47	(0.8%)	R: definite + probable relevance: 38%	1
2007-2016 USA	0.01% pet.	2313		(0.6%)	R: not stated	137
2011-2015 USA	0.1% pet.	2575	18	(0.7%)	R: not stated	2
2013-2014 NACDG	0.1% pet.	4859	42	(0.9%)	R: definite + probable relevance: 26%	3
2007-2014 NACDG	0.1% pet.	17,978	156	(0.9%)	R: definite + probable relevance: 40%	119
2011-2012 NACDG	0.1% pet.	4232	35	(0.8%)	R: definite + probable relevance: 54%	6
2009-2010 NACDG	0.1% pet.	4304		(0.8%)	R: definite + probable relevance: 43%	7
2006-2010 USA	0.1% pet.	975		(1.2%)	R: 67%	8
	0.01% pet.	2096		(1.2%)	R: 36%	
2007-2008 NACDG	0.1% pet.	5083		(0.9%)	R: definite + probable relevance: 40%	11
2005-2006 NACDG	0.1% pet.	4439		(1.5%)	R: definite + probable relevance: 26%	12
	0.01% pet.	4435		(1.5%)	R: definite + probable relevance: 23%	
2001-2005 USA	0.1% pet.	3843		(1.5%)	R: 47%	14
2003-2004 NACDG	0.1% pet.	5142	81	(1.6%)	R: not stated	16
2001-2002 NACDG	0.1% pet.	4901		(1.1%)	R: definite + probable relevance: 44%	18
1998-2000 USA	0.1% pet.	5806		(1.4%)	R: definite + probable relevance: 40%	21
1998-2000 USA	0.1% pet.	713		(3.2%)	R: not stated	22
Other countries						
2013-2019 Turkey	0.01% pet.	1309	7	(0.5%)	R: not stated	138
2006-2018 Thailand	0.01% pet.	2803	29	(1.0%)	R: not stated; C: decrease from 5.2% in 2006-2008 to 0.1% in the period 2016-2018	45
2009-2013 Singapore		2598		(0.3%)	R: present + past relevance: 0%; C: range of positive reactions per year 0% - 0.5%	5
2001-2010 Australia	0.1% pet.	5132	97	(1.9%)	R: 11%	9

[a] age-standardized and sex-standardized proportions; [b] study of the ESSCA (European Surveillance System on Contact Allergies); EECDRG: European Environmental and Contact Dermatitis Research Group; IVDK: Information Network of Departments of Dermatology, Germany, Austria, Switzerland; NACDG: North American Contact Dermatitis Group (USA, Canada)

Patch testing in consecutive patients suspected of contact dermatitis: routine testing

Budesonide has been present in the screening series of the North American Contact Dermatitis Group (NACDG, USA + Canada), the European baseline series and probably many national standard series for two decades, so data on routine testing with this corticosteroid are abundant (table 3.49.2). In Europe, prevalences of positive patch tests tobudesonide have generally been in the range of 0.4% to 1.5%. The highest scores were seen in Greece (2.8% [41]) and Poland (3.9% [123]), the latter being a very small study of 257 patients. Relevance was hardly ever addressed. In a study from Belgium, 96% of all reactions to topical drugs and antiseptics were considered to be relevant. However, as budesonide is used in nasal and pulmonary inhalation preparations only, it must be assumed that most reactions to budesonide are cross-reactions to other corticosteroids, attesting to its efficacy as a screening agent for cortico-steroid allergy. In the multi-national studies performed by the ESSCA (European Surveillance System on Contact Allergies), there was a tendency to lower rates of positive tests in more recent years. This was also noted – in a slightly different time frame – in a large center in Belgium, where there was a significant decrease in the frequency of positive reactions from 3.3% in 1990-1994 to 1.3% in 2010-2014 (48).

In the USA and Canada, frequencies of sensitization to budesonide generally ranged from 0.8% to 1.5%, with only one higher percentage of 3.2 in a smaller USA study performed in the period 1998 to 2000 (22). In the NACDG studies, percentages definite + probable relevance were mostly in the 25-50% range, which was also the case in a few other USA studies (8,14). Here, too, was a tendency to slightly lower frequencies in the second decade visible.

Data on routine patch testing budesonide from some other countries can be found in table 3.49.2.

Patch testing in groups of selected patients

In groups of patients with leg ulcers/stasis dermatitis, prevalences of positive patch tests to budesonide have ranged from 1.5% to 20%, the latter in two Polish studies (26,28). However, the number of patients in these investigations was very small (n=50), relevance was not discussed and the groups were probably the same.

In groups of patients with perianal, genital or anogenital dermatitis, the frequencies of sensitization ranged from 0.8% to 7.4%. In most studies with a control group, the rates were higher than in the control group. Quite baffling are the differences in an NACDG study between testing with budesonide 0.1% (2.7% positive reactions) and with 0.01% (7.4% positive reactions). It should be mentioned that the latter was in a group of 68 patients only (5 positive reactions) and it is uncertain whether this smaller group was also included in the group tested with budesonide 0.1%, or possibly had been selected in a different manner (31).

In groups of patients suspected of allergy to corticosteroids or with otitis externa, frequencies of positive reactions to budesonide were in the 2.5% to 3% range. Relevance rates, where mentioned, ranged from 18% (37) to 100% (117). However, it would seem to be impossible that all 36 patients reacting to budesonide in this study had allergic reactions to inhaled, nasal or oral budesonide pharmaceuticals (117). The high percentage of 9 in a Canadian study is an artifact, as patients were selected on the basis of previously established contact allergy to medicaments (50).

Table 3.49.3 Patch testing in groups of patients: Selected patient groups

Years and Country	Test conc. & vehicle	Number of patients tested	positive (%)	Selection of patients (S); Relevance (R); Comments (C)	Ref.
Patients with leg ulcers/stasis dermatitis					
<2017 India	0.01% pet.	172	3 (1.7%)	S: patients with venous leg ulcers of over 6 weeks duration; R: 'the majority of reactions were relevant'	30
2003-2014 IVDK	0.1% pet.	1133	(3.4%)	S: patients with stasis dermatitis/chronic leg ulcers; R: not stated	24
2005-2008 France	0.1% pet.	423	30 (7.1%)	S: patients with leg ulcers; R: not stated	25
<2008 Poland	0.01% pet.	50	10 (20%)	S: patients with venous leg ulcers; R: not stated	28
<2005 Poland	0.01% pet.	50	10 (20%)	S: patients with chronic venous leg ulcers; R: not stated	26
1997-2001 U.K.		200	3 (1.5%)	S: patients with venous or mixed venous/arterial leg ulcers; R: all reactions to topical drugs were considered to be of probable, past or current relevance	27
Patients with perianal, genital or anogenital dermatitis					
2005-2016 NACDG	0.1% or 0.01% pet.	449	7 (1.6%)	S: patients with only anogenital dermatitis; R: all positives represent relevant reactions; C: the frequency was significantly higher than in a control group	36
2004-2016 Spain		124	1 (0.8%)	S: patients with perianal dermatitis lasting >4 weeks; R: 100%	32
2003-2010 USA	0.1% pet.	55	1 (2%)	S: women with (ichty) vulvar dermatoses; R: not stated	33
2004-2008 IVDK			(2.1%)	S: patients with anogenital dermatoses tested with a medicament series; R: not stated; C: number of patients tested unknown	34

Table 3.49.3 Patch testing in groups of patients: Selected patient groups (continued)

Years and Country	Test conc. & vehicle	Number of patients tested | positive (%)		Selection of patients (S); Relevance (R); Comments (C)	Ref.
1994-2004 NACDG	0.1% pet.	339	9 (2.7%)	S: patients with anogenital signs or symptoms; R: only relevant reactions were included; C: the frequency in a control group was 1.1%	31
	0.01% pet.	68	5 (7.4%)	R: 26%; C: the frequency was significantly higher than in a control group (1.3%)	
1990-2003 Belgium	0.01% pet.	92	4 (4%)	S: women suffering from vulval complaints referred for patch testing; R: relevant as corticosteroid allergy marker	40
1990-2003 IVDK		193	3 (1.6%)	S: patients with perianal dermatoses; R: not stated; C: the frequency was not higher than in a control group	35
Other patient groups					
2000-2010 Canada		100	9 (9%)	S: charts reviewed and included in the study when there was at least one positive reaction to a topical drug; R: not stated; C: the high percentages to all drugs are obviously the result of the – rather unusual – selection procedure targeting at topical drug allergy	50
2000-2005 USA	0.1% pet.	1172	36 (3.1%)	S: patients suspected of corticosteroid allergy; R: 100%	117
2002-2004 IVDK	0.1% pet.	2349	(2.5%)	S: not stated; R: not stated	46
1995-2004 IVDK	0.1% pet.	2260	57 (2.6%)	S: patients tested with a corticosteroid series; R: not stated	122
1995-9 Netherlands	0.1% pet.	79	2 (2.5%)	S: patients with chronic otitis externa; R: 18% of all reactions to ingredients of topical medications were relevant	37

IVDK: Information Network of Departments of Dermatology, Germany, Austria, Switzerland; NACDG: North American Contact Dermatitis Group (USA, Canada)

Case series

In Leuven, Belgium, in the period 2007-2011, 81 patients have been diagnosed with occupational airborne allergic contact dermatitis. In 23 of them, drugs were the offending agents, including budesonide in one case. Another patient had allergic contact dermatitis of the face from budesonide aerosol used by a family member (49).

From January 1990 to June 2008, also in Leuven, Belgium, 315 patients were diagnosed with contact allergy to/allergic contact dermatitis from corticosteroids (CSs) from routine patch testing with a baseline series including tixocortol pivalate, budesonide, hydrocortisone butyrate and prednisone caproate, patch testing with patients' own CS preparations, and testing those with proven contact allergy to a corticosteroid or strongly suspected of CS allergy later with a series of 66 CSs, including two sex hormones (progesterone and testosterone). 71% of the patients had relevant reactions, but these were not specified. In this group of 315 CS allergic patients, 191 had positive patch tests to budesonide 0.01% pet./0.002% alc. (116). It is unknown how many of these reactions were caused by the use of a pharmaceutical product containing budesonide and how many were cross-reactions to other corticosteroids.

Again in Leuven, Belgium, between 1990 and 2008, 15 individuals (14 women and 1 man) out of 315 corticosteroid-allergic patients (5%) who were not themselves treated by aerosols, but who took care and/or lived together with patients using them regularly (son, daughter, father, mother, grand-daughter, husband), appeared to have been sensitized by airborne exposure to budesonide (sensitization 'by proxy', connubial contact dermatitis, consort contact dermatitis). Nine of them presented clinical signs of airborne contact dermatitis (mostly face and eyelids); the other 6 were sensitized (as shown by positive patch tests to budesonide), but had dermatitis unrelated to budesonide exposure. Two patients had occupational sensitization, a woman working with children at a day care center and a female nurse, who took care of patients and her father using budesonide. Avoidance of budesonide resulted in improvement or clearing of dermatitis. It was concluded that air exposure to inhalation corticosteroids used 'by proxy' and to budesonide, in particular, needs to be taken into account as a potential cause of primary sensitization and/or airborne allergic contact dermatitis, sometimes also in an occupational context (60).

Yet again in Leuven, Belgium, between 1990 and 2008, of 315 patients with contact allergy to corticosteroids, 11 subjects (3.5%) presented with allergic manifestations due to orally (n=5), nasally (n=4), or by both routes (n=2) inhaled budesonide. Nine of these patients presented with facial eczema most pronounced around the nose, mouth and eyes. One patient also presented with stomatitis and another with conjunctivitis. Three patients developed a generalized eczematous or maculo-papular eruption from absorption of budesonide through the mucosae. All patients tested positively to budesonide 0.01% pet., most of them with concomitant or cross-reactions to many other CSs (121).

Two patients out of a group of 30 with asthma, allergic rhinitis, or other conditions treated with inhaled or intranasal corticosteroids for at least 3 months were patch tested and reacted to budesonide. One used a budesonide-containing nasal spray. The other reacted to budesonide and a nasal spray containing budesonide, which she did not use at the moment and probably represented a cross-reaction (126).

From Japan, 5 cases of allergic contact dermatitis from an ointment or cream containing 0.05% budesonide were reported, 3 women and 2 men, ages ranging from 19-23 years. The pharmaceuticals had been prescribed for various forms of eczema, which worsened from budesonide after 3 applications to one month after initiation of treatment.Patch tests were positive in all individuals to the cream and ointment and to budesonide 0.05%, 0.005% and 0.0005% pet. at D2 and D4 (93).

Another case series of 5 similar patients with allergic contact dermatitis from the same brand of 0.05% budesonide-containing ointment and cream was reported from Japan. One had been treated for psoriasis, the others for eczema. In three, patch tests were positive to the cream, the ointment and budesonide 0.05% pet., the fourth reacted to budesonide (cream and ointment not tested) and in the fifth, who had suffered worsening of chronic leg dermatitis twice from budesonide cream, no patch tests were performed (94-96). The author also reviewed the Japanese literature and found 34 more reported cases of allergic contact dermatitis to budesonide in 15 articles published in the period 1991-1992 (94).

In the period 1982-1994, in a center in Tokyo, 11 patients were diagnosed with allergic contact dermatitis from topical corticosteroids, including five cases caused by budesonide (134). In reviewing the Japanese literature up to 1994, 43 patients with allergic contact dermatitis from topical corticosteroid were identified, including 13 caused by budesonide (134).

Case reports

A 51-year-old woman was treated with an inhalant spray containing budesonide for asthma. After 10 days of treatment, her lips and face became swollen overnight and she reported breathing difficulties and a hoarse voice. She stopped using budesonide and was treated with oral prednisolone. Over the following days, she developed urticaria on most of the body. One year later, the patient was treated with a different brand of inhalant spray containing budesonide and now developed within 24 hours swollen lips but no urticaria. Open tests and prick tests with different formulations of budesonide were negative after 20 minutes, but the prick tests became red and itchy after 8 hours. Patch tests were positive to budesonide 0.1% on D2, D5 and D7 (55). Although the authors were cautious on this aspect, the urticaria may well have been an expression of systemic contact dermatitis.

A 32-year-old man presented with a 3-month history of recurrent vesiculobullous lesions on the dorsum of his right hand, sometimes extending to the forearm and arm. Patch tests were positive to budesonide 0.01% pet. and various other corticosteroids. Relevance was not found at that moment. Some weeks later, the patient returned to the clinic with another outbreak and suggested a relationship with contact with budesonide spray during asthma treatment for his 2-year-old daughter. The patient began to avoid exposure to the budesonide-containing aerosol, and no more episodes of dermatitis have occurred (56).

A 63-year-old woman experienced 10 days after the initiation of inhalation treatment with budesonide a symmetrical and itching eczematous eruption, beginning on her décolleté chest and spreading over the trunk and the proximal extremities over 1-2 days. The eruption cleared completely with scaling 2 weeks after discontinuation of the drug. Budesonide was reintroduced after 6 weeks, owing to progression of the asthmatic symptoms, and after 1 day of use a similar, more severe eruption occurred. Patch testing was positive to budesonide 0.1% pet. (58). This was a case of systemic allergic contact dermatitis from absorption of budesonide through the pulmonary airways.

In a 38-year-old man, intradermal tests with budesonide and various other corticosteroids resulted in generalized erythema (erythroderma) after 24 hours. It was difficult to interpret the intradermal injections because of the confluent erythema. However, the budesonide injection site showed increased erythema and induration compared with the surrounds. Later, a patch test with budesonide cream was positive (budesonide itself was not patch tested) (115).

Short summaries of other case reports of allergic contact dermatitis from budesonide are shown in table 3.49.4 (nasal spray), table 3.49.5 (inhalation preparations), table 3.49.6 (cream and ointments) and table 3.49.7 (oral spray, occupational contact, contact 'by proxy'). In a patient with erythema multiforme-like allergic contact dermatitis from triamcinolone acetonide, budesonide may have played a causal or contributing role (72). Three cases of allergic contact dermatitis to budesonide, data of which are not available to the author, were reported from Portugal (82). Inhalation of budesonide for asthma in one patient resulted in generalized urticaria. It was uncertain whether the reaction was due to type-I, type IV or a combined-type hypersensitivity to budesonide (127). A patient with eczema unresponsive to budesonide cream had a positive patch test to the cream but negative to its ingredients including budesonide on 2 occasions. The authors diagnosed compound allergy (the formation of a new allergenic chemical in the cream) (114). However, there are more likely explanations, including problems with the concentration, vehicle and/or times of reading used in patch testing of budesonide.

Table 3.49.4 Short summaries of case reports of allergic contact dermatitis from budesonide in nasal spray

Year and country	Sex	Age	Positive patch tests	Clinical data and comments	Ref.
2010 Spain	M	40	budesonide	generalized exanthema 5 days after starting budesonide nasal spray; positive lymphocyte proliferation test	139
2010 Greece	F	37	BUD TRUE test and sol. 0.25% in pet.	swelling of the lips and nasal itching after 3 days BUD spray; dysphagia, edema of tongue and face after 2nd treatment	59
	F	15	as above	nasal congestion, pruritic papular erythema on the nose and cheeks, labial and facial edema after BUD nasal spray	
2007 Sweden	F	30	BUD nasal spray 'as is'; BUD 0.01% pet.	pain in the nose, papules around the nose	123
2003 Portugal	M	9	BUD nasal spray 'as is'; BUD 0.1% pet.	erythema, edema, scaling around the nose, pruritus after 3 weeks' treatment with budesonide nasal spray	66
2001 Australia	F	78	BUD 0.1% pet.	after 3 days' use of BUD nasal spray, itchy facial rash Rapidly progressing into generalized eczematous eruption	142
1997 Sweden	F	38	BUD 0.02%, 0.002% and 0.0002% alc.; neg. to BUD 0.1% pet.	nasal bleeding and crust formation leading to septum perforation; misuse of both corticosteroids and vasoconstrictive medications	73
1996 Switzerland	F	32	BUD nasal spray 'as is'; BUD 1% pet.	sneezing and pruritic burning vesicular eczema around the eyes, on the sides of the nose and at both wrists one week after initiating treatment with BUD nasal spray	129
1995 Spain	F	41	BUD 1% pet.	after 2 days' BUD spray for perennial rhinitis worsening, nasal burning and itching and erythematous edematous plaques with vesiculation on the upper lip, perinasal area and cheeks; after BUD oral inhaler: pharyngeal dryness and pruritus with intense erythema of the oral mucosa and orodynophagia; erythematous vesicular and papular eruption on the cheeks and peribuccal area	132
1994 Spain	F	43	BUD nasal spray 'as is'; BUD 1% and 0.1% in pet., alc./DMSO and alc./water	sore throat and worsening rhinitis after 10 days' BUD nasal spray; 3 days later pruritic erythematous papular eruption on the neck and chest; re-introduction: identical symptoms	74
1993 Spain	F	44	BUD 1% pet.	erythema and edema of the face and neck after 7 months' BUD nasal spray; erythema, edema and pruritus of the face, neck and chest 2 days after patch testing	78
	F	37	BUD 1% pet.	vesicular lesions in the lip commissures, eyelids, nasal edema after 1 month' treatment with BUD nasal spray	
1993 Finland	F	59	BUD 0.25% and 0.025% pet.	known with contact allergy to corticosteroids; after three application of BUD inhaler and nasal spray in one day, next morning erythema lips, nose, cheeks and neck and worsening of dermatitis on the arm; dubious IgE-related immediate type I mechanism also involved (?)	90
1992 Spain	F	29	BUD nasal spray 'as is'; BUD 0.25% and 1% pet.	itching in the nose and eczema with vesicles and edema around the nose after 2 months' use of BUD nasal spray	84
1992 Portugal	M	24	BUD nasal spray 'as is'; BUD 0.1% and 0.01% pet.	eczema around the nose, cheeks, upper lip and trunk after 6 months' use of BUD nasal spray	81
1991 Spain	F	46	BUD nasal spray 'as is'; BUD 1%, 0.5%, 0.25% and 0.025% pet.	dermatitis on the nose and upper lip; when BUD nasal spray was used again, perinasal dermatitis and worsening of rhinitis ensued	75
1991 Spain	F	38	BUD nasal spray 'as is'; BUD 0.2% pet.	vesicular eruption in the lip commissure, lip, eyelid and nasal edema; recurrence from renewed application	85
1990 Spain	F	24	BUD nasal spray 'as is'; BUD 0.2% and 0.05% pet.	after a month treatment with BUD nasal spray nasal dryness, congestion, burning, pruritus, erythematous and edematous plaques on the upper lip with vesicles; with continued treatment maculopapular eruption on the arms	88
1986 Sweden	F	37	BUD nasal spray 'as is'; BUD 0.25%-0.00025% alc.	after 2 months' treatment of BUD nasal spray, erythema with vesicles around the nose and on the cheeks; soreness of the nasal mucosa	89

BUD: budesonide

Table 3.49.5 Short summaries of case reports of allergic contact dermatitis from budesonide in inhalation preparations

Year and country	Sex	Age	Positive patch tests	Clinical data and comments	Ref.
2020 USA	F	47	budesonide	24 hours of starting budesonide inhalation for rhinitis and asthma, maculopapular rash developed	136
2016 Italy	M	3	BUD 0.01% pet.	erythema multiforme-like ACD of the face and arm after BUD inhalation for adenoid hypertrophy	140
2016 Spain	F	66	budesonide	contact stomatitis from inhaled budesonide	57
2008 Spain	M	49	budesonide 0.1% pet.	oropharyngeal itching, erythema, dysphonia, tongue edema, dysphagia; challenge inhalation test positive	62
2005 Spain	F	44	budesonide	shortness of breath and swelling of the upper lip 24 hours after budesonide inhalation; positive intradermal test to the BUD inhalation at D2	124
2003 Germany	F	43	BUD 1%, 0.1%, 0.01% and 0.001% pet.	sore throat, swelling of the lips and oral cavity 2 weeks after daily use of budesonide spray, starting 3-4 hours after inhalation, suggesting type-I allergy	65
2003 Finland	F	50	budesonide 0.1% pet.	after 4 BUD inhalations for asthma, eczematous, very itchy dermatitis appeared all over the trunk and thighs; her palms turned red	125
2001 Belgium	M	69	budesonide 0.1% pet.	generalized skin eruption after administration of a budesonide-containing inhalation product; prick and i.c. tests neg. after 20 minutes, infiltrated lesions after 1 day	69
2000 Switzerland	M	37	BUD 0.1% pet.	after 5 days' of BUD inhalation stomatitis, perioral eczema and swelling of the lips and eyelids; macular erythema of the trunk 2 days after patch testing	128
1995 Spain	F	41	BUD 1% pet.	within 12 hours after BUD oral inhaler: pharyngeal dryness and pruritus with intense erythema of the oral mucosa and orodynophagia; also erythematous vesicular and papular eruption on the cheeks and peribuccal area	132
1993 Finland	F	59	BUD 0.25% and 0.025% pet.	known with contact allergy to corticosteroids; after three application of BUD inhaler and nasal spray in one day, next morning erythema lips, nose, cheeks and neck and worsening of dermatitis on the arm; dubious IgE-related immediate type I mechanism also involved (?)	90
1992 Australia	M	40	budesonide	maculopapular exanthema on buttocks, trunk, arms and legs within 2 days of starting budesonide inhalation	141

BUD: budesonide

Table 3.49.6 Short summaries of case reports of allergic contact dermatitis from budesonide in creams/ointments

Year and country	Sex	Age	Positive patch tests	Clinical data and comments	Ref.
2000 Italy	M	6	BUD ointment or cream 'as is'; BUD 0.1% pet.	blanching in the pubic area surrounded by an erythematous halo ('edge effect') and massive edema of the preputium after application of budesonide 0.025%	70
1998 Italy	F	29	BUD ointment 'as is'; BUD 0.1% and 1% pet.	worsening of hand eczema	111
1996 Spain	F	35	BUD cream/ointment 'as is'; budesonide	acute vulvar and perianal edema after BUD cream or ointment	131
	M	13	BUD cream/ointment 'as is'; budesonide	erythema multiforme-like eruption in the inguinal area and on the thighs 2 days after starting BUD crem/ointment	
1996 Italy	F	19	BUD 0.1% ointment 'as is'; BUD 1% pet.	erythema multiforme-like eruption on the hands, forearms after 5 days' treatment of ACD from benzoyl peroxide on the hands and face with BUD ointment	76
	F	65	as above	generalized erythema multiforme-like eruption after a week treatment with BUD 0.1% ointment; spreading over the entire body after 4 days treatment with oral triamcinolone, to which the patch test was also positive	
1993 Italy	F	42	BUD cream, ointment and lotion 'as is'; BUD 1% pet.	diffuse eczema after 4 days' treatment of psoriasis with BUD 0.025% cream	79
1992 Japan	F	20	BUD ointment and cream 'as is'; BUD 1%, 0.5%, 0.05% and 0.025% pet.	worsening of atopic dermatitis on the arms after use of BUD ointment, which she had previously used without problems	80

Table 3.49.6 Short summaries of case reports of allergic contact dermatitis from budesonide in creams/ointments (continued)

Year and country	Sex	Age	Positive patch tests	Clinical data and comments	Ref.
1991 Japan	M	19	BUD ointment and cream 'as is'; BUD 0.05%, 0.02%, 0.005% and 0.002% pet.	worsening of eczema of the elbow after 3 applications of budesonide 0.05% ointment	83
1991 Japan	M	21	BUD ointment and cream 'as is'; BUD 0.1%, 0.05% and 0.025% pet.	itchy erythema and edema on the foot and limbs one day after application of BUD ointment for insect bites; previous use for contact dermatitis had been tolerated well	86
1991 Italy	F	32	BUD cream 'as is'; BUD 1% pet.	worsening of hand eczema from BUD cream	87
	F	26	as above	worsening of dermatitis on the arms	
1980 Belgium	F	39	budesonide 1% pet.	edematous and erythematous plaques on the inner aspect of the thighs, knees and elbows from BUD 0.025% ointment for the treatment of psoriasis	92

ACD: allergic contact dermatitis; BUD: budesonide

Table 3.49.7 Short summaries of case reports of allergic contact dermatitis from budesonide in oral spray, from occupational contact or 'by proxy' contact

Year and country	Sex	Age	Positive patch tests	Clinical data and comments	Ref.
Oral spray					
1995 Spain	F	13	BUD oral spray 'as is'; BUD 1% pet.	burning face with erythema and papules on the cheeks, nose, lips, forehead and behind the ears with dry cough after 5 days of budesonide oral spray; previously used budesonide nasal spray without problems	77
Occupational/by proxy					
2019 Belgium	F	?	budesonide	nurse; eczema of the face and around the eyes	135
	M	?	budesonide	nurse; eczema of the hands	
	F	?	budesonide	eczema of the face	
2008 Italy	F	44	budesonide 0.1% pet.	airborne allergic contact dermatitis from occupational exposure as a baby sitter to budesonide inhaler used by the child she was taking care of	61
2006 Sweden	F	48	BUD 0.01% pet.; down to 0.000000001% alc. pos.	urticaria-like lesions in the face, neck and on the arms, only at her work as nurse assisting patients with budesonide inhalation; airborne occupational allergic contact dermatitis	64
2005 France	F	?	BUD inhaler; budesonide	mother who gave son BUD inhalers for asthma; after 4 days she developed swollen itchy lesions on the face with conjunctivitis several hours after administration to child; ACD mimicking angioedema; worsening from desonide cream; contact allergy to the cream demonstrated; allergic contact dermatitis 'by proxy'	91
1999 Ireland	F	40	budesonide 0.1% pet. BUD inhaler in pet.	eczematous patches on the wrists, forearms and legs; contact dermatitis from budesonide inhaler and nebulized budesonide used by her asthmatic children ('by proxy')	71

ACD: allergic contact dermatitis; BUD: budesonide

Cross-reactions, pseudo-cross-reactions and co-reactions

Cross-reactions between corticosteroids are discussed in Chapter 2.8. No cross-reactivity between budesonide and desonide, despite their close chemical similarity (97). Cross-reactions to other corticosteroids in 46 patients with positive patch test reactions have been investigated in ref. 113.

Patch test sensitization

A man, who was patch tested with various corticosteroids including budesonide 0.1% pet. and an alcoholic corticosteroid mixture containing 0.05% budesonide, was seen at D17 because of 2 reactions on the back since 2 days, one of which was thought to be at the site of the budesonide patch test. In 2 more test sessions, the patient was patch tested with a dilution series of budesonide from 1% alc. – 0.000001% alc. and he had positive reactions at D2, D4 and D8 down to the 0.001% concentration. It was concluded that the patient had been sensitized by the patch tests to budesonide (and to p-phenylenediamine). The authors mentioned that a colleague had observed late positive reactions to budesonide in 2 patients, during a systematic late patch test reading (53,54).

Late reactions occurring at D10-D13 need not be a sign of active sensitization (133).

Provocation tests

In a randomized, double-blind, placebo-controlled study, fifteen non-asthmatic patients who had previously had a positive patch test to budesonide from less than 1 up to 8 years before the study, were provoked with budesonide or placebo (lactose) by inhalation 6 weeks after they had been patch tested with budesonide, its *R* and *S* diastereomers, and potentially cross-reacting substances (hydrocortisone butyrate, triamcinolone acetonide). Lung function was studied by using spirometry and repeated peak expiratory flow measurements. In 4 of 7 patients who inhaled budesonide with a total dose of 900-1700 µgr, reactivation of 1,2,4, and 5 of the previously positive patch test reactions was noted within 24 hours, in contrast to 0 of 8 patients who inhaled placebo. Two patients also had extensive maculopapular exanthema, one had flare-up of previous budesonide dermatitis from a repetitive use test and the 4th developed 2 dermal lesions on the left side of the back. No adverse pulmonary responses could be detected (67,68). This study proved the existence of systemic contact dermatitis from inhaled budesonide.

To determine the clinical relevance of contact allergy to topical budesonide, seven patients allergic to both budesonide and nickel repeatedly applied 0.025% budesonide cream, betamethasone valerate 0.1% cream or the common base for both corticosteroids to areas of experimentally induced nickel dermatitis in a controlled, randomized, double-blind repetitive usage test. Nineteen controls allergic to nickel, but not budesonide, went through the same procedure. Five of seven patients hypersensitive to budesonide (71%) had to stop treatment of the dermatitis with budesonide cream because of severe deterioration at the application site, i.e. increased erythema and homogeneous infiltration, an increased number of papules and developing vesicles, within three to eight applications. In two of these individuals, the arm treated with budesonide cream became swollen up to an increase of 5 cm in circumference. In addition, distinct, demarcated erythematous infiltrated plaques without visible signs of any epidermal changes on the ipsilateral trunk, arm (both upper and lower), neck and hip developed within the same time. These lesions ranged from one to six in number, measured from 2-5 cm in diameter, and no visible connection with the contact dermatitis site on the arm was noted. The areas of dermatitis in all of the 19 controls (not allergic to budesonide) healed uneventfully. This experiment substantiated the clinical relevance of a contact allergy to budesonide (98).

LITERATURE

1   DeKoven JG, Warshaw EM, Zug KA, Maibach HI, Belsito DV, Sasseville D, et al. North American Contact Dermatitis Group patch test results: 2015-2016. Dermatitis 2018;29:297-309
2   Veverka KK, Hall MR, Yiannias JA, Drage LA, El-Azhary RA, Killian JM, et al. Trends in patch testing with the Mayo Clinic standard series, 2011-2015. Dermatitis 2018;29:310-315
3   DeKoven JG, Warshaw EM, Belsito DV, Sasseville D, Maibach HI, Taylor JS, et al. North American Contact Dermatitis Group Patch Test Results: 2013-2014. Dermatitis 2017;28:33-46
4   Uter W, Amario-Hita JC, Balato A, Ballmer-Weber B, Bauer A, Belloni Fortina A, et al. European Surveillance System on Contact Allergies (ESSCA): results with the European baseline series, 2013/14. J Eur Acad Dermatol Venereol 2017;31:1516-1525
5   Ochi H, Cheng SWN, Leow YH, Goon ATJ. Contact allergy trends in Singapore – a retrospective study of patch test data from 2009 to 2013. Contact Dermatitis 2017;76:49-50
6   Warshaw EM, Maibach HI, Taylor JS, Sasseville D, DeKoven JG, Zirwas MJ, et al. North American Contact Dermatitis Group patch test results: 2011-2012. Dermatitis 2015;26:49-59
7   Warshaw EM, Belsito DV, Taylor JS, Sasseville D, DeKoven JG, Zirwas MJ, et al. North American Contact Dermatitis Group patch test results: 2009 to 2010. Dermatitis 2013;24:50-59
8   Wentworth AB, Yiannias JA, Keeling JH, Hall MR, Camilleri MJ, Drage LA, et al. Trends in patch-test results and allergen changes in the standard series: a Mayo Clinic 5-year retrospective review (January 1, 2006, to December 31, 2010). J Am Acad Dermatol 2014;70:269-275
9   Toholka R, Wang Y-S, Tate B, Tam M, Cahill J, Palmer A, Nixon R. The first Australian Baseline Series: Recommendations for patch testing in suspected contact dermatitis. Australas J Dermatol 2015;56:107-115
10 Uter W, Aberer W, Armario-Hita JC, , Fernandez-Vozmediano JM, Ayala F, Balato A, et al. Current patch test results with the European baseline series and extensions to it from the 'European Surveillance System on Contact Allergy' network, 2007-2008. Contact Dermatitis 2012;67:9-19
11 Fransway AF, Zug KA, Belsito DV, Deleo VA, Fowler JF Jr, Maibach HI, et al. North American Contact Dermatitis Group patch test results for 2007-2008. Dermatitis 2013;24:10-21
12 Zug KA, Warshaw EM, Fowler JF Jr, Maibach HI, Belsito DL, Pratt MD, et al. Patch-test results of the North American Contact Dermatitis Group 2005-2006. Dermatitis 2009;20:149-160
13 Uter W, Rämsch C, Aberer W, Ayala F, Balato A, Beliauskiene A, et al. The European baseline series in 10 European Countries, 2005/2006 – Results of the European Surveillance System on Contact Allergies (ESSCA). Contact Dermatitis 2009;61:31-38

14 Davis MD, Scalf LA, Yiannias JA, Cheng JF, El-Azhary RA, Rohlinger AL, et al. Changing trends and allergens in the patch test standard series. Arch Dermatol 2008;144:67-72

15 ESSCA Writing Group. The European Surveillance System of Contact Allergies (ESSCA): results of patch testing the standard series, 2004. J Eur Acad Dermatol Venereol 2008;22:174-181

16 Warshaw EM, Belsito DV, DeLeo VA, Fowler JF Jr, Maibach HI, Marks JG, et al. North American Contact Dermatitis Group patch-test results, 2003-2004 study period. Dermatitis 2008;19:129-136

17 Uter W, Hegewald J, Aberer W et al. The European standard series in 9 European countries, 2002/2003 – First results of the European Surveillance System on Contact Allergies. Contact Dermatitis 2005;53:136-145

18 Pratt MD, Belsito DV, DeLeo VA, Fowler JF Jr, Fransway AF, Maibach HI, et al. North American Contact Dermatitis Group patch-test results, 2001-2002 study period. Dermatitis 2004;15:176-183

19 Lindberg M, Edman B, Fischer T, Stenberg B. Time trends in Swedish patch test data from 1992 to 2000. A multi-centre study based on age- and sex-adjusted results of the Swedish standard series. Contact Dermatitis 2007;56:205-210

20 Britton JE, Wilkinson SM, English JSC, Gawkrodger DJ, Ormerod AD, Sansom JE, et al. The British standard series of contact dermatitis allergens: validation in clinical practice and value for clinical governance. Br J Dermatol 2003;148:259-264

21 Marks JG Jr, Belsito DV, DeLeo VA, Fowler JF Jr, Fransway AF, Maibach HI, et al. North American Contact Dermatitis Group patch-test results, 1998–2000. Am J Contact Dermat 2003;14:59-62

22 Wetter DA, Davis MDP, Yiannias JA, Cheng JF, Connolly SM, el-Azhary RA, et al. Patch test results from the Mayo Contact Dermatitis Group, 1998–2000. J Am Acad Dermatol 2005;53:416-421

23 Bruynzeel DP, Diepgen TL, Andersen KE, Brandão FM, Bruze M, Frosch PJ, et al (EECDRG). Monitoring the European Standard Series in 10 centres 1996–2000. Contact Dermatitis 2005;53:146-152

24 Erfurt-Berge C, Geier J, Mahler V. The current spectrum of contact sensitization in patients with chronic leg ulcers or stasis dermatitis - new data from the Information Network of Departments of Dermatology (IVDK). Contact Dermatitis 2017;77:151-158

25 Barbaud A, Collet E, Le Coz CJ, Meaume S, Gillois P. Contact allergy in chronic leg ulcers: results of a multicentre study carried out in 423 patients and proposal for an updated series of patch tests. Contact Dermatitis 2009;60:279-287

26   Zmudzinska M, Czarnecka-Operacz M, Silny W, Kramer L. Contact allergy in patients with chronic venous leg ulcers – possible role of chronic venous insufficiency. Contact Dermatitis 2006;54:100-105

27 Tavadia S, Bianchi J, Dawe RS, McEvoy M, Wiggins E, Hamill E, et al. Allergic contact dermatitis in venous leg ulcer patients. Contact Dermatitis 2003;48:261-265

28 Zmudzinska M, Czarnecka-Operacz M, SilnyW. Contact allergy to glucocorticosteroids in patients with chronic venous leg ulcers, atopic dermatitis and contact allergy. Acta Dermatovenerol Croat 2008;16:72-78

29 Uter W, Spiewak R, Cooper SM, Wilkinson M, Sánchez Pérez J, Schnuch A, et al. Contact allergy to ingredients of topical medications: results of the European Surveillance System on Contact Allergies (ESSCA), 2009-2012. Pharmacoepidemiol Drug Saf 2016;25:1305-1312

30 Rai R, Shenoy MM, Viswanath V, Sarma N, Majid I, Dogra S. Contact sensitivity in patients with venous leg ulcer: A multi-centric Indian study. Int Wound J 2018;15:618-622

31 Warshaw EM, Furda LM, Maibach HI, Rietschel RL, Fowler JF Jr, Belsito DV, et al. Anogenital dermatitis in patients referred for patch testing: retrospective analysis of cross-sectional data from the North American Contact Dermatitis Group, 1994-2004. Arch Dermatol 2008;144:749-755

32 Agulló-Pérez AD, Hervella-Garcés M, Oscoz-Jaime S, Azcona-Rodríguez M, Larrea-García M, Yanguas-Bayona JI. Perianal dermatitis. Dermatitis 2017;28:270-275

33 Lucke TW, Fleming CJ, McHenry P, Lever R. Patch testing in vulval dermatoses: how relevant is nickel? Contact Dermatitis 1998;38:111-112

34 Bauer A. Contact sensitization in the anal and genital area. Curr Probl Dermatol 2011;40:133-141

35 Kügler K, Brinkmeier T, Frosch PJ, Uter W. Anogenital dermatoses—allergic and irritative causative factors. Analysis of IVDK data and review of the literature. J Dtsch Dermatol Ges 2005;3:979-986

36 Warshaw EM, Kimyon RS, Silverberg JI, Belsito DV, DeKoven JG, Maibach HI, et al. Evaluation of patch test findings in patients with anogenital dermatitis. JAMA Dermatol 2019;156:85-91

37 Devos SA, Mulder JJ, van der Valk PG. The relevance of positive patch test reactions in chronic otitis externa. Contact Dermatitis 2000;42:354-355

38 Diepgen TL, Ofenloch RF, Bruze M, Bertuccio P, Cazzaniga S, Coenraads P-J, et al. Prevalence of contact allergy in the general population in different European regions. Br J Dermatol 2016;174:319-329

39   Mortz CG, Bindslev-Jensen C, Andersen KE. Prevalence, incidence rates and persistence of contact allergy and allergic contact dermatitis in The Odense Adolescence Cohort Study: a 15-year follow-up. Brit J Dermatol 2013;168:318-325

40   Nardelli A, Degreef H, Goossens A. Contact allergic reactions of the vulva: a 14-year review. Dermatitis 2004;15:131-136

41   Tagka A, Stratigos A, Lambrou GI, Nicolaidou E, Katsarou A, Chatziioannou A. Prevalence of contact dermatitis in the Greek population: A retrospective observational study. Contact Dermatitis 2019;81:460-462

42   Lagrelius M, Wahlgren CF, Matura M, Kull I, Lidén C. High prevalence of contact allergy in adolescence: results from the population-based BAMSE birth cohort. Contact Dermatitis 2016;74:44-51

43   van Amerongen CCA, Ofenloch R, Dittmar D, Schuttelaar MLA. New positive patch test reactions on day 7—The additional value of the day 7 patch test reading. Contact Dermatitis 2019;81:280-287

44   Madsen JT, Andersen KE. Outcome of a second patch test reading of TRUE Tests® on D6/7. Contact Dermatitis 2013;68:94-97

45   Sukakul T, Chaweekulrat P, Limphoka P, Boonchai W. Changing trends of contact allergens in Thailand: A 12-year retrospective study. Contact Dermatitis 2019;81:124-129

46   Menezes de Padua CA, Uter W, Schnuch A. Contact allergy to topical drugs: prevalence in a clinical setting and estimation of frequency at the population level. Pharmacoepidemiol Drug Saf 2007;16:377-384

47   Menezes de Padua CA, Schnuch A, Nink K, Pfahlberg A, Uter W. Allergic contact dermatitis to topical drugs – Epidemiological risk assessment. Pharmacoepidemiol Drug Saf 2008;17:813-821

48   Gilissen L, Goossens A. Frequency and trends of contact allergy to and iatrogenic contact dermatitis caused by topical drugs over a 25-year period. Contact Dermatitis 2016;75:290-302

49   Swinnen I, Goossens A. An update on airborne contact dermatitis: 2007-2011. Contact Dermatitis 2013;68:232-238

50   Spring S, Pratt M, Chaplin A. Contact dermatitis to topical medicaments: a retrospective chart review from the Ottawa Hospital Patch Test Clinic. Dermatitis 2012;23:210-213

51   Wilkinson M, Gonçalo M, Aerts O, Badulici S, Bennike NH, Bruynzeel D, et al. The European baseline series and recommended additions: 2019. Contact Dermatitis 2019;80:1-4

52   Baeck M, Goossens A. Immediate and delayed allergic hypersensitivity to corticosteroids: practical guidelines. Contact Dermatitis 2012;66:38-45

53   Le Coz CJ, El Bakali A, Untereiner F, Grosshans E. Active sensitization to budesonide and para-phenylenediamine from patch testing. Contact Dermatitis 1998;39:153-155

54   Le Coz C, El Bakali A, Untereiner F, Grosshans E. Sensibilisation active au budésonide et à la paraphénylènediami-ne par tests épicutanés. La Lettre du GERDA 1998;15:88-92 (Article in French)

55   Opstrup MS, Garvey LH, Johansen JD, Bregnbak DK, Thyssen JP. A contact allergic reaction to budesonide mimicking immediate-type allergy. Contact Dermatitis 2017;77:62-63

56   Teixeira V, Coutinho I, Gonçalo M. Budesonide allergic contact dermatitis "by proxy"? Dermatitis 2013;24:144-146

57   Vega F, Ramos T, Las Heras P, Blanco C. Concomitant sensitization to inhaled budesonide and oral nystatin presenting as allergic contact stomatitis and systemic allergic contact dermatitis. Cutis 2016;97:24-27

58   Salava A, Alanko K, Hyry H. A case of systemic allergic dermatitis caused by inhaled budesonide: cross-reactivity in patch tests with the novel inhaled corticosteroid ciclesonide. Contact Dermatitis 2012;67:244-246

59   Pitsios C, Stefanaki EC, Helbling A. Type IV delayed-type hypersensitivity of the respiratory tract due to budesonide use: report of two cases and a literature review. Prim Care Respir J 2010;19:185-188

60   Baeck M, Goossens A. Patients with airborne sensitization/contact dermatitis from budesonide-containing aerosols 'by proxy'. Contact Dermatitis 2009;61:1-8

61   Corazza M, Baldo F, Osti F, Virgili A. Airborne allergic contact dermatitis due to budesonide from professional exposure. Contact Dermatitis 2008;59:318-319

62   García AP, Tovar V, de Barrio M, Villanueva A, Tornero P. Contact allergy to inhaled budesonide. Contact Dermatitis 2008;59:60-61

63   Isaksson M. Skin reactions to inhaled corticosteroids. Drug Saf 2001;24:369-373

64   Pontén A. Airborne occupational contact dermatitis caused by extremely low concentrations of budesonide. Contact Dermatitis 2006;55:121-124

65   Pirker C, Misić A, Frosch PJ. Angioedema and dysphagia caused by contact allergy to inhaled budesonide. Contact Dermatitis 2003;49:77-79

66   Cunha AP, Mota AV, Barros MA, Bonito-Victor A, Resende C. Corticosteroid contact allergy from a nasal spray in a child. Contact Dermatitis 2003;48:277

67 Isaksson M, Bruze M. Allergic contact dermatitis in response to budesonide reactivated by inhalation of the allergen. J Am Acad Dermatol 2002;46:880-885

68 Isaksson M, Bruze M. Allergic contact dermatitis to budesonide reactivated by inhalation of the allergen. Am J Contact Dermat 2001;12:130 (Abstract)

69 Goossens A, Huygens S, Matura M, Degreef H. Fluticasone propionate: a rare contact sensitizer. Eur J Dermatol 2001;11:29-34

70 Auricchio L, Nino M, Suppa F. Contact sensitivity to budesonide in a child. Contact Dermatitis 2000;42:359

71 O'Hagan AH, Corbett JR. Contact allergy to budesonide in a breath-actuated inhaler. Contact Dermatitis 1999;41:53

72 Valsecchi R, Reseghetti A, Leghissa P, Cologni L, Cortinovis R. Erythema-multiforme-like lesions from triamcinolone acetonide. Contact Dermatitis 1998;38:362-363

73 Isaksson M, Bruze M, Wihl JA. Contact allergy to budesonide and perforation of the nasal septum. Contact Dermatitis 1997;37:133

74 Iglesias-Cadarso A, Diaz C, Laguna JJ, Hernandez-Weigand P. Allergic contact dermatomucositis to budesonide. J Allergy Clin Immunol 1994;94(3Pt.1):559-560

75 Peris-Tortajada A, Giner A, Perez C, Hernandez D, Basomba A. Contact allergy to topical budesonide. J Allergy Clin Immunol 1991;87:597-598

76 Stingeni L, Caraffini S, Assalve D, Lapomarda V, Lisi P. Erythema-multiforme-like contact dermatitis from budesonide. Contact Dermatitis 1996;34:154-155

77 Garcia-Abujeta JL, Fernandez L, Maquiera E, Picans I, Rodriguez F, Jerez J. Contact allergy to budesonide in an oral spray. Contact Dermatitis 1995;32:253

78 Jorro G, Rochina A, Morales C, Burchés E, Peláez A. Contact allergy to topical budesonide in nasal spray. Contact Dermatitis 1993;28:254

79 Veraldi S, Fallahdar D, Riboldi A. Allergic contact dermatitis from budesonide. Contact Dermatitis 1993;28:116

80 Hayakawa R, Suzuki M, Ukei C, Ogino Y, Arisu K, Hiramoto K, Hirose O. A case of budesonide contact allergy. Contact Dermatitis 1992;26:128-129

81 Faria A, Marote J, de Freitas C. Contact allergy to budesonide in a nasal spray. Contact Dermatitis 1992;27:57

82 Goncalo S, Reis JP. Dermite de Contacto ao Budesonide. Boletim Informativo do GPEDC 1989;no.3:31-33 (Article in Portuguese, data cited in ref. 81)

83 Noda H, Nishida T, Ihda Y, Fukaya Y, Abe M, Ueda H. Contact dermatitis due to budesonide. Contact Dermatitis 1991;25:72-73. Erratum in: Contact Dermatitis 1992;26:71

84 Sastre J, Ibañez MD. Contact allergy to budesonide contained in a nasal spray. Allergy 1992;47:661-662

85 Gamboa PM, Jáuregui I, Antépara I. Contact dermatitis from budesonide in a nasal spray without cross-reactivity to amcinonide. Contact Dermatitis 1991;24:227-228

86 Hayakawa R, Matsunaga K, Suzuki M, Ogino Y, Arisu K, Arima Y, Hirose O. Allergic contact dermatitis due to budesonide. Contact Dermatitis 1991;24:136-137

87 Piraccini BM, Bardazzi F, Morelli R, Tosti A. Contact dermatitis due to budesonide. Contact Dermatitis 1991;24:54-55

88 Jerez J, Rodríguez F, Garcés M, Martín-Gil D, Jiménez I, Antón E, Duque S. Allergic contact dermatitis from budesonide. Contact Dermatitis 1990;22:231-232

89 Meding B, Dahlberg E. Contact allergy to budesonide in a nasal spray. Contact Dermatitis 1986;14:253-254

90 Lauerma AI, Kiistala R, Makinen-Kiljunen S, Haahtela T. Allergic skin reaction after inhalation of budesonide. Clin Exp Allergy 1993;23:232-233

91 Raison-Peyron N, Co Minh HB, Vidal-Mazuy A, Guilhou JJ, Guillot B. Connubial contact dermatitis to an inhaled corticosteroid. Ann Dermatol Venereol 2005;132:143-146 (Article in French)

92 Van Hecke E, Temmerman L. Contact allergy to the corticosteroid budesonide. Contact Dermatitis 1980;6:509

93 Noda H, Matsunaga K, Noda T, Abe M, Ohtani T, Shimizu Y, et al. Contact sensitivity and cross-reactivity of budesonide. Contact Dermatitis 1993;28:212-215

94 Okano M. Contact dermatitis due to budesonide: report of five cases and review of the Japanese literature. Int J Dermatol 1994;33:709-715

95 Okano M. A case of contact dermatitis due to budesonide ointment. Jpn J Dermatol 1991;101:373 (Article in Japanese, cited in ref. 94)

96 Okano M. Contact dermatitis due to budesonide. The 14th International Congress of Allergology and Clinical Immunology, Kyoto, October 1991 (Abstract in Japanese, cited in ref. 94)

97 Foti C, Cassano N, Vena GA. Evaluation of cross-reactivity between budesonide and desonide. Contact Dermatitis 2002;47:109-110

98 Isaksson M, Bruze M. Repetitive usage testing with budesonide in experimental nickel-allergic contact dermatitis in individuals hypersensitive to budesonide. Br J Dermatol 2001;145:38-44

99 Isaksson M, Bruze M, Lepoittevin J-P, Goossens A. Patch testing with serial dilutions of budesonide, its *R* and *S* diastereomers, and potentially cross-reacting substances. Am J Contact Dermat 2001;12:170-176

100 Wilkinson SM, Beck MH. Corticosteroid contact hypersensitivity: what vehicle and concentration? Contact Dermatitis 1996;34:305-308

101 Dooms-Goossens A, Pauwels M, Bourda A, Degreef H, Kinget R. Patch testing with corticosteroids in xerogel formulations. Contact Dermatitis 1992;26:206-207

102 Isaksson M, Beck MH, Wilkinson SM. Comparative testing with budesonide in petrolatum and ethanol in a standard series. Contact Dermatitis 2002;47:123-124

103 Chowdhury MM, Statham BN, Sansom JE, Foulds IS, English JS, Podmore P, et al. Patch testing for corticosteroid allergy with low and high concentrations of tixocortol pivalate and budesonide. Contact Dermatitis 2002;46:311-312

104 Isaksson M, Andersen KE, Brandão FM, Bruynzeel DP, Bruze M, Diepgen T, et al. Patch testing with budesonide in serial dilutions. A multicentre study of the EECDRG. Contact Dermatitis 2000;42:352-354

105 Wilkinson SM, Beck MH. Patch testing for corticosteroid allergy using high and low concentrations. Contact Dermatitis 2000;42:350-351

106 Bjarnason B, Flosadóttir E, Fischer T. Assessment of budesonide patch tests. Contact Dermatitis 1999;41:211-217

107 Isaksson M, Bruze M, Goossens A, Lepoittevin JP. Patch testing with budesonide in serial dilutions: the significance of dose, occlusion time and reading time. Contact Dermatitis 1999;40:24-31

108 Isaksson M, Bruze M, Matura M, Goossens A. Patch testing with low concentrations of budesonide detects contact allergy. Contact Dermatitis 1997;37:241-242

109 Wilkinson SM, Beck MH. Patch testing with budesonide. Contact Dermatitis 1998;39:52-53

110 Isaksson M, Brandão FM, Bruze M, Goossens A. Recommendation to include budesonide and tixocortol pivalate in the European standard series. Contact Dermatitis 2000;43:41-42.

111 Corazza M, Virgili A. Allergic contact dermatitis from 6alpha-methylprednisolone aceponate and budesonide. Contact Dermatitis 1998;38:356-357

112 Ferguson AD, Emerson RM, English JS. Cross-reactivity patterns to budesonide. Contact Dermatitis 2002;47:337-340

113 Wilkinson M, Hollis S, Beck M. Reactions to other corticosteroids in patients with positive patch test reactions to budesonide. J Am Acad Dermatol 1995;33:963-968

114 Corazza M, Mantovani L, Romani I, Bettoli V, Virgili A. Compound allergy to topical budesonide. Contact Dermatitis 1994;30:246-247

115 Wilkinson SM, Smith AG, English JS. Erythroderma following the intradermal injection of the corticosteroid budesonide. Contact Dermatitis 1992;27:121-122

116 Baeck M, Chemelle JA, Terreux R, Drieghe J, Goossens A. Delayed hypersensitivity to corticosteroids in a series of 315 patients: clinical data and patch test results. Contact Dermatitis 2009;61:163-175

117 Davis MD, El-Azhary RA, Farmer SA. Results of patch testing to a corticosteroid series: a retrospective review of 1188 patients during 6 years at Mayo Clinic. J Am Acad Dermatol 2007;56:921-927

118 Mercader-García P, Pastor-Nieto MA, García-Doval I, Giménez-Arnau A, González-Pérez R, Fernández-Redondo V, et al. GEIDAC. Are the Spanish baseline series markers sufficient to detect contact allergy to corticosteroids in Spain? A GEIDAC prospective study. Contact Dermatitis 2018;78:76-82

119 Pratt MD, Mufti A, Lipson J, Warshaw EM, Maibach HI, Taylor JS, et al. Patch test reactions to corticosteroids: Retrospective analysis from the North American Contact Dermatitis Group 2007-2014. Dermatitis 2017;28:58-63

120 Vind-Kezunovic D, Johansen JD, Carlsen BC. Prevalence of and factors influencing sensitization to corticosteroids in a Danish patch test population. Contact Dermatitis 2011;64:325-329

121 Baeck M, Pilette C, Drieghe J, Goossens A. Allergic contact dermatitis to inhalation corticosteroids. Eur J Dermatol 2010;20:102-108

122 Uter W, de Pádua CM, Pfahlberg A, Nink K, Schnuch A, Lessmann H. Contact allergy to topical corticosteroids – results from the IVDK and epidemiological risk assessment. J Dtsch Dermatol Ges 2009;7:34-41

123 Isaksson M. Systemic contact allergy to corticosteroids revisited. Contact Dermatitis 2007;57:386-388

124 Reduta T, Laudanska H. Contact hypersensitivity to topical corticosteroids – frequency of positive reactions in patch-tested patients with allergic contact dermatitis. Contact Dermatitis 2005;52:109-110

124 Cadinha S, Malheiro D, Rodrigues J, Castro E, Castel-Branco MG. Delayed hypersensitivity reactions to corticosteroids. Allergol Immunopathol 2005;33:329-332

125 Kilpio K, Hannuksela M. Corticosteroid allergy in asthma. Allergy 2003;58:1131-1135

126 Bennett ML, Fountain JM, McCarty MA, Sherertz EF. Contact allergy to corticosteroids in patients using inhaled or intranasal corticosteroids for allergic rhinitis or asthma. Am J Contact Dermat 2001;12:193-196

127 Fuchs T, Uter W, Sprotte U. Generalisierte Urtikaria durch Budesonid -verzögerte, IgE-vermittelte Sofortreaktion? Allergologie 1991;14:234-238 (Article in German)

128 Bircher AJ, Bigliardi P, Zaugg T, Mäkinen-Kiljunen S. Delayed generalized allergic reactions to corticosteroids. Dermatology 2000;200:349-351

129 Bircher AJ, Pelloni F, Langauer Messmer S, Müller D. Delayed hypersensitivity reactions to corticosteroids applied to mucous membranes. Br J Dermatol 1996;135:310-313

130 Hürlimann AF, Schmid-Grendelmeier P, Elsner P, Wüthrich B. Allergische Kontaktdermatitis auf das Kortikosteroid Budesonid nach intranasaler Anwendung. Otorhinolaryngol Nova 1994;4:168-170 (Article in German)

131 Rodríguez-Serna M, Silvestre JF, Quecedo E, Martínez A, Miguel FJ, Gauchía R. Corticosteroid allergy: report of 3 unusually acute cases. Contact Dermatitis 1996;35:361-362

132 Gonzalo Garijo MA, Bobadilla González P. Cutaneous-mucosal allergic contact reaction due to topical corticosteroids. Allergy 1995;50:833-836

133 Isaksson M, Bruze M. Late patch-test reactions to budesonide need not be a sign of sensitization induced by test procedure. Am J Contact Derm 2003;14:154-156

134 Oh-i T. Contact dermatitis due to topical steroids with conceivable cross reactions between topical steroid preparations. J Dermatol 1996;23:200-208

135 Gilissen L, Boeckxstaens E, Geebelen J, Goossens A. Occupational allergic contact dermatitis from systemic drugs. Contact Dermatitis 2020;82:24-30

136 Al-Shaikhly T, Rosenthal JA, Chau AS, Ayars AG, Rampur L. Systemic contact dermatitis to inhaled and intranasal corticosteroids. Ann Allergy Asthma Immunol 2020;125:103-105

137 Tam I, Schalock PC, González E, Yu J. Patch testing results from the Massachusetts General Hospital Contact Dermatitis Clinic, 2007-2016. Dermatitis 2020;31:202-208

138 Boyvat A, Kalay Yildizhan I. Patch test results of the European baseline series among 1309 patients in Turkey between 2013 and 2019. Contact Dermatitis 2020 Jul 3. doi: 10.1111/cod.13653. Online ahead of print.

139 Lopez S, Torres MJ, Antunez C, Rodríguez-Pena R, Canto G, Blanca M, MayorgaC. Specific immunological response to budesonide in a patient with delayed-type hypersensitivity reaction. J Invest Dermatol 2010; 130:895-897

140 Neri I, Loi C, Magnano M, Vincenzi C, La Placa M PA. Erythema multiforme-like eruption in a 3-year-old boy. Arch Dis Child 2016;101:652

141 Holmes P, Cowen P. Spongiotic (eczematous-type) dermatitis after inhaled budesonide. Austr NZ J Med 1992;22:511.

142 Poon E, Fewings JM. Generalized eczematous reaction to budesonide in a nasal spray with cross-reactivity to triamcinolone. Australas J Dermatol 2001;42;36-37

Chapter 3.50   BUFEXAMAC

IDENTIFICATION

Description/definition   : Bufexamac is the benzeneacetamide that conforms to the structural formula shown
                            below
Pharmacological classes   : Anti-inflammatory agents, non-steroidal
IUPAC name   : 2-(4-Butoxyphenyl)-N-hydroxyacetamide
]Other names   : Bufexamic acid; 4-butoxy-N-hydroxybenzeneacetamide
CAS registry number   : 2438-72-4
EC number   : 219-451-1
Merck Index monograph   : 2749
Patch testing   : 5.0% pet. (Chemotechnique, SmartPracticeCanada, SmartPracticeEurope)
Molecular formula   : $C_{12}H_{17}NO_3$

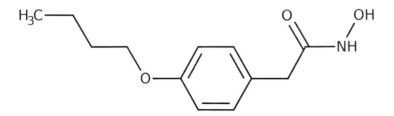

GENERAL

Bufexamac is a nonsteroidal anti-inflammatory drug (NSAID) with anti-inflammatory, analgesic, and antipyretic properties. It is typically administered topically for the treatment of subacute and chronic eczema of the skin, including atopic eczema and other inflammatory dermatoses, as well as sunburn and other minor burns, and itching. Bufexamac has also been used in suppositories in combination with local anesthetics indicated for hemorrhoids. Due to undetermined clinical efficacy and a high prevalence of contact sensitization, in April 2010, the drug was withdrawn by the European Medicines Agency (EMA) (20). One month later, Japanese pharmaceutical companies voluntarily recalled bufexamac-containing medicines (19), soon thereafter followed by some Australian producers (53). However, it was still registered there in 2019 for other-the-counter use (55). Currently (2020) it is not available in New Zealand, Japan, the European Union, the United States of America and Canada (55).

CONTACT ALLERGY

General

Bufexamac has caused many cases of sensitization, often with generalized eczematous eruptions, especially in Germany and Austria, where bufexamac-containing preparations were widely used. Sensitization not infrequently occurred after a few applications already. Many patients had (very) extensive eruptions, some with hematogenous spread from application inside and around the anus but also from percutaneous absorption of the drug elsewhere on the skin (systemic contact dermatitis). The course of allergic contact dermatitis was often protracted and refractory to treatment and many patients had to be hospitalized (1,42).

Atypical features have included erythema multiforme-like eruptions (26,27,35,41,55), generalized rash resembling the baboon-syndrome (37), allergic contact dermatitis (ACD) presenting as acute generalized exanthematous pustulosis (AGEP) (21), ACD 'by proxy' (19) and pigmented purpuric dermatosis (54). Because of the many cases of sensitization, bufexamac was included in the German standard series in June 1999 (37) and in 2015 in the Australian baseline series (55). After the drug was withdrawn in the EU in 2010 (20), the frequency of sensitization in Germany, Austria and Switzerland dropped considerably (7,58).

Problems of an allergic nature could have been anticipated: In France, in 1973, a trial of the treatment of various dermatoses with bufexamac cream and ointment was prematurely discontinued, as some 5% of the treated patients had developed symptoms highly suggestive of sensitization to bufexamac (48).

Contact allergy in the general population

With the CE-DUR approach, the incidence of sensitization to bufexamac in the German population was estimated to range from 8 to 33 cases/100,000/year in the period 2000-2004 (14). Also in Germany, for the period 1995-2004, the population-based relative incidence (RI) of contact sensitization to bufexamac (cases/100,000 defined daily doses per year) was estimated to be 76. In the group of all topical drugs, the RI ranged from 0.3 (dexamethasone sodium phosphate and pilocarpine) to 414 (benzocaine) (15).

Patch testing in consecutive patients suspected of contact dermatitis: routine testing

As bufexamac was added to the baseline series by the IVDK in 1999, most studies performing routine tested have come from Germany, Austria and Switzerland; two such studies have been performed in Australia and France (table 3.50.1). Prevalences of sensitization have ranged from 0.76% to 4.0%. The higher rates were found in the 1990s (22,24). Between 2000 and 2010, the rates were relatively stable in the IVDK studies between 1% and 1.6%. After bufexamac was prohibited in the EU in 2010, sensitization frequencies dropped to 0.8% (7,58). In Australia, 2.1% of patients tested between 2001 and 2010 were sensitized to bufexamac (2). In most studies, no data on the relevance of the positive patch tests were provided, but those authors who did, considered 42% to 'the majority' to 83% of current or past relevance (2,22,24).

Table 3.50.1 Patch testing in groups of patients: Routine testing

Years and Country	Test conc. & vehicle	Number of patients tested | positive (%)		Selection of patients (S); Relevance (R); Comments (C)	Ref.
2011-2014 IVDK	5% pet.	35,772	(0.76%)	R: not stated; the drug was withdrawn in April 2010	58
2009-12, Germany, Austria, Switzerland [a]	5% pet.	13,591	(0.8%)	R: not stated; C: range per country: 0.6% - 1.1%	7
2007-2010 IVDK	5% pet.	42,990	(1.02%)	R. not stated	58
2001-2010 Australia	5% pet.	899	19 (2.1%)	R: 42%	2
2007-2008 Germany, Austria, Switzerland [b]	5% pet.	5774	68 (1.2%) [a]	R: not stated; C: prevalences ranged from 0.9% to 1.8%	3
2001-2004 IVDK	5% pet.	31,033	474 (1.5%)	R: not stated	4
1999-2004 IVDK	5% pet.	39,392	560 (1.4%)	R: not stated	39
1995-2004 IVDK	5% pet.	38,128	(1.4%)	R: not stated	14
2002-2003 Europe [b]	5% pet.	3538	(1.6%)	R: not stated; C: only tested in Germany, Austria and Switzerland	5
<1998 Germany	5% pet.	371	12 (3.2%)	R: 10/12 (83%)	22
<1998 France	5% pet.	1008	11 (1.1%)	R: 9 had used bufexamac-containing creams before, but only 2 remembered intolerance	23
1995 Austria	5% pet.; comm prep	504	20 (4.0%)	R: the majority were of present or past relevance; patients reacted to both the commercial preparation and to bufexamac 5% pet.; 30/504 had definitely used the cream, 5 probably	24

[a] age-standardized and sex-standardized proportions; [b] study of the ESSCA (European Surveillance System on Contact Allergies); comm prep: commercial preparation; IVDK: Information Network of Departments of Dermatology, Germany, Austria, Switzerland

Patch testing groups of selected patients

Results of studies in which groups of selected patients (e.g. patients with leg ulcers/stasis dermatitis, individuals with perianal or anogenital dermatitis, patients suspected of medicament/bufexamac allergy, subjects with chronic otitis externa) were patch tested with bufexamac are shown in table 3.50.2. Prevalences of sensitization have ranged from 1.4% (partly performed in a period when bufexamac was already prohibited) (8) to 6.3%. High rates were found in patients with otitis externa (6.3% [13]; small study), patients with anogenital dermatoses (5.8% [11]; 3.5% [12]), and individuals suspected of medicament allergy (5.7% [25]; 4.2% [53]). The frequencies were mostly higher than in control groups (11,12) and   - as with most medicament allergies – patients >60 years were sensitized more frequently (6). Where relevance was mentioned, 63-100% of the positive reactions were relevant to the patients' present of past skin complaints (10,25,53).

Case series

In Leuven, Belgium, in the period 1990-2014, iatrogenic contact dermatitis was diagnosed in 2600 individuals (17% of the total patch test population). 96% of all positive patch test reactions to topical drugs and antiseptics were considered to be relevant. Bufexamac (5% pet.) was tested in 84 patients and there were 17 positive reactions to it (16). In Japan, before 1994, a group of 74 patients with severe refractory atopic dematitis involving the face were patch tested. There were 3 relevant reactions to bufexamac (4%). It is uncertain whether all patients were tested with this hapten and the test concentration and vehicle were not mentioned (17).

In a period of 1.5 year in 1991-1992, at a University clinic in Germany, 4 patients (3 women, 1 man, ages 28,29,34, and 62 years) were investigated who had developed an erythema multiforme-like reaction with urticarial papules and plaques following acute contact dermatitis from 2 different bufexamac-containing topical preparations. Bufexamac had been applied to discrete erythema (n=3) or circumscribed atopic dermatitis, and allergic contact dermatitis appeared after a few days to 3 weeks. Histologically, the lesions did not show changes typical of erythema multiforme. Patch tests were positive to bufexamac 5% in all 4, to 1% pet. in 2 and all 4 also reacted to their own bufexamac ointment/cream. The authors suggest that a systemic allergic reaction with spreading of the lesions may have occurred from percutaneous absorption of the allergen (26).

Table 3.50.2 Patch testing in groups of patients: Selected patient groups

Years and Country	Test conc. & vehicle	Number of patients tested | positive (%)		Selection of patients (S); Relevance (R); Comments (C)	Ref.
2003-2014 IVDK	5% pet.	5202	(1.4%)	S: patients with stasis dermatitis/chronic leg ulcers; R: not stated; C: percentage of reactions not significantly higher than in a control group of routine testing (1.0%)	8
2004-2013 Germany	5% pet.	52	1 (1.9%)	S: patients with lower leg dermatitis, chronic venous disease, and/or chronic leg ulcer; R: 100%	10
1993-2010 Australia	5% pet.	451	19 (4.2%)	S: patients suspected of bufexamac allergy, in 2009-2010 routinely tested; R: 13/19 (68%) present relevance	53
2005-2008 France	5% pet.	423	13 (3.1%)	S: patients with leg ulcers; R: not stated	9
2004-2008 IVDK		1373	(5.8%)	S: patients with anogenital dermatoses; R: not stated; C: the frequency was significantly higher than in a control group (1.3%)	11
1990-2003 IVDK		1168	(3.5%)	S: patients with perianal dermatoses; R: not stated; C: the frequency was significantly higher than in a control group	12
1996-1999 IVDK	5% pet.	18,899	383 (2.0%)	S: not stated; R: not stated; C: the frequency of sensitization rose from 1.8% in patients of 60 years or younger to 2.8% in the age group of 61-75 years and to 2.9% in patients over 75 years old	6
1992-1997 IVDK		48	3 (6.3%)	S: patients with chronic otitis externa; R: not stated	13
1994-1995 Austria	5% pet.	141	8 (5.7%)	S: patients suspected of medicament allergy; R: 5/8 (63%); most patients developed eczema after short-term application	25

IVDK: Information Network of Departments of Dermatology, Germany, Austria, Switzerland

In Italy, before 1993, the members of the GIRDCA Multicentre Study Group diagnosed 102 patients (49 men, 53 women), aged 16 to 66 years (mean 37 years), with (photo)dermatitis induced by systemic or topical NSAIDs. Bufexamac caused one contact allergic and zero photocontact allergic reactions (18). In 1987, a series of 13 patients allergic to bufexamac was presented from Germany, two of who had an erythema multiforme-like eruption (27).

In a university hospital in The Netherlands, 5 cases of allergic contact dermatitis to bufexamac were observed in the period 1984-1987. Indications for use were perianal eczema, vulval lichen simplex chronicus, and eczema (n=3). Sensitization caused acute dermatitis in 3 patients, subacute in one and worsening of atopic dermatitis in the 5th. They all had positive patch tests to their own cream or ointment and 4/5 (one was not tested with it because of strong reaction previously) to bufexamac 5% pet. (29).

Other case series include: 42 patients from Germany seen in the period 1983-1987 (42), 3 from Australia published in 2019 (55), three patients from France in 1991 with erythema multiforme-like contact dermatitis (35), 3 from The Netherlands and 2 from Belgium in 1973 (38,39; the first 2 publications on the subject),

Case reports

A 39-year-old woman presented with a rash on the right nipple. She was prescribed a corticosteroid-antibiotic ointment, but the lesions enlarged and increased in number on the face, trunk and left arm. Physical examination revealed light brownish patches on the right cheek, perioral and mandibular regions, and the neck. Dark reddish, edematous erythema with serous papules and vesicles was seen on the breasts, abdominal wall, and left upper arm. The patient now told that she would apply 5% bufexamac cream to the infantile eczema on the face of her nursing baby. Patch testing gave positive reactions to bufexamac cream 'as is' and bufexamac 5%, 1%, 0.2%, 0.05% and 0.01% pet. with negative reactions to the excipients of the cream (19). This was a case of allergic contact dermatitis 'by proxy' (formerly termed 'consort' or 'connubial' allergic contact dermatitis).

A 3-year old girl received topical bufexamac twice a day for mild atopic dermatitis of the cheeks. Two days later, she showed erythematous and pustular lesions on the face, rapidly spreading to the rest of the body, while developing fever. The parents denied the intake of other medications during the last weeks. Laboratory investigations showed a leucocytosis with 90% neutrophils. C-reactive protein, blood and urine tests were normal and pustule cultures were negative. The diagnosis of AGEP was suspected and confirmed by a skin biopsy. Later, patch tests were positive to the cream and to bufexamac (21).

Short summaries of other case reports of allergic contact dermatitis from bufexamac are shown in table 3.50.3. Other case reports of contact allergy to/allergic contact dermatitis from bufexamac, adequate data of which are not available to the author, can be found in refs. 43-47 and 49-52.

Table 3.50.3 Short summaries of case reports of allergic contact dermatitis from bufexamac

Year and country	Sex	Age	Positive patch tests	Clinical data and comments	Ref.
2019 Australia	F	41	BUF 5% pet.	facial edema and widespread polymorphic eruption within 2 hours applying BUF cream to abrasion on the right foot; hospitalization was required	56
2019 Australia	F	7	bufexamac	after 1 day application of BUF cream to the abdomen a rash started and developed into pruritic erythematous eruption affecting the face, trunk, limbs and genitals; edema of the eyelids, vesicles and bullae involving the nose, upper lip and palms	55
	M	68	bufexamac	pruritic erythematous and edematous eruption affecting the trunk, limbs and genitals a days after BUF cream to the arms 3 days previously; urticated plaques, some target-like, were seen at the periphery	
	F	78	bufexamac	2 days after BUF cream pruritic, erythematous and edematous eruption affecting the face, trunk and limbs	
2010 Japan	F	73	BUF 0.05%-5% pet.	erythema and edema at the application site at the trunk for herpes zoster, expanding to the surrounding area; the site affected by herpes zoster showed *no* erythema, presumably due to suppressed antigen-presenting function of the skin affected by herpes zoster	57
2009 Germany	F	56	bufexamac	4 days after application of BUF ointment for hemorrhoids generalized non-palpable purpura trunk and extremities; dermatitis with infiltration, hemorrhage and papules in anal region; histopathology: pigmented purpuric dermatosis	54
2004 Japan	F	47	BUF 0.1%, 1% and 5%	after 2 weeks' application of BUF ointment to the ankles symmetrical edematous erythema on the arms, abdomen and legs	56
	F	40	BUF 0.1%, 1% and 5%	erythema and edema after applying BUF ointment to the vulva for a week; spreading to the face, trunk, and limbs	
2003 Germany	F	48	bufexamac 5% pet.	fierce red maculopapular dermatitis anogenital area with disseminated maculae over the back several days after applying an ointment containing bufexamac to the anal region; the clinical picture was compared by the authors with the baboon syndrome	37
1999 Germany	F	48	bufexamac 5% pet.	acute perianal vesicular and bullous contact dermatitis, spreading to the trunk, face, neck and wrists after application of an ointment containing bufexamac to treat hemorrhoids; flare-up of previous dermatitis one week after patch tests	40
1998 Japan	M	52	BUF ointment 'as is'; BUF down to 0.01% pet.	after a few weeks BUF ointment around and inside the acute eczema on the buttocks, spreading to the abdomen, inguinal and genital region and neck resembling erythema multiforme; flare-up during patch testing; phototests provoked a similar EM-like eruption, which was called a 'photo Koebner phenomenon'	41
1993 Italy	F	39	BUF cream 'as is'; BUF 5% pet.	acute ACD from bufexamac cream applied to sunburn	28
1987 Italy	F	57	BUF ointment 'as is'; bufexamac	dermatitis on the face after treatment of seborrheic dermatitis with 5% BUF ointment for 2 months	30
1983 Japan	M	26	BUF cream and ointment 'as is'; BUF 5% pet.	exacerbation of seborrheic dermatitis immediately after applying BUF cream for a second course	31
	M	30	BUF ointment 'as is'; bufexamac 5% pet.	worsening and spreading of intertrigo in the axilla after using bufexamac ointment for 2 weeks	
1979 Italy	F	55	BUF cream 'as is'; bufexamac 5% pet.	allergic contact dermatitis several days after applying BUF cream to erythema induratum on the legs	32
1975 Belgium	F	26	BUF cream 'as is'; BUF 5% pet.	eczematous eruption after applying BUF cream for some months to the leg for superficial phlebitis	33
	F	47			
1974 United Kingdom	F	40	BUF cream 'as is'; BUF 5% paraffinum mole	vesicular allergic contact dermatitis of the back of the hands and wrist from 1 week BUF cream on mild eczema	34

ACD: allergic contact dermatitis; BUF: bufexamac

LITERATURE

1   Wong GN, Shen S, Nixon R. Severe allergic contact dermatitis caused by topical bufexamac requiring hospitalization. Contact Dermatitis 2019;80:391-393

2   Toholka R, Wang Y-S, Tate B, Tam M, Cahill J, Palmer A, Nixon R. The first Australian Baseline Series: Recommendations for patch testing in suspected contact dermatitis. Australas J Dermatol 2015;56:107-115

3   Uter W, Aberer W, Armario-Hita JC, , Fernandez-Vozmediano JM, Ayala F, Balato A, et al. Current patch test results with the European baseline series and extensions to it from the 'European Surveillance System on Contact Allergy' network, 2007-2008. Contact Dermatitis 2012;67:9-19

4   Worm M, Brasch J, Geier J, Uter W, Schnuch A. Epikutantestung mit der DKG-Standardreihe 2001-2004. Hautarzt 2005;56:1114-1124 (Article in German)

5   Uter W, Hegewald J, Aberer W et al. The European standard series in 9 European countries, 2002/2003 – First results of the European Surveillance System on Contact Allergies. Contact Dermatitis 2005;53:136-145

6   Uter W, Geier J, Pfahlberg A, Effendy I. The spectrum of contact allergy in elderly patients with and without lower leg dermatitis. Dermatology 2002;204:266-272

7   Uter W, Spiewak R, Cooper SM, Wilkinson M, Sánchez Pérez J, Schnuch A, et al. Contact allergy to ingredients of topical medications: results of the European Surveillance System on Contact Allergies (ESSCA), 2009-2012. Pharmacoepidemiol Drug Saf 2016;25:1305-1312

8   Erfurt-Berge C, Geier J, Mahler V. The current spectrum of contact sensitization in patients with chronic leg ulcers or stasis dermatitis - new data from the Information Network of Departments of Dermatology (IVDK). Contact Dermatitis 2017;77:151-158

9   Barbaud A, Collet E, Le Coz CJ, Meaume S, Gillois P. Contact allergy in chronic leg ulcers: results of a multicentre study carried out in 423 patients and proposal for an updated series of patch tests. Contact Dermatitis 2009;60:279-287

10   Erfurt-Berge C, Mahler V. Contact sensitization in patients with lower leg dermatitis, chronic venous insufficiency, and/or chronic leg ulcer: Assessment of the clinical relevance of contact allergens. J Investig Allergol Clin Immunol 2017;27:378-380

11   Bauer A. Contact sensitization in the anal and genital area. Curr Probl Dermatol 2011;40:133-141

12   Kügler K, Brinkmeier T, Frosch PJ, Uter W. Anogenital dermatoses—allergic and irritative causative factors. Analysis of IVDK data and review of the literature. J Dtsch Dermatol Ges 2005;3:979-986

13   Hillen U, Geier J, Goos M. Contact allergies in patients with eczema of the external ear canal. Results of the Information Network of Dermatological Clinics and the German Contact Allergy Group. Hautarzt 2000;51:239-243 (Article in German)

14   Menezes de Padua CA, Uter W, Schnuch A. Contact allergy to topical drugs: prevalence in a clinical setting and estimation of frequency at the population level. Pharmacoepidemiol Drug Saf 2007;16:377-384

15   Menezes de Padua CA, Schnuch A, Nink K, Pfahlberg A, Uter W. Allergic contact dermatitis to topical drugs – Epidemiological risk assessment. Pharmacoepidemiol Drug Saf 2008;17:813-821

16   Gilissen L, Goossens A. Frequency and trends of contact allergy to and iatrogenic contact dermatitis caused by topical drugs over a 25-year period. Contact Dermatitis 2016;75:290-302

17   Tada J, Toi Y, Arata J. Atopic dermatitis with severe facial lesions exacerbated by contact dermatitis from topical medicaments. Contact Dermatitis 1994;31:261-263

18   Pigatto PD, Mozzanica N, Bigardi AS, Legori A, Valsecchi R, Cusano F, et al. Topical NSAID allergic contact dermatitis. Italian experience. Contact Dermatitis 1993;29:39-41

19   Nakada T, Matsuzawa Y. Allergic contact dermatitis syndrome from bufexamac for nursing infant. Dermatitis 2012;23:185-186

20   Uter W, Schnuch A. EMA revokes marketing authorization for bufexamac. Contact Dermatitis 2011;64:235-236

21   Belhadjali H, Ghannouchi N, Njim L, Mohamed M, Moussa A, Bayou F, et al. Acute generalized exanthematous pustulosis induced by bufexamac in an atopic girl. Contact Dermatitis 2008;58:247-248

22   Gniazdowska B, Ruëff F, Przybilla B. Delayed contact hypersensitivity to non-steroidal anti-inflammatory drugs. Contact Dermatitis 1999;40:63-65

23   Barbaud A, Tréchot P, Aublet-Cuvelier A, Reichert-Penetrat S, Schmutz JL. Bufexamac and diclofenac: frequency of contact sensitization and absence of cross-reactions. Contact Dermatitis 1998;39:272-273

24   Kränke B, Szolar-Platzer C, Komericki P, Derhaschnig J, Aberer W. Epidemiological significance of bufexamac as a frequent and relevant contact sensitizer. Contact Dermatitis 1997;36:212-215

25   Kränke B, Derhaschnig J, Komericki P, Aberer W. Bufexamac is a frequent contact sensitizer. Contact Dermatitis 1996;34:63-64

26   Koch P, Bahmer FA. Erythema-multiforme-like, urticarial papular and plaque eruptions from bufexamac: report of 4 cases. Contact Dermatitis 1994;31:97-101

27   Frosch PJ, Raulin C. Kontaktallergie auf Bufexamac. Hautarzt 1987;38:331-334 (Article in German)

28 Corazza M, Roveggio C, Virgili A. Allergic contact dermatitis from bufexamac cream. Contact Dermatitis 1993;29:219

29 Perret CM, Happle R. Contact allergy to bufexamac. Contact Dermatitis 1989;20:307-308

30 Melino M, Bardazzi F, Manuzzi P. Contact dermatitis due to bufexamac. Contact Dermatitis 1987;16:287

31 Watanabe K, Yoshikawa K. Contact dermatitis due to bufexamac. Contact Dermatitis 1983;9:433

32 Meneghini CL, Angelini G. Contact allergy to antirheumatic drugs. Contact Dermatitis 1979;5:197-198

33 Lachapelle JM. Contact sensitivity to bufexamac. Contact Dermatitis 1975;1:261

34 Sneddon IB. Contact dermatitis to bufexamac (Parfenac). Contact Dermatitis Newsletter 1974;16:519

35 Poli F, Pouget F, Revuz J. Erythème polymorphe après application de bufexamac: 3 cas. Ann Dermatol Venereol 1991;118:901-902 (Article in French)

36 Schnuch A, Gefeller O, Uter W. Eine heimtückische und häufige Nebenwirkung: Das Ekzemtherapeutikum Bufexamac verursacht Kontaktallergien. Ergebnisse des IVDK. Dtsch Med Wochenschr 2005;130:2881-2886 (Article in German)

37 Proske S, Uter W, Schnuch A, Hartschuh W. Severe allergic contact dermatitis with generalized spread due to bufexamac presenting as the "baboon" syndrome (Schwere allergische Kontaktdermatitis mit generalisierter Streuung auf Bufexamac unter dem Bild eines 'Baboon'-Syndroms. Dtsch Med Wochenschr 2003;128:545-547 (Article in German)

38 Smeenk G. Contact allergy to bufexamac. Dermatologica 1973;147:334-337

39 Van Hecke E. Allergy to bufexamac. Arch Belg Dermatol 1973;29:301-303

40 Bauer A, Greif C, Gebhardt M, Elsner P. Schwere Epikutantestreaktion auf Bufexamac in einem Hämorrhoidal-Therapeutikum. Dtsch Med Wochenschr 1999;124:1168-1170 (Article in German)

41 Kurumaji Y. Photo Koebner phenomenon in erythema-multiforme-like eruption induced by contact dermatitis due to bufexamac. Dermatology 1998;197:183-186

42 Geier J, Fuchs T. Kontaktallergien auf Bufexamac. Med Klin (Münich) 1989;84:333-338 (Article in German)

43 Collet E, Lacroix M, Boulitrop Morvan C, Dalac S, Sallin J, Lambert D. Dermite de contact grave à la crème Parfenac. Ann Dermatol Venereol 1993;120:892-893 (Article in French)

44 Reiffers J. Contact dermatitis to bufexamac. Dermatologica 1982;164:354-356

45 Ebner H. Contact allergy to non-steroidal antiphlogistics and antirheumatics. Akt Dermatol 1986;12:197-199 (Article in German)

46 Anonymus. Kontaktallergie gegen Bufexamac. Dermatosen 1993;41:214 (Article in German)

47 Koch P, Bahmer FA. Akute atypische Kontaktdermatitis durch Bufexamac. Dermatosen 1996;44:72-76 (Article in German)

48 Achten G, Bourlond A, Haven E, Lapière CM, Piérard J, Reynaers H. Etude du bufuxamac crème et du bufexamac onguent dans le traitement de diverses dermatoses [Bufexamac cream and ointment for the treatment of various dermatoses]. Dermatologica 1973;146:1-7 (Article in French)

49 Senff H, Kunze J, Kreyes E, Hausen BM. Allergisches Kontaktekzem gegen Bufexamac und Amcinonid mit Kreuzreaktionen zu chemisch verwandten Glukokortikosteroiden. Dermatosen 1992;40:62-65 (Article in German)

50 Landrieux C, Estève E, Serpier H, Kalis B. Dermite de contact grave au bufexamac avec desquamation en grands lambeaux. Ann Dermatol Venereol 1996;123:198-199 (Article in French)

51 Yoon SH, Harada S, Nakagawa K. Contact dermatitis due to bufexamac (in Japanese). Hifubyoh Shinryoh 1986;8:273-274 (Article in Japanese)

52 Imamura K, Kase K, Urushibata O, Saito R. A case of contact dermatitis due to bufexamac. Skin Res 1993;35(suppl.16):243-246 (Article in Japanese)

53 Pan Y, Nixon R. Allergic contact dermatitis to topical preparations of bufexamac. Australas J Dermatol 2012;53:207-210

54 Waltermann K, Marsch WCh, Kreft B. Bufexamac-induced pigmented purpuric eruption. Hautarzt 2009; 60:424-427 (Article in German)

55 Harris AG, Saikal SL, Scurry J, Relic J, Nixon RL, Chee P. Severe cutaneous eruptions following the topical use of preparations containing bufexamac: Is it time to reconsider its registration in Australia? Australas J Dermatol 2019;60:53-56

56 Arikawa J, Okabe S, Kaneko T. Allergic contact dermatitis with spreading over extensive regions due to topical use of 5% bufexamac ointment. J Dermatol 2004;31:136-138

57 Fukuda H, Sato Y, Usami N, Yokouchi Y, Mukai H. Contact dermatitis caused by bufexamac sparing the eruption of herpes zoster. J Dermatol 2012;39:405-407

58 Uter W, Gefeller O, Mahler V, Geier J. Trends and current spectrum of contact allergy in Central Europe: results of the Information Network of Departments of Dermatology (IVDK) 2007-2018. Br J Dermatol 2020 Feb 27. doi: 10.1111/bjd.18946. Online ahead of print.

Chapter 3.51   BUPRENORPHINE

IDENTIFICATION

Description/definition : Buprenorphine is the synthetic opioid analgesic and thebaine derivative that conforms to the structural formula shown below

Pharmacological classes : Analgesics, opioid; narcotic antagonists; narcotics

IUPAC name : (1S,2S,6R,14R,15R,16R)-5-(Cyclopropylmethyl)-16-[(2S)-2-hydroxy-3,3-dimethylbutan-2-yl]-15-methoxy-13-oxa-5-azahexacyclo[13.2.2.1^{2,8}.0^{1,6}.0^{2,14}.0^{12,20}]icosa-8(20),9,11-trien-11-ol

CAS registry number : 52485-79-7

EC number : 257-950-6

Merck Index monograph : 2771

Patch testing : 10% water (4); 0.002% in petrolatum (8); 0.005% and 1% water (9)

Molecular formula : C$_{29}$H$_{41}$NO$_4$

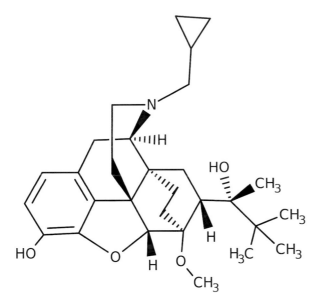

GENERAL

Buprenorphine is an orally available semisynthetic derivative of the opioid alkaloid thebaine that is a more potent (25-40 times) and longer lasting analgesic than morphine. It is indicated for the treatment of moderate to severe pain, peri-operative analgesia, and opioid dependence. In pharmaceutical products, both buprenorphine (in transdermal patches) and, in injection fluids, buprenorphine hydrochloride (CAS number 53152-21-9, EC number 258-396-8, molecular formula C$_{29}$H$_{42}$ClNO$_4$) may be employed (1).

CONTACT ALLERGY

Case series

In Leuven, Belgium, in the period 1990-2014, iatrogenic contact dermatitis was diagnosed in 2600 individuals (17% of the total patch test population). 96% of all positive patch test reactions to topical drugs and antiseptics were considered to be relevant. Buprenorphine (transdermal therapeutic system 'as is') was tested in 6 patients and there were 2 positive reactions to it. However, the active principle itself was not tested separately (2).

Three patients using transdermal buprenorphine developed signs of allergic contact dermatitis with erythema, vesicles and bullae under the patches. After healing of the dermatitis, 2 of these individuals were tested with the patch (the third was not tested because of a severe bullous reaction to the buprenorphine patch) and all with a placebo transdermal therapeutic system (= an identical system without buprenorphine). There were only positive reactions to buprenorphine, not to the placebo (3).

Three similar cases of buprenorphine allergy were reported from Finland (4). Two patients were tested with the buprenorphine part of the patch, the adhesive part and buprenorphine 10% in water and petrolatum. Both reacted to the buprenorphine patch and buprenorphine 10% water, one also to the drug 10% in petrolatum. The third patient was tested with the two parts of the patch only and reacted to the buprenorphine part. She was using oral buprenorphine 0.4 mg four to six times daily without signs of systemic contact dermatitis, either before or after patch testing (4).

Five older patients suffering from chronic pain had developed persistent, pruritic erythematous plaques at the contact sites of buprenorphine transdermal patches, and two of them also had a generalized skin eruption, considered to be systemic contact dermatitis. All five patients were tested with the buprenorphine patches and reacted positively; in four, placebo patches were also applied but were negative. The fifth patient had positive reactions to the active patch and 3 other forms of buprenorphine. Twenty-eight controls were negative to a buprenorphine patch test. The authors found a transdermal system (TDS) containing fentanyl to be a good alternative to control severe pain (7).

Case reports

A 52-year-old woman, suffering from joint pain resulting from Ehlers–Danlos syndrome, had developed allergic contact dermatitis on her shoulder after prolonged cutaneous exposure to buprenorphine patches. A patch test with buprenorphine 0.3 mg/ml in petrolatum was positive. The patient was prescribed oral treatment with tramadol retard. Within 24 hours, a severe systemic allergic dermatitis reaction was observed at the site of the patch, with homogeneous infiltration, vesicles, and bullae, as well as an orange-red discoloration, and systemic symptoms, including influenza-like symptoms, fever, and shivering. However, no 'baboon syndrome' was observed (5). A similar case of systemic contact dermatitis from tramadol in a patient sensitized to buprenorphine was presented in the same year from France (6).

An 81-year-old woman developed allergic contact dermatitis under her buprenorphine TDS. She had positive patch tests to the 35 μg/hour patch 'as is', a buprenorphine 0.3 mg vial 'as is', and buprenorphine at 0.001% and 0.002% in petrolatum. Fentanyl patches were well tolerated (8). Another woman who had developed itching and erythema in the contact zone of the buprenorphine patch had positive patch tests to the TDS and to buprenorphine 0.005% and 1% water (9).

LITERATURE

1   The data in the section 'General' may have been obtained from literature discussed in this chapter, but mostly also or exclusively from one or more of the following online sources: ChemIDPlus Advanced, PubChem, DrugBank, RxList, Drug Central, Drugs.com, and Wikipedia

2   Gilissen L, Goossens A. Frequency and trends of contact allergy to and iatrogenic contact dermatitis caused by topical drugs over a 25-year period. Contact Dermatitis 2016;75:290-302

3   Huilaja L, Riekki R, Immonen A, Tasanen K. Allergic contact dermatitis from buprenorphine and oral tolerance to other opioid derivatives in three patients. Dermatology 2014;228:130-131

4   Kyrklund C, Hyry H, Alanko K. Allergic contact dermatitis caused by transdermal buprenorphine. Contact Dermatitis 2013;69:60-61

5   Kaae J, Menné T, Thyssen JP. Systemic contact dermatitis following oral exposure to tramadol in a patient with allergic contact dermatitis caused by buprenorphine. Contact Dermatitis 2012;66:106-107

6   Schmutz JL, Trechot P. Systemic reactivation of contact eczema following tramadol administration in a patient with buprenorphine-induced eczema. Ann Dermatol Venereol 2012;139:335-336 (Article in French)

7   Van der Hulst K, Parera Amer E, Jacobs C, Dewulf V, Baeck M, Pujol Vallverdú RM, et al. Allergic contact dermatitis from transdermal buprenorphine. Contact Dermatitis 2008;59:366-369

8   Pérez-Pérez L, Cabanillas M, Loureiro M, Fernández-Redondo V, Labandeira J, Toribio J. Allergic contact dermatitis due to transdermal buprenorphine. Contact Dermatitis 2008;58:310-312

9   Callejo Melgosa AM, Martinez JC, Fuentes MJ, Martin C. Allergic contact dermatitis from buprenorphine. Allergy 2005;60:1217-1218

Chapter 3.52   BUTAMBEN

IDENTIFICATION

Description/definition   : Butamben is the benzoate ester and substituted aniline that conforms to the structural
                             formula shown below
Pharmacological classes   : Anesthetics, local
IUPAC name   : Butyl 4-aminobenzoate
Other names   : Butyl *p*-aminobenzoate; butyl PABA; Butesin
CAS registry number   : 94-25-7
EC number   : 202-317-1
Merck Index monograph   : 2785
Patch testing   : 5% pet.
Molecular formula   : $C_{11}H_{15}NO_2$

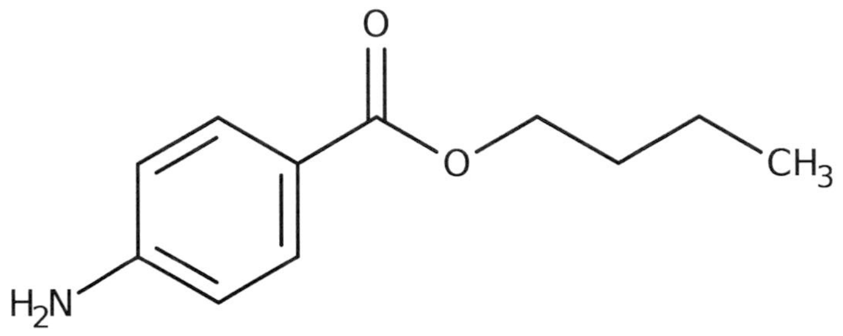

GENERAL

Butamben is a local anesthetic that has been used for surface anesthesia of the skin and mucous membranes, and for relief of pain and itching associated with some anorectal disorders. It was also used in epidural anesthesia for the treatment of chronic pain due to its long-duration effect. Although it has been stated that all butamben-containing products have been removed from the market under the belief that it is unsafe or ineffective (1), it is apparently still available in some countries (www.drugs.com).

CONTACT ALLERGY

Case series

In a period of 15 years before 1991, in a University hospital in Amsterdam, 10 patients (9 women, 1 man, average age 48 years) with contact allergy to butyl aminobenzoate (butamben) have been investigated, six before 1980 and the other 4 in the period 1988-1990.   They had all been sensitized from topical medicaments applied in the anogenital area and had positive patch tests to butyl aminobenzoate 2% pet. Six also reacted to ethyl aminobenzoate (benzocaine) 5% pet. Whether these were cross-reactions or co-reactions is not clear, as the patients might also have used topical pharmaceuticals with ethyl aminobenzoate and some pharmaceuticals contained, at that time, both butyl and ethyl aminobenzoate (2).

Case reports

A girl aged 12 with atopic dermatitis had an exacerbation after using an ointment containing butyl aminobenzoate and ethyl aminobenzoate on her face to relieve the pain and itching. Patch tests with the oil and its components yielded positive reactions to the oil (D2 ++) and butyl aminobenzoate 5% pet. (D2 ++), but there was no reaction to ethyl aminobenzoate (= benzocaine) (3).

    Case reports and case series of allergy to butamben published in early literature can be found in refs. 4 and 5.

Cross-reactions, pseudo-cross-reactions and co-reactions

Butamben is a *p*-aminobenzoic acid ester. Therefore, cross-reactions to other 'para-compounds' are possible (see Chapter 3.34 Benzocaine). No cross-reaction to tetracaine (2). Of 10 patients allergic to butyl aminobenzoate (6 of who were also sensitive to ethyl aminobenzoate = benzocaine), one reacted to the benzoate anesthetic procaine (2). A patient sensitized to butamben cross-reacted to a series of homologous PABA alkyl esters including the methyl, ethyl, propyl, isopropyl, isobutyl and amyl esters (5).

LITERATURE

1   The data in the section 'General' may have been obtained from literature discussed in this chapter, but mostly also or exclusively from one or more of the following online sources: ChemIDPlus Advanced, PubChem, DrugBank, RxList, Drug Central, Drugs.com, and Wikipedia
2   Van Ketel WG, Bruynzeel DP. A 'forgotten' topical anaesthetic sensitizer: butyl aminobenzoate. Contact Dermatitis 1991;25:131-132
3   Van Ketel WG. Allergic contact dermatitis from butyl aminobenzoate. Contact Dermatitis 1978;4:55-56
4   Lane CG, Luikart R. Dermatitis from local anesthetics, with a review of one hundred and seven cases from the literature. J Am Med Assoc 1951;146:717-720
5   Laden EL, Rubin L. Epidermal sensitization to butesin, an experimental study on the range of specificity. J Invest Dermatol 1948;11:119-125

Chapter 3.53   CAINE MIX III

The caine mix III (Chemotechnique) is a mixture of three local anesthetics used for patch testing. The mix contains 5% benzocaine, 2.5% dibucaine (= cinchocaine) hydrochloride and 2.5% tetracaine hydrochloride. The caine mix is also available as T.R.U.E. TEST ® (www.smartpracticecanada.com). It contains 630 microgram/cm^2 active ingredients, corresponding to 510 microgram/patch: 5 parts benzocaine, one part dibucaine hydrochloride and one part tetracaine hydrochloride.

In this chapter, only results of testing with the mix are presented. The individual local anesthetics are monographed in Chapter 3.34 (Benzocaine), Chapter 3.101 (Dibucaine) and Chapter 3.330 (Tetracaine). SmartPractice Canada provides the 'caine mix A', which contains the same 3 local anesthetics but in a different concentration: 5% benzocaine, 1% dibucaine hydrochloride and 1% tetracaine hydrochloride. The studies found for this chapter either used the caine mix III or the T.R.U.E. test.

CONTACT ALLERGY

General
The caine mix III was included in the European baseline series in 2019, replacing benzocaine, that had been present in this screening series for decades (17). That this change took place only very recently may be called somewhat surprising. Already in 1988 it was demonstrated in the United Kingdom that over half of the sensitizations to local anesthetics are missed with benzocaine as single screening agent in the baseline series (19). This was convincingly confirmed a decade later, again in the U.K. (18). Meanwhile, the EECDRG had already experimented with caine mixes containing benzocaine, dibucaine and tetracaine in 1990 (20). What probably stimulated discussions about replacing benzocaine with the caine mix III was a 2013 publication from Portuguese investigators (16). They had tested the caine mix III in the period 2000-2010 and demonstrated that almost 70% of allergic reactions to local anaesthetics would have been missed if benzocaine had been used as a screening allergen instead of the caine mix III (16).

Patch testing in the general population and in subgroups
Several investigations have been performed in European countries in which random samples of the population of certain age groups have been patch tested with the caine mix III (table 3.53.1, data back to 1995). In all, prevalences of sensitization have been low, ranging from 0% to 0.4% (0.6% in women in a study in 5 European countries [4]). Relevance was not addressed.

Table 3.53.1 Contact allergy in the general population and in subgroups

Year and country	Selection and number tested	Prevalence of contact allergy			Comments	Ref.
		Total	Women	Men		
General population						
2008-11 five European countries	general population, random sample, 18-74 years, n=3119	0.4%	0.6%	0.2%	TRUE test	4
<2007 Norway	general population, random sample, 18-69 years, n=531	0.4%	0.3%	0.4%	TRUE test	12
2006 Denmark	general population, random sample, 18-69 years, n=3460	0.1%	0.1%	0%	patch tests were read on day 2 only; TRUE test	6
2005 Norway	general population, random sample, 18-69 years, n=1236	0.1%	0.1%	0%	TRUE test	7
1998 Denmark	general population, random sample, 15-41 years, n=469		0.4%	0%	patch tests were read on day 2 only; TRUE test	8
Subgroups						
2011-2013 Sweden	adolescents from a birth cohort, 15.8-18.9 years, n=2285	0.13%	0.18%	0.09%	TRUE test; patch tests were read at day 2 only	13
2010 Denmark	unselected population of 8th grade schoolchildren in Denmark, 15 years later; n=442	0%	0%	0%	follow-up study; TRUE test	5
1997-1998 Denmark	twins aged 20-44 years, n=1076	0%	0%	0%	TRUE test; 449 had self-reported hand eczema	11
1995-1996 Denmark	8th grade school children, 12-16 years, n=1146	0.2%	0.2%	0.2%		9,10

In all studies, T.R.U.E. test materials have been used. It should be realized that, in a number of these investigations, the results were read at D2 only. This may have resulted in an underestimation of the true prevalence of sensitization, as positive reactions at D3 or D4 (the usual time for the second reading) with negative result at D2 commonly occur with most haptens. Moreover, a considerable number (up to 25%) of positive patch test reactions to the T.R.U.E. test caine mix may be missed when readings are not performed at day 7 (14).

Patch testing in groups of patients

Results of patch testing with the caine mix III in consecutive patients suspected of contact dermatitis (routine testing) back to 1998 are shown in table 3.53.2. In these 8 studies (with 10 results, two were performed in 2 different countries, the U.K. and Finland [1,15]), the frequencies of sensitization have ranged from 0.5% to 4.5%, with 7 of 9 scoring 1.7% or lower. High prevalences were found in Portugal (4.1%) and the USA (4.5%). Relevance rates were given in 3 investigations only and were 33% (2), 55% (18) and 60% (3).

Table 3.53.2 Patch testing in groups of patients: Routine testing

Years and Country	Test conc. & vehicle	Number of patients tested | positive (%)		Selection of patients (S); Relevance (R); Comments (C)	Ref.
2008-18 Netherlands [b]		3217	34 (1.1%)	R: not stated; C: >25% of the reactions may be missed when patch are not read at D7	14
2009-2012 Finland	10% pet.	544	3 (0.5%) [a]	R: not stated	15
2009-2012 U.K.	10% pet.	15,587	202 (1.3%) [a]	R: not stated	15
2000-2010 Portugal	10% pet.	2736	110 (4.1%)	R: specified for individual local anesthetics only	16
2007-2008 U.K	10% pet.	8909	106 (1.2%) [a]	R: not stated	1
2007-2008 Finland	10% pet.	760	12 (1.6%) [a]	R: not stated	1
1997-2007 Finland	10% pet.	7814	(1.1%)	R: not stated	21
2001-2005 USA	10% pet.	1669	(4.5%)	R: 33%	2
2000 United Kingdom	10% pet.	3063	(1.5%)	R: 60% (current and past relevance in one center); C: range of positive reactions per center 0.9% - 2.5%	3
1988 1998 U.K.	10% pet. [c]	5464	91 (1.7%)	R: 55% of reactions to the mix and other single local anesthetics were considered to be of present or past relevance	18

[a] age-standardized and sex-standardized proportions
[b] T.R.U.E. test
[c] from 1994 on, the concentrations of dibucaine HCl and tetracaine HCl were increased from 2.5% to 5%

LITERATURE

1   Uter W, Aberer W, Armario-Hita JC, , Fernandez-Vozmediano JM, Ayala F, Balato A, et al. Current patch test results with the European baseline series and extensions to it from the 'European Surveillance System on Contact Allergy' network, 2007-2008. Contact Dermatitis 2012;67:9-19

2   Davis MD, Scalf LA, Yiannias JA, Cheng JF, El-Azhary RA, Rohlinger AL, et al. Changing trends and allergens in the patch test standard series. Arch Dermatol 2008;144:67-72

3   Britton JE, Wilkinson SM, English JSC, Gawkrodger DJ, Ormerod AD, Sansom JE, et al. The British standard series of contact dermatitis allergens: validation in clinical practice and value for clinical governance. Br J Dermatol 2003;148:259-264

4   Diepgen TL, Ofenloch RF, Bruze M, Bertuccio P, Cazzaniga S, Coenraads P-J, et al. Prevalence of contact allergy in the general population in different European regions. Br J Dermatol 2016;174:319-329

5   Mortz CG, Bindslev-Jensen C, Andersen KE. Prevalence, incidence rates and persistence of contact allergy and allergic contact dermatitis in The Odense Adolescence Cohort Study: a 15-year follow-up. Brit J Dermatol 2013;168:318-325

6   Thyssen JP, Linneberg A, Menné T, Nielsen NH, Johansen JD. Contact allergy to allergens of the TRUE-test (panels 1 and 2) has decreased modestly in the general population. Br J Dermatol 2009;161:1124-1129

7   Dotterud LK, Smith-Sivertsen T. Allergic contact sensitization in the general adult population: a population-based study from Northern Norway. Contact Dermatitis 2007;56:10-15

8   Nielsen NH, Linneberg A, Menné T, Madsen F, Frølund L, Dirksen A, et al. Allergic contact sensitization in an adult Danish population: two cross-sectional surveys eight years apart (the Copenhagen Allergy Study). Acta Derm Venereol 2001;81:31-34

9   Mortz CG, Lauritsen JM, Bindslev-Jensen C, Andersen KE. Contact allergy and allergic contact dermatitis in adolescents: prevalence measures and associations. Acta Derm Venereol 2002;82:352-358

10 Mortz CG, Lauritsen JM, Bindslev-Jensen C, Andersen KE. Prevalence of atopic dermatitis, asthma, allergic rhinitis, and hand and contact dermatitis in adolescents. The Odense Adolescence Cohort Study on Atopic Diseases and Dermatitis. Br J Dermatol 2001;144:523-532

11   Bryld LE, Hindsberger C, Kyvik KO, Agner T, Menné T. Risk factors influencing the development of hand eczema in a population-based twin sample. Br J Dermatol 2003;149:1214-1220

12   Dotterud LK. The prevalence of allergic contact sensitization in a general population in Tromsø, Norway. Int J Circumpolar Health 2007;66:328-334

13   Lagrelius M, Wahlgren CF, Matura M, Kull I, Lidén C. High prevalence of contact allergy in adolescence: results from the population-based BAMSE birth cohort. Contact Dermatitis 2016;74:44-51

14   van Amerongen CCA, Ofenloch R, Dittmar D, Schuttelaar MLA. New positive patch test reactions on day 7—The additional value of the day 7 patch test reading. Contact Dermatitis 2019;81:280-287.

15   Uter W, Spiewak R, Cooper SM, Wilkinson M, Sánchez Pérez J, Schnuch A, et al. Contact allergy to ingredients of topical medications: results of the European Surveillance System on Contact Allergies (ESSCA), 2009-2012. Pharmacoepidemiol Drug Saf 2016;25:1305-1312

16   Brinca A, Cabral R, Gonçalo M. Contact allergy to local anaesthetics— value of patch testing with a caine mix in the baseline series. Contact Dermatitis 2013;68:156-162

17   Wilkinson M, Gonçalo M, Aerts O, Badulici S, Bennike NH, Bruynzeel D, et al. The European baseline series and recommended additions: 2019. Contact Dermatitis 2019;80:1-4

18   Sidhu SK, Shaw S, Wilkinson JD. A 10-year retrospective study on benzocaine allergy in the United Kingdom. Am J Cont Dermat 1999;10:57-61

19   Beck MH, Holden A. Benzocaine—an unsatisfactory indicator of topical local anaesthetic sensitization for the U.K. Br J Dermatol 1988;118:91-94

20   Wilkinson JD, Andersen KE, Lahti A, Rycroft RJ, Shaw S, White IR. Preliminary patch testing with 25% and 15% 'caine'-mixes. The EECDRG. Contact Dermatitis 1990;22:244-245

21   Jussi L, Lammintausta K. Sources of sensitization, cross-reactions, and occupational sensitization to topical anaesthetics among general dermatology patients. Contact Dermatitis 2009;60:150-154

Chapter 3.54 CALCIPOTRIOL

IDENTIFICATION

Description/definition : Calcipotriol is the vitamin D analog that conforms to the structural formula shown below
Pharmacological classes : Dermatological agents
IUPAC name : (1R,3S,5Z)-5-[(2E)-2-[(1R,3aS,7aR)-1-[(E,2R,5S)-5-Cyclopropyl-5-hydroxypent-3-en-2-yl]-7a-methyl-2,3,3a,5,6,7-hexahydro-1H-inden-4-ylidene]ethylidene]-4-methylidenecyclo-hexane-1,3-diol
Other names : Calcipotriene
CAS registry number : 112965-21-6
EC number : 601-218-4
Merck Index monograph : 2914
Patch testing : 2µg/ml calcipotriol in citrate-buffered isopropyl alcohol solution (4); a positive test should be repeated after a minimum period of 3 months to ensure its consistency over time; a repeated open application test may be indicated (4); patch testing with topical pharmaceutical preparations containing calcipotriol frequently induce false-positive, irritant patch test reactions
Molecular formula : $C_{27}H_{40}O_3$

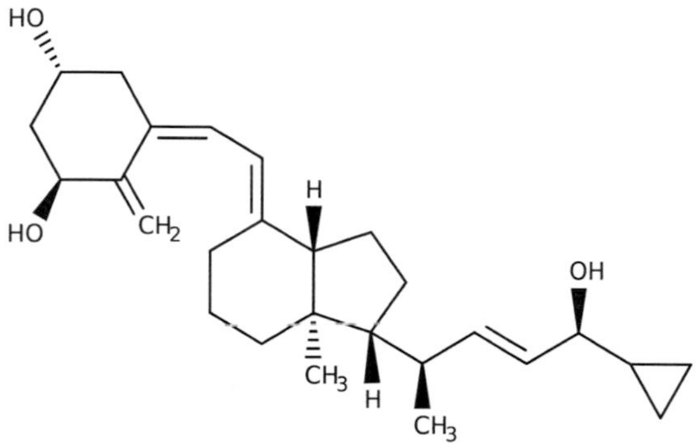

GENERAL

Calcipotriol is a synthetic vitamin D derivative usually formulated for topical dermatological use as antipsoriatic. It competes with vitamin D for vitamin D-receptors in regulating cell proliferation and differentiation. Calcipotriol thereby induces differentiation and suppresses proliferation of keratinocytes, reversing abnormal keratinocyte changes in psoriasis, and leads to normalization of epidermal growth. It is indicated as monotherapy or in a combination product with betamethasone dipropionate for the treatment of moderate plaque psoriasis. In pharmaceutical products, calcipotriol is usually employed as calcipotriol monohydrate (CAS number 147657-22-5, EC number not available, molecular formula $C_{27}H_{42}O_4$) (1).

CONTACT ALLERGY

General

Few cases of allergic contact dermatitis to calcipotriol have been reported (table 3.54.1). Preparations with calcipotriol are often irritant, both with patch testing and in a ROAT. Indeed, a positive ROAT with calcipotriol-containing preparations certainly does not always indicate sensitization (17). Calcipotriol itself should always be tested in a dilution series and positive reactions preferably be confirmed by retesting after some months.

Patch testing in groups of selected patients

In the period 1990-2014, in Leuven, Belgium, 139 patients suspected of iatrogenic contact dermatitis have been tested with calcipotriol 2µg/g alc. and there were 6 (4.3%) positive reactions. 96% of the positive patch test reactions to all topical drugs and antiseptics were considered to be relevant (2, possibly overlap with ref. 3).

Case series

In the period 2004-2016, in Leuven, Belgium, 6 patients were diagnosed with allergic contact dermatitis from calcipotriol (3, possibly overlap with the data from ref. 2). The group consisted of 4 women and two men, age range 10-59, mean age 38 years. They had been treated with calcipotriol cream (n=3), ointment (n=1), lotion (n=1) and a calcipotriol-corticosteroid preparation (n=1). Lesions suspected of allergic contact dermatitis were located at the feet (n=2), hands and feet (n=2), hands (n=1) and scalp (n=1). All had positive patch tests to calcipotriol 2µg/ml isopropyl alcohol and, when tested, with their calcipotriol-containing product. Quite curiously, some patients had been prescribed calcipotriol not for psoriasis, but for (atopic) dermatitis. It was stressed that when topical treatment with calcipotriol fails to improve or worsens the existing (psoriasis) lesions, contact allergy should be ruled out (3).

Researchers from Italy presented 3 patients with psoriasis who developed a severe eczematous eruption after the use of calcipotriol ointment. When patch tested, all revealed a strong allergic reaction to calcipotriol ointment and to a dilution series of calcipotriol in isopropyl alcohol (only one patient reacted to calcipotriol in petrolatum). Quite curiously, the allergic reactions always occurred on the legs, even if the patients had applied the ointment elsewhere. There was no cross-reactivity to tacalcitol and calcitriol, which are also vitamin D3 analogs (5).

Case reports
Short summaries of case reports of allergic contact dermatitis from calcipotriol are shown in table 3.54.1. Possibly, another case has been reported from France (16, data unknown)

Table 3.54.1 Short summaries of case reports of allergic contact dermatitis from calcipotriol

Year and country	Sex	Age	Positive patch tests	Clinical data and comments	Ref.
2001 South Korea	M	31	ointment and solution 'as is'; CAL 0.4-50µg/ml in IA	papulovesicular and bullous eruption; ROAT with ointment ++ after 3 days	6
1999 Germany	F	64	ointment 'as is'; CAL 2-10-50 µg/ml in IA	itchy papulovesicular dermatitis around psoriatic plaques; ROAT with ointment pos. at D4, with placebo ointment neg.; retesting after 6 months positive to CAL 0.4 µg/ml	7
1999 Switzerland	F	67	cream and ointment 'as is'; CAL 2µg/ml in IA	pronounced erythema	9
1997 Germany	M	54	ointment 'as is'; CAL 0.4-10 µg/ml	erythema and itching within 20 min., papulovesicles after 2-3 days; ROAT with ointment pos. D3; positive lymphocyte stimulation test	10
1996 Spain	F	54	cream 'as is'; negative to CAL 0.4 µg/ml in IA	acute dermatitis; repeat test with cream positive; ROAT 2x positive after 3 applications	11
	M	62	cream 'as is'; positive to CAL 0.4 µg/ml in IA	generalized exudative very itchy eczema on the psoriasis plaques; ROAT with the cream 2x negative	
1995 Germany	F	54	ointment 'as is'; CAL 0.5-50 µg/ml in IA	itchy dermatitis; ROAT ++ D5-14, placebo neg.; retest with ointment after 3 months ++	8
1994 The Netherlands	F	36	ointment 'as is'; CAL 2-10 µg/ml in IA (twice)	ROAT with ointment: ++ at D3; ROAT with 10 µg/ml in IA positive, with 2 µg neg.; erythema of psoriatic and surrounding skin with itching	12
1994 Norway	M	64	ointment 'as is'; CAL 2-10-50 µg/ml in IA (twice)	exudative dermatitis; ROAT: some erythema	15
1992 The Netherlands	F	50	ointment 'as is'; CAL 10 and 50 µg/ml in 67% alc. (twice)	dermatitis; cross-reaction to 25-hydroxyvitamin D (calcife-diol) and 1α-hydroxyvitamin D3 (alfacalcidol)	13
1991 Australia/U.K.	F	68	ointment 'as is'	acute dermatitis; 2 controls were negative; as the ointment is irritant, contact allergy was not proven	14

CAL: calcipotriol; IA: isopropyl alcohol (isopropanol)

Cross-reactions, pseudo-cross-reactions and co-reactions
No cross-reactivity to tacalcitol and calcitriol in 3 patients (5). Not to tacalcitol (10). A patient sensitized to calcipotriol cross-reacted to 25-hydroxyvitamin D3 (calcifediol) and 1α-hydroxyvitamin D3 (alfacalcidol) (13).

LITERATURE
1   The data in the section 'General' may have been obtained from literature discussed in this chapter, but mostly also or exclusively from one or more of the following online sources: ChemIDPlus Advanced, PubChem, DrugBank, RxList, Drug Central, Drugs.com, and Wikipedia
2   Gilissen L, Goossens A. Frequency and trends of contact allergy to and iatrogenic contact dermatitis caused by topical drugs over a 25-year period. Contact Dermatitis 2016;75:290-302 (overlap with ref. 3)
3   Gilissen L, Huygens S, Goossens A. Allergic contact dermatitis caused by calcipotriol. Contact Dermatitis 2018;78:139-142 (overlap with ref. 2)
4   Fullerton A, Benfeldt E, Petersen JR, Jensen SB, Serup J. The calcipotriol dose–irritation relationship. 48 hour occlusive testing in healthy volunteers using Finn Chambers. Br J Dermatol 1998;138:259-265
5   Foti C, Carnimeo L, Bonamonte D Conserva A, Casulli C, Angelini G. Tolerance to calcitriol and tacalcitol in three patients with allergic contact dermatitis to calcipotriol. J Drugs Dermatol 2005;4:756-759
6   Park YK, Lee JH, Chung WG. Allergic contact dermatitis from calcipotriol. Acta Derm Venereol 2002;82:71-72

7   Frosch PJ, Rustemeyer T. Contact allergy to calcipotriol does exist. Report of an unequivocal case and review of the literature. Contact Dermatitis 1999;40:66-71

8   Schmid P. Allergisches Kontaktekzem auf Calcipotriol. Akt Dermatol 1995;21:401-402 (Article in German)765

9   Krayenbühl BH, Elsner P. Allergic and irritant contact dermatitis to calcipotriol. Am J Contact Dermat 1999;10:78-80

10   Zollner TM, Ochsendorf FR, Hensel O, Thaci D, Diehl S, Kalveram CM, et al. Delayed-type reactivity to calcipotriol without cross-sensitization to tacalcitol. Contact Dermatitis 1997;37:251

11   Garcia-Bravo B, Camacho F. Two cases of contact dermatitis caused by calcipotriol cream. Am J Contact Dermat 1996;7:118-119

12   De Groot AC. Contact allergy to calcipotriol. Contact Dermatitis 1994;30:242-243

13   Bruynzeel DP, Hol CW, Nieboer C. Allergic contact dermatitis to calcipotriol. Br J Dermatol 1992;127:66

14   Yip J, Goodfield M. Contact dermatitis from MC 903, a topical vitamin D3 analogue. Contact Dermatitis 1991;25:139-140

15   Steinkjer B. Contact dermatitis from calcipotriol. Contact Dermatitis 1994;31:122

16   Giordano-Labadie F, Laplanche G, Bazex J. Contact eczema caused by calcipotriol. Ann Dermatol Venereol 1996;123:196-197 (Article in French)

17   Molin L. Contact dermatitis after calcipotriol and patch test evaluation. Acta Derm Venereol 1996;76:163-164

Chapter 3.55   CANRENONE

IDENTIFICATION

Description/definition   : Canrenone is the synthetic pregnadiene compound that conforms to the structural
                            formula shown below
Pharmacological classes   : Mineralocorticoid receptor antagonists
IUPAC name   : (8R,9S,10R,13S,14S,17R)-10,13-Dimethylspiro[2,8,9,11,12,14,15,16-octahydro-1H-
                            cyclopenta[a]phenanthrene-17,5'-oxolane]-2',3-dione
Other names   : 17-Hydroxy-3-oxo-17α-pregna-4,6-diene-21-carboxylic acid γ-lactone
CAS registry number   : 976-71-6
EC number   : 213-554-5
Merck Index monograph   : 3023
Patch testing   : 1% pet.
Molecular formula   : $C_{22}H_{28}O_3$

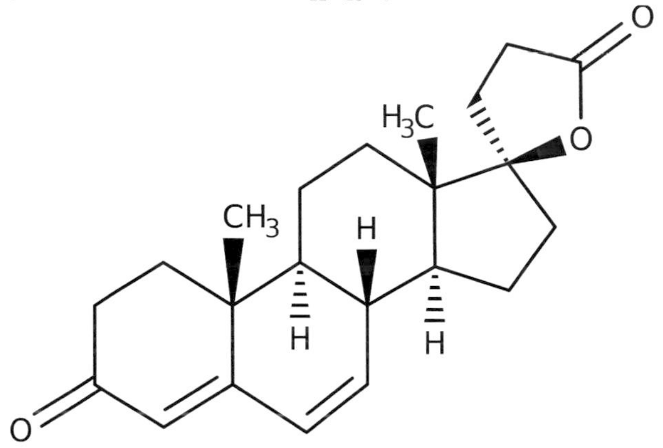

GENERAL

Canrenone is a synthetic pregnadiene compound and aldosterone antagonist with potassium-sparing diuretic activity. It is the major metabolite of spironolactone. The drug is indicated for the treatment of primary hyperaldosteronism, edematous conditions resulting from secondary hyperaldosteronism (congestive heart failure, ascitic liver cirrhosis, nephrotic syndrome) and therapy-resistant essential arterial hypertension (1). As canrenone inhibits 5α-dehydrotestosterone receptors in the pilosebaceous unit and thus has anti-androgenic properties, it has also been suggested for the treatment of acne, hirsutism and alopecia androgenetica (2). In pharmaceutical products, canrenone is usually employed as canrenoate potassium (CAS number 2181-04-6, EC number 218-554-9, molecular formula $C_{22}H_{29}KO_4$) (1).

CONTACT ALLERGY

Case report

A 38-year-old woman presented with an eczematous reaction in the face and on the scalp. She had been using a topical formulation made of minoxidil 2%, canrenone 2%, biotin 1%, and panthenol 1% in alcoholic solution since 5 days as a treatment for her androgenic alopecia. Patch tests were positive to the solution and – later -to canrenone 0.1%, 0.5%, 1%, and 2% pet. Ten controls were negative (2).

LITERATURE

1   The data in the section 'General' may have been obtained from literature discussed in this chapter, but mostly
    also or exclusively from one or more of the following online sources: ChemIDPlus Advanced, PubChem,
    DrugBank, RxList, Drug Central, Drugs.com, and Wikipedia
2   Blaya B, González M, Ratón JA, Díaz-Pérez JL. Allergic contact dermatitis produced by canrenone. Contact
    Dermatitis 2007;57:197

Chapter 3.56   CARBARSONE

IDENTIFICATION

Description/definition : Carbarsone is the arsenic containing urea compound that that conforms to the structural
formula shown below
Pharmacological classes : Anti-parasitic agents
IUPAC name : [4-(Carbamoylamino)phenyl]arsonic acid
Other names : 4-((Aminocarbonyl)amino)phenyl)arsonic acid; 4-ureidophenylarsonic acid; aminarsone
CAS registry number : 121-59-5
EC number : 204-484-6
Merck Index monograph : 3057
Patch testing : 1% and 5% pet.
Molecular formula : $C_7H_9AsN_2O_4$

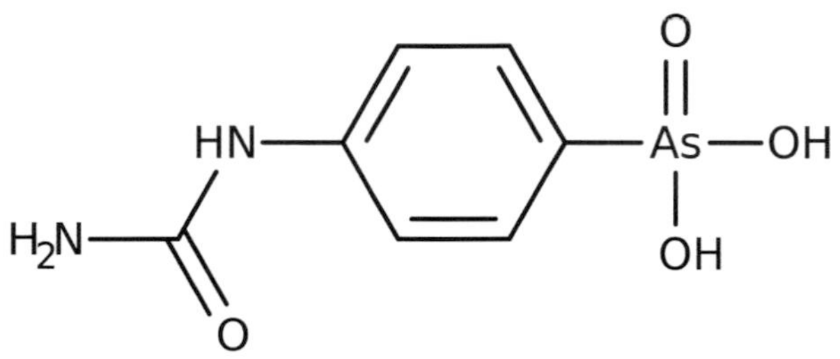

GENERAL

Carbarsone is an organoarsenic compound formerly used as an antiprotozoal drug for treatment of intestinal amebiasis and other infections. It was available for amebiasis in the United States as late as 1991. Thereafter, it remained available as a turkey feed additive for increasing weight gain and controlling histomoniasis. However, it is not used anymore in the USA since 2013, when the producer voluntarily withdrew FDA approval (1).

CONTACT ALLERGY

Case report

A 57-year-old woman had a generalized eczematous eruption which appeared after carbarsone-containing vaginal pessaries for 4 months. Patch tests showed a strong reaction to a piece of the pessary and – later – to its ingredient carbarsone 1% and 5% pet. Five controls were negative. This was probably a case of systemic contact dermatitis from resorption of carbarsone from the vaginal mucosa after sensitization (2,3).

LITERATURE

1   The data in the section 'General' may have been obtained from literature discussed in this chapter, but mostly also or exclusively from one or more of the following online sources: ChemIDPlus Advanced, PubChem, DrugBank, RxList, Drug Central, Drugs.com, and Wikipedia
2   Verburgh-van der Zwan N, van Ketel WG. Allergic drug eruption to carbason. Contact Dermatitis 1981;7:274-275
3   Verburgh-van der Zwan N, van Ketel WG. Contact allergy to an arsenic containing drug administered intravaginally. Ned Tijdschr Geneeskd 1981;125:1718-1719 (Article in Dutch)

Chapter 3.57   CARMUSTINE

IDENTIFICATION

Description/definition   : Carmustine is the nitrosurea that conforms to the structural formula shown below
Pharmacological classes : Antineoplastic agents, alkylating
IUPAC name   : 1,3-bis(2-Chloroethyl)-1-nitrosourea
CAS registry number   : 154-93-8
EC number   : 205-838-2
Merck Index monograph : 3115
Patch testing   : 0.1% water
Molecular formula   : $C_5H_9Cl_2N_3O_2$

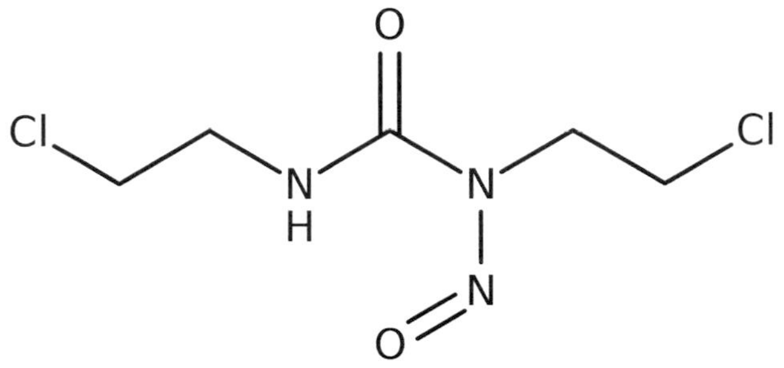

GENERAL

Carmustine is a nonspecific nitrosurea derivative and alkylating antineoplastic agent. It alkylates and cross-links DNA during all phases of the cell cycle, resulting in disruption of DNA function, cell cycle arrest, and apoptosis. This agent also carbamoylates proteins, including DNA repair enzymes, resulting in an enhanced cytotoxic effect. Carmustine is indicated for the treatment of brain tumors, multiple myeloma, Hodgkin's disease and non-Hodgkin's lymphomas. It may also be administered topically in the treatment of cutaneous T-cell lymphoma (1).

CONTACT ALLERGY

Case report

A 67-year-old woman was treated topically with a standard preparation of carmustine (2 mg/ml in 95% alc.) 2 times daily for intermittent periods of 4 weeks for stage 1 cutaneous T-cell lymphoma. Six months after an initially good response, a repeat course resulted in severe erosive inflammation at the treatment site. Patch tests were positive to carmustine 1%, 0.5% and 0.1% water (all D2 +++, D4 +++) and also to the related drug lomustine 1.0%, 0.5%, 0.1% and 0.05% water. Twenty-two controls tested to 0.1% carmustine and 15 tested to 0.1% lomustine were all negative (2).

Cross-reactions, pseudo-cross-reactions and co-reactions

One patient sensitized to carmustine also reacted to the related lomustine (2).

LITERATURE

1   The data in the section 'General' may have been obtained from literature discussed in this chapter, but mostly also or exclusively from one or more of the following online sources: ChemIDPlus Advanced, PubChem, DrugBank, RxList, Drug Central, Drugs.com, and Wikipedia
2   Thomson KF, Sheehan-Dare RA, Wilkinson SM. Allergic contact dermatitis from topical carmustine. Contact Dermatitis 2000;42:112

Chapter 3.58   CARTEOLOL

IDENTIFICATION

Description/definition : Carteolol is the quinolinone derivative that conforms to the structural formula shown
below

Pharmacological classes : β-Adrenergic antagonists; antihypertensive agents; anti-arrhythmia agents;
sympatholytics

IUPAC name : 5-[3-(*tert*-Butylamino)-2-hydroxypropoxy]-3,4-dihydro-1*H*-quinolin-2-one

Other names : 5-(3-((1,1-Dimethylethyl)amino)-2-hydroxypropoxy)-3,4-dihydro-2(1*H*)-quinolinone

CAS registry number : 51781-06-7

EC number : Not available

Merck Index monograph : 3136

Patch testing : 5% water

Molecular formula : $C_{16}H_{24}N_2O_3$

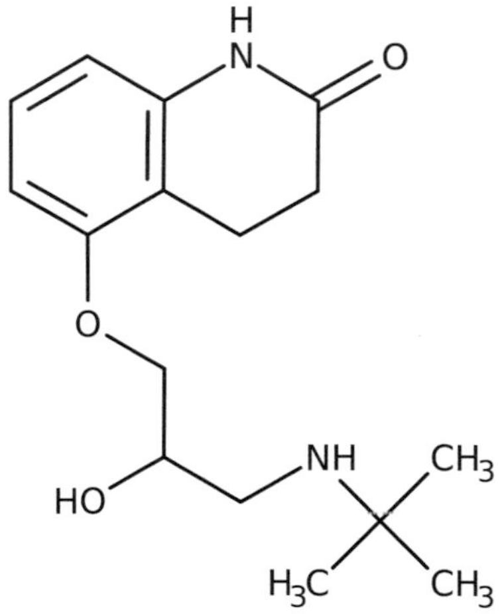

GENERAL

Carteolol is a synthetic quinolinone derivative and nonselective β-adrenoceptor blocking agent with anti-glaucoma activity. Upon topical administration to the eye, carteolol decreases aqueous humor production, thereby reducing both elevated and normal intraocular pressure. It is indicated for the treatment of intraocular hypertension and chronic open-angle glaucoma. According to some sources, carteolol may also be used as an anti-arrhythmia, anti-angina and antihypertensive agent. In pharmaceutical products (mostly eye drops), carteolol is employed as carteolol hydrochloride (CAS number 51781-21-6, EC number 257-415-7, molecular formula $C_{16}H_{25}ClN_2O_3$) (1).

CONTACT ALLERGY

General

For information on patch testing with commercial eye drops containing beta-blockers see Chapter 3.338 Timolol.

Case series

In a hospital in Slovenia, in an undefined period before 2015, 55 patients with suspected contact allergy to topical drugs used for treatment of glaucoma were retrospectively analyzed. Each patient had been tested to individual products in serial dilutions (1%-10%-50% and as is). Patch tests to vehicles, emulsifiers and preservatives(including benzalkonium chloride) were also performed. Three controls on healthy volunteers were always performed in case of a positive reaction. The causal relationship between the drug and dermatitis was always confirmed by a positive elimination test. Eight of 55 patients (5 women, 3 men) had positive patch tests to one or more products. Six patients were positive to beta-blockers (5x timolol and 1x carteolol), 2 to latanoprost, two to dorzolamide, and two to brinzolamide (8). It is highly likely that these patients have not been tested with the active drugs but only with the commercial eye drops and that the diagnosis of allergic contact dermatitis to the active ingredients was, at best, made *per exclusionem*.

In Leuven, Belgium, in the period 1990-2014, iatrogenic contact dermatitis was diagnosed in 2600 individuals (17% of the total patch test population). 96% of all positive patch test reactions to topical drugs and antiseptics were considered to be relevant. Carteolol (5% water) was tested in 26 patients and there were 2 positive reactions to it (2).

Case reports

A 60-year-old man had used carteolol and timolol-containing eye drops for the treatment of chronic open-angle glaucoma for several years, when he developed a severe symmetrical eyelid dermatitis with itching, swelling, hyperemia and conjunctival chemosis. Positive patch tests were observed to carteolol eye drops, timolol eye drops, and to their active principles (test concentrations not mentioned). There was also a reaction to levobunolol eye drops, possibly from cross-reactivity, as the patient had never used levobunolol medication before (3).

A 65-year-old man developed eczema localized on the upper and lower eyelids. Patch testing showed contact allergy to timolol and befunolol, both of which had been used in eye drops for the treatment of glaucoma. Later prescription of eye drops containing carteolol led to recurrence of the eczema, which was considered to be a cross-reaction (5).

A 61-year-old female patient had been treated with timolol and dorzolamide eye drops for the past few years for open-angle glaucoma and with various other ophthalmic preparations after she underwent bilateral iridectomy. All medications were stopped and she was started on 2% carteolol eye drops. A week later, the patient developed pruriginous bilateral erythematous, edematous lesions of the eyelids. The withdrawal of carteolol and replacement with timolol led to improvement within 10 days. Patch testing showed positive reactions to carteolol eye drops, and carteolol HCl 1% and 2% water (negative in 23 controls). Although not discussed by the authors, it seems likely. considering the short time between starting therapy and development of dermatitis of one week, that the patient was already allergic when treatment was started (6).

A 68-year-old woman presented with palpebral edema, pruritus, and periorbital and conjunctival hyperemia one day after the administration of carteolol 2% eye drops for glaucoma. The patient had previously used beta-blocker eye drops. Epicutaneous tests on the back with commercial eye drops with levobunolol 0.5%, betaxolol 0.5%, carteolol 2%, and timolol 0.5% yielded negative results. Patch tests on tape-stripped skin were also negative, as was a ROAT with carteolol 2%. Finally, a use test with carteolol 2% was performed, applying one drop to the right eye. A positive response with similar characteristics to those previously described by the patient was evident after one day. Betaxolol eye drops, containing the same preservative as the carteolol medication, were well tolerated (7).

Cross-reactions, pseudo-cross-reactions and co-reactions

Possible cross-reactivity to levobunolol (3). A woman sensitized to befunolol had co-reactivity to carteolol, which she had never used before (4). A male patient sensitized tot timolol and befunolol later had a recurrence of dermatitis from eye drops containing carteolol, which was considered to be the result of cross-sensitivity (5).

LITERATURE

1   The data in the section 'General' may have been obtained from literature discussed in this chapter, but mostly also or exclusively from one or more of the following online sources: ChemIDPlus Advanced, PubChem, DrugBank, RxList, Drug Central, Drugs.com, and Wikipedia
2   Gilissen L, Goossens A. Frequency and trends of contact allergy to and iatrogenic contact dermatitis caused by topical drugs over a 25-year period. Contact Dermatitis 2016;75:290-302
3   Quiralte J, Florido F, de San Pedro BS. Allergic contact dermatitis from carteolol and timolol in eyedrops. Contact Dermatitis 2000;42:245
4   Nino M, Suppa F, Ayala F, Balato N. Allergic contact dermatitis due to the beta-blocker befunolol in eyedrops, with cross-sensitivity to carteolol. Contact Dermatitis 2001;44:369
5   Giordano-Labadie F, Lepoittevin JP, Calix I, Bazex J. [Contact allergy to beta blockaders in eye drops: cross allergy?]. Ann Dermatol Venereol 1997;124:322-324 (Article in French)
6   Sánchez-Pérez J, Córdoba S, Bartolomé B, García-Díez A. Allergic contact dermatitis due to the beta-blocker carteolol in eyedrops. Contact Dermatitis 1999;41:298
7   Gonzalo-Garijo MA, Zambonino MA, Pérez-Calderón R, Pérez-Rangel I, Sánchez-Vega S. Allergic contact dermatitis due to carteolol, with good tolerance to betaxolol. Dermatitis 2011;22:232-233
8   Slavomir U, Kuklova Bielikova M. Allergic reactions due to antiglaucomatics. Dermatitis 2015;26(2):e12-13

Chapter 3.59 CHLORAL HYDRATE

IDENTIFICATION

Description/definition : Chloral hydrate is the synthetic monohydrate of chloral that conforms to the structural formula shown below
Pharmacological classes : Hypnotics and sedatives
IUPAC name : 2,2,2-Trichloroethane-1,1-diol
Other names : 1,1,1-Trichloro-2,2-dihydroxyethane; 1,1,1-trichloro-2,2-ethanediol
CAS registry number : 302-17-0
EC number : 206-117-5
Merck Index monograph : 3341
Patch testing : 5% water
Molecular formula : $C_2H_3Cl_3O_2$

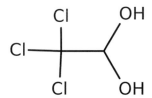

GENERAL

Chloral hydrate is a synthetic monohydrate of chloral with sedative, hypnotic, and anticonvulsive properties. Formerly it was used as a hypnotic and sedative in the treatment of insomnia, but is was effective for short time use only. It is no longer considered useful as an anti-anxiety medication. This agent has also been used as a routine sedative preoperatively to decrease anxiety and cause sedation and/or sleep with respiration depression or cough reflex. Chloral hydrate is probably not used anymore (1). Formerly, chloral hydrate was also used topically as a counter-Irritant (rubefacient) for the relief of itching and pain (2).

CONTACT ALLERGY

Case reports

A 35-year-old woman treated a dermatitis caused by her metal watch with several topical agents, including a counter-irritant ointment. The rash subsequently spread to her arms, neck and legs. Patch tests gave a positive reaction to nickel, cobalt and the counter-irritant. The ingredients of this ointment were chloral hydrate (4.9 mg/gr), camphor, iodinated oil, menthol, hyoscyamine, chloroform, zinc oxide and 'ointment base'. Patch testing to these constituents showed positive reactions to chloral hydrate 1% (++/++) and 5% (+++/+++) in water. Thirty controls were negative. The patient stopped using the ointment and wearing metal jewelry and she has had no more skin problems (2).

Contact allergy to chloral hydrate had some 50 years earlier been reported a few times (3,4,5). In 2 cases (4,5), the patients had an eruption from oral chloral hydrate. One of these patients had apparently been sensitized previously by an ointment containing 0.5% chloral hydrate, a case of systemic contact dermatitis (5). Another had become sensitized by handling chloral hydrate during his work as a pharmacist: occupational contact dermatitis (3). All patients had positive patch test reactions to 'chloral hydrate solutions' 1%-2.5%.

LITERATURE

1 The data in the section 'General' may have been obtained from literature discussed in this chapter, but mostly also or exclusively from one or more of the following online sources: ChemIDPlus Advanced, PubChem, DrugBank, RxList, Drug Central, Drugs.com, and Wikipedia
2 De Groot AC, Conemans J. Chloral hydrate. The contact allergen that fell asleep. Contact Dermatitis 1987;16:229-231
3 Abramowitz EW, Noun MH. Eczematous dermatitis due to exposure to chloral. J Allergy 1933;4:338-343
4 Flandin MM, Rabeau H, Ukrainczyk M. Dermite par intolérance au chloral. Test cutané. Bull Soc Franc Derm Syph 1936;43:1231-1234 (Article in French)
5 Baer RL, Sulzberger MB. Eczematous dermatitis due to chloral hydrate (following both oral administration and topical application). J Allergy 1938;9:519-520

Chapter 3.60   CHLORAMPHENICOL

IDENTIFICATION

Description/definition : Chloramphenicol is a broad-spectrum antibiotic that was derived from the bacterium *Streptomyces venezuelae* and is now produced synthetically; it conforms to the structural formula shown below
Pharmacological classes : Protein synthesis inhibitors; anti-bacterial agents
IUPAC name : 2,2-Dichloro-*N*-[(1*R*,2*R*)-1,3-dihydroxy-1-(4-nitrophenyl)propan-2-yl]acetamide
CAS registry number : 56-75-7
EC number : 200-287-4
Merck Index monograph : 3347
Patch testing : 5% pet. (Chemotechnique, SmartPracticeCanada, SmartPracticeEurope); 2% alc. (SmartPracticeCanada)
Molecular formula : $C_{11}H_{12}Cl_2N_2O_5$

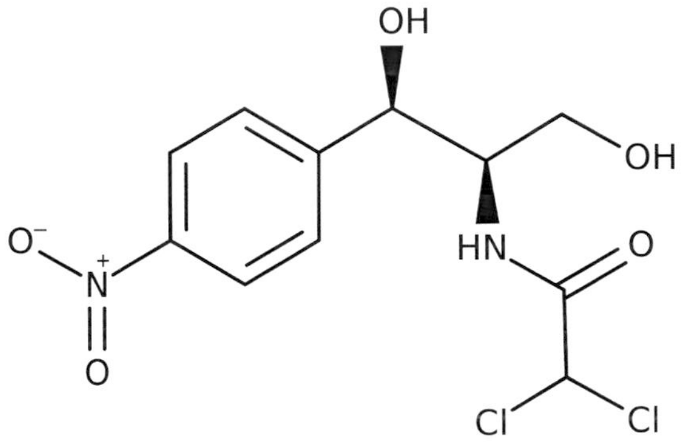

GENERAL

Chloramphenicol is a broad-spectrum antibiotic of the amphenicol class with primarily bacteriostatic activity, that was first isolated from cultures of *Streptomyces venezuelae* in 1947, but is now produced synthetically. It is effective against a wide variety of microorganisms, but due to serious adverse effects (e.g. damage to the bone marrow, including aplastic anemia) in humans, it is usually reserved for the treatment of serious and life-threatening infections. It is used for typhoid fever and for the treatment of cholera, as it destroys the vibrios and decreases the diarrhea. Chloramphenicol is mostly used in skin ointments, ear drops for the treatment of otitis externa and eye drops or ointment to treat bacterial conjunctivitis (1). Chloramphenicol was first reported as a contact sensitizer in 1951 (44).

CONTACT ALLERGY

General

Chloramphenicol sensitization is often said to be infrequent, because of limited use. Nevertheless, in one clinic in Belgium, in the period 1990-2014, 72 patients suspected of iatrogenic contact dermatitis had positive patch tests to this antibiotic (13) and in a group of 75 patients from Serbia with chronic venous leg ulcers, 9 (12%) reacted to chloramphenicol 5% pet. (5). In 1979, chloramphenicol was the 6th most frequent contact allergen in female hospitalized women in Budapest, Hungary, and the 5th most frequent in male hospitalized patients (45). Sources of sensitization are mostly ointments applied to leg ulcers or stasis dermatitis, eye drops and – to a lesser degree - ear drops. Much work on chloramphenicol allergy has been published in early German literature (46-50).

Contact allergy in the general population

In Germany, for the period 1995-2004, the population-based relative incidence (RI) of contact sensitization to chloramphenicol (cases/100,000 defined daily doses (DDDs) per year) was estimated to be 23.3. In the group of antibiotics, the RI ranged from 1.6 (oxytetracycline) to 86.2 (framycetin). For chloramphenicol used in ophthalmic drugs, the RI was 4.0 (12).

Patch testing in groups of selected patients

Studies in which groups of selected patients (e.g. patients with leg ulcers/stasis dermatitis, individuals with otitis externa, patients suspected of contact allergy to [ophthalmological] drugs) have been patch tested with chloramphenicol back to 1985 are shown in table 3.60.1. Prevalences of sensitization have ranged from 0.4% to 12%. Generally speaking, higher frequencies were observed in patients with leg ulcers/stasis dermatitis (12% [5], 4.1% [7], 3% [3]), but a low rate was observed in the United Kingdom (0.5% [2]). Such differences may relate to the selection of patients and the frequencies of usage of chloramphenicol pharmaceuticals in the investigated populations. Patients with chronic otitis externa or media had 3% resp. 4.2% positive reactions (7,8). A high rate of 9.8% was observed in a group of patients suspected of iatrogenic contact dermatitis, in Belgium, in a University clinic specialized in cosmetic and drug allergy (13). Relevance rates of the positive patch tests, where mentioned, were generally very high.

Table 3.60.1 Patch testing in groups of patients: Selected patient groups

Years and Country	Test conc. & vehicle	Number of patients tested | positive (%)		Selection of patients (S); Relevance (R); Comments (C)	Ref.
<2017 India	5% pet.	172	7 (4.1%)	S: patients with venous leg ulcers of over 6 weeks' duration; R: 'the majority of reactions were relevant'	4
1990-2014 Belgium	5% pet.	737	72 (9.8%)	S: patients suspected of iatrogenic contact dermatitis and tested with a pharmaceutical series and their own products; R: 96% of the positive patch test reactions to all topical drugs and antiseptics were considered to be relevant	13
2001-2010 IVDK	5% pet.	1125	9 (0.8%)	S: patients with periorbital dermatitis tested with an ophthalmic tray; R: not stated	15
1995-2004 IVDK	5% pet.	2920	(0.9%)	S: patients patch tested for suspected contact allergy to ophthalmological drugs; R: not stated; C: the estimated number of sensitizations per 100,000 prescriptions was 4.0	10
1995-2004 IVDK	5% pet.	5703	(3.0%)	S: not stated; R: not stated	11
2000-2002 Serbia	5% pet.	75	9 (12%)	S: patients with chronic venous leg ulcers; R: not stated; C: 0% reactions in a (small) control group (difference significant)	5
1997-2001 U.K.		200	1 (0.5%)	S: patients with venous or mixed venous/arterial leg ulcers; R: all reactions to topical drugs were considered to be of probable, past or current relevance	2
<1999 Croatia	5% pet.	100	3 (3%)	S: patients with leg ulcers; R: not stated	3
1994-1998 U.K.	5% pet.	232	1 (0.4%)	S: patients with periorbital contact dermatitis; R: 100%; C: the frequency in a control group was 0%	6
1993-4 Netherlands	20% pet.	34	1 (3%)	S: patients with chronic otitis externa or media; R: not stated	7
<1985 Finland		142	6 (4.2%)	S: patients with chronic otitis externa; C: details unknown	8

IVDK: Information Network of Departments of Dermatology (Germany, Austria, Switzerland)

Case series

In Leuven, Belgium, between 1990 and 2017, 16,065 patients were investigated for contact allergy and 118 (0.7%) showed positive patch test reactions to topical ophthalmic medications and/or to their ingredients. Eighty-four individuals (71%) reacted to an active principle. Chloramphenicol was tested in 779 patients and was the allergen in eye medications in 9. There were also 44 reactions to chloramphenicol in other types of medications (9).

During the 2-year period April 1992 to March 1994, in a hospital in the United Kingdom, 63 (7%) of 865 patients patch tested had dermatitis of the face and 25 (3%) eyelid dermatitis. In the latter group, 2 patients had relevant reactions to chloramphenicol versus zero in the group with facial dermatitis. The culprit products were not mentioned, but must likely have been eye drops or ointments (38).

In Pamplona, Spain, in one year's time (1992), 13 patients were diagnosed with contact allergy to ophthalmic medications, nearly 3% of all patients investigated for suspected contact dermatitis. There was one reaction to chloramphenicol 5% pet. in an individual who had previously used chloramphenicol eye drops (16).

In Bologna, Italy, before 1989, 136 patients (35 men, 101 women, range of age 14-84 years, average 36 years) suspected of contact conjunctivitis from eye drops (including contact lens solutions), were patch tested. There was one reaction to chloramphenicol eye drops and chloramphenicol itself (17).

In Rotterdam, The Netherlands, in the period January to August 1984, 8 patients were seen with a periocular (n=6) or periauricular (n=2) dermatitis of possible allergic origin. All reacted to chloramphenicol powder (100%; 20 controls were negative), but only 3 of them to chloramphenicol 1% pet. Two had used and reacted to an eye ointment containing chloramphenicol and 4 reacted to eye drops (which two had used in the ear) (31).

In the period 1971-1976 inclusive, at St. John's Hospital, London, 14 patients were diagnosed with contact allergy to chloramphenicol. In ten cases, the reaction was relevant, in 5 of these the patients had reacted to chloramphenicol eye medicaments (42).

Case reports

A 65-year-old non-atopic man presented with a generalized papular and nodular eruption, which had been diagnosed as prurigo nodularis and treated for 2 months with topical corticosteroids, without any improvement. The patient reported that the lesions appeared one month after he had started using topical therapy with an ointment containing chloramphenicol and clostridiopeptidase A for a leg ulcer, which was found surrounded by an eczematous reaction. Topical drug withdrawal and clobetasol dipropionate ointment (twice daily) cleared the eczematous dermatitis surrounding the leg ulcer. Moreover, within one month, healing of nodular lesions was achieved, with residual hypopigmentation. Histopathology of a nodular lesion showed hyperkeratosis, mild acanthosis, and focal spongiosis; the absence of cutaneous neural hyperplasia and an increase in the number of nerve fibers led the authors to exclude typical prurigo nodularis. Patch tests were positive to the ointment 'as is', to chloramphenicol 5% pet. (on 2 occasions), but negative to the ointment without chloramphenicol (21). This may have been a case of systemic contact dermatitis from absorption of chloramphenicol through the damaged ulcer skin.

A 40-year-old man, who had been treated with chloramphenicol eye drops as a young adult, underwent treatment with chloramphenicol 0.5% eye drops for bilateral conjunctivitis. After 2 days the patient developed profound bilateral infiltrative erythema and edema involving his eyelids with erythematous papules periorbitally and a generalized maculopapular exanthema. Patch testing was – according to the authors - contraindicated because of recent intense ultraviolet exposure. Intradermal skin testing was performed with the undiluted stock solution of the intravenous preparation of chloramphenicol in addition to serial dilutions (1, 10, and 100 mg/ml). At 15 minutes, there were no reactions. At D2, however, all 3 testing sites were strongly positive with papules, vesicles, edema, induration, and erythema. At the same time, the patient also developed significant bilateral eyelid edema, erythema, and ill-defined erythematous papules around the eyes with a mild generalized maculopapular exanthema (23). This was a case of systemic allergic contact dermatitis.

Short summaries of other case reports of allergic contact dermatitis from chloramphenicol are shown in table 3.60.2. Apparently, two cases of erythema multiforme-like contact dermatitis from chloramphenicol have been described (40,41). One or more patients with allergic contact dermatitis from chloramphenicol were reported from Germany in 1999; details are not available to the author (43). The first two cases of allergic contact dermatitis to chloramphenicol (termed chloromycetin) were described in 1951 (44).

Table 3.60.2 Short summaries of case reports of allergic contact dermatitis from chloramphenicol

Year and country	Sex	Age	Positive patch tests	Clinical data and comments	Ref.
2001 Germany	M	56	chloramphenicol 5% pet.	periorbital dermatitis from eye drops; cross-reaction to azidamphenicol but not to thiamphenicol; lymphocyte transformation tests positive to chlor- and azidamphenicol	39
1998 France	F	73	eye drops 'as is'; chloramphenicol 5% pet.	redness, edema and pruritus of the eyelids, occlusion of the eyelids; cross-reaction to thiamphenicol	25
1996 Spain	M	30	chloramphenicol 5% pet.	periorbital dermatitis from eye drops; occupational allergic contact dermatitis of the hands from veterinary medicines	26
1992 Spain	M	36	ChA succinate 20 mg/ml water; intradermal test pos. at D2 and D4	periorbital dermatitis; swelling of face and exudation after intramuscular injection; conjunctivitis 8 hours after patch and intradermal tests; possibly systemic contact dermatitis	27
	F	43	chloramphenicol 5% pet.	micropapular dermatitis on the thighs and genital area, itching, dysuria after vaginal pessary with chloramphenicol	
1991 Italy	M	54	leg ulcer ointment 'as is'; ChA 2%, 3% and 5% pet.	erythema, swelling and vesicles on the legs, hips and arms from application to leg ulcer; possibly systemic contact dermatitis	28
	F	53	leg ulcer ointment 'as is'; chloramphenicol 3% pet.	dermatitis of the hands, forearms and legs after application of chloramphenicol-containing ointment to leg ulcers	
1989 Japan	F	69	eye drops 'as is' and ChA	periorbital dermatitis; also allergy to colistimethate	19
1987 Japan	M	9	cream 'as is' and in dilution series; ChA 2% in water and alcohol	itchy red swelling and vesicles with erosions around an excoriation. ChA 2% pet. was positive only with 1% sodium lauryl sulfate or on abraded skin	30
1986 Poland	F	37	chloramphenicol 5% pet.	oculist with occupational allergic contact dermatitis; no clinical details provided	29
1985 Germany	M	43	eye ointment 'as is'; chloramphenicol 2% pet.	itchy dermatitis of the right periorbital region after therapy for chalazion	20
1975 Germany	?	?	ointment for leg ulcer, chloramphenicol	no clinical or patch testing data provided	18

ChA: chloramphenicol

Cross-reactions, pseudo-cross-reactions and co-reactions

A patient sensitized to chloramphenicol cross-reacted to azidamphenicol but not to thiamphenicol; lymphocyte transformation tests were positive to chlor- and to azidamphenicol (39). Cross-reaction in another patient to thiamphenicol (25). Fifteen patients sensitized to chloramphenicol and reacting to chloramphenicol 5% pet. were patch tested with chloramphenicol succinate and chloramphenicol palmitate (dilution series 0.5%-1%-2%-5%-10%) and all had positive reactions to both chloramphenicol esters (34).

Patients sensitized to chloramphenicol may cross-react to *p*-nitrobenzoic acid and *p*-dinitrobenzene (1,4-dinitrobenzene) (32,34). Both compounds are apparently intermediates in the chloramphenicol synthesis (34).

In an early study, 12 of 30 (40%) patients who developed contact dermatitis to chloramphenicol ointment, predominantly used for treating chronic stasis dermatitis and leg ulcers, also showed a positive patch test to 0.1% DNCB (dinitrochlorobenzene, 1-chloro-2,4-dinitrobenzene) in acetone, despite a negative history of previous exposure to DNCB (32). In several following studies, it has been well documented that patients sensitized to DNCB do *not* cross-react to chloramphenicol (33-36). However, the authors of the first study did not investigate whether patients sensitized to DNCB cross-react to chloramphenicol, but the reverse situation: patients sensitized to chloramphenicol cross-reacting to DNCB. In other words: the results of the later studies do *not* necessarily contradict the findings in ref. 32 and cannot invalidate them. However, in one of the later studies, 15 patients who had positive patch test reactions to chloramphenicol were tested with DNCB, and there were no positive reactions (34). It was suggested that the test material of DNCB used in the first study (0.1% in acetone) (32) may have caused irritant, false-positive reactions (as has been observed in ref. 36), interpreted by the authors as true allergic reactions (34).

Immediate contact reactions

Immediate contact reactions (contact urticaria) to chloramphenicol are presented in Chapter 5.

Cutaneous adverse drug reactions from systemic administration caused by type IV (delayed-type) hypersensitivity

Cutaneous adverse drug reactions from systemic administration of chloramphenicol caused by type IV (delayed-type) hypersensitivity, including a 'drug eruption' (37) and acute generalized exanthematous pustulosis (AGEP) (24), and in addition occupational allergic contact dermatitis (14), are planned to be discussed in Volume IV of the *Monographs in Contact Allergy* series on Systemic drugs.

LITERATURE

1   The data in the section 'General' may have been obtained from literature discussed in this chapter, but mostly also or exclusively from one or more of the following online sources: ChemIDPlus Advanced, PubChem, DrugBank, RxList, Drug Central, Drugs.com, and Wikipedia
2   Tavadia S, Bianchi J, Dawe RS, McEvoy M, Wiggins E, Hamill E, et al. Allergic contact dermatitis in venous leg ulcer patients. Contact Dermatitis 2003;48:261-265
3   Marasovic D, Vuksic I. Allergic contact dermatitis in patients with leg ulcers. Contact Dermatitis1999;41:107-109
4   Rai R, Shenoy MM, Viswanath V, Sarma N, Majid I, Dogra S. Contact sensitivity in patients with venous leg ulcer: A multi-centric Indian study. Int Wound J 2018;15:618-622
5   Jankićević J, Vesić S, Vukićević J, Gajić M, Adamic M, Pavlović MD. Contact sensitivity in patients with venous leg ulcers in Serbia: comparison with contact dermatitis patients and relationship to ulcer duration. Contact Dermatitis 2008;58:32-36
6   Cooper SM, Shaw S. Eyelid dermatitis: an evaluation of 232 patchtest patients over 5 years. Contact Dermatitis 2000;42:291-293
7   Van Ginkel CJ, Bruintjes TD, Huizing EH. Allergy due to topical medications in chronic otitis externa and chronic otitis media. Clin Otolaryngol Allied Sci 1995;20:326-328
8   Fräki JE, Kalimo K, Tuohimaa P, Aantaa E. Contact allergy to various components of topical preparations for treatment of external otitis. Acta Otolaryngol 1985;100:414-418
9   Gilissen L, De Decker L, Hulshagen T, Goossens A. Allergic contact dermatitis caused by topical ophthalmic medications: Keep an eye on it! Contact Dermatitis 2019;80:291-297
10 Uter W, Menezes de Pádua C, Pfahlberg A, Nink K, Schnuch A, Behrens-Baumann W. Contact allergy to topical ophthalmological drugs - epidemiological risk assessment. Klin Monbl Augenheilkd 2009;226:48-53 (Article in German)
11 Menezes de Padua CA, Uter W, Schnuch A. Contact allergy to topical drugs: prevalence in a clinical setting and estimation of frequency at the population level. Pharmacoepidemiol Drug Saf 2007;16:377-384

12 Menezes de Padua CA, Schnuch A, Nink K, Pfahlberg A, Uter W. Allergic contact dermatitis to topical drugs – Epidemiological risk assessment. Pharmacoepidemiol Drug Saf 2008;17:813-821

13 Gilissen L, Goossens A. Frequency and trends of contact allergy to and iatrogenic contact dermatitis caused by topical drugs over a 25-year period. Contact Dermatitis 2016;75:290-302

14 Rudzki E, Rebandel P, Grzywa Z. Contact allergy in the pharmaceutical industry. Contact Dermatitis 1989;21:121-122

15 Landeck L, John SM, Geier J. Topical ophthalmic agents as allergens in periorbital dermatitis. Br J Ophthalmol 2014;98:259-262

16 Tabar AI, García BE, Rodríguez A, Quirce S, Olaguibel JM. Etiologic agents in allergic contact dermatitis caused by eyedrops. Contact Dermatitis 1993;29:50-51

17 Tosti A, Tosti G. Allergic contact conjunctivitis due to ophthalmic solution. In: Frosch PJ, Dooms-Goossens A, Lachapelle JM, Rycroft RJG, Scheper RJ (eds). Current Topics in Contact Dermatitis. Berlin: Springer-Verlag, 1989: 269-272

18 Braun WP. Contact allergy to collagenase mixture (Iruxol). Contact Dermatitis 1975;1:241-242

19 Kubo Y, Tanaka M, Yoshida H. A case of contact dermatitis from Colymy C ophthalmic solution. Skin Research (Hifu) 1989;31(suppl.7):105-110 (Article in Japanese)

20 Frosch PJ, Olbert D, Weickel R. Contact allergy to tolazoline. Contact Dermatitis 1985;13:272

21 Romita P, Stingeni L, Hansel K, Ettorre G, Bosco A, Ambrogio F, et al. Allergic contact dermatitis caused by chloramphenicol with prurigo nodularis-like spreading. Contact Dermatitis 2019;80:251-252

22 Schwank R, Jirasek L. Contact eczema after chloramphenicol ointment. Cesk Dermatol 1961;36:4-11 (Article in Czech, data cited in ref. 21).

23 Watts TJ. Severe delayed-type hypersensitivity to chloramphenicol with systemic reactivation during intradermal testing. Ann Allergy Asthma Immunol 2017;118:644-645

24 Lee AY, Yoo SH. Chloramphenicol induced acute generalized exanthematous pustulosis proved by patch test and systemic provocation. Acta Derm Venereol 1999;79:412-413

25 Le Coz CJ, Santinelli F. Facial contact dermatitis from chloramphenicol with cross-sensitivity to thiamphenicol. Contact Dermatitis 1998;38:108-109

26 Moyano JC, Alvarez M, Fonseca JL, Bellido J, Munoz Bellido FJ. Allergic contact dermatitis to chloramphenicol. Allergy 1996;51:67-69

27 Urrutia I, Audícana M, Echechipía S, Gastaminza G, Bernaola G, Fernández de Corrès L. Sensitization to chloramphenicol. Contact Dermatitis 1992;26:66-67

28 Vincenzi C, Morelli R, Bardazzi F, Guerra L. Contact dermatitis from chloramphenicol in a leg ulcer cream. Contact Dermatitis 1991;25:64-65

29 Rebandel P. Rudzki. Occupational contact sensitivity in oculists. Contact Dermatitis 1986;15:92

30 Kubo Y, Nonaka S, Yoshida H. Contact sensitivity to chloramphenicol. Contact Dermatitis 1987;17:245-247, probably also published in Japanese in Skin Research 1987;29:418-423

31 Van Joost T, Dikland W, Stolz E, Prens E. Sensitization to chloramphenicol; a persistent problem. Contact Dermatitis 1986;14:176-178

32 Schwank R, Jirásek L. Kontaktallergie gegen Chloramphenicol mit besonderer Berücksichtigung der Gruppensensibilisierung. Hautarzt 1963;14:24-30 (Article in German)

33 Strick RA. Lack of cross-reaction between DNCB and chloramphenicol. Contact Dermatitis 1983;9:484-487

34 Eriksen K. Cross allergy between paranitro compounds with special reference to DNCB and chloramphenicol. Contact Dermatitis 1978;4:29-32

35 Palacios JS, Nemuth MG, Blaylock WK. Lack of cross sensitization between 2,4-dinitrobenzene and chloramphenicol. South Med J 1968;51:243-245

36 Catalona WJ, Taylor PT, Rabson AS, Chretien PB. A method for DNCB contact sensitization. N Engl J Med 1972;286:399-402

37 Rudzki E, Grzywa Z, Maciejowska E. Drug reaction with positive patch test to chloramphenicol. Contact Dermatitis 1976;2:181

38 Shah M, Lewis F M, Gawkrodger D J. Facial dermatitis and eyelid dermatitis: a comparison of patch test results and final diagnoses. Contact Dermatitis 1996;34:140-141

39 Sachs B, Erdmann S, al Masaoudi T, Merk HF. Molecular features determining lymphocyte reactivity in allergic contact dermatitis to chloramphenicol and azidamphenicol. Allergy 2001;56:69-72

40 Fisher AA. Erythema-multiforme-like eruptions due to topical medicaments: part II. Cutis 1986;37:158-161

41 Meneghini CL, Angelini G. Secondary polymorphic eruptions in allergic contact dermatitis. Dermatologica 1981;163:63-70

42 Cronin E. Contact Dermatitis. Edinburgh: Churchill Livingstone, 1980:204-206

43 Erdmann S, Sachs B, Merk HF. Allergische Kontaktdermatitis auf Chloramphenicol. Z Hautkr 1999;74:762-764 (Article in German)

44 Robinson HM Jr, Zeligman I, Shapiro A, Cohen MM. Acquired contact sensitivity to chloromycetin. J Invest Dermatol 1951;17:205-206

45 Korossy S, Nebenführer L, Vincze E. Frequency, relevance and latency of chemical allergy in hospitalized patients in Budapest. Derm Beruf Umwelt 1983;31:39-44 (Article in German)

46 Korossy S, Vincze E, Gózony M. Chloramphenicol (Chloracid)-allergy. Z Haut Geschlechtskr 1966;41:375-379 (Article in German)

47 Linss G, Arlt C, Schmidt G, Goethe A, Schmollack E. HLA typing and chloramphenicol allergy. Dermatol Monatsschr 1985;171:250-253 (Article in German)

48 Langer H. Sensitization of skin disease patients by the local use of the antibiotics xanthocillin and chloramphenicol. Z Haut Geschlechtskr 1962;33:210-215 (Article in German)

49 Adam H. The evidence of the chloramphenicol-sensibilisation by the usual lymphocyte transformation test. Dermatol Monatsschr 1979;165:270-273 (Article in German)

50 Eberhartinger C, Ebner H. Contribution to the research of contact allergy to chloramphenicol. Arch Klin Exp Dermatol 1966;224:463-470 (Article in German)

Chapter 3.61   CHLORDANTOIN

IDENTIFICATION

Description/definition   : Chlordantoin is the imidazole that conforms to the structural formula shown below
Pharmacological classes   : Antifungal agents
IUPAC name   : 5-Heptan-3-yl-3-(trichloromethylsulfanyl)imidazolidine-2,4-dione
Other names   : Clodantoin; 5-(1-ethylpentyl)-3-((trichloromethyl)thio)hydantoin
CAS registry number   : 5588-20-5
EC number   : 226-995-3
Merck Index monograph   : 639
Patch testing   : 1% pet.
Molecular formula   : $C_{11}H_{17}Cl_3N_2O_2S$

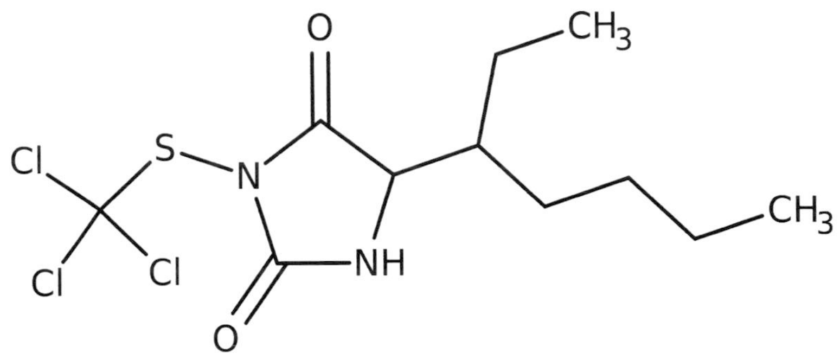

GENERAL

Chlordantoin is a topical imidazole antifungal agent with activity against *Candida albicans*. It is not used anymore (1).

CONTACT ALLERGY

Case reports

A woman developed vulvitis from using a vaginal cream containing 1% chlordantoin and 0.05% benzalkonium chloride for the treatment of vaginal candidiasis for 2 weeks. She had positive patch tests to the cream and to chlordantoin 0.1%, 0.5% and 1% pet., but 0.1% benzalkonium chloride was negative. None of five control patients showed positive reactions to chlordantoin (2). Previously, in the 1960s, two women had been described who developed vulvitis from a vaginal cream containing 1% chlordantoin for superficial candidiasis. Both reacted to patch tests with the cream, but not with the cream base. Chlordantoin itself was not tested (3).

LITERATURE

1   The data in the section 'General' may have been obtained from literature discussed in this chapter, but mostly also or exclusively from one or more of the following online sources: ChemIDPlus Advanced, PubChem, DrugBank, RxList, Drug Central, Drugs.com, and Wikipedia
2   Helander I, Hollmén A, Hopsu-Havu VK. Allergic contact dermatitis to chlordantoin. Contact Dermatitis 1979;5:54-55
3   Epstein E. Allergic dermatitis from chlordantoin vaginal cream. Report of 2 cases. Obstet Gynecol 1966;27:369-370

Chapter 3.62   CHLORPHENESIN

IDENTIFICATION

Description/definition	: Chlorphenesin is the glycol, member of propane-1,2-diols and of monochloroben-zenes that conforms to the structural formula shown below
Pharmacological classes	: Muscle relaxants, central; dermatological agents
IUPAC name	: 3-(4-Chlorophenoxy)propane-1,2-diol
Other names	: *p*-Chlorophenyl glyceryl ether
CAS registry number	: 104-29-0
EC number	: 203-192-6
Merck Index monograph	: 3454
Patch testing	: 1% and 0.5% pet; both concentrations may cause irritant reactions (11)
Molecular formula	: $C_9H_{11}ClO_3$

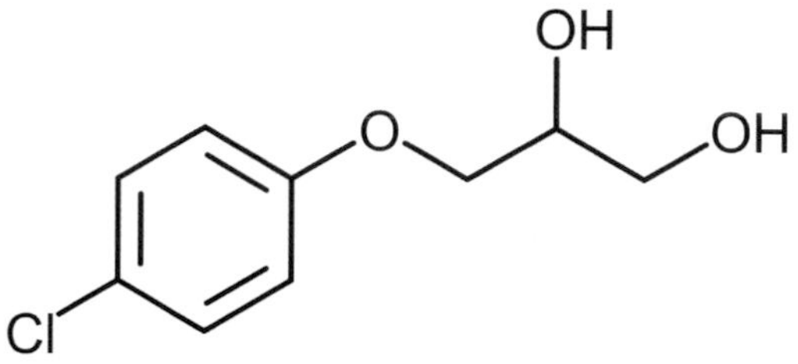

GENERAL

Chlorphenesin is an antimicrobial agent with antifungal, antibacterial and anticandidal activity. It has been included in preparations to treat dermatophytosis of the feet and vaginal infections for more than 70 years and may also be present in cosmetics as a biocide and preservative. Chlorphenesin is hardly used anymore in topical drugs (1).

CONTACT ALLERGY

Patch testing in groups of patients

In Korea, in 2010-2011, 584 patients suspected of allergic cosmetic dermatitis were patch tested with chlorphenesin 1% pet. and 24 (4.1%) had a positive reaction. The relevance of these reactions was not mentioned (11). It should be appreciated, however, that the test concentration of 1% pet. may result in false-positive, irritant, patch test reactions.

Case reports

Two patients (3) and two more individuals (4,5) had allergic contact dermatitis from chlorphenesin in antifungal preparations. Chlorphenesin has also been the cause of *cosmetic dermatitis* (2,6-10). Culprit products containing chlorphenesin have included deodorant (6), facial moisturizer (7), face and body cream (8), foundation make-up (10) and moisturizer (10).

LITERATURE

1   The data in the section 'General' may have been obtained from literature discussed in this chapter, but mostly also or exclusively from one or more of the following online sources: ChemIDPlus Advanced, PubChem, DrugBank, RxList, Drug Central, Drugs.com, and Wikipedia

2   Travassos AR, Claes L, Boey L, Drieghe J, Goossens A. Non-fragrance allergens in specific cosmetic products. Contact Dermatitis 2011;65:276-285

3   Cronin E. Contact Dermatitis. Edinburgh: Churchill Livingstone, 1980: 227

4   Brown R. Chlorphenesin sensitivity. Contact Dermatitis 1981;7:162

5   Burns DA. Allergic contact sensitivity to chlorphenesin. Contact Dermatitis 1986;14:246

6   Goh CL. Dermatitis from chlorphenesin in a deodorant. Contact Dermatitis 1987;16:287

7   Wakelin SH, White IR. Dermatitis from chlorphenesin in a facial cosmetic. Contact Dermatitis 1997;37:138-139

8   Dyring-Andersen B, Elberling J, Johansen JD, Zachariae C. Contact allergy to chlorphenesin. JEADV 2015;29:1019

9   Goossens A. Cosmetic contact allergens. Cosmetics 2016, 3, 5; doi:10.3390/cosmetics3010005

10   Brown VL, Orton DI. Two cases of facial dermatitis due to chlorphenesin in cosmetics. Contact Dermatitis 2005;52:48-49

11   Lee SS, Hong DK, Jeong NJ, Lee JH, Choi YS, Lee AY, et al. Multicenter study of preservative sensitivity in patients with suspected cosmetic contact dermatitis in Korea. J Dermatol 2012;39:677-681

Chapter 3.63   CHLORPHENIRAMINE

IDENTIFICATION

Description/definition   : Chlorpheniramine is the synthetic alkylamine derivative that conforms to the structural
                             formula shown below
Pharmacological classes : Histamine H1 antagonists; anti-allergic agents; antipruritics
IUPAC name   : 3-(4-Chlorophenyl)-*N,N*-dimethyl-3-pyridin-2-ylpropan-1-amine
Other names   : 2-Pyridinepropanamine, γ-(4-chlorophenyl)-*N,N*-dimethyl-; chlorphenamine
CAS registry number   : 132-22-9
EC number   : 205-054-0
Merck Index monograph : 3456
Patch testing   : Maleate, 5% pet. (SmartPracticeCanada)
Molecular formula   : $C_{16}H_{19}ClN_2$

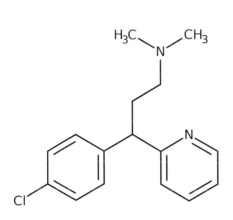

Chlorpheniramine

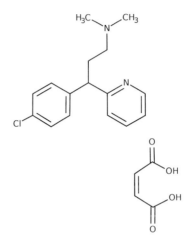

Chlorpheniramine maleate

GENERAL

Chlorpheniramine is a synthetic alkylamine derivative and a competitive histamine H1 receptor antagonist, and displays anticholinergic and mild sedative effects as well. It is indicated for the treatment of rhinitis, urticaria, allergy, common cold, asthma and hay fever. It has also been used in veterinary applications. In pharmaceutical products, chlorpheniramine is employed as chlorpheniramine maleate (CAS number 113-92-8, EC number 204-037-5, molecular formula $C_{20}H_{23}ClN_2O_4$) (1).

CONTACT ALLERGY

Patch testing in groups of selected patients

In the period 1996-2005, in Hungary, 133 patients with periorbital dermatitis were patch tested with chlorphenira-mine maleate 5% pet. and there was one (0.8%) positive reaction, which was probably relevant from its presence in an ophthalmic medication (2). In Italy, between 1968 and 1977, 251 selected patients (selection criteria not stated) were tested with chlorpheniramine 2% pet. and there were 3 (1.2%) positive reactions. The issue of relevance was not addressed (3). In the late 1960s, in Bari, 1.4% of patients tested for eczematous dermatitis (routine testing?) had positive reactions to chlorpheniramine maleate (12; no details available, data cited in ref. 11).

Case reports

An 89-year-old man presented with pruritic lesions over his left knee with linear extension down onto the lower leg, where a wound had been treated with an antiseptic containing chlorpheniramine maleate, benzalkonium chloride, dibucaine HCl, naphazoline HCl, and a mixture of fragrance ingredients for 2 days. Several years previously, he had developed eczema following the repeated use of an over-the-counter antiseptic. A patch test to the product was positive. In a second session, the patient reacted to chlorpheniramine maleate 1% pet., dibucaine HCl 1% pet., and naphazoline HCl 1% pet. Twenty controls were negative to chlorpheniramine maleate 1% pet. (5).

An almost identical case, also from Japan, and possibly from the same pharmaceutical product, had been reported 8 years previously (6). A 71-year-old man developed dermatitis on his right lower leg, where he had applied an over-the-counter medicament for skin wounds following a trauma. This contained chlorpheniramine maleate, dibucaine HCl and naphazoline HCl. Patch testing showed the patient to be allergic to chlorpheniramine maleate (1% pet.) and dibucaine HCl (1% pet.) (6).

A 45-year-old woman had suffered from eczema of the face for 4 years. She had not been using topical medicaments or cosmetics there, with the exception of eye drops for chronic conjunctivitis. Patch tests were positive to the eye drops 'as is' and its ingredient chlorpheniramine maleate, tested 5% pet. The patient had sporadically used a cream containing this antihistamine for the treatment of insect bites. The eczema cleared completely after stopping the use of the eye drops (7).

A 46-year-old woman presented with a 6-day history of facial swelling and itching. She had taken a multi-ingredient cold medication including chlorpheniramine maleate 11 days earlier and had applied an external preparation of chlorpheniramine maleate 7 days previously. Physical examination revealed edematous erythema, papules, and erosive lesions on the patient's face, especially on the eyelids and cheeks. The lesions were localized to the sites to which the external preparation had been applied. Epicutaneous tests revealed positive reactions to 1% and 5% chlorpheniramine maleate at D2. An 'intracutaneous type IV allergy test' was also positive (probably an intradermal test read at D2) (10). It is uncertain whether the oral preparation had initiated the dermatitis.

A 70-year-old man presented with itchy erythema and papules on the genitals and inguinal region, which had started 1 week earlier. The patient had applied a cream containing chlorpheniramine maleate to treat an itchy sensation there for 4 weeks. Patch tests were positive to chlorpheniramine maleate 1% and 0.1% pet. (negative at D2 and D3, positive at D7), but negative to the cream (which contained prednisolone acetate) and the other ingredients (11).

Cutaneous adverse drug reactions from systemic administration caused by type IV (delayed-type) hypersensitivity
Cutaneous adverse drug reactions from systemic administration of chlorpheniramine caused by type IV (delayed-type) hypersensitivity, including erythematous drug eruption (8) and systemic contact dermatitis (9), are planned to be discussed in Volume IV of the *Monographs in Contact Allergy* series on Systemic drugs.

Cross-reactions, pseudo-cross-reactions and co-reactions
Possible cross-reactivity between pheniramine, chlorpheniramine and dexchlorpheniramine (4). Two patients who developed allergic contact dermatitis from a cream containing dexchlorpheniramine maleate co-reacted to chlorpheniramine maleate 1% water (9).

LITERATURE

1   The data in the section 'General' may have been obtained from literature discussed in this chapter, but mostly also or exclusively from one or more of the following online sources: ChemIDPlus Advanced, PubChem, DrugBank, RxList, Drug Central, Drugs.com, and Wikipedia
2   Temesvári E, Pónyai G, Németh I, Hidvégi B, Sas A, Kárpáti S. Periocular dermatitis: a report of 401 patients. J Eur Acad Dermatol Venereol 2009;23:124-128
3   Angelini G, Vena GA, Meneghini CL. Allergic contact dermatitis to some medicaments. Contact Dermatitis 1985;12:263-269
4   Parente G, Pazzaglia M, Vincenzi C, Tosti A. Contact dermatitis from pheniramine maleate in eyedrops. Contact Dermatitis 1999;40:338
5   Yamadori Y, Oiso N, Hirao A, Kawara S, Kawada A. Allergic contact dermatitis from dibucaine hydrochloride, chlorpheniramine maleate, and naphazoline hydrochloride in an over-the-counter topical antiseptic. Contact Dermatitis 2009;61:52-53
6   Hayashi K, Kawachi S, Saida T. Allergic contact dermatitis due to both chlorpheniramine maleate and dibucaine hydrochloride in an over-the-counter medicament. Contact Dermatitis 2001;44:38-39
7   Tosti A, Bardazzi F, Piancastelli E. Contact dermatitis due to chlorpheniramine maleate in eyedrops. Contact Dermatitis 1990;22:55
8   Brown VL, Orton DI. Cutaneous adverse drug reaction to oral chlorphenamine detected with patch testing. Contact Dermatitis 2005;52:49-50
9   Santucci B, Cannistraci C, CristaudoA, Picardo M. Contact dermatitis from topical alkylamines. Contact Dermatitis 1992;27:200-201
10   Shikino K, Masuyama T, Yamashita T, Ikusaka M. Allergic contact dermatitis following drug rash due to chlorpheniramine maleate. Clin Case Rep 2017;5:718-719
11   Ueno T, Kawana S. Case of allergic contact dermatitis due to a topical agent containing chlorpheniramine maleate, with a false-negative patch test. J Dermatol 2008;35:46-48
12   Meneghini CL, Rantuccio F, Lomuto M. Additives, vehicles and active drugs of topical medicaments as cause of delayed-type allergic dermatitis. Dermatologica 1971;143:137-147

Chapter 3.64   CHLORPHENOXAMINE

IDENTIFICATION

Description/definition   : Chlorphenoxamine is the diphenylmethane that conforms to the structural formula
                             shown below
Pharmacological classes : Histamine antagonists; muscarinic antagonists
IUPAC name   : 2-[1-(4-Chlorophenyl)-1-phenylethoxy]-N,N-dimethylethanamine
Other names   : Phenoxene
CAS registry number   : 77-38-3
EC number   : Not available
Merck Index monograph : 3457
Patch testing   : 1.5% pet. or water
Molecular formula   : $C_{18}H_{22}ClNO$

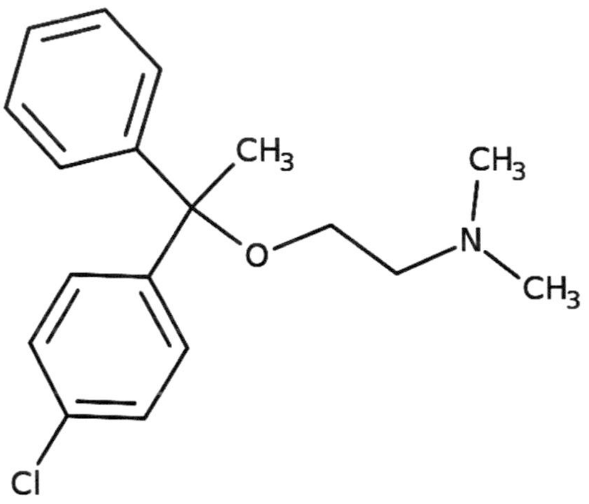

GENERAL

Chlorphenoxamine is a diphenylmethane antihistamine and anticholinergic agent. It is used as an antipruritic and antiparkinsonian agent. In pharmaceutical products, chlorphenoxamine may be employed as chlorphenoxamine hydrochloride (CAS number 562-09-4, EC number 209-227-1, molecular formula $C_{18}H_{23}Cl_2NO$) (1).

CONTACT ALLERGY

Case report

A 61-year-old man suffered an insect bite on the penis during his vacation. He applied an antihistamine emulsion containing 1.5% chlorphenoxamine hydrochloride on it and after 3 days developed a weeping dermatitis in the inguinal and perianal areas. Patch tests with the pharmaceutical and chlorphenoxamine HCl 1.5% in pet. and in water were all strongly positive (+++). It was not mentioned how the patients had acquired his already existing contact allergy to the drug (2).

LITERATURE

1   The data in the section 'General' may have been obtained from literature discussed in this chapter, but mostly
    also or exclusively from one or more of the following online sources: ChemIDPlus Advanced, PubChem,
    DrugBank, RxList, Drug Central, Drugs.com, and Wikipedia
2   Van Ketel WG. Sensitivity to chlorphenoxamine HCl. Contact Dermatitis 1976;2:121

Chapter 3.65   CHLORPROETHAZINE

IDENTIFICATION

Description/definition : Chlorproethazine is the phenothiazine that conforms to the structural formula shown below

Pharmacological classes : Muscle relaxants, central

IUPAC name : 3-(2-Chlorophenothiazin-10-yl)-*N*,*N*-diethylpropan-1-amine

Other names : 2-Chloro-10-(3-diethylaminopropyl)phenothiazine

CAS registry number : 84-01-5

EC number : 201-510-8

Merck Index monograph : 3459

Patch testing : 1% pet.; if negative and allergy or photoallergy is strongly suspected, use 10% pet. (risk of false-positives)

Molecular formula : $C_{19}H_{23}ClN_2S$

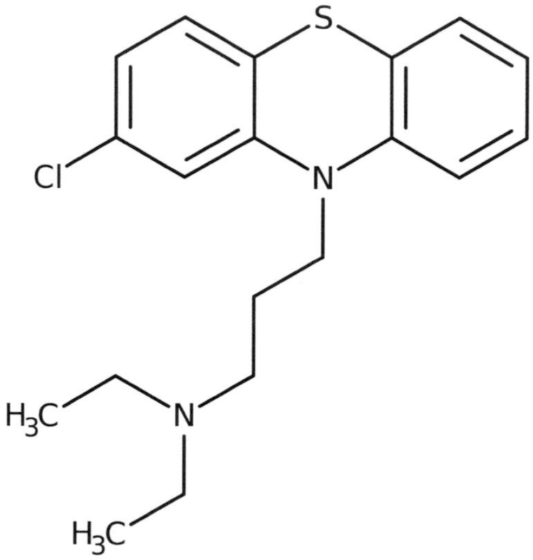

GENERAL

Chlorproethazine is a member of the phenothiazine group described as a muscle relaxant or tranquillizer. It has been marketed in Europe as a topical cream for the treatment of muscle pain. Chlorproethazine is probably hardly, if at all, used anymore, either orally or topically (1). It was withdrawn from the French market in 2007 (6).

CONTACT ALLERGY

Case series

In France, before 2001, 10 patients (5 women, 5 men, mean age 65 years) suspected to be allergic to a muscle-relaxing ointment containing chlorproethazine have been investigated in 2 hospitals. A severe contact dermatitis had been observed at the site of application in all 10 cases. There was secondary spread to distant sites in 7 of them, with photo-distribution in 4. During follow-up, a relapse occurred in 2 patients who used the pharmaceutical to treat another person. Two developed persistent light reactions. The patients were patch tested with the ointment as is (n=10) and chlorproethazine 10% pet. (n=10), 1% pet. (n=9) and 0.1% pet. (n=6). Additionally, photopatch tests with chlorproethazine 1% and 0.1% pet. were performed in 3 patients. Patch tests and photopatch tests were also carried out with the related phenothiazines chlorpromazine 0.1% pet. (n=5) and promethazine 0.1% pet. (n=4).

Patch tests were positive in all 10 patients to the ointment (which was irritant) as is and chlorproethazine 10% pet. (also irritant), causing a focal flare in two. The excipients were negative in all patients. Chlorproethazine 1% pet. was positive in 8 of the 9 tested. All 3 patients who were photopatch tested with the ointment and chlorproethazine 10% and 1% pet. showed photo-aggravation of the patch tests. Patch tests with the related phenothiazines chlorpromazine and promethazine were negative in all 5 patients tested. However, *photo*patch tests were positive to chlorpromazine in 4/4 cases and promethazine in 3/4 cases (2).

Cross-reactions, pseudo-cross-reactions and co-reactions

Contact allergy to chlorproethazine frequently results in *photo*cross-reactions to the related phenothiazines chlorpromazine and promethazine (2).

PHOTOSENSITIVITY

Chlorproethazine in a cream (Neuroplege ®) was responsible for drug-induced photosensitivity in 25/1239 cases (2%) reported to the French Agency for Drug Pharmacovigilance before 2000 (cited in ref. 2). It was shown that chlorproethazine and the cream with chlorproethazine had strong phototoxic and photocontact allergic potential (6). Indeed, two volunteers became photosensitized to chlorproethazine from patch and photopatch tests to chlorproethazine and chlorproethazine cream (6)

This phenothiazine can induce contact dermatitis which is photoaggravated and in addition result in persistent light reactions (2,3,4,7). See also the section 'Case series' above.

Immediate contact reactions

Immediate contact reactions (contact urticaria) to chlorproethazine are presented in Chapter 5.

LITERATURE

1   The data in the section 'General' may have been obtained from literature discussed in this chapter, but mostly also or exclusively from one or more of the following online sources: ChemIDPlus Advanced, PubChem, DrugBank, RxList, Drug Central, Drugs.com, and Wikipedia

2   Barbaud A, Collet E, Martin S, Granel F, Trechot P, Lambert D, et al. Contact sensitization to chlorproethazine can induce persistent light reaction and cross-photoreactions to other phenothiazines. Contact Dermatitis 2001;44:373-374

3.   Fenard-Naud D. La photosensibilité au Neuriplège® pommade á propos de 8 observations – Photoallergie croisée entre la chlorproéthazine et les autres phénothiazines. Medical Thesis, 1988, Lille (France): page 130 (in French)

4.   Jeanmougin M, Sigal-Nahum M, Manciet JR, Petit A, Flageul B, Dubertret L. Photosensibilité rémanente induite par la chlorproéthazine. Ann Dermatol Venereol 1993;120:840-843 (Article in French)

5   Cariou C, Droitcourt C, Osmont MN, Marguery MC, Dutartre H, Delaunay J, et al. Photodermatitis from topical phenothiazines: A case series. Contact Dermatitis 2020;83:19-24

6   Kerr A, Woods J, Ferguson J. Photocontact allergic and phototoxic studies of chlorproethazine. Photodermatol Photoimmunol Photomed 2008;24:11-15

7   Marguery MC. Contact allergy to chlorproethazine with photoaggravation. DermoFocus 2000;7:17-19

Chapter 3.66   CHLORQUINALDOL

IDENTIFICATION

Description/definition : Chlorquinaldol is the halogenated 8-hydroxyquinoline that conforms to the structural formula shown below
Pharmacological classes : Anti-infective agents
IUPAC name : 5,7-Dichloro-2-methylquinolin-8-ol
Other names : Hydroxydichloroquinaldine; 5,7-Dichloro-8-hydroxyquinaldine
CAS registry number : 72-80-0
EC number : 200-789-3
Merck Index monograph : 3466
Patch testing : 5% pet. (Chemotechnique, SmartPracticeCanada, SmartPracticeEurope)
Molecular formula : $C_{10}H_7Cl_2NO$

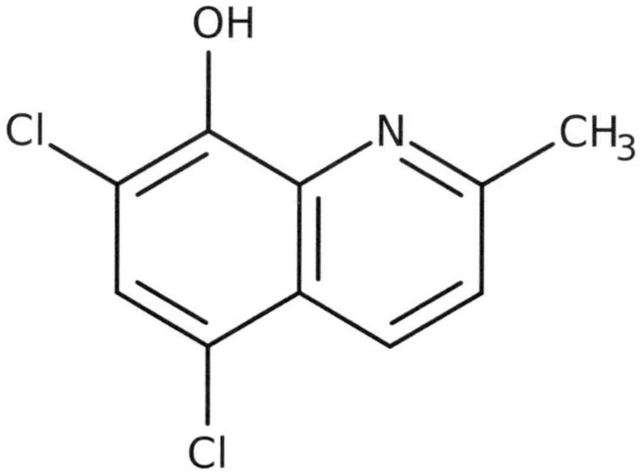

GENERAL

Chlorquinaldol is an organochlorine compound and a monohydroxyquinoline. An antifungal and antibacterial, chlorquinaldol was formerly used for topical treatment of skin, gastrointestinal, and vaginal infections with fungi, protozoa, and certain bacteria. It was formerly present in some routine series in the quinoline mix together with clioquinol, in which it caused far less positive reactions than the latter (15).

CONTACT ALLERGY

Patch testing in groups of patients

Results of patch testing chlorquinaldol in consecutive patients (routine testing) and in groups of selected patients are shown in table 3.66.1. Routine testing in the USA yielded low prevalences of 0.3% and 0.4% of sensitization (2,3) and somewhat higher (1.1%) in Germany in the period 1972-1983 (19). Higher rates were – as expected – observed in patients patch tested for leg ulcers: 2.8% in 1970-1973 in Italy (5) and even 17% in 1976-1978 in Finland. In those days, hydroxyquinolines were widely used on leg ulcers and stasis dermatitis. Very likely, many of the observed reactions represent cross-reactions to clioquinol, which was far more used than chlorquinaldol.

Table 3.66.1 Patch testing in groups of patients

Years and Country	Test conc. & vehicle	Number of patients tested	positive (%)	Selection of patients (S); Relevance (R); Comments (C)	Ref.
Routine testing					
2001-2005 USA	5% pet.	2665	(0.3%)	R: 22%	2
1998-2000 USA	5% pet.	713	(0.4%)	R: not stated	3
1972-1983 Germany			(1.1%)	R: unknown; there was an increase to 1.7% reactions in 1983 and it was concluded that inclusion of chlorquinaldol in the standard battery was justified	19
Testing in groups of selected patients					
1976-1978 Finland	5% pet.	78	13 (17%)	S: patients with leg ulcers or stasis dermatitis; R: not stated	4
1968-1977 Italy	5% pet.	869	8 (0.9%)	S: not stated; R: not stated	7
1970-1973 Italy	5% pet.	143	4 (2.8%)	S: patients with stasis dermatitis with or without leg ulcers; R: not stated	5

Case reports and case series

In Leuven, Belgium, in the period 1990-2014, iatrogenic contact dermatitis was diagnosed in 2600 individuals (17% of the total patch test population). 96% of all positive patch test reactions to topical drugs and antiseptics were considered to be relevant. Chlorquinaldol (3% pet.) was tested in 14 patients and there were 7 positive reactions to it (6). Two patients had relevant positive patch tests to chlorquinaldol. They both had used a combination preparation containing a corticosteroid and chlorquinaldol. Clinical data were not provided (9).

A 57-year-old woman presented with a long-standing severely itchy papular and pustular eruption on the arms and legs. The lesions had appeared after treatment for eczematoid tinea pedis with 1% chlorquinaldol cream and later with isoconazole nitrate cream. Patch tests were positive to both antifungal creams 'as is', to the quinoline mix (containing chlorquinaldol) and to isoconazole nitrate. Chlorquinaldol itself was not tested (8). A papulopustular reaction is unusual in allergic contact dermatitis.

A 31-year-old woman who had previously become sensitized to clioquinol from an antiseptic ointment to treat otitis externa, suffered vaginal erythema, itching, and edema, and eczematous lesions in abdominal and thoracic areas one day after application of a vaginal ovule containing chlorquinaldol and promestriene. Patch tests were positive to chlorquinaldol, clioquinol (both 5% pet.) and promestriene 0.1% alcohol. (17). This was a case of local and systemic contact dermatitis.

A man was sensitized to chlorquinaldol and clioquinol from several topical pharmaceuticals applied to allergic contact dermatitis caused by benzoyl peroxide used as catalyst in the material to make an arm prosthesis (18).

The first case of contact allergy to/allergic contact dermatitis from chlorquinaldol was described in 1944. Apparently, it concerned a cross-reaction to clioquinol (11).

Cross-reactions, pseudo-cross-reactions and co-reactions

Cross-reactions between halogenated hydroxyquinolines such as clioquinol (5-chloro-7-iodoquinolin-8-ol), chlorquinaldol (5,7-dichloro-2-methylquinolin-8-ol), cloxyquin (5-chloroquinolin-8-ol), oxyquinoline (8-hydroxyquinoline, non-halogenated), iodoquinol (5,7-diiodoquinolin-8-ol), halquinol (a mixture of 4 hydroxy-quinolines), 5,7-dichloro-8-quinolinol and 5-chloro-8-quinolinol may occur (9,10,12,13 [examples of references]). Of 16 patients sensitized to clioquinol, 8 (50%) cross-reacted to chlorquinaldol (14); this was also the case in 9 of 10 clioquinol-allergic individuals in another series (16).

LITERATURE

1   The data in the section 'General' may have been obtained from literature discussed in this chapter, but mostly also or exclusively from one or more of the following online sources: ChemIDPlus Advanced, PubChem, DrugBank, RxList, Drug Central, Drugs.com, and Wikipedia

2   Davis MD, Scalf LA, Yiannias JA, Cheng JF, El-Azhary RA, Rohlinger AL, et al. Changing trends and allergens in the patch test standard series. Arch Dermatol 2008;144:67-72

3   Wetter DA, Davis MDP, Yiannias JA, Cheng JF, Connolly SM, el-Azhary RA, et al. Patch test results from the Mayo Contact Dermatitis Group, 1998–2000. J Am Acad Dermatol 2005;53:416-421

4   Fräki JE, Peltonen L, Hopsu-Havu VK. Allergy to various components of topical preparations in stasis dermatitis and leg ulcer. Contact Dermatitis 1979;5:97-100

5   Angelini G, Rantuccio F, Meneghini CL. Contact dermatitis in patients with leg ulcers. Contact Dermatitis 1975;1:81-87

6   Gilissen L, Goossens A. Frequency and trends of contact allergy to and iatrogenic contact dermatitis caused by topical drugs over a 25-year period. Contact Dermatitis 2016;75:290-302

7   Angelini G, Vena GA, Meneghini CL. Allergic contact dermatitis to some medicaments. Contact Dermatitis 1985;12:263-269

8   Lazarov A, Ingber A. Pustular allergic contact dermatitis to isoconazole nitrate. Am J Contact Dermat 1997;8:229-230

9   Myatt AE, Beck MH. Contact sensitivity to chlorquinaldol. Contact Dermatitis 1983;9:523

10   Allenby CF. Skin sensitization to Remiderm and cross-sensitization to hydroxyquinoline compounds. Br Med J 1965;2(5455):208-209

11   Jadassohn W, Fierz HE, Pfanner E. Schweiz Med Wschr 1944;74:168 (Article in German, title unknown, cited in ref. 10)

12   Leifer W, Steiner K. Studies in sensitization to halogenated hydroxyquinolines and related compounds. J Invest Dermatol 1951;17:233-240

13 Wantke F, Götz M, Jarisch R. Contact dermatitis from cloxyquin. Contact Dermatitis 1995;32:112-113

14 Bielicky T, Novak M. Gruppensensibilisierung gegen Chinolinderivate. Dermatologica 1969;138:45-48

15 Agner T, Menné T. Sensitivity to clioquinol and chlorquinaldol in the quinoline mix. Contact Dermatitis 1993;29:163

16 Soesman-van Waadenoijen Kernekamp A, van Ketel WG. Persistence of patch test reactions to clioquinol (Vioform) and cross-sensitization. Contact Dermatitis 1980;6:455-460

17 Rodríguez A, Cabrerizo S, Barranco R, de Frutos C, de Barrio M. Contact cross-sensitization among quinolines. Allergy 2001;56:795

18 Vincenzi C, Cameli N, Vassilopoulou A, Tosti A. Allergic contact dermatitis due to benzoyl peroxide in an arm prosthesis. Contact Dermatitis 1991;24:66-67

19 Hutzler D, Pevny I. Allergies to 8-hydroxyquinoline derivatives. Derm Beruf Umwelt 1988;36:86-90

Chapter 3.67   CHLORTETRACYCLINE

IDENTIFICATION

Description/definition   : Chlortetracycline is a tetracycline with a 7-chloro substitution that conforms to the
                              structural formula shown below
Pharmacological classes   : Anti-bacterial agents; antiprotozoal agents; protein synthesis inhibitors
IUPAC name   : (4S,4aS,5aS,6S,12aR)-7-Chloro-4-(dimethylamino)-1,6,10,11,12a-pentahydroxy-6-methyl-
                              3,12-dioxo-4,4a,5,5a-tetrahydrotetracene-2-carboxamide
Other names   : Chlorotetracycline
CAS registry number   : 57-62-5
EC number   : 200-341-7
Merck Index monograph   : 3468
Patch testing   : Hydrochloride, 1% pet. (SmartPracticeCanada)
Molecular formula   : $C_{22}H_{23}ClN_2O_8$

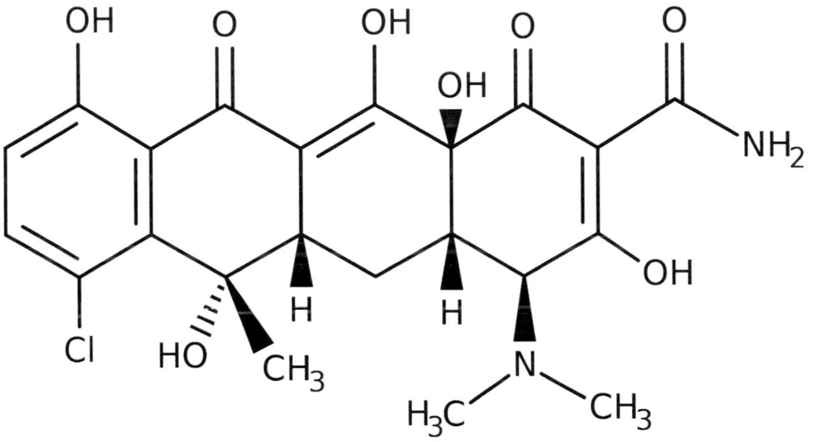

GENERAL

Chlortetracycline is a tetracycline antibiotic (the first tetracycline to be identified) isolated from an actinomycete named *Streptomyces aureofaciens*. The designated name of this microorganism and that of the isolated drug, Aureomycin, derive from their golden color. Chlortetracycline is currently used in the manufacturing of medicated animal feeds and as antibacterial agent in eye ointments. In pharmaceutical products, chlortetracycline is employed as chlortetracycline hydrochloride (CAS number 64-72-2, EC number 200-591-7, molecular formula $C_{22}H_{24}Cl_2N_2O_8$), probably as an ointment for eye infections only (1).

CONTACT ALLERGY

Patch testing in groups of patients

In the period 1997-2001, in the United Kingdom, 200 patients with venous or mixed venous/arterial leg ulcers were patch tested with chlortetracycline (test concentration and vehicle unknown) and 4 (2.0%) had a positive patch test. All reactions were considered to be of probable, past or current relevance (2). Also in the U.K., in 1988-1989, of 81 patients with leg ulcers patch tested with oxytetracycline (test concentration and vehicle unknown), three had a positive reaction, each being considered relevant (3).

Case reports

A woman had used chlortetracycline ointment on a dermabraded tattoo and developed dermatitis. Patch tests were positive to the ointment 'as is', chlortetracycline and demethylchlortetracycline, but not to oxytetracycline and tetracycline (4). Another woman had applied the ointment on varicose ulcers. She had a positive patch test to chlortetracycline 0.5% pet. (or the ointment 0.5% pet.?) (4).

   One case of occupational allergic contact dermatitis to chlortetracycline in a pharmaceutical worker was reported from Warsaw, Poland, in 1984 (5).

Cross-reactions, pseudo-cross-reactions and co-reactions
A woman who had developed allergic contact dermatitis from an ointment containing chlortetracycline also reacted to demethylchlortetracycline, but not to oxytetracycline and tetracycline (4).

LITERATURE

1   The data in the section 'General' may have been obtained from literature discussed in this chapter, but mostly also or exclusively from one or more of the following online sources: ChemIDPlus Advanced, PubChem, DrugBank, RxList, Drug Central, Drugs.com, and Wikipedia
2   Tavadia S, Bianchi J, Dawe RS, McEvoy M, Wiggins E, Hamill E, et al. Allergic contact dermatitis in venous leg ulcer patients. Contact Dermatitis 2003;48:261-265
3   Wilson CL, Cameron J, Powell SM, Cherry G, Ryan TJ. High incidence of contact dermatitis in leg-ulcer patients – implications for management. Clin Exp Dermatol 1991;16:250-253
4   Calnan CD. Chlortetracycline sensitivity. Contact Dermatitis Newsletter 1967;1:16
5   Rudzki E, Rebendel P. Contact sensitivity to antibiotics. Contact Dermatitis 1984;11:41-42

Chapter 3.68   CICLESONIDE

IDENTIFICATION

Description/definition   : Ciclesonide is the synthetic glucocorticoid that conforms to the structural formula shown
                             below
Pharmacological classes   : Anti-allergic agents; glucocorticoids
IUPAC name   : [2-[(1S,2S,4R,6R,8S,9S,11S,12S,13R)-6-Cyclohexyl-11-hydroxy-9,13-dimethyl-16-oxo-5,7-
                             dioxapentacyclo[10.8.0.0^{2,9}.0^{4,8}.0^{13,18}]icosa-14,17-dien-8-yl]-2-oxoethyl] 2-methyl-
                             propanoate
Other names   : Pregna-1,4-diene-3,20-dione, 16,17-(((R)-cyclohexylmethylene)bis(oxy))-11-hydroxy-21-
                             (2-methyl-1-oxopropoxy)-, (11β,16α)-
CAS registry number   : 126544-47-6
EC number   : Not available
Merck Index monograph   : 3536
Patch testing   : In general, corticosteroids may be tested at 0.1% and 1% in alcohol; late readings (6-10
                             days) are strongly recommended; 0.1% and 0.5% in alcohol (2)
Molecular formula   : C$_{32}$H$_{44}$O$_7$

GENERAL

General aspects of corticosteroids used on the skin and mucous membranes are discussed in Chapter 2.4. A practical guideline for diagnosing allergic reactions to corticosteroids is presented in ref. 1.

CONTACT ALLERGY

Patch testing in groups of patients

In the period 2009-2011, In Helsinki, Finland, 79 patients had undergone patch testing with a steroid series including ciclesonide 0.1% and 0.5% in pet. and 0.1% and 0.5% alcohol, mainly because of eczematous conditions and suspected contact allergy to topical corticosteroids. None of these patients had had previous treatment with inhalation pharmaceuticals containing ciclesonide. They were also tested with the commercial ciclesonide inhalation spray at 160 mg/dose 0.1% and 0.5% in alcohol and 0.1% and 0.5% in pet. Ten (13%) of the 79 patch tested patients showed positive reactions to inhaled ciclesonide; all were also positive to the other 4 ciclesonide test materials. Budesonide 0.1% in pet. gave a positive reaction in all four patients with positive patch tests to ciclesonide. Conversely, there were no patients with a positive reaction to budesonide but a negative reaction to ciclesonide. It was concluded that all reactions to ciclesonide were cross-reactions to budesonide (2).

LITERATURE

1   Baeck M, Goossens A. Immediate and delayed allergic hypersensitivity to corticosteroids: practical guidelines.
    Contact Dermatitis 2012;66:38-45
2   Salava A, Alanko K, Hyry H. A case of systemic allergic dermatitis caused by inhaled budesonide: cross-reactivity
    in patch tests with the novel inhaled corticosteroid ciclesonide. Contact Dermatitis 2012;67:244-246

Chapter 3.69   CICLOPIROX OLAMINE

IDENTIFICATION

Description/definition   : Ciclopirox olamine is the olamine salt form of ciclopirox that conforms to the structural formula shown below

Pharmacological classes : Antifungal agents

IUPAC name   : 2-Aminoethanol;6-cyclohexyl-1-hydroxy-4-methylpyridin-2-one

Other names   : 6-Cyclohexyl-1-hydroxy-4-methyl-2(1*H*)-pyridinone compd. with 2-aminoethanol (1:1); ciclopirox ethanolamine

CAS registry number   : 41621-49-2

EC number   : 255-464-9

Merck Index monograph : 3539 (Ciclopirox)

Patch testing   : 1% pet.

Molecular formula   : $C_{14}H_{24}N_2O_3$

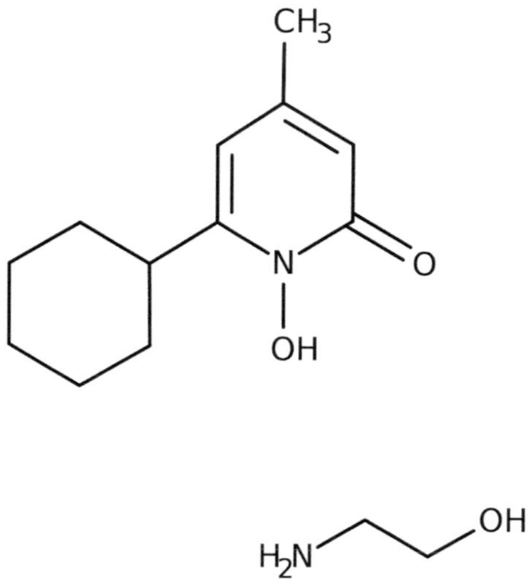

GENERAL

Ciclopirox olamine is the olamine salt form of ciclopirox, a synthetic, broad-spectrum antifungal agent with additional antibacterial and anti-inflammatory activities. It is used for topical dermatologic treatment of superficial mycoses. Ciclopirox olamine is especially effective in treating pityriasis (tinea) versicolor (1).

CONTACT ALLERGY

Patch testing in groups of selected patients

In the period 2001-2002, 20 members of the IVDK tested a series of antifungals in selected patients (selection procedure not specified, <10% - 22% of all routine patients were tested). 1094 patients were tested with ciclopirox olamine 1% pet. and there were 4 (0.4%) positive reactions. Relevance data were not provided (2).

Case reports

A 43-year-old woman had been treated for 14 days with a milk containing 1% ciclopirox olamine (CPO) for interdigital tinea pedis, when intensely itchy erythematovesicular and erythematosquamous lesions appeared in the interdigital spaces, spreading to the toes and upper surface of the feet. Patch tests revealed strongly positive reactions to the milk and to CPO 1% pet., but no reactions to the other constituents of the milk (3).

    A 50-year-old man developed dermatitis spreading from the toes to the lower shins after 4 weeks' treatment of interdigital mycosis with CPO cream. Patch tests were positive to the cream and to CPO 1% pet. Ten controls were negative (4). A 43-year-old patient had been treated for interdigital mycosis with a cream containing CPO for 3 days, when a massive bullous reaction was noticed between the toes and on the dorsum of the right foot. Patch tests were positive to the cream, CPO 1% in 40% alcohol (10 controls negative) and to 2 constituents of the cream, cetearyl alcohol and benzyl alcohol (5).

The first case of ACD from CPO was published in 1986 from Italy in a woman with *Microsporum canis* infection of the abdomen, developing after 12 days' treatment. Patch tests were positive to the 1% CPO milk and CPO 1% pet. (6)

LITERATURE

1   The data in the section 'General' may have been obtained from literature discussed in this chapter, but mostly also or exclusively from one or more of the following online sources: ChemIDPlus Advanced, PubChem, DrugBank, RxList, Drug Central, Drugs.com, and Wikipedia

2   Menezes de Pádua CA, Uter W, Geier J, Schnuch A, Effendy I; German Contact Dermatitis Research Group (DKG); Information Network of Departments of Dermatology (IVDK). Contact allergy to topical antifungal agents. Allergy 2008;63:946-947

3   Romano C, Ghilardi A, Calò D, Maritati E. Contact dermatitis due to ciclopiroxolamine. Mycoses 2006;49:338-339

4   Foti C, Diaferio A, Bonamonte D. Allergic contact dermatitis from ciclopirox olamine. Australas J Dermatol 2001;42:145

5   Jager SU, Pönninghaus JM, Koch P. Allergic contact dermatitis from cyclopiroxolamine? Contact Dermatitis 1995;33:349-350

6   Goitre M, Bedello PG, Cane D, Pulatti P, Forte M, Cervetti O. Contact dermatitis due to cyclopyroxolamine. Contact Dermatitis 1986;15:94-95

Chapter 3.70   CLEMIZOLE

IDENTIFICATION

Description/definition   : Clemizole is the benzimidazole that conforms to the structural formula shown below
Pharmacological classes : Histamine H1 antagonists
IUPAC name   : 1-[(4-Chlorophenyl)methyl]-2-(pyrrolidin-1-ylmethyl)benzimidazole
Other names   : 1-(p-Chlorobenzyl)-2-pyrrolidylmethylenebenzimidazole
CAS registry number   : 442-52-4
EC number   : 207-133-5
Merck Index monograph   : 3614
Patch testing   : 2% pet.
Molecular formula   : $C_{19}H_{20}ClN_3$

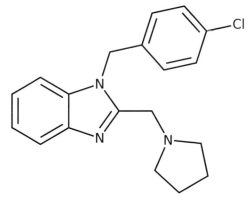

GENERAL

Clemizole is a member of the class of benzimidazoles and a histamine H1 blocker used to treat allergies. In pharmaceutical products (very rarely used), clemizole may be employed as clemizole hydrochloride (CAS number 1163-36-6, EC number 214-605-4, molecular formula $C_{19}H_{21}Cl_2N_3$) (1).

CONTACT ALLERGY

Case reports and case series

In Leuven, Belgium, in the period 1990-2014, iatrogenic contact dermatitis was diagnosed in 2600 individuals (17% of the total patch test population). 96% of all positive patch test reactions to topical drugs and antiseptics were considered to be relevant. Clemizole (1% pet.) was tested in 31 patients and there were 3 positive reactions to it (2, overlap with the data in ref. 3).

In the period 1978-1997, in the same University Clinic in Leuven, Belgium, 6 patients had positive patch test reactions to clemizole (patch test concentration and vehicle not mentioned). The relevance of these reactions was not mentioned, but it was stated that clemizole in Belgium was used in topical antihemorrhoidal preparations (3, overlap with the data in ref. 2).

A patient allergic to a corticosteroid cream containing fluocortolone and/or fluocortolone caproate reacted to the steroid, to clemizole 2% pet., to clemizole penicillin and to hexachlorophene 1% pet. The cream to which the patient was allergic did not contain these materials, but a corticosteroid ointment of the same brand, to which he also had a positive patch test, contained 'clemizolum hexachlorophenate' (unknown chemical, possibly clemizole AND hexachlorophene). It was uncertain whether the patient had ever used this ointment (4).

LITERATURE

1   The data in the section 'General' may have been obtained from literature discussed in this chapter, but mostly also or exclusively from one or more of the following online sources: ChemIDPlus Advanced, PubChem, DrugBank, RxList, Drug Central, Drugs.com, and Wikipedia
2   Gilissen L, Goossens A. Frequency and trends of contact allergy to and iatrogenic contact dermatitis caused by topical drugs over a 25-year period. Contact Dermatitis 2016;75:290-302
3   Goossens A, Linsen G. Contact allergy to antihistamines is not common. Contact Dermatitis 1998;39:38-39
4   Van Ketel WG. Allergy to Ultralan® preparations. Contact Dermatitis Newsletter 1974;15:427

Chapter 3.71   CLINDAMYCIN

IDENTIFICATION

Description/definition : Clindamycin is a semisynthetic antibiotic produced by chemical modification of the parent compound lincomycin; it conforms to the structural formula shown below
Pharmacological classes : Anti-bacterial agents; protein synthesis inhibitors
IUPAC name : (2S,4R)-N-[(1S,2S)-2-Chloro-1-[(2R,3R,4S,5R,6R)-3,4,5-trihydroxy-6-methylsulfanyloxan-2-yl]propyl]-1-methyl-4-propylpyrrolidine-2-carboxamide
Other names : 7-Chloro-7-deoxylincomycin
CAS registry number : 18323-44-9
EC number : 242-209-1
Merck Index monograph : 3624
Patch testing : Phosphate, 10.0% pet. (Chemotechnique)
Molecular formula : $C_{18}H_{33}ClN_2O_5S$

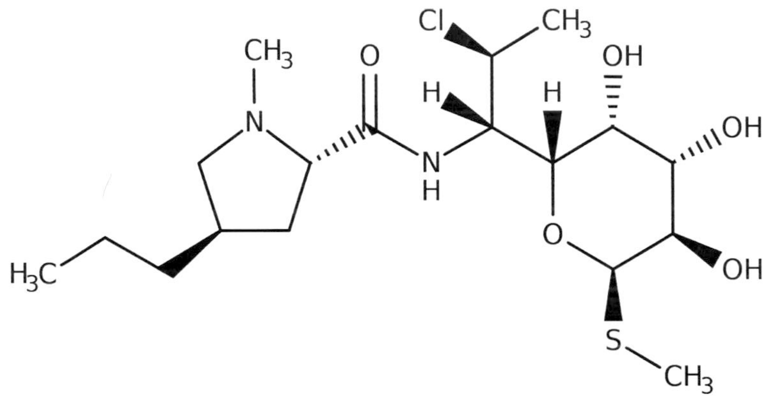

GENERAL

Clindamycin is a semisynthetic broad-spectrum antibiotic produced by chemical modification of the parent compound lincomycin. This agent dissociates peptidyl-tRNA from the bacterial ribosome, thereby disrupting bacterial protein synthesis. Clindamycin is indicated for the treatment of serious infections caused by susceptible anaerobic bacteria, including *Bacteroides* spp., *Peptostreptococcus*, anaerobic streptococci, *Clostridium* spp., and microaerophilic streptococci. The antibiotic may be useful in polymicrobial infections such as intra-abdominal or pelvic infections, osteomyelitis, diabetic foot ulcers, aspiration pneumonia and dental infections. Another use of clindamycin is vaginally to treat vaginosis caused by *Gardnerella vaginosa*. Clindamycin reduces the toxin-producing effects of *S. aureus* and *S. pyogenes* and as such, may be particularly useful for treating necrotizing fasciitis. In topical preparations, clindamycin is widely used in the treatment of inflammatory acne vulgaris (1).

In pharmaceutical products, clindamycin is usually employed as clindamycin phosphate (CAS number 24729-96-2, EC number 246-433-0, molecular formula $C_{18}H_{34}ClN_2O_8PS$), sometimes as clindamycin hydrochloride (CAS number 21462-39-5, EC number 244-398-6, molecular formula $C_{18}H_{34}Cl_2N_2O_5S$) (1).

CONTACT ALLERGY

General

Allergic contact dermatitis caused by topical clindamycin was first reported in 1978 (14). Since then, some 10 cases have been described, caused by clindamycin hydrochloride or clindamycin phosphate. In some cases, allergic contact dermatitis presented with atypical clinical pictures, such as rosacea-like rash (11) or mimicking a 'retinoid flare' (7). As clindamycin is often used on large areas of the skin and for long periods of time, its potential for contact sensitization is apparently very low (6).

Patch testing in groups of selected patients

In the period 2003-2010, in the USA, 73 women with predominantly vulvar symptoms were patch tested with clindamycin phosphate 2% pet and one (1%) had a positive patch test, which was considered to be relevant (4).

Case series

In Leuven, Belgium, in the period 1990-2014, iatrogenic contact dermatitis was diagnosed in 2600 individuals (17% of the total population). 96% of all positive patch test reactions to topical drugs and antiseptics were considered to be relevant. Clindamycin (10% pet.) was tested in 43 patients and there were 4 positive reactions to it (2).

Case reports

A 58-year-old woman was treated for hidradenitis suppurativa of both inguinal folds with oral amoxicillin and topical clindamycin phosphate 1% gel 2x daily. One week later, the patient returned with an acute very itchy rash at the application sites, characterized by erythema and vesicles. Allergic contact dermatitis was diagnosed and clindamycin treatment was stopped. Patch tests were positive to the gel and another clindamycin-containing pharmaceutical, to clindamycin hydrochloride 1% pet. and clindamycin phosphate 1% water, but were negative to the other ingredients of the gel. Eight controls were negative (5).

A 24-year-old woman presented with an eczematous and pustular eruption on her face associated with itching, burning, and stinging, resembling a 'retinoid-flare'. She had noted worsening of the facial lesions from using 1% clindamycin phosphate cream and 1% adapalene cream. Patch tests were positive to the clindamycin cream and – later – to its ingredient clindamycin 1% pet. A repeated open application test with clindamycin 1% pet. was positive after 2 applications already. Ten controls were negative to patch tests with clindamycin 1% pet. (7).

Short summaries of other case reports of allergic contact dermatitis from clindamycin are shown in table 3.71.1.

Table 3.71.1 Short summaries of case reports of allergic contact dermatitis from clindamycin

Year and country	Sex	Age	Positive patch tests	Clinical data and comments	Ref.
2020 USA	F	63	ClinP lotion 'as is'; Clin 10% pet. (2x)	eczema in the axillae and on lateral trunk from ClinP 1% lotion used as deodorant for bromhidrosis	22
1996 Italy	F	24	ClinP 1% pet.	ACD 2-3 days after using ClinP 1% lotion for acne vulgaris	8
1995 Denmark	F	36	ClinP 1% lotion 'as is'; ClinP 1% water	ACD in the axillae after using ClinP lotion for 2 months on hidradenitis suppurativa; subsequently generalized dermatitis unresponsive to topical corticosteroids	9
1991 Japan	M	21	ClinHCl 1% lotion 'as is'; ClinHCl 0.5% water; ClinP 0.1% water	facial erythema and papules after using ClinHCl lotion for one month for acne vulgaris	10
1989 Netherlands	F	43	ClinP 1% lotion 'as is'; ClinP 1% water	pustular and erythematous eruption after 3 days treatment of 'acne' with ClinP lotion; it was probably an exacerbation of previously existing rosacea from the allergic reaction	11
1983 Spain	M	55	ClinHCl and ClinP 1% water	not well-described reaction to oral clindamycin and topical 5% clindamycin solution for atopic dermatitis; cross-reaction to the related lincomycin	12
1982 Netherlands	F	31	ClinHCl 2% lotion 'as is'; ClinHCl 1% water	facial itching with a mild eruption of papules after 5 weeks' treatment of acne vulgaris with ClinHCl lotion	13
1978 United States	F	33	ClinHCl 1% water	itching and multiple excoriations from treatment of acne vulgaris with 1% ClinHCl in alcohol	14
1978 United States	F	23	ClinP caps. 1% water	12 days after initiating ClinP caps. 1% solution for acne papular and nodular eruption; reintroduction of ClinP in another base led to the same reaction within 36 hours	15

ACD: allergic contact dermatitis; ClinHCl: clindamycin hydrochloride; ClinP: clindamycin phosphate

Cross-reactions, pseudo-cross-reactions and co-reactions

A patient sensitized to lincomycin cross-reacted to the structurally closely related antibiotic clindamycin (3). An individual sensitized to clindamycin cross-reacted to lincomycin (12).

Cutaneous adverse drug reactions from systemic administration caused by type IV (delayed-type) hypersensitivity

Cutaneous adverse drug reactions from systemic administration of clindamycin caused by type IV (delayed-type) hypersensitivity, including maculopapular exanthemas (16-21,24,25), acute generalized exanthematous pustulosis (AGEP) (23,24) and drug reaction with eosinophilia and systemic symptoms (DRESS) (24) are planned to be discussed in Volume IV of the *Monographs in Contact Allergy* series on Systemic drugs.

LITERATURE

1   The data in the section 'General' may have been obtained from literature discussed in this chapter, but mostly also or exclusively from one or more of the following online sources: ChemIDPlus Advanced, PubChem, DrugBank, RxList, Drug Central, Drugs.com, and Wikipedia

2   Gilissen L, Goossens A. Frequency and trends of contact allergy to and iatrogenic contact dermatitis caused by topical drugs over a 25-year period. Contact Dermatitis 2016;75:290-302

3   Conde-Salazar L, Guimaraens D, Romero L, Gonzalez M, Yus S. Erythema multiforme-like contact dermatitis from lincomycin. Contact Dermatitis 1985;12:59-61

4   O'Gorman SM, Torgerson RR. Allergic contact dermatitis of the vulva. Dermatitis 2013;24:64-72

5   Veraldi S, Guanziroli E, Ferrucci S, Nazzaro G. Allergic contact dermatitis caused by clindamycin. Contact Dermatitis 2019;80:68-69

6   Veraldi S, Brena M, Barbareschi M. Allergic contact dermatitis caused by topical antiacne drugs. Expert Rev Clin Pharmacol 2015;8:377-381

7   Romita P, Ettorre G, Corazza M, Borghi A, Foti C. Allergic contact dermatitis caused by clindamycin mimicking 'retinoid flare'. Contact Dermatitis 2017;77:181-182

8   García R, Galindo PA, Feo F, Gómez E, Fernández F. Delayed allergic reactions to amoxycillin and clindamycin. Contact Dermatitis 1996;35:116-117

9   Vejlstrup E, Menné T. Contact dermatitis from clindamycin. Contact Dermatitis 1995;32:110

10 Yokoyama R, Mizuno E, Takeuchi M, Abe M, Ueda H. Contact dermatitis due to clindamycin. Contact Dermatitis 1991;25:125

11 De Kort WJ, de Groot AC. Clindamycin allergy presenting as rosacea. Contact Dermatitis 1989;20:72-73

12 Conde-Salazar L, Guimaraens D, Romero LV. Contact dermatitis from clindamycin. Contact Dermatitis 1983;9:225

13 De Groot AC. Contact allergy to clindamycin. Contact Dermatitis 1982;8:428

14 Coskey RJ. Contact dermatitis due to clindamycin. Arch Dermatol 1978;114:446

15 Herstoff JK. Sensitization to topical antibiotics. Arch Dermatol 1978;114:1402

16 Lammintausta K, Tokola R, Kalimo K. Cutaneous adverse reactions to clindamycin: results of skin tests and oral exposure. Br J Dermatol 2002;146:643-648

17 Papakonstantinou E, Müller S, Röhrbein JH, Wieczorek D, Kapp A, Jakob T, Wedi B. Generalized reactions during skin testing with clindamycin in drug hypersensitivity: a report of 3 cases and review of the literature. Contact Dermatitis 2018;78:274-280

18 Pereira N, Canelas MM, Santiago F, Brites MM, Gonçalo M. Value of patch tests in clindamycin-related drug eruptions. Contact Dermatitis 2011;65:202-207

19 Monteagudo B, Cabanillas M, Iriarte P, Ramírez-Santos A, León-Muinos E, González-Vilas D, Suárez-Amor Ó. Clindamycin-induced maculopapular exanthema with preferential involvement of striae distensae: A Koebner phenomenon? Acta Dermatovenerol Croat 2018;26:61-63

20 Seitz CS, Bröcker EB, Trautmann A. Allergy diagnostic testing in clindamycin-induced skin reactions. Int Arch Allergy Immunol 2009;149:246-250

21 Muñoz D, Del Pozo MD, Audícana M, Fernandez E, Fernandez De Corres LF. Erythema-multiforme-like eruption from antibiotics of 3 different groups. Contact Dermatitis 1996;34:227-228

22 Voller LM, Kullberg SA, Warshaw EM. Axillary allergic contact dermatitis to topical clindamycin. Contact Dermatitis 2020;82:313-314

23 Valois M, Phillips EJ, Shear NH, Knowles SR. Clindamycin-associated acute generalized exanthematous pustulosis. Contact Dermatitis 2003;48:169

24 Gilissen L, Huygens S, Goossens A, Breynaert C, Schrijvers R. Utility of patch testing for the diagnosis of delayed-type drug hypersensitivity reactions to clindamycin. Contact Dermatitis 2020;83:237-239

25 Vicente J, Fontela JL. Delayed reaction to oral treatment with clindamycin. Contact Dermatitis 1999;41:221

Chapter 3.72   CLIOQUINOL

IDENTIFICATION

Description/definition	: Clioquinol is the halogenated 8-hydroxyquinoline that conforms to the structural formula shown below
Pharmacological classes	: Dermatological drugs; antiseptics and disinfectants; anti-infective agents
IUPAC name	: 5-Chloro-7-iodoquinolin-8-ol
Other names	: Iodochlorhydroxyquin; 5-chloro-7-iodo-8-hydroxyquinoline
CAS registry number	: 130-26-7
EC number	: 204-984-4
Merck Index monograph	: 6345
Patch testing	: 5% pet. (Chemotechnique, SmartPracticeCanada, SmartPracticeEurope)
Molecular formula	: C_9H_5ClINO

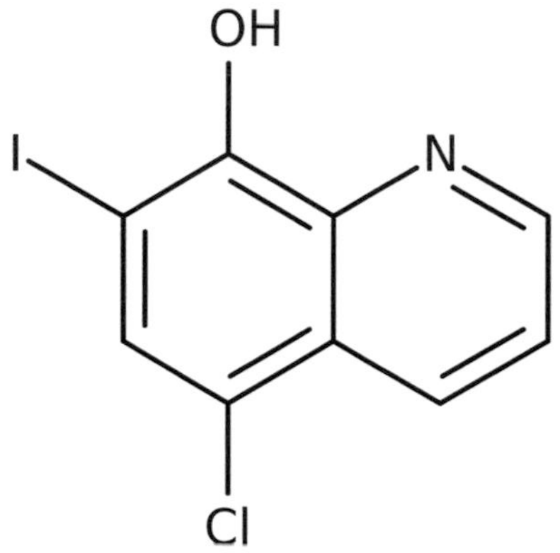

GENERAL

Clioquinol is a broad-spectrum antibacterial agent with antifungal properties. It is used as a topical antibacterial and antifungal treatment. However, topical absorption of this iodine-containing agent is rapid and extensive, especially when the skin is covered with an occlusive dressing or if the medication is applied to extensive or eroded areas of the skin, which may affect thyroid function tests (1).

CONTACT ALLERGY

General

Clioquinol was formerly used widely in topical medicaments for the treatment of leg ulcers and in combination preparations with corticosteroids for the treatment of (secondarily infected) eczema. Allergic reactions were mostly seen in patients with venous leg ulcers and stasis dermatitis, who are known to be highly susceptible to contact sensitization. Because of its yellow color and the availability of more effective anti-infective agents, it is infrequently used in European countries anymore. As this inevitably resulted in low frequencies of sensitization, clioquinol, which had been part of the European baseline series for decades, was removed from it in 2019 (48).

Several cases of systemic contact dermatitis to clioquinol have been reported, both from percutaneous/mucosal absorption (41,42,44) and from oral administration of the drug (40,43,49,50). Cases of irritant contact dermatitis from topical clioquinol preparations have been observed (24,45).

Contact allergy in the general population and subgroups

Estimates of the 10-year prevalence (1997-2006) of contact allergy to clioquinol in the general population of Denmark based on the CE-DUR method ranged from 0.13% to 0.18% (23). In Germany, for the period 1995-2004, the population-based relative incidence (RI) of contact sensitization to clioquinol (cases/100,000 defined daily doses (DDDs) per year) was estimated to be 64.5. In the group of all topical drugs, the RI ranged from 0.3 (dexamethasone sodium phosphate and pilocarpine) to 413.9 (benzocaine) (27).

In 2014, a group of 481 healthy Chinese university student volunteers aged 20-27 years (332 women, 149 men) were patch tested with a battery of 34 haptens and there was one reaction to clioquinol 0.2% pet. in a female volunteer (frequency of sensitization in the entire group 0.2%, in women 0.3%, in men 0%) (25).

Patch testing in groups of patients

Results of patch testing clioquinol in consecutive patients suspected of contact dermatitis (routine testing) back to 1998 are shown in table 3.72.1. As clioquinol until 2019 was included in the European baseline series, most studies with routine testing have been performed in European countries and other countries where the same baseline series is used, e.g. Israel and Turkey; there are also 2 investigations from the USA. Prevalences of sensitization to clioquinol were invariably low, ranging from 0.1% to 0.8%. Studies performed after 2005 tended to have lower scores than the older ones. The relevance of the positive patch tests was addressed in 3 studies only and ranged from 40% to 96%, but culprit products were not mentioned (5,21,28)

Table 3.72.1 Patch testing in groups of patients: Routine testing

Years and Country	Test conc. & vehicle	Number of patients tested	positive (%)		Selection of patients (S); Relevance (R); Comments (C)	Ref.
2013-2019 Turkey	5% pet.	1247	3	(0.2%)	R: not stated	54
2013-2014 12 European countries [b]	5% pet.	11,922		(0.3%)	R: not stated; C: range of positive reactions: 0% - 0.9%	2
1990-2014 Belgium	1% pet.	10,370	49	(0.5%)	R: 96% of the positive patch test reactions to all topical drugs and antiseptics were considered to be relevant; C: there was a significant decrease in the frequency of positive reactions from 0.5% in 1990-1994 to 0.1% in 2010-2014; before 1995, the quinoline mix 6% pet. was used for patch testing clioquinol; the 1% test concentration may be a mistake, 5% is the usual and commercially available concentration	28
2009-12, six European countries [a]	5% pet.	16,435		(0.3%)	R: not stated; C: range per country: 0.2% - 0.8%	17
2007-2008 11 European countries [b]	5% pet.	23,439	16	(0.1%) [a]	R: not stated; C: prevalences ranged from 0% to 0.5%	3
2005-2006 9 European countries [b]	5% pet.	4441	12	(0.3%)	R: not stated; C: prevalences were 0.1% in Central Europe, 0.5% in Northeast and 0.2% in South Europe; in Western Europe (U.K.), clioquinol was not tested	4
2001-2005 USA	5% pet.	2705		(0.6%)	R: 40%	5
1985-2005 Denmark	5% pet.	14,996		(0.7%)	R: not stated	6
2004, 11 European countries [b]	5% pet.	1629	5	(0.3%) [a]	R: not stated; C: range positives per center: 0.0% - 0.7%	7
1998-2004 Israel	5% pet.	2156	3	(0.1%)	R: not stated	8
2002-2003 Europe [b]	5% pet.	2792		(0.4%)	R: not stated	9
1997-2001 Czech Rep.	5% pet.	12,058	69	(0.6%)	R: not stated	10
1998-2000 USA	5% pet.	713		(0.4%)	R: not stated	11
1996-2000 Europe	5% pet.	26,210		(0.8%)	R: not stated; C: ten centers, seven countries, EECDRG study; an unspecified number of these patients had been tested with and reacted to the quinoline mix	12
1998 United Kingdom	5% pet.	1119	8	(0.7%)	R: the 'vast majority' of all reactions were relevant	21

[a] age-standardized and sex-standardized proportions; [b] study of the ESSCA (European Surveillance System on Contact Allergies); EECDRG: European Environmental and Contact Dermatitis Research Group

Results of testing clioquinol in groups of *selected* patients (e.g. patients with stasis dermatitis/leg ulcers, individuals with perianal, vulval dermatitis or otitis externa or media) back to 1982 are shown in table 3.72.2. In patients with stasis dermatitis/leg ulcers, prevalences of positive patch tests to clioquinol were higher, ranging from 1% to 10%. Here also, the more recent studies tended to have lower rates, but these are obviously dependent on the country and the extent of clioquinol usage there. However, the IVDK (Information Network of Departments of Dermatology: Germany, Austria, Switzerland) in 1992-1997 noted a sensitization rate of 2.4% (19), which had decreased to 1.0% in the period 2003-2014 (13).

Table 3.72.2 Patch testing in groups of patients: Selected patient groups

Years and Country	Test conc. & vehicle	Number of patients tested	positive (%)		Selection of patients (S); Relevance (R); Comments (C)	Ref.
Patients with leg ulcers/stasis dermatitis						
2003-2014 IVDK	5% pet.	1782		(1.0%)	S: patients with stasis dermatitis/chronic leg ulcers; R: not stated	13
2005-2008 France	5% pet.	423	6	(1.4%)	S: patients with leg ulcers; R: not stated	14
<2008 Poland	5% pet.	50	5	(10%)	S: patients with venous leg ulcers; R: not stated	16
<2005 Poland	5% pet.	50	5	(10%)	S: patients with chronic venous leg ulcers; R: not stated	15

Table 3.72.2 Patch testing in groups of patients: Selected patient groups (continued)

Years and Country	Test conc. & vehicle	Number of patients tested | positive (%)		Selection of patients (S); Relevance (R); Comments (C)	Ref.
1992-1997 IVDK	5% pet.	1429	40 (2.8%)	S: patients with leg ulcers and/or leg eczema; R: not stated	19
<1994 U.K.		85	8 (9%)	S: patients with longstanding venous ulceration or eczema complicating leg ulcers; R: not stated	18
Other patient groups					
2004-2016 Spain		124	1 (0.8%)	S: patients with perianal dermatitis lasting >4 weeks; R: 0%	22
1995-2004 IVDK	5% pet.	10,410	(1.1%)	S: not stated; R: not stated	26
1991-1995 U.K.		121	3 (2.5%)	S: women with pruritus vulvae and primary vulval dermatoses suspected of secondary allergic contact dermatitis; R: 49% for all positive reactions together	20
<1982 U.K.	5% pet.	40	5 (12%)	S: patients with otitis externa, chronic suppurative otitis media or discharging mastoid cavities >1 year; R: not stated	55

IVDK: Information Network of Departments of Dermatology (Germany, Austria, Switzerland)

Case reports

A woman aged 33 with anogenital pruritis had used a wide variety of local applications. Examination showed an extensive lichenified and excoriated dermatitis of the perianal and vulval skin. *Candida albicans* was grown from both sites. Patch tests were positive to clioquinol, clioquinol-hydrocortisone cream, 2 pharmaceuticals containing neomycin resp. polidocanol and their active ingredients (34). Several other patients have become sensitized to clioquinol from its presence in a clioquinol-corticosteroid preparation (35,36,37,46,51).

After 2 weeks treatment for axillary intertrigo with a cream containing clioquinol, tolnaftate, gentamicin and betamethasone valerate, a 47-year-old man developed acute eczema in both axillae. Patch tests were positive to the cream, to clioquinol 5% pet. and to tolnaftate 0.1% and 1% pet.; all other active ingredients and excipients tested negative (2). A teenage school girl developed allergic contact dermatitis of the cheeks and the sides of her right index and middle finger from sensitization to clioquinol in an ointment she used for her dog, containing 3% clioquinol and 0.5% hydrocortisone hemisuccinate (37).

Contact allergy to clioquinol has also caused an erythema multiforme-like allergic contact dermatitis (33). Other cases of allergic contact dermatitis from clioquinol (not a full literature review) have been described in refs. 52 and 53 (with photosensitivity).

Systemic contact dermatitis

Oral administration of clioquinol in patients previously sensitized by topical application of this drug will induce an eruption in some (some say: most) patients: systemic contact dermatitis. In some of them a flare up of the original dermatitis is seen (50), while in other cases a generalized dermatitis may be observed (40). Absorption of clioquinol through the skin and mucosae may also have induced systemic contact dermatitis (41,42,44).

A 3-year-old female child was prescribed clioquinol/hydrocortisone combination cream to be used topically to the vaginal area daily for burning in the urogenital area post voiding. After 2 days of topical therapy, she developed a rapid onset painful, erythematous, dermatitic eruption around her neck. The eruption then spread to involve her bilateral axillae, antecubital fossae, groin, and perioral area. She was diagnosed with a systemic contact dermatitis to topical clioquinol, which is unsubstantiated as patch testing was not performed (41,42). The clinical picture was classic for systemic contact dermatitis, though, and clioquinol was a very likely candidate. In this case, absorption of clioquinol through the genital skin and mucosae may have caused the systemic symptoms.

A 40-year-old woman started to take clioquinol tablets as prophylaxis against travellers' diarrhea. After 2 days, her palms began to itch. The next day, the itching spread to the whole body, and during the following 2 days a generalized eruption developed. At the time of admission, by which time she had taken four tablets, she had extremely itchy, edematous plaques with papules and vesicles especially on the breast, abdomen, back and hands. She also had tinea pedis and was by mistake treated with clioquinol powder. During the next 24 h she developed a severe bullous reaction on her feet. Patch tests were positive to clioquinol and the related chlorquinaldol (43).

A 68-year old woman developed generalized erythema after one application of a cream containing 3% clioquinol, nystatin and 0.5% hydrocortisone) for submammary intertrigo. The patient was known to be sensitive to oral quinine, and, because of structural similarities, the exanthema was (implicitly) ascribed to clioquinol, although patch tests were not performed (44). Percutaneous absorption of clioquinol from the moist and occluded skin under the mammae may have been responsible for the eruption.

The left side of the body of a 32-year-old man was treated with an ointment containing 1% clioquinol for pustular psoriasis. This resulted in a contact dermatitis localized at the sites of application. The treatment was discontinued and the reaction disappeared within a week. Later, to treat a slight enteritis, the patient took two tablets of 0.25 gram clioquinol. A few hours later, the left side of his body began to itch and he developed a severe, oozing dermatitis which was limited to the area which had been treated topically with clioquinol nine months earlier. Patch tests were not performed, but this case history is highly suggestive for systemic allergic contact dermatitis from clioquinol (50).

Two more patients sensitized to clioquinol had a generalized eruption after having taken clioquinol tablets (49).

Cross-reactions, pseudo-cross-reactions and co-reactions
Cross-reactions between halogenated hydroxyquinolines such as clioquinol (5-chloro-7-iodoquinolin-8-ol), chlorquinaldol (5,7-dichloro-2-methylquinolin-8-ol), cloxyquin (5-chloroquinolin-8-ol), oxyquinoline (8-hydroxyquinoline, non-halogenated), iodoquinol (5,7-diiodoquinolin-8-ol), halquinol (a mixture of 4 hydroxy-quinolines), 5,7-dichloro-8-quinolinol and 5-chloro-8-quinolinol may occur (29-32,38,47 [examples of references]).

Ten patients sensitized to clioquinol were patch tested with a number of structurally-related chemicals and there were 9 cross-reactions to broxyquinoline (5,7-dibromo-8-quinolinol), 9 to chlorquinaldol, 4 to quinine, one to chloroquine (possibly false-positive from excited skin syndrome), one to amodiaquine (ibid), and two to potassium iodide (10% pet. ++ and + , 5% pet. neg., possibly also false-positive) (39). Cross-reactions to chloroquine had been observed before (38). Possibly cross-sensitivity from quinine to clioquinol (44; patch tests not performed).

Immediate contact reactions
Immediate contact reactions (contact urticaria) to clioquinol are presented in Chapter 5.

LITERATURE

1   The data in the section 'General' may have been obtained from literature discussed in this chapter, but mostly also or exclusively from one or more of the following online sources: ChemIDPlus Advanced, PubChem, DrugBank, RxList, Drug Central, Drugs.com, and Wikipedia

2   Uter W, Amario-Hita JC, Balato A, Ballmer-Weber B, Bauer A, Belloni Fortina A, et al. European Surveillance System on Contact Allergies (ESSCA): results with the European baseline series, 2013/14. J Eur Acad Dermatol Venereol 2017;31:1516-1525

3   Uter W, Aberer W, Armario-Hita JC, , Fernandez-Vozmediano JM, Ayala F, Balato A, et al. Current patch test results with the European baseline series and extensions to it from the 'European Surveillance System on Contact Allergy' network, 2007-2008. Contact Dermatitis 2012;67:9-19

4   Uter W, Rämsch C, Aberer W, Ayala F, Balato A, Beliauskiene A, et al. The European baseline series in 10 European Countries, 2005/2006 – Results of the European Surveillance System on Contact Allergies (ESSCA). Contact Dermatitis 2009;61:31-38

5   Davis MD, Scalf LA, Yiannias JA, Cheng JF, El-Azhary RA, Rohlinger AL, et al. Changing trends and allergens in the patch test standard series. Arch Dermatol 2008;144:67-72

6   Carlsen BC, Menné T, Johansen JD. 20 Years of standard patch testing in an eczema population with focus on patients with multiple contact allergies. Contact Dermatitis 2007;57:76-83

7   ESSCA Writing Group. The European Surveillance System of Contact Allergies (ESSCA): results of patch testing the standard series, 2004. J Eur Acad Dermatol Venereol 2008;22:174-181

8   Lazarov A. European Standard Series patch test results from a contact dermatitis clinic in Israel during the 7-year period from 1998 to 2004. Contact Dermatitis 2006;55:73-76

9   Uter W, Hegewald J, Aberer W et al. The European standard series in 9 European countries, 2002/2003 – First results of the European Surveillance System on Contact Allergies. Contact Dermatitis 2005;53:136-145

10 Machovcova A, Dastychova E, Kostalova D, et al. Common contact sensitizers in the Czech Republic. Patch test results in 12,058 patients with suspected contact dermatitis. Contact Dermatitis 2005;53:162-166

11 Wetter DA, Davis MDP, Yiannias JA, Cheng JF, Connolly SM, el-Azhary RA, et al. Patch test results from the Mayo Contact Dermatitis Group, 1998–2000. J Am Acad Dermatol 2005;53:416-421

12 Bruynzeel DP, Diepgen TL, Andersen KE, Brandão FM, Bruze M, Frosch PJ, et al (EECDRG). Monitoring the European Standard Series in 10 centres 1996–2000. Contact Dermatitis 2005;53:146-152

13 Erfurt-Berge C, Geier J, Mahler V. The current spectrum of contact sensitization in patients with chronic leg ulcers or stasis dermatitis - new data from the Information Network of Departments of Dermatology (IVDK). Contact Dermatitis 2017;77:151-158

14 Barbaud A, Collet E, Le Coz CJ, Meaume S, Gillois P. Contact allergy in chronic leg ulcers: results of a multicentre study carried out in 423 patients and proposal for an updated series of patch tests. Contact Dermatitis 2009;60:279-287

15 Zmudzinska M, Czarnecka-Operacz M, Silny W, Kramer L. Contact allergy in patients with chronic venous leg ulcers – possible role of chronic venous insufficiency. Contact Dermatitis 2006;54:100-105

16 Zmudzinska M, Czarnecka-Operacz M, SilnyW. Contact allergy to glucocorticosteroids in patients with chronic venous leg ulcers, atopic dermatitis and contact allergy. Acta Dermatovenerol Croat 2008;16:72-78

17 Uter W, Spiewak R, Cooper SM, Wilkinson M, Sánchez Pérez J, Schnuch A, et al. Contact allergy to ingredients of topical medications: results of the European Surveillance System on Contact Allergies (ESSCA), 2009-2012. Pharmacoepidemiol Drug Saf 2016;25:1305-1312

18 Zaki I, Shall L, Dalziel KL. Bacitracin: a significant sensitizer in leg ulcer patients? Contact Dermatitis1994;31:92-94

19 Renner R, Wollina U. Contact sensitization in patients with leg ulcers and/or leg eczema: comparison between centers. Int J Low Extrem Wounds 2002;1:251-255

20 Lewis FM, Shah M, Gawkrodger DJ. Contact sensitivity in pruritus vulvae: patch test results and clinical outcome. Dermatitis 1997;8:137-140

21 Morris SD, Rycroft RJ, White IR, Wakelin SH, McFadden JP. Comparative frequency of patch test reactions to topical antibiotics. Br J Dermatol 2002;146:1047-1051

22 Agulló-Pérez AD, Hervella-Garcés M, Oscoz-Jaime S, Azcona-Rodríguez M, Larrea-García M, Yanguas-Bayona JI. Perianal dermatitis. Dermatitis 2017;28:270-275

23 Thyssen JP, Uter W, Schnuch A, Linneberg A, Johansen JD. 10-year prevalence of contact allergy in the general population in Denmark estimated through the CE-DUR method. Contact Dermatitis 2007;57:265-272

24 Kero M, Hannuksela M, Sothman A. Primary irritant dermatitis from topical clioquinol. Contact Dermatitis 1979;5:115-117

25 Zhao L, Li LF. Contact sensitization to 34 common contact allergens in University students in Beijing. Contact Dermatitis 2015;73:323-324

26 Menezes de Padua CA, Uter W, Schnuch A. Contact allergy to topical drugs: prevalence in a clinical setting and estimation of frequency at the population level. Pharmacoepidemiol Drug Saf 2007;16:377-384

27 Menezes de Padua CA, Schnuch A, Nink K, Pfahlberg A, Uter W. Allergic contact dermatitis to topical drugs – Epidemiological risk assessment. Pharmacoepidemiol Drug Saf 2008;17:813-821

28 Gilissen L, Goossens A. Frequency and trends of contact allergy to and iatrogenic contact dermatitis caused by topical drugs over a 25-year period. Contact Dermatitis 2016;75:290-302

29 Myatt AE, Beck MH. Contact sensitivity to chlorquinaldol. Contact Dermatitis 1983;9:523

30 Leifer W, Steiner K. Studies in sensitization to halogenated hydroxyquinolines and related compounds. J Invest Dermatol 1951;17:233-240

31 Allenby CF. Skin sensitization to Remiderm and cross-sensitization to hydroxyquinoline compounds. Br Med J 1965;2(5455):208-209

32 Wantke F, Götz M, Jarisch R. Contact dermatitis from cloxyquin. Contact Dermatitis 1995;32:112-113

33 Meneghini CL, Angelini G. Secondary polymorphic eruptions in allergic contact dermatitis. Dermatologica 1981;163:63-70

34 Calnan CD. Oxypolyethoxydodecane in an ointment. Contact Dermatitis 1978;4:168-169

35 González Pérez R, Aguirre A, Oleaga JM, Eizaguirre X, Díaz Pérez JL. Allergic contact dermatitis from tolnaftate. Contact Dermatitis 1995;32:173

36 Freeman S, Stephens R. Cheilitis: analysis of 75 cases referred to a contact dermatitis clinic. Am J Contact Dermat 1999;10:198-200

37 Malten KE. Therapeutics for pets as neglected causes of contact dermatitis in housewives. Contact Dermatitis 1978;4:296-299

38 Bielicky T, Novak M. Gruppensensibilisierung gegen Chinolinderivate. Dermatologica 1969;138:45-58 (Article in German, data cited in ref. 39)

39 Soesman-van Waadenoijen Kernekamp A, van Ketel WG. Persistence of patch test reactions to clioquinol (Vioform) and cross-sensitization. Contact Dermatitis 1980;6:455-460

40 Ekelund AG, Möller H. Oral provocation in eczematous contact allergy to neomycin and hydroxyquinolines. Acta Dermato-venereologica 1969;49:422-426

41 Mussani F, Poon D, Skotnicki-Grant S. Systemic contact dermatitis to topical clioquinol/hydrocortisone combination cream. Dermatitis 2013;24:196-197

42 Mussani F, Skotnicki S. Systemic contact dermatitis: Two Interesting cases of systemic eruptions following exposure to drugs clioquinol and metronidazole. Dermatitis 2013;24(4):e3

43 Skog E. Systemic eczematous contact-type dermatitis induced by iodochlorhydroxyquin and chloroquine phosphate. Contact Dermatitis 1975;1:187

44 Simpson JR. Reversed cross-sensitization between quinine and iodochlorhydroxyquinoline. Contact Dermatitis Newsletter 1974;15:431

45 Beck MH, Wilkinson SM. A distinctive irritant contact reaction to Vioform (clioquinol). Contact Dermatitis 1994;31:54-55

46 Newbold PC. Contact reaction to penicillamine in vaginal secretions. Lancet 1979;1(8130):1344

47 Van Ketel WG. Cross-sensitization to 5,7 dibromo-8-hydroxyquinoline (DBO). Contact Dermatitis 1975;1:385

48 Wilkinson M, Gonçalo M, Aerts O, Badulici S, Bennike NH, Bruynzeel D, et al. The European baseline series and recommended additions: 2019. Contact Dermatitis 2019;80:1-4

49 Cronin E. Contact Dermatitis. Edinburgh: Churchill Livingstone, 1980:220

50 Domar M, Juhlin L. Allergic dermatitis produced by oral clioquinol. Lancet 1967;1(7500):1165-1166

51 Khoo B-P, Leow Y-H, Ng S-K, Goh C-L. Corticosteroid contact hypersensitivity screening in Singapore. Am J Contact Dermat 1998;9:87-91

52 Goh CL. Cross-sensitivity to multiple topical corticosteroids. Contact Dermatitis 1989;20:65-67

53 Rivara G, Barile M, Guarrera M. Photosensitivity in a patient with contact allergic dermatitis from clioquinol. Photodermatol Photoimmunol Photomed 1991;8:225-226

54 Boyvat A, Kalay Yildizhan I. Patch test results of the European baseline series among 1309 patients in Turkey between 2013 and 2019. Contact Dermatitis 2020 Jul 3. doi: 10.1111/cod.13653. Online ahead of print.

55 Holmes RC, Johns AN, Wilkinson JD, Black MM, Rycroft RJ. Medicament contact dermatitis in patients with chronic inflammatory ear disease. J R Soc Med 1982;75:27-30

Chapter 3.73   CLOBETASOL PROPIONATE

IDENTIFICATION

Description/definition : Clobetasol propionate is the 17-*O*-propionate ester of the synthetic glucocorticoid clobetasol that conforms to the structural formula shown below

Pharmacological classes : Anti-inflammatory agents; glucocorticoids

IUPAC name : [(8*S*,9*R*,10*S*,11*S*,13*S*,14*S*,16*S*,17*R*)-17-(2-Chloroacetyl)-9-fluoro-11-hydroxy-10,13,16-trimethyl-3-oxo-6,7,8,11,12,14,15,16-octahydrocyclopenta[a]phenanthren-17-yl] propanoate

Other names : 21-Chloro-9-fluoro-11β,17-dihydroxy-16β-methylpregna-1,4-diene-3,20-dione 17-propionate; clobetasol 17-propionate

CAS registry number : 25122-46-7

EC number : 246-634-3

Merck Index monograph : 3628 (Clobetasol)

Patch testing : 1% pet. (Chemotechnique, SmartPracticeCanada, SmartPracticeEurope); 0.25% pet. (SmartPracticeCanada, SmartPracticeEurope); 0.1% alc. (SmartPracticeCanada)

Molecular formula : $C_{25}H_{32}ClFO_5$

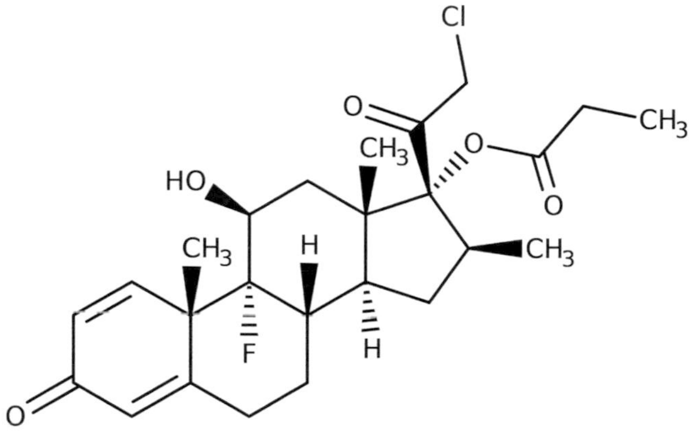

GENERAL

General aspects of corticosteroids used on the skin and mucous membranes are discussed in Chapter 2.4. A practical guideline for diagnosing allergic reactions to corticosteroids is presented in ref. 15. Clobetasol 17-propionate 1% pet. is included in the American core allergen series (www.smartpracticecanada.com).

CONTACT ALLERGY

Contact allergy in the general population

In Germany, for the period 1995-2004, the population-based relative incidence (RI) of contact sensitization to clobetasol propionate (cases/100,000 defined daily doses (DDDs) per year) was estimated to be 1.4. In the group of corticosteroids, the RI ranged from 0.3 (dexamethasone sodium phosphate) to 43.2 (budesonide) (14).

Patch testing in groups of patients

Results of patch testing clobetasol propionate in consecutive patients suspected of contact dermatitis (routine testing) are shown in table 3.73.1. Results of testing in groups of *selected* patients (individuals suspected of corticosteroid allergy, patients with anogenital dermatitis or leg ulcers, individuals with eczema or psoriasis unresponsive to topical corticosteroid treatment) are shown in table 3.73.2.

Patch testing in consecutive patients suspected of contact dermatitis: routine testing

Clobetasol propionate 1% pet. has been included in the screening series of the North American Contact Dermatitis Group (NACDG, USA + Canada) since 2003. Since then, the prevalences of sensitization in studies from the NACDG and other USA centers have been invariably low, 0.1%-0.8%. Previously, low rates had been reported from Israel, the United Kingdom, Belgium and 'Europe'. Only one study from Poland had a frequency of >1% positive patch test reactions, but the sample was very small (n=257) (34). Relevance rates, where mentioned, were 30-50%, and in the NACDG studies the combined definite + probable relevance ranged from 36% to 67% (table 3.73.1).

Table 3.73.1 Patch testing in groups of patients: Routine testing

Years and Country	Test conc. & vehicle	Number of patients tested | positive (%)		Selection of patients (S); Relevance (R); Comments (C)	Ref.
2015-2017 NACDG	1% pet.	5593	13 (0.2%)	R: definite + probable relevance: 61%	1
2007-2014 NACDG	1% pet.	17,978	58 (0.3%)	R: definite + probable relevance 45%	32
2015-2016 Spain	0.1% alc.	3699	13 (0.4%)	R: not stated	31
2011-2015 USA	1% pet.	2572	5 (0.2%)	R: not stated	2
2013-2014 NACDG	1% pet.	4859	6 (0.1%)	R: definite + probable relevance: 67%	3
2011-2012 NACDG	1% pet.	4230	11 (0.3%)	R: definite + probable relevance: 55%	4
2009-2010 NACDG	1% pet.	4304	(0.3%)	R: definite + probable relevance: 62%	5
2006-2010 USA	1% pet.	3083	(0.8%)	R: 31%	6
2007-2008 NACDG	1% pet.	5082	(0.6%)	R: definite + probable relevance: 36%	7
2005-2006 NACDG	1% pet.	4435	(0.8%)	R: definite + probable relevance: 45%	8
<2005 Poland	0.25% pet.	257	4 (1.6%)	R: 50%	34
2003-2004 NACDG	1% pet.	5140	38 (0.7%)	R: not stated	9
<2002 U.K	various	1526	13 (0.8%)	R: not stated; 1% alcohol was found to be the best test material (Abstract in ref. 16)	35
1995-1998 Israel		660	3 (0.5%)	R: not stated	38
1993-1994 Europe	1% pet.	7238	21 (0.3%)	R: 43% current relevance	40
1992-1993 U.K	1% alc.	2123	8 (0.4%)	R: not stated	41
1988-1990 Belgium	1% alc. 94%	1960	4 (0.2%)	R: not stated	43

IVDK: Information Network of Departments of Dermatology, Germany, Austria, Switzerland; NACDG: North American Contact Dermatitis Group (USA, Canada)

Patch testing in groups of selected patients

In most selected patient groups (table 3.73.2), the prevalences of positive patch tests reactions were below 1%, despite selection (anogenital dermatitis, leg ulcers, patients suspected of corticosteroid allergy). In only 2 investigations, >1% positive patch tests were observed (1.4% (30); 1.9% [46]). Relevance was hardly ever discussed, but in one study from the USA, all 17 reactions were scored as relevant (relevance rates from this group are always much higher than from other investigators) (30).

Table 3.73.2 Patch testing in groups of patients: Selected patient groups

Years and Country	Test conc. & vehicle	Number of patients tested | positive (%)		Selection of patients (S); Relevance (R); Comments (C)	Ref.
2005-2016 NACDG	1% pet.	449	2 (0.5%)	S: patients with only anogenital dermatitis; R: all positives represent relevant reactions; C: the frequency was not significantly different from a control group	12
2004-2008 IVDK			(0.9%)	S: patients with anogenital dermatoses tested with a medicament series; R: not stated; C: number of patients tested unknown	11
2000-2005 USA	1% pet.	1178	17 (1.4%)	S: patients suspected of corticosteroid allergy; R: 100%	30
1995-2004 IVDK	0.25% pet.	5882	(0.4%)	S: not stated; R: not stated	13
1995-2004 IVDK	0.25% pet.	5677	24 (0.4%)	S: patients tested with a corticosteroid series; R: not stated	33
1997-2001 U.K.		200	1 (0.5%)	S: patients with venous or mixed venous/arterial leg ulcers; R: all reactions to topical drugs were considered to be of probable, past or current relevance	10
1995-9 Netherlands	1% pet.	215	1 (0.5%)	S: not stated; R: probable 100%	37
1996-1997 IVDK	0.1% pet.	608	3 (0.5%)	S: patients tested with a corticosteroid series; R: not stated	36
1988-1991 U.K.	20% pet.	528	5 (0.9%)	S: patients with a positive patch test to tixocortol pivalate or suspected of corticosteroid allergy; R: not stated	42
1978-1980 Finland	0.05% alc.	105	2 (1.9%)	S: patients unresponsive to topical steroids or in whom eczema or psoriasis deteriorated; S: not stated	46

IVDK: Information Network of Departments of Dermatology, Germany, Austria, Switzerland; NACDG: North American Contact Dermatitis Group (USA, Canada)

Case series

From January 1990 to June 2008, in Leuven, Belgium, 315 patients were diagnosed with contact allergy to/allergic contact dermatitis from corticosteroids (CSs) from routine patch testing with a baseline series including tixocortol pivalate, budesonide, hydrocortisone butyrate and prednisone caproate, patch testing with patients' own CS preparations, and testing those with proven contact allergy to a corticosteroid or strongly suspected of CS allergy later with a series of 66 CSs, including two sex hormones (progesterone and testosterone). 71% of the patients had

relevant reactions, but these were not specified. In this group of 315 CS allergic patients, 16 had positive patch tests to clobetasol propionate 0.1% alc. (29). It is unknown how many of these reactions were caused by the use of a pharmaceutical product containing clobetasol propionate and how many were cross-reactions to other corticosteroids.

In the period 1982-1994, in a center in Tokyo, 11 patients were diagnosed with allergic contact dermatitis from topical corticosteroids, including one case caused by clobetasol propionate (47). One author in 1994 reviewed the Japanese literature and found 2 reported cases of allergic contact dermatitis to clobetasol propionate in 2 articles published in the period 1988-1991 (48).

Case reports

A 63-year-old man presented with severe scaly erythema on his scalp, face, and neck, and erythema on his trunk, arms, and legs after having been treated with clobetasol propionate ointment and lanoconazole cream. Patch tests were positive to clobetasol 0.05% ointment 'as is', clobetasol propionate 0.05% pet., lanoconazole cream 'as is' and lanoconazole 0.1% and 1% pet. (17). A 74-year-old-man, with eczema of the external auditory meatus for 6 months, was referred because of worsening of the otitis externa one day after applying clobetasol cream, together with spread to the thorax and axilla. He had also previously suffered a generalized erythema some hours after an intra-articular injection of paramethasone. The patient had a history of recurrent episodes of eczema in the axillae, flexural folds of the arm and the inguinal folds (highly suggestive of systemic contact dermatitis). Patch tests were positive to clobetasol propionate (CP) 0.05% cream, CP 1% pet., paramethasone acetate 2 mg/ml injection fluid and many other corticosteroids (39).

Short summaries of other case reports of allergic contact dermatitis from clobetasol propionate are shown in table 3.73.3. Additional case reports can be found in ref. 46 (cases 1 and 5) and ref. 49 (cases 3 and 9). Case reports of contact allergy to/allergic contact dermatitis from clobetasol propionate, adequate data of which are not available to the author, can be found in refs. 19,20,24 and 28 (2 cases).

Table 3.73.3 Short summaries of case reports of allergic contact dermatitis from clobetasol propionate

Year and country	Sex	Age	Positive patch tests	Clinical data and comments	Ref.
1991 Germany	F	41	CP comm prep 'as is'; CP 0.5% pet.	dermatitis around leg ulcers	44
	M	62	CP comm prep 'as is'; CP 0.5% pet.	worsening of stasis dermatitis	44
1990 United Kingdom	M	60	CP 20% pet.	worsening and spreading of dermatitis on the inner thighs; systemic contact dermatitis from i.m. triamcinolone acet.	45
1988 United kingdom	M	62	CP 5% pet.	worsening of long-standing eczema from CP preparations	27
	F	37	CP cream and ointment 'as is'; CP 5%, 0.5% and 0.05% pet.	facial edema, erythroderma, hand eczema; topical application of CP (cream or ointment?) caused severe local erythema and swelling of the skin	
1987 Italy	M	39	CP ointment 0.05% 'as is'; CP 0.5% pet.	contact dermatitis of the face and forearms	18
	M	38	CP ointment 0.05% 'as is'; CP 0.5% pet.	chronic dermatitis back of the hands; also contact allergy to gentamicin	
	M	33	CP ointment 0.05% 'as is'; CP 0.5% pet.	contact dermatitis of the hands that did not improve with clobetasol propionate ointment treatment	
1986 United Kingdom	F	27	CP cream and scalp lotion 'as is'; CP 5% pet.	exacerbation of flexural psoriasis suspected to have been caused by CP cream	21
1984 United Kingdom	M	69	CP cream and ointment 'as is'; CP 0.05% & 1% pet.	exacerbation of eczema on the face, neck and arms from CP ointment	22
1983 Belgium	F	40	CP 0.5% pet.	id-like spreading of dermatitis of the leg to face, neck, arms and trunk	23
	M	30	CP 5% pet.	very extensive dermatitis, mostly treated with CP cream	
1983 United Kingdom	F	?	CP 0.05% cream 'as is'; CP 20% pet.	recurrent eczema of the eyelids, the sides of the neck and the antecubital fossae	25
1981 Netherlands	M	60	CP cream and ointment 'as is'; CP 0.05% and 0.5% cetomacrogol cream	eczema of the face from contact allergy to aminoglycosides in eardrops for bilateral otitis externa; CP cream and ointment were not effective	26
	M	75	CP cream and ointment 'as is'; CP 10% pet.	exacerbation of nummular eczema from CP ointment	

CP: clobetasol propionate

Cross-reactions, pseudo-cross-reactions and co-reactions

Cross-reactions between corticosteroids are discussed in Chapter 2.8.

LITERATURE

1   DeKoven JG, Warshaw EM, Zug KA, Maibach HI, Belsito DV, Sasseville D, et al. North American Contact Dermatitis Group patch test results: 2015-2016. Dermatitis 2018;29:297-309

2   Veverka KK, Hall MR, Yiannias JA, Drage LA, El-Azhary RA, Killian JM, et al. Trends in patch testing with the Mayo Clinic standard series, 2011-2015. Dermatitis 2018;29:310-315

3   DeKoven JG, Warshaw EM, Belsito DV, Sasseville D, Maibach HI, Taylor JS, et al. North American Contact Dermatitis Group Patch Test Results: 2013-2014. Dermatitis 2017;28:33-46

4   Warshaw EM, Maibach HI, Taylor JS, Sasseville D, DeKoven JG, Zirwas MJ, et al. North American Contact Dermatitis Group patch test results: 2011-2012. Dermatitis 2015;26:49-59

5   Warshaw EM, Belsito DV, Taylor JS, Sasseville D, DeKoven JG, Zirwas MJ, et al. North American Contact Dermatitis Group patch test results: 2009 to 2010. Dermatitis 2013;24:50-59

6   Wentworth AB, Yiannias JA, Keeling JH, Hall MR, Camilleri MJ, Drage LA, et al. Trends in patch-test results and allergen changes in the standard series: a Mayo Clinic 5-year retrospective review (January 1, 2006, to December 31, 2010). J Am Acad Dermatol 2014;70:269-275

7   Fransway AF, Zug KA, Belsito DV, Deleo VA, Fowler JF Jr, Maibach HI, et al. North American Contact Dermatitis Group patch test results for 2007-2008. Dermatitis 2013;24:10-21

8   Zug KA, Warshaw EM, Fowler JF Jr, Maibach HI, Belsito DL, Pratt MD, et al. Patch-test results of the North American Contact Dermatitis Group 2005-2006. Dermatitis 2009;20:149-160

9   Warshaw EM, Belsito DV, DeLeo VA, Fowler JF Jr, Maibach HI, Marks JG, et al. North American Contact Dermatitis Group patch-test results, 2003-2004 study period. Dermatitis 2008;19:129-136

10   Tavadia S, Bianchi J, Dawe RS, McEvoy M, Wiggins E, Hamill E, et al. Allergic contact dermatitis in venous leg ulcer patients. Contact Dermatitis 2003;48:261-265

11   Bauer A. Contact sensitization in the anal and genital area. Curr Probl Dermatol 2011;40:133-141

12   Warshaw EM, Kimyon RS, Silverberg JI, Belsito DV, DeKoven JG, Maibach HI, et al. Evaluation of patch test findings in patients with anogenital dermatitis. JAMA Dermatol 2019;156:85-91

13   Menezes de Padua CA, Uter W, Schnuch A. Contact allergy to topical drugs: prevalence in a clinical setting and estimation of frequency at the population level. Pharmacoepidemiol Drug Saf 2007;16:377-384

14   Menezes de Padua CA, Schnuch A, Nink K, Pfahlberg A, Uter W. Allergic contact dermatitis to topical drugs – Epidemiological risk assessment. Pharmacoepidemiol Drug Saf 2008;17:813-821

15   Baeck M, Goossens A. Immediate and delayed allergic hypersensitivity to corticosteroids: practical guidelines. Contact Dermatitis 2012;66:38-45

16   Sommer S, Wilkinson SM, English J, Gawkrodger DJ, Green C, king CM. Type-IV hypersensitivity to betamethasone valerate and clobetasol propionate: results of a multicentre study. Am J Contact Dermat 2001;12:125

17   Kawai M, Sowa-Osako J, Omura R, Fukai K, Tsuruta D. Allergic contact dermatitis due to clobetasol propionate and lanoconazole. Contact Dermatitis 2020 May 28. doi: 10.1111/cod.13623. Online ahead of print.

18   Tosti A, Guerra L, Manuzzi P, Lama L. Contact dermatitis from clobetasol propionate. Contact Dermatitis 1987;17:256-257

19   Bachmann-Buffie B. Allergy to clobetasol-17-propionate (Dermovate). Dermatologica 1983;167:104

20   Calnan CD. Use and abuse of topical steroids. Dermatologica 1976:152(Suppl.1):247-251

21   Spiro JG, Lawrence CM. Contact sensitivity to clobetasol propionate. Contact Dermatitis 1986;14:116-117

22   Boyle J, Peachey RD. Allergic contact dermatitis to Dermovate and Eumovate. Contact Dermatitis 1984;11:50-51

23   Dooms-Goossens A, Vanhee J, Vanderheyden D, Gevers D, Willems L, Degreef H. Allergic contact dermatitis to topical corticosteroids: clobetasol propionate and clobetasone butyrate. Contact Dermatitis 1983;9:470-478

24   Camarasa JG, Serra-Baldrich E. Contact dermatitis caused by creams containing clobetasol propionate. Med Cutan Ibero Lat Am 1988;16:328-330 (Article in Spanish)

25   Chalmers RJ, Beck MH, Muston HL. Simultaneous hypersensitivity to clobetasone butyrate and clobetasol propionate. Contact Dermatitis 1983;9:317-318

26   Van Ketel WG, Swain AF. Allergy to clobetasol-17-propionate (Dermovate). Contact Dermatitis 1981;7:278-279

27   Cox NH. Contact allergy to clobetasol propionate. Arch Dermatol 1988;124:911-913

28   Kuhlwein A, Hausen BM, Hoting E: Contact allergy due to halogenated corticosteroids. Z Hautkr 1983;58:794-804 (Article in German)

29   Baeck M, Chemelle JA, Terreux R, Drieghe J, Goossens A. Delayed hypersensitivity to corticosteroids in a series of 315 patients: clinical data and patch test results. Contact Dermatitis 2009;61:163-175

30   Davis MD, El-Azhary RA, Farmer SA. Results of patch testing to a corticosteroid series: a retrospective review of 1188 patients during 6 years at Mayo Clinic. J Am Acad Dermatol 2007;56:921-927

31   Mercader-García P, Pastor-Nieto MA, García-Doval I, Giménez-Arnau A, González-Pérez R, Fernández-Redondo V, et al. GEIDAC. Are the Spanish baseline series markers sufficient to detect contact allergy to corticosteroids in Spain? A GEIDAC prospective study. Contact Dermatitis 2018;78:76-82

32 Pratt MD, Mufti A, Lipson J, Warshaw EM, Maibach HI, Taylor JS, et al. Patch test reactions to corticosteroids: Retrospective analysis from the North American Contact Dermatitis Group 2007-2014. Dermatitis 2017;28:58-63

33 Uter W, de Pádua CM, Pfahlberg A, Nink K, Schnuch A, Lessmann H. Contact allergy to topical corticosteroids – results from the IVDK and epidemiological risk assessment. J Dtsch Dermatol Ges 2009;7:34-41

34 Reduta T, Laudanska H. Contact hypersensitivity to topical corticosteroids – frequency of positive reactions in patch-tested patients with allergic contact dermatitis. Contact Dermatitis 2005;52:109-110

35 Sommer S, Wilkinson S M, English JS, et al. Type-IV hypersensitivity to betamethasone valerate and clobetasol propionate: results of a multicentre study. Br J Dermatol 2002;147:266-269

36 Uter W, Geier J, Richter G, Schnuch A; IVDK Study Group, German Contact Dermatitis Research Group. Patch test results with tixocortol pivalate and budesonide in Germany and Austria. Contact Dermatitis 2001;44:313-314

37 Devos SA, Van der Valk PG. Relevance and reproducibility of patch-test reactions to corticosteroids. Contact Dermatitis 2001;44:362-365

38 Weltfriend S, Marcus-Farber B, Friedman-Birnbaum R. Contact allergy to corticosteroids in Israeli patients. Contact Dermatitis 2000;42:47

39 Marcos C, Allegue F, Luna I, González R. An unusual case of allergic contact dermatitis from corticosteroids. Contact Dermatitis 1999;41:237-238

40 Dooms-Goossens A, Andersen KE, Brandão FM, Bruynzeel D, Burrows D, Camarasa J, et al. Corticosteroid contact allergy: an EECDRG multicentre study. Contact Dermatitis 1996;35:40-44

41 Boffa MJ, Wilkinson SM, Beck MH. Screening for corticosteroid contact hypersensitivity. Contact Dermatitis 1995;33:149-151

42 Burden AD, Beck MH. Contact hypersensitivity to topical corticosteroids. Br J Dermatol 1992;127:497-501

43 Dooms-Goossens A, Morren M. Results of routine patch testing with corticosteroid series in 2073 patients. Contact Dermatitis 1992;26:182-191

44 Dunkel FG, Elsner P, Burg G. Contact allergies to topical corticosteroids: 10 cases of contact dermatitis. Contact Dermatitis 1991;25:97-103

45 English JS, Ford G, Beck MH, Rycroft RJ. Allergic contact dermatitis from topical and systemic steroids. Contact Dermatitis 1990;23:196-197

46 Förström L, Lassus A, Salde L, Niemi KM. Allergic contact eczema from topical corticosteroids. Contact Dermatitis 1982;8:128-133

47 Oh-i T. Contact dermatitis due to topical steroids with conceivable cross reactions between topical steroid preparations. J Dermatol 1996;23:200-208

48 Okano M. Contact dermatitis due to budesonide: report of five cases and review of the Japanese literature. Int J Dermatol 1994;33:709-715

49 Reitamo S, Lauerma AI, Stubb S, Käyhkö K, Visa K, Förström L. Delayed hypersensitivity to topical corticosteroids. J Am Acad Dermatol 1986:14:582-589

Chapter 3.74   CLOBETASONE BUTYRATE

IDENTIFICATION

Description/definition   : Clobetasone butyrate is the butyrate ester of the synthetic glucocorticoid clobetasone
                             that conforms to the structural formula shown below
Pharmacological classes   : Anti-inflammatory agents; glucocorticoids
IUPAC name   : [(8S,9R,10S,13S,14S,16S,17R)-17-(2-Chloroacetyl)-9-fluoro-10,13,16-trimethyl-3,11-dioxo-
                             7,8,12,14,15,16-hexahydro-6H-cyclopenta[a]phenanthren-17-yl] butanoate
Other names   : 21-Chloro-9-fluoro-17-hydroxy-16-β-methylpregna-1,4-diene-3,11,20-trione butyrate;
                             clobetasone 17-butyrate
CAS registry number   : 25122-57-0
EC number   : 246-635-9
Merck Index monograph   : 3629 (Clobetasone)
Patch testing   : In general, corticosteroids may be tested at 0.1% and 1% in alcohol; late readings (6-10
                             days) are strongly recommended
Molecular formula   : $C_{26}H_{32}ClFO_5$

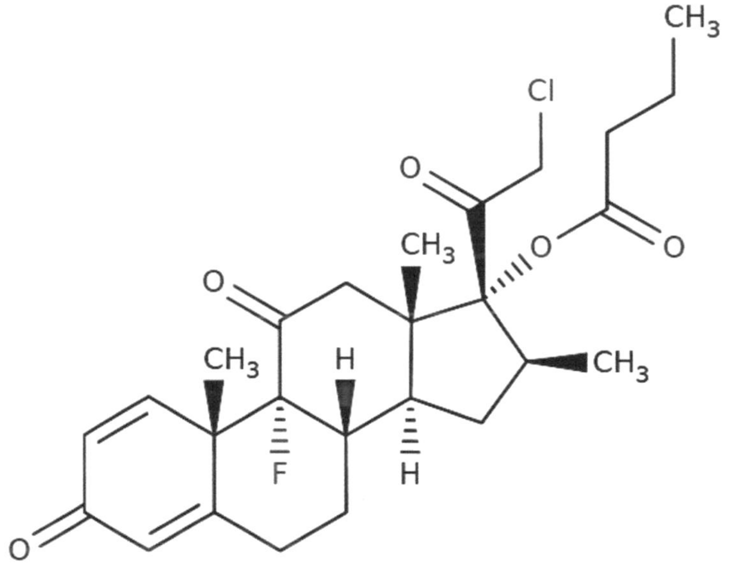

GENERAL

General aspects of corticosteroids used on the skin and mucous membranes are discussed in Chapter 2.4. A practical guideline for diagnosing allergic reactions to corticosteroids is presented in ref. 1.

CONTACT ALLERGY

Patch testing in groups of patients

In the United Kingdom, in the period 1992-1993, 2123 consecutive patients suspected of contact dermatitis were patch tested with clobetasone butyrate 1% alc. and there were 11 (0.5%) positive reactions, the relevance of which was not mentioned (10). Also in the U.K., from 1988 to 1991, 528 selected patients with a positive patch test to tixocortol pivalate or suspected of corticosteroid allergy were patch tested with clobetasone butyrate 20% pet. and 3 patients (0.6%) had a positive patch test. Their relevance was not addressed (11).

Case series

From January 1990 to June 2008, in Leuven, Belgium, 315 patients were diagnosed with contact allergy to/allergic contact dermatitis from corticosteroids (CSs) from routine patch testing with a baseline series including tixocortol pivalate, budesonide, hydrocortisone butyrate and prednisone caproate, patch testing with patients' own CS preparations, and testing those with proven contact allergy to a corticosteroid or strongly suspected of CS allergy later with a series of 66 CSs, including two sex hormones (progesterone and testosterone). 71% of the patients had relevant reactions, but these were not specified. In this group of 315 CS allergic patients, 25 had positive patch tests to clobetasone butyrate 0.1% alc. (9). It is unknown how many of these reactions were caused by the use of a pharmaceutical product containing clobetasone butyrate and how many were cross-reactions to other corticosteroids.

Case reports

A 36-year-old man, with a long history of atopic dermatitis of the neck, chest and arms, complained of an exacerbation on his neck after applying 0.05% clobetasone butyrate ointment. Patch tests were positive to the ointment and clobetasone butyrate 0.05%, 0.01% and 0.005% pet. at D7 only. It was emphasized that late readings at D7 are essential with testing corticosteroids (2). Another 36-year-old male patient suffered from pruritus ani for many years. He had been treated with various preparations, and the pruritus now had become worse when using a corticosteroid cream and a corticosteroid combination cream. When patch tested, he reacted to the combination product containing clobetasone butyrate, oxytetracycline and nystatin. Ingredient patch testing was positive only to clobetasone butyrate 2% pet. (3).

Three similar case reports had been published previously (6,7,8). The patients all noted an exacerbation of eczema while using clobetasone butyrate cream and/or ointment. Patch tests were positive to the topicals and to clobetasone butyrate, tested in various concentrations. All three patients also reacted to clobetasol propionate. These were more likely co-sensitizations than cross-reactions, as the patients all had used pharmaceuticals with this corticosteroid also (6,7,8).

Two more single case reports of allergic contact dermatitis from clobetasone butyrate have been reported in Japanese literature (4,5).

Cross-reactions, pseudo-cross-reactions and co-reactions

Cross-reactions between corticosteroids are discussed in Chapter 2.8.

LITERATURE

1   Baeck M, Goossens A. Immediate and delayed allergic hypersensitivity to corticosteroids: practical guidelines. Contact Dermatitis 2012;66:38-45
2   Murata T, Tanaka M, Dekio I, Tanikawa A, Nishikawa T. Allergic contact dermatitis due to clobetasone butyrate. Contact Dermatitis 2000;42:305
3   Corbett JR. Allergic contact dermatitis to Trimovate. Contact Dermatitis 1985;13:281
4   Kusunoki T, Kusunoki M. (Japanese title only). Jpn J Dermatol 1993;9:1208-1209 (Article in Japanese, data cited in ref. 2).
5   Takekawa K, Okajima M, Wada H, Kawaguchi T, Ohnuma S, Ohsawa J, et al. (Japanese title only). The Allergy in Practice 1997;4:58 (Article in Japanese, data cited in ref. 2)
6   Boyle J, Peachey RD. Allergic contact dermatitis to Dermovate and Eumovate. Contact Dermatitis 1984;11:50-51
7   Dooms-Goossens A, Vanhee J, Vanderheyden D, Gevers D, Willems L, Degreef H. Allergic contact dermatitis to topical corticosteroids: clobetasol propionate and clobetasone butyrate. Contact Dermatitis 1983;9:470-478
8   Chalmers RJ, Beck MH, Muston HL. Simultaneous hypersensitivity to clobetasone butyrate and clobetasol propionate. Contact Dermatitis 1983;9:317-318
9   Baeck M, Chemelle JA, Terreux R, Drieghe J, Goossens A. Delayed hypersensitivity to corticosteroids in a series of 315 patients: clinical data and patch test results. Contact Dermatitis 2009;61:163-175
10 Boffa MJ, Wilkinson SM, Beck MH. Screening for corticosteroid contact hypersensitivity. Contact Dermatitis 1995;33:149-151
11 Burden AD, Beck MH. Contact hypersensitivity to topical corticosteroids. Br J Dermatol 1992;127:497-501

Chapter 3.75   CLOCORTOLONE PIVALATE

IDENTIFICATION

Description/definition   : Clocortolone pivalate is the 21-*O*-pivalate ester of the synthetic glucocorticoid
                             clocortolone that conforms to the structural formula shown below
Pharmacological classes   : Glucocorticoids
IUPAC name   : [2-[(6*S*,8*S*,9*R*,10*S*,11*S*,13*S*,14*S*,16*R*,17*S*)-9-Chloro-6-fluoro-11-hydroxy-10,13,16-trimethyl-
                             3-oxo-7,8,11,12,14,15,16,17-octahydro-6*H*-cyclopenta[a]phenanthren-17-yl]-2-oxoethyl]
                             2,2-dimethylpropanoate
Other names   : 9-Chloro-6α-fluoro-11β,21-dihydroxy-16α-methylpregna-1,4-diene-3,20-dione 21-pivalate
CAS registry number   : 34097-16-0
EC number   : 251-826-5
Merck Index monograph   : 3633 (Clocortolone)
Patch testing   : In general, corticosteroids may be tested at 0.1% and 1% in alcohol; late readings (6-10
                             days) are strongly recommended
Molecular formula   : $C_{27}H_{36}ClFO_5$

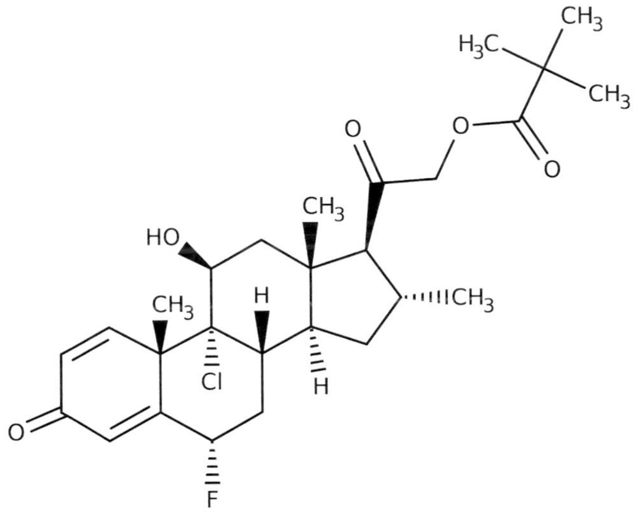

GENERAL

General aspects of corticosteroids used on the skin and mucous membranes are discussed in Chapter 2.4. A practical guideline for diagnosing allergic reactions to corticosteroids is presented in ref. 1.

CONTACT ALLERGY

Case report

A 75-year-old woman had a 20 year history of an intermittent, pruritic eruption of her arms and legs. Patch testing revealed a positive reaction to a cream she had used containing 0.1% clocortolone pivalate. In addition, she reacted to many other corticosteroids including clobetasol, triamcinolone, budesonide, hydrocortisone 17-butyrate, betamethasone 17-valerate and various commercial corticosteroid preparations. The authors concluded that this patient had contact allergies to multiple topical corticosteroid preparations including clocortolone pivalate. However, clocortolone pivalate itself was not patch tested, nor were the other constituents of the cream (2).

Cross-reactions, pseudo-cross-reactions and co-reactions

Cross-reactions between corticosteroids are discussed in Chapter 2.8.

LITERATURE

1   Baeck M, Goossens A. Immediate and delayed allergic hypersensitivity to corticosteroids: practical guidelines. Contact Dermatitis 2012;66:38-45
2   Clark T, Pedvis-Leftick A. Contact allergy to clocortolone pivalate (Cloderm) cream 0.1% in a patient with allergies to multiple topical corticosteroid preparations. Dermatitis 2008;19:167-168

Chapter 3.76   CLONIDINE

IDENTIFICATION

Description/definition : Clonidine is the imidazoline derivative that conforms to the structural formula shown below
Pharmacological classes : α_2-Adrenergic receptor agonists; analgesics; antihypertensive agents; sympatholytics
IUPAC name : *N*-(2,6-Dichlorophenyl)-4,5-dihydro-1*H*-imidazol-2-amine
Other names : 2-((2,6-Dichlorophenyl)imino)imidazolidine
CAS registry number : 4205-90-7
EC number : 224-119-4
Merck Index monograph : 3650
Patch testing : 9% pet.; not infrequently, patients are negative to clonidine 9% pet. but have a positive reaction to the clonidine-TTS, applied for 7 days (10)
Molecular formula : $C_9H_9Cl_2N_3$

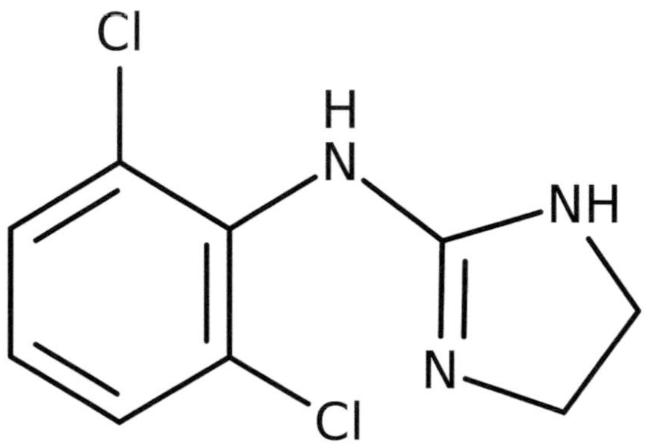

GENERAL

Clonidine is a centrally active α_2-adrenergic agonist. It is used predominantly as an antihypertensive agent, usually in combination with other drugs. Clonidine binds to and stimulates central α_2-adrenergic receptors, thereby decreasing sympathetic outflow to the heart, kidneys, and peripheral vasculature. This leads to decreased peripheral vascular resistance, decreased blood pressure, and decreased heart rate. In transdermal patches, clonidine base is used; in other pharmaceuticals, clonidine is employed as clonidine hydrochloride (CAS number 4205-91-8, EC number 224-121-5, molecular formula $C_9H_{10}Cl_3N_3$) (1). The patches and compounded topical formulations with clonidine are sometimes used to treat chronic pain, e.g. from diabetic neuropathy (16).

CONTACT ALLERGY

General

Clonidine was found to have such as low allergenicity that it impeded studies of sensitization mechanisms in guinea pig models (8). Nevertheless, in clinical trials, high percentages (up to nearly 50%) of sensitization to clonidine in clonidine-TTS have been observed (4,5,6,9,13). Apparently, the persistent skin contacts with sustained drug delivery and occlusion from the TTS device generate unique conditions favoring the development of allergic contact dermatitis. Indeed, using Draize repeat insult patch test assays in men, clonidine 9% pet. did not induce contact allergy, whereas 4 of 74 volunteers having undergone the procedure with the clonidine-TTS became sensitized to it (11). Another possible explanation given was that the small amount of acetaldehyde present in the device is critical and may induce a clonidine derivative, a reactive condensation product of acetaldehyde and two molecules of clonidine, which is the actual sensitizing hapten in a number of the patients (11,13).

Sensitization rates to clonidine have been cited to be highest in white women (34%) followed by white men (18%), black women (14%) and black men (8%) (14). Oral administration of clonidine in patients sensitized to the TTS are in most cases well tolerated (9,10,11). When contact allergy has appeared, pretreatment with a potent corticosteroid may be able to prevent the development of allergic contact dermatitis (17).

Case series

In a prospective study, 29 patients with hypertension were treated with clonidine-TTS. In the study period of 3 months, local skin reactions suspected to be allergic occurred in 11 (38%) of the patients. Patch testing with all components of clonidine-TTS in 7 of these individuals confirmed contact allergy to clonidine in 6 and a reaction to polyisobutylene in the TTS in the 7th patient (4). With continued treatment and inclusion of more patients, these authors reported 15 (47%) allergic reactions in 32 patients treated. Not all of them were patch tested, but those who were all reacted positively to clonidine (apart from the one reacting to polyisobutylene) (6).

In a similar study, 4 of 85 patients discontinued treatment after 5-15 weeks because of localised skin reactions (erythema, vesiculation, and/or infiltration); contact allergy was confirmed by patch testing. An additional patient first had a localised skin reaction and later developed a generalized papular rash (5). Of 20 patients on long-term clonidine-TTS treatment, 10 (50%) stopped the therapy because of localized skin rashes. Five of them were tested (probably some sort of patch testing, 'clonidine in petrolatum') and all developed a skin rash at the test site. Oral clonidine in 2 was well tolerated (9).

Of 52 patients who were previously withdrawn from clinical trials with clonidine-TTS because of the development of contact dermatitis, and who had confirmed sensitization to clonidine (positive patch test to clonidine 9% pet. or clonidine-TTS applied for 7 days), 29 agreed to oral challenge with clonidine HCl in a 4-day program (day 1 - 0.025 mg; day 2 - 0.05 mg; day 3 - 0.1 mg; day 4 - 0.2 mg). Only one patient, on the second day of clonidine treatment, developed a localized flare-up at the site of the original dermatitis 2 hours after receiving the 0.05 mg dose (10). In another group of such patients, a woman developed a generalized maculopapular rash and pruritus on the fifth day or oral treatment (10). These were cases of systemic contact dermatitis.

In a group of 21 patients treated with clonidine-TTS, 3 became sensitized after 17-28 weeks. Two had localized reactions, the third developed a generalized rash following continuing the TTS for 4 weeks after a localized reaction had developed. All 3 patients had positive patch tests to 'clonidine-containing patch-test materials' (13).

Case reports

A 64-year-old man developed localised eczema below the patches of clonidine-TTS after using them regularly for 3 months. Because systemic side effects were minimal and his blood pressure was well-controlled, the patient chose to continue treatment despite local irritation and was able to tolerate the patches for the recommended period of 1 week before changing to a new site. After 14 months of continuous use, the eczema remained localised to the sites of application. Patch tests to active and placebo patches provided by the manufacturer showed a ++ reaction to the active patch alone. Clonidine at 1% and 9% in pet. produced a ++ response. An oral challenge dose of 25 μg clonidine did not provoke any reaction (12).

In another patient, allergic contact dermatitis to the clonidine-TTS was associated with a sudden rise in blood pressure, comparable to rebound hypertension after stopping oral clonidine (also known as clonidine withdrawal syndrome). The authors suggested that clonidine was not absorbed from the patch secondary to the inflammation of the type IV allergic reactions (15). In yet another case, in a patient with a positive reaction to the TTS, patch tests with clonidine, a placebo device and a placebo system plus plaster cover were negative. It was suggested that the actual allergen was a chemical formed by the reaction of clonidine with acetaldehyde during manufacture (7).

Other case reports of contact allergy to/allergic contact dermatitis from clonidine, adequate data of which are not available to the author, can be found in refs. 2 and 3.

LITERATURE

1   The data in the section 'General' may have been obtained from literature discussed in this chapter, but mostly also or exclusively from one or more of the following online sources: ChemIDPlus Advanced, PubChem, DrugBank, RxList, Drug Central, Drugs.com, and Wikipedia
2   Crivellaro MA, Bonadonna P, Dama A, Senna G, Passalacqua G. Skin reactions to clonidine: not just a local problem. Case report. Allergol Immunopathol (Madr) 1999;27:318-319
3   Polster AM, Warner MR, Camisa C. Allergic contact dermatitis from transdermal clonidine in a patient with mycosis fungoides. Cutis 1999;63:154-155
4   Groth H, Vetter H, Knuesel J, Vetter W. Allergic skin reactions to transdermal clonidine. Lancet 1983;2(8354):850-851
5   McMahonn FG, Weber MA. Allergic skin reactions to transdermal clonidine. Lancet 1983;2(8354):851
6   Groth H, Vetter H, Knüsel J, Foerster E, Siegenthaler W, Vetter W. Transdermal clonidine application: long-term results in essential hypertension. Klin Wochenschr 1984;62:925-930 (Article in German)
7   Corazza M, Mantovani L, Virgili A, Strumia R. Allergic contact dermatitis from a clonidine transdermal delivery system. Contact Dermatitis 1995;32:246

8   Scheper RJ, Von Blomberg BME, De Groot J, Goeptar AR, Lang M, et al. Low allergenicity of clonidine impedes studies of sensitization mechanisms in guinea pig models. Contact Dermatitis 1990;23:81-89

9   Horning JR, Zawada ET Jr, Simmons JL, Williams L, McNulty R. Efficacy and safety of two-year therapy with transdermal clonidine for essential hypertension. Chest 1988;93:941-945

10   Maibach HI. Oral substitution in patients sensitized by transdermal clonidine treatment. Contact Dermatitis 1987;16:1-8

11   Maibach H. Clonidine: irritant and allergic contact dermatitis assays. Contact Dermatitis 1985;12:192-195

12   Grattan CE, Kennedy CT. Allergic contact dermatitis to transdermal clonidine. Contact Dermatitis 1985;12:225-226

13   Boekhorst JC. Allergic contact dermatitis with transdermal clonidine. Lancet 1983;2(8357):1031-1032

14   Berardesca E, Maibach HI. Sensitive and ethnic skin. A need for special skin-care agents? Dermatol Clin 1991;9:89-92

15   White TM, Guidry JR. Rebound hypertension associated with transdermal clonidine and contact dermatitis. Western J Med 1986;145:104

16   Cline AE, Turrentine JE. Compounded topical analgesics for chronic pain. Dermatitis 2016;27:263-271

17   McChesney JA. Preventing the contact dermatitis caused by a transdermal clonidine patch. West J Med 1991;154:736

Chapter 3.77   CLOPREDNOL

IDENTIFICATION

Description/definition   : Cloprednol is the synthetic glucocorticoid that conforms to the structural formula shown
                            below
Pharmacological classes   : Glucocorticoids; anti-inflammatory agents
IUPAC name   : (8S,9S,10R,11S,13S,14S,17R)-6-Chloro-11,17-dihydroxy-17-(2-hydroxyacetyl)-10,13-
                            dimethyl-9,11,12,14,15,16-hexahydro-8H-cyclopenta[a]phenanthren-3-one
Other names   : 6-Chloro-11β,17,21-trihydroxypregna-1,4,6-triene-3,20-dione
CAS registry number   : 5251-34-3
EC number   : 226-052-6
Merck Index monograph   : 3657
Patch testing   : In general, corticosteroids may be tested at 0.1% and 1% in alcohol; late readings (6-10
                            days) are strongly recommended
Molecular formula   : $C_{21}H_{25}ClO_5$

GENERAL
General aspects of corticosteroids used on the skin and mucous membranes are discussed in Chapter 2.4. A practical
guideline for diagnosing allergic reactions to corticosteroids is presented in ref. 1.

CONTACT ALLERGY

Case series
From January 1990 to June 2008, in Leuven, Belgium, 315 patients were diagnosed with contact allergy to/allergic
contact dermatitis from corticosteroids (CSs) from routine patch testing with a baseline series including tixocortol
pivalate, budesonide, hydrocortisone butyrate and prednisone caproate, patch testing with patients' own CS
preparations, and testing those with proven contact allergy to a corticosteroid or strongly suspected of CS allergy
later with a series of 66 CSs, including two sex hormones (progesterone and testosterone). 71% of the patients had
relevant reactions, but these were not specified. In this group of 315 CS allergic patients, 61 had positive patch tests
to cloprednol 0.1% alc. (2). As this corticosteroid has never been used in pharmaceuticals in Belgium, these positive
reactions must all be considered cross-reactions to other corticosteroids.

Cross-reactions, pseudo-cross-reactions and co-reactions
Cross-reactions between corticosteroids are discussed in Chapter 2.8.

LITERATURE
1   Baeck M, Goossens A. Immediate and delayed allergic hypersensitivity to corticosteroids: practical guidelines.
    Contact Dermatitis 2012;66:38-45
2   Baeck M, Chemelle JA, Terreux R, Drieghe J, Goossens A. Delayed hypersensitivity to corticosteroids in a series of
    315 patients: clinical data and patch test results. Contact Dermatitis 2009;61:163-175

Chapter 3.78   CLOTRIMAZOLE

IDENTIFICATION

Description/definition	: Clotrimazole is the phenylmethyl imidazole derivate that conforms to the structural formula shown below
Pharmacological classes	: 14α-Demethylase inhibitors; anti-infective agents, local; antifungal agents
IUPAC name	: 1-[(2-Chlorophenyl)-diphenylmethyl]imidazole
Other names	: 1H-Imidazole, 1-[(2-chlorophenyl)diphenylmethyl]-
CAS registry number	: 23593-75-1
EC number	: 245-764-8
Merck Index monograph	: 3671
Patch testing	: 1% pet. (SmartPracticeCanada, SmartPracticeEurope); generally, alcohol or MEK (methyl ethyl ketone) appear to be more suitable for patch testing imidazoles (7,8)
Molecular formula	: $C_{22}H_{17}ClN_2$

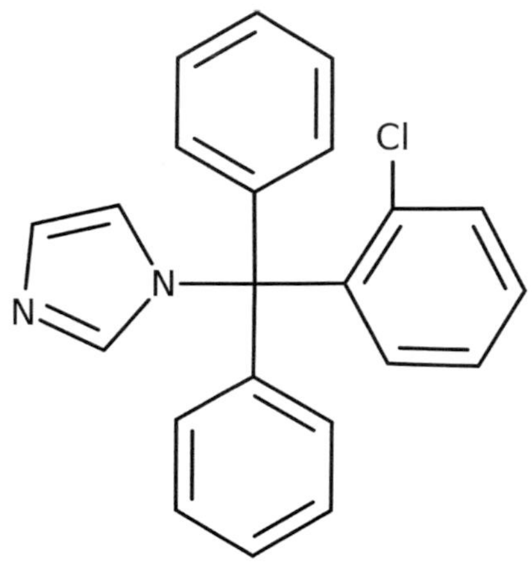

GENERAL

Clotrimazole is a synthetic imidazole derivative with a broad spectrum of antimycotic activity. Topical clotrimazole preparations are indicated for the treatment of superficial fungal infections caused by dermatophytes or *Candida albicans* (including diaper rash and candidiasis vaginalis), and for treating pityriasis versicolor. The oral preparation of clotrimazole (troche, slowly dissolving tablets) is indicated for the local treatment of oropharyngeal candidiasis, and as a prophylactic drug to reduce the incidence of oropharyngeal candidiasis in patients immunocompromised by conditions such as chemotherapy, radiotherapy, or steroid therapy utilized in the treatment of leukemia, solid tumors, or renal transplantation. Troche preparations are not indicated for the treatment of any systemic mycoses. Clotrimazole has also become a drug of interest in treating several other diseases such as sickle cell disease, malaria and some cancers (1).

CONTACT ALLERGY

Contact allergy in the general population

In Germany, for the period 1995-2004, the population-based relative incidence (RI) of contact sensitization to clotrimazole (cases/100,000 defined daily doses (DDDs) per year) was estimated to be 0.8. In the group of all topical drugs, the RI ranged from 0.3 (dexamethasone sodium phosphate and pilocarpine) to 413.9 (benzocaine) (3).

Patch testing in groups of selected patients

The results of patch testing groups of selected patients with clotrimazole are shown in table 3.78.1. Frequencies of sensitization in 4 studies have ranged from 0.3% to 2.4%. The highest prevalences were found in Belgium in patients tested for suspected iatrogenic contact dermatitis (2.4%) (4) and by the IVDK in patients with anogenital dermatoses (1.8%) (9).

Table 3.78.1 Patch testing in groups of patients: Selected patient groups

Years and Country	Test conc. & vehicle	Number of patients tested	positive (%)	Selection of patients (S); Relevance (R); Comments (C)	Ref.
1990-2014 Belgium	2% pet.	167	4 (2.4%)	S: patients suspected of iatrogenic contact dermatitis and tested with a pharmaceutical series and their own products; R: 96% of the positive patch test reactions to all topical drugs and antiseptics were considered to be relevant	4
2004-2008 IVDK			(1.8%)	S: patients with anogenital dermatoses tested with a medicament series; R: not stated; C: number of patients tested unknown	9
1995-2004 IVDK	5% pet.	10,400	(0.5%)	S: not stated; R: not stated	2
2001-2002 IVDK	5% pet.	1180	3 (0.3%)	S: patients tested with an imidazole antifungal series; the selection procedure was not well described; R: not stated	6

IVDK: Information Network of Departments of Dermatology (Germany, Austria, Switzerland)

Case series

In the period 2008-2011, 10 patients (age 40-75 years, 8 males) were referred to a university clinic in Bern, Switzerland, because of severe cutaneous drug eruptions following application of topical antimycotics with or without corticosteroids. Seven to 21 days after initiation of antifungal therapy for clinically or mycologically diagnosed infections with dermatophytes or intertrigo, the patients developed widespread eczema as well as erythematous, maculopapular, erythema multiforme-like and blistering eruptions, that occurred in addition to intense eczematous reactions at application sites associated with peripheral blood eosinophilia, suggesting drug reactions. After the eruptions had resolved, patch testing was performed. Two patients showed sensitization to both clotrimazole and corticosteroids, as well as to the culprit antifungal creams. Two patients were sensitized to clotrimazole but not to corticosteroids, and five patients were sensitized to tixocortol pivalate or prednisolone. In one individual, a positive patch test reaction exclusively to the culprit drug was found. The most severe, erythema multiforme-like and blistering eruptions were associated with corticosteroid allergy (unspecified which corticosteroids the antifungal creams contained), and eczematous and maculopapular exanthemas were associated with clotrimazole allergy (17).

In Japan, in the period 1984 to 1994, 3049 outpatients were patch tested for suspected contact dermatitis and 218 of these with topical antifungal preparations. Thirty-five were allergic to imidazoles, including clotrimazole in one individual (5). Between 1977 and 1986, 9 patients with contact allergy to imidazole antimycotics were investigated in the University clinic of Heidelberg, Germany. Three patients were allergic clotrimazole, 6 to miconazole, 3 to econazole, 3 to isoconazole, and one to oxiconazole (8).

In a literature review up to 1994, the authors identified 105 reported patients who had a positive patch test reaction to at least one imidazole derivative, ranging from 51 reactions to miconazole to one reaction each to bifonazole and enilconazole. The number of reported reactions to clotrimazole was 13 (7).

Case reports

A 56-year-old woman was treated for suspected candidiasis vaginalis with clotrimazole cream, which caused a worsening of the vulvovaginal discomfort. Next, she received another clotrimazole cream and solution (containing also metronidazole), resulting in severe erythematous, edematous, and itching vulvovaginitis after 3 days. She denied having used tampons, sanitary napkins, condoms, or other topical ointments. Patch tests were positive to all topical medicaments containing the antifungal and to clotrimazole 1% pet., but not to metronidazole (14).

A 44-year-old woman was prescribed 1% clotrimazole cream for suspected anogenital candidiasis. After she had applied it on and off for 2 months, her itch deteriorated, and she developed erythema in her groins. For this, her physician prescribed fluconazole (150 mg single dose), which resulted in a widespread maculopapular exanthema 2 days later. Patch testing showed positive reactions to clotrimazole 5%, methylisothiazolinone (MI), MCI/MI and the baby wipes she used on her anogenital skin before the start of the problem. The patient was not tested with fluconazole. The authors postulated – without corroborating evidence (such as a positive patch test to fluconazole) – that the maculopapular exanthema to oral fluconazole was the result of cross-sensitivity to primary clotrimazole sensitization (10).

Short summaries of other case reports of allergic contact dermatitis from clotrimazole are shown in table 3.78.2.

Table 3.78.2 Short summaries of case reports of allergic contact dermatitis from clotrimazole

Year and country	Sex	Age	Positive patch tests	Clinical data and comments	Ref.
2015 India	M	44	clotrimazole cream 1%	acute ACD scrotum and groins; clotrimazole itself not tested; also reaction to miconazole cream	16
2010 USA	F	47	clotrimazole cream 'as is'; 'purified clotrimazole' on 2 occasions	a large, confluent, and bright red lichenified plaque with linear excoriations, small erosions, and scale covering the entire mons pubis, labia majora, medial thighs, inguinal folds, and medial buttocks	18
2009 Germany	M	58	cream 'as is'; clotrimazole 5% pet.	weeping erosions in genital and inguinal area; also positive reactions to creams with croconazole, oxiconazole, and tioconazole, which may have been used before, and to croconazole and tioconazole 1% pet.	19
1999 U.K.	F	71	cream 1% 'as is'; clotrimazole 1% pet. (D2 +/ D4 -)	exacerbation of vulval dermatitis	21
1999 Germany	M	52	cream 'as is'; clotrimazole 1% alc., 1% and 5% pet.	contact dermatitis in the genital area, spreading to lower abdomen and upper thighs; possible cross-reaction to croconazole and itraconazole	22
1995 Belgium	M	47	cream 'as is'; clotrimazole 1% pet.	exacerbation of presternal seborrheic dermatitis	24
1995 Germany	M	13	cream 'as is'; clotrimazole 1% pet.	worsening of facial atopic dermatitis spreading to the neck and thorax; also contact allergy to prednisolone-21-acetate and 3 other ingredients of the cream; repeat tests positive	25
1994 Italy	M	35	cream 'as is'; clotrimazole 1% pet.	acute dermatitis in the genital area; connubial contact dermatitis: clotrimazole cream and suppositories had been used by his wife for vulvovaginitis	26
1987 Germany	M	58	clotrimazole 1% alcohol and methyl ethyl ketone	cream applied to stasis dermatitis; later contact allergy to oxiconazole, used for tinea pedis; at that moment allergy to clotrimazole was detected	27
1985 Italy	F	50	cream 'as is'; clotrimazole 1% pet.	dermatitis of the trunk from applying clotrimazole 1% cream to pityriasis versicolor	28
1978 U.K.	M	48	cream 'as is'; clotrimazole 1% MEK	oozing dermatitis in the genital and perianal area	29

ACD: allergic contact dermatitis

Other case reports of contact allergy to/allergic contact dermatitis from clotrimazole, adequate data of which are not available to the author, can be found in refs. 11-13,20, and 23.

Cross-reactions, pseudo-cross-reactions and co-reactions

Although there are some unique and sporadic reports of simultaneous reactions, cross-reactions are very unlikely between clotrimazole and other imidazoles (7, data up to 1994, before the introduction of other imidazoles). A patient sensitized to clotrimazole may have cross-reacted to croconazole and itraconazole (22). Possible cross-reaction to fluconazole (10).

The question has been raised (but not answered) whether contact allergy to clotrimazole and other (nitro)imidazoles may be overrepresented in patients allergic to methylchloroisothiazolinone/methylisothiazolinone (MCI/MI (15).

LITERATURE

1   The data in the section 'General' may have been obtained from literature discussed in this chapter, but mostly also or exclusively from one or more of the following online sources: ChemIDPlus Advanced, PubChem, DrugBank, RxList, Drug Central, Drugs.com, and Wikipedia

2   Menezes de Padua CA, Uter W, Schnuch A. Contact allergy to topical drugs: prevalence in a clinical setting and estimation of frequency at the population level. Pharmacoepidemiol Drug Saf 2007;16:377-384

3   Menezes de Padua CA, Schnuch A, Nink K, Pfahlberg A, Uter W. Allergic contact dermatitis to topical drugs – Epidemiological risk assessment. Pharmacoepidemiol Drug Saf 2008;17:813-821

4   Gilissen L, Goossens A. Frequency and trends of contact allergy to and iatrogenic contact dermatitis caused by topical drugs over a 25-year period. Contact Dermatitis 2016;75:290-302

5   Yoneyama E. [Allergic contact dermatitis due to topical imidazole antimycotics. The sensitizing ability of active ingredients and cross-sensitivity]. Nippon Ika Daigaku Zasshi 1996;63:356-364 (article in Japanese)

6   Menezes de Pádua CA, Uter W, Geier J, Schnuch A, Effendy I; German Contact Dermatitis Research Group (DKG); Information Network of Departments of Dermatology (IVDK). Contact allergy to topical antifungal agents. Allergy 2008;63:946-947

7   Dooms-Goossens A, Matura M, Drieghe J, Degreef H. Contact allergy to imidazoles used as antimycotic agents. Contact Dermatitis 1995;33:73-77

8   Raulin C, Frosch PJ. Contact allergy to imidazole antimycotics. Contact Dermatitis 1988;18:76-80

9   Bauer A. Contact sensitization in the anal and genital area. Curr Probl Dermatol 2011;40:133-141

10   Nasir S, Goldsmith P. Anogenital allergic contact dermatitis caused by methylchloroisothiazolinone, methylisothiazolinone and topical clotrimazole with subsequent generalized exanthem triggered by oral fluconazole. Contact Dermatitis 2016;74:296-297

11   Raulin C, Frosch PJ. Kontaktallergie auf Clotrimazol und Azidamfenicol. Dermatosen in Beruf und Umwelt 1987;35:64-66 (Article in German)

12   Kalb RE, Grossmann ME. Contact dermatitis to clotrimazole. Cutis 1985;36:240-242

13   Napolitano M, Cappello M, Festa B, Patruno C. Allergic contact dermatitis due to clotrimazole. G Ital Dermatol Venereol 2020;155:355-357 (Article in Italian)

14   Corazza M, Scuderi V, Toni G, Forconi R, Zedde P, Borghi A. Severe vulvovaginal allergic contact dermatitis due to clotrimazole contained in multiple topical products. Contact Dermatitis 2020;82:57-59

15   Stingeni L, Rigano L, Lionetti N, Bianchi L, Tramontana M, Foti C, et al. Sensitivity to imidazoles/nitroimidazoles in subjects sensitized to methylchloroisothiazolinone/methylisothiazolinone: A simple coincidence? Contact Dermatitis 2019;80:181-183

16   Abhinav C, Mahajan VK, Mehta KS, Chauhan PS. Allergic contact dermatitis due to clotrimazole with cross-reaction to miconazole. Indian J Dermatol Venereol Leprol 2015;81:80-82

17   Tang MM, Corti MA, Stirnimann R, Pelivani N, Yawalkar N, Borradori L, et al. Severe cutaneous allergic reactions following topical antifungal therapy. Contact Dermatitis 2013;68:56-67

18   Pullen SK, Warshaw EM. Vulvar allergic contact dermatitis from clotrimazole. Dermatitis 2010;21:59-60

19   Brans R, Wosnitza M, Baron JM, Merk HF. Contact sensitization to azole antimycotics. Hautarzt 2009;60:372-375 (Article in German)

20   Higashi M, Isonomaki M, Hazano S. 3 cases of contact dermatitis from topical imidazole antimycotics. Skin Res 1990;32(Suppl.):149-154 (article in Japanese, cited in ref. 19)

21   Cooper SM, Shaw S. Contact allergy to clotrimazole: an unusual allergen. Contact Dermatitis 1999;41:168

22   Erdmann S, Hertl M, Merk HF. Contact dermatitis from clotrimazole with positive patch-test reactions also to croconazole and itraconazole. Contact Dermatitis 1999;40:47-48

23   Nöhle M, Szliska C, Uter W, Schwanitz H-J. Spättyp-Sensibilisierung gegen ausgewählte Antimykotika. Z Hautkr 1997;72:934 (Article in German, cited in ref. 22)

24   Baes H. Contact dermatitis from clotrimazole. Contact Dermatitis 1995;32:187-188

25   Brand CU, Ballmer-Weber BK. Contact sensitivity to 5 different ingredients of a topical medicament (Imacort cream). Contact Dermatitis 1995;33:137

26   Valsecchi R, Pansera B, di Landro A, Cainelli T. Connubial contact sensitization to clotrimazole. Contact Dermatitis 1994;30:248

27   Raulin C, Frosch PJ. Contact allergy to oxiconazole. Contact Dermatitis 1987;16:39-40

28   Balato N, Lembo G, Nappa P, Ayala F. Contact dermatitis from clotrimazole. Contact Dermatitis 1985;12:110

29   Roller JA. Contact allergy to clotrimazole. Br Med J 1978;2(6139):737

Chapter 3.79   CLOXYQUIN

IDENTIFICATION

Description/definition : Cloxyquin is the quinoline and organochlorine compound that conforms to the structural formula shown below
Pharmacological classes : Dermatological agents; antiseptics and disinfectants; anti-infective agents
IUPAC name : 5-Chloroquinolin-8-ol
Other names : 5-Chloro-8-hydroxyquinoline; monochlorohydroxyquinoline; cloxyquine
CAS registry number : 130-16-5
EC number : 204-978-1
Merck Index monograph : 3675
Patch testing : 5% pet.
Molecular formula : C_9H_6ClNO

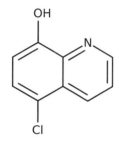

GENERAL

Cloxyquin is a monohalogenated 8-hydroxyquinoline with activity against bacteria, fungi, and protozoa. This agent is also reported to have activity against *Mycobacterium tuberculosis*, including strains that show resistance to common first-line antibacterial agents. Cloxyquin was formerly used as topical antifungal drug (1).

CONTACT ALLERGY

Case reports and case series

A 68-year-old woman was prescribed a cream containing cloxyquin 5.0%, fluprednidene acetate 1.0% and gentamicin sulfate 1.67% for a fungal infection. After 2 days of application, severe itchy dermatitis developed. Patch tests were positive to cloxyquin 5.0% pet. and to clioquinol 5.0%. As the patient developed dermatitis after 2 days already, she must previously have become sensitized to cloxyquin or have cross-reacted to clioquinol, although the patient's history suggested that she had never used antimycotics and had never developed contact dermatitis before (2).

Three patients applied a combination preparation of 0.75% halquinol (5,7-dichloro-8-quinolinol, 5-chloro-8-quinolinol [cloxyquin], 7-chloro-8-quinolinol and 8-hydroxyquinoline) and triamcinolone acetonide to stasis dermatitis (n=2) or venous ulcers (n=1), which resulted in (worsening of) dermatitis, in one case spreading to become generalized within 2 days. They all had used clioquinol (5-chloro-7-iodoquinolin-8-ol) ointment previously. Patch tests were positive to halquinol, its ingredients cloxyquin and 5,7-dichloro-8-quinolinol, and several other hydroxyquinolines (3).

Cross-reactions, pseudo-cross-reactions and co-reactions

Cross-reactions between halogenated hydroxyquinolines such as clioquinol (5-chloro-7-iodoquinolin-8-ol), chlorquinaldol (5,7-dichloro-2-methylquinolin-8-ol), cloxyquin (5-chloroquinolin-8-ol), oxyquinoline (8-hydroxy-quinoline, non-halogenated), iodoquinol (5,7-diiodoquinolin-8-ol), halquinol (a mixture of 4 hydroxyquinolines), 5,7-dichloro-8-quinolinol and 5-chloro-8-quinolinol may occur (2-5 [examples of references]).

LITERATURE

1 The data in the section 'General' may have been obtained from literature discussed in this chapter, but mostly also or exclusively from one or more of the following online sources: ChemIDPlus Advanced, PubChem, DrugBank, RxList, Drug Central, Drugs.com, and Wikipedia
2 Wantke F, Götz M, Jarisch R. Contact dermatitis from cloxyquin. Contact Dermatitis 1995;32:112-113
3 Allenby CF. Skin sensitization to Remiderm and cross-sensitization to hydroxyquinoline compounds. Br Med J 1965;2(5455):208-209
4 Myatt AE, Beck MH. Contact sensitivity to chlorquinaldol. Contact Dermatitis 1983;9:523
5 Leifer W, Steiner K. Studies in sensitization to halogenated hydroxyquinolines and related compounds. J Invest Dermatol 1951;17:233-240

Chapter 3.80   COAL TAR

IDENTIFICATION

Description/definition   : Coal tar is a thick liquid or semi-solid obtained as a by-product in the destructive
                               distillation of bituminous coal
Pharmaceutical classes   : Dermatological agents; keratolytic agents
Other names   : Tar, coal; pix lithanthracis; pix ex carbone
CAS registry number   : 8007-45-2
EC number   : 232-361-7
Merck Index monograph : 3830
Patch testing   : 5.0% pet. (Chemotechnique, SmartPracticeEurope, SmartPracticeCanada)

GENERAL

Coal tar is one of the by-products when coal is carbonized to make coke or gasified to make coal gas. It is a brown or black liquid of extremely high viscosity. Coal tars are complex and variable mixtures of phenols, polycyclic aromatic hydrocarbons, and heterocyclic compounds. Coal tar products are used to relief itch, irritation, redness, dryness, scaling, and flaking of the skin caused by certain skin diseases. Coal tar ointment is a keratolytic. It works by slowing bacterial growth and loosening and softening scales and crust. Indeed, topical coal tar products are widely used in the treatment of psoriasis and atopic dermatitis, although the level of evidence of their efficacy is not strong (3) and people dislike their odor, messy application, and staining of clothing. The short-term side effects of coal tar pharmaceutical preparations are folliculitis, irritation, phototoxicity and contact allergy. Occupational dermal exposure to coal tar and coal tar pitches can be the cause of phototoxic reactions, irritation and burn, allergic dermatitis, folliculitis, occupational acne, atrophy of the epidermis, hyperpigmentation, and malignancies (22).

Coal tar contains carcinogens. The carcinogenicity of coal tar has been shown in animal studies and studies in occupational settings. Yet, there is no clear evidence of an increased risk of skin tumors or internal tumors in patients treated with coal tar (4). Shampoos with coal tar were formerly widely used in the treatment of dandruff and seborrheic dermatitis of the scalp. Because of the known carcinogenicity of coal tar, however, its use in the EU is prohibited, but it is widely available in the USA (www.drugs.com).

CONTACT ALLERGY

Patch testing in groups of selected patients

Results of patch testing coal tar in groups of selected patients are shown in table 3.80.1. Prevalences of sensitization

Table 3.80.1 Patch testing in groups of patients: Selected patient groups

Years and Country	Test conc. & vehicle	Number of patients tested	positive (%)	Selection of patients (S); Relevance (R); Comments (C)	Ref.
2018 Croatia	5% pet.	88	8 (9%)	S: patients with psoriasis and palmar and/or plantar involvement; R: current relevance in 6 of 8 patients (75%); one reaction (2%) in a control group of 61 patients with psoriasis with no plantar and/or palmar psoriasis	17
1990-2014 Belgium	5% pet.	309	15 (4.9%)	S: patients suspected of iatrogenic contact dermatitis and tested with a pharmaceutical series and their own products; R: 96% of the positive patch test reactions to all topical drugs and antiseptics were considered to be relevant	10
1996-2005 Hungary		401	2 (5.3%)	S: patients with periorbital dermatitis; R: not stated	9
<1997 U.K.		75	3 (4%)	S: adult patients with psoriasis; R: not stated	18
1992-6 Netherlands	5% pet.	47	6 (13%)	S: patients hospitalized for psoriasis; R: not stated	11
<1994 U.K.		85	8 (9%)	S: patients with longstanding venous ulceration or eczema complicating leg ulcers; R: not stated	6
1993 U.K.	3% pet.	74	2 (3%)	S: consecutive patients with psoriasis admitted to the dermatology ward; R: 13 patients intolerant to coal tar, 2 had positive patch tests to a dilution series of coal tar; R: one current relevance, one past relevance	28
1976-1978 Finland	5% pet.	72	2 (3%)	S: patients with leg ulcers or stasis dermatitis; R: not stated	7
1968-1977 Italy	5% pet.	1283	14 (1.1%)	S: not stated; R: not stated	12
1970-1973 Italy	5% pet.	111	2 (1.8%)	S: patients with stasis dermatitis with or without leg ulcers; R: not stated	8
1967-1970 Poland	5% pet.	877	(7.9%)	S: not stated; R: not stated	29

ranged from 3% to 9% in patients with psoriasis, from 1.8% to 9% in individuals with leg ulcers/stasis dermatitis and from 1.1% to 7.9% in other patients. Data on relevance were largely lacking, no study had a control group and culprit products were never mentioned.

Case reports

A 57-year-old man used an ointment containing tar, salicylic acid and a corticosteroid for lichenification of the scrotum. After 3 weeks, dermatitis had developed on the scrotum and left upper thigh. Patch tests were positive to the ointment, coal tar (5% pet.) and wood tar mix (4x3%) (19). A 54-year-old woman with arthropathic psoriasis noticed itching, redness and serous exudation in the areas where she had applied tar preparations on psoriasis lesions. After she stopped using them, the new symptoms rapidly disappeared. Patch tests were positive to coal tar, plant tar and the commercial tar products used (23).

Two patients with psoriasis became allergic to a lotion containing coal tar extract (15). One patient who reported a flare of psoriasis after using tar products had positive patch test reactions to coal tar 1% and 5%. After avoiding tar products, the psoriasis improved (13). A woman had dermatitis of both hands, arms, back and the periorbital area of her face. Patch tests were positive to a shampoo containing coal tar and an 'udderfat'. When tested with their ingredients, the patient reacted to coal tar 0.1%, present in the shampoo and to Osmaron B 1% in the udderfat; there was a co-reaction to birch tar (5).

Two more patients may have become sensitized to coal tar from treatment of their psoriasis in the early 1960s (30). In one, the combination of coal tar and UV-light (Goeckerman therapy) apparently led to acute erythroderma followed by exfoliative dermatitis. Patch tests were strongly positive to coal tar and liquor carbonis detergens. As a photopatch test was slightly stronger, the authors diagnosed photocontact allergy. However, according to current standards this diagnosis cannot be accepted and aggravation of the positive patch test by phototoxicity is far more likely (30).

Non-pharmaceutical applications

Coal tar was responsible for 1 out of 399 cases of cosmetic allergy where the causal allergen was identified in a study of the NACDG, USA, 1977-1983 (1). Another case of non-pharmaceutical contact allergy to coal tar was a woman who became sensitized to the printing ink in newspapers. Such inks may contain 5% asphaltic material to produce a top quality, high-speed production ink. Asphalt or bitumen is the bottom residue obtained from the distillation of crude coal tar (26).

Occupational allergic contact dermatitis

Coal tar is one of the oldest treatments in dermatology but is also used in different industries such as coke production, aluminum reduction, pipe coating, roofing and paving. A 48-old-man worked in a factory where brushes for electrical engines are manufactured. He had been in contact with tars for 2 days when he suffered an outbreak of widespread eczema which began on the front of his trunk, arms and legs, but did not affect his hands which were protected by gloves. The patient had previously handled tars occasionally. Patch tests were strongly positive to products E30 and E40 (composed of lampblack, tar and artificial graphite) and to coal tar. The company confirmed that coal tar was present in the E compounds (21).

A female pharmacist suffered from a mild but persisting dermatitis, appearing exclusively around the eyes. The patient proved strongly positive to coal tar and to formaldehyde, cassia oil, bitter almond oil, cinnamyl alcohol and isoeugenol. She came daily in contact with coal tar, preparing on average two ointments, with formaldehyde (antiperspirant drugs and for rinsing the oral cavity) and with bitter almond oil. When the patient ceased to prepare drugs, the dermatitis healed and did not return (24). The authors ascribed the dermatitis partly to coal tar, but dermatitis around the eyes suggests airborne contact dermatitis, which is to be expected from formaldehyde and essential oils but not from coal tar. There could, of course, have been some contamination from the hands to the eyelids.

Occupational allergic contact dermatitis from coal tar has also been observed in an aluminum-processing worker (25). In one retrospective study from Bosnia and Herzegovina, it was found that coal tar was even the most frequent cause (14%) of allergic dermatitis in metal workers (27).

Cross-reactions, pseudo-cross-reactions and co-reactions.

Co-reactivity between coal tar and dithranol has been observed several times (2,14,15,16). Possible explanations are independent sensitization in psoriasis patients having used both coal tar and dithranol, a shared allergen, or the presence of minute quantities of dithranol in coal tar (14).

Some authors believe that combined allergy to wood and coal tars could be the consequence of cross-sensitization rather than due to long-term previous topical treatment with both tar derivatives (20). Co-reaction to birch tar (5); to wood tars (19,23).

PHOTOSENSITIVITY
An unconvincing case of photocontact allergy to coal tar was reported in 1961 (30). However, the phototoxic properties of coal tar are well-known.

LITERATURE
1   Adams RM, Maibach HI, Clendenning WE, Fisher AA, Jordan WJ, Kanof N, et al. A five-year study of cosmetic reactions. J Am Acad Dermatol 1985;13:1062-1069

2   De Groot AC, Nater JP, Bleumink E, de Jong MC. Does DNCB therapy potentiate epicutaneous sensitization to non-related contact allergens? Clin Exp Dermatol 1981;6:139-144

3   Slutsky JB, Clark RA, Remedios AA, Klein PA. An evidence-based review of the efficacy of coal tar preparations in the treatment of psoriasis and atopic dermatitis. J Drugs Dermatol 2010;9:1258-1264

4   Roelofzen JH, Aben KK, van der Valk PG, van Houtum JL, van de Kerkhof PC, Kiemeney LA. Coal tar in dermatology. J Dermatolog Treat 2007;18:329-334

5   Goldermann R, Scharffetter-Kochanek K, Brunner M, Merk H, Goerz G. 3 cases of contact dermatitis from alkylammonium amidobenzoate (Osmaron BA). Contact Dermatitis 1992;27:337-339

6   Zaki I, Shall L, Dalziel KL. Bacitracin: a significant sensitizer in leg ulcer patients? Contact Dermatitis1994;31:92-94

7   Fräki JE, Peltonen L, Hopsu-Havu VK. Allergy to various components of topical preparations in stasis dermatitis and leg ulcer. Contact Dermatitis 1979;5:97-100

8   Angelini G, Rantuccio F, Meneghini CL. Contact dermatitis in patients with leg ulcers. Contact Dermatitis 1975;1:81-87

9   Temesvári E, Pónyai G, Németh I, Hidvégi B, Sas A, Kárpáti S. Periocular dermatitis: a report of 401 patients. J Eur Acad Dermatol Venereol 2009;23:124-128

10 Gilissen L, Goossens A. Frequency and trends of contact allergy to and iatrogenic contact dermatitis caused by topical drugs over a 25-year period. Contact Dermatitis 2016;75:290-302

11 Heule F, Tahapary GJ, Bello CR, van Joost T. Delayed-type hypersensitivity to contact allergens in psoriasis. A clinical evaluation. Contact Dermatitis 1998;38:78-82

12 Angelini G, Vena GA, Meneghini CL. Allergic contact dermatitis to some medicaments. Contact Dermatitis 1985;12:263-269

13 Clark AR, Sherertz EF. The incidence of allergic contact dermatitis in patients with psoriasis vulgaris. Dermatitis 1998;9:96-99

14 Burden AD, Stapleton M, Beck MH. Dithranol allergy: fact or fiction? Contact Dermatitis 1992;27:291-293

15 Goncalo S, Sousa I, Moreno A. Contact dermatitis to coal tar. Contact Dermatitis 1984;10:57-58

16 Di Landro A, Valsecchi R, Cainelli T. Contact allergy to dithranol. Contact Dermatitis 1992;26:49-50

17 Žužul K, Kostović K, Čeović R, Ljubojević Hadžavdić S. Contact hypersensitivity in patients with psoriasis. Contact Dermatitis 2018;78:287-289

18 Fleming CJ, Burden AD. Contact allergy in psoriasis. Contact Dermatitis 1997;36:274-276

19 Cusano F, Capozzi M, Errico G. Allergic contact dermatitis from coal tar. Contact Dermatitis 1992;27:51-52

20 Roesyanto ID, van den Akker TW, van Joost TW. Wood tars allergy, cross-sensitization and coal tar. Contact Dermatitis 1990;22:95-98

21 Condé-Salazar L, Guimaraens D, Romero LV, Gonzalez MA. Occupational coal tar dermatitis. Contact Dermatitis 1987;16:231

22 Moustafa GA, Xanthopoulou E, Riza E, Linos A. Skin disease after occupational dermal exposure to coal tar: a review of the scientific literature. Int J Dermatol 2015;54:868-879

23 Riboldi A, Pigatto PD, Innocenti MO, Giacchetti A, Morelli M. Contact dermatitis to coal tar in psoriasis. Contact Dermatitis 1986;14:187-188

24 Rudzki E, Grzywa Z. Occupatonal dermatitis partly elicited by the coal tar. Contact Dermatitis 1977;3:54

25 Doré MA, Houle M-C. Occupational coal tar allergy: An old story revisited. Dermatitis 2015;26:e5

26 Illchyshyn A, Cartwright PH, Smith AG. Contact sensitivity to newsprint: a rare manifestation of coal tar allergy. Contact Dermatitis 1987;17:52-53

27 Sijercić N, Hadzigrahić N, Kamberović S, Suljagić E. Frequency of standard and occupational contact allergens in Tuzla area, Bosnia and Herzegovina: retrospective study. Acta Dermatovenerol Croat 2003;11:75-79

28 Burden AD, Muston H, Beck MH. Intolerance and contact allergy to tar and dithranol in psoriasis. Contact Dermatitis 1994;31:185-186

29 Rudzki E, Kleniewska D. The epidemiology of contact dermatitis in Poland. Br J Dermatol 1970;83:543-545

30 Starke JC, Jillson OF. Photosensitization to coal tar. A cause of psoriatic erythroderma. Arch Dermatol 1961;84:935-936

Chapter 3.81 COLISTIMETHATE

IDENTIFICATION

Description/definition : Colistimethate sodium, also called colistimethate, is the sodium salt form of colistinmethanesulfonic acid; it conforms to the structural formula shown below

Pharmacological classes : Anti-bacterial agents

IUPAC name : Pentasodium;[2-[17-(1-hydroxyethyl)-22-[[2-[[3-hydroxy-2-[[2-(6-methyloctanoylamino)-4-(sulfonatomethylamino)butanoyl]amino]butanoyl]amino]-4-(sulfonatomethylamino) butanoyl]amino]-5,8-bis(2-methylpropyl)-3,6,9,12,15,18,23-heptaoxo-11,14-bis[2-(sulfonatomethylamino)ethyl]-1,4,7,10,13,16,19-heptazacyclotricos-2-yl]ethylamino] methanesulfonate

Other names : Sodium colistimethate; colistin sodium methanesulfonate

CAS registry number : 8068-28-8

EC number : Not available

Merck Index monograph : 3733 (Colistin)

Patch testing : 1% water or pet.

Molecular formula : $C_{58}H_{105}N_{16}Na_5O_{28}S_5$

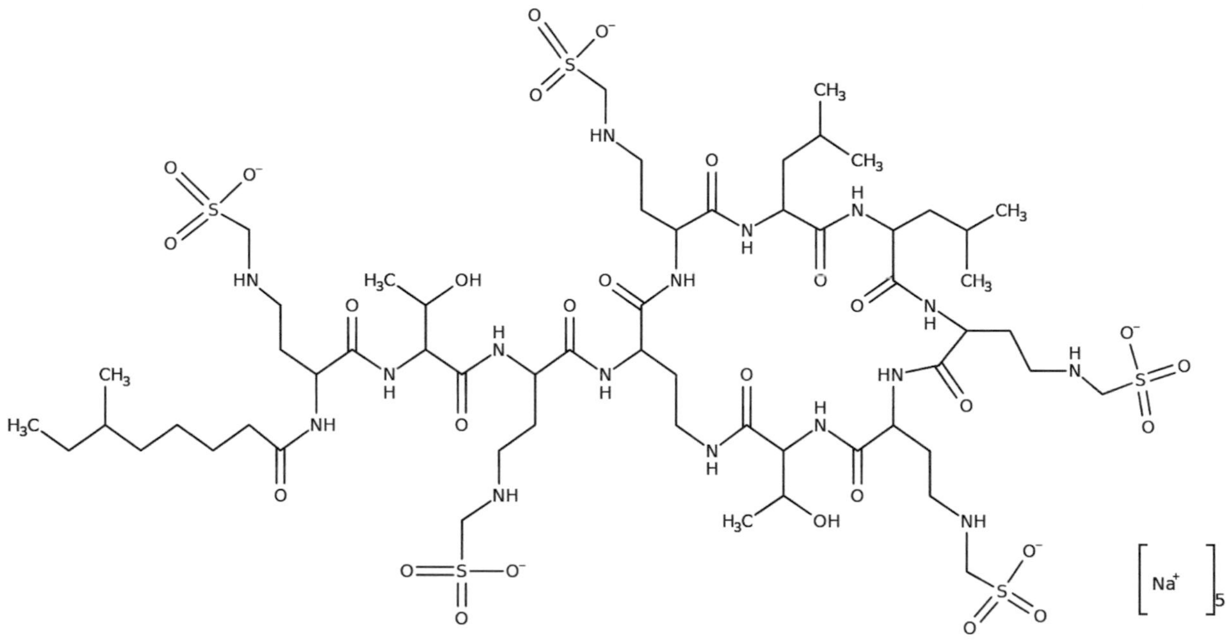

GENERAL

Colistimethate is a broad-spectrum polymyxin antibiotic active against most aerobic gram-negative bacteria except *Proteus* bacteria. This agent is a mixture of methanesulfonate derivatives of the cyclic polypeptides colistin A and B from *Bacillus colistinus or B. polymyxa*. Colistimethate is indicated for the treatment of acute or chronic infections due to sensitive strains of certain gram-negative bacilli, particularly *Pseudomonas aeruginosa*. It may also be employed for ear and ocular infections (1).

CONTACT ALLERGY

Case reports

A 4-year-old Japanese girl was referred because of an itchy eruption on the bilateral periorbital regions. She had undergone enucleations of both eye balls because of microphthalmia and wore ocular prostheses since then. Ofloxacin eye drops were applied to prevent ocular infection. At one point, the ophthalmologist substituted ofloxacin for an ophthalmic solution containing a combination of chloramphenicol and colistimethate. After 3 weeks, an itchy erythematous eruption developed on both periorbital areas. In spite of treatment with a steroid ointment, the eruption gradually extended to both temples and upper cheeks. No other symptoms such as lacrimation or conjunctival hyperemia were seen. The eye drops were discontinued and the lesions subsided within two weeks. Patch tests were positive to the chloramphenicol-colistimethate ophthalmic solution. Later, ingredient patch testing revealed negative reactions to chloramphenicol and to the excipients and the following reactions to colistimethate dilution series (D2/D3): 1% water +/+, 0.1% water ?+/?+, 0.01% water negative; 0.1% pet. +/+, 1% pet. ?+/?+ (2).

A 69-year-old Japanese woman used an ophthalmic solution containing chloramphenicol and colistimethate (the same as the previous patient) on her right eye after surgery for cataract. One year later, right periorbital dermatitis developed, and patch tests revealed positive reactions to the commercial drug and both active ingredients chloramphenicol and colistimethate. Details are not available (3, article in Japanese; data cited in ref. 2).

Another patient also developed allergic contact dermatitis on the eyelids due to colistimethate. Patch tests were positive to colistimethate and polymyxin B. Details are not available (4, article in Japanese, data cited in ref. 2; the title of the article mentions colistin).

LITERATURE

1   The data in the section 'General' may have been obtained from literature discussed in this chapter, but mostly also or exclusively from one or more of the following online sources: ChemIDPlus Advanced, PubChem, DrugBank, RxList, Drug Central, Drugs.com, and Wikipedia
2   Sasaki S, Mitsuhashi Y, Kondo S. Contact dermatitis due to sodium colistimethate. J Dermatol 1998;25:415-417
3   Kubo Y, Tanaka M, Yoshida H. A case of contact dermatitis from Colymy C ophthalmic solution. Skin Research (Hifu) 1989;31(suppl.7):105-110 (Article in Japanese) (data cited in ref. 2)
4   Akai Y, Masuda R, Akaeda T. A case of allergic contact dermatitis from colistin. Skin Research (Hifu) 1992;34(Special Issue):176-180 (Article in Japanese)

Chapter 3.82   COLISTIN

IDENTIFICATION

Description/definition   : Colistin is a cyclic polypeptide antibiotic obtained from *Bacillus colistinus*
Pharmacological classes   : Anti-bacterial agents
IUPAC name   : *N*-[(2*S*)-4-Amino-1-[[(2*S*,3*R*)-1-[[(2*S*)-4-amino-1-oxo-1-[[(6*R*,9*S*,12*R*,15*R*,18*S*,21*S*)-6,9,18-
                          tris(2-aminoethyl)-3-[(1*R*)-1-hydroxyethyl]-12,15-bis(2-methylpropyl)-2,5,8,11,14,17,20-
                          heptaoxo-1,4,7,10,13,16,19-heptazacyclotricos-21-yl]amino]butan-2-yl]amino]-3-hydroxy-
                          1-oxobutan-2-yl]amino]-1-oxobutan-2-yl]-5-methylheptanamide
Other names   : Colomycin; polymyxin E
CAS registry number   : 1066-17-7
EC number   : 213-907-3
Merck Index monograph : 3733
Patch testing   : 5% pet.
Molecular formula   : $C_{52}H_{98}N_{16}O_{13}$

GENERAL

Colistin is a cyclic polypeptide antibiotic derived from *Bacillus colistinus*; it is composed of polymyxins E1 and E2 (or colistins A, B, and C) which act as detergents on cell membranes. Colistin is less toxic than polymyxin B, but otherwise similar; the methanesulfonate (see Chapter 3.81 Colistimethate) is used orally. Colistin is indicated for the treatment of acute or chronic infections due to sensitive strains of certain gram-negative bacilli, particularly *Pseudomonas aeruginosa*. In pharmaceutical products, colistin is employed as colistin sulfate (CAS number 1264-72-8, EC number 215-034-3, molecular formula $C_{53}H_{102}N_{16}O_{17}S$) (1).

CONTACT ALLERGY

Case series

In the period 1993-1994, in The Netherlands, 34 patients (17 women and 17 men) with otorrhea for more than 3 months were referred for patch testing. Seven patients (23%) had a positive reaction to colistin 5% pet. The relevance of the reactions was not addressed (2).

Case reports

A 27-year-old woman had both ears pierced followed by prophylactic application of an ointment containing colistin sulfate $5x10^4$ U, bacitracin 250 U, liquid paraffin and white petrolatum to 1 gram for 1 week. Three years later, both ears were pierced again followed by application of the same ointment, which resulted in pruritic exudative erythema on both auricles. Patch testing with the ointment and its ingredients resulted in positive reactions to the ointment 'as is', colistin sulfate 5% and 0.5% pet. (negative to 0.05%) , and bacitracin 5% and 0.5% pet. (negative to 0.05%) (4). A 45-year-old man had used the same ointment for 10 days on stasis dermatitis, when edematous eczema developed. Patch tests were positive to the ointment and to colistin sulfate 20% and 10% pet. (+ to 1% on D7 only and negative to 0.5%). Five controls were negative (5).

Occupational allergic contact dermatitis

In a group of 107 workers in the pharmaceutical industry with dermatitis, investigated in Warsaw, Poland, before 1989, one reacted to colistin, tested 1.000.000 U/gr pet. (3). Also in Warsaw, Poland, in the period 1979-1983, 27 pharmaceutical workers, 24 nurses and 30 veterinary surgeons were diagnosed with occupational allergic contact dermatitis from antibiotics. The numbers that had positive patch tests to colistin (ampoule content) were 1, 0, and 0, respectively, total 1 (7).

Cross-reactions, pseudo-cross-reactions and co-reactions

Not to polymyxin B sulfate (colistin = polymyxin E) (5). A patient sensitized to polymyxin B sulfate co-reacted to colistin (polymyxin E), which was considered to be a cross-reaction (6).

LITERATURE

1   The data in the section 'General' may have been obtained from literature discussed in this chapter, but mostly also or exclusively from one or more of the following online sources: ChemIDPlus Advanced, PubChem, DrugBank, RxList, Drug Central, Drugs.com, and Wikipedia

2   Van Ginkel CJ, Bruintjes TD, Huizing EH. Allergy due to topical medications in chronic otitis externa and chronic otitis media. Clin Otolaryngol Allied Sci 1995;20:326-328

3   Rudzki E, Rebandel P, Grzywa Z. Contact allergy in the pharmaceutical industry. Contact Dermatitis 1989;21:121-122

4   Sowa J, Tsuruta D, Kobayashi H, Ishii M. Allergic contact dermatitis caused by colistin sulfate & bacitracin. Contact Dermatitis 2005;53:175-176

5   Inoue A, Shoji A. Allergic contact dermatitis from colistin. Contact Dermatitis 1995;33:200

6   Van Ketel WG. Polymixine B-sulfate and bacitracin. Contact Dermatitis Newsletter 1974;15:445

7   Rudzki E, Rebendel P. Contact sensitivity to antibiotics. Contact Dermatitis 1984;11:41-42

Chapter 3.83 COLLAGENASE CLOSTRIDIUM HISTOLYTICUM

IDENTIFICATION

Description/definition	: Collagenase clostridium histolyticum is a member of the matrix metalloproteinases that cleaves triple-helical collagen types I, II, and III
Pharmacological classes	: Enzymes
IUPAC name	: (5S,6S,9R,10S,13S,17S,23S,24S,27R,28S,31S,35S)-5,6,9,13,17,23,24,27,31,35-Decamethyl-10,28-dioctyl-2,20-diazanonacyclo[19.15.0.0^{3,19}.0^{5,17}.0^{6,14}.0^{9,13}.0^{23,35}.0^{24,32}.0^{27,31}]hexa-triaconta-1(21),2,19-triene
Other names	: Matrix metalloproteinase-1; collagen protease; collagenase A; clostridiopeptidase A
CAS registry number	: 9001-12-1
EC number	: 232-582-9
Merck Index monograph	: 3735
Patch testing	: 1% pet.
Molecular formula	: $C_{60}H_{100}N_2$

GENERAL

Collagenase clostridium histolyticum is an enzyme produced by the bacterium *Clostridium histolyticum* and a member of the matrix metalloproteinases that cleaves triple-helical collagen types I, II, and III. It is used as a powder-and-solvent injection kit for the treatment of Dupuytren's contracture, a condition where the fingers bend towards the palm and cannot be fully straightened, and Peyronie's disease, a connective tissue disorder involving the growth of fibrous plaques in the soft tissue of the penis. Used in the topical treatment of skin ulcers and severe burns, collagenase is able to digest collagen in necrotic tissue at physiological pH by hydrolysing the peptide bonds of undenatured and denatured collagen. Collagenase thus contributes towards the formation of granulation tissue and subsequent epithelization. The action of collagenase may remove substrates necessary for bacterial proliferation or may permit antibodies, leukocytes, and antibiotics better access to the infected area (1).

CONTACT ALLERGY

Case series

Four female patients aged 75, 83, 70 and 35 years were treated with an ointment containing clostridiopeptidase A for leg ulcers when they developed eczema around the wounds. Patch tests were performed with the ointment as is, clostridiopeptidase A 1%, 0.1%, 0.05% and 0.001% pet. and the excipients of the ointment. There were 3 doubtful and one positive reactions to the ointment. All 4 women had positive reactions to the active ingredient 1% pet., one of them also to all lower concentrations. Ten healthy controls were negative (2).

A 62-year old woman and a 53-year-old female patient were both treated with an ointment containing 0.6 U clostridiopeptidase A (collagenase) and 0.01 gr. chloramphenicol/gr. ointment. Clinical details were not provided, but both had positive patch tests to the ointment, collagenase A 1.2 mg/g pet. and various other ingredients of topical drugs, including chloramphenicol in one (5).

Case reports

A 77-year-old woman had treated her venous leg ulcers for 2 months with clostridiopeptidase A containing ointment, when she developed a very itchy erythematous, edematous, exudative and crusted dermatitis on the legs which later spread to the forearms and trunk. Patch tests were positive to the ointment and negative to its vehicle ingredients (soft paraffin, white petrolatum). Forty controls were negative. It was concluded (*per exclusionem*) that the active ingredient clostridiopeptidase A was the allergenic culprit (4).

Another case of contact allergy was reported in the German literature. The 75-year-old male patient had a post-traumatic ulcer on the right leg. He reacted to the ointment and many topical drug constituents, but clostridiopeptidase A itself was (probably) not tested (3; no further details known).

A 62-year-old woman was referred for an itchy eruption on the left leg that had appeared 2 weeks into a bandage treatment with an ointment containing a mixture of collagenase clostridiopeptidase A and proteases. Based on patient history and vesicular appearance of the lesions, allergic contact dermatitis was suspected and confirmed by a positive patch test to the ointment; the ingredients were not tested separately (6).

LITERATURE

1   The data in the section 'General' may have been obtained from literature discussed in this chapter, but mostly also or exclusively from one or more of the following online sources: ChemIDPlus Advanced, PubChem, DrugBank, RxList, Drug Central, Drugs.com, and Wikipedia
2   Foti C, Conserva A, Casulli C, Scrimieri V, Pepe ML, Quaranta D. Contact dermatitis with clostridiopeptidase A contained in Noruxol ointment. Contact Dermatitis 2007;56:361-362
3   Meichlbock A, Bayerl C. Kontakteczem auf Noruxol® Salbe bei posttraumatischem Ulkus. Akt Dermatol 1999;25:272-273
4   Lisi P, Brunelli L. Extensive allergic contact dermatitis from a topical enzymatic preparation (Noruxol). Contact Dermatitis 2001;45:186-187
5   Braun WP. Contact allergy to collagenase mixture (Iruxol). Contact Dermatitis 1975;1:241-242
6   Baroni A, Piccolo V, Russo T. A possible explanation for the high frequency of contact sensitisation in chronic venous ulcers. Int Wound J 2015;12:369-370

Chapter 3.84   CORTISONE ACETATE

IDENTIFICATION

Description/definition : Cortisone acetate is the acetate ester of cortisone, a synthetic or semisynthetic glucocorticoid; it conforms to the structural formula shown below

Pharmacological classes : Anti-inflammatory agents

IUPAC name : [2-[(8S,9S,10R,13S,14S,17R)-17-Hydroxy-10,13-dimethyl-3,11-dioxo-1,2,6,7,8,9,12,14,15, 16-decahydrocyclopenta[a]phenanthren-17-yl]-2-oxoethyl] acetate

Other names : 17,21-Dihydroxypregn-4-ene-3,11,20-trione 21-acetate; cortisone 21-acetate

CAS registry number : 50-04-4

EC number : 200-006-5

Merck Index monograph : 3795 (Cortisone)

Patch testing : In general, corticosteroids may be tested at 0.1% and 1% in alcohol; late readings (6-10 days) are strongly recommended

Molecular formula : $C_{23}H_{30}O_6$

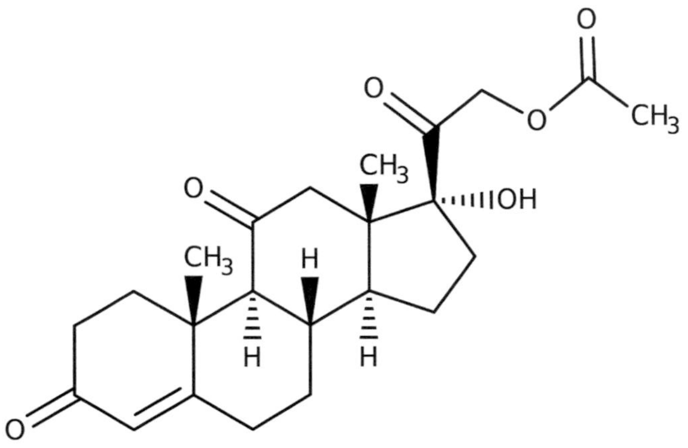

GENERAL

General aspects of corticosteroids used on the skin and mucous membranes are discussed in Chapter 2.4. A practical guideline for diagnosing allergic reactions to corticosteroids is presented in ref. 1.

CONTACT ALLERGY

Case series

From January 1990 to June 2008, in Leuven, Belgium, 315 patients were diagnosed with contact allergy to/allergic contact dermatitis from corticosteroids (CSs) from routine patch testing with a baseline series including tixocortol pivalate, budesonide, hydrocortisone butyrate and prednisone caproate, patch testing with patients' own CSs preparations, and testing those with proven contact allergy to a corticosteroid or strongly suspected of CS allergy later with a series of 66 CSs, including two sex hormones (progesterone and testosterone). 71% of the patients had relevant reactions, but these were not specified. In this group of 315 CS allergic patients, 67 had positive patch tests to cortisone acetate 0.5% alc./DMSO (2). It is unknown how many of these reactions were caused by the use of a pharmaceutical product containing cortisone acetate and how many were cross-reactions to other corticosteroids.

Cross-reactions, pseudo-cross-reactions and co-reactions

Cross-reactions between corticosteroids are discussed in Chapter 2.8.

LITERATURE

1   Baeck M, Goossens A. Immediate and delayed allergic hypersensitivity to corticosteroids: practical guidelines. Contact Dermatitis 2012;66:38-45

2   Baeck M, Chemelle JA, Terreux R, Drieghe J, Goossens A. Delayed hypersensitivity to corticosteroids in a series of 315 patients: clinical data and patch test results. Contact Dermatitis 2009;61:163-175

Chapter 3.85   CROCONAZOLE

IDENTIFICATION

Description/definition   : Croconazole is the phenylmethyl imidazole that conforms to the structural formula shown
                              below
Pharmacological classes : Antifungal agents
IUPAC name   : 1-[1-[2-[(3-Chlorophenyl)methoxy]phenyl]ethenyl]imidazole
Other names   : 1-(1-(o-((m-Chlorobenzyl)oxy)phenyl)vinyl)imidazole; cloconazole
CAS registry number   : 77175-51-0
EC number   : Not available
Merck Index monograph : 3850
Patch testing   : 1% pet.; 1% alcohol or MEK (methyl ethyl ketone) is usually preferred for imidazoles
Molecular formula   : $C_{18}H_{15}ClN_2O$

GENERAL

Croconazole is an imidazole antifungal drug used for the topical treatment of superficial fungal infections. In
pharmaceutical products, croconazole is usually employed as croconazole hydrochloride (CAS number 77174-66-4,
EC number not available, molecular formula $C_{18}H_{16}Cl_2N_2O$) (1).

CONTACT ALLERGY

General

Croconazole is (or was) an important sensitizer especially in East Asia (Japan, Korea). It was developed in Japan and
released in 1986. Soon, cases of allergic contact dermatitis were published. In guinea pigs, croconazole was found to
be a strong sensitizer (12).

Patch testing in groups of selected patients

In the period 2001-2002, 20 members of the IVDK tested a series of antifungals in selected patients (selection
procedure not specified, <10% - 22% of all routine patients were tested). 1094 patients were tested with croconazole
HCl 1% pet. and there were 2 (0.2%) positive reactions. Relevance data were not provided (2).

Case series

In Japan, in the period 1984 to 1994, 3049 outpatients were patch tested for suspected contact dermatitis and 218
of these with topical antifungal preparations. Thirty-five were allergic to imidazoles, including croconazole in 11
individuals. In 60% of the cases, there were cross-reactions between imidazoles. Based on the limited use of
croconazole and the short period from first use to sensitization, it was concluded that croconazole has a stronger
sensitizing ability than other imidazoles (3).

From Japan, 6 cases of allergic contact dermatitis from croconazole were reported, 3 years after its introduction
in 1986 (8). Three of the patients were seen in the first year and had been reported in Japanese literature before (9).
The group consisted of 3 women and three men, age range 38-61 years, mean age 51. Four of them had been
treated with a gel and the other 2 with a cream containing 1% croconazole HCl for tinea pedis. After 1-4 months (in
one case after a week), signs of sensitization appeared: vesicular dermatitis, erosions, worsening of lesions, pruritic
erythema, itchy infiltrated lesions. Patch tests were positive to the gel and cream in all; in a dilution series of 5% pet.
to 0.01% pet., all reacted down to 0.5% and 2 to 0.1%. Ten controls were negative to all concentrations. There were
also 3 reactions to sulconazole, which were considered to be cross-reactions (8).

Other small case series reporting allergic contact dermatitis from croconazole in 2 patients (11,16), 3 individuals (10) and 4 patients (14), details of which are not available to the author, have been reported in Japanese literature (cited in ref. 12). In a literature review up to 1994, the authors identified 105 reported patients who had a positive patch test reaction to at least one imidazole derivative, ranging from 51 reactions to miconazole to one reaction each to bifonazole and enilconazole. The number of reported reactions to croconazole was 12 (5).

Case reports
A 74-year-old man presented with a 3-month history of interdigital mycosis of both feet. Treatment included the topical imidazole antifungals clotrimazole, bifonazole and croconazole. Four weeks of treatment with croconazole led, after initial improvement, to an eczema spreading from the toes to the lower shins. Patch testing revealed positive reactions only to croconazole cream and croconazole (test concentration and vehicle not mentioned) (7).

A 58-year-old man presented with weeping erosions in the genital and inguinal area, where a tinea cruris had been treated with various topical imidazoles including croconazole cream. There were positive reactions to the cream as is, croconazole 1% pet., creams with clotrimazole, oxiconazole, and tioconazole (which may have been used before), and to clotrimazole 5% pet. and tioconazole 1% pet. (4).

Other single case reports of allergic contact dermatitis from croconazole, details of which are not available to the author, have been reported in Japanese literature (13 [patient also presented in ref. 11], 15).

Cross-reactions, pseudo-cross-reactions and co-reactions
A patient sensitized to clotrimazole very likely cross-reacted to croconazole (also a phenylmethyl imidazole) and itraconazole (a triazole) (6). Three patients sensitized to croconazole may have cross-reacted to sulconazole (8). Possible cross-sensitivity between croconazole, isoconazole, bifonazole and clotrimazole (11).

LITERATURE
1   The data in the section 'General' may have been obtained from literature discussed in this chapter, but mostly also or exclusively from one or more of the following online sources: ChemIDPlus Advanced, PubChem, DrugBank, RxList, Drug Central, Drugs.com, and Wikipedia
2   Menezes de Pádua CA, Uter W, Geier J, Schnuch A, Effendy I; German Contact Dermatitis Research Group (DKG); Information Network of Departments of Dermatology (IVDK). Contact allergy to topical antifungal agents. Allergy 2008;63:946-947
3   Yoneyama E. Allergic contact dermatitis due to topical imidazole antimycotics. The sensitizing ability of active ingredients and cross-sensitivity. Nippon Ika Daigaku Zasshi 1996;63:356-364 (Article in Japanese)
4   Brans R, Wosnitza M, Baron JM, Merk HF. Contact sensitization to azole antimycotics. Hautarzt 2009;60:372-375 (Article in German)
5   Dooms-Goossens A, Matura M, Drieghe J, Degreef H. Contact allergy to imidazoles used as antimycotic agents. Contact Dermatitis 1995;33:73-77
6   Erdmann S, Hertl M, Merk HF. Contact dermatitis from clotrimazole with positive patch-test reactions also to croconazole and itraconazole. Contact Dermatitis 1999;40:47-48
7   Steinmann A, Mayer G, Breit R, Agathos M. Allergic contact dermatitis from croconazole without cross-sensitivity to clotrimazole and bifonazole. Contact Dermatitis 1996;35:255-256
8   Shono M, Hayashi K, Sugimoto R. Allergic contact dermatitis from croconazole hydrochloride. Contact Dermatitis 1989;21:225-227
9   Shono M, Hayashi K. Contact dermatitis to croconazole hydrochloride. Skin Research 1987;29(suppl.):231-236 (Article in Japanese, data cited in ref. 8)
10 Matsumura E, Iizumi Y, Hata M, Yajima J, Hattori S, Honda M. Allergic contact dermatitis to imidazole antymycotics. Rinsho Derma 1987;29:673-677 (Article in Japanese, data cited in ref. 12)
11 Higashi N. Contact dermatitis from imidazole derivatives. Hifubyoh-Shinryoh 1988;10:713-716 (Article in Japanese, data cited in ref. 12)
12 Hausen BM, Angel M. Studies on the sensitizing capacity of imidazole and triazole derivatives. Part II. Amer J Contact Dermat 1992;3: 95-101
13 Higashi N, Matsumura T, Iwasa M. Contact dermatitis from topical antimycotics containing imidazole derivatives. Report of 7 cases. Skin Res 1988;30(Suppl.5):55-61 (Article in Japanese, data cited in ref. 12)
14 Fujimoto K, Hashimoto S. Contact dermatitis due to cloconazole hydrochloride. Skin Res 1989;31(Suppl.):111-114 (Article in Japanese, data cited in ref. 12)
15 Higashi M, Isonomaki M, Hazano S. Three cases of contact dermatitis from topical imidazole antimycotics. Skin Res 1990;32:(Suppl.):149-154 (Article in Japanese, data cited in ref. 12)
16 Akai Y, Masuda R, Akaeda T. Two cases of contact dermatitis from croconazole hydrochloride. Skin Res 1990;32(Suppl.):144-148 (Article in Japanese, data cited in ref. 12)

Chapter 3.86   CROMOGLICIC ACID

IDENTIFICATION

Description/definition   : Cromoglicic acid is the chromone complex that conforms to the structural formula shown
                           below
Pharmacological classes : Anti-asthmatic agents
IUPAC name   : 5-[3-(2-Carboxylato-4-oxochromen-5-yl)oxy-2-hydroxypropoxy]-4-oxochro-mene-2-
                           carboxylic acid
Other names   : 5,5'-(2-Hydroxytrimethylenedioxy)bis(4-oxochromene-2-carboxylic acid); cromolyn;
                           cromoglycic acid
CAS registry number   : 16110-51-3
EC number   : 240-279-8
Merck Index monograph   : 3851 (Cromolyn)
Patch testing   : 2% water
Molecular formula   : $C_{23}H_{16}O_{11}$

GENERAL

Cromoglicic acid (better known as cromolyn) is a mast cell stabilizer with anti-inflammatory activity. It probably interferes with the antigen-stimulated calcium transport across the mast cell membrane, thereby inhibiting mast cell release of histamine, leukotrienes, and other substances that cause hypersensitivity reactions. Cromoglicic acid also inhibits eosinophil chemotaxis. It is used as inhalation aerosol in the prophylactic treatment of both allergic and exercise-induced asthma, but does not affect an established asthmatic attack. In eye drops, it has been used to treat allergic conjunctivitis and vernal keratoconjunctivitis and it is useful as nasal spray in patients with allergic rhinitis. In oral form, sodium cromoglycate may prevent allergic reactions to foods which cannot be avoided. In pharmaceutical products, cromoglicic acid is employed as its sodium salt cromolyn sodium (sodium cromoglycate, disodium cromoglycate) (CAS number 15826-37-6, EC number 239-926-7, molecular formula $C_{23}H_{14}Na_2O_{11}$) (1).

CONTACT ALLERGY

Case reports

A 32-year-old woman was prescribed eye drops containing 4% sodium cromoglycate for complaints of ocular discomfort, with severe itching and burning sensations, photophobia and conjunctival injection, which were ascribed to the cleaning product of her contact lenses. She continued to use the same contact lenses and cleaning solution. A few weeks later, a severe, itchy, erythematous edematous periorbital dermatitis appeared, spreading over the cheeks. Positive patch test were observed to the eye drops, the cleaning solution and thimerosal, which was the preservative in the cleaning solution. Sodium cromoglycate was tested at 1% in water and petrolatum, but the results were negative. Avoiding the cleaning solution and cromolyn eye drops resulted in disappearance of her complaints. Next, a ROAT was performed with the commercial eye drops, which elicited a severe vesicular eczema after 4 days. The patient was patch tested with all ingredients including benzalkonium chloride, which were all negative. However, a patch test with sodium cromoglycate 1% combined with benzalkonium chloride 0.1% in petrolatum was strongly positive (D2++, D4++). Ten controls were negative to this material (3).

The authors called this phenomenon 'compound allergy', but this is probably incorrect. The more likely explanation is that the test concentration of 1% for sodium cromoglycate was too low to elicit a positive reaction, and that the irritant properties of benzalkonium chloride and its effect on the skin facilitated percutaneous absorption and the elicitation of a contact allergic reaction. False-negative patch test reactions with eye drops in patients with ocular and periocular symptoms are well-known.

A 59-year-old woman with allergic conjunctivitis had been treated with eye drops containing 2% sodium cromoglycate for 4 months, when she developed erythema around both eyes. Patch tests showed a positive reaction to the eye drops. Later, sodium cromoglycate was tested at 2% water, and there were positive reactions at D2 and D3. Twenty controls were negative (4).

Immediate contact reactions

Immediate contact reactions (contact urticaria) to cromolyn sodium (cromoglycate disodium) are presented in Chapter 5.

LITERATURE

1   The data in the section 'General' may have been obtained from literature discussed in this chapter, but mostly also or exclusively from one or more of the following online sources: ChemIDPlus Advanced, PubChem, DrugBank, RxList, Drug Central, Drugs.com, and Wikipedia

2   Temesvári E, Pónyai G, Németh I, Hidvégi B, Sas A, Kárpáti S. Periocular dermatitis: a report of 401 patients. J Eur Acad Dermatol Venereol 2009;23:124-128

3   Camarasa JG, Serra-Baldrich E, Monreal P, Soller J. Contact dermatitis from sodium-cromoglycate-containing eyedrops. Contact Dermatitis 1997;36:160-161

4   Kudo H, Tanaka T, Miyachi Y, Imamura S. Contact dermatitis from sodium cromoglycate eyedrops. Contact Dermatitis 1988;19:312

Chapter 3.87   CROTAMITON

IDENTIFICATION

Description/definition   : Crotamiton is the enamide and tertiary carboxamide that conforms to the structural
                            formula shown below
Pharmacological classes : Pesticides
IUPAC name   : (*E*)-*N*-Ethyl-*N*-(2-methylphenyl)but-2-enamide
Other names   : *N*-Ethyl-*o*-crotonotoluidide
CAS registry number   : 483-63-6
EC number   : 207-596-3
Merck Index monograph   : 3855
Patch testing   : 5% pet.
Molecular formula   : $C_{13}H_{17}NO$

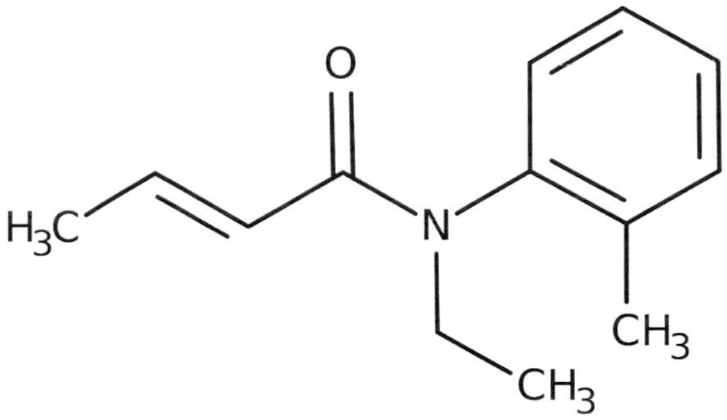

GENERAL

Crotamiton is an enamide and tertiary carboxamide resulting from the formal condensation of crotonic acid with *N*-ethyl-2-methylaniline. It is used in the treatment of pruritus by producing a counter-irritation: as it evaporates from the skin, it produces a cooling effect that diverts attention away from the itching. It has also been used as an acaricide in the treatment of scabies, though more effective drugs are usually preferred (1). Crotamiton is also used as solubilizer and may be included in topical NSAIDs, antimycotics, corticosteroids and others (5).

CONTACT ALLERGY

Case reports and case series

At St. John's Hospital, London, in the period 1971 to 1976, 5 women and one man were thought to have been sensitized by an ointment containing 10% crotamiton, which they had used for treatment of various forms of dermatitis. They all had positive patch tests to crotamiton 1% MEK (methyl ethyl ketone) (12). In Leuven, Belgium, in

Table 3.87.1 Short summaries of case reports of allergic contact dermatitis from crotamiton

Year and country	Sex	Age	Positive patch tests	Clinical data and comments	Ref.
2014 Japan	F	67	CRO 0.1%, 1%, 10% pet. and pure	erythema and yellow crusts on right sole from application to fungus; present in antifungal cream for anti-itch	9
2003 Japan	F	48	CRO ointment and tape; CRO 5% and 10% pet.	acute dermatitis on the right ankle first from CRO tape on sprain, which was subsequently treated with CRO ointment	5
1997 Japan	F	48	CRO 1% and 10% pet.	pruritic scaly erythema on the toes and dorsum of the right foot; 5 controls negative; also allergy to lidocaine	8
1992 Portugal	M	51	CRO 5% and 10% pet.	erythema, itch and small papules on abdomen and limbs 2 days after application of anti-itch cream; 20 controls neg.	10
1986 Italy	M	40	CRO 10% water; CRO cream 'as is'	cream applied to legs; after 1 month acute dermatitis of the hands, face and trunk	13
1972 Netherlands	F	67	CRO ointment 'as is'; CRO 5% pet.	ointment applied to stasis dermatitis; acute vesicular dermatitis of legs, arms and back; erythema and scaling of the face	14
1952 USA	F	?	CRO ointment 'as is'; CRO (conc./vehicle ?)	treatment of otitis externa resulted within 24 hours in burning, itching and edema extending down the cheeks to the chin and anterior neck	15

CRO: crotamiton

the period 1990-2014, iatrogenic contact dermatitis was diagnosed in 2600 individuals (17% of the total population patch test). 96% of all positive patch test reactions to topical drugs and antiseptics were considered to be relevant. Crotamiton (5% pet.) was tested in 8 patients and there was one positive reaction to it (2).

A 65-year-old man developed an extremely pruritic and diffusely spreading allergic contact dermatitis from crotamiton (tested 1% pet.), the NSAID felbinac and the excipient diisopropanolamine in a compress (3). A 60-year-old woman presented with pruritic erythematous macules on the neck and in the left cubital fossa, where she had previously applied an anti-itch ointment. A patch test with the ointment was positive at D2 and D3 with a flare-up of the previous symptoms. When tested later with all ingredients, the patient reacted positively to crotamiton 5% pet. and cetyl alcohol (30% pet.) (4).

Short summaries of other case reports of allergic contact dermatitis from crotamiton are shown in table 3.87.1. Other case reports of contact allergy to/allergic contact dermatitis from crotamiton, adequate data of which are not available to the author, can be found in refs. 6 and 7 (articles in Japanese, cited in ref. 5). Early reports of sensitization (1948-1968) have been reviewed in ref. 11.

Cross-reactions, pseudo-cross-reactions and co-reactions
Crotamiton and lidocaine have butanoyl toluidine in common in their chemical structure, suggesting the possibility of cross-sensitization (8).

LITERATURE

1   The data in the section 'General' may have been obtained from literature discussed in this chapter, but mostly also or exclusively from one or more of the following online sources: ChemIDPlus Advanced, PubChem, DrugBank, RxList, Drug Central, Drugs.com, and Wikipedia
2   Gilissen L, Goossens A. Frequency and trends of contact allergy to and iatrogenic contact dermatitis caused by topical drugs over a 25-year period. Contact Dermatitis 2016;75:290-302
3   Oiso N, Fukai K, Ishii M. Triple allergic contact sensitivities due to ferbinac, crotamiton and diisopropanolamine. Contact Dermatitis 2003;49:261-263
4   Oiso N, Fukai K, Ishii M. Concomitant allergic reaction to cetyl alcohol and crotamiton. Contact Dermatitis 2003;49:261
5   Hara H, Masuda T, Yokoyama A, Asaki H, Okada T, Suzuki H. Allergic contact dermatitis due to crotamiton. Contact Dermatitis 2003;49:219
6   Morita C, Ito M, Wakita M, Shimizu A, Takeuchi Y, Tsuyuki S. Two cases of allergic contact dermatitis due to crotamiton. Environ Dermatol 1997;4:277-282 (Article in Japanese, cited in ref. 5)
7   Onodera H, Mori Y, Akasaka T. Contact dermatitis due to Damalin-L: report of two cases. Rinsho Derma (Tokyo) 2002;44:417-420 (Article in Japanese, cited in ref. 5)
8   Kawada A, Hiruma M, Fujioka A, Tajima S, Akiyama M, Ishibashi A. Simultaneous contact sensitivity due to lidocaine and crotamiton. Contact Dermatitis 1997;37:45
9   Mori N, Mizawa M, Hara H, Norisugi O, Makino T, Nakano H, et al. Hailey-Hailey disease diagnosed based on an exacerbation of contact dermatitis with topical crotamiton. Eur J Dermatol 2014;24:263-264
10 Baptista A, Barros MA. Contact dermatitis from crotamiton. Contact Dermatitis 1992;27:59
11 Hausen BM, Kresken J. The sensitizing capacity of crotamiton. Contact Dermatitis 1988;18:298-299
12 Cronin E. Contact dermatitis. Edinburgh: Churchill Livingstone, 1980:254-255
13 Lama L, Bardazzi F, Vanni D, Vassiloupou A. Contact dermatitis due to crotamiton. Contact Dermatitis 1986;15:255-256
14 Van Dijk TJA. Allergic contact dermatitis from Eurax® (containing 10% crotonyl-N-aethyl-o-toluidinum). Contact Dermatitis Newsletter 1972;12:344
15 Bereston ES. Contact dermatitis due to N-ethyl-o-crotonotoluide ointment (Eurax®). Arch Derm 1952;65:100-101

Chapter 3.88   CYCLOBENZAPRINE

IDENTIFICATION

Description/definition   : Cyclobenzaprine is the tricyclic compound that conforms to the structural formula shown
                           below
Pharmacological classes : Muscle relaxants, central; tranquilizing agents; antidepressive agents, tricyclic
IUPAC name   : *N*,*N*-Dimethyl-3-(2-tricyclo[9.4.0.0^{3,8}]pentadeca-1(15),3,5,7,9,11,13-heptaenylidene)pro-
                           pan-1-amine
Other names   : 5-(3-Dimethylaminopropyliden)-5*H*-dibenzo-(a,d)-cyclopentene
CAS registry number   : 303-53-7
EC number   : 206-145-8
Merck Index monograph   : 3976
Patch testing   : 2% pet.
Molecular formula   : C$_{20}$H$_{21}$N

GENERAL

Cyclobenzaprine, a chemical closely related to the antidepressant amitriptyline, is a centrally acting skeletal muscle
relaxant with antidepressant activity. It is indicated for use as an adjunct to rest and physical therapy for relief of
muscle spasm associated with acute, painful musculoskeletal conditions. In pharmaceutical products, cyclobenzaprine is employed as cyclobenzaprine hydrochloride (CAS number 6202-23-9, EC number 228-264-4, molecular
formula C$_{20}$H$_{22}$ClN) (1).

CONTACT ALLERGY

Case report

A 39-year-old man with degenerative disc disease complicated by cervical radiculopathy was treated topically with a
compounded pain medication containing ketamine 10%, diclofenac 5%, baclofen 2%, bupivacaine 1%, cyclobenzaprine 2%, gabapentin 6%, ibuprofen 3%, and pentoxifylline 3% in a cream base. The patient benefitted from
substantial pain relief, but after several weeks of use, he developed an itchy rash at the site of application. Patch
testing revealed a positive reaction to the cream and to cyclobenzaprine 2% (vehicle not mentioned, probably the
cream base). Later, a serial dilution test with cyclobenzaprine (2%, 1%, 0.2%, 0.02%, 0.002%) gave a response in a
graded manner to all concentrations (photo shown, but strength of reactions not specified, highest concentration
probably ++ or +++, lowest concentration ?+ or negative). The topical compounded formulation without cyclobenzaprine was subsequently well tolerated by the patient (2).

LITERATURE

1   The data in the section 'General' may have been obtained from literature discussed in this chapter, but mostly
    also or exclusively from one or more of the following online sources: ChemIDPlus Advanced, PubChem,
    DrugBank, RxList, Drug Central, Drugs.com, and Wikipedia
2   Turrentine JE, Marrazzo G, Cruz PD Jr. Novel use of patch testing in the first report of allergic contact dermatitis
    to cyclobenzaprine. Dermatitis 2015;26:60-61 (also published as Abstract: Dermatitis 2015;26:e3)

Chapter 3.89   CYCLOPENTOLATE

IDENTIFICATION

Description/definition   : Cyclopentolate is the aromatic homomonocyclic compound that conforms to the structural formula shown below
Pharmacological classes   : Mydriatics and cycloplegics
IUPAC name   : 2-(Dimethylamino)ethyl 2-(1-hydroxycyclopentyl)-2-phenylacetate
CAS registry number   : 512-15-2
EC number   : 208-136-4
Merck Index monograph   : 4011
Patch testing   : 0.5% water
Molecular formula   : $C_{17}H_{25}NO_3$

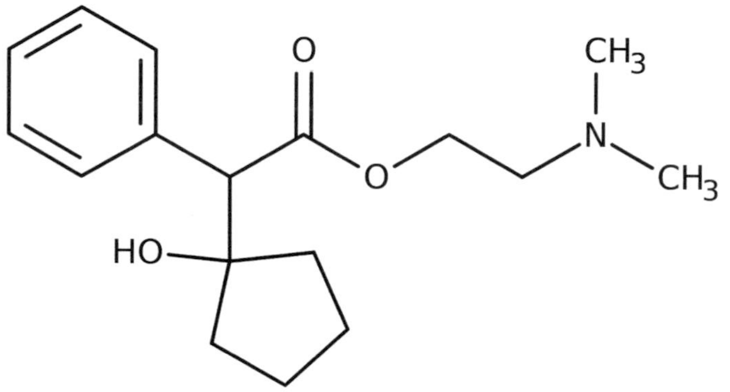

GENERAL

Cyclopentolate is a parasympatholytic anticholinergic drug. Administered in the eye, cyclopentolate blocks the acetylcholine receptor in the sphincter muscle of the iris and the ciliary muscle, thereby preventing contraction. This produces mydriasis (excessive dilation of the pupil) and cycloplegia (paralysis of the ciliary muscle of the eye), which facilitates ophthalmic diagnostic procedures. Cyclopentolate acts more quickly than atropine and has a shorter duration of action (1). In pharmaceutical products, cyclopentolate is employed as cyclopentolate hydrochloride (CAS number 5870-29-1, EC number 227-521-8, molecular formula $C_{17}H_{26}ClNO_3$) (1).

CONTACT ALLERGY

Case report

A 49-year-old man presented with a few months' history of severe relapses of erythema, edema, itching and burning of both eyes, eyelids and upper cheeks. The patient was under ophthalmological investigation for progressive loss of sight. Before each examination, at home he applied one eye drop containing cyclopentolate in each eye, every 15 minutes for one hour. Afterwards, when arriving at the consultation, the ophthalmologist would apply one drop of a phenylephrine solution. Patch tests gave positive reactions to cyclopentolate hydrochloride 0.5% and 0.1% water and to phenylephrine 1% water. Control tests in 20 healthy individuals with the same solutions were negative (2).

Immediate contact reactions

Immediate contact reactions (contact urticaria) to cyclopentolate are presented in Chapter 5.

LITERATURE

1   The data in the section 'General' may have been obtained from literature discussed in this chapter, but mostly also or exclusively from one or more of the following online sources: ChemIDPlus Advanced, PubChem, DrugBank, RxList, Drug Central, Drugs.com, and Wikipedia
2   Camarasa JG, Pla C. Allergic contact dermatitis from cyclopentolate. Contact Dermatitis 1996;35:368-369

Chapter 3.90   CYPROHEPTADINE

IDENTIFICATION

Description/definition   : Cyproheptadine is the synthetic methylpiperidine derivative that conforms to the
                              structural formula shown below
Pharmacological classes   : Anti-allergic agents; gastrointestinal agents; histamine H1 antagonists; serotonin
                              antagonists; antipruritics
IUPAC name   : 1-Methyl-4-(2-tricyclo[9.4.0.0^{3,8}]pentadeca-1(15),3,5,7,9,11,13-heptaenylidene)pi-
                              peridine
Other names   : 4-(5-Dibenzo(a,e)cycloheptatrienylidene)piperidine
CAS registry number   : 129-03-3
EC number   : 204-928-9
Merck Index monograph   : 4040
Patch testing   : 2% pet.
Molecular formula   : $C_{21}H_{21}N$

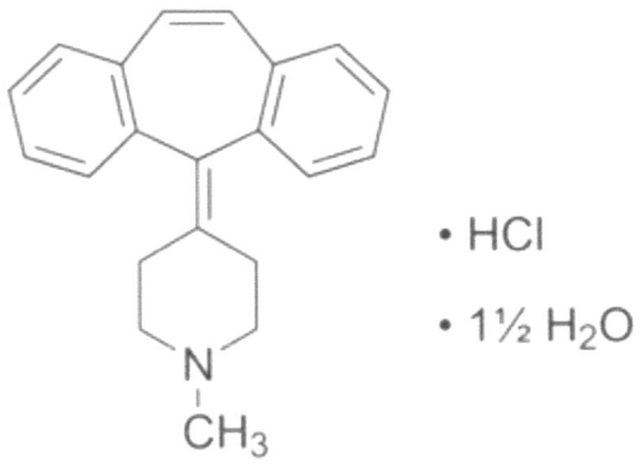

• HCl

• 1½ H_2O

Cyproheptadine hydrochloride sesquihydrate

GENERAL

Cyproheptadine is a synthetic methylpiperidine derivative with antihistaminic (H1-blocking) and anti-serotoninergic properties. This agent exhibits anticholinergic and sedative properties and has been shown to stimulate appetite and weight gain. It is indicated for treatment of perennial and seasonal allergic rhinitis, vasomotor rhinitis, allergic conjunctivitis due to inhalant allergens and foods, mild uncomplicated allergic skin manifestations of urticaria and angioedema, amelioration of allergic reactions to blood or plasma, cold urticaria, dermographism, and as therapy for anaphylactic reactions adjunctive to epinephrine. In China, this drug is (or was) also used in topical preparations to relieve itching (2). In pharmaceutical products, cyproheptadine is employed as cyproheptadine hydrochloride (sesquihydrate) (CAS number 41354-29-4, EC number not available, molecular formula $C_{42}H_{50}Cl_2N_2O_3$) (1).

CONTACT ALLERGY

Case report

A 26-year-old woman developed exudative edematous erythematous lesions on her face a week after having used vitamin B$_6$ cream, cyproheptadine hydrochloride cream and tetracycline cream simultaneously for facial dryness with itching and fine scaling. Patch testing revealed positive reactions to the cyproheptadine hydrochloride cream and to cyproheptadine HCl 1% and 2% pet. (+++). Twenty controls were negative (2). Previously, a case of contact dermatitis from cyproheptadine HCl cream had been reported in Chinese literature (3; no details known, data cited in ref. 2).

LITERATURE

1   The data in the section 'General' may have been obtained from literature discussed in this chapter, but mostly
    also or exclusively from one or more of the following online sources: ChemIDPlus Advanced, PubChem,
    DrugBank, RxList, Drug Central, Drugs.com, and Wikipedia
2   Li LF, Sun XY, Li SY. Allergic contact dermatitis from cyproheptadine hydrochloride. Contact Dermatitis
    1995;33:50
3   Lang FM. Cyproheptadine hydrochloride cream contact dermatitis, a case report. Chin J Clin Dermatol
    1988;17:74 (Article in Chinese)

Chapter 3.91   DESONIDE

IDENTIFICATION

Description/definition   : Desonide is the synthetic glucocorticoid that conforms to the structural formula shown below

Pharmacological classes   : Anti-inflammatory agents

IUPAC name   : (1S,2S,4R,8S,9S,11S,12S,13R)-11-Hydroxy-8-(2-hydroxyacetyl)-6,6,9,13-tetramethyl-5,7-dioxapentacyclo[10.8.0.0^{2,9}.0^{4,8}.0^{13,18}]icosa-14,17-dien-16-one

Other names   : 11β,16α,17,21-Tetrahydroxypregna-1,4-diene-3,20-dione cyclic 16,17-acetal with acetone; 16α-hydroxyprednisolone 16,17-acetonide

CAS registry number   : 638-94-8

EC number   : 211-351-6

Merck Index monograph   : 4199

Patch testing   : In general, corticosteroids may be tested at 0.1% and 1% in alcohol; late readings (6-10 days) are strongly recommended

Molecular formula   : $C_{24}H_{32}O_6$

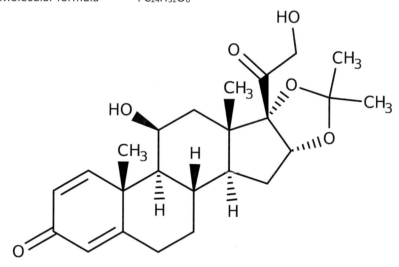

GENERAL

General aspects of corticosteroids used on the skin and mucous membranes are discussed in Chapter 2.4. A practical guideline for diagnosing allergic reactions to corticosteroids is presented in ref. 1.

CONTACT ALLERGY

Patch testing in groups of selected patients

In the period 1988-1991, in the United Kingdom, 528 patients with a positive patch test to tixocortol pivalate or suspected of corticosteroid allergy were tested with desonide 20% pet. and there was one (0.2%) positive reaction; its relevance was not mentioned (7).

Case series

From January 1990 to June 2008, in Leuven, Belgium, 315 patients were diagnosed with contact allergy to/allergic contact dermatitis from corticosteroids (CSs) from routine patch testing with a baseline series including tixocortol pivalate, budesonide, hydrocortisone butyrate and prednisone caproate, patch testing with patients' own CS preparations, and testing those with proven contact allergy to a corticosteroid or strongly suspected of CS allergy later with a series of 66 CSs, including two sex hormones (progesterone and testosterone). 71% of the patients had relevant reactions, but these were not specified. In this group of 315 CS allergic patients, 59 had positive patch tests to desonide 0.1% alc. (6). It is unknown how many of these reactions were caused by the use of a pharmaceutical product containing desonide and how many were cross-reactions to other corticosteroids.

Three cases of allergic contact dermatitis from desonide were reported from France. A 49-year-old man with leg ulcers was treated with several topical corticosteroids. 5 months later, dermatitis occurred on his face, trunk and arms. Patch tests were positive to a preparation containing desonide and to desonide 0.1% pet. (8). A 46-year-old man had been using in the perianal area several proprietary corticosteroid creams including one with desonide, which tended to aggravate the condition. Patch testing with desonide 0.1% pet. revealed a ++ allergic reaction to this molecule (8). A 30-year-old woman suffering from leg ulcers had papulovesicular dermatitis in the area where she had applied several ointments, including one containing desonide. She reacted to the ointment, and to desonide 0.1% pet., but was negative to the other ingredients (8).

Case reports

A 55-year-old woman presented with facial dermatitis of 4 weeks duration. She did not use any cosmetics. Physical examination showed erythematous, scaly plaques involving the eyelids, perioral area, and inner arms. The eruption cleared completely after treatment with desonide ointment, but recurred after its discontinuation. Patch tests to the routine series and a corticosteroid series were negative. When she later had a recurrence, the patient was patch tested again and now reacted to the desonide ointment and desonide 0.05% pet. (3). She was diagnosed with allergic contact dermatitis from desonide, but it is doubtful that the ointment was the cause of the dermatitis. During treatment, the patient may have become sensitized to it, continued use thus contributing to persistence.

A 21-year-old woman presented with intertrigo of the groin. She was treated with 0.05% desonide cream three times a day and the intertrigo cleared after several days of treatment. Several months later, the patient had a recurrence of intertrigo and was again given desonide cream, but now, the dermatitis became worse. An open patch test to the desonide cream was strongly positive. Patch tests were positive to desonide cream, desonide ointment and desonide 0.05% pet. (4).

A child was treated for asthma with budesonide aerosol inhaler. From the fourth day of treatment onwards, his mother would develop swollen and itchy lesions on the face with conjunctivitis several hours after each administration of the corticosteroid to her son. The patient reported worsening of her eruption when she was treated with a cream containing desonide. Prick-tests (??) conducted later confirmed contact allergy to budesonide, budesonide inhaler, and the desonide cream (5).

A 35-year-old man had a 10-year history of a recurrent non-specific dermatitis occurring on various aspects of the extremities, trunk and genitals. Each exacerbation was treated with a different corticosteroid preparation and they all aggravated the condition, necessitating short courses of oral prednisolone. He was patch test positive to all commercial preparations, their active ingredients hydrocortisone, desonide, amcinonide, fluocinolone acetonide, fluocinonide and various base ingredients (10). A 16-year-old girl was treated in the course of several years with various corticosteroids for seborrheic dermatitis of the axillae, face or scalp and became sensitized to fluocinonide, desonide, and triamcinolone acetonide in these topical pharmaceuticals (11).

Another case of allergic contact dermatitis from desonide may have been presented in ref. 9.

Cross-reactions, pseudo-cross-reactions and co-reactions

Cross-reactions between corticosteroids are discussed in Chapter 2.8. Twenty patients with allergic contact dermatitis from budesonide were patch tested with desonide 0.1% and 1% pet. to investigate cross-reactivity and only one reacted to desonide. This patient had used topical desonide before, so cross-sensitivity could not be verified. The authors suggested a lack of cross-reactivity between budesonide and desonide, despite their close chemical similarity (2).

LITERATURE

1   Baeck M, Goossens A. Immediate and delayed allergic hypersensitivity to corticosteroids: practical guidelines. Contact Dermatitis 2012;66:38-45
2   Foti C, Cassano N, Vena GA. Evaluation of cross-reactivity between budesonide and desonide. Contact Dermatitis 2002;47:109-110
3   Garner LA, Cruz PD Jr. Contact puzzle: Worsening of a recurrent facial eruption despite treatment. Diagnosis: allergic contact dermatitis caused by desonide. Am J Contact Dermat 1999;10:81-82
4   Sturz RP, Rau RC. Contact dermatitis to desonide. Arch Dermatol 1983;119:1023
5   Raison-Peyron N, Co Minh HB, Vidal-Mazuy A, Guilhou JJ, Guillot B. Connubial contact dermatitis to an inhaled corticosteroid. Ann Dermatol Venereol 2005;132:143-146 (Article in French)
6   Baeck M, Chemelle JA, Terreux R, Drieghe J, Goossens A. Delayed hypersensitivity to corticosteroids in a series of 315 patients: clinical data and patch test results. Contact Dermatitis 2009;61:163-175
7   Burden AD, Beck MH. Contact hypersensitivity to topical corticosteroids. Br J Dermatol 1992;127:497-501
8   Rivara G, Tomb RR, Foussereau J. Allergic contact dermatitis from topical corticosteroids. Contact Dermatitis 1989;21:83-91
9   Fisher AA. Allergic reactions to intralesional and multiple topical corticosteroids. Cutis 1979;23:564,708-709
10 Feldman SB, Sexton FM, Buzas J, Marks JG Jr. Allergic contact dermatitis from topical steroids. Contact Dermatitis 1988;19:226-228
11 Coskey RJ. Contact dermatitis due to multiple corticosteroid creams. Arch Dermatol 1978;114:115-117

Chapter 3.92   DESOXIMETASONE

IDENTIFICATION

Description/definition : Desoximetasone is the synthetic glucocorticoid that conforms to the structural formula shown below

Pharmacological classes : Glucocorticoids; anti-inflammatory agents

IUPAC name : (8S,9R,10S,11S,13S,14S,16R,17S)-9-Fluoro-11-hydroxy-17-(2-hydroxyacetyl)-10,13,16-trimethyl-7,8,11,12,14,15,16,17-octahydro-6H-cyclopenta[a]phenanthren-3-one

Other names : Desoxymethasone; 9-fluoro-11β,21-dihydroxy-16α-methylpregna-1,4-diene-3,20-dione

CAS registry number : 382-67-2

EC number : 206-845-3

Merck Index monograph : 4201

Patch testing : 1.0% pet. (Chemotechnique, SmartPracticeCanada)

Molecular formula : $C_{22}H_{29}FO_4$

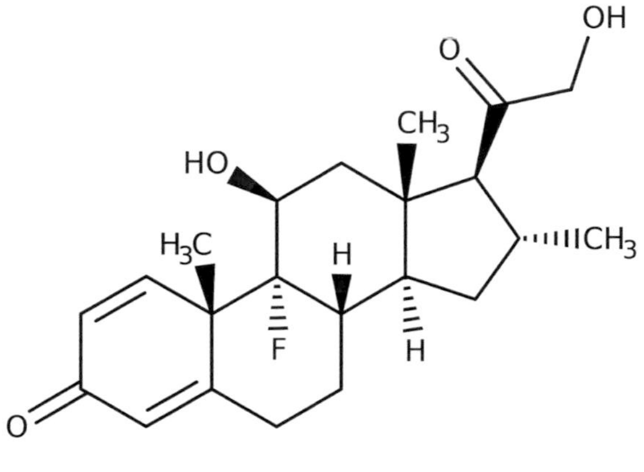

GENERAL

General aspects of corticosteroids used on the skin and mucous membranes are discussed in Chapter 2.4. A practical guideline for diagnosing allergic reactions to corticosteroids is presented in ref. 6.

CONTACT ALLERGY

Patch testing in groups of patients

The results of patch testing with desoximetasone in consecutive patients suspected of contact dermatitis (routine testing) and of testing in groups of *selected* patients (individuals suspected of corticosteroid allergy, patients previously reacting to tixocortol pivalate) are shown in table 3.92.1. Desoximetasone has been included in the screening series of the North American Contact Dermatitis group since 2007, and prevalences of sensitization since then have been invariably very low, 0.1% or 0.2%. Definite + probable relevance rates varied widely, ranging from 0% to 83% (1-5,13). In 2 studies in groups of selected patients, the frequencies of positive patch tests were also low: 0.4% (12) and 1.0% (14).

Table 3.92.1 Patch testing in groups of patients

Years and Country	Test conc. & vehicle	Number of patients tested	positive (%)	Selection of patients (S); Relevance (R); Comments (C)	Ref.
Routine testing					
2015-2017 NACDG	1% pet.	5588	8 (0.1%)	R: definite + probable relevance: 50%	1
2013-2014 NACDG	1% pet.	4859	6 (0.1%)	R: definite + probable relevance: 83%	2
2007-2014 NACDG	1% pet.	17,978	29 (0.2%)	R: definite + probable relevance: 17%	13
2011-2012 NACDG	1% pet.	4230	10 (0.2%)	R: definite + probable relevance: 20%	3
2009-2010 NACDG	1% pet.	4304	(0.2%)	R: definite + probable relevance: 29%	4
2007-2008 NACDG	1% pet.	5070	(0.2%)	R: definite + probable relevance: 0%	5
Testing in groups of selected patients					
2000-2005 USA	0.05% pet.	727	3 (0.4%)	S: patients suspected of corticosteroid allergy; R: 100%	12
1988-1991 U.K.	20% pet.	192	2 (1.0%)	S: patients with a positive patch test to tixocortol pivalate or suspected of corticosteroid allergy; R: not stated	14

NACDG: North American Contact Dermatitis Group (USA, Canada)

Case series
From January 1990 to June 2008, in Leuven, Belgium, 315 patients were diagnosed with contact allergy to/allergic contact dermatitis from corticosteroids (CSs) from routine patch testing with a baseline series including tixocortol pivalate, budesonide, hydrocortisone butyrate and prednisone caproate, patch testing with patients' own CS preparations, and testing those with proven contact allergy to a corticosteroid or strongly suspected of CS allergy later with a series of 66 CSs, including two sex hormones (progesterone and testosterone). 71% of the patients had relevant reactions, but these were not specified. In this group of 315 CS allergic patients, 18 had positive patch tests to desoximetasone 0.1% alc. (11). It is unknown how many of these were cross-reactions.

Case reports
A 61-year-old man with chronic venous insufficiency had a 2-month history of stasis dermatitis of the right leg, for which he was treated with 2.5% desoximetasone emulsion. A week later, he presented with itching, erythema, and miliary papules on the right ankle and dorsum of the foot. Over the next 2 days, a morbilliform and erythema multiforme-like eruption appeared, mainly in the skin folds. Patch tests were positive to the commercial cream and to 5 other corticosteroids. The patient was diagnosed with contact allergy to desoximetasone, but the corticosteroid itself was not tested and probably not al of the ingredients of the emulsion (7).

A 32-year-old man applied desoximetasone cream (0.25%) to an itchy rash on the left forearm. He had previously used this cream for insect bites. After a few days an eczematous reaction developed at the site of application of the cream. At his doctor's suggestion, the patient stopped the topical treatment and began to take 8 mg/day of triamcinolone orally. The day after, a slight red papulovesicular eruption appeared on almost his entire skin surface, rapidly worsening. Patch tests were positive to desoximetasone 1% pet., desoximetasone acetate 1% pet. and triamcinolone acetonide 1% pet. Intradermal tests were positive at D2. Repetition of patch and intradermal tests 3 months later gave the same results (8). This was a case of systemic contact dermatitis.

A 63-year-old woman developed an acute itchy, maculopapular eruption of the trunk, arms and legs after taking deflazacort as an adjuvant for chemotherapy. She had suffered from eczematous dermatitis in the past after using various topical corticosteroids for lichen planus, including desoximetasone. Patch tests were positive to deflazacort 1% alc., desoximetasone 1% alc. and 8 other corticosteroids (9). This also was a case of systemic contact dermatitis.

Cross-reactions, pseudo-cross-reactions and co-reactions
Cross-reactions between corticosteroids are discussed in Chapter 2.8.

LITERATURE

1   DeKoven JG, Warshaw EM, Zug KA, Maibach HI, Belsito DV, Sasseville D, et al. North American Contact Dermatitis Group patch test results: 2015-2016. Dermatitis 2018;29:297-309
2   DeKoven JG, Warshaw EM, Belsito DV, Sasseville D, Maibach HI, Taylor JS, et al. North American Contact Dermatitis Group Patch Test Results: 2013-2014. Dermatitis 2017;28:33-46
3   Warshaw EM, Maibach HI, Taylor JS, Sasseville D, DeKoven JG, Zirwas MJ, et al. North American Contact Dermatitis Group patch test results: 2011-2012. Dermatitis 2015;26:49-59
4   Warshaw EM, Belsito DV, Taylor JS, Sasseville D, DeKoven JG, Zirwas MJ, et al. North American Contact Dermatitis Group patch test results: 2009 to 2010. Dermatitis 2013;24:50-59
5   Fransway AF, Zug KA, Belsito DV, Deleo VA, Fowler JF Jr, Maibach HI, et al. North American Contact Dermatitis Group patch test results for 2007-2008. Dermatitis 2013;24:10-21
6   Baeck M, Goossens A. Immediate and delayed allergic hypersensitivity to corticosteroids: practical guidelines. Contact Dermatitis 2012;66:38-45
7   Stingeni L, Hansel K, Lisi P. Morbilliform erythema-multiforme-like eruption from desoxymethasone. Contact Dermatitis 1996;35:363-364
8   Brambilla L, Boneschi V, Chiappino G, Fossati S, Pigatto PD. Allergic reactions to topical desoxymethasone and oral triamcinolone. Contact Dermatitis 1989;21:272-274.
9   Bianchi L, Hansel K, Antonelli E, Bellini V, Rigano L, Stingeni L. Deflazacort hypersensitivity: a difficult-to-manage case of systemic allergic dermatitis and literature review. Contact Dermatitis 2016;75:54-56
10 Donovan JC, Dekoven JG. Cross-reactions to desoximetasone and mometasone furoate in a patient with multiple topical corticosteroid allergies. Dermatitis 2006 ;17:147-151
11 Baeck M, Chemelle JA, Terreux R, Drieghe J, Goossens A. Delayed hypersensitivity to corticosteroids in a series of 315 patients: clinical data and patch test results. Contact Dermatitis 2009;61:163-175
12 Davis MD, El-Azhary RA, Farmer SA. Results of patch testing to a corticosteroid series: a retrospective review of 1188 patients during 6 years at Mayo Clinic. J Am Acad Dermatol 2007;56:921-927
13 Pratt MD, Mufti A, Lipson J, Warshaw EM, Maibach HI, Taylor JS, et al. Patch test reactions to corticosteroids: Retrospective analysis from the North American Contact Dermatitis Group 2007-2014. Dermatitis 2017;28:58-63
14 Burden AD, Beck MH. Contact hypersensitivity to topical corticosteroids. Br J Dermatol 1992;127:497-501

Chapter 3.93   DEXAMETHASONE

IDENTIFICATION

Description/definition : Dexamethasone is the synthetic glucocorticoid that conforms to the structural formula shown below

Pharmacological classes : Antineoplastic agents, hormonal; antiemetics; anti-inflammatory agents; glucocorticoids

IUPAC name : (8S,9R,10S,11S,13S,14S,16R,17R)-9-Fluoro-11,17-dihydroxy-17-(2-hydroxyacetyl)-10,13,16-trimethyl-6,7,8,11,12,14,15,16-octahydrocyclopenta[a]phenanthren-3-one

Other names : (11β,16α)-9-Fluoro-11,17,21-trihydroxy-16-methylpregna-1,4-diene-3,20-dione

CAS registry number : 50-02-2

EC number : 200-003-9

Merck Index monograph : 4215

Patch testing : 0.5% pet. (SmartPracticeCanada); late readings (6-10 days) are strongly recommended

Molecular formula : $C_{22}H_{29}FO_5$

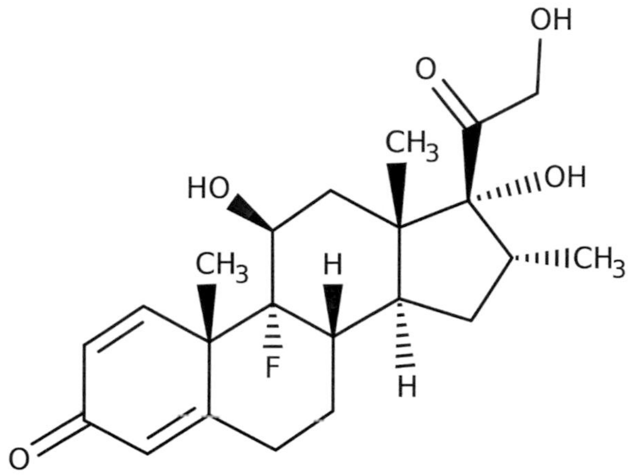

GENERAL

General aspects of corticosteroids used on the skin and mucous membranes are discussed in Chapter 2.4. A practical guideline for diagnosing allergic reactions to corticosteroids is presented in ref. 3. In pharmaceutical products, dexamethasone may be employed as base, as acetate (Chapter 3.94), phosphate (Chapter 3.95) or as dexamethasone sodium phosphate (Chapter 3.96). Dexamethasone *base* is used as tablet and as elixir for oral use only, which implies that by far most allergic reactions to 'dexamethasone' have in fact been the result of sensitization to a salt or ester of dexamethasone or of cross-reactivity to another corticosteroid. It is also likely that there has been confusion in some publications on the correct forms of the drugs used, e.g. that dexamethasone was mentioned where a salt or ester form should have been mentioned.

CONTACT ALLERGY

Patch testing in groups of selected patients

The results of patch testing with dexamethasone in groups of selected patients (patients tested with a corticosteroid series, individuals with leg ulcers) are shown in table 3.93.1. Low prevalences of sensitization ranging from 0.4% to 1.0% were found with no cases of certain current relevance, i.e. allergic reaction caused by oral dexamethasone (1,8,9).

Table 3.93.1 Patch testing in groups of patients: Selected patient groups

Years and Country	Test conc. & vehicle	Number of patients tested	positive (%)	Selection of patients (S); Relevance (R); Comments (C)	Ref.
1995-2004 IVDK	1% pet.	2212	8 (0.4%)	S: patients tested with a corticosteroid series; R: not stated	8
1997-2001 U.K.		200	2 (1.0%)	S: patients with venous or mixed venous/arterial leg ulcers; R: all reactions to topical drugs were considered to be of probable, past or current relevance	1
1995-9 Netherlands	1% alc. 97%	213	1 (0.5%)	S: not stated; R: probable 100%	9

IVDK: Information Network of Departments of Dermatology (Germany, Austria, Switzerland)

Case series

In Leuven, Belgium, between 1990 and 2017, 16,065 patients were investigated for contact allergy and 118 (0.7%) showed positive patch test reactions to topical ophthalmic medications and/or to their ingredients. Eighty-four individuals (71%) reacted to an active principle. Dexamethasone was tested in 744 patients and was the allergen in eye medications in 6. There were also two reactions to dexamethasone in other types of medications (2, overlap with refs. 6 and 7).

From January 1990 to June 2008, in the same clinic in Leuven, Belgium, 315 patients were diagnosed with contact allergy to/allergic contact dermatitis from corticosteroids (CSs) from routine patch testing with a baseline series including tixocortol pivalate, budesonide, hydrocortisone butyrate and prednisone caproate, patch testing with patients' own CS preparations, and testing those with proven contact allergy to a corticosteroid or strongly suspected of CS allergy later with a series of 66 CSs, including two sex hormones (progesterone and testosterone). 71% of the patients had relevant reactions, but these were not specified. In this group of 315 CS allergic patients, 27 had positive patch tests to dexamethasone 0.1% alc. (6, overlap with refs. 2 and 7). It is unknown how many of these reactions were caused by the use of a pharmaceutical product containing dexamethasone and how many were cross-reactions to other corticosteroids.

In the period 1990-2008, in Leuven, Belgium, 315 patients were diagnosed with contact allergy to/allergic contact dermatitis from corticosteroids. Eighteen subjects (5.7%) presented with allergic manifestations (conjunctivitis, eczema of the face, periocular skin or eyelids) caused by the use of CS-containing ocular preparations. Five patients had used ophthalmic preparations containing dexamethasone. All 18 patients were patch test positive to the 'group A' CSs, that is, hydrocortisone, prednisone, prednisolone, methylprednisolone, and/or particularly to tixocortol pivalate (a marker for allergy to group A CSs), and to many other CSs, including dexamethasone in 3 of the five that had used dexamethasone eye drops (7, overlap with refs. 2 and 6).

In Bologna, Italy, before 1989, 136 patients (35 men, 101 women, range of age 14-84 years, average 36 years) suspected of contact conjunctivitis from eye drops (including contact lens solutions), were patch tested. In sixty, there was also eyelid contact dermatitis. In 75 individuals, the causative agents were found. The great majority was caused by thimerosal (n=52), followed by benzalkonium chloride (n=7), both present in contact lens solutions. There were 2 reactions to dexamethasone eye drops and dexamethasone itself (4).

It is highly likely that, in the studies mentioned above, patients reacting to 'dexamethasone' in eye medications, had in fact used products containing dexamethasone sodium phosphate.

Case reports

A 38-year-old man was treated for facial pyoderma with an ointment containing virginiamycin, polymyxin B and dexamethasone. A few days later, contact dermatitis developed in the treated area. Patch testing with the ointment was positive and in the corticosteroid series, dexamethasone 0.1% pet. revealed a ++ allergic reaction (10). Two patients apparently sensitized to dexamethasone in topical preparations were reported in 1969 (11).

Cutaneous adverse drug reactions from systemic administration caused by type IV (delayed-type) hypersensitivity

Cutaneous adverse drug reactions from systemic administration of dexamethasone caused by type IV (delayed-type) hypersensitivity, including generalized maculopapular exanthema (5) and exanthema/delayed urticaria (12), are planned to be discussed in Volume IV of the *Monographs in Contact Allergy* series on Systemic drugs.

Cross-reactions, pseudo-cross-reactions and co-reactions

Cross-reactions between corticosteroids are discussed in Chapter 2.8.

LITERATURE

1   Tavadia S, Bianchi J, Dawe RS, McEvoy M, Wiggins E, Hamill E, et al. Allergic contact dermatitis in venous leg ulcer patients. Contact Dermatitis 2003;48:261-265
2   Gilissen L, De Decker L, Hulshagen T, Goossens A. Allergic contact dermatitis caused by topical ophthalmic medications: Keep an eye on it! Contact Dermatitis 2019;80:291-297
3   Baeck M, Goossens A. Immediate and delayed allergic hypersensitivity to corticosteroids: practical guidelines. Contact Dermatitis 2012;66:38-45
4   Tosti A, Tosti G. Allergic contact conjunctivitis due to ophthalmic solution. In: Frosch PJ, Dooms-Goossens A, Lachapelle JM, Rycroft RJG, Scheper RJ (eds). Current Topics in Contact Dermatitis. Berlin: Springer-Verlag, 1989: 269-272

5   Watts TJ, Thursfield D, Haque R. Cutaneous adverse drug reaction induced by oral dexamethasone with possible cross-reactivity to Group 1 corticosteroids confirmed by patch testing and intradermal testing. Contact Dermatitis 2019;81:384-386

6   Baeck M, Chemelle JA, Terreux R, Drieghe J, Goossens A. Delayed hypersensitivity to corticosteroids in a series of 315 patients: clinical data and patch test results. Contact Dermatitis 2009;61:163-175

7   Baeck M, De Potter P, Goossens A. Allergic contact dermatitis following ocular use of corticosteroids. J Ocul Pharmacol Ther 2011;27:83-92

8   Uter W, de Pádua CM, Pfahlberg A, Nink K, Schnuch A, Lessmann H. Contact allergy to topical corticosteroids – results from the IVDK and epidemiological risk assessment. J Dtsch Dermatol Ges 2009;7:34-41

9   Devos SA, Van der Valk PG. Relevance and reproducibility of patch-test reactions to corticosteroids. Contact Dermatitis 2001;44:362-365

10 Rivara G, Tomb RR, Foussereau J. Allergic contact dermatitis from topical corticosteroids. Contact Dermatitis 1989;21:83-91

11 Wiegel O. Kontaktallergie durch Kortikosteroidhaltige externa (Triamcinolon-Acetonid und Dexamethason). Med Welt 1968;13:828-829 (Article in German)

12 Padial A, Posadas S, Alvarez J, Torres M-J, Alvarez JA, Mayorga C, Blanca M. Nonimmediate reactions to systemic corticosteroids suggest an immunological mechanism. Allergy 2005;60:665-670

Chapter 3.94   DEXAMETHASONE ACETATE

There are two forms of dexamethasone acetate, the (mono)hydrate form and the anhydrous form, both of which are termed dexamethasone acetate.

IDENTIFICATION	DEXAMETHASONE ACETATE (HYDRATE)
Description/definition	: Dexamethasone acetate hydrate is the acetate ester of the synthetic glucocorticoid dexamethasone that conforms to the structural formula shown below
Pharmacological classes	: Glucocorticoids; anti-inflammatory agents
IUPAC name	: [2-[(8S,9R,10S,11S,13S,14S,16R,17R)-9-Fluoro-11,17-dihydroxy-10,13,16-trimethyl-3-oxo-6,7,8,11,12,14,15,16-octahydrocyclopenta[a]phenanthren-17-yl]-2-oxoethyl] acetate;hydrate
Other names	: 9-Fluoro-11β,17,21-trihydroxy-16α-methylpregna-1,4-diene-3,20-dione 21-acetate, monohydrate
CAS registry number	: 55812-90-3
EC number	: 611-316-9
Patch testing	: Generally, corticosteroids may be tested at 0.1% and 1% in alcohol; late readings (6-10 days) are strongly recommended
Molecular formula	: $C_{24}H_{33}FO_7$

IDENTIFICATION	DEXAMETHASONE ACETATE (ANHYDROUS)
Description/definition	: Dexamethasone acetate anhydrous is the acetate ester of the synthetic glucocorticoid dexamethasone that conforms to the structural formula shown below
Pharmacological classes	: Glucocorticoids; anti-inflammatory agents
IUPAC name	: [2-[(8S,9R,10S,11S,13S,14S,16R,17R)-9-Fluoro-11,17-dihydroxy-10,13,16-trimethyl-3-oxo-6,7,8,11,12,14,15,16-octahydrocyclopenta[a]phenanthren-17-yl]-2-oxoethyl]acetate
Other names	: 9-Fluoro-11β,17,21-trihydroxy-16α-methylpregna-1,4-diene-3,20-dione 21-acetate; dexamethasone 21-acetate
CAS registry number	: 1177-87-3
EC number	: 214-646-8
Merck Index monograph	: 4215 (Dexamethasone)
Patch testing	: Generally, corticosteroids may be tested at 0.1% and 1% in alcohol; late readings (6-10 days) are strongly recommended
Molecular formula	: $C_{24}H_{31}FO_6$

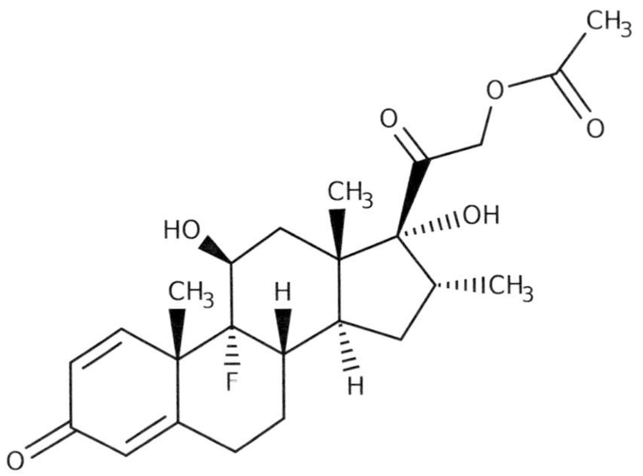

Dexamethasone acetate anhydrous 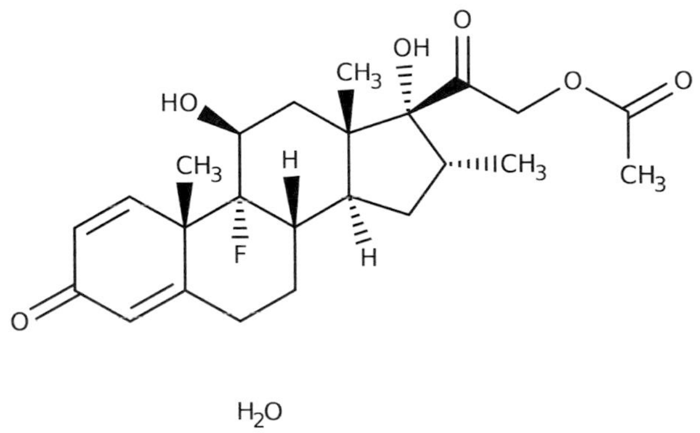   Dexamethasone acetate hydrate

CONTACT ALLERGY

General aspects of corticosteroids used on the skin and mucous membranes are discussed in Chapter 2.4. A practical guideline for diagnosing allergic reactions to corticosteroids is presented in ref. 1. See also dexamethasone (Chapter 3.93), dexamethasone phosphate (Chapter 3.95), and dexamethasone sodium phosphate (Chapter 3.96).

Case series

In a period of 11 years (2000-2010), the French Pharmacovigilance received 53 reports of adverse skin reactions to a pharmaceutical product containing as active ingredients dexamethasone acetate, hydroxyethyl (glycol) salicylate, and salicylamide marketed for treatment of benign joint conditions such as mild tendinitis, small joint arthritis and sprains. The main cutaneous side effect (n=41) was contact dermatitis with secondary extension in 15 cases. Onset was immediate in 12 cases, delayed in 32 cases and unspecified in eight cases. Twelve patients were hospitalized. Allergological tests were performed in 14 cases and were positive for the drug itself (eight cases), dexamethasone acetate (n=3), glycol salicylate (n=7), salicylamide (n=6), and propylene glycol (n=2) (2).

From January 1990 to June 2008, in Leuven, Belgium, 315 patients were diagnosed with contact allergy to/allergic contact dermatitis from corticosteroids (CSs) from routine patch testing with a baseline series including tixocortol pivalate, budesonide, hydrocortisone butyrate and prednisone caproate, patch testing with patients' own CS preparations, and testing those with proven contact allergy to a corticosteroid or strongly suspected of CS allergy later with a series of 66 CSs, including two sex hormones (progesterone and testosterone). 71% of the patients had relevant reactions, but these were not specified. In this group of 315 CS allergic patients, 18 had positive patch tests to dexamethasone acetate 0.1% alc. (3). It is unknown how many of these reactions were caused by the use of a pharmaceutical product containing dexamethasone acetate and how many were cross-reactions to other corticosteroids.

Cross-reactions, pseudo-cross-reactions and co-reactions

Cross-reactions between corticosteroids are discussed in Chapter 2.8.

LITERATURE

1   Baeck M, Goossens A. Immediate and delayed allergic hypersensitivity to corticosteroids: practical guidelines. Contact Dermatitis 2012;66:38-45
2   Remy C, Barbaud A, Lebrun-Vignes B, Perrot JL, Beyens MN, Mounier G, et al. Skin toxicity related to Percutalgine (®): analysis of the French pharmacovigilance database. Ann Dermatol Venereol 2012;139:350-354 (Article in French)
3   Baeck M, Chemelle JA, Terreux R, Drieghe J, Goossens A. Delayed hypersensitivity to corticosteroids in a series of 315 patients: clinical data and patch test results. Contact Dermatitis 2009;61:163-175

Chapter 3.95 DEXAMETHASONE PHOSPHATE

IDENTIFICATION

Description/definition : Dexamethasone phosphate is the 21-O-phospho derivative of the synthetic glucocorticoid dexamethasone that conforms to the structural formula shown below

Pharmacological classes : Glucocorticoids

IUPAC name : [2-[(8S,9R,10S,11S,13S,14S,16R,17R)-9-Fluoro-11,17-dihydroxy-10,13,16-trimethyl-3-oxo-6,7,8,11,12,14,15,16-octahydrocyclopenta[a]phenanthren-17-yl]-2-oxoethyl] dihydrogen phosphate

Other names : Dexamethasone-21-phosphate; 9-fluoro-11β,17,21-trihydroxy-16α-methylpregna-1,4-diene-3,20-dione 21-(dihydrogen phosphate)

CAS registry number : 312-93-6

EC number : 206-232-0

Merck Index monograph : 4215 (Dexamethasone)

Patch testing : 1% pet. (SmartPracticeCanada); late readings (6-10 days) are strongly recommended

Molecular formula : $C_{22}H_{30}FO_8P$

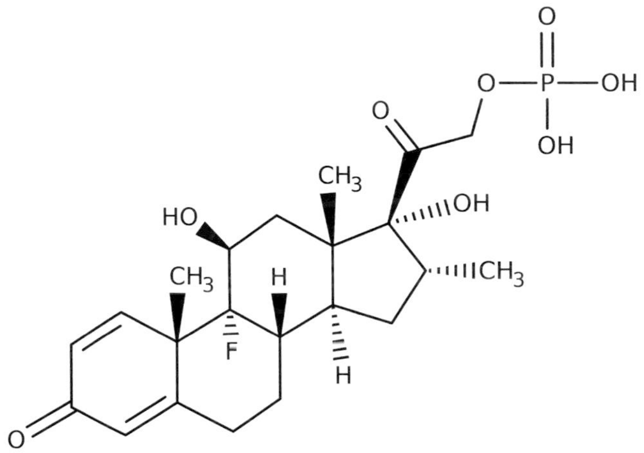

GENERAL

General aspects of corticosteroids used on the skin and mucous membranes are discussed in Chapter 2.4. A practical guideline for diagnosing allergic reactions to corticosteroids is presented in ref. 1. See also dexamethasone (Chapter 3.93), dexamethasone sodium phosphate (Chapter 3.96) and dexamethasone acetate (Chapter 3.94).

CONTACT ALLERGY

Case series

In The Netherlands, in the period 1993-1994, 34 patients with chronic otitis externa or media were patch tested with a battery of medicaments and there was one (3%) reaction to dexamethasone-21-phosphate; the relevance of this positive patch test was not mentioned (2).

Cutaneous adverse drug reactions from systemic administration caused by type IV (delayed-type) hypersensitivity

Cutaneous adverse drug reactions from systemic administration of dexamethasone phosphate caused by type IV (delayed-type) hypersensitivity, including generalized eczematous exanthema (3), are planned to be discussed in Volume IV of the *Monographs in Contact Allergy* series on Systemic drugs.

LITERATURE

1 Baeck M, Goossens A. Immediate and delayed allergic hypersensitivity to corticosteroids: practical guidelines. Contact Dermatitis 2012;66:38-45
2 Van Ginkel CJ, Bruintjes TD, Huizing EH. Allergy due to topical medications in chronic otitis externa and chronic otitis media. Clin Otolaryngol Allied Sci 1995;20:326-328
3 Plaza T, Nist G, Von den Driesch P. Type IV-allergy due to corticosteroids. Rare and paradoxical. Dtsch Med Wochenschr 2007;132:1692-1695 (Article in German)

Chapter 3.96   DEXAMETHASONE SODIUM PHOSPHATE

IDENTIFICATION

Description/definition	: Dexamethasone sodium phosphate is the sodium phosphate salt form of the synthetic glucocorticoid dexamethasone, that conforms to the structural formula shown below
Pharmacological classes	: Glucocorticoids
IUPAC name	: Disodium;[2-[(8S,9R,10S,11S,13S,14S,16R,17R)-9-fluoro-11,17-dihydroxy-10,13,16-trimethyl-3-oxo-6,7,8,11,12,14,15,16-octahydrocyclopenta[a]phenanthren-17-yl]-2-oxoethyl] phosphate
Other names	: Dexamethasone phosphate disodium salt; dexamethasone 21-(disodium phosphate); 9-fluoro-11β,17,21-trihydroxy-16α-methylpregna-1,4-diene-3,20-dione 21-(dihydrogen phosphate) disodium salt
CAS registry number	: 2392-39-4
EC number	: 219-243-0
Merck Index monograph	: 4215 (Dexamethasone)
Patch testing	: 1.0% pet. (Chemotechnique, SmartPracticeEurope); late readings (6-10 days) are strongly recommended
Molecular formula	: $C_{22}H_{28}FNa_2O_8P$

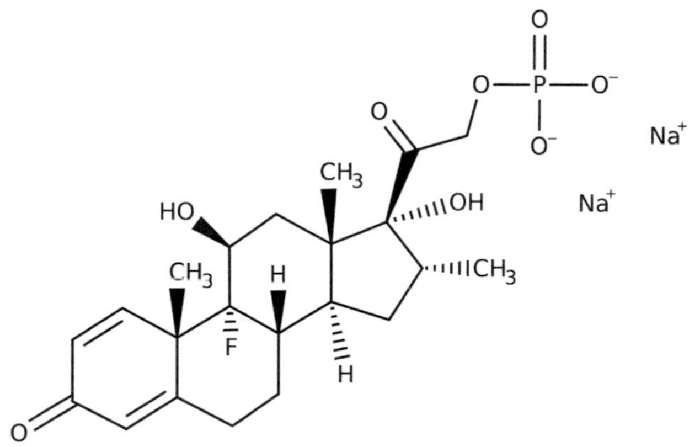

GENERAL

General aspects of corticosteroids used on the skin and mucous membranes are discussed in Chapter 2.4. A practical guideline for diagnosing allergic reactions to corticosteroids is presented in ref. 3. See also dexamethasone (Chapter 3.93), dexamethasone phosphate (Chapter 3.95), and dexamethasone acetate (Chapter 3.94).

CONTACT ALLERGY

Contact allergy in the general population

In Germany, for the period 1995-2004, the population-based relative incidence (RI) of contact sensitization to dexamethasone sodium phosphate (cases/100,000 defined daily doses (DDDs) per year) was estimated to be 0.3. In the group of corticosteroids, the RI ranged from 0.3 (dexamethasone sodium phosphate) to 43.2 (budesonide) (2).

Patch testing in groups of patients

The results of patch testing dexamethasone sodium phosphate in groups of selected patients (individuals with anogenital dermatoses, patients suspected of corticosteroid allergy) are shown in table 3.96.1. Prevalences of positive patch tests ranged from 0.4% to 5%, the latter in a very small series of 41 patients (8). In one study, 10/11 reactions were considered to be relevant (7), which seems high for a corticosteroid which is used little in topical preparations.

Case reports and case series

From January 1990 to June 2008, in Leuven, Belgium, 315 patients were diagnosed with contact allergy to/allergic contact dermatitis from corticosteroids (CSs) from routine patch testing with a baseline series including tixocortol pivalate, budesonide, hydrocortisone butyrate and prednisone caproate, patch testing with patients' own CS preparations, and testing those with proven contact allergy to a corticosteroid or strongly suspected of CS allergy later with a series of 66 CSs, including two sex hormones (progesterone and testosterone). 71% of the patients had

Table 3.96.1 Patch testing in groups of patients: Selected patient groups

Years and Country	Test conc. & vehicle	Number of patients tested	positive (%)	Selection of patients (S); Relevance (R); Comments (C)	Ref.
2004-2008 IVDK			(0.9%)	S: patients with anogenital dermatoses tested with a medicament series; R: not stated; C: number of patients tested unknown	4
2000-2005 USA	1% pet.	1177	11 (0.9%)	S: patients suspected of corticosteroid allergy; R: 91%	7
<2005 Turkey		41	2 (5%)	S: patients not reacting to treatment with corticosteroids or with dermatoses made worse by corticosteroids; R: not stated	8
2002-2004 IVDK	1% pet.	2302	(0.4%)	S: not stated; R: not stated	1

IVDK: Information Network of Departments of Dermatology (Germany, Austria, Switzerland)

relevant reactions, but these were not specified. In this group of 315 CS allergic patients, 9 had positive patch tests to dexamethasone sodium phosphate 0.1% alc. (6). It is unknown how many of these reactions were caused by the use of a pharmaceutical product containing dexamethasone sodium phosphate and how many were cross-reactions to other corticosteroids.

See also the section 'Case series' on ophthalmic medications in Chapter 3.93. Dexamethasone. It is highly likely that the patients reacting to 'dexamethasone' in ophthalmic medications in fact had used eye drops containing dexamethasone sodium phosphate.

Cutaneous adverse drug reactions from systemic administration caused by type IV (delayed-type) hypersensitivity
Cutaneous adverse drug reactions from systemic administration of dexamethasone sodium phosphate caused by type IV (delayed-type) hypersensitivity, including maculopapular eruption (5), are planned to be discussed in Volume IV of the *Monographs in Contact Allergy* series on Systemic drugs.

Cross-reactions, pseudo-cross-reactions and co-reactions
Cross-reactions between corticosteroids are discussed in Chapter 2.8.

LITERATURE

1   Menezes de Padua CA, Uter W, Schnuch A. Contact allergy to topical drugs: prevalence in a clinical setting and estimation of frequency at the population level. Pharmacoepidemiol Drug Saf 2007;16:377-384
2   Menezes de Padua CA, Schnuch A, Nink K, Pfahlberg A, Uter W. Allergic contact dermatitis to topical drugs – Epidemiological risk assessment. Pharmacoepidemiol Drug Saf 2008;17:813-821
3   Baeck M, Goossens A. Immediate and delayed allergic hypersensitivity to corticosteroids: practical guidelines. Contact Dermatitis 2012;66:38-45
4   Bauer A. Contact sensitization in the anal and genital area. Curr Probl Dermatol 2011;40:133-141
5   Bianchi L, Hansel K, Antonelli E, Bellini V, Rigano L, Stingeni L. Deflazacort hypersensitivity: a difficult-to-manage case of systemic allergic dermatitis and literature review. Contact Dermatitis 2016;75:54-56
6   Baeck M, Chemelle JA, Terreux R, Drieghe J, Goossens A. Delayed hypersensitivity to corticosteroids in a series of 315 patients: clinical data and patch test results. Contact Dermatitis 2009;61:163-175
7   Davis MD, El-Azhary RA, Farmer SA. Results of patch testing to a corticosteroid series: a retrospective review of 1188 patients during 6 years at Mayo Clinic. J Am Acad Dermatol 2007;56:921-927
8   Gönül M, Gül U. Detection of contact hypersensitivity to corticosteroids in allergic contact dermatitis patients who do not respond to topical corticosteroids. Contact Dermatitis 2005;53:67-70

Chapter 3.97   DEXCHLORPHENIRAMINE

IDENTIFICATION

Description/definition : Dexchlorpheniramine is the pheniramine that conforms to the structural formula shown below

Pharmacological classes : Histamine H1 antagonists

IUPAC name : (3S)-3-(4-Chlorophenyl)-N,N-dimethyl-3-pyridin-2-ylpropan-1-amine

Other names : 2-(p-Chloro-α-(2-(dimethylamino)ethyl)benzyl)pyridine

CAS registry number : 25523-97-1

EC number : 247-073-7

Merck Index monograph : 3456

Patch testing : 1% water; 5% pet.

Molecular formula : $C_{16}H_{19}ClN_2$

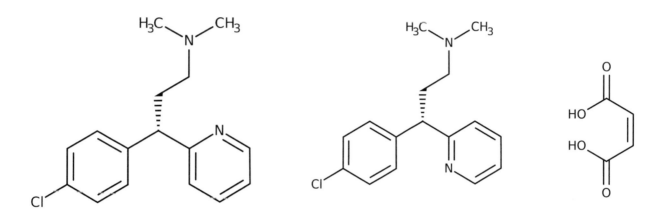

Dexchlorpheniramine   Dexchlorpheniramine maleate

GENERAL

Dexchlorpheniramine is an alkylamine first-generation histamine H1 antagonist with anti-allergic activity. It is the S-enantiomer of chlorpheniramine; this agent has more pharmacological activity than the R-enantiomer and therefore is more potent than the racemic mixture. In pharmaceutical products, dexchlorpheniramine is employed as dexchlorpheniramine maleate (CAS number 2438-32-6, EC number 219-450-6, molecular formula $C_{20}H_{23}ClN_2O_4$) (1).

CONTACT ALLERGY

Case series

In Leuven, Belgium, in the period 1990-2014, iatrogenic contact dermatitis was diagnosed in 2600 individuals (17% of the total population). 96% of all positive patch test reactions to topical drugs and antiseptics were considered to be relevant. Dexchlorpheniramine (5% pet.) was tested in 4 patients and there was one positive reaction to it (2).

Case reports

A 45-year-old woman, successfully treated for 2 years with oral dexchlorpheniramine maleate for allergic rhinitis, applied dexchlorpheniramine maleate cream to insect bites on her right leg. A few hours later, she developed intensely itchy eczema at the application site, for which she was treated with injections of chlorpheniramine maleate. Her dermatitis became generalized over the next 24 hours (4). A second patient, a 52-year-old man with a 3-month history of itchy eczema on the abdomen, was prescribed dexchlorpheniramine maleate cream, which he had safely used in the past. After 3 hours, an acute exacerbation of the eczema appeared at the application site. Chlorpheniramine maleate injections were prescribed for the pruritus. After one hour, a generalized pruritic dermatitis, spreading from the injection site, appeared (4). Both patients were patch tested with the cream, its ingredients, dexchlorpheniramine maleate and related alkylamine antihistamines and both had ++ reactions to the cream 'as is', and to dexchlorpheniramine maleate, pheniramine maleate, chlorpheniramine maleate and brompheniramine maleate, all tested 1% water (4). Thus, these patients had allergic contact dermatitis from dexchlorpheniramine maleate and systemic contact dermatitis from chlorpheniramine maleate injections. Surprisingly, both could take oral dexchlorpheniramine without problems.

A 16-year-old girl had hand eczema due to the glues that she handled in a pottery factory. Treatment with many different topical medicaments resulted in periodic exacerbations of the dermatitis, with spread to the forearms and face. Patch tests gave a +++ reaction to epoxy resin (1% pet.) and a cream containing dexchlorpheniramine maleate 1%. Patch tests with the pure drug and some other antihistamines were positive to dexchlorpheniramine maleate, pheniramine (2% pet.) and 2 antihistamines of another class, mepyramine HCl (tripelennamine 2% pet.) and diphenhydramine HCl (2% pet.). The authors considered these latter two reactions also (just as pheniramine) to be cross-reactions, but, belonging to another class, this is less likely. In addition, the patient had been treated with many topical medicaments before, which may well have included these antihistamines (5).

Cross-reactions, pseudo-cross-reactions and co-reactions
Possible cross-reactivity between pheniramine, chlorpheniramine and dexchlorpheniramine (3). Two patients who had developed allergic contact dermatitis from a cream containing dexchlorpheniramine maleate, cross-reacted to pheniramine maleate, chlorpheniramine maleate and brompheniramine maleate 1% water (4) and another one to pheniramine 2% pet. (5).

LITERATURE
1   The data in the section 'General' may have been obtained from literature discussed in this chapter, but mostly also or exclusively from one or more of the following online sources: ChemIDPlus Advanced, PubChem, DrugBank, RxList, Drug Central, Drugs.com, and Wikipedia
2   Gilissen L, Goossens A. Frequency and trends of contact allergy to and iatrogenic contact dermatitis caused by topical drugs over a 25-year period. Contact Dermatitis 2016;75:290-302
3   Parente G, Pazzaglia M, Vincenzi C, Tosti A. Contact dermatitis from pheniramine maleate in eyedrops. Contact Dermatitis 1999;40:338
4   Santucci B, Cannistraci C, Cristaudo A, Picardo M. Contact dermatitis from topical alkylamines. Contact Dermatitis 1992;27:200-201
5   Cusano F, Capozzi M, Errico G. Contact dermatitis from dexchlorpheniramine. Contact Dermatitis 1989;21:340

Chapter 3.98   DEXKETOPROFEN

IDENTIFICATION

Description/definition   : Dexketoprofen is the propionic acid derivative and isomer of ketoprofen that conforms to the structural formula shown below
Pharmacological classes : Anti-inflammatory agents, non-steroidal
IUPAC name   : (2S)-2-(3-Benzoylphenyl)propanoic acid
Other names   : (S)-(+)-Ketoprofen
CAS registry number   : 22161-81-5
EC number   : 606-944-5
Patch testing   : 1.0% pet. (Chemotechnique, SmartPracticeCanada)
Molecular formula   : $C_{16}H_{14}O_3$

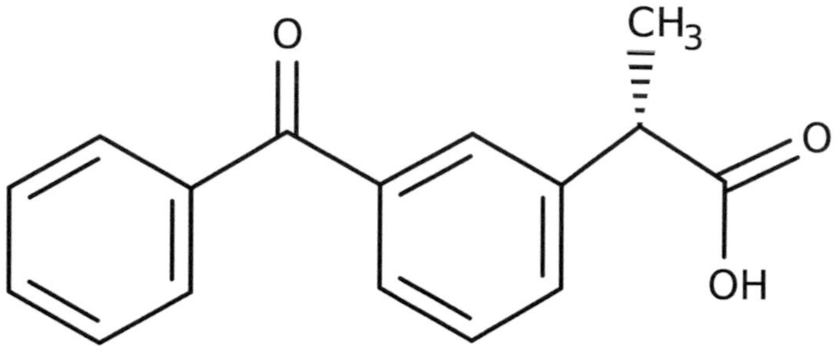

GENERAL

Dexketoprofen is a propionic acid derivative and nonsteroidal anti-inflammatory drug (NSAID) with analgesic, anti-inflammatory, and antipyretic properties. It is the (S)-enantiomer and active isomer of ketoprofen and works by blocking the action of cyclooxygenase. Dexketoprofen is indicated for short-term treatment of mild to moderate pain, including dysmenorrhea, musculoskeletal pain and toothache. It is available in topical and oral formulations (1).

CONTACT ALLERGY

General

One case series of 6 patients (4), one of 2 (5) and a few single case reports (6,7,8) of photoallergic contact dermatitis from a topical gel containing 1.25% dexketoprofen have been published. All reports came from Spain and all patients had used the same brand of dexketoprofen gel. Cross-reactions were noted to or from ketoprofen in all individuals: dexketoprofen is the (S)-enantiomer of ketoprofen. One case of photoallergy from oral dexketoprofen has been observed in a man previously photosensitized to ketoprofen (3).

Patch testing in groups of selected patients

In Spain, in the period 2014-2016, 116 selected patients (selection criteria not mentioned, but elected for photopatch testing) were patch and photopatch tested with dexketoprofen (concentration and vehicle not mentioned) and there was one positive patch test to this NSAID, the relevance of which was not mentioned (10).

PHOTOSENSITIVITY

Photopatch testing in groups of patients

In Spain, in the period 2014-2016, 116 selected patients (selection criteria not mentioned) were patch and photopatch tested with dexketoprofen (concentration and vehicle not mentioned) and there were 16 (14%) positive photopatch test to this NSAID. The relevance of these reactions was not specified for individual allergens, but the rate was 57% for all NSAID reactions together (10).

Case series

In a hospital in Spain, in the period 2003-2006, 6 cases of photoallergic contact dermatitis from dexketoprofen 12.5 mg/gr gel have been investigated. The group consisted of 4 men and 2 women, ages ranging from 23 to 72 years.

They had applied the gel to various parts of the body, mostly the (joints of the) arms of legs. After 2-10 days, a strong eczematous picture with bullous lesions would develop, restricted to the application areas. Five had been exposed to sunlight, the 6th to artificial UVA rays. Two patients had a history of photocontact dermatitis to topical ketoprofen. Patch and photopatch tests (with UVA irradiation at a dose of 5 J/cm^2) were performed in all patients with the dexketoprofen gel, its components and a NSAID series. All 6 patients had negative patch tests, but positive photo-patch tests to dexketoprofen (6 to 2% pet., 5 to 1% pet.) and to ketoprofen 1% pet. Eight controls were negative (4).

Two women presented with a characteristic clinical picture of contact photodermatitis and had used topical dexketoprofen gel in the days before the rashes appeared. Photopatch tests were positive to the gel and to dexketoprofen in both patients. One of the women had also photosensitivity to other NSAIDs and to several excipients of the gel. The investigators advise to use dexketoprofen 0.1-1% in petrolatum for photopatch testing. Further details are not available (5).

Case reports

A 65-year-old woman developed a pruritic erythema and blisters on her left ankle 1 week after topical application of dexketoprofen 1.25% gel twice daily for joint pain. She had also applied this gel on her right shoulder but no lesions appeared there. Patch tests were positive to piketoprofen (++) but negative to the gel and dexketoprofen 1% and 5% pet. (in the table a ++++ reaction is given for the dexketoprofen patch test at D4, which is probably a mistake considering the title of the article, the text and the fact that the patient did not develop allergic contact dermatitis on the non-irradiated shoulder). Photopatch tests (with UVA 5J/cm^2 at D2) were strongly positive to the gel 'as is', dexketoprofen 1% and 5% pet., ketoprofen 1% pet, and piketoprofen 2% pet. The reaction to the latter was far stronger (++++) than the patch test (++), indicating both contact and photocontact allergy to piketoprofen. Fifteen controls had negative photopatch tests to dexketoprofen 1% and 5% pet. (6).

A 27-year-old woman presented with an itchy lesion on her right hand after applying dexketoprofen 1.25% gel for joint pain and after 2 days of sun exposure. Patch tests were negative, photopatch tests (with UVA 7.5 J/cm^2) were positive to dexketoprofen 1% and 5% pet. and to ketoprofen 1% pet., but negative to the other ingredients of the gel (7). A 61-year-old woman presented with an acute eczematous plaque over the upper 1/3 of her left leg, of 10 days' duration, where she had applied dexketoprofen 1.25% gel 2× to 3× daily for several days prior to the onset. She admitted to moderate sun exposure of the treated area. Patch tests were negative, photopatch tests positive to the gel, dexketoprofen 1% pet., ketoprofen 1% pet. and piketoprofen 2% pet. (8).

A woman who had suffered photoallergic contact dermatitis from dexketoprofen, 7 years later developed allergic contact dermatitis from the fragrances cinnamal and hexyl cinnamal. The authors try to link this to the possible cross-reactivity between ketoprofen (which always cross-reacts with dexketoprofen) and cinnamal (11). However, the association is with cinnamyl alcohol, not with cinnamal (cinnamaldehyde) (see Chapter 3.193 Ketoprofen).

One case of photoallergy from *oral* dexketoprofen has been observed in a man previously photosensitized to ketoprofen (3).

Cross-reactions, pseudo-cross-reactions and co-reactions

Dexketoprofen is the (S)-enantiomer of ketoprofen and photocross-reactions between these NSAIDs (in both directions) nearly always occur (3,4,6,7,8). Of 16 patients with photocontact allergy to ketoprofen and dexketoprofen, 11 cross-reacted to piketoprofen (9). Photocross-sensitivity to or from ketoprofen or piketoprofen (6,8). A patient who developed allergic photocontact dermatitis from piketoprofen had a photocross-reaction to dexketoprofen (2).

LITERATURE

1   The data in the section 'General' may have been obtained from literature discussed in this chapter, but mostly also or exclusively from one or more of the following online sources: ChemIDPlus Advanced, PubChem, DrugBank, RxList, Drug Central, Drugs.com, and Wikipedia
2   Fernández-Jorge B, Goday Buján JJ, Paradela S, Mazaira M, Fonseca E. Consort photocontact dermatitis from piketoprofen. Contact Dermatitis 2008;58:113-115
3   Asensio T, Sanchís ME, Sánchez P, Vega JM, García JC. Photocontact dermatitis because of oral dexketoprofen. Contact Dermatitis 2008;58:59-60

4   Goday-Bujan JJ, Rodríguez-Lozano J, Martínez-González MC, Fonseca E. Photoallergic contact dermatitis from dexketoprofen: study of 6 cases. Contact Dermatitis 2006;55:59-61

5   González-Pérez R, Trébol I, García-Ríos I, Arregui MA, Soloeta-Arechavala R. Photocontact dermatitis due to dexketoprofen. Report of 2 cases. Actas Dermosifiliogr 2006;97:456-459 (Article in Spanish)

6   López-Abad R, Paniagua MJ, Botey E, Gaig P, Rodriguez P, Richart C. Topical dexketoprofen as a cause of photocontact dermatitis. J Investig Allergol Clin Immunol 2004;14:247-249

7   Cuerda Galindo E, Goday Buján JJ, del Pozo Losada J, García Silva J, Peña Penabad C, Fonseca E. Photocontact dermatitis due to dexketoprofen. Contact Dermatitis 2003;48:283-284

8   Valenzuela N, Puig L, Barnadas MA, Alomar A. Photocontact dermatitis due to dexketoprofen. Contact Dermatitis 2002;47:237

9   Subiabre-Ferrer D, Esteve-Martínez A, Blasco-Encinas R, Sierra-Talamantes C, Pérez-Ferriols A, Zaragoza-Ninet V. European photopatch test baseline series: A 3-year experience. Contact Dermatitis 2019;80:5-8

10   Subiabre-Ferrer D, Esteve-Martínez A, Blasco-Encinas R, Sierra-Talamantes C, Pérez-Ferriols A, Zaragoza-Ninet V. European photopatch test baseline series: A 3-year experience. Contact Dermatitis 2019;80:5-8

11   Rubio-González B, Ortiz-de Frutos FJ. Allergic contact dermatitis to hexyl cinnamaldehyde, cinnamaldehyde, and 3,4 methylbenzylidene camphor in a patient with previous photoallergic contact dermatitis to dexketoprofen. Actas Dermosifiliogr 2015;106:146-148

Chapter 3.99   DIBEKACIN

IDENTIFICATION

Description/definition	: Dibekacin is the 4,6-disubstituted 2-deoxystreptamine that conforms to the structural formula shown below
Pharmacological classes	: Anti-bacterial agents
IUPAC name	: (2S,3R,4S,5S,6R)-4-Amino-2-[(1S,2S,3R,4S,6R)-4,6-diamino-3-[(2R,3R,6S)-3-amino-6-(aminomethyl)oxan-2-yl]oxy-2-hydroxycyclohexyl]oxy-6-(hydroxymethyl)oxane-3,5-diol
Other names	: 3',4'-Dideoxykanamycin B
CAS registry number	: 34493-98-6
EC number	: 252-064-6
Merck Index monograph	: 4278
Patch testing	: No data available; suggested: 10% pet.
Molecular formula	: $C_{18}H_{37}N_5O_8$

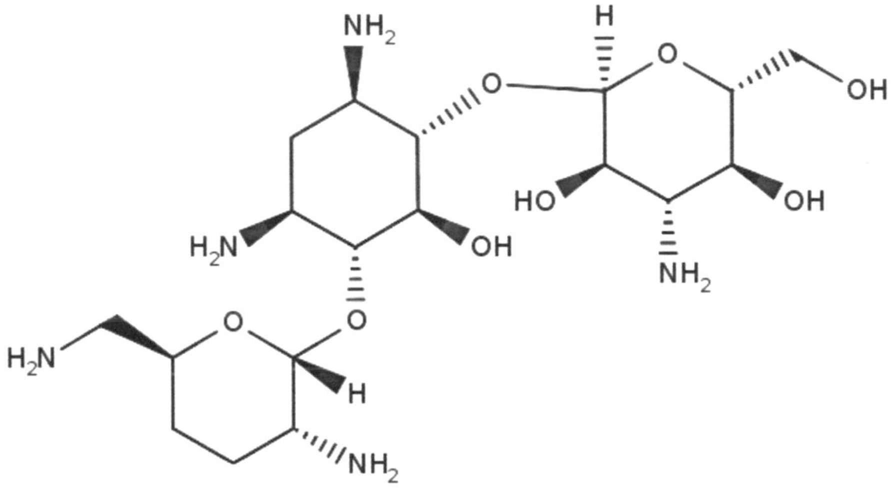

GENERAL

Dibekacin is an aminoglycoside antibiotic derived from kanamycin B, active against bacterial strains that are resistant to kanamycin. In pharmaceutical products, dibekacin is employed as dibekacin sulfate (CAS numbers 58580-55-5 and 64070-13-9, EC numbers 261-341-0 and 264-651-4, molecular formula $C_{18}H_{39}N_5O_{12}S$) (1).

CONTACT ALLERGY

Case series

In Tokyo, Japan, out of 3903 patients who were patch tested in the period January 1987 to December 1995, 141 (3.6%) were patch tested with eye drops and 49 individuals (35%) reacted positively and were diagnosed with allergic contact dermatitis. In 36 cases, ingredient patch testing was performed and there were three reactions to dibekacin sulfate (2).

LITERATURE

1   The data in the section 'General' may have been obtained from literature discussed in this chapter, but mostly also or exclusively from one or more of the following online sources: ChemIDPlus Advanced, PubChem, DrugBank, RxList, Drug Central, Drugs.com, and Wikipedia
2   Aoki J. Allergic contact dermatitis due to eye drops. Their clinical features and the patch test results. Nihon Ika Daigaku Zasshi 1997;64:232-237 (article in Japanese)

Chapter 3.100 DIBROMPROPAMIDINE

IDENTIFICATION

Description/definition   : Dibrompropamidine is the organobromine compound that conforms to the structural formula shown below
Pharmacological classes : Anti-infective agents, local
IUPAC name   : 3-Bromo-4-[3-(2-bromo-4-carbamimidoylphenoxy)propoxy]benzenecarboximidamide
Other names   : 4,4'-(Trimethylenedioxy)bis(3-bromobenzamide); dibromopropamidine
CAS registry number   : 496-00-4
EC number   : Not available
Merck Index monograph : 4299
Patch testing   : 5% pet.
Molecular formula   : $C_{17}H_{18}Br_2N_4O_2$

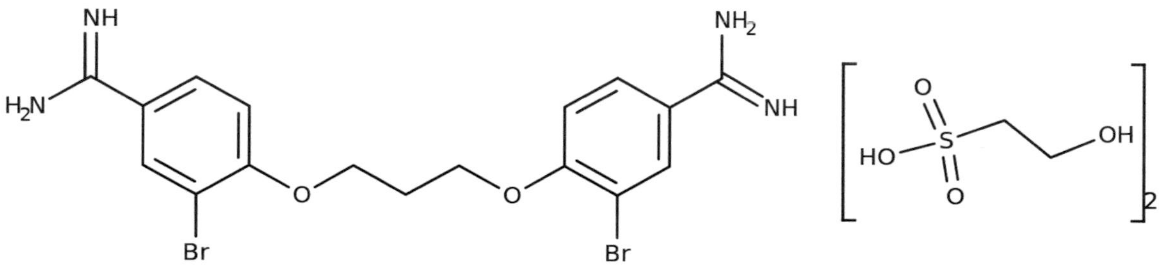

Dibrompropamidine   isethionate

GENERAL

Dibrompropamidine is an antibacterial agent used for topical therapy. It is active against streptococci and staphylococci, including penicillin-resistant strains. It also has some activity against gram-negative microorganisms. In pharmaceutical products, dibrompropamidine is usually employed as dibrompropamidine isethionate (CAS number 614-87-9, EC number 210-399-5, molecular formula $C_{21}H_{30}Br_2N_4O_{10}S_2$) (1).

CONTACT ALLERGY

Case series

Three patients sensitized to dibrompropamidine were reported from Norway. The first was a 43-year-old woman with a post-thrombotic chronic leg ulcer, who noticed aggravation of the ulcer accompanied by local skin infection, which was treated with a cream containing 0.15% dibrompropamidine isethionate. Exudation and spreading erythema was followed by an acute itchy maculopapular dermatitis on the upper thorax and distal parts of the arms, with progressive dissemination to proximal parts of the legs. The second patient, an 18-year-old woman, had treated a superinfection of her atopic dermatitis with the same cream, which seemed to aggravate the eczema rather than to ameliorate it. Patient 3 was a 26-year old male physician with longstanding atopic dermatitis, who, during surgical internship, developed hand eczema with spreading to the face and neck. Flares of erythema followed the use of dibrompropamidine cream. The latter patient had a positive patch test to the cream, but the ingredients could not be tested. The first and second patient both reacted to the cream and were tested with its ingredients, resulting in positive reactions to dibrompropamidine isethionate 5% and 10% pet. in the first patient (negative to 0.2%) and to 0.2% and 5.0% in the second (10% not tested). Both patients performed a usage test on the volar forearm resulting in an erythematous and papulovesicular reaction. Twenty controls were negative to the cream and to dibrompropamidine 5% pet. (4).

After having reported on these 3 patients allergic to dibrompropamidine cream (4), the author from Norway stated about a year later to have seen a further 9 patients with clinically relevant reactions to the cream. All 9 patients were patch tested with the components of the cream and showed positive reactions, ++ or +++ at D2 and D3, to dibrompropamidine isethionate 5% and 10% pet. (3).

Case reports

A 40-year-old man had developed an acute edematous vesiculobullous dermatitis of the penile shaft, one week after he had treated a new outbreak of herpes genitalis with a cream containing dibrompropamidine isethionate 0.15%, as he had done many times before. Patch testing with the cream and its ingredients showed positive reactions to the cream 'as is' and to dibrompropamidine 5%, 1%, 0.5% and 0.15% in petrolatum (2).

A 60-year-old female patient grazed her knee during a fall and applied dibrompropamidine cream. The next morning, she noticed a circular patch of intensely irritating blisters around the wound. Patch tests were positive to the cream 'as is' and to dibrompropamidine 0.1%, 0.5%, 1.0% and 5.0% pet. (all ++/++), but negative to the other ingredients of the cream. Twelve controls were negative (5).

At least two (6) and possibly one or more additional patients sensitized to dibrompropamidine (7) have been reported in the Norwegian literature (6,7).

LITERATURE

1   The data in the section 'General' may have been obtained from literature discussed in this chapter, but mostly also or exclusively from one or more of the following online sources: ChemIDPlus Advanced, PubChem, DrugBank, RxList, Drug Central, Drugs.com, and Wikipedia
2   Selvaag E. Contact allergy to dibromopropamidine cream. Contact Dermatitis 1999;40:58
3   Lützow-Holm C. Allergic contact dermatitis from dibrompropamidine cream. Contact Dermatitis 1989;20:160
4   Lützow-Holm C, Rønnevig JR. Allergic contact dermatitis from dibrompropamidine cream. Contact Dermatitis 1988;18:100-101
5   Wright S. Contact allergy to dibrompropamidine cream. Contact Dermatitis 1983;9:226
6   Løvold Berents T, Nodenes K. Graze with unusual progress. Tidsskr Nor Laegeforen 2009;129:305-307 (Article in Norwegian)
7   Holsen DS. When the treatment is the problem. Tidsskr Nor Laegeforen 2009;129:307 (Article in Norwegian)

Chapter 3.101 DIBUCAINE

IDENTIFICATION

Description/definition	: Dibucaine is the aminoalkylamide local anesthetic and quinoline derivative that conforms to the structural formula shown below
Pharmacological classes	: Anesthetics, local
IUPAC name	: 2-Butoxy-*N*-(2-(diethylamino)ethyl)quinoline-4-carboxamide
Other names	: Cinchocaine
CAS registry number	: 85-79-0
EC number	: 201-632-1
Merck Index monograph	: 4306
Patch testing	: Hydrochloride 5% pet. (Chemotechnique, SmartPracticeCanada, SmartPracticeEurope); 2.5% pet. (Chemotechnique, SmartPracticeCanada)
Molecular formula	: $C_{20}H_{29}N_3O_2$

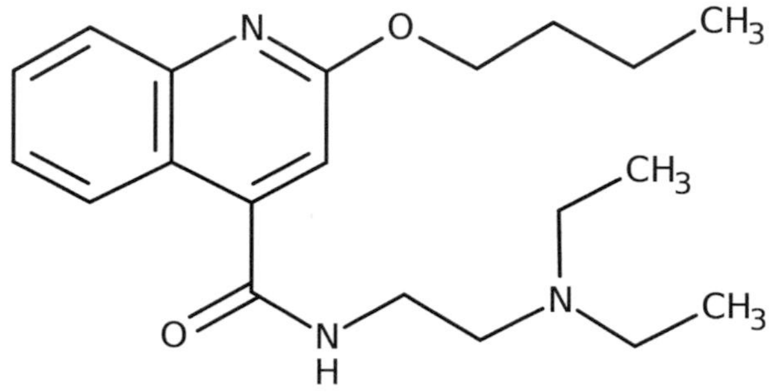

GENERAL

Dibucaine is a local anesthetic of the aminoalkylamide type now generally used for surface anesthesia. It is one of the most potent and toxic of the long-acting local anesthetics and its parenteral use is restricted to spinal anesthesia. In pharmaceutical products, dibucaine may either be used as dibucaine base or employed as dibucaine hydrochloride (CAS number 61-12-1, EC number 200-498-1, molecular formula $C_{20}H_{30}ClN_3O_2$) (1).

CONTACT ALLERGY

General

Dibucaine was a frequent cause of allergic contact dermatitis in the 1980s and 1990s in Japan, from its widespread use in OTC medicaments (66 products available in 1981, 33 in 2000), e.g. antiseptics, antipruritics and antifungals (29,30,31). It is probably the most important local anesthetic contact allergen in countries where benzocaine is no longer widely used, including the United Kingdom, Finland and the IVDK countries (Germany, Austria, Switzerland), especially from its presence in topical pharmaceuticals for the treatment of hemorrhoids and anal pruritus. Several cases of systemic contact dermatitis from dibucaine in antihemorrhoidal ointments containing dibucaine have been reported, mostly in the form of the baboon syndrome (21-25) and rarely as a maculopapular eruption (26) or acute generalized exanthematous pustulosis (AGEP) (47). Dibucaine also caused 2 cases of photocontact dermatitis (39,46).

Case reports and case series of allergy to dibucaine published in early literature can be found in ref. 43.

Contact allergy in the general population

In Germany, for the period 1995-2004, the population-based relative incidence (RI) of contact sensitization to dibucaine (cases/100,000 defined daily doses (DDDs) per year) was estimated to be 5.9. In the group of local anesthetics, the RI ranged from 1.5 (lidocaine) to 413.9 (benzocaine) (17).

Patch testing in consecutive patients suspected of contact dermatitis: routine testing

The results of studies in which dibucaine was tested in consecutive patients suspected of contact dermatitis (routine testing) are shown in table 3.101.1 Dibucaine 2.5% was added to the NACDG screening series in 2001. Prevalences of sensitization were relatively low and constant, ranging from 0.5% to 1.1%. Definite + probable relevance was always lower than 20%, which is probably why the local anesthetic was removed again from the screening series in 2010.

Table 3.101.1 Patch testing in groups of patients: Routine testing

Years and Country	Test conc. & vehicle	Number of patients tested | positive (%)		Selection of patients (S); Relevance (R); Comments (C)	Ref.
2009-2010 NACDG	2.5% pet.	4299	(0.5%)	R: definite + probable relevance: 18%	2
2007-2008 NACDG	2.5% pet.	5074	(1.1%)	R: definite + probable relevance: 16%	3
2005-2006 NACDG	2.5% pet.	4426	(1.1%)	R: definite + probable relevance: 11%	4
2003-2004 NACDG	2.5% pet.	5137	51 (1.0%)	R: not stated	5
2001-2004 USA		10,061	96 (1.0%)	R: definite + probable relevance : 16%	42
2001-2002 NACDG	2.5% pet.	4891	(0.9%)	R: definite + probable relevance: 9%	6

Patch testing in groups of selected patients

Results of patch testing dibucaine in groups of selected patients (patients with perianal, genital or anogenital dermatitis, individuals suspected of medicament allergy) are shown in table 3.101.2. In the first category, prevalences of sensitization ranged from 3% to 10.7%. In 5/7 investigations, the rates were 4% or higher. The highest frequency of sensitization to dibucaine (10.7%) was found in a study by the IVDK (Information Network of Departments of Dermatology: Germany, Austria, Switzerland) performed in the period 2004-2008 in patients with anogenital dermatoses. The relevance was not mentioned. In other patient groups, the frequency of sensitization ranged from 3% to 7.7%, the latter in a large study from Belgium in patients suspected of iatrogenic contact dermatitis (18). In general, relevance rates – where mentioned – were higher than 50% (table 3.101.2).

Table 3.101.2 Patch testing in groups of patients: Selected patient groups

Years and Country	Test conc. & vehicle	Number of patients tested | positive (%)		Selection of patients (S); Relevance (R); Comments (C)	Ref.
Patients with perianal, genital or anogenital dermatitis					
2005-2016 NACDG	2.5% pet.	250	10 (4.0%)	S: patients with only anogenital dermatitis; R: all positives represent relevant reactions; C: the frequency was significantly higher than in a control group	14
2003-2010 USA	5% pet.	55	3 (6%)	S: women with (ichty) vulvar dermatoses; R: not stated	10
2004-2008 IVDK			(10.7%)	S: patients with anogenital dermatoses tested with a medicament series; R: not stated; C: number of patients tested unknown	11
1994-2004 NACDG	2.5% pet.	146	6 (4.2%)	S: patients with anogenital signs or symptoms; R: 14%; C: the frequency was significantly higher than in a control group	9
1990-2003 IVDK			(6.6%)	S: patients with perianal dermatoses; R: not stated; C: the frequency was significantly higher than in a control group; number of patients tested uncertain	12
1992-1997 IVDK		1008		S: patients evaluated for allergic anogenital contact dermatitis; C: the frequency is unknown to the author, but dibucaine was the 4th most common allergen in the patient group	13
1991-1995 U.K.		121	4 (3.3%)	S: women with pruritus vulvae and primary vulval dermatoses suspected of secondary allergic contact dermatitis; R: 49% for all positive reactions together	7
1992-1994 U.K.	5% pet.	69	2 (3%)	S: women with 'vulval problems'; R: 58% for all allergens together	8
Other patient groups					
1990-2014 Belgium	5% pet.	169	13 (7.7%)	S: patients suspected of iatrogenic contact dermatitis and tested with a pharmaceutical series and their own products; R: 96% of the positive patch test reactions to all topical drugs and antiseptics were considered to be relevant	18
2000-2010 Canada		100	7 (7%)	S: charts reviewed and included in the study when there was at least one positive reaction to a topical drug; R: not stated C: the high percentage may (or may not) be the result of the – rather unusual – selection procedure targeting at previously demonstrated topical drug allergy	19
1997-2007 Finland	5% pet.	624	20 (3.2%)	S: patients suspected of contact allergy to topical therapies; R: 12 had perianal dermatitis from dibucaine and 7 formerly had suffered from perianal dermatitis	49
1995-2004 IVDK	3% pet.	7580	(1.5%)	S: not stated; R: not stated	16

IVDK: Information Network of Departments of Dermatology, Germany, Austria, Switzerland
NACDG: North American Contact Dermatitis Group (USA, Canada)

Case series

Between 1990 and 2003, in Leuven, Belgium, 92 patients were patch tested for chronic vulval complaints. Fifteen of these women were tested with a series of topical drugs; there was one reaction to dibucaine 2.5% pet. which was relevant (15). In Japan, in the period 1988-1995, 138 patients were patch tested with their own topical medicaments including local anesthetics, and 70 (51%) showed positive reactions. In 49 of the 60 cases who reacted positively to local anesthetics, the individual ingredients were tested separately. The local anesthetics causing positive reactions were dibucaine HCl (35 patients), benzocaine (n=12), lidocaine HCl (n=2), and procaine HCl (n=1) (29).

In a study in the United Kingdom in the period 1988-1998, 63 patients reacting to a mix of benzocaine, tetracaine and dibucaine were tested with the three active ingredients and there were 28 reactions to dibucaine, 22 to benzocaine, and 23 to tetracaine. Of the entire group, 55% were interpreted as either of current or of past relevance (41). In a study performed by members of the EECDRG, of 13 patients reacting to an investigational 'caine mix', tested with its anesthetic components, 7 reacted to dibucaine 2.5% pet., 5 to benzocaine 5% pet., and 2 to amethocaine (tetracaine) 2.5% pet. Relevance was not specified (38).

In a similar study in the United Kingdom, before 1988, 40 patients reacting to a 'caine mix' containing 5% benzocaine, 1% dibucaine and 1% tetracaine were tested with its ingredients and there were 12 reactions to dibucaine, 19 reactions to benzocaine, and 16 to tetracaine. The most involved primary sites were the legs (29%) and the anogenital region (27%). Relevance for the entire group was a little over 50%, past relevance 19% (40).

In the period 1971-1976 inclusive, at St. John's Hospital, London, 18 patients were found to have been sensitized by dibucaine, of who 8 through a rectal ointment containing 0.5% dibucaine HCl (45). In a 1966 U.K. series of 20 patients sensitized to local anesthetic ointments, nine of 16 with anogenital pruritus and two of four with eczema elsewhere reacted to dibucaine 1% pet. (44).

Case reports

Short summaries of case reports of allergic contact dermatitis from dibucaine are shown in table 3.101.3.

Table 3.101.3 Short summaries of case reports of allergic contact dermatitis from dibucaine

Year and country	Sex	Age	Positive patch tests	Clinical data and comments	Ref.
2009 Japan	M	89	antiseptic ointment 'as is'; dibucaine HCl 1% pet.	eczema on knee from antiseptic ointment on wound; also contact allergy to ingredients naphazoline and chlorpheniramine	20
2008 New Zealand	F	45	dibucaine 5% pet.	perianal dermatitis from a hemorrhoid ointment containing dibucaine	48
2007 Spain	F	32	AH ointment 'as is'; dibucaine 5% pet.	eczema perianally and intergluteal fold from AH ointment	27
2001 Japan	M	71	antiseptic ointment 'as is'; dibucaine HCl 1% pet.	eczema on right lower leg from OTC ointment for skin wounds; also contact allergy to ingredient chlorpheniramine	28
2001 Australia	M	79	AH ointment 'as is'; dibucaine 5% pet.	weeping dermatitis of the perianal skin, buttocks and proximal thighs from antihemorrhoidal ointment	36
2000 Japan	F	23	antipruritic jelly 'as is'; dibucaine HCl 0.1% pet; caine mix with 1% dibucaine	eczema on right forearm from an antipruritic jelly against itch from insect bites	30
1998 Korea	F	36	AH ointment 'as is'; dibucaine	perianal eczema from antihemorrhoidal ointment	34
1983 Netherlands	F	36	AH ointment 'as is'; dibucaine 2% water	perianal dermatitis from antihemorrhoidal ointment; also contact allergy to the ingredient soframycin	35

AH; Antihemorrhoidal; OTC: over-the-counter

Other case reports or information on contact allergy to dibucaine, adequate data of which are not available to the author, can be found in refs. 31,32,33, and 37.

Systemic contact dermatitis

Several cases of systemic contact dermatitis from dibucaine have been reported, mostly in the form of the baboon syndrome (21-25), and as a maculopapular eruption (26) and acute generalized exanthematous pustulosis (AGEP) in one patient each. All patients had used antihemorrhoidal ointments containing dibucaine, applying them to the anus and the rectal mucosa. This mucosal surface, due to its high vascularization and the liver metabolism by-pass, and helped by the anatomical occlusion in the gluteal fold, allows a considerable systemic absorption of the topicals there applied, transforming them in systemic drugs as if they were taken orally and so expanding their desirable but also undesirable effects.

Baboon syndrome

A 69-year-old woman had a strongly positive patch test reaction to dibucaine HCl (probably from sensitization in antihemorrhoidal ointments) and weaker positive reactions to multiple corticosteroids. She was prescribed mometasone furoate cream for the patch test reactions, but two days later, she presented because of pruritus and erythematous exanthema of the back and arms. This was interpreted as dermatitis, and another cream (unspecified) was advised. However, three days later, eczema involved the antecubital fossae, anterior and posterior axillary folds, wrist, inner sides of both thighs, and anal fold. This was diagnosed as systemic contact dermatitis in the form of the baboon syndrome from percutaneous absorption of dibucaine HCl (because of the fierce reaction), the corticosteroids from the patch tests and the creams applied to suppress these reactions (21).

A 61-year-old woman suffered from a chronic anal fissure for which she applied both an ointment with dibucaine 1% and a cream with 2% diltiazem hydrochloride. The patient reported erythema of the perianal area in the first days, and 5 days later, she presented with a symmetric intertriginous and flexural exanthema. She stopped the medications with complete recovery within 10 days. Patch tests were strongly positive (+++) to the cream and ointment 'as is', dibucaine 5% pet. and diltiazem 10% pet. The clinical presentation was that of the baboon syndrome, one of the main presentations of systemic contact dermatitis. The cause was most likely systemic absorption of dibucaine and diltiazem through the anal mucosa and skin (22).

An almost identical case was reported from Portugal in 2011. A 43-year-old woman had the classic clinical picture of the baboon syndrome, while using dibucaine HCl ointment. She had positive patch tests to the ointment, dibucaine 5% pet. and the caine mix (containing 2.5% dibucaine) (23). In 2000, systemic contact dermatitis had been reported from Germany (24). A 62-year-old woman presented with erythematovesicular lesions of the perianal area, and an erythematous, edematous rash of the face, axillae, elbow flexures and upper inner thighs after several days' application of dibucaine ointment to the perianal skin and rectal mucosa for hemorrhoids. Patch tests were positive to the ointment and dibucaine 5% pet. (24).

The first patient with systemic contact dermatitis due to dibucaine was reported in 1995 from Portugal (25). He had used an ointment containing dibucaine (the same as in most other publications) because he had undergone a transrectal prostatic ultrasound, which caused pruriginous erythematous exudative lesions in the perianal area. His family physician then prescribed another ointment which also contained dibucaine, resulting in systemic contact dermatitis with erythematovesicular exudative lesions of the intergluteal fold, buttocks and scrotum, and a diffuse symmetrical erythematous and papular eruption, with small vesicles and pustules, mainly on flexural areas (neck, axillae, antecubital fossae, popliteal fossae), pubic region and upper inner thighs (inverted triangle), lateral trunk and abdomen (25).

Possibly, a case of systemic contact dermatitis from the local anesthetic in a lubricant gel for digital rectal examination was reported in French literature (50), but details are unknown.

Other eruptions

A 60-year-old woman presented with a pruriginous centrifugally spreading maculopapular erythematous eruption affecting her trunk and arms, which had been evolving over the previous week. Ibuprofen and rabeprazole were suspected, but patch tests were negative. However, a positive reaction was observed (++) at D2 and D4 to caine mix III, which contains benzocaine 5%, tetracaine HCl 2.5%, and dibucaine HCl 2.5%. Upon a more thorough questioning, the patient now reported that she had used two ointments for hemorrhoids, which both caused local reactions in the perianal area controlled by corticosteroids. Both ointments contained dibucaine HCl and a patch test to this anesthetic 5% pet. was positive. Ibuprofen and rabeprazole were re-introduced without problems (26).

A 48-year old woman who developed a generalized cutaneous eruption without fever had attended several times in the emergency department and was finally admitted. Physical examination revealed diffuse edematous erythema with non-follicular pustules all over the body, resembling acute generalized exanthematous pustulosis (AGEP). A skin biopsy was compatible with AGEP. All oral drugs were stopped and with treatment she recovered slowly. Patch tests were positive to the caine mix. The patient now reported having used a compound for hemorrhoids containing ruscogenin, prednisolone, cinchocaine (dibucaine), menthol, and zinc oxide some days before and during hospitalization. The patient was diagnosed with drug-induced AGEP from dibucaine, which was contained in both the compound and the caine-mix (47).

Cross-reactions, pseudo-cross-reactions and co-reactions

Dibucaine is often considered to be an amide-type local anesthetic, just as articaine, bupivacaine, lidocaine, mepivacaine, prilocaine, and ropivacaine. However, whereas all others are aminoacylamides, dibucaine is an aminoalkylamide. Dibucaine does neither cross-react to the ester-type local anesthetics (esters of p-aminobenzoic acid [PABA] such as benzocaine [ethyl aminobenzoate], butamben [butyl aminobenzoate] or procaine), nor to the other amide-type anesthetics.

PHOTOSENSITIVITY

Case reports

A 39-year-old woman presented with pruritic skin lesions in the perianal area, extending sometimes to the legs and arms, appearing at irregular intervals more in the summer, and responding well to topical corticosteroids. She suffered from hemorrhoids, which she treated with an antihemorrhoidal ointment containing dibucaine. However, patch tests to the ointment, dibucaine and the caine mix, were negative. On the 5th day, the patient was exposed to UV radiation on a sunbed for skin tanning. Some hours later, 3 unknown patches became positive. Photopatch tests with UVA radiation were now performed and there were positive reactions to the ointment and dibucaine. After avoiding dibucaine, the patient became asymptomatic (39).

In a 13-year-old girl, an itchy eruption suddenly developed in light-exposed areas. Examination of the skin showed a sharply demarcated eczematous eruption with erythema, edema, and vesiculation on both forearms, extending to the dorsa of the hands. The finger webs were spared. The perioral region also showed vesiculation and crusting. The oral mucous membranes were not involved. Prior to this episode, she had been at the dentist twice who had anesthetized her with dibucaine (surface anesthesia or by injection?). Two days after the first dental treatment, she noted a swelling of her lips that subsided in a few days. A week later she was treated again, and then had extensive sun exposure at a swimming pool. That night she noted severe pruritus on her forearms and dorsa of the hands. By the next day, vesicle formation had developed in these areas. Patch and photopatch tests (using UVA 4 mW/cm^2, target distance 15 cm, irradiation for 15 minutes) were carried out with 0.3% dibucaine, 2% procaine HCl and 1% lidocaine HCl in physiologic saline solution. The photopatch test site with dibucaine showed papules and erythema with slight edema. Only a few papules were observed at the patch test site with dibucaine. Four controls were negative. The author managed to induce experimental photocontact allergy in guinea pigs. It was concluded that the patient had both photocontact allergy and contact allergy to dibucaine (46). As no dibucaine can have touched the skin of the hands and forearms, this must have been a case of systemic photoallergic contact dermatitis from dibucaine absorbed into the blood stream.

LITERATURE

1   The data in the section 'General' may have been obtained from literature discussed in this chapter, but mostly also or exclusively from one or more of the following online sources: ChemIDPlus Advanced, PubChem, DrugBank, RxList, Drug Central, Drugs.com, and Wikipedia

2   Warshaw EM, Belsito DV, Taylor JS, Sasseville D, DeKoven JG, Zirwas MJ, et al. North American Contact Dermatitis Group patch test results: 2009 to 2010. Dermatitis 2013;24:50-59

3   Fransway AF, Zug KA, Belsito DV, Deleo VA, Fowler JF Jr, Maibach HI, et al. North American Contact Dermatitis Group patch test results for 2007-2008. Dermatitis 2013;24:10-21

4   Zug KA, Warshaw EM, Fowler JF Jr, Maibach HI, Belsito DL, Pratt MD, et al. Patch-test results of the North American Contact Dermatitis Group 2005-2006. Dermatitis 2009;20:149-160

5   Warshaw EM, Belsito DV, DeLeo VA, Fowler JF Jr, Maibach HI, Marks JG, et al. North American Contact Dermatitis Group patch-test results, 2003-2004 study period. Dermatitis 2008;19:129-136

6   Pratt MD, Belsito DV, DeLeo VA, Fowler JF Jr, Fransway AF, Maibach HI, et al. North American Contact Dermatitis Group patch-test results, 2001-2002 study period. Dermatitis 2004;15:176-183

7   Lewis FM, Shah M, Gawkrodger DJ. Contact sensitivity in pruritus vulvae: patch test results and clinical outcome. Dermatitis 1997;8:137-140

8   Lewis FM, Harrington CI, Gawkrodger DJ. Contact sensitivity in pruritus vulvae: a common and manageable problem. Contact Dermatitis 1994;31:264-265

9   Warshaw EM, Furda LM, Maibach HI, Rietschel RL, Fowler JF Jr, Belsito DV, et al. Anogenital dermatitis in patients referred for patch testing: retrospective analysis of cross-sectional data from the North American Contact Dermatitis Group, 1994-2004. Arch Dermatol 2008;144:749-755

10 Lucke TW, Fleming CJ, McHenry P, Lever R. Patch testing in vulval dermatoses: how relevant is nickel? Contact Dermatitis 1998;38:111-112

11 Bauer A. Contact sensitization in the anal and genital area. Curr Probl Dermatol 2011;40:133-141

12 Kügler K, Brinkmeier T, Frosch PJ, Uter W. Anogenital dermatoses—allergic and irritative causative factors. Analysis of IVDK data and review of the literature. J Dtsch Dermatol Ges 2005;3:979-986

13 Bauer A, Geier J, Elsner P. Allergic contact dermatitis in patients with anogenital complaints. J Reprod Med 2000;45:649-654

14 Warshaw EM, Kimyon RS, Silverberg JI, Belsito DV, DeKoven JG, Maibach HI, et al. Evaluation of patch test findings in patients with anogenital dermatitis. JAMA Dermatol 2019;156:85-91

15 Nardelli A, Degreef H, Goossens A. Contact allergic reactions of the vulva: a 14-year review. Dermatitis 2004;15:131-136

16 Menezes de Padua CA, Uter W, Schnuch A. Contact allergy to topical drugs: prevalence in a clinical setting and estimation of frequency at the population level. Pharmacoepidemiol Drug Saf 2007;16:377-384

17 Menezes de Padua CA, Schnuch A, Nink K, Pfahlberg A, Uter W. Allergic contact dermatitis to topical drugs – Epidemiological risk assessment. Pharmacoepidemiol Drug Saf 2008;17:813-821

18 Gilissen L, Goossens A. Frequency and trends of contact allergy to and iatrogenic contact dermatitis caused by topical drugs over a 25-year period. Contact Dermatitis 2016;75:290-302

19 Spring S, Pratt M, Chaplin A. Contact dermatitis to topical medicaments: a retrospective chart review from the Ottawa Hospital Patch Test Clinic. Dermatitis 2012;23:210-213

20 Yamadori Y, Oiso N, Hirao A, Kawara S, Kawada A. Allergic contact dermatitis from dibucaine hydrochloride, chlorpheniramine maleate, and naphazoline hydrochloride in an over-the-counter topical antiseptic. Contact Dermatitis 2009;61:52-53

21 Alves da Silva C, Paulsen E. Systemic allergic dermatitis after patch testing with cinchocaine and topical corticosteroids. Contact Dermatitis 2019;81:301-303

22 Santiago L, Moura AL, Coutinho I, Gonçalo M. Systemic allergic contact dermatitis associated with topical diltiazem and/or cinchocaine. J Eur Acad Dermatol Venereol 2018;32:e284-e285

23 Oliveira A, Rosmaninho A, Lobo I, Selores M. Intertriginous and flexural exanthema after application of a topical anesthetic cream: a case of baboon syndrome. Dermatitis 2011;22:360-362

24 Erdmann SM, Sachs B, Merk HF. Systemic contact dermatitis from cinchocaine. Contact Dermatitis 2001;44:260-261

25 Marques C, Faria E, Machado A, Gonçalo M, Gonçalo S. Allergic contact dermatitis and systemic contact dermatitis from cinchocaine. Contact Dermatitis 1995;33:443

26 Matos D, Serrano P, Brandão FM. Maculopapular rash of unsuspected cause: systemic contact dermatitis to cinchocaine. Cutan Ocul Toxicol 2015;34:260-261

27 González Mahave I, Lobera T, Blasco A, Del Pozo MD. Allergic contact dermatitis caused by cinchocaine. Contact Dermatitis 2008;58:55-58

28 Hayashi K, Kawachi S, Saida T. Allergic contact dermatitis due to both chlorpheniramine maleate and dibucaine hydrochloride in an over-the-counter medicament. Contact Dermatitis 2001;44:38-39

29 Amano K. Clinical study of patients with positive reactions in patch tests with local anesthetics. Nippon Ika Daigaku Zasshi 1997;64:139-146 (Article in Japanese).

30 Nakada T, Iijima M. Allergic contact dermatitis from dibucaine hydrochloride. Contact Dermatitis 2000;42:283

31 Watanabe K, Sugai T. Contact dermatitis from dibucaine. Skin Research 1981;23:537-543 (article in Japanese, cited in ref. 30)

32 Kurumaji Y. A case of allergic contact dermatitis due to nupercaine (dibucaine hydrochloride) in an antipruritic plaster. Skin Research 1993;35(suppl.16):349-352 (article in Japanese, cited in ref. 30)

33 Yamamoto M, Onuma S, Mukawa R, Kawaguchi T, Kitamura K, Ikezawa Z. Amikacin rash and contact dermatitis syndrome patients with contact dermatitis induced by amino-glycosidal antibiotics and dibucaine HCl. The Clinical Journal of Dermatology 1998;52:705-708 (article in Japanese, cited in ref. 30)

34 Lee AY. Allergic contact dermatitis from dibucaine in Proctosedyl ointment without cross-sensitivity. Contact Dermatitis 1998;39:261

35 Van Ketel WG. Contact allergy to different antihaemorrhoidal anaesthetics. Contact Dermatitis 1983;9:512-513

36 Kearney CR, Fewings J. Allergic contact dermatitis to cinchocaine. Australas J Dermatol 2001;42:118-119

37 Edman B, Möller H. Medicament contact allergy. Derm Beruf Umwelt 1986;34:139-143 (Article in German)

38 Wilkinson JD, Andersen KE, Lahti A, Rycroft RJ, Shaw S, White IR. Preliminary patch testing with 25% and 15% 'caine'-mixes. The EECDRG. Contact Dermatitis 1990;22:244-245

39 Urrutia I, Jáuregui I, Gamboa P, González G, Antépara I. Photocontact dermatitis from cinchocaine (dibucaine). Contact Dermatitis 1998;39:139-140

40 Beck MH, Holden A. Benzocaine—an unsatisfactory indicator of topical local anaesthetic sensitization for the U.K. Br J Dermatol 1988;118:91-94

41 Sidhu SK, Shaw S, Wilkinson JD. A 10-year retrospective study on benzocaine allergy in the United Kingdom. Am J Cont Dermat 1999;10:57-61

42 Warshaw EM, Schram SE, Belsito DV, DeLeo VA, Fowler JF Jr, Maibach HI, et al. Patch-test reactions to topical anesthetics: retrospective analysis of cross-sectional data, 2001 to 2004. Dermatitis 2008;19:81-85

43 Lane CG, Luikart R. Dermatitis from local anesthetics, with a review of one hundred and seven cases from the literature. J Am Med Assoc 1951;146:717-720

44 Wilson HTH. Dermatitis from anaesthetic ointments. Practitioner 1966;197:673 (cited in ref. 45)

45 Cronin E. Contact Dermatitis. Edinburgh: Churchill Livingstone, 1980:198-199

46 Horio T. Photosensitivity reaction to dibucaine. Case report and experimental induction. Arch Dermatol 1979;115:986-987

47 Cubero JL, Garcés MM, Segura N, Sobrevía MT, Fraj J, Lezaun A, Colás C. Topical drug-induced acute generalized exanthematous pustulosis misdiagnosed as an oral drug-related eruption. J Investig Allergol Clin Immunol 2010;20: 620-621

48 Gunson TH, Greig DE. Allergic contact dermatitis to all three classes of local anaesthetic. Contact Dermatitis 2008;59:126-127

49 Jussi L, Lammintausta K. Sources of sensitization, cross-reactions, and occupational sensitization to topical anaesthetics among general dermatology patients. Contact Dermatitis 2009;60:150-154

50 Schmutz JL. Generalised contact eczema: Care is needed in selecting a lubricant for digital rectal examination. Ann Dermatol Venereol 2015;142:624-625

Chapter 3.102   DICLOFENAC

IDENTIFICATION

Description/definition	: Diclofenac is the nonsteroidal benzeneacetic acid derivative that conforms to the structural formula shown below
Pharmacological classes	: Anti-inflammatory agents, non-steroidal; cyclooxygenase inhibitors
IUPAC name	: 2-[2-(2,6-Dichloroanilino)phenyl]acetic acid
Other names	: 2-((2,6-Dichlorophenyl)amino)benzeneacetic acid
CAS registry number	: 15307-86-5
EC number	: 239-348-5
Merck Index monograph	: 4361
Patch testing	: Sodium salt, 2.5% pet. (SmartPracticeCanada, SmartPracticeEurope); sodium salt, 5% pet. (Chemotechnique, SmartPracticeCanada); sodium salt, 1% pet. (Chemotechnique)
Molecular formula	: $C_{14}H_{11}Cl_2NO_2$

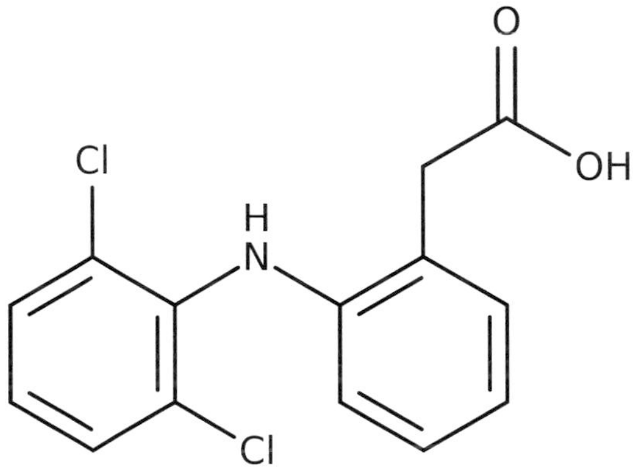

GENERAL

Diclofenac is a benzeneacetic acid-derived nonsteroidal anti-inflammatory drug (NSAID) with antipyretic, anti-inflammatory and analgesic actions. Its mechanism of action is as a cyclooxygenase inhibitor. Oral diclofenac is generally used to treat pain from dysmenorrhea, osteoarthritis, rheumatoid arthritis, ankylosing spondylitis, and from other causes. In pharmaceutical products, diclofenac is employed as diclofenac sodium (CAS number 15307-79-6, EC number 239-346-4, molecular formula $C_{14}H_{10}Cl_2NNaO_2$) (1). Diclofenac is also available in topical formulations including ointment, gel, suppositories, and as transdermal therapeutic system (TTS) for treatment of pain due to minor sprains, strains, and contusions (7,9). Lipid nanoemulsions of diclofenac may be used for parenteral applications.

CONTACT ALLERGY

Patch testing in groups of selected patients

The results of patch testing with diclofenac in groups of selected patients are shown in table 3.102.1. Only a few positive reactions have been observed.

Table 3.102.1 Patch testing in groups of patients: Selected patient groups

Years and Country	Test conc. & vehicle	Number of patients tested	positive (%)	Selection of patients (S); Relevance (R); Comments (C)	Ref.
2014-2016 Spain		116	2 (1.7%)	S: not stated; R: 3/4 patch and photopatch test reactions were relevant; C: there were also two positive photopatch tests to diclofenac	24
1990-2014 Belgium	5% pet.	336	3 (0.9%)	S: patients suspected of iatrogenic contact dermatitis and tested with a pharmaceutical series and their own products; R: 96% of the positive patch test reactions to all topical drugs and antiseptics were considered to be relevant	2
2001-2007 Belgium	5% pet.	34	1 (3%)	S: patients suspected of photoallergic contact dermatitis to ketoprofen; R: not stated	23

Case series

A case series of 4 patients allergic to diclofenac was reported in 2016 (9). A 62-year-old female patient with rheumatoid arthritis applied the gel twice daily on her left wrist under occlusive dressing. She noticed erythema with slight edema at the site of drug application on the third day of use. A 70-year-old woman with right gonarthrosis applied the gel to this knee for 5 days. She was seen with a large desquamating erythematous lesion, well delineated, associated with slight edema and scratch signs on the right knee. Scattered erythematous papules coved by scales and diffuse erythema were noticed on the left knee and left lower leg. A 51-year-old man presented with intense erythema, edema, and pruritus on the lumbosacral area that had developed after self-application of 1% diclofenac gel several times per day for 4 consecutive days for acute lumbago. A 23-year old man with psoriatic arthritis developed redness, swelling and itching in the lumbar area where he had applied the gel for 7 days. Patch testing showed positive reactions to the gel and to diclofenac sodium 1% pet. in all 4 patients. The excipients were not tested (9).

In the period 1996-2001, in 2 hospitals in Spain, 3 patients were diagnosed with contact allergy and with zero photocontact allergy to diclofenac. The accumulated incidence per million inhabitants (catchment population of the hospitals) of both side effects together was 3.6 (27). In Italy, before 1993, the members of the GIRDCA Multicentre Study Group diagnosed 102 patients (49 men, 53 women), aged 16 to 66 years (mean 37 years), with (photo)dermatitis induced by systemic or topical NSAIDs. Diclofenac caused one contact allergic and zero photocontact allergic reactions (3).

Case reports

A 25-year-old woman developed perioral, upper eyelid and hand dermatitis. She was a massage therapist and for several years had handled three drugs including diclofenac (probably topical). A patch test with diclofenac 2.5% pet. was positive (+++) after 3 days. A patch test with one of the other drugs (benzydamine) was negative (6). This was a case of occupational allergic contact dermatitis of the hands with eczema of the face from contamination by the hands.

Following cataract surgery, a 70-year-old woman was treated with various eye drops, including diclofenac sodium 0.1% eye drops. Two weeks later, she developed itchy erythema and edema of her eyelids, without conjunctivitis. She had a previous history of contact dermatitis from − unknown - topical non-steroidal anti-inflammatory drugs (NSAIDs). Patch tests were positive to the eye drops 'as is', diclofenac sodium 2% pet., 1% indomethacin cream 'as is' and indomethacin 2% pet. The authors suggested that the patient had previously become sensitized to indomethacin from frequent use of topical NSAIDs (10).

Short summaries of other case reports of allergic contact dermatitis from diclofenac are shown in table 3.102.2. One more patient had allergic contact dermatitis from diclofenac and co-reacted to various other NSAIDs (not cross-reactions); details are not available to the author (8).

Table 3.102.2 Short summaries of case reports of allergic contact dermatitis from diclofenac

Year and country	Sex	Age	Positive patch tests	Clinical data and comments	Ref.
2011 Japan	M	82	DICLO eye solution 'as is'; DICLO 5% pet.	erythema and edema on the upper eyelids while treated with various eye drops; other eye drops patch test negative; concentrations of DICLO <5% gave ?+ reactions	11
2004 United Kingdom	F	65	DICLO gel 5% and 10% pet.; DICLO 1% water	contact dermatitis after applying 3% DICLO gel for a few weeks on actinic porokeratosis	12
2004 Italy	F	37	DICLO 1% and 2% water	itchy, erythematous, edematous and vesicular lesions on the lumbar region after iontophoresis with diclofenac ampoules 2x/week for 1 week	18
2002 United Kingdom	M	56	DICLO gel 'as is'; DICLO 2.5% and 5% pet.	pain and inflammation after 2 weeks' application of DICLO 3% gel to actinic keratoses on the forehead and the scalp	13
1996 Italy	M	36	DICLO eye drops 'as is'; DICLO 1% and 2% pet.	itchy erythematous edematous vesicular dermatitis of the eyelids after using diclofenac eye drops	16
1994 Germany	M	58	DICLO ointment 'as is'; DICLO 10% pet.	subacute eczema after DICLO gel on both knees for gonarthrosis; ROAT pos. at D5; also pos. patch test to DICLO diethylamine 10% pet.; 10 controls were negative	17

DICLO: diclofenac

PHOTOSENSITIVITY

Photopatch testing in groups of patients

The results of studies in which groups of selected patients, mostly those with suspected photoallergic contact dermatitis, were photopatch tested with diclofenac are shown in table 3.102.3. Only a few positive photopatch test reactions have been observed. Some were relevant, in other studies relevancy was not addressed.

Table 3.102.3 Photopatch testing in groups of patients

Years and Country	Test conc. & vehicle	Number of patients tested	positive (%)	Selection of patients (S); Relevance (R); Comments (C)	Ref.
2014-2016 Spain		116	2 (1.7%)	S: not stated; R: 3/4 patch and photopatch test reactions were relevant; C: there were also two positive 'plain' patch tests to diclofenac	24
<2013 Poland		36	1 (3%)	S: patients with a history of dermatitis induced or aggravated by light; R: the reaction was relevant	21
2008-2011 Europe	5% pet.	1031	2 (0.2%)	S: patients suspected or photoallergic contact dermatitis; R: not specified for individual allergens	33
2003-2007 Portugal	10% pet.	30	1 (3%)	S: subgroup of 83 patients with suspected photoaggravated facial dermatitis or systemic photosensitivity; R: attributed to systemic photosensitivity	25
2001-2007 Belgium	5% pet.	34	1 (3%)	S: patients suspected of photoallergic contact dermatitis to ketoprofen; R: not stated	23

Case reports and case series
In a multicenter study in Italy, performed in the period 1985-1994, 3 photopatch test reactions were seen to diclofenac; the patch test concentration used was not mentioned and relevance was not specified (78% for all photoallergens together) (26).

A 63-year-old man presented with subacute eczema on his hands, neck, and face. Two days before, he had been prescribed diclofenac gel for joint pains of the hands. The day before the eruption started, he had sun exposure in his garden. A 61-year-old woman presented with an eczematous rash on her legs which subsequently spread to the face and neck. For several days, she had been applying diclofenac gel on her knee because of osteoarthritic pain. The skin reaction appeared several hours after sun exposure. A year later, a similar eczematous eruption developed on her right knee after the application of a cream containing aceclofenac. Patch tests with diclofenac were negative, but both patients had positive photopatch tests, one to 5% and 10% pet. (negative to 1% pet.) and the other to all three concentrations. Both had photocross-reactions to aceclofenac 10% pet. (negative to 5% and 1% pet. (19).

A 77-year-old female patient with a single superficial actinic keratosis at her right cheek developed an infiltrated and exudative, stinging and itching erythema at this site in early summer following 7 weeks of treatment with a gel containing 3% diclofenac. Patch tests with the gel and its ingredients were negative. However, photopatch tests using 10 J/cm^2 UVA at D2 were positive to the gel 'as is', 5% and 10% pet. and to diclofenac 1% water, with negative reactions to the other ingredients (20).

A 40-year-old man was treated with 1% diclofenac gel twice daily for elbow trauma. At D3, acute vesicular eczema appeared at the application area. It was summer and the patient wore a short-sleeved shirt. One year before, he had complained of the same symptoms after application of the same gel on a finger. Patch and photopatch tests (5 J/cm^2 UVA) were negative to the plain patch tests but positive to photopatch tests with the gel 'as is' and diclofenac 1% pet. Five controls were negative (22).

A 32-year-old woman applied diclofenac gel to her left knee which was exposed to sunlight; after 3 days, dermatitis developed. Patch tests were negative, but photopatch tests positive to the gel and its ingredient diclofenac 1% alc. (negative to 5% pet.) (31).

Cross-reactions, pseudo-cross-reactions and co-reactions
Two patients photosensitized to diclofenac had photocross-reactions to aceclofenac (19). A patient who had a fixed drug eruption with a positive patch test to aceclofenac also reacted to diclofenac (4). Another patient, who developed allergic contact dermatitis from aceclofenac, had a weaker positive reaction to diclofenac (5). Spanish authors considered a positive patch test reaction to diclofenac cream to be a cross-reaction of diclofenac to indomethacin. However, only the creams were tested, not the active ingredients themselves; in addition, as they noticed themselves, these two drugs belong to different classes of NSAIDs not likely to cross-react (14).

Cutaneous adverse drug reactions from systemic administration caused by type IV (delayed-type) hypersensitivity
Cutaneous adverse drug reactions from systemic administration of diclofenac caused by type IV (delayed-type) hypersensitivity, including maculopapular rash (28,29,32), drug reaction with eosinophilia and systemic symptoms (DRESS) (30), and urticaria (6,15), and in addition occupational allergic contact dermatitis (6), are planned to be discussed in Volume IV of the *Monographs in Contact Allergy* series on Systemic drugs.

Immediate contact reactions

Immediate contact reactions (contact urticaria) to diclofenac sodium are presented in Chapter 5.

LITERATURE

1 The data in the section 'General' may have been obtained from literature discussed in this chapter, but mostly also or exclusively from one or more of the following online sources: ChemIDPlus Advanced, PubChem, DrugBank, RxList, Drug Central, Drugs.com, and Wikipedia

2 Gilissen L, Goossens A. Frequency and trends of contact allergy to and iatrogenic contact dermatitis caused by topical drugs over a 25-year period. Contact Dermatitis 2016;75:290-302

3 Pigatto PD, Mozzanica N, Bigardi AS, Legori A, Valsecchi R, Cusano F, et al. Topical NSAID allergic contact dermatitis. Italian experience. Contact Dermatitis 1993;29:39-41

4 Linares T, Marcos C, Gavilan MJ, Arenas L. Fixed drug eruption due to aceclofenac. Contact Dermatitis 2007;56:291-292

5 Pitarch Bort G, de la Cuadra Oyanguren J, Torrijos Aguilar A, García-Melgares Linares ML. Allergic contact dermatitis due to aceclofenac. Contact Dermatitis 2006;55:365-366

6 Schiavino D, Papa G, Nucera E, Schinco G, Fais G, Pirrotta LR, Patriarca G. Delayed allergy to diclofenac. Contact Dermatitis 1992;26:357-358

7 Bershow A, Warshaw E. Cutaneous reactions to transdermal therapeutic systems. Dermatitis 2011;22:193-203

8 Gonzalo MA, Revenga F. Multiple cutaneous sensitization to non-steroidal anti-inflammatory drugs. Dermatology 1996;193:59-60

9 Gulin SJ, Chiriac A. Diclofenac-induced allergic contact dermatitis: A series of four patients. Drug Saf Case Rep 2016;3(1):15

10 Ueda K, Higashi N, Kume A, Ikushima-Fujimoto M, Ogiwara S. Allergic contact dermatitis due to diclofenac and indomethacin. Contact Dermatitis 1998;39:323

11 Miyazato H, Yamaguchi S, Taira K, Asato Y, Yamamoto Y, Hagiwara K, Uezato H. Allergic contact dermatitis due to diclofenac sodium in eye drops. J Dermatol 2011;38:276-279

12 Kleyn CE, Bharati A, King CM. Contact dermatitis from 3 different allergens in Solaraze gel. Contact Dermatitis 2004;51:215-216

13 Kerr OA, Kavanagh G, Horn H. Allergic contact dermatitis from topical diclofenac in Solaraze gel. Contact Dermatitis 2002;47:175

14 Gómez A, Florido JF, Quiralte J, Martín AE, Sáenz de San Pedro B. Allergic contact dermatitis due to indomethacin and diclofenac. Contact Dermatitis 2000;43:59

15 Barbaud A, Tréchot P, Aublet-Cuvelier A, Reichert-Penetrat S, Schmutz JL. Bufexamac and diclofenac: frequency of contact sensitization and absence of cross-reactions. Contact Dermatitis 1998;39:272-273

16 Valsecchi R, Pansera B, Leghissa P, Reseghetti A. Allergic contact dermatitis of the eyelids and conjunctivitis from diclofenac. Contact Dermatitis 1996;34:150-151

17 Gebhardt M, Reuter A, Knopf B. Allergic contact dermatitis from topical diclofenac. Contact Dermatitis 1994;30:183-184

18 Foti C, Cassano N, Conserva A, Vena GA. Allergic contact dermatitis due to diclofenac applied with iontophoresis. Clin Exp Dermatol 2004;29:91

19 Fernández-Jorge B, Goday-Buján JJ, Murga M, Molina FP, Pérez-Varela L, Fonseca E. Photoallergic contact dermatitis due to diclofenac with cross-reaction to aceclofenac: two case reports. Contact Dermatitis 2009;61:236-237

20 Kowalzick L, Ziegler H. Photoallergic contact dermatitis from topical diclofenac in Solaraze gel. Contact Dermatitis 2006;54:348-349

21 Spiewak R. The frequency and causes of photoallergic contact dermatitis among dermatology outpatients. Acta Dermatovenerol Croat 2013;21:230-235

22 Montoro J, Rodríguez M, Díaz M, Bertomeu F. Photoallergic contact dermatitis due to diclofenac. Contact Dermatitis 2003;48:115

23 Devleeschouwer V, Roelandts R, Garmyn M, Goossens A. Allergic and photoallergic contact dermatitis from ketoprofen: results of (photo) patch testing and follow-up of 42 patients. Contact Dermatitis 2008;58:159-166

24 Subiabre-Ferrer D, Esteve-Martínez A, Blasco-Encinas R, Sierra-Talamantes C, Pérez-Ferriols A, Zaragoza-Ninet V. European photopatch test baseline series: A 3-year experience. Contact Dermatitis 2019;80:5-8

25 Cardoso J, Canelas MM, Gonçalo M, Figueiredo A. Photopatch testing with an extended series of photoallergens: a 5-year study. Contact Dermatitis 2009;60:325-329

26   Pigatto PD, Legori A, Bigardi AS, Guarrera M, Tosti A, Santucci B, et al. Gruppo Italiano recerca dermatiti da contatto ed ambientali Italian multicenter study of allergic contact photodermatitis: epidemiological aspects. Am J Contact Dermatitis 1996;17:158-163

27   Diaz RL, Gardeazabal J, Manrique P, Ratón JA, Urrutia I, Rodríguez-Sasiain JM, Aguirre C. Greater allergenicity of topical ketoprofen in contact dermatitis confirmed by use. Contact Dermatitis 2006;54:239-243

28   Romano A, Quaratino D, Papa G, Di Fonso M, Artesani MC, Venuti A. Delayed hypersensitivity to diclofenac: a report on two cases. Ann Allergy Asthma Immunol 1998;81:373-375

29   Romano A, Pietrantonio F, Di Fonso M, Garcovich A, Chiarelli C, Venuti A, Barone C. Positivity of patch tests in cutaneous reaction to diclofenac. Two case reports. Allergy 1994;49:57-59

30   Klingenberg RD, Bassukas ID, Homann N, Stange EF, Ludwig D. Delayed hypersensitivity reactions to diclofenac. Allergy 2003;58:1076-1077

31   Adamski H, Benkalfate L, Delaval Y, Ollivier I, le Jean S, Toubel G, et al. Photodermatitis from non-steroidal anti-inflammatory drugs. Contact Dermatitis 1998;38:171-174

32   Alonso R, Enrique E, Cisteró A. Positive patch test to diclofenac in Stevens-Johnson syndrome. Contact Dermatitis 2000;42:367

33   The European Multicentre Photopatch Test Study (EMCPPTS) Taskforce. A European multicentre photopatch test study. Br J Dermatol 2012;166:1002-1009

Chapter 3.103 DIETHYLSTILBESTROL

IDENTIFICATION

Description/definition : Diethylstilbestrol is the synthetic nonsteroidal estrogen that conforms to the structural formula shown below
Pharmaceutical classes : Estrogens, non-steroidal; carcinogens
Chemical/IUPAC name : 4-[(E)-4-(4-Hydroxyphenyl)hex-3-en-3-yl]phenol
CAS registry number : 56-53-1
EC number : 200-278-5
Merck Index monograph : 4418
Patch testing : 1% pet.; 0.1% alc.
Molecular formula : $C_{18}H_{20}O_2$

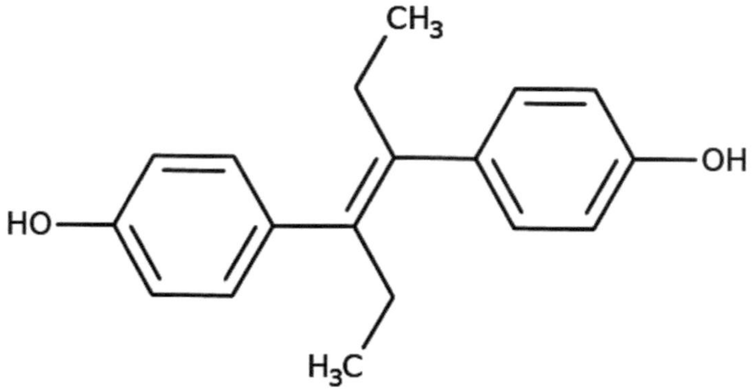

CONTACT ALLERGY

GENERAL

Diethylstilbestrol (DES) is a synthetic nonsteroidal form of estrogen that was developed to supplement a woman's natural estrogen production. It was used in the prevention of miscarriage or premature delivery in pregnant women prone to miscarriage or premature delivery. However, In 1971, the Food and Drug Administration advised physicians to stop prescribing DES to pregnant women because it was linked to a rare vaginal cancer in female offspring. It is apparently still used in the treatment of prostate cancer. Diethylstilbestrol inhibits the hypothalamic-pituitary-gonadal axis, thereby blocking the testicular synthesis of testosterone, lowering plasma testosterone, and inducing a chemical castration (1).

Case report

One patient had allergic contact dermatitis from diethylstilbestrol used as a therapeutic agent against hair loss in a hair lotion (2).

Cross-reactions, pseudo-cross-reactions and co-reactions

Benzestrol (3); dienestrol, hexestrol (2).

Immediate contact reactions

Immediate contact reactions (contact urticaria) to diethylstilbestrol are presented in Chapter 5.

LITERATURE

1   The data in the section 'General' may have been obtained from literature discussed in this chapter, but mostly also or exclusively from one or more of the following online sources: ChemIDPlus Advanced, PubChem, DrugBank, RxList, Drug Central, Drugs.com, and Wikipedia
2   Fregert S, Rorsman H. Hypersensitivity to diethyldtilbestrol. Cross-sensitization to dienestrol, hexestrol, bisphenol A, p-benzylphenol, hydroquinone-monobenzylether, and p-hydroxybenzoic-benzylester. Acta Derm Venereol 1960;40:206-219
3   Fregert S, Rorsman H. Hypersensitivity to diethylstilbestrol with cross-sensitization to benzestrol. Acta Derm Venereol 1962;42:290-293

Chapter 3.104 DIFLORASONE DIACETATE

IDENTIFICATION

Description/definition   : Diflorasone diacetate is the 17,21 diacetate derivative of the synthetic glucocorticoid
                            diflorasone, that conforms to the structural formula shown below
Pharmacological classes   : Glucocorticoids; anti-inflammatory agents
IUPAC name   : [2-[(6S,8S,9R,10S,11S,13S,14S,16S,17R)-17-Acetyloxy-6,9-difluoro-11-hydroxy-10,13,16-
                            trimethyl-3-oxo-6,7,8,11,12,14,15,16-octahydrocyclopenta[a]phenanthren-17-yl]-2-
                            oxoethyl] acetate
Other names   : 6α,9-Difluoro-11β,17,21-trihydroxy-16β-methylpregna-1,4-diene-3,20-dione 17,21-
                            diacetate
CAS registry number   : 33564-31-7
EC number   : 251-575-1
Merck Index monograph   : 4429 (Diflorasone)
Patch testing   : In general, corticosteroids may be tested at 0.1% and 1% in alcohol; late readings (6-10
                            days) are strongly recommended
Molecular formula   : $C_{26}H_{32}F_2O_7$

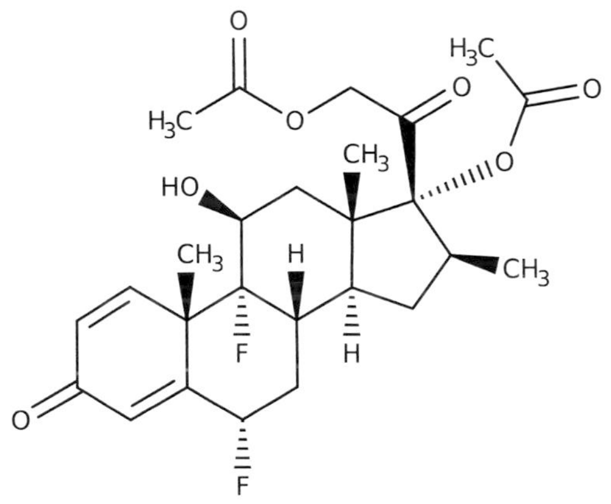

GENERAL

General aspects of corticosteroids used on the skin and mucous membranes are discussed in Chapter 2.4. A practical
guideline for diagnosing allergic reactions to corticosteroids is presented in ref. 1.

CONTACT ALLERGY

Case series

From January 1990 to June 2008, in Leuven, Belgium, 315 patients were diagnosed with contact allergy to/allergic
contact dermatitis from corticosteroids (CSs) from routine patch testing with a baseline series including tixocortol
pivalate, budesonide, hydrocortisone butyrate and prednisone caproate, patch testing with patients' own CS
preparations, and testing those with proven contact allergy to a corticosteroid or strongly suspected of CS allergy
later with a series of 66 CSs, including two sex hormones (progesterone and testosterone). 71% of the patients had
relevant reactions, but these were not specified. In this group of 315 CS allergic patients, 10 had positive patch tests
to diflorasone diacetate 0.1% alc. (2). As this corticosteroid has never been used in pharmaceuticals in Belgium,
these positive reactions must all be considered cross-reactions to other corticosteroids.

In reviewing the Japanese literature up to 1994, 43 patients with allergic contact dermatitis from topical
corticosteroid were identified, including one caused by diflorasone diacetate (3).

Cross-reactions, pseudo-cross-reactions and co-reactions
Cross-reactions between corticosteroids are discussed in Chapter 2.8.

LITERATURE

1   Baeck M, Goossens A. Immediate and delayed allergic hypersensitivity to corticosteroids: practical guidelines.
     Contact Dermatitis 2012;66:38-45
2   Baeck M, Chemelle JA, Terreux R, Drieghe J, Goossens A. Delayed hypersensitivity to corticosteroids in a series of
     315 patients: clinical data and patch test results. Contact Dermatitis 2009;61:163-175
3   Oh-i T. Contact dermatitis due to topical steroids with conceivable cross reactions between topical steroid
     preparations. J Dermatol 1996;23:200-208

Chapter 3.105 DIFLUCORTOLONE VALERATE

IDENTIFICATION

Description/definition : Diflucortolone valerate is the valerate ester of the synthetic glucocorticoid diflucortolone that conforms to the structural formula shown below

Pharmacological classes : Glucocorticoids; anti-inflammatory agents

IUPAC name : [2-[(6S,8S,9R,10S,11S,13S,14S,16R,17S)-6,9-Difluoro-11-hydroxy-10,13,16-trimethyl-3-oxo-7,8,11,12,14,15,16,17-octahydro-6H-cyclopenta[a]phenanthren-17-yl]-2-oxoethyl] pentanoate

Other names : 6α,9-Difluoro-11β,21-dihydroxy-16α-methylpregna-1,4-diene-3,20-dione 21-valerate

CAS registry number : 59198-70-8

EC number : 261-655-8

Merck Index monograph : 4432 (Diflucortolone)

Patch testing : In general, corticosteroids may be tested at 0.1% and 1% in alcohol; late readings (6-10 days) are strongly recommended

Molecular formula : $C_{27}H_{36}F_2O_5$

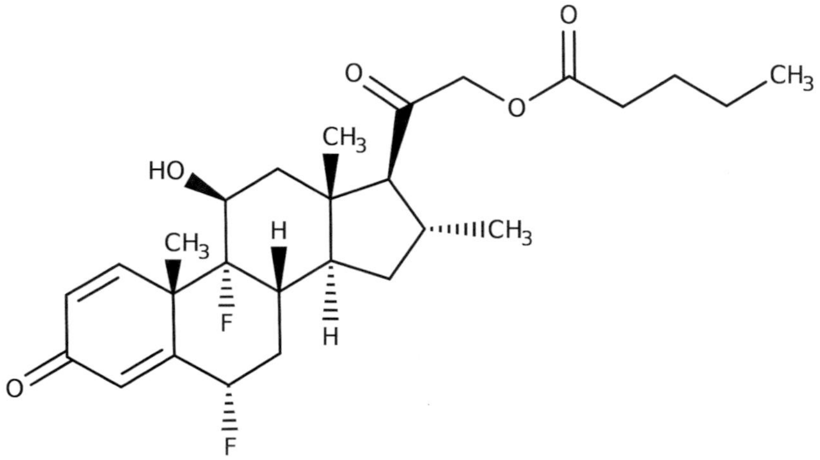

GENERAL

General aspects of corticosteroids used on the skin and mucous membranes are discussed in Chapter 2.4. A practical guideline for diagnosing allergic reactions to corticosteroids is presented in ref. 1.

CONTACT ALLERGY

Case series

From January 1990 to June 2008, in Leuven, Belgium, 315 patients were diagnosed with contact allergy to/allergic contact dermatitis from corticosteroids (CSs) from routine patch testing with a baseline series including tixocortol pivalate, budesonide, hydrocortisone butyrate and prednisone caproate, patch testing with patients' own CS preparations, and testing those with proven contact allergy to a corticosteroid or strongly suspected of CS allergy later with a series of 66 CSs, including two sex hormones (progesterone and testosterone). 71% of the patients had relevant reactions, but these were not specified. In this group of 315 CS allergic patients, 6 had positive patch tests to diflucortolone valerate 0.1% alc. (2). It is unknown how many of these reactions were caused by the use of a pharmaceutical product containing diflucortolone valerate and how many were cross-reactions to other corticosteroids.

In reviewing the Japanese literature up to 1994, 43 patients with allergic contact dermatitis from topical corticosteroid were identified, including one caused by diflucortolone valerate (5). Another Japanese author in 1994 reviewed the Japanese literature and found 2 reported cases of allergic contact dermatitis to diflucortolone valerate in 2 articles published in the period 1990-1991 (3).

Cross-reactions, pseudo-cross-reactions and co-reactions
Cross-reactions between corticosteroids are discussed in Chapter 2.8.

LITERATURE

1   Baeck M, Goossens A. Immediate and delayed allergic hypersensitivity to corticosteroids: practical guidelines.
     Contact Dermatitis 2012;66:38-45
2   Baeck M, Chemelle JA, Terreux R, Drieghe J, Goossens A. Delayed hypersensitivity to corticosteroids in a series of
     315 patients: clinical data and patch test results. Contact Dermatitis 2009;61:163-175
3   Okano M. Contact dermatitis due to budesonide: report of five cases and review of the Japanese literature.
     Int J Dermatol 1994;33:709-715
4   Burden AD, Beck MH. Contact hypersensitivity to topical corticosteroids. Br J Dermatol 1992;127:497-501
5   Oh-i T. Contact dermatitis due to topical steroids with conceivable cross reactions between topical steroid
     preparations. J Dermatol 1996;23:200-208

Chapter 3.106 DIFLUPREDNATE

IDENTIFICATION

Description/definition : Difluprednate is the synthetic glucocorticoid that conforms to the structural formula shown below

Pharmacological classes : Glucocorticoids

IUPAC name : [(6S,8S,9R,10S,11S,13S,14S,17R)-17-(2-Acetyloxyacetyl)-6,9-difluoro-11-hydroxy-10,13-dimethyl-3-oxo-6,7,8,11,12,14,15,16-octahydrocyclopenta[a]phenanthren-17-yl]-butanoate

Other names : Difluoroprednisolone butyrate acetate; 6α,9-difluoro-11β,17,21-trihydroxypregna-1,4-diene-3,20-dione 21-acetate 17-butyrate

CAS registry number : 23674-86-4

EC number : 245-815-4

Merck Index monograph : 4441

Patch testing : In general, corticosteroids may be tested at 0.1% and 1% in alcohol; late readings (6-10 days) are strongly recommended

Molecular formula : $C_{27}H_{34}F_2O_7$

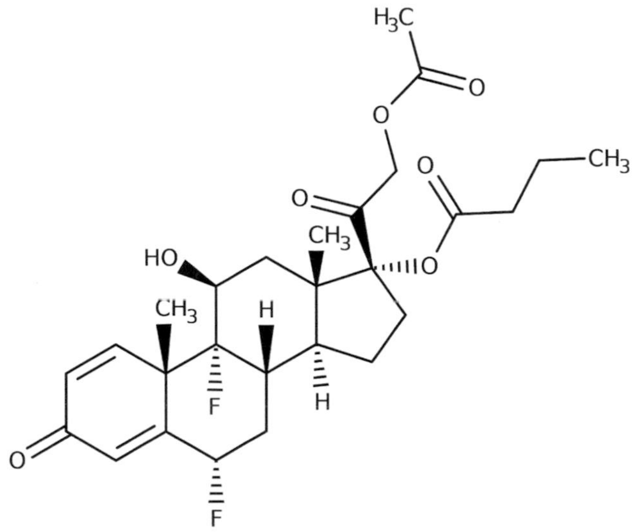

GENERAL

General aspects of corticosteroids used on the skin and mucous membranes are discussed in Chapter 2.4. A practical guideline for diagnosing allergic reactions to corticosteroids is presented in ref. 1.

CONTACT ALLERGY

Case series

From January 1990 to June 2008, in Leuven, Belgium, 315 patients were diagnosed with contact allergy to/allergic contact dermatitis from corticosteroids (CSs) from routine patch testing with a baseline series including tixocortol pivalate, budesonide, hydrocortisone butyrate and prednisone caproate, patch testing with patients' own CS preparations, and testing those with proven contact allergy to a corticosteroid or strongly suspected of CS allergy later with a series of 66 CSs, including two sex hormones (progesterone and testosterone). 71% of the patients had relevant reactions, but these were not specified. In this group of 315 CS allergic patients, 15 had positive patch tests to difluprednate 0.1% alc. (2). As this corticosteroid is not used in pharmaceuticals in Belgium, these positive reactions must all be considered cross-reactions to other corticosteroids.

Cross-reactions, pseudo-cross-reactions and co-reactions

Cross-reactions between corticosteroids are discussed in Chapter 2.8.

LITERATURE

1   Baeck M, Goossens A. Immediate and delayed allergic hypersensitivity to corticosteroids: practical guidelines. Contact Dermatitis 2012;66:38-45

2   Baeck M, Chemelle JA, Terreux R, Drieghe J, Goossens A. Delayed hypersensitivity to corticosteroids in a series of 315 patients: clinical data and patch test results. Contact Dermatitis 2009;61:163-175

Chapter 3.107 4,5-DIHYDRO-1H-IMIDAZOLE MONOHYDROCHLORIDE

IDENTIFICATION

Description/definition   : 4,5-Dihydro-1H-imidazole monohydrochloride is the imidazole derivative that conforms to
                             the structural formula shown below
Pharmacological classes   : Sympathomimetic agents
IUPAC name   : 4,5-Dihydro-1H-imidazole;hydrochloride
Other names   : Imidazoline hydrochloride; imidazoline chlorohydrate (in publication); 1H-imidazole, 4,5-
                             dihydro-, monohydrochloride; imidazolinium chloride
CAS registry number   : 34301-57-0
EC number   : Not available
Patch testing   : 0.1% pet.
Molecular formula   : $C_3H_7ClN_2$

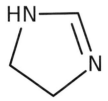

GENERAL

4,5-Dihydro-1H-imidazole monohydrochloride is a sympathomimetic agent with marked α-adrenergic activity used as a vasoconstrictor with rapid and prolonged action to relieve nasal congestion caused by rhinitis and sinusitis (1).

CONTACT ALLERGY

Case report

A 31-year-old man noticed that itchy eczematous lesions appeared around his mouth and nose, with itching spreading to the rest of his face and neck. The condition was chronic with alternating periods of remission and relapse, but did not clear up completely. When he first presented for patch testing, the appearance was similar to impetigo contagiosa with red scaly and crusting lesions localized to the anterior nares, the nasolabial folds, the ala nasi, bridge of the nose, upper lip, lips and cheeks. The patient said that he had only used nasal sprays for his rhinitis. Patch testing with various cosmetics and nasal sprays and ointments showed a positive reaction to three nasal sprays and imidazoline chlorohydrate (4,5-dihydro-1H-imidazole monohydrochloride) 0.1% pet. The causative ingredient of the other sprays was not found. The patient stopped using the nasal sprays, but the symptoms recurred, which was shown to be caused by contact allergy to amyl nitrite, which he used intranasally as a sexual stimulant (2).

LITERATURE

1   The data in the section 'General' may have been obtained from literature discussed in this chapter, but mostly
    also or exclusively from one or more of the following online sources: ChemIDPlus Advanced, PubChem,
    DrugBank, RxList, Drug Central, Drugs.com, and Wikipedia
2   Romaguera C, Grimalt F. Contact dermatitis from nasal sprays and amyl nitrite. Contact Dermatitis 1982;8:266-
    267

Chapter 3.108 DILTIAZEM

IDENTIFICATION

Description/definition	: Diltiazem is the benzothiazepine that conforms to the structural formula shown below
Pharmacological classes	: Antihypertensive agents; calcium channel blockers; cardiovascular agents; vasodilator agents
IUPAC name	: [(2S,3S)-5-[2-(Dimethylamino)ethyl]-2-(4-methoxyphenyl)-4-oxo-2,3-dihydro-1,5-benzothiazepin-3-yl] acetate
CAS registry number	: 42399-41-7
EC number	: 255-796-4
Merck Index monograph	: 4494
Patch testing	: Hydrochloride, 10% pet. (Chemotechnique); because patch tests are often very strongly positive, it has been suggested to lower the concentration of pure diltiazem to 1% or a maximum of 3% (18)
Molecular formula	: $C_{22}H_{26}N_2O_4S$

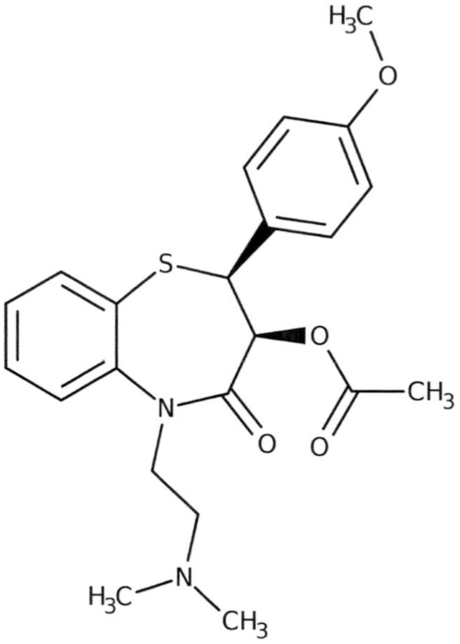

GENERAL

Diltiazem is a benzothiazepine calcium channel blocking agent and vasodilator, which is used in the management of angina pectoris and hypertension. Topical diltiazem creams are used to treat anal fissures and Raynaud's syndrome. They are antipruritic and relieve ischemia by relaxing the muscles (anal sphincter in the case of fissures) and dilating small blood vessels in the affected areas of skin and mucosa (9). In pharmaceutical products, diltiazem is employed as diltiazem hydrochloride (CAS number 33286-22-5, EC number 251-443-3, molecular formula $C_{22}H_{27}ClN_2O_4S$) (1).

CONTACT ALLERGY

General

A few cases of allergic contact dermatitis to diltiazem cream used for anal fissures have been described (2,3,7,8, 9,10). The resorption of diltiazem through the perianal skin and the mucosa has led to systemic contact dermatitis presenting as generalized (maculo-)papular rashes (2,7) and the baboon syndrome (3).

Case reports

47-year-old man developed a generalized papular rash starting from the anogenital area a few days after commencing topical treatment of an anal fissure with a pharmacy preparation containing 2% lidocaine and 2% diltiazem hydrochloride. The rash improved with topical glucocorticoid treatment and cessation of the diltiazem cream. Two weeks later, the patient restarted treatment of the anal fissure with the same cream, and the rash reappeared. An identical cream without diltiazem was well tolerated. Patch tests were positive to the cream 'as is', diltiazem 1% and 10% pet. and to other calcium channel blockers nifedipine and verapamil hydrochloride (both 1% and 10% pet.). One control was negative. This was a case of systemic contact dermatitis (2).

A 61-year-old woman suffered from a chronic anal fissure for which she applied both a cream with 2% diltiazem hydrochloride and an ointment with dibucaine 1%. The patient reported erythema of the perianal area in the first days, and 5 days later, she presented with a symmetric intertriginous and flexural exanthema. Patch tests were strongly positive (+++) to the cream and ointment 'as is', diltiazem 10% pet. and dibucaine 5% pet. The clinical presentation was that of the baboon syndrome, one of the main presentations of systemic contact dermatitis. The cause was most likely systemic absorption of diltiazem and dibucaine through the anal mucosa and perianal skin (3).

A 53-year-old man was troubled by pain and bleeding during defecation. Anal fissures were diagnosed and a gel containing diltiazem hydrochloride 2% was prescribed. After 1 day of treatment, the patient suffered local pain around the anus but he continued the treatment. Four days later, he developed a very pruritic maculopapular rash on most of the body, including the face, the neck, most of the trunk and arms, the flexures and dorsal parts of the hands, the groin and the buttocks, and both legs. Patch tests were positive to diltiazem gel (D2 +++, D4 +++, D7 +) and negative to its preservative methylparaben; the excipients sorbitol and hypromellose were not tested, as they were not considered possible allergens (7). This was also a case of systemic allergic contact dermatitis from percutaneous and mucosal absorption of – probably - diltiazem (which itself was not tested) in the (peri)anal area.

A 36-year-old woman underwent a limited hemorrhoidectomy for anal fissures. Postoperatively, to relieve painful muscular spasms, she applied topical diltiazem cream. Perianal erythema and itch occurred after she had used the product for 1 month, and improved following its avoidance. Patch tests were positive to the cream and to diltiazem 0.5%, 1% and 5% pet. Fifty controls were negative (8).

A 53-year-old woman had anal fissures, which were repeatedly treated with 2% diltiazem cream for 2 years. At one point she had developed localized skin irritation and eczema in the anogenital area. Over several months, the skin symptoms worsened and spread to the lower body. A ROAT with the cream was strongly positive but negative with plain cream without diltiazem. Patch tests were positive to diltiazem 0.1% and 0.01% (vehicle not mentioned), but not to 0.001%. Twenty-one controls were negative (9).

A 37-year-old man had anal fissures, which he treated with 2% diltiazem hydrochloride cream; however, this caused increasing symptoms of perianal itch. Patch tests were positive to the cream and - later - to the powder of diltiazem HCl capsules 50% pet. Thirty controls were negative. There were no common ingredients in the cream and the powder other than diltiazem (10).

Cross-reactions, pseudo-cross-reactions and co-reactions
A patient sensitized to diltiazem co-reacted to 2 other calcium channel blockers, nifedipine and verapamil HCl. Although these reactions were suggestive of cross-reactivity (the patients had never used them before), the chemical structures of the 3 different drugs do not show similarities that may easily explain these findings (2). Nevertheless, other authors have also found co-reactions to verapamil (4,5,6) and nifedipine (4) in patients with delayed-type hypersensitivity to diltiazem, but negative reactions also occur.

Cutaneous adverse drug reactions from systemic administration caused by type IV (delayed-type) hypersensitivity
Cutaneous adverse drug reactions from systemic administration of diltiazem caused by type IV (delayed-type) hypersensitivity, including erythema multiforme-like eruption (4,12), erythroderma (4,12), exfoliative dermatitis (12), maculopapular rash (5,15,18), acute generalized exanthematous pustulosis (AGEP) (6,14,17,18), photocontact dermatitis (13), and psoriasiform eruptions (16), are planned to be discussed in Volume IV of the *Monographs in Contact Allergy* series on Systemic drugs.

PHOTOSENSITIVITY
In the period 2004-2005, in a Spanish multicenter study performing photopatch testing, 2 relevant positive photopatch tests were observed to diltiazem. It was not mentioned how many patients had been tested with this chemical (11).

LITERATURE
1   The data in the section 'General' may have been obtained from literature discussed in this chapter, but mostly also or exclusively from one or more of the following online sources: ChemIDPlus Advanced, PubChem, DrugBank, RxList, Drug Central, Drugs.com, and Wikipedia

2   Forkel S, Baltzer AB, Geier J, Buhl T. Contact dermatitis caused by diltiazem cream and cross-reactivity with other calcium channel blockers. Contact Dermatitis 2018;79:244-246

3   Santiago L, Moura AL, Coutinho I, Gonçalo M. Systemic allergic contact dermatitis associated with topical diltiazem and/or cinchocaine. J Eur Acad Dermatol Venereol 2018;32:e284-e285

4   Gonzalo MA, Pérez R, Argila D, Rangel JF. Cutaneous reactions due to diltiazem and cross reactivity with other calcium channel blockers. Allergol Immunopathol (Madr) 2005;33:238-240

5   Cholez C, Trechot P, Schmutz JL, Faure G, Bene MC, Barbaud A. Maculopapular rash induced by diltiazem: allergological investigations in four patients and cross reactions between calcium channel blockers. Allergy 2003;58:1207-1209

6   Sáenz de Santa María García M, Noguerado-Mellado B, Perez-Ezquerra PR, Hernandez-Aragües I, De Barrio Fernández M. Acute generalized exanthematous pustulosis due to diltiazem: investigation of cross-reactivity with other calcium channel blockers. J Allergy Clin Immunol Pract 2016;4:765-766

7   Opstrup MS, Guldager S, Zachariae C, Thyssen JP. Systemic allergic dermatitis caused by diltiazem. Contact Dermatitis 2017;76:364-365

8   Wong TH, Horn HM. Allergic contact dermatitis caused by topical diltiazem. Contact Dermatitis 2010;63:228

9   Leinonen PT, Riekki R, Oikarinen A. Contact allergy to diltiazem cream. Contact Dermatitis 2010;63:228-230

10   Rose RF, Wilkinson SM. Contact sensitization to topical diltiazem. Contact Dermatitis 2009;60:347-348

11   De La Cuadra-Oyanguren J, Perez-Ferriols A, Lecha-Carrelero M, et al. Results and assessment of photopatch testing in Spain: towards a new standard set of photoallergens. Actas DermoSifiliograficas 2007;98:96-101

12   Sousa-Basto A, Azenha A, Duarte ML, Pardal-Oliveira F. Generalized cutaneous reaction to diltiazem. Contact Dermatitis 1993;29:44-45

13   Ramírez A, Pérez-Pérez L, Fernández-Redondo V, Toribio J. Photoallergic dermatitis induced by diltiazem. Contact Dermatitis 2007;56:118-119

14   Wakelin SH, James MP. Diltiazem-induced acute generalised exanthematous pustulosis. Clin Exp Dermatol 1995;20:341-344

15   Romano A, Pietrantonio F, Garcovich A, Rumi C, Bellocci F, Caradonna P, Barone C. Delayed hypersensitivity to diltiazem in two patients. Ann Allergy 1992;69:31-32

16   Kitamura K, Kanasashi M, Suga C, Saito S, Yoshida S, Ikezawa Z. Cutaneous reactions induced by calcium channel blocker: high frequency of psoriasiform eruptions. J Dermatol 1993;20:279-286

17   Vicente-Calleja JM, Aguirre A, Landa N, Crespo V, González-Pérez R, Diaz-Pérez JL. Acute generalized exanthematous pustulosis due to diltiazem: confirmation by patch testing. Br J Dermatol 1997;137:837-839

18   Assier H, Ingen-Housz-Oro S, Zehou O, Hirsch G, Chosidow O, Wolkenstein P. Strong reactions to diltiazem patch tests: Plea for a low concentration. Contact Dermatitis 2020;83:224-225

Chapter 3.109 DIMETHINDENE

IDENTIFICATION

Description/definition   : Dimethindene is the indene that conforms to the structural formula shown below
Pharmacological classes : Histamine H1 antagonists; antipruritics; anti-allergic agents
IUPAC name   : N,N-Dimethyl-2-[3-(1-pyridin-2-ylethyl)-1H-inden-2-yl]ethanamine
Other names   : Dimetindene; dimethpyrindene
CAS registry number   : 5636-83-9
EC number   : 227-083-8
Merck Index monograph   : 4508
Patch testing   : 5% pet.
Molecular formula   : $C_{20}H_{24}N_2$

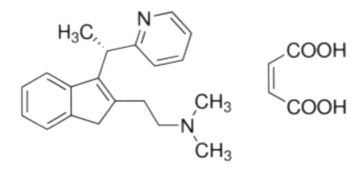

Dimethindene maleate

GENERAL

Dimethindene is an indene histamine H1 antagonist. It is indicated as symptomatic treatment of allergic reactions such as urticaria, allergies of the upper respiratory tract such as hay fever and perennial rhinitis, and food and drug allergies. It is also used for pruritus of various origins, e.g. from insect bites, varicella, eczema and other pruriginous dermatoses on account of its sedative properties. In pharmaceutical products, dimethindene is employed as dimethindene maleate (CAS number 3614-69-5, EC number 222-789-2 [Dimetindene hydrogen maleate], molecular formula $C_{24}H_{28}N_2O_4$) (1).

CONTACT ALLERGY

Case series

In Leuven, Belgium, in the period 1990-2014, iatrogenic contact dermatitis was diagnosed in 2600 individuals (17% of the total patch test population). 96% of all positive patch test reactions to topical drugs and antiseptics were considered to be relevant. Dimethindene (5% pet.) was tested in 83 patients and there was one positive reaction to it (2). In the same clinic, in the period 1978-1997, one patient with contact allergy to dimethindene was observed (3). As the time periods of the studies have an overlap (1990-1997), it cannot be excluded that the patients in the studies were the same one.

Case reports

A 12-year-old boy presented with an acute eczematous eruption on one leg, after having applied a gel containing dimethindene maleate on insect bites. In order to calm the pruritus, drops containing dimethindene maleate were taken orally, and 24 hr later the patient developed a diffuse, itchy, maculopapular and vesicular rash. Patch testing showed positive reactions to the gel and to the drops 'as is' and diluted 30%. Later, patch tests were conducted with the individual constituents of the gel and the drops, which yielded positive reactions to dimethindene maleate (contained in both preparations), to benzalkonium chloride (present in the gel) and to benzoic acid (present in the drops)®. Five normal controls were negative, test concentrations and vehicles were not mentioned (4). This was a case of systemic contact dermatitis.

A 55-year-old woman, with a long history of atopic dermatitis of the flexures, hands and legs, presented with a severe, edematous, vesiculobullous and intensely itchy eruption of the legs, with spread to the thighs. The rash had appeared two weeks after starting treatment of atopic dermatitis with a dimethindene maleate containing gel. Patch tests gave a +++ reaction to the topical medicament containing dimethindene maleate at 2 and 4 days (5 controls were negative). Subsequently, she was patch tested with the individual constituents of the gel (acrylic acid, methylparaben, propylene glycol), except for dimethindene maleate (which was not available), with negative results. Thus, the diagnosis of allergic contact dermatitis from dimethindene maleate was made *per exclusionem* (5).

LITERATURE

1   The data in the section 'General' may have been obtained from literature discussed in this chapter, but mostly also or exclusively from one or more of the following online sources: ChemIDPlus Advanced, PubChem, DrugBank, RxList, Drug Central, Drugs.com, and Wikipedia

2   Gilissen L, Goossens A. Frequency and trends of contact allergy to and iatrogenic contact dermatitis caused by topical drugs over a 25-year period. Contact Dermatitis 2016;75:290-302

3   Goossens A, Linsen G. Contact allergy to antihistamines is not common. Contact Dermatitis 1998;39:38-39

4   Leroy A, Baeck M, Tennstedt D. Contact dermatitis and secondary systemic allergy to dimethindene maleate. Contact Dermatitis 2011;64:170-171

5   Valsecchi R, di Landro A, Pansera B, Cainelli T. Contact dermatitis from a gel containing dimethindene maleate. Contact Dermatitis 1994;30:248-249

Chapter 3.110 DIOXOPROMETHAZINE

IDENTIFICATION

Description/definition   : Dioxopromethazine is the phenothiazine that conforms to the structural formula shown
                             below
Pharmacological classes   : Dermatological agents; anti-allergic agents; antipruritics
IUPAC name   : 1-(5,5-Dioxophenothiazin-10-yl)-N,N-dimethylpropan-2-amine
Other names   : 5,5-Dioxo-10-(2-(dimethylamino)propyl)phenothiazine
CAS registry number   : 13754-56-8
EC number   : Not available
Patch testing   : 0.1% and 0.5% pet.
Molecular formula   : $C_{17}H_{20}N_2O_2S$

GENERAL

Dioxopromethazine is a phenothiazine that was formerly used in Germany in a topical preparation for the treatment of dermatitis solaris and insect bites. Currently, it is not available anymore there, but it may still be used in China (www.drugs.com). In pharmaceutical products, dioxopromethazine is or was employed as dioxopromethazine hydrochloride (CAS number 15374-15-9, EC number not available, molecular formula $C_{17}H_{21}ClN_2O_2S$) (1).

PHOTOSENSITIVITY

Case report

A 63-year-old woman presented with severe dermatitis of the lower eyelids that spread to sun-exposed areas. She had used a gel containing 0.5% dioxopromethazine against periocular pruritus as a result of high grade allergy to cat dander. Patch and photopatch tests showed positive photoreactions to promethazine HCl 0.1% pet., the gel 'as is' and dioxopromethazine hydrochloride in concentrations between 0.005% and 0.5%. Twenty controls were negative. After the gel was discontinued, the photocontact dermatitis subsided. However, later she developed burning erythema and itching papules at the back of the hands when driving a car during sunny weather. Erythema and papules would develop at the neck after gardening. She was therefore diagnosed with persistent light reaction. The reaction to promethazine was considered to be photocross-reactivity to dioxopromethazine (2).

Cross-reactions, pseudo-cross-reactions and co-reactions

A patient photosensitized to dioxopromethazine photocross-reacted to the parent compound promethazine (2).

LITERATURE

1   The data in the section 'General' may have been obtained from literature discussed in this chapter, but mostly also or exclusively from one or more of the following online sources: ChemIDPlus Advanced, PubChem, DrugBank, RxList, Drug Central, Drugs.com, and Wikipedia
2   Schauder S. Dioxopromethazine-induced photoallergic contact dermatitis followed by persistent light reaction. Dermatitis 1998;9:182-187

Chapter 3.111 DIPERODON

IDENTIFICATION

Description/definition : Diperodon is the carbamate that conforms to the structural formula shown below
Pharmacological classes : Anesthetics, local
IUPAC name : [2-(Phenylcarbamoyloxy)-3-piperidin-1-ylpropyl] N-phenylcarbamate
Other names : 3-(1-Piperidyl)-1,2-propanediol dicarbanilate
CAS registry number : 101-08-6
EC number : 202-913-1
Merck Index monograph : 148
Patch testing : 1% pet.
Molecular formula : $C_{22}H_{27}N_3O_4$

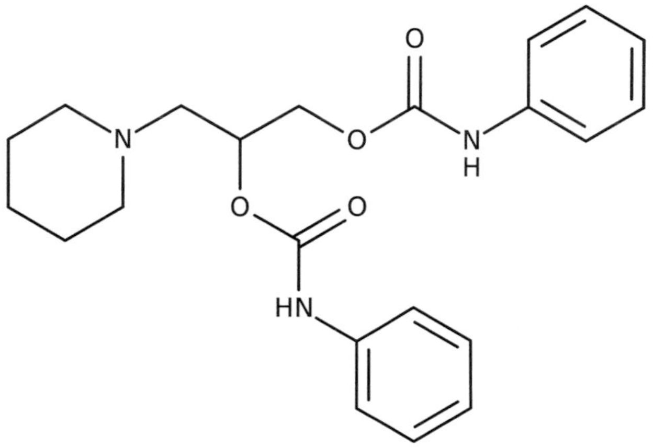

GENERAL

Diperodon is a carbamate that was formerly used as local anesthetic. Very few data can be found on this compound. In the publication discussed below, diperodon was used as diperodon hydrochloride (CAS number 537-12-2, EC number 208-659-8, molecular formula $C_{22}H_{28}ClN_3O_4$). According to www.drugs.com, diperodon may still be available in bacitracin/diperodon/neomycin/polymyxin B topical preparations (1).

CONTACT ALLERGY

Case report

A man aged 30 who had psoriasis for many years, developed pruritus ani. He had used an antihemorrhoidal ointment containing the local anesthetic diperodon. Patch tests were positive to the ointment and to diperodon HCl 1% and negative to the other ingredients (2).

LITERATURE

1 The data in the section 'General' may have been obtained from literature discussed in this chapter, but mostly also or exclusively from one or more of the following online sources: ChemIDPlus Advanced, PubChem, DrugBank, RxList, Drug Central, Drugs.com, and Wikipedia
2 Calnan CD. Allergy to the local anaesthetic diperodon. Contact Dermatitis 1980;6:367

Chapter 3.112 DIPHENHYDRAMINE

IDENTIFICATION

Description/definition   : Diphenhydramine is the diphenylmethane that conforms to the structural formula shown
                              below
Pharmacological classes   : Anesthetics, local; hypnotics and sedatives; anti-allergic agent; antiemetics; histamine H1
                              antagonists; sleep aids, pharmaceutical
IUPAC name   : 2-Benzhydryloxy-N,N-dimethylethanamine
Other names   : 2-Diphenylmethoxy-N,N-dimethylethylamine
CAS registry number   : 58-73-1
EC number   : 200-396-7
Merck Index monograph   : 4609
Patch testing   : Hydrochloride, 1.0% pet. (Chemotechnique)
Molecular formula   : $C_{17}H_{21}NO$

GENERAL

Diphenhydramine is an ethanolamine first-generation histamine H1 receptor antagonist with anti-allergic, antiemetic, antitussive, antimuscarinic and sedative effects. As an over-the-counter (OTC) medication, diphenhydramine is typically formulated as tablets and creams indicated for use in treating sneezing, runny nose, itchy/watery eyes, itching of nose or throat, insomnia, pruritus, urticaria, insect bites/stings, allergic rashes, and nausea. It may apparently also be used as an antiparkinsonian agent. In pharmaceutical products, both diphenhydramine base and – far more often – diphenhydramine hydrochloride (CAS number 147-24-0, EC number 205-687-2, molecular formula $C_{17}H_{22}ClNO$) may be employed, and possibly - rarely - diphenhydramine citrate or methylbromide (1).

CONTACT ALLERGY

General

Currently, contact allergy to diphenhydramine (better known under one of its trade names, Benadryl) is rarely reported. Nevertheless, it was stated to be a common cause of allergic contact dermatitis in the USA in the 1980s (15), although another American author, in 1983, presenting 3 patients sensitized to diphenhydramine, could find only a few other case reports (24). However, 2 series of 19 (26) and 12 (27) cases of diphenhydramine hydrochloride dermatitis had previously been reported from the United Kingdom (data cited in ref. 24). Diphenhydramine has also caused some cases of photocontact dermatitis (see the section 'Photosensitivity' below) and systemic contact dermatitis (12,25).

Patch testing in groups of selected patients

In Leuven, Belgium, in the period 1990-2014, iatrogenic contact dermatitis was diagnosed in 2600 individuals (17% of the total patch test population). 96% of all positive patch test reactions to topical drugs and antiseptics were considered to be relevant. Diphenhydramine (1-5% pet.) was tested in 177 patients and there were 9 (5.1%) positive reactions to it (2, overlap with the data in ref. 4).

In the period 1978-1997, in the same University Clinic in Leuven, Belgium, 8 patients had positive patch test reactions to diphenhydramine (patch test concentration and vehicle not mentioned). The relevance of these reactions was not mentioned (4, overlap with the data in ref. 2).

In the period 1971-1976 inclusive, at St. John's Hospital, London, 7 patients were diagnosed with contact allergy to diphenhydramine and another 6 with probable sensitization. The diphenhydramine containing medicaments had been used to treat a variety of pruritic dermatoses. All 13 patients had positive patch tests to their proprietary medications containing 1-2% diphenhydramine HCl; 7 of these were patch tested with the antihistamine 10% in water or 5% pet. and each reacted positively (23).

Case series

Three cases of contact allergy to diphenhydramine HCl were reported from the USA in 1983. A 10-year-old child had developed an eruption from diphenhydramine HCl lotion and cream used on an insect bite. Patch tests were positive to these pharmaceuticals at D2. A 60-year-old man had a weeping dermatitis involving the hands for 1 week, which had been treated with 5 (!) topical drugs, 3 of which contained diphenhydramine HCL. These 3 were positive on patch testing at D2. The third patient was a 54-year-old man who presented with generalized dermatitis on the trunk and extremities. According to his history, he developed localized dermatitis 2 days after being exposed to a large vine. He was treated with injections in an emergency department, a capsule containing diphenhydramine HCl (Benadryl) to take orally and he applied a lotion containing the antihistamine topically. This resulted in a temperature of 38° C and worsening of the eruption. Patch tests were performed with the diphenhydramine HCl and injectable diphenhydramine HCl. Both patch tests were read as positive 3 days later. This was a case of allergic contact dermatitis and systemic contact dermatitis. Neither the active ingredient nor the excipients were patch tested and the reactions were read at D2 only in the first 2 patients (24).

Two series of 19 (26) and 12 (27) cases of diphenhydramine hydrochloride dermatitis have been reported from the United Kingdom in the 1950s (data cited in ref. 24).

Case reports

A 45-year-old woman had a 15-day history of an acute, itchy, vesicular and erythematous eruption around the mouth. For rosacea she had been treated with a cream containing chlorophyll, kamillosan, erythromycin, metronidazole and diphenhydramine, the concentrations of which were not recorded. Patch testing showed positive reactions to the cream at D2 (++) and D4 (+++) and – later – to its ingredients diphenhydramine 1% pet. and metronidazole 5% and 10% pet. Photopatch tests gave a reaction to diphenhydramine equal to the patch test. Twelve controls were negative (3).

A 58-year-old woman presented with a severe rash progressing since 4 days. On clinical examination, there was a fierce facial edema with exudation and erythemato-vesicular plaques with serous and sero-hemorrhagic confluent vesicles, distributed along the posterior neck, upper extremities, inner thighs and knees, with a linear pattern in some areas. The hands were not affected. Pruritus was intense. In the night before this outbreak she had applied on her husband's back a cream containing 2% diphenhydramine HCl. Patch tests were positive to the cream, promethazine HCl cream (unknown relevance) and diphenhydramine 1% pet. Six weeks later, a patch test with diphenhydramine was again positive. As the patient had no direct contact with the cream, except in the palmar skin which was not affected, and some of the localizations of her dermatitis were not very easily accessible by her own hands, the authors proposed that this was due to the transfer of the allergen from her husband's skin, and therefore a case of connubial contact dermatitis (5).

A 59-year old woman developed allergic contact dermatitis from diphenhydramine (1% pet.) in a cream used on her upper arm for the treatment of insect bites (6).

According to the author of ref. 6, contact allergic reactions to diphenhydramine have been reported in older German literature (7-11). Details are unknown, but is it likely that refs. 7,8 and 9 refer to contact allergy to topical preparations, whereas ref. 10 and 11 may deal with occupational allergic contact dermatitis, e.g. in health care providers (10).

PHOTOSENSITIVITY

Photopatch testing in groups of patients

In a multicenter study in Italy, performed in the period 1985-1994, one photopatch test reactions was seen to diphenhydramine; the patch test concentration used was not mentioned and relevance was not specified (78% for all photoallergens together (19). The result of photopatch testing diphenhydramine in groups of selected patients (patients suspected of photosensitivity/photodermatitis/sun-related skin diseases) are shown in table 3.112.1. The prevalences of sensitization were low with only a few patients reacting. There were 5 patients with photocontact allergy in Greece, but these were observed in a period of 15 years (21).

Table 3.112.1 Photopatch testing in groups of patients

Years and Country	Test conc. & vehicle	Number of patients		Selection of patients (S); Relevance (R); Comments (C)	Ref.
		tested	positive (%)		
1993-2006 USA	1% pet.	76	1 (1%)	S: not stated; R: 21% of all reactions to medications were considered 'of possible relevance'	22
1992-2006 Greece	1% pet.	207	5 (2.5%)	S: patients suspected of photosensitivity; R: not stated	21
2000-2005 USA	1% pet.	178	2 (1.1%)	S: patients photopatch tested for suspected photodermatitis; R: 50%	18
1980-1981 4 Scandinavian countries	1% pet.	745	1 (0.1%)	S: patients suspected of sun-related skin disease; R: not specified	20

Case reports

A 56-year-old woman had an itchy rash on the face and backs of the hands for several months. On examination, she had diffuse erythema with papules and scaling, exclusively on sun-exposed areas. She had been applying 1% diphenhydramine ointment for 6 months. She stopped using the ointment, but her condition persisted. Photopatch testing showed a positive reaction to diphenhydramine HCl 1% pet., whereas a patch test was negative. The patient stopped the use of the ointment and avoided UV exposure as much as possible. Over the next 2 years, her skin lesions gradually subsided with application of a corticosteroid ointment (13).

A 52-year-old man had a pruritic eruption of five years' duration that was confined to the light-exposed areas. The lesions appeared initially on the scalp following several applications of hair dye. Despite the use of a number of prescribed therapeutic modalities, the eruption spread to the forehead, cheeks, nape and V of the neck, extensor aspects of the forearms, and dorsa of the hands. He had been receiving no medication other than those prescribed for the skin eruption. Diphenhydramine had been given orally, topically, and by injection. Patch and photopatch tests (using UVA) were positive to p-phenylenediamine (hair dye) and to diphenhydramine 1% and 5% pet. The reactions to the diphenhydramine photopatch tests were slightly stronger than those of the patch tests. Five controls were negative (14). Two years earlier, another case of photosensitivity induced by topical application of diphenhydramine, elicited by ultraviolet light in the 290-320 nm range (UVB), had been described (16).

Cross-reactions, pseudo-cross-reactions and co-reactions

In 2 patients sensitized to diphenhydramine and tested with chemically-related derivatives, 2 positive reactions were observed to bromazine (bromodiphenhydramine), 2 to medrylamine (4-methoxydiphenhydramine) and 1 to dimenhydrinate (diphenhydramine + 8-chlorotheophylline) and p-methyldiphenhydramine, respectively (4).

Cutaneous adverse drug reactions from systemic administration caused by type IV (delayed-type) hypersensitivity

A 51-year-old woman had developed a widespread vesiculobullous eruption affecting the feet, hands, trunk, thighs and face, 2 days after starting a cough medicine containing diphenhydramine. Twice previously she had developed a similar eruption. On both occasions she had used another expectorant containing diphenhydramine. The patient had never used topical antihistamines. Patch tests yielded a ++ reaction to diphenhydramine 2% pet. (12). Probably, the patient had previously become sensitized to diphenhydramine on the oral mucosa and now had systemic contact dermatitis (12). Another patient also developed systemic contact dermatitis from oral diphenhydramine following sensitization from a topical diphenhydramine product (25, data cited in ref. 24).

Photoallergy to oral diphenhydramine in 2 patients was reported in 1962 (17).

LITERATURE

1   The data in the section 'General' may have been obtained from literature discussed in this chapter, but mostly also or exclusively from one or more of the following online sources: ChemIDPlus Advanced, PubChem, DrugBank, RxList, Drug Central, Drugs.com, and Wikipedia

2   Gilissen L, Goossens A. Frequency and trends of contact allergy to and iatrogenic contact dermatitis caused by topical drugs over a 25-year period. Contact Dermatitis 2016;75:290-302

3   Fernández-Jorge B, Goday Buján J, Fernández-Torres R, Rodríguez-Lojo R, Fonseca E. Concomitant allergic contact dermatitis from diphenhydramine and metronidazole. Contact Dermatitis 2008;59:115-116

4   Goossens A, Linsen G. Contact allergy to antihistamines is not common. Contact Dermatitis 1998;39:38

5   Teixeira V, Cabral R, Gonçalo M. Exuberant connubial allergic contact dermatitis from diphenhydramine. Cut Ocular Toxicol 2013;33:82-84

6   Heine A. Diphenhydramine: a forgotten allergen? Contact Dermatitis 1996;35:311-312

7   Fleck M. Zur Sensibilisierungsneigung lokal angewendeter Antihistaminika. Dtsch Ges Wes 1959;14:1384-1385 (Article in German)

8   Richter G. Zur Kritik der externen Dermatika unter dem Aspekt ihrer allergischen Nebenwirkungen. Dermatol Monatsschr 1975;161:384-387 (Article in German)

9   Sönnichsen N, Stäps R. Zur Problematik allergischer Reaktionen durch Antihistaminika. Z Haut-Geschlskrht 1960;29:283-288 (Article in German)

10   Gertler H, Laubstein H. Berufsdermatosen bei Angehörigen der medizinischen Berufe. Zschr Ärztl Fortbild 1969;59:251-255 (Article in German)

11   Laubstein H, Mönnich HT. Zur Epidemiologie der Berufsdermatosen. Dermatol Wochenschr 1968;154:649-667 (Article in German)

12   Lawrence CM, Byrne JP. Eczematous eruption from oral diphenhydramine. Contact Dermatitis 1981;7:276-277

13   Yamada S, Tanaka M, Kawahara Y, Inada M, Ohata Y. Photoallergic contact dermatitis due to diphenhydramine hydrochloride. Contact Dermatitis 1998;38:282

14   Horio T. Allergic and photoallergic dermatitis from diphenhydramine. Arch Dermatol 1976;112:1124- 1126

15   Bigby M, Stern RS, Arndt KA. Allergic cutaneous reactions to drugs. Prim Care 1989;16:713-727

16   Emmett EA. Diphenhydramine photoallergy. Arch Dermatol 1974;110:249-252

17   Schreiber MM, Naylor LZ. Antihistamine photosensitivity. Arch Dermatol 1962;86:58-62

18   Scalf LA, Davis MDP, Rohlinger AL, Connolly SM. Photopatch testing of 182 patients: A 6-year experience at the Mayo Clinic. Dermatitis 2009;20:44-52

19   Pigatto PD, Legori A, Bigardi AS, Guarrera M, Tosti A, Santucci B, et al. Gruppo Italiano recerca dermatiti da contatto ed ambientali Italian multicenter study of allergic contact photodermatitis: epidemiological aspects. Am J Contact Dermatitis 1996;17:158-163

20   Wennersten G, Thune P, Brodthagen H, Jansen C, Rystedt I, Crames M, et al. The Scandinavian multicenter photopatch study. Contact Dermatitis 1984;10:305-309

21   Katsarou A, Makris M, Zarafonitis G, Lagogianni E, Gregoriou S, Kalogeromitros D. Photoallergic contact dermatitis: the 15-year experience of a tertiary referral center in a sunny Mediterranean city. Int J Immunopathol Pharmacol 2008;21:725-727

22   Victor FC, Cohen DE, Soter NA. A 20-year analysis of previous and emerging allergens that elicit photoallergic contact dermatitis. J Am Acad Dermatol 2010;62:605-610

23   Cronin E. Contact Dermatitis. Edinburgh: Churchill Livingstone, 1980:236

24   Coskey RJ. Contact dermatitis caused by diphenhydramine hydrochloride. J Am Acad Dermatol 1983;8:204-206

25   Shelley WB, Bennett RG. Primary contact sensitization site: A determination for localization of a diphenhydramine eruption. Acta Derm Venereol (Stockh) 1972;52:376-378

26   Calnan CD. Contact dermatitis from drugs. Proc R Soc Med 1962;44:39-42 (data cited in ref. 24)

27   Vickers CFH. Dermatitis medicamentosa. Br Med J 1961;1:1366-1367 (data cited in ref. 24)

Chapter 3.113 DIPIVEFRIN

IDENTIFICATION

Description/definition   : Dipivefrin is the dipivalate ester of (+-)-epinephrine (racepinephrine); it conforms to the structural formula shown below

Pharmacological classes : Adrenergic agonists

IUPAC name   : [2-(2,2-Dimethylpropanoyloxy)-4-[1-hydroxy-2-(methylamino)ethyl]phenyl] 2,2-dimethylpropanoate

Other names   : Dipivalyl epinephrine; 3,4-dihydroxy-α-((methylamino)methyl)benzyl alcohol 3,4-dipivalate

CAS registry number   : 52365-63-6

EC number   : Not available

Merck Index monograph : 4648

Patch testing   : 1% and 5% water

Molecular formula   : $C_{19}H_{29}NO_5$

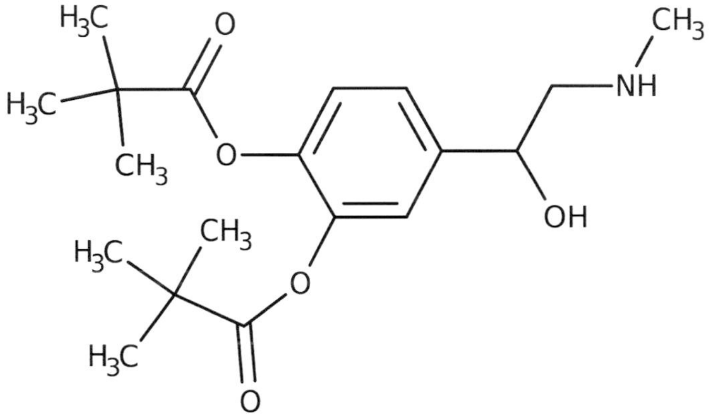

GENERAL

Dipivefrin is the dipivalate ester of epinephrine (adrenaline) and a prodrug of epinephrine, used topically as eye drops to reduce intra-ocular pressure in the treatment of open-angle glaucoma or ocular hypertension. Due to its lipophilicity, dipivefrin penetrates more easily to the anterior chamber of the eye and is hydrolyzed to the pharmacologically active form epinephrine there. Epinephrine, an adrenergic agonist, enhances the outflow of aqueous humor and decreases the production of aqueous humor by vasoconstriction. The overall result is a reduction in intraocular pressure. In pharmaceutical products, dipivefrin is employed as dipivefrin hydrochloride (CAS number 64019-93-8, EC number 264-609-5, molecular formula $C_{19}H_{30}ClNO_5$) (1).

CONTACT ALLERGY

Case series

In Leuven, Belgium, between 1990 and 2017, 16,065 patients were investigated for contact allergy and 118 (0.7%) showed positive patch test reactions to topical ophthalmic medications and/or to their ingredients. Eighty-four individuals (71%) reacted to an active principle. Dipivefrin was tested in 4 patients and was the allergen in eye medications in one. There were no reactions to dipivefrin in other types of medications (2, overlap with ref. 3).

In Leuven, Belgium, in the period 1990-2014, iatrogenic contact dermatitis was diagnosed in 2600 individuals (17% of the total patch test population). 96% of all positive patch test reactions to topical drugs and antiseptics were considered to be relevant. Dipivefrin (1% water) was tested in 4 patients and there was one positive reaction to it (3, overlap with ref. 2).

Five men, ranging in age from 35 to 76 years, with signs of ocular hypersensitivity reactions to dipivefrin eye drops were examined. The adverse reactions included follicular conjunctivitis in each patient and eyelid pruritus, erythema, and/or scaling in four of the subjects. Patch tests were performed with the 5 components of dipivefrin ophthalmic solution 0.1%, including the drug at 0.5% water, but no positive reactions were observed. Next, challenge tests were performed with sterile solutions each containing one of the five constituents at the same concentration as in the commercial dipivefrin eye drops. Ocular reactions were noted only to dipivefrin in each of the subjects after ocular challenge. Adverse reactions began 3 to 96 hours after ocular challenge with dipivefrin (mean, 44 hours). All patients developed conjunctival injection; follicular conjunctivitis was noted in two patients. Four of the five patients also developed significant eyelid symptoms, including pruritus and erythema. The reactions resolved within 24 to 72 hours after stopping the dipivefrin-containing eye drops (7).

Comments: It is uncertain whether these patients had irritant reactions (follicular conjunctival changes are considered to be a hallmark of the 'long-term drug irritation syndrome') or allergic reactions to dipivefrin; the patch test concentration of 0.5% may have been too low to detect sensitization.

Case reports

A 62-year-old man presented with a 3-year history of glaucoma, for which he was treated with several topical ophthalmic solutions, most recently dipivefrin eye drops. Four months after initiating the latter therapy, the patient presented with bilateral pruritus, conjunctivitis and periocular eczema. Patch tests were positive to the eye drops and – in a second session – to dipivefrin hydrochloride 1% water. Epinephrine 1% water was negative (4).

A 68-year-old man with glaucoma had been treated for 3 years with timolol and dipivefrin eye drops in his right eye and timolol in the left. He presented with a 12-month history of eczema around the right eye, which had temporarily responded to topical corticosteroids. On examination, he showed a right-sided hyperemic bulbar conjunctiva and edematous and hyperpigmented periocular skin, with fine scaling. His left eye was not involved. After discontinuing dipivefrin eye drops and treatment with topical corticosteroids, all lesions gradually cleared. Patch tests showed a weakly positive reaction to the dipivefrin eye drops and – in a second session – strong reactions to dipivefrin 1% and 5% water, but only a weak reaction to 5% pet. Five controls were negative to all tests. Epinephrine also reacted strongly positive at 1% water (5).

A 65-year-old man presented with a 3-month history of redness of both eyes, increased tear production and pruritus. He was noted to have periocular eczema as well as moderate bulbar conjunctivitis. The patient had a 3-year history of glaucoma, for which he was being treated with timolol maleate eye drops, pilocarpine ophthalmic solution 1% and dipivefrin eye drops 0.1%. Patch tests were positive to the commercial dipivefrin product and – later – to dipivefrin hydrochloride 0.5% water. Within one week of discontinuing the dipivefrin eye drops, the eyes and skin improved dramatically (6).

Cross-reactions, pseudo-cross-reactions and co-reactions

No cross-reactivity to epinephrine (4). Possible cross-reaction to epinephrine in a patient sensitized to dipivefrin (5).

LITERATURE

1 The data in the section 'General' may have been obtained from literature discussed in this chapter, but mostly also or exclusively from one or more of the following online sources: ChemIDPlus Advanced, PubChem, DrugBank, RxList, Drug Central, Drugs.com, and Wikipedia

2 Gilissen L, De Decker L, Hulshagen T, Goossens A. Allergic contact dermatitis caused by topical ophthalmic medications: Keep an eye on it! Contact Dermatitis 2019;80:291-297

3 Gilissen L, Goossens A. Frequency and trends of contact allergy to and iatrogenic contact dermatitis caused by topical drugs over a 25-year period. Contact Dermatitis 2016;75:290-302

4 Vilaplana J, Zaballos P, Romaguera C. Contact dermatitis by dipivefrine. Contact Dermatitis 2005;52:169-170

5 Parra A, Casas L, Tombo MJ, Yanez M, Vila L. Contact dermatitis from dipivefrine with possible cross-reaction to epinephrine. Contact Dermatitis 1998;39:325-326

6 Gaspari AA. Contact allergy to ophthalmic dipivalyl epinephrine hydrochloride: demonstration by patch testing. Contact Dermatitis 1993;28:35-37

7 Petersen PE, Evans RB, Johnstone MA, Henderson WR Jr. Evaluation of ocular hypersensitivity to dipivalyl epinephrine by component eye-drop testing. J Allergy Clin Immunol 1990;85:954-958

Chapter 3.114 DITHRANOL

IDENTIFICATION

Description/definition   : Dithranol is the natural anthraquinone derivative that conforms to the structural formula shown below
Pharmacological classes : Dermatological agents
IUPAC name   : 1,8-Dihydroxy-10*H*-anthracen-9-one
Other names   : Anthralin; 1,8-dihydroxy-9-anthrone
CAS registry number   : 1143-38-0
EC number   : 214-538-0
Merck Index monograph   : 1946
Patch testing   : 0.001% pet.; test in higher concentration if negative and contact allergy is strongly suspected, but at 0.005% pet., 1/3 of patients may show irritant reactions (9)
Molecular formula   : $C_{14}H_{10}O_3$

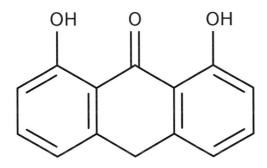

GENERAL

Dithranol is a natural anthraquinone derivative with anti-psoriatic and anti-inflammatory properties. It controls skin growth by reducing DNA synthesis and mitotic activity in the hyperplastic epidermis, restoring a normal rate of cell proliferation and keratinization. Dithranol is indicated for treatment of stable plaque psoriasis of the skin and scalp. Salicylic acid is frequently added to dithranol preparations to augment the stability of the drug and to increase its penetration and efficacy. Dithranol has also been studied in the treatment of warts, showing promising results (1).

CONTACT ALLERGY

General

Related to its widespread use, contact allergy to dithranol appears to be very infrequent. The irritant properties of dithranol are well known, and it is likely that a number of cases considered to be allergic on the basis of a positive patch test reaction, in fact had shown a – wrongly interpreted - false-positive, irritant patch test reaction, especially with test concentrations of 0.005% or higher. Conversely, as all physicians are well aware of irritation and irritant dermatitis from dithranol, cases of real allergic contact dermatitis may be interpreted as irritant; consequently, no patch tests are performed and sensitization goes unnoticed.

Patch testing in groups of selected patient

In a 4-month-period in 1993, in the United Kingdom, 74 consecutive patients with psoriasis admitted to the dermatology ward were patch tested to dithranol 0.005% pet. A positive reaction was followed by a dilution series down to 0.0005%. Thirteen patient were intolerant to dithranol of who only one was contact allergic to the material (8). In Rotterdam, The Netherlands, 47 patients hospitalized for psoriasis were patch tested with an extended baseline series and a topical treatment series. There was one reactions to dithranol 0.1% or 0.2% pet. Relevance was not addressed (2).

   In India, between 1999 and 2001, 200 patients with psoriasis were tested with dithranol 0.01% pet. and there were 13 (6.5%) positive reactions. Only four of these had a positive history of exposure to dithranol, with three developing intolerance to dithranol in the past. Twelve had a positive ROAT to a cream (diluted 1:4) containing 1.15% dithranol, 5.3% decolorized coal tar and 1.15% salicylic acid. This makes the authors suggest that the patients were indeed allergic to dithranol, although they could not exclude irritant reactions (17). This author thinks it's likely that a

– considerable - number of these patients had false-positive, irritant reactions to dithranol 0.01% pet. and certainly – given its composition - to the ROAT (17).

Case series

In Groningen, The Netherlands, nineteen patients (13 women, 6 men, mean age 27 years, range 7-55 years) with persistent alopecia areata were treated concomitantly with DNCB after sensitization and with dithranol 1% in acetone and 2% in chloroform-benzene (as an irritant control) once a week to separate areas of the scalp. After 8 to 10 successive treatments, four patients developed a pruritic vesicular spreading dermatitis at the dithranol treatment sites, suggestive of an allergic reaction. All 19 patients were patch tested with dithranol 0.05% and the solvents and 7 (including the 4 with signs of ACD) had positive patch tests to dithranol: 3x +++, 3x ++, 1x +. There were no reactions to chemically related compounds. After 4 months the patch tests were repeated with essentially the same results. Previously, the authors had tested hundreds of patients, including individuals with psoriasis, with dithranol 0.05% without ever having noticed any clearly positive skin test reaction (6). Another strong indication that the reactions to dithranol were allergic rather than irritant was that 4 of the 7 patients who reacted to dithranol had hair growth not only at the DNCB-treated sites but also on the dithranol sites, which did not occur in any of the 12 patients who only had irritant reactions to dithranol (7). It was concluded that DNCB therapy may increase the susceptibility of the treated individual to contact allergens (6).

Case reports

A 60-year-old man with psoriasis developed erythema with vesiculation and exudation on resistant psoriasis plaques treated with dithranol 0.01% cream. There was a past history of dithranol intolerance. Patch tests were positive to dithranol 0.01% (++), 0.005% (+) and 0.0025% pet. (+). Five controls were negative (4).

A 39-year-old woman with psoriasis was treated with short contact treatment with dithranol cream 0.1% for 15 minutes, when she developed a generalized severe exanthematous reaction after 2 days of widespread application. A biopsy showed subacute dermatitis. She recovered in a few days and after 2 weeks, the majority of her psoriasis lesions had disappeared. Patch tests were negative to the constituents of the cream base. Patch testing with a dilution series of dithranol cream ranging from 0.00025% to 0.01% gave + reactions at D2 and D3 with the 0.005 and 0.01 concentrations and ?+ reactions to all lower concentrations. Thereafter, she was treated with low but increasing concentrations of dithranol and started to develop irritation at 0.03% with an application time of 15 minutes (5).

Short summaries of other case reports of allergic contact dermatitis from dithranol are shown in table 3.114.1.

Table 3.114.1 Short summaries of case reports of allergic contact dermatitis from dithranol

Year and country	Sex	Age	Positive patch tests	Clinical data and comments	Ref.
1998 USA	?	?	dithranol 0.02%	no details provided, but after avoiding products with dithranol the psoriasis improved	3
1992 U.K.	M	53	dithranol 0.05-0.0005% pet.	burning from dithranol 0.1% in Lassar's paste; also contact allergy to coal tar 3% pet.	9
1992 Italy	M	29	dithranol 0.02% and 0.05% pet.	erythema and vesicles with pruritus after dithranol; also contact allergy to coal tar and wood tar; one positive patch test reaction turned into psoriasis (Köbner phenomenon)	13
1990 Italy	M	58	dithranol 0.05-0.005% pet.	intense pruritus after dithranol followed by erythematous and later squamous pigmented lesions following healing of psoriasis lesions	14
1982 U.K.	F	?	dithranol 0.02% in aceto-	acute eczematous lesions after dithranol	15
	F	?	ne and pet.	irritation from dithranol	
1981 Netherlands	F	31	dithranol 0.05% pet.	papuloversicular eczematous lesions on both arms, legs and abdomen after treatment of many warts on her hands with a collodion containing 3% dithranol; oozing vesicular dermatitis of the hands; recurrence of hand dermatitis during patch testing; primary source of sensitization unknown, possibly cross-reaction to unknown chemical; complete cure of all warts after 3 weeks	16

Two more cases of contact allergy to dithranol have been reported in the German literature. The first was of a psoriatic patient intolerant of dithranol who reacted on patch testing to dithranol 0.0005% (vehicle not stated) (10). The second also involved a patient with psoriasis who had an adverse reaction to dithranol. Patch tests revealed a marked erythema and bullous reaction to dithranol 0.02% pet. and 0.1% acetone (11). More details are not available to the author (data cited in ref. 9).

Cross-reactions, pseudo-cross-reactions and co-reactions.
Co-reactivity with coal tar has been observed several times (6,9,12,13). Possible explanations are independent sensitization in patients having used both coal tar and dithranol, a shared allergen, or the presence of minute quantities of dithranol in coal tar (9).

LITERATURE

1   The data in the section 'General' may have been obtained from literature discussed in this chapter, but mostly also or exclusively from one or more of the following online sources: ChemIDPlus Advanced, PubChem, DrugBank, RxList, Drug Central, Drugs.com, and Wikipedia
2   Heule F, Tahapary GJ, Bello CR, van Joost T. Delayed-type hypersensitivity to contact allergens in psoriasis. A clinical evaluation. Contact Dermatitis 1998;38:78-82
3   Clark AR, Sherertz EF. The incidence of allergic contact dermatitis in patients with psoriasis vulgaris. Dermatitis 1998;9:96-99
4   Chadha V, Shenoi SD. Allergic contact dermatitis from dithranol. Contact Dermatitis 1999;41:166
5   Prins M, Swinkels OQ, Mommers JM, Gerritsen MJ, van der Valk PG. Dithranol treatment of psoriasis in dithranol-sensitive patients. Contact Dermatitis 1999;41:116-117
6   De Groot AC, Nater JP, Bleumink E, de Jong MC. Does DNCB therapy potentiate epicutaneous sensitization to non-related contact allergens? Clin Exp Dermatol 1981;6:139-144
7   De Groot AC, Nater JP, De Jong MCJM. De behandeling van alopecia areata met dinitrochloorbenzeen. Nederlands Tijdschr Geneeskd 1980;124:1634-1640 (Article in Dutch)
8   Burden AD, Muston H, Beck MH. Intolerance and contact allergy to tar and dithranol in psoriasis. Contact Dermatitis 1994;31:185-186
9   Burden AD, Stapleton M, Beck MH. Dithranol allergy: fact or fiction? Contact Dermatitis 1992;27:291-293
10 Schauder S. Allergisches Kontaktekzem auf Cignolin (Dithranol). Z Hautkr 1984;59:773-774 (Article in German)
11 Bratzke B, Albrecht G, Orfanos CE. Hautreaktion auf Dithranol und ihre Beeinflussung durch Teerzusatz (LCD). Hautarzt 1987;38:356-360 (Article in German)
12 Goncalo S, Sousa I, Moreno A. Contact dermatitis to coal tar. Contact Dermatitis 1984;10:57-58
13 Di Landro A, Valsecchi R, Cainelli T. Contact allergy to dithranol. Contact Dermatitis 1992;26:49-50
14 Romaguera C, Grimalt F, Vilaplana J. Acute contact dermatitis by dithranol. Am J Cont Dermat 1990;1:186-188
15 Lawlor F, Hindson C. Allergy to dithranol. Contact Dermatitis 1982;8:137-138
16 De Groot AC, Nater JP. Contact allergy to dithranol. Contact Dermatitis 1981;7:5-8.
17 Malhotra V, Kaur I, Saraswat A, Kumar B. Frequency of patch-test positivity in patients with psoriasis: a prospective controlled study. Acta Derm Venereol 2002;82:432-435

Chapter 3.115 DORZOLAMIDE

IDENTIFICATION

Description/definition : Dorzolamide is the sulfonamide derivative that conforms to the structural formula shown below

Pharmacological classes : Antihypertensive agents; carbonic anhydrase inhibitors

IUPAC name : (4S,6S)-4-(Ethylamino)-6-methyl-7,7-dioxo-5,6-dihydro-4H-thieno[2,3-b]thiopyran-2-sulfonamide

Other names : 4H-Thieno(2,3-b)thiopyran-2-sulfonamide, 4-(ethylamino)-5,6-dihydro-6-methyl-, 7,7-dioxide, (4S,6S)-

CAS registry number : 120279-96-1

EC number : Not available

Merck Index monograph : 4745

Patch testing : 10% water or pet.; it may be necessary to test on tape-stripped or scarified skin or perform a ROAT

Molecular formula : $C_{10}H_{16}N_2O_4S_3$

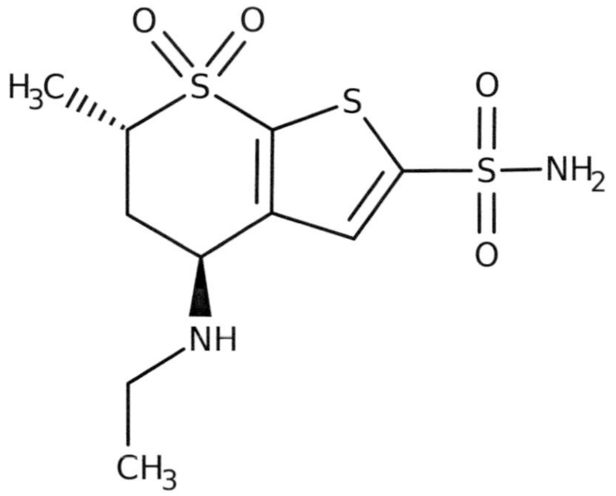

GENERAL

Dorzolamide is a topical sulfonamide carbonic anhydrase inhibitor that is indicated for the reduction of elevated intraocular pressure in patients with open-angle glaucoma or ocular hypertension who are insufficiently responsive to β-blockers. In pharmaceutical products, dorzolamide is employed as dorzolamide hydrochloride (CAS number 130693-82-2, EC number not available, molecular formula $C_{10}H_{17}ClN_2O_4S_3$) (1).

CONTACT ALLERGY

General

Only few cases of proven allergic contact dermatitis to dorzolamide from its presence in eye drops have been reported. However, in one eye hospital in the U.K., 14 patients who developed periorbital dermatitis while on dorzolamide eye drops were seen in a 3-year-period (9). In addition, false-negative patch tests to commercial 2% dorzolamide eye drops may occur not infrequently (7,8,9,11). Sometimes, patch testing with dorzolamide is positive only when applied to scarified (8) or tape-stripped skin (11). Test concentrations of 5-10% may be false-negative and higher concentrations have to be used to obtain a positive patch test reaction (2). A ROAT (7) and conjunctival challenge (8) may aid in reaching the correct diagnosis, but ROATs may also be false-negative (2).

Finally, ophthalmologists diagnose 'allergy' on the basis of clinical manifestations and rarely appear to refer these individuals to a dermatologist for patch testing. Thus, the few cases of contact allergy to dorzolamide reported may be an underestimation of the actual contact allergy problem to this anti-glaucoma drug.

Case series

In a hospital in Slovenia, in an undefined period before 2015, 55 patients with suspected contact allergy to topical drugs used for treatment of glaucoma were retrospectively analyzed. Each patient had been tested to individual products in serial dilutions (1%-10%-50% and as is). Patch tests to vehicles, emulsifiers and preservatives (including benzalkonium chloride) were also performed. Three controls on healthy volunteers were always performed in case of a positive reaction. The causal relationship between the drug and dermatitis was always confirmed by a positive elimination test. Eight of 55 patients (5 women, 3 men) had positive patch tests to one or more products. Two patients were positive to dorzolamide, six to beta-blockers (5x timolol and 1x carteolol), 2 to latanoprost, and two to brinzolamide (13). It is highly likely that these patients have not been tested with the active drugs but only with the commercial eye drops and that the diagnosis of allergic contact dermatitis to the active ingredients was, at best, made *per exclusionem*.

In Ferrara, Italy, over a 65-month period before 2005, 50 patients affected by periorbital dermatitis while using topical ocular products were patch tested, including with their own ophthalmic medications (n=210). Only 15 positive reactions were detected in 12 subjects, including 14 reactions to commercial eye drops. There was one reaction to dorzolamide. The active ingredient was not tested separately, but contact allergy to the excipients and preservatives was excluded by patch testing. The authors concluded that patch testing with commercial eye drops has doubtful value (4).

Between 1996 and 1998, 14 patients were referred to an eye hospital in the U.K. for periorbital dermatitis while on dorzolamide hydrochloride eye drops. The periorbital dermatitis occurred after a mean period of 20 weeks of commencing dorzolamide therapy. Thirteen patients had used preserved topical β-blocker treatment for a mean period of 34 months without complication before the introduction of dorzolamide. In eight (57%) the dermatitis resolved completely after discontinuing dorzolamide but in six (43%) resolution of the dermatitis did not occur until the concomitant preserved β-blocker was stopped and substituted with preservative-free β-blocker drops. Six of the 14 patients were patch tested to a standard series of allergens, to benzalkonium chloride 0.1% in water, to dorzolamide eye drops, and to pure dorzolamide hydrochloride 1%, both in petrolatum and water. These patients were also prick tested to the eye drops with results read at 15 minutes and 48 hours. Three patients later underwent patch testing to dorzolamide hydrochloride 5%, applying one series to normal skin and another series to cellotape-stripped skin, to aid penetration of dorzolamide through the epidermis. However, all tests were negative (9).

Case reports
A 64-year-old man had bilateral palpebral redness and swelling for 2 months with intense itching. He had recently been diagnosed with open-angle glaucoma which was treated with dorzolamide 2% eye drops (1 drop t.i.d.). Patch tests were positive to dorzolamide 2% eye drops in paraffin (??) and its preservative benzalkonium chloride. Despite the fact that the eye drops contain benzalkonium chloride and dorzolamide itself was not tested separately, all 6 authors conclude that this patient had contact allergy to dorzolamide, which obviously is a unsubstantiated conclusion (5).

An 80-year-old woman presented with conjunctival inflammation and severe periorbital eczematous dermatitis of the right eye, which had been evolving for 6 months. Since then, she had also a few round patchy 'fixed' dermatitis-like lesions on the abdomen. She used latanoprost eye drops and eye drops containing dorzolamide hydrochloride and timolol maleate. All eye drops were withdrawn and the dermatitis was successfully treated with topical corticosteroid ointment and anti-allergic eye drops. Patch testing was negative to all eye drops used and benzalkonium chloride 0.1% pet, the preservative in the eye drops. A ROAT was positive to the eye drops containing dorzolamide and timolol, but negative to timolol eye drops. During the ROAT, a discrete dermatitis of the right eye

Table 3.115.1 Short summaries of case reports of allergic contact dermatitis from dorzolamide

Year	Sex	Age	Clinical picture	Positive patch tests	Comments	Ref.
2015	F	77	periorbital eczematous edema both eyes, pruritus, burning	dorzolamide HCl 1% (vehicle not stated)		6
2011	?	?	blepharoconjunctivitis	dorzolamide eye drops; negative to all excipients of the eye drops	diagnosis *per exclusionem*	3
2006	F	80	conjunctival inflammation, eczema of eyelids and cheeks	dorzolamide 40% pet., negative to 20% and 10%	patch test and ROAT to eye-drops with 2% dorzolamide had been negative	2
2005	M	71	dermatitis on the eyelids	negative to eye drops on tape-stripped and scarified skin; dorzolamide 5% aq. negative, 5% in pet. positive only on scarified skin	two patients, both M 71, same histories and patch test results; conjunctival provocation test positive in both patients	8
2001	M	72	edema and exudation of the eyelids and upper cheeks	dorzolamide 2% eye drops, dorzolamide HCl 2%, 1%, 0.1% and 0.01% water	no cross-sensitivity to other sulfonamides	10
2001	M	61	eyelid dermatitis	negative to dorzolamide 1% eye drops; positive to dorzolamide 10% and 5% water and pet., but only on skin stripped with tape	five controls were negative	11
1998	F	56	eyelid eczema	dorzolamide 2% eye drops, as is and 50% water; dorzolamide 10%, 5%, 1% and 0.5% water, and 10%, 5% and 1% pet.	twenty controls were negative to dorzolamide 0.5% water and to 1% pet.	12

recurred, despite no eye drop being started again, as well as a reactivation of the fixed eczematous lesions of the abdomen. Then, 1 month later, a new ROAT was performed with eye drops containing only dorzolamide, which was positive already after 2 days. Severe recurrence of the right periocular dermatitis and of the eczematous lesions of the abdomen was observed at the same time. It was concluded that this was a case of systemic contact dermatitis from dorzolamide allergy (7).

Short summaries of other case reports of allergic contact dermatitis from dorzolamide are shown in table 3.115.1.

Cross-reactions, pseudo-cross-reactions and co-reactions
No cross-reactions to other sulfonamides (sulfacetamide and sulfanilamide) (10).

LITERATURE

1 The data in the section 'General' may have been obtained from literature discussed in this chapter, but mostly also or exclusively from one or more of the following online sources: ChemIDPlus Advanced, PubChem, DrugBank, RxList, Drug Central, Drugs.com, and Wikipedia

2 Kalavala M, Statham BN. Allergic contact dermatitis from timolol and dorzolamide eye drops. Contact Dermatitis 2006;54:345

3 Urbancek S, Kuklova M. Allergic blepharoconjunctivitis caused by antiglaucomatics—Central Slovakia experiences. American Contact Dermatitis Society 22nd Annual Meeting, New Orleans, USA, February 3, 2011. Meeting program, poster presentations, pages 32-33. Available at: https://www.contactderm.org/files/2011_ACDS_Abstracts.pdf

4 Corazza M, Massieri LT, Virgili A. Doubtful value of patch testing for suspected contact allergy to ophthalmic products. Acta Derm Venereol 2005;85:70-71

5 Orsini D, D'Arino A, Pigliacelli F, Assorgi C, Latini A, Cristaudo A. Allergic contact dermatitis to dorzolamide and benzalkonium chloride. Postepy Dermatol Alergol 2018;35:538-539

6 Lee SJ, Kim M. Allergic contact dermatitis caused by dorzolamide eyedrops. Clin Ophthalmol 2015;9:575-577

7 Kluger N, Guillot B, Raison-Peyron N. Systemic contact dermatitis to dorzolamide eye drops. Contact Dermatitis 2008;58:167-168

8 Linares Mata T, Pardo Sánchez J, de la Cuadra Oyanguren J. Contact dermatitis caused by allergy to dorzolamide. Contact Dermatitis 2005;52:111-112

9 Delaney YM, Salmon JF, Mossa F, Gee B, Beehne K, Powell S. Periorbital dermatitis as a side effect of topical dorzolamide. Br J Ophthalmol 2002;86:378-380

10 Mancuso G, Berdondini RM. Allergic contact blepharoconjunctivitis from dorzolamide. Contact Dermatitis 2001;45:243

11 Shimada M, Higaki Y, Kawashima M. Allergic contact dermatitis due to dorzolamide eyedrops. Contact Dermatitis 2001;45:52

12 Aalto-Korte K. Contact allergy to dorzolamide eyedrops. Contact Dermatitis 1998;39:206

13 Slavomir U, Kuklova Bielikova M. Allergic reactions due to antiglaucomatics. Dermatitis 2015;26(2):e12-13

Chapter 3.116 DOXEPIN

IDENTIFICATION

Description/definition   : Doxepin is the dibenzoxepin tricyclic compound that conforms to the structural formula
                                         shown below
Pharmacological classes   : Antidepressive agents, tricyclic; sleep aids, pharmaceutical; histamine antagonists
IUPAC name   : (3E)-3-(6H-Benzo[c][1]benzoxepin-11-ylidene)-N,N-dimethylpropan-1-amine
Other names   : 11-(3-Dimethylaminopropylidene)-6,11-dihydrodibenz(b,e)oxipin
CAS registry number   : 1668-19-5
EC number   : Not available
Merck Index monograph : 4753
Patch testing   : 1% and 5% pet.
Molecular formula   : $C_{19}H_{21}NO$

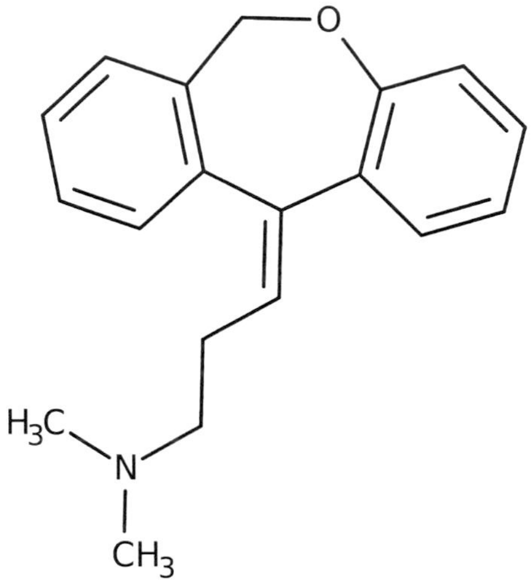

GENERAL

Doxepin is a dibenzoxepin derivative and tricyclic antidepressant-like drug with antipruritic, anti-depressive, sedative and anxiolytic activities with structural similarities to phenotheazines. Oral doxepin is approved for treatment of depression and/or anxiety associated with different conditions, including alcoholism, organic disease and manic-depressive disorders and for treatment of insomnia characterized by difficulties with sleep maintenance. Topical doxepin is also approved for short-term (up to 8 days) management of moderate pruritus in adult patients with atopic dermatitis, pruritus or lichen simplex chronicus. In pharmaceutical products, doxepin is employed as doxepin hydrochloride (CAS number 1229-29-4, EC number 214-966-8, molecular formula $C_{19}H_{22}ClNO$) (1).

CONTACT ALLERGY

General

Doxepin appears to be a frequent cause of allergic contact dermatitis, with a possible rate of sensitization of >10% (11). It is certainly the most frequent sensitizer in the group of the antihistamines.

Case series

In the year 2002, 26 post-marketing cases of allergic contact dermatitis to doxepin 5% cream were reported to the Food and Drug Administration (2). In 13 of the 20 cases (65%) with use information, duration of use exceeded the recommended 8 days. The manifestations of hypersensitivity included eczema, urticaria, purpura, and papulo-vesicular lesions on the body. Of the 13 cases for which data on treatment were provided, 10 were severe enough to warrant treatment with systemic corticosteroids, with one hospitalization. In 7 cases, the presenting symptom was acute worsening of their original dermatitis after topical doxepin. All 23 patients for whom information was provided recovered on discontinuation of doxepin cream. Patch testing to doxepin 5% cream was positive in all patients for whom that information was provided (21/26), but it is unknown whether doxepin itself had been patch tested. The

FDA authors conclude that ACD to doxepin 5% cream is not uncommon. They emphasize that the true number is likely to be much higher, as underreporting is a well-recognized limitation of passive surveillance systems such as the FDA's adverse-event reporting system, only 1-10% of adverse reactions usually being reported (2).

Six patients developed severe allergic contact dermatitis to doxepin 5% cream after using it for 2 weeks to 7 months. The reactions included photodermatitis, angioedema-like swelling, and generalized weeping vesicular dermatitis. In some cases these reactions were mistaken for exacerbations of the original dermatitis. Patch tests with the doxepin cream were positive in all six patients. Three individuals were patch tested to the ingredients and had strongly positive reactions to doxepin 5% pet. but not to the other ingredients of the cream. The package insert with doxepin cream suggests use of the cream for pruritus in adults with atopic dermatitis, with limitation of use to 8 days. According to the authors, it is probably unrealistic to expect patients with chronic pruritic skin diseases, such as atopic dermatitis, to stop using a product on the ninth day when it is helpful in relieving their chronic itch. They also suggest that doxepin augments the development of delayed hypersensitivity by interfering with the immunoinhibittory role of histamine, actually amplifying T-cell activity in cell-mediated hypersensitivity (10).

In the USA, 97 consecutive patients were routinely patch tested with doxepin 5% cream for various forms of dermatitis in the period November 1994 - November 1995, half a year after it was approved by the FDA. Seventeen patients (10 women, 7 men, age range 29-76, mean age 52 years) had relevant positive patch test reactions to doxepin cream, which all of them used or had used. In 13 of the 17 patients with positive reactions, the diagnosis of allergic contact dermatitis to doxepin cream was confirmed by positive patch test reactions to both the active ingredient and the whole formulation of doxepin cream, by an observed positive repeated open application test (ROAT) reaction to doxepin cream, or by both. Of 14 patients who completed testing with doxepin cream ingredients, all had positive reactions to the whole formulation, and 12 had positive reactions to doxepin hydrochloride. None reacted to any of the other ingredients. ROATs with doxepin cream on normal skin resulted in positive eczematous responses in eight of 10 patients. Eight of the 17 patients had concurrent, relevant positive reactions to other allergens. Many had long-standing dermatitis, and each had used doxepin cream for several days to 1 year. Two patients appeared to have had, according to the authors, systemic contact dermatitis from oral administration of doxepin after topical sensitization (11). However, the evidence for that was very weak.

Shortly thereafter, these results were criticized and it was suggested that doxepin cannot be a sensitizer because of the negative results of predictive patch tests with doxepin cream and that the excited skin syndrome may have played a role (13). In their rebuttal, the authors stated that they had seen another 6 patients with contact allergy to doxepin, that many colleagues had observed such patients, and that they had tested another 120 control patients to doxepin cream with negative results (13).

Case reports
Short summaries of case reports of allergic contact dermatitis from doxepin are shown in table 3.116.1. One patient who was contact allergic to doxepin hydrochloride 5% cream later developed systemic contact dermatitis from oral intake of doxepin (3).

Table 3.116.1 Short summaries of case reports of allergic contact dermatitis from doxepin

Year and country	Sex	Age	Positive patch tests	Clinical data and comments	Ref.
2001 U.K.	M	51	DHCl 1% and 0.5% pet.	patient with localized recessive dystrophic epidermolysis bullosa	4
2000 U.K.	M	59	5% cream 'as is'; DHCl 0.05%-5%	sensitization after 3 weeks; 20 controls negative to cream	5
1999 U.K.	M	40	5% cream 'as is'; DHCl 0.5%, 1% and 5% pet.	patient with epidermolysis bullosa pruriginosa	7
1996 Spain	F	53	5% cream 'as is'; DHCl 2.5% and 1% pet.	long duration of use (2 months); 14 controls negative to doxepin 5% (probably the cream, not the neat drug)	8
1995 USA	M	68	5% cream 'as is'; DHCL 5% pet.	ACD after 12 days; application to the scalp led to severe dermatitis with swollen and later closed eyelids	9

ACD: allergic contact dermatitis; DHCl: doxepin hydrochloride

Two more patients allergic to doxepin (or doxepin cream?) were reported from India. Details are not available to the author (6). Another patient had a positive patch test reaction to both the cream and doxepin, but details are not available (12).

LITERATURE

1   The data in the section 'General' may have been obtained from literature discussed in this chapter, but mostly also or exclusively from one or more of the following online sources: ChemIDPlus Advanced, PubChem, DrugBank, RxList, Drug Central, Drugs.com, and Wikipedia

2   Bonnel RA, La Grenade L, Karwoski CB, Beitz JG. Allergic contact dermatitis from topical doxepin: Food and Drug Administration's postmarketing surveillance experience. J Am Acad Dermatol 2003;48:294-296

3   Brancaccio RR, Weinstein S. Systemic contact dermatitis to doxepin. J Drugs Dermatol 2003;2:409-410

4   Horn HM, Tidman MJ, Aldridge RD. Allergic contact dermatitis due to doxepin cream in a patient with dystrophic epidermolysis bullosa. Contact Dermatitis 2001;45:115

5   Buckley DA. Contact allergy to doxepin. Contact Dermatitis 2000;43:231-232

6   Reddy G, Shenoi SD, Pai BS, Sandra A, Deepak S. Allergic contact dermatitis to doxepin. Indian J Dermatol Venereol Leprol 1999;65:277-278

7   Wakelin SH, Rycroft RJG. Allergic contact dermatitis from doxepin. Contact Dermatitis 1999;40:214

8   Bilbao I, Aguirre A, Vicente JM, Raton JA, Zabala R, Diaz Perez JL. Allergic contact dermatitis due to 5% doxepin cream. Contact Dermatitis 1996;35:254-255

9   Greenberg JH. Allergic contact dermatitis from topical doxepin. Contact Dermatitis 1995;33:281

10 Shelly WB, Shelly ED, Talanin NY. Self-potentiating allergic contact dermatitis caused by doxepin hydrochloride cream. J Am Acad Dermatol 1996;34:143-144

11 Taylor JS, Praditsuwan P, Handel D, Kuffner G. Allergic contact dermatitis from doxepin cream. Arch Dermatol 1996:132:515-518

12 Porres J. Dermatitis from Zonalon cream. Schoch Lett. October 1995;45:39. Item 143 (data cited in ref. 11)

13 Rapaport MJ. Allergic contact dermatitis from doxepin cream. Arch Dermatol 1996;132:1516-1518

Chapter 3.117 DYCLONINE

IDENTIFICATION

Description/definition	: Dyclonine is the alkyl-phenylketone that conforms to the structural formula shown below
Pharmacological classes	: Anesthetics, local
IUPAC name	: 1-(4-Butoxyphenyl)-3-piperidin-1-ylpropan-1-one
Other names	: 4'-Butoxy-3-piperidinopropiophenone
CAS registry number	: 586-60-7
EC number	: Not available
Merck Index monograph	: 4790
Patch testing	: 1% pet. or water
Molecular formula	: $C_{18}H_{27}NO_2$

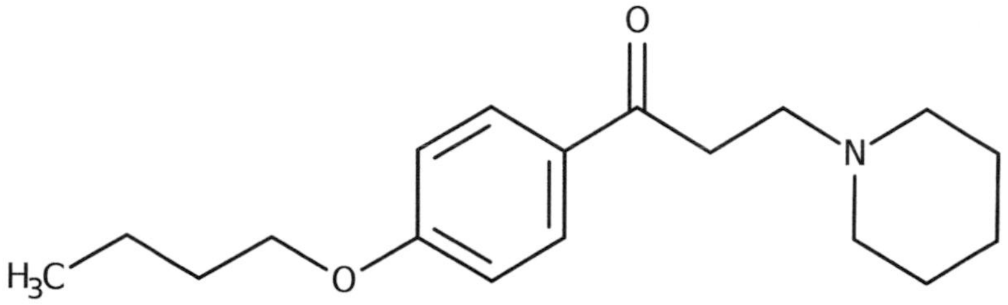

GENERAL

Dyclonine is a member of the piperidines with local anesthetic effect. This agent is used to provide topical anesthesia of accessible mucous membranes prior to examination, endoscopy or instrumentation, or other procedures involving the esophagus, larynx, mouth, pharynx or throat, respiratory tract or trachea, urinary tract, or vagina. Dyclonine is also used to suppress the gag reflex and/or other laryngeal and esophageal reflexes to facilitate dental examination or procedures (including oral surgery), endoscopy, or intubation. Another indication is for relief of canker sores (aphthosis stomatis) or cold sores/fever blisters (herpes simplex infection). Preparations containing dyclonine may be available over-the-counter (1,2). In pharmaceutical products, dyclonine is employed as dyclonine hydrochloride (CAS number 536-43-6, EC number 208-633-6, molecular formula $C_{18}H_{28}ClNO_2$) (1).

CONTACT ALLERGY

Case reports

A 56-year-old woman presented with a 2.5-year history of a pruritic, papular facial rash that had recently spread to her bilateral armpits and shins. She had tried several over-the-counter treatments, including a cream containing 0.5% dyclonine HCl. Patch tests read at D7 were strongly positive to the cream, tested as is. Because she tested negative to other ingredients in the cream (including allantoin, benzyl alcohol, EDTA, triethanolamine, mineral oil, paraben mix, stearyl alcohol, and mineral oil), allergy to the active ingredient, dyclonine HCl was suspected. Purified dyclonine powder was obtained from the manufacturer, tested to 1% aqueous and 1% petrolatum preparations, with positive reactions to the aqueous preparation (++) and to the petrolatum preparation (+). Five controls were negative (2).

A 24-year-old man with recurrent lip herpes simplex vesicles used an OTC topical agent containing dyclonine for symptomatic relief, but the lesions spread rather than healed in the customary manner. He had used the product earlier without difficulty. Physical examination revealed a spreading eczematous eruption surrounding the original vesicles. Patch testing revealed a ++ reaction to the final formulation and 1% dyclonine in white petrolatum. Ten controls were negative to both (3).

Two more patients had become sensitized to dyclonine gel used for herpes labialis (4). A ROAT was positive in both individuals after 2 resp. 3 days. Patch tests were positive (++) to the gel 'as is' and 1% pure dyclonine hydrochloride solution (4).

Cross-reactions, pseudo-cross-reactions and co-reactions
No cross-reactivity to the related pramoxine (2,4).

LITERATURE
1   The data in the section 'General' may have been obtained from literature discussed in this chapter, but mostly also or exclusively from one or more of the following online sources: ChemIDPlus Advanced, PubChem, DrugBank, RxList, Drug Central, Drugs.com, and Wikipedia
2   Kimyon RS, Schlarbaum JP, Liou YL, Warshaw EM, Hylwa SA. Allergic dermatitis to dyclonine (Dyclocaine). Dermatitis 2019;30:372-373
3   Maibach HI. Dyclonine hydrochloride, local anaesthetic allergic contact dermatitis. Contact Dermatitis 1986;14:114
4   Purcell SM, Dixon SL. Allergic contact dermatitis to dyclonine hydrochloride simulating extensive herpes simplex labialis. J Am Acad Dermatol 1985;12(2 Pt.1):231-234

Chapter 3.118 ECHOTHIOPHATE

IDENTIFICATION

Description/definition   : Echothiophate is the tetraalkylammonium salt that conforms to the structural formula
                                       shown below
Pharmacological classes : Miotics; cholinesterase inhibitors; parasympathomimetics
IUPAC name   : 2-Diethoxyphosphorylsulfanylethyl(trimethyl)azanium
Other names   : Ecothiophate
CAS registry number   : 6736-03-4
EC number   : Not available
Merck Index monograph : 4818
Patch testing   : 1% and 2.5% water
Molecular formula   : $C_9H_{23}NO_3PS$

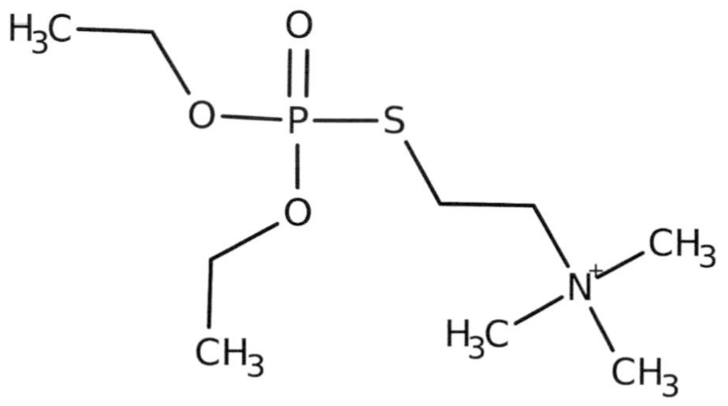

GENERAL

Echothiophate is a potent, long-acting cholinesterase inhibitor with parasympathomimetic activity. Its iodide salt echothiophate iodide (CAS number 513-10-0, EC number 208-152-1, molecular formula $C_9H_{23}INO_3PS$) potentiates the action of endogenous acetylcholine by inhibiting acetylcholinesterase that hydrolyzes acetylcholine. When applied topically to the eye, this agent prolongs stimulation of the parasympathetic receptors at the neuromuscular junctions of the longitudinal muscle of the ciliary body. Contraction of longitudinal muscle pulls on the scleral spur, and opens the trabecular meshwork, thereby increasing aqueous humor outflow from the eye and reducing intraocular pressure. Echothiopate iodide is indicated for use in the treatment of subacute or chronic angle-closure glaucoma after iridectomy or where surgery is refused or contraindicated (1).

CONTACT ALLERGY

Case report

A 75-year-old woman with a 20-year history of open-angle glaucoma had a two year history of chronic conjunctivitis and periorbital erythema; contact dermatitis was clinically suspected. Ophthalmic medications at the time of evaluation included 4% pilocarpine hydrochloride, 0.25% echothiophate iodide, 10% phenylephrine, and gentamicin sulfate ophthalmic solution. In patch testing, echothiophate iodide, tested as the 0.25% commercial product, gave a positive reaction. Next, echothiophate iodide powder, diluted in water and tested at 0.25%, 1.0%, and 2.5% concentrations, were patch tested and all gave positive reactions. Twenty controls were negative. The patient also proved to be allergic to phenylephrine (2).

LITERATURE

1   The data in the section 'General' may have been obtained from literature discussed in this chapter, but mostly
     also or exclusively from one or more of the following online sources: ChemIDPlus Advanced, PubChem,
     DrugBank, RxList, Drug Central, Drugs.com, and Wikipedia
2   Mathias CGT, Maibach HI. Allergic contact dermatitis to echothiophate iodide and phenylephrine. Arch
     Ophthalmol 1979;97:286-287

Chapter 3.119 ECONAZOLE

IDENTIFICATION

Description/definition   : Econazole is the imidazole that conforms to the structural formula shown below
Pharmacological classes : 14α-Demethylase Inhibitors; antifungal agents
IUPAC name   : 1-[2-[(4-Chlorophenyl)methoxy]-2-(2,4-dichlorophenyl)ethyl]imidazole
CAS registry number   : 27220-47-9
EC number   : 248-341-6
Merck Index monograph   : 4819
Patch testing   : Nitrate, 1.0% alc. (Chemotechnique)
Molecular formula   : $C_{18}H_{15}Cl_3N_2O$

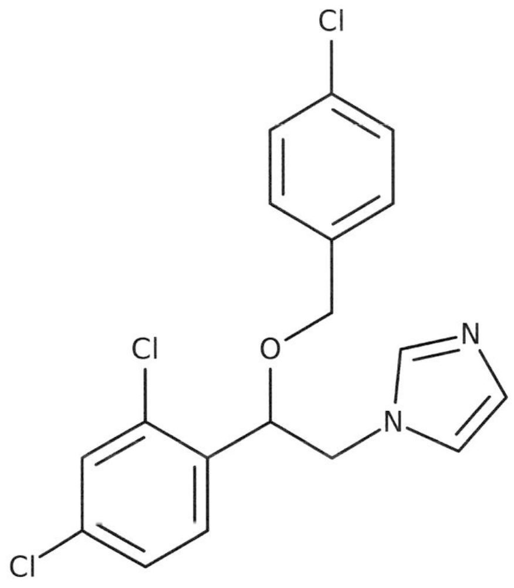

GENERAL

Econazole is a broad-spectrum imidazole antimycotic with fungistatic properties and some action against gram-positive bacteria. It is indicated for topical application in the treatment of superficial fungal infections caused by dermatophytes, in the treatment of cutaneous candidiasis, and in the treatment of pityriasis (tinea) versicolor. In pharmaceutical products, econazole is employed as econazole nitrate (CAS number 24169-02-6, EC number 246-053-5, molecular formula $C_{18}H_{16}Cl_3N_3O_4$) (1).

CONTACT ALLERGY

Case series

In the period 1990 to 2014, in Leuven, Belgium, 164 patients suspected of iatrogenic contact dermatitis were tested with econazole 1% pet. and there were 9 (5.5%) positive reactions. Relevance was not specified for individual drugs, but 96% of the positive patch test reactions to all topical drugs and antiseptics were considered to be relevant (3).

Between 1990 and 2003, in Leuven, Belgium, 92 patients were patch tested for chronic vulval complaints. Fifteen of these women were tested with a series of topical drugs; there was one reaction to econazole nitrate 2% pet. which was relevant (2). In Bologna, Italy, in 'the few years' before 1995, 20 cases of allergic contact dermatitis due to imidazole preparations were found. However, patch tests with the relevant imidazoles were positive in only 12, including 4 to econazole (4).

In the period 1976-1988, 13 cases of allergic contact dermatitis to imidazole antifungal preparations were seen in a private dermatological practice in Saverne, France, and 2 in the Université Catholique de Louvain, Belgium. Most had been prescribed for tinea cruris, pedis or corporis, some for anal pruritus, dermatitis and even a leg ulcer. The responsible antifungals (tested 2% pet.) were econazole in 2 cases, miconazole in 6 cases, isoconazole in 5 cases, and tioconazole in one case; one patient reacted to a base ingredient (13).

Between 1977 and 1986, 9 patients with contact allergy to imidazole antimycotics were investigated in Heidelberg, Germany. Three patients were allergic to econazole, 6 to miconazole, 3 to clotrimazole, 3 to isoconazole, and one to oxiconazole (6).

Case reports

A 26-year-old woman had a persistent foot dermatitis, which had improved in recent months following the wearing of new shoes and the use of a corticosteroid ointment. Patch tests were positive to econazole nitrate 1% in alcohol, miconazole 1% in alcohol, a 1% econazole prescription cream that the patient had previously used and a sample from a shoe that the patient had had for the previous 5 years. The foot dermatitis had persisted after stopping econazole cream from retention of this imidazole in the patient's shoes (9).

A 43-year-old woman was prescribed miconazole cream for a rash which appeared to be ringworm at the site of an injury. When this did not work, she was given econazole cream, but this aggravated the condition. Subsequently she was given clotrimazole cream, but the condition had by now deteriorated to a severe eczema. Patch tests were positive to miconazole and econazole cream and the active imidazoles 2% pet. She had also positive reactions to isoconazole cream, sulconazole 1% and tioconazole solution, which were considered to be cross-reactions, as the patient had never used them before. She did not react to clotrimazole (11).

A 27-year-old man suffering from a fungal infection of the genital area was treated with a cream containing econazole nitrate. After 15-20 days, he developed intensely erythemato-edematous urticarial-like lesions with pruritus at the application site of the cream, spreading to the trunk and arms. Patch tests were positive to the cream and econazole nitrate 1% pet., but negative to the cream base (12).

In a literature review up to 1994, the authors identified 105 reported patients who had a positive patch test reaction to at least one imidazole derivative, ranging from 51 reactions to miconazole to one reaction each to bifonazole and enilconazole. The number of reported reactions to econazole was 37 (10).

Other cases of contact allergy to/allergic contact dermatitis from econazole, details of which are not available to the author, have been presented in refs. 7 and 8 (cross-reaction to miconazole).

Cross-reactions, pseudo-cross-reactions and co-reactions

Statistically significant associations have been found in patient data between miconazole, econazole, and isoconazole; between sulconazole, miconazole, and econazole (10,14). A patient sensitized to econazole (or to miconazole cross-reacting to econazole) had co-reactions to isoconazole cream, sulconazole 1% and tioconazole solution, which were considered to be cross-reactions, as the patient had never used them before (11). A patient sensitized to sertaconazole, who had never used antifungal preparations before, cross-reacted to econazole and miconazole (5).

LITERATURE

1   The data in the section 'General' may have been obtained from literature discussed in this chapter, but mostly also or exclusively from one or more of the following online sources: ChemIDPlus Advanced, PubChem, DrugBank, RxList, Drug Central, Drugs.com, and Wikipedia

2   Nardelli A, Degreef H, Goossens A. Contact allergic reactions of the vulva: a 14-year review. Dermatitis 2004;15:131-136

3   Gilissen L, Goossens A. Frequency and trends of contact allergy to and iatrogenic contact dermatitis caused by topical drugs over a 25-year period. Contact Dermatitis 2016;75:290-302

4   Guidetti MS, Vincenzi C, Guerra L, Tosti A. Contact dermatitis due to imidazole antimycotics. Contact Dermatitis 1995;33:282

5   Goday JJ, Yanguas I, Aguirre A, Ilardia R, Soloeta R. Allergic contact dermatitis from sertaconazole with cross-sensitivity to miconazole and econazole. Contact Dermatitis 1995;32:370-371

6   Raulin C, Frosch PJ. Contact allergy to imidazole antimycotics. Contact Dermatitis 1988;18:76-80

7   Böttger EM, Mücke C, Tronnier H. Kontaktdermatitis auf neuere Antimykotika und Kontakturticaria. Aktuelle Dermatologie 1981;7:70-74 (Article in German)

8   Mücke C. Ein Fall von Überempfindlichkeit gegen Econazol und Tolciclat. Dermatosen in Beruf und Umwelt 1980;28:118 (Article in German)

9   Zirwas MJ. Allergy to imidazole antifungals retained in shoes. Dermatitis 2009;20:172-173

10   Dooms-Goossens A, Matura M, Drieghe J, Degreef H. Contact allergy to imidazoles used as antimycotic agents. Contact Dermatitis 1995;33:73-77

11   Motley RJ, Reynolds AJ. Contact allergy to 2,4-dichlorophenylethyl imidazole derivatives. Contact Dermatitis 1988;19:381-382

12   Valsecchi R, Tornaghi A, Tribbia G, Cainelli T. Contact dermatitis from econazole. Contact Dermatitis 1982;8:422

13   Jelen G, Tennstedt D. Contact dermatitis from topical imidazole antifungals: 15 new cases. Contact Dermatitis 1989;21:6-11

14   Carmichael AJ, Foulds IS. Imidazole cross-sensitivity to sulconazole. Contact Dermatitis 1988;19:237-238

Chapter 3.120 EFINACONAZOLE

IDENTIFICATION

Description/definition   : Efinaconazole is the azole antifungal that conforms to the structural formula shown below
Pharmacological classes   : Antifungal agents; 14α-demethylase inhibitors
IUPAC name   : (2R,3R)-2-(2,4-Difluorophenyl)-3-(4-methylidenepiperidin-1-yl)-1-(1,2,4-triazol-1-yl)butan-2-ol
CAS registry number   : 164650-44-6
EC number   : Not available
Merck Index monograph   : 11729
Patch testing   : 10% pet. or alcohol
Molecular formula   : $C_{18}H_{22}F_2N_4O$

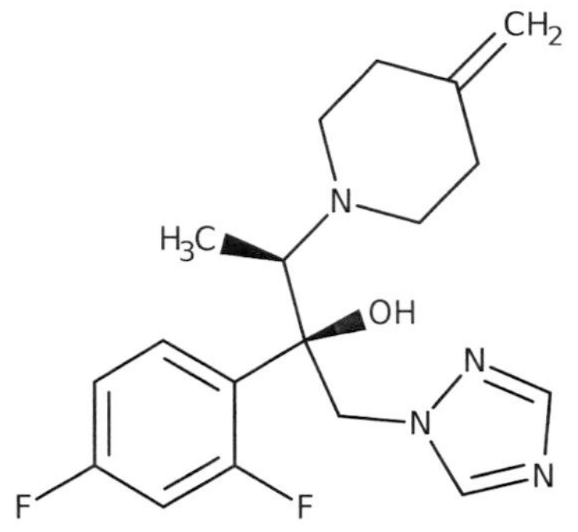

GENERAL

Efinaconazole is a triazole antifungal and 14α-demethylase inhibitor used in the topical treatment of fungal infections of the nail (onychomycosis). It was introduced in Japan in 2014 (1,4).

CONTACT ALLERGY

Case reports

A 74-year-old man was using efinaconazole 10% solution, bifonazole 1% cream and terbinafine 1% cream to treat fungal nail infection, when he developed pruritic erythematous macules on the bilateral toes. Patch tests were positive to the efinaconazole solution and a cream containing 1% luliconazole. Later, patch tests were positive to efinaconazole 10%, 1% and 0.1% pet., and negative to the excipients of the antifungal solution. Controls were not mentioned (2).

A 66-year-old man was prescribed a 10% topical solution of efinaconazole for fungal infection of the toenails; he had used this solution before. One week later, the patient presented with redness, swelling and onycholysis affecting all toes. All toenails were macerated and softened, and the nail plates had separated from the nail beds. Patch testing was performed with several previously used topical agents, including efinaconazole solution, tested 'as is', and all constituents of efinaconazole solution (at concentrations as in the product). There were positive reactions to efinaconazole solution and efinaconazole 10% alc. and a diagnosis of allergic contact dermatitis leading to onychomadesis ('nail shedding') was made (3).

Previously, a 64-year-old man, also from Japan, had been diagnosed with allergic contact dermatitis from efinaconazole solution. He had positive patch test reactions to the 10% solution, efinaconazole 10% pet., efinaconazole 10% alcohol, and one of the excipients in the antifungal preparation, alkyl lactate (test concentration not stated). Nine controls were negative (4).

LITERATURE

1   The data in the section 'General' may have been obtained from literature discussed in this chapter, but mostly also or exclusively from one or more of the following online sources: ChemIDPlus Advanced, PubChem, DrugBank, RxList, Drug Central, Drugs.com, and Wikipedia
2   Oiso N, Tatebayashi M, Kawada A. Allergic contact dermatitis caused by efinaconazole: positive patch test reactions up to 0.1% pet. Contact Dermatitis 2017;76:53-54
3   Yamaguchi Y, Hoshina D, Furuya K. Onychomadesis caused by efinaconazole. Contact Dermatitis 2017;76:57-58
4   Hirohata A, Hanafusa T, Mabuchi-Kiyohara E, Ikegami R. Contact dermatitis caused by efinaconazole solution for treatment of onychomycosis. Contact Dermatitis 2015;73:190-192

Chapter 3.121 ENILCONAZOLE

IDENTIFICATION

Description/definition	: Enilconazole is the phenylethyl imidazole derivative that conforms to the structural formula shown below
Pharmacological classes	: Fungicides, industrial
IUPAC name	: 1-[2-(2,4-Dichlorophenyl)-2-prop-2-enoxyethyl]imidazole
Other names	: 1-[2-(Allyloxy)-2-(2,4-dichlorophenyl)ethyl]-1H-imidazole
CAS registry number	: 35554-44-0
EC number	: 252-615-0
Merck Index monograph	: 4907
Patch testing	: 2% pet.; 1% alcohol or MEK (methyl ethyl ketone) is usually preferred for imidazole antifungal drugs
Molecular formula	: $C_{14}H_{14}Cl_2N_2O$

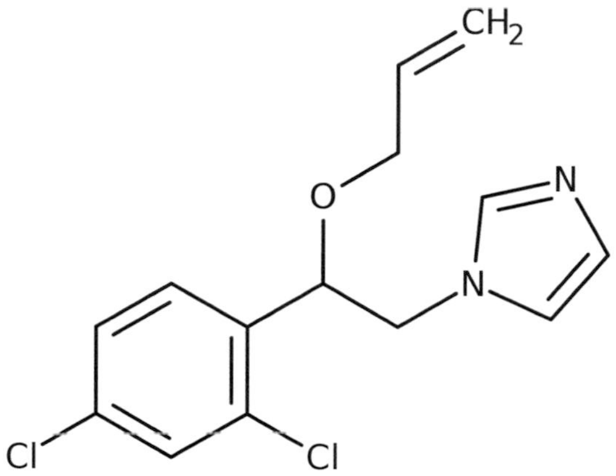

GENERAL

Enilconazole is an imidazole antifungal agent used in veterinary medicine and as agricultural fungicide, particularly in the growing of citrus fruits. In veterinary medicine, it is employed in topical preparations for the treatment of fungal infections in cattle, horses and dogs caused by *Trichophyton* and *Microsporum* species (1).

CONTACT ALLERGY

Case reports

A 43-year-old female technician in a veterinarian's laboratory developed a scaly red patch on the dorsal aspect of the right hand. Her employer considered the lesion as a dermatophytic infection of the skin and gave her the veterinary antimycotic enilconazole 0.2% solution. Two weeks later, she had developed an acute eczematous contact dermatitis of the right hand and forearm. Patch tests revealed contact sensitivity to econazole, enilconazole, isoconazole and miconazole, all tested 2% pet. The patient also reacted to the commercial enilconazole solution 0.2% and (again) to enilconazole 2% pet. (3).

There is also one case report of occupational allergic contact dermatitis to enilconazole in a metalworker from its presence in a water-based metalworking fluid (2).

In a literature review up to 1994, the authors identified 105 reported patients who had a positive patch test reaction to at least one imidazole derivative, ranging from 51 reactions to miconazole to one reaction each to enilconazole and bifonazole (4).

LITERATURE

1 The data in the section 'General' may have been obtained from literature discussed in this chapter, but mostly also or exclusively from one or more of the following online sources: ChemIDPlus Advanced, PubChem, DrugBank, RxList, Drug Central, Drugs.com, and Wikipedia
2 Piebenga WP, van der Walle HB. Allergic contact dermatitis from 1-[2-(2,4-dichlorophenyl)-2-(2-propenyloxy) ethyl]-1H-imidazole in a water-based metalworking fluid. Contact Dermatitis 2003;48:285-286
3 Van Hecke E, de Vos L. Contact sensitivity to enilconazole. Contact Dermatitis 1983;9:144
4 Dooms-Goossens A, Matura M, Drieghe J, Degreef H. Contact allergy to imidazoles used as antimycotic agents. Contact Dermatitis 1995;33:73-77

Chapter 3.122 ENOXOLONE

IDENTIFICATION

Description/definition   : Enoxolone is the pentacyclic triterpenoid aglycone metabolite of glycyrrhizin that conforms to the structural formula shown below

Pharmacological classes : Anti-inflammatory agents

UPAC name   : (2S,4aS,6aR,6aS,6bR,8aR,10S,12aS,14bR)-10-Hydroxy-2,4a,6a,6b,9,9,12a-hepta-methyl-13-oxo-3,4,5,6,6a,7,8,8a,10,11,12,14b-dodecahydro-1H-picene-2-carboxylic acid

Other names   : Glycyrrhetinic acid; glycyrrhetic acid; olean-12-en-29-oic acid, 3-hydroxy-11-oxo-, (3β,20β)-; uralenic acid; glycyrrhetin

CAS registry number   : 471-53-4

EC number   : 207-444-6

Merck Index monograph : 4914

Patch testing   : 10% pet.

Molecular formula   : $C_{30}H_{46}O_4$

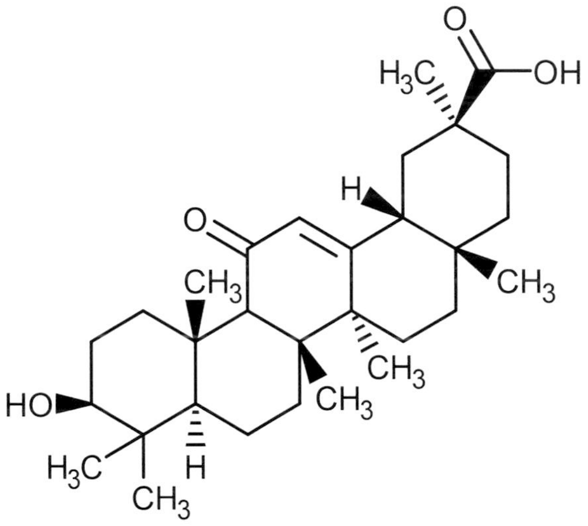

GENERAL

Enoxolone (glycyrrhetinic acid) is a pentacyclic triterpene derived from the hydrolysis of glycyrrhizic acid (glycyrrhizin), one of the main components of licorice root *Glycyrrhiza glabra*. Because of its steroid-like chemical structure, it possesses anti-inflammatory activity, as well as antibacterial, antiviral and antifungal activities. It is used for its anti-inflammatory and antipruritic effects in the treatment of atopic dermatitis (2).

CONTACT ALLERGY

Case reports

A 7-year-old girl with atopic dermatitis developed pruritic erythematous macules on the bilateral cubital fossa where she had applied an over-the-counter medicament for eczema. Patch tests were positive to the ointment and its ingredient enoxolone 1% pet. (5). A 43-year-old man developed a perianal eruption 2 hours after applying an ointment containing enoxolone 0.7%, benzocaine 0.25%, procaine HCl 1%, allantoin 0.8%, menthol 0.2% and zinc oxide 20%, which he had previously used for hemorrhoids for 2 years without problems. Patch tests were positive to the ointment 'as is' and 10%, and to enoxolone 10% pet., benzocaine 1% pet. and procaine HCl 1% pet. The other ingredients were all negative (4).

A 62-year-old man with a tracheostomy was treated with an ointment containing enoxolone 1.5%, mafenide 7%, neomycin 0.35% and sulfanilamide 3% for a skin fissure adjacent to the tracheostomy. After 3 days, he developed itchy erythematous and vesicular exudative lesions at the application site. Nonetheless, the patient continued the treatment for a further week, with spreading of the lesions to the upper chest and shoulders. Later, after healing of the dermatitis, he applied the same ointment around his nostrils, because of irritation related to an upper respiratory tract infection, and developed vesicular lesions around the nose within a day. Patch tests were positive to the ointment, to enoxolone 10% pet. and mafenide 10% pet. Ten controls were negative (3).

Enoxolone has also caused allergic *cosmetic* dermatitis from its presence in a deodorant and an after-sun product. The patient had positive patch tests to glycyrrhetinic acid (enoxolone) 0.02% and 0.2% alcohol/water (1).

A 53-yer-old woman developed facial angioedema 12 hours after taking 2 tablets (to remain in the mouth for temporary relief of inflammation and pain in the oral cavity) containing tyrothricin, enoxolone, sulfaguanidine and benzocaine. Examination revealed severe swelling of the lips and the left side of her face (the side on which the tablets had been left to dissolve), accompanied by inflammation of the oral mucosa. Patch testing with the tablet (probably crushed) was positive, as was the caine mix. Oral administration of tyrothricin and sulfaguanidine gave no reaction, benzocaine was not tested (because of the positive patch test to the caine mix), but an oral provocation with 10 mg of enoxolone resulted in discomfort in the mouth after 10 hours and erythema and edema on the left side of the oral mucosa a few hours later. Patch testing with enoxolone showed minimal erythema at 10% pet. but a positive patch test reaction at 20% pet. Ten controls were negative (6).

LITERATURE

1   Sasseville D, Desjardins M, Almutawa F. Allergic contact dermatitis caused by glycyrrhetinic acid and castor oil. Contact Dermatitis 2011;64:168-169
2   Veraldi S, De Micheli P, Schianchi R, Lunardon L. Treatment of pruritus in mild-to-moderate atopic dermatitis with a topical non-steroidal agent. J Drugs Dermatol 2009;8:537-359
3   Fernàndez JC, Gamboa P, Jáuregui I, González G, Antépara I. Concomitant sensitization to enoxolone and mafenide in a topical medicament. Contact Dermatitis 1992;27:262
4   Tanaka S, Otsuki T, Matsumoto Y, Hayakawa R, Sugiura M. Allergic contact dermatitis from enoxolone. Contact Dermatitis 2001;44:192
5   Oiso N, Ota T, Yoshinaga E, Endo H, Kawara S, Kawada A. Allergic contact dermatitis mimicking atopic dermatitis due to enoxolone in a topical medicament. Contact Dermatitis 2006;54:351
6   Villas-Martinez F, Joral Badas A, Garmendia Goitia JF, Aguirre I. Sensitization to oral enoxolone. Contact Dermatitis 1994;30:124

Chapter 3.123 EPHEDRINE

IDENTIFICATION

Description/definition   : Ephedrine is the phenethylamine alkaloid that conforms to the structural formula shown
                             below
Pharmacological classes   : Vasoconstrictor agents; central nervous system stimulants; adrenergic agents;
                             sympathomimetics
IUPAC name   : (1R,2S)-2-(Methylamino)-1-phenylpropan-1-ol
Other names   : Benzenemethanol, α-[(1S)-1-(methylamino)ethyl]-, (αR)-
CAS registry number   : 299-42-3
EC number   : 206-080-5
Merck Index monograph   : 4933
Patch testing   : 10% pet. or water
Molecular formula   : $C_{10}H_{15}NO$

GENERAL

Ephedrine is an alkaloid and hydroxylated form of phenethylamine which is found in the plant *Ephedra sinica* and various other plants in the genus *Ephedra*. It is an α- and β-adrenergic agonist that may also enhance release of norepinephrine. Following administration, ephedrine activates post-synaptic noradrenergic receptors. Activation of α-adrenergic receptors in the vasculature induces vasoconstriction, and activation of β-adrenergic receptors in the lungs leads to bronchodilation. Ephedrine is commonly used as a stimulant, appetite suppressant, concentration aid, decongestant, and to treat hypotension associated with anesthesia. In pharmaceutical products, ephedrine is employed as ephedrine sulfate (CAS number 134-72-5, EC number 205-154-4, molecular formula $C_{20}H_{32}N_2O_6S$) or ephedrine hydrochloride (CAS number 50-98-6, EC number 200-074-6, molecular formula $C_{10}H_{16}ClNO$) (1).

CONTACT ALLERGY

Case reports and case series

Formerly, ephedrine was used in nasal sprays and, in older literature, many cases and some small case series of allergic contact dermatitis (7-15) and a few of systemic contact dermatitis (10,11) have been reported. In one case, this presented as dermatitis of the penis from contamination by the patient's hands (13), in another it was used as eye drops and caused ACD of the eyelids and the face (11). The literature up to 1933 has been reviewed in ref. 7.

A 64-year-old woman had a recurrent generalized maculopapular eruption following the 'intake' (meant is probably the use in the nasal cavity) of nasal decongestants containing ephedrine, pseudoephedrine and norephedrine. Patch tests were positive to phenylephrine hydrochloride 10% pet. and phenylephrine eye drops, but negative to the other drugs, including ephedrine. However, oral challenges with ephedrine and norephedrine 50 mg every 8 hours resulted in a generalized eruption after 2 days and this was apparently enough evidence for the authors to conclude that the generalized eruptions were caused by ephedrine (5). Nevertheless, patients sensitized to phenylephrine may cross-react to ephedrine (16) and it is quite possible that this patient had systemic contact dermatitis from ephedrine with a false-negative patch test.

Cutaneous adverse drug reactions from systemic administration caused by type IV (delayed-type) hypersensitivity

Cutaneous adverse drug reactions from systemic administration of ephedrine caused by type IV (delayed-type) hypersensitivity, including erythroderma (3), maculopapular exanthema (2), fixed drug eruption (17) and 'itchy dermatosis' (6), are planned to be discussed in Volume IV of the *Monographs in Contact Allergy* series on Systemic drugs.

Cross-reactions, pseudo-cross-reactions and co-reactions
Epinephrine (adrenaline) and phenylephrine cross-reacted in a patient with an allergic reaction to ephedrine (3). Of 9 patients sensitized to phenylephrine, 5 cross-reacted to ephedrine 20% DMSO (16). Of 3 patients sensitized to pseudoephedrine, 2 cross-reacted to ephedrine 20% DMSO (16). A patient sensitized by topical fepradinol cross-reacted to ephedrine (16).

LITERATURE

1   The data in the section 'General' may have been obtained from literature discussed in this chapter, but mostly also or exclusively from one or more of the following online sources: ChemIDPlus Advanced, PubChem, DrugBank, RxList, Drug Central, Drugs.com, and Wikipedia
2   Maul LV, Streit M, Grabbe J. Ephedrine-induced maculopapular rash. Contact Dermatitis 2018;79:193-194
3   Tanno LK, Fillard A, Landry Q, Ramdane C, Bourrain JL, Demoly P, et al. Ephedrine-induced erythrodermia: Clinical diagnostic procedure and cross-sensitivity. Contact Dermatitis 2018;79:43-44
4   Buzo-Sánchez G, Martín-Muñoz MR, Navarro-Pulido AM, Orta-Cuevas JC. Stereoisomeric cutaneous hypersensitivity. Ann Pharmacother 1997;31:1091 (cited in ref. 3)
5   Villas Martínez F, Badas AJ, Garmendia Goitia JF, Aguirre I. Generalized dermatitis due to oral ephedrine. Contact Dermatitis 1993;29:215-216
6   Audicana M, Urrutia I, Echechipia S, Muñoz D, Fernández de Corres L. Sensitization to ephedrine in oral anticatarrhal drugs. Contact Dermatitis 1991;24:223
7   Abramowitz EW, Noun MH. Ephedrine dermatoses: Clinical and experimental study of a personal case with a review of the literature. Br J Derm Syph 1933;45:225-237
8   Scheer M. A case of dermatitis venenata due to ephedrine. Arch Derm Syph 1929;20:641
9   Ramirez MA, Eller JJ. The 'patch test' in 'contact dermatitis' (dermatitis venenata). J Allergy 1930;1:489
10 Ayres S Jr, Anderson NP. Dermatitis medicamentosa due to ephedrine. JAMA 1931;97:437-440
11 Bullen S, Francis N, Parker JM. Dermatitis medicamentosa due to ephedrine. J Allergy 1932;3:485
12 Lewis GM. Contact dermatitis-like lesions following intranasal application of ephedrine. Arch Derm Syph 1944;49:379-380
13 Hollander L. Dermatitis of the penis caused by ephedrine. JAMA 1936;106:706
14 Spencer GA. Hypersensitivity to ephedrine. Arch Derm Syph 1945;51:48-49
15 Spencer CC, Summit NJ. Pompholyx produced by sensitivity to ephedrine. Arch Derm Syph 1936;34:1028-1030
16 Barranco R, Rodríguez A, de Barrio M, Trujillo MJ, de Frutos C, Matheu V, et al. Sympathomimetic drug allergy: cross-reactivity study by patch test. Am J Clin Dermatol 2004;5:351-355
17 Tanimoto K, Shimakage T, Ayabe Y, Yamakawa T, Kumekawa M, Moriyama S. A case of fixed drug eruption due to ephedrine hydrochloride. Masui 2000;49:1374-1376 (Article in Japanese)

Chapter 3.124 EPINEPHRINE

IDENTIFICATION

Description/definition   : Epinephrine is the amine that conforms to the structural formula shown below
Pharmacological classes : β-Adrenergic agonists; bronchodilator agents; mydriatics; sympathomimetics;
                            vasoconstrictor agents; α-adrenergic agonists
IUPAC name   : 4-[(1R)-1-Hydroxy-2-(methylamino)ethyl]benzene-1,2-diol
Other names   : Adrenaline; (-)-3,4-dihydroxy-α-((methylamino)methyl)benzyl alcohol
CAS registry number   : 51-43-4
EC number   : 200-098-7
Merck Index monograph   : 4944
Patch testing   : Bitartrate or hydrochloride, 1% water
Molecular formula   : $C_9H_{13}NO_3$

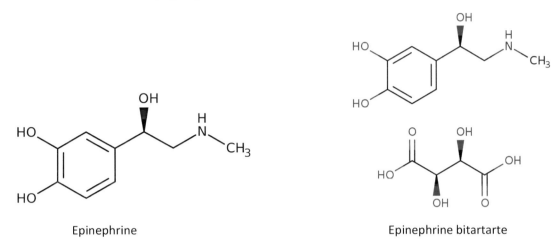

Epinephrine   Epinephrine bitartarte

GENERAL

Epinephrine (adrenaline) is the active sympathomimetic hormone from the adrenal medulla. It stimulates both the α- and β-adrenergic systems, causes systemic vasoconstriction and gastrointestinal relaxation, stimulates the heart, and dilates bronchi and cerebral vessels. Epinephrine injections are used in the emergency treatment of allergic reactions (Type I) including anaphylaxis to insects, allergen immunotherapy, foods, drugs, diagnostic testing substances (e.g. radiocontrast media) and other allergens, as well as idiopathic or exercise-induced anaphylaxis. This agent is also used in hay fever, rhinitis, acute sinusitis, bronchial asthmatic paroxysms, syncope due to complete heart block or carotid sinus hypersensitivity, serum sickness, urticaria, angioedema, for resuscitation in cardiac arrest following anesthetic accidents, in simple (open-angle) glaucoma, for relaxation of uterine musculature and to inhibit uterine contractions. Epinephrine can also be utilized to prolong the action of local anesthetics, for the maintenance of mydriasis during intraocular surgery and as a hemostatic agent. In addition, epinephrine is used as an over-the-counter agent for the intermittent symptoms of asthma, such as wheezing, tightness of the chest and shortness of breath.

In most pharmaceuticals, epinephrine base is used, sometimes epinephrine hydrochloride (CAS number 55-31-2, EC number 200-230-3, molecular formula $C_9H_{14}ClNO_3$) or epinephrine bitartrate (CAS number 51-42-3, EC number 200-097-1, molecular formula $C_{13}H_{19}NO_9$) are employed (1).

CONTACT ALLERGY

Case reports

A 53-year-old woman with chronic conjunctivitis developed edema of the eyelids followed by red, scaly itchy lesions after using a brand of epinephrine eye drops. Patch tests were positive to the eye drops and – later – to epinephrine bitartrate 1% water. Twenty controls were negative (3).

A 53-year-old man had been using an epinephrine-containing ophthalmic solution for one year for chronic wide-angle glaucoma, when he developed redness, swelling, and itching of both eyelids and eyes. Examination showed erythema, scaling, and some edema of the periorbital area. Patch testing revealed +++ reactions to L-epinephrine (test concentration and vehicle not stated, bitartrate or hydrochloride not specified) on two occasions, but not to D-epinephrine. Later, the patient reacted within hours to trial installation of epinephrine eye drops with an exacerbation of his original acute allergic periorbital contact dermatitis. Quite surprisingly, patch tests to epinephrine were now negative (4).

A 74-year-old man had been treated with epinephrine bitartrate eye drops for open-angle glaucoma, when contact dermatitis was noted, which promptly receded when the drops were stopped. Later, after a recurrence from treatment with epinephrine borate, patch tests showed positive reactions to epinephrine chloride (= hydrochloride) 1.0% and 0.1% (vehicle not mentioned, probably water), which persisted for nine days after removal of the patches. The eye drops were discontinued and the dermatitis healed spontaneously within the next few days (5).

One patient reported in an early publication apparently had contact allergy to adrenaline as shown by 'skin tests' (6, data cited in ref. 5).

Cross-reactions, pseudo-cross-reactions and co-reactions
Possible cross-reaction to primary dipivefrin contact allergy (2). A patient with an allergic reaction to ephedrine cross-reacted to epinephrine (adrenaline) 1% and 5%, both in pet. and in water (7).

LITERATURE

1   The data in the section 'General' may have been obtained from literature discussed in this chapter, but mostly also or exclusively from one or more of the following online sources: ChemIDPlus Advanced, PubChem, DrugBank, RxList, Drug Central, Drugs.com, and Wikipedia
2   Parra A, Casas L, Tombo MJ, Yanez M, Vila L. Contact dermatitis from dipivefrine with possible cross-reaction to epinephrine. Contact Dermatitis 1998;39:325-326
3   Romaguera C, Grimalt F. Contact dermatitis from epinephrine. Contact Dermatitis 1980;6:364
4   Gibbs RC. Allergic contact dermatitis to epinephrine. Arch Dermatol 1970;101:92-94
5   Alani SD, Alani MD. Allergic contact dermatitis and conjunctivitis from epinephrine. Contact Dermatitis 1976;2:147-150
6   Colldahl H, Fagerberg E. Conjunctivitis and eyelid eczema due to hypersensitiveness to adrenalin solution employed in spray-treatment of asthma. Acta Allergol 1956;10:77-81
7   Maul LV, Streit M, Grabbe J. Ephedrine-induced maculopapular rash. Contact Dermatitis 2018;79:193-194

Chapter 3.125 ERYTHROMYCIN

IDENTIFICATION

Description/definition   : Erythromycin a macrolide antibiotic produced by *Saccharopolyspora erythraea* (formerly
                              Streptomyces erythraeus); the structural formula of erythromycin A, its major active
                              component, is shown below
Pharmacological classes   : Gastrointestinal agents; protein synthesis inhibitors; anti-bacterial agents
IUPAC name   : (3R,4S,5S,6R,7R,9R,11R,12R,13S,14R)-6-[(2S,3R,4S,6R)-4-(Dimethylamino)-3-hydroxy-6-
                              methyloxan-2-yl]oxy-14-ethyl-7,12,13-trihydroxy-4-[(2R,4R,5S,6S)-5-hydroxy-4-methoxy-
                              4,6-dimethyloxan-2-yl]oxy-3,5,7,9,11,13-hexamethyl-oxacyclotetradecane-2,10-dione
Other names   : Erythromycin A
CAS registry number   : 114-07-8
EC number   : 204-040-1
Merck Index monograph   : 5009
Patch testing   : 2% pet. (SmartPracticeCanada, SmartPracticeEurope); 1% pet. (SmartPracticeCanada);
                              10.0% pet. (Chemotechnique); 1% may sometimes be too low (6,7)
Molecular formula   : $C_{37}H_{67}NO_{13}$

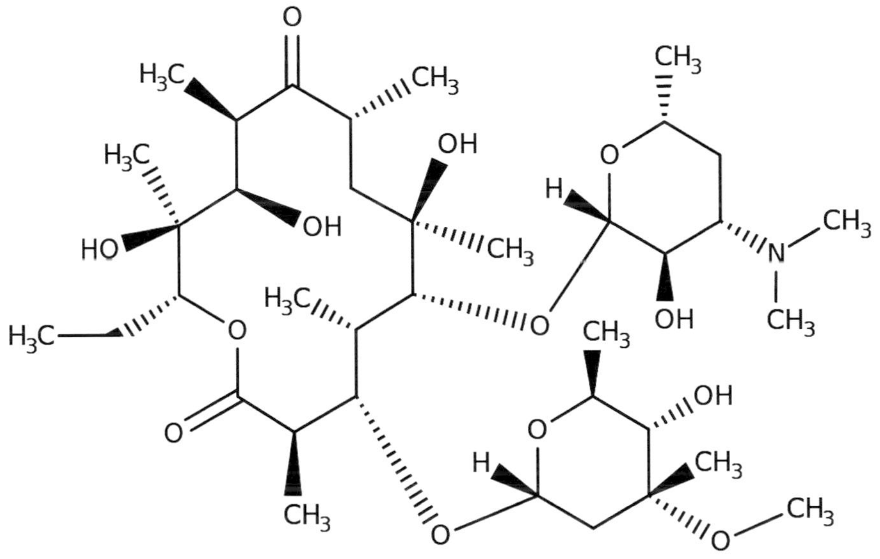

GENERAL

Erythromycin is a broad-spectrum antibiotic drug produced by a strain of *Saccharopolyspora erythraea* (formerly
Streptomyces erythraeus) and belongs to the macrolide group of antibiotics. Erythromycin may be bacteriostatic or
bactericidal in action, depending on the concentration of the drug at the site of infection and the susceptibility of the
organism involved. Erythromycin is widely used for treating a variety of infections caused by gram-positive bacteria,
gram-negative bacteria and many other organisms of the respiratory tract, skin, gastrointestinal and genital tracts
including sexually transmitted diseases (syphilis, gonorrhea, *Chlamydia* infections). In pharmaceutical products,
mostly erythromycin base is used, but many other forms (salts and esters) are possible (1).

CONTACT ALLERGY

General

Formerly, it was thought that erythromycin (at least as erythromycin base) is non-sensitizing (9,10). This has proven
to be incorrect, albeit the number of reported cases of sensitization to erythromycin is indeed extremely small.

Patch testing in groups of selected patients

In Leuven, Belgium, in the period 1990-2014, erythromycin 5% pet. was patch tested in 194 patients suspected of
iatrogenic contact dermatitis and there were 5 positive reactions to it. Relevance was not specified for individual
allergens, but 96% of all positive patch test reactions to topical drugs and antiseptics were considered to be relevant
(2).

Case reports and case series

A 35-year-old man had been treated with 1% metronidazole cream and 3% erythromycin (base) cream for rosacea for 12 days, when the patient presented with acute, severe dermatitis of the face, eyelids and forehead. Patch tests showed a strong reaction (+++) to the erythromycin cream; subsequent patch tests with its other individual ingredients (paraffin and petrolatum) were negative. Erythromycin base itself was not tested (4).

A 21-year-old woman developed pruriginous, erythematous papules and vesicles on her face, 8 days after treatment of acne with a topical preparation containing erythromycin base. Patch tests were positive to the fragrance mix I and erythromycin base 2% pet. (++). Among the other constituents of the topical medicament, there was a positive reaction to the perfume. Later, patch tests with erythromycin ethylsuccinate in ethanol were positive at 2%, 5% and 10%, but no reaction was observed when this salt of erythromycin was tested in pet. (5).

A woman aged 72 with stasis ulcers and stasis dermatitis had been treated with many different topicals drugs including erythromycin and had a positive patch test to erythromycin sulfate 25% pet. Two years later the positive result was reproduced. No mention was made of control testing (11). The authors refer to an Italian article in which 3 patients in a group of 32 with leg ulcers and surrounding eczema were found to be allergic to erythromycin, including this woman (12).

A 52-year-old woman suffering from a venous leg ulcer had been treated with 5% erythromycin stearate in yellow petrolatum for 3 weeks when the patient complained of pain in the ulcer and an eczematous eruption appeared around it. Patch tests were positive to the 5% ointment (+++) and to erythromycin stearate 0.1%, 1% and 5% pet. (13).

Cutaneous adverse drug reactions from systemic administration caused by type IV (delayed-type) hypersensitivity

Cutaneous adverse drug reactions from systemic administration of erythromycin caused by type IV (delayed-type) hypersensitivity, including fixed drug eruption (3), systemic contact dermatitis (6,7,14), acute generalized exanthematous pustulosis (AGEP) (15) and toxic epidermal necrolysis (TEN) (8), are planned to be discussed in Volume IV of the *Monographs in Contact Allergy* series on Systemic drugs.

LITERATURE

1   The data in the section 'General' may have been obtained from literature discussed in this chapter, but mostly also or exclusively from one or more of the following online sources: ChemIDPlus Advanced, PubChem, DrugBank, RxList, Drug Central, Drugs.com, and Wikipedia
2   Gilissen L, Goossens A. Frequency and trends of contact allergy to and iatrogenic contact dermatitis caused by topical drugs over a 25-year period. Contact Dermatitis 2016;75:290-302
3   Florido Lopez JF, Lopez Serrano MC, Belchi Hernandez J, Estrada Rodriguez JL. Fixed eruption due to erythromycin. Allergy 1991;46:77-78
4   Valsecchi R, Pansera B, Reseghetti A. Contact allergy to erythromycin. Contact Dermatitis 1996;34:428
5   Martins C, Freitas JD, Gonçalo M, Gonçalo S. Allergic contact dermatitis from erythromycin. Contact Dermatitis 1995;33:360
6   Fernandez Redondo V, Casas L, Taboada M, Toribio J. Systemic contact dermatitis from erythromycin. Contact Dermatitis 1994;30:311 (same as ref. 7)
7   Fernandez Redondo V, Casas L, Taboada M, Toribio J. Systemic contact dermatitis from erythromycin. Contact Dermatitis 1994;30:43-44 (same as ref. 6)
8   Lund Kofoed M, Oxholm A. Toxic epidermal necrolysis due to erythromycin. Contact Dermatitis 1985;13:273
9   Fisher AA. Is topical erythromycin base non-allergenic? Contact Dermatitis 1983;9:243
10 Fisher AA. The safety of topical erythromycin. Contact Dermatitis 1976;2:43-44
11 Lombardi P, Campolmo P, Spallanzani P, Sertoli A. Delayed hypersensitivity to erythromycin. Contact Dermatitis 1982;8:416
12 Marliani A, Lombardi P, Spallanzani P, Sertoli A. Eczema allergico da contatto iatrogeno neele ulcere degli aranti inferiori. Annali Italiani di Dermatologia Clinica e Sperimentale 1981;35:341-345 (Article in Italian, data cited in ref. 11)
13 Van Ketel WG. Immediate- and delayed-type allergy to erythromycin. Contact Dermatitis 1976;2:363-364
14 Goossens C, Sass U, Song M. Baboon syndrome. Dermatology 1997;194:421-422 aanvragen
15 Moreau A, Dompmartin A, Castel B, Remond B, Leroy D. Drug-induced acute generalized exanthematous pustulosis with positive patch tests. Int J Dermatol 1995;34:263-266

Chapter 3.126 ESTRADIOL

IDENTIFICATION

Description/definition   : Estradiol is the estrogenic steroid that conforms to the structural formula shown below
Pharmacological classes : Estrogens
IUPAC name   : (8R,9S,13S,14S,17S)-13-Methyl-6,7,8,9,11,12,14,15,16,17-decahydrocyclopenta[a]-
                           phenanthrene-3,17-diol
Other names   : 1,3,5-Estratriene-3,17β-diol; 3,17-epidihydroxyoestratriene; 17β-estradiol;
                           dihydrofolliculin; oestradiol
CAS registry number   : 50-28-2
EC number   : 200-023-8
Merck Index monograph   : 5028
Patch testing   : 5% alc. 96%
Molecular formula   : $C_{18}H_{24}O_2$

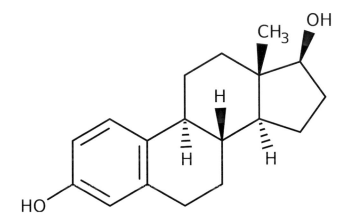

GENERAL

Estradiol is a naturally occurring hormone that circulates endogenously within the human body. It is the most potent form of mammalian estrogenic steroids and acts as the major female sex hormone. As such, estradiol plays an essential role in the regulation of the menstrual cycle, in the development of puberty and secondary female sex characteristics, as well as in ageing and several hormonally-mediated disease states. Estradiol and estradiol esters (esterification of estradiol aims to improve absorption and bioavailability after oral administration or to sustain release from depot intramuscular injections) are commercially available in oral, transdermal, and injectable hormone therapy products for managing conditions associated with reduced estrogen production such as menopausal and peri-menopausal symptoms as well as hypoestrogenism. It is also used in transgender hormone therapy, as a component of oral contraceptive pills for preventing pregnancy (most commonly as ethinylestradiol), and is sometimes used for the palliative treatment of some hormone-sensitive cancers like breast and prostate cancer (1).

CONTACT ALLERGY

General

Fifteen cases of allergic contact dermatitis from estradiol in transdermal patches have been described (2-6, 8-15). In some, oral administration of estradiol (or another estrogen derivative [12]) induced systemic allergic contact dermatitis (6,9,12), in another, it was tolerated well (13). In one sensitized patient, intravaginal application of estradiol tablets did cause local nor systemic side effects (8). Alcohol appears to be a more suitable vehicle for patch testing than petrolatum (3,9).

Case reports and case series

Between 1990 and 2003, in Leuven, Belgium, 92 patients were patch tested for chronic vulval complaints. Fifteen of these women were tested with a series of topical drugs; there was one reaction to estradiol 0.5% pet. which was relevant (2).

A 50-year-old postmenopausal woman, using 2x weekly transdermal patches containing 5 mg estradiol and 15 mg norethisterone, developed itchy discoid indurated erythema under the 3rd patch applied, followed by a bullous reaction. Because of this, treatment was switched to the estradiol-containing gel, but acute eczema also developed on sites of application of this gel. Patch tests yielded strongly positive reactions to 17β-estradiol 1% and 0.1% alc. and to norethisterone acetate 0.1% and 1% alcohol (also present in the patches). Twenty controls were negative. The patient did not react to estradiol 5% pet. (3).

A 61-year-old woman had a past history of allergic contact dermatitis to a hormone replacement patch containing both estrogen and progesterone. Positive patch test reactions were observed to estradiol 1% alc., hydroxyprogesterone 1% alc., and progesterone 1% alc. (?+/?+). There were also multiple corticosteroid allergies. This case demonstrated, according to the authors, that patients who develop contact allergies to sex steroids are at risk of developing multiple corticosteroid allergies (4).

Short summaries of other case reports of allergic contact dermatitis from estradiol in transdermal therapeutic systems are shown in table 3.126.1.

Table 3.126.1 Short summaries of case reports of allergic contact dermatitis from estradiol

Year and country	Sex	Age	Positive patch tests	Clinical data and comments	Ref.
2014 USA	F	43	TTS and spray 'as is'; EST (conc. ?) in alc.	ACD hours after first application from TTS; ACD from spray; oral estrogens had just before been well tolerated	5
2002 Italy	F	47	EST (details unknown)	ACD from TTS and gel; pruritic rash from oral estradiol (systemic contact dermatitis)	6
2000 Germany	F	29	TSS (n=3); EST 1%, 2% and 4% alc. 96%; tablet (powder?)	ACD within a few days; oral estradiol had just before been been well tolerated; estradiol tablets for vaginal application caused local nor systemic side effects; estriol was tolerated	8
1999 Portugal	F	42	TTS and gel 'as is'; EST 1%, 5% and 10% alc. 96%; 1% and 10% acetone	ACD after 6 applications; ACD to gel; systemic contact dermatitis from oral application; pet. unsuitable for PT; estradiol 5% alc. negative in 10 controls	9
	F	52	TTS and gel 'as is'; EST 1%, 5% and 10% alc. 96%	ACD after 8 applications; ACD from gel; negative or weak patch test reactions in petrolatum and acetone; systemic contact dermatitis from oral estradiol	
1996 Spain	F	36	TTS and gel 'as is'	ACD after 4 months; estradiol itself was not patch tested but reaction to identical patch without estradiol negative	11
	F	49	TTS and gel 'as is'	ACD after 7 months; estradiol itself was not patch tested but reaction to identical patch without estradiol negative	
1996 France	F	52	EST (details unknown)	ACD from TTS; generalized eczema (systemic contact dermatitis) from an estrogen derivative (details unknown)	12
1996 Germany	F	55	TTS 'as is'; EST 1%, 2% and 4% alc. 96%	ACD followed by generalization after 9th application becoming more urticarial; oral drug with estradiol and other female hormones was tolerated well; 10 controls negative	13
1992 United Kingdom	F	51	TTS 'as is'	ACD after 8 applications; estradiol itself was not patch tested, negative reaction to placebo patch (identical TTS without estradiol)	14
1989 USA	?	?	estradiol (test conc. and vehicle not provided)	pruritic papular reaction underneath the TTS, reproduced at each new application; 8 other patients with similar eruptions did not react to estradiol, but may have been sensitized by an unidentified hapten in the TSS	15

ACD: Allergic contact dermatitis; EST: Estradiol; PT: Patch testing; TTS: Transdermal therapeutic system (transdermal patch)

Another case report of contact allergy to/allergic contact dermatitis from estradiol, adequate data of which are not available to the author, can be found in refs. 10, 17 and 18.

Some women with premenstrual (flares of) urticaria, eczema, or generalized pruritus may have type-IV sensitivity to *endogenous* estrogens, as shown by positive delayed intradermal skin tests. The authors termed this autoimmune estrogen dermatitis (16).

Cross-reactions, pseudo-cross-reactions and co-reactions
A patient sensitized to testosterone had positive reactions to 2 patches containing estradiol. *In vivo*, testosterone is metabolized to estradiol by the enzyme complex aromatase, particularly in the liver and adipose tissue. Therefore, cross-reactivity does not come as a surprise (7). According to some authors, patients who develop contact allergies to sex steroids are at risk of developing multiple corticosteroid allergies (4).

LITERATURE

1   The data in the section 'General' may have been obtained from literature discussed in this chapter, but mostly also or exclusively from one or more of the following online sources: ChemIDPlus Advanced, PubChem, DrugBank, RxList, Drug Central, Drugs.com, and Wikipedia

2   Nardelli A, Degreef H, Goossens A. Contact allergic reactions of the vulva: a 14-year review. Dermatitis 2004;15:131-136

3   Koch P. Allergic contact dermatitis from estradiol and norethisterone acetate in a transdermal hormonal patch. Contact Dermatitis 2001;44:112-113

4   Lamb SR, Wilkinson SM. Contact allergy to progesterone and estradiol in a patient with multiple corticosteroid allergies. Dermatitis 2004;15:78-81

5   Ta V, Chin WK, White AA. Allergic contact dermatitis to testosterone and estrogen in transdermal therapeutic systems. Dermatitis 2014;25:279

6   Corazza M, Mantovani L, Montanari A, Virgili A. Allergic contact dermatitis from transdermal estradiol and systemic contact dermatitis from oral estradiol. A case report. J Reprod Med 2002;47:507-509

7   Shouls J, Shum KW, Gadour M, Gawkrodger DJ. Contact allergy to testosterone in an androgen patch: control of symptoms by pre-application of topical corticosteroid. Contact Dermatitis 2001;45:124-125

8   Panhans-Gross A, Gall H, Dziuk M, Peter RU. Contact dermatitis from estradiol in a transdermal therapeutic system. Contact Dermatitis 2000;43:368-369

9   Gonçalo M, Oliveira HS, Monteiro C, Clerins I, Figueiredo A. Allergic and systemic contact dermatitis from estradiol. Contact Dermatitis 1999;40:58-59

10 Fisher AA. Contact Dermatitis: Highlights from the 1987 Meeting of the American Academy of Dermatology, San Antonio, Texas. Cutis 1988;41:87-88 (cited in ref. 14)

11 Quirce S, Garde A, Baz G, Gonzalez P, Alonso Diaz de Durana MD. Allergic contact dermatitis from estradiol in a transdermal therapeutic system. Allergy 1996;51:62-63

12 El Sayed F, Bayle-Lebey P, Marguery MC, Bazex J. Systemic sensitization to 17-beta estradiol induced by transcutaneous administration. (Sensibilisation systémique au 17-β-oestradiol induite par voie transcutanée). Ann Dermatol Venereol 1996;123:26-28 (Article in French)

13 Boehncke WH, Gall H. Type-IV hypersensitivity to topical estradiol in a patient tolerant to it orally. Contact Dermatitis 1996;35:187-188

14 Carmichael AJ, Foulds IS. Allergic contact dermatitis from oestradiol in oestrogen patches. Contact Dermatitis 1992;26:194-195

15 McBurney E, Boustany Noel S, Collins J A. Contact dermatitis to transdermal estradiol system. J Am Acad Dermatol 1989;20:508-510

16 Shelley WB, Shelley ED, Talanin NY, Santoso-Pham J. Estrogen dermatitis. J Am Acad Dermatol 1995;32:25-31

17 Moureaux P. Contact eczema to 17 beta estradiol. Allerg Immunol (Paris) 1998;30:53-54

18 Coustou D, Gautier C, Ducombs G, Barbaud A, Geniaux M. Dermatitis caused by estrogens. Ann Dermatol Venereol 1998;125:505-508 (Article in French)

Chapter 3.127 ESTRADIOL BENZOATE

IDENTIFICATION

Description/definition : Estradiol benzoate is the synthetic benzoate ester of the steroid sex hormone estradiol that conforms to the structural formula shown below

Pharmacological classes : Contraceptive agents

IUPAC name : [(8R,9S,13S,14S,17S)-17-Hydroxy-13-methyl-6,7,8,9,11,12,14,15,16,17-decahydro-cyclopenta[a]phenanthren-3-yl] benzoate

Other names : 17β-Estradiol 3-benzoate

CAS registry number : 50-50-0

EC number : 200-043-7

Merck Index monograph : 5028 (Estradiol)

Patch testing : 0.01-0.05% MEK (methyl ethyl ketone)

Molecular formula : $C_{25}H_{28}O_3$

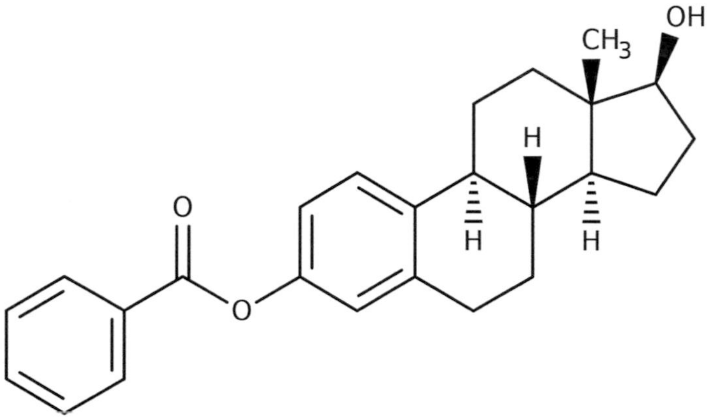

GENERAL

Estradiol benzoate is a prodrug ester of estradiol, a naturally occurring hormone that circulates endogenously within the human body. Estradiol is the most potent form of all mammalian estrogenic steroids and acts as the major female sex hormone. Following absorption, the ester is cleaved, resulting in the release of estradiol bioidentical to endogenous estradiol (1). Indications for the use of estradiol (esters) are presented in Chapter 3.126 Estradiol.

CONTACT ALLERGY

Case reports

Three female patients were prescribed a cream (one also a solution) containing 0.05% estradiol benzoate for dermatitis resp. an undefined eruption resp. seborrheic dermatitis of the face. After a few days (patients 1 and 3) to two weeks, dermatitis developed in all three. Patch tests were positive to the estradiol preparations (++). In a second patch test session, they were all positive to estradiol benzoate 0.01% MEK (++), but there were no cross-reactions to estradiol valerate, estradiol, ethinylestradiol, or diethylstilbestrol (all tested 1% alc.). Ten controls were negative.

As estradiol was negative in all three patients, the author hypothesized that the sensitizing capacity of estradiol benzoate resides in the benzoate part of the molecule rather than in the hormone portion. Accordingly, two of the patients reacted to Myroxylon pereirae resin (balsam of Peru, which contains benzoic acid and benzoates) and all 3 to resorcinol monobenzoate. The author suggested that estradiol benzoate was the primary sensitizer in these patients, although he admitted that the induction period in the present cases seems rather short. An alternate explanation would, according to the author, be that the primary sensitization was due to resorcinol monobenzoate (2).

Cross-reactions, pseudo-cross-reactions and co-reactions

No cross-reactions to estradiol valerate, estradiol, ethinylestradiol, or diethylstilbestrol in 1-3 patients sensitized to estradiol benzoate (2).

LITERATURE

1   The data in the section 'General' may have been obtained from literature discussed in this chapter, but mostly also or exclusively from one or more of the following online sources: ChemIDPlus Advanced, PubChem, DrugBank, RxList, Drug Central, Drugs.com, and Wikipedia

2   Ljunggren B. Contact dermatitis to estradiol benzoate. Contact Dermatitis 1981;7:141-144

Chapter 3.128 ESTROGENS, CONJUGATED

IDENTIFICATION

Description/definition   : Estrogens, conjugated, is a pharmaceutical preparation containing a mixture of water-
                            soluble, conjugated estrogens derived wholly or in part from urine of pregnant mares or
                            synthetically from estrone and equilin
Pharmacological classes : Estrogens
Other names   : Conjugated estrogenic hormones
CAS registry number   : 12126-59-9
EC number   : 235-199-5
Merck Index monograph   : 3762
Patch testing   : 0.625 mg/g pet.

GENERAL

Estrogens, conjugated, is a pharmaceutical preparation containing a mixture of water-soluble, conjugated estrogens derived wholly or in part from urine of pregnant mares or synthetically from estrone and equilin. It contains a sodium-salt mixture of estrone sulfate (52-62%) and equilin sulfate (22-30%) with a total of the two between 80 and 88 per cent. Other concomitant conjugates include 17α-dihydroequilin, 17α-estradiol, and 17β-dihydroequilin. The product is indicated for use as treatment of vasomotor symptoms or vulvar and vaginal atrophy due to menopause, hypoestrogenism due to hypogonadism, castration or primary ovarian failure, as palliative treatment of breast cancer with metastatic disease, as palliative treatment of androgen-dependent carcinoma of the prostate, and as preventive therapy for postmenopausal osteoporosis (1).

CONTACT ALLERGY

Case series
In the USA, in the period 2003-2010, 74 women with (predominantly) vulvar symptoms were patch tested with 'conjugate estrogen' 0.625 mg/g pet., and one patient had a relevant positive patch test. The culprit product was not mentioned (2).

LITERATURE

1   The data in the section 'General' may have been obtained from literature discussed in this chapter, but mostly
    also or exclusively from one or more of the following online sources: ChemIDPlus Advanced, PubChem,
    DrugBank, RxList, Drug Central, Drugs.com, and Wikipedia
2   O'Gorman SM, Torgerson RR. Allergic contact dermatitis of the vulva. Dermatitis 2013;24:64-72

Chapter 3.129 ETHACRIDINE

IDENTIFICATION

Description/definition : Ethacridine is the acridine-derived compound that conforms to the structural formula shown below

Pharmacological classes : Anti-infective agents, local

IUPAC name : 7-Ethoxyacridine-3,9-diamine

Other names : 2,5-Diamino-7-ethoxyacridine; Rivanol ® (ethacridine lactate)

CAS registry number : 442-16-0

EC number : 207-130-9

Merck Index monograph : 5041

Patch testing : 1% pet.

Molecular formula : $C_{15}H_{15}N_3O$

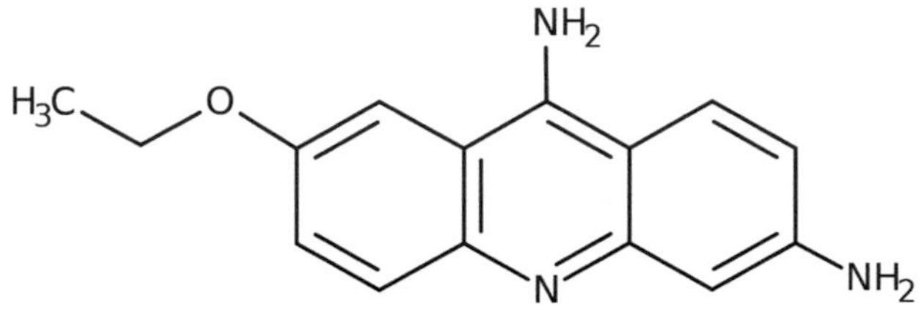

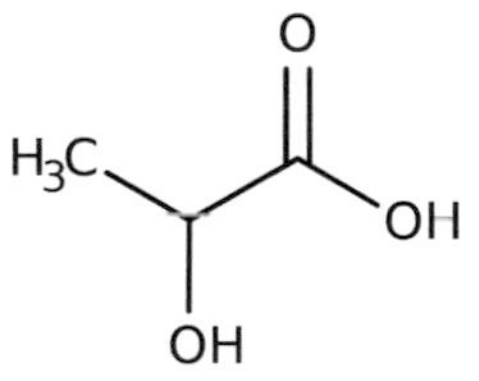

Ethacridine lactate

GENERAL

Ethacridine is an aromatic organic compound topically applied as anti-infective agent. In solutions of 0.1%, it is effective against mostly gram-positive bacteria, such as streptococci and staphylococci, but ineffective against gram-negative bacteria such as *Pseudomonas aeruginosa*. Ethacridine is or was apparently also used as an agent for second trimester abortion. Up to 150 ml of a 0.1% solution was instilled extra-amniotically using a foley catheter. After 20 to 40 hours, 'mini labor' ensues. In China, an intra-amniotic method has also been used (1).

In topical anti-infectives and antiseptics, ethacridine is used as ethacridine lactate monohydrate (CAS number 6402-23-9, EC number not available, molecular formula $C_{18}H_{23}N_3O_5$). There is also a non-hydrate ethacridine lactate with CAS number 1837-57-6. The monohydrate form is still available in local anti-infective preparations in several countries (1).

CONTACT ALLERGY

Patch testing in groups of selected patients

In Croatia, in an undefined period before 2007, 60 patients with venous leg ulcers with or without (allergic contact) dermatitis were patch tested with ethacridine lactate 1% water and there were two (3%) positive reactions; their relevance was not specified (3). Also in Croatia, in an undefined period before 1999, a group of 100 patients with leg ulcers was patch tested with ethacridine lactate 1% pet. and 20 subjects (20%) had positive reactions; relevance was not mentioned (2).

In Warsaw, Poland, in the period 1967-1970, 316 selected patients (selection criteria unknown) were patch tested with ethacridine lactate (Rivanol) 1% pet. and there were 2.9% positive reactions; their relevance was not mentioned (6).

Case reports
A male patient aged 50 had been a veterinary surgeon for 24 years and had worked all that time with ethacridine lactate. Several times per week he used to prepare 2% solutions of this drug for application to the wounds of animals. When he suffered extensive burns of both forearms, the patient was advised to use compresses of 1% ethacridine solution. The next day, he noted periorbital and palpebral edema and felt breathless. Within several hours, dyspnoea and palpebral edema increased, and dermatitis developed on both forearms at the site of ethacridine compress application. Since that time, the patient observed that after contact with ethacridine, periorbital edema of variable intensity developed, requiring, sometimes, administration of corticosteroids. After several such incidents, he stopped using ethacridine. When patch tested, he reacted to ethacridine lactate 1% pet. A diagnosis of occupational airborne contact dermatitis was made (4).

Contact allergy to ethacridine lactate (Rivanol) was also described in an early publication (5).

LITERATURE

1 The data in the section 'General' may have been obtained from literature discussed in this chapter, but mostly also or exclusively from one or more of the following online sources: ChemIDPlus Advanced, PubChem, DrugBank, RxList, Drug Central, Drugs.com, and Wikipedia
2 Marasovic D, Vuksic I. Allergic contact dermatitis in patients with leg ulcers. Contact Dermatitis1999;41:107-109
3 Tomljanović-Veselski M, Lipozencić J, Lugović L. Contact allergy to special and standard allergens in patients with venous ulcers. Coll Antropol 2007;31:751-756
4 Rudzki E, Rebandel P. Airborne contact dermatitis due to ethacridine lactate in a veterinary surgeon. Contact Dermatitis 2001;45:234
5 Epstein S. Dermal contact dermatitis; sensitivity to Rivanol and gentian violet. Dermatologica 1958;117:287-296
6 Rudzki E, Kleniewska D. The epidemiology of contact dermatitis in Poland. Br J Dermatol 1970;83:543-545

Chapter 3.130 ETISAZOLE

IDENTIFICATION

Description/definition : Etisazole is the benzothiazole that conforms to the structural formula shown below
Pharmacological classes : Antifungal agents, veterinary
IUPAC name : *N*-Ethyl-1,2-benzothiazol-3-amine
Other names : 3-(Ethylamino)-1,2-benzisothiazole; etisazol
CAS registry number : 7716-60-1; 703-83-3
EC number : 231-739-9
Merck Index monograph : 178
Patch testing : 2% pet.
Molecular formula : $C_9H_{10}N_2S$

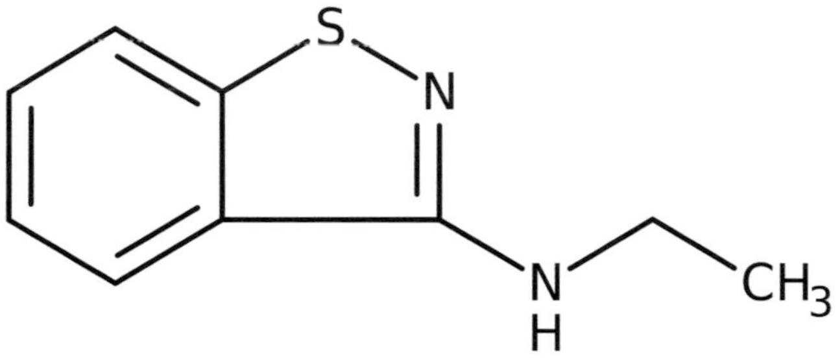

GENERAL

Etisazole is an antifungal agent used in veterinary medicine. In pharmaceutical products, etisazole may be employed as etisazole hydrochloride (CAS number 7716-59-8, EC number 231-738-3, molecular formula $C_9H_{11}ClN_2S$) (1).

CONTACT ALLERGY

Case reports

A 55-year-old farm hand had applied a local antimycotic veterinary liniment to an itchy patch on his wrist 2-3 times per day for 2 days, when an acute contact dermatitis with pruritus, swelling and bullae appeared on his wrist and arm. A few days later, a severe id-like spread developed, which persisted for 3 weeks, despite intensive local and systemic corticosteroid therapy. Patch tests with the ingredients of the liniment (etisazole, DMSO, glycerin) gave a positive reaction to etisazole 2% pet. The prolonged course of the dermatitis was ascribed to the penetration enhancing effect of DMSO. Quite curiously, the authors did not comment on the fact that the reaction already developed after 2 days, suggesting that the patient had been sensitized before. He had used miconazole earlier, but the chemical structures of these two chemicals are very dissimilar (2).

A kennel owner with relapsing vesicular hand dermatitis was examined because of acute worsening. He treated dogs with a drug containing the antifungal chemical etisazole. A patch test with etisazole 1% pet. was positive. However, the patient also had a strong patch test reaction to buclosamide, which he used as powder in his protective rubber gloves (4).

One year later, this same medicament was positive in a patch test in a woman who had used it for bilateral vesicular hand dermatitis, but the ingredients were not tested separately (3).

LITERATURE

1   The data in the section 'General' may have been obtained from literature discussed in this chapter, but mostly also or exclusively from one or more of the following online sources: ChemIDPlus Advanced, PubChem, DrugBank, RxList, Drug Central, Drugs.com, and Wikipedia
2   Vanhee J, Ceuterick A, Dooms M, Dooms-Goossens A. Etisazole: an animal antifungal agent with skin sensitizing properties in man. Contact Dermatitis 1980;6:443
3   Malten KE. Therapeutics for pets as neglected causes of contact dermatitis in housewives. Contact Dermatitis 1978;4:296-299
4   Dahlquist I. Contact allergy to 3-ethylamino-1,2-benzisothiazol-hydrochloride, a veterinary fungicide. Contact Dermatitis 1977;3:277

Chapter 3.131 ETOFENAMATE

IDENTIFICATION

Description/definition   : Etofenamate is the non-steroidal anti-inflammatory drug that conforms to the structural formula shown below
Pharmacological classes : Anti-inflammatory agents, non-steroidal
IUPAC name   : 2-(2-Hydroxyethoxy)ethyl 2-[3-(trifluoromethyl)anilino]benzoate
Other names   : 2-[3-(Trifluoromethyl)anilino]benzoic acid 2-(2-hydroxyethoxy)ethyl ester
CAS registry number   : 30544-47-9
EC number   : 250-231-8
Merck Index monograph   : 5191
Patch testing   : 2.0% pet. (Chemotechnique, SmartPracticeCanada)
Molecular formula   : $C_{18}H_{18}F_3NO_4$

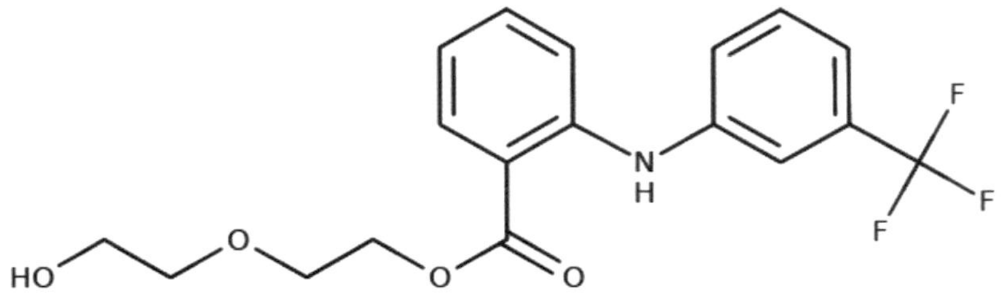

GENERAL

Etofenamate is a benzoate ester and anthranilic acid derivative with analgesic, antirheumatic, antipyretic and anti-inflammatory properties. It is a non-selective COX inhibitor affecting also the lipo-oxygenase pathway. This nonsteroidal anti-inflammatory drug (NSAID) is commonly used in Mediterranean countries (Spain, Italy, Portugal) in topical preparations for the treatment of joint and muscular pain (1).

CONTACT ALLERGY

General

Allergic contact dermatitis, photoallergic contact dermatitis and photoaggravated allergic contact dermatitis to etofenamate have been observed and reported frequently. Of the NSAIDs, etofenamate appears to be the most frequent cause of contact allergy after ketoprofen and of photocontact allergy after ketoprofen and possibly benzydamine.

Patch testing in groups of patients

In Germany, before 1998, 371 consecutive patients suspected of contact dermatitis (routine testing) were patch tested with etofenamate 5% pet. and there were 2 (0.5%) positive reactions. The relevance of one was 'probable', the other had unknown relevance (4).

The result of patch testing etofenamate in groups of *selected* patients are shown in table 3.131.1. In 3 studies, prevalences of sensitization to etofenamate ranged from 1.0% to 12.9%, likely dependent on the selection procedure. Relevance was either not stated, nearly 100%, or mostly unknown. In all 3 studies, there were also an equal number or many more cases of *photo*sensitization to etofenamate (2,23,26).

Table 3.131.1 Patch testing in groups of patients

Years and Country	Test conc. & vehicle	Number of patients tested	positive (%)	Selection of patients (S); Relevance (R); Comments (C)	Ref.
2014-2016 Spain		116	6 (5.2%)	S: not stated; R: not stated; C: there were also 7 positive photopatch tests to etofenamate	23
1990-2014 Belgium	2% pet.	295	38 (12.9%)	S: patients suspected of iatrogenic contact dermatitis; nearly all were relevant; C: 10/38 had photoaggravated contact allergy; there were also 32 cases of photocontact allergy	2
2008-2011 Europe	2% pet.	1031	10 (1.0%)	S: patients suspected or photoallergic contact dermatitis; R: most had unknown relevance; C: there were also 59 positive photopatch tests	26

Case series

In a hospital in Spain, in a period of 8 years before 2009, 14 cases of allergic and/or photoallergic contact dermatitis due to etofenamate have been identified. There were 10 women and 4 men, ages ranging from 21 to 70 years. The time of onset from commencement of therapy to first signs of (photo)contact dermatitis was less than 10 days in 12, of whom 4 within 4 days. The most common sites of involvement of the reactions were the feet, legs and neck. In some cases, bullous lesions were seen. All patients had a history of using one or more NSAID topical products containing etofenamate. All 14 individuals were patch tested with an NSAID series (not containing related anthranilic acid derivatives such as meclofenamate, niflumic acid and mefenamic acid), etofenamate 1%, 5% and 10% pet., as well as with the actual commercial medicament used by the individual patient when possible. Photopatch tests were performed in 10 patients. Seven patients had allergic contact dermatitis, five photoallergic contact dermatitis and two combined allergic and photoallergic. In only 2 patients, there were positive photopatch tests to other NSAIDs (ketoprofen, dexketoprofen, piketoprofen), probably independent photosensitizations. In all cases where they were tested, the commercial products showed positive (photo)patch tests (6).

Another four patients from Spain had allergic contact dermatitis from using an anti-inflammatory gel containing etofenamate. Patch tests were positive to the commercial preparation and etofenamate 2% pet. Only one cross-reacted to the related flufenamic acid 2% in an NSAIDs series (7).

In Italy, before 1993, the members of the GIRDCA Multicentre Study Group diagnosed 102 patients (49 men, 53 women, aged 16 to 66 years, mean 37 years), with (photo)dermatitis induced by systemic or topical NSAIDs. Etofenamate caused one contact allergic and zero photocontact allergic reactions (3).

In the period 1996-2001, in 2 hospitals in Spain, 10 patients were diagnosed with contact allergy and zero with photocontact allergy to etofenamate. The accumulated incidence per million inhabitants (catchment population of the hospitals) of both side effects together was 12.1 (28).

Between 1980 and 1994, in a hospital in Madrid, Spain, 9 patients (5 women, 4 men; ages ranging from 14 to 71 years) were investigated who had developed eczema from using etofenamate gel or solution. The preparations had been used in 3 patients for longer than 10 days before symptoms started, but in 4, the latent period was only 1-4 days. None of the patients had – apparently – used etofenamate preparations previously. The pharmaceuticals had mostly been applied to the legs, hands and feet. All 9 patients had positive patch tests to the commercial preparation and to etofenamate 2% pet. In two, photopatch tests were performed, but the reactions were not stronger than in the conventional patch tests (12).

Three patients with allergic contact dermatitis from etofenamate were reported from Germany in 1989 (14). Clinical details were not provided except that all 3 were women and they aged 39, 47 and 49 years (14). Three more cases of contact dermatitis following topical application of etofenamate were reported from Germany in 1988. Each patient had positive patch test reactions to etofenamate 1% and 10% pet. In one, there was a flare-up of previous dermatitis after patch testing (15).

Case reports

A 44-year-old man had been treated with an anti-inflammatory gel containing etofenamate for soreness on his right thigh, when, on the fourth day, itchy erythema, vesicles and exudation developed. He had not been exposed to sunlight. Patch tests were positive to the gel and to etofenamate 2% pet., but not to other NSAIDs including related anthranilic acid derivatives (8).

A 67-year-old woman had applied etofenamate 5% gel on her right ankle for arthralgias for one month, when she developed itching erythematous, bilateral asymmetric plaques on her ankles and legs after exposure to the sun. Patch and photopatch tests were both positive (+++, equal strength) to the gel and to etofenamate 0.5% and 5% pet. To etofenamate 0.05% pet. there was a +/++ reaction on D2 and D4 in the patch test and a far stronger (+++) reaction to the photopatch test. Controls were negative (9). The authors conclude – only based on the difference in the 0.05% patch/photopatch – that this was a case of combined contact and photocontact dermatitis, which could be challenged.

A 70-year-old woman had been treated daily with etofenamate gel to a hematoma on the right leg for one week when she developed erythema, edema and vesicles at the application site. Patch testing revealed a positive reaction to etofenamate 2% pet. but she was negative to the topical product itself and its excipients (13). A 53-year-old woman presented with exudative erythematous dermatitis localized to the arms and face. For rheumatic elbow pain, she had been applying etofenamate cream for 3 days. Patch tests were positive to the cream and to etofenamate 5% in lanolin; lanolin itself tested negative. Five controls were negative (17).

A 55-year-old woman had arthrosis in the right hand and was receiving topical treatment with a product containing 5% etofenamate. About 20 minutes after the 3rd application, severe itching followed by erythematous papular lesions appeared in the area of application of the aerosol, but disappeared spontaneously 3 days after discontinuing the medication. Patch tests were positive to etofenamate 5% pet. at D2 and D4 (18). Although the authors did not comment on this, appearance of lesions after 20 minutes is incompatible with allergic contact dermatitis and suggests contact urticaria (immediate contact reaction). On the other hand, the patch test was positive and immediate contact reactions should disappear in a matter of hours, not 3 days.

A 13-year-old girl developed allergic contact dermatitis with erythema, vesicles and exudation on her thigh, after 3 days' application of a gel containing etofenamate. A use test was strongly positive and patch testing yielded positive reactions to the gel and etofenamate, but negative to the ingredients of the gel (19). A 20-year-old woman presented with a vesicular weeping eczema on her right ankle. An intensely pruritic and severe id-like spread occurred simultaneously on all her limbs and in her groin. She had been treated with etofenamate gel for a sprained ankle. Patch testing showed positive reactions to the gel and to etofenamate 2% pet., whereas the other ingredients of the pharmaceutical remained negative (20).

A 31-year-old-man complained of lumbosacral pain and was given etofenamate gel and diclofenac tablets for the first time. Four days later, he developed generalized pruritic papuloerythematous lesions, beginning at the site of application, with thin scaling. Axillary and inguinal lymphadenopathy, fever (38°C), leukocytosis (12.2 X 10^9/l) with marked eosinophilia (3.5 X 10^9/l) and elevated serum glutamic pyruvic transaminase (57 U/l) were also present. Skin biopsy revealed moderate parakeratosis, exocytosis and a perivascular mononuclear infiltrate. Patch tests with the Portuguese standard series were negative, positive to etofenamate 0.5%, 1% and 2% pet. at D2 and D4, while diclofenac at 1%, 5% and 10% pet. was negative (22). The authors ascribe the 'exfoliative dermatitis' to contact allergy to etofenamate. However, it is far more likely that, although the patient had contact allergy to etofenamate, the lymphadenopathy and other systemic symptoms and possibly also (part of) the skin exanthema were caused by diclofenac. A cross-reaction between etofenamate and diclofenac is unlikely, as they are NSAIDs from different structural groups.

A 66-year-old woman presented with acute eczema at the contact area of an elbow strap on the right distal elbow she had been using for relief of her epicondylitis for 6 weeks. The elbow strap had a neoprene pressure pad in the center. Topical nonsteroidal anti-inflammatory drugs were used concomitantly at the beginning, but not any longer during the last 2-3 weeks. Patch testing showed a +++ positive reaction to the inner part of the used neoprene pad and +++ positive reactions to the previously used anti-inflammatory creams 'as is', containing etofenamate or thiocolchicoside. Patch testing with a new piece of pressure band of an unused elbow strap of the same brand showed no reaction, suggesting the role of the retained NSAIDs in the used pad. Patch testing could not be performed with etofenamate or thiocolchicoside separately, but was negative with the tested inactive ingredients: parabens, lanolin alcohol and lavender oil (29).

One patient had allergic contact dermatitis from etofenamate and diclofenac (11) and another to etofenamate (16); details are not available to the author.

PHOTOSENSITIVITY

The action spectrum for etofenamate photoallergic contact dermatitis was found to be 335-430 ± 27 nm, supporting the use of UVA irradiation for photopatch testing with this NSAID (5).

Photopatch testing in groups of selected patients

In a group of 38 patients with photocontact allergy, one with photoaggravated contact allergy and 3 with contact allergy to topical ketoprofen pharmaceuticals, 36 were patch tested and 35 photopatch tested with etofenamate 2% pet. and there were 4 positive patch tests, 15 positive photopatch tests and 5 photoaggravated reactions (25).

Table 3.131.2 Photopatch testing in groups of patients

Years and Country	Test conc. & vehicle	Number of patients tested | positive (%)		Selection of patients (S); Relevance (R); Comments (C)	Ref.
2014-2016 Spain		116	7 (6.0%)	S: not stated; R: 57% for all NSAID reactions together; C: there were also 6 positive 'plain' patch tests to etofenamate	23
1990-2014 Belgium	2% pet.	295	32 (10.8%)	S: patients suspected of iatrogenic contact dermatitis; nearly all were relevant; C: there were also 38 cases of contact allergy to etofenamate, 10 of which were photoaggravated	2
2008-2011 Europe	2% pet.	1031	59 (5.7%)	S: patients suspected or photoallergic contact dermatitis; R: most had unknown relevance; C: it was suggested that a significant number or reactions were phototoxic; there were also 10 positive conventional patch tests	26
2004-2008 Italy	10% pet.	1082	1 (0.1%)	S: patients with histories and clinical features suggestive of photoallergic contact dermatitis; R: the reaction was relevant; C: multicenter study	24

Why there were so many reactions to etofenamate is unknown; the differences in structural formulas make cross-reactivity to ketoprofen unlikely. However, it is well known that photocontact allergy to ketoprofen is often associated with other photocontact sensitizations to unrelated chemicals (Chapter 1.193 Ketoprofen).

The results of photopatch testing with etofenamate are summarized in table 3.131.2. Prevalences of photocontact allergy have ranged from 0.1% to 10.8%, likely depending on the selection criteria. Many (23) or most (2) reactions were considered to be relevant, but in another European study, most of 59 positive photopatch tests had unknown relevance and it was suggested that a significant number of these were phototoxic rather than photoallergic (26).

Case reports and case series

In Krakow, Poland, before 2013, 36 patients with a history of dermatitis induced or aggravated by exposure to light and were photopatch tested. Nine patients were photoallergic to at least one non-steroidal anti-inflammatory drug, with ketoprofen photoallergy being most frequent (5 patients, in each case clinically relevant), followed by etofenamate (4 non-relevant reactions) and diclofenac (1 relevant reaction) (27).

In a hospital in Spain, in a period of 8 years before 2009, five patients with photoallergic contact dermatitis and two with combined allergic and photoallergic dermatitis were seen (6). These are discussed in more detail in the section 'Case reports' (under Contact allergy) above.

In the period 2004-2005, in seven hospitals in Spain, 5 patients had a positive photopatch test to etofenamate 1% pet., the highest number after ketoprofen (n=43) and benzydamine (n=7); It was not mentioned how many patients had been tested with this chemical (21).

A 17-year-old man presented in the summer with sharply outlined acute eczematous lesions on his knees. He played football in shorts and used etofenamate spray for muscle stiffness. He had occasionally used the spray without adverse effects for approximately 2 months previously. Patch tests were negative, but photopatch tests were positive to the spray and to etofenamate 2% pet. Fifteen controls were negative (10).

A possible case of combined photoallergic contact dermatitis and contact dermatitis to etofenamate is discussed in the section 'Case reports' (under Contact allergy) above (9). In one patient, photoaggravated contact allergy and allergic contact dermatitis to etofenamate led to prolonged photosensitivity of the affected site (30).

Cross-reactions, pseudo-cross-reactions and co-reactions

Patients who are allergic to etofenamate do not cross-react to non-structurally-related NSAIDs such as the arylpropionic acid group (ketoprofen, dexketoprofen, piketoprofen, tiaprofenic acid, ibuprofen, fenoprofen) (6,8), and infrequently to other anthranilic acid derivatives such as sodium meclofenamate, mefenamic acid, flufenamic acid (7) and niflumic acid (8). A patient photosensitized to flufenamic acid photocross-reacted to etofenamate (10).

Immediate contact reactions

Immediate contact reactions (contact urticaria) to etofenamate are presented in Chapter 5.

LITERATURE

1   The data in the section 'General' may have been obtained from literature discussed in this chapter, but mostly also or exclusively from one or more of the following online sources: ChemIDPlus Advanced, PubChem, DrugBank, RxList, Drug Central, Drugs.com, and Wikipedia
2   Gilissen L, Goossens A. Frequency and trends of contact allergy to and iatrogenic contact dermatitis caused by topical drugs over a 25-year period. Contact Dermatitis 2016;75:290-302
3   Pigatto PD, Mozzanica N, Bigardi AS, Legori A, Valsecchi R, Cusano F, et al. Topical NSAID allergic contact ` dermatitis. Italian experience. Contact Dermatitis 1993;29:39-41
4   Gniazdowska B, Rueff F, Przybilla B. Delayed contact hypersensitivity to non-steroidal anti-inflammatory drugs. Contact Dermatitis 1999;40:63-65
5   Kerr A, Becher G, Ibbotson S, Ferguson J. Action spectrum for etofenamate photoallergic contact dermatitis. Contact Dermatitis 2011;65:117-118
6   Goday Buján JJ, Pérez Varela L, Piñeyro Molina F, Díaz Román T, Fonseca E. Allergic and photoallergic contact dermatitis from etofenamate: study of 14 cases. Contact Dermatitis 2009;61:118-120
7   Alcántara Villar M, Pagan JA, Palacios L, Quiralte J, Ramirez M. Allergic contact dermatitis to etofenamate. Cross-reaction to other nonsteroidal anti-inflammatory drugs. Contact Dermatitis 2008;58:118-119
8   Chu CY, Chen YL, Lin LJ, Sun CC. Allergic contact dermatitis from etofenamate without cross-sensitization to other anthranilic acid derivatives. Dermatology 2003;206:341-342
9   Sánchez-Pérez J, Sánchez TS, García-Díez A. Combined contact and photocontact allergic dermatitis to etofenamate in flogoprofen gel. Am J Contact Dermat 2001;12:215-216

10 Montoro J, Rodriguez-Serna M, Liñana JJ, Ferré MA, Sanchez-Motilla JM. Photoallergic contact dermatitis due to flufenamic acid and etofenamate. Contact Dermatitis 1997;37:139-140

11 Gonzalo MA, Revenga F. Multiple cutaneous sensitization to non-steroidal anti-inflammatory drugs. Dermatology 1996;193:59-60

12 Hergueta JP, Ortiz FJ, Iglesias L. Allergic contact dermatitis from etofenamate: report of 9 cases. Contact Dermatitis 1994;31:60-62

13 Guerra L, Piraccini BM, Adami F, Venturo N, Tardio M. Contact dermatitis due to etofenamate. Contact Dermatitis 1992;26:199

14 Beller U. Cross-allergy to non-steroidal antiphlogistic drugs. In: Frosch P, Dooms-Goossens A, Lachapelle J-M, Rycroft RJG, Scheper RJ, Eds. Current topics in contact dermatitis. Berlin: Springer-Verlag, 1989:4248-249

15 Degenhardt A, Zick C, Hausen BM. Allergic contact dermatitis caused by etofenamate. Z Hautkr 1988;63:475-478 (Article in German)

16 Knoll R, Ulrich R, Spallek W. Allergic contact dermatitis to etofenamate and dwarf pine oil. Sportverletz Sportschaden 1990;4:96-98 (Article in German)

17 Götze A, Teikemeier G, Goerz G. Contact dermatitis from etofenamate. Contact Dermatitis 1992;26:209

18 Piñol J, Navarro M, Carapeto FJ. Allergic contact dermatitis to etophenamate. Contact Dermatitis 1985;13:193

19 Balato N, Lembo G, Cantelli V, Ayala F. Allergic dermatitis from etofenamate. Contact Dermatitis 1984;11:190

20 Vanhee J, Gevers D, Dooms-Goossens A. Contact dermatitis from an antirheumatic gel containing etofenamate. Contact Dermatitis 1981;7:50-51

21 Cuadra-Oyanguren J, Pérez-Ferriols A, Lecha-Carrelero M, Giménez-Arnau AM, Fernández-Redondo V, Ortiz de Frutos FJ, et al. Results and assessment of photopatch testing in Spain: towards a new standard set of photoallergens. Actas Dermosifilogr 2006;98:96-101

22 Correia O, Barros M A. Exfoliative dermatitis with etofenamate. Contact Dermatitis 1990;23:264

23 Subiabre-Ferrer D, Esteve-Martínez A, Blasco-Encinas R, Sierra-Talamantes C, Pérez-Ferriols A, Zaragoza-Ninet V. European photopatch test baseline series: A 3-year experience. Contact Dermatitis 2019;80:5-8

24 Pigatto PD, Guzzi G, Schena D, Guarrera M, Foti C, Francalanci, S, Cristaudo A, et al. Photopatch tests: an Italian multicentre study from 2004 to 2006. Contact Dermatitis 2008;59:103-108

25 Devleeschouwer V, Roelandts R, Garmyn M, Goossens A. Allergic and photoallergic contact dermatitis from ketoprofen: results of (photo) patch testing and follow-up of 42 patients. Contact Dermatitis 2008;58:159-166

26 The European Multicentre Photopatch Test Study (EMCPPTS) Taskforce. A European multicentre photopatch test study. Br J Dermatol 2012;166:1002-1009

27 Spiewak R. The frequency and causes of photoallergic contact dermatitis among dermatology outpatients. Acta Dermatovenerol Croat 2013;21:230-235

28 Diaz RL, Gardeazabal J, Manrique P, Ratón JA, Urrutia I, Rodríguez-Sasiain JM, Aguirre C. Greater allergenicity of topical ketoprofen in contact dermatitis confirmed by use. Contact Dermatitis 2006;54:239-243

29 Özkaya E. Patch testing with used and unused personal products : a practical way to show contamination with contact allergens. Contact Dermatitis 2016;75:328-330

30 Espasandín-Arias M, Vázquez-Osorio I, Salgado-Boquete L, Rodríguez-Granados MT, Toribio J. Prolonged localized photosensitivity following allergic and photo-aggravated contact dermatitis from etofenamate. Photodermatol Photoimmunol Photomed 2014;30:340-342

Chapter 3.132 EUCAINE

IDENTIFICATION

Description/definition : Eucaine is the piperidine derivative and benzoate ester that conforms to the structural formula shown below
Pharmacological classes : Anesthetics, local
IUPAC name : (2,2,6-Trimethylpiperidin-4-yl) benzoate
Other names : Benzamine lactate (in publication, wrong name); 2,2,6-trimethyl-4-piperidinol benzoate (ester); beta-eucaine; β-eucaine
CAS registry number : 500-34-5
EC number : Not available
Merck Index monograph : 5208
Patch testing : 1% and 5% pet.
Molecular formula : $C_{15}H_{21}NO_2$

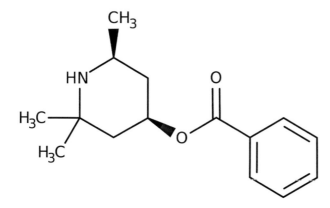

GENERAL
Eucaine is an analog of cocaine that was formerly used as a local anesthetic (1).

CONTACT ALLERGY

Case report
A man aged 57 used a local anesthetic ointment for treatment of hemorrhoids. He applied it two or three times a day and after 2 days found that it caused soreness. In addition he used suppositories and a corticosteroid cream. A generalized eruption on the hands, arms, chest and thighs ensued. Examination showed a widespread dermatitis and a macerated perianal area. Treatment with a topical steroid and clioquinol produced complete resolution in 2 weeks. Patch test were positive to the cream (+/+) and – later – to its ingredient eucaine (wrongly termed benzamine lactate in the publication) 1% pet. (2). This was probably a case of systemic contact dermatitis from percutaneous absorption in the skin and mucosa of the anal area.

LITERATURE
1   The data in the section 'General' may have been obtained from literature discussed in this chapter, but mostly also or exclusively from one or more of the following online sources: ChemIDPlus Advanced, PubChem, DrugBank, RxList, Drug Central, Drugs.com, and Wikipedia
2   Calnan CD. Sensitivity to benzamine lactate. Contact Dermatitis 1975;12:56

Chapter 3.133 FELBINAC

IDENTIFICATION
Description/definition   : Felbinac is the monocarboxylic acid that conforms to the structural formula shown below
Pharmacological classes : Anti-inflammatory agents, non-steroidal
IUPAC name   : 2-(4-Phenylphenyl)acetic acid
Other names   : (1,1'-Biphenyl)-4-acetic acid; 4-biphenylacetic acid
CAS registry number   : 5728-52-9
EC number   : 227-233-2
Merck Index monograph : 5255
Patch testing   : 1% and 5% pet.
Molecular formula   : $C_{14}H_{12}O_2$

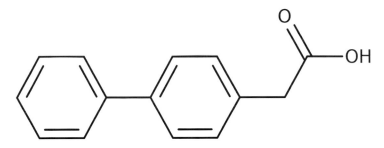

GENERAL
Felbinac is a nonsteroidal anti-inflammatory drug (NSAID) with analgesic, antipyretic and anti-inflammatory properties. It is used in topical pharmaceutical preparations to treat muscle inflammation and arthritis (1).

CONTACT ALLERGY

Case report
A 65-year-old man developed an extremely pruritic eruption following the use of compresses containing felbinac (in the article erroneously termed 'ferbinac') for one day. Examination showed a rectangular-shaped, severely pruritic exudative erythematous macular area on the right shoulder and upper arm and diffuse erythema on the chest, abdomen, back and extremities. After discontinuing the compress and following medical treatment, the patient had no recurrence of the dermatitis. Patch testing demonstrated positive reactions to felbinac 1% pet., crotamiton 1% pet., and diisopropanolamine 1% pet. Later, the patient was patch tested again with felbinac, 0.1%, 1% and 5% pet., yielding positive reactions to the 1% and 5% test materials. Twenty controls were negative (2).

LITERATURE
1   The data in the section 'General' may have been obtained from literature discussed in this chapter, but mostly also or exclusively from one or more of the following online sources: ChemIDPlus Advanced, PubChem, DrugBank, RxList, Drug Central, Drugs.com, and Wikipedia
2   Oiso N, Fukai K, Ishii M. Triple allergic contact sensitivities due to ferbinac, crotamiton and diisopropanolamine. Contact Dermatitis 2003;49:261-263

Chapter 3.134 FENOTEROL

IDENTIFICATION

Description/definition : Fenoterol is the aralkylamine that conforms to the structural formula shown below
Pharmacological classes : Bronchodilator agents; β_2-adrenergic receptor agonists; sympathomimetics; tocolytic agents
IUPAC name : 5-[1-Hydroxy-2-[1-(4-hydroxyphenyl)propan-2-ylamino]ethyl]benzene-1,3-diol
Other names : 1-(3,5-Dihydroxyphenyl)-2-((1-(4-hydroxybenzyl)ethyl)amino)ethanol
CAS registry number : 130156-24-0; 69421-37-0
EC number : Not available
Merck Index monograph : 5282
Patch testing : No data available
Molecular formula : $C_{17}H_{21}NO_4$

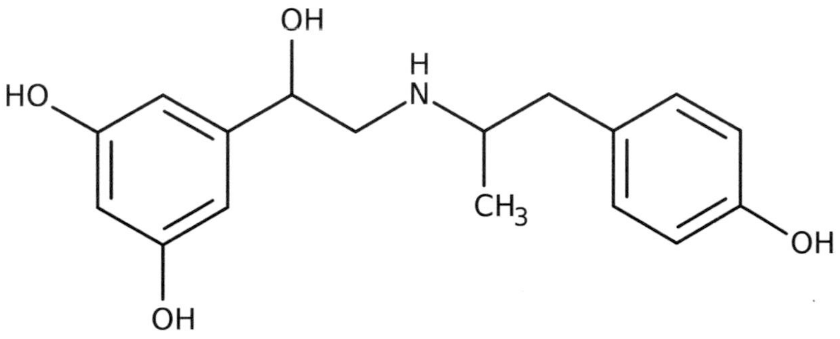

GENERAL

Fenoterol is a β_2-adrenergic receptor agonist and a short-acting sympathomimetic agent with bronchodilator and tocolytic activities. It is used for the treatment of asthma. In pharmaceutical products, fenoterol is usually employed as fenoterol hydrobromide (CAS number 1944-12-3, EC number 217-742-8, molecular formula $C_{17}H_{22}BrNO_4$) (1).

CONTACT ALLERGY

Case report

A man aged 66 years had suffered from asthma for 20 years, when he reported sore lips, mouth, tongue and throat that had started 4 weeks earlier, and 2 weeks' duration of swelling of the upper lip. Examination revealed crusted vermilion of the lips, angular cheilitis, swollen upper lip, hyperemic buccal mucosa and pharynx, increased rugae of swollen tongue and swollen uvula. There was no itching. For his asthma he inhaled fenoterol hydrobromide, salbutamol and beclomethasone. Spray of these agents to the skin of the upper back with plastic tape occlusion of the test sites revealed a ++ reaction to fenoterol and a negative reaction to a fenoterol placebo (the same spray but without fenoterol). One of 6 controls had a + reaction to fenoterol spray. The patient was advised to discontinue inhalation of fenoterol but to continue inhalation of salbutamol and beclomethasone. The symptoms disappeared within 2 weeks. The tests were repeated and were identical, but 4 months later, when the patient was still symptom free, fenoterol had become negative in a third test session (2). Contact allergy to fenoterol was not shown convincingly in this case.

LITERATURE

1 The data in the section 'General' may have been obtained from literature discussed in this chapter, but mostly also or exclusively from one or more of the following online sources: ChemIDPlus Advanced, PubChem, DrugBank, RxList, Drug Central, Drugs.com, and Wikipedia
2 Groot D, Mitchell JC. Oro-pharyngeal mucosal reaction to fenoterol. Contact Dermatitis 1981;7:48

Chapter 3.135 FENTICONAZOLE

IDENTIFICATION

Description/definition   : Fenticonazole is the member of the imidazoles that conforms to the structural formula
                                           shown below
Pharmacological classes : Antifungal agents
IUPAC name   : 1-[2-(2,4-Dichlorophenyl)-2-[(4-phenylsulfanylphenyl)methoxy]ethyl]imidazole
CAS registry number   : 72479-26-6
EC number   : Not available
Merck Index monograph : 5302
Patch testing   : 1% alc.
Molecular formula   : $C_{24}H_{20}Cl_2N_2OS$

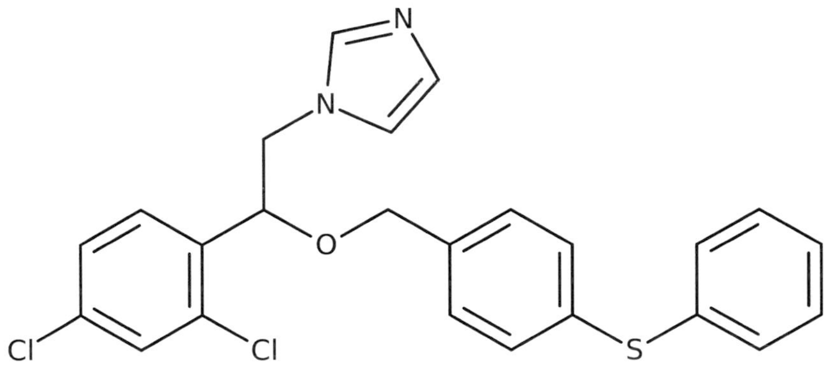

GENERAL

Fenticonazole is an imidazole antifungal agent, used both topically (especially for vulvovaginitis caused by *Candida albicans*) and systemically. In pharmaceutical products, fenticonazole is usually employed as fenticonazole nitrate (CAS number 73151-29-8, EC number 277-302-6, molecular formula $C_{24}H_{21}Cl_2N_3O_4S$) (1).

CONTACT ALLERGY

Case series

In Bologna, Italy, in 'the few years' before 1995, 20 cases of allergic contact dermatitis due to imidazole preparations were found. However, patch tests with the relevant imidazoles were positive in only 12, including one to fenticonazole (2).

LITERATURE

1   The data in the section 'General' may have been obtained from literature discussed in this chapter, but mostly
     also or exclusively from one or more of the following online sources: ChemIDPlus Advanced, PubChem,
     DrugBank, RxList, Drug Central, Drugs.com, and Wikipedia
2   Guidetti MS, Vincenzi C, Guerra L, Tosti A. Contact dermatitis due to imidazole antimycotics. Contact Dermatitis
     1995;33:282

Chapter 3.136 FEPRADINOL

IDENTIFICATION

Description/definition : Fepradinol is the aralkylamine that conforms to the structural formula shown below
Pharmacological classes : Anti-inflammatory agents, non-steroidal
IUPAC name : 2-[(2-Hydroxy-2-phenylethyl)amino]-2-methylpropan-1-ol
Other names : α-(((2-Hydroxy-1,1-dimethylethyl)amino)methyl)benzyl alcohol
CAS registry number : 63075-47-8
EC number : Not available
Merck Index monograph : 5308
Patch testing : 1-2% alcohol
Molecular formula : $C_{12}H_{19}NO_2$

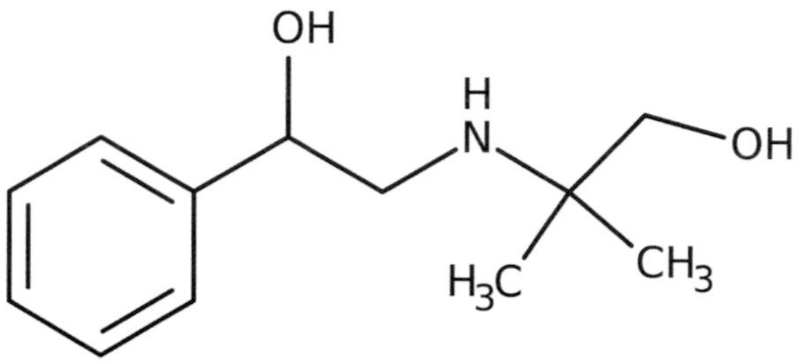

GENERAL

Fepradinol is a phenylethanolamine derivative and nonsteroidal anti-inflammatory drug (NSAID) with analgesic, antipyretic and anti-inflammatory properties. In pharmaceutical products, fepradinol is usually employed as fepradinol hydrochloride (CAS number 67704-50-1, EC number not available, molecular formula $C_{12}H_{20}ClNO_2$) (1). Fepradinol is considered an adrenergic compound, due to its structural similarity with other α-adrenergic drugs, and is unrelated to the 'classic' NSAIDs such as ibuprofen and ketoprofen (5).

CONTACT ALLERGY

General

Several case reports and small case series of allergic contact dermatitis (3,6,7,8,9,10,11,12,13) and one of photoallergic contact dermatitis (4) from topical use of fepradinol have been reported. All came from Spain, where two commercial sprays containing 6% fepradinol are frequently used for muscle and joint pains (3).

Case series

Three men and a woman, ages ranging from 48-70 years, developed allergic contact dermatitis after having used 6% fepradinol spray to various parts of the body for 6 days to 10 weeks. All had positive patch tests to fepradinol 1% pet., alcohol and water. Three reacted to 0.1% in all these vehicles and one to 0.01%. Fifteen controls were negative to fepradinol 1% pet., alcohol and water (9).

In the period 1996-2001, in 2 hospitals in Spain, 3 patients were diagnosed with contact allergy and zero with photocontact allergy to fepradinol. The accumulated incidence per million inhabitants (catchment population of the hospitals) of both side effects together was 3.6 (14).

Case reports

A 75-year-old man presented with a one-week history of purpuric papulonecrotic lesions in the lumbar region. Two days before the onset, he had begun to apply a fepradinol-containing spray intermittently for muscle pain. Patch tests were strongly positive to the spray and to fepradinol 0.1%, 1% and 2% alc. Thirty controls were negative. Biopsies of the initial lesion and the patch test reaction to the spray showed a thrombotic vasculopathy with epidermal necrosis and a lymphocytic perivascular infiltrate with eosinophils in the adjacent dermis. The case was diagnosed as vascular-occlusive contact dermatitis, a not previously described form of allergic contact dermatitis (3).

A 50-year-old man had applied fepradinol spray 2-3x/day for muscular pain on the left side of his back for 3 days, when itchy dermatitis developed. Patch tests were positive to the spray, and to fepradinol 0.01%, 0.1% and 1% in water and in alcohol and to fepradinol 1% pet. There were no reactions to the other constituents (6). The same authors also describe a 71-year-old woman with rheumatoid arthritis. After application of fepradinol spray on her knees and dorsum of her left foot due to joint pain for 8 days, very pruriginous erythematous and edematous patches with a tendency to scaling appeared. Patch tests were positive to the spray and to fepradinol 0.01% and

0.1% in alcohol, but negative to these concentrations of fepradinol in water. There were no cross-reactions to the related phenylethanolamines phenylephrine, epinephrine, ephedrine, salbutamol and terbutaline (6).

Three similar single case reports (10,11,12) and a small case series of two patients (13) were reported between 1992 and 1994. One was in a child (11); in another individual, localized allergic contact dermatitis on the right arm was followed by severe generalized eczema (13). Other case reports of contact allergy to/allergic contact dermatitis from fepradinol, adequate data of which are not available to the author, can be found in refs. 7 and 8.

Cross-reactions, pseudo-cross-reactions and co-reactions
Three patients sensitized by oral pseudoephedrine (causing generalized dermatitis) and two from phenylephrine in eye drops cross-reacted to fepradinol (2). Conversely, a patient sensitized by topical fepradinol cross-reacted to pseudoephedrine, ephedrine, and phenylpropanolamine (2). Fepradinol is not a classic NSAID but is considered an adrenergic compound, due to its structural similarity with other α-adrenergic drugs (5). This explains the cross-reactivity to pseudoephedrine, phenylephrine, ephedrine and phenylpropanolamine (2).

No cross-reactions to phenylephrine, epinephrine, ephedrine, salbutamol and terbutaline (6).

PHOTOSENSITIVITY
A 71-year-old woman had an itchy rash suggestive of contact photodermatitis for 3 days, which had begun one week after using meloxicam and fepradinol spray to relieve pain in the right knee. She had used fepradinol 2 months before without any problem. Examination revealed, on sun-exposed areas of the left cheek, neck and nape, erythematous edematous itchy plaques. Patch and photopatch tests with meloxicam 1% pet. and fepradinol 1% were positive only to the irradiated fepradinol patch test. The absence of reaction in the application areas and development of dermatitis on sun-exposed areas suggests transport of the allergen by contaminated hands (4).

LITERATURE
1   The data in the section 'General' may have been obtained from literature discussed in this chapter, but mostly also or exclusively from one or more of the following online sources: ChemIDPlus Advanced, PubChem, DrugBank, RxList, Drug Central, Drugs.com, and Wikipedia
2   Barranco R, Rodríguez A, de Barrio M, Trujillo MJ, de Frutos C, Matheu V, et al. Sympathomimetic drug allergy: cross-reactivity study by patch test. Am J Clin Dermatol 2004;5:351-355
3   Santos-Briz A, Antúnez P, Muñoz E, Morán M, Fernández E, Unamuno P. Vascular-occlusive contact dermatitis from fepradinol. Contact Dermatitis 2004;50:44-46
4   Rodríguez Granados T, Piñeiro G, de la Torre C, Cruces Prado MJ. Photoallergic contact dermatitis from fepradinol. Contact Dermatitis 1998;39:194-195
5   Schnuch A. Fepradinol allergy: possibly a case of unnoticed cross-reaction due to misclassification. Contact Dermatitis 1994;30:243-245
6   Goday JJ, Yanguas I, González-Güemes M, Oleaga JM, Ilardia R, Soloeta R. No evidence of cross-reaction between fepradinol and other phenylethanolamines. Contact Dermatitis 1997;36:170-171
7   Gonzalo MA, Revenga F. Multiple cutaneous sensitization to nonsteroidal anti-inflammatory drugs. Dermatology 1996;193:59-60
8   Rodriguez Granados T, Pineiro Corrales G, De la Torre Fraga C, Cruces Prado M J. Alergia de contacto al fepradinol. XXII Congreso Nacional de Dermatologia, Granada, Spain. Mayo 1993, oral communication (in Spanish, cited in ref. 6)
9   Ortiz-Frutos FJ, Hergueta JP, Quintana I, Zarco C, Iglesias L. Allergic contact dermatitis from fepradinol: report of 4 cases and review of the literature. Contact Dermatitis 1994;31:193-195
10 Aranzabal A, De Barrio M, Rodriguez V, Baeza ML, Tornero P. Allergic contact dermatitis from fepradinol. Contact Dermatitis 1994;31:121
11 Gomez A, Martorell A, de la Cuadra J. Allergic contact dermatitis from fepradinol in a child. Contact Dermatitis 1994;30:44
12 Goday JJ, Yanguas I, Ilardia R, Soloeta R. Subacute contact dermatitis from fepradinol. Contact Dermatitis 1993;29:160
13 Izu R, Aguirre A, Irazabal B, Goday J, Díaz-Pérez JL. Allergic contact dermatitis from fepradinol. Contact Dermatitis 1992;27:266-267
14 Diaz RL, Gardeazabal J, Manrique P, Ratón JA, Urrutia I, Rodríguez-Sasiain JM, Aguirre C. Greater allergenicity of topical ketoprofen in contact dermatitis confirmed by use. Contact Dermatitis 2006;54:239-243

Chapter 3.137 FEPRAZONE

IDENTIFICATION

Description/definition : Feprazone is the pyrazole that conforms to the structural formula shown below
Pharmacological classes : Anti-inflammatory agents, non-steroidal
IUPAC name : 4-(3-Methylbut-2-enyl)-1,2-diphenylpyrazolidine-3,5-dione
Other names : Phenylprenazone; 1,2-diphenyl-4-(3-methyl-2-butenyl)-3,5-pyrazolidinedione
CAS registry number : 30748-29-9
EC number : 250-324-3
Merck Index monograph : 5309
Patch testing : 5% pet.
Molecular formula : $C_{20}H_{20}N_2O_2$

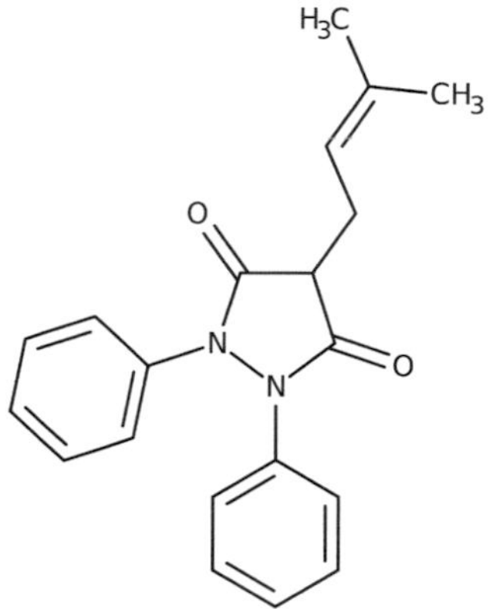

GENERAL

Feprazone is a pyrazole derivative and nonsteroidal anti-inflammatory drug (NSAID) that has analgesic, anti-inflammatory and antipyretic properties. It is or has been used to treat mild to moderate pain, fever, and inflammation associated with musculoskeletal and joint disorders (1).

CONTACT ALLERGY

Case series

In a hospital in Italy, in a period of 5 years before 1993, three patients had been diagnosed with allergic contact dermatitis from feprazone, 2 men aged 41 and 45 and a woman aged 45. One patient had dermatitis on the legs and knees, the second on the back and patient 3 on the back and in the neck. In all, the dermatitis had been preceded by 2-3 weeks of use of feprazone cream, followed by iontophoresis and other physiotherapeutic modalities for rheumatic pains. One patient reacted to feprazone 5% pet., in the other2 feprazone itself was not tested, but they had positive reactions to the feprazone-containing cream and negative reactions to its excipients. All patients denied having previously used feprazone either topically or systemically (3, possibly overlap with ref. 2).

In Italy, before 1993, the members of the GIRDCA Multicentre Study Group diagnosed 102 patients (49 men, 53 women), aged 16 to 66 years (mean 37 years), with (photo)dermatitis induced by systemic or topical NSAIDs. Feprazone caused two contact allergic and zero photocontact allergic reactions (2, possibly overlap with ref. 3).

Cross-reactions, pseudo-cross-reactions and co-reactions

Two patients who had allergic contact dermatitis from topical oxyphenbutazone co-reacted (cross-reacted?) to feprazone, phenylbutazone, and suxibuzone (4). One patient with allergic contact dermatitis from feprazone had a positive patch test reaction to phenylbutazone 10% pet. and a ?+ reaction to oxyphenbutazone 10% pet. (3).

LITERATURE

1 The data in the section 'General' may have been obtained from literature discussed in this chapter, but mostly also or exclusively from one or more of the following online sources: ChemIDPlus Advanced, PubChem, DrugBank, RxList, Drug Central, Drugs.com, and Wikipedia
2 Pigatto PD, Mozzanica N, Bigardi AS, Legori A, Valsecchi R, Cusano F, et al. Topical NSAID allergic contact dermatitis. Italian experience. Contact Dermatitis 1993;29:39-41
3 Cusano F, Luciano S, Iannazzone S, Adamo F. Contact dermatitis from feprazone. Contact Dermatitis 1993;28:109
4 Figueiredo A, Gonçalo S, Freitas JD. Contact sensitivity to pyrazolone compounds. Contact Dermatitis 1985;13:271

Chapter 3.138 FLUCLOROLONE ACETONIDE

IDENTIFICATION

Description/definition   : Fluclorolone acetonide is the acetonide ester of the synthetic glucocorticoid fluclorolone
                                that conforms to the structural formula shown below
Pharmacological classes   : Glucocorticoids
IUPAC name   : (1S,2S,4R,8S,9S,11S,12R,13S,19S)-11,12-Dichloro-19-fluoro-8-(2-hydroxyacetyl)-6,6,9,13-
                                tetramethyl-5,7-dioxapentacyclo[10.8.0.0^{2,9}.0^{4,8}.0^{13,18}]icosa-14,17-dien-16-one
Other names   : Flucloronide; 9,11β-dichloro-6α-fluoro-16α,17,21-trihydroxypregna-1,4-diene-3,20-dione
                                cyclic 16,17-acetal with acetone
CAS registry number   : 3693-39-8
EC number   : 223-010-9
Merck Index monograph : 1166
Patch testing   : In general, corticosteroids may be tested at 0.1% and 1% in alcohol; late readings (6-10
                                days) are strongly recommended
Molecular formula   : C$_{24}$H$_{29}$Cl$_2$FO$_5$

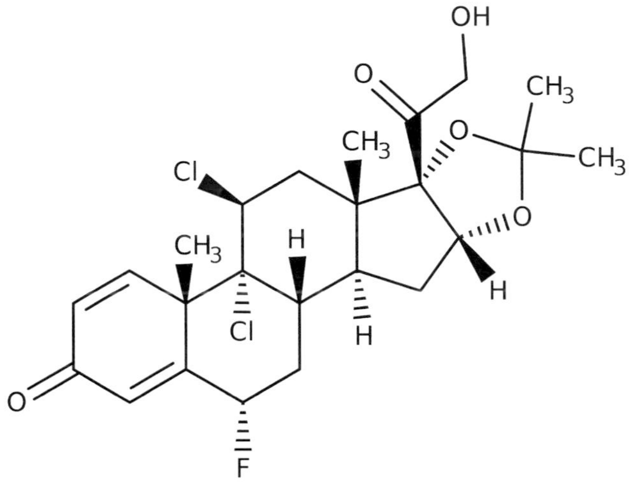

GENERAL

General aspects of corticosteroids used on the skin and mucous membranes are discussed in Chapter 2.4. A practical guideline for diagnosing allergic reactions to corticosteroids is presented in ref. 1.

CONTACT ALLERGY

Case series

From January 1990 to June 2008, in Leuven, Belgium, 315 patients were diagnosed with contact allergy to/allergic contact dermatitis from corticosteroids (CSs) from routine patch testing with a baseline series including tixocortol pivalate, budesonide, hydrocortisone butyrate and prednisone caproate, patch testing with patients' own CS preparations, and testing those with proven contact allergy to a corticosteroid or strongly suspected of CS allergy later with a series of 66 CSs, including two sex hormones (progesterone and testosterone). 71% of the patients had relevant reactions, but these were not specified. In this group of 315 CS allergic patients, 15 had positive patch tests to flucloronide (fluclorolone acetonide) 0.1% alc. (2). As this corticosteroid has – very likely – not been used in pharmaceuticals in Belgium, these positive reactions must all be considered cross-reactions to other corticosteroids.

Cross-reactions, pseudo-cross-reactions and co-reactions

Cross-reactions between corticosteroids are discussed in Chapter 2.8.

LITERATURE

1   Baeck M, Goossens A. Immediate and delayed allergic hypersensitivity to corticosteroids: practical guidelines. Contact Dermatitis 2012;66:38-45
2   Baeck M, Chemelle JA, Terreux R, Drieghe J, Goossens A. Delayed hypersensitivity to corticosteroids in a series of 315 patients: clinical data and patch test results. Contact Dermatitis 2009;61:163-175

Chapter 3.139 FLUDROCORTISONE ACETATE

IDENTIFICATION

Description/definition : Fludrocortisone acetate is the acetate ester of the synthetic glucocorticoid fludrocortisone that conforms to the structural formula shown below

Pharmacological classes : Glucocorticoids

IUPAC name : [2-[(8S,9R,10S,11S,13S,14S,17R)-9-Fluoro-11,17-dihydroxy-10,13-dimethyl-3-oxo-1,2,6,7,8,11,12,14,15,16-decahydrocyclopenta[a]phenanthren-17-yl]-2-oxoethyl] acetate

Other names : Fludrocortisone 21-acetate; 9-fluoro-11β,17,21-trihydroxypregn-4-ene-3,20-dione 21-acetate

CAS registry number : 514-36-3

EC number : 208-180-4

Merck Index monograph : 5431 (Fludrocortisone)

Patch testing : In general, corticosteroids may be tested at 0.1% and 1% in alcohol; late readings (6-10 days) are strongly recommended

Molecular formula : $C_{23}H_{31}FO_6$

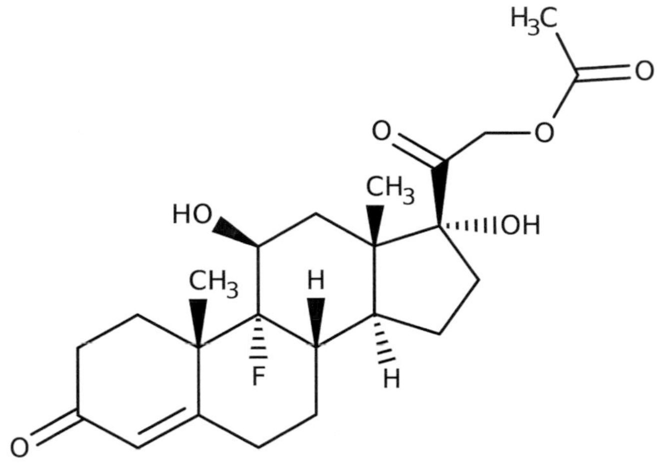

GENERAL

General aspects of corticosteroids used on the skin and mucous membranes are discussed in Chapter 2.4. A practical guideline for diagnosing allergic reactions to corticosteroids is presented in ref. 1.

CONTACT ALLERGY

Case series

From January 1990 to June 2008, in Leuven, Belgium, 315 patients were diagnosed with contact allergy to/allergic contact dermatitis from corticosteroids (CSs) from routine patch testing with a baseline series including tixocortol pivalate, budesonide, hydrocortisone butyrate and prednisone caproate, patch testing with patients' own CS preparations, and testing those with proven contact allergy to a corticosteroid or strongly suspected of CS allergy later with a series of 66 CSs, including two sex hormones (progesterone and testosterone). 71% of the patients had relevant reactions, but these were not specified. In this group of 315 CS allergic patients, 25 had positive patch tests to fludrocortisone acetate 0.1% alc. (2). It is unknown how many of these reactions were caused by the use of a pharmaceutical product containing fludrocortisone acetate and how many were cross-reactions to other corticosteroids.

Cross-reactions, pseudo-cross-reactions and co-reactions

Cross-reactions between corticosteroids are discussed in Chapter 2.8.

LITERATURE

1   Baeck M, Goossens A. Immediate and delayed allergic hypersensitivity to corticosteroids: practical guidelines. Contact Dermatitis 2012;66:38-45
2   Baeck M, Chemelle JA, Terreux R, Drieghe J, Goossens A. Delayed hypersensitivity to corticosteroids in a series of 315 patients: clinical data and patch test results. Contact Dermatitis 2009;61:163-175

Chapter 3.140 FLUFENAMIC ACID

IDENTIFICATION

Description/definition   : Flufenamic acid is the anthranilic acid derivative that conforms to the structural formula
                              shown below
Pharmacological classes   : Anti-inflammatory agents
IUPAC name   : 2-[3-(Trifluoromethyl)anilino]benzoic acid
Other names   : 2-((3-(Trifluoromethyl)phenyl)amino)benzoic acid
CAS registry number   : 530-78-9
EC number   : 208-494-1
Merck Index monograph   : 5433
Patch testing   : 2% pet.
Molecular formula   : $C_{14}H_{10}F_3NO_2$

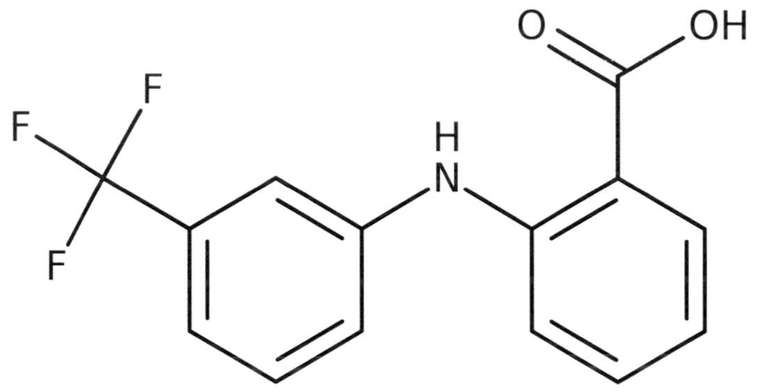

GENERAL

Flufenamic acid is an anthranilic acid derivative and nonsteroidal anti-inflammatory drug (NSAID) with analgesic, anti-inflammatory, and antipyretic properties. It is used in musculoskeletal and joint disorders and can be administered both by mouth and topically (1).

CONTACT ALLERGY

Patch testing in groups of patients
Before 1998, in Germany, 371 consecutive patients were patch tested with flufenamic acid 5% pet. and there was one (0.3%) positive reaction; its relevance remained unknown (5).

Case reports and case series
In Italy, before 1993, the members of the GIRDCA Multicentre Study Group diagnosed 102 patients (49 men, 53 women), aged 16 to 66 years (mean 37 years), with (photo)dermatitis induced by systemic or topical NSAIDs. Flufenamic acid caused one contact allergic and zero photocontact allergic reactions (2).

PHOTOSENSITIVITY

Case report
A 21-year-old woman complained of sharply-outlined acute exudative eczema localized to her knees and ankles. She practiced body-building and had applied a spray containing salicylic acid and flufenamic acid at these sites because of mild joint pain for 15 days before the appearance of lesions. Eight days before the onset of symptoms, she had had 6 daily UVA sessions after training for 30 minutes. Patch and photopatch tests to the spray, salicylic acid, flufenamic acid 0.5%, 1% and 2% pet., salicylic acid, the other ingredients of the spray and etofenamate 2% pet. showed positive photopatch test reactions to the spray, all three concentrations of flufenamic acid and etofenamate; reactions were negative to salicylic acid and the other ingredients. Fifteen controls were negative (4).

Cross-reactions, pseudo-cross-reactions and co-reactions

Of 4 patients sensitized to etofenamate, one had a positive reaction to a commercial cream containing flufenamic acid. As both are anthranilic compounds and etofenamate is the 2-(2-hydroxyethoxy)ethyl ester of flufenamic acid, the authors suggested that this was a case of cross-reactivity (3). A patient photosensitized to flufenamic acid may have photocross-reacted to etofenamate (4).

LITERATURE

1   The data in the section 'General' may have been obtained from literature discussed in this chapter, but mostly also or exclusively from one or more of the following online sources: ChemIDPlus Advanced, PubChem, DrugBank, RxList, Drug Central, Drugs.com, and Wikipedia

2   Pigatto PD, Mozzanica N, Bigardi AS, Legori A, Valsecchi R, Cusano F, et al. Topical NSAID allergic contact dermatitis. Italian experience. Contact Dermatitis 1993;29:39-41

3   Alcántara Villar M, Pagan JA, Palacios L, Quiralte J, Ramirez M. Allergic contact dermatitis to etofenamate. Cross-reaction to other nonsteroidal anti-inflammatory drugs. Contact Dermatitis 2008;58:118-119

4   Montoro J, Rodriguez-Serna M, Liñana JJ, Ferré MA, Sanchez-Motilla JM. Photoallergic contact dermatitis due to flufenamic acid and etofenamate. Contact Dermatitis 1997;37:139-140

5   Gniazdowska B, Ruëff F, Przybilla B. Delayed contact hypersensitivity to non-steroidal anti-inflammatory drugs. Contact Dermatitis 1999;40:63-65

Chapter 3.141 FLUMETHASONE PIVALATE

IDENTIFICATION

Description/definition   : Flumethasone pivalate is the 21-pivalate ester of the synthetic glucocorticoid
                            flumethasone that conforms to the structural formula shown below
Pharmacological classes   : Glucocorticoids; anti-inflammatory agents
IUPAC name   : [2-[(6S,8S,9R,10S,11S,13S,14S,16R,17R)-6,9-Difluoro-11,17-dihydroxy-10,13,16-trimethyl-
                            3-oxo-6,7,8,11,12,14,15,16-octahydrocyclopenta[a]phenanthren-17-yl]-2-oxoethyl] 2,2-
                            dimethylpropanoate
Other names   : 6α,9-Difluoro-11β,17,21-trihydroxy-16α-methylpregna-1,4-diene-3,20-dione 21-pivalate;
                            flumethasone 21-pivalate
CAS registry number   : 2002-29-1
EC number   : 217-901-1
Merck Index monograph   : 5439 (Flumethasone)
Patch testing   : In general, corticosteroids may be tested at 0.1% and 1% in alcohol; late readings (6-10
                            days) are strongly recommended
Molecular formula   : $C_{27}H_{36}F_2O_6$

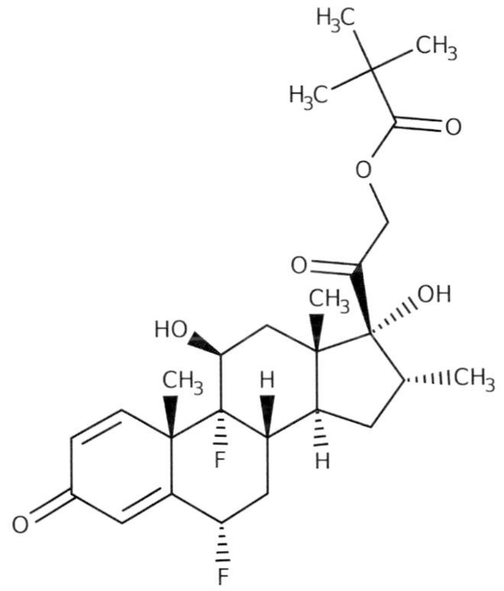

GENERAL

General aspects of corticosteroids used on the skin and mucous membranes are discussed in Chapter 2.4. A practical guideline for diagnosing allergic reactions to corticosteroids is presented in ref. 1.

CONTACT ALLERGY

Case series

From January 1990 to June 2008, in Leuven, Belgium, 315 patients were diagnosed with contact allergy to/allergic contact dermatitis from corticosteroids (CSs) from routine patch testing with a baseline series including tixocortol pivalate, budesonide, hydrocortisone butyrate and prednisone caproate, patch testing with patients' own CS preparations, and testing those with proven contact allergy to a corticosteroid or strongly suspected of CS allergy later with a series of 66 CSs, including two sex hormones (progesterone and testosterone). 71% of the patients had relevant reactions, but these were not specified. In this group of 315 CS allergic patients, 6 had positive patch tests to flumethasone pivalate 0.1% alc. (2). It is unknown how many of these reactions were caused by the use of a pharmaceutical product containing flumethasone pivalate and how many were cross-reactions to other corticosteroids.

Cross-reactions, pseudo-cross-reactions and co-reactions

Cross-reactions between corticosteroids are discussed in Chapter 2.8.

LITERATURE

1   Baeck M, Goossens A. Immediate and delayed allergic hypersensitivity to corticosteroids: practical guidelines. Contact Dermatitis 2012;66:38-45
2   Baeck M, Chemelle JA, Terreux R, Drieghe J, Goossens A. Delayed hypersensitivity to corticosteroids in a series of 315 patients: clinical data and patch test results. Contact Dermatitis 2009;61:163-175

Chapter 3.142 FLUNISOLIDE

IDENTIFICATION

Description/definition	: Flunisolide is the synthetic glucocorticoid that conforms to the structural formula shown below
Pharmacological classes	: Anti-asthmatic agents; anti-inflammatory agents
IUPAC name	: (1S,2S,4R,8S,9S,11S,12S,13R,19S)-19-Fluoro-11-hydroxy-8-(2-hydroxyacetyl)-6,6,9,13-tetramethyl-5,7-dioxapentacyclo[10.8.0.0^{2,9}.0^{4,8}.0^{13,18}]icosa-14,17-dien-16-one
Other names	: 6α-Fluoro-11β,16α,17,21-tetrahydroxypregna-1,4-diene-3,20-dione cyclic 16,17-acetal with acetone
CAS registry number	: 3385-03-3
EC number	: 222-193-2
Merck Index monograph	: 5446
Patch testing	: In general, corticosteroids may be tested at 0.1% and 1% in alcohol; late readings (6-10 days) are strongly recommended
Molecular formula	: C$_{24}$H$_{31}$FO$_6$

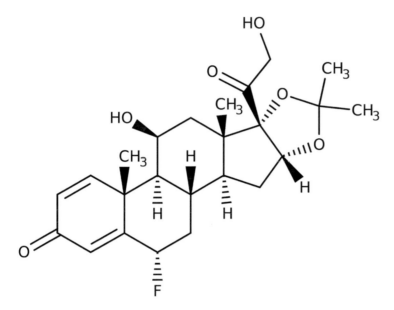

GENERAL

General aspects of corticosteroids used on the skin and mucous membranes are discussed in Chapter 2.4. A practical guideline for diagnosing allergic reactions to corticosteroids is presented in ref. 1. In pharmaceutical products, flunisolide may both be employed as such and as flunisolide hemihydrate (CAS number 77326-96-6, EC number not available, molecular formula C$_{48}$H$_{64}$F$_2$O$_{13}$).

CONTACT ALLERGY

Case reports and case series

From January 1990 to June 2008, in Leuven, Belgium, 315 patients were diagnosed with contact allergy to/allergic contact dermatitis from corticosteroids (CSs) from routine patch testing with a baseline series including tixocortol pivalate, budesonide, hydrocortisone butyrate and prednisone caproate, patch testing with patients' own CS preparations, and testing those with proven contact allergy to a corticosteroid or strongly suspected of CS allergy later with a series of 66 CSs, including two sex hormones (progesterone and testosterone). 71% of the patients had relevant reactions, but these were not specified. In this group of 315 CS allergic patients, 35 had positive patch tests to flunisolide 0.1% alc. (2). It is unknown how many of these reactions were caused by the use of a pharmaceutical product containing flunisolide and how many were cross-reactions to other corticosteroids.

A 42-year-old woman presented with periocular dermatitis from contact allergy to flunisolide and budesonide, both used in oral inhalation therapy (3).

Cross-reactions, pseudo-cross-reactions and co-reactions
Cross-reactions between corticosteroids are discussed in Chapter 2.8.

LITERATURE
1   Baeck M, Goossens A. Immediate and delayed allergic hypersensitivity to corticosteroids: practical guidelines. Contact Dermatitis 2012;66:38-45
2   Baeck M, Chemelle JA, Terreux R, Drieghe J, Goossens A. Delayed hypersensitivity to corticosteroids in a series of 315 patients: clinical data and patch test results. Contact Dermatitis 2009;61:163-175
3   Baeck M, Pilette C, Drieghe J, Goossens A. Allergic contact dermatitis to inhalation corticosteroids. Eur J Dermatol 2010;20:102-108

Chapter 3.143 FLUOCINOLONE ACETONIDE

IDENTIFICATION

Description/definition : Fluocinolone acetonide is the acetonide ester of the synthetic glucocorticoid fluocinolone that conforms to the structural formula shown below

Pharmacological classes : Anti-inflammatory agents; glucocorticoids

IUPAC name : (1S,2S,4R,8S,9S,11S,12R,13S,19S)-12,19-Difluoro-11-hydroxy-8-(2-hydroxyacetyl)-6,6,9,13-tetramethyl-5,7-dioxapentacyclo[10.8.0.0^{2,9}.0^{4,8}.0^{13,18}]icosa-14,17-dien-16-one

Other names : 6α,9-Difluoro-11β,21-dihydroxy-16α,17α-(isopropylidenedioxy)pregna-1,4-diene-3,20-dione

CAS registry number : 67-73-2

EC number : 200-668-5

Merck Index monograph : 5451

Patch testing : In general, corticosteroids may be tested at 0.1% and 1% in alcohol; late readings (6-10 days) are strongly recommended

Molecular formula : C$_{24}$H$_{30}$F$_2$O$_6$

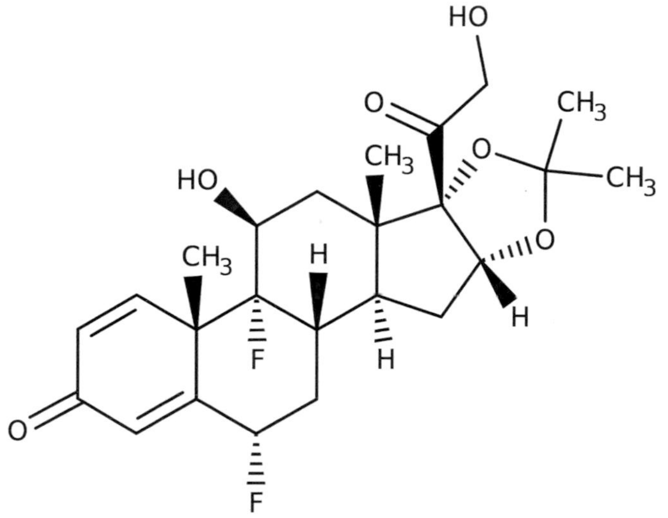

GENERAL

General aspects of corticosteroids used on the skin and mucous membranes are discussed in Chapter 2.4. A practical guideline for diagnosing allergic reactions to corticosteroids is presented in ref. 1.

CONTACT ALLERGY

Patch testing in groups of selected patients

In the period 2000 to 2005, in the USA, 1172 patients suspected of corticosteroid allergy were patch tested with fluocinolone acetonide 1% alc. and there were 12 (1.0%) positive reactions, of which 11 (92%) were considered to be relevant (5).

Case series

From January 1990 to June 2008, in Leuven, Belgium, 315 patients were diagnosed with contact allergy to/allergic contact dermatitis from corticosteroids (CSs) from routine patch testing with a baseline series including tixocortol pivalate, budesonide, hydrocortisone butyrate and prednisone caproate, patch testing with patients' own CS preparations, and testing those with proven contact allergy to a corticosteroid or strongly suspected of CS allergy later with a series of 66 CSs, including two sex hormones (progesterone and testosterone). 71% of the patients had relevant reactions, but these were not specified. In this group of 315 CS allergic patients, 27 had positive patch tests to fluocinolone acetonide 0.1% alc. (4). It is unknown how many of these reactions were caused by the use of a pharmaceutical product containing fluocinolone acetonide and how many were cross-reactions to other corticosteroids.

Case reports

A 58-year-old man had itchy lichenified lesions on both legs for 12 years. He used fluocinolone acetonide 0.1% ointment, but 18 days later, had increased itching and redness and stopped therapy. Six months later, he again applied this ointment and soon after noted itching, erythema and vesiculation on both legs. Patch tests were positive to various commercial fluocinolone acetonide ointments and to fluocinolone acetonide 0.025% water. He also – weakly – reacted to the base of the ointment and to propyl- and benzylparaben, but it was not mentioned whether these were present in the ointment (2).

In a report on contact allergy to nystatin, the patient had a negative reaction to fluocinolone, present in the combination product containing nystatin. However, the authors stated that they had observed a case of contact allergy to fluocinolone acetonide a few years previously. Clinical and patch testing details were not provided (3).

A 56-year-old man, who was allergic to neomycin, lanolin alcohol and clioquinol, first reacted well to betamethasone valerate cream and fluocinolone acetonide cream. After 4 years, however, both preparations seemed to worsen the condition. Hydrocortisone cream too aggravated the eczema. Oral prednisolone, previously well-tolerated and effective, now caused a generalized maculopapular exanthema (systemic contact dermatitis). Patch tests were positive to the three commercial topical preparations and their ingredients betamethasone valerate, fluocinolone acetonide and hydrocortisone. Prednisolone was not patch tested (7).

A 35-year-old man had a 10-year history of a recurrent non-specific dermatitis occurring on various aspects of the extremities, trunk and genitals. Each exacerbation was treated with a different corticosteroid preparation and they all aggravated the condition, necessitating short courses of oral prednisolone. He was patch test positive to all commercial preparations, their active ingredients fluocinolone acetonide, hydrocortisone, desonide, amcinonide, fluocinonide and various base ingredients (6).

Cross-reactions, pseudo-cross-reactions and co-reactions

Cross-reactions between corticosteroids are discussed in Chapter 2.8.

LITERATURE

1   Baeck M, Goossens A. Immediate and delayed allergic hypersensitivity to corticosteroids: practical guidelines. Contact Dermatitis 2012;66:38-45
2   Pasricha JS, Gupta R. Contact sensitivity to betamethasone 17-valerate and fluocinolone acetonide. Contact Dermatitis 1983;9:330-331
3   Foussereau J, Limam-Mestiri S, Khochnevis A. Contact allergy to nystatin. Contact Dermatitis Newsletter 1971;10:221
4   Baeck M, Chemelle JA, Terreux R, Drieghe J, Goossens A. Delayed hypersensitivity to corticosteroids in a series of 315 patients: clinical data and patch test results. Contact Dermatitis 2009;61:163-175
5   Davis MD, El-Azhary RA, Farmer SA. Results of patch testing to a corticosteroid series: a retrospective review of 1188 patients during 6 years at Mayo Clinic. J Am Acad Dermatol 2007;56:921-927
6   Feldman SB, Sexton FM, Buzas J, Marks JG Jr. Allergic contact dermatitis from topical steroids. Contact Dermatitis 1988;19:226-228
7   Goh CL. Cross-sensitivity to multiple topical corticosteroids. Contact Dermatitis 1989;20:65-67

Chapter 3.144 FLUOCINONIDE

IDENTIFICATION

Description/definition : Fluocinonide is the synthetic glucocorticoid that conforms to the structural formula shown below

Pharmacological classes : Anti-inflammatory agents

IUPAC name : [2-[(1S,2S,4R,8S,9S,11S,12R,13S,19S)-12,19-Difluoro-11-hydroxy-6,6,9,13-tetramethyl-16-oxo-5,7-dioxapentacyclo[10.8.0.0^{2,9}.0^{4,8}.0^{13,18}]icosa-14,17-dien-8-yl]-2-oxoethyl]

CAS registry number : 356-12-7

EC number : 206-597-6

Merck Index monograph : 5452

Patch testing : In general, corticosteroids may be tested at 0.1% and 1% in alcohol; late readings (6-10 days) are strongly recommended

Molecular formula : $C_{26}H_{32}F_2O_7$

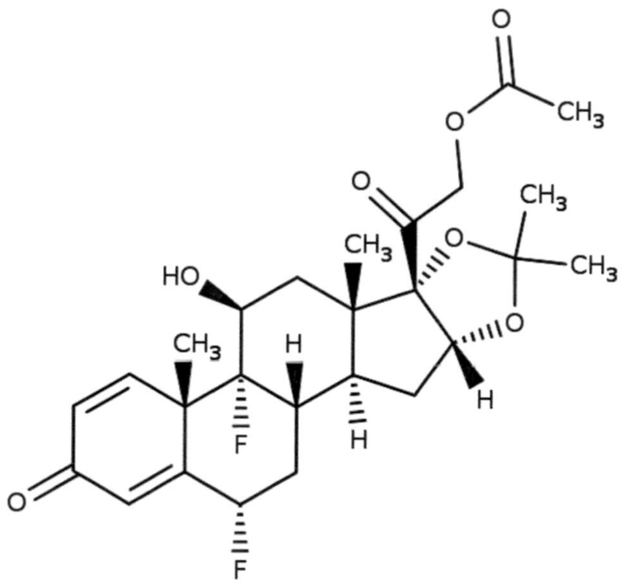

GENERAL

General aspects of corticosteroids used on the skin and mucous membranes are discussed in Chapter 2.4. A practical guideline for diagnosing allergic reactions to corticosteroids is presented in ref. 1.

CONTACT ALLERGY

Patch testing in groups of selected patients

In the period 2000 to 2005, in the USA, 1179 patients suspected of corticosteroid allergy were patch tested with fluocinonide 1% pet. and there were 12 (1.0%) positive reactions, which were all considered to be relevant (5). Also in the U.K., from 1988 to 1991, 528 selected patients with a positive patch test to tixocortol pivalate or suspected of corticosteroid allergy were patch tested with fluocinonide 20% pet. and 1 patient (0.2%) had a positive patch test. Its relevance was not addressed (6).

Case series

From January 1990 to June 2008, in Leuven, Belgium, 315 patients were diagnosed with contact allergy to/allergic contact dermatitis from corticosteroids (CSs) from routine patch testing with a baseline series including tixocortol pivalate, budesonide, hydrocortisone butyrate and prednisone caproate, patch testing with patients' own CS preparations, and testing those with proven contact allergy to a corticosteroid or strongly suspected of CS allergy later with a series of 66 CSs, including two sex hormones (progesterone and testosterone). 71% of the patients had relevant reactions, but these were not specified. In this group of 315 CS allergic patients, 29 had positive patch tests to fluocinonide 1% alc. (4). It is unknown how many of these reactions were caused by the use of a pharmaceutical product containing fluocinonide and how many were cross-reactions to other corticosteroids.

In reviewing the Japanese literature up to 1994, 43 patients with allergic contact dermatitis from topical corticosteroid were identified, including one caused by fluocinonide (9).

Case reports
A 60-year-old man had a small area of dermatitis on his lower leg. He was initially treated with betamethasone dipropionate. The dermatitis persisted, and his topical steroid was changed to fluocinonide, which led to increased erythema, pruritus, and (ultimately) bullae formation. Subsequent administration of hydrocortisone valerate worsened the erythema. A punch biopsy specimen showed subepidermal bullae with a large number of eosinophils, compatible with a diagnosis of bullous contact dermatitis. Patch tests were positive to budesonide 0.1% pet., commercial 0.05% fluocinonide cream, 2 hydrocortisone valerate commercial creams and many other corticosteroids (3).

A 35-year-old man had a 10-year history of a recurrent non-specific dermatitis occurring on various aspects of the extremities, trunk and genitals. Each exacerbation was treated with a different corticosteroid preparation and they all aggravated the condition, necessitating short courses of oral prednisolone. He was patch test positive to all commercial preparations, their active ingredients fluocinonide, hydrocortisone, desonide, amcinonide, fluocinolone acetonide, and various base ingredients (7).

A 16-year-old girl was treated in the course of several years with various corticosteroids for seborrheic dermatitis of the axillae, face or scalp and became sensitized to fluocinonide, desonide, and triamcinolone acetonide in these topical pharmaceuticals (8).

Apparently, in 1981, one or more cases of contact allergy to fluocinonide have been reported in Japanese literature (2). Details are not available to the author.

Cross-reactions, pseudo-cross-reactions and co-reactions
Cross-reactions between corticosteroids are discussed in Chapter 2.8.

LITERATURE
1 Baeck M, Goossens A. Immediate and delayed allergic hypersensitivity to corticosteroids: practical guidelines. Contact Dermatitis 2012;66:38-45
2 Kawatzu T, Tokizane K. Contact dermatitis due to fluocinonide. Jap J Dermatol 1981;35:953-955 (Article in Japanese)
3 Donovan JC, Dekoven JG. Cross-reactions to desoximetasone and mometasone furoate in a patient with multiple topical corticosteroid allergies. Dermatitis 2006 ;17:147-151
4 Baeck M, Chemelle JA, Terreux R, Drieghe J, Goossens A. Delayed hypersensitivity to corticosteroids in a series of 315 patients: clinical data and patch test results. Contact Dermatitis 2009;61:163-175
5 Davis MD, El-Azhary RA, Farmer SA. Results of patch testing to a corticosteroid series: a retrospective review of 1188 patients during 6 years at Mayo Clinic. J Am Acad Dermatol 2007;56:921-927
6 Burden AD, Beck MH. Contact hypersensitivity to topical corticosteroids. Br J Dermatol 1992;127:497-501
7 Feldman SB, Sexton FM, Buzas J, Marks JG Jr. Allergic contact dermatitis from topical steroids. Contact Dermatitis 1988;19:226-228
8 Coskey RJ. Contact dermatitis due to multiple corticosteroid creams. Arch Dermatol 1978;114:115-117
9 Oh-i T. Contact dermatitis due to topical steroids with conceivable cross reactions between topical steroid preparations. J Dermatol 1996;23:200-208

Chapter 3.145 FLUOCORTOLONE

IDENTIFICATION

Description/definition : Fluocortolone is the synthetic glucocorticoid that conforms to the structural formula shown below

Pharmacological classes : Glucocorticoids; anti-inflammatory agents

IUPAC name : (6S,8S,9S,10R,11S,13S,14S,16R,17S)-6-Fluoro-11-hydroxy-17-(2-hydroxyacetyl)-10,13,16-trimethyl-6,7,8,9,11,12,14,15,16,17-decahydrocyclopenta[a]phenanthren-3-one

Other names : 6α-Fluoro-11β,21-dihydroxy-16α-methylpregna-1,4-diene-3,20-dione

CAS registry number : 152-97-6

EC number : 205-811-5

Merck Index monograph : 5454

Patch testing : In general, corticosteroids may be tested at 0.1% and 1% in alcohol; late readings (6-10 days) are strongly recommended

Molecular formula : $C_{22}H_{29}FO_4$

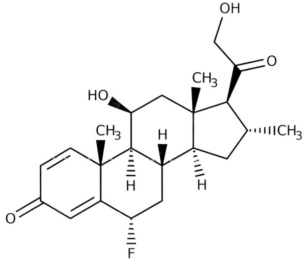

GENERAL

General aspects of corticosteroids used on the skin and mucous membranes are discussed in Chapter 2.4. A practical guideline for diagnosing allergic reactions to corticosteroids is presented in ref. 1. See also Chapter 3.146 Fluocortolone caproate and Chapter 3.147 Fluocortolone pivalate.

CONTACT ALLERGY

Patch testing in groups of selected patients

In the United Kingdom, in the period 1988-1991, 528 patients with a positive patch test to tixocortol pivalate or suspected of corticosteroid allergy were tested with fluocortolone 20% pet. and there was one (0.2%) positive reaction, the relevance of which was not mentioned (6).

Case series

From January 1990 to June 2008, in Leuven, Belgium, 315 patients were diagnosed with contact allergy to/allergic contact dermatitis from corticosteroids (CSs) from routine patch testing with a baseline series including tixocortol pivalate, budesonide, hydrocortisone butyrate and prednisone caproate, patch testing with patients' own CS preparations, and testing those with proven contact allergy to a corticosteroid or strongly suspected of CS allergy later with a series of 66 CSs, including two sex hormones (progesterone and testosterone). 71% of the patients had relevant reactions, but these were not specified. In this group of 315 CS allergic patients, 7 had positive patch tests to fluocortolone 0.1% alc. (5). It is unknown how many of these reactions were caused by the use of a pharmaceutical product containing fluocortolone and how many were cross-reactions to other corticosteroids.

Case reports

A 68-year old woman had intermittent eczema of the hands and buttocks. She had been treated with various topical corticosteroids without any adverse effect. After changing to a fatty ointment containing 0.25% fluocortolone and 0.25% fluocortolone caproate, the eczema became worse a few days later. Patch tests revealed positive reactions to the fatty ointment and 2 other commercial preparations of the same brand containing either fluocortolone and fluocortolone caproate or fluocortolone pivalate and fluocortolone pivalate. In addition, there were positive reactions to fluocortolone and its 2 esters, tested at 1% pet. The vehicles of the preparations tested negative, as did other corticosteroids (3).

The same authors report on a 53-year-old female patient with recurrent genitofemoral eczema for 3 years. She had suffered from severe spreading eczema twice after using a corticosteroid preparation, at least one of which was of the brand containing either fluocortolone and fluocortolone caproate or fluocortolone caproate and fluocortolone pivalate. Patch tests were positive to fluocortolone but not to the esters. When retested later, the patient now reacted to fluocortolone and its pivalate ester, but not fluocortolone caproate (all tested 1% pet.) (3). The authors suggested that a higher concentration must possibly be used to detect corticosteroid allergy. However, the second test was read at D2 only, and late positive reactions (D6-D10) are frequent with corticosteroids.

A 40-year-old man presented with an acute contact dermatitis of the hands and neck, which had developed after he had applied a steroid cream containing fluocortin butyl ester to the atopic dermatitis of his daughter. Patch tests were positive to the cream, to an ointment of the same brand, to fluocortin butyl 1% in water and in pet. and to fluocortolone 1% water and pet. In the past, the patient had occasionally used a steroid cream containing fluocortolone without any side effects. Probably, he had become sensitized then and now cross-reacted to fluocortin butyl, which is derived from fluocortolone (2).

A man aged 26 was treated with a corticosteroid cream containing fluocortolone and/or fluocortolone caproate for dermatitis of the face. After 3 weeks' treatment, the face became swollen, red and itchy. Patch tests were positive to the cream and 3 ointments and a lotion of the same brand. Patch testing with crystallized powder of the corticosteroid(s) in the cream also yielded a positive reaction. The author could not determine whether the allergy was due to fluocortolone, fluocortolone caproate or both (4).

A female patient had a widespread papular, angry, irritable rash which had started under the base metal of her watch strap. She had used various corticosteroids. There were positive patch tests to a number of them including an ointment containing fluocortolone hydrate. The steroid itself at 1% and 10% pet. gave positive responses (8). A man of 60 and a woman aged 45 both treated chronic dermatitis with fluocortolone cream/ointment and noticed worsening of the skin condition. Patch tests were positive to the commercial preparation and to fluocortolone 1% pet. in both patients (7).

Cross-reactions, pseudo-cross-reactions and co-reactions
Cross-reactions between corticosteroids are discussed in Chapter 2.8.

LITERATURE

1   Baeck M, Goossens A. Immediate and delayed allergic hypersensitivity to corticosteroids: practical guidelines. Contact Dermatitis 2012;66:38-45
2   Tosti A, Peluso AM, Morelli R. Allergic contact dermatitis from fluocortin butylester. Contact Dermatitis 1990;22:123-124
3   Menné T, Andersen KE. Allergic contact dermatitis from fluocortolone, fluocortolone pivalate and fluocortolone caproate. Contact Dermatitis 1977;3:337-340
4   Van Ketel WG. Allergy to Ultralan® preparations. Contact Dermatitis Newsletter 1974;15:427
5   Baeck M, Chemelle JA, Terreux R, Drieghe J, Goossens A. Delayed hypersensitivity to corticosteroids in a series of 315 patients: clinical data and patch test results. Contact Dermatitis 2009;61:163-175
6   Burden AD, Beck MH. Contact hypersensitivity to topical corticosteroids. Br J Dermatol 1992;127:497-501
7   Dunkel FG, Elsner P, Burg G. Contact allergies to topical corticosteroids: 10 cases of contact dermatitis. Contact Dermatitis 1991;25:97-103
8   Brown R. Simultaneous hypersensitivity to 3 topical corticosteroids. Contact Dermatitis 1982;8:339-340

Chapter 3.146 FLUOCORTOLONE CAPROATE

IDENTIFICATION

Description/definition : Fluocortolone caproate is the caproate ester of the synthetic glucocorticoid fluocortolone that conforms to the structural formula shown below

Pharmacological classes : Anti-inflammatory agents; glucocorticoids

IUPAC name : [2-[(6S,8S,9S,10R,11S,13S,14S,16R,17S)-6-Fluoro-11-hydroxy-10,13,16-trimethyl-3-oxo-6,7,8,9,11,12,14,15,16,17-decahydrocyclopenta[a]phenanthren-17-yl]-2-oxoethyl] hexanoate

Other names : Fluocortolone 21-hexanoate; 6α-fluoro-11β,21-dihydroxy-16α-methylpregna-1,4-diene-3,20-dione 21-hexanoate

CAS registry number : 303-40-2

EC number : 206-140-0

Merck Index monograph : 5454 (Fluocortolone)

Patch testing : In general, corticosteroids may be tested at 0.1% and 1% in alcohol; late readings (6-10 days) are strongly recommended

Molecular formula : $C_{28}H_{39}FO_5$

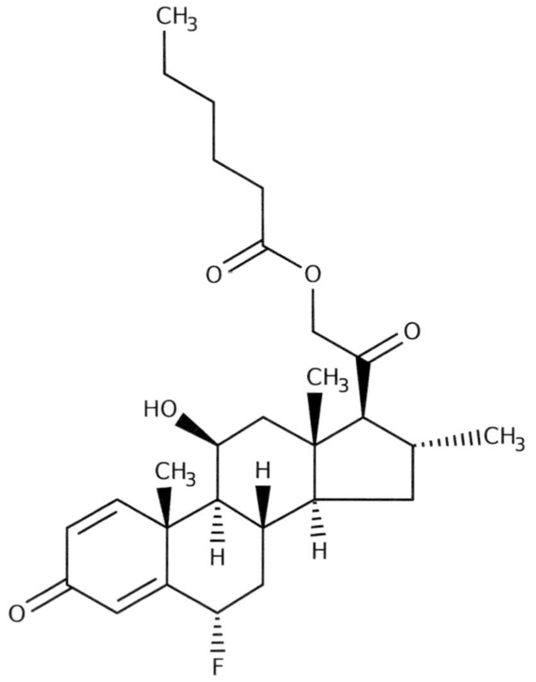

GENERAL

General aspects of corticosteroids used on the skin and mucous membranes are discussed in Chapter 2.4. A practical guideline for diagnosing allergic reactions to corticosteroids is presented in ref. 1. See also Chapter 3.145 Fluocortolone and Chapter 3.147 Fluocortolone pivalate.

CONTACT ALLERGY

Case series

From January 1990 to June 2008, in Leuven, Belgium, 315 patients were diagnosed with contact allergy to/allergic contact dermatitis from corticosteroids (CSs) from routine patch testing with a baseline series including tixocortol pivalate, budesonide, hydrocortisone butyrate and prednisone caproate, patch testing with patients' own CS preparations, and testing those with proven contact allergy to a corticosteroid or strongly suspected of CS allergy later with a series of 66 CSs, including two sex hormones (progesterone and testosterone). 71% of the patients had relevant reactions, but these were not specified. In this group of 315 CS allergic patients, 13 had positive patch tests to fluocortolone caproate 0.1% alc. (4). It is unknown how many of these reactions were caused by the use of a pharmaceutical product containing fluocortolone caproate and how many were cross-reactions to other corticosteroids.

Case reports

A 68-year old woman had intermittent eczema of the hands and buttocks. She had been treated with various topical corticosteroids without any adverse effect. After changing to a fatty ointment containing 0.25% fluocortolone and 0.25% fluocortolone caproate, the eczema became worse a few days later. Patch tests revealed positive reactions to the fatty ointment and 2 other commercial preparations of the same brand containing either fluocortolone and fluocortolone caproate or fluocortolone pivalate and fluocortolone pivalate. In addition, there were positive reactions to fluocortolone and its 2 esters, tested at 1% pet. The vehicles of the preparations tested negative, as did other corticosteroids (2).

The same authors report on a 53-year-old female patient with recurrent genitofemoral eczema for 3 years. She had suffered from severe spreading eczema twice after using a corticosteroid preparation, at least one of which was of the brand containing either fluocortolone and fluocortolone caproate or fluocortolone caproate and fluocortolone pivalate. Patch tests were positive to fluocortolone but not to the esters. When retested later, the patient now reacted to fluocortolone and its pivalate ester, but not fluocortolone caproate (all tested 1% pet.) (2). The authors suggested that a higher concentration must possibly be used to detect corticosteroid allergy. However, the second test was read at D2 only, and late positive reactions (D6-D10) are frequent with corticosteroids.

A man aged 26 was treated with a corticosteroid cream containing fluocortolone and/or fluocortolone caproate for dermatitis of the face. After 3 weeks treatment, the face became swollen, red and itchy. Patch tests were positive to the cream and 3 ointments and a lotion of the same brand. Patch testing with crystallized powder of the corticosteroid(s) in the cream also yielded a positive reaction. The author could not determine whether the allergy was due to fluocortolone, fluocortolone caproate or both (3).

Cross-reactions, pseudo-cross-reactions and co-reactions

Cross-reactions between corticosteroids are discussed in Chapter 2.8.

LITERATURE

1   Baeck M, Goossens A. Immediate and delayed allergic hypersensitivity to corticosteroids: practical guidelines. Contact Dermatitis 2012;66:38-45
2   Menné T, Andersen KE. Allergic contact dermatitis from fluocortolone, fluocortolone pivalate and fluocortolone caproate. Contact Dermatitis 1977;3:337-340
3   Van Ketel WG. Allergy to Ultralan® preparations. Contact Dermatitis Newsletter 1974;15:427
4   Baeck M, Chemelle JA, Terreux R, Drieghe J, Goossens A. Delayed hypersensitivity to corticosteroids in a series of 315 patients: clinical data and patch test results. Contact Dermatitis 2009;61:163-175

Chapter 3.147 FLUOCORTOLONE PIVALATE

IDENTIFICATION

Description/definition : Fluocortolone pivalate is the pivalate ester of the synthetic glucocorticoid fluocortolone that conforms to the structural formula shown below

Pharmacological classes : Anti-inflammatory agents; glucocorticoids

IUPAC name : [2-[(6S,8S,9S,10R,11S,13S,14S,16R,17S)-6-Fluoro-11-hydroxy-10,13,16-trimethyl-3-oxo-6,7,8,9,11,12,14,15,16,17-decahydrocyclopenta[a]phenanthren-17-yl]-2-oxoethyl] 2,2-dimethylpropanoate

Other names : 6α-Fluoro-11β,21-dihydroxy-16α-methylpregna-1,4-diene-3,20-dione 21-pivalate; fluocortolone 21-pivalate

CAS registry number : 29205-06-9

EC number : 249-504-4

Merck Index monograph : 5454 (Fluocortolone)

Patch testing : In general, corticosteroids may be tested at 0.1% and 1% in alcohol; late readings (6-10 days) are strongly recommended

Molecular formula : $C_{27}H_{37}FO_5$

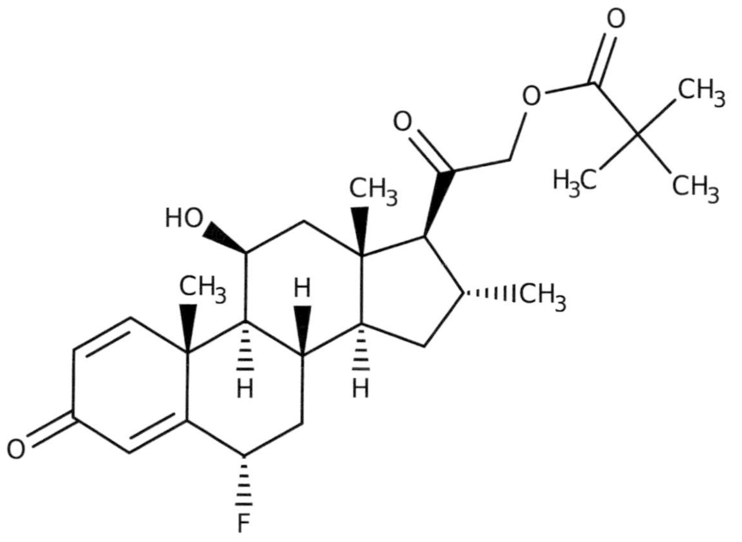

GENERAL

General aspects of corticosteroids used on the skin and mucous membranes are discussed in Chapter 2.4. A practical guideline for diagnosing allergic reactions to corticosteroids is presented in ref. 1. See also Chapter 3.145 Fluocortolone and Chapter 3.146 Fluocortolone caproate.

CONTACT ALLERGY

Case series

From January 1990 to June 2008, in Leuven, Belgium, 315 patients were diagnosed with contact allergy to/allergic contact dermatitis from corticosteroids (CSs) from routine patch testing with a baseline series including tixocortol pivalate, budesonide, hydrocortisone butyrate and prednisone caproate, patch testing with patients' own CS preparations, and testing those with proven contact allergy to a corticosteroid or strongly suspected of CS allergy later with a series of 66 CSs, including two sex hormones (progesterone and testosterone). 71% of the patients had relevant reactions, but these were not specified. In this group of 315 CS allergic patients, 9 had positive patch tests to fluocortolone pivalate 0.1% alc. (3). It is unknown how many of these reactions were caused by the use of a pharmaceutical product containing fluocortolone pivalate and how many were cross-reactions to other corticosteroids.

Case reports

A 68-year old woman had intermittent eczema of the hands and buttocks. She had been treated with various topical corticosteroids without any adverse effect. After changing to a fatty ointment containing 0.25% fluocortolone and 0.25% fluocortolone caproate, the eczema became worse a few days later. Patch tests revealed positive reactions to the fatty ointment and 2 other commercial preparations of the same brand containing either fluocortolone and fluocortolone caproate or fluocortolone pivalate and fluocortolone pivalate. In addition, there were positive reactions to fluocortolone and its 2 esters, tested at 1% pet. The vehicles of the preparations tested negative, as did other corticosteroids (2).

The same authors report on a 53-year-old female patient with recurrent genitofemoral eczema for 3 years. She had suffered from severe spreading eczema twice after using a corticosteroid preparation, at least one of which was of the brand containing either fluocortolone and fluocortolone caproate or fluocortolone caproate and fluocortolone pivalate. Patch tests were positive to fluocortolone but not to the esters. When retested later, the patient now reacted to fluocortolone and its pivalate ester, but not fluocortolone caproate (all tested 1% pet.) (2). The authors suggested that a higher concentration must possibly be used to detect corticosteroid allergy. However, the second test was read at D2 only, and late positive reactions (D6-D10) are frequent with corticosteroids.

Cross-reactions, pseudo-cross-reactions and co-reactions
Cross-reactions between corticosteroids are discussed in Chapter 2.8.

LITERATURE
1   Baeck M, Goossens A. Immediate and delayed allergic hypersensitivity to corticosteroids: practical guidelines. Contact Dermatitis 2012;66:38-45
2   Menné T, Andersen KE. Allergic contact dermatitis from fluocortolone, fluocortolone pivalate and fluocortolone caproate. Contact Dermatitis 1977;3:337-340
3   Baeck M, Chemelle JA, Terreux R, Drieghe J, Goossens A. Delayed hypersensitivity to corticosteroids in a series of 315 patients: clinical data and patch test results. Contact Dermatitis 2009;61:163-175

Chapter 3.148 FLUOROMETHOLONE

IDENTIFICATION

Description/definition   : Fluorometholone is the synthetic glucocorticoid that conforms to the structural formula shown below

Pharmacological classes   : Anti-allergic agents; anti-inflammatory agents; glucocorticoids

IUPAC name   : (6*S*,8*S*,9*R*,10*S*,11*S*,13*S*,14*S*,17*R*)-17-Acetyl-9-fluoro-11,17-dihydroxy-6,10,13-trimethyl-6,7,8,11,12,14,15,16-octahydrocyclopenta[a]phenanthren-3-one

Other names   : 21-Desoxy-9α-fluoro-6-methylprednisolone; 6α-methyl-9α-fluoro-21-desoxyprednisolone; 9-fluoro-11β,17-dihydroxy-6α-methylpregna-1,4-diene-3,20-dione

CAS registry number   : 426-13-1

EC number   : 207-041-5

Merck Index monograph   : 5477

Patch testing   : In general, corticosteroids may be tested at 0.1% and 1% in alcohol; late readings (6-10 days) are strongly recommended

Molecular formula   : $C_{22}H_{29}FO_4$

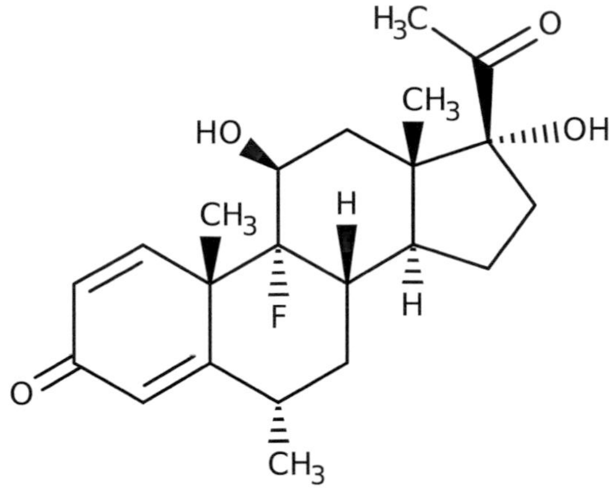

GENERAL

General aspects of corticosteroids used on the skin and mucous membranes are discussed in Chapter 2.4. A practical guideline for diagnosing allergic reactions to corticosteroids is presented in ref. 1.

CONTACT ALLERGY

Case reports and case series

From January 1990 to June 2008, in Leuven, Belgium, 315 patients were diagnosed with contact allergy to/allergic contact dermatitis from corticosteroids (CSs) from routine patch testing with a baseline series including tixocortol pivalate, budesonide, hydrocortisone butyrate and prednisone caproate, patch testing with patients' own CS preparations, and testing those with proven contact allergy to a corticosteroid or strongly suspected of CS allergy later with a series of 66 CSs, including two sex hormones (progesterone and testosterone). 71% of the patients had relevant reactions, but these were not specified. In this group of 315 CS allergic patients, 13 had positive patch tests to fluorometholone 0.1% alc. (3). It is unknown how many of these reactions were caused by the use of a pharmaceutical product containing fluorometholone and how many were cross-reactions to other corticosteroids.

In a group of 203 patients with eyelid dermatitis, one reacted to fluorometholone (test concentration and vehicle not mentioned); the relevance was not specified and it is unknown how many individuals were tested with this material (2).

Cross-reactions, pseudo-cross-reactions and co-reactions

Cross-reactions between corticosteroids are discussed in Chapter 2.8.

LITERATURE

1   Baeck M, Goossens A. Immediate and delayed allergic hypersensitivity to corticosteroids: practical guidelines. Contact Dermatitis 2012;66:38-45
2   Guin JD. Eyelid dermatitis: experience in 203 cases. J Am Acad Dermatol 2002;47:755-765
3   Baeck M, Chemelle JA, Terreux R, Drieghe J, Goossens A. Delayed hypersensitivity to corticosteroids in a series of 315 patients: clinical data and patch test results. Contact Dermatitis 2009;61:163-175

Chapter 3.149 FLUOROURACIL

IDENTIFICATION

Description/definition   : Fluorouracil is the fluoropyrimidine analog of the nucleoside pyrimidine that conforms to
                                the structural formula shown below
Pharmacological classes   : Antimetabolites; antimetabolites, antineoplastic; immunosuppressive agents
IUPAC name   : 5-Fluoro-1*H*-pyrimidine-2,4-dione
Other names   : 5-Fluorouracil; 5-fluoro-2,4(1*H*,3*H*)-pyrimidinedione
CAS registry number   : 51-21-8
EC number   : 200-085-6
Merck Index monograph   : 5483
Patch testing   : commercial 5% cream 'as is'; 5-fluorouracil 50 mg/ml (5%) and 10 mg/ml (1%) in saline or
                                5% pet.; if negative and contact allergy is strongly suspected, or with ?+ patch tests,
                                perform intradermal tests with 5-FU 10 mg/ml (1%) saline with reading at D2
Molecular formula   : $C_4H_3FN_2O_2$

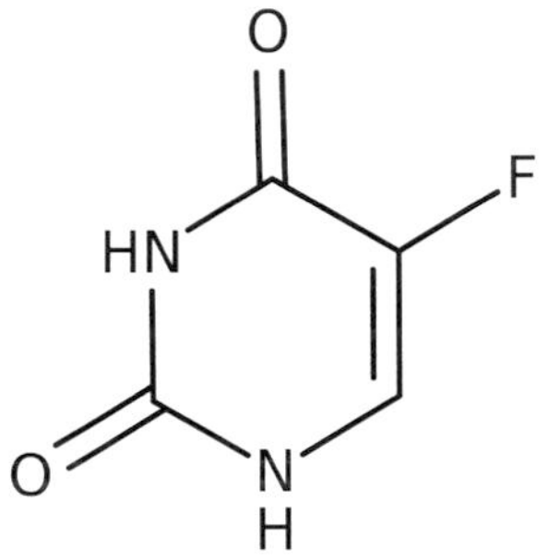

GENERAL

Fluorouracil (5-FU) is an antimetabolite fluoropyrimidine analog of the nucleoside pyrimidine with antineoplastic activity. It interferes with DNA synthesis by blocking the thymidylate synthetase conversion of deoxyuridylic acid to thymidylic acid. Fluorouracil is indicated for the topical treatment of multiple actinic (solar) keratoses. In the 5% strength it is also useful in the treatment of superficial basal cell carcinomas when conventional methods are impractical, such as with multiple lesions or difficult treatment sites. Fluorouracil injection is indicated in the palliative management of some types of cancer, including of the colon, esophagus, rectum, breast, biliary tract, stomach, head and neck, cervix, pancreas, renal cell cancer and carcinoid (1).

CONTACT ALLERGY

General

Related to its widespread use, contact allergy to 5-fluorouracil (5-FU) is often believed to be infrequent. However, in a period of 3 years (2004-2006), 8 cases were observed in a clinic in The Netherlands (3). In 1980-1984, one investigator from the USA had already presented 10 patients 'who were clearly allergic to 5-FU' (7,8) and in 1977, a case series of 6 allergic patients with positive patch and intradermal tests (of 35 investigated!) had been published (9). This clearly indicates that contact allergy to and allergic contact dermatitis from 5-FU may be far from rare.

Moreover, in another study from the USA, sensitization rates to 5-FU of 68% (retrospective study in 25 patients previously treated with 5-FU) and 74% (prospective study in 15 patients, sensitization studies performed before and after therapy with 5-FU) were reported. However, only intradermal tests and not patch tests were performed (10).

Possibly, cases of allergic contact dermatitis are not recognized and are mistaken for the irritant dermatitis that usually results from treatment with 5-FU cream. In addition, patch tests may be false-negative and intradermal tests with 5-FU with late readings sometimes have to be performed to reach the correct diagnosis of contact allergy (4,7,8). Application of 5-FU under occlusion may enhance and the former use of high concentrations of 5-FU (up to 25%) may have increased the risk of sensitization (7).

Case series

In The Netherlands, between 2004 and 2006, 14 patients (7 women and 7 men, aged 51-80 years) presenting with symptoms of a possible allergic contact dermatitis during topical use of 5-FU cream for of actinic keratoses, were investigated. Symptoms consisted of erythema, severe itching, swelling and ulcerations at the site of application of the cream and skin beyond, more severe and extensive than the irritant reaction expected from the topical use of this cream. The application sites involved were the face, the vertex, the lower legs and/or the dorsa of the hands. The duration of the initial treatment was 3-4 weeks with a frequency of 2 daily applications. The earliest reaction was observed in one patient 10 days after initial treatment with the cream. In the remaining patients, the extensive skin reactions developed during a second or third treatment after an unsatisfactory response to previous therapy with 5-FU. Patch tests were performed with the cream 'as is' and diluted 0.1% and 1% pet., as well as with all the ingredients individually: propylene glycol (20% pet.), methyl- and propylparaben (both 3% pet.), stearyl alcohol (30% pet.), polysorbate 80 (5% pet.), sorbitan monooleate (5% pet.), and 5-FU in serial dilutions of 0.1%, 1% and 5% in saline. Eight of the 15 patients had positive patch tests to the 5-fluorouracil cream. In one, the reaction was caused by stearyl alcohol and not by 5-FU. The others also had positive patch tests to the cream 1% pet. (n=4), the cream diluted 0.1% pet. (n=1), to 5-FU in saline 50 mg/ml (5%) (n=7), to 1% (n=6) and to 0.1% (n=1). Two of these patients also reacted to propylene glycol, and the same individuals also to stearyl alcohol (3).

In Leuven, Belgium, in the period 1990-2014, iatrogenic contact dermatitis was diagnosed in 2600 individuals (17% of the total patch test population). 96% of all positive patch test reactions to topical drugs and antiseptics were considered to be relevant. Fluorouracil (5% pet.) was tested in 16 patients and there were 4 positive reactions to it (2).

In the USA, in the period 1980-1984, a dermatologist from California presented 10 patients 'who were clearly allergic to 5-FU' (7,8). Only 6 or 7 had positive patch tests with 5-FU 2%, 5% and 25% pet., but all had positive intradermal tests to 0.1 ml of 1% and 4% 5-FU solution at D2 and D3 with erythematous and edematous reactions more than 10 mm in diameter. The author suggested that 1 in 4 patients with allergy to 5-FU will be missed when only patch tests are performed (7,8).

In the USA, in 1976, 35 patients who had been treated for actinic keratoses with topical 5-FU in the preceding three months to three years on one or more occasions, received intracutaneous tests with 0.1 ml 5-FU 5%, 2.5% and 1% in sterile water (reading at D2) and 20 of these also received patch tests with 5-FU 1%, 5% and 10% in petrolatum. Six patients had positive intracutaneous tests to all 3 concentrations and all 6 individuals also had positive patch tests. They had previously experienced either a moderate local flare reaction from 5-FU treatment or a moderate to severe pruritic, widespread dermatitis. Biopsy specimens of positive intra- and epicutaneous tests were consistent with allergic contact dermatitis. Control tests in 35 patients who had never used 5-FU before were negative with the exception of one patient who showed asymptomatic macular erythema at the intracutaneous test sites (9).

Case reports

A 64-year-old man with androgenetic alopecia and actinic damage presented with multiple actinic keratoses on the scalp, which were successfully treated with 5-FU ointment 2x daily for 21 days. 18 months later, the patient began to apply 5-FU ointment again for further such lesions. Five days later, erythematous and edematous plaques appeared on and around the sites of application. Pruriginous vesicles and exudation were also noted, with eyelid edema. Patch tests with the cream and 5-FU 5% pet. were ?+ on 2 occasions. Intradermal testing with 0.1 ml 5-FU in saline in concentrations of 10 mg/ml (1%), 25 mg/ml (2.5%) and 50 mg/ml (5%) resulted in plaques of more of 10 mm diameter of erythema, edema and vesicles to all concentrations (4).

Pustular allergic contact dermatitis from fluorouracil with rosacea-like sequelae has been observed in a 51-year-old woman from treatment of actinic keratoses on the face. A patch test with the fluorouracil cream was negative, but ingredient patch testing showed a papulopustular eruption to 1% fluorouracil in water in a wide area around the test site. Nine months later, a standard patch test with the fluorouracil cream was performed and the pustular response again occurred (15).

Short summaries of other case reports of allergic contact dermatitis from 5-fluorouracil are shown in table 3.149.1.

Table 3.149.1 Short summaries of case reports of allergic contact dermatitis from 5-fluorouracil

Year and country	Sex	Age	Positive patch tests	Clinical data and comments	Ref.
1987 Belgium	M	24	5% FU ointment 'as is'; 5-FU 1% pet. (both 2x)	treatment of plane warts on the hands; after a week pruritic vesicular eruption at the sites of application; healing of warts from allergic contact dermatitis	6
1980 USA	F	52	intradermal tests pos. with 1, 2 and 4% 5-FU in saline; patch tests neg.	sensitized by 25% 5-FU in petrolatum under occlusion for basal cell carcinomas; later, 2 days after starting 5% 5-FU cream for actinic keratoses, acute dermatitis of the face	8
1968 USA	M	60	5-FU 5% in Aquaphor base, in hydrophilic petrolatum and in water	widespread eczema involving the face, neck and ears with moderate edema and crusting from treatment of actinic keratoses of the ears; cross-reaction to 5-bromouracil	12

FU: fluorouracil

According to the authors of ref. 6, one or more cases of sensitization to 5-FU have been reported in ref. 13. Details are not available to the author. Erythema multiforme-like allergic contact dermatitis from fluorouracil has been cited in ref. 14.

A 61-year-old man had undergone several courses of topical 5-FU cream for actinic keratoses. In a following course, the patient noticed severe inflammation at sites of 5-FU application, developing 2-3 days hours after the onset of treatment. Patch tests were positive to 5-FU 0.5% and 1% pet. Four years later, the patient was diagnosed as having rectal adenocarcinoma, for which chemotherapy with i.v. 5-FU was scheduled. One day later, the patient developed acute dermatitis with severe head and neck edema. An erythematovesicular eruption involving the scalp, face, neck and the lateral aspects of both hands was also present. The eruption was more severe on the right upper arm, where the i.v. perfusion had been administered. A diagnosis of systemic contact dermatitis was made (5).

Cross-reactions, pseudo-cross-reactions and co-reactions
A patient sensitized to 5-fluorouracil cross-reacted to 5-bromouracil (12).

LITERATURE
1   The data in the section 'General' may have been obtained from literature discussed in this chapter, but mostly also or exclusively from one or more of the following online sources: ChemIDPlus Advanced, PubChem, DrugBank, RxList, Drug Central, Drugs.com, and Wikipedia
2   Gilissen L, Goossens A. Frequency and trends of contact allergy to and iatrogenic contact dermatitis caused by topical drugs over a 25-year period. Contact Dermatitis 2016;75:290-302
3   Meijer BU, de Waard-van der Spek FB. Allergic contact dermatitis because of topical use of 5-fluorouracil (Efudix cream). Contact Dermatitis 2007;57:58-60
4   Sánchez-Pérez J, Bartolomé B, del Río MJ, García-Díez A. Allergic contact dermatitis from 5-fluorouracil with positive intradermal test and doubtful patch test reactions. Contact Dermatitis 1999;41:106-107
5   Nadal C, Pujol RM, Randazzo L, Marcuello E, Alomar A. Systemic contact dermatitis from 5-fluorouracil. Contact Dermatitis 1996;35:124-125
6   Tennstedt D, Lachapelle JM. Allergic contact dermatitis to 5-fluorouracil. Contact Dermatitis 1987;16:279-280
7   Epstein E. Testing for 5 fluorouracil allergy: patch and intradermal tests. Contact Dermatitis 1984;10:311-312
8   Epstein E. Contact dermatitis to 5-fluorouracil with false negative patch tests. Contact Dermatitis 1980;6:220-221
9   Goette DK, Odom RB. Allergic contact dermatitis to topical fluorouracil. Arch Dermatol 1977;113:1058-1061
10 Mansell PWA, Litwin MS, Ichinose H, Krementz ET. Delayed hypersensitivity to 5-fluorouracil following topical chemotherapy of cutaneous cancers. Cancer Res 1975;35:1288-1294
11 Goette DK, Odom RB, Owens R. Allergic contact dermatitis from topical fluorouracil. J Am Acad Dermatol 1977;113:196-198
12 Sams WM. Untoward response with topical fluorouracil. Arch Dermatol 1968;97:14-23
13 Leyden JJ, Kligman A. Studies on the allergenicity of 5-fluorouracil. J Derm Surg Oncol 1977;3:518-519
14 De Groot AC, Weyland JW, Nater JP. Unwanted effects of cosmetics and drugs used in dermatology, 3rd edition. Amsterdam: Elsevier Science BV, 1994:43
15 Sevadjian CM. Pustular contact hypersensitivity to fluorouracil with rosacealike sequelae. Arch Dermatol 1985;121:240-242

Chapter 3.150 FLUPREDNIDENE ACETATE

IDENTIFICATION

Description/definition : Fluprednidene acetate is the acetate ester of the synthetic glucocorticoid fluprednidene that conforms to the structural formula shown below

Pharmacological classes : Glucocorticoids

IUPAC name : [2-[(8S,9R,10S,11S,13S,14S,17R)-9-Fluoro-11,17-dihydroxy-10,13-dimethyl-16-methyl-idene-3-oxo-7,8,11,12,14,15-hexahydro-6H-cyclopenta[a]phenanthren-17-yl]-2-oxo-ethyl] acetate

Other names : 9-Fluoro-11β,17,21-trihydroxy-16-methylenepregna-1,4-diene-3,20-dione 21-acetate

CAS registry number : 1255-35-2

EC number : 215-013-9

Merck Index monograph : 5492

Patch testing : In general, corticosteroids may be tested at 0.1% and 1% in alcohol; late readings (6-10 days) are strongly recommended

Molecular formula : $C_{24}H_{29}FO_6$

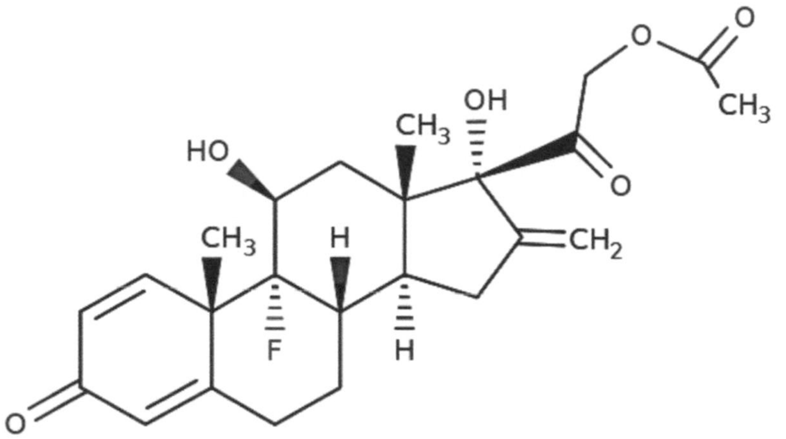

GENERAL

General aspects of corticosteroids used on the skin and mucous membranes are discussed in Chapter 2.4. A practical guideline for diagnosing allergic reactions to corticosteroids is presented in ref. 1.

CONTACT ALLERGY

Case series

From January 1990 to June 2008, in Leuven, Belgium, 315 patients were diagnosed with contact allergy to/allergic contact dermatitis from corticosteroids (CSs) from routine patch testing with a baseline series including tixocortol pivalate, budesonide, hydrocortisone butyrate and prednisone caproate, patch testing with patients' own CS preparations, and testing those with proven contact allergy to a corticosteroid or strongly suspected of CS allergy later with a series of 66 CSs, including two sex hormones (progesterone and testosterone). 71% of the patients had relevant reactions, but these were not specified. In this group of 315 CS allergic patients, 10 had positive patch tests to fluprednidene acetate 0.1% alc. (2). It is unknown how many of these reactions were caused by the use of a pharmaceutical product containing fluprednidene acetate and how many were cross-reactions to other corticosteroids (probably all).

Cross-reactions, pseudo-cross-reactions and co-reactions

Cross-reactions between corticosteroids are discussed in Chapter 2.8.

LITERATURE

1 Baeck M, Goossens A. Immediate and delayed allergic hypersensitivity to corticosteroids: practical guidelines. Contact Dermatitis 2012;66:38-45

2 Baeck M, Chemelle JA, Terreux R, Drieghe J, Goossens A. Delayed hypersensitivity to corticosteroids in a series of 315 patients: clinical data and patch test results. Contact Dermatitis 2009;61:163-175

Chapter 3.151 FLUPREDNISOLONE

IDENTIFICATION

Description/definition   : Fluprednisolone is the synthetic glucocorticoid that conforms to the structural formula
                               shown below
Pharmacological classes   : Glucocorticoids; anti-inflammatory agents
IUPAC name   : (6S,8S,9S,10R,11S,13S,14S,17R)-6-Fluoro-11,17-dihydroxy-17-(2-hydroxyacetyl)-10,13-
                               dimethyl-7,8,9,11,12,14,15,16-octahydro-6H-cyclopenta[a]phenanthren-3-one
Other names   : 6α-Fluoro-11β,17,21-trihydroxypregna-1,4-diene-3,20-dione; 6α-fluoroprednisolone
CAS registry number   : 53-34-9
EC number   : 200-170-8
Merck Index monograph : 5493
Patch testing   : Generally, corticosteroids may be tested at 0.1% and 1% in alcohol; late readings (6-10
                               days) are strongly recommended
Molecular formula   : $C_{21}H_{27}FO_5$

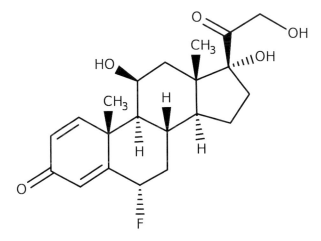

GENERAL

General aspects of corticosteroids used on the skin and mucous membranes are discussed in Chapter 2.4. A practical
guideline for diagnosing allergic reactions to corticosteroids is presented in ref. 1.

CONTACT ALLERGY

Case series

From January 1990 to June 2008, in Leuven, Belgium, 315 patients were diagnosed with contact allergy to/allergic
contact dermatitis from corticosteroids (CSs) from routine patch testing with a baseline series including tixocortol
pivalate, budesonide, hydrocortisone butyrate and prednisone caproate, and from patch testing with patients' own
CS preparations. Individuals with proven contact allergy to a corticosteroid or strongly suspected of CS allergy were
later tested with a series of 66 CSs, including two sex hormones (progesterone and testosterone). 71% of the
patients had relevant reactions, but relevancy was not specified for individual corticosteroids. In this group of 315 CS
allergic patients, 15 had positive patch tests to fluprednisolone 0.1% alc. As this corticosteroid has never been used
in pharmaceuticals in Belgium, these positive reactions must all be considered cross-reactions to other
corticosteroids.

LITERATURE

1   Baeck M, Goossens A. Immediate and delayed allergic hypersensitivity to corticosteroids: practical guidelines.
    Contact Dermatitis 2012;66:38-45
2   Baeck M, Chemelle JA, Terreux R, Drieghe J, Goossens A. Delayed hypersensitivity to corticosteroids in a series of
    315 patients: clinical data and patch test results. Contact Dermatitis 2009;61:163-175

Chapter 3.152 FLURANDRENOLIDE

IDENTIFICATION

Description/definition   : Flurandrenolide is the synthetic glucocorticoid that conforms to the structural formula shown below

Pharmacological classes : Anti-inflammatory agents; glucocorticoids

IUPAC name   : (1S,2S,4R,8S,9S,11S,12S,13R,19S)-19-Fluoro-11-hydroxy-8-(2-hydroxyacetyl)-6,6,9,13-tetramethyl-5,7-dioxapentacyclo[10.8.0.0^{2,9}.0^{4,8}.0^{13,18}]icos-17-en-16-one

Other names   : Flurandrenolone

CAS registry number   : 1524-88-5

EC number   : 216-196-8

Merck Index monograph : 5497

Patch testing   : Generally, corticosteroids may be tested at 0.1% and 1% in alcohol; late readings (6-10 days) are strongly recommended

Molecular formula   : C$_{24}$H$_{33}$FO$_6$

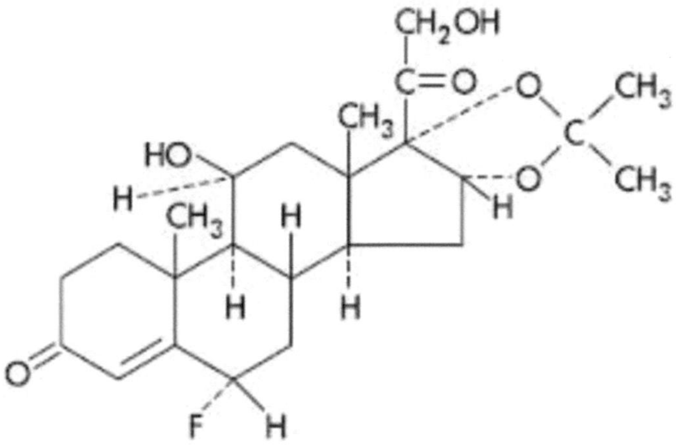

GENERAL

General aspects of corticosteroids used on the skin and mucous membranes are discussed in Chapter 2.4. A practical guideline for diagnosing allergic reactions to corticosteroids is presented in ref. 1.

CONTACT ALLERGY

Patch testing in groups of selected patients

In the United Kingdom, from 1988 to 1991, 528 selected patients with a positive patch test to tixocortol pivalate or suspected of corticosteroid allergy were patch tested with flurandrenolide 20% pet. and 2 patients (0.4%) had a positive reaction. Their relevance was not addressed (2).

Case report

A 66-year-old female patient developed increasing eczematous reactions after treatment of her leg ulcer with a corticosteroid ointment. When patch tested with the constituents, she reacted positively to fludroxycortide (flurandrenolide) and cetyl alcohol (3).

Cross-reactions, pseudo-cross-reactions and co-reactions

Cross-reactions between corticosteroids are discussed in Chapter 2.8.

LITERATURE

1   Baeck M, Goossens A. Immediate and delayed allergic hypersensitivity to corticosteroids: practical guidelines. Contact Dermatitis 2012;66:38-45

2   Burden AD, Beck MH. Contact hypersensitivity to topical corticosteroids. Br J Dermatol 1992;127:497-501

3   Hausen BM, Kulenkamp D. Contact allergy to fludroxycortid and cetyl alcohol. Derm Beruf Umwelt 1985;33:27-28 (Article in German)

Chapter 3.153 FLURBIPROFEN

IDENTIFICATION

Description/definition   : Flurbiprofen is the arylpropionic acid derivative that conforms to the structural formula
                             shown below
Pharmacological classes : Anti-Inflammatory agents, non-steroidal; cyclooxygenase inhibitors; analgesics
IUPAC name   : 2-(3-Fluoro-4-phenylphenyl)propanoic acid
Other names   : 2-(2-Fluoro-4-biphenylyl)propionic acid
CAS registry number   : 5104-49-4
EC number   : 225-827-6
Merck Index monograph : 5499
Patch testing   : 1% and 5% pet.
Molecular formula   : $C_{15}H_{13}FO_2$

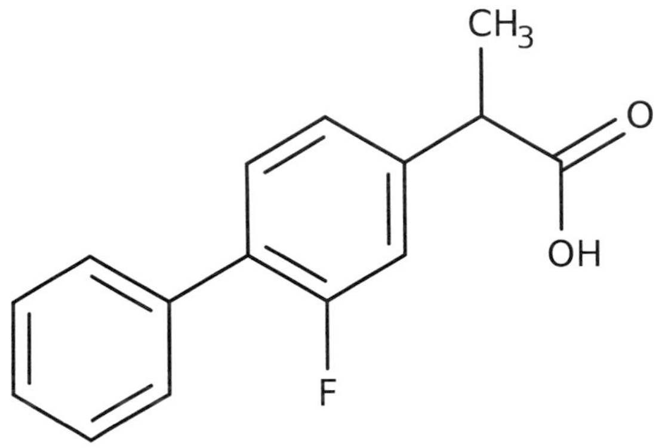

GENERAL

Flurbiprofen is a derivative of propionic acid and a nonsteroidal anti-inflammatory drug (NSAID) with analgesic, anti-inflammatory and antipyretic effects. Flurbiprofen non-selectively binds to and inhibits cyclooxygenase (COX). This results in a reduction of arachidonic acid conversion into prostaglandins that are involved in the regulation of pain, inflammation and fever. Upon ocular administration, flurbiprofen may reduce bicarbonate ion concentrations leading to a decrease in the production of aqueous humor, thereby lowering intraocular pressure (1).

Flurbiprofen tablets are indicated for the acute or long-term symptomatic treatment of rheumatoid arthritis, osteoarthritis and ankylosing spondylitis. It may also be used to treat pain associated with dysmenorrhea and mild to moderate pain accompanied by inflammation (e.g. bursitis, tendonitis, soft tissue trauma). Topical ophthalmic formulations may be used pre-operatively to prevent intraoperative miosis. In pharmaceutical products, both flurbiprofen base (in tablets) and – in eye drops – flurbiprofen sodium (CAS number 56767-76-1, EC number 260-373-2, molecular formula $C_{15}H_{12}FNaO_2$) may be employed (1).

CONTACT ALLERGY

Case series

In 3 hospitals in Italy, in a 7-year-period before 1996, 123 patients with contact or photocontact allergic reactions to topical NSAIDs were investigated. Only one patient had allergic contact dermatitis from flurbiprofen (3).

Case report

A 22-year-old woman had developed pruritic erythema over her right wrist joint and forearm, one day after having applied a poultice containing flurbiprofen for pain in the right wrist joint. She had not been exposed to sunlight when or after applying the poultice. Patch and photopatch tests with flurbiprofen 1% and 5% pet. and a series of other NSAIDs showed positive patch test reactions to both concentrations of flurbiprofen. Five controls were negative (2).

Cross-reactions, pseudo-cross-reactions and co-reactions

One patient sensitized to ibuprofen may have cross-reacted to flurbiprofen and fenoprofen (3). An individual who had photoallergic contact dermatitis from ketoprofen showed cross-photosensitivity to flurbiprofen and ibuproxam (4). A patient with photocontact allergy to tiaprofenic acid photocross-reacted to flurbiprofen (5).

Cutaneous adverse drug reactions from systemic administration caused by type IV (delayed-type) hypersensitivity
Cutaneous adverse drug reactions from systemic administration of flurbiprofen caused by type IV (delayed-type) hypersensitivity, including maculopapular rash (6), are planned to be discussed in Volume IV of the *Monographs in Contact Allergy* series on Systemic drugs.

LITERATURE

1   The data in the section 'General' may have been obtained from literature discussed in this chapter, but mostly also or exclusively from one or more of the following online sources: ChemIDPlus Advanced, PubChem, DrugBank, RxList, Drug Central, Drugs.com, and Wikipedia

2   Kawada A, Aragane Y, Maeda A, Yudate T, Tezuka T. Contact dermatitis due to flurbiprofen. Contact Dermatitis 2000;42:167-168

3   Pigatto P, Bigardi A, Legori A, Valsecchi R, Picardo M. Cross-reactions in patch testing and photopatch testing with ketoprofen, tiaprophenic acid, and cinnamic aldehyde. Am J Cont Dermat 1996;7:220-223

4   Mozzanica N, Pigatto PD. Contact and photocontact allergy to ketoprofen: clinical and experimental study. Contact Dermatitis 1990;23:336-340

5   Valsecchi R, Landro AD, Pigatto P, Cainelli T. Tiaprofenic acid photodermatitis. Contact Dermatitis 1989;21:345-346

6   Romano A, Pietrantonio F. Delayed hypersensitivity to flurbiprofen. J Intern Med 1997;241:81-83

Chapter 3.154 FLUTICASONE PROPIONATE

IDENTIFICATION

Description/definition : Fluticasone propionate is the 17-propionate ester of the synthetic glucocorticoid fluticasone that conforms to the structural formula shown below

Pharmacological classes : Anti-allergic agents; anti-inflammatory agents; bronchodilator agents; dermatologic agents

IUPAC name : [(6S,8S,9R,10S,11S,13S,14S,16R,17R)-6,9-Difluoro-17-(fluoromethylsulfanylcarbonyl)-11-hydroxy-10,13,16-trimethyl-3-oxo-6,7,8,11,12,14,15,16-octahydrocyclopenta[a]-phenanthren-17-yl] propanoate

Other names : S-(Fluoromethyl) 6α,9-difluoro-11β,17-dihydroxy-16α-methyl-3-oxoandrosta-1,4-diene-17β-carbothioate, 17-propionate

CAS registry number : 80474-14-2

EC number : 617-082-4

Merck Index monograph : 5510

Patch testing : In general, corticosteroids may be tested at 0.1% and 1% in alcohol; late readings (6-10 days) are strongly recommended

Molecular formula : $C_{25}H_{31}F_3O_5S$

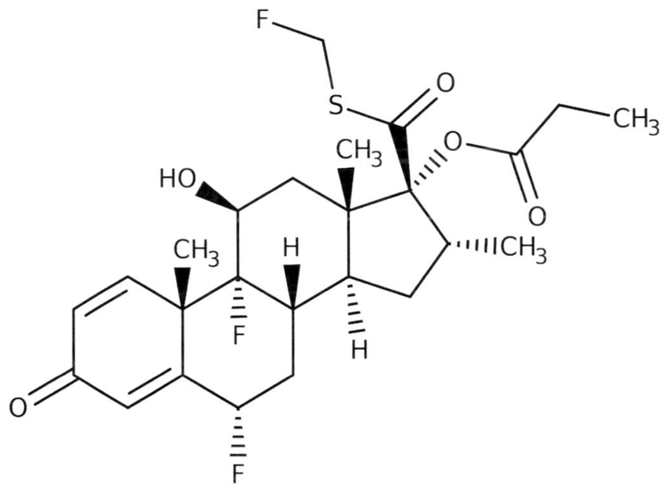

GENERAL

General aspects of corticosteroids used on the skin and mucous membranes are discussed in Chapter 2.4. A practical guideline for diagnosing allergic reactions to corticosteroids is presented in ref. 1.

CONTACT ALLERGY

Patch testing in consecutive patients suspected of contact dermatitis: Routine testing

In the United Kingdom, in the period 1994-1995, 1312 patients were routinely tested with fluticasone propionate 1% in acetone. There were 67 steroid-allergic patients, but none reacted to fluticasone propionate. It was concluded that this corticosteroid has a low risk of sensitization (3). It may, however, cross-react in patients with multiple corticosteroid sensitivities (4).

Case series

From January 1990 to June 2008, in Leuven, Belgium, 315 patients were diagnosed with contact allergy to/allergic contact dermatitis from corticosteroids (CSs) from routine patch testing with a baseline series including tixocortol pivalate, budesonide, hydrocortisone butyrate and prednisone caproate, patch testing with patients' own CS preparations, and testing those with proven contact allergy to a corticosteroid or strongly suspected of CS allergy later with a series of 66 CSs, including two sex hormones (progesterone and testosterone). 71% of the patients had relevant reactions, but these were not specified. In this group of 315 CS allergic patients, 14 had positive patch tests to fluticasone propionate 0.1% alc. (5, overlap with ref. 2). It is unknown how many of these reactions were caused by the use of a pharmaceutical product containing fluticasone propionate and how many were cross-reactions to other corticosteroids.

In the period 1990-1999, in Belgium, fluticasone propionate 1% alc. was patch tested in an extended corticosteroid series in 206 patients. 118 of them had actually used the product (89 used the cream and 29 the ointment) and 155 subjects were tested with the corticosteroid series. There were positive patch test reactions in only 7 patients, all of who also reacted to multiple other corticosteroids. Only one of these 7 individual had actually used a cream containing fluticasone propionate (2, overlap with ref. 5).

Cross-reactions, pseudo-cross-reactions and co-reactions
Cross-reactions between corticosteroids are discussed in Chapter 2.8.

LITERATURE

1   Baeck M, Goossens A. Immediate and delayed allergic hypersensitivity to corticosteroids: practical guidelines. Contact Dermatitis 2012;66:38-45
2   Goossens A, Huygens S, Matura M, Degreef H. Fluticasone propionate: a rare contact sensitizer. Eur J Dermatol 2001;11:29-34
3   Wilkinson SM, Beck MH. Fluticasone propionate and mometasone furoate have a low risk of contact sensitization. Contact Dermatitis 1996;34:365-366
4   Venning VA. Fluticasone propionate sensitivity in a patient with contact allergy to multiple corticosteroids. Contact Dermatitis 1995;33:48-49
5   Baeck M, Chemelle JA, Terreux R, Drieghe J, Goossens A. Delayed hypersensitivity to corticosteroids in a series of 315 patients: clinical data and patch test results. Contact Dermatitis 2009;61:163-175

Chapter 3.155 FRAMYCETIN

Framycetin is an aminoglycoside antibiotic. It is apparently used only in topical pharmaceutical preparations, notably for the eyes and ears, most often as framycetin sulfate. Chemical databases are not always consistent with nomenclature, synonyms, CAS numbers (many numbers provided) and EC numbers. Framycetin consists mainly of neomycin B, the main ingredient of neomycin. Consequently, framycetin, neomycin and neomycin B are often used as synonyms. See also Chapter 3.235 Neomycin.

IDENTIFICATION

Description/definition : Framycetin is an aminoglycoside antibiotic isolated from – according to some sources – *Streptomyces decaris*, mainly containing neomycin B; its structural formula (neomycin B) is shown below

Pharmacological classes : Anti-bacterial agents; protein synthesis inhibitors

IUPAC name : (2*R*,3*S*,4*R*,5*R*,6*R*)-5-Amino-2-(aminomethyl)-6-[(1*R*,2*R*,3*S*,4*R*,6*S*)-4,6-diamino-2-[(2*S*,3*R*,4*S*,5*R*)-4-[(2*R*,3*R*,4*R*,5*S*,6*S*)-3-amino-6-(aminomethyl)-4,5-dihydroxyoxan-2-yl]oxy-3-hydroxy-5-(hydroxymethyl)oxolan-2-yl]oxy-3-hydroxycyclohexyl]oxyoxane-3,4-diol

Other names : Neomycin; neomycin B; Soframycin

CAS registry number : 119-04-0

EC number : 204-292-2

Merck Index monograph : 8909 (Neomycin)

Patch testing : Sulfate, 20.0% pet. (Chemotechnique); sulfate, 10% pet. (SmartPracticeCanada, SmartPracticeEurope)

Molecular formula : $C_{23}H_{46}N_6O_{13}$

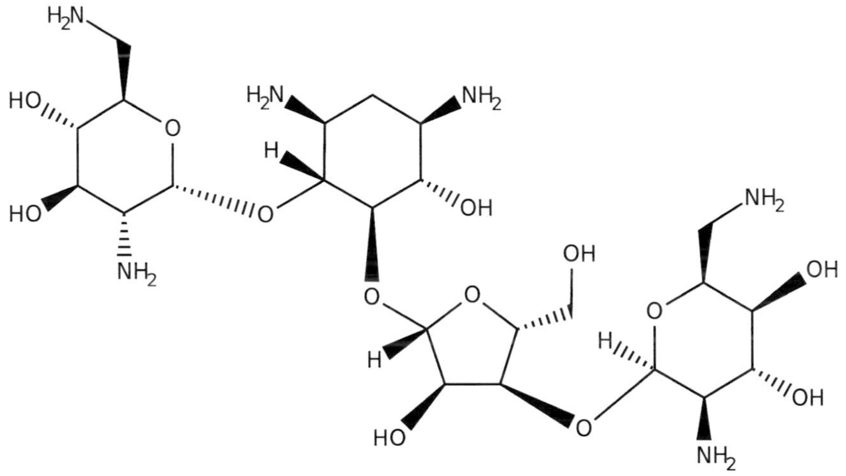

GENERAL

Framycetin is an aminoglycoside antibiotic isolated from *Streptomyces decaris*, mainly containing neomycin B, with broad-spectrum antibacterial activity. It is indicated for the topical treatment of bacterial blepharitis, bacterial conjunctivitis, corneal injuries, corneal ulcers and meibomianitis. Framycetin is also used for the prophylaxis of ocular infections following foreign body removal. It is also widely used in ointment and tulle for the treatment of wounds and leg ulcers, in a combination preparation with a corticosteroid in an ointment frequently used for pruritus ani and hemorrhoids and in eardrops and nose spray. In pharmaceutical products, framycetin is employed as framycetin sulfate (CAS number 4146-30-9, EC number 223-969-3, molecular formula $C_{23}H_{52}N_6O_{25}S_3$) (1).

CONTACT ALLERGY

General

Contact allergy/allergic contact dermatitis from framycetin is not infrequent, especially in patients with leg ulcers/ stasis dermatitis and otitis externa. Frequent neomycin co-reactivity can make interpretation of relevance difficult.

Contact allergy in the general population

With the CE-DUR approach, the incidence of sensitization to framycetin sulfate in the German population was estimated to range from 3 to 11 cases/100,000/year in the period 1995-1999 and from 3 to 13 cases/100,000/year in the period 2000-2004 (18). Also In Germany, for the period 1995-2004, the population-based relative incidence (RI) of contact sensitization to framycetin (cases/100,000 defined daily doses (DDDs) per year) was estimated to be 86.2. In the group of antibiotics, the RI ranged from 1.6 (oxytetracycline) to 86.2 (framycetin) (19).

Patch testing in groups of selected patients

The results of patch testing with framycetin in groups of selected patients (e.g. patients with leg ulcers/stasis dermatitis, individuals with (chronic) otitis externa, patients with perianal or genital dermatitis) are shown in table 3.155.1. In 7 studies patch testing patients with stasis dermatitis or leg ulcers, prevalences of sensitization ranged from 5% to 20%. Relevance figures were high ('the majority' to 100%) (4,5,8), but in an American study, none of 4

Table 3.155.1 Patch testing in groups of patients: Selected patient groups

Years and Country	Test conc. & vehicle	Number of patients tested	positive (%)		Selection of patients (S); Relevance (R); Comments (C)	Ref.
Patients with leg ulcers/stasis dermatitis						
<2017 India	20% pet.	172	16	(9.3%)	S: patients with venous leg ulcers of over 6 weeks' duration; R: 'the majority of reactions were relevant'	8
2003-2014 IVDK	10% pet.	2029		(5.0%)	S: patients with stasis dermatitis/chronic leg ulcers; R: not stated	2
2006-2008 Germany	20% pet.	95	10	(17%)	S: patients with chronic leg ulcers; R: 87% of all reactions were considered to be relevant	4
<2004 USA, Canada	10% pet.	54	4	(7%)	S: patients with past or present leg ulcers with or without dermatitis; R: definite + probable 0%	3
1996-1997 U.K.		109	12	(11%)	S: mostly patients with leg ulcers and stasis dermatitis; R: not stated	6
<1994 U.K.	20% pet.	85		(20%)	S: patients with longstanding venous ulceration or eczema complicating leg ulcers; R: not stated	7
1988-1989 U.K.	20% pet.	815	18	(22%)	S: patients with leg ulcers; R: all reactions were considered to be relevant	5
Patients with otitis externa						
<2002 U.K.		149	7	(4.7%)	S: patients with chronic otitis media or otosclerosis scheduled for middle ear surgery; R: not stated	20
1985-2002 U.K	20% pet.	179	29	(16.2%)	S: patients with chronic inflammatory ear disease; R: 97%; C: significant reduction in the frequency of contact allergy in the second half of the study period	14
1993-4 Netherlands	20% pet.	34	12	(35%)	S: patients with chronic otitis externa or media; R: not stated	34
1986-1988 U.K.		37	10	(27%)	S: patients with chronic otitis externa; R: not stated; C: it was suggested that all were cross-reactions to neomycin	36
<1982 U.K.	20% pet.	40	6	(15%)	S: patients with otitis externa, chronic suppurative otitis media or discharging mastoid cavities >1 year; R: most were considered to be cross-reactions to neomycin	33
Other patient groups						
2013-2015 Ireland	10% pet.	99	3	(3%)	S: patients patch tested for perianal and/or genital symptoms; R: all reactions to medicaments were relevant	12
2003-2010 USA	20% pet.	55	3	(6%)	S: women with (itchy) vulvar dermatoses; R: not stated	10
2000-2010 Canada		100	8	(8%)	S: charts reviewed and included in the study when there was at least one positive reaction to a topical drug; R: not stated; C: the high percentages to all drugs are obviously the result of the – rather unusual – selection procedure targeting at topical drug allergy	23
2004-2008 IVDK				(2.4%)	S: patients with anogenital dermatoses tested with a medicament series; R: not stated; C: number of patients tested unknown	11
2000-6 Netherlands		41	3	(7%)	S: patients patch tested for suspected contact allergy to ophthalmological drugs; R: 69% for all reactions to drugs together	16
1995-2004 IVDK	10% pet.	8822		(4.9%)	S: not stated; R: not stated	18
1995-1999 IVDK	10% pet.	106		(9.3%)	S: patients with *allergic* periorbital contact dermatitis; the frequency was not significantly higher than in a control group	9
1986-1990 U.K.	20% pet.	135	8	(5.9%)	S: patients referred to a vulva clinic for patch testing; R: 7/8	13

IVDK: Information Network of Departments of Dermatology (Germany, Austria, Switzerland)

positive patch tests to framycetin were of definite or probable relevance (3). In patients with otitis externa, rates of sensitization to framycetin ranged from 4.7% to 35%, and were 15% or higher in 4/5 studies. In one investigation, 28 of 29 positive patch test reactions were scored as relevant (14), in three no relevance data were provided (20,34,36) and in two most or all reactions were considered to be cross-reactions to neomycin (33,36).

In other patient groups, lower rates of sensitization, ranging from 2.4% to 9.3%, were observed. In a study performed by the IVDK (Information Network of Departments of Dermatology: Germany, Austria, Switzerland) 9.3% positive reactions were seen in individuals with allergic periorbital contact dermatitis (suggesting contact allergy to eye drops or eye ointments), but this frequency was not statistically higher than in a control group (9). Relevance rates, where mentioned, were generally 70% or higher (12,13,16).

Case series
In Leuven, Belgium, between 1990 and 2017, 16,065 patients were investigated for contact allergy and 118 (0.7%) showed positive patch test reactions to topical ophthalmic medications and/or to their ingredients. Eighty-four individuals (71%) reacted to an active principle. Framycetin was tested in 73 patients and was the allergen in eye medications in one. There was also one reaction to framycetin in other types of medications (15).

Also in Leuven, Belgium, in the period 1990-2014, iatrogenic contact dermatitis was diagnosed in 2600 individuals (17% of the total population). 96% of all positive patch test reactions to topical drugs and antiseptics were considered to be relevant. Framycetin (20% pet.) was tested in 71 patients and there were 31 positive reactions to it (21). In Tokyo, Japan, out of 3903 patients who were patch tested in the period January 1987 to December 1995, 141 (3.6%) were patch tested with eye drops and 49 individuals (35%) reacted positively and were diagnosed with allergic contact dermatitis. In 36 cases ingredient patch testing was performed and there were 14 reactions to framycetin (17). In a Finnish study performed in the first half of the 1980s, framycetin and neomycin were the most frequent contact allergens in a patch test study among 142 patients suffering from chronic otitis externa; details are not available to the author (35).

Case reports
A female patient aged 36 had a relapsing perianal dermatitis from antihemorrhoidal ointments. It recurred when she used an ointment containing hydrocortisone, dibucaine (cinchocaine) and 1% framycetin (Soframycin). Patch tests were positive to framycetin 20% pet., neomycin 20% pet. and dibucaine (24).

An 80-year-old woman was referred from the ophthalmology department and presented with an acute, weeping dermatitis predominantly affecting the lower legs, forearms and hands. Diffuse subacute dermatitis was noted on the trunk, but the face, including the conjunctivae, was spared. She had undergone a left cataract extraction under general anesthesia. A subconjunctival injection of framycetin (500 mg ophthalmic powder suspended in 0.5 ml sterile water for injection) was administered immediately after the operation as routine prophylaxis against endophthalmitis. Patch tests 8 years previously had been positive to neomycin. An intradermal test was now performed with framycetin (1 mg in 0.1 ml water). At D2, there was a 3-cm diameter area of erythema at the injection site (25). There is evidence of significant systemic absorption of framycetin following this route of administration (26) and it can be concluded that this patient had systemic contact dermatitis (25).

Cross-reactions, pseudo-cross-reactions and co-reactions
In patients sensitized to neomycin, about 70% cross-reacts to framycetin (Chapter 3.235 Neomycin; 15,22,27,30,31). A high degree of co-reactivity between framycetin and neomycin can be expected, as framycetin consists mostly of neomycin B, which is also the main constituent of neomycin (27,32). However, neomycin B is not the only sensitizer either in framycetin or neomycin, as a considerable number of patients allergic to neomycin do not react to framycetin and vice versa (28,29).

LITERATURE
1   The data in the section 'General' may have been obtained from literature discussed in this chapter, but mostly also or exclusively from one or more of the following online sources: ChemIDPlus Advanced, PubChem, DrugBank, RxList, Drug Central, Drugs.com, and Wikipedia
2   Erfurt-Berge C, Geier J, Mahler V. The current spectrum of contact sensitization in patients with chronic leg ulcers or stasis dermatitis - new data from the Information Network of Departments of Dermatology (IVDK). Contact Dermatitis 2017;77:151-158
3   Saap L, Fahim S, Arsenault E, Pratt M, Pierscianowski T, Falanga V, Pedvis-Leftick A. Contact sensitivity in patients with leg ulcerations: a North American study. Arch Dermatol 2004;140:1241-1246
4   Kulozik M, Powell SM, Cherry G, Ryan T J. Contact sensitivity in community-based leg ulcer patients. Clin Exp Dermatol 1988;13:82-84
5   Wilson CL, Cameron J, Powell SM, Cherry G, Ryan TJ. High incidence of contact dermatitis in leg-ulcer patients – implications for management. Clin Exp Dermatol 1991;16:250-253

6   Gooptu C, Powell SM. The problems of rubber hypersensitivity (types I and IV) in chronic leg ulcer and stasis eczema patients. Contact Dermatitis 1999;41:89-93

7   Zaki I, Shall L, Dalziel KL. Bacitracin: a significant sensitizer in leg ulcer patients? Contact Dermatitis1994;31:92-94

8   Rai R, Shenoy MM, Viswanath V, Sarma N, Majid I, Dogra S. Contact sensitivity in patients with venous leg ulcer: A multi-centric Indian study. Int Wound J 2018;15:618-622

9   Herbst RA, Uter W, Pirker C, Geier J, Frosch PJ. Allergic and non-allergic periorbital dermatitis: patch test results of the Information Network of the Departments of Dermatology during a 5-year period. Contact Dermatitis 2004;51:13-19

10   Lucke TW, Fleming CJ, McHenry P, Lever R. Patch testing in vulval dermatoses: how relevant is nickel? Contact Dermatitis 1998;38:111-112

11   Bauer A. Contact sensitization in the anal and genital area. Curr Probl Dermatol 2011;40:133-141

12   Foley CC, White S, Merry S, Nolan U, Moriarty B, et al. Understanding the role of cutaneous allergy testing in anogenital dermatoses: a retrospective evaluation of contact sensitization in anogenital dermatoses. Int J Dermatol 2019;58:806-810

13   Marren P, Wojnarowska F, Powell S. Allergic contact dermatitis and vulvar dermatoses. Br J Dermatol 1992;126:52-56

14   Millard TP, Orton DI. Changing patterns of contact allergy in chronic inflammatory ear disease. Contact Dermatitis 2004;50:83-86

15   Gilissen L, De Decker L, Hulshagen T, Goossens A. Allergic contact dermatitis caused by topical ophthalmic medications: Keep an eye on it! Contact Dermatitis 2019;80:291-297

16   Wijnmaalen AL, van Zuuren EJ, de Keizer RJ, Jager MJ. Cutaneous allergy testing in patients suspected of an allergic reaction to eye medication. Ophthalmic Res 2009;41:225-229

17   Aoki J. [Allergic contact dermatitis due to eye drops. Their clinical features and the patch test results]. Nihon Ika Daigaku Zasshi 1997;64:232-237 (Article in Japanese)

18   Menezes de Padua CA, Uter W, Schnuch A. Contact allergy to topical drugs: prevalence in a clinical setting and estimation of frequency at the population level. Pharmacoepidemiol Drug Saf 2007;16:377-384

19   Menezes de Padua CA, Schnuch A, Nink K, Pfahlberg A, Uter W. Allergic contact dermatitis to topical drugs – Epidemiological risk assessment. Pharmacoepidemiol Drug Saf 2008;17:813-821

20   Yung MW, RajendraT. Delayed hypersensitivity reaction to topical aminoglycosides in patients undergoing middle ear surgery. Clin Otolaryngol Allied Sci 2002;27:365-368

21   Gilissen L, Goossens A. Frequency and trends of contact allergy to and iatrogenic contact dermatitis caused by topical drugs over a 25-year period. Contact Dermatitis 2016;75:290-302

22   Bajaj AK, Gupta SC, Chatterjee AK. Contact sensitivity to topical aminoglycosides in India. Contact Dermatitis 1992;27:204-205

23   Spring S, Pratt M, Chaplin A. Contact dermatitis to topical medicaments: a retrospective chart review from the Ottawa Hospital Patch Test Clinic. Dermatitis 2012;23:210-213

24   Van Ketel WG. Contact allergy to different antihaemorrhidal anaesthetics. Contact Dermatitis 1983;9:512-513

25   Morton CA, Evans CD, Douglas WS. Allergic contact dermatitis following subconjunctival injection of framycetin. Contact Dermatitis 1993;29:42-43

26   Bron AJ, Richards AB, Knight-Jones D, Easty DE, Ainslie D. Systemic absorption of Soframycin and subconjunctival injection. Br J Ophthalmol 1970;56:615-620

27   Samsoen M, Metz R, Melchior E, Foussereau J. Cross-sensitivity between aminoside antibiotics. Contact Dermatitis 1980;6:141

28   Carruthers JA, Cronin E. Incidence of neomycin and framycetin sensitivity. Contact Dermatitis 1976;2:269-270

29   Kirton, V, Munro-Ashman D. Contact dermatitis from neomycin and framycetin. Lancet 1965;i:138-139

30   Förström L, Pirilä V, Pirilä L. Cross-sensitivity within the neomycin group of antibiotics. Acta Derm Venereol (Stockh) 1979;59(Suppl.):67-69

31   Pirilä V, Pirilä L. Sensitization to the neomycin group of antibiotics. Patterns of cross-sensitivity as a function of polyvalent sensitization to different portions of the neomycin molecule. Acta Derm Venereol (Stockh) 1966;46:489-496

32   Malten KE. Soframycin - neomycin reactivity. Contact Dermatitis Newsletter 1973;13:358

33   Holmes RC, Johns AN, Wilkinson JD, Black MM, Rycroft RJ. Medicament contact dermatitis in patients with chronic inflammatory ear disease. J R Soc Med 1982;75:27-30

34   Van Ginkel CJ, Bruintjes TD, Huizing EH. Allergy due to topical medications in chronic otitis externa and chronic otitis media. Clin Otolaryngol Allied Sci 1995;20:326-328

35   Fräki JE, Kalimo K, Tuohimaa P, Aantaa E. Contact allergy to various components of topical preparations for treatment of external otitis. Acta Otolaryngol 1985;100:414-418

36   Smith IM, Keay MG, Buxton PK. Contact hypersensitivity in patients with chronic otitis externa. Clin Otolaryngol Allied Sci 1990;15:155-158

Chapter 3.156 FURALTADONE

IDENTIFICATION

Description/definition   : Furaltadone is the nitrofuran antibiotic that conforms to the structural formula shown
                            below
Pharmacological classes   : Anti-infective agents, urinary
IUPAC name   : 5-(Morpholin-4-ylmethyl)-3-[(*E*)-(5-nitrofuran-2-yl)methylideneamino]-1,3-oxazolidin-2-
                            one
Other names   : Nitrofurmethone
CAS registry number   : 139-91-3
EC number   : 205-384-5
Merck Index monograph   : 5593
Patch testing   : 1% pet.
Molecular formula   : $C_{13}H_{16}N_4O_6$

GENERAL

Furaltadone is a nitrofuran antibiotic that is used in veterinary medicine and is effective against bacterial infections in birds when added to feed or drinking water. It was formerly used in humans orally but was withdrawn due to toxicity. However, furaltadone hydrochloride (CAS number 3759-92-0, EC number 223-169-4, molecular formula $C_{13}H_{17}ClN_4O_6$) may still be used topically in some countries for treatment of ear disorders (1).

CONTACT ALLERGY

Case report

A 70-year-old woman had been treated for some time with eardrops for seborrheic dermatitis in the right ear canal, when she developed worsening of itchy erythematous, edematous and vesicular dermatitis in the ear canal and on the right ear. She stopped the use of the eardrops, with clearance of the lesions within a week. Patch tests with the eardrops and all its ingredients showed positive (++) patch test reactions at 2 and 4 days to the eardrops and its active ingredients furaltadone hydrochloride 1% pet. and neomycin sulfate 20% pet. Twenty controls were negative to furaltadone HCl 1% pet. (2).

A 58-year-old farmhand presented with a 6-month history of pruritus and papules on his forearms, face, neck and legs. When he worked preparing and distributing feed for pigs and chickens, the rash rapidly deteriorated. Patch testing with all ingredients of the feeds showed a positive reaction to furaltadone tartrate 1% pet. at D2 (+) and D4 (++). Twenty controls were negative. The patient was diagnosed with occupational allergic contact dermatitis (3).

LITERATURE

1   The data in the section 'General' may have been obtained from literature discussed in this chapter, but mostly
     also or exclusively from one or more of the following online sources: ChemIDPlus Advanced, PubChem,
     DrugBank, RxList, Drug Central, Drugs.com, and Wikipedia
2   Sánchez-Pérez J, Córdoba S, del Río MJ, García-Díes A. Allergic contact dermatitis from furaltadone in eardrops.
     Contact Dermatitis 1999;40:222
3   Vilaplana J, Grimalt F, Romaguera C. Contact dermatitis from furaltadone in animal feed. Contact Dermatitis
     1990;22:232-323

Chapter 3.157 FUSIDIC ACID

IDENTIFICATION

Description/definition	: Fusidic acid is an antibiotic isolated from the fermentation broth of *Fusidium coccineum* that conforms to the structural formula shown below
Pharmacological classes	: Anti-bacterial agents; protein synthesis inhibitors
IUPAC name	: (2Z)-2-[(3R,4S,5S,8S,9S,10S,11R,13R,14S,16S)-16-Acetyloxy-3,11-dihydroxy-4,8,10,14-tetramethyl-2,3,4,5,6,7,9,11,12,13,15,16-dodecahydro-1H-cyclopenta[a]phenanthren-17-ylidene]-6-methylhept-5-enoic acid
CAS registry number	: 6990-06-3
EC number	: 230-256-0
Merck Index monograph	: 5616
Patch testing	: Sodium fusidate, 2% pet. (Chemotechnique, SmartPracticeCanada, SmartPracticeEurope)
Molecular formula	: $C_{31}H_{48}O_6$

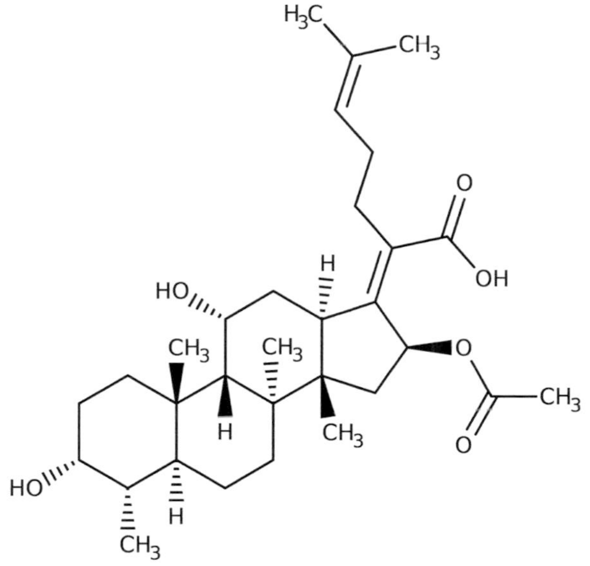

GENERAL

Fusidic acid is a bacteriostatic antibiotic derived from the fungus *Fusidium coccineum* that is used mostly as a topical medication to treat bacterial skin infections caused by *Staphylococcus aureus*, but is also given systemically as a tablet or injection. In pharmaceutical products, fusidic acid is employed as sodium fusidate (CAS number 751-94-0, EC number 212-030-3, molecular formula $C_{31}H_{47}NaO_6$) (1).

CONTACT ALLERGY

General

Related to its widespread use, allergic contact dermatitis from topical sodium fusidate is infrequent and is most often the result of application to leg ulcers, stasis dermatitis or atopic dermatitis. Co-reactivity to other antibiotics and lanolin (derivatives) was high in patients with leg ulcers/stasis dermatitis (12). Fusidic acid has a high molecular weight of 500 kDa and has a unique structure, different from those of other antibiotics.

Contact allergy in the general population

In Germany, for the period 1995-2004, the population-based relative incidence (RI) of contact sensitization to fusidic acid (cases/100,000 defined daily doses (DDDs) per year) was estimated to be 7.4. In the group of antibiotics, the RI ranged from 1.6 (oxytetracycline) to 86.2 (framycetin) (16).

Patch testing in groups of patients

Results of patch testing fusidic acid in consecutive patients suspected of contact dermatitis (routine testing) and of testing in groups of *selected* patients (e.g. patients with leg ulcers/stasis dermatitis, individuals with perianal dermatitis, patients suspected of allergy to [ophthalmological] drugs) are shown in table 3.157.1. In routine testing, low prevalences of 0.3% and 0.5% positive reactions to fusidic acid were found in the United Kingdom (2,12). In 9 studies in which patients with leg ulcers/stasis dermatitis were tested, the frequencies of sensitization ranged from 0.9% to 27% (9). The highest rates were found in Croatia (27% [9]) and Serbia (17% [11]), presumably because the sodium fusidate ointment and tulle were often used there. Relevance data were mostly not provided, but when they were, the rates of relevant reactions was usually high. In other groups of patients, rates of positive patch tests ranged from 0.8% to 4.1%, depending on selection criteria and exposure. Rates were 1.5%-4% in patients with perianal dermatitis (13) and those suspected of medicament allergy (12,14,17). Most reactions were considered to

Table 3.157.1 Patch testing in groups of patients

Years and Country	Test conc. & vehicle	Number of patients tested | positive (%)		Selection of patients (S); Relevance (R); Comments (C)	Ref.
Routine testing					
2000 United Kingdom	2% pet.	3063	(0.5%)	R: 100% (current and past relevance in one center); C: range of positive reactions per center 0% - 3.2%	2
1998 United Kingdom	2% pet.	1119	3 (0.3%)	R: the 'vast majority' of all reactions were relevant	12
Testing in groups of selected patients					
Patients with leg ulcers and/or stasis dermatitis					
<2017 India	2% pet.	172	4 (2.3%)	S: patients with venous leg ulcers of over 6 weeks' duration; R: 'the majority of reactions were relevant'	10
2003-2014 IVDK	2% pet.	2029	(2.3%)	S: patients with stasis dermatitis/chronic leg ulcers; R: not stated	3
2005-2008 France	2% pet.	423	4 (0.9%)	S: patients with leg ulcers; R: not stated	4
2006-2007 Canada		100	7 (7%)	S: patients with leg ulcers or venous disease; R: not stated	8
<2004 NACDG	2% pet.	54	2 (4%)	S: patients with past or present leg ulcers with or without dermatitis; R: definite + probable 100%	5
2000-2002 Serbia	2% pet.	75	13 (17%)	S: patients with chronic venous leg ulcers; R: not stated; C: 0% reactions in a (small) control group (difference significant)	11
1997-2001 U.K.	2% pet.	200	9 (4.5%)	S: patients with venous or mixed venous/arterial leg ulcers; R: all reactions to topical drugs were considered to be of probable, past or current relevance	6
1980-2000 U.K.	2% pet.	3307	48 (1.5%)	S: patients tested with a medicament series; R: not specified; C: after 1984, the rate dropped, despite increased use; 26 (54%) had stasis ulcers; >50% co-reactivity to lanolin and Amerchol-L101 (lanolin was present in a fusidic acid gauze dressing frequently used for leg ulcers in the early 1980s in the U.K) and 65% to other antibiotics	12
<1999 Croatia	2% pet.	100	27 (27%)	S: patients with leg ulcers; R: not stated	9
<1988 U.K.	2% pet.	59	2 (3%)	S: patients with chronic leg ulcers; R: 87% of all reactions were considered to be relevant	7
Other patient groups					
2004-2016 Spain		25	1 (4%)	S: patients with perianal dermatitis lasting >4 weeks; R: 100%	13
1990-2014 Belgium	2% pet.	1035	42 (4.1%)	S: patients suspected of iatrogenic contact dermatitis and tested with a pharmaceutical series and their own products; R: 96% of the positive patch test reactions to all topical drugs and antiseptics were considered to be relevant	17
2000-6 Netherlands		65	2 (3%)	S: patients patch tested for suspected contact allergy to ophthalmological drugs; R: 69% for all reactions to drugs together	14
1978-2005 Taiwan		603	5 (0.8%)	S: patients suspected of contact allergy to medicaments; R: 65% of the reactions to all medicaments were considered to be relevant	18
2002-2004 IVDK	2% pet.	3031	(1.1%)	S: not stated; R: not stated	15
1980-2000 U.K.	2% pet.	3307	48 (1.5%)	S: patients tested with a medicament series; R: not specified; C: after 1984, the rate dropped, despite increased use; 26 (54%) had stasis ulcers; >50% co-reactivity to lanolin and Amerchol-L101 (lanolin was present in a fusidic acid gauze dressing frequently used for leg ulcers in the early 1980s in the U.K) and 65% to other antibiotics	12

IVDK: Information Network of Departments of Dermatology (Germany, Austria, Switzerland); NACDG: North American Contact Dermatitis Group (USA, Canada)

be relevant. In one study from the United Kingdom, 50% of the sensitized patients co-reacted to lanolin and the lanolin-derivative Amerchol-L101 (lanolin was present in a fusidic acid gauze dressing frequently used for leg ulcers in the early 1980s in the U.K) and 65% to other antibiotics (12).

Case series

Fourteen cases of sensitization to sodium fusidate were seen in a 6-year period (1971-1976) at the St. John's Hospital in London. Ten of the patients had stasis eczema, 2 otitis externa, one patient had used fusidic acid ointment for herpes simplex, and another one had treated a graze on the leg with this topical drug. All patients showed a positive reaction to sodium fusidate 2% in petrolatum (29). The data of a small case series of 3 patients, seen in Switzerland in a period of 3 months, are shown in table 3.157.2.

Case reports

A 37-year-old woman had an 8-month history of widespread eczema, thought to be endogenous. She had been using fusidic acid in combination with a potent topical corticosteroid, which had not cleared the rash. There was a positive patch test reaction at day 4 only to fusidic acid 2% pet. A 39-year-old woman had widespread atopic eczema since the age of 3 years which had worsened over the last year. Patch tests showed a positive reaction at days 2 and 4 to fusidic acid 2% pet., which was confirmed on repeated testing. Both patients noticed a marked improvement following discontinuation of fusidic acid use (12).

Short summaries of other case reports of allergic contact dermatitis from fusidic acid are shown in table 3.157.2.

Table 3.157.2 Short summaries of case reports of allergic contact dermatitis from fusidic acid

Year and country	Sex	Age	Positive patch tests	Clinical data and comments	Ref.
2005 Korea	F	25	FA ointment 'as is'; SF 2%, 1%, 0.5% pet.	ACD from 2% FA ointment to unspecified papular dermatosis	19
2000 Korea	F	26	FA ointment 'as is'; SF 1% and 2% pet.	ACD from 2% FA ointment to abrasion; probably previous occupational sensitization while working as a nurse	21
1996 France	M	44	FA ointment 'as is'; SF 2% water, pet.	ACD of the perianal area, buttocks and backs of the thighs from FA ointment applied to hemorrhoids	22
1993 Korea	?	?	Sodium fusidate	two patients; FA ointment applied to a biopsy wound and after laser treatment of vascular nevi	23
1990 Portugal	M	24	FA cream 'as is'; SF 0.5%, 1% and 2% pet.	dermatitis of the face after 30 days of treatment with FA cream for superficial folliculitis	24
1990 Switzerland	F	79	FA cream and ointment 'as is'; FA and SF 2% alc.	one patient had a leg ulcer and developed acute eczema on the 10th day of application; the other 2 developed ACD superimposed on atopic dermatitis; these 3 cases were seen in one clinic in a period of 3 months	25
	M	17			
	M	30			
1988 Canada	M	33	FA 2% in 'cream' and in petrolatum	ACD around traumatic leg ulcer spreading as papular dermatitis to the arms, upper body and face	27
1985 Italy	M	68	FA ointment 'as is'; SF 2%, 1% and 0.1% pet.	'generalized' erythema on the face, neck, nape and upper chest after treatment of impetigo on the face	26
1982 The Netherlands	F	64	FA ointment 'as is'; SF 0.1% and 1% pet.	mild eczema arounds traumatic ulcer from FA ointment; repeated patch tests again positive; also reaction to the excipient lanolin	28
1973 United Kingdom	F	41	SF 2% water	mild eczema around a stasis ulcer from FA tulle; acute dermatitis affecting the forehead, eyelids and cheeks	30
1970 United Kingdom	F	?	SF 2% water	ACD around stasis ulcer; previously she had had dermatitis of the face and eyelids which recurred during patch testing	31

ACD: allergic contact dermatitis; FA: Fusidic acid (FA ointment and cream contain 2% sodium fusidate); SF: sodium fusidate

Cutaneous adverse drug reactions from systemic administration caused by type IV (delayed-type) hypersensitivity

A 51-year-old man developed a pruritic micropapular generalized exanthema, 4 hours after a first oral 250 mg dose of fusidic acid. Four days before, the patient had been treated with topical fusidic acid 2% ointment to an Impetiginized skin lesion with good tolerance. It was the first time he was exposed to fusidic acid. A patch test with the ointment was positive and with its excipients (lanolin alcohol and petrolatum) negative; fusidic acid itself was not tested. The patient was diagnosed with systemic contact dermatitis from fusidic acid (20).

LITERATURE

1 The data in the section 'General' may have been obtained from literature discussed in this chapter, but mostly also or exclusively from one or more of the following online sources: ChemIDPlus Advanced, PubChem, DrugBank, RxList, Drug Central, Drugs.com, and Wikipedia

2   Britton JE, Wilkinson SM, English JSC, Gawkrodger DJ, Ormerod AD, Sansom JE, et al. The British standard series of contact dermatitis allergens: validation in clinical practice and value for clinical governance. Br J Dermatol 2003;148:259-264

3   Erfurt-Berge C, Geier J, Mahler V. The current spectrum of contact sensitization in patients with chronic leg ulcers or stasis dermatitis - new data from the Information Network of Departments of Dermatology (IVDK). Contact Dermatitis 2017;77:151-158

4   Barbaud A, Collet E, Le Coz CJ, Meaume S, Gillois P. Contact allergy in chronic leg ulcers: results of a multicentre study carried out in 423 patients and proposal for an updated series of patch tests. Contact Dermatitis 2009;60:279-287

5   Saap L, Fahim S, Arsenault E, Pratt M, Pierscianowski T, Falanga V, Pedvis-Leftick A. Contact sensitivity in patients with leg ulcerations: a North American study. Arch Dermatol 2004;140:1241-1246

6   Tavadia S, Bianchi J, Dawe RS, McEvoy M, Wiggins E, Hamill E, et al. Allergic contact dermatitis in venous leg ulcer patients. Contact Dermatitis 2003;48:261-265

7   Kulozik M, Powell SM, Cherry G, Ryan T J. Contact sensitivity in community-based leg ulcer patients. Clin Exp Dermatol 1988;13:82-84

8   Smart V, Alavi A, Coutts P, Fierheller M, Coelho S, Holness LD, et al. Contact allergens in persons with leg ulcers: a Canadian study in contact sensitization. Int J Low Extrem Wounds 2008;7:120-125

9   Marasovic D, Vuksic I. Allergic contact dermatitis in patients with leg ulcers. Contact Dermatitis1999;41:107-109

10   Rai R, Shenoy MM, Viswanath V, Sarma N, Majid I, Dogra S. Contact sensitivity in patients with venous leg ulcer: A multi-centric Indian study. Int Wound J 2018;15:618-622

11   Jankićević J, Vesić S, Vukićević J, Gajić M, Adamic M, Pavlović MD. Contact sensitivity in patients with venous leg ulcers in Serbia: comparison with contact dermatitis patients and relationship to ulcer duration. Contact Dermatitis 2008;58:32-36

12   Morris SD, Rycroft RJ, White IR, Wakelin SH, McFadden JP. Comparative frequency of patch test reactions to topical antibiotics. Br J Dermatol 2002;146:1047-1051

13   Agulló-Pérez AD, Hervella-Garcés M, Oscoz-Jaime S, Azcona-Rodríguez M, Larrea-García M, Yanguas-Bayona JI. Perianal dermatitis. Dermatitis 2017;28:270-275

14   Wijnmaalen AL, van Zuuren EJ, de Keizer RJ, Jager MJ. Cutaneous allergy testing in patients suspected of an allergic reaction to eye medication. Ophthalmic Res 2009;41:225-229

15   Menezes de Padua CA, Uter W, Schnuch A. Contact allergy to topical drugs: prevalence in a clinical setting and estimation of frequency at the population level. Pharmacoepidemiol Drug Saf 2007;16:377-384

16   Menezes de Padua CA, Schnuch A, Nink K, Pfahlberg A, Uter W. Allergic contact dermatitis to topical drugs – Epidemiological risk assessment. Pharmacoepidemiol Drug Saf 2008;17:813-821

17   Gilissen L, Goossens A. Frequency and trends of contact allergy to and iatrogenic contact dermatitis caused by topical drugs over a 25-year period. Contact Dermatitis 2016;75:290-302

18   Shih Y-H, Sun C-C, Tseng Y-H, Chu C-Y. Contact dermatitis to topical medicaments: a retrospective study from a medical center in Taiwan. Dermatol Sinica 2015;33:181-186

19   Kim CS, Kim TH, Park TH, Yoo JH, Kim KJ. A case of allergic contact dermatitis to sodium fusidate. Ann Dermatol 2005;17:95-97

20   De Castro Martinez FJ, Ruiz FJ, Tornero P, De Barrio M, Prieto A. Systemic contact dermatitis due to fusidic acid. Contact Dermatitis 2006;54:169

21   Lee AY, Joo HJ, Oh JG, Kim YG. Allergic contact dermatitis from sodium fusidate with no underlying dermatosis. Contact Dermatitis 2000;42:53

22   Giordano-Labadie F, Pelletier N, Bazex J. Contact dermatitis from sodium fusidate. Contact Dermatitis 1996;34:159

23   Kim CH, Oh YS, Seo S-H, Kye YC, Moon KC. Two cases of allergic contact dermatitis to sodium fusidate. Korean Journal of Dermatology 1993;31: 944-947

24   Baptista A, Barros MA. Contact dermatitis from sodium fusidate. Contact Dermatitis 1990;23:186-187

25   Riess CE, Bruckner-Tuderman L. Delayed type hypersensitivity to fusidic acid in patients with chronic dermatitis. Lancet 1990;335(8704):1525-1526

26   Romaguera C, Grimalt F. Contact dermatitis to sodium fusidate. Contact Dermatitis 1985;12:176-177

27   Hogan DJ. Widespread dermatitis after topical treatment of chronic leg ulcers and stasis dermatitis. Can Med Assoc J 1988;138:336-338

28   De Groot AC. Contact allergy to sodium fusidate. Contact Dermatitis 1982;8:429

29   Cronin E. Contact dermatitis. Edinburgh: Churchill Livingstone, 1980:208

30   Dave VK, Main RA. Contact sensitivity to sodium fusidate. Contact Dermatitis Newsletter 1973;14:398

31   Verbov JL. Sensitivity to sodium fusidate. Contact Dermatitis Newsletter 1970;7:153

Chapter 3.158 GENTAMICIN

IDENTIFICATION

Description/definition : Gentamicin is an antibiotic mixture of closely related aminoglycosides obtained from *Micromonospora purpurea* and related species

Pharmacological classes : Anti-bacterial agents; protein synthesis inhibitors

IUPAC name : 2-[4,6-Diamino-3-[3-amino-6-[1-(methylamino)ethyl]oxan-2-yl]oxy-2-hydroxycyclo-hexyl]oxy-5-methyl-4-(methylamino)oxane-3,5-diol

CAS registry number : 1403-66-3

EC number : 215-765-8

Merck Index monograph : 5697

Patch testing : Sulfate, 20.0% pet. (Chemotechnique, SmartPracticeCanada, SmartPracticeEurope); late readings (D6-D8) are necessary to avoid missing positive patch tests evolving after D3-4

Molecular formula : $C_{21}H_{43}N_5O_7$

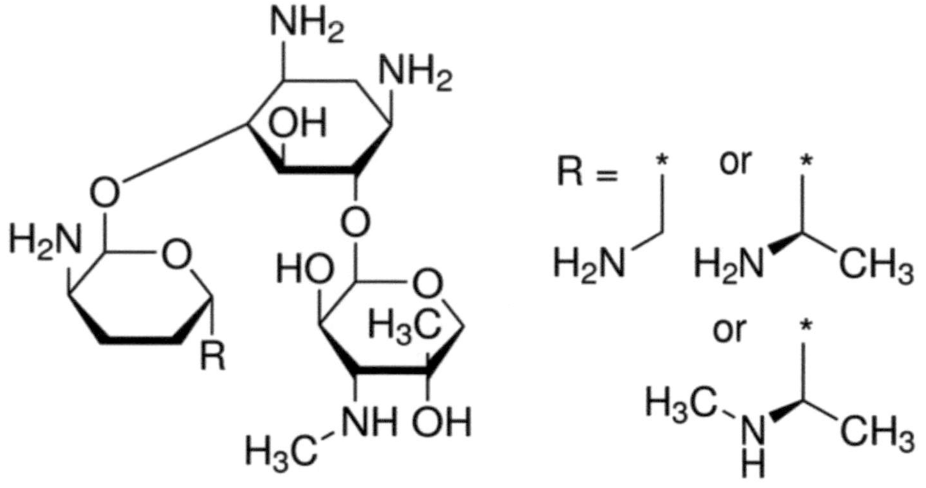

Gentamicin C_{1A}: R = CH$_2$NH$_2$
Gentamicin C_1: R = CH(CH$_3$)NHCH$_3$
Gentamicin C_2: R = CH(CH$_3$)NH$_2$

GENERAL

Gentamicin is a broad-spectrum aminoglycoside antibiotic produced by fermentation of *Micromonospora purpurea* or *M. echinospora*. It is an antibiotic complex consisting of four major (C1, C1a, C2, and C2a) and several minor components. Gentamicin is indicated for treatment of serious infections caused by susceptible strains of *Pseudomonas aeruginosa*, *Proteus* species (indole-positive and indole-negative), *E. coli*, *Klebsiella-Enterobacter-Serratia* species, *Citrobacter* species and *Staphylococcus* species (coagulase-positive and coagulase-negative). This antibiotic is also used in topical pharmaceuticals for the treatment of superficial skin and eye infections. In pharmaceutical products, gentamicin is employed as gentamicin sulfate (CAS number 1405-41-0, EC number 215-778-9, molecular formula $C_{19}H_{40}N_4O_{10}S$) (1).

CONTACT ALLERGY

General

Contact allergy to gentamicin is far from rare. High prevalences of sensitization (table 3.158.1 and table 3.158.2) have been observed in patients with otitis externa/media (5.9%-26%), individuals with periorbital dermatitis /eyelid eczema (up to 10.4%), patients with leg ulcers/stasis dermatitis (up to 10%), patients suspected of allergic contact dermatitis from (ophthalmic) medicaments (19.5% [28], 26% [24]) and even in routine testing in India (14.2% [36]). Unfortunately, in the great majority of these studies, no specific data on the relevance of the positive patch tests were provided. It cannot be excluded, and is in fact likely, that a great number of such reactions, probably the majority, represent cross-reactions to primary neomycin sensitization. Indeed, in previous studies, the great majority

of patients with positive patch tests to gentamicin were also allergic to neomycin (44 [62%], 45 (91%), 46 [97%], 47 [80%]), 48 [72%], 53 [100%]).

Nevertheless, primary sensitization to gentamicin certainly occurs, as has convincingly been shown in patients who have undergone a hip or knee arthroplasty, where gentamicin was added to bone cement (49). In a case series from Finland, of 29 patients reacting to gentamicin, 11 were allergic to gentamicin alone and not to neomycin (44). Also, of 10 patients treated for a month with gentamicin-containing eardrops, 4 proved to be allergic afterwards to gentamicin (unfortunately, no patch tests had been performed before the start of therapy) (41). Patients sensitized by gentamicin tend also to display other reactions to (ingredients of) topical drugs such as benzalkonium chloride (41) and corticosteroids (44). However, cross-reactions to neomycin in these patients are very infrequent (49).

Late reading of gentamicin patch tests is imperative, as up to 65% of positive reactions first appear after D3 (49).

Contact allergy in the general population

With the CE-DUR approach, the incidence of sensitization to gentamicin sulfate in the German population was estimated to range from 2 to 7 cases/100,000/year in the period 1995-1999 and from 3 to 10 cases/100,000/year in the period 2000-2004 (25). Also in Germany, for the period 1995-2004, the population-based relative incidence (RI) of contact sensitization to gentamicin (cases/100,000 defined daily doses [DDDs] per year) was estimated to be 29.0. In the group of antibiotics, the RI ranged from 1.6 (oxytetracycline) to 86.2 (framycetin). For gentamicin used in ophthalmic drugs, the RI was 4.8 (26).

Patch testing in groups of patients

Results of patch testing gentamicin in consecutive patients suspected of contact dermatitis (routine testing) are shown in table 3.158.1. Results of testing in groups of *selected* patients (e.g. patients with leg ulcers/stasis dermatitis, individuals with periorbital dermatitis/eyelid eczema, patients with perianal, genital or anogenital dermatitis, individuals with ear diseases, patients suspected of allergy to [ophthalmological] drugs) are shown in table 3.158.2.

Patch testing in consecutive patients suspected of contact dermatitis: routine testing

In 4 studies in which routine testing was performed with gentamicin, of which 3 were from India, prevalences of sensitization have ranged from 0.2% to an amazing 14.2%. In the study with this high percentage, relevance was not stated. The reactions were ascribed to the increasing use of gentamicin-corticosteroid preparations, but cross-allergy to neomycin (which scored 8.5% positive reactions), was also considered a possibility (36). Presumably, patients in India are highly selected for patch tests to be performed.

Table 3.158.1 Patch testing in groups of patients: Routine testing

Years and Country	Test conc. & vehicle	Number of patients tested | positive (%)		Relevance (R); Comments (C)	Ref.
1997-2006 India	20% pet.	1000	25 (2.5%)	R: 100% present relevance	35
<2004 India	20%pet.	220	3 (1.4%)	R: 82% for all allergens together	34
1992-1993 India		212	30 (14.2%)	R: not stated; C: the reactions were ascribed to the increasing use of gentamicin-corticosteroid preparations, but cross- allergy to neomycin (which scored 8.5% positive reactions), was also considered a possibility	36
1985-1991 Singapore	20% pet.	3145	7 (0.2%)	R: not stated; the low figure is somewhat surprising, as there were 193 patients with positive reactions to neomycin and, generally, 50% of such patients cross-react to gentamicin (see Chapter 3.235 Neomycin)	55

Patch testing in groups of selected patients

The results of patch testing gentamicin in groups of selected patients are shown in table 3.158.2. Prevalences of sensitization have varied widely and ranged from 0.4% to 26%, probably depending on the mode of selection and the extent of exposure in the populations investigated. Generally speaking, the rates were the highest in patients suspected of allergy to (eye) medications, individuals with otitis externa/media, patients with periorbital dermatitis/ eyelid eczema and individuals with leg ulcers/stasis dermatitis. Unfortunately, in the great majority of these studies, no or non-specific data on the relevance of the positive patch tests were provided. It may be assumed that a large number of the positive patch tests have represented cross-reactivity to primary neomycin sensitization.

Table 3.158.2 Patch testing in groups of patients: Selected patient groups

Years and Country	Test conc. & vehicle	Number of patients tested | positive (%)		Selection of patients (S); Relevance (R); Comments (C)	Ref.
Patients with leg ulcers/stasis dermatitis					
2003-2014 IVDK	20% pet.	2029	(3.1%)	S: patients with stasis dermatitis/chronic leg ulcers; R: not stated	3
2006-2008 Germany	20% pet.	95	(5.3%)	S: patients with leg ulcers; R: not stated	2
2005-2008 France	10% water	423	11 (2.6%)	S: patients with leg ulcers; R: not stated	4
<2008 Poland	20% pet.	50	2 (4%)	S: patients with venous leg ulcers; R: not stated	6
<2005 Poland	20% pet.	50	2 (4%)	S: patients with chronic venous leg ulcers; R: not stated	5
2000-2002 Serbia	20% pet.	75	4 (5%)	S: patients with chronic venous leg ulcers; R: not stated; C: 1.2% reactions in a (small) control group (difference not significant)	9
<1999 Croatia	20% pet.	100	10 (10%)	S: patients with leg ulcers; R: not stated	8
1996-1997 U.K.		109	5 (4.6%)	S: mostly patients with leg ulcers and stasis dermatitis; R: not stated	7
Patients with periorbital dermatitis/eyelid eczema					
2000-2010 IVDK		503	33 (5.6%)	S: patients with periorbital dermatitis; R: not stated; C: the frequency in patients suspected of eye medicament allergy was 5.2% and in a control group of dermatitis patients 3.0%	10
2001-2010 IVDK	20% pet.	258	21 (8.1%)	S: patients with periorbital dermatitis tested with an ophthalmic tray; R: not stated	30
1999-2004 Germany		88	2 (2%)	S: patients with periorbital eczema; R: not stated	11
1995-1999 IVDK	20% pet.	618	(4.5%)	S: patients with *allergic* periorbital contact dermatitis; the frequency was significantly higher than in a control group	13
1994-1998 U.K.	2% pet. (?)	232	1 (0.4%)	S: patients with periorbital contact dermatitis; R: 100%; C: the frequency in a control group was 0%	14
1990-1994 IVDK		269	28 (10.4%)	S: patients with periorbital eczema; R: not stated; C: frequency of sensitization in dermatitis controls: 5%	12
Patients with perianal, genital or anogenital dermatitis					
2013-2015 Ireland	20% pet.	99	2 (2%)	S: patients patch tested for perianal and/or genital symptoms; R: all reactions to medicaments were relevant	18
2005-2015 U.K.		123	3 (2.4%)	S: patients with perianal dermatoses and/or pruritus ani; R: 100%; C: number of patients tested uncertain	16
2004-2008 IVDK			(2.8%)	S: patients with anogenital dermatoses tested with a medicament series; R: not stated; C: number of patients tested unknown	15
1990-2003 IVDK		393	15 (3.8%)	S: patients with perianal dermatoses; R: not stated; C: the frequency was not higher than in a control group	17
Patients with ear diseases					
1985-2002 U.K	20% pet.	146	12 (8.2%)	S: patients with chronic inflammatory ear disease; R: 92%	19
<2002 U.K.		149	20 (13.4%)	S: patients with chronic otitis media or otosclerosis scheduled for middle ear surgery; R: not stated	27
1992-1997 IVDK		51	3 (5.9%)	S: patients with chronic otitis externa; R: not stated	20
1993-4 Netherlands	20% pet.	34	9 (26%)	S: patients with chronic otitis externa or media; R: not stated	21
<1982 U.K	10% pet.	40	4 (10%)	S: adults with otitis externa, chronic suppurative otitis media or discharging mastoid cavities of greater than one year's duration; R: at least one was relevant, the others may have shown cross-reactivity to neomycin or framycetin	42
Other patient groups					
1990-2014 Belgium	3% pet.	257	50 (19.5%)	S: patients suspected of iatrogenic contact dermatitis and tested with a pharmaceutical series and their own products; R: 96% of the positive patch test reactions to all topical drugs and antiseptics were considered to be relevant	28
2000-2010 Canada		100	10 (10%)	S: charts reviewed and included in the study when there was at least one positive reaction to a topical drug; R: not stated; C: the high percentages to all drugs are obviously the result of the – rather unusual – selection procedure targeting at topical drug allergy previously demonstrated by patch testing	31

Table 3.158.2 Patch testing in groups of patients: Selected patient groups (continued)

Years and Country	Test conc. & vehicle	Number of patients tested | positive (%)		Selection of patients (S); Relevance (R); Comments (C)	Ref.
2000-6 Netherlands		23	6 (26%)	S: patients patch tested for suspected contact allergy to ophthalmological drugs; R: 69% for all reactions to drugs together	24
1995-2004 IVDK	20% pet.	10,576	(3.3%)	S: not stated; R: not stated	25
1995-2002 IVDK	20% pet.	285	(12.5%)	S: children 6-12 years; R: not stated; C: in adolescents (13-18 years) and a control group of adults (60-66 years), far lower frequencies were found of 2.9% resp. 2.6%; there were far fewer reactions to neomycin (3.7%); this author concludes that gentamicin 20% pet. must have caused many false-positive, irritant reactions in children	56
1995-2001 IVDK	20% pet.	3113	(3.5%)	S: patients patch tested for suspected contact allergy to ophthalmological drugs; R: not stated; C: the estimated number of sensitizations per 100,000 prescriptions was 4.8	23
1995-1999 IVDK	20% pet.	7332	182 (2.5%)	S: patients tested with an antibiotic series; R: not stated; C: the prevalence was not significantly different between atopics and non-atopic patients	54

IVDK: Information Network of Departments of Dermatology (Germany, Austria, Switzerland)

Case series

In Leuven, Belgium, between 1990 and 2017, 16,065 patients were investigated for contact allergy and 118 (0.7%) showed positive patch test reactions to topical ophthalmic medications and/or to their ingredients. Eighty-four individuals (71%) reacted to an active principle. Gentamicin was tested in 274 patients and was the allergen in eye medications in 10. There were also 12 reactions to gentamicin in other types of medications (22).

In the university hospital of Turku, Finland, between 1997 and 2007, 620 patients were patch tested, depending on the individual exposure and suspicion of sensitization, with a medicament series including gentamicin sulfate 20% pet. and there were 29 (4.6%) positive reactions to the antibiotic. Mean age of the patients was 62 years (range 17-85 years). Eighteen individuals co-reacted to neomycin and 7 to kanamycin. There were also 13 co-reactivities to bacitracin (11/13 neomycin-allergic, see the chapters on neomycin and bacitracin) and 7 to tixocortol pivalate (marker for corticosteroid allergy). In Finland, gentamicin had been an ingredient in preparations only available with prescription (after 1995 in one cream only, combined with a potent corticosteroid), whereas a neomycin-containing ointment had been a bestseller among the over-the-counter products for 50 years. Therefore, most reactions to gentamicin were considered to be cross-reactions to neomycin. However, 11 patients reacted to gentamicin only. Four had allergic contact dermatitis from the cream containing gentamicin and a corticosteroid (all 4 also positive to tixocortol pivalate); in some patients, gentamicin may have contributed to leg ulcer dermatitis or otitis externa, and three nurses had hand dermatitis from occupational contact with gentamicin-containing bone cement. One patient had previously suffered a drug exanthema from netilmicin, which is structurally related to gentamicin (44).

Ten patients with active chronic suppurative otitis media were treated with gentamicin (0.3%) – hydrocortisone (1%) ear drops. Response to treatment was assessed on the basis of otoscopic appearance after four weeks of treatment. The patients were subsequently patch tested with the various constituents of the ear drops. Four patients showed a positive reaction to gentamicin at 4 days. Three of these patients and one other reacted to benzalkonium chloride, which is used as an antimicrobial in the eardrops. None of the five patients with positive skin tests (unknown to which the 5th reacted) had a satisfactory response to treatment (41).

Case reports

A 38-year-old man presented with chronic dermatitis of the back of the hands for 2 years. Many different topical drugs had been used with little benefit. Patch tests with the topical preparations used gave a positive reaction to an ointment containing clobetasol propionate and a cream containing gentamicin sulfate. Patch testing with their ingredients were positive to both the corticosteroid and gentamicin sulfate 20% pet. (33).

A 55-year-old woman presented with a 3-week history of pruritic, erythematous, scaly plaques on the eyelids, spreading in a few days periorbitally. The condition had developed one day after starting treatment with gentamicin eye drops. Patch tests were positive to the eye drops 'as is' and to its active principal gentamicin sulfate 20% pet. There was a cross-reaction to kanamycin but not to neomycin (38).

A 50-year-old man was treated for conjunctivitis with gentamicin eye drops. One day later, erythema and edema involving the eyelids and orbicular and nasal areas developed, followed by worse itching and conjunctival hyperemia. Skin prick tests and intradermal tests with gentamicin sulfate 1 and 5 mg/ml were negative after 20 minutes, but positive after 2 days with erythema, infiltration, and vesicles. Patch tests confirmed the contact allergy to gentamicin, but neomycin was negative. An intramuscular challenge test with gentamicin (an 80-mg dose of gentamicin sulfate was reached) was negative (43).

A patient using gentamicin cream on leg ulcers was tested because of stasis dermatitis. Patch tests were positive to gentamicin sulfate 30% pet.; the cream itself was not tested (45). In 1970 a case of contact sensitivity to gentamicin 0.1% cream was reported in a 49-year-old male, with no previous exposure to gentamicin or neomycin, who was applying gentamicin cream 3 times daily to leg ulcers. On the thirty-seventh day of treatment, the patient experienced itching, redness, and swelling around the ulcers. He was patch test positive to gentamicin cream 0.1% and to neomycin sulfate 20% pet. (57).

Allergic reactions to gentamicin in bone cement for hip or knee arthroplasty

In patients with complications after arthroplasty such as persistent pain and swelling, reduced range of motion, recurrent effusions, loosening of the prosthesis, or implant-related eczema, common potential causes, such as malpositioning, mechanical problems, or infections, are usually investigated first. After exclusion of such causes, or in cases of local eczema, implant allergy may be suspected. Today, most knee and hip endoprostheses are cemented. Antibiotic-loaded (e.g. gentamicin-containing) bone cement is frequently used to prevent infection. Correspondingly, prolonged local gentamicin release is intended, and can be observed for years (49,50).

Between 2010 and 2013, in Munich and Erlangen, Germany, 250 consecutive cemented arthroplasty-bearing patients suspected of having allergic reactions to the implant materials were analyzed (49). 'Classic' complication elicitors such as infection and mechanical causes had been previously excluded by the referring orthopedic surgeons. All patients were patch tested with the European baseline series and a bone cement component series, containing 7 haptens including gentamicin sulfate 20%. Patch tests were read at D2, D3 and D6. Twenty-five patients (10%) had positive reactions to gentamicin, of which 17 were positive at D6 only. Clinical relevance was addressed as follows: the patients were given a medical report indicating that – in the absence of alternative 'problem elicitors' – the gentamicin allergy might be a potential cause of the complication. Any decision to perform implant exchange was left to the treating orthopedic surgeon (49).

Three months after a 79-year-old male patient had received a right cemented total knee arthroplasty, pain, swelling and a reduced range of motion were noted after an extended walk. Physical examination by the orthopedic surgeon showed joint effusion. Subsequent computed tomography showed a correct implant position without any signs of loosening. Diagnostic joint aspiration and microbiological analysis showed no signs of infection. Scintigraphy was suggestive of local synovitis. Thus, synovitis caused by 'excessive walking' was diagnosed, and oral diclofenac was prescribed. However, the patient complained of increasing pain, and presented with local eczema of the right knee a few weeks later. Patch tests were positive to gentamicin and neomycin, but an antibiotic-free bone cement had been used for the operation. However, it was found that, while performing the diagnostic joint aspiration, the orthopedic surgeon had injected gentamicin solution to prevent infection. Taking this detail into account, the authors diagnosed 'synovitis and allergic contact dermatitis' resulting from intra-articular gentamicin application. In the course of the next 10 months, the patient's symptoms, including his eczema, completely resolved (50).

A case report of probable systemic contact dermatitis from gentamicin in bone cement is described below in the section 'Systemic contact dermatitis from topical application and presence in bone cement' (51) and another one was already described in Spanish literature in 1986 (52).

Systemic contact dermatitis from topical application and presence in bone cement

A 62-year-old woman developed a superficial and partial thickness thermal burn on her left leg and foot after accidental contact with boiling water. She was treated with various topicals including gentamicin 0.1% cream. After 3 weeks the burn wound worsened and perilesional eczematization appeared. An itchy eczema on her scalp and on the extensor part of her limbs and hands developed as well. Patch tests were positive to the gentamicin cream (D2 ++, D4 +++, D7 +++). During patch testing the patient developed a widespread flare of itchy eczematous dermatitis involving the whole body except the face. Systemic corticosteroids were prescribed and healing of the eczema and of the burn wound was achieved. Six weeks later, a semi-open test with gentamicin 25% pet. on the right arm was performed, which showed a positive, localized reaction on D2 and D4 readings. Systemic allergic dermatitis due to gentamicin was diagnosed (39).

A 74-year-old woman who had suffered from venous leg ulcers and stasis dermatitis for more than 20 years developed a pruritic eczematous rash 3 days after a left knee replacement. The dermatitis initially involved the left lower leg, but within 2 days, it had extended over the whole left leg, the left hip, the right lower leg, and both arms.

The rash cleared over the next 2 weeks without the need for any therapy. The patient reported that she had developed a similar rash 2 years previously when she had received a total right knee prosthesis. Patch tests showed many positives including gentamicin sulfate and framycetin sulfate. Gentamicin sulfate was present in the bone cement. Systemic contact dermatitis was diagnosed (51)

Cross-reactions, pseudo-cross-reactions and co-reactions
Of patients sensitized to neomycin, about 50% cross-reacts to gentamicin (Chapter 3.235 Neomycin). The cross-sensitivity pattern between aminoglycoside antibiotics in patients primarily sensitized to gentamicin has not been well investigated. In previous studies, the majority of patients with positive patch tests to gentamicin are also allergic to neomycin (44 [62%], 45 (91%), 46 [97%], 47 [80%]), 48 [72%], 53 [100%]) and may represent cross-reactions. However, primary sensitization to gentamicin does exist and may infrequently be accompanied by cross-sensitization to neomycin. Indeed, in a group of 25 patients sensitized to gentamicin present in bone cement used for arthroplasty, only 2 (8%) co-reacted to neomycin (49).

In one patient sensitized to gentamicin, netilmicin may have cross-reacted (44).

Cutaneous adverse drug reactions from systemic administration caused by type IV (delayed-type) hypersensitivity
Cutaneous adverse drug reactions from systemic administration of gentamicin caused by type IV (delayed-type) hypersensitivity, including erythroderma (29), systemic contact dermatitis (32,40), and 'severe drug reaction' (47), and in addition occupational allergic contact dermatitis (37,44,48) are planned to be discussed in Volume IV of the *Monographs in Contact Allergy* series on Systemic drugs.

Immediate contact reactions
Immediate contact reactions (contact urticaria) to gentamicin are presented in Chapter 5.

LITERATURE
1   The data in the section 'General' may have been obtained from literature discussed in this chapter, but mostly also or exclusively from one or more of the following online sources: ChemIDPlus Advanced, PubChem, DrugBank, RxList, Drug Central, Drugs.com, and Wikipedia
2   Reich-Schupke S, Kurscheidt J, Appelhans C Kreuter A, Altmeyer P, Stücker M. Patch testing in patients with leg ulcers with special regard to modern wound products. Hautarzt 2010;61:593-597 (Article in German)
3   Erfurt-Berge C, Geier J, Mahler V. The current spectrum of contact sensitization in patients with chronic leg ulcers or stasis dermatitis - new data from the Information Network of Departments of Dermatology (IVDK). Contact Dermatitis 2017;77:151-158
4   Barbaud A, Collet E, Le Coz CJ, Meaume S, Gillois P. Contact allergy in chronic leg ulcers: results of a multicentre study carried out in 423 patients and proposal for an updated series of patch tests. Contact Dermatitis 2009;60:279-287
5   Zmudzinska M, Czarnecka-Operacz M, Silny W, Kramer L. Contact allergy in patients with chronic venous leg ulcers – possible role of chronic venous insufficiency. Contact Dermatitis 2006;54:100-105
6   Zmudzinska M, Czarnecka-Operacz M, SilnyW. Contact allergy to glucocorticosteroids in patients with chronic venous leg ulcers, atopic dermatitis and contact allergy. Acta Dermatovenerol Croat 2008;16:72-78
7   Gooptu C, Powell SM. The problems of rubber hypersensitivity (types I and IV) in chronic leg ulcer and stasis eczema patients. Contact Dermatitis 1999;41:89-93
8   Marasovic D, Vuksic I. Allergic contact dermatitis in patients with leg ulcers. Contact Dermatitis1999;41:107-109
9   Jankićević J, Vesić S, Vukićević J, Gajić M, Adamic M, Pavlović MD. Contact sensitivity in patients with venous leg ulcers in Serbia: comparison with contact dermatitis patients and relationship to ulcer duration. Contact Dermatitis 2008;58:32-36
10 Landeck L, John SM, Geier J. Periorbital dermatitis in 4779 patients – patch test results during a 10-year period. Contact Dermatitis 2014;70:205-212
11 Feser A, Plaza T, Vogelgsang I, Mahler V. Periorbital dermatitis – a recalcitrant disease: causes and differential diagnoses. Brit J Dermatol 2008;159:858-863
12 Ockenfels H, Seemann U, Goos M. Contact allergy in patients with periorbital eczema: an analysis of allergens. Dermatology 1997;195:119-124
13 Herbst RA, Uter W, Pirker C, Geier J, Frosch PJ. Allergic and non-allergic periorbital dermatitis: patch test results of the Information Network of the Departments of Dermatology during a 5-year period. Contact Dermatitis 2004;51:13-19
14 Cooper SM, Shaw S. Eyelid dermatitis: an evaluation of 232 patchtest patients over 5 years. Contact Dermatitis 2000;42:291-293

15 Bauer A. Contact sensitization in the anal and genital area. Curr Probl Dermatol 2011;40:133-141

16 Abu-Asi MJ, White IR, McFadden JP, White JM. Patch testing is clinically important for patients with peri-anal dermatoses and pruritus ani. Contact Dermatitis 2016;74:298-300

17 Kügler K, Brinkmeier T, Frosch PJ, Uter W. Anogenital dermatoses—allergic and irritative causative factors. Analysis of IVDK data and review of the literature. J Dtsch Dermatol Ges 2005;3:979-986

18 Foley CC, White S, Merry S, Nolan U, Moriarty B, et al. Understanding the role of cutaneous allergy testing in anogenital dermatoses: a retrospective evaluation of contact sensitization in anogenital dermatoses. Int J Dermatol 2019;58:806-810

19 Millard TP, Orton DI. Changing patterns of contact allergy in chronic inflammatory ear disease. Contact Dermatitis 2004;50:83-86

20 Hillen U, Geier J, Goos M. Contact allergies in patients with eczema of the external ear canal. Results of the Information Network of Dermatological Clinics and the German Contact Allergy Group. Hautarzt 2000;51:239-243 (Article in German)

21 Van Ginkel CJ, Bruintjes TD, Huizing EH. Allergy due to topical medications in chronic otitis externa and chronic otitis media. Clin Otolaryngol Allied Sci 1995;20:326-328

22 Gilissen L, De Decker L, Hulshagen T, Goossens A. Allergic contact dermatitis caused by topical ophthalmic medications: Keep an eye on it! Contact Dermatitis 2019;80:291-297

23 Uter W, Menezes de Pádua C, Pfahlberg A, Nink K, Schnuch A, Behrens-Baumann W. Contact allergy to topical ophthalmological drugs - epidemiological risk assessment. Klin Monbl Augenheilkd 2009;226:48-53 (Article in German)

24 Wijnmaalen AL, van Zuuren EJ, de Keizer RJ, Jager MJ. Cutaneous allergy testing in patients suspected of an allergic reaction to eye medication. Ophthalmic Res 2009;41:225-229

25 Menezes de Padua CA, Uter W, Schnuch A. Contact allergy to topical drugs: prevalence in a clinical setting and estimation of frequency at the population level. Pharmacoepidemiol Drug Saf 2007;16:377-384

26 Menezes de Padua CA, Schnuch A, Nink K, Pfahlberg A, Uter W. Allergic contact dermatitis to topical drugs – Epidemiological risk assessment. Pharmacoepidemiol Drug Saf 2008;17:813-821

27 Yung MW, RajendraT. Delayed hypersensitivity reaction to topical aminoglycosides in patients undergoing middle ear surgery. Clin Otolaryngol Allied Sci 2002;27:365-368

28 Gilissen L, Goossens A. Frequency and trends of contact allergy to and iatrogenic contact dermatitis caused by topical drugs over a 25-year period. Contact Dermatitis 2016;75:290-302

29 Guin JD, Phillips D. Erythroderma from systemic contact dermatitis: a complication of systemic gentamicin in a patient with contact allergy to neomycin. Cutis 1989;43:564-567

30 Landeck L, John SM, Geier J. Topical ophthalmic agents as allergens in periorbital dermatitis. Br J Ophthalmol 2014;98:259-262

31 Spring S, Pratt M, Chaplin A. Contact dermatitis to topical medicaments: a retrospective chart review from the Ottawa Hospital Patch Test Clinic. Dermatitis 2012;23:210-213

32 Ghadially R, Ramsay CA. Gentamicin: systemic exposure to a contact allergen. J Am Acad Dermatol 1988;19(2Pt. 2):428-430

33 Tosti A, Guerra L, Manuzzi P, Lama L. Contact dermatitis from clobetasol propionate. Contact Dermatitis 1987;17:256-257

34 Sharma VK, Sethuraman G, Garg T, Verma KK, Ramam M. Patch testing with the Indian standard series in New Delhi. Contact Dermatitis 2004;51:319-321

35 Bajaj AK, Saraswat A, Mukhija G, Rastogi S, Yadav S. Patch testing experience with 1000 patients. Indian J Dermatol Venereol Leprol 2007;73:313-318

36 Shenoy SD, Srinivas CR, Balachandran C. Results of patch testing with standard series of allergens at Manipal. Indian J Dermatol Venereol Leprol 1994;60:133-135

37 Gielen K, Goossens A. Occupational allergic contact dermatitis from drugs in healthcare workers. Contact Dermatitis 2001;45:273-279

38 Sánchez-Pérez J, López MP, De Vega Haro JM, García-Díez A. Allergic contact dermatitis from gentamicin in eyedrops, with cross-reactivity to kanamycin but not neomycin. Contact Dermatitis 2001;44:54

39 Corazza M, Forconi R, Toni G, Scuderi V, Mantovani L, Borghi A. Systemic allergic dermatitis due to gentamicin. Contact Dermatitis 2019;81:402-403

40 Paniagua MJ, Garcia-Ortega P, Tella R, Gaig P, Richart C. Systemic contact dermatitis to gentamicin. Allergy 2002;57:1086-1087

41 Robinson M. Contact sensitivity to gentamicin-hydrocortisone ear drops. J Laryngol Otol 1988;102:577-578

42 Holmes RC, Johns AN, Wilkinson JD, Black MM, Rycroft RJ. Medicament contact dermatitis in patients with chronic inflammatory ear disease. JR Soc Med 1982;75:27-30

43 Muñoz Bellido FJ, Moyano JC, Alvarez M, Juan JL, Bellido J. Contact sensitivity to gentamicin with tolerance of systemic exposure. Allergy 1996;51:758-759

44 Liippo J, Lammintausta K. Positive patch test reactions to gentamicin show sensitization to aminoglycosides from topical therapies, bone cements, and from systemic medication. Contact Dermatitis 2008;59:268-272

45 Van Ketel WG, Bruynzeel DP. Sensitization to gentamicin alone. Contact Dermatitis 1989;20:303-304

46 Rudzki E, Zakrzewski Z, Rebandel P, Grzywa Z, Hudymowicz W. Cross reactions between aminoglycoside antibiotics. Contact Dermatitis 1988;18:314-316

47 Braun W, Schütz R. Beitrag zur Gentamycin Allergie. Hautarzt 1969;20:108 ((Article in German, data cited in ref. 45)

48 Bandmann H-J, Mutzeck E. Contact allergy to gentamycin sulfate. Contact Dermatitis Newsletter 1973;13:371

49 Thomas B, Kulichova D, Wolf R, Summer B, Mahler V, Thomas P. High frequency of contact allergy to implant and bone cement components, in particular gentamicin, in cemented arthroplasty with complications: usefulness of late patch test reading. Contact Dermatitis 2015;73:343-349

50 Wittmann D, Summer B, Thomas B, Halder A, Thomas P. Gentamicin allergy as an unexpected 'hidden' cause of complications in knee arthroplasty. Contact Dermatitis 2018;78:293-294

51 Haeberle M, Wittner B. Is gentamicin-loaded bone cement a risk for developing systemic allergic dermatitis? Contact Dermatitis 2009;60:176-177

52 Romaguera C, Grimalt F, Villaplana J, Mascaró JM. Contact dermatitis caused by Septopal. Med Cutan Ibero Lat Am 1986;14:43-47 (Article in Spanish)

53 Breit R. Allergen change in stasis dermatitis. Contact Dermatitis 1977;3:309-311

54 Jappe U, Schnuch A, Uter W. Frequency of sensitization to antimicrobials in patients with atopic eczema compared with nonatopic individuals: analysis of multicentre surveillance data, 1995-1999. Brit J Dermatol 2003;149:87-93

55 Goh CL. Contact sensitivity to topical antimicrobials. I. Epidemiology in Singapore. Contact Dermatitis 1989;21:46-48

56 Heine G, Schnuch A, Uter W, Worm M. Frequency of contact allergy in German children and adolescents patch tested between 1995 and 2002: results from the Information Network of Departments of Dermatology and the German Contact Dermatitis Research Group. Contact Dermatitis 2004;51:111-117

57 Lynfield YL. Allergic contact sensitization to gentamicin. NY State J Med 1970;70:2235-2236

Chapter 3.159 GLYCOL SALICYLATE

IDENTIFICATION

Description/definition : Glycol salicylate is the benzoate ester that conforms to the structural formula shown below

Pharmacological classes : Non-steroidal anti-inflammatory drugs

IUPAC name : 2-Hydroxyethyl 2-hydroxybenzoate

Other names : 2-Hydroxyethyl salicylate; ethylene glycol salicylate

CAS registry number : 87-28-5

EC number : 201-737-2

Merck Index monograph : 5805

Patch testing : 5% pet.

Molecular formula : $C_9H_{10}O_4$

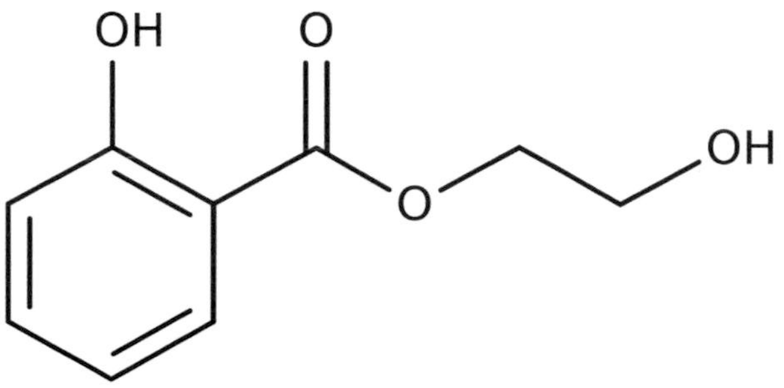

GENERAL

Glycol salicylate is a benzoate ester obtained by the formal condensation of the carboxy group of salicylic acid with one of the hydroxy groups of ethylene glycol. It is a nonsteroidal anti-inflammatory drug and an ingredient of many topical creams and sprays used for the relief of aches, pains, and stiffness of the muscles, joints, and tendons (1).

CONTACT ALLERGY

Case series

In a period of 11 years (2000-2010), the French Pharmacovigilance received 53 reports of adverse skin reactions to a pharmaceutical product containing hydroxyethyl (glycol) salicylate, dexamethasone acetate, and salicylamide marketed for treatment of benign joint conditions such as mild tendinitis, small joint arthritis and sprains. The main cutaneous side effect (n=41) was contact dermatitis with secondary extension in 15 cases. Onset was immediate in 12 cases, delayed in 32 cases and unspecified in eight cases. Twelve patients were hospitalized. Allergological tests were performed in 14 cases and were positive for the topical drug itself (eight cases), glycol salicylate (n=7), salicylamide (n=6), dexamethasone (n=3), and propylene glycol (n=2) (5).

Case reports

A 64-year-old man with cervical pain applied a gel containing 2-hydroxyethyl salicylate, menthol, and heparin sodium to the left shoulder and neck. The gel was a veterinary medicine indicated for the treatment of musculoskeletal inflammation in horses. After 2 weeks, the patient developed a pruritic erythematous vesicular eruption at the sites of application. Patch tests were positive to the gel 'as is', to 2-hydroxyethyl salicylate (glycol salicylate) 1%, 2% and 5% pet., but negative to heparin sodium. Fifteen controls were negative to glycol salicylate 5% pet. (2).

A 62-year-old woman with muscle and back pain had applied an antirheumatic and anti-inflammatory gel and cream to her back and thighs. Two days later, she developed acute contact dermatitis at the sites of application. She remembered that previous use of these medicaments had also caused dermatitis. Patch tests were positive to the gel and cream 'as is' and to glycol salicylate 5% water, present in both medicaments (3).

A 47-year-old man, with osteochondrosis dissecans of the right talotibial joint, received local treatment with a gel containing hydroxyethyl (glycol) salicylate. After 2 weeks, he developed acute contact dermatitis at the site of application. Eight months later, after an arthroscopy of the same joint, the gel was again applied and, over the next 24 hours, he developed acute contact dermatitis once more. Patch tests were positive to the gel 'as is' and to glycol salicylate 0.1%, 0.5%, 1%, 2%, 5% and 10% pet. Patch tests with acetylsalicylic acid and sodium salicylate were negative. Orally administered acetylsalicylic acid in a cumulative dose of 1000 mg and sodium salicylate 500 mg were well-tolerated (4).

A 19-year-old man had used a rubefacient cream for 2 days and a balsam for 14 days to treat a sprained muscle in his right thigh. He then changed to another rubefacient gel. Very soon after the first application, the patient developed a severe contact eczema on his thigh with an extensive id-like spread on his other leg and on his face with some erythema multiforme-like lesions. He was patch tested and reacted to the gel, its ingredient glycol salicylate 1% pet., the balsam, and its ingredient mephenesin. The cream he had used first also contained glycol salicylate, but a patch test to it was negative (6).

Cross-reactions, pseudo-cross-reactions and co-reactions
Not to acetylsalicylic acid and sodium salicylate (4).

LITERATURE

1   The data in the section 'General' may have been obtained from literature discussed in this chapter, but mostly also or exclusively from one or more of the following online sources: ChemIDPlus Advanced, PubChem, DrugBank, RxList, Drug Central, Drugs.com, and Wikipedia
2   Córdoba S, García-Donoso C, Villanueva CA, Borbujo J. Allergic contact dermatitis from a veterinary antiinflammatory gel containing 2-hydroxyethyl salicylate. Dermatitis 2011;22:171-172
3   Horak J, Hemmer W, Focke M, Götz M, Jarisch R. Contact dermatitis from anti-inflammatory gel containing hydroxyethyl salicylate. Contact Dermatitis 2002;47:120-121
4   Reichert C, Gall H. Contact dermatitis from hydroxyethyl salicylate. Contact Dermatitis 1995;33:275-276
5   Remy C, Barbaud A, Lebrun-Vignes B, Perrot JL, Beyens MN, Mounier G, et al. Skin toxicity related to Percutalgine (®): analysis of the French pharmacovigilance database. Ann Dermatol Venereol 2012;139:350-354 (Article in French)
6   Degreef H, Bonamie A, van Derheyden D, Dooms-Goossens A. Mephenesin contact dermatitis with erythema multiforme features. Contact Dermatitis 1984;10:220-223

Chapter 3.160 GRAMICIDIN

IDENTIFICATION

Description/definition : Gramicidin is a heterogeneous mixture of 6 peptide antibiotics obtained from the soil
                          bacterium *Bacillus brevis*
Pharmacological classes : Anti-bacterial agents
CAS registry number   : 1405-97-6; 1393-88-0 (Gramicidin D)
EC number   : 215-790-4; 215-738-0 (Gramicidin D)
Merck Index monograph : 5838 (Gramicidins)
Patch testing   :

GENERAL

Gramicidin is a heterogeneous mixture of 6 peptide antibiotics (PubChem) obtained from the soil bacterium *Bacillus brevis*. Gramicidin C or S is a cyclic, ten-amino acid polypeptide and gramicidins A, B, D are linear. Gramicidin is one of the two principal components of tyrothricin (ChemIDplus). It is active against most gram-positive bacteria and some gram-negative organisms. Gramicidin D is used primarily as a topical antibiotic for the treatment of skin lesions, surface wounds and eye infections (1).

CONTACT ALLERGY

Patch testing in groups of selected patients

In Leuven, Belgium, in the period 1990-2014, gramicidin 2% pet. was patch tested in 343 patients suspected of iatrogenic contact dermatitis and there was one positive reaction to it. Relevance was not specified for individual allergens, but 96% of all positive patch test reactions to topical drugs and antiseptics were considered to be relevant (2).

In the period 1985-2002, in the United Kingdom, 179 patients with chronic inflammatory ear disease were patch tested with gramicidin 10% pet. and there were 11 (16%) positive reactions, all of which were considered to be relevant (3).

LITERATURE

1   The data in the section 'General' may have been obtained from literature discussed in this chapter, but mostly also or exclusively from one or more of the following online sources: ChemIDPlus Advanced, PubChem, DrugBank, RxList, Drug Central, Drugs.com, and Wikipedia
2   Gilissen L, Goossens A. Frequency and trends of contact allergy to and iatrogenic contact dermatitis caused by topical drugs over a 25-year period. Contact Dermatitis 2016;75:290-302
3   Millard TP, Orton DI. Changing patterns of contact allergy in chronic inflammatory ear disease. Contact Dermatitis 2004;50:83-86

Chapter 3.161 HALCINONIDE

IDENTIFICATION

Description/definition	: Halcinonide is the synthetic glucocorticoid that conforms to the structural formula shown below
Pharmacological classes	: Anti-inflammatory agents
IUPAC name	: (1S,2S,4R,8S,9S,11S,12R,13S)-8-(2-Chloroacetyl)-12-fluoro-11-hydroxy-6,6,9,13-tetramethyl-5,7-dioxapentacyclo[10.8.0.0^{2,9}.0^{4,8}.0^{13,18}]icos-17-en-16-one
Other names	: 21-Chloro-9-fluoro-11β,16α,17-trihydroxypregn-4-ene-3,20-dione cyclic 16,17-acetal with acetone
CAS registry number	: 3093-35-4
EC number	: 221-439-6
Merck Index monograph	: 5896
Patch testing	: In general, corticosteroids may be tested at 0.1% and 1% in alcohol; late readings (6-10 days) are strongly recommended
Molecular formula	: $C_{24}H_{32}ClFO_5$

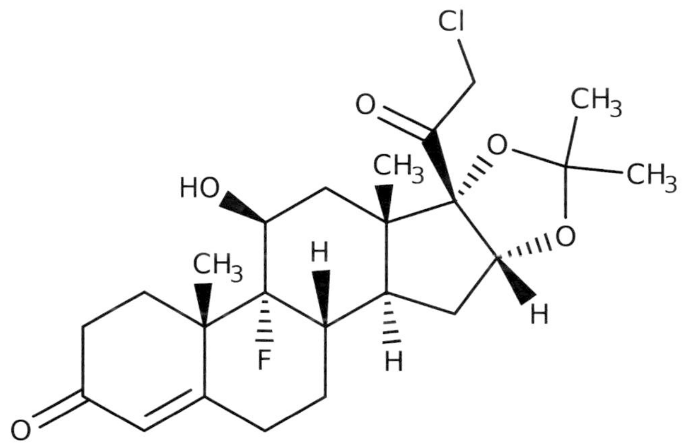

GENERAL

General aspects of corticosteroids used on the skin and mucous membranes are discussed in Chapter 2.4. A practical guideline for diagnosing allergic reactions to corticosteroids is presented in ref. 1.

CONTACT ALLERGY

Patch testing in groups of selected patients

In the period 2000 to 2005, in the USA, 1187 patients suspected of corticosteroid allergy were patch tested with halcinonide 1% pet. and there were 5 (0.4%) positive reactions, all of which were considered to be relevant (3). In Finland, in the period 1985-1990, 66 patients very likely to be corticosteroid-allergic were patch tested with halcinonide 0.1% and 1% pet., which yielded 3 (5%) positive reactions; their relevance was not mentioned (4).

Case series

From January 1990 to June 2008, in Leuven, Belgium, 315 patients were diagnosed with contact allergy to/allergic contact dermatitis from corticosteroids (CSs) from routine patch testing with a baseline series including tixocortol pivalate, budesonide, hydrocortisone butyrate and prednisone caproate, patch testing with patients' own CS preparations, and testing those with proven contact allergy to a corticosteroid or strongly suspected of CS allergy later with a series of 66 CSs, including two sex hormones (progesterone and testosterone). 71% of the patients had relevant reactions, but these were not specified. In this group of 315 CS allergic patients, 11 had positive patch tests to halcinonide 0.1% alc. (2). It is unknown how many of these reactions were caused by the use of a pharmaceutical product containing halcinonide and how many were cross-reactions to other corticosteroids (possibly all).

Cross-reactions, pseudo-cross-reactions and co-reactions
Cross-reactions between corticosteroids are discussed in Chapter 2.8.

LITERATURE

1   Baeck M, Goossens A. Immediate and delayed allergic hypersensitivity to corticosteroids: practical guidelines. Contact Dermatitis 2012;66:38-45
2   Baeck M, Chemelle JA, Terreux R, Drieghe J, Goossens A. Delayed hypersensitivity to corticosteroids in a series of 315 patients: clinical data and patch test results. Contact Dermatitis 2009;61:163-175
3   Davis MD, El-Azhary RA, Farmer SA. Results of patch testing to a corticosteroid series: a retrospective review of 1188 patients during 6 years at Mayo Clinic. J Am Acad Dermatol 2007;56:921-927
4   Lauerma AI. Contact hypersensitivity to glucocorticosteroids. Am J Contact Dermat 1992;3:112-132

Chapter 3.162 HALOMETASONE

IDENTIFICATION

Description/definition : Halometasone is the synthetic glucocorticoid that conforms to the structural formula shown below
Pharmacological classes : Anti-Inflammatory agents; dermatologic agents; glucocorticoids
IUPAC name : (6S,8S,9R,10S,11S,13S,14S,16R,17R)-2-Chloro-6,9-difluoro-11,17-dihydroxy-17-(2-hydroxyacetyl)-10,13,16-trimethyl-6,7,8,11,12,14,15,16-octahydrocyclopenta-[a]phenanthren-3-one
Other names : 2-Chloro-6α,9-difluoro-11β,17,21-trihydroxy-16α-methylpregna-1,4-diene-3,20-dione; 2-chloroflumethasone
CAS registry number : 50629-82-8
EC number : 256-664-9
Merck Index monograph : 5903
Patch testing : In general, corticosteroids may be tested at 0.1% and 1% in alcohol; late readings (6-10 days) are strongly recommended
Molecular formula : $C_{22}H_{27}ClF_2O_5$

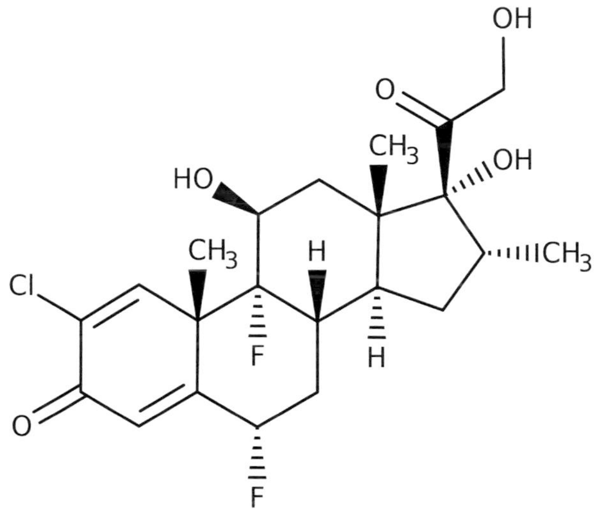

GENERAL

General aspects of corticosteroids used on the skin and mucous membranes are discussed in Chapter 2.4. A practical guideline for diagnosing allergic reactions to corticosteroids is presented in ref. 1.

CONTACT ALLERGY

Case series

From January 1990 to June 2008, in Leuven, Belgium, 315 patients were diagnosed with contact allergy to/allergic contact dermatitis from corticosteroids (CSs) from routine patch testing with a baseline series including tixocortol pivalate, budesonide, hydrocortisone butyrate and prednisone caproate, patch testing with patients' own CS preparations, and testing those with proven contact allergy to a corticosteroid or strongly suspected of CS allergy later with a series of 66 CSs, including two sex hormones (progesterone and testosterone). 71% of the patients had relevant reactions, but these were not specified. In this group of 315 CS allergic patients, 6 had positive patch tests to halometasone 0.1% alc. (2). It is unknown how many of these reactions were caused by the use of a pharmaceutical product containing halometasone and how many were cross-reactions to other corticosteroids.

Cross-reactions, pseudo-cross-reactions and co-reactions

Cross-reactions between corticosteroids are discussed in Chapter 2.8.

LITERATURE

1 Baeck M, Goossens A. Immediate and delayed allergic hypersensitivity to corticosteroids: practical guidelines. Contact Dermatitis 2012;66:38-45
2 Baeck M, Chemelle JA, Terreux R, Drieghe J, Goossens A. Delayed hypersensitivity to corticosteroids in a series of 315 patients: clinical data and patch test results. Contact Dermatitis 2009;61:163-175

Chapter 3.163 HALOPROGIN

IDENTIFICATION

Description/definition : Haloprogin is the halogenated phenolic ether that conforms to the structural formula shown below
Pharmacological classes : Antifungal agents
IUPAC name : 1,2,4-Trichloro-5-(3-iodoprop-2-ynoxy)benzene
CAS registry number : 777-11-7
EC number : 212-286-6
Merck Index monograph : 1184
Patch testing : 1% pet.
Molecular formula : $C_9H_4Cl_3IO$

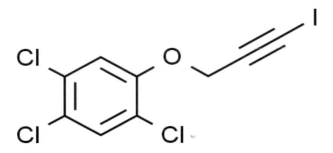

GENERAL

Haloprogin is an antifungal agent formerly used for the treatment of superficial dermatophyte infections and infections caused by *Candida albicans*. It was discontinued in favor of newer antifungals with fewer side effects (1).

CONTACT ALLERGY

Case report

A 21-year-old woman was given a prescription for a 1% haloprogin cream to be applied three times a day on a fungal infection on her right shoulder. The patient returned to the clinic two weeks later and exhibited a 9 cm round area of erythema, edema, and vesiculation at the site of application. The area had become pruritic after one week's use of the topical medication, and vesicles appeared shortly thereafter. Nevertheless, the patient had continued to apply the haloprogin cream. Patch tests were positive to haloprogin cream, haloprogin lotion and haloprogin 1% pet., but negative to 0.1% pet. (2).

LITERATURE

1 The data in the section 'General' may have been obtained from literature discussed in this chapter, but mostly also or exclusively from one or more of the following online sources: ChemIDPlus Advanced, PubChem, DrugBank, RxList, Drug Central, Drugs.com, and Wikipedia
2 Rudolph RL. Allergic contact dermatitis caused by haloprogin. Arch Dermatol 1975;111:1487-1488

Chapter 3.164 HALQUINOL

IDENTIFICATION

Description/definition   : Halquinol is a combination of hydroxyquinolines containing 5,7-dichloro-8-quinolinol,
                             5-chloro-8-quinolinol, 7-chloro-8-quinolinol and 8-hydroxyquinoline (PubChem);
                             halquinol is a mixture of 5,7-dichloro-8-quinolinol with 5-chloro-8-quinolinol and
                             7-chloro-8-quinolinol (ChemIDPlus)
Pharmacological classes   : Topical anti-infectives
Other names   : Halquinols
CAS registry number   : 8067-69-4
EC number   : Not available
Merck Index monograph   : 227
Patch testing   : 5% pet.

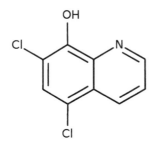 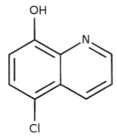 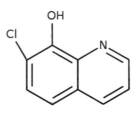

Halquinol according to ChemIDPlus

GENERAL

Halquinol is a combination of hydroxyquinolines currently described as 5,7-dichloro-8-quinolinol 14%, 5-chloro-8-quinolinol 42%, 7-chloro-8-quinolinol 10% and 8-hydroxyquinoline 34% (PubChem). ChemIDPlus describes it as 5,7-dichloro-8-quinolinol, mixture with 5-chloro-8-quinolinol and 7-chloro-8-quinolinol. It is or was used in antiseptic preparations, sometimes also in topical corticosteroid preparations (1).

CONTACT ALLERGY

Case series and case reports

Three patients applied a combination preparation of halquinol 0.75% and triamcinolone acetonide 0.025% to stasis dermatitis (n=2) or venous ulcers (n=1), which resulted in (worsening of) dermatitis, in one case spreading to become generalized within 2 days. They all had used clioquinol (5-chloro-7-iodoquinolin-8-ol) ointment previously. Patch tests were positive to halquinol, its ingredients 5,7-dichloro-8-quinolinol and 5-chloro-8-quinolinol, chlorquinaldol, clioquinol, and diiodoquin (5,7-diiodoquinolin-8-ol) (1).

Occupational allergic contact dermatitis to halquinol has been observed in a worker in an animal feed mill (6).

Cross-reactions, pseudo-cross-reactions and co-reactions

Cross-reactions between halogenated hydroxyquinolines such as clioquinol (5-chloro-7-iodoquinolin-8-ol), chlorquinaldol (5,7-dichloro-2-methylquinolin-8-ol), cloxyquin (5-chloroquinolin-8-ol), oxyquinoline (8-hydroxy-quinoline, non-halogenated), iodoquinol (5,7-diiodoquinolin-8-ol), halquinol (a mixture of 4 hydroxyquinolines), 5,7-dichloro-8-quinolinol and 5-chloro-8-quinolinol may occur (2-5 [examples of references]).

LITERATURE

1   The data in the section 'General' may have been obtained from literature discussed in this chapter, but mostly also or exclusively from one or more of the following online sources: ChemIDPlus Advanced, PubChem, DrugBank, RxList, Drug Central, Drugs.com, and Wikipedia
2   Allenby CF. Skin sensitization to Remiderm and cross-sensitization to hydroxyquinoline compounds. Br Med J 1965;2(5455):208-209
3   Myatt AE, Beck MH. Contact sensitivity to chlorquinaldol. Contact Dermatitis 1983;9:523
4   Leifer W, Steiner K. Studies in sensitization to halogenated hydroxyquinolines and related compounds. J Invest Dermatol 1951;17:233-240
5   Wantke F, Götz M, Jarisch R. Contact dermatitis from cloxyquin. Contact Dermatitis 1995;32:112-113
6   Burrows D. Contact dermatitis in animal feed mill workers. Br J Dermatol 1975;92:167-170

Chapter 3.165 HEXAMIDINE

IDENTIFICATION

Description/definition   : Hexamidine is the aromatic ether, member of guanidines and polyether that conforms to the structural formula shown below

Pharmacological classes   : Anti-infective agents

IUPAC name   : 4-[6-(4-Carbamimidoylphenoxy)hexoxy]benzenecarboximidamide

Other names   : Benzenecarboximidamide, 4,4'-(1,6-hexanediylbis(oxy))bis-

CAS registry number   : 3811-75-4

EU number   : Not available

Merck Index monograph : 6000

Patch testing   : 0.15% water or alcohol; 0.1% pet.

Molecular formula   : $C_{20}H_{26}N_4O_2$

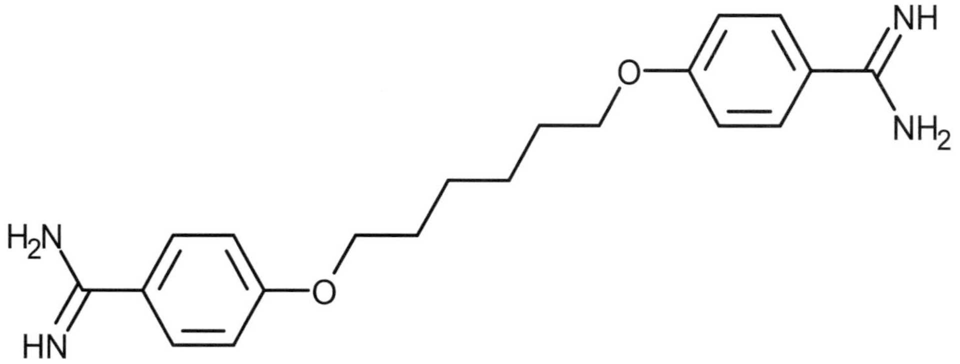

GENERAL

Hexamidine is a polyether that is the bis(4-guanidinophenyl) ether of hexane-1,6-diol. It has a role as an antimicrobial agent and an antiseptic drug. It is not always clear in literature (or data are unknown) whether 'hexamidine' means hexamidine base, or its salt hexamidine diisethionate (Chapter 3.166). Hexamidine diisethionate is present in the antiseptic and anti-infective preparation Hexomedine ®, which is widely used, especially in France. It cannot be excluded and is probably likely, that some of the data below refer to the diisethionate salt rather than to hexamidine itself. See also Chapter 3.166 Hexamidine diisethionate.

CONTACT ALLERGY

Patch testing in groups of selected patients

In the period 1990-2014, in a university clinic in Leuven, Belgium, 1174 patients suspected of iatrogenic contact dermatitis were patch tested with hexamidine 0.15% alc. and there were 16 (1.4%) positive reactions. Relevance was not specified for individual drugs, but 96% of the positive patch test reactions to all topical drugs and antiseptics were considered to be relevant (6, overlap with ref. 3).

In the period 2004-2008, in France, 423 patients with leg ulcers were patch tested with hexamidine 0.15% water and 5 (1.2%) had a positive reaction. The relevance of these reactions was not mentioned (2). In 1981, in France, 465 patients suspected of allergy to cosmetics, drugs, industrial products or clothes were tested with hexamidine and 1 (0.2%) reacted positively; its relevance was not mentioned (4).

Case series

In a 2-year retrospective study (2003-2004), the members of the French Dermato-Allergology Vigilance network Revidal together found 20 cases of contact allergy to 'hexamidine', possibly hexamidine diisethionate. Most were therapy-related, i.e. from its presence in antiseptics (5). In the period 1985-1997, in a university clinic in Leuven, Belgium, 8521 patients were patch tested, and 17 positive reactions were observed to hexamidine. It was not stated, however, how many patients had been tested with this allergen and how many of the reactions were relevant (3, overlap with ref. 6).

In France, 20 patients with contact dermatitis to hexamidine, collected in a 7 months period, were reported in 1984 (7). The clinical features were described as 'very peculiar' with more frequently papular semispheric and papulovesicular lesions than classical vesicular eczema lesions. The histopathology was also unusual, most often with dermal vasculitis and infrequently with epidermal spongiosis. Contact dermatitis had usually a long standing evolution despite removal of the allergen and topical corticosteroid treatment. Patch tests with hexamidine were frequently followed by exacerbation of the cutaneous lesions (7).

One (5) and 4 (1) cases of sensitization to hexamidine have been ascribed to cosmetic allergy.

LITERATURE

1   Kohl L, Blondeel A, Song M. Allergic contact dermatitis from cosmetics: retrospective analysis of 819 patch-tested patients. Dermatology 2002;204:334-337
2   Barbaud A, Collet E, Le Coz CJ, Meaume S, Gillois P. Contact allergy in chronic leg ulcers: results of a multicentre study carried out in 423 patients and proposal for an updated series of patch tests. Contact Dermatitis 2009;60:279-287
3   Goossens A, Claes L, Drieghe J, Put E. Antimicrobials, preservatives, antiseptics and disinfectants. Contact Dermatitis 1998;39:133-134
4   Meynadier JM, Meynadier J, Colmas A, Castelain PY, Ducombs G, Chabeau G, et al. Allergy to preservatives. Ann Dermatol Venereol 1982;109:1017-1023 (Article in French)
5   Barbaud A, Vigan M, Delrous JL, et al. Contact allergy to antiseptics: 75 cases analyzed by the dermato-allergovigilance network (Revidal). Ann Dermatol Venereol 2005;132:962-965
6   Gilissen L, Goossens A. Frequency and trends of contact allergy to and iatrogenic contact dermatitis caused by topical drugs over a 25-year period. Contact Dermatitis 2016;75:290-302
7   Revuz J, Poli F, Wechsler J, Dubertret L. Contact dermatitis from hexamidine. Ann Dermatol Venereol 1984;111:805-810 (Article in French)

Chapter 3.166 HEXAMIDINE DIISETHIONATE

IDENTIFICATION

Description/definition : Hexamidine diisethionate is the diisethionate salt form of hexamidine that conforms to the structural formula shown below

Pharmaceutical classes : Anti-infective agents

IUPAC name : 2-Hydroxyethanesulphonic acid, compound with 4,4'-[hexane-1,6-diylbis(oxy)]bis-[benzenecarboxamidine] (2:1)

Other names : Hexamidine isethionate

CAS registry number : 659-40-5

EC number : 211-533-5

Merck Index monograph : 6000 (Hexamidine)

Patch testing : 0.15% hydro-alcoholic solution

Molecular formula : $C_{24}H_{38}N_4O_{10}S_2$

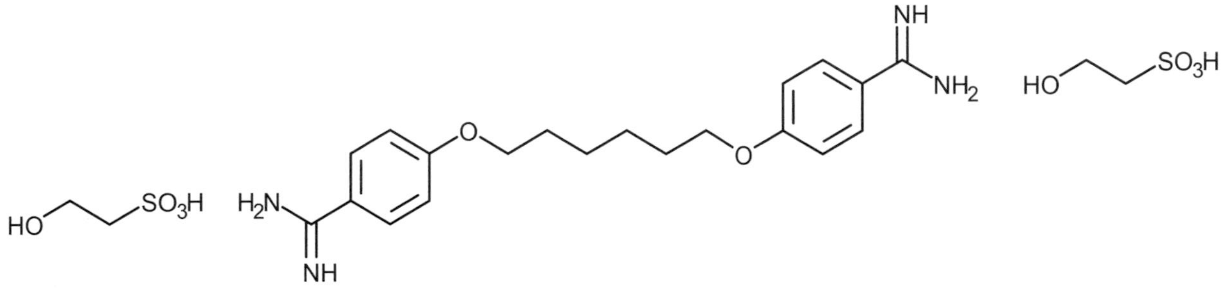

GENERAL

Hexamidine diisethionate is the diisethionate salt form of hexamidine, a polyether that is the bis(4-guanidinophenyl) ether of hexane-1,6-diol. Hexamidine diisethionate is the active ingredient of the antiseptic and anti-infective preparation Hexomedine ®, which is widely used, especially in France. See also Chapter 3.165 Hexamidine.

CONTACT ALLERGY

Case series

In a 2-year retrospective study (2003-2004), the members of the French Dermato-Allergology Vigilance network Revidal together found 20 cases of contact allergy to 'hexamidine', probably hexamidine diisethionate. Most were therapy-related, i.e. from its presence in antiseptics (9).

In a clinic in Paris, in the period 1976-1978, eight patients with positive patch test reactions to an antiseptic solution containing hexamidine diisethionate were observed, 6 men and 2 women, age range 25-81 years. Four patients had localized allergic contact dermatitis, the other four, in whom the lotion had been applied to stasis dermatitis or (surgical) wounds, had a disseminated eruption. The author mentioned that 'three or four cases of allergy to Hexomedine have been observed each year since 1969' (4). Although the title 'Contact dermatitis to hexamidine' suggests that the active ingredient was the allergen, it is very likely that only the commercial solution was tested.

In early French literature, 147 cases of allergic contact dermatitis from hexamidine diisethionate have been reported (5,6, data cited in ref. 4).

Case reports

A 13-year-old boy with mild atopic eczema developed 2 solitary erythematous lesions on his face. While being treated with a cream containing hexamidine diisethionate, clotrimazole and prednisolone acetate, the eczema worsened and spread to the neck and upper thorax. Patch tests were positive to the cream, all three active ingredients (hexamidine diisethionate tested 1% pet.) and two excipients (10).

A female patient aged 34 was suffering from eczema localized on the amputation stump of the right thigh. In the clinic of revalidation, often hexamidine diisethionate solution was used as disinfectant. Strongly positive patch tests were observed to the lotion (containing 0.15% active ingredient) and hexamidine 0.15% water (2). One patient (the first reported with contact allergy to hexamidine diisethionate) had generalized dermatitis from allergy to hexamidine diisethionate in an antiseptic lotion (3). One case of 'systemic allergy' from topical hexamidine has been reported from Australia; this was a case of contact urticaria syndrome (8).

A few cases of sensitization to or allergic contact dermatitis from hexamidine diisethionate have been caused by cosmetic products (1 [hair lotion], 9 [cosmetic unspecified]).

PHOTOSENSITIVITY
One patient had photoallergic contact dermatitis from hexamidine diisethionate in an antiseptic preparation (7).

Immediate contact reactions
Immediate contact reactions (contact urticaria) to hexamidine diisethionate are presented in Chapter 5.

LITERATURE
1   Dooms-Goossens A, Vandaele M, Bedert R, Marien K. Hexamidine isethionate: a sensitizer in topical pharmaceutical products and cosmetics. Contact Dermatitis 1989;21:270
2   Van Ketel WG. Allergic contact eczema by Hexomedine®. Contact Dermatitis 1975;1:332
3   Gougerot H, Tabernat J, Mlle. Raufast & Gascoin. Eczema généralisé par sensibilisation à l' hexomedine. Bulletin de la Societé Française de Dermatologie et Syphiligraphie 1950;57:271 (Article in French, cited in ref. 2)
4   Robin J. Contact dermatitis to hexamidine. Contact Dermatitis 1978;4:375-376
5   Sidi E, Bourgeois-Spinasse J, Arouète, J. Quelques causes d'eczema d'origine medicamenteuses. Revue Francaise d'Allergie 1969;9:179-182 (Article in French, data cited in ref. 4)
6   Sidi E, Bourgeois-Spinasse J, Arouète, J. Gazette des hôpitaux, 20 Decembre 1968: 1143-1144 (Article in French, data cited in ref. 4)
7   Boulitrop-Morvan C, Collet E, Dalac S, Bailly N, Jeanmougin M, Lambert D. Photoallergy to hexamidine. Photodermatol Photoimmunol Photomed 1993;9:154-155
8   Mullins RJ. Systemic allergy to topical hexamidine. Med J Aust 2006;185:177
9   Barbaud A, Vigan M, Delrous JL, Assier H, Avenel-Audran M, Collet E, et al Membres du Groupe du REVIDAL. Contact allergy to antiseptics: 75 cases analyzed by the dermato-allergovigilance network (Revidal). Ann Dermatol Venereol 2005;132(12 Part 1):962-965
10 Brand CU, Ballmer-Weber BK. Contact sensitivity to 5 different ingredients of a topical medicament (Imacort cream). Contact Dermatitis 1995;33:137

Chapter 3.167 HIRUDOID ® CREAM

IDENTIFICATION

Hirudoid ® cream (CAS 11097-84-0) is a pharmaceutical cream that contains, as active ingredient, a heparinoid (heparin-derivative) termed 'mucopolysaccharide polysulfate'. It is widely used for the treatment of phlebitis, leg ulcers, contusions, swelling and hematoma's, although its therapeutic efficacy is not well-documented. Hirudoid cream has caused many cases of allergic contact dermatitis. However, it was not the active drug itself, but a chemical formed in the cream by the interaction between 2 ingredients, that was the allergenic culprit. For that reason, the commercial preparation is discussed here instead of its active drug component. This is also the reason that this chapter differs considerably in format from all others.

CONTACT ALLERGY

Contact allergy to Hirudoid cream was first described in the early 1970s. At least 17 patients with allergic contact dermatitis from the cream were seen in a private practice in The Netherlands (1,2). Twelve per cent of all patients with leg ulcers were sensitized. In most cases, the allergenic ingredient proved to be the preservative chloroacetamide, which consequently was removed from the formulation by the manufacturer.

However, new cases of contact allergy kept being published and were by no means rare (3-9). The same Dutch investigators saw another 23 patients with allergic contact dermatitis from Hirudoid cream in the period 1975-1985. They also added the cream to their standard series between 1983 and 1985 and patch tested 490 consecutive patients suspected of contact dermatitis. Fifteen individuals (3.1%) had positive reactions to Hirudoid cream (3). In 3 hospitals in Portugal, in the period 1982-1985, thirty-one patients (17 women and 14 men) were seen with contact sensitization to Hirudoid cream (4). These authors noted that 14% of their patients patch tested for leg ulcers were allergic to the cream; all had used the cream in the past (5).

Most of the patients were elderly women, who had used the cream for leg disorders caused by chronic venous insufficiency such as stasis dermatitis, ulcers, thrombophlebitis and indurations (3). Others used it for trauma or were hemodialysis patients using it on the shunt area at the hands or arms to prevent inflammation (4). In about half of the patients, contact dermatitis was confined to the area of application, and the other half showed spreading to other skin areas. Some individuals required hospitalization (3).

The allergen

In many investigations, patients who had a positive patch test to Hirudoid cream were subsequently tested with its ingredients, provided by the manufacturer. In 31 such patients seen in Portugal, the following (inactive) ingredients reacted positively: myristyl alcohol (n=7), cetearyl alcohol (n=6), methylparaben (n=6), anhydrous lanolin (n=5), and 1,3,5-trihydroxyethylhexahydrotriazine (n=3; 21 patients tested). There was also one reaction to thymol and to the active heparinoid, but the authors suggested that these were false-positive due to the excited skin syndrome (4). Others found sensitization to lanolin alcohol (3) and myristyl alcohol (9). However, most patients allergic to Hirudoid cream also reacted to the cream base, but not to any ingredient. Thus, compound allergy was suspected, i.e. that 2 or more ingredients interact chemically to form a new compound, which is the actual allergenic ingredient (3,4,6,7,8)

To test this hypothesis and try to identify the allergen, 16 patients reacting to Hirudoid cream and the cream base but not any ingredient were further investigated (3). By testing with the cream base (prepared in-house) and leaving out individual ingredients, it was found that thymol and the preservative 1,3,5-trihydroxyethylhexahydrotriazine (triazinetriethanol; CAS number 4719-04-4) (THT) were necessary to obtain positive reactions; however, these individual components themselves always tested negative. Therefore, patients were tested further with a mixture of THT and thymol in alcohol 30%. This combination provoked strongly positive reactions, whereas in 10 non-allergic control subjects patch tests with this mixture gave negative results. Patch tests with thymol combined with one of the degradation products of THT (formaldehyde and ethanolamine) were negative, but positive when thymol was combined with *both* degradation products, indicating that a newly formed chemical from thymol, formaldehyde and ethanolamine had to be the allergenic culprit. This allergen was identified by nuclear magnetic resonance spectroscopy and infrared spectroscopy as 3-(hydroxyethyl)-5-methyl-8-(2-methylethyl)-3,4-dihydro-2H-1,3-benzoxazine (IUPAC name 2-(5-methyl-8-propyl-2,4-dihydro-1,3-benzoxazin-3-yl)ethanol, CAS number 117652-03-6). Its structural formula is shown below.

Currently, Hirudoid cream still contains thymol but not 1,3,5-trihydroxyethylhexahydrotriazine anymore. However, it does contains lanolin alcohol, cetearyl alcohol, myristyl alcohol and methyl- and propylparaben, which will certainly induce contact allergy in a number of patients, especially those who use it on stasis dermatitis or leg ulcers.

3-(Hydroxyethyl)-5-methyl-8-(2-methylethyl)-3,4-dihydro-2H-1,3-benzoxazine

LITERATURE

1   Prins FJ, Smeenk G. Contact eczema caused by Hirudoid ointment. Ned Tijdschr Geneeskd 1971;115:1934-1938 (Article in Dutch)
2   Smeenk G, Prins FJ. Allergic contact eczema due to chloracetamide. Dermatologica 1972;144:108-114
3   Smeenk G, Kerckhoffs HP, Schreurs PH. Contact allergy to a reaction product in Hirudoid cream: an example of compound allergy. Br J Dermatol 1987;116:223-231
4   Pecegueiro M, Brandão M, Pinto J, Conçalo S. Contact dermatitis to Hirudoid cream. Contact Dermatitis 1987;17:290-293
5   Pecegueiro M, Brandao M. Alergia de contacto em doentes com ulcera de perna. Trab Soc Port Derm Vener 1983: XLI:37-46 (Article in Portuguese, data cited in ref. 4).
6   Manuzzi P, Lama L, Barone M, Patrone P. Contact dermatitis to Hirudoid cream. Contact Dermatitis 1986;15:42
7   Dooms-Goossens A. Allergic contact dermatitis to ingredients used in topically applied pharmaceutical products and cosmetics. Doctoral Thesis. University of Leuven, Belgium, 1982
8   Kellett JK, King CM, Beck MH. Compound allergy to medicaments. Contact Dermatitis 1986;14:45
9   Edman B, Möller H. Medicament contact allergy. Dermatosen 1986;34:139-142

Chapter 3.168 HOMATROPINE

IDENTIFICATION

Description/definition   : Homatropine is the synthetic tertiary amine alkaloid that conforms to the structural
                             formula shown below
Pharmacological classes : Parasympatholytics
IUPAC name   : [(1R,5S)-8-Methyl-8-azabicyclo[3.2.1]octan-3-yl] 2-hydroxy-2-phenylacetate
Other names   : Mandelyltropine
CAS registry number   : 87-00-3
EC number   : 201-716-8
Merck Index monograph   : 6038
Patch testing   : 1% water
Molecular formula   : $C_{16}H_{21}NO_3$

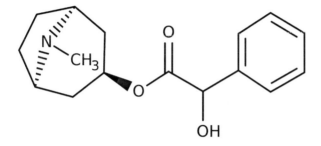

GENERAL

Homatropine is a synthetic tertiary amine alkaloid and an anticholinergic drug that acts as an antagonist at muscarinic acetylcholine receptors, blocking parasympathetic nerve stimulation. It is present in antitussives in combination with hydrocodone (dihydrocodeinone) bitartrate indicated for the symptomatic relief of cough. Homatropine is also administered as ophthalmic solution as a cycloplegic to temporarily paralyze accommodation and to induce mydriasis. In most pharmaceutical preparations, homatropine is present as homatropine hydrobromide (CAS number 51-56-9, EC number 200-105-3, molecular formula $C_{16}H_{22}BrNO_3$) (1).

CONTACT ALLERGY

Patch testing in groups of selected patients

In The Netherlands, in the period 2000-2006, 3 patients with suspected contact allergy to ophthalmological drugs were patch tested with homatropine hydrobromide and 2 had a positive reaction (test concentration and vehicle not mentioned). Relevance was not given for individual drugs, but 69% of all reactions to drugs together were considered to be relevant (3).

Case series

In Leuven, Belgium, between 1990 and 2017, 16,065 patients were investigated for contact allergy and 118 (0.7%) showed positive patch test reactions to topical ophthalmic medications and/or to their ingredients. Eighty-four individuals (71%) reacted to an active principle. Homatropine was tested in 12 patients and was the allergen in eye medications in one. There were no reactions to homatropine in other types of medications (2, overlap with ref. 4).

Also in Leuven, Belgium, in the period 1990-2014, iatrogenic contact dermatitis was diagnosed in 2600 individuals (17% of the total patch test population). 96% of all positive patch test reactions to topical drugs and antiseptics were considered to be relevant. Homatropine (1% water) was tested in 12 patients and there were 4 positive reactions to it (4, overlap with ref. 2).

Case report

A 43-year-old nurse, working in both the ophthalmology and otolaryngology departments of a hospital, over the last 3 years had developed an itchy erythematous dermatitis on her face, especially the eyelids, related to her occupational duty of applying topical ophthalmic drugs, mainly homatropine and phenylephrine. Lesions improved when she worked in otolaryngology, and almost disappeared on holidays. Patch tests revealed positive reactions to homatropine 1% water and phenylephrine 10% water. Ten controls were negative. Occupational allergic contact dermatitis was diagnosed. The nurse left her work in ophthalmology and has since remained symptomless (6).

Cross-reactions, pseudo-cross-reactions and co-reactions
Possible cross-reactivity to homatropine, scopolamine, and belladonna in a patient with allergic contact dermatitis from atropine (5). Another individual sensitized to atropine co-reacted to homatropine, to which he had not previously been exposed (7).

LITERATURE

1   The data in the section 'General' may have been obtained from literature discussed in this chapter, but mostly also or exclusively from one or more of the following online sources: ChemIDPlus Advanced, PubChem, DrugBank, RxList, Drug Central, Drugs.com, and Wikipedia

2   Gilissen L, De Decker L, Hulshagen T, Goossens A. Allergic contact dermatitis caused by topical ophthalmic medications: Keep an eye on it! Contact Dermatitis 2019;80:291-297

3   Wijnmaalen AL, van Zuuren EJ, de Keizer RJ, Jager MJ. Cutaneous allergy testing in patients suspected of an allergic reaction to eye medication. Ophthalmic Res 2009;41:225-229

4   Gilissen L, Goossens A. Frequency and trends of contact allergy to and iatrogenic contact dermatitis caused by topical drugs over a 25-year period. Contact Dermatitis 2016;75:290-302

5   Decraene T, Goossens A. Contact allergy to atropine and other mydriatic agents in eye drops. Contact Dermatitis 2001;45:309-310

6   Marcos ML, Garcés MM, Alonso L, Juste S, Carretero P, Blanco J, et al. Occupational allergic contact dermatitis from homatropine and phenylephrine eyedrops. Contact Dermatitis 1997;37:189

7   Yoshikawa K, Kawahara S. Contact allergy to atropine and other mydriatic agents. Contact Dermatitis 1985;12:56-57

Chapter 3.169 HYDROCORTISONE

IDENTIFICATION

Description/definition	: Hydrocortisone is the main glucocorticoid secreted by the adrenal cortex that conforms to the structural formula shown below
Pharmacological classes	: Glucocorticoids; anti-inflammatory agents
IUPAC name	: (8S,9S,10R,11S,13S,14S,17R)-11,17-Dihydroxy-17-(2-hydroxyacetyl)-10,13-dimethyl-2,6,7,8,9,11,12,14,15,16-decahydro-1H-cyclopenta[a]phenanthren-3-one
Other names	: Cortisol; 11β,17,21-trihydroxypregn-4-ene-3,20-dione; hydrocortisone alcohol
CAS registry number	: 50-23-7
EC number	: 200-020-1
Merck Index monograph	: 6094
Patch testing	: 1% pet. (SmartPracticeCanada, SmartPracticeEurope); late readings (6-10 days) are strongly recommended; testing with this preparation probably results in many false-negative reactions; in general, corticosteroids can be tested 0.1% and 1% alcohol
Molecular formula	: $C_{21}H_{30}O_5$

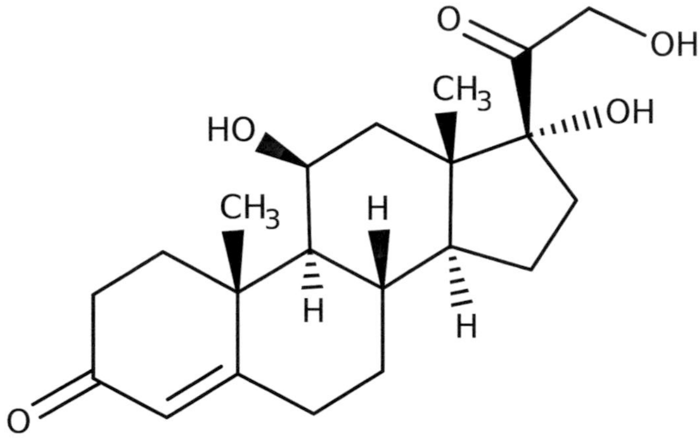

GENERAL

General aspects of corticosteroids used on the skin and mucous membranes are discussed in Chapter 2.4. A practical guideline for diagnosing allergic reactions to corticosteroids is presented in ref. 6. Hydrocortisone *base* (hydrocortisone alcohol, hydrocortisone free alcohol) is used in tablets only, which implies that by far most allergic reactions to 'hydrocortisone' have in fact been the result of sensitization to an ester of hydrocortisone or of cross-reactivity to another corticosteroid. It is also likely that there has been confusion in some publications on the correct forms of the drugs used, e.g. that hydrocortisone was mentioned where an ester form should have been mentioned.

See also hydrocortisone acetate (Chapter 3.171), hydrocortisone butyrate (Chapter 3.172), hydrocortisone aceponate (Chapter 3.170), hydrocortisone hemisuccinate (Chapter 3.173), hydrocortisone probutate (Chapter 3.174), and hydrocortisone valerate (Chapter 3.175).

CONTACT ALLERGY

General: problems with patch testing hydrocortisone
Patch testing with hydrocortisone has shown to be problematic. Patch tests with HC (acetate) in alcohol, pet. or DMSO are frequently false-negative (15,17,21,22), whereas hydrocortisone acetate in 70% propylene glycol causes many false-positive, irritant, reactions (17), which also applies to hydrocortisone alcohol 15% in 100% DMSO (20). Intradermal testing with hydrocortisone (HC) sodium succinate is a sensitive way to diagnose contact allergy to hydrocortisone (12,15,16,17,20,28,40). However, tixocortol pivalate, which is routinely tested in most baseline/routine/standard series (0.1% or 1% pet.) is a very sensitive marker and almost always points at contact allergy to hydrocortisone (12,15,16,17,20,27,33).

Contact allergy in the general population
In Germany, for the period 1995-2004, the population-based relative incidence (RI) of contact sensitization to hydrocortisone (cases/100,000 defined daily doses (DDDs) per year) was estimated to be 0.9. In the group of corticosteroids, the RI ranged from 0.3 (dexamethasone sodium phosphate) to 43.2 (budesonide) (5).

Patch testing in groups of patients

Results of patch testing hydrocortisone in consecutive patients suspected of contact dermatitis (routine testing) back to 2001 are shown in table 3.169.1. Results of testing in groups of *selected* patients (individuals with eyelid dermatitis, patient suspected of corticosteroid allergy) are shown in table 3.169.1. In routine testing in 2 studies (of which one in 2 countries with separate results shown), prevalences of sensitization were 0.2%, 0.5% and 1.5%. In Belgium, the rate was 1.5% in 1989-1990 (41), which, in the same clinic, only one year later, had dropped to 0.5% (37). A possible explanation is that the DMSO in the test material had caused a number of false-positive, irritant reactions (41).

In selected patient groups, high rates of positive reactions (16.2% and 18%) were observed in the United Kingdom (38) and in Finland (39), which can be attributed to the fact that the patients were strongly suspected of allergy to corticosteroids and that the U.K. study included individuals who had a positive patch test to tixocortol pivalate, which nearly always point at hydrocortisone hypersensitivity (38). The low rate of positive patch test reactions of 0.9% to hydrocortisone 20% pet. and the high percentage of 16.2 when the patients were tested with hydrocortisone 2% in alcohol, clearly showed that petrolatum is not a suitable vehicle for patch diagnosing contact allergy to hydrocortisone (38).

Table 3.169.1 Patch testing in groups of patients

Years and Country	Test conc. & vehicle	Number of patients tested | positive (%)		Selection of patients (S); Relevance (R); Comments (C)	Ref.
Routine testing					
1991 Belgium	1% alc.	610	3 (0.5%)	R: not stated	37
1991 The Netherlands	1% alc.	533	1 (0.2%)	R: not stated	37
1989-1990 Belgium	0.5% alc. 94%/DMSO	1150	17 (1.5%)	R: not stated	41
Testing in groups of selected patients					
1995-2004 IVDK	1% pet.	6145	(0.3%)	S: not stated; R: not stated	4
1995-2004 IVDK	1% pet.	5940	19 (0.3%)	S: patients tested with a corticosteroid series; R: not stated	26
1997-2001 USA		203	6 (3%)	S: patients with eyelid dermatitis; R: not stated; C: unknown in how many patients this material was tested	1
1995-9 Netherlands	1% pet.	218	2 (0.9%)	S: not stated; R: probable 100%	30
1994-1998 U.K.	2% alc.	232	1 (0.4%)	S: patients with periorbital contact dermatitis; R: 100%; C: the frequency in a control group was 0%	2
1996-1997 IVDK	1% pet.	608	2 (0.3%)	S: patients tested with a corticosteroid series; R: not stated	29
1988-1991 U.K.	20% pet. 2% alc.	528 68	5 (0.9%) 11 (16.2%)	S: patients with a positive patch test to tixocortol pivalate or suspected of corticosteroid allergy; R: not stated	38
1985-1990 Finland	1% and 10% alc. and pet.	66	12 (18%)	S: patients very likely to be corticosteroid-allergic; R: not stated	39

IVDK: Information Network of Departments of Dermatology (Germany, Austria, Switzerland)

Case series

From January 1990 to June 2008, in Leuven, Belgium, 315 patients were diagnosed with contact allergy to/allergic contact dermatitis from corticosteroids (CSs) from routine patch testing with a baseline series including tixocortol pivalate, budesonide, hydrocortisone butyrate and prednisone caproate, patch testing with patients' own CS preparations, and testing those with proven contact allergy to a corticosteroid or strongly suspected of CS allergy later with a series of 66 CSs, including two sex hormones (progesterone and testosterone). 71% of the patients had relevant reactions, but these were not specified. In this group of 315 CS allergic patients, 99 had positive patch tests to hydrocortisone 0.5% alc./DMSO (24, overlap with refs. 3 and 25). As hydrocortisone base (free alcohol) is used in tablets only, it may be assumed that virtually all positive patch tests were the result of cross-reactivity to hydrocortisone esters or other corticosteroids.

In the period 1990-2008, also in Leuven, Belgium, 315 patients were diagnosed with contact allergy to/allergic contact dermatitis from corticosteroids. Eighteen subjects (5.7%) presented with allergic manifestations (conjunctivitis, eczema of the face, periocular skin or eyelids) caused by the use of CS-containing ocular preparations. Fourteen patients had used ophthalmic preparations containing hydrocortisone (25, overlap with refs. 3 and 24). However, hydrocortisone as such is not used in eye drops, but usually its salt hydrocortisone sodium phosphate.

Again in Leuven, Belgium, between 1990 and 2017, 16,065 patients were investigated for contact allergy and 118 (0.7%) showed positive patch test reactions to topical ophthalmic medications and/or to their ingredients. Eighty-four individuals (71%) reacted to an active principle. Hydrocortisone was tested in 15,659 patients and was the allergen in eye medications in 9. There were also 105 reactions to hydrocortisone in other types of medications (3, overlap with refs. 24 and 25).

In a hospital in the United Kingdom, before 1992, 59 consecutive patients with positive patch tests to a nasal spray containing tixocortol pivalate (at that moment an often used screening agent for corticosteroid, notably hydrocortisone, allergy) were all positive to an intradermal test with hydrocortisone sodium succinate 1 mg in a volume of 0.1 ml, read at D2, confirming hydrocortisone sensitivity. Stasis, perineal, and chronic actinic dermatitis occurred significantly more frequent than in a control group of 199 consecutive patients. Patients with hydrocortisone sensitivity also had significantly more other positive patch test reactions than controls (12). A use test on eczematous skin with hydrocortisone was positive in 9 of 11 patients who were patch test positive to tixocortol pivalate (unpublished observations by the authors, possibly in an earlier study) (12).

In the same clinic in the United Kingdom, between June and December 1989, 497 consecutive patients with suspected allergic contact dermatitis were patch tested to tixocortol pivalate 1% nasal spray. All patients positive to the spray were given an intradermal injection of 1 mg hydrocortisone sodium phosphate and were patch tested to hydrocortisone acetate (1% in alcohol) and pure tixocortol pivalate (1% in white soft paraffin), to confirm tixocortol pivalate sensitivity. 24 patients (4.8%) were allergic to the spray, all of whom were also allergic to tixocortol pivalate 1% pet. Most patients had co-reactivities, mostly to the fragrance mix, quinoline mix, lanolin alcohol and neomycin. Of the 24 allergic patients, only 2 (8.3%) had a positive patch test reaction to hydrocortisone acetate 1% alc. However, all 24 patients had indurated erythema at the site of intradermal hydrocortisone injection within 24-48 hours, which was negative in 80 controls not reacting to tixocortol spray. This study showed that patch testing with hydrocortisone (acetate) 1% alc. will miss >90% of hydrocortisone sensitivities (15).

Four patients (2 women, 2 men, ages 31-70 years) treated hand eczema (n=2), vasculitis of the inner thighs (n=1) and perianal eczema (n=1) with hydrocortisone cream and developed worsening of eczema or allergic contact dermatitis (the patient with vasculitis). Patch tests were positive to hydrocortisone and to hydrocortisone butyrate. Oral provocation with hydrocortisone 100-250 mg resulted in reactivation of positive patch tests (both patients with hand eczema), indurated papules on the inner sides and reactivation of previous perianal eczema (systemic contact dermatitis) (13,14).

In Sydney, Australia, from 1988 to 1994, 19 cases of corticosteroid allergy were diagnosed: 11 to hydrocortisone (1% pet.), 7 to alclometasone dipropionate (1% alc. and pet.), 4 to betamethasone valerate (1% alc.), 2 to betamethasone dipropionate (1% alc.) and 2 to triamcinolone acetonide (cream and ointment). 15 patients showed positive patch tests to the corticosteroid in the cream base, but were negative to the same corticosteroid in the ointment base. Of 8 patients performing a ROAT with the commercial cream, 7 were strongly positive (36).

In a hospital in the USA, between 1981 and 1991 (before the introduction of screening markers), 19 patients were found to have positive patch test reactions to commercial topical steroid preparations. Eleven were allergic to the corticosteroid ingredients: 5 to hydrocortisone, 3 to amcinonide, 1 to hydrocortisone butyrate and 2 to multiple CSs. Ten co-reacted or reacted alone to another active ingredient (neomycin) or one or more excipients (35).

From a clinic in Sweden, early 1970s, 21 patients with contact allergy to/allergic contact dermatitis from hydrocortisone and/or hydrocortisone acetate were reported. Despite the large number, only 0.3% of consecutive patients reacted to one or more corticosteroids (detailed data are not available to the author) (51).

Case reports
A 68-year-old man, with pruritus and a burning sensation of the penis and scrotum, was treated with a cream containing hydrocortisone and miconazole nitrate, to which he developed an eczematous reaction in the genital and inguinal region. Prednicarbate ointment was then applied, and 3 days later, an intense edema of the penis and scrotum with erythema and vesicles was observed. Patch tests were positive to both creams, hydrocortisone 1% alcohol (negative to pet.), prednicarbate 1% pet., hydrocortisone butyrate and tixocortol pivalate (7). A 44-year-old woman had allergic contact dermatitis from hydrocortisone in an antifungal-corticosteroid combination preparation and from hydrocortisone acetate in a cream and in an ophthalmic ointment. It proved very difficult to clearly demonstrate contact allergy to hydrocortisone (23). One case of occupational sensitization to hydrocortisone has been reported in a pharmacist, whose work had included handling and manufacturing of ex tempore corticosteroid creams and ointments for over 10 years. She had never used topical corticosteroids, but when she applied hydrocortisone cream for a mild sunburn, worsening and spreading of skin lesions were noted. She had positive patch and prick tests to hydrocortisone and related corticosteroids (32).

Short summaries of other case reports of allergic contact dermatitis from hydrocortisone are shown in table 3.169.2.

Early (<1975) case reports of allergic contact dermatitis from hydrocortisone and cases in non-English literature can be found in refs. 42-47 and 50.

Table 3.169.2 Short summaries of case reports of allergic contact dermatitis from hydrocortisone

Year and country	Sex	Age	Positive patch tests	Clinical data and comments	Ref.
2002 Sweden	M	8	tixocortol pivalate (marker for HC sensitivity)	bilateral secretory otitis media; also allergy to neomycin and lanolin alcohol; after avoidance otitis healed fully	19
1998 Sweden	F	61	HC 0.5% alc./DMSO	longstanding use led to worsening of facial eczema with erythema and lichenification around the mouth and eyes; exacerbation and spreading to previously unaffected skin from oral prednisolone (systemic contact dermatitis)	8
1998 United Kingdom	F	52	tixocortol pivalate (marker for HC sensitivity)	periorbital eczema and allergic conjunctivitis not responding to HC ointment	31
	F	43	tixocortol pivalate (marker for HC sensitivity)	exacerbation of periorbital inflammation and edema from topical hydrocortisone	
1996 Spain	M	44	tixocortol pivalate (marker for HC sensitivity)	edema of the penis a few days after using HC cream for genital pruritus	34
1992 United Kingdom	M	35	HC acetate saturated solution in alcohol	acute swelling and erythema of the groin after using a cream containing hydrocortisone (free alcohol)	9
1989 Singapore	M	56	HC cream 'as is'; HC 12.5% pet.	worsening of eczema from hydrocortisone cream; maculopapular rash from prednisolone (not tested but very likely cross-reacting to hydrocortisone)	48
1988 Canada	M	35	HC cream 'as is'; HC 5% pet.	worsening of non-specific eczema	49
1988 Sweden	M	53	HC 25% pet.; HC 2.5%, 1.25%, 0.63% and 0.31% alcohol	worsening of hand eczema; also positive reactions to 3 commercial HC creams/ointment and tixocortol pivalate	11

HC: hydrocortisone (base, free alcohol)

Cross-reactions, pseudo-cross-reactions and co-reactions
Cross-reactions between corticosteroids are discussed in Chapter 2.8. Patients reacting to tixocortol pivalate are almost always also allergic to hydrocortisone (12,15,16,17,20).

PHOTOSENSITIVITY
A 59-year-old woman had a facial eruption thought at various times to be seborrheic dermatitis, rosacea, or contact dermatitis. She was treated with 1% hydrocortisone cream, which caused a worsening of the facial eruption. Patch tests with two commercial preparations containing 1% and 2.5% hydrocortisone were positive. Patch tests with all ingredients were positive to HC 2% in propylene glycol and HC 20% pet. at D4. The patch test was repeated and combined with a photopatch test with HC 20%. The photopatch test was positive at D4, whereas the 'plain' patch test became positive at D8 only. It was concluded that the patient had polymorphic light eruption and photoallergic contact dermatitis to hydrocortisone (18). The latter may be incorrect, as 'plain' contact allergy, next to the alleged photocontact allergy, had already been established, late reactions to corticosteroids – as we now know – are frequent, and patch testing with hydrocortisone is highly unreliable.

OTHER INFORMATION
Eleven patients, positive to tixocortol pivalate on patch testing, were tested intradermally with a variety of hydrocortisone (HC) analogs. Substitution at the C_{21} position (sodium phosphate, acetate, succinate, aldehyde hydrate) had no effect on the occurrence of positive reactions to hydrocortisone (43/44 positive), whereas alteration of any of the carbon rings (prednisolone acetate, cortisone acetate, tetrahydrocortisone) significantly reduced the number of positive reactions (4/33 positive), suggesting that these are the allergenic parts of the molecule. The authors also suggested that it is the hydrocortisone molecule itself that is important to allergenicity and not an industrial precursor (aldehyde hydrate) or a product of enzymatic metabolism (cortisone, tetrahydrocortisone) (16).

Cutaneous adverse drug reactions from systemic administration caused by type IV (delayed-type) hypersensitivity
Cutaneous adverse drug reactions from systemic administration of hydrocortisone caused by type IV (delayed-type) hypersensitivity, including systemic contact dermatitis (9,13,14,53), and in addition occupational allergic contact dermatitis (52), are planned to be discussed in Volume IV of the Monographs in Contact Allergy series on Systemic drugs.

LITERATURE

1   Guin JD. Eyelid dermatitis: experience in 203 cases. J Am Acad Dermatol 2002;47:755-765
2   Cooper SM, Shaw S. Eyelid dermatitis: an evaluation of 232 patch test patients over 5 years. Contact Dermatitis 2000;42:291-293
3   Gilissen L, De Decker L, Hulshagen T, Goossens A. Allergic contact dermatitis caused by topical ophthalmic medications: Keep an eye on it! Contact Dermatitis 2019;80:291-297
4   Menezes de Padua CA, Uter W, Schnuch A. Contact allergy to topical drugs: prevalence in a clinical setting and estimation of frequency at the population level. Pharmacoepidemiol Drug Saf 2007;16:377-384
5   Menezes de Padua CA, Schnuch A, Nink K, Pfahlberg A, Uter W. Allergic contact dermatitis to topical drugs – Epidemiological risk assessment. Pharmacoepidemiol Drug Saf 2008;17:813-821
6   Baeck M, Goossens A. Immediate and delayed allergic hypersensitivity to corticosteroids: practical guidelines. Contact Dermatitis 2012;66:38-45
7   Miranda-Romero A, Sánchez-Sambucety P, Bajo C, Martinez M, Garcia-Munõz M. Genital oedema from contact allergy to prednicarbate. Contact Dermatitis 1998;38:228-229
8   Isaksson M, Persson L-M. Contact allergy to hydrocortisone and systemic contact dermatitis from prednisolone with tolerance of betamethasone. Am J Contact Dermat 1998;9:136-138
9   Torres V, Tavares-Bello R, Melo H, Soares AP. Systemic contact dermatitis from hydrocortisone. Contact Dermatitis 1993;29:106
10 Wilkinson SM, Beck MH. Allergic contact dermatitis from dibutyl phthalate, propyl gallate and hydrocortisone in Timodine. Contact Dermatitis 1992;27:197
11 Lindelöf B. Contact allergy to hydrocortisone. Contact Dermatitis 1988;18:309
12 Wilkinson SM, English JSC. Hydrocortisone sensitivity: clinical features of 59 cases. J Am Acad Dermatol 1992;27:683-687
13 Lauerma AI, Reitamo S, Maibach HI. Systemic hydrocortisone/cortisol induces allergic skin reactions in presensitized subjects. J Am Acad Dermatol 1991;24:182-185
14 Lauerma AI, Reitamo S, Maibach HI. Systemic hydrocortisone/cortisol induces allergic skin reactions in presensitized subjects. Am J Contact Dermat 1991;2:68 (Abstract)
15 Wilkinson SM, Cartwright PH, English JSC. Hydrocortisone: an important cutaneous allergen. Lancet 1991;337:761-762
16 Wilkinson SM, English JSC. Hydrocortisone sensitivity: an investigation into the nature of the allergen. Contact Dermatitis 1991;25:178-181
17 Wilkinson SM, English JSC. Hydrocortisone sensitivity: a prospective study into the value of tixocortol pivalate and hydrocortisone acetate as patch test markers. Contact Dermatitis 1991;25:132-133
18 Rietschel RL. Photocontact dermatitis to hydrocortisone. Contact Dermatitis 1978;4:334-337
19 Isaksson M. Triple sensitization in a child with chronic otitis externa. Contact Dermatitis 2002;47:172
20 Lauerma AI, Tarvainen K, Forström L, Reitamo S. Contact hypersensitivity to hydrocortisone-free-alcohol in patients with allergic patch test reactions to tixocortol pivalate. Contact Dermatitis 1993;28:10-14
21 Wilkinson M, Cartwright P, English JS. The significance of tixocortol-pivalate-positive patch tests in leg ulcer patients. Contact Dermatitis 1990;23:120-121
22 Foussereau J, Jelen G. Tixocortol pivalate – an allergen closely related to hydrocortisone. Contact Dermatitis 1986;15:37-38
23 Dooms-Goossens A, Verschaeve H, Degreef H, van Berendoncks J. Contact allergy to hydrocortisone and tixocortol pivalate: problems in the detection of corticosteroid sensitivity. Contact Dermatitis 1986;14:94-102
24 Baeck M, Chemelle JA, Terreux R, Drieghe J, Goossens A. Delayed hypersensitivity to corticosteroids in a series of 315 patients: clinical data and patch test results. Contact Dermatitis 2009;61:163-175
25 Baeck M, De Potter P, Goossens A. Allergic contact dermatitis following ocular use of corticosteroids. J Ocul Pharmacol Ther 2011;27:83-92
26 Uter W, de Pádua CM, Pfahlberg A, Nink K, Schnuch A, Lessmann H. Contact allergy to topical corticosteroids – results from the IVDK and epidemiological risk assessment. J Dtsch Dermatol Ges 2009;7:34-41
27 Mimesh S, Pratt M. Allergic contact dermatitis from corticosteroids: reproducibility of patch testing and correlation with intradermal testing. Dermatitis 2006;17:137-142
28 Wilkinson SM, Heagerty AHM, English JSC. A prospective study into the value of patch and intradermal tests in identifying topical corticosteroid allergy. Br J Dermatol 1992;12:22-25
29 Uter W, Geier J, Richter G, Schnuch A; IVDK Study Group, German Contact Dermatitis Research Group. Patch test results with tixocortol pivalate and budesonide in Germany and Austria. Contact Dermatitis 2001;44:313-314

30 Devos SA, Van der Valk PG. Relevance and reproducibility of patch-test reactions to corticosteroids. Contact Dermatitis 2001;44:362-365

31 Lyon CC, Beck MH. Allergic contact dermatitis reactions to corticosteroids in periorbital inflammation and conjunctivitis. Eye (Lond) 1998;12(Pt.1):148-149

32 Lauerma AI. Occupational contact sensitization to corticosteroids. Contact Dermatitis 1998;39:328-329

33 Seukeran DC, Wilkinson SM, Beck MH. Patch testing to detect corticosteroid allergy: is it adequate? Contact Dermatitis 1997;36:127-130

34 Rodríguez-Serna M, Silvestre JF, Quecedo E, Martínez A, Miguel FJ, Gauchía R. Corticosteroid allergy: report of 3 unusually acute cases. Contact Dermatitis 1996;35:361-362

35 Jagodzinski LJ, Taylor JS, Oriba H. Allergic contact dermatitis from topical corticosteroid preparations. Am J Contact Dermat 1995;6:67-74

36 Freeman S. Corticosteroid allergy. Contact Dermatitis 1995;33:240-242

37 Dooms-Goossens A, Meinardi MM, Bos JD, Degreef H. Contact allergy to corticosteroids: the results of a two-centre study. Br J Dermatol 1994;130:42-47

38 Burden AD, Beck MH. Contact hypersensitivity to topical corticosteroids. Br J Dermatol 1992;127:497-501

39 Lauerma AI. Contact hypersensitivity to glucocorticosteroids. Am J Contact Dermat 1992;3:112-132

40 Räsänen L, Tuomi ML. Cross-sensitization to mometasone furoate in patients with corticosteroid contact allergy. Contact Dermatitis 1992;27:323-325

41 Dooms-Goossens A, Morren M. Results of routine patch testing with corticosteroid series in 2073 patients. Contact Dermatitis 1992;26:182-191

42 Edwards M, Rudner E J. Dermatitis venenata due to hydrocortisone alcohol. Cutis 1970;6:757-758

43 Wilkinson RD, McGarry EM, Solomon S. Allergic contact dermatitis to hydrocortisone. J Invest Dermatol 1967;43:295

44 Coskey RJ. Contact dermatitis due to topical hydrocortisone and prednisolone. Michigan Med 1965;64:669-670

45 Sönnichsen N. Beitrag zur Hydrocortison Überempfindlichkeit. Hautarzt 1962;13:226-227 (Article in German)

46 Krook G. Contact dermatitis due to Ficortril® (hydrocortisone 1% ointment, Pfizer). Contact Dermatitis Newsletter 1974;15:460-461

47 Kooij R. Hypersensitivity to hydrocortisone. Br J Dermatol 1959;71:392-394

48 Goh CL. Cross-sensitivity to multiple topical corticosteroids. Contact Dermatitis 1989;20:65-67

49 Feldman SB, Sexton FM, Buzas J, Marks JG Jr. Allergic contact dermatitis from topical steroids. Contact Dermatitis 1988;19:226-228

50 Jelen G, Dooms-Goossens A. A propos de l'allergie à l'hydrocortisone. La lettre du GERDA 1986:3:76 (Article in French)

51 Alani MS, Alani SD. Allergic contact dermatitis to corticosteroids. Ann Allergy 1972;30:181-185

52 Gilissen L, Boeckxstaens E, Geebelen J, Goossens A. Occupational allergic contact dermatitis from systemic drugs. Contact Dermatitis 2020;82:24-30

53 Malik M, Tobin AM, Shanahan F, O'Morain C, Kirby B, Bourke J. Steroid allergy in patients with inflammatory bowel disease. Br J Dermatol 2007;157:967-969

Chapter 3.170 HYDROCORTISONE ACEPONATE

IDENTIFICATION

Description/definition : Hydrocortisone aceponate is the aceponate (acetate propionate) ester of the synthetic glucocorticoid hydrocortisone that conforms to the structural formula shown below

Pharmacological classes : Glucocorticoids

IUPAC name : [(8S,9S,10R,11S,13S,14S,17R)-17-(2-Acetyloxyacetyl)-11-hydroxy-10,13-dimethyl-3-oxo-2,6,7,8,9,11,12,14,15,16-decahydro-1H-cyclopenta[a]phenanthren-17-yl] propanoate

Other names : 21-(Acetyloxy)-11β-hydroxy-17α-(propionyloxy)-4-pregnene-3,20-dione

CAS registry number : 74050-20-7

EC number : Not available

Patch testing : In general, corticosteroids may be tested at 0.1% and 1% in alcohol; late readings (6-10 days) are strongly recommended

Molecular formula : $C_{26}H_{36}O_7$

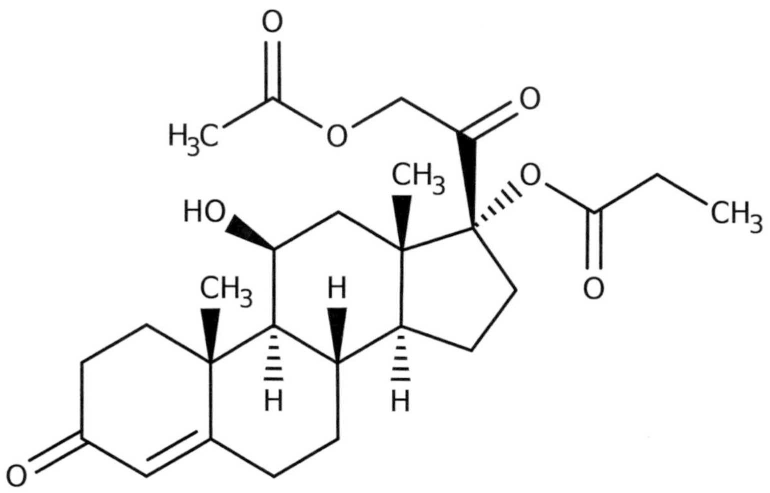

GENERAL

General aspects of corticosteroids used on the skin and mucous membranes are discussed in Chapter 2.4. A practical guideline for diagnosing allergic reactions to corticosteroids is presented in ref. 1. See also hydrocortisone (Chapter 3.169), hydrocortisone acetate (Chapter 3.171), hydrocortisone butyrate (Chapter 3.172), hydrocortisone hemisuccinate (Chapter 3.173), hydrocortisone probutate (Chapter 3.174), and hydrocortisone valerate (Chapter 3.175).

CONTACT ALLERGY

Case series

From January 1990 to June 2008, in Leuven, Belgium, 315 patients were diagnosed with contact allergy to/allergic contact dermatitis from corticosteroids (CSs) from routine patch testing with a baseline series including tixocortol pivalate, budesonide, hydrocortisone butyrate and prednisone caproate, patch testing with patients' own CS preparations, and testing those with proven contact allergy to a corticosteroid or strongly suspected of CS allergy later with a series of 66 CSs, including two sex hormones (progesterone and testosterone). 71% of the patients had relevant reactions, but these were not specified. In this group of 315 CS allergic patients, 86 had positive patch tests to hydrocortisone aceponate 1% alc. (4). As this corticosteroid has never been used in pharmaceuticals in Belgium, these positive reactions must all be considered cross-reactions to other corticosteroids.

Case reports

A 37-year-old woman presented with chronic hand dermatitis of 14 years duration. She was treated with several corticosteroids without success. When she started to apply 0.127% hydrocortisone aceponate cream, within 15 days she developed widespread plaques of eczema on her face, arms, trunk and thighs, in addition to an acute eczema of the dorsa of her hands. Patch tests were positive to hydrocortisone-17-butyrate, budesonide and the commercial cream. When tested with all its ingredients, there was a positive reaction to hydrocortisone aceponate 0.127% pet. (2).

A 50-year-old woman developed eczema on the backs of her legs, mainly in the popliteal fossae, while applying hydrocortisone aceponate cream to psoriatic lesions strictly limited to the lower back. There was no eczema on or around the site of application itself. An identical reaction occurred when betamethasone dipropionate was used to treat the psoriasis on the back. Extensive patch testing, including a corticosteroids series with dilution testing and the 2 commercial corticosteroid preparations used by the patient, revealed only negative reactions at day 2, 4 and 7. A repeated open application test with hydrocortisone aceponate cream and with hydrocortisone aceponate 0.127% elicited an annular positive reaction surrounding the site of application at D4. Some other corticosteroids were tested in a ROAT, showing a positive result to tixocortol pivalate 1% pet. on day 15 (3). This case was highly unusual in the sense that eczema only appeared at distant sites and not at the application site. Systemic contact dermatitis (localization in the popliteal fossae is suggestive) with suppression of the allergic reaction at the application site (and the patch tests) from the anti-inflammatory action of the corticosteroid is a possibility.

Cross-reactions, pseudo-cross-reactions and co-reactions
Cross-reactions between corticosteroids are discussed in Chapter 2.8.

LITERATURE

1   Baeck M, Goossens A. Immediate and delayed allergic hypersensitivity to corticosteroids: practical guidelines. Contact Dermatitis 2012;66:38-45
2   El Sayed F, Ammoury A, Launais F, Bazex J. Contact dermatitis to hydrocortisone aceponate in Efficort cream inducing widespread reaction. Contact Dermatitis 2005;53:242-243
3   Weber F, Barbaud A, Reichert-Penetrat S, Danchin A, Schmutz JL. Unusual clinical presentation in a case of contact dermatitis due to corticosteroids diagnosed by ROAT. Contact Dermatitis 2001;44:105-106
4   Baeck M, Chemelle JA, Terreux R, Drieghe J, Goossens A. Delayed hypersensitivity to corticosteroids in a series of 315 patients: clinical data and patch test results. Contact Dermatitis 2009;61:163-175

Chapter 3.171 HYDROCORTISONE ACETATE

IDENTIFICATION

Description/definition	: Hydrocortisone acetate is acetate ester of hydrocortisone that conforms to the structural formula shown below
Pharmacological classes	: Anti-inflammatory agents
IUPAC name	: [2-[(8S,9S,10R,11S,13S,14S,17R)-11,17-Dihydroxy-10,13-dimethyl-3-oxo-2,6,7,8,9,11,12, 14,15,16-decahydro-1H-cyclopenta[a]phenanthren-17-yl]-2-oxoethyl] acetate
Other names	: 21-Acetoxy-11β,17α-dihydroxypregn-4-ene-3,20-dione
CAS registry number	: 50-03-3
EC number	: 200-004-4
Merck Index monograph	: 6094 (Hydrocortisone)
Patch testing	: 1.0% pet. (Chemotechnique; SmartPracticeCanada); late readings (6-10 days) are strongly recommended; hydrocortisone acetate 1% in alc./DMSO is irritant (15)
Molecular formula	: $C_{23}H_{32}O_6$

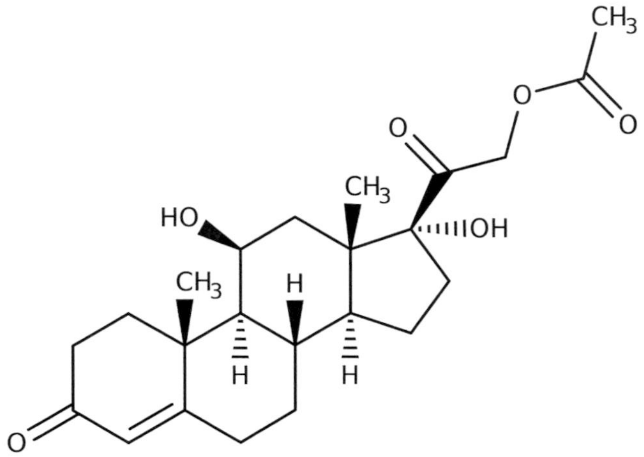

GENERAL

General aspects of corticosteroids used on the skin and mucous membranes are discussed in Chapter 2.4. A practical guideline for diagnosing allergic reactions to corticosteroids is presented in ref. 3. Patch testing wit hydrocortisone acetate is not a sensitive method to detect hypersensitivity to hydrocortisone (5,7). See also hydrocortisone (Chapter 3.169), hydrocortisone butyrate (Chapter 3.172), hydrocortisone aceponate (Chapter 3.170), hydrocortisone hemisuccinate (Chapter 3.173), hydrocortisone probutate (Chapter 3.174), and hydrocortisone valerate (Chapter 3.175).

CONTACT ALLERGY

Patch testing in groups of selected patients

In the period 2000 to 2005, in the USA, 1186 patients suspected of corticosteroid allergy were patch tested with hydrocortisone acetate 1% pet. and there were 8 (0.7%) positive reactions, all of which were considered to be relevant (14).

Case series

In Leuven, Belgium, between 1990 and 2017, 16,065 patients were investigated for contact allergy and 118 (0.7%) showed positive patch test reactions to topical ophthalmic medications and/or to their ingredients. Eighty-four individuals (71%) reacted to an active principle. Hydrocortisone acetate was tested in 1967 patients and was the allergen in eye medications in 6 (0.3%). There were also 28 reactions to hydrocortisone acetate in other types of medications (1; possibly overlap with refs. 2 and 13).

From January 1990 to June 2008, in Leuven, Belgium, 315 patients were diagnosed with contact allergy to/ allergic contact dermatitis from corticosteroids (CSs) from routine patch testing with a baseline series including tixocortol pivalate, budesonide, hydrocortisone butyrate and prednisone caproate, patch testing with patients' own

CS preparations, and testing those with proven contact allergy to a corticosteroid or strongly suspected of CS allergy later with a series of 66 CSs, including two sex hormones (progesterone and testosterone). 71% of the patients had relevant reactions, but these were not specified. In this group of 315 CS allergic patients, 48 had positive patch tests to hydrocortisone acetate 0.1% alc. It is unknown how many of these reactions were caused by the use of a pharmaceutical product containing hydrocortisone acetate and how many were cross-reactions to other corticosteroids (13; possibly overlap with refs. 1 and 2).

Between 1990 and 2003, in Leuven, Belgium, 92 patients were patch tested for chronic vulval complaints. Fifteen of these women were tested with a series of topical drugs; there were 4 reactions to hydrocortisone acetate 1% alc. of which 2 were relevant (2; possibly overlap with refs. 1 and 13).

From a clinic in Sweden, early 1970s, 21 patients with contact allergy to/allergic contact dermatitis from hydrocortisone and/or hydrocortisone acetate were reported. Despite the large number, only 0.3% of consecutive patients reacted to one or more corticosteroids (detailed data are not available to the author) (12).

Case reports

An 18-year-old woman presented with a pruritic eczematous eruption on the dorsum of the 4th proximal interdigital joint area and face, which developed after topically applying an ointment containing neomycin, *Centella asiatica* extract and hydrocortisone acetate. A provisional diagnosis of contact dermatitis and autosensitization dermatitis was made, and she was treated with topical methylprednisolone aceponate cream, which worsened the dermatitis. Patch tests were positive to hydrocortisone acetate 2% alc., methylprednisolone aceponate 1% alc., and also to neomycin and *Centella asiatica* extract (4).

A 44-year-old woman had allergic contact dermatitis from hydrocortisone in an antifungal-corticosteroid combination preparation and from hydrocortisone acetate in a cream and in an ophthalmic ointment. It proved very difficult to clearly demonstrate contact allergy to hydrocortisone (8).

In early literature (1959-1960), a series of five cases (6) and several single case reports (9,10,11) have described allergic contact dermatitis to hydrocortisone acetate ointment.

Cross-reactions, pseudo-cross-reactions and co-reactions
Cross-reactions between corticosteroids are discussed in Chapter 2.8.

LITERATURE
1   Gilissen L, De Decker L, Hulshagen T, Goossens A. Allergic contact dermatitis caused by topical ophthalmic medications: Keep an eye on it! Contact Dermatitis 2019;80:291-297
2   Nardelli A, Degreef H, Goossens A. Contact allergic reactions of the vulva: a 14-year review. Dermatitis 2004;15:131-136
3   Baeck M, Goossens A. Immediate and delayed allergic hypersensitivity to corticosteroids: practical guidelines. Contact Dermatitis 2012;66:38-45
4   Seok Oh C, Young Lee J. Contact allergy to various ingredients of topical medicaments. Contact Dermatitis 2003;49:49-50
5   Wilkinson SM, Cartwright PH, English JSC. Hydrocortisone: an important cutaneous allergen. Lancet 1991;337:761-762
6   Church R. Sensitivity to hydrocortisone acetate ointment. Br J Derm 1960;72:341-344
7   Wilkinson SM, English JSC. Hydrocortisone sensitivity: a prospective study into the value of tixocortol pivalate and hydrocortisone acetate as patch test markers. Contact Dermatitis 1991;25:132-3
8   Dooms-Goossens A, Verschaeve H, Degreef H, van Berendoncks J. Contact allergy to hydrocortisone and tixocortol pivalate: problems in the detection of corticosteroid sensitivity. Contact Dermatitis 1986;14:94-102
9   Burckhardt W. Kontaktekzem durch Hydrocortison. Hautarzt 1959;10:42-43 (Article in German)
10 Kooij R. Hypersensitivity to hydrocortisone. Br J Dermatol 1959;71:392-394
11 Dorn H. Kontaktallergie gegenüber Salben-Konservierungsmitteln und Hydrocortison. Z Hautkrankheiten 1959;27:305-310 (Article in German)
12 Alani MS, Alani SD. Allergic contact dermatitis to corticosteroids. Ann Allergy 1972;30:181-185
13 Baeck M, Chemelle JA, Terreux R, Drieghe J, Goossens A. Delayed hypersensitivity to corticosteroids in a series of 315 patients: clinical data and patch test results. Contact Dermatitis 2009;61:163-175
14 Davis MD, El-Azhary RA, Farmer SA. Results of patch testing to a corticosteroid series: a retrospective review of 1188 patients during 6 years at Mayo Clinic. J Am Acad Dermatol 2007;56:921-927
15 Devos SA, Van der Valk PG. Relevance and reproducibility of patch-test reactions to corticosteroids. Contact Dermatitis 2001;44:362-365

Chapter 3.172 HYDROCORTISONE BUTYRATE

IDENTIFICATION

Description/definition	: Hydrocortisone butyrate is the butyrate ester of the glucocorticoid hydrocortisone, that conforms to the structural formula shown below
Pharmacological classes	: Glucocorticoids; dermatologic agents
IUPAC name	: [(8S,9S,10R,11S,13S,14S,17R)-11-Hydroxy-17-(2-hydroxyacetyl)-10,13-dimethyl-3-oxo-2,6,7,8,9,11,12,14,15,16-decahydro-1H-cyclopenta[a]phenanthren-17-yl] butanoate
Other names	: 11β,17,21-Trihydroxypregn-4-ene-3,20-dione 17-butyrate; hydrocortisone 17-butyrate
CAS registry number	: 13609-67-1
EC number	: 237-093-4
Merck Index monograph	: 6094 (Hydrocortisone)
Patch testing	: 1.0% pet. (Chemotechnique, SmartPracticeCanada); 1% alc. (Chemotechnique, SmartPracticeCanada); 0.1% pet. (SmartPracticeCanada, SmartPracticeEurope); alcohol may be a better vehicle for testing hydrocortisone butyrate than petrolatum (37,38,53); late readings (6-10 days) are strongly recommended
Molecular formula	: $C_{25}H_{36}O_6$

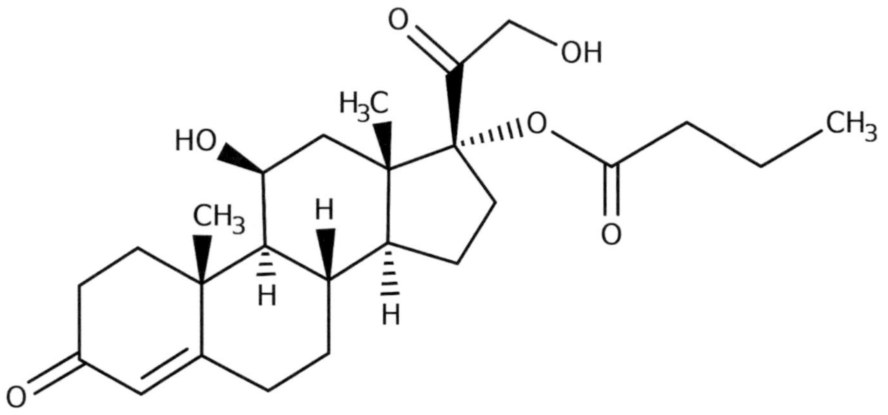

GENERAL

General aspects of corticosteroids used on the skin and mucous membranes are discussed in Chapter 2.4. A practical guideline for diagnosing allergic reactions to corticosteroids is presented in ref. 28. Hydrocortisone butyrate 1% pet. is included in the American core allergen series (www.smartpracticecanada.com) as screening agent for corticosteroid hypersensitivity (37). A considerable number of positive patch test reactions to hydrocortisone butyrate may be missed when readings are not performed at day 7 (22,23). MEK (methyl ethyl ketone) may be a suitable vehicle for patch testing hydrocortisone butyrate (32) and alcohol is superior over petrolatum (37,38,53). Intradermal testing with hydrocortisone butyrate may reveal additional cases of sensitization not picked up by patch testing (50,51). Hydrocortisone butyrate was suggested as a useful screening agent for corticosteroid allergy in 1986 (53), but was later replaced with tixocortol pivalate and budesonide.

See also hydrocortisone (Chapter 3.169), hydrocortisone acetate (Chapter 3.171), hydrocortisone aceponate (Chapter 3.170), hydrocortisone hemisuccinate (Chapter 3.173), hydrocortisone probutate (Chapter 3.174), and hydrocortisone valerate (Chapter 3.175).

CONTACT ALLERGY

Patch testing in the general population and in subgroups

The results of patch testing hydrocortisone butyrate in (subgroups of) the general population are shown in table 3.172.1. In adolescents from Sweden, the prevalence of positive patch tests was <0.1% (21) and in 8th grade school children from Denmark, not a single positive reaction was observed (19). Of adults from 5 European countries, 0.4% had positive reactions to hydrocortisone butyrate, 0.2% of the women and 0.6% of the men (18). It should be appreciated that the reactions were read at D3 or even once at D2. It is well known that many patch test reactions to corticosteroids first develop after D3-4, which means that a (possibly considerable) number of positive patch test reactions may have gone unnoticed.

Table 3.172.1 Contact allergy in the general population and in subgroups

Year and country	Selection and number tested	Prevalence of contact allergy			Comments	Ref.
		Total	Women	Men		
General population						
2008-11 five European countries	general population, random sample, 18-74 years, n=3119	0.4%	0.2%	0.6%	TRUE test	18
Subgroups						
2011-2013 Sweden	adolescents from a birth cohort, 15.8-18.9 years, n=2285	0.09%	0.08%	0.09%	TRUE test; patch tests were read at day 2 only	21
2010 Denmark	unselected population of 8th grade schoolchildren in Denmark, 15 years later; n=442	0%	0%	0%	follow-up study; TRUE test	19

With the CE-DUR approach, the incidence of sensitization to hydrocortisone butyrate in the German population was estimated to range from 1 to 2 cases/100,000/year in the period 1995-1999 and from 1 to 3 cases/100,000/year in the period 2000-2004 (25). Also in Germany, for the period 1995-2004, the population-based relative incidence (RI) of contact sensitization to hydrocortisone butyrate (cases/100,000 defined daily doses (DDDs) per year) was estimated to be 10.7. In the group of corticosteroids, the RI ranged from 0.3 (dexamethasone sodium phosphate) to 43.2 (budesonide) (26).

Patch testing in groups of patients
Results of patch testing hydrocortisone butyrate in consecutive patients suspected of contact dermatitis (routine testing) back to 2001 are shown in table 3.172.2. Results of testing in groups of *selected* patients (individuals with perianal dermatitis or leg ulcers, patient suspected of corticosteroid allergy) are shown in table 3.172.3.

Patch testing in consecutive patients suspected of contact dermatitis: routine testing
Hydrocortisone butyrate 1% has been present in the screening series of the North American Contact Dermatitis Group (NACDG, USA + Canada) since 2001, and 13 of the 19 studies shown in table 3.172.2 are from those countries. In these investigations, low prevalences of positive patch test reactions have been observed, ranging from 0.1% to 0.9%. Low rates were also reported from the Netherlands (0.4% [22]), Thailand (0.1% [24]), and Spain (0.3% [41]), with slightly higher frequencies in Belgium (1.2% [27]) and Denmark (1.0% [43]). A study from Poland had the highest

Table 3.172.2 Patch testing in groups of patients: Routine testing

Years and Country	Test conc. & vehicle	Number of patients tested	positive (%)	Selection of patients (S); Relevance (R); Comments (C)	Ref.
2008-18 Netherlands	0.1% alc.	3250	14 (0.4%)	R: not stated; C: >25% of the reactions may be missed when patch are not read at D7	22
2006-2018 Thailand	1% pet.	2803	2 (0.1%)	R: not stated	24
2015-2017 NACDG	1% pet.	5595	9 (0.2%)	R: definite + probable relevance: 33%	1
2015-2016 Spain		3699	12 (0.3%)	R: not stated; TRUE test or 1% alc. tested	41
2007-2016 USA	1% pet.	2312	(0.3%)	R: not stated	55
2011-2015 USA	1% alc.	2566	7 (0.3%)	R: not stated	2
2013-2014 NACDG	1% pet.	4858	5 (0.1%)	R: definite + probable relevance: 20%	3
2007-2014 NACDG		17,978	77 (0.4%)	R: definite + probable relevance 43%; C: tested 1% alc. or pet.	42
1990-2014 Belgium	0.1% alc.	12,311	151 (1.2%)	R: 96% of the positive patch test reactions to all topical drugs and antiseptics were considered to be relevant	27
2011-12 NACDG	1% pet.	4231	14 (0.3%)	R: definite + probable relevance: 43%	4
2009-10 NACDG	1% pet.	4304	(0.7%)	R: definite + probable relevance: 53%	5
2006-2010 USA	1% alc.	3058	(0.9%)	R: 55%	6
2007-2008 NACDG	1% pet.	4937	(0.5%)	R: definite + probable relevance: 39%	7
2005-2008 Denmark		3594	34 (1.0%)	R: not stated	43
2005-200 NACDG	1% pet.	4438	(0.6%)	R: definite + probable relevance: 27%	8
<2005 Poland	0.1% pet.	257	7 (2.7%)	R: 100%	46
2001-2005 USA	1% alc.	2796	(0.8%)	R: 59%	9
2003-2004 NACDG	1% pet.	5140	25 (0.3%)	R: not stated	10
2001-2002 NACDG	1% pet.	4888	(0.5%)	R: definite + probable relevance: 23%	11

IVDK: Information Network of Departments of Dermatology (Germany, Austria, Switzerland); NACDG: North American Contact Dermatitis Group (USA, Canada)

score (2.7%), but the number of patients (n=257) was very small (46). In the NACDG studies, percentages definite + probable relevance were mostly in the 30-40% range. In other investigations, relevance rates, where mentioned, were 55-100% (6,9,46).

Patch testing in groups of selected patients
In groups of selected patients, frequencies of positive reactions to hydrocortisone butyrate have ranged from 0.5% to 3.4%. The highest rate was seen in patients with anogenital signs or symptoms with 5 positive reactions in 146 patients. This percentage was significantly higher than in a control group, but only one of the 5 patch test reactions was relevant. Possibly, hydrocortisone butyrate acted as an indicator for other relevant corticosteroids here (15). Relevance data were hardly ever provided, but in one study from the USA, 24 of 25 positive patch tests were considered to be relevant (relevance rates from this group are always much higher than from other investigators) (40).

Table 3.172.3 Patch testing in groups of patients: Selected patient groups

Years and Country	Test conc. & vehicle	Number of patients tested | positive (%)		Selection of patients (S); Relevance (R); Comments (C)	Ref.
2003-2014 IVDK	0.1% pet.	1133	(1.2%)	S: patients with stasis dermatitis/chronic leg ulcers; R: not stated	12
2005-2008 France	0.1% pet.	423	7 (1.7%)	S: patients with leg ulcers; R: not stated	13
2004-2008 IVDK			(0.9%)	S: patients with anogenital dermatoses tested with a medicament series; R: not stated; C: number of patients tested unknown	17
2000-2005 USA	1% alc.	1171	25 (2.1%)	S: patients suspected of corticosteroid allergy; R: 96%	40
1995-2004 IVDK	0.1% pet.	6022	(2.5%)	S: not stated; R: not stated	25
1995-2004 IVDK	0.1% pet.	5817	87 (1.5%)	S: patients tested with a corticosteroid series; R: not stated	44
1994-2004 NACDG	1% pet.	146	5 (3.4%)	S: patients with anogenital signs or symptoms; R: 20%; C: the frequency was significantly higher than in a control group	15
1990-2003 IVDK		193	3 (1.6%)	S: patients with perianal dermatoses; R: not stated; C: the frequency was not higher than in a control group	16
1997-2001 U.K.		200	1 (0.5%)	S: patients with venous or mixed venous/arterial leg ulcers; R: all reactions to topical drugs were considered to be of probable, past or current relevance	14
1995-9 Netherlands	0.1% pet.	216	1 (0.5%)	S: not stated; R: possible	48
1996-1997 IVDK	0.1% pet.	608	11 (1.8%)	S: patients tested with a corticosteroid series; R: not stated	47

IVDK: Information Network of Departments of Dermatology (Germany, Austria, Switzerland); NACDG: North American Contact Dermatitis Group (USA, Canada)

Case series
From January 1990 to June 2008, in Leuven, Belgium, 315 patients were diagnosed with contact allergy to/allergic contact dermatitis from corticosteroids (CSs) from routine patch testing with a baseline series including tixocortol pivalate, budesonide, hydrocortisone butyrate and prednisone caproate, patch testing with patients' own CS preparations, and testing those with proven contact allergy to a corticosteroid or strongly suspected of CS allergy later with a series of 66 CSs, including two sex hormones (progesterone and testosterone). 71% of the patients had relevant reactions, but these were not specified. In this group of 315 CS allergic patients, 96 had positive patch tests to hydrocortisone butyrate 0.1% alc. (39, overlap with ref. 20). It is unknown how many of these reactions were caused by the use of a pharmaceutical product containing hydrocortisone butyrate and how many were cross-reactions to other corticosteroids.

Between 1990 and 2003, also in Leuven, Belgium, 92 patients were patch tested for chronic vulval complaints. Fifteen of these women were tested with a series of topical drugs; there were 3 reactions to hydrocortisone butyrate of which 2 were relevant (20, overlap with ref. 39). In the period 1982-1994, in a center in Tokyo, 11 patients were diagnosed with allergic contact dermatitis from topical corticosteroids, including one case caused by hydrocortisone butyrate (54). In reviewing the Japanese literature up to 1994, 43 patients with allergic contact dermatitis from topical corticosteroids were identified, including 5 caused by hydrocortisone butyrate (54).

In a period of 2.5 years (1985-1987), in Helsinki, Finland, 36 patients (24 women, age range 13-74 years) with contact allergy to hydrocortisone butyrate (HB) were collected by routine testing, testing in patients previously having shown a positive or ?+ reaction to HB, patients suspected of corticosteroid allergy and patients with known allergy to other steroids. 1% alcohol was the best test concentration and vehicle; using pet. as vehicle resulted in fewer positive results. Seventeen individuals (47%) had actually used HB preparations and also 17 had noted irritation from steroid creams or ointments. Twenty-eight (78%) also reacted to one or more of a series of 8 other corticosteroids, most often betamethasone valerate (n=8). Upon retesting in 13 patients, the allergy to HB was reproduced in 11 (85%). A ROAT with 3 times daily application of HB 0.1% ointment and 1% or 0.1% HB in alcohol for a maximum of 10-12 days was positive in 8 of 20 patients tested (40%). In all but 3 patients, the clinical condition improved or healed after discontinuation of corticosteroids to which contact allergy had been demonstrated (37).

Case reports

A 17-year-old female patient presented with acute erythematous and edematous patches on both cheeks characterized by well-defined borders and by the presence of a few papulopustules and vesicles, that had appeared about 24 hours earlier. The patient had applied and massaged hydrocortisone 17-butyrate cream on her cheeks to treat a mild erythrosis about 48 hours earlier. Patch tests were positive to a corticosteroid mix containing 1% hydrocortisone butyrate and to hydrocortisone butyrate 1% pet. It was not mentioned how the patient had become sensitized to the corticosteroid (29).

A 62-year-old man presented with widespread numerous symmetrically distributed non-follicular pustules on an erythematous base, primarily on his forearms and antecubital fossae and a crusted dermatitis affecting his right foot, thighs and both hands. No bacteria or fungi were cultured from the pustules. Histopathology was consistent with acute generalize exanthematous pustulosis (AGEP) and it was suspected that this had been caused by the antibiotics that the patient had received for the lesions on his foot. He was treated with topical betamethasone valerate and oral prednisolone and 4 days later the skin had recovered almost completely. Patch tests were positive to budesonide and hydrocortisone butyrate, pustules were not observed. The patient now told that, in addition to the oral antibiotics, he had used a topical corticosteroid containing hydrocortisone butyrate from the early onset of the skin symptoms. Tests for penicillin allergy were negative and it was concluded that the patient's AGEP-like eruption was a manifestation of allergic contact dermatitis to hydrocortisone butyrate (45).

Short summaries of other case reports of allergic contact dermatitis from hydrocortisone butyrate are shown in table 3.172.4.

Table 3.172.4 Short summaries of case reports of allergic contact dermatitis from hydrocortisone butyrate

Year and country	Sex	Age	Positive patch tests	Clinical data and comments	Ref.
1998 Poland	M	23	HB 0.1%	acute dermatitis of penis and scrotum with phimosis from HB cream; provocation test on previously affected skin positive within hours	49
1997 Italy	F	15	HB cream 'as is'; HB 1% alc.	worsening of atopic dermatitis; cross-reaction to 6α-methylprednisolone aceponate	31
1995 Mexico	F	25	HB 1% alc.	dermatitis of the face, earlobes and neck with erythema, vesicles, oozing and itching; also allergy to alclometasone	30
1991 Germany	M	48	HB comm prep 'as is'; HB 10% pet.	worsening of chronic dermatitis	52
1985 Japan	M	52	HB cream 'as is'; HB 0.1%	erythema and vesicles from HB cream applied to a burn; also contact allergy to parabens in the cream	33
1980 United Kingdom	F	51	HB cream and ointment 'as is'; vehicles negative	worsening and spreading of stasis dermatitis	34
	F	48	HB cream and ointment 'as is'; vehicles negative	exacerbation of eczema surrounding post-thrombotic leg ulcer	
1979 Spain	F	52	HB cream and ointment 'as is'; HB 0.1% pet.	erythema and pruritus of the ankle and foot from HB ointment on stasis dermatitis	35
	F	61	HB 0.1% pet.	worsening of stasis dermatitis	
1979 The Netherlands	F	28	HB cream, ointment and liniment 'as is'; HB 1% and 0.1% pet.	severe eczema complicating atopic dermatitis	36

HB: hydrocortisone butyrate; comm prep: commercial preparation

Cross-reactions, pseudo-cross-reactions and co-reactions
Cross-reactions between corticosteroids are discussed in Chapter 2.8.

LITERATURE

1   DeKoven JG, Warshaw EM, Zug KA, Maibach HI, Belsito DV, Sasseville D, et al. North American Contact Dermatitis Group patch test results: 2015-2016. Dermatitis 2018;29:297-309
2   Veverka KK, Hall MR, Yiannias JA, Drage LA, El-Azhary RA, Killian JM, et al. Trends in patch testing with the Mayo Clinic standard series, 2011-2015. Dermatitis 2018;29:310-315
3   DeKoven JG, Warshaw EM, Belsito DV, Sasseville D, Maibach HI, Taylor JS, et al. North American Contact Dermatitis Group Patch Test Results: 2013-2014. Dermatitis 2017;28:33-46

4   Warshaw EM, Maibach HI, Taylor JS, Sasseville D, DeKoven JG, Zirwas MJ, et al. North American Contact Dermatitis Group patch test results: 2011-2012. Dermatitis 2015;26:49-59

5   Warshaw EM, Belsito DV, Taylor JS, Sasseville D, DeKoven JG, Zirwas MJ, et al. North American Contact Dermatitis Group patch test results: 2009 to 2010. Dermatitis 2013;24:50-59

6   Wentworth AB, Yiannias JA, Keeling JH, Hall MR, Camilleri MJ, Drage LA, et al. Trends in patch-test results and allergen changes in the standard series: a Mayo Clinic 5-year retrospective review (January 1, 2006, to December 31, 2010). J Am Acad Dermatol 2014;70:269-275

7   Fransway AF, Zug KA, Belsito DV, Deleo VA, Fowler JF Jr, Maibach HI, et al. North American Contact Dermatitis Group patch test results for 2007-2008. Dermatitis 2013;24:10-21

8   Zug KA, Warshaw EM, Fowler JF Jr, Maibach HI, Belsito DL, Pratt MD, et al. Patch-test results of the North American Contact Dermatitis Group 2005-2006. Dermatitis 2009;20:149-160

9   Davis MD, Scalf LA, Yiannias JA, Cheng JF, El-Azhary RA, Rohlinger AL, et al. Changing trends and allergens in the patch test standard series. Arch Dermatol 2008;144:67-72

10 Warshaw EM, Belsito DV, DeLeo VA, Fowler JF Jr, Maibach HI, Marks JG, et al. North American Contact Dermatitis Group patch-test results, 2003-2004 study period. Dermatitis 2008;19:129-136

11 Pratt MD, Belsito DV, DeLeo VA, Fowler JF Jr, Fransway AF, Maibach HI, et al. North American Contact Dermatitis Group patch-test results, 2001-2002 study period. Dermatitis 2004;15:176-183

12 Erfurt-Berge C, Geier J, Mahler V. The current spectrum of contact sensitization in patients with chronic leg ulcers or stasis dermatitis - new data from the Information Network of Departments of Dermatology (IVDK). Contact Dermatitis 2017;77:151-158

13 Barbaud A, Collet E, Le Coz CJ, Meaume S, Gillois P. Contact allergy in chronic leg ulcers: results of a multicentre study carried out in 423 patients and proposal for an updated series of patch tests. Contact Dermatitis 2009;60:279-287

14 Tavadia S, Bianchi J, Dawe RS, McEvoy M, Wiggins E, Hamill E, et al. Allergic contact dermatitis in venous leg ulcer patients. Contact Dermatitis 2003;48:261-265

15 Warshaw EM, Furda LM, Maibach HI, Rietschel RL, Fowler JF Jr, Belsito DV, et al. Anogenital dermatitis in patients referred for patch testing: retrospective analysis of cross-sectional data from the North American Contact Dermatitis Group, 1994-2004. Arch Dermatol 2008;144:749-755

16 Kügler K, Brinkmeier T, Frosch PJ, Uter W. Anogenital dermatoses—allergic and irritative causative factors. Analysis of IVDK data and review of the literature. J Dtsch Dermatol Ges 2005;3:979-986

17 Bauer A. Contact sensitization in the anal and genital area. Curr Probl Dermatol 2011;40:133-141

18 Diepgen TL, Ofenloch RF, Bruze M, Bertuccio P, Cazzaniga S, Coenraads P-J, et al. Prevalence of contact allergy in the general population in different European regions. Br J Dermatol 2016;174:319-329

19 Mortz CG, Bindslev-Jensen C, Andersen KE. Prevalence, incidence rates and persistence of contact allergy and allergic contact dermatitis in The Odense Adolescence Cohort Study: a 15-year follow-up. Br J Dermatol 2013;168:318-325

20 Nardelli A, Degreef H, Goossens A. Contact allergic reactions of the vulva: a 14-year review. Dermatitis 2004;15:131-136

21 Lagrelius M, Wahlgren CF, Matura M, Kull I, Lidén C. High prevalence of contact allergy in adolescence: results from the population-based BAMSE birth cohort. Contact Dermatitis 2016;74:44-51

22 van Amerongen CCA, Ofenloch R, Dittmar D, Schuttelaar MLA. New positive patch test reactions on day 7—The additional value of the day 7 patch test reading. Contact Dermatitis 2019;81:280-287.

23 Madsen JT, Andersen KE. Outcome of a second patch test reading of TRUE Tests® on D6/7. Contact Dermatitis 2013;68:94-97

24 Sukakul T, Chaweekulrat P, Limphoka P, Boonchai W. Changing trends of contact allergens in Thailand: A 12-year retrospective study. Contact Dermatitis 2019;81:124-129

25 Menezes de Padua CA, Uter W, Schnuch A. Contact allergy to topical drugs: prevalence in a clinical setting and estimation of frequency at the population level. Pharmacoepidemiol Drug Saf 2007;16:377-384

26 Menezes de Padua CA, Schnuch A, Nink K, Pfahlberg A, Uter W. Allergic contact dermatitis to topical drugs – Epidemiological risk assessment. Pharmacoepidemiol Drug Saf 2008;17:813-821

27 Gilissen L, Goossens A. Frequency and trends of contact allergy to and iatrogenic contact dermatitis caused by topical drugs over a 25-year period. Contact Dermatitis 2016;75:290-302

28 Baeck M, Goossens A. Immediate and delayed allergic hypersensitivity to corticosteroids: practical guidelines. Contact Dermatitis 2012;66:38-45

29 D'Erme AM, Gola M. Allergic contact dermatitis induced by topical hydrocortisone-17-butyrate mimicking papular rosacea. Dermatitis 2012;23:95-96

30 Blancas R, Ancona A, Arévalo A. Allergic contact dermatitis from alclomethasone dipropionate. Am J Contact Dermat 1995;6:115-116

31 Balato N, Patruno C, Lembo G, Cuccurullo FM, Ayala F. Contact sensitization to 6α-methylprednisolone aceponate. Am J Contact Dermat 1997;8:24-25

32 Wilkinson SM, Beck MH, Steel I. Results of patch testing with hydrocortisone butyrate in different vehicles. Contact Dermatitis 2000;42:299-300

33 Yoshikawa K, Watanabe K, Mizuno N. Contact allergy to hydrocortisone 17-butyrate and pyridoxine hydrochloride. Contact Dermatitis 1985;12:55-56

34 Brown R. Allergy to hydrocortisone-17-butyrate. Contact Dermatitis 1980;6:504-505

35 Brandao FM, Camarasa FM. Contact allergy to hydrocortisone 17-butyrate. Contact Dermatitis 1979;5:354-356

36 Van Waadenoijen Kernekamp AS, van Ketel WG. Contact allergy to hydrocortisone-17-butyrate. Contact Dermatitis 1979;5:268-269

37 Reitamo S, Lauerma AI, Förström L. Detection of contact hypersensitivity to topical corticosteroids with hydrocortisone-17-butyrate. Contact Dermatitis 1989;21:159-165

38 Wilkinson SM, Beck M H. Corticosteroid contact hypersensitivity: what vehicle and concentration? Contact Dermatitis 1996;34:305-308

39 Baeck M, Chemelle JA, Terreux R, Drieghe J, Goossens A. Delayed hypersensitivity to corticosteroids in a series of 315 patients: clinical data and patch test results. Contact Dermatitis 2009;61:163-175

40 Davis MD, El-Azhary RA, Farmer SA. Results of patch testing to a corticosteroid series: a retrospective review of 1188 patients during 6 years at Mayo Clinic. J Am Acad Dermatol 2007;56:921-927

41 Mercader-García P, Pastor-Nieto MA, García-Doval I, Giménez-Arnau A, González-Pérez R, Fernández-Redondo V, et al. GEIDAC. Are the Spanish baseline series markers sufficient to detect contact allergy to corticosteroids in Spain? A GEIDAC prospective study. Contact Dermatitis 2018;78:76-82

42 Pratt MD, Mufti A, Lipson J, Warshaw EM, Maibach HI, Taylor JS, et al. Patch test reactions to corticosteroids: Retrospective analysis from the North American Contact Dermatitis Group 2007-2014. Dermatitis 2017;28:58-63

43 Vind-Kezunovic D, Johansen JD, Carlsen BC. Prevalence of and factors influencing sensitization to corticosteroids in a Danish patch test population. Contact Dermatitis 2011;64:325-329

44 Uter W, de Pádua CM, Pfahlberg A, Nink K, Schnuch A, Lessmann H. Contact allergy to topical corticosteroids – results from the IVDK and epidemiological risk assessment. J Dtsch Dermatol Ges 2009;7:34-41

45 Broesby-Olsen S, Clemmensen O, Andersen KE. Allergic contact dermatitis from a topical corticosteroid mimicking acute generalized exanthematous pustulosis. Acta Derm Venereol 2005;85:444-445

46 Reduta T, Laudanska H. Contact hypersensitivity to topical corticosteroids – frequency of positive reactions in patch-tested patients with allergic contact dermatitis. Contact Dermatitis 2005;52:109-110

47 Uter W, Geier J, Richter G, Schnuch A; IVDK Study Group, German Contact Dermatitis Research Group. Patch test results with tixocortol pivalate and budesonide in Germany and Austria. Contact Dermatitis 2001;44:313-314

48 Devos SA, Van der Valk PG. Relevance and reproducibility of patch-test reactions to corticosteroids. Contact Dermatitis 2001;44:362-365

49 Rudzki E, Rebandel P. Patch tests with corticosteroids in Poland. Contact Dermatitis 1998;39:269

50 Seukeran DC, Wilkinson SM, Beck MH. Patch testing to detect corticosteroid allergy: is it adequate? Contact Dermatitis 1997;36:127-130

51 Wilkinson SM, English JS. Patch tests are poor detectors of corticosteroid allergy. Contact Dermatitis 1992;26:67-68

52 Dunkel FG, Elsner P, Burg G. Contact allergies to topical corticosteroids: 10 cases of contact dermatitis. Contact Dermatitis 1991;25:97-103

53 Reitamo S, Lauerma AI, Stubb S, Käyhkö K, Visa K, Förström L. Delayed hypersensitivity to topical corticosteroids. J Am Acad Dermatol 1986;14:582-589

54 Oh-i T. Contact dermatitis due to topical steroids with conceivable cross reactions between topical steroid preparations. J Dermatol 1996;23:200-208

55 Tam I, Schalock PC, González E, Yu J. Patch testing results from the Massachusetts General Hospital Contact Dermatitis Clinic, 2007-2016. Dermatitis 2020;31:202-208

Chapter 3.173 HYDROCORTISONE HEMISUCCINATE

IDENTIFICATION

Description/definition : Hydrocortisone hemisuccinate is the hemisuccinate ester of the synthetic glucocorticoid hydrocortisone that conforms to the structural formula shown below

Pharmacological classes : Glucocorticoids; anti-inflammatory agents

IUPAC name : 4-[2-[[(8S,9S,10R,11S,13S,14S,17R)-11,17-Dihydroxy-10,13-dimethyl-3-oxo-2,6,7,8,9,11,12,14,15,16-decahydro-1H-cyclopenta[a]phenanthren-17-yl]-2-oxoethoxy]-4-oxobutanoic acid

Other names : Hydrocortisone hemisuccinate anhydrous; (11β)-21-(3-carboxy-1-oxopropoxy)-11,17-dihydroxypreg-4-ene-3,20-dione; cortisol hemisuccinate; hydrocortisone succinate

CAS registry number : 2203-97-6

EC number : 218-612-3 (hydrogen hemisuccinate)

Patch testing : In general, corticosteroids may be tested at 0.1% and 1% in alcohol; late readings (6-10 days) are strongly recommended

Molecular formula : $C_{25}H_{34}O_8$

In pharmaceutical preparations, hydrocortisone hemisuccinate is present as hydrocortisone sodium succinate (CAS number 125-04-2, EC number 204-725-5, molecular formula $C_{25}H_{33}NaO_8$).

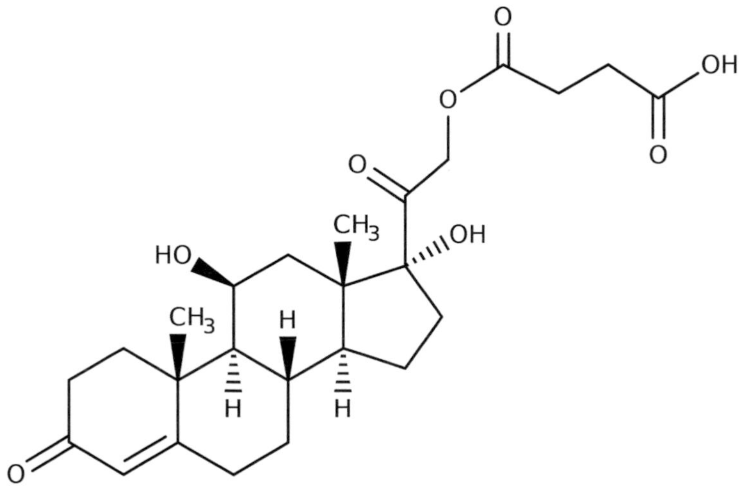

GENERAL

General aspects of corticosteroids used on the skin and mucous membranes are discussed in Chapter 2.4. A practical guideline for diagnosing allergic reactions to corticosteroids is presented in ref. 1. See also hydrocortisone (Chapter 3.169), hydrocortisone acetate (Chapter 3.171), hydrocortisone butyrate (Chapter 3.172), hydrocortisone aceponate (Chapter 3.170), hydrocortisone probutate (Chapter 3.174), and hydrocortisone valerate (Chapter 3.175).

CONTACT ALLERGY

Case reports and case series

From January 1990 to June 2008, in Leuven, Belgium, 315 patients were diagnosed with contact allergy to/allergic contact dermatitis from corticosteroids (CSs) from routine patch testing with a baseline series including tixocortol pivalate, budesonide, hydrocortisone butyrate and prednisone caproate, patch testing with patients' own CS preparations, and testing those with proven contact allergy to a corticosteroid or strongly suspected of CS allergy later with a series of 66 CSs, including two sex hormones (progesterone and testosterone). 71% of the patients had relevant reactions, but these were not specified. In this group of 315 CS allergic patients, 15 had positive patch tests to hydrocortisone hemisuccinate 0.1% alc. (3). It is unknown how many of these reactions were caused by the use of a pharmaceutical product containing hydrocortisone hemisuccinate and how many were cross-reactions to other corticosteroids.

Cross-reactions, pseudo-cross-reactions and co-reactions
Cross-reactions between corticosteroids are discussed in Chapter 2.8.

OTHER INFORMATION
Eleven patients, positive to tixocortol pivalate on patch testing, were tested intradermally with the hydrocortisone (HC) analogs HC acetate, HC sodium phosphate, HC succinate and HC aldehyde hydrate. 43/44 reactions were positive, there was only one negative reaction to HC acetate. It was concluded that substitution at the C_{21} position had no effect on the occurrence of positive reactions to hydrocortisone (2).

LITERATURE
1   Baeck M, Goossens A. Immediate and delayed allergic hypersensitivity to corticosteroids: practical guidelines. Contact Dermatitis 2012;66:38-45
2   Wilkinson SM, English JSC. Hydrocortisone sensitivity: an investigation into the nature of the allergen. Contact Dermatitis 1991;25:178-181
3   Baeck M, Chemelle JA, Terreux R, Drieghe J, Goossens A. Delayed hypersensitivity to corticosteroids in a series of 315 patients: clinical data and patch test results. Contact Dermatitis 2009;61:163-175

Chapter 3.174 HYDROCORTISONE PROBUTATE

IDENTIFICATION

Description/definition : Hydrocortisone probutate is the probutate (butyrate propionate) ester of hydrocortisone that conforms to the structural formula shown below

Pharmacological classes : Anti-inflammatory agents

IUPAC name : [(8S,9S,10R,11S,13S,14S,17R)-11-Hydroxy-10,13-dimethyl-3-oxo-17-(2-propanoyl-oxyacetyl)-2,6,7,8,9,11,12,14,15,16-decahydro-1H-cyclopenta[a]phenanthren-17-yl] butanoate

Other names : Hydrocortisone buteprate; hydrocortisone butyrate propionate; 11β,17,21-trihydroxy-pregn-4-ene-3,20-dione 17-butyrate 21-propionate; hydrocortisone 17-butyrate 21-propionate

CAS registry number : 72590-77-3

EC number : 276-726-9

Patch testing : In general, corticosteroids may be tested at 0.1% and 1% in alcohol; late readings (6-10 days) are strongly recommended

Molecular formula : $C_{28}H_{40}O_7$

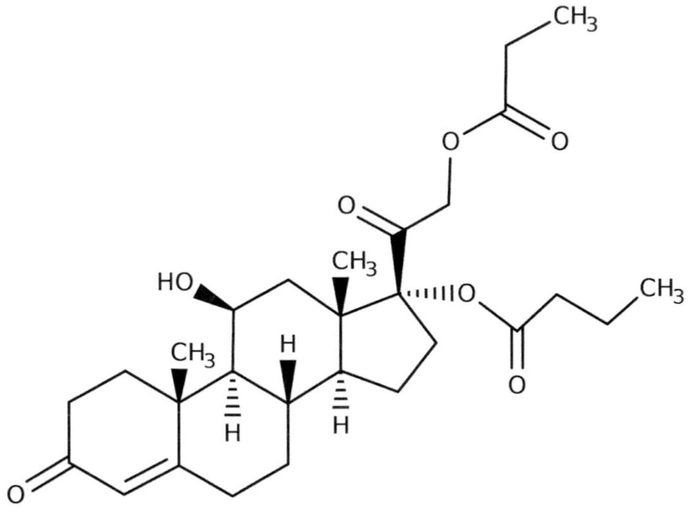

GENERAL

General aspects of corticosteroids used on the skin and mucous membranes are discussed in Chapter 2.4. A practical guideline for diagnosing allergic reactions to corticosteroids is presented in ref. 1. See also hydrocortisone (Chapter 3.169), hydrocortisone acetate (Chapter 3.171), hydrocortisone butyrate (Chapter 3.172), hydrocortisone aceponate (Chapter 3.170), hydrocortisone hemisuccinate (Chapter 3.173) and hydrocortisone valerate (Chapter 3.175).

CONTACT ALLERGY

Patch testing in groups of patients

In a group of 203 patients with eyelid dermatitis, one patient reacted to hydrocortisone probutate (test concentration and vehicle not mentioned); the relevance was not specified and it is unknown how many individuals were tested with this material (5).

Case reports

A 31-year-old Japanese woman with atopic dermatitis was treated by topical application of hydrocortisone butyrate propionate (hydrocortisone probutate) ointment. Soon after its start, she developed edematous erythema with pustules at the sites of application, which spread rapidly within two days to involve large parts of the body, associated with high-grade fever. At the first examination, erythematous papules and irregular erythematous lesions with purpura were noted on the back and legs. Numerous pustules were also observed on the neck, axilla and arms.

Laboratory studies revealed elevated white blood cell and neutrophil counts with a slight increase in the eosinophil count; bacterial cultures of blood, tonsillar swabs and material obtained from the pustules were negative. Patch tests were positive to hydrocortisone probutate 0.01% pet., hydrocortisone butyrate 0.1% pet. and prednisolone valerate acetate 0.1% pet. (all 3 have an ester substitution at the C17 position). The patient was diagnosed with allergic contact dermatitis from hydrocortisone probutate mimicking acute generalized exanthematous pustulosis (3).

In reviewing the Japanese literature up to 1994, 43 patients with allergic contact dermatitis from topical corticosteroid were identified, including 3 caused by hydrocortisone probutate (hydrocortisone butyrate propionate) (2).

Cross-reactions, pseudo-cross-reactions and co-reactions
Cross-reactions between corticosteroids are discussed in Chapter 2.8. A patient allergic to prednisolone valerate acetate may have cross-reacted to hydrocortisone butyrate propionate (hydrocortisone probutate (4).

LITERATURE
1   Baeck M, Goossens A. Immediate and delayed allergic hypersensitivity to corticosteroids: practical guidelines. Contact Dermatitis 2012;66:38-45
2   Oh-i T. Contact dermatitis due to topical steroids with conceivable cross reactions between topical steroid preparations. J Dermatol 1996;23:200-208
3   Tohgi N, Eto H, Maejima H, Saito N, Nakamura M, Katsuoka K. Allergic contact dermatitis induced by topical hydrocortisone butyrate propionate mimicking acute generalized exanthematous pustulosis. Eur J Dermatol 2009;19:518-519
4   Ogura, Natsuaki M, Hirano A, Yasugi Y, Miyata A, Yamanishi K. A case of contact dermatitis due to prednisolone valerate acetate. Hifu no kagaku 2005;4:111-115 (Article in Japanese)
5   Guin JD. Eyelid dermatitis: experience in 203 cases. J Am Acad Dermatol 2002;47:755-765

Chapter 3.175 HYDROCORTISONE VALERATE

IDENTIFICATION

Description/definition : Hydrocortisone valerate is the valerate ester of the synthetic glucocorticoid hydrocortisone that conforms to the structural formula shown below

Pharmacological classes : Anti-inflammatory agents; glucocorticoids

IUPAC name : [(8S,9S,10R,11S,13S,14S,17R)-11-Hydroxy-17-(2-hydroxyacetyl)-10,13-dimethyl-3-oxo-2,6,7,8,9,11,12,14,15,16-decahydro-1H-cyclopenta[a]phenanthren-17-yl] pentanoate

Other names : Cortisol 17-valerate; hydrocortisone 17-valerate; 11β,17,21-trihydroxypregn-4-ene-3,20-dione 17-valerate

CAS registry number : 57524-89-7

EC number : 260-786-8

Merck Index monograph : 6094 (Hydrocortisone)

Patch testing : In general, corticosteroids may be tested at 0.1% and 1% in alcohol; late readings (6-10 days) are strongly recommended

Molecular formula : $C_{26}H_{38}O_6$

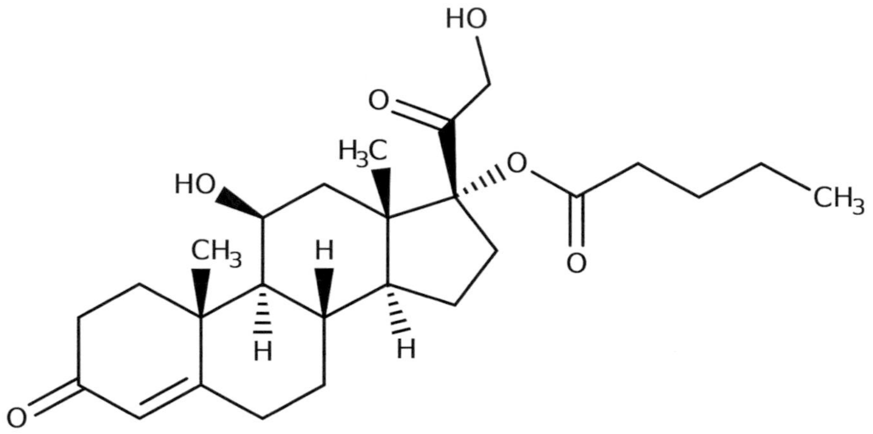

GENERAL

General aspects of corticosteroids used on the skin and mucous membranes are discussed in Chapter 2.4. A practical guideline for diagnosing allergic reactions to corticosteroids is presented in ref. 1. See also hydrocortisone (Chapter 3.169), hydrocortisone acetate (Chapter 3.171), hydrocortisone butyrate (Chapter 3.172), hydrocortisone aceponate (Chapter 3.170), hydrocortisone hemisuccinate (Chapter 3.173), and hydrocortisone probutate (Chapter 3.174).

CONTACT ALLERGY

Case reports and case series

In a group of 203 patients with eyelid dermatitis, two patients had positive patch tests to hydrocortisone valerate (test concentration and vehicle not mentioned); the relevance was not specified and it is unknown how many individuals were tested with hydrocortisone valerate (4).

A 60-year-old man had a small area of dermatitis on his lower leg. He was initially treated with betamethasone dipropionate. The dermatitis persisted, and his topical steroid was changed to fluocinonide, which led to increased erythema, pruritus, and (ultimately) bullae formation. Subsequent administration of hydrocortisone valerate worsened the erythema. A punch biopsy specimen showed subepidermal bullae with a large number of eosinophils, compatible with a diagnosis of bullous contact dermatitis. Patch tests were positive to budesonide 0.1% pet., commercial 0.05% fluocinonide cream, two hydrocortisone valerate commercial creams and many other corticosteroids. However, hydrocortisone valerate itself was not tested (9).

Another case of contact allergy to/allergic contact dermatitis from hydrocortisone valerate was published in 1979; data are not available to the author (5).

Cross-reactions, pseudo-cross-reactions and co-reactions
Cross-reactions between corticosteroids are discussed in Chapter 2.8. In a trial of the anti-inflammatory activity of some cream preparations containing steroids in volunteers, a 29-year-old woman developed some vesicles and redness on the 3rd day, only on the area of the forearm treated with 0.1% betamethasone valerate cream. Patch tests were positive to the cream and to betamethasone valerate 5% pet. Betamethasone itself was negative. There was also a positive reaction to hydrocortisone valerate (commercial preparation) but not to hydrocortisone base 5% pet. The author concluded that the contact allergic reactions were attributable to the side chains of the steroid compounds. Hydrocortisone valerate itself, however, was not tested (2).

LITERATURE

1   Baeck M, Goossens A. Immediate and delayed allergic hypersensitivity to corticosteroids: practical guidelines. Contact Dermatitis 2012;66:38-45
2   Koch EM. Contact allergic reaction to valerate esters of betamethasone and hydrocortisone. Contact Dermatitis. 1985;12:58
3   Donovan JC, Dekoven JG. Cross-reactions to desoximetasone and mometasone furoate in a patient with multiple topical corticosteroid allergies. Dermatitis 2006 ;17:147-151
4   Guin JD. Eyelid dermatitis: experience in 203 cases. J Am Acad Dermatol 2002;47:755-765
5   Fisher AA. Allergic reactions to intralesional and multiple topical corticosteroids. Cutis 1979;23:564,708-709

Chapter 3.176 HYDROQUINONE

IDENTIFICATION

Description/definition   : Hydroquinone is the aromatic organic compound that conforms to the formula shown
                             below
Pharmacological classes : Radiation-protective agents; antioxidants; mutagens; dermatological agents
IUPAC name   : Benzene-1,4-diol
Other names   : 1,4-Dihydroxybenzene; p-hydroxyphenol
CAS registry number   : 123-31-9
EC number   : 204-617-8
Merck Index monograph : 6115
Patch testing   : 1.0% pet. (Chemotechnique, SmartPracticeEurope, SmartPracticeCanada); this test
                           concentration may cause irritant reactions (37)
Molecular formula   : $C_6H_6O_2$

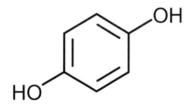

GENERAL

Hydroquinone is a phenolic compound widely used in industry as reducing agent, photographic developer, stabilizer in paints, varnishes, motor fuels and oils, as antioxidant for fats, and as an antioxidant and stabilizer for certain materials that polymerize in the presence of oxidizing agents such as resins and plastics. Hydroquinone may also be present in cosmetics, medicaments, and as feed additive for cattle. Finally, hydroquinone is present in cigarette smoke, some herbal remedies and in common foods such as cranberries, blueberries, wheat, pears, tea, coffee, red wine, rice, and onions (41).

Hydroquinone is also a depigmenting agent, well-known in cosmetic and pharmaceutical treatment of cutaneous hyperpigmentation disorders such as melasma, post-inflammatory hyperpigmentation, sunspots, and freckles in a concentration of 2-4%. Hydroquinone reduces melanin pigment production through inhibition of the tyrosinase enzyme, which is involved in the initial step of the melanin pigment biosynthesis pathway. These activities do not 'bleach the skin' but gradually suppress melanin pigment production (41).

Since 2001, hydroquinone has been banned from cosmetic products (except in artificial nail systems for professional use) in the European Union due to concerns about potential carcinogenicity. It is, however, available in OTC products in the USA.

Potential side effects of hydroquinone are diverse and include dermatitis (irritant, allergic), pigmentary disorders (hypo-, de- and hyperpigmentation, exogenous ochronosis, discoloration of the nails), corneal damage and the fish-odor syndrome. The subject has been extensively presented in several review articles (21,22,27,28,29,30,33). In this chapter, only allergic contact dermatitis to hydroquinone is discussed. Other side effects have recently been reviewed in ref. 41.

CONTACT ALLERGY

General population and subgroups
In 2014, a group of 481 healthy Chinese university student volunteers aged 20-27 years (332 women, 149 men) were patch tested with a battery of 34 haptens and there was one reaction to hydroquinone 1% pet. in a female volunteer (frequency of sensitization in the entire group 0.2%, in women 0.7%, in men 0%) (39).

Patch testing in consecutive patients suspected of contact dermatitis: routine testing
In the period 2015-2016, in the USA, the members of the North American Contact Dermatitis Group tested 5594 patients suspected of contact dermatitis with hydroquinone 1% pet. and there were 14 (0.3%) positive reactions. 42% of these were considered to be of definite or probable relevance, but culprit products were not mentioned (38).

Patch testing in groups of selected patients

Results of patch testing hydroquinone in groups of *selected* patients (mostly patients tested with a hairdressing series) are shown in table 3.176.1. In 14 investigations, frequencies of sensitization have ranged from 0.2% to 13%, but most scored below 2%. The highest frequency of 13% was found in a very small study from India, where 8 patients had a positive patch test to hydroquinone 1% pet.; the relevance of these reactions was not mentioned (20). In two studies performed by the IVDK, low rates of 0.4% and 0.9% were observed in female hairdressers with current or previous occupational contact dermatitis (3,17). Surprisingly, the frequencies of sensitization were far higher (4.2% and 5.4%) in clients of hairdressers, in who hair cosmetics were regarded as a cause of dermatitis; in these studies also, relevance was not mentioned (3,17).

Table 3.176.1 Patch testing in groups of patients: Selected patient groups

Years and Country	Test conc. & vehicle	Number of patients tested	positive (%)		Selection of patients (S); Relevance (R); Comments (C)	Ref.
2007-2012 IVDK	1% pet.	709	3	(0.4%)	S: female hairdressers with current or previous occupational contact dermatitis; R: not stated	17
		1703	71	(4.2%)	S: female patients, clients of hairdressers, in who hair cosmetics were regarded as a cause of dermatitis, and who had never worked as hairdressers; R: not stated	
2002-2011 Denmark		284	4	(1.4%)	S: hairdressers with contact dermatitis; R: not stated	16
2005-2010 Germany		66	3	(4.5%)	S: patients with complications from knee or hip arthroplasty; R: not stated	35
2000-2008 USA	1% pet.	209		(1.9%)	S: patients tested with a hairdresser's series; R: 50%	14
<2008 Germany		113	3	(2.7%)	S: patients with intolerance to endoprostheses; R: not stated	34
1997-2007 U.K.	1% pet.	80	1	(1.3%)	S; patients suspected of hair dye allergy; R: not stated	1
1980-2007 U.K.	1% pet.	538	1	(0.2%)	S: hairdressers tested with a hairdressers series; R: not specified	18
2003-2006 IVDK	1% pet.	431		(0.9%)	S: female hairdressers with suspected occupational contact dermatitis; R: not stated	3
	1% pet.	612		(5.4%)	S: women with suspected reactions to hair cosmetics; R: not stated	
1978-2005 Taiwan		603	7	(1.2%)	S: patients suspected of contact allergy to medicaments; R: 65% of the reactions to all medicaments were considered to be relevant	40
1999-2004 U.K.		518	7	(1.4%)	S: patients tested with the hairdressing series; R: only reactions that were of current or past relevance were collected	19
2000-2002 Finland		894		(0.3%)	S: patients tested with a hairdressing series; R: not stated	12
1995-2002 IVDK		884		(0.7%)	S: female hairdressers with present or past occupational contact dermatitis; R: not specified	15
		1217		(1.5%)	S: clients of hairdressers suspected to react to hairdressing cosmetics or hair care products; R: not specified	
1995-1996 Finland		438		(0.5%)	S: patients tested with a hairdressing series; R: not stated	12
<1990 India	1% pet.	63	8	(13%)	S: patients with features suggestive of allergic contact dermatitis and histories of prolonged use of numerous topical preparations, or of exacerbation or spread of dermatitis following the use of such preparations; R: the reactions were ascribed to the presence of hydroquinone in rubber footwear	20

IVDK: Information Network of Departments of Dermatology, Germany, Austria, Switzerland

Case series

Of 56 patients treated with 2% or 5% hydroquinone cream (exact composition not mentioned), one (using the 5% cream) developed local reactions suggestive of sensitization. This patient was patch tested with both the 2% and the 5% cream. After 48 hours, the patch tests were positive and a generalized eczematous eruption had appeared, but hydroquinone itself was not tested (24). Of 39 patients treated for a variety of pigmentary disorders with 5% hydroquinone ointment (exact composition not mentioned), 2 developed local reactions suggestive of sensitization. Both later reacted to the patch test with the ointment, but hydroquinone itself was not tested (23).

Four (13) and 6 (2) positive patch tests to hydroquinone were ascribed to cosmetic allergy.

Case reports

A patient became allergic to hydroquinone from a bleaching cream; she also had post-inflammatory hyperpigmentation, incorrectly called 'exogenous ochronosis' by the authors (4). Another patient also had allergic contact dermatitis from hydroquinone in a skin-lightening cream (11). A woman developed allergic contact dermatitis from hydroquinone present at a 5% concentration in a cream used to prevent post-inflammatory hyperpigmentation after a glycolic acid peel (32). Another woman had allergic contact dermatitis of the face from hydroquinone in a bleaching cream with reticulate post-inflammatory hyperpigmentation (10). Two patients with probable allergic contact dermatitis from hydroquinone presenting as leukomelanoderma were reported from Japan in 2017 (42).

Contact allergy to hydroquinone in non-pharmaceutical and non-cosmetic products

One patient had cheilitis and stomatitis from contact allergy to hydroquinone in acrylic dentures (5). One or more similar patients had already been identified in 1958 (6). Hydroquinone has also sensitized one or more individuals from its presence in cattle food (25, cited in ref. 26).

Cross-reactions, pseudo-cross-reactions and co-reactions

Pyrocatechol (7); resorcinol (8,9,36); monobenzone (monobenzyl ether of hydroquinone) (31). Hydroquinone may cross-react with other structurally related chemicals, notably those with a related para-structure.

LITERATURE

1   Basketter DA, English J. Cross-reactions among hair dye allergens. Cut Ocular Toxicol 2009;28:104-106
2   Kohl L, Blondeel A, Song M. Allergic contact dermatitis from cosmetics: retrospective analysis of 819 patch-tested patients. Dermatology 2002;204:334-337
3   Uter W, Lessmann H, Geier J, Schnuch A. Contact allergy to hairdressing allergens in female hairdressers and clients – current data from the IVDK, 2003-2006. J Dtsch Dermatol Ges 2007;5:993-1001
4   Camarasa JG, Serra-Baldrich E. Exogenous ochronosis with allergic contact dermatitis from hydroquinone. Contact Dermatitis 1994;31:57-58
5   Torres V, Cristina Mano-Azul A, Correia T, Pinto Soares A. Allergic contact cheilitis and stomatitis from hydroquinone in an acrylic dental prosthesis. Contact Dermatitis 1993;29:102-103
6   Magnusson B. Excerpta Medica. III Internal Congress Allergol 1958;131 (cited in ref. 5)
7   Andersen KE, Carisen L. Pyrocatechol contact allergy from a permanent cream dye for eyelashes and eyebrows. Contact Dermatitis 1988;18:306-307
8   Caron GA, Calnan CD. Studies in contact dermatitis. XIV. Resorcin. Trans St John's Hosp Dermatol Soc 1962;48:149-156 (data cited in ref. 15)
9   Keil H. Group reactions in contact dermatitis due to resorcinol. Arch Dermatol 1962;86:212-216 (data cited in ref. 15)
10   Tatebayashi M, Oiso N, Wada T, Suzuki K, Matsunaga K, Kawada A. Possible allergic contact dermatitis with reticulate postinflammatory pigmentation caused by hydroquinone. J Dermatol 2014;41:669-670
11   Romaguera C, Grimalt F. Dermatitis from PABA and hydroquinone. Contact Dermatitis 1983;9:226
12   Hasan T, Rantanen T, Alanko K, Harvima RJ, Jolanki R, Kalimo K, et al. Patch test reactions to cosmetic allergens in 1995–1997 and 2000–2002 in Finland –a multicentre study. Contact Dermatitis 2005;53:40-45
13   Goossens A. Cosmetic contact allergens. Cosmetics 2016, 3, 5; doi:10.3390/cosmetics3010005
14   Wang MZ, Farmer SA, Richardson DM, Davis MDP. Patch-testing with hairdressing chemicals. Dermatitis 2011;22:16-26
15   Uter W, Lessmann H, Geier J, Schnuch A. Contact allergy to ingredients of hair cosmetics in female hairdressers and clients: an 8-year analysis of IVDK data. Contact Dermatitis 2003;49:236-240
16   Schwensen JF, Johansen JD, Veien NK, Funding AT, Avnstorp C, Østerballe M, et al. Occupational contact dermatitis in hairdressers: an analysis of patch test data from the Danish Contact Dermatitis Group, 2002–2011. Contact Dermatitis 2014;70:233-237
17   Uter W, Gefeller O, John SM, Schnuch A, Geier J. Contact allergy to ingredients of hair cosmetics – a comparison of female hairdressers and clients based on IVDK 2007–2012 data. Contact Dermatitis 2014;71:13-20
18   O'Connell RL, White IR, McFadden JP, White JML. Hairdressers with dermatitis should always be patch tested regardless of atopy status. Contact Dermatitis 2010;62:177-181
19   Katugampola RP, Statham BN, English JSC, Wilkinson MM, Foulds IS, Green CM, Ormerod AD, et al. A multicentre review of the hairdressing allergens tested in the UK. Contact Dermatitis 2005;53:130-132
20   George ND, Srinivas CR, Balachandran C, Shenoi SD. Sensitivity to various ingredients of topical preparations following prolonged use. Contact Dermatitis 1990;23:367-368

21 Jow T, Hantash B. Hydroquinone-induced depigmentation: Case report and review of the literature. Dermatitis 2014;25:e1–e5

22 Levitt J. The safety of hydroquinone: a dermatologist's response to the 2006 Federal Register. J Am Acad Dermatol 2007;57:854-872

23 Spencer MS. Topical use of hydroquinone for depigmentation. JAMA 1965;194:962-964

24 Arndt KA, Fitzpatrick TB. Topical use of hydroquinone as a depigmenting agent. JAMA 1965;194:965

25 Jirasek L, Kalensky J. Kontakni alergicky ekzema z Krmnychsmesi v zivocisne vyrobe. Csekoslovenskti Dermatologie 1975; 50:217 (article in Czech) (data cited in ref. 47)

26 Van der Walle HB, Delbressine LPC, Seutter K. Concomitant sensitization to hydroquinone and p-methoxyphenol in the guinea pig; inhibitors in acrylic monomers. Contact Dermatitis 1982;8:147-154

27 Tse TW. Hydroquinone for skin lightening: safety profile, duration of use and when should we stop? J Dermatolog Treat 2010;21:272-275

28 Ruocco V, Florio M. Fish-odor syndrome: An olfactory diagnosis. Int J Dermatol 1995;34:92-93

29 Olumide YM, Akinkugbe AO, Altraide D, Mohammed T, Ahamefule N, Ayanlono S, et al. Complications of chronic use of skin lightening cosmetics. Int J Dermatol 2008;47:344-353

30 Ladizinski B, Mistry N, Kundu RV. Widespread use of toxic skin lightening compounds: medical and psychosocial aspects. Dermatol Clin 2011;29:111-123

31 Van Ketel WG. Sensitization to hydroquinone and the monobenzyl ether of hydroquinone. Contact Dermatitis 1984;10:253

32 Barrientos N, Ortiz-Frutos J, Gómez E, Iglesias L. Allergic contact dermatitis from a bleaching cream. Am J Cont Derm 2001;12:33-34

33 Nordlund JJ, Grimes PE, Ortonne JP. The safety of hydroquinone. J Eur Acad Dermatol Venereol 2006;20:781-787

34 Thomas P, Schuh A, Eben R, Thomsen M. Allergy to bone cement components. Orthopäde 2008;37:117-120 (Article in German)

35 Eben R, Dietrich KA, Nerz C, Schneider S, Schuh A, Banke IJ, Mazoochian F, Thomas P. Contact allergy to metals and bone cement components in patients with intolerance of arthroplasty. Dtsch Med Wochenschr 2010;135:1418-1322 (article in German)

36 Barbaud A, Reichert-Penetrat S, Trechot P, Granel F, Schmutz JL. Sensitization to resorcinol in a prescription verrucide preparation: unusual systemic clinical features and prevalence. Ann Dermatol Venereol 2001;128:615-618 (article in French)

37 Kanerva L, Jolanki R, Alanko K, Estlander T. Patch-test reactions to plastic and glue allergens. Acta Derm Venereol 1999;79:296-300

38 DeKoven JG, Warshaw EM, Zug KA, Maibach HI, Belsito DV, Sasseville D, et al. North American Contact Dermatitis Group patch test results: 2015-2016. Dermatitis 2018;29:297-309

39 Zhao L, Li LF. Contact sensitization to 34 common contact allergens in University students in Beijing. Contact Dermatitis 2015;73:323-324

40 Shih Y-H, Sun C-C, Tseng Y-H, Chu C-Y. Contact dermatitis to topical medicaments: a retrospective study from a medical center in Taiwan. Dermatol Sinica 2015;33:181-186

41 De Groot AC. Monographs in Contact Allergy Volume I. Non-Fragrance Allergens in Cosmetics (Part I and Part 2). Boca Raton, Fl, USA: CRC Press Taylor and Francis Group, 2018:626-633

42 Yanagishita-Nakatsuji S, Fukai K, Ohyama A, Umekoji A, Sowa-Osako J, Tsuruta D. Probable allergic contact dermatitis from hydroquinone presenting as leukomelanoderma: Report of two cases. J Dermatol 2017;44:e330-e331

Chapter 3.177 IBACITABINE

IDENTIFICATION

Description/definition   : Ibacitabine is the pyrimidine 2'-deoxyribonucleoside that conforms to the structural
formula shown below
Pharmacological classes   : Antiviral agents
IUPAC name   : 4-Amino-1-[(2R,4S,5R)-4-hydroxy-5-(hydroxymethyl)oxolan-2-yl]-5-iodopyrimidin-2-one
Other names   : 5-Iodo-2'-deoxycytidine; 2'-deoxy-5-iodocytidine
CAS registry number   : 611-53-0
EC number   : 210-269-8
Patch testing   : 1% and 5% pet.
Molecular formula   : $C_9H_{12}IN_3O_4$

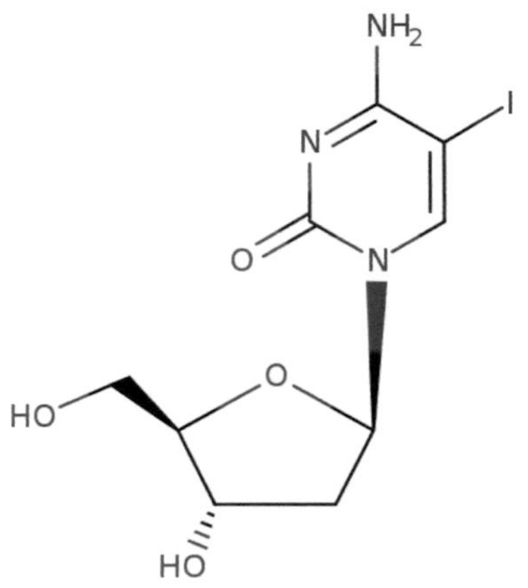

GENERAL

Ibacitabine is an iodinated analog of deoxycytidine used topically in the treatment of herpes labialis (1).

CONTACT ALLERGY

Case reports and case series

Two patients developed allergic contact dermatitis from a cream containing 1% ibacitabine (5-iodo-2'-deoxycytidine) (2). Clinical details and patch test results are unknown (cited in ref. 3). Another author has seen 7 patients allergic to the same cream containing 1% ibacitabine, but provides no clinical details (3). A 39 year-old woman developed eczema surrounding the mouth each time she used a cream containing 1% ibacitabine for the treatment of herpes simplex infections of the skin. Patch tests were positive to the commercial cream and ibacitabine 1% and 10% pet. (5). Contact allergy to ibacitabine has also been mentioned in ref. 4 (no specific data available).

Cross-reactions, pseudo-cross-reactions and co-reactions

Two patients allergic to ibacitabine (5-iodo-2'-deoxycytidine) also had positive patch tests to idoxuridine. The authors considered the latter reactions to be cross-reactions to ibacitabine. In one case this was highly likely, as the patient had never used pharmaceuticals with idoxuridine before. The other patient, however, had used a cream containing 0.24% idoxuridine, which may have caused sensitization (2). Three other patients sensitized to ibacitabine cross-reacted to idoxuridine (1% and 5% pet.) but not to trifluridine (3).

LITERATURE

1   The data in the section 'General' may have been obtained from literature discussed in this chapter, but mostly also or exclusively from one or more of the following online sources: ChemIDPlus Advanced, PubChem, DrugBank, RxList, Drug Central, Drugs.com, and Wikipedia
2   Cardinaud F, Janaud M, Michel F, Bernard P, Bonnetblanc JM. Sensibilisation croisée entre topiques antiviraux. Nouv Dermatol 1985;4:263-264
3   Foussereau J, Tomb R. Cross-allergy between 5-iodo-2'-deoxycytidine and idoxuridine. J Am Acad Dermatol 1987;1:145-147
4   Chabeau G. Quoted by Ducombs G, Chabeau G. Abrégé de dermato-allergologie de contact. Paris: Masson, 1979 (Book in French, data cited in ref. 3)
5   Serpentier-Daude A, Collet E, Didier AF, Touraud JP, Sgro C, Lambert D. Contact dermatitis to topical antiviral drugs. Ann Dermatol Venereol 2000;127:191-193 (Article in French)

Chapter 3.178 IBUPROFEN PICONOL

IDENTIFICATION

Description/definition   : Ibuprofen piconol is the pyridyl ester of ibuprofen that conforms to the structural formula shown below
Pharmacological classes : Anti-inflammatory agents, non-steroidal
IUPAC name   : Pyridin-2-ylmethyl 2-[4-(2-methylpropyl)phenyl] propanoate
Other names   : 2-(p-Isobutylphenyl)propionic acid 2-pyridylmethyl ester; 2-pyridylmethyl 2-(4-isobutylphenyl)propionate
CAS registry number   : 64622-45-3
EC number   : 264-979-8
Patch testing   : 1% and 5% pet.
Molecular formula   : $C_{19}H_{23}NO_2$

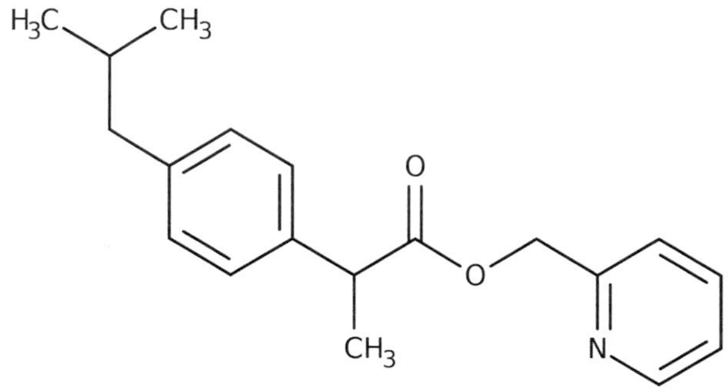

GENERAL

Ibuprofen piconol is a non-steroidal anti-inflammatory drug (NSAID) with analgesic and antipyretic properties. It is a 2-pyridinemethanol ester of ibuprofen and was designed to be more soluble in fat than ibuprofen, so it can remain in the tissue longer (2).

CONTACT ALLERGY

Case series

In Japan, before 1994, a group of 74 patients with severe refractory atopic dermatitis involving the face were patch tested. There were 10 relevant reactions to ibuprofen piconol (14%). It is uncertain whether all patients were tested with this hapten and the test concentrations and vehicles were not mentioned (1).

Case reports

An 18-year-old woman with atopic dermatitis had been treated with a cream containing ibuprofen piconol for a facial eruption. After 2 weeks, her face became hot, red and swollen. Patch tests were positive to the cream, an ointment of the same brand, and to their active ingredient ibuprofen piconol 0.1% (+), 1% (++) and 5% pet. (++), but not to the excipients (2). The authors also present a 16-year-old man with an almost identical history and patch test results, but in this individual, the cream base was also positive (2).

    Previously, (at least) 6 cases of allergic contact dermatitis from ibuprofen piconol had been reported from Japan (3-5; no details available, articles in Japanese, data cited in ref. 2).

LITERATURE

1   Tada J, Toi Y, Arata J. Atopic dermatitis with severe facial lesions exacerbated by contact dermatitis from topical medicaments. Contact Dermatitis 1994;31:261-263
2   Kubo K, Shirai K, Akaeda T, Oguchi M. Contact dermatitis from ibuprofen piconol. Contact Dermatitis 1988;18:188-189
3   Matsunaga K, Hayakawa R. Four cases of allergic contact dermatitis due to ibuprofen piconol. Skin Research 1986;28(suppl.):168-176 (Article in Japanese, data cited in ref. 2)
4   Ota M, Toda K. Contact dermatitis to nonsteroidal anti-inflammatory topical agents. Hifurinsho 1986;28:39-43 (Article in Japanese, data cited in ref. 2)
5   Suzuki M, Hosokawa K, Matsunaga K, Hayakawa R. Allergic contact dermatitis due to non-steroidal anti-inflammatory topical agents. Skin Research 1987;29(suppl.):211-217 (Article in Japanese, data cited in ref. 2)

Chapter 3.179 IBUPROXAM

IDENTIFICATION

Description/definition   : Ibuproxam is the hydroxamic acid derivative that conforms to the structural formula shown below

Pharmacological classes   : Anti-inflammatory agents, non-steroidal

IUPAC name   : *N*-Hydroxy-2-[4-(2-methylpropyl)phenyl]propanamide

Other names   : *p*-Isobutylhydratropohydroxamic acid

CAS registry number   : 53648-05-8

EC number   : 258-683-8

Merck Index monograph : 6190

Patch testing   : 5% pet.

Molecular formula   : $C_{13}H_{19}NO_2$

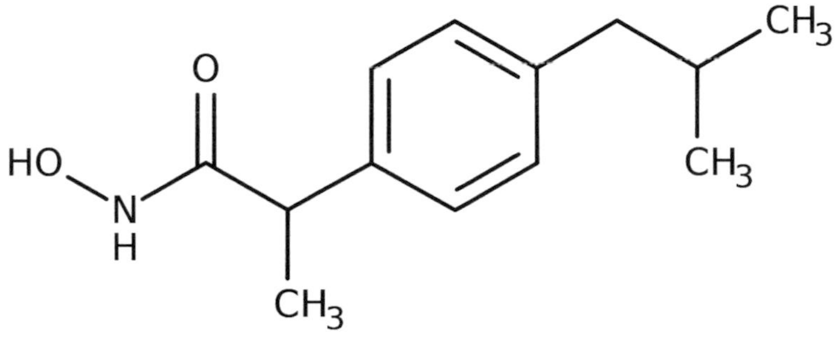

GENERAL

Ibuproxam is a hydroxamic acid obtained by formal condensation of the carboxy group of ibuprofen with the amino group of hydroxylamine. It is a nonsteroidal anti-inflammatory drug (NSAID) used for treatment of pain and inflammation associated with musculoskeletal and joint disorders (1).

CONTACT ALLERGY

Patch testing in groups of patients

In the period 2004-2008, in a multicenter Italian study, 1082 patients with histories and clinical features suggestive of photoallergic contact dermatitis were patch and photopatch tested with ibuproxam 5% pet. and there was one contact allergic reaction to the NSAID, which was photoaggravated; its relevance was not specified (8).

Case series

In 3 centers in Italy, in the period 1989-1995, 23 patients with contact allergy to ibuproxam were seen and 9 with photocontact allergy. Clinical data were not provided. After ketoprofen, ibuproxam was by far the most frequent NSAID sensitizer and photosensitizer (10).

Also in Italy, before 1993, the members of the GIRDCA Multicentre Study Group diagnosed 102 patients (49 men, 53 women), aged 16 to 66 years (mean 37 years), with (photo)dermatitis induced by systemic or topical NSAIDs. Ibuproxam caused 20 contact allergic and zero photocontact allergic reactions (2).

Case reports

A 39-year-old man, who had used a cream containing ibuproxam 2x daily for slight trauma to the left foot, developed erythema, edema and itching on the palm of his right hand, left ankle, glans and lips on the 3rd day of treatment. Later, vesicles turning into bullae appeared on his palm after a few days. Patch tests were positive to the cream and to ibuproxam 2.5% pet. but not to other NSAIDs (3).

Ten days after a 49-year-old woman started applying a cream containing ibuproxam for pain in the right foot after musculoskeletal trauma, an erythematous, edematous, vesicular dermatitis had developed, later spreading to the legs. Patch tests were positive to the cream, ibuproxam 5% pet. and ketoprofen 2.5% pet., the latter reaction being interpreted as due to cross-sensitivity. Five controls were negative (4).

A 19-year-old boy had used a cream containing ibuproxam for pain in the right knee for 2 weeks, when dermatitis with erythema, edema and vesicular lesions with severe itching appeared at the site of application. The area had not been exposed to sunlight. Patch test were positive to the cream and ibuproxam 5% pet. but not to ibuprofen 5% pet. (5). A second virtually identical case was also reported by the authors (5).

A 32-year old woman developed an eczematous eruption with severe itching on the left knee, after having used ibuproxam cream there for 2 weeks to treat pain after a musculoskeletal trauma. Patch tests were positive to the cream and ibuproxam 5% pet., but negative to other NSAIDs (6). The same authors describe a 22-year old man who used a cream containing ibuproxam for right elbow pain. After 15 days he stopped the treatment because an eczematous eruption with vesiculation and itching had developed at the site of application. He had not exposed the elbow to sunlight. Patch tests were strongly positive to the cream and to ibuproxam 5% pet. All other NSAIDs tested (ketoprofen, ibuprofen, naproxen, flurbiprofen, fenoprofen, tiaprofenic acid) were negative (6).

PHOTOSENSITIVITY

Photopatch testing in groups of patients
In the period 2004-2008, in a multicenter Italian study, 1082 patients with histories and clinical features suggestive of photoallergic contact dermatitis were patch and photopatch tested with ibuproxam 5% pet. and there were four photoallergic reactions to the NSAID; their relevance was not specified (8).

Case series
In 3 centers in Italy, in the period 1989-1995, 9 patients with photocontact allergy to ibuproxam were seen and 23 with contact allergy. Clinical data were not provided. After ketoprofen, ibuproxam was by far the most frequent NSAID photosensitizer and sensitizer (10).

In a multicenter study, also in Italy, performed in the period 1985-1994, one photopatch test reaction was seen to ibuproxam; the patch test concentration used was not mentioned and relevance was not specified (78% for all photoallergens together) (7).

Cross-reactions, pseudo-cross-reactions and co-reactions
A patient sensitized to ibuproxam may have cross-reacted to ketoprofen (4). One patient photosensitized to ketoprofen may have cross-reacted (contact allergy, not photocontact allergy) to ibuproxam (9). Co-reactivities to ketoprofen and ibuproxam (both contact and photocontact) have also been observed in ref. 10. True cross-sensitivity appears to be unlikely.

LITERATURE
1   The data in the section 'General' may have been obtained from literature discussed in this chapter, but mostly also or exclusively from one or more of the following online sources: ChemIDPlus Advanced, PubChem, DrugBank, RxList, Drug Central, Drugs.com, and Wikipedia
2   Pigatto PD, Mozzanica N, Bigardi AS, Legori A, Valsecchi R, Cusano F, et al. Topical NSAID allergic contact dermatitis. Italian experience. Contact Dermatitis 1993;29:39-41
3   Molinini R. Contact allergy to ibuproxam. Contact Dermatitis 1991;24:302-303
4   Valsecchi R, Cainelli T. Contact dermatitis from ibuproxam. A case with cross-reactivity with ketoprofen. Contact Dermatitis 1990;22:51
5   Mozzanica N, Pucci M, Negri M, Pigatto P. Contact dermatitis to ibuproxam. Contact Dermatitis 1987;16:281-282
6   Mozzanica N, Pucci M, Pigatto PD. Contact and photoallergic dermatitis to topical nonsteroidal anti-inflammatory drugs (propionic acid derivatives): a study of 8 cases. In: Frosch P, Dooms-Goossens A, Lachapelle J-M, Rycroft RJG, Scheper RJ, Eds. Current topics in contact dermatitis. Berlin: Springer-Verlag, 1989:499-506
7   Pigatto PD, Legori A, Bigardi AS, Guarrera M, Tosti A, Santucci B, et al. Gruppo Italiano recerca dermatiti da contatto ed ambientali Italian multicenter study of allergic contact photodermatitis: epidemiological aspects. Am J Contact Dermatitis 1996;17:158-163
8   Pigatto PD, Guzzi G, Schena D, Guarrera M, Foti C, Francalanci, S, Cristaudo A, et al. Photopatch tests: an Italian multicentre study from 2004 to 2006. Contact Dermatitis 2008;59:103-108
9   Cusano F, Capozzi M. Photocontact dermatitis from ketoprofen with cross-reactivity to ibuproxam. Contact Dermatitis 1992;27:50-51
10 Pigatto P, Bigardi A, Legori A, Valsecchi R, Picardo M. Cross-reactions in patch testing and photopatch testing with ketoprofen, tiaprophenic acid, and cinnamic aldehyde. Am J Contact Dermat 1996;7:220-223

Chapter 3.180 ICHTHAMMOL

IDENTIFICATION

Description/definition	: Ichthammol is a complex product obtained by the sulfonation and ammoniation of the distillation product from bituminous schists; it may contain saturated and unsaturated hydrocarbons, nitrogen bases and thiophene derivatives
Pharmacological classes	: Dermatological agents
Other names	: Ichthyol; ammonium bituminosulfonate
CAS registry number	: 8029-68-3
EC number	: 232-439-0
Merck Index monograph	: 6194
Patch testing	: 5% water; control tests are necessary

GENERAL

Ichthammol is a complex product obtained by the sulfonation and ammoniation of shale oil, the distillation product of bituminous schists. It consists of 10% sulfur, 5-7% ammonium sulfate, hydrocarbons, nitrogenous bases, acids and derivatives of thiophene. The allergen is said to be present in the water- and cyclohexane- soluble fraction (7). Ichthammol is available in OTC products as an active ingredient for the treatment of skin disorders such as eczema, psoriasis and boils as it is thought to exhibit anti-inflammatory, antibacterial and antimycotic properties (1).

CONTACT ALLERGY

Testing in groups of selected patients

In Leuven, Belgium, in the period 1990-2014, iatrogenic contact dermatitis was diagnosed in 2600 individuals (17% of the total patch test population). Ichthammol 5% pet. was tested in 309 patients and there were 3 (1%) positive reactions to it. Relevance was not specified for individual allergens, but 96% of all positive patch test reactions to topical drugs and antiseptics were considered to be relevant (2).

In Warsaw, Poland, in the period 1967-1970, 836 selected patients (selection criteria unknown) were patch tested with ichthammol 5% pet. and there were 2.7% positive reactions; their relevance was not mentioned (9).

Case reports

A 40-year-old geriatric nurse developed an acute eczematous eruption localized to her right lower leg one day after applying a glycerin and ichthammol dressing, used to alleviate the discomfort caused by an acute episode of super-ficial thrombophlebitis. Previous contact with glycerin and ichthammol had only occurred at work, approximately five times in four years, where the mixture was used as a soothing agent on bed sores. Patch tests were positive to the mixture and to ichthammol 1%, 5% and 10% water (3). The patients had developed occupational sensitization to ichthammol but not occupational allergic contact dermatitis.

A 33-year-old man thought to be responsive to local medicaments reacted to glycerin and ichthammol (D2 +++, D7 +); he had far weaker reactions to ichthammol 1% in zinc oxide ointment and 2% in zinc oxide cream. A 60-year-old woman was said to be responsive to ichthammol and indeed reacted strongly to ichthammol in pet., ichthammol 5% and 10% water and ichthammol/glycerin (4). A man had previously used glycerin and ichthammol solution to treat phlebitis and when he subsequently applied it to a leg injury, dermatitis developed. Patch tests were positive to the solution and to ichthammol 5% water (6).

A 49-year-old woman developed acute allergic contact dermatitis after applying pure ichthammol to a furuncle near the right eye. A patch test with ichthammol was positive (5). A 45-year-old woman treated eczema of the feet with an ointment containing ichthammol, which aggravated the eczema. Patch tests were positive to parabens and the ointment and a cream containing ichthammol, but negative to wood and coal tars. The author suggested that the reactions were caused by concomitant sensitization to butylparaben, to glycol (??) and to purified ichthammol, but presented no supporting evidence (5).

One or more cases of sensitization to ichthammol were also reported from Germany in 1985 (7). A 'severe allergic reaction to ichthyol' has been reported in Russian literature, but details are unknown to the author (8).

LITERATURE

1   The data in the section 'General' may have been obtained from literature discussed in this chapter, but mostly also or exclusively from one or more of the following online sources: ChemIDPlus Advanced, PubChem, DrugBank, RxList, Drug Central, Drugs.com, and Wikipedia

2   Gilissen L, Goossens A. Frequency and trends of contact allergy to and iatrogenic contact dermatitis caused by topical drugs over a 25-year period. Contact Dermatitis 2016;75:290-302

3   Lawrence CM, Smith AG. Ichthammol sensitivity. Contact Dermatitis 1981;7:335

4   Cooke MA, Senter GW, Hocken Robertson DE. Ichthammol dermatitis. Contact Dermatitis Newsletter 1972;11:299

5   Bandmann H-J. Ichthyol dermatitis. Contact Dermatitis Newsletter 1971;10:224

6   Calnan CD. Ichthyol. Contact Dermatitis Newsletter 1971;9:218

7   Schwale M, Frosch PJ. Kontaktallergie auf Ammoniumbituminosulfonat. Dermatosen 1983;31:183-186 (Article in German)

8   Patlan BD, Arefiev IuL. Severe allergic reaction to ichthyol. Klin Med (Mosk) 1973;51:140 (Article in Russian)

9   Rudzki E, Kleniewska D. The epidemiology of contact dermatitis in Poland. Br J Dermatol 1970;83:543-545

Chapter 3.181 IDOXURIDINE

IDENTIFICATION

Description/definition	: Idoxuridine is the iodinated analog of deoxyuridine that conforms to the structural formula shown below
Pharmacological classes	: Antiviral agents; nucleic acid synthesis inhibitors
IUPAC name	: 1-[(2R,4S,5R)-4-Hydroxy-5-(hydroxymethyl)oxolan-2-yl]-5-iodopyrimidine-2,4-dione
Other names	: Uridine, 2'-deoxy-5-iodo-; 5-iodo-2'-deoxyuridine; 1-(2-dehydroxy-β-D-ribofuranosyl)-5-iodouracil
CAS registry number	: 54-42-2
EC number	: 200-207-8
Merck Index monograph	: 6201
Patch testing	: 1% pet. (SmartPracticeCanada)
Molecular formula	: $C_9H_{11}IN_2O_5$

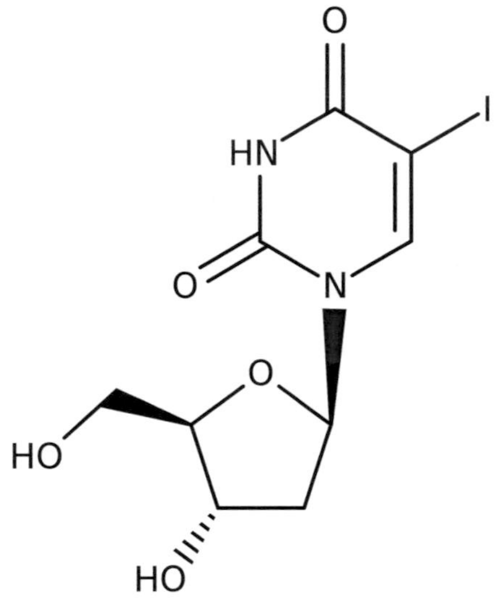

GENERAL

Idoxuridine is a iodinated analog of deoxyuridine, with antiviral activity against *Herpes simplex* virus (HSV) and potential radiosensitizing activities. In chemical structure idoxuridine closely approximates the configuration of thymidine, one of the four building blocks of DNA, the genetic material of the *Herpes* virus. As a result, idoxuridine is able to replace thymidine in the enzymatic step of viral replication or 'growth'. The consequent production of faulty DNA results in a pseudostructure which cannot infect or destroy tissue. Idoxuridine is indicated for use in keratoconjunctivitis and keratitis caused by *herpes simplex* virus. Formerly, the drug has been used in ointments and creams for the treatment of herpes labialis (cold sore), but these were soon found to be ineffective (1).

Most literature on contact allergy to idoxuridine is at least 30 years old and it is doubtful whether this drug is still widely used.

CONTACT ALLERGY

Patch testing in groups of patients

In Hungary, in the period 1996-2005, 133 patients with periorbital dermatitis were patch tested with 1% idoxuridine and there were 2 (1.5%) positive reactions; both were considered to be relevant (2).

Case series

In Bologna, Italy, before 1989, 136 patients (35 men, 101 women, range of age 14-84 years, average 36 years) suspected of contact conjunctivitis from eye drops (including contact lens solutions), were patch tested. In sixty, there was also eyelid contact dermatitis. In 75 individuals, the causative agents were found. The great majority was caused by thimerosal (n=52), followed by benzalkonium chloride (n=7), both present in contact lens solutions. There was one reaction to idoxuridine eye drops and idoxuridine itself (7).

In Japan, before 1987, 132 patients were treated with IDU for herpes keratitis. Patch tests were routinely done when patients exhibited contact dermatitis. Of the patients treated with IDU, 3 (2.3%) showed allergic contact dermatitis confirmed by patch testing (22).

In a hospital in Bari, Italy, before 1986, nine patients (5 women, 4 men, 18-45 years of age) have been diagnosed with contact sensitization to idoxuridine. They all had used an ointment containing either 5% or 40% IDU for recurrent herpes simplex labialis. Six had used the pharmaceutical for various recurrences before they developed acute, red oozing dermatitis of the face. The others had treated only one flare of herpes simplex, and after 8-10 days developed dermatitis spreading beyond the original site; in one case the dermatitis became generalized. Patch tests were positive to the ointments with IDU and to IDU 1% pet., but negative to the ointment base (8; possible overlap with ref. 3).

In Bari, Italy, in the period 1968-1983, 6 selected patients (selection procedure unknown) were patch tested with idoxuridine 0.5% pet. and there were 2 positive reactions; relevance was not discussed (3; possible overlap with ref. 8).

In 1976, in Denmark, 20 hospitalized patients with herpes zoster were treated with 40% IDU in DMSO (dimethyl sulfoxide) solution, 20 with 5% IDU in 60% DMSO ointment with propylene glycol and another 20 with 40% IDU in 60% DMSO ointment, applied locally twice daily for 4 days. Almost all patients were observed daily until the crusts had fallen off and then at intervals for 6 months. In one of the 60 patients was a rash with widespread erythema and papules observed 17 days after discontinuation of the treatment. Two years later, these 60 patients were recalled for examination and 45 responded. They were patch tested with IDU 5% and 0.5% pet. and propylene glycol, but not with DMSO. Three patients (7%) had positive patch tests to both concentrations of IDU, and had been sensitized by the treatment, but only one had developed skin symptoms (15). The authors consider 7% sensitization acceptable, but it should be realized that the patients had been treated for 4 days only! Application to damaged skin, high concentrations of IDU (5-40%) and using an irritant and penetration enhancer such as DMSO will have facilitated sensitization.

In 1975, from Portland, USA, 4 patients were presented with allergic contact dermatitis from IDU, who had been seen in a period of one year (20); their details are shown in table 3.181.1.

Case reports

A 70-year-old man had allergic contact cheilitis from idoxuridine in a topical pharmaceutical cream for treatment of cold sore (4). A 66-year old man was treated for herpes zoster with oral acyclovir and a solution of idoxuridine (IDU) containing DMSO for 4 days; it was stopped when the symptoms improved. Ten days later, the patient again observed swelling and pain in the lumbar region and treated himself again with the IDU solution. After a few hours, a skin reaction occurred with erythema, severe pruritic papulovesicles and scaling, not only at the site of dermatitis, but also on the left knee and scrotum. Patch tests were positive to the commercial IDU solution, IDU 1% in DMSO, and IDU 5%, 1% and 0.1% pet. DMSO 40% water remained negative. The author also describe a second patient allergic to IDU with a similar history and patch test results (5).

One patient had erythema multiforme/erythema multiforme-like contact dermatitis with erythematous urticarial papules, plaques and target-lesions from contact allergy to idoxuridine. Details are unknown (6).

Short summaries of other case reports of allergic contact dermatitis from idoxuridine are shown in table 3.181.1.

Table 3.181.1 Short summaries of case reports of allergic contact dermatitis from idoxuridine

Year	Sex	Age	Clinical picture	Positive patch tests	Comments	Ref.
1985	M	43	Dermatitis of the lips	IDU in DMSO, IDU 0.2% water	when investigated, the patient had periorbital ACD from chlor-amphenicol and tolazoline in eye ointment	25
1985	F	30	eczema of the lips, perioral region and the cheeks	IDU ointment and solution, IDU 0.5% pet.	treatment of herpes simplex labialis	9
1982	F	23	acute dermatitis of the face	IDU ointment and solution, IDU 0.2% - 5% pet.	treatment of verrucae planae	12
1982	M	65	dermatitis of the left thigh spreading to leg and face	IDU ointment and solution, IDU 0.2% - 5% pet.	treatment of herpes simplex; unknown why applied to thigh	12
1982	M	59	dermatitis around the right eye and on the cheek	IDU 0.2% in pet. and in DMSO/cetomacrogolis cream	cross-allergy to trufluridine	27
1981	M	35	swelling of the genitals, painful phimosis	IDU ointment 25%, IDU in DMSO-alc. 1:4 (concentration not stated)	treatment of genital herpes	13
1981	F	30	dermatitis of the left check	IDU ointment 25%, IDU 1% (vehicle ?)	treatment of herpes simplex	13
1979	M	25	secondarily infected herpes simplex (?) in the groin	IDU 5% in DMSO; IDU ointment 0.5%; IDU itself was not tested	in the lotion and the ointment IDU is the only common constituent	16

Table 3.181.1 Short summaries of case reports of allergic contact dermatitis from idoxuridine (continued)

Year	Sex	Age	Clinical picture	Positive patch tests	Comments	Ref.
1977	M	59	dermatitis around the right orbita spreading to the face	IDU eye drops 0.1%, IDU 0.2% pet. and IDU 0.2% DMSO/cream base	treatment of herpes keratitis of the right eye	17
1976	F	26	dermatitis around herpes on the back of left thigh	IDU in DMSO, IDU 0.5% pet.		19
1975	M	31	ACD of the genital area, spreading to hands, perioral area and neck	0.5% IDU ointment; IDU 0.1% (vehicle?), IDU 5% in dimethylacetamide	treatment of herpes simplex; generalized dermatitis from patch testing	20
1975	M	33	erythema and edema of the scrotal skin	0.1% IDU ointment, IDU 5% in dimethylacetamide	treatment of herpes genitalis	20
1975	M	63	acute ACD around the eyes and on the face	IDU 5% in dimethylacetamide	treatment of herpes keratitis; also allergy to scopolamine	20
1975	M	34	acute ACD of the genital region	IDU 5% in dimethylacetamide; IDU 0.5% pet.	treatment of herpes genitalis	20
1975	F	65	severe oozing dermatitis on the lower abdomen	IDU ointment, IDU 1% pet.	treatment of herpes simplex with IDU ophthalmic ointment	21
1975	F	53	very severe edematous dermatitis of the perianal area	IDU ointment, IDU 1% pet.	treatment of condylomata acuminata with IDU ointment; ROAT with IDU 1% was positive already after 3 days	21

ACD : allergic contact dermatitis; IDU : idoxuridine

One of 50 patients treated with 5% or 40% idoxuridine in DMSO became sensitized with acute allergic contact dermatitis; he later had positive patch tests to 1% and 5% idoxuridine in DMSO but not to DMSO alone (26). Other case reports of contact allergy to idoxuridine, adequate data of which are not available to the author, can be found in refs. 10,11 and 18. Calnan mentioned that, at a convention in Copenhagen in 1974 (?), 7 patients with allergic contact dermatitis from IDU were mentioned by 3 participants (19).

Cross-reactions, pseudo-cross-reactions and co-reactions
Two patients developed allergic contact dermatitis from a cream containing 1% ibacitabine (5-iodo-2'-deoxycytidine). They were also tested with the related idoxuridine and both had a positive patch test to it. The authors considered this to be cross-reactions. In one case this was highly likely, as the patient had never used pharmaceuticals with idoxuridine before. The other patient, however, had used a cream containing 0.24% idoxuridine, which may have caused sensitization (23).

Three other patients sensitized to the antiviral agent ibacitabine in a cream were patch tested with idoxuridine to detect cross-sensitization. All reacted to the commercial cream with ibacitabine that had sensitized them, ibacitabine 1% and 5% pet., a commercial gel containing 0.5% idoxuridine (IDU) and IDU 1% and 5% pet. (24). In patients sensitized to idoxuridine (5-iodo-2'-deoxyuridine), cross-reactivity has been observed to the pyrimidine nucleosides 5-bromodeoxyuridine, 5-chlorouridine, and 5-chlorocytidine and to the substituted pyrimidine bases 5-iodouracil and 5-bromouracil. Fluorine substituted compounds did not cross-react (20). A patient sensitized to idoxuridine may have cross-sensitized to trifluorthymidine (14).

LITERATURE
1 The data in the section 'General' may have been obtained from literature discussed in this chapter, but mostly also or exclusively from one or more of the following online sources: ChemIDPlus Advanced, PubChem, DrugBank, RxList, Drug Central, Drugs.com, and Wikipedia
2 Temesvári E, Pónyai G, Németh I, Hidvégi B, Sas A, Kárpáti S. Periocular dermatitis: a report of 401 patients. J Eur Acad Dermatol Venereol 2009;23:124-128
3 Angelini G, Vena GA, Meneghini CL. Allergic contact dermatitis to some medicaments. Contact Dermatitis 1985;12:263-269
4 Freeman S, Stephens R. Cheilitis: analysis of 75 cases referred to a contact dermatitis clinic. Am J Contact Dermat 1999;10:198-200
5 Senff H, Engelmann L, Kunze J, Hausen BM. Allergic contact dermatitis from idoxuridine. Contact Dermatitis 1990;23:43-45
6 Seidenari S, Di Nardo A, Motolese A, Pincelli C. Erythema multiforme associated with contact sensitization. Description of 6 clinical cases. G Ital Dermatol Venereol 1990;125:35-40 (article in Italian)
7 Tosti A, Tosti G. Allergic contact conjunctivitis due to ophthalmic solution. In: Frosch PJ, Dooms-Goossens A, Lachapelle JM, Rycroft RJG, Scheper RJ (eds). Current Topics in Contact Dermatitis. Berlin: Springer-Verlag, 1989: 269-272

8   Angelini G, Vena GA, Meneghini CL. Contact allergy to antiviral agents. Contact Dermatitis 1986;15:114-115

9   Balato N, Nappa P, Lembo G, Ayala F. Dermatitis from idoxuridine. Contact Dermatitis 1985;13:338-339

10   Frosch PJ. Kontakallergie auf Zostrum ® (Idoxuridin). Allergologie 1984;7:344 (article in German)

11   Elkjaer P, Wadskov S. Allergic contact eczema provoked by idoxuridine. Ugeskr Laeger 1982;144:877-878 (article in Danish)

12   Lombardi P, Spallanzani P, Giorgini S, Seroli A. Allergic contact dermatitis from idoxuridine. Contact Dermatitis 1982;8:350-351

13   Reiffers J. Allergy to 5-iodo-2'desoxyuridine. Contact Dermatitis 1981;7:125

14   Cirkel PK, van Ketel WG. Allergic contact dermatitis to trifluorothymidine eyedrops. Contact Dermatitis 1981;7:49-50

15   Thormann J, Wildenhoff KE. Contact allergy to idoxuridine. Sensitization following treatment of herpes zoster. Contact Dermatitis 1980;6:170-171

16   Calnan CD. Allergy to idoxuridine ointment. Contact Dermatitis 1979;5:194-195

17   Van Ketel WG. Allergy to idoxuridine eyedrops. Contact Dermatitis 1977;3:106-107

18   Amon RB, Hanifin JM. Allergic contact dermatitis due to idoxuridine (letter). N Engl J Med 1976;294(17):956-957

19   Calnan CD. Contact dermatitis to idoxuridine. Contact Dermatitis 1976;2:58

20   Amon RB, Lis AW, Hanifin JM. Allergic contact dermatitis caused by idoxuridine. Patterns of cross reactivity with other pyrimidine analogues. Arch Dermatol 1975;111:1581-1584

21   Osmundsen PE. Allergic contact dermatitis from idoxuridine. Contact Dermatitis 1975;1:251

22   Naito T, Shiota H, Mimura Y. Side effects in the treatment of herpetic keratitis. Curr Eye Res 1987;6:237-239

23   Cardinaud F, Janaud M, Michel F, Bernard P, Bonnetblanc JM. Sensibilisation croisée entre topiques antiviraux. Nouv Dermatol 1985;4:263-264 (Article in French)

24   Foussereau J, Tomb R. Cross-allergy between 5-iodo-2'-deoxycytidine and idoxuridine. J Am Acad Dermatol 1987;1:145-147

25   Frosch PJ, Olbert D, Weickel R. Contact allergy to tolazoline. Contact Dermatitis 1985;13:272

26   Simpson JR. Idoxuridine in herpes zoster. Br Med J 1974;3(5929):523

27   Cirkel PK, van Ketel WG. A patient with contact allergy for various virostatic agents used in herpetic keratitis. Ned Tijdschr Geneeskd 1982;126:1453-1454 (Article in Dutch)

Chapter 3.182 INDOMETHACIN

IDENTIFICATION

Description/definition   : Indomethacin is the synthetic nonsteroidal indole derivative that conforms to the structural formula shown below

Pharmacological classes   : Tocolytic agents; anti-inflammatory agents, non-steroidal; cardiovascular agents; cyclooxygenase inhibitors; gout suppressants

IUPAC name   : 2-[1-(4-Chlorobenzoyl)-5-methoxy-2-methylindol-3-yl]acetic acid

CAS registry number   : 53-86-1

EC number   : 200-186-5

Merck Index monograph   : 6279

Patch testing   : 1% pet. (SmartPracticeCanada, SmartPracticeEurope)

Molecular formula   : $C_{19}H_{16}ClNO_4$

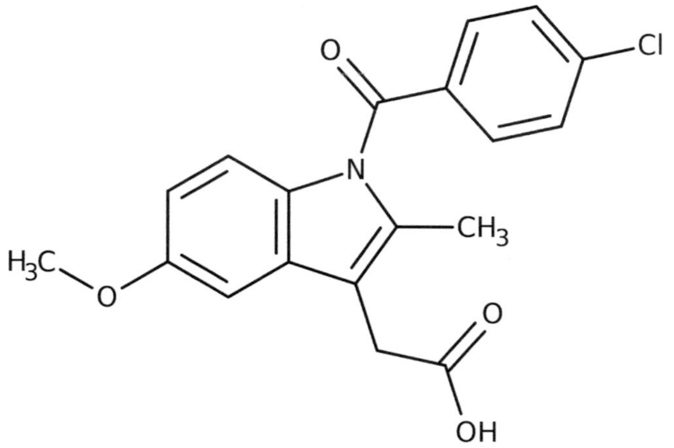

GENERAL

Indomethacin is an indole derivative and nonsteroidal anti-inflammatory drug (NSAID) with anti-inflammatory activity and chemopreventive properties. It inhibits the enzyme cyclooxygenase, which is necessary for the formation of prostaglandins. Indomethacin may also inhibit the expression of multidrug-resistant protein type 1, resulting in increased efficacies of some antineoplastic agents in treating multi-drug resistant tumors. Oral indomethacin is indicated for symptomatic management of (moderate to severe) rheumatoid arthritis including acute flares of chronic disease, ankylosing spondylitis, severe osteoarthritis, acute painful shoulder (bursitis and/or tendinitis) and acute gouty arthritis. In powder for injection solutions, indomethacin is present as indomethacin sodium trihydrate (CAS number 74252-25-8, EC number 200-186-5, molecular formula $C_{19}H_{21}ClNNaO_7$), in other preparations indomethacin base is employed. In some countries indomethacin it is also available in topical preparations for the relief of pain and inflammation of the muscles and joints (1).

CONTACT ALLERGY

Patch testing in groups of patients

In Germany, before 1998, 371 consecutive patients suspected of contact dermatitis were patch tested with indomethacin 5% pet. and there were 2 (0.5%) positive reactions. One of them was relevant, the relevance of the other unknown (4).

Case series

In Leuven, Belgium, in the period 1990-2014, iatrogenic contact dermatitis was diagnosed in 2600 individuals (17% of the total patch test population). 96% of all positive patch test reactions to topical drugs and antiseptics were considered to be relevant. Indomethacin (2% pet.) was tested in 44 patients and there was one positive reaction to it (2).

In the period 1996-2001, in 2 hospitals in Spain, 2 patients were diagnosed with contact allergy and zero with photocontact allergy to indomethacin. The accumulated incidence per million inhabitants (catchment population of the hospitals) of both side effects together was 2.4 (11).

In Italy, before 1993, the members of the GIRDCA Multicentre Study Group diagnosed 102 patients (49 men, 53 women), aged 16 to 66 years (mean 37 years), with (photo)dermatitis induced by systemic or topical NSAIDs. Indomethacin caused one contact allergic and zero photocontact allergic reactions (3).

Case reports

A 53-year-old man was treated for an acutely inflamed right elbow with 1% indomethacin gel. One day later, he developed increasing itching, erythema and vesiculation limited to the treatment area. Patch tests were positive to the indomethacin gel 'as is' and to 0.1% diclofenac gel 'as is'. It was concluded that the patient had 'Allergic contact dermatitis due to indomethacin and diclofenac' and that the patient cross-reacted to diclofenac. However, neither of the active drugs were tested, the gel bases were not tested nor the other ingredients. Also, the structural formulas of both NSAIDs are quite different, so not one of the conclusions of the authors was well substantiated (5).

Following cataract surgery, a 70-year-old woman was treated with various eye drops, including diclofenac sodium 0.1% eye drops. Two weeks later, she developed itchy erythema and edema of her eyelids, without conjunct-tivitis. She had a previous history of contact dermatitis from − unknown - topical non-steroidal anti-inflammatory drugs (NSAIDs). Patch tests were positive to the eye drops 'as is', diclofenac sodium 2%, 1% indomethacin cream 'as is' and indomethacin 2% pet. The authors suggested that she had previously become sensitized to indomethacin from frequent use of topical NSAIDs (6).

A 14-year-old boy presented with a pruriginous, erythematous, papulovesicular eruption on the right ankle that appeared 10 days after he had started therapy with indomethacin gel for a sprained ankle. Previously, he had taken NSAIDs orally but never topically. Patch tests were positive to the gel 'as is', indomethacin 1% pet. but negative to the excipients of the gel and 5 other NSAIDs. Oral administration of indomethacin up to the therapeutic dose was well tolerated (7).

A tendonitis of the left arm of a 23-year-old man was treated with indomethacin 5% spray for 14 days. Within that period, he developed increased itching and an acute papular and vesicular dermatitis on his left elbow which disappeared after the application was discontinued. Patch tests were strongly positive to the spray 'as is' and indomethacin 5% pet. at D2 and D3. When the patient was re-exposed to the spray applied to the other elbow for 2 days, an identical dermatitis appeared. Ten controls were negative (9).

PHOTOSENSITIVITY

In the period 2004-2005, in a Spanish multicenter study performing photopatch testing, one relevant positive photopatch tests was observed to indomethacin 10% pet. It was not mentioned how many patients had been tested with this chemical (10).

Cross-reactions, pseudo-cross-reactions and co-reactions

A patient sensitized to benzydamine may have cross-reacted to indomethacin; both are indolacetic acid derivatives (8).

LITERATURE

1   The data in the section 'General' may have been obtained from literature discussed in this chapter, but mostly also or exclusively from one or more of the following online sources: ChemIDPlus Advanced, PubChem, DrugBank, RxList, Drug Central, Drugs.com, and Wikipedia
2   Gilissen L, Goossens A. Frequency and trends of contact allergy to and iatrogenic contact dermatitis caused by topical drugs over a 25-year period. Contact Dermatitis 2016;75:290-302
3   Pigatto PD, Mozzanica N, Bigardi AS, Legori A, Valsecchi R, Cusano F, et al. Topical NSAID allergic contact dermatitis. Italian experience. Contact Dermatitis 1993;29:39-41
4   Gniazdowska B, Ruëff F, Przybilla B. Delayed contact hypersensitivity to non-steroidal anti-inflammatory drugs. Contact Dermatitis 1999;40:63-65
5   Gómez A, Florido JF, Quiralte J, Martín AE, Sáenz de San Pedro B. Allergic contact dermatitis due to indomethacin and diclofenac. Contact Dermatitis 2000;43:59
6   Ueda K, Higashi N, Kume A, Ikushima-Fujimoto M, Ogiwara S. Allergic contact dermatitis due to diclofenac and indomethacin. Contact Dermatitis 1998;39:323
7   Pulido Z, González E, Alfaya T, Alvarez JA, Ceña M, de la Hoz B. Allergic contact dermatitis from indomethacin. Contact Dermatitis 1999;41:112
8   Goday Buján JJ, Ilardia Lorentzen R, Soloeta Arechavala R. Allergic contact dermatitis from benzydamine with probable cross-reaction to indomethacin. Contact Dermatitis 1993;28:111-112
9   Beller U, Kaufmann R. Contact dermatitis to indomethacin. Contact Dermatitis 1987;17:121
10 De La Cuadra-Oyanguren J, Perez-Ferriols A, Lecha-Carrelero M, et al. Results and assessment of photopatch testing in Spain: towards a new standard set of photoallergens. Actas DermoSifiliograficas 2007;98:96-101
11 Diaz RL, Gardeazabal J, Manrique P, Ratón JA, Urrutia I, Rodríguez-Sasiain JM, Aguirre C. Greater allergenicity of topical ketoprofen in contact dermatitis confirmed by use. Contact Dermatitis 2006;54:239-243

Chapter 3.183 INTERFERONS

IDENTIFICATION   **Interferon alfa-2c**

Description/definition	: Interferon alfa-2c is a subtype of interferon α
Pharmacological classes	: Antineoplastic agents; antiviral agents
CAS registry number	: 135669-44-2; 142192-09-4
EC number	: 232-710-3 (Interferons)
Merck Index monograph	: 6304 (Interferon-α)
Patch testing	: Commercial preparation
Molecular formula	: Unspecified

IDENTIFICATION   **Interferon beta**

Description/definition	: Interferon beta is one of the type I interferons
Pharmacological classes	: Antineoplastic agents; antiviral agents
IUPAC name	: Not available
Other names	: Interferon alphaB (human leukocyte protein moiety reduced); beta-interferon
CAS registry number	: 77238-31-4
EC number	: 232-710-3 (Interferons)
Merck Index monograph	: 6305 (Interferon-β)
Patch testing	: Commercial preparation
Molecular formula	: Unspecified

IDENTIFICATION   **Interferon beta-1b**

Description/definition	: Interferon beta-1b is a non-glycosylated form of interferon β1 that has a serine at position 17
Pharmacological classes	: Adjuvants, immunologic
Other names	: 17-L-Serine-2-166-interferon beta-1b (human fibroblast reduced); interferon beta-1b, recombinant
CAS registry number	: 145155-23-3
EC number	: 232-710-3 (Interferons)
Merck Index monograph	: 6305 (Interferon-β)
Patch testing	: Commercial preparation
Molecular formula	: $C_{903}H_{1399}N_{245}O_{252}S_5$

GENERAL

Interferon alfa-2c is a subtype of interferon-alfa that was used in trials in the 1980s and 1990s for the treatment of specific forms of leukemia. It may currently be used for viral hepatitis and hiv-infection and may be investigated for the treatment of progressive metastatic renal cell carcinoma. Interferon-beta is one of the type I interferons produced by fibroblasts in response to stimulation by live or inactivated virus or by double-stranded RNA. It is a cytokine with antiviral, antiproliferative, and immunomodulating activity. Human interferon-beta may be investigated for treatment of metastatic Merkel cell carcinoma. Interferon beta-1b is a non-glycosylated form of interferon beta-1 that has a serine at position 17. It is used in the treatment of both relapsing-remitting multiple sclerosis and chronic progressive multiple sclerosis (1).

CONTACT ALLERGY

Case report: topical administration

A 57-year-old painter presented with acute facial eczema following the use of eye drops for conjunctivitis, containing beta-interferon (interferon-β; IFN-β). There was considerable facial swelling and erythema, with vesicles and, in places, serous scaling. The face was intensely itchy and excoriated. Patch tests showed a positive reactions to the commercial eye drops, even after 1:32 dilution. Ten control subjects were negative. Negative results were also obtained when each of the excipients was patch tested separately (PVP, human albumin, mannite, buffered salts).

IFN-beta itself could not be patch tested because it was available commercially only in bulk, but there were no reactions to interferon-alfa and interferon-gamma. Next, an anti-IFN-beta antibody was added to the eye drops at a concentration of 10 mg/ml, sufficient to inhibit the IFN-beta. After precipitation, centrifugation and preparation of the supernatant, retesting with this inhibited preparation, unlike the original preparation, did not provoke a positive patch test reaction (3).

Case reports: subcutaneous injections

A 36-year old woman was treated with subcutaneous injections of interferon beta-1b for multiple sclerosis. By the 14th of 22 injections, she noticed a transient erythema (appearing after 3-5 days), and after 5 non-consecutive injections (4 in the arm, 1 in the abdomen), she showed infiltrated, red, vesicular, itchy lesions at all 5 injection sites, which healed in 3-4 weeks. A patch test was negative to the commercial beta-interferon material. An intradermal test was negative after 20 minutes, but positive at D2 (2).

A 55-year-old man with hairy cell leukemia was treated with subcutaneous injections of recombinant interferon alfa-2c in a clinical study. At one point during treatment, he began to notice itching erythema at the injection site developing about 2 days after each injection. Patch tests were positive to the commercial interferon alfa-2c used, tested undiluted. A group of 10 patients, with no previous interferon treatment, showed no positive reactions when tested as controls (4).

LITERATURE

1   The data in the section 'General' may have been obtained from literature discussed in this chapter, but mostly also or exclusively from one or more of the following online sources: ChemIDPlus Advanced, PubChem, DrugBank, RxList, Drug Central, Drugs.com, and Wikipedia
2   Van Rengen A, Goossens A. Local reactions after subcutaneous injections of beta-interferon. Contact Dermatitis 1998;39:209
3   Pigatto PD, Bigardi A, Legori A, Altomare GF, Riboldi A. Allergic contact dermatitis from beta-interferon in eye drops. Contact Dermatitis 1991;25:199-200
4   Detmar U, Agathos M, Nerl C. Allergy of delayed type to recombinant interferon alpha 2c. Contact Dermatitis 1989;20:149-150

Chapter 3.184 IODINE

IDENTIFICATION

Description/definition   : Iodine is the non-metallic element of the halogen group that is represented by the atomic
                             symbol I and has atomic number 53
Pharmacological classes : Trace elements; anti-infective agents, local
IUPAC name   : Molecular iodine
CAS registry number   : 7553-56-2
EC number   : 231-442-4
Merck Index monograph : 6327
Patch testing   : 0.25% alcohol; this may cause false-positive, irritant reactions, but also allergic *and*
                           irritant patch test reactions; 0.5% pet. (9)
Molecular formula   : I_2

GENERAL

Iodine is a nonmetallic element of the halogen group that is represented by the atomic symbol I, atomic number 53, and atomic weight of 126.90. It is a nutritional element essential for growth and development and especially important in thyroid hormone synthesis. Iodine deficiency causes goitre and hypothyroidism in children and adults, and cretinism if present during fetal development. Since 1839, iodine compounds have been used as antiseptics and disinfectants. It has a broad germicidal action, being effective against bacteria, fungi, viruses, and protozoa. Iodine has traditionally been used in its tincture form, which consists of 2% iodine and 2.4% sodium iodide diluted in alcohol. Iodine preparations are currently available in aqueous solution, tincture of alcohol, aerosol, ointment, antiseptic gauze pad, foam and swab sticks (1,9). Iodine is also present in povidone-iodine (Chapter 3.270 Povidone-iodine).

CONTACT ALLERGY

General

Contact allergy to and allergic contact dermatitis from iodine are well-known but have not been frequent. In the 1930s iodine sensitivity was found in about 1 per cent of all patients tested at the Finsen Institute in Copenhagen. In the 1950s tincture of iodine was gradually replaced with other antiseptics and numbers of sensitization consequently dropped, to almost zero in 1953 in Denmark (12). Iodine and iodine hydro-alcoholic solutions are well-known local irritants which can cause iodine burns. Patch testing with iodine 1% pet. and 0.75% isopropyl alcohol will cause irritant reactions (9). It may be assumed that, especially in the older publications (but also in the more recent ones [2,3]) and when tested with hydro-alcoholic solutions, many cases of 'sensitization' to iodine have in fact been irritant patch test reactions. As iodine is – apart from some cases of sensitization to it in povidone-iodine – a historical allergen, the early literature is not discussed here. Contact allergy to iodine from its presence in povidone-iodine is presented in Chapter 3.270 Povidone-iodine.

Patch testing in groups of selected patients

In Leuven, Belgium, in the period 1990-2014, iatrogenic contact dermatitis was diagnosed in 2600 individuals (17% of the total patch test population). Iodine 0.5% alc. was tested in 1446 patients and there were 196 (13.6%) positive reactions to it. Relevance was not specified for individual allergens, but 96% of all positive patch test reactions to topical drugs and antiseptics were considered to be relevant (3, overlap with refs. 2 and 4).

In The Netherlands, in the period 1995-1999, 79 patients with chronic otitis externa were patch tested with iodine 1% alc. 96% and 4 patients (5.1%) had a positive reaction. Relevance was not specified for individual allergens, but 18% of all positive patch test reactions to ingredients of topical medications were considered to be relevant (5).

Case series

In Leuven, Belgium, between 1990 and 2017, 16,065 patients were investigated for contact allergy. Iodine was tested in 1538 patients and there were 176 reactions (11.4%) to iodine in various types of medications (2, overlap with refs. 3 and 4). In the period 1985-1997, in the same university clinic, 8521 patients were patch tested, and 89 positive reactions were observed to iodine. It was not stated, however, how many patients had been tested with this allergen and how many of the reactions were relevant (4, overlap with refs. 2 and 3).

Seven patients were sensitized to iodine (tested 0.25% alc.) and/or povidone-iodine, causing acute dermatitis during surgical interventions in two of them, perilesional dermatitis resulting from the treatment of a wound in four of them, and occupational dermatitis in one of them (a midwife with hand dermatitis sensitized to PVP-iodine in the hospital) (6). The data from iodine and povidone-iodine were not well-separated in this study.

Case report
A 41-year-old man was treated for a yellowish necrotic ulcer on the left ankle with an ointment containing 0.9% iodine and cadexomer. Although the topical medicament was effective for the necrotic ulcer, an itchy erythema suddenly developed around the ulcer 6 months later. Patch testing showed a positive reaction to the ointment itself and to 0.33% iodine/10% glycerin w/v but negative to glycerin and the other ingredients of the ointment. Six controls were negative (8).

Systemic contact dermatitis
A 60-year-old woman allergic to iodine developed erythroderma from an iodine-containing X-ray contrast medium (10), a manifestation of systemic contact dermatitis. Patients sensitized to iodine may also develop systemic contact dermatitis with a maculopapular eruption (11), generalized pruritus and urticaria-like erythema (10) from foods rich in iodine, notably seaweed, such as kelp (*konbu*) and *wakame*, which are generally consumed by the Japanese (10).

Cross-reactions, pseudo-cross-reactions and co-reactions
In 7 patients sensitized to iodine (tested 0.25% alc.) and/or povidone-iodine (tested 5% water or with the commercial Betadine solution), co-reactions were found to iodopropynyl butylcarbamate (IPBC), for which no relevance could be found. The authors suspected that free iodine, released from IPBC, had caused the positive patch test reactions to this preservative (6).

Six of seven patients with positive patch tests to iodinated contrast media (ICMs) had additional positive patch tests to 'one or more iodine formulations' (not further specified). Oral provocation with iodine was positive in only 2. Quite unexpectedly (probably from the low number of positive oral provocation tests), the authors concluded that in the majority of allergic reactions to ICMs, iodine is not the eliciting agent (7).

Patients sensitized to iodine may cross-react (or more accurately: pseudo-cross-react) to povidone-iodine and vice versa (see Chapter 3.270 Povidone-iodine).

LITERATURE

1   The data in the section 'General' may have been obtained from literature discussed in this chapter, but mostly also or exclusively from one or more of the following online sources: ChemIDPlus Advanced, PubChem, DrugBank, RxList, Drug Central, Drugs.com, and Wikipedia
2   Gilissen L, De Decker L, Hulshagen T, Goossens A. Allergic contact dermatitis caused by topical ophthalmic medications: Keep an eye on it! Contact Dermatitis 2019;80:291-297
3   Gilissen L, Goossens A. Frequency and trends of contact allergy to and iatrogenic contact dermatitis caused by topical drugs over a 25-year period. Contact Dermatitis 2016;75:290-302
4   Goossens A, Claes L, Drieghe J, Put E. Antimicrobials, preservatives, antiseptics and disinfectants. Contact Dermatitis 1998;39:133-134
5   Devos SA, Mulder JJ, van der Valk PG. The relevance of positive patch test reactions in chronic otitis externa. Contact Dermatitis 2000;42:354-355
6   Vanhoutte C, Goossens A, Gilissen L, Huygens S, Vital-Durand D, Dendooven E, et al. Concomitant contact-allergic reactions to iodopropynyl butylcarbamate and iodine. Contact Dermatitis 2019;81:17-23
7   Bircher AJ, Harr T, Bach S, Scherer K. Iodine is rarely the elicitor in hypersensitivity reactions to iodinated contrast media. Dermatitis 2008;19:347
8   Oiso N, Fukai K, Ishii M. Allergic contact dermatitis from iodine in a topical ointment. Contact Dermatitis 2006;54:347-348
9   Lee SK, Zhai H, Maibach HI. Allergic contact dermatitis from iodine preparations: a conundrum. Contact Dermatitis 2005;52:184-187
10   Kubota Y, Koga T, Nakayama J. Iodine allergy induced by consumption of iodine-containing food. Contact Dermatitis 2000;42:286-287
11   Adachi A, Horikawa T, Shimizu R. A case of iodine allergy. Hifu 1989;31(Suppl.):25-30 (Article in Japanese; data cited in ref. 10).
12   Cronin E. Contact Dermatitis. Edinburgh: Churchill Livingstone, 1980:257-258

Chapter 3.185 IODOFORM

IDENTIFICATION

Description/definition	: Iodoform is an organo-iodine compound with the formula CHI_3 that conforms to the structural formula shown below
Pharmacological classes	: Antiseptics; disinfectants; dermatological agents
IUPAC name	: Iodoform
Other names	: Triiodomethane; carbon triiodide; Jodoform
CAS registry number	: 75-47-8
EC number	: 200-874-5
Merck Index monograph	: 6346
Patch testing	: 5% pet. (SmartPracticeCanada); 10% pet.; also test iodine 0.5% pet. or 0.25% alc. (risk of false-positive, irritant reactions)
Molecular formula	: CHI_3

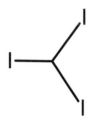

GENERAL

Iodoform is an organo-iodine compound with the formula CHI_3 and a tetrahedral molecular geometry. It is a relatively water-insoluble yellow solid powder or crystals with penetrating odor and unctuous feel, that is chemically reactive in free-radical reactions. Since the beginning of the 20th century, iodoform has been commonly used as a healing and antiseptic dressing or powder for wounds and sores, but currently such clinical use is limited, although bismuth iodoform paraffin paste (BIPP) may still be one of the most commonly used packs after middle and external ear surgery (4,7). Iodoform has also been found in dental paste and root canal filling materials in combination with other intracanal medications for its radiopacity. It may also still be used for veterinary purposes. When iodoform decomposes it releases iodine in nascent state (96.7% of iodine) when in contact with secretions, endodontic infections or granulation tissue (1).

CONTACT ALLERGY

General

A study from the United Kingdom showed that 12% of patients previously exposed to bismuth iodoform paraffin paste (BIPP) may have become sensitized, which makes it a frequent allergen. Unfortunately, the patients were tested with very high concentrations of iodoform of 50%, whereas controls were not performed. Nevertheless, in the group of patients with similar ear diseases that had *not* previously been exposed to BIPP, only 1% had positive reactions to the paste, which makes irritant reactions in the former group less likely (4). It appears that both iodine and iodoform itself may act as the sensitizer (4).

Case series

In Ipswitch, United Kingdom, in the period 2000-2007, 237 patients previously undergoing surgery, were patch tested with BIPP (tested pure). Twenty-eight (12%) were found to have positive reactions. Seventeen of these subjects had further tests with components: 50% iodoform paste (50% iodoform in liquid paraffin), 25% bismuth, 0.5% or 2% iodine solution and paraffin. All reacted to the iodoform paste, there were no allergies to bismuth. Of the 17 patients allergic to the iodoform paste, 10 (59%) did *not* react to iodine solution, but 7 did. In patients not previously exposed to BIPP, the rate of sensitization was 1% (4).

In 2002, from the U.K., 3 cases of allergic contact otitis externa due to bismuth subnitrate and iodoform paraffin paste (BIPP), seen at one hospital over a 3-year period, were reported. All 3 patients were women (aged 13, 16 and 52) who had had their external auditory meatus and concha packed with BIPP-impregnated gauze following surgery (myringoplasty n=2, mastoidectomy n=1). Clinical presentation was variable. One patient had mild erythema and swelling of the concha 3 weeks postoperatively after removal of the BIPP pack. In another patient, however, a florid eczematous reaction developed, which extended to the forehead, crossing the midline, one day after surgery, the patient having been previously exposed to BIPP. In all 3 individuals, patch tests were positive to iodoform 10% pet.

and negative to bismuth subnitrate 10% pet. One patient was tested with povidone-iodine but was negative; none were patch tested with iodine itself (5).

In the UK, between 1985 and 2002, 179 patients with chronic inflammatory ear disease were patch tested with iodoform 10% pet. and there were 3 (1.7%) positive reactions. All were considered to be relevant (3).

Case reports

A 50-year-old man had a radical mastoidectomy right for chronic suppurative otitis media. Post-operatively, a bismuth iodoform and paraffin paste-impregnated gauze was packed into the external auditory canal, left *in situ* for 2 weeks and changed biweekly. Four weeks after the operation, the patient developed acute dermatitis on his external auditory canal, right pinna and right side of the face, where the bismuth iodoform paraffin pack was applied. His dermatitis cleared after removal of the pharmaceutical and with topical steroid treatment. Patch testing showed positive reactions to the bismuth iodoform paraffin pack, iodoform 5%, 10% and 25% pet., potassium iodide 25% pet. and bismuth subnitrate 'as is'. Twenty controls were negative to iodoform 10% and 25% pet. and bismuth subnitrate pure. The authors suggested that the patient had probably become sensitized to iodine which was released from iodoform, as he reacted to potassium iodide, to which he had not been exposed before (2).

A 36-year-old man was referred with a 4-day history of ulceration in the mouth and soreness and crusting of the lips and surrounding skin. Twenty days previously, his lower left first premolar tooth had been extracted under general anaesthesia because of a periapical abscess. After 15 days, alveolar osteitis was diagnosed and the socket was packed with ribbon gauze impregnated with bismuth subnitrate and iodoform paste (iodoform 33%, bismuth subnitrate 17% and liquid paraffin 50%, abbreviated BIPP). On the following day, the patient developed mouth ulceration, swelling of the tongue and inflammation of the lips and perioral skin. He remembered having had a rash from iodine antiseptic previously. On examination, the patient had a severe left-sided stomatitis with ulceration of the left side of the tongue with an acute cheilitis and dermatitis of the perioral skin. Cervical lymphadenopathy was present. After successful treatment, an open patch test with BIPP gauze produced an area of erythema and edema 8x7 centimeter with central vesiculation after 2 days, persisting at D4, but without the vesiculation. Later, patch tests were positive to potassium iodide 5% pet. (6).

LITERATURE

1   The data in the section 'General' may have been obtained from literature discussed in this chapter, but mostly also or exclusively from one or more of the following online sources: ChemIDPlus Advanced, PubChem, DrugBank, RxList, Drug Central, Drugs.com, and Wikipedia
2   Goh CL, Ng SK. Contact allergy to iodoform and bismuth subnitrate. Contact Dermatitis 1987;16:109-110
3   Millard TP, Orton DI. Changing patterns of contact allergy in chronic inflammatory ear disease. Contact Dermatitis 2004;50:83-86
4   Bennett AMD, Bartle J, Yung MW. Avoidance of BIPP allergy hypersensitivity reactions following ear surgery. Clin Otolaryngol 2008;33:32-34
5   Roest MA, Shaw S, Orton DI. Allergic contact otitis externa due to iodoform in BIPP cavity dressings. Contact Dermatitis 2002;46:360
6   Maurice PD, Hopper C, Punnia-Moorthy A, Rycroft RJ. Allergic contact stomatitis and cheilitis from iodoform used in a dental dressing. Contact Dermatitis 1988;18:114-116
7   Crossland GJ, Bath AP. Bismuth iodoform paraffin paste: a review. J Laryngol Otol 2011;125:891-895

Chapter 3.186 IODOQUINOL

IDENTIFICATION

Description/definition	: Iodoquinol is the quinoline derivative that conforms to the structural formula shown below
Pharmacological classes	: Amebicides
IUPAC name	: 5,7-Diiodoquinolin-8-ol
Other names	: Diiodohydroxyquinoline; diiodoquin
CAS registry number	: 83-73-8
EC number	: 201-497-9
Merck Index monograph	: 6355
Patch testing	: 5% pet.
Molecular formula	: $C_9H_5I_2NO$

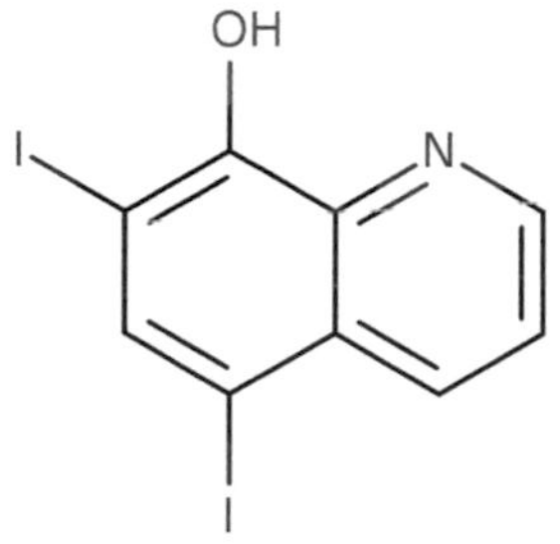

GENERAL

Iodoquinol is one of the halogenated 8-quinolinols, which is widely used as an intestinal antiseptic, especially as an anti-amebic agent. It is also used topically in other infections, often in combination with hydrocortisone acetate.

CONTACT ALLERGY

Case reports and case series

A 25-year-old man was prescribed 3% iodoquinol (diiodoquin) in Aquaphor base for an extensive infectious eczematoid dermatitis superimposed on a seborrheic dermatitis. By the following morning the patient had developed a severe eczematous contact dermatitis in all areas to which the ointment had been applied. The patient recalled that he suffered a similar episode of dermatitis about one year previously, after treatment with 3 per cent clioquinol ointment. Another patient treated chromate dermatitis of the feet with iodoquinol 3% in Aquaphor for a week when contact dermatitis developed. Both patients were patch test positive to iodoquinol 3-10% in Aquaphor and to related hydroxyquinolines (3). The first patient, after his skin had cleared, ingested a single tablet of 210 mg iodoquinol. A severe recrudescence of the eczematous dermatitis appeared within 12 hours after ingestion of the tablet (systemic contact dermatitis) (3).

In a group of 150 patients treated for infectious skin diseases or secondarily infected eczematous dermatoses with iodoquinol 3-15% in Aquaphor, one developed allergic contact dermatitis as shown by a positive patch test to iodoquinol with a cross-reaction to clioquinol. However, in a subgroup, 6 of 70 patients showed 'flare-ups' in the course of the therapy, who were probably not patch tested (4).

An early publication may have reported on systemic allergic dermatitis from vaginal absorption of iodoquinol. Floraquin is the trade name of vaginal tablets containing 100 mg di-iodohydroxyquinoline (iodoquinol) (5).

Cross-reactions, pseudo-cross-reactions and co-reactions
Cross-reactions between iodoquinol and other halogenated hydroxyquinolines may occur (2,3,4).

LITERATURE

1   Swinny B. Allergic dermatitis from vaginal absorption of sensitizers (floraquin and verazeptol) report of cases. Ann Allergy 1947;5:490
2   Allenby CF. Skin sensitization to Remiderm and cross-sensitization to hydroxyquinoline compounds. Br Med J 1965;2(5455):208-209
3   Leifer W, Steiner K. Studies in sensitization to halogenated hydroxyquinolines and related compounds. J Invest Dermatol 1951;17:233-240
4   Leifer W, Steiner K. Diiodohydroxyquinoline in dermatologic therapy. Arch Derm Syphilol 1950;62:46-53

Chapter 3.187 IOTHION

IDENTIFICATION

Description/definition	: Iothion is the alcohol and organoiodine compound that conforms to the structural formula shown below
Pharmacological classes	: Dermatological agents, antiseptics and disinfectants
IUPAC name	: 1,3-Diiodopropan-2-ol
Other names	: Diiodohydroxypropane
CAS registry number	: 534-08-7
EC number	: 208-586-1
Merck Index monograph	: 775
Patch testing	: 0.05% alc.
Molecular formula	: $C_3H_6I_2O$

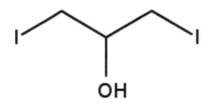

GENERAL

Iothion was formerly used in Denmark in 'White tincture of iodine', containing 10% iothion, 10% glycerin and 80% alcohol (1).

CONTACT ALLERGY

Case reports

Iothion containing tincture sometimes caused a blistering streaky contact dermatitis resembling plant dermatitis (phytophotodermatitis, dermatitis bullosa striata pratensis). One such patient had a positive patch test to iothion 0.05% alcohol. Twenty controls were negative (1). Apparently, a case of allergic contact dermatitis with concomitant urticarial reaction had already been described in 1928 in Germany; details are unknown (2).

LITERATURE

1   Hjorth N. Contact dermatitis from 1,3-diiodo-2-hydroxypropane. Contact Dermatitis Newsletter 1972;12:322
2   Löwenfeld W. Überempfindlichkeit gegen Iodthion mit gleichzeitiger urtikarieller Reaktion. Derm Wschr 1928;78:502 (Article in German).

Chapter 3.188 ISOCONAZOLE

IDENTIFICATION

Description/definition : Isoconazole is the phenylethyl imidazole compound that conforms to the structural formula shown below

Pharmacological classes : Antifungal agents

IUPAC name : 1-[2-(2,4-Dichlorophenyl)-2-[(2,6-dichlorophenyl)methoxy]ethyl]imidazole

Other names : 1-(2,4-Dichloro-β-(2,6-dichlorobenzyloxy)phenethyl)imidazole

CAS registry number : 27523-40-6

EC number : 248-508-3

Merck Index monograph : 6475

Patch testing : 1% alcohol or MEK (methyl ethyl ketone)

Molecular formula : $C_{18}H_{14}Cl_4N_2O$

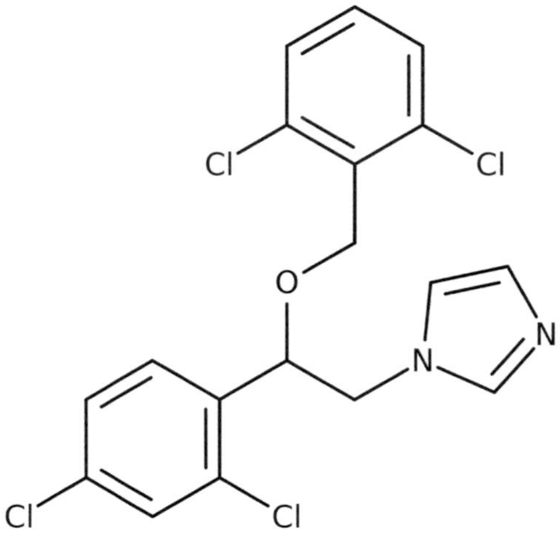

GENERAL

Isoconazole is an azole antifungal drug that is used in the treatment of dermatomycoses and vaginal *Candida* infections. In pharmaceutical products, both isoconazole and its salt isoconazole nitrate (CAS number 24168-96-5, EC number 246-051-4, molecular formula 246-051-4) may be employed (1).

CONTACT ALLERGY

Case series

In the period 1990 to 2014, in Leuven, Belgium, 148 patients suspected of iatrogenic contact dermatitis were tested with isoconazole 2% pet. and there were 8 (5.4%) positive reactions. Relevance was not specified for individual drugs, but 96% of the positive patch test reactions to all topical drugs and antiseptics were considered to be relevant (3). Between 1990 and 2003, in the same university hospital in Leuven, Belgium, 92 patients were patch tested for chronic vulval complaints. Fifteen of these women were tested with a series of topical drugs; there was one reaction to isoconazole 2% pet. which was relevant (2).

In the period 1976-1988, 13 cases of allergic contact dermatitis to imidazole antifungal preparations were seen in a private dermatological practice in Saverne, France, and 2 in the Université Catholique de Louvain, Belgium. Most had been prescribed for tinea cruris, pedis or corporis, some for anal pruritus, dermatitis and even a leg ulcer. The responsible antifungals (tested 2% pet.) were isoconazole in 5 cases, miconazole in 6 cases, econazole in 2 cases, and tioconazole in one case; one patient reacted to a base ingredient (7).

Between 1977 and 1986, 9 patients with contact allergy to imidazole antimycotics were investigated in the University clinic of Heidelberg, Germany. Three patients were allergic to isoconazole, 6 to miconazole, 3 to clotrimazole, 3 to econazole, and one to oxiconazole (5).

In a literature review up to 1994, the authors identified 105 reported patients who had a positive patch test reaction to at least one imidazole derivative, ranging from 51 reactions to miconazole to one reaction each to bifonazole and enilconazole. The number of reported reactions to isoconazole was 22 (4).

Case reports

A 65-year-old woman had itchy, erythematous and edematous diffuse dermatitis involving the arms and legs, abdomen, chest, and neck. Two weeks before, eczematous dermatitis began on the right thigh, where she had applied a cream containing diflucortolone valerate 0.1% and isoconazole nitrate 1% for suspected tinea corporis. She denied previous use of topical imidazoles. Patch tests were strongly positive to the cream and to isoconazole nitrate 2%. There was no reaction to the corticosteroid, but the patient did have cross-reactivity to miconazole (8).

A 48-year-old woman treated a fungal infection on her right hand with a solution containing isoconazole nitrate. After 4 weeks, she developed an itchy erythematous eruption which spread over the whole dorsum of her right hand and forearm. Patch tests were positive to isoconazole nitrate 5% and 10% pet. (negative in 10 controls) (9).

A 57-year-old woman presented with a long-standing severely itchy papular and pustular eruption on the arms and legs. The lesions had appeared after treatment for eczematoid tinea pedis with 1% chlorquinaldol cream and later with isoconazole nitrate cream and a cream containing isoconazole nitrate and a corticosteroid for a month. Patch tests were positive to both antifungal creams 'as is', to isoconazole nitrate 0.5% and 1% pet. (negative in 10 controls) and to the quinoline mix (containing chlorquinaldol). The patient did not react to the vehicle cream and other imidazoles (10). Another patient allergic to isoconazole (and probably miconazole) was reported from France in 1982 ; details are unknown to the author (6).

Isoconazole that had been retained in polyurethane and PVC slippers caused persistent allergic contact dermatitis of the dorsal feet of a 35-year-old man more than a month after he had stopped using isoconazole antifungal cream. Patch tests were positive to a piece of the inner upper part of the slippers (but negative to identical but *new* unused slippers), isoconazole cream and isoconazole nitrate 1% pet. and 1% alc. (12).

Cross-reactions, pseudo-cross-reactions and co-reactions

Statistically significant associations have been found in patient data between miconazole, econazole, and isoconazole; between isoconazole and tioconazole (4,5,7). Isoconazole is an isomer of miconazole which differs from econazole only by the addition of one chlorine atom (7). No cross-reactions to phenylmethyl imidazoles (bifonazole, clotrimazole), triazoles (fluconazole, itraconazole) or nitroimidazoles (albendazole, mebendazole, metronidazole, tinidazole) (8).

The question has been raised (but not answered) whether contact allergy to isoconazole and other (nitro)imidazoles may be overrepresented in patients allergic to methylchloroisothiazolinone/methylisothiazolinone (MCI/MI (11).

LITERATURE

1   The data in the section 'General' may have been obtained from literature discussed in this chapter, but mostly also or exclusively from one or more of the following online sources: ChemIDPlus Advanced, PubChem, DrugBank, RxList, Drug Central, Drugs.com, and Wikipedia
2   Nardelli A, Degreef H, Goossens A. Contact allergic reactions of the vulva: a 14-year review. Dermatitis 2004;15:131-136
3   Gilissen L, Goossens A. Frequency and trends of contact allergy to and iatrogenic contact dermatitis caused by topical drugs over a 25-year period. Contact Dermatitis 2016;75:290-302
4   Dooms-Goossens A, Matura M, Drieghe J, Degreef H. Contact allergy to imidazoles used as antimycotic agents. Contact Dermatitis 1995;33:73-77
5   Raulin C, Frosch PJ. Contact allergy to imidazole antimycotics. Contact Dermatitis 1988;18:76-80
6   Jelen G. Allergie gegen Imidazolhaltige Antimykotika: Kreuzallergie? Dermatosen im Beruf und Umwelt 1982;30:53-55 (Article in German)
7   Jelen G, Tennstedt D. Contact dermatitis from topical imidazole antifungals: 15 new cases. Contact Dermatitis 1989;21:6-11
8   Bianchi L, Hansel K, Antonelli E, Bellini V, Stingeni L. Contact allergy to isoconazole nitrate with unusual spreading over extensive regions. Contact Dermatitis 2017;76:243-245
9   Frenzel UH, Gutekunst A. Contact dermatitis to isoconazole nitrate. Contact Dermatitis 1983;9:74
10   Lazarov A, Ingber A. Pustular allergic contact dermatitis to isoconazole nitrate. Am J Contact Dermat 1997;8:229-230
11   Stingeni L, Rigano L, Lionetti N, Bianchi L, Tramontana M, Foti C, et al. Sensitivity to imidazoles/nitroimidazoles in subjects sensitized to methylchloroisothiazolinone/methylisothiazolinone: A simple coincidence? Contact Dermatitis 2019;80:181-183
12   Özkaya E. Patch testing with used and unused personal products: a practical way to show contamination with contact allergens. Contact Dermatitis 2016;75:328-330

Chapter 3.189 ISOFLUPREDONE ACETATE

IDENTIFICATION

Description/definition : Isoflupredone acetate is the acetate ester of the synthetic glucocorticoid isoflupredone that conforms to the structural formula shown below

Pharmacological classes : Glucocorticoids

IUPAC name : [2-[(8S,9R,10S,11S,13S,14S,17R)-9-Fluoro-11,17-dihydroxy-10,13-dimethyl-3-oxo-6,7,8,11,12,14,15,16-octahydrocyclopenta[a]phenanthren-17-yl]-2-oxoethyl] acetate

Other names : Isoflupredone 21-acetate; 9-fluoro-11β,17,21-trihydroxypregna-1,4-diene-3,20-dione 21-acetate; 9-fluoroprednisolone acetate

CAS registry number : 338-98-7

EC number : 206-423-9

Merck Index monograph : 6490 (Isoflupredone)

Patch testing : In general, corticosteroids may be tested at 0.1% and 1% in alcohol; late readings (6-10 days) are strongly recommended

Molecular formula : $C_{23}H_{29}FO_6$

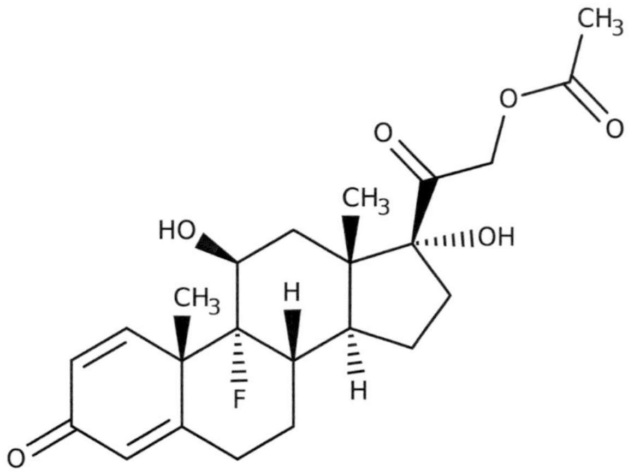

GENERAL

General aspects of corticosteroids used on the skin and mucous membranes are discussed in Chapter 2.4. A practical guideline for diagnosing allergic reactions to corticosteroids is presented in ref. 1. Isoflupredone acetate is used in some countries in veterinary medicine, both as isoflupredone and as its ester isoflupredone acetate.

CONTACT ALLERGY

Case series

From January 1990 to June 2008, in Leuven, Belgium, 315 patients were diagnosed with contact allergy to/allergic contact dermatitis from corticosteroids (CSs) from routine patch testing with a baseline series including tixocortol pivalate, budesonide, hydrocortisone butyrate and prednisone caproate, patch testing with patients' own CS preparations, and testing those with proven contact allergy to a corticosteroid or strongly suspected of CS allergy later with a series of 66 CSs, including two sex hormones (progesterone and testosterone). 71% of the patients had relevant reactions, but these were not specified. In this group of 315 CS allergic patients, 14 had positive patch tests to isoflupredone acetate 0.1% alc. (2). Very likely, most of these have been cross-reactions to other corticosteroid.

Cross-reactions, pseudo-cross-reactions and co-reactions

Cross-reactions between corticosteroids are discussed in Chapter 2.8.

LITERATURE

1 Baeck M, Goossens A. Immediate and delayed allergic hypersensitivity to corticosteroids: practical guidelines. Contact Dermatitis 2012;66:38-45

2 Baeck M, Chemelle JA, Terreux R, Drieghe J, Goossens A. Delayed hypersensitivity to corticosteroids in a series of 315 patients: clinical data and patch test results. Contact Dermatitis 2009;61:163-175

Chapter 3.190 ISOTHIPENDYL

IDENTIFICATION

Description/definition	: Isothipendyl is the azaphenothiazine that conforms to the structural formula shown below
Pharmacological classes	: Histamine antagonists
IUPAC name	: *N,N*-Dimethyl-1-pyrido[3,2-b][1,4]benzothiazin-10-ylpropan-2-amine
Other names	: 10-(2-Dimethylamino-2-methylethyl)-10*H*-pyrido(3,2-b)(1,4)benzothiazine
CAS registry number	: 482-15-5
EC number	: 207-578-5
Merck Index monograph	: 6543
Patch testing	: 1% pet.
Molecular formula	: $C_{16}H_{19}N_3S$

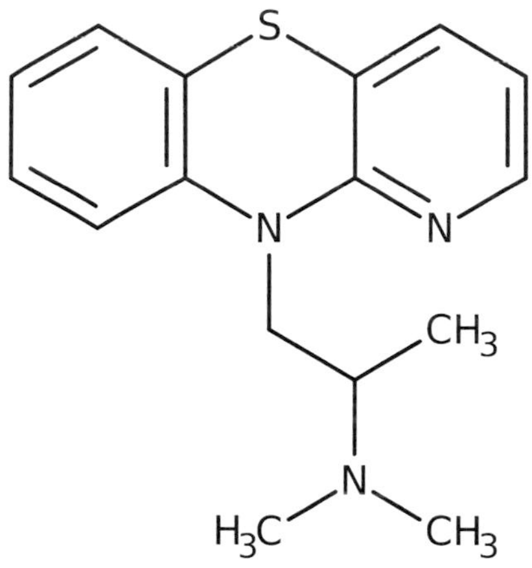

GENERAL

Isothipendyl is an azaphenothiazine H1 receptor antagonist with antihistaminic and anticholinergic properties. It is indicated for the topical treatment of itching associated with allergic reactions. In pharmaceutical products, isothipendyl is employed as isothipendyl hydrochloride (CAS number 1225-60-1, EC number 214-957-9, molecular formula $C_{16}H_{20}ClN_3S$) (1).

CONTACT ALLERGY

Case report

A 41-year-old woman with hand dermatitis used isothipendyl gel; the next day she experienced an exacerbation of the lesions as well as a new lesion on her face. Patch tests were positive to the gel and to isothipendyl 1% pet. Ten controls were negative. Patch tests with structurally related phenothiazines (trimeprazine, methotrimeprazine, chlorpromazine, prochlorpromazine, perphenazine, thioridazine) were negative (4).

Cross-reactions, pseudo-cross-reactions and co-reactions

Photocross-reactions between the phenothiazines isothipendyl, promethazine and chlorpromazine may occur (2,5). No cross-reactivity from isothipendyl to the phenothiazines trimeprazine, methotrimeprazine, chlorpromazine, prochlorpromazine, perphenazine, and thioridazine (4).

PHOTOSENSITIVITY

Case series

In a retrospective study performed in 21 French photodermatology centers from January 1, 2007 to October 31, 2019, 14 patients were diagnosed with photocontact dermatitis from topical phenothiazines: 6 women, 8 men, mean age 62 years, range 27-88. One patient had been reported previously (2). All individuals developed eczema at the site of application of the phenothiazine during sunny seasons. Distant sites were involved secondarily, including photodistributed areas, in 13 of 14 patients. The duration of eczema was longer than 3 months in 9 of 14, often with recurrence in spring and summer. The causative phenothiazine was isothipendyl in 13 of 14 cases. One patient used both isothipendyl gel and chlorproethazine ointment. Three patients were hospitalized, one developed persistent light reactions (5). In all of the nine cases tested, photopatch testing to the topical phenothiazine used 'as is' was positive. However, isothipendyl, chlorproethazine, and the excipients were not tested separately. Photopatch tests to chlorpromazine 0.1% pet. and promethazine 0.1% pet. were positive in 8 of 12 and 7 of 13 tested, respectively, and this was taken as proof, that the reactions to the isothipendyl gel were caused by the isothipendyl itself (photocross-reactivity). The authors acknowledge the phototoxic potential of isothipendyl and note that 'some positive photopatch tests with a + reaction to isothipendyl gel and promethazine may represent just a phototoxic reaction', but at the same time consider them to be photoallergic because the patients had prior eczema (5).

Case report

A 56-year-old woman with chronic eczema on the dorsum of the right foot associated with venous insufficiency developed eczematous lesions on sun-exposed areas (face, neck, nape, forearms, dorsum of the hands) progressing within 2 months to erythroderma, exacerbated on sun-exposed areas. She did not take any oral drugs. The patient was a farmer and handled the insecticide deltamethrin. Photopatch tests showed photocontact allergy to chlorpromazine with UVA at D2 and D3, but were negative to deltamethrin. The patient was informed by her chemist about the various topical drugs containing a phenothiazine and recognized the pharmaceutical gel containing 0.75% isothipendyl that she applied on her foot to reduce the pruritus. Next, a photopatch test with the gel was strongly positive (D1 ++, D2 +++) but was negative in the unirradiated patch test. After stopping use of the gel, the lesions disappeared within 1 month with symptomatic treatment and with strict external photoprotection (2).

The authors found that, between 1993 and 2010, 10 cases of photocontact allergy and one case of phototoxicity to isthipendyl-containing pharmaceuticals had been reported to the French Agency for Drug Pharmacovigilance (2). A phototoxic potential of isothipendyl has been reported (3).

LITERATURE

1   The data in the section 'General' may have been obtained from literature discussed in this chapter, but mostly also or exclusively from one or more of the following online sources: ChemIDPlus Advanced, PubChem, DrugBank, RxList, Drug Central, Drugs.com, and Wikipedia
2   Bibas N, Sartor V, Bulai Livideanu C, Bagheri H, Nougué J, Giordano-Labadie F, et al. Contact photoallergy to isothipendyl chlorhydrate. Dermatology 2012;224:289-291
3   Moreau A, Dompmartin A, Dubreuil A, Leroy D. Phototoxic and photoprotective effects of topical isothipendyl. Photodermatol Photoimmunol Photomed 1995;11:50-54
4   Takashima A, Yoshikawa K. Contact allergy to isothipendyl. Contact Dermatitis 1983;9:429-430
5   Cariou C, Droitcourt C, Osmont MN, Marguery MC, Dutartre H, Delaunay J, et al. Photodermatitis from topical phenothiazines: A case series. Contact Dermatitis 2020;83:19-24

Chapter 3.191 KANAMYCIN

IDENTIFICATION

Description/definition : Kanamycin is the aminoglycoside bactericidal antibiotic that conforms to the structural formula shown below

Pharmacological classes : Anti-bacterial agents; protein synthesis inhibitors

IUPAC name : (2R,3S,4S,5R,6R)-2-(Aminomethyl)-6-[(1R,2R,3S,4R,6S)-4,6-diamino-3-[(2S,3R,4S,5S,6R)-4-amino-3,5-dihydroxy-6-(hydroxymethyl)oxan-2-yl]oxy-2-hydroxycyclohexyl]oxyoxane-3,4,5-triol

Other names : 4,6-Diamino-2-hydroxy-1,3-cyclohexane 3,6'diamino-3,6'-dideoxydi-α-D-glucoside; kanamycin A

CAS registry number : 59-01-8

EC number : 200-411-7

Merck Index monograph : 6599

Patch testing : Sulfate, 10% pet. (Chemotechnique, SmartPracticeCanada, SmartPracticeEurope)

Molecular formula : $C_{18}H_{36}N_4O_{11}$

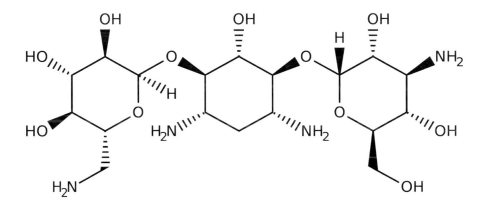

GENERAL

Kanamycin is an aminoglycoside bactericidal antibiotic isolated from the bacterium *Streptomyces kanamyceticus*. It is a complex comprising three components: kanamycin A, the major component, and kanamycins B and C. Kanamycin is indicated for treatment of infections where one or more of the following are the known or suspected pathogens: *E. coli*, *Proteus* species (both indole-positive and indole-negative), *E. aerogenes*, *K. pneumoniae*, *S. marcescens*, and *Acinetobacter* species. In pharmaceutical products, kanamycin is most commonly employed as kanamycin sulfate (CAS number 25389-94-0, EC number 246-933-9, molecular formula $C_{18}H_{38}N_4O_{15}S$) (1).

CONTACT ALLERGY

General

Kanamycin appears to be an infrequent sensitizer; no case reports of allergic contact dermatitis from topical pharmaceuticals have been found. Positive patch tests to kanamycin 10% pet. are not rare in patients with periorbital dermatitis and individuals suspected of contact allergy to ophthalmological medications, but in two studies, none of 6 reactions (7) and none of 10 positive patch tests (5) were relevant and all represented cross-reactions to other aminoglycosides. In not a single other such study (table 3.191.1) were relevant reactions reported, leaving the possibility open that most positive patch tests are cross-reactions rather than the result of primary sensitization from topical drugs.

Contact allergy in the general population

In Germany, for the period 1995-2004, the population-based relative incidence (RI) of contact sensitization to kanamycin (cases/100,000 defined daily doses (DDDs) per year) was estimated to be 8.7. In the group of ophthalmic drugs, the RI ranged from 0.3 (pilocarpine) to 21.6 (phenylephrine) (9).

Patch testing in groups of selected patients

The results of studies in which groups of selected patients were patch tested with kanamycin (notably patients with periorbital dermatitis and individuals suspected of contact allergy to ophthalmological drugs) are shown in table 3.191.1. In 8 investigations, 6 of which were performed by the IVDK (Information Network of Departments of Dermatology: Germany, Austria, Switzerland), prevalences of positive reactions to kanamycin have ranged from 2.9% to 9%. In not a single study were relevance data provided. In three of the investigations, a control group was included (2,3,4). In two, the rate of sensitization in the target population was higher than in the control group: 3.5% versus 1.4% (2) and 8.1% versus 3.8% (3); in the third, the frequency of reactions to kanamycin in patients with allergic periorbital dermatitis was not significantly higher than in a control group (4). Quite interestingly, the highest frequency (9%) was found in The Netherlands in a small group of 64 patients suspected of contact allergy to ophthalmological drugs, a country where kanamycin is not used in eye drops or other eye medications. All reactions were cross-reactions to other aminoglycoside antibiotics (7). As this was also the case in 10 kanamycin-allergic patients from Belgium (5) and not a single study reported relevant reactions, it cannot be excluded that most of these reactions represent cross-reactions rather than primary sensitization by kanamycin-containing drugs used by the patients.

Table 3.191.1 Patch testing in groups of patients: Selected patient groups

Years and Country	Test conc. & vehicle	Number of patients tested	positive (%)	Selection of patients (S); Relevance (R); Comments (C)	Ref.
2001-2010 IVDK	10% pet.	1142	56 (4.9%)	S: patients with periorbital dermatitis tested with an ophthalmic tray; R: not stated	10
2000-2010 IVDK		1819	69 (3.5%)	S: patients with periorbital dermatitis; R: not stated; C: the frequency in patients suspected of eye medicament allergy was 3.9% and in a control group of dermatitis patients 1.4%	2
2000-2010 Canada		100	5 (5%)	S: charts reviewed and included in the study when there was at least one positive reaction to a topical drug; R: not stated; C: the high percentages to all drugs are obviously the result of the – rather unusual – selection procedure targeting at previously diagnosed topical drug allergy	11
2000-6 Netherlands		64	6 (9%)	S: patients patch tested for suspected contact allergy to ophthalmological drugs; R: 69% for all reactions to drugs together; C: kanamycin is not used in eye medication in The Netherlands, the reactions were cross-reactions to other aminoglycoside antibiotics	7
1995-2004 IVDK	10% pet.	3875	(2.9%)	S: patients patch tested for suspected contact allergy to ophthalmological drugs; R: not stated; C: the estimated number of sensitizations per 100,000 prescriptions was 8.7	6
1995-2004 IVDK	10% pet.	4282	(3.1%)	S: not stated; R: not stated	8
1995-1999 IVDK	10% pet.	582	(3.7%)	S: patients with *allergic* periorbital contact dermatitis; the frequency was not significantly higher than in a control group	4
1990-1994 IVDK		283	23 (8.1%)	S: patients with periorbital eczema; R: not stated; C: frequency of sensitization in dermatitis controls: 3.8%	3

IVDK: Information Network of Departments of Dermatology (Germany, Austria, Switzerland)

Case series

In Leuven, Belgium, between 1990 and 2017, 16,065 patients were investigated for contact allergy and 118 (0.7%) showed positive patch test reactions to topical ophthalmic medications and/or to their ingredients. Eighty-four individuals (71%) reacted to an active principle. There were 10 reactions to kanamycin, but these were all the result of cross-sensitivity to other aminoglycosides (5).

Cross-reactions, pseudo-cross-reactions and co-reactions

In patients sensitized to neomycin, about 60% cross-reacts to kanamycin (Chapter 3.235 Neomycin). Four of 5 patients who had developed allergic contact dermatitis from paromomycin also had positive reactions to 20% neomycin sulfate and the same four also reacted to kanamycin sulfate 10% pet. It was not mentioned whether these patients had used either aminoglycoside before (12). A patient sensitized to gentamicin in eye drops co-reacted to kanamycin, which was considered to be a cross-reaction (13). Of 29 patients with a positive patch test reaction to gentamicin, 7 co-reacted to kanamycin (14).

The cross-sensitivity pattern between aminoglycoside antibiotics in patients primarily sensitized to kanamycin has not been well investigated.

LITERATURE

1   The data in the section 'General' may have been obtained from literature discussed in this chapter, but mostly also or exclusively from one or more of the following online sources: ChemIDPlus Advanced, PubChem, DrugBank, RxList, Drug Central, Drugs.com, and Wikipedia

2   Landeck L, John SM, Geier J. Periorbital dermatitis in 4779 patients – patch test results during a 10-year period. Contact Dermatitis 2014;70:205-212

3   Ockenfels H, Seemann U, Goos M. Contact allergy in patients with periorbital eczema: an analysis of allergens. Dermatology 1997;195:119-124

4   Herbst RA, Uter W, Pirker C, Geier J, Frosch PJ. Allergic and non-allergic periorbital dermatitis: patch test results of the Information Network of the Departments of Dermatology during a 5-year period. Contact Dermatitis 2004;51:13-19

5   Gilissen L, De Decker L, Hulshagen T, Goossens A. Allergic contact dermatitis caused by topical ophthalmic medications: Keep an eye on it! Contact Dermatitis 2019;80:291-297

6   Uter W, Menezes de Pádua C, Pfahlberg A, Nink K, Schnuch A, Behrens-Baumann W. Contact allergy to topical ophthalmological drugs - epidemiological risk assessment. Klin Monbl Augenheilkd 2009;226:48-53 (Article in German)

7   Wijnmaalen AL, van Zuuren EJ, de Keizer RJ, Jager MJ. Cutaneous allergy testing in patients suspected of an allergic reaction to eye medication. Ophthalmic Res 2009;41:225-229

8   Menezes de Padua CA, Uter W, Schnuch A. Contact allergy to topical drugs: prevalence in a clinical setting and estimation of frequency at the population level. Pharmacoepidemiol Drug Saf 2007;16:377-384

9   Menezes de Padua CA, Schnuch A, Nink K, Pfahlberg A, Uter W. Allergic contact dermatitis to topical drugs – Epidemiological risk assessment. Pharmacoepidemiol Drug Saf 2008;17:813-821

10 Landeck L, John SM, Geier J. Topical ophthalmic agents as allergens in periorbital dermatitis. Br J Ophthalmol 2014;98:259-262

11 Spring S, Pratt M, Chaplin A. Contact dermatitis to topical medicaments: a retrospective chart review from the Ottawa Hospital Patch Test Clinic. Dermatitis 2012;23:210-213

12 Veraldi S, Benzecry V, Faraci AG, Nazzaro G. Allergic contact dermatitis caused by paromomycin. Contact Dermatitis 2019;81:393-394

13 Sánchez-Pérez J, López MP, De Vega Haro JM, García-Díez A. Allergic contact dermatitis from gentamicin in eyedrops, with cross-reactivity to kanamycin but not neomycin. Contact Dermatitis 2001;44:54.

14 Liippo J, Lammintausta K. Positive patch test reactions to gentamicin show sensitization to aminoglycosides from topical therapies, bone cements, and from systemic medication. Contact Dermatitis 2008;59:268-272

Chapter 3.192 KETOCONAZOLE

IDENTIFICATION

Description/definition : Ketoconazole is a phenylethyl imidazole and synthetic derivative of phenylpiperazine that conforms to the structural formula shown below

Pharmacological classes : 14α-Demethylase inhibitors; antifungal agents; cytochrome P-450 CYP3A inhibitors

IUPAC name : 1-[4-[4-[[(2R,4S)-2-(2,4-Dichlorophenyl)-2-(imidazol-1-ylmethyl)-1,3-dioxolan-4-yl]methoxy]phenyl]piperazin-1-yl]ethanone

CAS registry number : 65277-42-1

EC number : 265-667-4

Merck Index monograph : 6619

Patch testing : 1% alcohol

Molecular formula : $C_{26}H_{28}Cl_2N_4O_4$

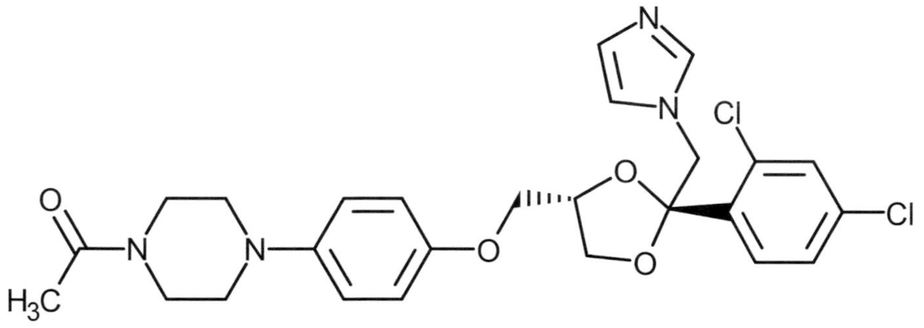

GENERAL

Ketoconazole is a synthetic derivative of phenylpiperazine and an imidazole-type antifungal with broad fungicidal properties and potential antineoplastic activity. It inhibits sterol 14α-demethylase, a microsomal cytochrome P450-dependent enzyme, thereby disrupting synthesis of ergosterol, an important component of the fungal cell wall.

Ketoconazole is indicated for the treatment of superficial fungal infections and the following systemic fungal infections: candidiasis, chronic mucocutaneous candidiasis, oral thrush, candiduria, blastomycosis, coccidioido-mycosis, histoplasmosis, chromomycosis, and paracoccidioidomycosis (1).

CONTACT ALLERGY

General

Contact allergy to ketoconazole appears to be rare, as only a few case reports have been published. However, a case series of 12 patient with allergic contact dermatitis and 5 with photoallergic contact dermatitis has been reported from Brazil, where topical preparations with ketoconazole are widely prescribed by medical practitioners and available as OTC preparations. Yet, this was in a period of 7 years (7). Many cases of reactions to ketoconazole cream are not caused by the antifungal drug itself but by excipients, e.g. propylene glycol (10), cetyl alcohol (14), polysorbate 80 (14) or sodium sulfite (14).

Patch testing in groups of selected patients

The results of patch testing in groups of selected patients (patients suspected of iatrogenic contact allergy, individuals tested with an antifungal series, patients with eyelid dermatitis) are shown in table 3.192.1. The prevalences of positive patch tests were low (0% - 1%) and only a few ketoconazole-sensitized patients were identified.

Case series

In a university hospital in São Paulo, Brazil, 749 patients were routinely patch tested between January 2010 and March 2017. 78 of them (10.4%) had suspected allergic contact dermatitis (ACD) or photoallergic contact dermatitis (PACD) caused by topical ketoconazole and they were patch tested (or photopatch tested, n=5) with ketoconazole 1% pet. Of these individuals, 17 (22%) had positive patch tests and 5 positive photopatch tests. Among the patients with a confirmed diagnosis of ACD/PACD, 12 were women. The mean age of the patients was 57 years, and the median age was 61 years. The face was affected in 11 patients, the legs in 9 and the arms in 7. Among the patients with facial involvement, 5 had PACD. The other body regions were related to the application areas of the medication.

There are at least 25 ketoconazole preparations available over-the-counter in Brazil and the authors suggest that it is their 'indiscriminate use' that makes ketoconazole a frequent allergen in their country (7).

Table 3.192.1 Patch testing in groups of patients: Selected patient groups

Years and Country	Test conc. & vehicle	Number of patients tested	positive (%)		Selection of patients (S); Relevance (R); Comments (C)	Ref.
1990-2014 Belgium	2% pet.	276	2	(0.7%)	S: patients suspected of iatrogenic contact dermatitis and tested with a pharmaceutical series and their own products; R: 96% of the positive patch test reactions to all topical drugs and antiseptics were considered to be relevant	3
2001-2002 IVDK	2% pet.	1104	0	(0%)	S: patients tested with an imidazole antifungal series; the selection procedure was not well described; R: not stated	4
1997-2001 USA		203	2	(1%)	S: patients with eyelid dermatitis; R: not stated; C: unknown in how many patients this material was tested	2

Spanish investigators reported 4 cases of hypersensitivity to ketoconazole preparations, but in only one was contact allergy to ketoconazole itself demonstrated (14). In a literature review up to 1994, the authors identified 105 reported patients who had a positive patch test reaction to at least one imidazole derivative, ranging from 51 reactions to miconazole to one reaction each to bifonazole and enilconazole. The number of reported reactions to ketoconazole was 6 (5).

Case reports

A 70-year-old woman had been treated for a week with ketoconazole lotion for seborrheic dermatitis of the scalp, when the dermatitis worsened greatly, with exudates and crust appearing on the face, scalp and neck. The patient revealed that she had started using an antidandruff shampoo containing miconazole, after which she got the scalp symptoms. Patch testing with the shampoo and all its ingredients and with ketoconazole lotion showed positive reactions to the shampoo 0.01% water, miconazole nitrate in pet. in the same concentration as in the shampoo (not specified) and 0.01% pet. and ketoconazole lotion. Ketoconazole itself was not tested. The authors considered it likely that that the patient had been sensitized first to miconazole in the shampoo and had then developed cross-reactive allergy to ketoconazole (8).

Short summaries of other case reports of allergic contact dermatitis from ketoconazole are shown in table 3.192.2.

Table 3.192.2 Short summaries of case reports of allergic contact dermatitis from ketoconazole

Year and country	Sex	Age	Positive patch tests	Clinical data and comments	Ref.
1993 Italy	M	36	cream 'as is'; ketocona-zole 1% and 2% alcohol	erythematous, edematous, vesicular dermatitis with severe itching on the abdomen, inguinal region and legs	9
	F	31	scalp fluid 'as is'; ketoco-zazole 1% and 2% alcohol	acute dermatitis with micropapules and erythema with pruritus from scalp lotion used for pityriasis versicolor	
1992 Italy	M	41	cream 'as is'; ketocona-zole 1% alcohol	vesicular dermatitis on facial seborrheic dermatitis; also allergy to propylene glycol, miconazole and sulconazole	10
	M	29	cream 'as is'; ketocona-zole 1% alcohol	acute dermatitis in the groins spreading to the legs and forearms; also allergy to miconazole and sulconazole, which had both been used before by the patient	
1992 Netherlands	M	41	cream 'as is'; ketocona-zole 2% alcohol	edema and aggravation of erythema from seborrheic dermatitis on the face; 17 controls negative to 2% alc.	11

The first case of allergic contact dermatitis to ketoconazole was reported in 1990 from Italy (12). A patient with a history of undiagnosed recurrent dermatitis developed acute facial swelling and pruritus after using ketoconazole cream and shampoo for the treatment of seborrheic dermatitis. Patch testing revealed 'true contact allergy' to ketoconazole with no cross-reactivity to 4 other imidazoles (14). Details are not available to the author.

Cross-reactions, pseudo-cross-reactions and co-reactions

Although there are some unique and sporadic reports of simultaneous reactions (8,10,16), cross-reactions are very unlikely between ketoconazole and other imidazoles (5). The question has been raised (but not answered) whether contact allergy to ketoconazole and other (nitro)imidazoles may be overrepresented in patients allergic to methylchloroisothiazolinone/methylisothiazolinone (MCI/MI) (6).

PHOTOSENSITIVITY

Five patients were diagnosed in Brazil with photoallergic contact dermatitis of the face from topical ketoconazole preparations. They had positive photopatch tests to 1% ketoconazole in petrolatum with exposure to ultraviolet A 10 J/cm^2 (7).

A 41-year-old man with atopic dermatitis and seborrheic dermatitis presented with eczematous, desquamative lesions involving his face, neck and dorsum of the hands. The affected skin was easily demarcated, exhibiting a photodistributed pattern. The patient was taking fenofibrate and ramipril daily, and occasionally used ketoconazole 2% gel. The eruption started as pruritic erythema hours after being exposed to sunlight at work, worsening during the following days. Earlier that day and a few days before, the patient had washed his face and hands with ketoconazole gel. Photopatch tests were positive to ketoconazole 2% pet., the 'plain' patch test being negative. Ramipril was negative, fenofibrate (a well-known cause of photocontact allergy) apparently not tested, but this drug could be continued without side effects (15).

Cutaneous adverse drug reactions from systemic administration caused by type IV (delayed-type) hypersensitivity

A 63-year-old man developed eczema in the groin after applying 1% econazole nitrate cream to treat tinea cruris. Later, miconazole 2% cream was applied and the patient took one tablet ketoconazole 200 mg per day. Two days later, generalized eczema developed. Patch tests with ketoconazole 2% cream and powdered tablets were positive. The components of the ketoconazole cream were tested later and proved positive for sodium sulfite, polysorbate 80 and ketoconazole 1% pet. (14). It was not discussed whether there may have been a cross-sensitivity from econazole and/or miconazole to ketoconazole.

LITERATURE

1   The data in the section 'General' may have been obtained from literature discussed in this chapter, but mostly also or exclusively from one or more of the following online sources: ChemIDPlus Advanced, PubChem, DrugBank, RxList, Drug Central, Drugs.com, and Wikipedia
2   Guin JD. Eyelid dermatitis: experience in 203 cases. J Am Acad Dermatol 2002;47:755-765
3   Gilissen L, Goossens A. Frequency and trends of contact allergy to and iatrogenic contact dermatitis caused by topical drugs over a 25-year period. Contact Dermatitis 2016;75:290-302
4   Menezes de Pádua CA, Uter W, Geier J, Schnuch A, Effendy I; German Contact Dermatitis Research Group (DKG); Information Network of Departments of Dermatology (IVDK). Contact allergy to topical antifungal agents. Allergy 2008;63:946-947
5   Dooms-Goossens A, Matura M, Drieghe J, Degreef H. Contact allergy to imidazoles used as antimycotic agents. Contact Dermatitis 1995;33:73-77
6   Stingeni L, Rigano L, Lionetti N, Bianchi L, Tramontana M, Foti C, et al. Sensitivity to imidazoles/nitroimidazoles in subjects sensitized to methylchloroisothiazolinone/methylisothiazolinone: A simple coincidence? Contact Dermatitis 2019;80:181-183
7   Lazzarini R, Hafner MFS, Miguel BAF, Kawakami NT, Nakagome BHY. Allergic contact dermatitis caused by topical ketoconazole: a relevant issue? Review of ketoconazole-positive patch tests. Contact Dermatitis 2018;78:234-236
8   Imafuku S, Nakayama J. Contact allergy to ketoconazole cross-sensitive to miconazole. Clin Exp Dermatol 2009;34:411-412
9   Valsecchi R, Pansera B, di Landro A, Cainelli T. Contact dermatitis from ketoconazole. Contact Dermatitis 1993 ;29 :162
10 Santucci B, Cannistraci C, Cristaudo A, Picardo M. Contact dermatitis from ketoconazole cream. Contact Dermatitis 1992;27:274-275
11 Verschueren GL, Bruynzeel DP. Hypersensitivity to ketoconazole. Contact Dermatitis 1992;26:47-48
12 Foti C, Filotico R, Grandolfo M, Vena GA. Dermatite allergica da contatto con ketaconazolo. Boll Derm All Profess 1990;5:251-253 (Article in Italian, cited in refs. 9 and 11)
13 Liu J, Warshaw E M. Allergic contact dermatitis from ketoconazole. Cutis 2014;94:112-114
14 Garcia-Bravo B, Mazuecos J, Rodriguez-Pichardo A, Navas J, Camacho F. Hypersensitivity to ketoconazole preparations: study of 4 cases. Contact Dermatitis 1989;21:346-348
15 Martínez-Doménech A, García-Legaz-Martínez M, Valenzuela-Oñate C, Magdaleno-Tapial J, Zaragoza-Ninet V, Sánchez-Carazo JL, et al. Photoallergic contact dermatitis to topical ketoconazole. J Eur Acad Dermatol Venereol 2020;34:e499-e501
16 Jones SK, Kennedy CT. Contact dermatitis from tioconazole. Contact Dermatitis 1990;22:122-123

Chapter 3.193 KETOPROFEN

IDENTIFICATION

Description/definition   : Ketoprofen is the arylpropionic acid derivate that conforms to the structural formula
                            shown below
Pharmacological classes : Anti-inflammatory agents, non-steroidal; cyclooxygenase inhibitors
IUPAC name   : 2-(3-Benzoylphenyl)propanoic acid
Other names   : Benzeneacetic acid, 3-benzoyl-α-methyl-; 2-(3-benzoylphenyl)propionic acid
CAS registry number   : 22071-15-4
EC number   : 244-759-8
Merck Index monograph   : 6622
Patch testing   : 1% pet. (Chemotechnique, SmartPracticeCanada); 2.5% pet. (SmartPracticeCanada);
                            perform photopatch tests; late readings at D5-D7 are advisable; as photopatch testing
                            may induce severe reactions, one-hour patch test occlusion with ketoprofen has been
                            proposed, which was successful for the detection of photosensitivity and simplifies the
                            photopatch test procedure by eliminating one visit to the clinic (98)
Molecular formula   : $C_{16}H_{14}O_3$

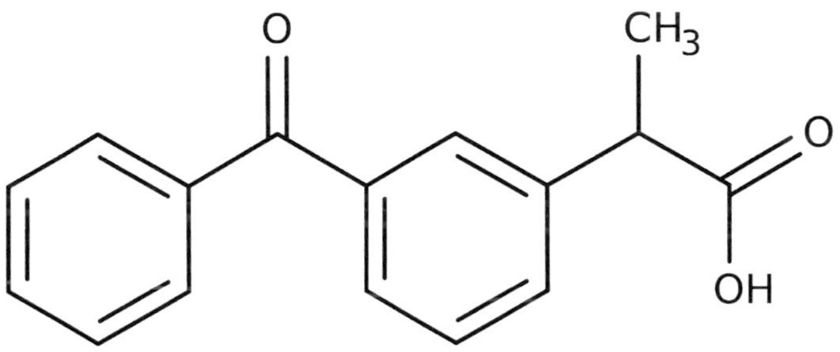

GENERAL

Ketoprofen is a propionic acid derivate and nonsteroidal anti-inflammatory drug (NSAID) with anti-inflammatory, analgesic and antipyretic effects. Ketoprofen inhibits the activity of the enzymes cyclo-oxygenase I and II, resulting in a decreased formation of precursors of prostaglandins and thromboxanes. The resulting decrease in prostaglandin synthesis, by prostaglandin synthase, is responsible for the therapeutic effects of this NSAID. Ketoprofen is indicated for symptomatic treatment of acute and chronic rheumatoid arthritis, osteoarthritis, ankylosing spondylitis, primary dysmenorrhea and mild to moderate pain associated with musculotendinous trauma (sprains and strains), postoperative (including dental surgery) or postpartum pain. It is also, and far more frequently, used in topical formulations (gel, cream, foam, ointment, tape) for musculoskeletal diseases and injuries for its analgesic and anti-inflammatory effects (1).

CONTACT ALLERGY

General

Topical ketoprofen can induce both contact allergy and photocontact allergy and cause allergic and photoallergic contact dermatitis. In almost all studies presenting case series of hypersensitivity to this NSAID, photoallergic reactions are far more frequent than contact allergic reactions (e.g. 2,5,10,42,43,48). It has been suggested that (some) contact allergies to ketoprofen may have been false-positive and in fact represent photocontact allergic reactions (49).

Patch testing in groups of selected patients

The results of patch testing with ketoprofen in groups of selected patients (patients suspected of iatrogenic contact dermatitis, patients with histories and clinical features suggestive of photoallergic contact dermatitis) are shown in table 3.193.1. Prevalences of positive patch tests were rather low, ranging from 0.9% to 2.7%. The majority of the contact allergic reactions were photoaugmented and, in the same studies, 7-16 times more photocontact allergies were demonstrated (2,42,43).

Table 3.193.1 Patch testing in groups of patients: Selected patient groups

Years and Country	Test conc. & vehicle	Number of patients tested	positive (%)	Selection of patients (S); Relevance (R); Comments (C)	Ref.
2014-2016 Spain		116	1 (0.9%)	S: not stated; R: not stated; C: there were also 16 positive photopatch test to ketoprofen	43
1990-2014 Belgium	1% pet.	294	8 (2.7%)	S: patients suspected of iatrogenic contact dermatitis and tested with a pharmaceutical series and their own products; R: 96% of the positive patch test reactions to all topical drugs and antiseptics were considered to be relevant; C: 5 of these patients had photo-aggravation of contact allergy; there were also 90 cases of *photo*contact allergy to ketoprofen (overlap with ref. 10)	2
2004-2008 Italy	1%, 2.5% or 5% pet.	1082	13 (1.2%)	S: patients with histories and clinical features suggestive of photoallergic contact dermatitis; R: not specified; all reactions were photoaugmented; there were also 97 (pure) photoallergic reactions	42

Case series

In the period 1993-2007, in a university hospital in Leuven, Belgium, 42 patients with a suspected (photo)allergic contact dermatitis from ketoprofen were investigated. 36 of 42 (photo)contact allergy tests took place between November 2000 and June 2007. Three patients showed contact allergic reactions, one photoaggravated allergic contact dermatitis, and 38 had photoallergic reactions to ketoprofen (10, overlap with ref. 2).

In the period 2001-2002, the members of the Belgian Contact & Environmental Dermatitis Group investigated 20 patients with suspected reactions to ketoprofen gels. Patch and photopatch tests with ketoprofen 2% in petrolatum showed contact allergy in 1 patient, photoaggravated contact allergy in 2 and contact photoallergy in 17 individuals (5). In the period 1996-2001, in 2 hospitals in Spain, 16 patients were diagnosed with contact allergy and 84 with photocontact allergy to ketoprofen. The accumulated incidence per million inhabitants (catchment population of the hospitals) of both side effects together was 101.4 (48).

In 3 centers in Italy, in the period 1989-1995, 26 patients with contact allergy to ketoprofen were seen and 25 with photocontact allergy (8). Also in Italy, before 1993, the members of the GIRDCA Multicentre Study Group diagnosed 102 patients (49 men, 53 women), aged 16 to 66 years (mean 37 years), with (photo)dermatitis induced by systemic or topical NSAIDs. Ketoprofen caused 21 contact allergic and 20 photocontact allergic reactions (3). These two studies are probably the only ones presenting more contact allergies than photocontact reactions (3,8).

A 43-year-old man applied ketoprofen 2.5% gel to his 'tennis-elbow' right, a 32-year-old woman did the same to her knee for postoperative pain and a 39-year-old man treated an occupational injury on this thighs with this gel. They all developed 'sensitivity reactions' and were patch test positive to the gel tested 'as is' and to ketoprofen pure, 5% water, 1% water (all +++) and 0.1% water (+, +, and ++, resp.) (89).

Small case series (always with more photo- than contact allergies) have also been presented in refs. 6,9,70,79, 82, and 87.

Case reports

An elderly married couple, the man aged 82 and the woman 80 years, were treated for diffuse rheumatic pains with a topical gel containing 2.5% ketoprofen. Each denied having taken NSAIDs systemically during the last month. They both presented with pruritic, erythematovesicular dermatitis on the hands and the sites where they had applied the gel to each other (back, shoulders and buttocks) for 10 days. In addition, the husband showed an erythema multiforme-like eruption extensively involving the trunk and limbs. Both patients had positive patch tests to the gel and to ketoprofen 1% pet. (37). A 21-year-old man had applied ketoprofen anti-inflammatory gel for cervical pain, until burning erythema appeared at the site of application. Patch tests were positive to the gel and to ketoprofen 10% pet. but negative to photopatch tests (84).

A 29-year-old man's right foot was put into a plaster cast because he had torn lateral ligaments on the dorsum of the right foot. On removing the cast 6 weeks later, irritation caused by friction from the cast was treated with a topical fluorinated steroid ointment, applied 3 x daily for 6 weeks. After these 6 weeks, and while still applying the steroid ointment, he began to use a ketoprofen gel to treat pain in the affected area. After 3 weeks, extremely itchy erythematous lesions and blisters appeared on the lower right leg and dorsum of the right foot. Curiously, this eruption spared the center of this area, where the skin was atrophied, probably due to excess application of the steroid. Patch tests were positive to ketoprofen gel and ketoprofen 1% pet. (85).

A 49-year-old man with 'tennis elbow' (epicondylitis lateralis humeri) applied an ointment containing ketoprofen to the right elbow for 2 weeks, when dermatitis developed. Patch tests were very strongly positive (+++) to the ointment and to ketoprofen 2.5% pet. (86). A 20-year-old man was treated with 2.5% ketoprofen gel for rheumatic pains in the left leg. After 10 days he developed contact dermatitis, which spread to the right leg. The eruption was characterized by confluent erythema multiforme-like patches. Mild vesiculation was also present. Patch tests were positive to the gel and to ketoprofen 1% pet., but negative to the other ingredients of the gel (90).

A 59-year-old woman had a long history of rheumatic pains in the feet, which she started to treat with ketoprofen cream. After 10 days' use, she noted erythema with itching at the application site; she stopped the use of the cream and was treated and cleared with wet dressings and a steroid ointment. One week later, she used the cream again, and after a few days had erythema, edema and vesicles at the application site and subsequently on the legs. Patch tests were positive to the cream and to ketoprofen 1% pet. (91).

A 40-year-old ballroom dance teacher presented with a figurate, acute eczematous reaction involving her right hand, back waist and left foot. She reported giving tango lessons and suspected that the dermatitis was related to close body contact while dancing. She reported past skin reactions to perfumes and deodorants, and summer recurrences of dermatitis at the dorsum of her left foot, where, several years before, she had developed a vesicobullous reaction after applying a ketoprofen-containing gel (possibly persistent light reaction). On direct questioning, she admitted sun bed use 24 hours prior to the abrupt onset of her present symptoms. One of her tango students, in turn, admitted using ketoprofen topical medications. The patient reacted strongly to ketoprofen (+++, photo-aggravated) and the fragrance mix. This represented a peculiar example of consort dermatitis, in which the allergen was transferred to the patient's skin by the hands of a tango partner, and skin penetration in areas covered by her dress was favored by close contact, friction and perspiration while dancing (36).

PHOTOSENSITIVITY

General

Photoallergic contact dermatitis from topical pharmaceuticals containing ketoprofen, notably gel and plasters (not the cream, which is surprising, as it was in the late 1980s as popular as the gel [82]), is well known and has been extensively documented in case series and case reports since 1985. Most patients were reported from Italy, France, Japan, and Belgium, where these products are widely used, but also from Sweden, Croatia and Spain. Currently, ketoprofen is by far the most frequent photosensitizer in (a number of countries in) Europe and is considered a potent photosensitizer (49).

Patients may have used the product for a period of 7-14 days or longer before becoming photosensitized, depending on the frequency of application and the intensity of exposure to ultraviolet light. Often, however, photodermatitis developed within 1-2 days of application and exposure to either sunlight or artificial UV-light from sunbed or solarium. These patients frequently remember having used the same product, a similar product containing ketoprofen or a pharmaceutical with a cross-reacting substance before, which was either well tolerated at that time or had given a rash.

Photoallergic contact dermatitis from topical ketoprofen is often acute and fierce with erythema, edema and vesicles or bullae (5,9,10) and begins at the sites of application exposed to ultraviolet light. It may be mistaken for (bullous) cellulitis (10,31). The lesions may be confined to the body sites where the ketoprofen product was applied, such as elbows, knees, ankles, forearms, and thighs, but dissemination is more frequent (5,10,47). Erythema multiforme-like eruptions have been observed (5,10,37,40).

Photodermatitis can also develop at other sites by transfer by hands or clothing. It may also affect family members or others in close contact (36,39; connubial contact dermatitis, consort dermatitis, dermatitis 'by proxy'). Moreover, ketoprofen contaminates clothing, shoes, bandages, etc., explaining some of the persistent reactions (38) and relapses (51). Not infrequently, the reactions to topical ketoprofen are severe and patients are hospitalized and/or treated with systemic in addition to topical corticosteroids (5,10,31,43,75). Usually, however, cessation of the causative agents and avoidance of sun exposure in combination with topical application of glucocorticosteroids improve the symptoms in 2 weeks. Residual post-inflammatory hyperpigmentation may occur (38,39). However, many patients have reported prolonged photosensitivity, sometimes up to one year or longer, after having stopped ketoprofen application, photodermatitis recurring in relation to sun exposure (9,10,49,67,93). One explanation for this is that benzophenone-3 in cosmetics and sunscreen products, to which many patients photocross-react, play a role (10). Contaminated clothing and other objects may sometimes contribute to persistence or relapses (51) and the persistent presence of ketoprofen in the skin after discontinuing its application, as shown in a skin biopsy specimen by chemical analysis (62), may be an alternative explanation.

When patients are patch and photopatch tested (mostly using ketoprofen 1% or 2.5% pet. and irradiation one day after application with UVA 5-10 J/cm^2), patch tests are usually negative and photopatch tests positive. As the intrinsic anti-inflammatory action of ketoprofen suppresses or delays the cutaneous response, late readings at D5 or D7 are advisable. At the first reading, the reaction may appear only at the edges of the test area, while it can be completely absent in its central portion, where the anti-inflammatory effect of ketoprofen is more evident because of the accumulation at higher concentrations. This phenomenon named 'edge effect' or 'border effect' fades away on successive readings a few days later (38). However, some patients have photoaggravated allergic contact dermatitis from combined sensitization and photosensitization and others may have (plain) allergic contact dermatitis to ketoprofen.

Many individuals with photosensitization to ketoprofen show photocross-reactions and co-reactions to related and non-related chemicals. This subject is discussed in the section 'Cross-reactions, pseudo-cross-reactions and co-reactions' below. Some patients with photosensitization to ketoprofen may also have contact or photocontact allergic reactions to excipients of the topical ketoprofen preparations, including parabens (87), menthol (41), hydrogenated glyceryl rosinate (41; colophonium derivative), orange flower oil (10), lavender oil/essence (5,10,47,49,75), eucalyptus oil (10), neroli oil (47), diisopropanolamine (76) and pine needle oil (10).

A (limited) review of ketoprofen photodermatitis was provided in 2016 (95).

Photopatch testing in groups of selected patients

The results of photopatch testing with ketoprofen in groups of selected patients (patients suspected of photoallergic contact dermatitis, individuals with possible iatrogenic contact dermatitis) are shown in table 3.193.2. Prevalences of photosensitization ranged from 2.8% to 30.6%, presumably depending on the mode of selection and the extent of exposure of the population investigated. Very high rates of 30.6% and 29% were observed in resp. Belgium (2) and Spain (46), where topical ketoprofen pharmaceuticals are frequently used. Relevance rates were (very) high. In most studies, ketoprofen was the number 1 cause of photocontact sensitizations. The relatively low rate in Portugal of 6.7% is explained by the fact that ketoprofen is little used in that country; however, piroxicam showed 30% positive photocontact reactions (45).

Table 3.193.2 Photopatch testing in groups of patients

Years and Country	Test conc. & vehicle	Number of patients tested | positive (%)		Selection of patients (S); Relevance (R); Comments (C)	Ref.
2014-2016 Spain		116	16 (14%)	S: not stated; R: 57% for all NSAID reactions together	43
1990-2014 Belgium	1% pet.	294	90 (30.6%)	S: patients suspected of iatrogenic contact dermatitis and tested with a pharmaceutical series and their own products; R: 96% of the positive patch test reactions to all topical drugs and antiseptics were considered to be relevant; C: there were also 8 cases of 'plain' contact allergy to ketoprofen, 5 of who had photo-aggravation	2
2008-2011 Europe	1% pet.	1031	128 (12.4%)	S: patients suspected of photoallergic contact dermatitis; R: 'commonly assigned current or old relevance'	14
2004-2008 Italy	1%, 2.5% or 5% pet.	1082	97 (9.0%)	S: patients with histories and clinical features suggestive of photoallergic contact dermatitis; R: 90%; C: there were also 13 photoaugmented contact allergic reactions	42
2003-2007 Portugal	1% pet.	30	2 (6.7%)	S: subgroup of 83 patients with suspected photoaggravated facial dermatitis or systemic photosensitivity; R: not specified; C: ketoprofen is little used in Portugal, but in this study, piroxicam, which is widely used there, gave 30% positive reactions	45
2004-2005 Spain	2.5% pet.	224	65 (29%)	S: not stated; R: 95%	46
1985-1994 Italy		1050	29 (2.8%)	S: patients with histories or clinical pictures suggestive of allergic contact photodermatitis; R: 78% for all photo-allergens together	44

Case series

Many case series of patients photosensitized to/with photoallergic contact dermatitis from topical ketoprofen have been reported, with numbers of affected patients ranging from 4 to 337. These are summarized in table 3.193.3. From October 1995 to December 1997, 42 cases of photocontact dermatitis attributable to ketoprofen have been collected by the Swedish Adverse Drug Reactions Advisory Committee (97).

Table 3.193.3 Case series of patients photosensitized to/with photoallergic contact dermatitis from topical ketoprofen

Years and country	Nr. Pat.	Test conc.	Irradiation	Additional data and comments	Ref.
<2013 Poland	5			from a group of 36 patients with suspected photosensitivity	33
1993-2007 Belgium	38	1% -5% pet.	UVA 5 J/cm²	also 3 patients with contact allergy and one with photoaggravated contact allergy; important article for clinical picture, cross- and co-reactivity; many patients had persistent photosensitivity; overlap with ref. 5	10
2004-2005 Sweden	35	1%-0.0001% alcohol	UVA 5 J/cm²	all reacted to 1%, 32/34 to 0.1%, 27/34 to 0.01%, 13/34 to 0.001% (10 ppm) and one to 0.0001% (1 ppm); all reactions were relevant; many cross-reactions and co-reactivities	49
2001-2002 Belgium	19	2% pet.	UVA 5 J/cm²	2 had photoaugmented contact allergy; overlap with ref. 10	5
1998-2002 France	18	2.5% pet.	UVA 10J/cm²	11 of these patients were formerly presented in ref. 52	9
<2001 France	22			no details available	93
1996-2001 Spain	68		UVA 10J/cm²	there were also 16 cases of 'plain' contact allergy	48
1995-2000 Japan	5	0.1% and 1% pet.	UVA 10J/cm²		56
<2000 Japan	4	1% pet.	UVA 5 J/cm²	17 days after cessation of treatment, ketoprofen was still present in the skin at a concentration of 312 ng/gr	62
1994-1999 France	19	2% pet.		6 photoallergy, 9 photoaggravated allergy, 2 contact allergy	6
1996-1997 France	12	1% pet.	UVA 5 J/cm²	mainly cross-reaction study	24
1993-1997 France	10	1% pet.	UVA 100 mW/cm²		70
<1997 France	7			photopatch tests performed with the products and crushed ketoprofen tablet 20% water/pet.; 2 were photoaugmented	73
<1997 France	5			details not available	92
1989-1995 Italy	25	2.5% or 5% pet.		this is one of 2 studies in which the number of photocontact allergic reactions (n=25) was lower than the number of 'plain' contact allergy reactions (n=26)	8
1993 France	337			the topical gel had been available at that moment since 4 years only; the large number of photocontact allergic reactions led to modifications for the prescriptions of ketoprofen gels	47
<1993 Italy	20		UVA 10J/cm²	also 21 patients with contact allergy; data from 6 centers in Italy; relevance not specified, no clinical data provided	3
<1992 Italy	5			no details available	72
<1989 Spain	20			no details available; data cited in ref. 61	96
1988-1989 Italy	4	10% pet.	UVA 10J/cm²	very surprising, these patients had positive patch tests to ketoprofen gel but not to ketoprofen 10% pet. and positive photopatch tests to ketoprofen 10% pet. but not to the gel	82
1987-1989 Italy	8	2.5% pet.	UVA 10J/cm²	also 2 patients with 'plain' contact dermatitis; partly also described in refs. 80 and 81	79
1985-1986 Italy	6	2% pet.	UVA 9 J/cm²	one co-reaction to parabens, present as preservative in the gel	87

Nr. Pat.: number of patients

Case reports

'Classic' case reports of photoallergic contact dermatitis from ketoprofen have been published in refs. 31,55,68,74 (2 cases), 75,77,84, and 88 (3 cases); prolonged photosensitivity has been described in refs. 16,67 and 93. Case reports in Italian literature can be found in refs. 71 and 72 and in Japanese literature in refs. 57-60 and 63-65, with additional cases, published before 2000, being cited in ref. 54.

Case reports with atypical/special features

A 40-year old woman had an acute eczematous reaction affecting the right lip commissure, extending to the upper and lower lip and chin region. The patient reported that about 7 days before, during summer holiday at the seaside, she had consumed oral granulated ketoprofen for a toothache. By dissolving directly into her mouth, and in taking, the drug was drained on the perioral skin. She had previously had a reaction to ketoprofen gel and foam and had a positive photopatch test to ketoprofen (33).

A case of 'connubial' or 'consort' photocontact dermatitis, also termed 'dermatitis by proxy', caused by ketoprofen, was in a 42-year-old housewife who presented with extremely pruritic erythematovesicular lesions on the inner left thigh, right thigh, left buttock and left shoulder. During the following days lesions progressed, leaving marked residual pigmentation. The lesions on the inside of the left thigh formed a hand. For the last 2 weeks, the patient had received UVA rays in a suntan cubicle, and her husband had been using ketoprofen gel to himself for the last few days, leaving ketoprofen to the palm of his hands (39).

Erythema multiforme-like photoallergic contact dermatitis has been reported by several authors (10,37,40). In one case, patch and photopatch tests with ketoprofen could not be performed, because the patient's consent was not given. However, lymphocyte stimulation tests with ketoprofen-photomodified peripheral blood mononuclear cells revealed that he had circulating lymphocytes reactive with a photohaptenic moiety of ketoprofen (40).

A 56-year-old man presented with patchy pigmentation and depigmentation on the left extensor forearm. A year ago, he had applied a 3% ketoprofen gel to this arm for 3 days. On the 4th day, he had felt sunburn-like pain on his arm soon after a few hours exposure to strong sunlight. There had been an erythematous macule with vesicles at the application site. The inflammation soon cleared with topical corticosteroids, but leukomelanoderma developed and remained. Patch and photopatch tests showed photocontact allergy to ketoprofen and contact allergy to the ingredient diisopropanolamine (76).

A 59-year-old woman, who was known with photopatch-proven photocontact allergy to ketoprofen, developed a pruritic, erythematous papulovesicular eruption on the face, neck, and legs, which had appeared on the third day of taking fenofibrate 100 mg/day to treat hypertriglyceridemia. Photopatch tests were positive to ketoprofen and fenofibrate. This was a case of systemic contact dermatitis to fenofibrate, cross-reacting to ketoprofen photosensitization (78).

Cross-reactions, pseudo-cross-reactions and co-reactions

In patients with photocontact allergy to ketoprofen, there are usually many photo-co-reactivities to both structurally similar (e.g. tiaprofenic acid, suprofen, benzophenone-3, fenofibrate) and structurally unrelated compounds (octocrylene, halogenated salicylanilides). A stable photoproduct of ketoprofen – 3-ethylbenzophenone – is responsible for T cell activation, photoallergy (102) and cross-reactions only with arylpropionic acid derivatives that share a similar benzophenone structure (tiaprofenic acid, suprofen, piketoprofen and dexketoprofen, but not naproxen or ibuprofen). This also explains cross-reactions with other benzophenone-derived chemicals such as benzophenone-3, benzophenone-10 and fenofibrate (10,69).

The mechanisms for photoreactions to *unrelated* compounds are largely unclear. An earlier hypothesis was that ketoprofen photosensitivity induces hyper-photo-susceptibility to non-relevant allergens. The mechanism was thought to involve a high photoreactivity induced by the association of a benzene ring with an oxygen group. More recently, it has been suggested that the formation of a specific immunogenic complex that does *not* contain a ketoprofen moiety but that can still induce photocontact allergy, would explain the large number of photo-co-reactivities with ketoprofen. These allergens do not have to be structurally similar as long as they can generate singlet oxygen (32).

Concomitant (conventional) contact allergic reactions to the fragrance mix I, Myroxylon pereirae resin and cinnamyl alcohol are frequent.

Other NSAIDs

Dexketoprofen is the (S)-enantiomer of ketoprofen and photocross-reactions between these NSAIDs (in both directions) nearly always occur (26-29). Patients photosensitized to ketoprofen may photocross-react to tiaprofenic acid (6,8,9 [61%],15,24 [12/12,100%], 53,54,56 [4/4, 100%], 67,79), piketoprofen (43,68), suprofen (54,56 [4/4, 100%]) and photo-coreact to etofenamate (10).

There have also been co-reactivities with ibuproxam (8,77,79), some of which were considered to be cross-reactivities, but the latter appears to be unlikely. Conversely, a patient sensitized to ibuproxam was thought to have cross-reacted to ketoprofen (83).

Benzophenones

Photoallergy to ketoprofen leads in 17-64% of the patients to photocontact allergy to the UV-absorber benzophenone-3 (6,9,10,13,15,16,18,19,43,49,53,73). This is probably due to photocross-reactivity (25): when irradiated with sunlight, ketoprofen is broken down into various benzophenones, of which 3-ethylbenzophenone, a stable photoproduct, is responsible for the photocontact allergy, and is structurally related to benzophenone-3 (17). Conversely, of 37 patients with a positive photopatch test to benzophenone-3, 22 (59%) co-reacted to ketoprofen (14). Photocross-reactivity has also been observed to unsubstituted benzophenone (24), benzophenone-10 (49), but it appears that there is no photocross-sensitivity to benzophenone-4 (5,49).

Octocrylene

Some patients in early reports on photoallergic contact dermatitis from sunscreens containing the UV-absorber octocrylene (20,21) reported a history of reactions to ketoprofen gel. The possible relationship was first discussed in 2008 (10) and was later confirmed (19). From the data of various studies (9,10,14,19,22,23,24,43) it has been concluded (20) that ketoprofen photosensitivity in a considerable number of the patients (27-80%) leads to octocrylene photocontact allergy. Conversely, octocrylene photocontact allergy is in the great majority (probably >80%) of cases the result of ketoprofen photosensitization (13,14,19,20). The pathomechanism of co-reactivity is not clear (20), possibly octocrylene may contain benzophenone residues from the manufacturing process (13).

Cinnamyl alcohol, fragrance mix I and Myroxylon pereirae resin

In various studies, a significant association between photocontact allergy to ketoprofen and contact sensitization to the fragrance mix I (FM I) and Myroxylon pereirae resin (balsam of Peru) (5,10,49) has been observed (4-10). In the studies in which the ingredients of the FM I were tested, by far most reactions were observed to cinnamyl alcohol (4,9,10,49). It has been suggested that this may be due to cross-reactivity, as computerized conformational analysis has shown that there is a strong similarity between the standard structure of cinnamyl alcohol and the UVA-excited ketoprofen structure (4). Indeed, in patients with a reaction to the FM I, who were patch and photopatch tested with ketoprofen and a number of other NSAIDs, there were 23% positive photopatch tests to ketoprofen and 31% to the related NSAID tiaprofenic acid. There was a clear association with contact allergy to cinnamyl alcohol but not with any other ingredient of the FM I. It was concluded that sensitization to cinnamyl alcohol may be considered a marker of or even a risk factor for photocontact allergy to ketoprofen and tiaprofenic acid (11). In the period 1990-2011, in Belgium, 48 ketoprofen-photopositive patients were also tested with cinnamyl alcohol, and 37 (77%) reacted positively, whereas only 6 of 45 patients (13%) tested with cinnamal, which was originally thought to be the co-reacting ingredient of the FM I (5,8), reacted to it; this establishes cinnamyl alcohol clearly as a potential marker for ketoprofen photosensitization (12).

Other co-reactivities

A structure similar to ketoprofen is present in fenofibrate, a systemic hypolipemic agent causing systemic photosensitivity with cross-reactions with ketoprofen (99-101). Fenofibrate intake was shown as a risk factor for more severe photoallergic contact dermatitis from ketoprofen (24,53). Patients with ketoprofen photosensitization very often photocross-react to fenofibrate (9 [9/13,72%], 24 [8/12, 67%],43,49 [73%], 53,67,70,73 [7/7, 100%],74,78).

Of 35 patients with photoallergic reactions to ketoprofen, simultaneous photoallergy to fentichlor and tetrachloro-salicylanilide was registered in 74% and 40% of the patients, respectively. Simultaneous photocontact allergies to chlorpromazine, bithionol, and promethazine were also overrepresented in this study. The higher the degree of ketoprofen reactivity (reacting to lower concentrations), the more likely was simultaneous occurrence of photocon-tact allergy to these structurally non-related chemicals (49). Such co-reactivities, sometimes in high percentages, to fentichlor, tetrachlorosalicylanilide, triclosan, bithionol, and/or tribromsalan have (also) been observed in refs. 6,9,52, and 94.

Ketoprofen-photoallergic patients in one study showed an overrepresentation of contact allergy to *p-tert*-butylphenolformaldehyde resin, with 5 of 18 (28%) patients testing positively (50).

Immediate contact reactions

Immediate contact reactions (contact urticaria) to ketoprofen are presented in Chapter 5.

Cutaneous adverse drug reactions from systemic administration caused by type IV (delayed-type) hypersensitivity

Cutaneous adverse drug reactions from systemic administration of ketoprofen caused by type IV (delayed-type) hypersensitivity including photodermatitis (35), as well as occupational photoallergic contact dermatitis (30) are planned to be discussed in Volume IV of the *Monographs in Contact Allergy* series on Systemic drugs.

LITERATURE

1   The data in the section 'General' may have been obtained from literature discussed in this chapter, but mostly also or exclusively from one or more of the following online sources: ChemIDPlus Advanced, PubChem, DrugBank, RxList, Drug Central, Drugs.com, and Wikipedia
2   Gilissen L, Goossens A. Frequency and trends of contact allergy to and iatrogenic contact dermatitis caused by topical drugs over a 25-year period. Contact Dermatitis 2016;75:290-302
3   Pigatto PD, Mozzanica N, Bigardi AS, Legori A, Valsecchi R, Cusano F, et al. Topical NSAID allergic contact dermatitis. Italian experience. Contact Dermatitis 1993 ;29 :39-41
4   Foti C, Bonamonte D, Conserva A, Stingeni L, Lisi P, Lionetti N, et al. Allergic and photoallergic contact dermatitis from ketoprofen: evaluation of cross-reactivities by a combination of photopatch testing and computerized conformational analysis. Curr Pharm Des 2008;14:2833-2839
5   Matthieu L, Meuleman L, Van Hecke E, Blondeel A, Dezfoulian B, Constandt L, Goossens A. Contact and photocontact allergy to ketoprofen. The Belgian experience. Contact Dermatitis 2004 ;50 :238-241

6 Durieu C, Marguery M-C, Giordano-Labadie F, 459nti-in F, Loche F, Bazex J. Allergies de contact photoaggravées et photoallergies de contact au kétoprofène : 19 Cas. Ann Dermatol Venereol 2001;128:1020-1024 (Article in French)

7 Girardin P, Vigan M, Humbert P, Aubin F. Cross-reactions in patch testing with ketoprofen, fragrance mix and cinnamic derivatives. Contact Dermatitis 2006;55:126-128

8 Pigatto P, Bigardi A, Legori A, Valsecchi R, Picardo M. Cross-reactions in patch testing and photopatch testing with ketoprofen, tiaprophenic acid, and cinnamic aldehyde. Am J Contact Dermat 1996;7:220-223

9 Durbize E, Vigan M, Puzenat E, Girardin P, Adessi B, Desprez PH, et al. Spectrum of cross-photosensitization in 18 consecutive patients with contact photoallergy to ketoprofen: associated photoallergies to non-benzophenone-containing molecules. Contact Dermatitis 2003;48:144-149

10 Devleeschouwer V, Roelandts R, Garmyn M, Goossens A. Allergic and photoallergic contact dermatitis from ketoprofen: results of (photo) patch testing and follow-up of 42 patients. Contact Dermatitis 2008;58:159-166

11 Stingeni L, Foti C, Cassano N, Bonamonte D, Vonella M, Vena GA, et al. Photocontact allergy to arylpropionic acid non-steroidal anti-inflammatory drugs in patients sensitized to fragrance mix I. Contact Dermatitis 2010;63:108-110

12 Nardelli A, Carbonez A, Drieghe J, Goossens A. Results of patch testing with fragrance mix 1, fragrance mix 2, and their ingredients, and Myroxylon pereirae and colophonium, over a 21-year period. Contact Dermatitis 2013;68:307-313

13 Karlsson I, VandenBroecke K, Martensson J, Goossens A, Börje A. Clinical and experimental studies of octocrylene's allergenic potency. Contact Dermatitis 2011;64:343-352

14 The European Multicentre Photopatch Test Study (EMCPPTS) Taskforce. A European multicentre photopatch test study. Br J Dermatol 2012;166:1002-1009

15 Kawada A, Aragane Y, Asai M, Tezuka T. Simultaneous photocontact sensitivity to ketoprofen and oxybenzone. Contact Dermatitis 2001;44:370

16 Horn HM, Humphreys F, Aldridge RD. Contact dermatitis and prolonged photosensititivity induced by ketoprofen and associated with sensitivity to benzophenone-3. Contact Dermatitis 1998;38:353-354

17 Bosca F, Miranda MA, Carganico G, Mauleon D. Photochemical and photobiological properties of ketoprofen associated with the benzophenone chromophore. Photochem Photobiol 1994 ;60 :96-101

18 Avenel-Audran M, Dutartre H, Goossens A, Jeanmougin M, Comte C, Bernier C, et al. Octocrylene, an emerging photoallergen. Arch Dermatol 2010;146:753-757

19 De Groot AC, Roberts DW. Contact and photocontact allergy to octocrylene: a review. Contact Dermatitis 2014;70:193-204

20 Delplace D, Blondeel A. Octocrylene: really non-allergenic? Contact Dermatitis 2006;54:295

21 Carrotte-Lefebvre I, Bonnevalle A, Segard M, Delaporte E, Thomas P. Contact allergy to octocrylene – first 2 cases. Contact Dermatitis 2003 ;48 :46-47

22 Veyrac G, Leroux A, Ruellan AL, Bernier C, Jolliet P. Kétoprofène gel et octocrylène : étude des réactions photoallergiques associées à partir des cas nantais de la base nationale de pharmacovigilance (Abstract). Rev Franç d'Allergol 2012;52:277 (Article in French)

23 Bonnevalle A, Thomas P. Cross-reactions between ketoprofen and octocrylene (abstract). Nouvelles Dermatologiques 2008 ;27(Suppl. 5) :64 (Article in French)

24 Le Coz CJ, Bottlaender A, Scrivener JN, Santinelli F, Cribier BJ, Heid E, et al. Photocontact dermatitis from ketoprofen and tiaprofenic acid: cross-reactivity study in 12 consecutive patients. Contact Dermatitis 1998;38:245-252

25 Sugiura M, Hayakawa R, Xie Z, Sugiura K, Hiramoto K, Shamoto M. Experimental study on phototoxicity and the anti-inflammatory potential of ketoprofen, suprofen, tiaprofenic acid and benzophenone and the photocross-reactivity in guinea pigs. Photodermatol Photoimmunol Photomed 2002;18:82-89

26 Goday-Bujan JJ, Rodríguez-Lozano J, Martínez-González MC, Fonseca E. Photoallergic contact dermatitis from dexketoprofen: study of 6 cases. Contact Dermatitis 2006;55:59-61

27 López-Abad R, Paniagua MJ, Botey E, Gaig P, Rodriguez P, Richart C. Topical dexketoprofen as a cause of photocontact dermatitis. J Investig Allergol Clin Immunol 2004;14:247-249

28 Cuerda Galindo E, Goday Buján JJ, del Pozo Losada J, García Silva J, Peña Penabad C, Fonseca E. Photocontact dermatitis due to dexketoprofen. Contact Dermatitis 2003;48:283-284

29 Valenzuela N, Puig L, Barnadas MA, Alomar A. Photocontact dermatitis due to dexketoprofen. Contact Dermatitis 2002;47:237

30 Maurel DT, Durand-Moreau Q, Pougnet R, Dewitte JD, Roguedas-Contios AM, Bensefa-Colas L, Loddé B. Why is occupational photocontact allergic dermatitis caused by ketoprofen rarely reported in the literature? Contact Dermatitis 2018;78:92-94

31   Aerts O, Goossens A, Bervoets A, Lambert J. Almost missed it! Photo-contact allergy to octocrylene in a ketoprofen-sensitized subject. Dermatitis 2016;27:33-34

32   Karlsson I, Persson E, Ekebergh A, Mårtensson J, Börje A. Ketoprofen-induced formation of amino acid photoadducts: possible explanation for photocontact allergy to ketoprofen. Chem Res Toxicol 2014;27:1294-1303

33   Conti R, Bassi A, Difonzo EM, Moretti S, Francalanci S. A case of photoallergic contact dermatitis caused by unusual exposure to ketoprofen. Dermatitis 2012;23:295-296

34   Foti C, Romita P, Antelmi A. Sunscreen allergy due to cinnamyl alcohol in a ketoprofen-sensitized patient. Eur J Dermatol 2011;21:295

35   Foti C, Cassano N, Vena GA, Angelini G. Photodermatitis caused by oral ketoprofen: two case reports. Contact Dermatitis 2011;64:181-183

36   Gallo R, Paolino S, Marcella G, Parodi A. Consort allergic contact dermatitis caused by ketoprofen in a tango dancer. Contact Dermatitis 2010;63:172-174

37   Mastrolonardo M, Loconsole F, Rantuccio F. Conjugal allergic contact dermatitis from ketoprofen. Contact Dermatitis 1994;30:110

38   Cantisani C, Grieco T, Faina V, Mattozzi C, Bohnenberger H, Silvestri E, Calvieri S. Ketoprofen allergic reactions. Recent Pat Inflamm Allergy Drug Discov 2010;4:58-64

39   Mirande-Romero A, González-López A, Esquivias JI, Bajo C, García-Muñoz M. Ketoprofen-induced connubial photodermatitis. Contact Dermatitis 1997;37:242

40   Izu K, Hino R, Isoda H, Nakashima D, Kabashima K, Tokura Y. Photocontact dermatitis to ketoprofen presenting with erythema multiforme. Eur J Dermatol 2008;18:710-713

41   Ota T, Oiso N, Iba Y, Narita T, Kawara S, Kawada A. Concomitant development of photoallergic contact dermatitis from ketoprofen and allergic contact dermatitis from menthol and rosin (colophony) in a compress. Contact Dermatitis 2007;56:47-48

42   Pigatto PD, Guzzi G, Schena D, Guarrera M, Foti C, Francalanci, S, Cristaudo A, et al. Photopatch tests: an Italian multicentre study from 2004 to 2006. Contact Dermatitis 2008;59:103-108

43   Subiabre-Ferrer D, Esteve-Martínez A, Blasco-Encinas R, Sierra-Talamantes C, Pérez-Ferriols A, Zaragoza-Ninet V. European photopatch test baseline series: A 3-year experience. Contact Dermatitis 2019;80:5-8

44   Pigatto PD, Legori A, Bigardi AS, Guarrera M, Tosti A, Santucci B, et al. Gruppo Italiano recerca 460nti-infla da contatto ed ambientali Italian multicenter study of allergic contact photodermatitis: epidemiological aspects. Am J Contact Dermatitis 1996;17:158-163

45   Cardoso J, Canelas MM, Gonçalo M, Figueiredo A. Photopatch testing with an extended series of photoallergens: a 5-year study. Contact Dermatitis 2009 ;60 :325-329

46   De La Cuadra-Oyanguren J, Perez-Ferriols A, Lecha-Carrelero M, et al. Results and assessment of photopatch testing in Spain: towards a new standard set of photoallergens. Actas DermoSifiliograficas 2007;98:96-101

47   Baudot S, Milpied B, Larousse C. Cutaneous side effects of ketoprofen gels: results of a study based on 337 cases. Therapie 1993;53:137-144 (Article in French)

48   Diaz RL, Gardeazabal J, Manrique P, Ratón JA, Urrutia I, Rodríguez-Sasiain JM, Aguirre C. Greater allergenicity of topical ketoprofen in contact dermatitis confirmed by use. Contact Dermatitis 2006;54:239-243

49   Hindsén M, Zimerson E, Bruze M. Photoallergic contact dermatitis from ketoprofen in southern Sweden. Contact Dermatitis 2006;54:150-157

50   Hindsén M, Zimerson E, Garnemark V, Bruze M. Simultaneous contact allergies in patients with photocontact allergy to ketoprofen. Dermatitis 2006;17:95

51   Hindsén M, Isaksson M, Persson L, Zimersson E, Bruze M. Photoallergic contact dermatitis from ketoprofen induced by drug-contaminated personal objects. J Am Acad Dermatol 2004;50:215-219

52   Vigan M, Girardin P, Desprez P, Adessi B, Aubin F, Laurent R. Photocontact dermatitis due to ketoprofen and photosensitization to tetrachlorosalicylanide and to Fenticlorl. Ann Dermatol Venereol 2002;129(10Pt.1):1125-1127 (Article in French)

53   Veyrac G, Paulin M, Milpied B, Bourin M, Jolliet P. Results of a French nationwide survey of cutaneous side effects of ketoprofen gel reported between September 1996 and August 2000. Therapie 2002;57(1):55-64 (Article in French)

54   Sugiyama M, Nakada T, Hosaka H, Sueki H, Iijima M. Photocontact dermatitis to ketoprofen. Am J Contact Dermat 2001;12:180-181

55   Miralles JC, Negro JM, Sánchez-Gascón F, García M. Contact dermatitis due to ketoprofen with good tolerance to piketoprofen. Allergol Inmunol Clin 2001;16:105-108

56   Matsushita T, Kamide R. Five cases of photocontact dermatitis due to topical ketoprofen: photopatch testing and cross-reaction study. Photodermatol Photoimmunol Photomed 2001;17:26-31

57 Sugiura M, Hayakawa R, Suzuki M. A case of photocontact dermatitis due to ketoprofen. Environ Dermatol 1996: 3: 370-377 (Article in Japanese)

58 Kataoka Y, Tashiro M. A case of photoallergic dermatitis from ketoprofen. Skin Res 1988;30:167-170 (Article in Japanese)

59 Hosokawa K, Mitsuya K, Nishijima S, Horio T, Asada Y. Photocontact dermatitis and contact dermatitis from a non-steroidal 461nti-inflammatory drugs (Sector Lotion ®). Skin Res 1993;35:26-32 (Article in Japanese)

60 Osada A, Tsukamoto K, Shibagaki N, Shimada S. Photocontact dermatitis due to ketoprofen. Rinsho Derma (Tokyo) 1999;41:1405-1407 (Article in Japanese)

61 Bagheri H, Lhiamber V, Montastruc JL, Chouni-Lahanne N. Photosensitivity to ketoprofen: mechanisms and pharmacoepidemiological data. Drug Saf 2000;22:339-349

62 Sugiura M, Hayakawa R, Kato Y, Sugiura K, Ueda H. 4 cases of photocontact dermatitis due to ketoprofen. Contact Dermatitis 2000;43:16-19

63 Isonokami M, Matsumura T, Higashi N. Two cases of contact dermatitis due to ketoprofen. Skin Research 1989;31:121-125 (Article in Japanese)

64 Arisu K, Hayakawa R, Matsunaga K, Suzuki M, Ogino Y. A case of photocontact dermatitis due to ketoprofen. Annu Rep Nagoya Univ Br Hosp 1990;24:48-54 (Article in Japanese)

65 Ohtsu A. Mechanism of ketoprofen photosensitivity. Rinsho Derma 1990;32:1039-1046 (Article in Japanese)

66 Uchida T, Ota M, Matsusita T, Izumi H, Kamide R. Ketoprofen poulitice photoallergic dermatitis developed in a soler saloon customer. Jpn J Dermatol 1995;105:1008-1009 (Article in Japanese)

67 Albès B, Marguery MC, Schwarze HP, Journé F, Loche F, Bazex J. Prolonged photosensitivity following contact photoallergy to ketoprofen. Dermatology 2000;201:171-174

68 García Bara MT, Matheu V, Pérez A, Díaz MP, Martínez MI, Zapatero L. Contact photodermatitis due to ketoprofen and piketoprofen. Allergol Inmunol Clin 1999;14:148-150

69 Bosca F, Miranda MA. Photosensitizing drugs containing the benzophenone chromophore. J Photochem Photobiol B 1998;43:1-26

70 Adamski H, Benkalfate L, Delaval Y, Ollivier I, le Jean S, Toubel G, et al. Photodermatitis from non-steroidal anti-inflammatory drugs. Contact Dermatitis 1998;38:171-174

71 Lembo G, Patruno C, Mozzariello C, Aurricchio L Ayala F. Fotodermatiti allergiche do contatto da chetoprofene. Ann Ital Dermatol Clin Sper 1991;45:103-104 (Article in Italian)

72 Catrani S, Calista D, Arcangeli F, et al. Dermatite fotoallergica ad un prodotto topico a base di ketoprofene. Presentazione di 5 casi. G Ital Dermatol Venereol 1992 ;127 :167-168 (Article in Italian)

73 Leroy D, Dompmartin A, Szczurko C, Michel M, Louvet S. Photodermatitis from ketoprofen with cross-reactivity to fenofibrate and benzophenones. Photodermatol Photoimmunol Photomed 1997;13:93-97

74 Jeanmougin M, Petit A, Manciet JR, Sigal M, Dubertret L. Contact photoallergic eczema caused by ketoprofen. Ann Dermatol Venereol 1996;123:251-255 (Article in French)

75 Fernández de Corrès L, Díez JM, Audicana M, García M, Muñoz D, Fernández E, Etxenagusía M. Photodermatitis from plant derivatives in topical and oral medicaments. Contact Dermatitis 1996;35:184-185

76 Nabeya RT, Kojima T, Fujita M. Photocontact dermatitis from ketoprofen with an unusual clinical feature. Contact Dermatitis 1995;32:52-53

77 Cusano F, Capozzi M. Photocontact dermatitis from ketoprofen with cross-reactivity to ibuproxam. Contact Dermatitis 1992 ;27 :50-51

78 Serrano G, Fortea JM, Latasa JM, et al. Photosensitivity induced by fibric acid derivatives and its relation to photocontact dermatitis to ketoprofen. J Am Acad Dermatol 1992;27(2Pt.1):204-208

79 Mozzanica N, Pigatto PD. Contact and photocontact allergy to ketoprofen: clinical and experimental study. Contact Dermatitis 1990;23:336-340

80 Mozzanica N, Pucci M, Pigatto PD. Contact and photoallergic dermatitis to topical nonsteroidal anti-inflammatory drugs (propionic acid derivatives): a study of 8 cases. In: Frosch P, Dooms-Goossens A, Lachapelle J-M, Rycroft RJG, Scheper RJ, Eds. Current topics in contact dermatitis. Berlin: Springer-Verlag, 1989:499-506

81 Mozzanica N, Pigatto PD. Contact and photoallergic dermatitis to topical ketoprofen: a study of 4 cases. Bollettino di Dermatologia Allergologica e Professionale 1989;4:105-110 (Article in Italian, cited in ref. 82)

82 Tosti A, Gaddoni G, Valeri F, Bardazzi F. Contact allergy to ketoprofen: report of 7 cases. Contact Dermatitis 1990;23:112-113

83 Valsecchi R, Cainelli T. Contact dermatitis from ibuproxam. A case with cross-reactivity with ketoprofen. Contact Dermatitis 1990;22:51

84 Lanzarini M, Bardezzi F, Morelli R, Reggiani M. Contact allergy to ketoprofen. Contact Dermatitis 1989 ;21 :51

85 Romaguera C, Grimalt F, Vilaplana J, Palou J. Subclinical contact dermatitis from ketoprofen. Contact Dermatitis 1989;20:310-311

86 Mozzanica N, Pucci M, Pigatto P. Contact allergy from ketoprofen. Contact Dermatitis 1987;17:325-326

87 Cusano F, Rafenelli A, Bacchilega R, Errico G. Photo-contact dermatitis from ketoprofen. Contact Dermatitis 1987;17:108-109

88 Alomar A. Ketoprofen photodermatitis. Contact Dermatitis 1985;12:112-113

89 Camarasa JG. Contact dermatitis to ketoprofen. Contact Dermatitis 1985;12:120-122

90 Angelini G, Vena GA. Contact allergy to ketoprofen. Contact Dermatitis =9:234

91 Valsecchi R, Falgheri G, Cainelli T. Contact dermatitis from ketoprofen. Contact Dermatitis 1983;9:163-164

92 Bastien M, Milpied-Homsi B, Baudot S, Dutartre H, Litoux P. Photosensibilisation de contact au kétoprofène. 5 Observations. Ann Dermatol Venereol 1997;124:523-526 (Article in French)

93 Offidani AM, Cellini A, Amerio P, Simonetti O, Bossi G. A case of persistent light reaction phenomenon to ketoprofen? Eur J Dermatol 2000 ;10:153-154

94 Martin S, Barbaud A, Trechot P, Reichert-Penetrat S, Schmutz JL. Existence de photoallergies associées entre le kétoprofène et des molécules ne possédant pas de structure benzophénone: 22 explorations de photoallergies au Kétum gel. Ann Dermatol Venereol 2001;128:829-830 (Article in French)

95 Loh TY, Cohen PR. Ketoprofen-induced photoallergic dermatitis. Indian J Med Res 2016;144:803-806

96 Romaguera C, Alomar A, Lecha M. Dermatoses de contact aux anti-inflammatoires non stéroidiens. 10ème Cours du GERDA 1989;227-228 (Article in French)

97 Swedish Adverse Drug Reactions Advisory Committee (SADRAC). Ketoprofen gel contact dermatitis and photosensitivity. Bulletin Swedish Adverse Drug Reactions Advisory Committee (SADRAC) 1998 Oct;67:4

98 Marmgren V, Hindsén M, Zimerson E, Bruze M. Successful photopatch testing with ketoprofen using one-hour occlusion. Acta Derm Venereol 2011;91:131-136

99 Kuwatsuka S, Kuwatsuka Y, Takenaka M, Utani A. Case of photosensitivity caused by fenofibrate after photosensitization to ketoprofen. J Dermatol 2016;43:224-225

100 Rato M, Gil F, Monteiro AF, Parente J. Fenofibrate photoallergy – relevance of patch and photopatch testing. Contact Dermatitis 2018;78:413-414

101 Tsai K, Yang J, Hung S. Fenofibrate-induced photosensitivity – a case series and literature review. Photoderm Photoimmunol Photomed 2017;33:213-219

102 Imai S, Atarashi K, Ikesue K, Akiyama K, Tokura Y. Establishment of murine model of allergic photocontact dermatitis to ketoprofen and characterization of pathogenic T cells. J Dermatol Sci 2005;41:127-136

Chapter 3.194 KETOROLAC TROMETHAMINE

IDENTIFICATION

Description/definition : Ketorolac tromethamine is the combination of tromethamine with ketorolac, a synthetic pyrrolizine carboxylic acid derivative; their structural formulas are shown below
Pharmacological classes : Anti-inflammatory agents, non-steroidal; cyclooxygenase inhibitors
IUPAC name : 2-Amino-2-(hydroxymethyl)propane-1,3-diol;5-benzoyl-2,3-dihydro-1*H*-pyrrolizine-1-carboxylic acid
Other names : Ketorolac tris salt
CAS registry number : 74103-07-4
EC number : Not available
Merck Index monograph : 6623 (Ketorolac)
Patch testing : No data available; suggestion: 1% and 5% pet.
Molecular formula : $C_{19}H_{24}N_2O_6$

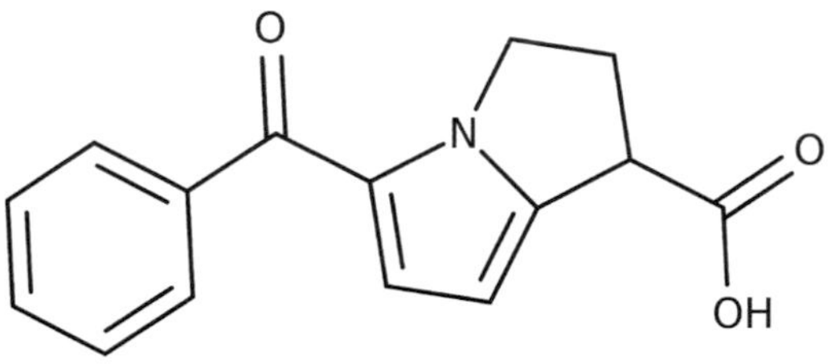

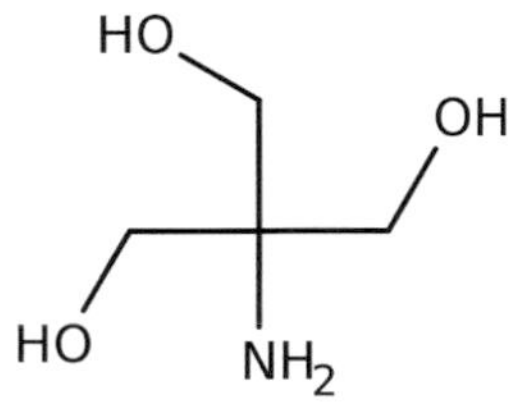

GENERAL

Ketorolac tromethamine is a pyrrolizine carboxylic acid derivative and nonsteroidal anti-inflammatory drug (NSAID) with antipyretic, analgesic and anti-inflammatory properties. It is a non-selective inhibitor of the cyclooxygenases (COX), inhibiting both COX-1 and COX-2 enzymes. This agent exerts its anti-inflammatory effect by preventing conversion of arachidonic acid to prostaglandins at inflammation sites mediated through inhibition of COX-2. Ketorolac tromethamine is indicated for short term management of acute pain that requires the caliber of pain management offered by opioids, e.g. for the management of post-operative pain, spinal and soft tissue pain, rheumatoid arthritis, osteoarthritis, ankylosing spondylitis, menstrual disorders and headaches (1).

CONTACT ALLERGY

Case report

A 65-year-old man was treated with eye-drops containing 0.5% ketorolac tromethamine, when he developed conjunctival injection, edematous swelling of the eyelids and periorbital dermatitis, apparently due to contact allergy to the NSAID (2, no details available, article in Spanish).

LITERATURE

1   The data in the section 'General' may have been obtained from literature discussed in this chapter, but mostly also or exclusively from one or more of the following online sources: ChemIDPlus Advanced, PubChem, DrugBank, RxList, Drug Central, Drugs.com, and Wikipedia
2   Rodríguez NA, Abarzuza R, Cristóbal JA, Sierra J, Mínguez E, Del Buey MA. Eyelid contact allergic eczema caused by topical ketorolac tromethamine 0.5%. Arch Soc Esp Oftalmol 2006;81:213-216 (Article in Spanish)

Chapter 3.195 KETOTIFEN

IDENTIFICATION

Description/definition   : Ketotifen the cycloheptathiophene derivative that conforms to the structural formula shown below

Pharmacological classes   : Anti-allergic agents; histamine H1 antagonists; antipruritics

IUPAC name   : 10-(1-Methylpiperidin-4-ylidene)-5H-benzo[1,2]cyclohepta[3,4-b]thiophen-4-one

Other names   : 4,9-Dihydro-4-(1-methylpiperidin-4-ylidene)-10H-benzo[4,5]cyclohepta[1,2-b]thiophene-10-one

CAS registry number   : 34580-13-7

EC number   : 252-099-7

Merck Index monograph   : 6624

Patch testing   : 2.5% pet. and 0.7% water

Molecular formula   : $C_{19}H_{19}NOS$

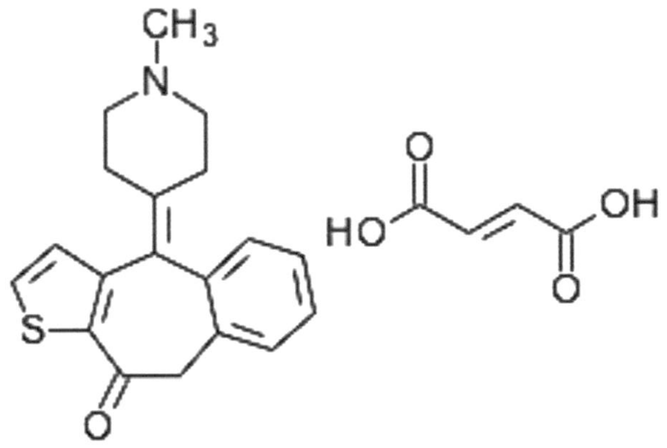

Ketotifen fumarate

GENERAL

Ketotifen is a cycloheptathiophene derivative that selectively blocks histamine H1 receptors and prevents the typical symptoms caused by histamine release. This agent also interferes with the release of inflammatory mediators from mast cells involved in hypersensitivity reactions, thereby decreasing chemotaxis and activation of eosinophils. Ketotifen is indicated as an add-on or prophylactic oral medication in the chronic treatment of mild atopic asthma in children. It is also used in eye drops for the temporary relief of itching of the eye due to allergic conjunctivitis (ketotifen ophthalmic) (1). It has also proven useful in urticaria (4). In pharmaceutical products, ketotifen is employed as ketotifen fumarate (CAS number 34580-14-8, EC number 252-100-0, molecular formula $C_{23}H_{23}NO_5S$) (1).

CONTACT ALLERGY

Case series

In Leuven, Belgium, between 1990 and 2017, 16,065 patients were investigated for contact allergy and 118 (0.7%) showed positive patch test reactions to topical ophthalmic medications and/or to their ingredients. Eighty-four individuals (71%) reacted to an active principle. Ketotifen hydrogen fumarate was tested in 16 patients and was the allergen in eye medications in one. There were no reactions to ketotifen hydrogen fumarate in other types of medications (2).

In Tokyo, Japan, out of 3903 patients who were patch tested in the period January 1987 to December 1995, 141 (3.6%) were patch tested with eye drops and 49 individuals (35%) reacted positively and were diagnosed with allergic contact dermatitis. In 36 cases ingredient patch testing was performed and there were 6 reactions to ketotifen fumarate (3).

Case reports

A 69-year-old woman presented with eczema on her eyelids associated with itching and burning, which had appeared about 1 month after starting therapy with ketotifen fumarate 0.05 mg/ml eye drops for seasonal allergic conjunctivitis. Patch tests were positive to the eye drops 'as is' (negative in 20 controls) and ketotifen hydrogen fumarate 2.5% pet. She also reacted to benzalkonium chloride present in the eye drops (7).

A 32-year-old woman suffered from chronic conjunctivitis with a burning sensation in the eyes and on the eyelid margins, sometimes accompanied by a feeling of pressure in the eyes; however, she had never had eczematous lesions. For the past 1.5 years, there had been a continuous need to use local corticosteroids. When she stopped applying these for 1 week, watery eyes and some irritation on the cheeks, most probably caused by tears, were noted. The personal history for atopy was strongly positive. Currently, the patient was using eye drops containing ketotifen fumarate and two other eye drops. Clinical investigation showed hyperaemic conjunctivae, but no eczema on the eyelids. Patch tests showed a positive reaction to the ketotifen eye drops. The active ingredient was obtained from the manufacturer and additional patch tests resulted in positive reactions to ketotifen hydrogen fumarate 2.5% pet. and 0.7% water. Twenty control subjects were negative. Avoiding contact with the eye drops containing ketotifen resulted in complete clearance of the conjunctivitis (4).

A 54-year-old woman with seasonal allergies presented with a several year history of intermittently pruritic eyes and eyelids with periorbital rash. Multiple brands of eye drops were of minimal benefit. Patch testing revealed clinically relevant positive reactions to 2 eye drops, both of which contained ketotifen, benzalkonium chloride, and glycerin. Pertinent negatives included purified benzalkonium chloride (tested twice) and purified glycerin. Purified ketotifen was not commercially available for testing. A complete resolution of symptoms was noted after discontinuation of the ophthalmic drops (5).

A 37-year-old man had been treated with ketotifen fumarate and several other eye drops for allergic conjunctivitis for 8 months, when he developed redness around his eyes. Examination showed erythema with well-circumscribed margins around both eyes. Patch tests gave a positive reaction to the ketotifen eye drops only. The skin lesions subsided rapidly after stopping all eye drops and applying methylprednisolone. The eye drops contained 0.069% ketotifen fumarate and benzalkonium chloride (0.1 mg/ml). Further patch tests showed a strongly positive reaction to 0.069% water at 2, 3 and 7 days, and an irritant reaction to benzalkonium chloride 0.1 mg/ml and 0.025 mg/ml water (6).

A 55-year-old woman with chronic allergic rhinoconjunctivitis had a 2-year history of eye redness, later progressing into an itchy rash around both eyes. Patch tests with her 3 different eye drops and cosmetics yielded a positive reaction to ketotifen-containing eye drops. The use of these drops was stopped and all symptoms completely resolved within 2 weeks. The authors presented this as a case of contact allergy to ketotifen, but this active principle was not tested separately and neither were the excipients (8).

A 61-year-old woman had a several-year history of episodic red eyes and periorbital rash. A month before presentation, she discontinued use of contact lenses, antihistamine eye drops, sunscreen, and cosmetics but continued to use shea butter on the face. Patch testing was positive to eye drops containing ketotifen and negative to its ingredient benzalkonium chloride. Ketotifen itself was not tested and neither was the other ingredient, glycerin (9).

LITERATURE

1   The data in the section 'General' may have been obtained from literature discussed in this chapter, but mostly also or exclusively from one or more of the following online sources: ChemIDPlus Advanced, PubChem, DrugBank, RxList, Drug Central, Drugs.com, and Wikipedia
2   Gilissen L, De Decker L, Hulshagen T, Goossens A. Allergic contact dermatitis caused by topical ophthalmic medications: Keep an eye on it! Contact Dermatitis 2019;80:291-297
3   Aoki J. [Allergic contact dermatitis due to eye drops. Their clinical features and the patch test results]. Nihon Ika Daigaku Zasshi 1997;64:232-237 (article in Japanese) (data cited in refs. 4 and 5)
4   Smets K, Werbrouck J, Goossens A, Gilissen L. Sensitization from ketotifen fumarate in eye drops presenting as chronic conjunctivitis. Contact Dermatitis 2017;76:124-126
5   Aschenbeck KA, Warshaw EM. Periorbital allergic contact dermatitis due to ketotifen. Dermatitis 2017;28:164-165
6   Niizeki H, Inamoto N, Nakamura K, Nakanoma J, Nakano T. Contact dermatitis from ketotifen fumarate eyedrops. Contact Dermatitis 1994;31:266
7   Romita P, Stingeni L, Barlusconi C, Hansel K, Foti C. Allergic contact dermatitis in response to ketotifen fumarate contained in eye drops. Contact Dermatitis 2020;83:35-37
8   Blackwell W, Cruz P Jr. Ketotifen in over-the-counter products is a contact allergen. Dermatitis 2020;31:e39-e40
9   DeBord L, Enos T, Cruz P Jr. A subtle historical clue unlocks a contact puzzle. Dermatitis 2019;30:77

Chapter 3.196 LACTIC ACID

IDENTIFICATION

Description/definition   : Lactic acid is the 2-hydroxy monocarboxylic acid that conforms to the structural formula
                             shown below
Pharmacological classes : Dermatological agents; gynaecological anti-infectives and antiseptics
IUPAC name   : 2-Hydroxypropanoic acid
Other names   : 2-Hydroxypropionic acid
CAS registry number   : 50-21-5
EC number   : 200-018-0
Merck Index monograph : 6654
Patch testing   : 3% water
Molecular formula   : $C_3H_6O_3$

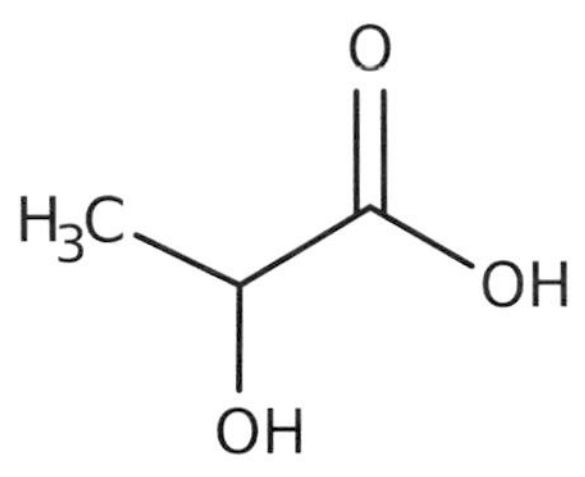

Lactic acid 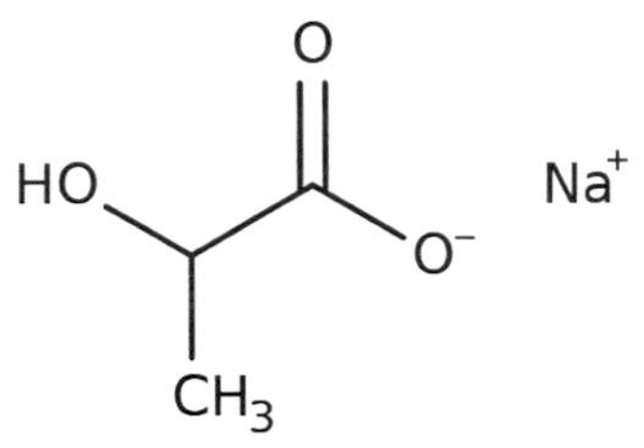   Sodium lactate 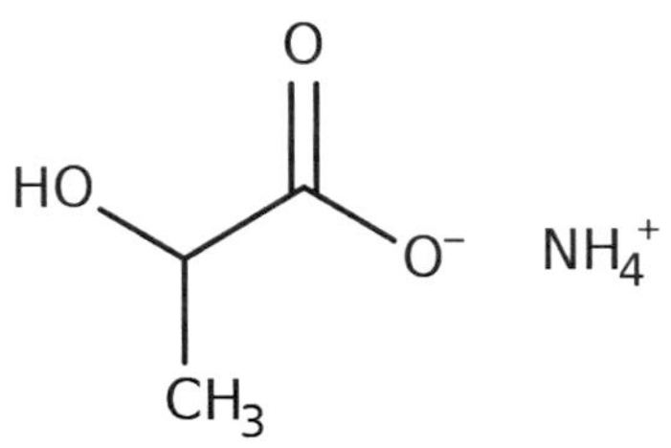   Ammonium lactate

GENERAL

Lactic acid is a normal intermediate in the fermentation (oxidation, metabolism) of sugar. It may be used in topical drugs, notably with salicylic acid in collodion for the treatment of warts. It is also applied with calcium chloride, magnesium chloride, dextrose monohydrate, sodium chloride, and sodium bicarbonate in replacement solutions in Continuous Renal Replacement Therapy to replace plasma volume removed by ultrafiltration. The sodium salt of racemic or inactive lactic acid (sodium lactate) is a hygroscopic agent used intravenously as a systemic and urinary alkalizer. Topical preparations containing (mostly) 12% ammonium lactate are indicated for the treatment of dry, scaly skin (xerosis) and ichthyosis vulgaris and for temporary relief of itching associated with these conditions. Lactic acid is also used to make cultured dairy products, as a food preservative, and to produce other chemicals (1).

In pharmaceutical preparations, lactic acid is also employed as sodium lactate (CAS number 72-17-3, EC number 200-772-0, molecular formula $C_3H_5NaO_3$) and as ammonium lactate (CAS number 515-98-0, EC number 208-214-8, molecular formula $C_3H_9NO_3$).

CONTACT ALLERGY

Case report

A 51-year-old woman, on the 20th day of treatment with an anti-wart solution, developed 2 roundish erythematous vesicular plaques, 30 and 10 cm in diameter, respectively, on areas of application in the left gluteal and right anterolateral neck regions. These remitted in a week on discontinuing the wart treatment and administering a topical corticosteroid. Patch testing was performed with the wart remover solution 'as is' and its components, which yielded positive reactions to the solution (D2 ++, D4 ++), lactic acid 3% water (D2 +, D4 +) and castor oil (Ricinus communis seed oil) as is and 20% pet (D2 -, D4 +) (2).

LITERATURE

1   The data in the section 'General' may have been obtained from literature discussed in this chapter, but mostly
    also or exclusively from one or more of the following online sources: ChemIDPlus Advanced, PubChem,
    DrugBank, RxList, Drug Central, Drugs.com, and Wikipedia
2   Tabar AI, Muro MD, Quirce S, Olaguibel JM. Contact dermatitis due to sensitization to lactic acid and castor oil in
    a wart remover solution. Contact Dermatitis 1993;29:49-50

Chapter 3.197 LANOCONAZOLE

IDENTIFICATION

Description/definition : Lanoconazole is the vinyl imidazole that conforms to the structural formula shown below
Pharmacological classes : Antifungal agents
IUPAC name : (2E)-2-[4-(2-Chlorophenyl)-1,3-dithiolan-2-ylidene]-2-imidazol-1-ylacetonitrile
CAS registry number : 101530-10-3
EC number : Not available
Merck Index monograph : 6678
Patch testing : 1% pet.; 1% alcohol or MEK (methyl ethyl ketone) is often preferred for testing imidazoles
Molecular formula : $C_{14}H_{10}ClN_3S_2$

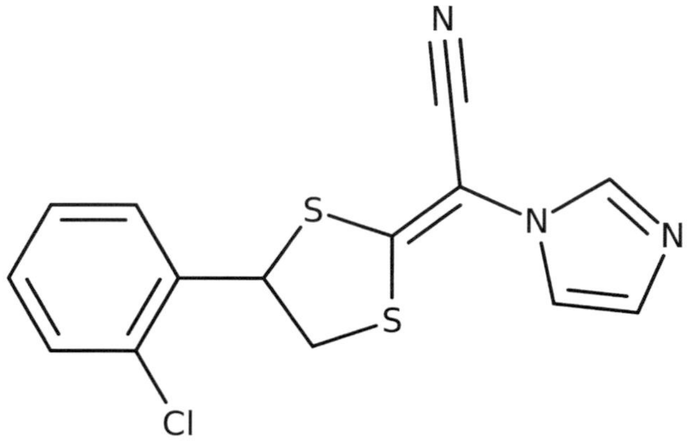

GENERAL

Lanoconazole is a vinyl imidazole antifungal drug that has been used since 1994 in Japan in topical preparations to treat superficial fungal infections (1).

CONTACT ALLERGY

Case reports

A 63-year-old man presented with severe scaly erythema on his scalp, face, and neck, and erythema on his trunk, arms, and legs after having been treated with lanoconazole cream and clobetasol propionate ointment. Patch tests were positive to lanoconazole cream 'as is', lanoconazole 0.1% and 1% pet., clobetasol 0.05% ointment 'as is' and clobetasol propionate 0.05% pet. The patient also reacted to commercial luliconazole cream 'as is'(10).

A 60-year-old man had been treated with 1% lanoconazole cream for tinea pedis for one year, when he developed a pruritic eczematous eruption on the treated lesions. Patch tests were positive to the cream and – later – to its ingredients lanoconazole (1% pet.), diethyl sebacate (5% pet.) and cetyl alcohol (5% pet.). He also reacted to commercial preparations of other – related and unrelated – antifungals, but these were caused by the presence therein of diethyl sebacate or cetyl alcohol (4).

A 65-year-old man presented with dermatitis on the feet after having used lanoconazole ointment to treat tinea pedis for 10 months. Patch testing with lanoconazole (10%, 1%, 0.1% pet.) was positive. Patch testing with other imidazoles (isoconazole, sulconazole, miconazole, neticonazole, ketoconazole) was negative at that time. He then started using neticonazole ointment and, 4 months later, the patient again developed dermatitis on the feet. Patch testing now was positive to neticonazole ointment and neticonazole (1%, 0.3%, 0.1% pet.; in the table concentrations were given as 10%, 1% and 0.1% pet.) and diethyl sebacate, an emulsifier in neticonazole ointment (5).

A 51-year-old woman developed erythematous scaly dermatitis on the vulva and in the groins starting after having used 1% lanoconazole cream for one week. Patch tests were positive to lanoconazole 1% and 10% pet. and negative to the base of the cream and many other imidazoles (6). A 72-year-old man developed a pruritic erythematous dermatitis of the dorsa of the feet after having used 1% lanoconazole cream for 2 weeks for tinea pedis. Patch testing with lanoconazole, oxiconazole, croconazole, bifonazole, ketoconazole and neticonazole gave positive reactions only to lanoconazole 1% and 10% pet. Fifteen controls were negative (7).

A 79-year-old man developed severe eczema on the dorsa of the feet after having treated tinea pedis for 6 weeks with lanoconazole solution. Patch tests were positive to the solution and to lanoconazole 20% pet. (negative in 20 controls) (8). Another case report on allergic contact dermatitis from lanoconazole in Japanese literature, adequate data of which are not available to the author, can be found in ref. 9.

Cross-reactions, pseudo-cross-reactions and co-reactions
A patient sensitized to lanoconazole later developed allergic contact dermatitis from luliconazole resulting from cross-sensitization. The authors suggested that the dithioacetal structure was essential for inducing contact allergy to both antifungals (2). Conversely, a patient sensitized to luliconazole cross-reacted to lanoconazole (3). In most patients sensitized to lanoconazole, no cross-reactions to other imidazoles have been observed (5,6,7).

LITERATURE

1   The data in the section 'General' may have been obtained from literature discussed in this chapter, but mostly also or exclusively from one or more of the following online sources: ChemIDPlus Advanced, PubChem, DrugBank, RxList, Drug Central, Drugs.com, and Wikipedia
2   Tanaka T, Satoh T, Yokozeki H. Allergic contact dermatitis from luliconazole: implication of the dithioacetal structure. Acta Derm Venereol 2007;87:271-722
3   Shono M. Allergic contact dermatitis from luliconazole. Contact Dermatitis 2007;56:296-297
4   Soga F, Katoh N, Kishimoto S. Contact dermatitis due to lanoconazole, cetyl alcohol and diethyl sebacate in lanoconazole cream. Contact Dermatitis 2004;50:49-50
5   Umebayashi Y, Ito S. Allergic contact dermatitis due to both lanoconazole and neticonazole ointments. Contact Dermatitis 2001;44:48-49
6   Tanaka N, Kawada A, Hiruma M, Tajima S, Ishibashi A. Contact dermatitis from lanoconazole. Contact Dermatitis 1996;35:256-257
7   Taniguchi S, Kono T. Allergic contact dermatitis due to lanoconazole with no cross-reactivity to other imidazoles. Dermatology 1998;196:366
8   Nakano R, Miyoshi H, Kanzaki T. Allergic contact dermatitis from lanoconazole. Contact Dermatitis 1996;35:63
9   Tani A, Hamada T, Kanzaki T. A case of allergic contact dermatitis due to lanoconazole and sulconazole. Environ Dermatol 1997;4:148 (Article in Japanese)
10   Kawai M, Sowa-Osako J, Omura R, Fukai K, Tsuruta D. Allergic contact dermatitis due to clobetasol propionate and lanoconazole. Contact Dermatitis 2020 May 28. doi: 10.1111/cod.13623. Online ahead of print.

Chapter 3.198 LATANOPROST

IDENTIFICATION

Description/definition   : Latanoprost is the prostaglandin F2α analog that conforms to the structural formula shown below

Pharmacological classes : Antihypertensive agents

IUPAC name   : Propan-2-yl (Z)-7-[(1R,2R,3R,5S)-3,5-dihydroxy-2-[(3R)-3-hydroxy-5-phenylpentyl]-cyclopentyl]hept-5-enoate

Other names   : Isopropyl *(Z)*-7-((1R,2R,3R,5S)-3,5-dihydroxy-2-((3R)-3-hydroxy-5-phenylpentyl) cyclopentyl)-5-heptenoate

CAS registry number   : 130209-82-4

EC number   : Not available

Merck Index monograph   : 6701

Patch testing   : No data available; all patients have been patch tested with latanoprost 0.005% solution, but this will without any doubt result in many false-negatives; higher concentrations (x 10-25) recommended

Molecular formula   : $C_{26}H_{40}O_5$

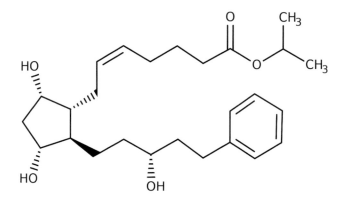

GENERAL

Latanoprost is a prostaglandin F2α analog and a prostanoid selective FP receptor agonist with an ocular antihypertensive effect. It increases outflow of aqueous fluid from the eyes and thereby reduces intraocular pressure. Latanoprost is indicated for the reduction of elevated intraocular pressure in patients with open-angle glaucoma or ocular hypertension (1).

CONTACT ALLERGY

Case series

In Ferrara, Italy, over a 65-month period before 2005, 50 patients affected by periorbital dermatitis while using topical ocular products were patch tested, including with their own ophthalmic medications (n=210). Only 15 positive reactions were detected in 12 subjects, including 14 reactions to commercial eye drops. There were two reactions to latanoprost eye drops. The active ingredients were not tested separately, but contact allergy to the excipients and preservatives was excluded by patch testing. The authors concluded that patch testing with commercial eye drops has doubtful value (2).

In a hospital in Slovenia, in an undefined period before 2015, 55 patients with suspected contact allergy to topical drugs used for treatment of glaucoma were retrospectively analyzed. Each patient had been tested to individual products in serial dilutions (1%-10%-50% and as is). Patch tests to vehicles, emulsifiers and preservatives(including benzalkonium chloride) were also performed. Three controls on healthy volunteers were always performed in case of a positive reaction. The causal relationship between the drug and dermatitis was always confirmed by a positive elimination test. Eight of 55 patients (5 women, 3 men) had positive patch tests to one or more products. Two patients were positive to latanoprost, 6 to beta-blockers (5x timolol and 1x carteolol), 2 to dorzolamide, and two to brinzolamide (10). It is highly likely that these patients have not been tested with the active drugs but only with the commercial eye drops and that the diagnosis of allergic contact dermatitis to the active ingredients was, at best, made *per exclusionem*.

Case reports

A 58-year-old woman was referred with itchy eczema of her scalp, face, and neck. The dermatitis appeared after 2 months of topical treatment with 2 ml daily of a galenic preparation composed of minoxidil and latanoprost (minoxidil 5%, latanoprost 0.02%, propylene glycol 15%, and ethanol/water 70:30 79.98%) for diffuse multifocal alopecia areata. The patient reported that she had used minoxidil 2% solution for 5 months previously, without any skin reaction. Patch testing revealed a positive reaction to the galenic solution, but not to propylene glycol 5% pet., ethanol 100% or freshly prepared minoxidil 5% ethanol. Later, patch tests were performed with 0.005% latanoprost eye drops with a positive (+) result and a negative reaction to its preservative benzalkonium chloride. A repeated open application test (ROAT), performed with the galenic preparation, resulted in a pruritic, eczematous rash after 3 days. Although latanoprost itself was not tested, contact allergy to it in this case is highly likely (3).

Short summaries of other case reports of allergic contact dermatitis from latanoprost, all from its application in eye drops for the treatment of elevated intraocular pressure and glaucoma, are shown in table 3.198.1. Ref. 6 is often quoted as presenting cases of allergic contact dermatitis from latanoprost, but the authors did not perform patch tests, so their claim 'allergic contact dermatitis' is invalid (6).

Table 3.198.1 Short summaries of case reports of allergic contact dermatitis from latanoprost in eye drops

Year	Sex	Age	Clinical picture	Positive patch tests	Comments	Ref.
2013	M	70	erythematous oozing patches on both eyelids, severe pruritus and burning	latanoprost 0.005% eye drops	contact allergy to latanoprost not proven	4
2008	F	48	ocular itching, erythema, tearing, and eyelid eczema	latanoprost 0.005% and levobunolol eye drops; not to 'additives' such as benzalkonium chloride	eyelid provocation test caused recurrence of symptoms in 2-3 days; allergy to latanoprost likely but not proven	5
2004	F	24	eczema of the eyelids	none, not to latanoprost ophthalmic	despite the negative patch tests the authors concluded that 'latanoprost was the antigen of the contact dermatitis'	7
2002	M	85	tearing, red eyes, pruritic edematous eczematous eyelids	latanoprost 0.005% eye drops; negative to the vehicle without latanoprost; negative to benzalkonium chloride, the eye drops' preservative	ROAT with the eye drops were strongly positive after 4 days; diagnosis *per exclusionem*, but latanoprost allergy is likely	8, 9

LITERATURE

1   The data in the section 'General' may have been obtained from literature discussed in this chapter, but mostly also or exclusively from one or more of the following online sources: ChemIDPlus Advanced, PubChem, DrugBank, RxList, Drug Central, Drugs.com, and Wikipedia

2   Corazza M, Massieri LT, Virgili A. Doubtful value of patch testing for suspected contact allergy to ophthalmic products. Acta Derm Venereol 2005;85:70-71

3   Napolitano M, Cantelli M, Vastarella M, Nappa P, Fabbrocini G, Patruno C. Allergic contact dermatitis probably caused by latanoprost during treatment for alopecia areata. Contact Dermatitis 2019;81:67-68

4   Lee JH, Kim TH, Kim SC. Allergic contact dermatitis caused by topical eye drops containing latanoprost. Ann Dermatol 2014;26:269-270

5   Pérez-Rodríguez E, González-Pérez R, Poza P, Feliciano L, López-Correcher B, Matheu V. Contact dermatitis caused by latanoprost-containing eye drops with good tolerance to bimatoprost eye drops. Contact Dermatitis 2008;58:370-371

6   Lai CH, Lai IC, Chi CC. Allergic contact dermatitis caused by latanoprost ophthalmic solution. Eur J Ophthalmol 2006;16:627-629

7   Pires MC, dos Santos Rodrigues RN. Contact dermatitis to latanoprost. Dermatitis 2005;16:104

8   Jerstad KM, Warshaw E. Allergic contact dermatitis to latanoprost. Am J Contact Dermat 2002;13:39-41

9   Jerstad K, Warshaw E. First known reported case of allergic contact dermatitis to latanoprost, the active ingredient in Xalatan ophthalmic solution. Dermatitis 2001;12:123

Chapter 3.199 LEVOBUNOLOL

IDENTIFICATION

Description/definition   : Levobunolol is the L-isomer of bunolol and conforms to the structural formula shown
                             below
Pharmacological classes   : Sympatholytics; β-adrenergic antagonists
IUPAC name   : 5-[(2S)-3-(tert-Butylamino)-2-hydroxypropoxy]-3,4-dihydro-2H-naphthalen-1-one
CAS registry number   : 47141-42-4
EC number   : Not available
Merck Index monograph   : 6784
Patch testing   : 10% pet.
Molecular formula   : $C_{17}H_{25}NO_3$

GENERAL

Levobunolol, the L-isomer of bunolol, is a naphthalenone and non-cardioselective β-adrenergic receptor antagonist with anti-glaucoma activity. Upon administration in the eye, this agent blocks β-adrenergic receptors, thereby causing vasoconstriction. Levobunolol also decreases the ciliary body's production of aqueous humor, which, in turn, results in a lower intraocular pressure. Levobunolol is indicated for lowering intraocular pressure in patients with chronic open-angle glaucoma or ocular hypertension. In pharmaceutical products, levobunolol is employed as levobunolol hydrochloride (CAS number 27912-14-7, EC number 248-725-3, molecular formula $C_{17}H_{26}ClNO_3$) (1).

CONTACT ALLERGY

General

For information on patch testing with commercial eye drops containing beta-blockers see Chapter 3.338 Timolol.

Case series

Out of 112,430 patients patch tested by the IVDK between 1993 and 2004, 332 had been tested with their own topical anti-glaucoma eye drops containing different β-blockers because of suspected allergic contact dermatitis. Eight-four subjects were tested with levobunolol eye drops and there were eleven (13%) positive reactions. The patients were not tested with the active substance, but reactions to the (possible) adjuvants benzalkonium chloride, sodium EDTA and sodium disulfite were excluded (2).

Case reports

A 45-year-old woman, affected by a 6-year primary open-angle glaucoma and treated for years with levobunolol eye drops, presented with bilateral itching erythematosquamous edematous plaques on the upper and lower eyelids. Patch testing with the constituents of the eye drops gave a positive reaction to levobunolol 1% water. Levobunolol was discontinued and the clinical manifestations completely resolved within 4 weeks. Afterwards, the patient used an ophthalmic solution containing befunolol, which had not been used previously, and she developed similar lesions 7 days later. Patch testing with a series of β-blockers showed a positive reaction to befunolol 1% water, which was considered to be a cross-reaction to levobunolol (8).

A 55-year-old woman presented with itchy erythematous plaques with scales surrounding both eyelids and periorbital areas. Patch tests with 3 types of eye drops she used showed a positive reaction to eye drops containing 0.5% levobunolol HCl. In a second session, the patient was tested with all ingredients and she reacted to levobunolol

10% and 1% pet. (negative to 0.1% pet.). The levobunolol preparation had already been withdrawn, followed by rapid healing within a week (10).

A 59-year-old woman presented with a 6-month history of bilateral symmetrical eyelid dermatitis. Glaucoma had been diagnosed 4 years earlier and treated with levobunolol 0.5% eye drops throughout. Patch tests with the eye drops and their excipients, however, were negative, as was a ROAT. In view of the continuing dermatitis and the clinical suspicion of allergic contact dermatitis from medicaments, the eye drops were changed to a preparation containing latanoprost, and there was complete resolution of the dermatitis. A supply of pure levobunolol was then obtained, and patch testing performed with levobunolol 10% pet. was positive (D2 ?+/D4 ++) (11).

A woman aged 72 had dermatitis of both eyelids and periorbital areas, while using various medications for her glaucoma. Patch tests were positive to a commercial preparation containing levobunolol and – in a second patch test session - to levobunolol 1% in water, but only on tape-stripped skin (negative on normal skin). Possibly, she was also allergic to pilocarpine (4). In another case, a positive patch test to levobunolol was only be obtained by pricking the skin 10x with a prick-test lancet before applying the test material (negative on normal skin) (12).

Short summaries of other case reports of allergic contact dermatitis from levobunolol are shown in table 3.199.1.

Table 3.199.1 Short summaries of other case reports of allergic contact dermatitis from levobunolol

Year	Sex	Age	Clinical picture	Positive patch tests	Comments	Ref.
2008	F	48	itching, erythema, tearing, eyelid eczema	levobunolol eye drops; negative to benzalkonium chloride	levobunolol not tested; diagnosis made *per exclusionem*	7
2008	F	48	ocular itching, erythema, tearing, and eyelid eczema	levobunolol eye drops; not to 'additives' such as benzalkonium chloride	eyelid provocation test caused recurrence of symptoms in 2-3 days; allergy to levobunolol likely but not proven	5
2006	M	66	erythema, edema, lichenification and excoriations eyelids	levobunolol eye drops, levobunolol (test conc. and vehicle not mentioned)		9
1999	F	68	blepharitis and dermatitis of the eyelids and periorbitally	levobunolol eye drops 0.5%; levobunolol 0.1%, 1% and 2% water and petrolatum	the materials were placed on tape-stripped skin	13
1998	F	80	erythema, edema; scattered papules around the left eye, on cheek and in eyebrow	levobunolol eye drops 0.5% (2x)	active principle not tested, but no positive patch test reaction to excipients	14
	F	62	redness and scaling of both upper and lower eyelids	levobunolol eye drops 0.5%	active principle not tested, but no positive patch test reaction to excipients	
1997	F	56	conjunctivitis and eczema of the right eyelids and periorbital swelling	levobunolol 0.5% eye drops; levobunolol in various concentrations and vehicles (not specified)	possibly cross-reaction to timolol	15
1997	F	32	bilateral dermatitis of the eyelids	levobunolol eye drops 0.5%	active principle not tested, but no positive patch test reactions to excipients; no cross-reactions to large number of other beta-blockers	16
1995	M	70	symmetrical eyelid dermatitis and conjunctival hyperemia	levobunolol eye drops 0.5%; levobunolol 0.25%, 0.5% and 1% water	later, the patient became sensitized to befunolol	17
1995	M	70	symmetrical eyelid dermatitis	levobunolol eye drops 0.5% ; levobunolol 0.25%, 0.5% and 1% water	5 controls were negative	18
1995	M	29	blepharoconjunctivitis	levobunolol eye drops 0.5%; levobunolol 0.5%, 2% and 5% water	co-, but probably not cross-reactivity to befunolol	19
1993	M	35	symmetrical eyelid dermatitis	levobunolol eye drops 0.5% ; levobunolol 0.5%, 1% and 3% pet.	vehicle ingredients were negative in patch testing	20
1989			periocular allergic contact dermatitis	eye drops with befunolol and befunolol itself	details not available	21

Cross-reactions, pseudo-cross-reactions and co-reactions

Cross-reactions between beta-blockers appear to be infrequent (2). Levobunolol has cross-reacted in a patient sensitized to timolol (3). A woman sensitized to levobunolol also had a positive patch test to befunolol, to which she had never been exposed (8). Timolol may have cross-reacted in one patient to levobunolol sensitization (15). Two patients who had developed allergic contact dermatitis from befunolol cross-reacted to levobunolol (6,10).

LITERATURE

1   The data in the section 'General' may have been obtained from literature discussed in this chapter, but mostly also or exclusively from one or more of the following online sources: ChemIDPlus Advanced, PubChem, DrugBank, RxList, Drug Central, Drugs.com, and Wikipedia

2   Jappe U, Uter W, Menezes de Pádua CA, Herbst RA, Schnuch A. Allergic contact dermatitis due to beta-blockers in eye drops: a retrospective analysis of multicentre surveillance data 1993-2004. Acta Derm Venereol 2006;86:509-514

3   Horcajada-Reales C, Rodríguez-Soria VJ, Suárez-Fernández R1. Allergic contact dermatitis caused by timolol with cross-sensitivity to levobunolol. Contact Dermatitis 2015;73:368-369

4   Koch P. Allergic contact dermatitis due to timolol and levobunolol in eyedrops, with no cross-sensitivity to other ophthalmic beta-blockers. Contact Dermatitis 1995;33:140-141

5   Pérez-Rodríguez E, González-Pérez R, Poza P, Feliciano L, López-Correcher B, Matheu V. Contact dermatitis caused by latanoprost-containing eye drops with good tolerance to bimatoprost eye drops. Contact Dermatitis 2008;58:370-371

6   Corazza M, Virgili A, Mantovani L, Masieri LT. Allergic contact dermatitis from cross-reacting beta-blocking agents. Contact Dermatitis 1993;28:188-189

7   Pérez-Rodríguez E, González-Pérez R, Poza P, Feliciano L, López-Correcher B, Matheu V. Contact dermatitis caused by latanoprost-containing eye drops with good tolerance to bimatoprost eye drops. Contact Dermatitis 2008;58:370-371

8   Nino M, Balato A, Ayala F, Balato N. Allergic contact dermatitis due to levobunolol with cross-sensitivity to befunolol. Contact Dermatitis 2007;56:53-54

9   Blumetti B, Brodell RT, Helms SE, Brodell LP, Bredle DL. Contact dermatitis to levobunolol eyedrops superimposed on IgE-mediated rhinoconjunctivitis. Ann Allergy Asthma Immunol 2006;97:817-818

10 Hashimoto Y, Aragane Y, Kawada A. Allergic contact dermatitis due to levobunolol in an ophthalmic preparation. J Dermatol 2006;33:507-509

11 Statham BN. Failure of patch testing with levobunolol eyedrops to detect contact allergy. Contact Dermatitis 2000;43:365-366

12 Wilkinson SM. False-negative patch test with levobunolol. Contact Dermatitis 2001;44:264

13 Erdmann S, Hertl M, Merk HF. Contact dermatitis from levobunolol eyedrops. Contact Dermatitis 1999;41:44-46

14 De Groot AC, van Ginkel CJ, Bruynzeel DP, Smeenk G, Conemans JM. [Contact allergy to eyedrops containing beta-blockers]. Ned Tijdschr Geneeskd. 1998;142:1034-1036 (Article in Dutch)

15 Förster W. Allergisches Kontaktekzem auf Levobunolol und Timolol in der Glaukombehandlung. Derm (Praktische Dermatologie) 1997;3:130-131 (Article in German)

16 Garcia F, Blanco J, Juste S, Garces MM, Alonso L, Marcos ML, et al. Contact dermatitis due to levobunolol in eyedrops. Contact Dermatitis 1997;36:230

17 Morelli R, Arcangeli F, Brunelli D, Vincenzi C, Landi G. Contact allergy to β-blocking agents in eyedrops. Dermatitis 1995;6:172-173

18 di Lernia V, Albertini G, Bisighini G. Allergic contact dermatitis from levobunolol eyedrops. Contact Dermatitis 1995;33:57

19 Zucchelli V, Silvani S, Vezzani C, Lorenzi S, Tosti A. Contact dermatitis from levobunolol and befunolol. Contact Dermatitis 1995;33:66-67

20 van der Meeren HL, Meurs PJ. Sensitization to levobunolol eyedrops. Contact Dermatitis 1993;28:41-42

21 Schultheiss E. Hypersensitivity to levobunolol. Derm Beruf Umwelt 1989;37:185-186 (Article in German)

Chapter 3.200 LEVOCABASTINE

IDENTIFICATION

Description/definition   : Levocabastine is the phenylpiperidine that conforms to the structural formula shown
                              below
Pharmacological classes : Histamine H1 antagonists, non-sedating
IUPAC name   : (3S,4R)-1-[4-Cyano-4-(4-fluorophenyl)cyclohexyl]-3-methyl-4-phenylpiperidine-4-
                             carboxylic acid
CAS registry number   : 79516-68-0
EC number   : Not available
Merck Index monograph   : 6785
Patch testing   : No data available; in general, active ingredients in ophthalmic medications should be
                             tested in (far) higher concentrations than in the medication itself to avoid false-negative
                             reactions
Molecular formula   : $C_{26}H_{29}FN_2O_2$

GENERAL

Levocabastine is a synthetic piperidine derivative with antihistamine properties. It is a second generation histamine-1 receptor antagonist. When applied locally into the eye as a topical solution, this agent reduces itching, rhinorrhea and symptoms of allergic rhinitis or conjunctivitis. Levocabastine can also reduce symptoms of allergic rhinitis by preventing an increase in vascular permeability of nasal mucosa. In pharmaceutical products, it is employed as levocabastine hydrochloride (CAS number 79547-78-7, EC number not available, molecular formula $C_{26}H_{30}ClFN_2O_2$) (1).

CONTACT ALLERGY

Case reports and case series

In Ferrara, Italy, over a 65-month period before 2005, 50 patients affected by periorbital dermatitis while using topical ocular products were patch tested, including with their own ophthalmic medications (n=210). Only 15 positive reactions were detected in 12 subjects, including 14 reactions to commercial eye drops. There was one reaction to levocabastine hydrochloride. The active ingredient was not tested separately, but contact allergy to the excipients and preservatives was excluded. The authors concluded that patch testing with commercial eye drops has doubtful value (2).

LITERATURE

1   The data in the section 'General' may have been obtained from literature discussed in this chapter, but mostly also or exclusively from one or more of the following online sources: ChemIDPlus Advanced, PubChem, DrugBank, RxList, Drug Central, Drugs.com, and Wikipedia
2   Corazza M, Massieri LT, Virgili A. Doubtful value of patch testing for suspected contact allergy to ophthalmic products. Acta Derm Venereol 2005;85:70-71

Chapter 3.201 LIDOCAINE

IDENTIFICATION

Description/definition : Lidocaine is the aminoethylamide that conforms to the structural formula shown below
Pharmacological classes : Anesthetics, local; anti-arrhythmia agents; voltage-gated sodium channel blockers
IUPAC name : 2-(Diethylamino)-N-(2,6-dimethylphenyl)acetamide
Other names : Lignocaine; Xylocaine ®
CAS registry number : 137-58-6
EC number : 205-302-8
Merck Index monograph : 6805
Patch testing : Hydrochloride, 5% pet. (Chemotechnique); hydrochloride, 15% pet. (Chemotechnique, SmartPracticeCanada, SmartPracticeEurope)
Molecular formula : $C_{14}H_{22}N_2O$

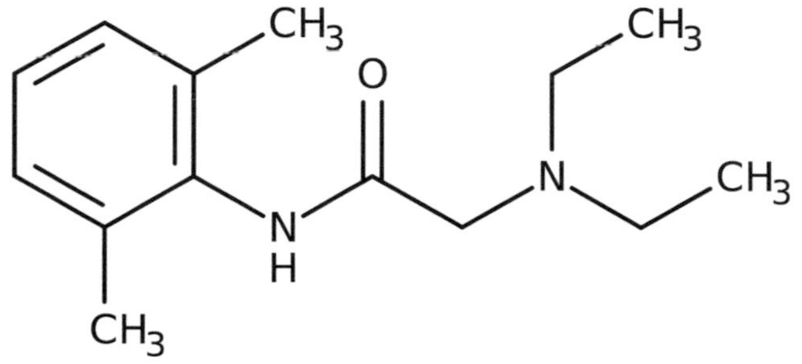

GENERAL

Lidocaine is an aminoethylamide and the prototypical member of the amide class anesthetics. It also exhibits class IB antiarrhythmic effects. Lidocaine is indicated for production of local or regional anesthesia by infiltration techniques such as percutaneous injection and intravenous regional anesthesia, by peripheral nerve block techniques such as brachial plexus and intercostal blocks, and by central neural techniques such as lumbar and caudal epidural blocks. It is also present in products for surface anesthesia, e.g. in a commonly used preparation combined with prilocaine and as adhesive patch (transdermal therapeutic system) (22). According to some sources, lidocaine may also be used as an anti-arrythmia agent. In pharmaceutical products, both lidocaine and lidocaine hydrochloride (CAS number 73-78-9, EC number 200-803-8, molecular formula $C_{14}H_{23}ClN_2O$) may be employed (1).

CONTACT ALLERGY

General

Formerly, lidocaine allergy was very infrequent. Up to 1998, only 33 cases had been reported worldwide in the English literature, of which 19 came from Australia (51). That year, 29 patients were reported from a hospital in Adelaide, Australia; thus, at that moment, 48 of 62 reported cases originated from this country. This was explained by the fast growth of the number of over-the-counter available topical pharmaceuticals containing lidocaine in Australia. Yet, these 29 cases had been collected in a period of 22 years and partly by routine testing (51).

According to some authors, there are indications that lidocaine sensitization is becoming more frequent (36,49). This is attributed to increasing use of lidocaine in topical pharmaceuticals, e.g. in creams and ointments used for pruritus ani or hemorrhoids (these are the most frequent sources of lidocaine sensitization [39,44,48,54,55,57, 60,61,66,67,68]), oral ulcer care, athlete's foot remedies, corn and callus treatments, a cream containing 2.5% lidocaine and 2.5% prilocaine for anesthesia before superficial dermatological interventions (41,52) and antibiotic ointments (37). Yet, in the USA, no increased frequency is seen in routine testing (table 3.201.1) and such data from other countries are not available. This author thinks that, related to the widespread use of topical and subcutaneous administration, lidocaine allergy is still very infrequent.

Cross-reactions between lidocaine and other amide class anesthetics such as bupivacaine, mepivacaine and prilocaine in lidocaine-sensitized individuals appear to be (fairly) common.

As lidocaine is also used subcutaneously, intramuscularly (59), intra-articularly (70) and intravenously (23) and can cause localized (24-32,40,49,58) and rarely generalized (33,70) allergic reactions in sensitized individuals, it has been advised that, following a positive patch test, patients undergo intradermal testing (also to exclude type-I allergy [44,52]), followed by a subcutaneous challenge with lidocaine and other local anesthetic agents in the 'amide' family to identify – if necessary – a safe alternative (36).

Contact allergy in the general population

In Germany, for the period 1995-2004, the population-based relative incidence (RI) of contact sensitization to lidocaine (cases/100,000 defined daily doses (DDDs) per year) was estimated to be 1.5. In the group of local anesthetics, the RI ranged from 1.5 (lidocaine) to 413.9 (benzocaine) (16).

Patch testing in groups of patients

Results of patch testing lidocaine in consecutive patients suspected of contact dermatitis (routine testing) back to 2000 are shown in table 3.201.1. Results of testing in groups of *selected* patients (e.g. patients with anogenital dermatoses, individuals suspected of allergy to medicaments) are shown in table 3.201.2.

Patch testing in consecutive patients suspected of contact dermatitis: routine testing

As lidocaine 15% pet. has been present in the screening series of the North American Contact Dermatitis Group (NACDG) since 2001, most data on routine testing with lidocaine come from the USA and Canada. In the NACDG studies, prevalences of sensitization have been fairly constant, ranging from 0.6% to 1.1%. 'Definite + probable relevance' was below 20% in the period 2001-2008, but recently rose to 40-50%. In most other studies from the USA, equally low rates of sensitization were observed with the exception of a very small 2001 study, which scored 2.2% positive patch tests (n=4) in 183 consecutive patients, 2 of which were relevant because of reactions to injected lidocaine (49). Somewhat higher frequencies of sensitization were also reported from Canada (1.7% [36]) and Australia 1.6% [8]).

Table 3.201.1 Patch testing in groups of patients: Routine testing

Years and Country	Test conc. & vehicle	Number of patients tested | positive (%)		Selection of patients (S); Relevance (R); Comments (C)	Ref.
2015-17 NACDG	15% pet.	5593	42 (0.8%)	R: definite + probable relevance: 45%	2
2011-2015 USA	15% pet.	2572	11 (0.4%)	R: not stated	3
2013-2014 Canada	15% pet.	756	13 (1.7%)	R: 9/13 (69%)	36
2013-14 NACDG	15% pet.	4859	49 (1.0%)	R: definite + probable relevance: 43%	4
2011-12 NACDG	15% pet.	4230	33 (0.8%)	R: definite + probable relevance: 39%	5
2009-10 NACDG	15% pet.	4300	(0.7%)	R: definite + probable relevance: 55%	6
2006-2010 USA	5% pet.	3083	(0.3%)	R: 0%	7
2001-2010 Australia	5% pet.	1208	19 (1.6%)	R: 47%	8
2007-2009 Denmark	15% pet.	1360	2 (0.1%)	R: current 1, unknown 1	38
2007-8 NACDG	15% pet.	5077	(0.9%)	R: definite + probable relevance: 18%	9
2005-6 NACDG	15% pet.	4429	(1.1%)	R: definite + probable relevance: 19%	10
2001-2005 USA	5% pet.	1895	(0.8%)	R: 60%	11
2001-2004 NACDG	15% pet.	10,061	66 (0.7%)	R: definite + probable relevance : 14%	72
2003-4 NACDG	15% pet.	5137	44 (0.6%)	R: not stated	12
2001-02 NACDG	15% pet.	4892	(0.7%)	R: definite + probable relevance: 12%	13
2001 USA	15% pet.	183	4 (2.2%)	R: 50%, reactions to injected lidocaine	49
1994-2001 Denmark	15% pet.	6265	18 (0.3%)	R: one current, one past relevance	38

NACDG: North American Contact Dermatitis Group (USA, Canada)

Patch testing in groups of selected patients

In groups of selected patients, prevalences of sensitization to lidocaine varied widely and ranged from 0.3% to 12%, but were mostly (very) low. The 12% rate can be explained by the mode of data selection, targeted at previously established contact allergy to topical medications in patient charts (18). In a University clinic in Leuven, Belgium, which is specialized in allergy to medicaments and cosmetics, 19 patients (5.5%) tested positive to lidocaine in a group of 345 individuals suspected of iatrogenic contact dermatitis, indicating proper selection for testing a pharmaceutical series (17).

Case series

In the period 2001-2005, in Cleveland, USA, 1143 consecutive patients were patch tested with lidocaine 15% and there were 16 (1.4%) positive reactions. The patients ranged in age from 28 to 77 years. The dermatitis involved the hands or the hands and feet in 8 patients, the arms in 3 cases, and the face and the groin in one case each; it was disseminated in 5 patients. Patch tests with lidocaine dilutions (in petrolatum) gave the following results: 3 of 4, 4 of 6, and 3 of 6 patients reacted positively to 10%, 5%, and 1% dilutions, respectively. Relevance was definite in 2 cases, probable in 1, possible in 11, past in 1 and unknown in 1. The 2 patients with definite relevancy had a history of allergic reactions to subcutaneous resp. intra-articular lidocaine injections. There were 10 co-reactions to neomycin

Table 3.201.2 Patch testing in groups of patients: Selected patient groups

Years and Country	Test conc. & vehicle	Number of patients tested | positive (%)		Selection of patients (S); Relevance (R); Comments (C)	Ref.
2005-2016 NACDG	15% pet.	449	5 (1.1%)	S: patients with only anogenital dermatitis; R: all positives represent relevant reactions; C: the frequency was significantly higher than in a control group	21
1990-2014 Belgium	5% pet.	345	19 (5.5%)	S: patients suspected of iatrogenic contact dermatitis and tested with a pharmaceutical series and their own products; R: 96% of the positive patch test reactions to all topical drugs and antiseptics were considered to be relevant	17
2000-2010 Canada		100	12 (12%)	S: charts reviewed and included in the study when there was at least one positive reaction to a topical drug; R: not stated; C: the high percentages to all drugs are obviously the result of the – rather unusual – selection procedure targeting at topical drug allergy	18
2004-2008 IVDK			(2.8%)	S: patients with anogenital dermatoses tested with a medicament series; R: not stated; C: number of patients tested unknown	20
1997-2007 U.K.	5% pet.	620	2 (0.3%)	S: patient with suspected topical medicament allergy; R: one patient was sensitized by repeated ocular anesthesia	69
1995-2004 IVDK	15% pet.	1692	(1.1%)	S: not stated; R: not stated	15
1992-1994 U.K.	15% pet.	69	1 (1%)	S: women with 'vulval problems'; R: 58% for all allergens together	19

IVDK: Information Network of Departments of Dermatology (Germany, Austria, Switzerland); NACDG: North American Contact Dermatitis Group (USA, Canada)

and 9 to bacitracin (70). Quite curiously, the authors of this study did not comment on this, e.g. if topical pharmaceuticals containing these antibiotics and lidocaine are available in the USA. An internet search learned that there are indeed – at this moment – at least 2 such products (e.g. Lanabiotic ®).

Between 1990 and 2003, in Leuven, Belgium, 92 patients were patch tested for chronic vulval complaints. Fifteen of these women were tested with a series of topical drugs; there was one reaction to lidocaine 5% pet. which was relevant (14).

In a hospital in Adelaide, Australia, where all patients were routinely patch tested with lidocaine (2% pet., later 15% pet.), 18 cases of sensitization were observed in the period 1972-1985 and another 11 between 1990 and 1998. In the initial group of 18, twelve were patch tested to their lidocaine-containing creams, lotions, ointments or eye drops and 11 had positive reactions. In the 2nd group of 11 patients, only 3 had a history of using a lidocaine-containing product (51).

In a hospital in Victoria, Australia, 10 patients were seen during the years 1978-1983 who showed delayed contact allergy to 2 proprietary products containing lidocaine sold in Australia. One product is used for minor skin irritation, the other for hemorrhoids and perineal itch. Five of the 10 patients were applying the cream in the perineal region and 4 were applying the medicament for hand dermatitis. Patch tests showed marked positive reactions (+++) to the product concerned in all 10 patients. Eight were subsequently patch tested to the individual components of the creams and all had marked positive reactions to 2% lidocaine base in soft paraffin. Four were also tested to 1% lidocaine HCl in aqueous solution, with strongly positive reactions in each case (60).

An unknown number of patients with contact allergy to/allergic contact dermatitis from lidocaine was reported from Australia in 1973 and 1980 (62,65) and from Germany in 1974 (66).

Case reports

A 54-year-old woman developed a severe perianal eczematous reaction a few days after applying lidocaine cream for hemorrhoids. She was prescribed a cream containing lidocaine and triamcinolone acetonide, but this worsened the dermatitis with bullae, swelling, erosions, and extending it to the genital region. An internal cause was suspected and the patient was referred to a proctologist after the dermatitis resolved. The patient was examined twice, using a lidocaine-containing lubricant and after 2 days she developed redness, itching, and blisters. Patch tests showed a positive reaction to lidocaine 1% pet. with cross-reactions to bupivacaine, mepivacaine and prilocaine. A provocation test with a subcutaneous injection of 2 ml lidocaine 2% showed erythema, swelling and vesicles after 5 days (39).

A man aged 54 was treated for hemorrhoids on two separate occasions with a combination ointment containing 2.5% lidocaine HCl, metaoxedrine (phenylephrine) HCl 0.1% and a corticosteroid. On the second occasion he developed a widespread eczema. There was no eczema in the treated area, possibly from the presence of the corticosteroid. Patch tests were positive to lidocaine HCl 1% and 2% pet. and to phenylephrine HCl 0.5% and 1%

water (68). This may have been a case of systemic contact dermatitis from lidocaine, phenylephrine or both from percutaneous and transmucosal absorption.

Short summaries of other case reports of allergic contact dermatitis from lidocaine are shown in table 3.201.3.

Table 3.201.3 Short summaries of case reports of allergic contact dermatitis from lidocaine

Year and country	Sex	Age	Positive patch tests	Clinical data and comments	Ref.
2017 Canada	F	28	lidocaine (not specified)	acute dermatitis of the lips with edema, vesicles and oozing 3 days after applying an ointment containing lidocaine	35
2009 The Netherlands	F	80	LIDO 1% pet.; PRILO 5% pet.; LIDO-PRILO cream 'as is'	after 10 days LIDO-PRILO cream 'erythema and inflamma-tion' of the left thorax	41
2008 Spain	M	66	LIDO eardrops 'as is'; LIDO 5% pet.	acute dermatitis of the pinna and left cheek after applying lidocaine eardrops	42
2006 Serbia	F	47	2 LIDO comm prep 'as is'; LIDO 2,5,15% pet., 1% aq.	recurrent dermatitis of perianal region; also type-I hypersensitivity with generalized urticaria	44
2004 Germany	M	65	LIDO lubricant 'as is'; LIDO 15% pet.	30 minutes after lido lubricant in urethra, edema of the penis and scrotum; immediate type skin tests to LIDO neg.; IgE-reaction to latex, but clinically not relevant	45
1999 Italy	M	47	LIDO ointment 'as is'; LIDO 5% pet.	after using LIDO combination ointment for a few days, eczema in the (peri)anal region	48
1999 USA	F	26	LIDO (concentration and vehicle not mentioned)	redness and itching from Aloe vera gel with lidocaine	50
1998 United Kingdom	F	35	LIDO 1% pet.	worsening of sunburn from LIDO-PRILO cream resulting in blistering; also contact allergy to prilocaine and type-I anaphylactic reaction to lidocaine buccal injection	52
1997 Japan	F	48	LIDO 2% pet.	pruritic scaly erythema on the toes and foot from lotion containing lidocaine and crotamiton; co- or cross-sensitization to crotamiton	53
1996 Australia	F	43	2 LIDO creams 'as is'; LIDO 1% pet.	localized acute weeping dermatitis of the right inner ankle after various topical pharmaceuticals including lidocaine	31
1994 United Kingdom	M	71	LIDO ointment 'as is'; LIDO 1% pet. and 2% aq.	acute perianal inflammation after LIDO ointment for pruritus ani	54
1992 United Kingdom	M	47	LIDO 1% aqueous gel	lichenified eczema peri-anally after LIDO ointment	55
1990 United Kingdom	F	42	LIDO ointment 'as is'; LIDO 1% pet.	acute contact dermatitis of the perianal area, hips, face and neck after 2 days LIDO ointment for hemorrhoids	57
1983 The Netherlands	F	36	LIDO ointment 'as is'; LIDO 2% water	perianal dermatitis from LIDO ointment; also contact allergy to dibucaine and pramocaine (*not* cross-reactivity)	61
1979 The Netherlands	F	78	LIDO 2% water	allergic contact dermatitis around a leg ulcer from an 5% aqueous solution of lidocaine and lidocaine cream	63
1979 Sweden	F	33	LIDO ointment 'as is'; LIDO 1% water	dermatitis of the perianal region, later spreading to the the face and body from several weeks 5% LIDO ointment	64
	F	43	LIDO ointment 'as is'; LIDO 1% water	after one month' application of 5% LIDO ointment for pruritus ani, dermatitis of the anal region and the thighs	
1977 Australia	M	45	LIDO spray 'as is'; LIDO 2% water	widespread rash in the treated area after application of a spray-on relief agent containing lidocaine	67

aq.: water; comm prep: commercial preparation(s); LIDO: lidocaine; PRILO: prilocaine

Fixed drug eruption
A 35-year-old woman presented with a slightly pigmented lesion on the dorsum of the left thumb. A week earlier, she had used on herself a suppository containing lidocaine HCl and diflucortolone valerate for the treatment of hemorrhoids for 2 days, which was followed by erythema and blistering on the dorsum of the left thumb with pain and itching. The patient was patch tested with both chemicals on unaffected skin of the upper arm 20% and 10% pet., and on involved skin with 20% pet. Only lidocaine HCl 20% pet. was positive and only on the involved thumb (34).

Cross-reactions, pseudo-cross-reactions and co-reactions
Patients sensitized to lidocaine fairly frequently cross-react to one or more of the other amide-type (aminoacyl-amides) local anesthetics bupivacaine, mepivacaine and prilocaine (32,39,40,43,50,51,54,56,57,58,61). One cross-reaction to ropivacaine 1% pet. (40). In one case, a cross-reaction from lidocaine sensitization to pseudoephedrine has been suggested (52) and in another to crotamiton (53). Cross-reaction from mepivacaine allergy (46) and from prilocaine allergy (47) to lidocaine.

Provocation tests
In eleven patients who had positive patch tests to lidocaine 15% pet., intradermal tests with 0.02 ml of 0.01% lidocaine solution (to exclude concomitant type-I allergy) and provocative challenge tests with 1.5 ml of subcutaneous lidocaine 1% solution were performed. Intradermal tests were negative in all. Subcutaneous provocation caused allergic reactions after 24 hours with (a combination of) pruritus, erythema, papules and vesicles in 4 subjects (36).

Cutaneous adverse drug reactions from systemic administration caused by type IV (delayed-type) hypersensitivity
Cutaneous adverse drug reactions from systemic administration of lidocaine caused by type IV (delayed-type) hypersensitivity, including extensive eczematous eruption from intravenous injection (23) and allergic reactions from intramuscular (59), intra-articular (70) or subcutaneous injections (mostly localized reactions [24-32,40,49,58,71], sometimes generalized [33,70]), are planned to be discussed in Volume IV of the *Monographs in Contact Allergy* series on Systemic drugs.

Immediate contact reactions
Immediate contact reactions (contact urticaria) to lidocaine are presented in Chapter 5.

LITERATURE

1   The data in the section 'General' may have been obtained from literature discussed in this chapter, but mostly also or exclusively from one or more of the following online sources: ChemIDPlus Advanced, PubChem, DrugBank, RxList, Drug Central, Drugs.com, and Wikipedia
2   DeKoven JG, Warshaw EM, Zug KA, Maibach HI, Belsito DV, Sasseville D, et al. North American Contact Dermatitis Group patch test results: 2015-2016. Dermatitis 2018;29:297-309
3   Veverka KK, Hall MR, Yiannias JA, Drage LA, El-Azhary RA, Killian JM, et al. Trends in patch testing with the Mayo Clinic standard series, 2011-2015. Dermatitis 2018;29:310-315
4   DeKoven JG, Warshaw EM, Belsito DV, Sasseville D, Maibach HI, Taylor JS, et al. North American Contact Dermatitis Group Patch Test Results: 2013-2014. Dermatitis 2017;28:33-46
5   Warshaw EM, Maibach HI, Taylor JS, Sasseville D, DeKoven JG, Zirwas MJ, et al. North American Contact Dermatitis Group patch test results: 2011-2012. Dermatitis 2015;26:49-59
6   Warshaw EM, Belsito DV, Taylor JS, Sasseville D, DeKoven JG, Zirwas MJ, et al. North American Contact Dermatitis Group patch test results: 2009 to 2010. Dermatitis 2013;24:50-59
7   Wentworth AB, Yiannias JA, Keeling JH, Hall MR, Camilleri MJ, Drage LA, et al. Trends in patch-test results and allergen changes in the standard series: a Mayo Clinic 5-year retrospective review (January 1, 2006, to December 31, 2010). J Am Acad Dermatol 2014;70:269-275
8   Toholka R, Wang Y-S, Tate B, Tam M, Cahill J, Palmer A, Nixon R. The first Australian Baseline Series: Recommendations for patch testing in suspected contact dermatitis. Australas J Dermatol 2015;56:107-115
9   Fransway AF, Zug KA, Belsito DV, Deleo VA, Fowler JF Jr, Maibach HI, et al. North American Contact Dermatitis Group patch test results for 2007-2008. Dermatitis 2013;24:10-21
10 Zug KA, Warshaw EM, Fowler JF Jr, Maibach HI, Belsito DL, Pratt MD, et al. Patch-test results of the North American Contact Dermatitis Group 2005-2006. Dermatitis 2009;20:149-160
11 Davis MD, Scalf LA, Yiannias JA, Cheng JF, El-Azhary RA, Rohlinger AL, et al. Changing trends and allergens in the patch test standard series. Arch Dermatol 2008;144:67-72
12 Warshaw EM, Belsito DV, DeLeo VA, Fowler JF Jr, Maibach HI, Marks JG, et al. North American Contact Dermatitis Group patch-test results, 2003-2004 study period. Dermatitis 2008;19:129-136
13 Pratt MD, Belsito DV, DeLeo VA, Fowler JF Jr, Fransway AF, Maibach HI, et al. North American Contact Dermatitis Group patch-test results, 2001-2002 study period. Dermatitis 2004;15:176-183
14 Nardelli A, Degreef H, Goossens A. Contact allergic reactions of the vulva: a 14-year review. Dermatitis 2004;15:131-136
15 Menezes de Padua CA, Uter W, Schnuch A. Contact allergy to topical drugs: prevalence in a clinical setting and estimation of frequency at the population level. Pharmacoepidemiol Drug Saf 2007;16:377-384
16 Menezes de Padua CA, Schnuch A, Nink K, Pfahlberg A, Uter W. Allergic contact dermatitis to topical drugs – Epidemiological risk assessment. Pharmacoepidemiol Drug Saf 2008;17:813-821
17 Gilissen L, Goossens A. Frequency and trends of contact allergy to and iatrogenic contact dermatitis caused by topical drugs over a 25-year period. Contact Dermatitis 2016;75:290-302
18 Spring S, Pratt M, Chaplin A. Contact dermatitis to topical medicaments: a retrospective chart review from the Ottawa Hospital Patch Test Clinic. Dermatitis 2012;23:210-213

19 Lewis FM, Harrington CI, Gawkrodger DJ. Contact sensitivity in pruritus vulvae: a common and manageable problem. Contact Dermatitis 1994;31:264-265

20 Bauer A. Contact sensitization in the anal and genital area. Curr Probl Dermatol 2011;40:133-141

21 Warshaw EM, Kimyon RS, Silverberg JI, Belsito DV, DeKoven JG, Maibach HI, et al. Evaluation of patch test findings in patients with anogenital dermatitis. JAMA Dermatol 2019;156:85-91

22 Bershow A, Warshaw E. Cutaneous reactions to transdermal therapeutic systems. Dermatitis 2011;22:193-203

23 Hickey JR, Cook SD, Gutteridge G, Sansom JE. Delayed hypersensitivity following intravenous lidocaine. Contact Dermatitis 2006;54:215-216

24 Trautmann A, Stoevesandt J. Differential diagnosis of late-type reactions to injected local anaesthetics: inflammation at the injection site is the only indicator of allergic hypersensitivity. Contact Dermatitis 2019;80:118-124

25 Voorberg AN, Schuttelaar MLA. A case of postoperative bullous allergic contact dermatitis caused by injection with lidocaine. Contact Dermatitis 2019;81:304-306

26 Halabi-Tawil M, Kechichian E, Tomb R. An unusual complication of minor surgery: contact dermatitis caused by injected lidocaine. Contact Dermatitis 2016;75:253-255

27 Bircher AJ, Messmer SL, Surber C, Rufli T. Delayed-type hypersensitivity to subcutaneous lidocaine with tolerance to articaine: confirmation by in vivo and in vitro tests. Contact Dermatitis 1996;34:387-389

28 Breit S, Ruëff F, Przybilla B. 'Deep impact' contact allergy after subcutaneous injection of local anesthetics. Contact Dermatitis 2001;45:296-297

29 Whalen J D, Dufresne RG. Delayed-type hypersensitivity after subcutaneous administration of amide anesthetic. Arch Dermatol 1996;132:1256-1257

30 Kaufmann JM, Hale EK, Ashinoff RA, Cohen DE. Cutaneous lidocaine allergy confirmed by patch testing. J Drugs Dermatol 2002;1:192-194

31 Bassett I, Delaney T, Freeman S. Can injected lignocaine cause allergic contact dermatitis? Australas J Dermatol 1996;37:155-156

32 Duque S, Fernández L. Delayed-type hypersensitivity to amide local anesthetics. Allergol Immunopathol (Madr) 2004;32:233-234

33 Hofmann H, Maibach HI, Prout E. Presumed generalised exfoliative dermatitis to lidocaine. Arch Dermatol 1975;111:266

34 Kawada A, Noguchi H, Hiruma M, Tajima S, Ishibashi A, Marshall J. Fixed drug eruption induced by lidocaine. Contact Dermatitis 1996;35:375

35 Colantonio S, Kirshen C. Severe allergic contact dermatitis due to Polysporin. CMAJ 2017 Aug 8;189(31):E1018

36 Corbo MD, Weber E, DeKoven J. Lidocaine allergy: Do positive patch results restrict future use? Dermatitis 2016;27:68-71

37 To D, Kossintseva I, de Gannes G. Lidocaine contact allergy is becoming more prevalent. Dermatol Surg 2014;40:1367-1372

38 Thyssen JP, Engkilde K, Menné T, Johansen JD. Prevalence of benzocaine and lidocaine patch test sensitivity in Denmark: temporal trends and relevance. Contact Dermatitis 2011;65:76-80

39 Yuen WY, Schuttelaar ML, Barkema LW, Coenraads PJ. Bullous allergic contact dermatitis to lidocaine. Contact Dermatitis 2009;61:300-301

40 Gunson TH, Greig DE. Allergic contact dermatitis to all three classes of local anaesthetic. Contact Dermatitis 2008;59:126-127

41 Timmermans MW, Bruynzeel DP, Rustemeyer T. Allergic contact dermatitis from EMLA cream: concomitant sensitization to both local anesthetics lidocaine and prilocaine. J Dtsch Dermatol Ges 2009;7:237-238

42 Gómez-de la Fuente E, Rosado A, Alvarez JG, Vicente FJ. Allergic contact dermatitis from lidocaine in ear drops. Actas Dermosifiliogr 2008;99:407-410 (Article in Spanish)

43 Langan SM, Collins P. Photocontact allergy to oxybenzone and contact allergy to lignocaine and prilocaine. Contact Dermatitis 2006;54:173-174

44 Jovanović M, Karadaglić D, Brkić S. Contact urticaria and allergic contact dermatitis to lidocaine in a patient sensitive to benzocaine and propolis. Contact Dermatitis 2006;54:124-126

45 Lippert U, Lessmann H, Niedenfuhr S, Fuchs T. Allergic contact dermatitis caused by lidocaine and latex gloves? Urologe A 2004;43:580-583 (Article in German)

46 Sanchez-Morillas L, Martinez JJ, Martos MR, Gomez-Tembleque P, Andres ER. Delayed-type hypersensitivity to mepivacaine with cross-reaction to lidocaine. Contact Dermatitis 2005;53:352-353

47 Suhonen R, Kanerva L. Contact allergy and cross-reactions caused by prilocaine. Am J Contact Dermat 1997;8:231-234

48 Lodi A, Ambonati M, Coassini A, Kouhdari Z, Palvarini M, Crosti C. Contact allergy to 'caines' caused by anti-hemorrhoidal ointments. Contact Dermatitis 1999;41:221-222

49 Mackley CL, Marks JG, Anderson BE. Delayed-type hypersensitivity to lidocaine. Arch Dermatol 2003;139:343-346

50 Redfern DC. Contact sensitivity to multiple local anesthetics. J Allergy Clin Immunol 1999;104(4Pt.1):890-891

51 Weightman W, Turner T. Allergic contact dermatitis from lignocaine: report of 29 cases and review of the literature. Contact Dermatitis 1998;39:265-266

52 Downs AM, Lear JT, Wallington TB, Sansom JE. Contact sensitivity and systemic reaction to pseudoephedrine and lignocaine. Contact Dermatitis 1998;39:33

53 Kawada A, Hiruma M, Fujioka A, Tajima S, Akiyama M, Ishibashi A. Simultaneous contact sensitivity due to lidocaine and crotamiton. Contact Dermatitis 1997;37:45

54 Hardwick N, King CM. Contact allergy to lignocaine with cross-reaction to bupivacaine. Contact Dermatitis 1994;30:245-246

55 Handfield-Jones SE, Cronin E. Contact sensitivity to lignocaine. Clin Exp Dermatol 1993;18:342-343

56 Klein CE, Gall H. Type IV allergy to amide-type local anesthetics. Contact Dermatitis 1991;25:45-48

57 Black RJ, Dawson TA, Strang WC. Contact sensitivity to lignocaine and prilocaine. Contact Dermatitis 1990;23:117-118

58 Curley RK, Macfarlane AW, King CM. Contact sensitivity to the amide anesthetics lidocaine, prilocaine and mepivacine. Case report and review of the literature. Arch Dermatol 1986;122:924-926

59 Fernándes de Corres L, Leanizbarrutia I. Dermatitis from lignocaine. Contact Dermatitis 1985;12:114-115

60 Nurse DS, Rosner SA. Contact dermatitis due to lignocaine. Contact Dermatitis 1983;9:513

61 Van Ketel WG. Contact allergy to different antihaemorrhoidal anaesthetics. Contact Dermatitis 1983;9:512-513

62 Burry JN, Kirk J, Reid JG, Turner T. Environmental dermatitis: changing patterns. Med J Aust 1980;1:183-184

63 Soesman-Van Waadenooyen Kernekamp AS, Van Ketel WG. Contact allergy to lidocaine (Xylocaine, Lignocaine). Contact Dermatitis 1979;5:403

64 Fregert S, Tegner E, Thelin I. Contact allergy to lidocaine. Contact Dermatitis 1979;5:185-188

65 Burry JH, Kirk J, Reid JC, Turner T. Environmental dermatitis: Patch test in 1,000 cases of allergic contact dermatitis. Med J Austral 1973;2:681-685 (Data cited in ref. 64).

66 Wozniak KD, Lübbe D. Zur Diagnostik von Arzneimittel - Intoleranzreaktionen der Haut. Dermatologische Monatsschrift 1974;160:451-459 (Article in German, data cited in ref. 64)

67 Turner TW. Contact dermatitis due to lignocaine. Contact Dermatitis 1977;3:210-211

68 Roed-Petersen J. Contact sensitivity to metaoxedrine. Contact Dermatitis 1976;2:235-236

69 Jussi L, Lammintausta K. Sources of sensitization, cross-reactions, and occupational sensitization to topical anaesthetics among general dermatology patients. Contact Dermatitis 2009;60:150-154

70 Amado A, Sood A, Taylor JS. Contact allergy to lidocaine: a report of sixteen cases. Dermatitis 2007;18:215-220

71 Fuzier R, Lapeyre-Mestre M, Mertes PM, Nicolas JF, Benoit Y, Didier A, et al. Immediate- and delayed-type allergic reactions to amide local anesthetics: clinical features and skin testing. Pharmacoepidemiol Drug Saf 2009;18:595-601

72 Warshaw EM, Schram SE, Belsito DV, DeLeo VA, Fowler JF Jr, Maibach HI, et al. Patch-test reactions to topical anesthetics: retrospective analysis of cross-sectional data, 2001 to 2004. Dermatitis 2008;19:81-85

Chapter 3.202 LINCOMYCIN

IDENTIFICATION

Description/definition   : Lincomycin is a lincosamide antibiotic derived from the bacillus *Streptomyces lincolnensis* that conforms to the structural formula shown below

Pharmacological classes   : Anti-bacterial agents; protein synthesis inhibitors

IUPAC name   : (2S,4R)-N-[(1R,2R)-2-Hydroxy-1-[(2R,3R,4S,5R,6R)-3,4,5-trihydroxy-6-methylsulfanyloxan-2-yl]propyl]-1-methyl-4-propylpyrrolidine-2-carboxamide

Other names   : Cillimycin

CAS registry number   : 154-21-2

EC number   : 205-824-6

Merck Index monograph   : 6825

Patch testing   : Hydrochloride, 1% water (3); hydrochloride, 5% pet. (2)

Molecular formula   : $C_{18}H_{34}N_2O_6S$

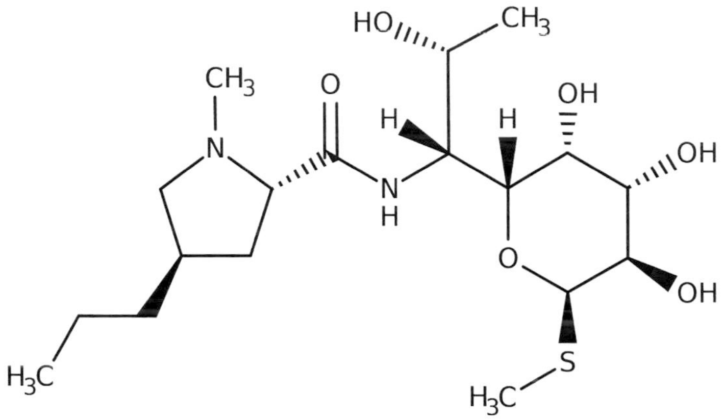

GENERAL

Lincomycin is a lincosamide antibiotic produced by *Streptomyces lincolnensis*. Lincomycin is indicated for the treatment of staphylococcal, streptococcal, and *Bacteroides fragilis* infections caused by susceptible microorganisms. In pharmaceutical products, lincomycin is employed as lincomycin hydrochloride hydrate (CAS number 7179-49-9, EC number 615-424-7, molecular formula $C_{18}H_{37}ClN_2O_7S$).

CONTACT ALLERGY

Case reports

A 37-year-old woman had chronic otitis in her left ear. She was treated with a topical application of a 50-50 mixture of lincomycin hydrochloride and 6-methylprednisolone. Twelve days later, the patient developed an erythematous-exudative lesion and intense pruritus on her left ear lobule, later extending and disseminating over her face, neck, trunk and extremities. Many lesions had a red violaceous color and the central area was depressed. Histopathology was consistent with erythema multiforme. The patient was treated with systemic steroids, and recovered in 10 days. Patch tests were positive to lincomycin HCl and clindamycin HCl (1% water) at D2 and D4. 25 controls were negative. The diagnosis was erythema multiforme-like allergic contact dermatitis (1).

Two months after using a mixture of lincomycin and spectinomycin for the first time, a 27-year-old female chicken vaccinator developed dermatitis of the hands and forearms. Patch tests were positive to the antibiotic mixture, lincomycin HCl 5% pet. and spectinomycin sulfate 1%, 5% and 20% pet. (2). This was a case of occupational allergic contact dermatitis to lincomycin.

Cross-reactions, pseudo-cross-reactions and co-reactions

A patient sensitized to lincomycin cross-reacted to the structurally closely related antibiotic clindamycin (1).

LITERATURE

1   Conde-Salazar L, Guimaraens D, Romero L, Gonzalez M, Yus S. Erythema multiforme-like contact dermatitis from lincomycin. Contact Dermatitis 1985;12:59-61

2   Vilaplana J, Romaguera C, Grimalt F. Contact dermatitis from lincomycin and spectinomycin in chicken vaccinators. Contact Dermatitis 1991;24:225-226

Chapter 3.203 LULICONAZOLE

IDENTIFICATION

Description/definition : Luliconazole is the azole and dichlorobenzene that conforms to the structural formula shown below
Pharmacological classes : Antifungal agents
IUPAC name : (2E)-2-[(4R)-4-(2,4-Dichlorophenyl)-1,3-dithiolan-2-ylidene]-2-imidazol-1-ylacetonitrile
Other names : 4-(2,4-Dichlorophenyl)-1,3-dithiolan-2-ylidene-1-imidazolylacetonitrile
CAS registry number : 187164-19-8
EC number : Not available
Merck Index monograph : 6925
Patch testing : 2% pet.
Molecular formula : $C_{14}H_9Cl_2N_3S_2$

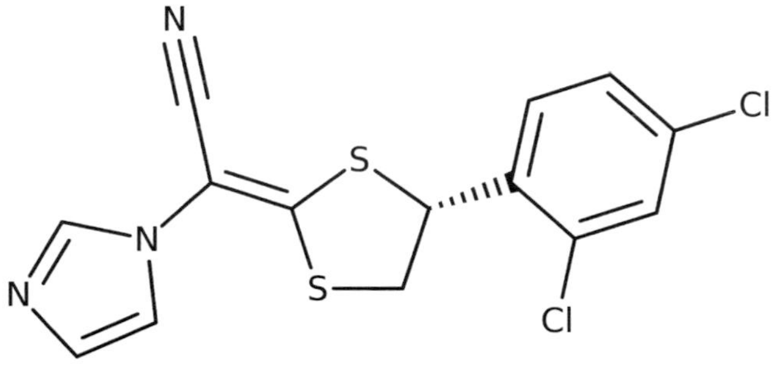

GENERAL

Luliconazole is an azole and dichlorobenzene antifungal agent. Its mechanism of action is as a cytochrome P450 2C19 inhibitor, altering the synthesis of fungal cell membranes. Luliconazole is indicated for the topical treatment of fungal infections caused by *Trichophyton rubrum* and *Epidermophyton floccosum*, specifically tinea pedis, cruris, and corporis.

CONTACT ALLERGY

Case reports

A 65-year-old woman was treated for chromomycosis due to *Phialophora verrucosa* with systemic antifungal treatment supplemented with topical luliconazole 1% cream. Two days later, itchy erythema and papules appeared at the application site. Patch testing showed a positive reaction to luliconazole 1% pet. at D2 (++) and D3 (++). Since the patient also reported a history of contact dermatitis from lanoconazole cream used for the treatment of tinea pedis, she was patch tested with lanoconazole 1% and 10% pet. and a battery of other antimycotic creams, which yielded ++ reactions to both concentrations of lanoconazole only. Because luliconazole and lanoconazole have a similar chemical structure, the patient was considered to have been sensitized with lanoconazole and to have cross-reacted with luliconazole, resulting in the allergic contact dermatitis. The authors suggest that the dithioacetal structure was essential for inducing contact allergy to luliconazole and lanoconazole (2).

A 59-year-old woman suffered from vesicular and squamous lesions on her left foot, which was diagnosed as tinea pedis and treated with 1% luliconazole cream. The lesion soon improved, but after the 27th day of treatment, itchy oozing erythematous dermatitis developed where the cream had been applied. Patch testing revealed positive reactions to luliconazole cream and solution and luliconazole 1% and 2% pet.; all other ingredients were negative. The patient was also patch tested with a range of other antimycotic creams 'as is' and there were weak-positive (+) reactions to 1% lanoconazole cream and ointment (= petrolatum), interpreted as cross-reactions (1).

Cross-reactions, pseudo-cross-reactions and co-reactions

A patient sensitized to lanoconazole later developed allergic contact dermatitis from luliconazole resulting from cross-sensitization (2). Conversely, a patient sensitized to luliconazole cross-reacted to lanoconazole (1).

LITERATURE

1 Shono M. Allergic contact dermatitis from luliconazole. Contact Dermatitis 2007;56:296-297
2 Tanaka T, Satoh T, Yokozeki H. Allergic contact dermatitis from luliconazole: implication of the dithioacetal structure. Acta Derm Venereol 2007;87:271-722

Chapter 3.204 MABUPROFEN

IDENTIFICATION

Description/definition : Mabuprofen is the propionic acid derivative that conforms to the structural formula shown below
Pharmacological classes : Anti-inflammatory agents, non-steroidal; cyclooxygenase inhibitors
IUPAC name : N-(2-Hydroxyethyl)-2-[4-(2-methylpropyl)phenyl]propanamide
Other names : Aminoprofen; ibuprofen aminoethanol
CAS registry number : 82821-47-4
EC number : 280-048-9
Patch testing : No data available: suggested: 1%, 2% and 5% pet.
Molecular formula : $C_{15}H_{23}NO_2$

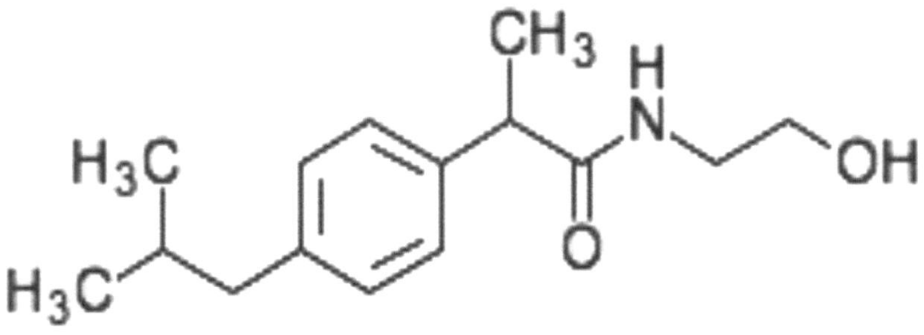

GENERAL

Mabuprofen is a non-steroidal anti-inflammatory drug (NSAID) and propionic acid derivative with anti-inflammatory, analgesic and anti-inflammatory properties. It is used in a spray for musculoskeletal diseases and injuries in certain countries, including Spain (1).

CONTACT ALLERGY

Case reports and case series

In the period 1996-2001, in 2 hospitals in Spain, one patient was diagnosed with contact allergy to mabuprofen. The accumulated incidence per million inhabitants (catchment population of the hospitals) of this side effect was 1.2. Clinical nor patch testing details were provided (2).

LITERATURE

1 The data in the section 'General' may have been obtained from literature discussed in this chapter, but mostly also or exclusively from one or more of the following online sources: ChemIDPlus Advanced, PubChem, DrugBank, RxList, Drug Central, Drugs.com, and Wikipedia
2 Diaz RL, Gardeazabal J, Manrique P, Ratón JA, Urrutia I, Rodríguez-Sasiain JM, Aguirre C. Greater allergenicity of topical ketoprofen in contact dermatitis confirmed by use. Contact Dermatitis 2006;54:239-243

Chapter 3.205 MAFENIDE

IDENTIFICATION

Description/definition   : Mafenide is a methylated sulfonamide that conforms to the structural formula shown
                             below
Pharmacological classes   : Anti-bacterial agents; anti-infective agents, local; carbonic anhydrase inhibitors
IUPAC name   : 4-(Aminomethyl)benzenesulfonamide
Other names   : 4-Homosulfanilamide; α-amino-p-toluenesulfonamide; Sulfamylon ®
CAS registry number   : 138-39-6
EC number   : 205-326-9
Merck Index monograph   : 6982
Patch testing   : 10% pet. (SmartPracticeCanada, SmartPracticeEurope)
Molecular formula   : $C_7H_{10}N_2O_2S$

GENERAL

Mafenide is a sulfonamide-type topical medication used to treat severe burns. It acts by reducing the bacterial population present in the burn tissue, with particular efficacy against *Pseudomonas aeruginosa*, and is said to promote healing of deep burns. In pharmaceutical products, mafenide is employed as mafenide acetate (CAS number 13009-99-9, EC number 235-855-0, molecular formula $C_9H_{14}N_2O_4S$) (1).

CONTACT ALLERGY

General

Mafenide was a frequently used topical chemotherapeutic in Germany from 1941 to 1974, when the marketing of the main mafenide product, a powder, was stopped. The first cases of allergic contact dermatitis were reported in 1949, after the extensive use of mafenide in hospitals of the German army at the end of the second world war (19). Following this, it became a notorious sensitizer, with 1.5%-5.8% positive patch test reactions in routine testing in various German studies (10,12-16). Indeed, in the years 1960-1965 mafenide held the second place in the ranking order of drug allergies; with 437 cases in the Munich dermatological department, it was outnumbered only by benzocaine with 765 positives (18).

The most recent report on allergic contact dermatitis from mafenide acetate stems from 2007 (3,4). That there are no recent publications on sensitization, although the product is still available in many countries including the USA (1), either means that it is very little used, or that allergic reactions to mafenide acetate are so well-known, that it need not be reported anymore.

Patch testing in groups of patients

In Munich, Germany, where mafenide was a popular topical product for treating infected wounds (e.g. leg ulcers), 6.4% of the men and 5.4% of the women reacted positively to mafenide acetate 10% pet.; details are not available (10). Of 3914 consecutive patients tested in the years 1964-1966 in Munich, 162 (4.1%) reacted to mafenide (20). In the period 1970-1972, 242 of 4394 tested eczematous patients (5.5%) had positive patch tests to mafenide acetate 10% pet. in the same clinic (14). Again in Munich, a few years later, in the period 1976 to 1987, 4270 consecutive patients were tested with mafenide hydrochloride 10% pet. and there were 149 (3.5%) positive reactions. Although the main mafenide product had been taken off the market in 1974, there was only a small decline in sensitizations compared with earlier studies from the same clinic. The largest clinical category was (still) dermatitis of the lower leg (42%). 27% of the patients reported an onset of dermatitis less than one year previously. The authors found that, up to 1984, mafenide had still been used for burns, and it was also available in eye drops, wound dressings, in veterinary pharmaceuticals and even as the powder which had apparently been banned in 1974 (12).

Case series

A 61-year-old man was treated with mafenide acetate for burns of his hands, left foot, and face, resulting from a gas explosion. After the agent was applied twice a day for three weeks the patient developed a diffuse papulovesicular cutaneous eruption over the areas treated. A second patient, a 70-year-old man, was treated with mafenide acetate for a burn on his left leg caused by hot food. After one month of therapy, he developed a papulovesicular eruption mainly on the left leg, but also on the trunk and arms. The third patient was a 70-year-old woman who burned her right foot with hot grease, and was treated as an outpatient with mafenide acetate. After three weeks of therapy, she developed a generalized papulovesicular eruption with periorbital and facial edema. When patch tested, all three patients had strongly positive patch tests to mafenide acetate 8.5% and mafenide HCl 5% solution (negative in 15 controls); the reactions to the excipients of the mafenide cream and to a mixture of sulfadiazine, sulfamerazine, and sulfamethazine were negative (17). The generalized eruption with periorbital and facial edema may well have been a case of systemic contact dermatitis from absorption of mafenide from the burn.

Four male soldiers who sustained extensive burns during Operation Iraqi Freedom/Operation Enduring Freedom were treated with mafenide acetate and developed contact dermatitis to this burn medication. Three of 4 patients who were patch tested to mafenide acetate 7% solution were positive. A challenge with mafenide acetate resulted in recurrence of the eruption in 2 out of the 4 patients (3,4).

Four hundred burn patients, ambulatory and hospitalized, were treated with mafenide acetate 11.2% in a hydrophilic cream base. Thirty-nine (9.5%) developed pruritic local maculopapular and sometimes vesicular cutaneous reactions after 8 to 44 days very suspicious for allergic contact dermatitis. Two patients developed erythema multiforme-like eruptions. Eleven of these patients were 'patch tested': the test material was placed on a band-aid and applied to the upper inner aspect of the arm. The tests were removed at 48 hours and read twice, once in a 30 minute period and again two days later. Materials tested were mafenide acetate, 11.2% in a cream base, the cream base, and the band-aid alone. Surprisingly, there were no positive reactions (21). Possibly, application under a band-aid was not occlusive enough (unlikely) or the tackifying material interfered with absorption of mafenide.

Case reports

A 62-year-old man with a tracheostomy was treated with an ointment containing mafenide 7%, enoxolone 1.5%, neomycin 0.35% and sulfanilamide 3% for a skin fissure adjacent to the tracheostomy. After 3 days, he developed itchy erythematous and vesicular exudative lesions at the application site. Nonetheless, the patient continued the treatment for a further week, with spreading of the lesions to the upper chest and shoulders. Later, after healing of the dermatitis, he applied the same ointment around his nostrils, because of irritation related to an upper respiratory tract infection, and developed vesicular lesions around the nose within a day. Patch tests were positive to the ointment, to mafenide 10% pet. and to enoxolone 10% pet. Ten controls were negative (2). A 60-year-old man was treated with the same ointment as the previous patient, for perioral erosions. After 2 days treatment, he developed acute eczema on the cheeks, eyelids and forehead. Patch tests were positive to the ointment and mafenide 7% pet. but not to sulfanilamide or the other ingredients. Fifteen controls were negative (9).

A 38-year-old man was burned by a piece of steel while soldering at work. His injury was a 2 x 4 cm partial thickness burn to the left posterior auricle and adjacent scalp. Mafenide acetate twice daily was prescribed because of the burn location over a cartilaginous area. After 2 weeks the patient developed erythema around the burn site. Mafenide acetate and dry dressings were continued with added oral antibiotics to prevent chondritis. Five days later, the patient returned with complaints of marked increase in swelling and tenderness of his left ear. At this time the edema extended down the angle of the jaw onto the lower face and side of the neck. The entire area was quite indurated. There was a considerable amount of watery drainage but no purulence or fluctuance. The patient did not have any adenopathy, he was afebrile, and his white blood cell count was 5900/mm without a left shift. Within 2 days of stopping mafenide acetate there was significant improvement in the amount of erythema and edema. A patch test and a repeat patch test one year later to the mafenide acetate product were positive (5).

Mafenide acetate is frequently used in auricular burns to prevent acute suppurative chondritis, which may result in cartilage necrosis with subsequent deformity. There have been several case reports in which chondritis (or at least the clinical picture of chondritis) was ascribed to 'allergy' of 'hypersensitivity' to mafenide acetate in cases where a bacterial infection was unlikely (no suppuration, no pain, no fever, no leukocytosis, no reaction to antibiotics) and the clinical picture improved rapidly after suspending the use of mafenide acetate. However, in these case reports, patch tests with mafenide were not performed (6,7,8).

Cross-reactions, pseudo-cross-reactions and co-reactions

Most sulfonamides may cross-react to each other. However, the structure of mafenide is not strictly that of the sulfonamides, as its amino group is not directly in para position on the benzene ring, but is separated from this ring by a methyl group. Therefore, mafenide has an aliphatic and the sulfonamides an aromatic quality. The chemical properties of an aliphatic amino group differ essentially from those of an amino group directly attached to the benzene ring. Therefore, mafenide in one study did not cross-react to the sulfonamides sulfadiazine, sulfamerazine, and sulfamethazine (17). Indeed, although in many countries mafenide was marketed in combination with sulfonamides in a powder for topical application, the very first studies on sensitization had already revealed the fact that in many cases patch tests were positive only to mafenide and not to sulfonamides (19).

However, over half of the patients sensitized to mafenide cross-react to other para-compounds including benzocaine and related esters of PABA (14).

LITERATURE

1   The data in the section 'General' may have been obtained from literature discussed in this chapter, but mostly also or exclusively from one or more of the following online sources: ChemIDPlus Advanced, PubChem, DrugBank, RxList, Drug Central, Drugs.com, and Wikipedia

2   Fernàndez JC, Gamboa P, Jáuregui I, González G, Antépara I. Concomitant sensitization to enoxolone and mafenide in a topical medicament. Contact Dermatitis 1992;27:262

3   Firoz EF, Firoz BF, Williams JF, Henning JS. Allergic contact dermatitis to mafenide acetate: a case series and review of the literature. J Drugs Dermatol 2007;6:825-828

4   Goksel D, Henning JS. Allergic contact dermatitis to mafenide acetate (Sulfamylon™). Dermatitis 2007;18:114

5   McKenna SR, Latenser BA, Jones LM, Barrette RR, Sherman HF, Varcelotti JR. Serious silver sulphadiazine and mafenide acetate dermatitis. Burns 1995;21:310-312

6   Pickus EJ, Lionelli GT, Charles EW 3rd, Korentager RA. Mafenide acetate allergy presenting as recurrent chondritis. Ann Plast Surg 2002;48:202-204

7   Perry AW, Gottlieb LJ, Krizek TJ, Parsons RW, Goodwin CW, Finkelstein JL, et al. Mafenide-induced pseudochondritis. J Burn Care Rehabil 1988;9:145-147

8   Kroll SS, Gerow FJ. Sulfamylon allergy simulating chondritis. Plast Reconstr Surg 1987;80:298-299

9   Sanz de Galdeano C, Aguirre A, Oleaga JM, Goday J, Diaz Perez JL. Allergic contact dermatitis from topical mafenide. Contact Dermatitis 1993;28:249

10 Bandmann H-J, Breit R. Die medikamentöse allergische Kontactdermatitis. Internist 1974;15:47- ? (Article in German,cited in ref. 11)

11 Cronin E. Contact Dermatitis. Edinburgh: Churchill Livingstone, 1980:223

12 Breit R, Seifert P. Mafenide – Still an allergen of importance? In: Frosch PJ, Dooms-Goossens A, Lachapelle JM, Rycroft RJG, Scheper RJ (eds). Current Topics in Contact Dermatitis. Berlin: Springer-Verlag, 1989:222-225

13 Bandmann H-J. Die Kontaktallergie durch Arnzneimittel. Pharm Zeit 1966;111:1470-1479 (Article in German, data cited in ref. 12)

14 Bandmann H-J, Breit R. The mafenide story. Br J Dermatol 1973;89:219-221

15 Braun W. Iatrogene Sensibilisierung und Beinleiden. Med Klin 1970;65:506-509 (Article in German, data cited in ref. 12)

16 Lämmer D. Testergebnisse von 1008 Patienten mit Kontaktallergie. Z Hautkr 1979;54:571-579 (Article in German, data cited in ref. 12)

17 Velasco JE, Africk JA. Contact dermatitis to mafenide acetate. Arch Dermatol 1971;103:61-63

18 Bandmann H-J. Die Kontaktallergie durch Arzneimitteln. Pharmazeutische Zeitung 1966;iii:1470 (Article in German, data cited in ref. 14)

19 Koch F. Über allergene Wirkung von Sulfonamiden. Archiv für Dermatologie und Syphilis 1949;187:213 (Article in German, data cited in refs. 3 and 14)

20 Bandmann H-J. Kontaktekzem und Ekzematogene. Münch Med Wochenschr 1967;109:157 (Article in German, data cited in ref. 14)

21 Yaffee HS, Dressler DP. Topical application of mafenide acetate. Its association with erythema multiforme and cutaneous reactions. Arch Dermatol 1969;100:277-281

Chapter 3.206 MAZIPREDONE

IDENTIFICATION

Description/definition : Mazipredone is the synthetic glucocorticoid that conforms to the structural formula
shown below
Pharmacological classes : Glucocorticoids
IUPAC name : (8S,9S,10R,11S,13S,14S,17R)-11,17-Dihydroxy-10,13-dimethyl-17-[2-(4-methylpiperazin-1-
yl)acetyl]-7,8,9,11,12,14,15,16-octahydro-6H-cyclopenta[a]phenanthren-3-one
Other names : 11β,17-Dihydroxy-21-(4-methyl-1-piperazinyl)pregna-1,4-diene-3,20-dione
CAS registry number : 13085-08-0
EC number : Not available
Merck Index monograph : 7102
Patch testing : In general, corticosteroids may be tested at 0.1% and 1% in alcohol; late readings (6-10
days) are strongly recommended
Molecular formula : $C_{26}H_{38}N_2O_4$

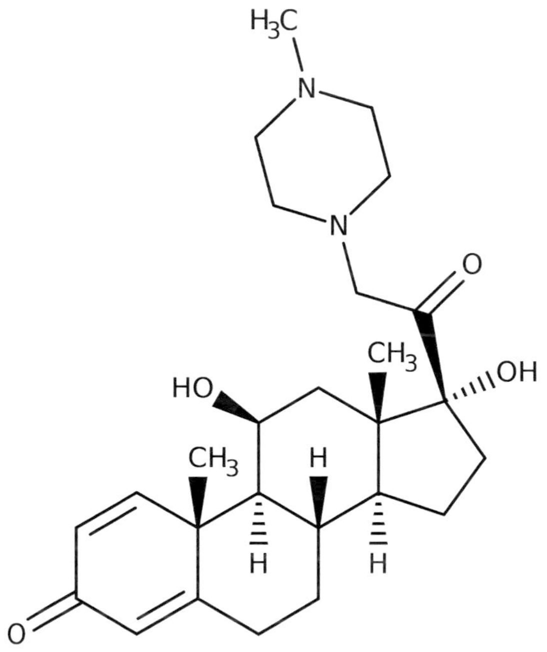

GENERAL
General aspects of corticosteroids used on the skin and mucous membranes are discussed in Chapter 2.4. A practical
guideline for diagnosing allergic reactions to corticosteroids is presented in ref. 1.

CONTACT ALLERGY

Case reports and case series
From January 1990 to June 2008, in Leuven, Belgium, 315 patients were diagnosed with contact allergy to/allergic
contact dermatitis from corticosteroids (CSs) from routine patch testing with a baseline series including tixocortol
pivalate, budesonide, hydrocortisone butyrate and prednisone caproate, patch testing with patients' own CS
preparations, and testing those with proven contact allergy to a corticosteroid or strongly suspected of CS allergy
later with a series of 66 CSs, including two sex hormones (progesterone and testosterone). 71% of the patients had
relevant reactions, but these were not specified. In this group of 315 CS allergic patients, 67 had positive patch tests
to mazipredone 0.1% alc. (2). As this corticosteroid has never been used in pharmaceuticals in Belgium, these
positive reactions must all be considered cross-reactions to other corticosteroids.

In Hungary, in the period 1994-1997, when mazipredone was widely used in that country, 8 patients reacted to mazipredone, of whom 2 *only* to this corticosteroid. Unfortunately, no clinical details were provided and it was not mentioned whether these reactions were relevant (3).

Cross-reactions, pseudo-cross-reactions and co-reactions
Cross-reactions between corticosteroids are discussed in Chapter 2.8.

LITERATURE
1   Baeck M, Goossens A. Immediate and delayed allergic hypersensitivity to corticosteroids: practical guidelines. Contact Dermatitis 2012;66:38-45
2   Baeck M, Chemelle JA, Terreux R, Drieghe J, Goossens A. Delayed hypersensitivity to corticosteroids in a series of 315 patients: clinical data and patch test results. Contact Dermatitis 2009;61:163-175
3   Matura M. Corticosteroid contact allergy in Hungary. Contact Dermatitis 1998;38:225-226

Chapter 3.207 MECHLORETHAMINE

IDENTIFICATION

Description/definition   : Mechlorethamine is the nitrogen mustard compound that conforms to the structural
                              formula shown below
Pharmacological classes   : Antineoplastic agents, alkylating; irritants; alkylating agents; chemical warfare agents
IUPAC name   : 2-Chloro-N-(2-chloroethyl)-N-methylethanamine
Other names   : Chlormethine; nitrogen mustard; 2,2'-dichlorodiethyl-methylamine
CAS registry number   : 51-75-2
EC number   : 200-120-5
Merck Index monograph   : 7116
Patch testing   : 0.02% water; perform an open test first
Molecular formula   : $C_5H_{11}Cl_2N$

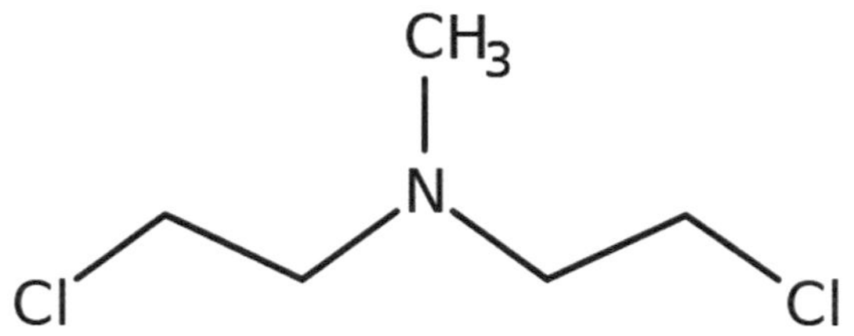

$$CH_3$$
$$|$$
$$N$$

Cl   Cl

GENERAL

Mechlorethamine is a nitrogen mustard with antineoplastic and immunosuppressive activities. This agent alkylates DNA, resulting in DNA base pair mismatching, DNA interstrand crosslinking, the inhibition of DNA repair and synthesis, cell-cycle arrest, and apoptosis. It is a vesicant and necrotizing irritant destructive to mucous membranes, which was formerly used as a war gas. Mechlorethamine is indicated for the palliative treatment of Hodgkin's disease (stages III and IV), lymphosarcoma, chronic myelocytic or chronic lymphocytic leukemia, polycythemia vera, mycosis fungoides, and bronchogenic carcinoma. As a topical medication, since 1959 it is commonly used in cutaneous mycosis fungoides and some other dermatoses such as parapsoriasis (1).

In pharmaceutical products, mechlorethamine is employed as mechlorethamine hydrochloride (CAS number 55-86-7, EC number 200-246-0, molecular formula $C_5H_{12}Cl_3N$) (1). Although mechlorethamine HCl is a known carcinogen, most studies have not found a strong association between topical treatment with it and cutaneous skin cancers, but the subject is controversial (21).

CONTACT ALLERGY

General

Allergic contact dermatitis is a frequent side effect of topical treatment of mycosis fungoides with mechlorethamine hydrochloride, that usually begins after 2-4 weeks (5-20,28; not a full literature search performed). In some studies, >50-60% of the patients apparently became sensitized (8,9,12,17,19,20). One group of investigators diagnosed more than 60 such patients (14). However, it should be realized that, with a few exceptions (13,14,16,24), patch tests (open) to confirm sensitization, sometimes in a dilution series, were hardly ever performed. Therefore, in a number of cases, irritant contact dermatitis from mechlorethamine HCl, which is also a frequent side effect, may well have been mistaken for allergic contact dermatitis. Indeed, in some studies, low rates of sensitization have been reported (9,10,11) and in a number of investigations, only few patients with a history of hypersensitivity to mechlorethamine developed allergic contact dermatitis from renewed treatment (10,11).

The risk of sensitization may both depend on patient factors (age, immune status, skin disease, stage of malignant skin disease and characteristics) and treatment factors such as concentration of mechlorethamine, vehicle for administering the drug (far more sensitizations from aqueous solutions than from ointment preparations: 10,11, 19), treatment protocol, surface treated, and immunosuppressive therapies, e.g. prior electron beam treatment (22). Interestingly, in quite a few studies, patients who became sensitized to the drug generally had better treatment results (12,14,15,16,19,23,29). In a number of investigations, it has been tried to desensitize allergic patients by using lower concentrations of mechlorethamine HCl (6,8,12,14,15,19,25) or by administering small doses of the drug intravenously (13,14) with incomplete success (7).

A recent review of mechlorethamine HCl treatment and its side effects was published in 2018 (21).

Cross-reactions, pseudo-cross-reactions and co-reactions
Patients with psoriasis or mycosis fungoides sensitized to mechlorethamine HCl have cross-reacted to *N*-methyl-2-chloroethyl-2-hydroxyethylamine, diethyl(2-chloroethyl)amine, chloroethyl piperidine, chloroethyl morpholine, chloroethyl hexamethyleneimine and methoxyethyl benzylchloroethylamine (26).

Patch test sensitization
Patch testing with mechlorethamine HCl is said to carry the risk of patch test sensitization (27).

Irritant contact reactions
Cutaneous irritation and itch are very frequent adverse effects of topical mechlorethamine treatment of mycosis fungoides. Mechlorethamine HCl is also well-known vesicant (1). Treatment of vitiligo with a tincture of 0.05% mechlorethamine HCl led to erythema, vesicles and bullae within one day, which healed with a burn scar in the area of application (2).

An 81-year-old man had had 11 applications of mechlorethamine HCl 0.02% for mycosis fungoides. After the 12th, he developed a local bullous reaction within one hour, localized exactly to the area of topical application. An open patch test with mechlorethamine 0.02% gave erythema and edema within one hour, but no bullous lesions. Twelve controls were negative. Histopathology of the bullous lesion and of the patch test showed an inflammatory reaction in the dermis and minimal changes in the epidermis. The authors interpreted this unusual reaction as a severe toxic contact effect that may probably be included in the immediate contact reaction spectrum (3).

In an 82-year-old woman with moderate peripheral arterial disease, treatment with 0.02% mechlorethamine HCl may have contributed to the development of necrotic ulcers of the legs following local trauma (4).

Immediate contact reactions
Immediate contact reactions (contact urticaria) to mechlorethamine HCl are presented in Chapter 5.

LITERATURE
1   The data in the section 'General' may have been obtained from literature discussed in this chapter, but mostly also or exclusively from one or more of the following online sources: ChemIDPlus Advanced, PubChem, DrugBank, RxList, Drug Central, Drugs.com, and Wikipedia
2   Ngai KY, Chan HY, Ng F. Topical chlormethine hydrochloride causing bullous reaction. Clin Toxicol (Phila) 2009;47:834-835
3   Goday JJ, Aguirre A, Ratón JA, Díaz-Pérez JL. Local bullous reaction to topical mechlorethamine (mustine). Contact Dermatitis 1990;22:306-307
4   Gary C, Gautier V, Lazareth I, Bagot M, Asgari R, Priollet P. Necrotic leg ulcers after topical application of chlormethine. Ann Dermatol Venereol 2019;146:226-231 (Article in French)
5   Vonderheid EC, Tan ET, Kantor AF, Shrager L, Micaily B, Van Scott EJ. Long-term efficacy, curative potential and carcinogenicity of topical mechlorethamine chemotherapy in cutaneous T-cell lymphoma. J Am Acad Dermatol 1989;20:416-428
6   Ramsay DL, Halperin PS, Zeleniuch-Jacquotte A. Topical mechlorethamine therapy for early stage mycosis fungoides. J Am Acad Dermatol 1988;19:684-691
7   Vonderheid EC. Topical mechlorethamine chemotherapy. Considerations on its use in mycosis fungoides. Int J Dermatol 1984;23:180-186
8   Ramsay DL, Parnes RE, Dubin N. Response of mycosis fungoides to topical chemotherapy with mechlorethamine. Arch Dermatol 1984;120:1585-1590
9   Hamminga B, Noordijk EM, Van Vloten WA: Treatment of mycosis fungoides: Total skin electron-beam irradiation vs topical mechlorethamine therapy. Arch Dermatol 1982;118:150-153
10 Price NM, Hoppe RT, Deneau DC. Ointment-based mechlorethamine treatment for mycosis fungoides. Cancer 1983;52:2214-2219
11 Price NM, Deneau DG, Hoppe RT. The treatment of mycosis fungoides with ointment-based mechlorethamine. Arch Dermatol 1982;118:234-237
12 Price NM, Constantine VS, Hoppe RT, Fuks ZY, Farber EM. Topical mechlorethamine therapy for mycosis fungoides. Br J Dermatol 1977;97:547-550
13 Leshaw S, Simon RS, Baer RL. Failure to induce tolerance to mechlorethamine hydrochloride. Arch Dermatol 1977;113:1406-1408
14 Vonderheid EC, VanScott EJ, Johnson WC, Grekin DA, Asbell SO. Topical chemotherapy and immunotherapy of mycosis fungoides. Arch Dermatol 1977;113:454-462

15 Van Scott EJ, Kalmanson JD. Complete remission of mycosis fungoides lymphoma induced by topical nitrogen mustard (HN2). Control of delayed hypersensitivity to HN2 by desensitization and by induction of specific immunologic tolerance. Cancer 1973;32:18-30

16 Van Scott EJ, Winters PL. Responses of mycosis fungoides to intensive external treatment with nitrogen mustard. Arch Dermatol 1970;102:507-514

17 Waldorf DS, Haynes HA, Van Scott EJ. Cutaneous hypersensitivity and desensitization to mechlorethamine in patients with mycosis fungoides lymphomas. Ann Intern Med 1967;67:282-290

18 Zachariae H, Thestrup-Pedersen K, Sogaard H. Topical nitrogen mustard in early mycosis fungoides. A 12-year experience. Acta Derm Venereol 1985;65:53-58

19 Kim YH, Martinez G, Varghese A, Hoppe RT. Topical nitrogen mustard in the management of mycosis fungoides: update of the Stanford experience. Arch Dermatol 2003;139:165-173

20 Lindahl LM, Fenger-Gron M, Iversen L. Topical nitrogen mustard therapy in patients with mycosis fungoides or parapsoriasis. J Eur Acad Dermatol Venereol 2013;27:163-168

21 Liner K, Brown C, McGirt LY. Clinical potential of mechlorethamine gel for the topical treatment of mycosis fungoides-type cutaneous T-cell lymphoma: a review on current efficacy and safety data. Drug Des Devel Ther 2018;12:241-254

22 Price NM, Hoppe RT, Constantine VS, Fuks ZY, Farber EM. The treatment of mycosis fungoides: adjuvant topical mechlorethamine after electron beam therapy. Cancer 1977;40:2851-2853

23 De Quatrebarbes J, Esteve E, Bagot M, Bernard P, Beylot-Barry M, Delaunay M, et al. Treatment of early-stage mycosis fungoides with twice-weekly applications of mechlorethamine and topical corticosteroids: a prospective study. Arch Dermatol 2005;141:1117-1120

24 Talpur R, Venkatarajan S, Duvic M. Mechlorethamine gel for the topical treatment of stage IA and IB mycosis fungoides-type cutaneous T-cell lymphoma. Expert Rev Clin Pharmacol 2014;7:591-597

25 Constantine VA, Fuks ZY, Farber EM. Mechlorethamine desensitization in therapy for mycosis fungoides. Arch Dermatol 1975;113:484-488

26 Van Scott EJ, Yu RJ. Antimitotic, antigenic, and structural relationships of nitrogen mustard and its homologues. J Invest Dermatol 1974;62:378-383

27 De Groot AC, Nater JP, Weijland JW. Unwanted effects of cosmetics and drugs used in dermatology, 3rd Edition. Amsterdam: Elsevier Science, 1994:29

28 Epstein E Jr, Ugel AR. Effects of topical mechlorethamine on skin lesions of psoriasis. Arch Dermatol 1970;102:504-506

29 Vonderheid EC, Ekbote SK, Kerrigan K, Kalmanson JD, Van Scott EJ, Rook AH, Abrams JT. The prognostic significance of delayed hypersensitivity to dinitrochlorobenzene and mechlorethamine hydrochloride in cutaneous T cell lymphoma. J Invest Dermatol 1998;110:946-950

Chapter 3.208 MEDRYSONE

IDENTIFICATION

Description/definition : Medrysone is the synthetic glucocorticoid that conforms to the structural formula shown below

Pharmacological classes : Glucocorticoids

IUPAC name : (6S,8S,9S,10R,11S,13S,14S,17S)-17-Acetyl-11-hydroxy-6,10,13-trimethyl-1,2,6,7,8,9,11,12,14,15,16,17-dodecahydrocyclopenta[a]phenanthren-3-one

Other names : 11β-Hydroxy-6α-methylpregn-4-ene-3,20-dione; hydroxymesterone

CAS registry number : 2668-66-8

EC number : 220-208-7

Merck Index monograph : 7135

Patch testing : In general, corticosteroids may be tested at 0.1% and 1% in alcohol; late readings (6-10 days) are strongly recommended

Molecular formula : $C_{22}H_{32}O_3$

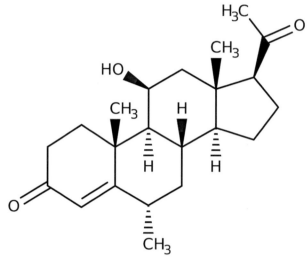

GENERAL

General aspects of corticosteroids used on the skin and mucous membranes are discussed in Chapter 2.4. A practical guideline for diagnosing allergic reactions to corticosteroids is presented in ref. 1.

CONTACT ALLERGY

Case reports and case series

From January 1990 to June 2008, in Leuven, Belgium, 315 patients were diagnosed with contact allergy to/allergic contact dermatitis from corticosteroids (CSs) from routine patch testing with a baseline series including tixocortol pivalate, budesonide, hydrocortisone butyrate and prednisone caproate, patch testing with patients' own CS preparations, and testing those with proven contact allergy to a corticosteroid or strongly suspected of CS allergy later with a series of 66 CSs, including two sex hormones (progesterone and testosterone). 71% of the patients had relevant reactions, but these were not specified. In this group of 315 CS allergic patients, 23 had positive patch tests to medrysone 0.1% alc. (2). Most, if not all of these positive reactions, are probably the result of cross-reactivity to other corticosteroids.

Cross-reactions, pseudo-cross-reactions and co-reactions

Cross-reactions between corticosteroids are discussed in Chapter 2.8.

LITERATURE

1 Baeck M, Goossens A. Immediate and delayed allergic hypersensitivity to corticosteroids: practical guidelines. Contact Dermatitis 2012;66:38-45
2 Baeck M, Chemelle JA, Terreux R, Drieghe J, Goossens A. Delayed hypersensitivity to corticosteroids in a series of 315 patients: clinical data and patch test results. Contact Dermatitis 2009;61:163-175

Chapter 3.209 MEPHENESIN

IDENTIFICATION

Description/definition   : Mephenesin is the synthetic cresol glyceryl ether that conforms to the structural formula
                              shown below
Pharmacological classes : Muscle relaxants, central
IUPAC name   : 3-(2-Methylphenoxy)propane-1,2-diol
Other names   : 1,2-Dihydroxy-3-(2-methylphenoxy)propane
CAS registry number   : 59-47-2
EC number   : 200-427-4
Merck Index monograph : 7187
Patch testing   : 5% pet.
Molecular formula   : $C_{10}H_{14}O_3$

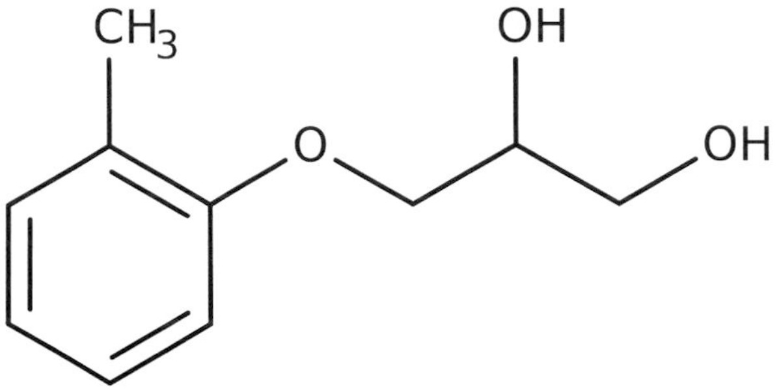

GENERAL

Mephenesin is a synthetic cresol glyceryl ether which produces transient muscle relaxation and paralysis via central nervous system depression. It was formerly used for the treatment of muscle spasticity in Parkinson's disease and multiple sclerosis. Mephenesin is also applied locally, used primarily by young people after sports injuries because of its analgesic and anti-inflammatory properties. With local application, a vasodilating agent is generally added (methyl nicotinate, glycol salicylate) to promote skin penetration (1).

CONTACT ALLERGY

General

Several cases of allergic contact dermatitis to mephenesin have been reported, mostly from Belgium and France, most of which had an erythema multiforme-like aspect (3,4,8). Possibly, it caused acute generalized exanthematous pustulosis (AGEP) from topical application in 2 patients (6,7).

Case series and case reports

In Leuven, Belgium, in the period 1990-2014, iatrogenic contact dermatitis was diagnosed in 2600 individuals (17% of the total patch test population). 96% of all positive patch test reactions to topical drugs and antiseptics were considered to be relevant. Mephenesin (5% pet.) was tested in 55 patients and there were 10 positive reactions to it (2).

A 30-year-old man presented with a very acute dermatitis on his upper right leg. His thigh was warm, violaceous, edematous, indurated, and painful to the touch. The limits of the dermatitis were sharp and map-like. The lesions looked very much like erysipelas, but there were no general symptoms, no fever, and no leukocytosis. Some target or cocardiform lesions were situated in the periphery of the eczema. Treatment with wet dressings and hydrocortisone butyrate cream was initiated. The acute dermatitis regressed, but the erythema multiforme lesions spread out over his arms and legs over the next few days. Ten days before the onset of the acute dermatitis, the patient had a muscle injury and had been treated with a balsam under a rubber bandage. Patch testing revealed contact allergy to rubber ingredients and the balsam. Ingredient patch testing showed a strongly positive reaction to mephenesin 5% pet. (3).

The authors describe an additional four patients who developed extensive contact dermatitis with erythema multiforme-like spreading after applying mephenesin-containing pharmaceuticals to sprained joints of muscles. Three were patch tested and they reacted to the products used and mephenesin 5% pet. There were also reactions to 2 other active ingredients in one patient each, methyl nicotinate (1% pet.) and glycol salicylate (1% pet.) (3).

One patient suffered purpuric contact dermatitis associated with urticaria and an erythema multiforme-like eruption from a cream containing mephenesin and phenylbutazone. Patch tests were strongly positive to the cream and mephenesin and positive to phenylbutazone, the other active principle in the cream (4). Another case of erythema multiforme-like allergic contact dermatitis caused by an ointment containing mephenesin was reported from Germany. Patch testing showed marked sensitization to mephenesin, with positive reactions to different dilutions down to 0.1% (8).

One or more cases of mephenesin allergic contact dermatitis were reported from France in 1986, but details are not available to the author (5).

Acute generalized exanthematous pustulosis
A 51-year-old man presented with a skin rash with fever which started 3 days earlier. His personal history consisted only of lumbar pain and he was not under any oral medication nor had he used any tablets. The skin rash was initially limited to the left hand and the back but quickly spread to the entire skin with multiple non-follicular pustules emerging on widespread and inflamed erythema. Three days before the skin rash appeared, the patient had applied mephenesin cream to the lumbar area with his left hand. The whole clinical presentation led the investigators to the diagnose AGEP induced by mephenesin cream. Patch tests performed 3 months later were positive for the pharmaceutical product 'as is'. According to the authors, this confirmed their diagnosis (which is incorrect, as it only indicates contact allergy, not AGEP). Mephenesin itself was not tested (6).

Possibly, a similar case was reported 5 years later, also in France, but it is unknown whether patch tests have been performed (7).

LITERATURE

1   The data in the section 'General' may have been obtained from literature discussed in this chapter, but mostly also or exclusively from one or more of the following online sources: ChemIDPlus Advanced, PubChem, DrugBank, RxList, Drug Central, Drugs.com, and Wikipedia
2   Gilissen L, Goossens A. Frequency and trends of contact allergy to and iatrogenic contact dermatitis caused by topical drugs over a 25-year period. Contact Dermatitis 2016;75:290-302
3   Degreef H, Bonamie A, van Derheyden D, Dooms-Goossens A. Mephenesin contact dermatitis with erythema multiforme features. Contact Dermatitis 1984;10:220-223
4   Bachmeyer C, Blum L, Picard O, Cabane I, Imbert JC. Dermite de contact au Traumalgyl ®. Ann Derm Venereol 1994;121:93 (Article in French)
5   Drain JP, Bouvier C, Brouet JF, Martin P. Contact dermatitis caused by mephenesin. Allerg Immunol (Paris) 1986;18:36,38 (Article in French)
6   Beltran C, Vergier B, Doutre M-S, Beylot C, Beylot-Barry M. Acute generalized exanthematous pustulosis induced by topical application of Algipan. Ann Dermatol Venereol 2009;136:709-712 (Article in French)
7   Lakhoua G, El Aidli S, Zaïem A, Ben Mously R, Ben Brahim E, Daghfous R. Acute generalized exanthematous pustulosis with mephenesin balm. Presse Med 2014;43:867-869 (Article in French)
8   Schulze-Dirks A, Frosch PJ. Contact allergy to mephenesin. Hautarzt 1993;44:403-406 (Article in German)

Chapter 3.210 MEPRYLCAINE

IDENTIFICATION

Description/definition   : Meprylcaine is the 2-methyl-2-(propylamino)propyl ester of benzoic acid that conforms
                             to the structural formula shown below
Pharmacological classes : Local anesthetics
IUPAC name   : [2-Methyl-2-(propylamino)propyl] benzoate
Other names   : Benzoic acid [2-methyl-2-(propylamino)propyl] ester
CAS registry number   : 495-70-5
EC number   : Not available
Merck Index monograph : 307
Patch testing   : 5% pet.
Molecular formula   : $C_{14}H_{21}NO_2$

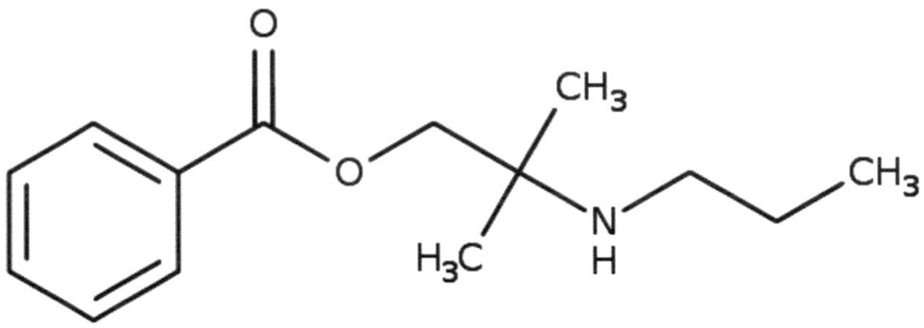

GENERAL

Meprylcaine is a local anesthetic that was used as surface anesthetic for the oral mucosa. It appears not to be employed as pharmaceutical anymore (1).

CONTACT ALLERGY

Case series

In Leuven, Belgium, in the period 1990-2014, iatrogenic contact dermatitis was diagnosed in 2600 individuals (17% of the total patch test population). 96% of all positive patch test reactions to topical drugs and antiseptics were considered to be relevant. Meprylcaine (5% pet.) was tested in 69 patients and there were 2 positive reactions to it (2).

LITERATURE

1   The data in the section 'General' may have been obtained from literature discussed in this chapter, but mostly also or exclusively from one or more of the following online sources: ChemIDPlus Advanced, PubChem, DrugBank, RxList, Drug Central, Drugs.com, and Wikipedia
2   Gilissen L, Goossens A. Frequency and trends of contact allergy to and iatrogenic contact dermatitis caused by topical drugs over a 25-year period. Contact Dermatitis 2016;75:290-302

Chapter 3.211 MESULFEN

IDENTIFICATION

Description/definition : Mesulfen is the thianthrene that conforms to the structural formula shown below
Pharmacological classes : Dermatological agents; antiparasitic agents
IUPAC name : 2,7-Dimethylthianthrene
Other names : Mesulphen
CAS registry number : 135-58-0
EC number : 205-202-4
Merck Index monograph : 7258
Patch testing : 5% pet.
Molecular formula : $C_{14}H_{12}S_2$

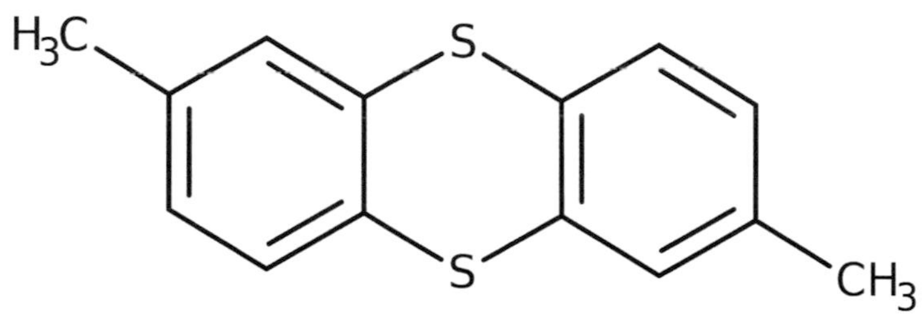

GENERAL

Mesulfen is a thianthrene that was formerly used in the treatment of scabies and dermatomycosis. Topical preparations with mesulfen are apparently still available in some countries, e.g. as a parasiticide and antipruritic in a range of skin disorders including acne, scabies, and seborrhoea (1).

CONTACT ALLERGY

Case reports and case series

In Bari, Italy, in the period 1968-1983 (or a 5-year period before 1982? [3]), 73 patients with widespread eczematous dermatitis following scabies were examined. They all had a history of treatment with topical scabicides over periods of not less than 10 days, mostly mesulfen. They were patch tested with mesulfen 5% pet. and there were 4 (5%) positive reactions; relevance was not discussed but likely (2,3).

Previously, 9 patients (of who 6 in The Netherlands [6]) had become sensitized to mesulfen in an antimycotic ointment containing 8% mesulfen. Most had treated tinea pedis with the ointment and developed allergic contact dermatitis. Patch test concentrations were 5% (4,7), 8% (6) and 1% and 8% (5). These concentrations appear not to be irritant (3).

LITERATURE

1   The data in the section 'General' may have been obtained from literature discussed in this chapter, but mostly also or exclusively from one or more of the following online sources: ChemIDPlus Advanced, PubChem, DrugBank, RxList, Drug Central, Drugs.com, and Wikipedia
2   Angelini G, Vena GA, Meneghini CL. Allergic contact dermatitis to some medicaments. Contact Dermatitis 1985;12:263-269
3   Meneghini CL, Vena GA, Angelini G. Contact dermatitis to scabicides. Contact Dermatitis 1982;8:285-286
4   Calnan CD. Contact dermatitis to mesulphen. Contact Dermatitis Newsletter 1972;11:283
5   Connor BL. Mesulphen in Tineafax ointment. Contact Dermatitis Newsletter 1973;14:417
6   Van Ketel WG. Allergic dermatitis caused by Tineafax ointment. Dermatologica 1967;135:121-125
7   Cronin E. Contact Dermatitis. Edinburgh: Churchill Livingstone, 1980:229

Chapter 3.212 METHOXSALEN

IDENTIFICATION

Description/definition   : Methoxsalen is the naturally occurring furocoumarin compound that conforms to the
                              structural formula shown below
Pharmacological classes : Photosensitizing agents; cross-linking reagents
IUPAC name   : 9-Methoxyfuro[3,2-g]chromen-7-one
Other names   : 8-Methoxypsoralen; 7H-furo(3,2-g)(1)benzopyran-7-one, 9-methoxy-; xanthotoxin
CAS registry number   : 298-81-7
EC number   : 206-066-9
Merck Index monograph   : 7329
Patch testing   : For patch testing: 0.15% alcohol; for photopatch testing a dilution series (0.15%, 0.015%,
                              0.0015%) may be necessary to differentiate photoallergy from phototoxicity (2)
Molecular formula   : $C_{12}H_8O_4$

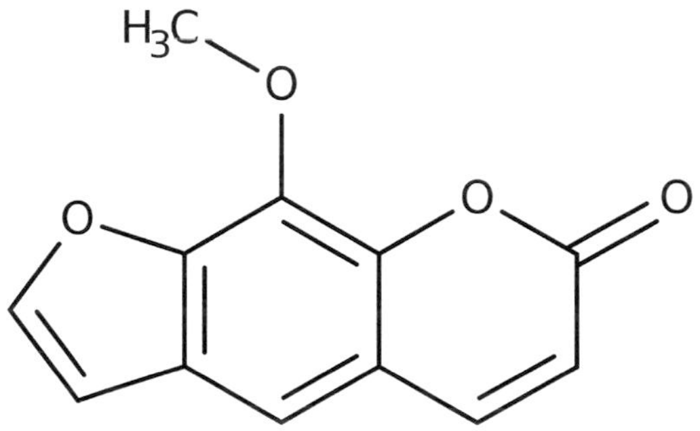

GENERAL

Methoxsalen is a naturally occurring substance isolated from the seeds of the plant *Ammi majus* with photoactivating properties. As a member of the family of compounds known as psoralens or furocoumarins, its exact mechanism of action is unknown. Upon photoactivation by ultraviolet A irradiation, methoxsalen has been observed to bind covalently to and crosslink DNA. This drug is indicated, together with UVA-irradiation, as photochemotherapy (synonym: PUVA, Psoralen – UVA), for the treatment of psoriasis, vitiligo and other skin disorders and can be used both topically and systemically (1). Because of its carcinogenic properties, PUVA has been largely replaced with UVB phototherapy. Methoxsalen and other furocoumarins such as 5-methoxypsoralen are phototoxic substances which are naturally present in many plants.

CONTACT ALLERGY

Case reports and case series

In a group of 371 patients treated with topical PUVA for psoriasis, 3 patients developed acute dermatitis within the treated areas. Patch tests and photopatch tests showed that one patient had contact allergy to methoxsalen and the other two photocontact allergy (5).

A 36-year-old female patient was treated with topical PUVA for severe dyshidrotic hand eczema, but noticed acute deterioration of her current eczema in the treated areas, with erythema and papulovesicles after the 6th treatment. Patch tests were positive to the 0.15% commercial methoxsalen solution tested 'as is' (+++) and diluted to 0.015% (+++), 0.0015% (++) and 0.00015% (+). The other, non-psoralen, ingredients of the commercial solution were negative (4).

A 37-year-old woman was treated for psoriasis of the soles and medial aspects of the heals with local photochemotherapy (PUVA) using methoxsalen 0.01% in alcohol irradiated with UVA in dosages of 4-5 J/cm². The therapy was given 3x per week for a month, when it had to be stopped because of itching dermatitis with blistering. Eight years later, the patient was treated with oral PUVA, which again resulted in burning, swelling and vesicles. Patch and photopatch tests were positive to methoxsalen 0.1%, 0.01% and 0.001% alcohol on both the exposed and unexposed sites. There was a cross-reaction to trimethylpsoralen (14, abstracts in refs. 6 and 7).

A 20-year-old man was treated with topical photochemotherapy for alopecia areata when, after the 6th treatment, a papulovesicular itching dermatitis in the treated areas developed. Patch tests were positive to 0.15% methoxsalen solution (apparently the pure material) and its 1:10 dilution. Photopatch tests were only slightly stronger. The reactions to the excipients of the commercial methoxsalen solution were negative. Histology of the patch test showed typical signs of allergic contact dermatitis. Ten controls were negative (9).

The first 2 cases of contact allergy to methoxsalen were reported in 1979 after topical PUVA treatment of psoriasis of the palms. An erythematous rash and increased irritation were observed in the treated areas after 5 and 6 weeks, respectively. Contact allergy was verified in both cases by patch testing with the 0.15% methoxsalen emulsion and methoxsalen 1% and 10% pet. (10). Another case of allergic contact dermatitis from methoxsalen, data of which are not available to the author, was reported from Italy (8).

PHOTOSENSITIVITY

General
Methoxsalen and other furocoumarins such as 5-methoxypsoralen are present in many plants and have phototoxic properties. In contact with the skin, furocoumarins bind to the DNA. Exposure to UV light causes cross-linking of the DNA that blocks cell division, DNA repair, DNA synthesis, and eventually causes cell death. This occurs mainly in epidermal DNA, leading to vesicle formation and blistering (2).

Case reports and case series
In a group of 371 patients treated with topical PUVA for psoriasis, 3 patients developed acute dermatitis within the treated areas. Patch tests and photopatch tests showed that one patient had contact allergy to methoxsalen and the other two photocontact allergy. In one of the cases of photocontact allergy, methoxsalen may have photocross-reacted to primary 3-carbethoxypsoralen photosensitization (5).

A 62-year-old woman with pustulosis palmoplantaris was treated with topical photochemotherapy (PUVA) using methoxsalen 0.01% in alcohol irradiated with UVA in dosages of 4-5 J/cm^2. However, the skin condition became worse and therapy was discontinued. Six years later, oral PUVA resulted immediately in a severe eczematous dermatitis of the hands and feet (systemic photocontact dermatitis). Patch and photopatch testing yielded positive photopatch tests to methoxsalen 0.1%, 0.01% and 0.01% in alcohol only. The patient was diagnosed with photocontact allergy to methoxsalen induced by local PUVA , later elicited by systemic PUVA. There was no cross-reaction to trimethylpsoralen (14, abstracts in refs. 6 and 7).

Three cases of experimentally induced photoallergic contact dermatitis with positive photopatch tests to 0.1% methoxsalen in hydrophilic ointment were reported in an early study (13). Photocontact dermatitis to methoxsalen used for the treatment of vitiligo was already described in 5 patients in France in 1953, but the test procedures and the results were not given in detail, so phototoxic reactions cannot be excluded (12, data cited in ref. 5).

Photocontact allergy to methoxsalen from non-medical products
Photocontact allergy to methoxsalen (8-methoxypsoralen) present in figs (*Ficus carica* L.) has been reported in cultivators of these plants with erythematovesicular or bullous contact dermatitis both on photo-exposed and covered skin and is by no means rare. Twelve such patients were reported from Italy. They all had positive photopatch test reactions to methoxsalen in alcohol down to a concentration of 0.001%. The authors emphasized that serial dilution testing is necessary to differentiate between phototoxic and photoallergic reactions (2).

Cross-reactions, pseudo-cross-reactions and co-reactions
A patient with contact allergy to methoxsalen cross-reacted to trimethylpsoralen [trioxsalen]) (14). One patient photosensitized to methoxsalen had a photocross-reaction to 3-carbethoxypsoralen; another individual may have had photocross-reactivity to methoxsalen from primary 3-carbethoxypsoralen photosensitization (5).

Cutaneous adverse drug reactions from systemic administration caused by type IV (delayed-type) hypersensitivity
Cutaneous adverse drug reactions from systemic administration of methoxsalen caused by type IV (delayed-type) hypersensitivity, including eczematous eruption (3), systemic photocontact dermatitis (14), and generalized photoallergic dermatitis (11), are planned to be discussed in Volume IV of the *Monographs in Contact Allergy* series on Systemic drugs.

LITERATURE

1   The data in the section 'General' may have been obtained from literature discussed in this chapter, but mostly also or exclusively from one or more of the following online sources: ChemIDPlus Advanced, PubChem, DrugBank, RxList, Drug Central, Drugs.com, and Wikipedia

2   Bonamonte D, Foti C, Lionetti N, Rigano L, Angelini G. Photoallergic contact dermatitis to 8-methoxypsoralen in *Ficus carica*. Contact Dermatitis 2010;62:343-348

3   Ravenscroft J, Goulden V, Wilkinson M. Systemic allergic contact dermatitis to 8-methoxypsoralen (8-MOP). J Am Acad Dermatol 2001;45(6 Suppl.):S218-S219

4   Korffmacher H, Hartwig R, Matthes U, Dirschka T, Albassam A, Weindorf N, et al. Contact allergy to 8-methoxypsoralen. Contact Dermatitis 1994;30:283-285

5   Takashima A, Yamamoto K, Kimura S, Takakuwa Y, Mizuno N. Allergic contact and photocontact dermatitis due to psoralens in patients with psoriasis treated with topical PUVA. Br J Dermatol 1991;124:37-42

6   Möller H. Contact and photocontact allergy to psoralens. Am J Contact Dermat 1990;1:254

7   Möller H. Contact and photocontact allergy to psoralens. Am J Contact Dermat 1990;1:202

8   Angelini G, D'Ovidio R, Vena GA. Allergia da contatto con 8-metossipsoralene. Boll Dermatol Allergol Prof 1987;3:69-74 (Article in Italian)

9   Weissmann I, Wagner G, Plewig G. Contact allergy to 8-methoxypsoralen. Br J Dermatol 1980;102:113-115

10 Saihan EM. Contact allergy to methoxalen. Br Med J 1979;2:20

11 Plewig G, Hofmann C, Braun-Falco O. Photoallergic contact dermatitis from 8-methoxypsoralen. Arch Dermatol Res 1978:261:201-211

12 Sidi E, Bourgeois-Gavardin J. Mise au point du traitement du vitiligo par l'Ammi majus. La Presse Medicale 1953;61:436-440 (Article in French, data cited in ref. 5)

13 Fulton JE, Willis I. Photoallergy to methoxsalen. Arch Dermatol 1968;98:445-450

14 Möller H. Contact and photocontact allergy to psoralens. Photodermatol Photoimmunol Photomed 1990;7:43-44

Chapter 3.213 METHYL AMINOLEVULINATE

IDENTIFICATION

Description/definition : Methyl aminolevulinate is the porphyrin precursor that conforms to the structural formula shown below

Pharmacological classes : Antineoplastic agents

IUPAC name : Methyl 5-amino-4-oxopentanoate

Other names : Aminolevulinic acid methyl ester

CAS registry number : 33320-16-0

EC number : Not available

Merck Index monograph : 7360

Patch testing : Inadequate data for pure methyl aminolevulinate; 21% pet. has been used; mostly, methyl aminolevulinate (MAL) is not available and the commercial cream with 16% MAL can be used for patch testing in a dilution series, e.g. 100%, 50%, 20%, 10% and 5% pet.

Molecular formula : $C_6H_{11}NO_3$

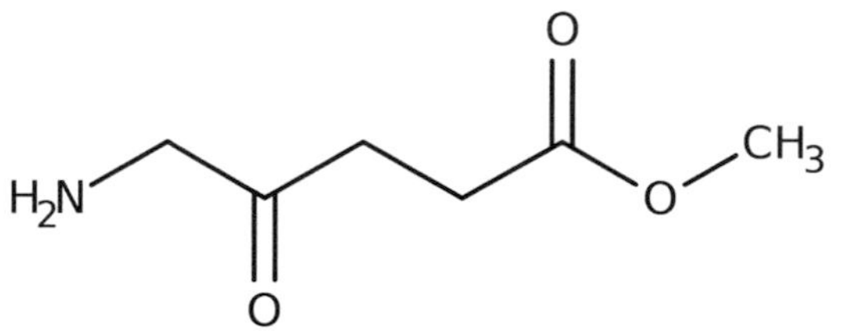

GENERAL

Methyl aminolevulinate (MAL) is the methyl ester of 5-aminolevulinic acid. A prodrug, it is metabolised to protoporphyrin IX, a photosensitizer, and is used in the photodynamic treatment (PDT) of non-melanoma skin cancer (including basal cell carcinoma) and solar keratoses. Topical application as methyl aminolevulinate hydrochloride results in an accumulation of protoporphyrin IX in the skin lesions to which the cream has been applied. Subsequent illumination with 570 to 670 nm wavelength red light results in the generation of toxic singlet oxygen that destroys cell membranes and thereby kills the tumor cells. In pharmaceutical products, methyl aminolevulinate is employed as methyl aminolevulinate hydrochloride (CAS number 79416-27-6, EC number 279-151-1, molecular formula $C_6H_{12}ClNO_3$) (1), nearly always in a cream containing 16% MAL (here termed '16% MAL-cream')

CONTACT ALLERGY

General

Photodynamic treatment with 16% MAL-cream bears a considerable risk of sensitization to the cream (and probably to its active ingredient methyl aminolevulinate) in patients treated at least five times (3). Severe or moderate local clinical reactions with redness, crusting, and scaling a few days after treatment are often considered to be local infection or simply a good response to the treatment, whereas, in fact, it may indicate sensitization. Therefore, it is advisable that all patients with repeated strong inflammatory reactions after PDT are offered patch testing (3).

Dermal tests by the manufacturer have shown that the cream has a considerable sensitizing potential (sensitization in 14-52% of subjects previously exposed to the cream on at least 4 occasions) (6). Methyl aminolevulinate (MAL) has a far higher risk of sensitization than aminolevulinic acid; it is suspected that the methyl group in MAL may be responsible for this (9).

Patch testing in groups of selected patients

In the period 2007-2008, in Aarhus, Denmark, twenty patients previously treated with PDT with 16% MAL-cream at least five times and 60 unexposed controls were patch tested with 16% MAL-cream 'as is', with 50% and 20% dilutions of the cream in petrolatum, and with the cream vehicle 'as is'. Pure methyl aminolevulinate was not available. Patch tests to 16% MAL-cream were positive in 7 (35%) patients on D3 and D7; 5/7 reacted to all dilutions of the cream. Four were considered to be relevant, in their medical files there were notes of strong local skin reactions. In the control group, one patient (1.6%) had a positive (+) reaction. All tests with the vehicle were negative (3).

Case series

In Aarhus, Denmark, from January 2002 to the end of January 2008, 500 patients were treated with PDT using 16% MAL-cream with an average of 2.8 treatments. Fifteen of these had a positive patch test to cream, giving a 3.0%

calculated risk of sensitization. However, only a minority of the patients had been offered a patch test and, therefore, the true number of patients with allergic contact dermatitis towards the cream is undoubtedly higher (3).

A 64-year-old man with actinic keratoses on the dorsal side of both hands, another 64-year-old man with actinic keratoses and a superficial basal cell carcinoma, and a 62-year-old woman with actinic keratoses on arms and legs were treated with PDT with 16% MAL-cream, when after 4, 9 resp. 6 treatments (severe) dermatitis developed. They all reacted strongly to patch tests with the cream but were negative to the cream vehicle. They were also tested with (one or more concentrations of) aminolevulinic acid (ALA) 1%, 5%, 10% and 20% pet. and 2 reacted to 20% ALA, but not to lower concentrations. However, irritant reactions were observed to the 10% concentrations and also in 2 of 4 controls, so the authors wondered whether the 2 'positive' reactions to ALA 20% were in fact allergic, or rather false-positive, irritant (4).

Six women and 3 men, aged 31-70 years, who were treated with PDT using 16% MAL-cream, suddenly developed an unexpected aggravated and spreading local inflammatory reaction at the treated skin site compatible with acute severe spreading dermatitis after 2-21 (medium 4) treatments. Some patients developed symptoms within 1-2 days and were treated with antibiotics by their general physician, who suspected a local infection. When patch tested, all patients reacted to the cream (8/9 ++/+++) but not to the cream base; methyl aminolevulinate itself was not available for patch testing. Of 25 controls, 5 had a ?+ macular erythema but no positive or obviously irritant reactions (5).

In Dundee, United Kingdom, over a 17-year period, 14 patients were identified who experienced severe blistering and/or prolonged eczematous reactions following PDT, suggestive of allergic contact dermatitis. There were nine women, 5 men, median age 71 (range 53-82) years. All of the patients had multiple areas of actinic keratosis, Bowen's disease or superficial basal cell carcinoma (medium 8, range 3-25), which had been treated with multiple PDT sessions (medium 7, range 3-21 sessions). Eight patients had undergone PDT exclusively with 16% MAL-cream, two patients exclusively with 20% ALA cream, two patients with both these preparations, one with 16% MAL-cream and 7.8% ALA gel and one patient with all three of the topical prodrugs.

Patch testing was performed to 16% MAL-cream in all 14 patients in a dilution series (10%, 5%, 1%, 0.5% and 0.1% MAL in petrolatum). Ten patients were also patch tested to a dilution series of 20% ALA cream (10%, 5%, 1%, 0.5% and 0.1% ALA in pet.) and 10 patients were patch tested to 7.8% ALA gel ('as is', 3.9%, 1.6% and 0.78% ALA). Ten patients had one or more positive patch test reactions. The results were not adequately specified and it was not shown which of the preparations had been used by the individual patients. All 10 reacted to 16% MAL-cream, 7 (of 10 tested) to 20% ALA cream and 5 (of 10 tested) to the 7.8% ALA gel. Results of dilution testing were not shown. Nine reacted both to MAL and one or two ALA preparations. Of the 4 patients who were tested with the 3 preparations, 3 reacted to all 3, indicating, according to the authors, cross-reactivity. There were 9 reactions to ALA (one or both preparations), whereas only 6 had been treated with ALA (12). When one assumes that the reactions were really allergic (and not irritant), this may indeed indicate cross-sensitization to ALA in patients sensitized to MAL. It should be realized, that not a single patient was tested with either ALA, MAL or the cream or gel base or its excipients. Also, control testing was inadequate and 3 of the 6 controls tested with the 7.8% gel had irritant reactions

Case reports
A 53-year-old woman with keratosis–ichthyosis–deafness (KID) syndrome presented with a long history of extensive hypertrophic lesions on both legs, which were treated with 15 sessions of photodynamic therapy (PDT) with aminolevulinic acid (ALA) as a photosensitizer. Seven years later, biopsies revealed multiple in situ squamous cell carcinomas. PDT with methyl aminolevulinate (MAL) cream was initiated. The topical drug was applied exclusively on the legs. The first session was unremarkable. However, 2 hours after the second session, performed 1 week later on the same skin area, the patient developed a generalized symmetric eruption with multiple erythematous and edematous pruritic papules and plaques on the trunk, arms, and legs. No obvious eczematous lesions were noted on treated sites. A similar eruption accompanied by intense pruritus occurred after the three following sessions, despite premedication with oral steroids. Older lesions later became hyperkeratotic and pigmented. Patch and photopatch tests showed UVA- and UVB-aggravated contact allergy to the cream. Patch tests showed weakly positive reactions (+) reactions on D2 and D3 with MAL 21% pet. and a negative reaction to the cream base and all ingredients. A diagnosis of systemic allergic contact dermatitis caused by MAL was made (2).

A 79-year-old woman developed, within a few days of the fourth PDT treatment for basal cell carcinoma on the left leg, an eczematous rash over the treatment area, which rapidly became generalized over much of her body. She had a positive patch test to 16%MAL-cream, but not to the cream diluted to 10% in pet.; the base was not tested (10). A 30-year-old woman with extensive necrobiosis lipoidica of her lower legs presented with eczema after PDT with both ALA and MAL (on different parts of the skin). Patch tests with 20% ALA and MAL were both positive. However, 6 weeks later, she reacted to MAL 10% (+++), 5% (+++) and 1% (++) in the cream base, but only had a ?+ reaction to 10% ALA. The cream bases were negative (11).

A 43-year-old woman with segmental Darier's disease on the lateral aspect of the trunk had been on treatment with PDT for 2 years. After receiving the fifteenth session she presented with erythematous plaques confined to the area of application of MAL 16% cream. Patch tests were positive (+++) to the cream and to MAL, as provided by the manufacturer (concentration unknown). Ten controls were negative (13).

Occupational contact dermatitis has been observed in 2 women with hand dermatitis working in the same department, whose current daily work included application and removal of ALA, MAL, and occlusive dressings on several patients per day, with gloves worn on a regular basis. Patch tests were positive to 16% MAL-cream 50% pet. and its active ingredient methyl aminolevulinate HCl (tested 21% pet.) but negative to the excipients (7). Previously, a similar case of occupational sensitization had been observed in a 49-year old nurse's aide whose job also consisted of applying and removing the MAL cream and the occlusive dressing. She did not wear gloves regularly. Patch tests were positive to the cream 'as is' (15 controls negative) but negative to the base of the cream and some of the excipients (8).

Cross-reactions, pseudo-cross-reactions and co-reactions
Most patients sensitized to MAL do not cross-react to aminolevulinic acid (ALA) (3,4,5,7,11). In one case, it was stated that 'a minor cross-reaction to ALA seems to be present'. Quite curiously, the authors gave their article the title 'Allergic contact dermatitis to 5-aminolaevulinic acid methylester but **NOT** to 5-aminolaevulinic acid after photodynamic therapy (11). Nevertheless, there are – weak – indications that some patients sensitized to MAL may cross-react to ALA (12, see the section Case series).

PHOTOSENSITIVITY
In one patient, patch and photopatch tests showed UVA- and UVB-aggravated contact allergy to MAL 16% cream (2). This patient is presented in the section 'Case reports' above.

Immediate contact reactions
Immediate contact reactions (contact urticaria) to methyl aminolevulinate are presented in Chapter 5.

LITERATURE
1   The data in the section 'General' may have been obtained from literature discussed in this chapter, but mostly also or exclusively from one or more of the following online sources: ChemIDPlus Advanced, PubChem, DrugBank, RxList, Drug Central, Drugs.com, and Wikipedia
2   Al Malki A, Marguery MC, Giordano-Labadie F, Konstantinou MP, Mokeddem L, Lamant L, et al. Systemic allergic contact dermatitis caused by methyl aminolaevulinate in a patient with keratosis-ichthyosis-deafness syndrome. Contact Dermatitis 2017;76:190-192
3   Korshøj S, Sølvsten H, Erlandsen M, Sommerlund M. Frequency of sensitization to methyl aminolaevulinate after photodynamic therapy. Contact Dermatitis 2009;60:320-324
4   Jungersted JM, Dam TN, Bryld LE, Agner T. Allergic reactions to Metvix (ALA-ME). Contact Dermatitis 2008;58:184-186
5   Hohwy T, Andersen KE, Sølvsten H, Sommerlund M. Allergic contact dermatitis to methyl aminolevulinate after photodynamic therapy in 9 patients. Contact Dermatitis 2007;57:321-323
6   Product monograph METVIX®. Methyl aminolevulinate topical cream 168 mg/g (as methyl aminolevulinate hydrochloride). Thornhill, Ontario, Canada: Galderma Canada Inc., May 2019. Available at: https://www.galderma.com/sites/g/files/jcdfhc196/files/inline-files/Metvix_Cream_PM_E_May_17_2019.pdf
7   Hartmann J, Enk AH, Gholam P. Sensitization following occupational exposure to methyl aminolevulinate: Report of two cases. Contact Dermatitis 2020;83:50-52
8   Pastor-Nieto AM, Olivares M, Sánchez-Herreros C, Belmar P, De Eusebio E. Occupational allergic contact dermatitis from methyl aminolevulinate. Dermatitis 2011;22:216-219
9   Roberts DW, Goodwin BF, Basketter D. Methyl groups as antigenic determinants in skin sensitisation. Contact Dermatitis 1988;18:219-225
10   Harries MJ, Street G, Gilmour E, Rhodes LE, Beck MH. Allergic contact dermatitis to methyl aminolevulinate (Metvix) cream used in photodynamic therapy. Photodermatol Photoimmunol Photomed 2007;23:35-36
11   Wulf HC, Phillipsen P. Allergic contact dermatitis to 5-aminolaevulinic acid methylester but not to 5-aminolaevulinic acid after photodynamic therapy. Br J Dermatol 2004;150:143-145
12   Cordey H, Ibbotson S. Allergic contact dermatitis to topical prodrugs used in photodynamic therapy. Photodermatol Photoimmunol Photomed 2016;32:320-322
13   Pastor-Nieto M, Jiménez-Blázquez E, Sánchez-Herreros C, Belmar-Flores P. Allergic contact dermatitis caused by methyl aminolevulinate. Actas Dermosifiliogr 2013;104:168-170

Chapter 3.214 METHYLDIPHENHYDRAMINE

It is uncertain which chemical is meant in the publication below (2), it was listed under the heading 'Antihistamines' of topical drugs. Most likely it is 2- (or *o*-) methyldiphenhydramine, INN name orphenadrine. This pharmaceutical is used in some compounded topical pain medications (www.drugs.com, https://ipscompounding.com/ips-pain-compounding-2/)

IDENTIFICATION

Description/definition   : Methyldiphenhydramine (INN name orphenadrine) is the tertiary amino compound that
                                conforms to the structural formula shown below
Pharmacological classes   : Antiparkinson agents; parasympatholytics; cytochrome P-450 CYP2B6 inhibitors;
                                muscle relaxants, central; muscarinic antagonists
IUPAC name   : *N,N*-Dimethyl-2-[(2-methylphenyl)-phenylmethoxy]ethanamine
Other names   : Orphenadrine; beta-dimethylaminoethyl 2-methylbenzhydryl ether
CAS registry number   : 83-98-7
EC number   : 201-509-2
Merck Index monograph   : 8245
Patch testing   : 1% pet.
Molecular formula   : $C_{18}H_{23}NO$

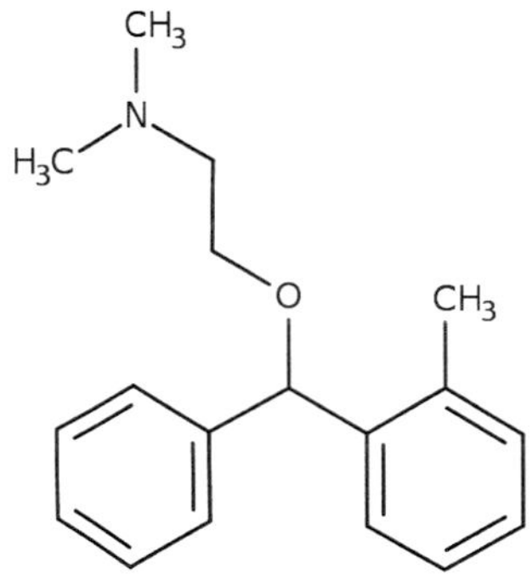

Methyldiphenhydramine 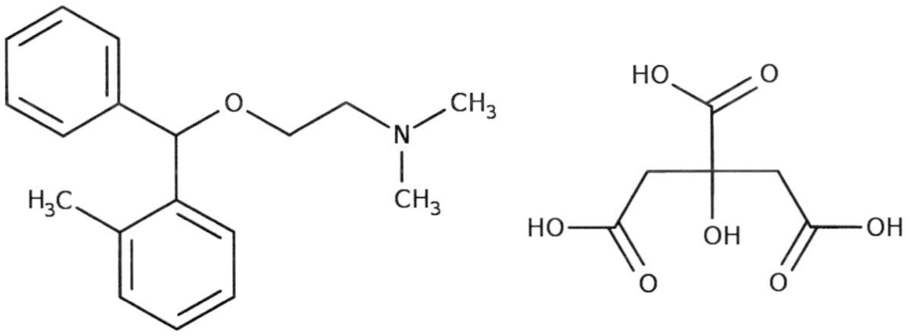   Methyldiphenhydramine citrate

GENERAL

Orphenadrine (2-methyldiphenhydramine) is a tertiary amino compound which is the phenyl-*o*-tolylmethyl ether of 2-(dimethylamino)ethanol. It has a role as a NMDA receptor antagonist, a H1-receptor antagonist, an antiparkinson drug, a parasympatholytic, a muscle relaxant, a muscarinic antagonist and an antidyskinesia agent. Orphenadrine is indicated as an adjunct to rest, physical therapy, and other measures for the relief of discomfort associated with acute painful musculoskeletal conditions. In pharmaceutical products, orphenadrine is employed as orphenadrine citrate (CAS number 4682-36-4, EC number 225-137-5, molecular formula $C_{24}H_{31}NO_8$) or as orphenadrine hydrochloride (CAS number 341-69-5, EC number 206-435-4, molecular formula $C_{18}H_{24}ClNO$) (1).

Case series

In Leuven, Belgium, in the period 1990-2014, iatrogenic contact dermatitis was diagnosed in 2600 individuals (17% of the total patch test population). 96% of all positive patch test reactions to topical drugs and antiseptics were considered to be relevant. Methyldiphenhydramine (1% pet.) was tested in 15 patients and there was one positive reaction to it (2).

LITERATURE

1   The data in the section 'General' may have been obtained from literature discussed in this chapter, but mostly
    also or exclusively from one or more of the following online sources: ChemIDPlus Advanced, PubChem,
    DrugBank, RxList, Drug Central, Drugs.com, and Wikipedia
2   Gilissen L, Goossens A. Frequency and trends of contact allergy to and iatrogenic contact dermatitis caused by
    topical drugs over a 25-year period. Contact Dermatitis 2016;75:290-302

Chapter 3.215 METHYL NICOTINATE

IDENTIFICATION

Description/definition : Methyl nicotinate is the ester of methyl alcohol and nicotinic acid that conforms to the structural formula shown below
Pharmaceutical classes : Rubefacients
IUPAC name : Methyl pyridine-3-carboxylate
Other names : Nicotinic acid methyl ester
CAS registry number : 93-60-7
EC number : 202-261-8
Merck Index monograph : 7443
Patch testing : 1% pet.
Molecular formula : $C_7H_7NO_2$

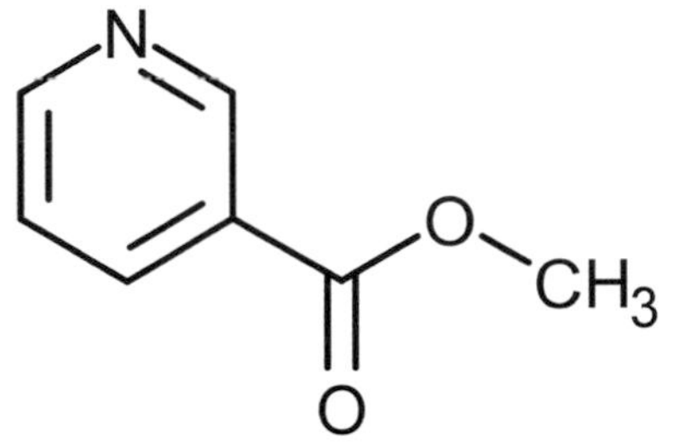

GENERAL

Methyl nicotinate is the methyl ester of nicotinic acid that is used as an active ingredient, a rubefacient and counter-irritant in over-the-counter topical preparations indicated for the temporary relief of aches and pains in muscles, tendons, and joints. The action of methyl nicotinate as a rubefacient is thought to involve peripheral vasodilation. For veterinary purposes, methyl nicotinate is used to treat respiratory diseases, vascular disorders, rheumatoid arthritis, and muscle and joint pains. Methyl nicotinate can be found in in guava fruit, papaya, strawberry, soursop (*Annona muricata*), beer, grape brandy, coffee, roasted filbert, roasted peanut and Bourbon vanilla (1).

CONTACT ALLERGY

Case reports and case series

In Leuven, Belgium, in the period 1990-2014, iatrogenic contact dermatitis was diagnosed in 2600 individuals (17% of the total patch test population). 96% of all positive patch test reactions to topical drugs and antiseptics were considered to be relevant. Methyl nicotinate (2% pet.) was tested in 66 patients and there were 12 positive reactions to it (2).

A 23-year-old woman had applied a balsam for 8 days for a sprained foot. Two days after stopping this medication, a bullous contact eczema developed with an extensive erythema multiforme id-like spread. Patch tests with the balsam and its constituents were positive to the pharmaceutical 'as is', methyl nicotinate 1% pet. and mephenesin 5% pet. (3).

OTHER SIDE EFFECTS

Other side effects of methyl nicotinate (notably non-immunological immediate contact reactions [contact urticaria]) have been presented in ref. 4. See also Chapter 5 Immediate contact reactions (contact urticaria).

LITERATURE

1   The data in the section 'General' may have been obtained from literature discussed in this chapter, but mostly also or exclusively from one or more of the following online sources: ChemIDPlus Advanced, PubChem, DrugBank, RxList, Drug Central, Drugs.com, and Wikipedia
2   Gilissen L, Goossens A. Frequency and trends of contact allergy to and iatrogenic contact dermatitis caused by topical drugs over a 25-year period. Contact Dermatitis 2016;75:290-302
3   Degreef H, Bonamie A, Vanderheyden D, Dooms-Goossens A. Mephenesin contact dermatitis with erythema multiforme features. Contact Dermatitis 1984;10:220-223
4   De Groot AC. Monographs in Contact Allergy Volume I. Non-Fragrance Allergens in Cosmetics (Part I and Part 2). Boca Raton, Fl, USA: CRC Press Taylor and Francis Group, 2018: 843-844

Chapter 3.216 METHYLPHENIDATE

IDENTIFICATION

Description/definition   : Methylphenidate is the synthetic central nervous system stimulant that conforms to the
                             structural formula shown below
Pharmacological classes : Central nervous system stimulants; dopamine uptake inhibitors
IUPAC name   : Methyl 2-phenyl-2-piperidin-2-ylacetate
Other names   : 2-Piperidineacetic acid, α-phenyl-, methyl ester
CAS registry number   : 113-45-1
EC number   : 204-028-6
Merck Index monograph   : 7453
Patch testing   : 10% pet.; concentrations up to 20% in water and pet. are non-irritating (3)
Molecular formula   : $C_{14}H_{19}NO_2$

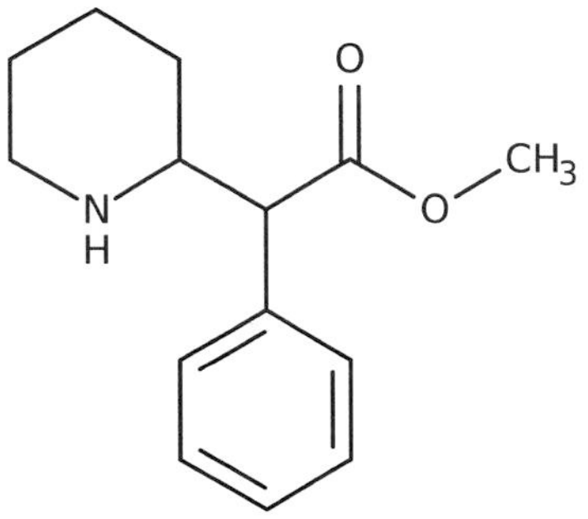

GENERAL

Methylphenidate is a synthetic central nervous system stimulant. It appears to activate the brain stem arousal system and cortex to produce its stimulant effect and, in some clinical settings, may improve cognitive function. Methylphenidate is indicated for the treatment of attention deficit hyperactivity disorder (ADHD) in patients 6 years of age and older and for the treatment of narcolepsy. It is available in a transdermal therapeutic system (1). In pharmaceutical products, both methylphenidate and methylphenidate hydrochloride (CAS number 298-59-9, EC number 206-065-3, molecular formula $C_{14}H_{20}ClNO_2$) may be employed (1).

CONTACT ALLERGY

Case reports

A 9-year-old female patient developed itchy, burning, red lesions under a methylphenidate (MPH) patch she used for attention-deficit/hyperactivity disorder. Her symptoms initially began on her hip at the area of patch placement, and then progressively spread to her arms, legs, abdomen, and back. Her symptoms continued for 2 months, despite the fact that she stopped using the patch. The patient was patch tested with 1%, 5% and 10% MPH, obtained with the use of oral MPH tablets suspended in sterile water and she reacted to the 10% suspension only. Later, patch tests with neat MPH in petrolatum yielded positive reactions to 0.5% (+), 1% (+), 5% (+), 10% (++) and 20% pet. (+++). Nine days after initial patch placement, the patient presented with a recall reaction, characterized by a return of the original pruritic dermatitis to her entire back, similar to the eruption that had occurred months previously after therapeutic use of the MPH patch, a manifestation of systemic contact dermatitis (2).

In a group of 305 patients participating in a methylphenidate dose-finding study, one patient discontinued treatment because of skin reactions at the TSS patch site. Patch tests were positive to 10% and 0.1% in pet. and water (4).

LITERATURE

1   The data in the section 'General' may have been obtained from literature discussed in this chapter, but mostly also or exclusively from one or more of the following online sources: ChemIDPlus Advanced, PubChem, DrugBank, RxList, Drug Central, Drugs.com, and Wikipedia
2   Vashi NA, Souza A, Cohen N, Franklin B, Cohen DE. Allergic contact dermatitis caused by methylphenidate. Contact Dermatitis 2011;65:183-185
3   Fowler JF, Warshaw EM, Squires L. Methylphenidate patch-test protocol and irritancy threshold determination in healthy adult subjects. Dermatitis 2009;20:271-274
4   Warshaw EM, Squires L, Li Y, Civil R, Paller AS. Methylphenidate transdermal system: a multisite, open-label study of dermal reactions in pediatric patients diagnosed with ADHD. Prim Care Companion J Clin Psychiatry 2010;12:E1–E9

Chapter 3.217 METHYLPREDNISOLONE

IDENTIFICATION

Description/definition	: Methylprednisolone is the synthetic glucocorticoid that conforms to the structural formula shown below
Pharmacological classes	: Glucocorticoids; neuroprotective agents; anti-inflammatory agents; antiemetics
IUPAC name	: (6S,8S,9S,10R,11S,13S,14S,17R)-11,17-Dihydroxy-17-(2-hydroxyacetyl)-6,10,13-trimethyl-7,8,9,11,12,14,15,16-octahydro-6H-cyclopenta[a]phenanthren-3-one
Other names	: 11β,17,21-Trihydroxy-6α-methylpregna-1,4-diene-3,20-dione
CAS registry number	: 83-43-2
EC number	: 201-476-4
Merck Index monograph	: 7454
Patch testing	: In general, corticosteroids may be tested at 0.1% and 1% in alcohol; late readings (6-10 days) are strongly recommended
Molecular formula	: $C_{22}H_{30}O_5$

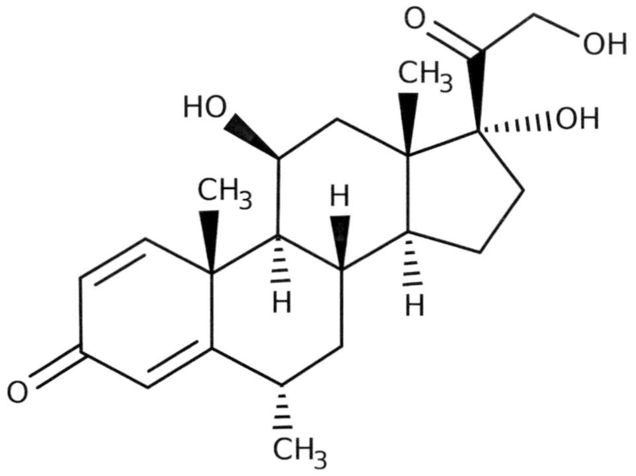

GENERAL
General aspects of corticosteroids used on the skin and mucous membranes are discussed in Chapter 2.4. A practical guideline for diagnosing allergic reactions to corticosteroids is presented in ref. 2. Methylprednisolone base is used in oral forms only. Esters used in other applications include methylprednisolone acetate (Chapter 3.119), methylprednisolone hemisuccinate (Chapter 3.220), and methylprednisolone aceponate (Chapter 3.218). As methylprednisolone *base* is used as tablet only, this implies that by far most allergic reactions to 'methylprednisolone' have in fact been the result of sensitization to an ester of methylprednisolone or of cross-reactivity to another corticosteroid. It is also likely that there has been confusion in some publications on the correct forms of the drugs used, e.g. that methylprednisolone was mentioned where in fact an ester form should have been mentioned (1,3).

CONTACT ALLERGY

Patch testing in groups of patients
In the period 2001-2010, in Australia, 3966 patients were patch tested with 'methylprednisolone' 1% pet. and there were 52 (1.3%) positive reactions, of which 9 (17%) were considered to be relevant (1). As a commercial ointment and cream containing methylprednisolone aceponate were also tested and reference was made to a supplier of patch test haptens, that provides methylprednisolone only as aceponate, it is likely that this ester was tested and not methylprednisolone base.

Cross-reactions, pseudo-cross-reactions and co-reactions
Cross-reactions between corticosteroids are discussed in Chapter 2.8.

Cutaneous adverse drug reactions from systemic administration caused by type IV (delayed-type) hypersensitivity
Cutaneous adverse drug reactions from systemic administration of methylprednisolone caused by type IV (delayed-type) hypersensitivity including systemic contact dermatitis (4; baboon syndrome]), generalized erythematous rash (3), and also occupational allergic contact dermatitis (5), are planned to be discussed in Volume IV of the *Monographs in Contact Allergy* series on Systemic drugs.

LITERATURE

1   Toholka R, Wang Y-S, Tate B, Tam M, Cahill J, Palmer A, Nixon R. The first Australian Baseline Series: Recommendations for patch testing in suspected contact dermatitis. Australas J Dermatol 2015;56:107-115
2   Baeck M, Goossens A. Immediate and delayed allergic hypersensitivity to corticosteroids: practical guidelines. Contact Dermatitis 2012;66:38-45
3   Fernández de Corres L, Bernaola G, Urrutia I, Muñoz D. Allergic dermatitis from systemic treatment with corticosteroids. Contact Dermatitis 1990;22:104-106
4   Treudler R, Simon J. Symmetric, drug-related, intertriginous, and flexural exanthema in a patient with polyvalent intolerance to corticosteroids. J Allergy Clin Immunol 2006;118:965-967
5   Gielen K, Goossens A. Occupational allergic contact dermatitis from drugs in healthcare workers. Contact Dermatitis 2001;45:273-279

Chapter 3.218 METHYLPREDNISOLONE ACEPONATE

IDENTIFICATION

Description/definition : Methylprednisolone aceponate is the aceponate (acetate propionate) ester of the synthetic glucocorticoid methylprednisolone that conforms to the structural formula shown below

Pharmacological classes : Anti-inflammatory agents

IUPAC name : [(6S,8S,9S,10R,11S,13S,14S,17R)-17-(2-Acetyloxyacetyl)-11-hydroxy-6,10,13-trimethyl-3-oxo-7,8,9,11,12,14,15,16-octahydro-6H-cyclopenta[a]phenanthren-17-yl] propanoate

Other names : Pregna-1,4-diene-3,20-dione, 21-(acetyloxy)-11-hydroxy-6-methyl-17-(1-oxopropoxy)-, (6α,11β)-

CAS registry number : 86401-95-8

EC number : Not available

Merck Index monograph : 7454 (Methylprednisolone)

Patch testing : 1% pet. (Chemotechnique); 0.1% alc. (SmartPracticeCanada); late readings (6-10 days) are strongly recommended

Molecular formula : $C_{27}H_{36}O_7$

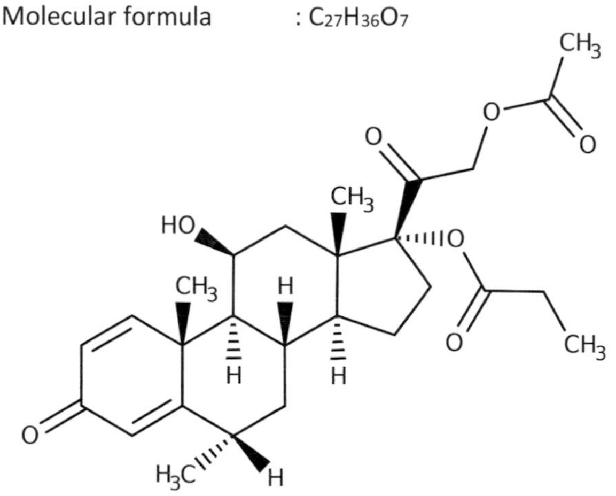

GENERAL

General aspects of corticosteroids used on the skin and mucous membranes are discussed in Chapter 2.4. A practical guideline for diagnosing allergic reactions to corticosteroids is presented in ref. 1. See also methylprednisolone (Chapter 3.217), methylprednisolone acetate (Chapter 3.219), and methylprednisolone hemisuccinate (Chapter 3.220).

CONTACT ALLERGY

Patch testing in consecutive patients suspected of contact dermatitis: Routine testing

In the period 2001-2010, in Australia, 4353 patients were patch tested with a commercial ointment containing 0.1% methylprednisolone aceponate and there were 52 (1.2%) positive reactions, of which 12 (23%) were considered to be relevant. Methylprednisolone (base? or aceponate?) 1% pet. was tested in 3966 patients and there were also 52 (1.3%) positive reactions, of which 9 (17%) were relevant (9).

In Spain, in the years 2015-2016, 3699 consecutive patients were patch tested with methylprednisolone aceponate and there were 18 (0.5%) positive reactions; their relevance was not mentioned (11).

Case series

From January 1990 to June 2008, in Leuven, Belgium, 315 patients were diagnosed with contact allergy to/allergic contact dermatitis from corticosteroids (CSs) from routine patch testing with a baseline series including tixocortol pivalate, budesonide, hydrocortisone butyrate and prednisone caproate, patch testing with patients' own CS preparations, and testing those with proven contact allergy to a corticosteroid or strongly suspected of CS allergy later with a series of 66 CSs, including two sex hormones (progesterone and testosterone). 71% of the patients had relevant reactions, but these were not specified. In this group of 315 CS allergic patients, 105 had positive patch tests to methylprednisolone aceponate 0.1% alc. (10). It is unknown how many of these reactions were caused by the use of a pharmaceutical product containing methylprednisolone aceponate and how many were cross-reactions to other corticosteroids.

Case reports

A 27-year-old man developed a disseminated eczema from a chronic dermatitis located on the dorsal aspect of his left foot, which had been treated with methylprednisolone aceponate (MPA) cream over the last 4 months. He presented with erythematous and scaly papules and plaques involving the trunk, arms and legs, associated with a violaceous, dry, clearly outlined plaque on his left foot. Patch tests were positive to MPA cream, to MPA 1% pet., to budesonide and 2 other corticosteroids. Methylprednisolone base was negative (2). The authors also presented the case of a 43-year-old woman, previously diagnosed with atopic dermatitis and nickel allergy, who presented with a chronic dermatitis on her hands and forearms, consisting of vesicles, erythematous and scaly papules and also fissures. She had been treated with various corticosteroid creams with uncomplete improvement. When the patient applied methylprednisolone aceponate cream for 3 weeks, worsening of the dermatitis was noted. Patch tests were positive to MPA and hydrocortisone-17-butyrate (2).

A patient with multiple corticosteroid allergies reacted to MPA 0.1% cream and ointment. A ROAT with the cream was positive at D8. The reactions were considered to be highly relevant clinically (6). However, it was not stated that the patient had used these preparations before, and therefore they were more likely cross-reactions.

Short summaries of other case reports of allergic contact dermatitis from methylprednisolone aceponate are shown in table 3.218.1. One case – of which no details are available – has been reported in Italian literature (12).

Table 3.218.1 Short summaries of case reports of allergic contact dermatitis from methylprednisolone aceponate

Year and country	Sex	Age	Positive patch tests	Clinical data and comments	Ref.
2004 Australia	F	42	MPA fatty ointment 'as is'; MPA 1% pet.	worsening of hand eczema with severe and extensive spreading after using MPA fatty ointment for a month	5
2003 Korea	F	18	MPA ointment 'as is'; MPA 1% alc.	worsening or iatrogenic allergic contact dermatitis; also contact allergy to neomycin and *Centella asiatica*	8
2000 Italy	F	26	MPA	patient with hand eczema; no details provided	7
1998 Italy	F	26	MPA cream 'as is'; MPA 1% alc.	hand eczema in a nurse worsened by various cortico-steroid preparations including MPA	3
1997 Italy	F	15	3 MPA preparations 'as is'; MPA 1% and 0.1% alc.	worsening of atopic dermatitis from MPA cream	4

MPA: methylprednisolone aceponate

Cross-reactions, pseudo-cross-reactions and co-reactions

Cross-reactions between corticosteroids are discussed in Chapter 2.8.

LITERATURE

1   Baeck M, Goossens A. Immediate and delayed allergic hypersensitivity to corticosteroids: practical guidelines. Contact Dermatitis 2012;66:38-45

2   Calzado L, Ortiz-Frutos FJ, Galera C, Sánchez-Caminero P, Vanaclocha F. Allergic contact dermatitis caused by 6alpha-methylprednisolone aceponate. Contact Dermatitis 2005;53:62-63

3   Corazza M, Virgili A. Allergic contact dermatitis from 6alpha-methylprednisolone aceponate and budesonide. Contact Dermatitis 1998;38:356-357

4   Balato N, Patruno C, Lembo G, Cuccurullo FM, Ayala F. Contact sensitization to 6α-methylprednisolone aceponate. Am J Contact Dermat 1997;8:24-25

5   Cahill J, Nixon R. Allergic contact dermatitis to methylprednisolone aceponate in a topical corticosteroid. Australas J Dermatol 2004;45:192-193

6   Chow ET. Multiple corticosteroid allergies. Australas J Dermatol 2001;42:62-63

7   Corazza M, Mantovani L, Maranini C, Bacilieri S, Virgili A. Contact sensitization to corticosteroids: increased risk in long-term dermatoses. Eur J Dermatol 2000;10:533-535

8   Seok Oh C, Young Lee J. Contact allergy to various ingredients of topical medicaments. Contact Dermatitis 2003;49:49-50

9   Toholka R, Wang Y-S, Tate B, Tam M, Cahill J, Palmer A, Nixon R. The first Australian Baseline Series: Recommendations for patch testing in suspected contact dermatitis. Australas J Dermatol 2015;56:107-115

10 Baeck M, Chemelle JA, Terreux R, Drieghe J, Goossens A. Delayed hypersensitivity to corticosteroids in a series of 315 patients: clinical data and patch test results. Contact Dermatitis 2009;61:163-175

11 Mercader-García P, Pastor-Nieto MA, García-Doval I, Giménez-Arnau A, González-Pérez R, Fernández-Redondo V, et al. GEIDAC. Are the Spanish baseline series markers sufficient to detect contact allergy to corticosteroids in Spain? A GEIDAC prospective study. Contact Dermatitis 2018;78:76-82

12 D'Erme AM, Milanesi N, Gola M. Allergic contact dermatitis to 6α-methylprednisolone aceponate. G Ital Dermatol Venereol 2013;148:307-308 (Article in Italian)

Chapter 3.219 METHYLPREDNISOLONE ACETATE

IDENTIFICATION

Description/definition : Methylprednisolone acetate is the acetate ester of the synthetic glucocorticoid methylprednisolone that conforms to the structural formula shown below

Pharmacological classes : Anti-inflammatory agents

IUPAC name : [2-[(6S,8S,9S,10R,11S,13S,14S,17R)-11,17-Dihydroxy-6,10,13-trimethyl-3-oxo-7,8,9,11,12,14,15,16-octahydro-6H-cyclopenta[a]phenanthren-17-yl]-2-oxoethyl] acetate

Other names : Methylprednisolone 21-acetate; 11β,17,21-trihydroxy-6α-methylpregna-1,4-diene-3,20-dione 21-acetate

CAS registry number : 53-36-1

EC number : 200-171-3

Merck Index monograph : 7454 (Methylprednisolone)

Patch testing : In general, corticosteroids may be tested at 0.1% and 1% in alcohol; late readings (6-10 days) are strongly recommended

Molecular formula : $C_{24}H_{32}O_6$

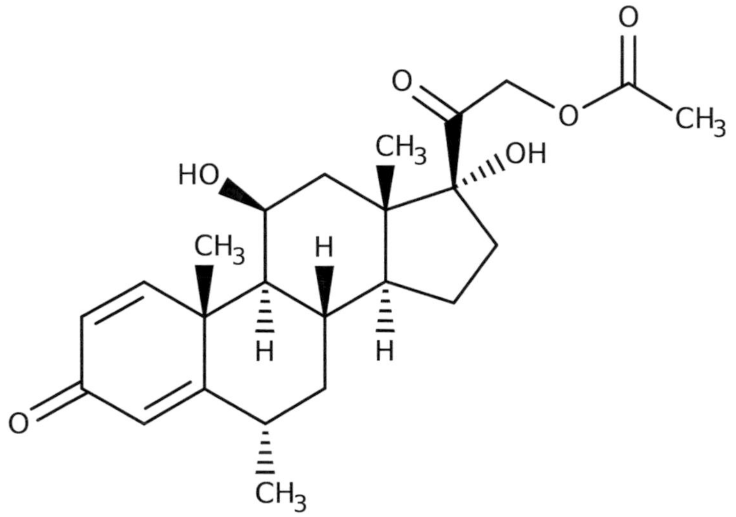

GENERAL

General aspects of corticosteroids used on the skin and mucous membranes are discussed in Chapter 2.4. A practical guideline for diagnosing allergic reactions to corticosteroids is presented in ref. 1. Methylprednisolone acetate is used for injection (epidural; Infiltration; Intra-articular; Intralesional; Intramuscular; Intravenous; soft tissue; subcutaneous) and in topical preparations for acne. See also methylprednisolone (Chapter 3.217), methylprednisolone hemisuccinate (Chapter 3.220) and methylprednisolone aceponate (Chapter 3.218).

CONTACT ALLERGY

Patch testing in groups of selected patients

In the United Kingdom, from 1988 to 1991, 528 selected patients with a positive patch test to tixocortol pivalate or suspected of corticosteroid allergy were patch tested with methylprednisolone acetate 20% pet. 1 patient (0.2%) had a positive reaction, the relevance of which was not mentioned (3).

Case reports and case series

From January 1990 to June 2008, in Leuven, Belgium, 315 patients were diagnosed with contact allergy to/allergic contact dermatitis from corticosteroids (CSs) from routine patch testing with a baseline series including tixocortol pivalate, budesonide, hydrocortisone butyrate and prednisone caproate, patch testing with patients' own CS preparations, and testing those with proven contact allergy to a corticosteroid or strongly suspected of CS allergy later with a series of 66 CSs, including two sex hormones (progesterone and testosterone). 71% of the patients had relevant reactions, but these were not specified. In this group of 315 CS allergic patients, 58 had positive patch tests to methylprednisolone acetate 0.1% alc. (2). It is unknown how many of these reactions were caused by the use of a pharmaceutical product containing methylprednisolone acetate and how many were cross-reactions to other corticosteroids.

A 49-year-old woman with chronic idiopathic, unilateral iridocyclitis for five years, had been treated with various topical corticosteroid ophthalmic drops, cryotherapy, and periodic (one to three per year) retrobulbar injections of methylprednisolone acetate (MPA). Approximately 24 hours after the last retrobulbar injection, the patient noted the onset of severe pain and conjunctival swelling in the treated eye. Patch tests were negative at D2 and D4, but subcutaneous and intradermal injections of 0.1% MPA suspension produced 60- and 30-mm indurated, erythematous reactions, respectively, at 48 hours. Additional intradermal testing with 0.025 ml of 0.1% MPA powder suspended in normal saline produced a 12-mm indurated erythematous nodule at 48 hours. Histopathology was compatible with a dermal hypersensitivity reaction. The patient had also a positive patch test to the preservative in the solution, myristyl γ-picolinium chloride (miripirium chloride) 0.1% (negative to 0.01%; vehicle not mentioned). Thirty controls were negative (9). The authors also reported a virtually identical case of a 30-year-old woman with Ehlers-Danlos syndrome and chronic idiopathic, unilateral iridocyclitis of approximately three years' duration, who developed severe left upper eyelid edema and conjunctival inflammation within six hours of a retrobulbar injection of methylprednisolone acetate. In this case, too, all patch tests, even to MPA 25% pet. were negative, but an intradermal test with MPA 0.1% was strongly positive at D2. Allergy to the preservative was demonstrated by an intradermal injection of 0.025 ml of a 0.00025% myristyl γ-picolinium chloride (9). Seven years earlier, in a similar case, allergy to MPA could also be demonstrated by intradermal testing only (10).

Cases of allergic contact dermatitis from methylprednisolone acetate have apparently been reported in early publications (4-7; specific data not available to the author).

Cross-reactions, pseudo-cross-reactions and co-reactions
Cross-reactions between corticosteroids are discussed in Chapter 2.8.

Cutaneous adverse drug reactions from systemic administration caused by type IV (delayed-type) hypersensitivity
Cutaneous adverse drug reactions from systemic administration of methylprednisolone acetate caused by type IV (delayed-type) hypersensitivity, including maculopapular eruption (8) and a local skin rash after intra-articular injection (11,12), are planned to be discussed in Volume IV of the *Monographs in Contact Allergy* series on Systemic drugs.

LITERATURE

1   Baeck M, Goossens A. Immediate and delayed allergic hypersensitivity to corticosteroids: practical guidelines. Contact Dermatitis 2012;66:38-45
2   Baeck M, Chemelle JA, Terreux R, Drieghe J, Goossens A. Delayed hypersensitivity to corticosteroids in a series of 315 patients: clinical data and patch test results. Contact Dermatitis 2009;61:163-175
3   Burden AD, Beck MH. Contact hypersensitivity to topical corticosteroids. Br J Dermatol 1992;127:497-501
4   Zina G, Bonu G. Contact sensitivity to corticosteroids. Contact Dermatitis Newsletter 1967;2:26 (same as ref. 5)
5   Bonu G, Zina G. Sensibilizzazione da contatto a cortisonici. Minerva Dermatologica 1967;42:513-514 (same as ref. 4) (Article in Italian)
6   Gutzwiller P. Zum Problem der Kortikosteroid-Allergie. Dermatologica 1974; 148: 253-256.
7   Coskey RJ, Bryan MG. Contact dermatitis to methylprednisolone. JAMA 1967;199:136 (bibliography incorrect)
8   Bursztejn AC, Tréchot P, Cuny JF, Schmutz JL, Barbaud A. Cutaneous adverse drug reactions during chemotherapy: consider non-antineoplastic drugs. Contact Dermatitis 2008;58:365-368
9   Mathias CGT, Robertson DB. Delayed hypersensitivity to a corticosteroid suspension containing methylprednisolone. Two cases of conjunctival inflammation after retrobulbar injection. Arch Dermatol 1985;121:258-261
10   Mathias CGT, Maibach HI, Ostler HB, Conant MA, Nelson W. Delayed hypersensitivity to retrobulbar injections of methylprednisolone acetate. Am J Ophthalmol 1978;86:816-819
11   Konttinen YT, Friman C, Tolvanen E, Reitamo S, Johansson E. Local skin rash after intraarticular methyl prednisolone acetate injection in a patient with rheumatoid arthritis. Arthritis Rheum 1983;26:231-233
12   De Boer EM, van den Hoogenband HM, van Ketel WG. Positive patch test reactions to injectable corticosteroids. Contact Dermatitis 1984;11:261-262

Chapter 3.220 METHYLPREDNISOLONE HEMISUCCINATE

IDENTIFICATION

Description/definition : Methylprednisolone hemisuccinate is the hemisuccinate ester of the synthetic glucocorticoid methylprednisolone that conforms to the structural formula shown below

Pharmacological classes : Antineoplastic agents, hormonal; anti-inflammatory agents

IUPAC name : 4-[2-[(6S,8S,9S,10R,11S,13S,14S,17R)-11,17-Dihydroxy-6,10,13-trimethyl-3-oxo-7,8,9,11,12,14,15,16-octahydro-6H-cyclopenta[a]phenanthren-17-yl]-2-oxoethoxy]-4-oxobutanoic acid

Other names : Methylprednisolone succinate; 11β,17,21-trihydroxy-6α-methylpregna-1,4-diene-3,20-dione 21-(hydrogen succinate)

CAS registry number : 2921-57-5

EC number : 220-863-9

Patch testing : In general, corticosteroids may be tested at 0.1% and 1% in alcohol; late readings (6-10 days) are strongly recommended

Molecular formula : $C_{26}H_{34}O_8$

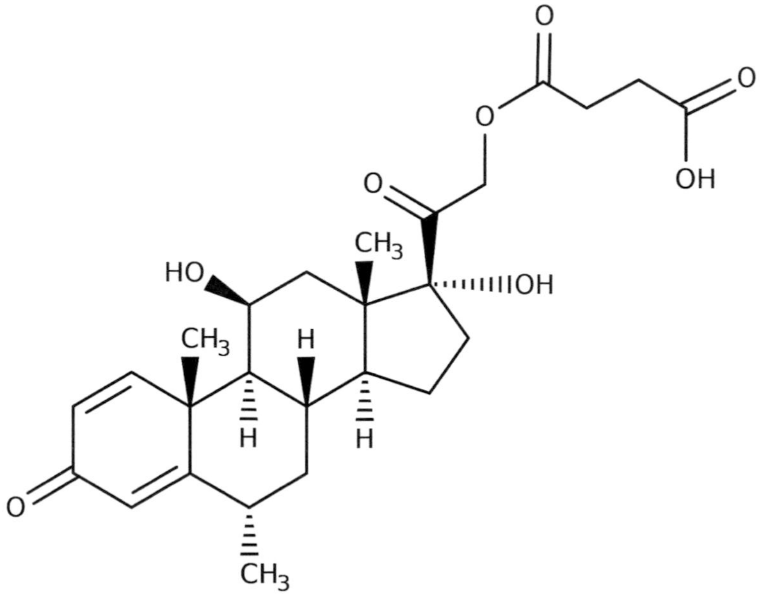

GENERAL

General aspects of corticosteroids used on the skin and mucous membranes are discussed in Chapter 2.4. A practical guideline for diagnosing allergic reactions to corticosteroids is presented in ref. 1. In pharmaceutical products, methylprednisolone hemisuccinate is employed as methylprednisolone sodium (hemi)succinate (CAS number 2375-03-3, EC number 219-156-8, molecular formula $C_{26}H_{33}NaO_8$). It is available only for intravenous and intramuscular injections. As methylprednisolone hemisuccinate is not used topically, most positive patch tests to it are the result of cross-sensitivity (e.g. 3).

See also Chapter 3.217 (Methylprednisolone), Chapter 3.219 (Methylprednisolone acetate), and Chapter 3.218 (Methylprednisolone aceponate).

CONTACT ALLERGY

Case series

From January 1990 to June 2008, in Leuven, Belgium, 315 patients were diagnosed with contact allergy to/allergic contact dermatitis from corticosteroids (CSs) from routine patch testing with a baseline series including tixocortol pivalate, budesonide, hydrocortisone butyrate and prednisone caproate, patch testing with patients' own CS preparations, and testing those with proven contact allergy to a corticosteroid or strongly suspected of CS allergy later with a series of 66 CSs, including two sex hormones (progesterone and testosterone). 71% of the patients had relevant reactions, but these were not specified. In this group of 315 CS allergic patients, 17 had positive patch tests to methylprednisolone hemisuccinate 0.1% alc. (2). As this corticosteroid is not used topically, it may be assumed that all these reactions were the result of cross-reactivity to other corticosteroids.

Cross-reactions, pseudo-cross-reactions and co-reactions
Cross-reactions between corticosteroids are discussed in Chapter 2.8.

Cutaneous adverse drug reactions from systemic administration caused by type IV (delayed-type) hypersensitivity
Cutaneous adverse drug reactions from systemic administration of methylprednisolone hemisuccinate caused by type IV (delayed-type) hypersensitivity, including generalized erythematous exanthema (4), maculopapular eruption (6,7) erythroderma (5), and 'generalized skin rashes' (8), are planned to be discussed in Volume IV of the *Monographs in Contact Allergy* series on Systemic drugs.

LITERATURE

1   Baeck M, Goossens A. Immediate and delayed allergic hypersensitivity to corticosteroids: practical guidelines. Contact Dermatitis 2012;66:38-45

2   Baeck M, Chemelle JA, Terreux R, Drieghe J, Goossens A. Delayed hypersensitivity to corticosteroids in a series of 315 patients: clinical data and patch test results. Contact Dermatitis 2009;61:163-175

3   Bianchi L, Hansel K, Antonelli E, Bellini V, Rigano L, Stingeni L. Deflazacort hypersensitivity: a difficult-to-manage case of systemic allergic dermatitis and literature review. Contact Dermatitis 2016;75:54-56

4   Fernández de Corres L, Bernaola G, Urrutia I, Muñoz D. Allergic dermatitis from systemic treatment with corticosteroids. Contact Dermatitis 1990;22:104-106

5   Fernández de Corrés L, Urrutia I, Audicana M, Echechipia S, Gastaminza G. Erythroderma after intravenous injection of methylprednisolone. Contact Dermatitis 1991;25:68-70

6   Bursztejn AC, Tréchot P, Cuny JF, Schmutz JL, Barbaud A. Cutaneous adverse drug reactions during chemotherapy: consider non-antineoplastic drugs. Contact Dermatitis 2008;58:365-368

7   Hotta E, Tamagawa-Mineoka R, Katoh N. Delayed-type hypersensitivity to 6-methyl-prednisolone sodium succinate. J Dermatol 2014;41:754-755

8   Kuga A, Futamura N, Funakawa I, Jinnai K. Allergic skin rashes by methylprednisolone in a case with multiple sclerosis. Rinsho Shinkeigaku 2004;44:691-694 (Article in Japanese)

Chapter 3.221 METHYL SALICYLATE

IDENTIFICATION

Description/definition   : Methyl salicylate is the ester of methyl alcohol and salicylic acid that conforms to the
                                 structural formula shown below
Pharmacological classes   : Fixatives; antirheumatic agents
IUPAC name   : Methyl 2-hydroxybenzoate
Other names   : Synthetic wintergreen oil; synthetic sweet birch oil
CAS registry number   : 119-36-8
EC number   : 204-317-7
Merck Index monograph : 7463
Patch testing   : 1% pet. (SmartPracticeEurope, SmartPracticeCanada)
Molecular formula   : $C_8H_8O_3$

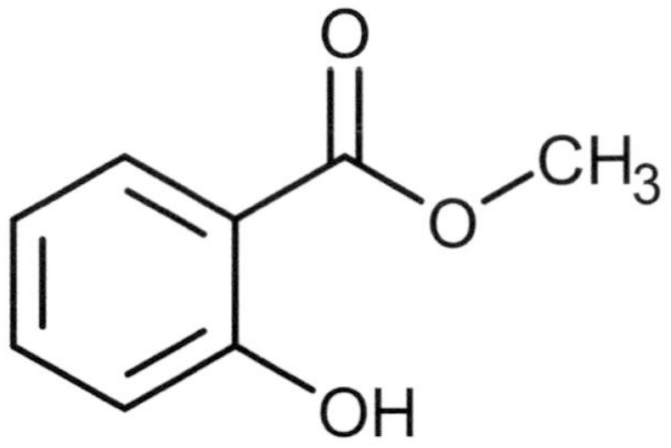

GENERAL

In many countries, methyl salicylate is available as an analgesic, anti-inflammatory agent, rubefacient and counterirritant in a wide range of over-the-counter liniments, ointments, lotions and medical oils for muscle pains (7,10). Other uses include or have included in perfumery, (veterinary and human) medications, UV-absorber in sunburn lotions, flavor in foods and beverages, solvent for insecticides, polishes, and inks, and as chemical intermediate (U.S. National Library of Medicine). Cutaneous and non-cutaneous (systemic) side effects of over-the-counter pharmaceutical preparations containing methyl salicylate have been reviewed up to 1996 (8), 2007 (11) and 2019 (12). In this chapter, only contact allergy to methyl salicylate is discussed.

CONTACT ALLERGY

Patch testing in groups of patients
Results of studies testing methyl salicylate in consecutive patients suspected of contact dermatitis (routine testing) and those of testing in groups of *selected* patients (patients suspected of oral and lip contact allergy, women with vulval pruritus) are shown in table 3.221.1.

Patch testing in consecutive patients suspected of contact dermatitis: routine testing
In three studies in which methyl salicylate 2% pet. was tested in consecutive patients suspected of contact dermatitis, low frequencies of sensitization (0.1%, 0.4% and 0.5%) have been observed (1,6,9). The relevance of the positive patch test reactions was not mentioned. In routine testing in the late 1970s and early 1980s, the percentage of positive reactions to methyl salicylate was decreasing in Hamburg, Germany (15).

Patch testing in groups of selected patients
In three studies in which methyl salicylate was tested in selected patient groups, low frequencies of sensitization of 1.3% (2), 1.6% (3) and 2% (very small group [13]) have been observed. The relevance of the positive patch test reactions was not mentioned.

Case reports
A man developed acute dermatitis of the neck, upper back, shoulders and dorsa of the hands. The patient had been applying an analgesic ointment to his neck and back containing menthol, camphor and 12% methyl salicylate. Patch testing to the constituents of the ointment gave a positive reaction to methyl salicylate 2% in olive oil. Three months later, the patient returned with a reappearance of his eczema at the previous sites. He denied using any topical applications but he had taken oral Aspirin (acetylsalicylic acid). Patch tests with acetylsalicylic acid 0.3% and 5% aqua were negative, but when he took 500 mg of Aspirin, the patient noticed pruritus and erythema again in the previously affected areas (4). Two cases of contact allergy to methyl salicylate have been reported from a liniment

Table 3.221.1 Patch testing in groups of patients

Years and Country	Test conc. & vehicle	Number of patients tested	positive (%)		Selection of patients (S); Relevance (R); Comments (C)	Ref.
Routine testing						
1998-9 Netherlands	2% pet.	1825	7	(0.4%)	R: not stated	1
1978-1980 USA	2% pet.	585	3	(0.5%)	R: not stated; C: testing with it was abandoned	9
1973-1977 Spain	2% pet.	4600	6	(0.1%)	R: not stated	6
Testing in groups of selected patients						
2014-2015 USA		149	2	(1.3%)	S: patients suspected of oral and lip contact allergy; R: not stated	2
2005-2007 Turkey		50	1	(2%)	S: women with vulval pruritus; R: relevant, but the culprit product was not mentioned	13
1975-1976 USA	2% pet.	183	3	(1.6%)	S: not stated; R: not stated	3

containing 25% methyl salicylate in arachis oil (5). A woman had a rectangular pruritic erythematous macule on the hip following the use of a compress. The manufacturer provided samples of the ingredients and patch testing showed a positive reaction to methyl salicylate 2% in olive oil (7).

Possibly, a case of allergic contact dermatitis from methyl salicylate was reported in 1979 (14).

Cross-reactions, pseudo-cross-reactions and co-reactions
Cross-reactions (or pseudo-cross-reactions from the same allergenic moiety) have been observed to phenyl salicylate (2), benzyl salicylate (2) and sodium salicylate (4).

Immediate contact reactions
Immediate contact reactions (contact urticaria) to methyl salicylate are presented in Chapter 5.

LITERATURE
1   De Groot AC, Coenraads PJ, Bruynzeel DP, Jagtman BA, Van Ginkel CJW, Noz K, et al. Routine patch testing with fragrance chemicals in The Netherlands. Contact Dermatitis 2000;42:184-185
2   Scheman A, Te R. Contact allergy to salicylates and cross-reactions. Dermatitis 2017;28:291
3   Rudner EJ. North American group results. Contact Dermatitis 1977;3:208-209
4   Hindson C. Contact eczema from methyl salicylate reproduced by oral aspirin (acetyl salicylic acid). Contact Dermatitis 1977;3:348-349
5   Morgan JK. British Journal of Clinical Practice 1968;22:261. Data cited in ref. 8
6   Romaguera C, Grimalt F. Statistical and comparative study of 4600 patients tested in Barcelona (1973–1977). Contact Dermatitis 1980;6:309-315
7   Oiso N, Fukai K, Ishii M. Allergic contact dermatitis due to methyl salicylate in a compress. Contact Dermatitis 2004;51:34-35
8   Chan TY. Potential dangers from topical preparations containing methyl salicylate. Hum Exp Toxicol 1996;15:747-750
9   Mitchell JC, Adams RM, Glendenning WE, Fisher A, Kanof N, Larsen W, et al. Results of standard patch tests with substances abandoned. Contact Dermatitis 1982;8:336-337
10   Anderson A, McConville A, Fanthorpe L, Davis J. Salicylate poisoning potential of topical pain relief agents: From age old remedies to engineered smart patches. Medicines (Basel) 2017 June 30;4(3). pii: E48. doi: 10.3390/medicines4030048
11   Davis JE. Are one or two dangerous? Methyl salicylate exposure in toddlers. J Emerg Med 2007;32:63-69
12   De Groot AC. Monographs in Contact Allergy, Volume II. Fragrances and Essential Oils. Boca Raton, Fl, USA: CRC Press Taylor and Francis Group, 2019: 500-503
13   Utas S, Ferahbas, A, Yildiz S. Patients with vulval pruritus: patch test results. Contact Dermatitis 2008;58:296-298
14   Speer F. Allergy to methyl salicylate. Ann Allergy 1979;43:36-37
15   Kuhlwein A, Hausen BM. Standard epicutaneous tests: interpretation of the statistical findings of the last five years. Z Hautkr 1982;57:1179-1186 (Article in German)

Chapter 3.222 METHYL SALICYLATE 2-ETHYLBUTYRATE

IDENTIFICATION

Description/definition : Methyl salicylate 2-ethylbutyrate is the salicylic acid derivative that conforms to the structural formula shown below
Pharmacological classes : Dermatologic agents; anti-inflammatory agents; topical analgesics
IUPAC name : Methyl 2-(2-ethylbutanoyloxy)benzoate
Other names : Methyl butetisalicylate; methyl 2-O-(ethylbutyryl)salicylate; methyl 2-[(2-ethylbutanoyl)oxy]benzoate; methyl diethylacetylsalicylate
CAS registry number : 85005-92-1
EC number : 285-023-6
Patch testing : 30% pet.
Molecular formula : $C_{14}H_{18}O_4$

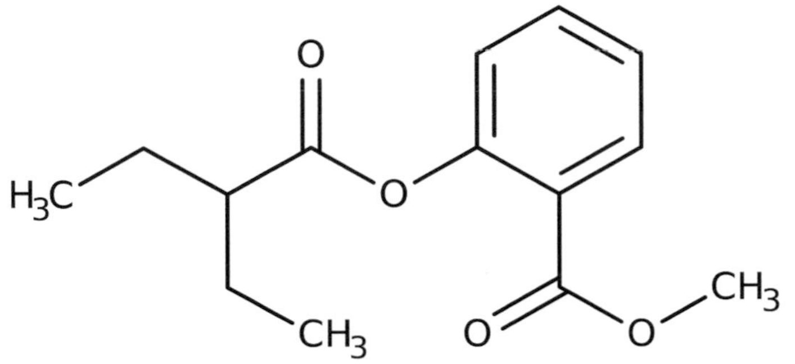

GENERAL

Methyl salicylate 2-ethylbutyrate is a salicylic acid derivative and nonsteroidal anti-inflammatory drug (NSAID) with analgesic, antipyretic and anti-inflammatory properties. It may be present in topical pharmaceutical preparations for the treatment of muscular and joint pain (1).

CONTACT ALLERGY

Case report

A 40-year-old woman presented with itchy, erythematous vesiculobullous lesions on her right knee and thigh, with erythema and micropapules spreading to the abdomen, after having used an analgesic cream containing 30% methyl butetisalicylate (methyl salicylate 2-ethylbutyrate) to treat a painful right knee for 10 days. Patch tests with the cream and – later – its ingredients were positive to the cream 'as is' and to methyl butetisalicylate 30% pet. Fifteen controls were negative. There were no positive patch tests to other NSAIDs: ibuproxam 5%, ketoprofen 2.5%, ibuprofen 5%, naproxen 5%, piroxicam 1%, flurbiprofen 5%, tiaprofenic acid 5% and fenoprofen 5%. Oral administration of a cumulative dose of 1000 mg acetylsalicylic acid (Aspirin) gave no cutaneous or systemic reaction (2).

Cross-reactions, pseudo-cross-reactions and co-reactions

Not to ibuproxam 5%, ketoprofen 2.5%, ibuprofen 5%, naproxen 5%, piroxicam1%, flurbiprofen 5%, tiaprofenic acid 5% and fenoprofen 5% (2).

LITERATURE

1   The data in the section 'General' may have been obtained from literature discussed in this chapter, but mostly also or exclusively from one or more of the following online sources: ChemIDPlus Advanced, PubChem, DrugBank, RxList, Drug Central, Drugs.com, and Wikipedia
2   Valsecchi R, Aiolfi M, Leghissa P, Cologni L, Cortinovis R. Contact dermatitis from methyl butetisalicylate. Contact Dermatitis 1998;38:360-361

Chapter 3.223 METIPRANOLOL

IDENTIFICATION

Description/definition   : Metipranolol is the propanolamine, acetate ester, aromatic ether and secondary amino
                             compound that conforms to the structural formula shown below
Pharmacological classes   : Sympatholytics; β-adrenergic antagonists; antihypertensive agents; anti-arrhythmia
                             agents
IUPAC name   : [4-[2-Hydroxy-3-(propan-2-ylamino)propoxy]-2,3,6-trimethylphenyl] acetate
Other names   : 1-(4-Hydroxy-2,3,5-trimethylphenoxy)-3-(isopropylamino)-2-propanol 4-acetate
CAS registry number   : 22664-55-7
EC number   : 245-151-5
Merck Index monograph   : 7486
Patch testing   : 5%-10% pet.
Molecular formula   : $C_{17}H_{27}NO_4$

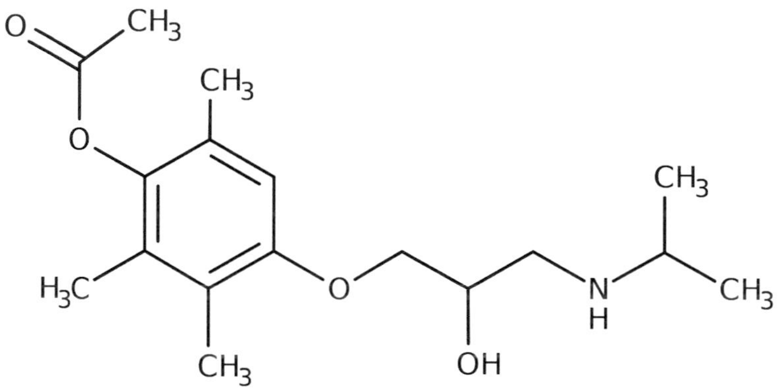

GENERAL

Metipranolol is a β- adrenergic antagonist effective for both β1- and β2- receptors that has antiarrhythmic, antihypertensive, and antiglaucoma properties. Metipranolol is indicated in the treatment of elevated intraocular pressure in patients with ocular hypertension or open-angle glaucoma. According to some sources, it is also used as an antihypertensive and anti-arrhythmia drug. Metipranolol is used in eye drops as metipranolol hydrochloride (CAS number 36592-77-5, EC number not available, molecular formula $C_{17}H_{28}ClNO_4$) (1).

CONTACT ALLERGY

General

For information on patch testing with commercial eye drops containing beta-blockers see Chapter 3.338 Timolol.

Case series

In Leuven, Belgium, in the period 1990-2014, iatrogenic contact dermatitis was diagnosed in 2600 individuals (17% of the total patch test population). 96% of all positive patch test reactions to topical drugs and antiseptics were considered to be relevant. Metipranolol (5% water) was tested in 13 patients and there were 3 positive reactions to it (3).

Out of 112,430 patients patch tested by the IVDK between 1993 and 2004, 332 had been tested with their own topical anti-glaucoma eye drops containing different β-blockers because of suspected allergic contact dermatitis. Eighty-six subjects were tested with metipranolol eye drops and there were 13 (15%) positive reactions. The patients were not tested with the active substance, but reactions to the (possible) adjuvants benzalkonium chloride, sodium EDTA and sodium disulfite were excluded (2).

In Graz, Austria, before 1990, 7 patients with periorbital dermatitis had allergic contact dermatitis from topical beta-blockers used for treatment of glaucoma. Treatment preceded the appearance of the skin lesions for one month up to one year. The beta-blockers were patch tested as hydrous solutions with the same concentrations as present in therapeutic preparations. Four patients were suspected with metipranolol allergy, but they had negative patch tests. Only after enlargement of the test area to 2x2 centimeter did metipranolol give reproducible positive test reactions. There were no cross-reactions between metipranolol, timolol, and befunolol (7).

Case reports

A 70-year-old man was referred for recurrent bilateral upper and lower eyelid eczema, which proved to be caused by contact allergy to timolol and betaxolol in eye drops used for glaucoma. He was then treated with metipranolol eye drops but experienced a further relapse; contact allergy to metipranolol was shown by a positive patch test to metipranolol 0.6% water but not 2% MEK (methyl ethyl ketone). Thus, it appears that the patient developed successive independent sensitizations to three beta-blockers. There were no cross-reactions to a battery of beta-blockers to which he had not been exposed (4).

A 51-year-old man had used 0.3% metipranolol-containing eye drops for treatment of glaucoma, when itchy, red and scaly skin changes of the eyelids and periorbital skin developed. The same eye drops without preservative (benzalkonium chloride) did not provide relief. Upon patch testing, no positive reactions to either of these solutions were noted. In a second session, metipranolol was tested in a dilution series ranging from 0.1% to 10% in petrolatum. There were positive reactions to metipranolol 10%, 5% and 2.5%, a dubious positive reaction to 1% and no reactions to the lower concentrations. Twenty controls were negative to metipranolol 10% pet. After ceasing the use of metipranolol, the patient has been free of skin complaints (5).

The authors of the latter case (5) also presented a similar case. A 43-year old woman developed an acute weeping dermatitis around the right eye while on metipranolol eye drops, which disappeared after cessation of the use of this medication. A patch test with the eye drops containing 0.1% metipranolol was negative, but she had positive patch tests to all concentrations of metipranolol in petrolatum in a dilution series (0.1%, 0.25%, 1%, 2.5%, 5% and 10%). Twenty controls were negative to the highest concentration. No recurrence of dermatitis has occurred after cessation of metipranolol (5).

A 52-year-old woman had dermatitis around both eyes. She used eye drops containing 0.6% metipranolol for glaucoma. In addition, 2 cosmetic lotions were applied daily to the face and eyelids. When patch tested, she reacted to nickel sulphate, p-tert-butylphenolformaldehyde resin, one of her cosmetic lotions and its main ingredient oleamidopropyl dimethylamine, but not to the metipranolol eye drops. Later, she was patch tested with the ingredients of this medication and several other beta-blockers. The patient now showed positive patch test reactions to metipranolol and l-penbutolol sulfate2% water; 15 controls were negative to both preparations (6).

Cross-reactions, pseudo-cross-reactions and co-reactions

A patient sensitized to metipranolol was patch tested with a series of other beta-blockers and had a positive reaction only to l-penbutolol sulfate, to which she had most likely not been exposed before and consequently may represent a cross-reaction (6).

LITERATURE

1 The data in the section 'General' may have been obtained from literature discussed in this chapter, but mostly also or exclusively from one or more of the following online sources: ChemIDPlus Advanced, PubChem, DrugBank, RxList, Drug Central, Drugs.com, and Wikipedia
2 Jappe U, Uter W, Menezes de Pádua CA, Herbst RA, Schnuch A. Allergic contact dermatitis due to beta-blockers in eye drops: a retrospective analysis of multicentre surveillance data 1993-2004. Acta Derm Venereol 2006;86:509-514
3 Gilissen L, Goossens A. Frequency and trends of contact allergy to and iatrogenic contact dermatitis caused by topical drugs over a 25-year period. Contact Dermatitis 2016;75:290-302
4 O'Donnell BF, Foulds IS. Contact allergy to beta-blocking agents in ophthalmic preparations. Contact Dermatitis 1993;28:121-122
5 De Groot AC, van Ginkel CJ, Bruynzeel DP, Smeenk G, Conemans JM. [Contact allergy to eyedrops containing beta-blockers]. Ned Tijdschr Geneeskd 1998;142:1034-1036 (Article in Dutch)
6 De Groot AC, Conemans J. Contact allergy to metipranolol. Contact Dermatitis 1988;18:107-108
7 Gailhofer G, Ludvan M. 'Beta-blockers': sensitizers in periorbital allergic contact dermatitis. Contact Dermatitis 1990;23:262

Chapter 3.224 METOPROLOL

IDENTIFICATION

Description/definition   : Metoprolol is the secondary alcohol that conforms to the structural formula shown below
Pharmacological classes   : β_1-Adrenergic receptor antagonists; anti-arrhythmia agents; sympatholytics;
                           antihypertensive agents
IUPAC name   : 1-[4-(2-Methoxyethyl)phenoxy]-3-(propan-2-ylamino)propan-2-ol
CAS registry number   : 51384-51-1
EC number   : 253-483-7
Merck Index monograph   : 7498
Patch testing   : 10% pet.
Molecular formula   : $C_{15}H_{25}NO_3$

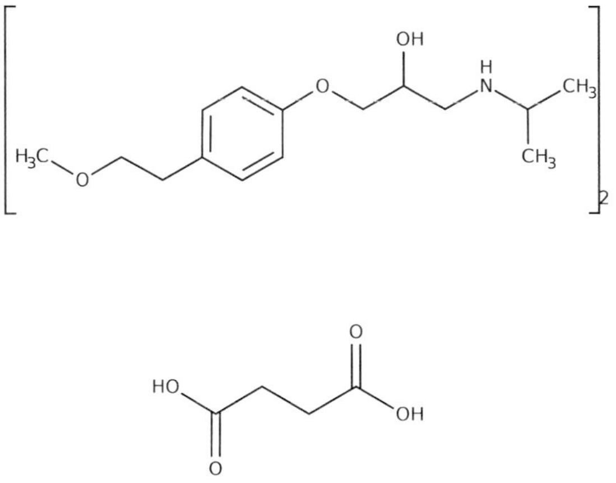

Metoprolol succinate 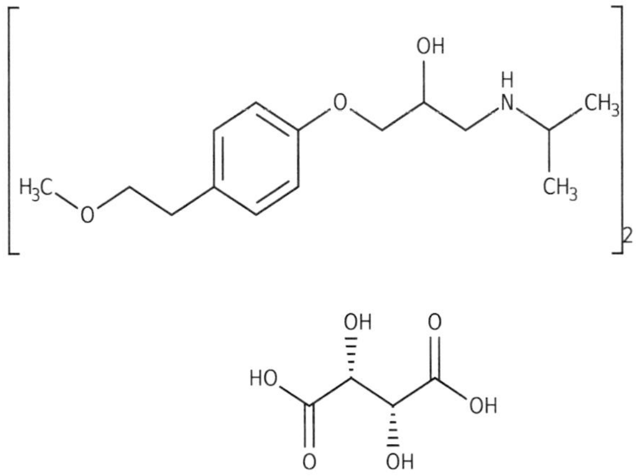   Metoprolol tartrate

GENERAL

Metoprolol is a cardioselective competitive β_1-adrenergic receptor antagonist with antihypertensive properties. This agent antagonizes β_1-adrenergic receptors in the myocardium, thereby reducing the rate and force of myocardial contraction, leading to a reduction in cardiac output. Metoprolol is indicated for the treatment of angina, heart failure, myocardial infarction, atrial fibrillation, atrial flutter and hypertension. Off-label uses of metoprolol include supraventricular tachycardia and thyroid storm. In pharmaceutical products, metoprolol is most often employed as metoprolol succinate (CAS number 98418-47-4, EC number not available, molecular formula $C_{34}H_{56}N_2O_{10}$) or as metoprolol tartrate (CAS number 56392-17-7, EC number 260-148-9, molecular formula $C_{34}H_{56}N_2O_{12}$) (1).

CONTACT ALLERGY

Case series
A group of 28 patients was treated with metoprolol eye drops 1% 3-4 times daily or metoprolol 3% or 4% two to three times a day for glaucoma. The mean duration of treatment was 3.8 months (range 0.5 to 6 months). Eleven of the twenty-eight patients (39%) developed ocular (conjunctival edema and/or hyperemia) and/or periocular adverse reactions (dermatitis) with subjective symptoms such as itching, burning and smarting 2 weeks to 5 months after beginning of treatment. The clinical symptoms disappeared 2-6 weeks after treatment was stopped. Patch tests were performed with metoprolol tartrate 3% in water and there were 5 positive reactions. Repeated patch tests after 3-6 months were again positive, albeit weaker. 'A group' of controls was negative (4).

Cross-reactions, pseudo-cross-reactions and co-reactions
Of 7 patients sensitized to alprenolol, 2 (28%) reacted to metoprolol (3). Three patients sensitized to metoprolol also had positive skin tests to propranolol 1% water, practolol 20% water, and timolol 0.5% water, one each (4).

Cutaneous adverse drug reactions from systemic administration caused by type IV (delayed-type) hypersensitivity
Cutaneous adverse drug reactions from systemic administration of metoprolol caused by type IV (delayed-type) hypersensitivity including a psoriasiform eruption (5) and occupational allergic contact dermatitis (2) are planned to be discussed in Volume IV of the *Monographs in Contact Allergy* series on Systemic drugs.

LITERATURE

1 The data in the section 'General' may have been obtained from literature discussed in this chapter, but mostly also or exclusively from one or more of the following online sources: ChemIDPlus Advanced, PubChem, DrugBank, RxList, Drug Central, Drugs.com, and Wikipedia
2 Swinnen I, Ghys K, Kerre S, Constandt L, Goossens A. Occupational airborne contact dermatitis from benzodiazepines and other drugs. Contact Dermatitis 2014;70:227-232
3 Ekenvall L, Forsbeck M. Contact eczema produced by a beta-adrenergic blocking agent (alprenolol). Contact Dermatitis 1978;4:190-194
4 Van Joost T, Middelkamp Hup J, Ros FE. Dermatitis as a side effect of long-term topical treatment with certain beta-blocking agents. Br J Dermatol 1979;101:171-176
5 Neumann HA, Van Joost T, Westerhof W. Dermatitis as side-effect of long-term metoprolol. Lancet 1979;2(8145):745

Chapter 3.225 METRONIDAZOLE

IDENTIFICATION

Description/definition   : Metronidazole is a synthetic nitroimidazole derivative that conforms to the structural
                              formula shown below
Pharmacological classes : Anti-infective agents; anti-bacterial agents; antiprotozoal agents
IUPAC name   : 2-(2-Methyl-5-nitroimidazol-1-yl)ethanol
CAS registry number   : 443-48-1
EC number   : 207-136-1
Merck Index monograph : 7506
Patch testing   : 1% pet. (SmartPracticeCanada, SmartPracticeEurope)
Molecular formula   : $C_6H_9N_3O_3$

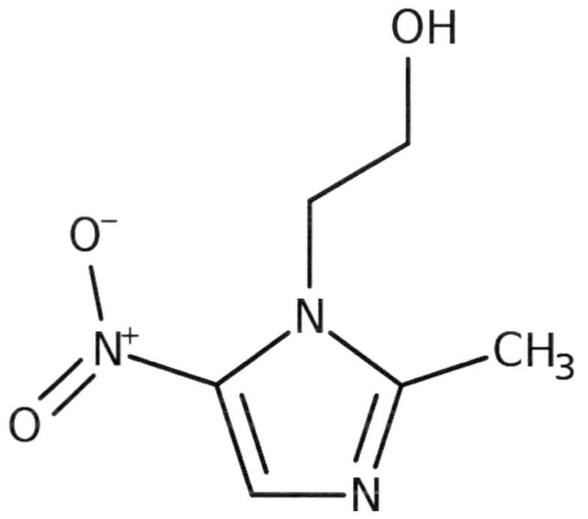

GENERAL

Metronidazole is a synthetic nitroimidazole derivative with antiprotozoal and antibacterial activities; it is extremely effective against anaerobic bacterial infections. Metronidazole is indicated for the treatment of anaerobic infections and mixed infections, surgical prophylaxis requiring anaerobic coverage, *Clostridium difficile*-associated diarrhea and colitis, *Helicobacter pylori* infection and duodenal ulcer disease, bacterial vaginosis, *Giardia lamblia* gastro-enteritis, amebiasis caused by *Entamoeba histolytica*, and *Trichomonas* infections. In topical formulations, it is used in the treatment of rosacea (1).

CONTACT ALLERGY

Patch testing in groups of patients

In Leuven, Belgium, in the period 1990-2014, 209 patients suspected of iatrogenic contact dermatitis were patch tested with a pharmaceutical series and with their own products; 4 (1.9%) had a positive reaction to metronidazole 1% pet. 96% of the positive patch test reactions to all topical drugs and antiseptics were considered to be relevant (2). Of 361 patients with rosacea who were patch tested in the departments of dermatology of the IVDK between 1995 and 2002, only one reacted to metronidazole cream; metronidazole itself was not tested. It was not specified how many patients had been tested with metronidazole (10).

Case series

A small case series of 3 patients with allergic contact dermatitis to metronidazole was reported in 2007, probably one from Denmark and the other 2 from Belgium (7). One is discussed below (female nurse, aged 40 [5]). The second patient was a 47-year-old woman who developed an infected weeping dermatitis on her chin and both cheeks, with spreading of erythematous papules to the neck, following the application of 0.75% metronidazole gel. Patch tests were positive to the gel and to metronidazole 2% pet. (D2 ?+, D4 ++) (7). The third patient, a 57-year-old woman, presented with facial dermatitis that had been present for about 2 years. Previous therapy with metronidazole cream was not effective and had been replaced with sodium fusidate cream and a corticosteroid, which had improved the dermatitis considerably. She had been advised to stop this treatment and use the metronidazole cream again, but this resulted in sharply demarcated papular eczema around the nose, upper lip and chin. Patch tests were positive to the cream and to metronidazole 2% pet. Other azole derivatives (miconazole, econazole, isoconazole, bifonazole, ketoconazole, clotrimazole, and sulconazole, all diluted 2% pet.) were negative (7).

Case reports

A 67-year-old female community nurse was referred with rosacea. After 2 days of treatment with topically applied 0.75% metronidazole cream, she developed a weeping, vesicular, and erythematous facial dermatitis. Patch tests were strongly positive to the cream and to metronidazole 5% pet. at D3 and D7. She also reacted to MCI/MI. It was suggested that the patient may have become sensitized from prior occupational exposure to metronidazole tablets. However, she had also previously had a metronidazole pessary, albeit only once (5).

Another female nurse, aged 40, whose history was previously already published (7), was referred with rosacea and developed facial dermatitis with erythema, swelling, and itching after a few days of treatment with 1% metronidazole cream. Use tests with metronidazole 1% cream and 0.75% gel showed acute vesicular dermatitis after six applications on D3. Patch tests were positive to the cream and to metronidazole 5% pet. Also in this case, the authors suggested occupational sensitization from previous exposure to metronidazole at work. However, this patient had taken metronidazole tablets herself several times, albeit without any adverse reactions (5).

A 45-year-old woman had a 15-day history of an acute, itchy, vesicular and erythematous eruption around the mouth. For rosacea she had been treated with a cream containing chlorophyll, kamillosan, erythromycin, metronidazole and diphenhydramine, the concentrations of which were not recorded. Patch testing showed positive reactions to the cream at D2 (++) and D4 (+++) and – later – to its ingredients metronidazole 5% and 10% pet. and diphenhydramine 1% pet. Twelve controls were negative (6).

A 68-year-old woman presented with an erythematous itchy eruption of 5 days' duration on her face after using metronidazole topical gel 0.75% for the treatment of mild rosacea for 6 days. Examination showed a severe bright red erythema around the mouth, on the nose reaching the nasal parts of the cheeks and on the upper and lower eyelids, with multiple pustules around the mouth, 'characteristic of acute contact dermatitis' (which is incorrect, as multiple pustules are not characteristic of allergic contact dermatitis). Patch tests were positive to the gel, to a solution of metronidazole 0.5% (0.5 gr/100 ml) for intravenous injection and to MCI/MI. The authors suggested that the possibility of a cross-reaction between these substances is very plausible (which it is most likely not) (8).

A 26-year-old girl had a 6-month' history of an itchy erythematovesicular eruption of the central face and forehead. She had been treated with topical corticosteroids and antibiotics, without benefit. She had also been treated with 1% topical metronidazole gel. Patch tests were positive to the gel and to metronidazole 1% pet., but negative to tioconazole, miconazole, econazole, clotrimazole and bifonazole (9).

Cross-reactions, pseudo-cross-reactions and co-reactions

The question has been raised (but not answered) whether contact allergy to metronidazole and other (nitro)imidazoles may be overrepresented, from cross-reactivity or otherwise, in patients allergic to methylchloroisothiazolinone /methylisothiazolinone (MCI/MI) (3,5,8).

Of 5 patients with delayed-type hypersensitivity to benznidazole from oral administration, 2 (40%) cross-reacted to metronidazole; both are nitroimidazoles (4). One patient sensitized to tioconazole may have cross-reacted to metronidazole (11).

Cutaneous adverse drug reactions from systemic administration caused by type IV (delayed-type) hypersensitivity

Cutaneous adverse drug reactions from systemic administration of metronidazole caused by type IV (delayed-type) hypersensitivity, including fixed drug eruption (13-15), acute generalized exanthematous pustulosis (AGEP) (12), linear IgA bullous dermatosis (8), systemic contact dermatitis (16), TEN (toxic epidermal necrolysis), and DRESS (drug reaction with eosinophilia and systemic symptoms) (3), are planned to be discussed in Volume IV of the *Monographs in Contact Allergy* series on Systemic drugs.

LITERATURE

1 The data in the section 'General' may have been obtained from literature discussed in this chapter, but mostly also or exclusively from one or more of the following online sources: ChemIDPlus Advanced, PubChem, DrugBank, RxList, Drug Central, Drugs.com, and Wikipedia
2 Gilissen L, Goossens A. Frequency and trends of contact allergy to and iatrogenic contact dermatitis caused by topical drugs over a 25-year period. Contact Dermatitis 2016;75:290-302
3 Stingeni L, Rigano L, Lionetti N, Bianchi L, Tramontana M, Foti C, et al. Sensitivity to imidazoles/nitroimidazoles in subjects sensitized to methylchloroisothiazolinone/methylisothiazolinone: A simple coincidence? Contact Dermatitis 2019;80:181-183

4 Noguerado-Mellado B, Rojas-Pérez-Ezquerra P, Calderón-Moreno M, Morales-Cabeza C, Tornero-Molina P. Allergy to benznidazole: cross-reactivity with other nitroimidazoles. J Allergy Clin Immunol Pract 2017;5:827-828

5 Madsen JT, Lorentzen HF, Paulsen E. Contact sensitization to metronidazole from possible occupational exposure. Contact Dermatitis 2009;60:117-118

6 Fernández-Jorge B, Goday Buján J, Fernández-Torres R, Rodríguez-Lojo R, Fonseca E. Concomitant allergic contact dermatitis from diphenhydramine and metronidazole. Contact Dermatitis 2008;59:115-116

7 Madsen JT, Thormann J, Kerre S, Andersen KE, Goossens A. Allergic contact dermatitis to topical metronidazole - 3 cases. Contact Dermatitis 2007;56:364-366

8 Wolf R, Orion E, Matz H. Co-existing sensitivity to metronidazole and isothiazolinone. Clin Exp Dermatol 2003;28:506-507

9 Vincenzi C, Lucente P, Ricci C, Tosti A. Facial contact dermatitis due to metronidazole. Contact Dermatitis 1997;36:116-117

10 Jappe U, Schnuch A, Uter W. Rosacea and contact allergy to cosmetics and topical medicaments – retrospective analysis of multicentre surveillance data 1995-2002. Contact Dermatitis 2005;52:96-101

11 Izu R, Aguirre A, Gonzales M, Diaz-Perez J L. Contact dermatitis from tioconazole with cross-sensitivity to other imidazoles. Contact Dermatitis 1992;26:130

12 Girardi M, Duncan KO, Tigelaar RE, Imaeda S, Watsky KL, McNiff JM. Cross-comparison of patch test and lymphocyte proliferation responses in patients with a history of acute generalized exanthematous pustulosis. Am J Dermatopathol 2005;27:343

13 Vila JB, Bernier MA, Gutierrez JV, Gómez MT, Polo AM, Harrison JM, Miranda-Romero A, et al. Fixed drug eruption caused by metronidazole. Contact Dermatitis 2002;46:122

14 Gastaminza G, Anda M, Audicana MT, Fernandez E, Muñoz D. Fixed-drug eruption due to metronidazole with positive topical provocation. Contact Dermatitis 2001;44:36

15 Hermida MD, Consalvo L, Lapadula MM, Della Giovanna P, Cabrera HN. Bullous fixed drug eruption induced by intravaginal metronidazole ovules, with positive topical provocation test findings. Arch Dermatol 2011;147:250-251

16 Mussani F, Skotnicki S. Systemic contact dermatitis: Two Interesting cases of systemic eruptions following exposure to drugs clioquinol and metronidazole. Dermatitis 2013;24(4):e3

Chapter 3.226 MICONAZOLE

IDENTIFICATION

Description/definition : Miconazole is a synthetic phenethyl imidazole that conforms to the structural formula shown below

Pharmacological classes : Antifungal agents; 14α-demethylase inhibitors; cytochrome P-450 CYP2C9 inhibitors; cytochrome P-450 CYP3A inhibitors

IUPAC name : 1-[2-(2,4-Dichlorophenyl)-2-[(2,4-dichlorophenyl)methoxy]ethyl]imidazole

CAS registry number : 22916-47-8

EC number : 245-324-5

Merck Index monograph : 7527

Patch testing : 1% alc. (Chemotechnique)

Molecular formula : $C_{18}H_{14}Cl_4N_2O$

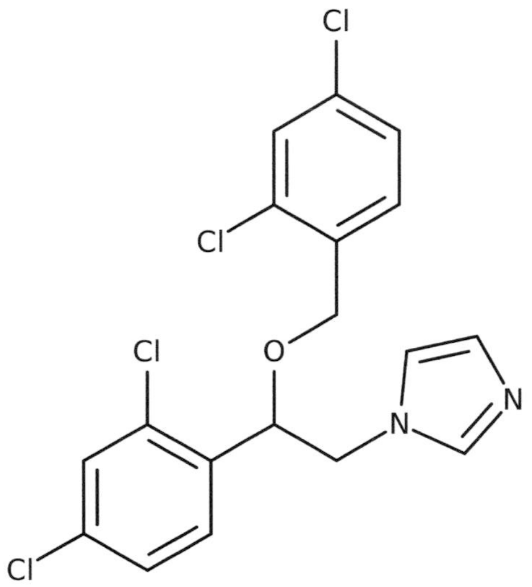

GENERAL

Miconazole is a synthetic phenethyl imidazole antifungal agent that can be used topically and by intravenous infusion. This agent selectively affects the integrity of fungal cell membranes, high in ergosterol content and different in composition from mammalian cell membranes. Miconazole is indicated for topical application in the treatment of tinea pedis, tinea cruris, and tinea corporis caused by *Trichophyton rubrum*, *Trichophyton mentagrophytes*, and *Epidermophyton floccosum*, in the treatment of cutaneous candidiasis and in the treatment of pityriasis (tinea) versicolor. In pharmaceutical products, miconazole is nearly always employed as miconazole nitrate (CAS number 22832-87-7, EC number 245-256-6, molecular formula $C_{18}H_{15}Cl_4N_3O_4$); buccal tablets may contain miconazole base (1).

CONTACT ALLERGY

Patch testing in groups of selected patients

In the period 1990 to 2014, in Leuven, Belgium, 1011 patients suspected of iatrogenic contact dermatitis were tested with miconazole 1% alc. and there were 21 (2.1%) positive reactions. Relevance was not specified for individual drugs, but 96% of the positive patch test reactions to all topical drugs and antiseptics were considered to be relevant (4).

Between 1990 and 2003, in Leuven, Belgium, 92 patients were patch tested for chronic vulval complaints. Fifteen of these women were tested with a series of topical drugs; there was one reaction to miconazole 2% pet. which was relevant (3). In Serbia, in the period 2000-2002, 75 patients with chronic venous leg ulcers were patch tested with miconazole 5% pet. and there were 2 (3%) positive reactions. Relevance was not addressed (2).

Case series

In Bologna, Italy, in 'the few years' before 1995, 20 cases of allergic contact dermatitis due to imidazole preparations were found. However, patch tests with the relevant imidazoles were positive in only 12, including 2 to miconazole (6). In Japan, in the period 1984 to 1994, 3049 outpatients were patch tested for suspected contact dermatitis and 218 of these with topical antifungal preparations. Thirty-five were allergic to imidazoles, including miconazole in 3 individuals. In 60% of the cases, there were cross-reactions between imidazoles (5).

In the period 1976-1988, 13 cases of allergic contact dermatitis to imidazole antifungal preparations were seen in a private dermatological practice in Saverne, France, and 2 in the Université Catholique de Louvain, Belgium. Most had been prescribed for tinea cruris, pedis or corporis, some for anal pruritus, dermatitis and even a leg ulcer. The responsible antifungals (tested 2% pet.) were miconazole in 6 cases, isoconazole in 5 cases, econazole in 2 cases, and tioconazole in one case; one patient reacted to a base ingredient (9).

Between 1977 and 1986, 9 patients with contact allergy to imidazole antimycotics were investigated in the University clinic of Heidelberg, Germany. Six patients were allergic to miconazole, 3 to clotrimazole, 3 to econazole, 3 to isoconazole, and one to oxiconazole. The active ingredient at 1% in ethanol seemed to be the most suitable choice for routine patch testing, petrolatum is less effective (12).

In a literature review up to 1994, the authors identified 105 reported patients who had a positive patch test reaction to at least one imidazole derivative. Miconazole headed the listing with 51 reactions (10).

Case reports

A 70-year old woman had started using an antidandruff shampoo containing miconazole, after which she developed scalp symptoms suggestive of seborrheic dermatitis. Later, after having been treated for a week with ketoconazole lotion on the scalp, the dermatitis worsened greatly, with exudates and crust appearing on the face, scalp and neck. Patch testing with the shampoo and all of its ingredients and with ketoconazole lotion showed positive reactions to the shampoo 0.01% water, miconazole nitrate in pet. in the same concentration as in the shampoo (not specified) and 0.01% pet. and to ketoconazole lotion. Ketoconazole itself was not tested. The authors considered it likely that that the patient had been sensitized first to miconazole in the shampoo and had then developed cross-reactive allergy to ketoconazole (7).

A 40-year-old woman presented after her third episode of acute facial edema within 4 months. Each episode occurred 1–2 days after visiting her mother's house. Patch testing showed positive reactions to miconazole and econazole nitrate 1% alc. The patient denied use of any antifungal creams, but it transpired that her mother's dog was being shampooed weekly with an antifungal shampoo containing 2% miconazole. On her arrival, the patient would, after kissing her mother, greet the dog. This involved petting the dog's coat and rubbing her face against the dog's head. This led to both direct, and indirect (via her hands), contact with the dog's hair, thereby transferring miconazole to her face. No further episodes were recorded after discontinuing the antifungal shampoo (18).

Short summaries of other case reports of allergic contact dermatitis from miconazole are shown in table 3.226.1.

Table 3.226.1 Short summaries of case reports of allergic contact dermatitis from miconazole

Year and country	Sex	Age	Positive patch tests	Clinical data and comments	Ref.
1991 Belgium	F	55	cream and vaginal cream 'as is'; miconazole 2% pet. and MEK	acute vesicular eczematous eruption on the inner thighs, vulva, perineum, pelvic region from cream and vaginal cream; also cross-reactions to iso-, tio- and oxiconazole	20
1988 Netherlands	M	56	cream 'as is'; miconazole 2% MEK; in pet.: negative	acute vesicular dermatitis with superinfection on the dorsa of the feet with papulovesicular dermatitis on arms and legs	21
1982 Denmark	M	13	2 creams 'as is'; negative to miconazole 2% pet., but in MEK not tested	miconazole cream on fungus right arm; widespread tricho-phytide reaction, eruption on the face, primary lesion became eczematous	22
1981 Belgium	F	37	miconazole 2% pet.	eczematous dermatitis of the toe webs and back of the right foot, spreading to the legs, the neck and periorbital	23
	F	42	miconazole 2% pet.	cream on scaly patch right thigh resulting in red, oozing dermatitis over the thighs, abdominal wall, chest and arms	
1977 France	F	23	gel 'as is'; miconazole nitrate 1% pet.	acute disseminated dermatitis after treatment of mycosis of the left hand with miconazole gel for 2 weeks	24
1975 Belgium	F	20	cream 'as is'; micona-zole 2% pet.	spreading vesicular dermatitis after miconazole cream on tinea corporis	25
1974 The Netherlands	F	40	cream and powder 'as is'; miconazole 2% water, 2% pet. and pure	dermatitis of the feet after 6 weeks' treatment with miconazole cream for tinea pedis	17

Other case reports of contact allergy to/allergic contact dermatitis from miconazole, adequate data of which are not available to the author, can be found in refs. 14,15, 16 and 26.

Cross-reactions, pseudo-cross-reactions and co-reactions

Statistically significant associations have been found in patient data between miconazole, econazole, and isoconazole; between sulconazole, miconazole, and econazole (10,13). Possible cross-reactivity from miconazole to ketoconazole (8). A patient sensitized to sertaconazole, who had never used antifungal preparations before, cross-reacted to miconazole and econazole (11).The question has been raised (but not answered) whether contact allergy to miconazole and other (nitro)imidazoles may be overrepresented in patients allergic to methylchloroisothiazolinone/methylisothiazolinone (MCI/MI (8).

Cutaneous adverse drug reactions from systemic administration caused by type IV (delayed-type) hypersensitivity
A 35-year-old man with oral candidiasis was treated with miconazole oral solution. 7 days later, he presented with a generalized, itchy, maculopapular eruption. Patch tests were positive to the solution (as is) and miconazole 2% pet. (19). It was not mentioned whether the patient had used topical imidazoles before.

LITERATURE

1   The data in the section 'General' may have been obtained from literature discussed in this chapter, but mostly also or exclusively from one or more of the following online sources: ChemIDPlus Advanced, PubChem, DrugBank, RxList, Drug Central, Drugs.com, and Wikipedia
2   Jankićević J, Vesić S, Vukićević J, Gajić M, Adamic M, Pavlović MD. Contact sensitivity in patients with venous leg ulcers in Serbia: comparison with contact dermatitis patients and relationship to ulcer duration. Contact Dermatitis 2008;58:32-36
3   Nardelli A, Degreef H, Goossens A. Contact allergic reactions of the vulva: a 14-year review. Dermatitis 2004;15:131-136
4   Gilissen L, Goossens A. Frequency and trends of contact allergy to and iatrogenic contact dermatitis caused by topical drugs over a 25-year period. Contact Dermatitis 2016;75:290-302
5   Yoneyama E. [Allergic contact dermatitis due to topical imidazole antimycotics. The sensitizing ability of active ingredients and cross-sensitivity]. Nippon Ika Daigaku Zasshi 1996;63:356-364 (article in Japanese)
6   Guidetti MS, Vincenzi C, Guerra L, Tosti A. Contact dermatitis due to imidazole antimycotics. Contact Dermatitis 1995;33:282
7   Imafuku S, Nakayama J. Contact allergy to ketoconazole cross-sensitive to miconazole. Clin Exp Dermatol 2009;34:411-412
8   Stingeni L, Rigano L, Lionetti N, Bianchi L, Tramontana M, Foti C, et al. Sensitivity to imidazoles/nitroimidazoles in subjects sensitized to methylchloroisothiazolinone/methylisothiazolinone: A simple coincidence? Contact Dermatitis 2019;80:181-183
9   Jelen G, Tennstedt D. Contact dermatitis from topical imidazole antifungals: 15 new cases. Contact Dermatitis 1989;21:6-11
10 Dooms-Goossens A, Matura M, Drieghe J, Degreef H. Contact allergy to imidazoles used as antimycotic agents. Contact Dermatitis 1995;33:73-77
11 Goday JJ, Yanguas I, Aguirre A, Ilardia R, Soloeta R. Allergic contact dermatitis from sertaconazole with cross-sensitivity to miconazole and econazole. Contact Dermatitis 1995;32:370-371
12 Raulin C, Frosch PJ. Contact allergy to imidazole antimycotics. Contact Dermatitis 1988;18:76-80
13 Bianchi L, Hansel K, Antonelli E, Bellini V, Stingeni L. Contact allergy to isoconazole nitrate with unusual spreading over extensive regions. Contact Dermatitis 2017;76:243-245
14 Nishioka K, Hisamoto K, Ogasawara M, Asagami C. Contact dermatitis due to miconazole nitrate: a case with cross sensitivity to multiple imidazole derivatives. Skin Res 1987;29(Suppl.):227-230 (Article in Japanese)
15 Böttger EM, Mücke C, Tronnier H. Kontaktdermatitis auf neuere Antimykotika und Kontakturticaria. Aktuelle Dermatologie 1981;7:70-74 (Article in German)
16 Jelen G. Allergie gegen Imiazolhaltige Antimykotika: Kreuzallerg1e? Dermatosen im Beruf und Umwelt 1982;30:53-55 (Article in German)
17 Van Ketel WG. Allergy to miconazole nitrate (Daktarin gel). Contact Dermatitis Newsletter 1974;8:517
18 Rademaker M, Barker S. Contact dermatitis to a canine anti-dandruff shampoo. Australas J Dermatol 2007;48:62-63
19 Fernandez L, Maquiera E, Rodriguez F, Picans I, Duque S. Systemic contact dermatitis from miconazole. Contact Dermatitis 1996;34:217
20 Baes H. Contact sensitivity to miconazole with ortho-chloro cross-sensitivity to other imidazoles. Contact Dermatitis 1991;24:89-93
21 Perret CM, Happle R. Contact allergy to miconazole. Contact Dermatitis 1988;19:75
22 Foged EK, Hammershøy O. Contact dermatitis due to miconazole nitrate. Contact Dermatitis 1982;8:284
23 Van Hecke E, van Brabandt S. Contact sensitivity to imidazole derivatives. Contact Dermatitis 1981;7:348-349
24 Samsoen M, Jelen G. Allergy to Daktarin gel. Contact Dermatitis 1977;3:351-352
25 Degreef H, Verhoeve L. Contact dermatitis to miconazole nitrate. Contact Dermatitis 1975;1:269-270
26 Mücke C. Ein Fall von Überempfindlichkeit gegen Econazol und Tolciclat. Dermatosen in Beruf und Umwelt 1980;28:118 (Article in German)

Chapter 3.227 MINOXIDIL

IDENTIFICATION

Description/definition   : Minoxidil is the dialkylarylamine that conforms to the structural formula shown below
Pharmacological classes   : Antihypertensive agents; vasodilator agents
IUPAC name   : 3-Hydroxy-2-imino-6-piperidin-1-ylpyrimidin-4-amine
Other names   : 2,4-Pyrimidinediamine, 6-(1-piperidinyl)-, 3-oxide; 2,4-diamino-6-piperidinopyrimidine 3-
                            N-oxide
CAS registry number   : 38304-91-5
EC number   : 253-874-2
Merck Index monograph   : 7555
Patch testing   : 5% alcohol; petrolatum is not a suitable vehicle for patch testing minoxidil (6,11,17,35,38)
Molecular formula   : $C_9H_{15}N_5O$

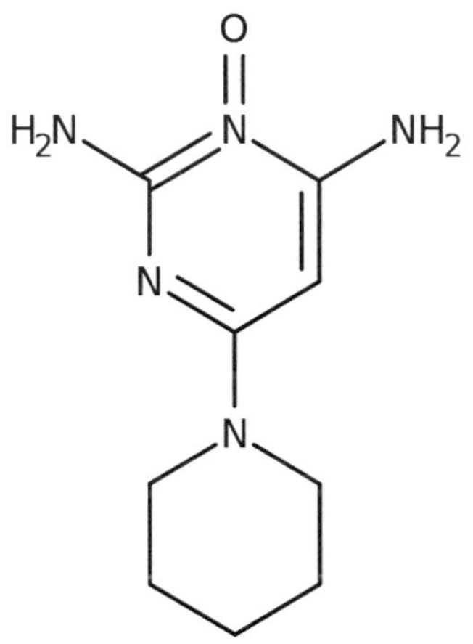

GENERAL

Minoxidil is an orally administered vasodilator with hair growth stimulatory and antihypertensive effects. Minoxidil is converted into its active metabolite minoxidil sulfate by sulfotransferase enzymes. This agent's hair growth stimulatory effect may be mediated through its vasodilatory activity, thereby increasing cutaneous blood flow, or due to its direct stimulatory effect on hair follicle cells and forcing them from their resting phase into their active growth phase. Minoxidil is indicated for the treatment of severe hypertension and in the topical treatment (regrowth) of androgenic alopecia in males and females and stabilization of hair loss in patients with male or female pattern hair loss. In several countries including the USA, topical formulations with 2% or 5% minoxidil are available as OTC pharmaceuticals. Excipients are nearly always water, alcohol and propylene glycol (1).

CONTACT ALLERGY

General

Contact allergy to and allergic contact dermatitis from topical minoxidil lotion used for androgenic alopecia or alopecia areata have been reported frequently, sometimes in the form of single case reports (4,5,6,8,9,11,14,16, 17,34,36), but also in case series of 2 (3,35,37), 4 (13,33), 6 (38), 7 (7), 8 (29), 22 (2) and 25 (30) patients. In a number of sensitized individuals, the allergic contact dermatitis was not due to the active ingredient itself, but was caused by contact allergy to the excipient propylene glycol (3,12,13,15,26,29,30,31,35,36,39). In one study (13), the majority of allergic reactions were stated to be due to propylene glycol, but the test concentration (unspecified, but certainly 50% was tested), was too high, risking irritant, false-positive reactions. In another investigation, deemed to be unreliable by some authors (30), all 13 cases of contact allergy to minoxidil lotion were ascribed to propylene glycol (31). Conversely, in an Italian study, of 25 patients reacting to minoxidil lotion, propylene glycol co-reacted in one patient only. However, propylene glycol was tested at 2% in petrolatum, which must very likely have resulted in false-negative reactions (30). Combined allergy to minoxidil and propylene glycol has also been observed (3,13,29,39).

Special forms of contact dermatitis have included pustular allergic contact dermatitis (8,16), psoriasiform dermatitis (3), photocontact dermatitis (30,37) and persisting pseudolymphoma-like patch test reactions (4,5). A case of pigmented contact dermatitis was reported from Israel, but proof for it was definitely lacking (10). In one instance, allergic contact dermatitis was followed by acute telogen effluvium (3). Some patients with persistent and severe itching but without signs of inflammation such as erythema, vesicles or scaling proved to be allergic to minoxidil when patch tested (7). Videodermoscopy is not helpful in distinguishing scalp contact dermatitis due to topical minoxidil from other conditions that cause severe scalp itching (7).

The data presented in this chapter suggest that contact allergy to/allergic contact dermatitis from minoxidil is relatively frequent. The fact that in most clinical trials only few cases of sensitization were reported (22-28,32), may partly be related to the observation that most patients who become sensitized had been using the drug for a long period of time, in one large series an average of 10 months (30). However, shorter treatment periods before sensitization develops are by no means rare (35). On the other hand, in a trial study from the United Kingdom, 6 of 161 subjects (3.7%) developed allergic contact dermatitis (38). It may well be that in other studies, patients with mild dermatitis were interpreted as having irritant contact dermatitis and were not patch tested and thus, allergic contact dermatitis remained unrecognized.

The use of minoxidil in dermatology and its side effects were reviewed in 2012 (40).

Case series

In Leuven, Belgium, in the period 1990-2014, iatrogenic contact dermatitis was diagnosed in 2600 individuals (17% of the total patch test population). 96% of all positive patch test reactions to topical drugs and antiseptics were considered to be relevant. Minoxidil (2% alcohol) was tested in 62 patients and there were 22 positive reactions to it (2).

In a university hospital in New York, 11 patients suspected of allergic contact dermatitis from minoxidil lotion were patch tested. Four were allergic to minoxidil. It was not stated how it was tested, possibly 1% in isopropyl alcohol. Nine were allergic to propylene glycol, of who two co-reacted to minoxidil. However, is was not stated whether the minoxidil lotion itself was tested and what the test concentration for propylene glycol was. It can be deducted that, at least in some patients, a 50% concentration was used, which entails a high risk of false-positive, irritant reactions (13).

In a university clinic in Coimbra, Portugal, in 1990 and 1991, 7 women and one man, with ages ranging from 24 to 41 years (mean 33 years), who had been using minoxidil solution for the treatment of androgenetic alopecia for periods of 3 weeks to 1 year, were seen with pruriginous erythema, papules and scaling of the scalp and forehead. In 2 patients, there were also vesicular lesions extending to the auricular area, retro-auricular folds and neck. They all had positive patch tests to at least 3 minoxidil test preparations and – except one, who was also allergic to propylene glycol – negative reactions to the placebo solution (29).

Between 1985 and 1990, in Bologna, Italy, 67 patients who complained of burning, itching, or the onset of papules and vesicles after using a topical minoxidil solution for alopecia, were patch tested with their personal minoxidil solution, a freshly prepared sample (2% minoxidil, 70% alcohol, 10% propylene glycol, 18% water) and propylene glycol 2% in petrolatum. In 18 patients who reported the onset of their symptoms after exposure to sunlight, photopatch tests were also performed using UVA. Twenty-five patients had positive patch test reactions to the lotion, of whom only one reacted to propylene glycol. Two individuals were negative to the patch tests but positive to photopatch tests to the solutions. It should be realized that 2% for patch testing propylene glycol is too low and will certainly have missed some cases of propylene glycol sensitization (30).

Four patients sensitized to minoxidil were reported from Spain. Three had an eczematous eruption, the 4th complained of pruritus only. They all reacted to minoxidil 1% and/or 2% in alcohol (33). In a clinical study in the Netherlands, 3 patients of 95 treated with 2% minoxidil lotion became sensitized with itchy dermatitis on the scalp and ears and periocular edema in one subject. One was allergic to propylene glycol. The other two reacted to patch tests with the 2% lotion (negative to 0.5%), but negative to the base without minoxidil and negative to propylene glycol 10% and 20% pet. (35). Minoxidil 2% in pet. was also negative: petrolatum would later be confirmed as unsuitable for patch testing minoxidil (6,11,17,38).

Of 161 patients participating in a clinical trial of treating common baldness with 2% minoxidil lotion for at least 2 years, 6 (3.7%) developed allergic contact dermatitis. They reacted to the lotion, but not to the base and not to propylene glycol. One was tested with and reacted to 2% minoxidil in propylene glycol and in methyl ethyl ketone. Petrolatum proved to be unsuitable for patch testing, resulting in false-negative reactions (38). Six patients with contact allergy to minoxidil were reported from Spain. They all had positive patch tests to the 2% solution, some of who also reacted to 1% and 0.5%. One patient was also allergic to propylene glycol (39).

Case reports

A 46-year-old woman with severe female pattern hair loss had been under treatment with 2% topical minoxidil for several years, when symptoms deteriorated and a scaly eruption developed. Examination showed diffuse alopecia with large areas of scaly psoriatic plaques involving almost the entire scalp. The pull test was strongly positive. Patch tests were positive to a 2% minoxidil lotion with propylene glycol (PG), 5% solution without PG but negative to propylene glycol 20% water. Minoxidil treatment was subsequently discontinued, and topical therapy with clobetasol ointment under occlusion was started. The psoriasiform dermatitis and telogen effluvium improved within a few weeks (3). In a second patient presented by the authors, a similar picture with psoriasiform ACD and telogen effluvium was caused by contact allergy to propylene glycol in the minoxidil solution (3).

A 59-year-old woman presented with a 15-day history of eczematous lesions affecting her face and scalp. For the previous 3 months she had been applying minoxidil 5% solution. Patch tests were positive to the lotion and negative to the other ingredients (alcohol and propylene glycol). A ROAT produced an eczematous response at D3. Two months later, the positive reaction to the lotion persisted, showing an atypical clinical presentation. A skin biopsy was therefore performed. Histopathological examination revealed a polymorphous infiltrate involving the superficial and deep dermis and extending close to the subcutaneous fat tissue, predominantly consisting of T cells (lymphomatoid hyperplasia), with some degree of interface dermatitis (4).

Short summaries of other case reports of allergic contact dermatitis from minoxidil are shown in table 3.227.1.

Table 3.227.1 Short summaries of case reports of allergic contact dermatitis from minoxidil

Year and country	Sex	Age	Positive patch tests	Clinical data and comments	Ref.
2010 Italy	F	60	5% lotion 'as is'; MIN 5% alcohol	itchy eczematous exudative lesions scalp; ROAT pos. at day 3	6
2007 Spain	M	22	MIN 5% alcohol and propylene glycol	multiple millimetric papulovesicles and papulopustules with erythema and edema frontal scalp and forehead; pustular ACD	8
2005 Germany	F	72	MIN 2% and 5% lotion; MIN 5% alc. and pet.	pruritus, scaling, weeping dermatitis; positive lymphocyte transformation test to the lotion and minoxidil	9
2002 Japan	M	54	MIN 1% lotion; MIN 1% alc. (neg. in pet.)	itchy erythematous plaques on the scalp and in the neck	11
2002 Australia	F	24	MIN lotion 'as is; MIN 1%, 2% and 4% (vehicle?)	erythema, scaling, erosions on the scalp	14
1998 Spain	F	26	2% MIN lotion; MIN 2% alcohol	vesicular and pustular lesions over an erythematous and edematous area; pustular contact dermatitis and patch test	16
1995 Austria	M	27	MIN 2% lotion; MIN 2% water and alcohol	increasing redness, swelling and scaling on the scalp and ears	17
1994 Spain	F	44	MIN 1% and 2% water	acute eczema of the scalp after 1 month' treatment with 2% MIN hydroalcoholic solution	42
1992 Italy	M	45	MIN 2% lotion 'as is'; MIN 2% water	erythematous, edematous desquamative dermatitis on the fingers from applying MIN lotion to customer; hairdresser, occupational allergic contact dermatitis	18
1992 USA	M	34	MIN 2% lotion; MIN 2% propylene glycol	irritation, redness and scaling on the scalp and forehead; patch test 2% minoxidil in petrolatum and alcohol negative	41
1987 Italy	M	22	MIN 2% solution	burning, pruritus, diffuse red papules and vesicles; the solution contained only MIN, alcohol and water	34
1985 Belgium	F	25	MIN lotion; ?+ to MIN 1% alcohol	burning, itching, diffuse red papules on the scalp, neck and eyebrows (alopecia totalis); ROAT ++ after 2 days; also allergy to propylene glycol	36
1985 Italy	M	42	MIN 1% lotion	erythema and edema on the scalp and face	37
	M	38	MIN 1% lotion	itchy vesicular rash behind the ears	

MIN: Minoxidil

One (19), one (20) and 4 cases (21) of allergic contact dermatitis from minoxidil have been reported in Italian journals (cited in ref. 18). In several clinical trials of treatment of alopecia androgenetica or alopecia areata with topical minoxidil, one or 2 cases of sensitization have briefly been mentioned (22-28,32).

Cross-reactions, pseudo-cross-reactions and co-reactions
Not to piperidine, pyrimidine or diaminopyrimidine (11).

PHOTOSENSITIVITY

Case reports
A 22-year old man and another male patient, aged 38, had developed signs suggestive of contact dermatitis on the scalp, beginning after exposure to sunlight, whilst using minoxidil lotion for androgenetic alopecia. Patch tests to minoxidil lotion (minoxidil 2%, alcohol 70%, propylene glycol 10%, water 20%) were negative, but photopatch tests using irradiation with 4J/cm^2 UVA were strongly positive. Minoxidil itself was not tested (30). A 34-year-old man developed a severe eczematous eruption of the scalp, forehead and periorbital area which appeared after exposure

to sunlight. He had been using 1% minoxidil lotion for 3 months. Patch tests to the lotion and propylene glycol were negative, but photopatch tests, using irradiation with 4J/cm^2 UVA were positive to the minoxidil lotion (37).

LITERATURE

1   The data in the section 'General' may have been obtained from literature discussed in this chapter, but mostly also or exclusively from one or more of the following online sources: ChemIDPlus Advanced, PubChem, DrugBank, RxList, Drug Central, Drugs.com, and Wikipedia

2   Gilissen L, Goossens A. Frequency and trends of contact allergy to and iatrogenic contact dermatitis caused by topical drugs over a 25-year period. Contact Dermatitis 2016;75:290-302

3   La Placa M, Balestri R, Bardazzi F, Vincenzi C. Scalp psoriasiform contact dermatitis with acute telogen effluvium due to topical minoxidil treatment. Skin Appendage Disord 2016;1:141-143

4   García-Rodiño S, Espasandín-Arias M, Suárez-Peñaranda JM, Rodríguez-Granados MT, Vázquez-Veiga H, Fernández-Redondo V. Persisting allergic patch test reaction to minoxidil manifested as cutaneous lymphoid hyperplasia. Contact Dermatitis 2015;72:413-416

5   Schmutz JL. Persistent reaction to minoxidil skin patch testing mimicking pseudolymphoma. Ann Dermatol Venereol 2015;142:724-725 (Article in French)

6   Corazza M, Borghi A, Ricci M, Sarno O, Virgili A. Patch testing in allergic contact dermatitis from minoxidil. Dermatitis 2010;21:217-218

7   Tosti A, Donati A, Vincenzi C, Fabbrocini G. Videodermoscopy does not enhance diagnosis of scalp contact dermatitis due to topical minoxidil. Int J Trichology 2009;1:134-137

8   Rodríguez-Martín M, Sáez-Rodríguez M, Carnerero-Rodríguez A, Cabrera de Paz R, Sidro-Sarto M, Pérez-Robayna N, et al. Pustular allergic contact dermatitis from topical minoxidil 5%. J Eur Acad Dermatol Venereol 2007;21:701-702

9   Hagemann T, Schlütter-Böhmer B, Allam JP, Bieber T, Novak N. Positive lymphocyte transformation test in a patient with allergic contact dermatitis of the scalp after short-term use of topical minoxidil solution. Contact Dermatitis 2005;53:53-55

10 Trattner A, David M. Pigmented contact dermatitis from topical minoxidil 5%. Contact Dermatitis 2002;46:246

11 Suzuki K, Suzuki M, Akamatsu H, Matsungaga K. Allergic contact dermatitis from minoxidil: study of the cross-reaction to minoxidil. Am J Contact Dermat 2002;13:45-46

12 Fisher AA. Use of glycerin in topical minoxidil solutions for patients allergic to propylene glycol. Cutis 1990;45:81-82

13 Friedman ES, Friedman PM, Cohen DE, Washenik K. Allergic contact dermatitis to topical minoxidil solution: etiology and treatment. J Am Acad Dermatol 2002;46:309-312

14 Sinclair RD, Mallari RS, Tate B. Sensitization to saw palmetto and minoxidil in separate topical extemporaneous treatments for androgenetic alopecia. Australas J Dermatol 2002;43:311-312

15 Scheman AJ, West DP, Hordinsky MK, Osburn AH, West LE. Alternative formulation for patients with contact reactions to topical 2% and 5% minoxidil vehicle ingredients. Contact Dermatitis 2000;42:241

16 Sánchez-Motilla JM, Pont V, Nagore E, Rodríguez-Serna M, Sánchez JL, Aliaga A. Pustular allergic contact dermatitis from minoxidil. Contact Dermatitis 1998;38:283-284

17 Ebner H, Müller E. Allergic contact dermatitis from minoxidil. Contact Dermatitis 1995;32:316-317

18 Veraldi S, Benelli C, Pigatto PD. Occupational allergic contact dermatitis from minoxidil. Contact Dermatitis 1992 ;26 :211-212

19 De Padova MP, Lama L, Montagnani A. Correlazione tra dose/effetto nella terapia topica con minoxidil nell'alopecia androgenica. Giorn It Derm Vener 1988;123:59-61 (Article in Italian, data cited in ref. 18)

20 Riboldi A, Pigatto PD, Pucci M, Morelli M. Dermatite da contatto al minoxidil 2% lozione: descrizione di 5 casi. Boll Dermatol Allergol Prof 1989;4:119-122 (Article in Italian, data cited in ref. 18)

21 Foti C, Filotico R, Vena GA, Grandolfo M, Angelini G. Dermatite da contatto con minoxidil. Boll Dermatol Allergol Prof 1991;6:33-36 (Article in Italian, data cited in ref. 18)

22 Weiss VC, West DP, Fu TS, Robinson LA, Cook B, Cohen RL, et al. Alopecia areata treated with topical minoxidil. Arch Dermatol 1984;120:457-463

23 Fiedler-Weiss VC. Topical minoxidil solution (1% and 5%) in the treatment of alopecia areata. J Am Acad Dermatol 1987;16(3Pt.2):745-748

24 Weiss VC, West DP. Topical minoxidil therapy and hair regrowth. Arch Dermatol 1985;121:191-192

25 Fiedler-Weiss VC, West DP, Buys CM, Rumsfield JA. Topical minoxidil dose-response effect in alopecia areata. Arch Dermatol 1986;122:180-182

26 Rietschel RL, Duncan SH. Safety and efficacy of topical minoxidil in the management of androgenetic alopecia. J Am Acad Dermatol 1987;16(3Pt.2):677-685

27 Price VH. Topical minoxidil in extensive alopecia areata, including 3-year follow-up. Dermatologica 1987;175(Suppl.2):36-41 (cited in ref. 18)

28 Roenigk HH, Pepper E, Kuruvilla S. Topical minoxidil therapy for hereditary male pattern alopecia. Cutis 1987;39:337-342 (cited in ref. 18)

29 Ruas E, Gonçalo M, Figueiredo A, Gonçalo S. Allergic contact dermatitis from minoxidil. Contact Dermatitis 1992;26:57-58

30 Tosti A, Guerra L, Bardazzi F. Contact dermatitis caused by topical minoxidil: Case reports and review of the literature. Am J Cont Dermat 1991;2:56-59

31 Castillo, Moshell AN, Jorgensen H, et al. Contact sensitization to topically applied minoxidil solution: possible mechanisms of action. Clin Res 1985;33:629A (data cited in ref. 30)

32 Pestana A, Olsen EA, Delong ER, Murray JC. Effect of ultraviolet light on topical minoxidil-induced hair growth in advanced male pattern baldness. J Am Acad Dermatol 1987;16:971-976

33 Alomar A, Smandia JA. Allergic contact dermatitis from minoxidil. Contact Dermatitis 1988;18:51-52

34 Valsecchi R, Cainelli T. Allergic contact dermatitis from minoxidil. Contact Dermatitis 1987;17:58-59

35 Van der Willigen AH, Dutrée-Meulenberg RO, Stolz E, Geursen-Reitsma AM, van Joost T. Topical minoxidil sensitization in androgenic alopecia. Contact Dermatitis 1987;17:44-45

36 Degreef H, Hendrickx I, Dooms-Goossens A. Allergic contact dermatitis to minoxidil. Contact Dermatitis 1985;13:194-195

37 Tosti A, Bardazzi F, De Padova MP, Caponeri GM, Melino M, Veronesi S. Contact dermatitis to minoxidil. Contact Dermatitis 1985;13:275-276

38 Wilson C, Walkden V, Powell S, Shaw S, Wilkinson J, Dawber R. Contact dermatitis in reaction to 2% minoxidil solution. J Am Acad Dermatol 1991;24:661-662

39 Camarasa JG, Serra-Baldrich E, Garcia-Bravo B, Vozmediano JF. Contact dermatitis to minoxidil. In: Frosch PJ, Dooms-Goossens A, Lachapelle J-M, Rycroft RJG, Scheper RJ (eds): Current topics in contact dermatitis. Berlin: Springer-Verlag, 1989: 261-263

40 Rossi A, Cantisani C, Melis L, Iorio A, Scali E, Calvieri S. Minoxidil use in dermatology, side effects and recent patents. Recent Pat Inflamm Allergy Drug Discov 2012;6:130-136

41 Whitmore SE. The importance of proper vehicle selection in the detection of minoxidil sensitivity. Arch Dermatol 1992;128:653-656

42 Aguirre A, Manzano D, Zabala R, Eizaguirre X, Díaz-Pérez JL. Allergic contact dermatitis from spironolactone. Contact Dermatitis 1994;30:312

Chapter 3.228 MISCELLANEOUS TOPICAL DRUG ALLERGENS

Drugs that have caused contact allergy/allergic contact dermatitis from topical application, but are not discussed in a separate monograph (chapter), are shown in table 3.228.1 Most are 'historical' allergens, that are currently not in use anymore. Various sulfonamides such as sulfanilamide (Chapter 3.323), sulfapyridine, sulfathiazole, and sulfadiazine were used on a large scale during and after the second world war on wounds and to treat cutaneous infections, but they caused many cases of sensitization and the topical use of most was gradually abandoned in many countries (23,24). Nevertheless, in Bari, Italy, in the period 1968-1977, 8.2% of 3758 consecutive patients still reacted to 'sulfonamide' 5% pet., although the frequency steeply dropped to 0.2% in the following years 1978-1983 (21). Some sulfonamides (notably sulfanilamide) also caused photosensitization; most patients recovered but some developed persistent light reactions (22). Sulfonamide is stated to have cause erythema multiforme-like allergic contact dermatitis (25).

Table 3.228.1 Miscellaneous topical drug allergens

Name	IUPAC name	CAS number	Pharmaceutical class	PTC/veh.	References
Butacaine	3-dibutylaminopropyl p-amino-benzoate	149-16-6	local anesthetic	5% pet.	1
Buclosamide [b] (Jadit)	N-butyl-4-chloro-2-hydroxy-benzamide	575-74-6	antifungal agent	5% pet.	13-18
Butethamine	isobutylaminoethyl p-amino-benzoate	2090-89-3	local anesthetic	5% pet.	2 (page 74)
Castellani's solution	see Chapter 3.298 Resorcinol		antifungal agent		
Chlorcyclizine	1-[(4-chlorophenyl)-phenyl-methyl]-4-methylpiperazine	82-93-9	histamine H1 antagonist	1% pet.	5
Clofenoxyde	2-chloro-1-[4-[4-(2-chloroace-tyl)phenoxy]phenyl]ethanone	3030-53-3	antifungal agent	1% ethyl acetate	2 (page 87)
Diacetazotol (Pellidol)	N-acetyl-N-[2-methyl-4-[(2-methylphenyl)diazenyl]phenyl]acetamide	83-63-6	dermatological agent; stimulant of epithelialization of wounds	2% pet.	2 (page 102), 19
Dimethisoquin hydrochloride	(3-butyl-1-(2-(dimethylamino)-ethoxy) isoquinoline HCl (Quotane)	2773-92-4	local anesthetic	2% pet.	3,4
Ethyl lactate	Ethyl 2-hydroxypropanoate	97-64-3	acne treatment	1% pet.	11
Hexetidine	1,3-bis(2-ethylhexyl)-5-methyl-1,3-diazinan-5-amine	141-94-6	bactericidal and fungicidal antiseptic	0.1% pet.	9
Methapyrilene	N,N-dimethyl-N'-pyridin-2-yl-N'-(thiophen-2-ylmethyl)-ethane-1,2-diamine	91-80-5	histamine H1 antagonist	2% pet.	6
Mycanodin [a]	combination of 3-(2-oxychloro-phenyl)pyrazole and bamipine		antifungal agent	1-2% ?	10
Orthocaine	methyl 3-amino-p-hydroxy-benzoate (Orthoform, old)	536-25-4	local anesthetic	1% pet.	1
	methyl 3-amino-m-hydroxy-benzoate (Orthoform, new)	17672-21-8	local anesthetic	1% pet.	1
Phenindamine tartrate	2,3-dihydroxybutanedioic acid; 2-methyl-9-phenyl-1,3,4,9-te-trahydroindeno[2,1-c]pyridine	569-59-5	histamine H1 antagonist	5% pet.	7,8
Sulfathiourea	(4-aminophenyl)sulfonyl-thiourea	515-49-1	anti-infective agent	5% pet.	20
Triaziquone	2,3,5-tris(aziridin-1-yl)cyclo-hexa-2,5-diene-1,4-dione	68-76-8	antineoplastic agent (used topically)	0.05% eucerin	12

[a] photoallergic contact dermatitis; terms Mycanodin and 3-(2-oxychlorophenyl)pyrazole unknown; [b] by far most often photoallergic contact dermatitis; HCl: hydrochloride; PTC/veh.: patch test concentration and vehicle

LITERATURE

1 Lane CG, Luikart R. Dermatitis from local anesthetics, with a review of one hundred and seven cases from the literature. J Am Med Assoc 1951;146:717-720
2 De Groot AC, Weyland JW, Nater JP. Unwanted effects of cosmetics and drugs used in dermatology, 3rd edition. Amsterdam: Elsevier Science BV, 1994:

3   Daly JF. Contact dermatitis due to 'Quotane'. Arch Dermatol Syph 1952;66:393-394
4   Grolnick M. Case reports; sensitization to Quotane and to para-aminobenzoic acid derivatives; report of a case
     with cross-sensitization reactions. NY State J Med 1953;53:442-444
5   Ayres S 3rd, Ayres S Jr. Contact dermatitis from chlorcyclizine hydrochloride (Perazil) cream; report of four cases.
     AMA Arch Derm Syphilol 1954;69:502-503
6   Loveman AB, Fliegelman MT. Local cutaneous sensitivity to methapyrilene; report of a case. AMA Arch Derm
     Syphilol 1951;63:250-251
7   Ellis FA, Bundick WR. Reactions to the local use of thephorin. J Invest Dermatol 1949;13:25-28
8   Stritzler C. Studies on topical thephorin therapy; index of sensitization and effectiveness as an antipruritic. J
     Allergy 1950;21:432-441
9   Merk H, Ebert L, Goerz G. Allergic contact dermatitis due to the fungicide hexetidine. Contact Dermatitis
     1982;8:216
10   Burckhardt W, Mahler F, Schwarz-Speck M. Photoallergische Kontaktekzem durch Mycanodin. Dermatologica
     (Basel) 1968;137:208 (cited in ref. 2, page 88) (original article in German)
11   Marot L, Grosshans E. Allergic contact dermatitis to ethyl lactate. Contact Dermatitis 1987;17:45-46
12   Helm F, Klein E. Effects of allergic contact dermatitis on basal cell epitheliomas. Arch Dermatol 1965;91:142-144
13   Burry JN, Hunter GA. Photocontact dermatitis from Jadit. Br J Dermatol 1960;82:224-229
14   Jung EG, Schwartz K. Photo-allergic Jadit eczema. Dermatologica (Basel) 1964;129:401-404
15   Jung EG, Schwarz K. Photoallergy to Jadit with photo cross-reactions to derivative of sulfanilamide. Int Arch
     Allergy Appl Immunol 1965;27:313-317
16   Nurse DS. Allergic contact sensitivity to Jadit. Med J Austr 1973;1(13):651-652
17   Gritiyarangsan P. A three-year photopatch study in Thailand. J Dermatol Sci 1991;2:371-375
18   Burry JN. Persistent light reactions from buclosamide. Arch Dermatol 1970;101:95-97
19   Rudzki E, Kleniewska D. The epidemiology of contact dermatitis in Poland. Br J Dermatol 1970;83:543-545
20   Degreef H, Dooms-Goossens A. Patch testing with silver sulfadiazine cream. Contact Dermatitis 1985;12:33-37
21   Angelini G, Vena GA, Meneghini CL. Allergic contact dermatitis to some medicaments. Contact Dermatitis
     1985;12:263-269
22   Peterkin GAG. Skin eruptions due to the local application of sulphonamides. Br J Dermatol 1945;57:1-9
23   Sulzberger MB, Kanof A, Baer RL, Lowenberg C. Sensitization by topical application of sulfonamides. J Allergy
     1947;18:92-103
24   Kooij R, Van Vloten TJ. Epidermal sensitization due to sulphonamide drugs; groups- and cross-sensitivity;
     photosensitivity; passive transfer of antibodies; leftwich reaction. Dermatologica 1952;104:151-167
25   Meneghini CL, Angelini G. Contact dermatitis from pyrrolnitrin. Contact Dermatitis 1982;8:55-58

Chapter 3.229 MITOMYCIN C

IDENTIFICATION

Description/definition : Mitomycin is a methylazirinopyrroloindoledione antineoplastic antibiotic isolated from the bacterium *Streptomyces caespitosus* and other *Streptomyces* bacterial species; its structural formula is shown below

Pharmacological classes : Antibiotics, antineoplastic; cross-linking reagents; nucleic acid synthesis inhibitors; alkylating agents

IUPAC name : Azirino(2',3':3,4)pyrrolo(1,2-a)indole-4,7-dione, 6-amino-8-(((aminocarbonyl)oxy)methyl)-1,1a,2,8,8a,8b-hexahydro-8a-methoxy-5-methyl-, (1aS-(1aα,8β,8aα,8bα))-

CAS registry number : 50-07-7

EC number : 200-008-6

Merck Index monograph : 7570 (Mitomycins)

Patch testing : 0.1% pet.

Molecular formula : $C_{15}H_{18}N_4O_5$

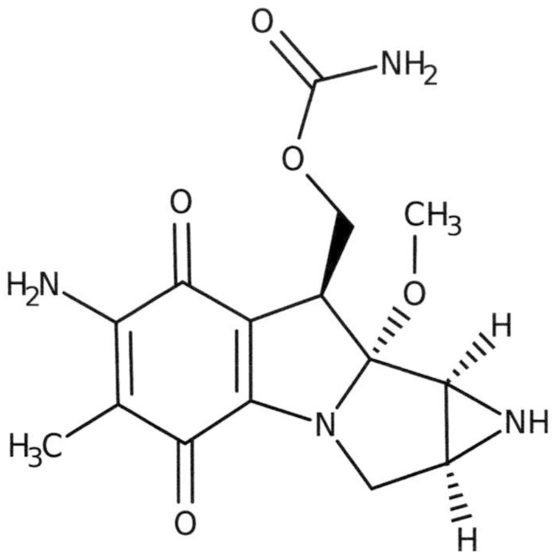

GENERAL

Mitomycin is an antineoplastic antibiotic isolated from the bacterium *Streptomyces caespitosus* and other *Streptomyces* bacterial species. Bioreduced mitomycin C generates oxygen radicals, alkylates DNA, and produces inter-strand DNA cross-links, thereby inhibiting DNA synthesis. Preferentially toxic to hypoxic cells, this agent also inhibits RNA and protein synthesis at high concentrations. Mitomycin C is indicated for treatment of malignant neoplasm of lip, oral cavity, pharynx, digestive organs, peritoneum, female breast, and urinary bladder (1).

CONTACT ALLERGY

Case report

A 63-year-old woman was treated for a squamous cell carcinoma on the left conjunctiva with topical 0.2 mg/ml mitomycin C four times a day. Two days later, she had developed a severe dermatitis involving her eyelids. A patch test was positive to mitomycin C 0.01%. The question how the patient had previously become sensitized to mitomycin C was not addressed (2).

Cutaneous adverse drug reactions from systemic administration caused by type IV (delayed-type) hypersensitivity

Cutaneous adverse drug reactions from systemic administration of mitomycin C caused by type IV (delayed-type) hypersensitivity and occupational allergic contact dermatitis (7) are planned to be discussed in Volume IV of the *Monographs in Contact Allergy* series on Systemic drugs. These mostly concern systemic contact dermatitis reactions from intravesical instillation for bladder cancer, usually presenting as dermatitis of the palms, soles and genitals (e.g. 3,5,6), sometimes as symmetrical drug-related intertriginous and flexural exanthema (SDRIFE) (4,8).

LITERATURE

1 The data in the section 'General' may have been obtained from literature discussed in this chapter, but mostly also or exclusively from one or more of the following online sources: ChemIDPlus Advanced, PubChem, DrugBank, RxList, Drug Central, Drugs.com, and Wikipedia

2 Cumurcu T, Sener S, Cavdar M. Periocular allergic contact dermatitis following topical Mitomycin C eye drop application. Cutan Ocul Toxicol 2011;30:239-240

3 Colomer Gallardo A, Martínez Rodríguez R, Castillo Pacheco C, González Satue C, Ibarz Servio L. Dermatological side effects of intravesical Mitomycin C: Delayed hypersensitivity. Arch Esp Urol 2016;69:89-91

4 Tan SC, Tan JW. Symmetrical drug-related intertriginous and flexural exanthema. Curr Opin Allergy Clin Immunol 2011;11:313-318

5 Fernandez DG, Fernández AF, Vallejo MS, Osorio JL, Anguita MJ, Jimenez AL. Delayed hypersensitivity to Mitomycin C. Contact Dermatitis 2009;61:237-238

6 Peitsch WK, Klemke CD, Michel MS, Goerdt S, Bayerl C. Hematogenous contact dermatitis after intravesicular instillation of mitomycin C. Hautarzt 2007;58:246-249 (Article in German)

7 Rogers B. Health hazards to personnel handling antineoplastic agents. Occup Med 1987;2:513-515

8 De Groot AC, Conemans JM. Systemic allergic contact dermatitis from intravesical instillation of the antitumor antibiotic mitomycin C. Contact Dermatitis 1991;24:201-209

Chapter 3.230 MOMETASONE FUROATE

IDENTIFICATION

Description/definition : Mometasone furoate is the furoate ester of the synthetic glucocorticoid mometasone that conforms to the structural formula shown below
Pharmacological classes : Anti-allergic agents; anti-inflammatory agents; dermatologic agents
IUPAC name : [(8S,9R,10S,11S,13S,14S,16R,17R)-9-Chloro-17-(2-chloroacetyl)-11-hydroxy-10,13,16-trimethyl-3-oxo-6,7,8,11,12,14,15,16-octahydrocyclopenta[a]phenanthren-17-yl] furan-2-carboxylate
Other names : 9,21-Dichloro-11β,17-dihydroxy-16α-methylpregna-1,4-diene-3,20-dione 17-(2-furoate)
CAS registry number : 83919-23-7
EC number : 617-501-0
Merck Index monograph : 7597
Patch testing : 0.1% alc. (SmartPracticeCanada)' late readings (6-10 days) are strongly recommended
Molecular formula : $C_{27}H_{30}Cl_2O_6$

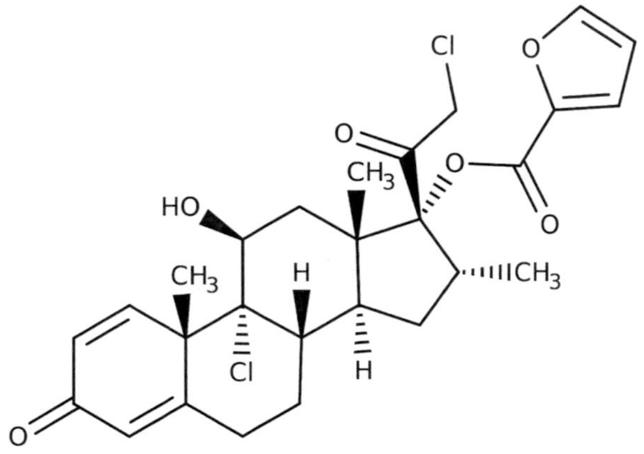

GENERAL

General aspects of corticosteroids used on the skin and mucous membranes are discussed in Chapter 2.4. A practical guideline for diagnosing allergic reactions to corticosteroids is presented in ref. 1.

CONTACT ALLERGY

Patch testing in consecutive patients suspected of contact dermatitis: Routine testing

In Spain, during 2015-2016, 3699 consecutive patients were patch tested with mometasone furoate 0.1% alc. and there were 7 (0.2%) positive reactions (12). Their relevance was not mentioned (12). In a French-Belgian study performed in the mid-1990s, none of 628 routinely tested patients reacted to mometasone furoate (7).

Patch testing in groups of selected patients

In Croatia, before 2007, 60 patients with venous leg ulcers with or without (allergic contact) dermatitis were patch tested with 'mometasone' (no details provided) and there was one (2%) positive reaction. The relevance of this positive patch test was not mentioned (2). In the period 2000 to 2005, in the USA, 1186 patients suspected of corticosteroid allergy were patch tested with mometasone furoate 0.1% pet. and there were 13 (1.1%) positive reactions, of which 12 (92%) were considered to be relevant (11).

Among a French-Belgian group of 38 glucocorticoid-sensitive patients, two reacted to mometasone 0.5% in alcohol. However, both patients also reacted to alcohol alone (7). In the United Kingdom, in the period 1993-1994, mometasone furoate 1% alcohol in a corticosteroid series was tested in 100 patients found to be allergic to a screening corticosteroid in the baseline series or suspected of corticosteroid hypersensitivity. There was not a single reaction to mometasone furoate and it was concluded that this corticosteroid has a low risk of sensitization (4).

Case series

From January 1990 to June 2008, in Leuven, Belgium, 315 patients were diagnosed with contact allergy to/allergic contact dermatitis from corticosteroids (CSs) from routine patch testing with a baseline series including tixocortol pivalate, budesonide, hydrocortisone butyrate and prednisone caproate, patch testing with patients' own CS preparations, and testing those with proven contact allergy to a corticosteroid or strongly suspected of CS allergy later with a series of 66 CSs, including two sex hormones (progesterone and testosterone). 71% of the patients had

relevant reactions, but these were not specified. In this group of 315 CS allergic patients, 15 had positive patch tests to mometasone furoate 0.1% alc. (10). It is unknown how many of these reactions were caused by the use of a pharmaceutical product containing mometasone furoate and how many were cross-reactions to other corticosteroids.

Case reports

A 65-year-old woman presented with an itchy rash of the arms and hand dermatitis, for which she had previously used various topical corticosteroids. Three weeks before consultation, while applying mometasone furoate ointment twice daily, the skin lesions worsened and the rash spread. The dermatitis was first erythematosquamous but later began to ooze and to spread from the hands and wrists to the popliteal fossae. Patch testing showed positive reactions to budesonide (which she had never used before) and mometasone furoate 0.01% at D3. When the patient inadvertently used the mometasone furoate ointment again, the rash was reproduced (13).

A 70-year-old patient presented with contact dermatitis on the right forearm and hands 7 days after application of mometasone furoate 0.1% cream because of granuloma annulare. Patch testing was positive to the cream and several corticosteroids (which he had apparently never used). A ROAT with the cream was strongly positive after 3 days. Further patch testing yielded positive reactions to mometasone furoate at 0.01% and 0.1% pet. (3).

A 60-year-old woman presented with a 3-year history of recalcitrant vulval dermatitis thought to be due to lichen simplex chronicus with or without psoriasis with intermittent *Candida* infection. This was treated with various topical corticosteroids and antifungals with very little improvement. She had used different combinations of hydrocortisone 1% cream, mometasone furoate 0.1% cream, betamethasone dipropionate 0.05% cream, and betamethasone valerate 0.1% cream/ointment. One month prior to presentation, her treatment was changed to strictly ichthammol in zinc cream and hydrocortisone ointment only and this dramatically cleared her vulval dermatitis. Patch tests were negative to hydrocortisone and the antifungals, but positive to budesonide and the commercial mometasone furoate and betamethasone dipropionate preparations. ROATs were positive after 8 days. The corticosteroids themselves were not patch tested (8).

Cross-reactions, pseudo-cross-reactions and co-reactions

Cross-reactions between corticosteroids are discussed in Chapter 2.8. A few patients with multiple corticosteroid sensitizations have shown cross-reactivity to mometasone furoate (5,6,9).

LITERATURE

1   Baeck M, Goossens A. Immediate and delayed allergic hypersensitivity to corticosteroids: practical guidelines. Contact Dermatitis 2012;66:38-45
2   Tomljanović-Veselski M, Lipozencić J, Lugović L. Contact allergy to special and standard allergens in patients with venous ulcers. Coll Antropol 2007;31:751-756
3   Seyfarth F, Elsner P, Tittelbach J, Schliemann S. Contact allergy to mometasone furoate with cross-reactivity to group B corticosteroids. Contact Dermatitis 2008;58:180-181
4   Wilkinson SM, Beck MH. Fluticasone propionate and mometasone furoate have a low risk of contact sensitization. Contact Dermatitis 1996;34:365-366
5   Räsänen L, Tuomi M. Cross-sensitization to mometasone furoate in patients with corticosteroid contact allergy. Contact Dermatitis 1992;27:323-324
6   Venning VA. Fluticasone propionate sensitivity in a patient with contact allergy to multiple corticosteroids. Contact Dermatitis 1995;33:48-49
7   Dooms-Goossens A, Lepoittevin J-P. Studies on the contact allergenic potential of momethasone furoate: a clinical and molecular study. Eur J Dermatol 1996;6:339-340
8   Chow ET. Multiple corticosteroid allergies. Australas J Dermatol 2001;42:62-63
9   Donovan JC, Dekoven JG. Cross-reactions to desoximetasone and mometasone furoate in a patient with multiple topical corticosteroid allergies. Dermatitis 2006 ;17:147-151
10   Baeck M, Chemelle JA, Terreux R, Drieghe J, Goossens A. Delayed hypersensitivity to corticosteroids in a series of 315 patients: clinical data and patch test results. Contact Dermatitis 2009;61:163-175
11   Davis MD, El-Azhary RA, Farmer SA. Results of patch testing to a corticosteroid series: a retrospective review of 1188 patients during 6 years at Mayo Clinic. J Am Acad Dermatol 2007;56:921-927
12   Mercader-García P, Pastor-Nieto MA, García-Doval I, Giménez-Arnau A, González-Pérez R, Fernández-Redondo V, et al. GEIDAC. Are the Spanish baseline series markers sufficient to detect contact allergy to corticosteroids in Spain? A GEIDAC prospective study. Contact Dermatitis 2018;78:76-82
13   Tatu AL, Ionescu MA, Nwabudike LC. Contact allergy to topical mometasone furoate confirmed by rechallenge and patch test. Am J Ther 2018;25:e497-e498

Chapter 3.231 MONOBENZONE

IDENTIFICATION

Description/definition	: Monobenzone is the monobenzyl ether of hydroquinone that conforms to the structural formula shown below
Pharmacological classes	: Depigmenting agent
IUPAC name	: 4-Phenylmethoxyphenol
Other names	: Monobenzyl ether of hydroquinone; 4-benzyloxyphenol
CAS registry number	: 103-16-2
EC number	: 203-083-3
Merck Index monograph	: 7604
Patch testing	: 1.0% pet. (Chemotechnique, SmartPracticeEurope, SmartPracticeCanada)
Molecular formula	: $C_{13}H_{12}O_2$

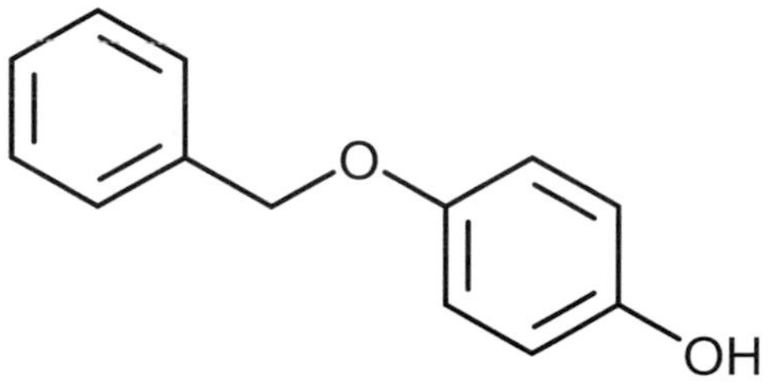

GENERAL

Monobenzone is a monobenzyl ether of hydroquinone with topical depigmentation activity. Although the exact mechanism of action of depigmentation is unknown, the metabolites of monobenzone appear to have a cytotoxic effect on melanocytes. Furthermore, the depigmentation effect might be mediated through the inhibition of tyrosinase, which is essential in the synthesis of melanin pigments, thereby causing permanent depigmentation of the skin. Monobenzone has been used since the beginning of the 1950s to treat several forms of hyperpigmentation. After some years, many cases were being reported of dermatitis, depigmentation (long-lasting and sometimes permanent) and hyperpigmentation (9). From the 1970s on, it was, and still is used mainly to depigment residual pigmented areas in patients with very extensive therapy-resistant vitiligo (7). In the USA and the EU, its use in cosmetics has long been banned.

In this chapter, only allergic contact dermatitis to monobenzone is discussed. Other side effects have been reviewed in ref. 9.

CONTACT ALLERGY

Case reports and case series
A woman with a 16-year history of extensive vitiligo was given a depigmenting cream containing 20% monobenzone to clear residual areas of normal pigmentation. Over the subsequent 6 weeks, she developed dermatitis at the sites of application. The patient undertook a ROAT with monobenzone 5% cream at an unaffected site and this produced a similar dermatitis. Patch tests were positive to 1% monobenzone in pet. and in two other vehicles; she did not co-react to hydroquinone (1).

A man had pigmented areas on the face; no precise diagnosis was made. Treatment with a 2% hydroquinone cream for 3 months did not improve the situation. He was then advised to use a cream containing 5% monobenzone, but 2 days after the first application, an acute dermatitis developed. Patch tests were positive to the cream, to monobenzone 2% pet. and to hydroquinone 2% pet. It was suggested that the patient had previously become sensitized to hydroquinone and now had a reaction from cross-sensitization to monobenzone (2).

It has been suggested that 5% of all patients treated with monobenzone cream become sensitized (6). Indeed, contact dermatitis from a 20% preparation was frequent, up to 13% of those treated (3,4). This was usually termed 'sensitization', but without patch tests being performed or on the basis of a positive patch test with the 20% monobenzone cream without proper controls.

As part of a clinical study to determine the effectiveness of monobenzone as a depigmenting agent, simulated use tests were done on a group of 43 middle-aged and older dark-skinned subjects with normal skin. Monobenzone in a vanishing cream base 1%, 4%, 7% and 10% was applied for 4 months plus the base itself. Inflammatory reactions during the course of treatment were observed in 23. No reactions were observed until after the 4th week of treatment. In general, erythematous reactions were observed on all areas except where the base was applied. In all cases where these reactions were observed, complete subsidence usually took about 2 weeks even though regular applications were continued. Several subjects experienced subsequently 2 or 3 similar episodes of erythema. In every case, with the exception of one, these reactions subsided despite continued applications (which would virtually exclude contact allergy). At the end of 4 months of the simulated use test, 26 of the 42 that completed the study showed depigmentation. This depigmentation was not uniform but was mottled and was more pronounced in those subjects who had experienced the more severe inflammatory reactions. After 2 years, 33 of the subjects were re-examined. Persistent leukoderma was present in 13 of this group. A patch test with the 10% monobenzone cream was applied to 12 who had leukoderma. All of these had a positive patch test which after 2 months revealed an early hypopigmentation at the patch test site. One can argue that the test concentration of 10% is too high (currently advised: 1% pet.), but the same patch test was applied on 16 of the remaining 20 who did *not* show any evidence of leukoderma and all had negative patch tests (5).

Cross-reactions, pseudo-cross-reactions and co-reactions
Not to hydroquinone (1). Cross-reactivity from hydroquinone (2); cross-reactivity from diethylstilbestrol (8).

LITERATURE

1 Lyon CC, Beck MH. Contact hypersensitivity to monobenzyl ether of hydroquinone used to treat vitiligo. Contact Dermatitis 1998;39:132-133
2 Van Ketel WG. Sensitization to hydroquinone and the monobenzyl ether of hydroquinone. Contact Dermatitis 1984;10:253
3 Dorsey CS. Dermatitis and pigmentary reactions to monobenzyl ether of hydroquinone: Report of two cases. Arch Dermatol 1960;81:245-248
4 Lerner AB, Fitzpatrick TB. Treatment of melanin hyperpigmentation. JAMA 1953;152:577-582
5 Spencer MC. Leukoderma following monobenzyl ether of hydroquinone bleaching. Arch Dermatol 1962;86:615-618
6 Fisher AA. Contact Dermatitis. Philadelphia: Lea and Febiger, 1973:235
7 Tan ES, Sarkany R. Topical monobenzyl ether of hydroquinone is an effective and safe treatment for depigmentation of extensive vitiligo in the medium term: a retrospective cohort study of 53 cases. Br J Dermatol 2015;172:1662-1664
8 De Groot AC, Weyland JW, Nater JP. Unwanted effects of cosmetics and drugs used in dermatology, 3rd Edition. Amsterdam – London – New York – Tokyo: Elsevier, 1994:23
9 De Groot AC. Monographs in Contact Allergy Volume I. Non-Fragrance Allergens in Cosmetics (Part I and Part 2). Boca Raton, Fl, USA: CRC Press Taylor and Francis Group, 2018:855-857

Chapter 3.232 MUPIROCIN

IDENTIFICATION

Description/definition : Mupirocin is the natural crotonic acid derivative extracted from *Pseudomonas fluorescens* that conforms to the structural formula shown below

Pharmacological classes : Anti-bacterial agents; protein synthesis inhibitors

IUPAC name : 9-[(E)-4-[(2S,3R,4R,5S)-3,4-Dihydroxy-5-[[(2S,3S)-3-[(2S,3S)-3-hydroxybutan-2-yl]oxiran-2-yl]methyl]oxan-2-yl]-3-methylbut-2-enoyl]oxynonanoic acid

Other names : (E)-(2S,3R,4R,5S)-5-((2S,3S,4S,5S)-2,3-Epoxy-5-hydroxy-4-methylhexyl)tetrahydro-3,4-dihydroxy-β-methyl-2H-pyran-2-crotonic acid, ester with 9-hydroxynonanoic acid

CAS registry number : 12650-69-0

EC number : Not available

Merck Index monograph : 7658

Patch testing : 2% pet.

Molecular formula : $C_{26}H_{44}O_9$

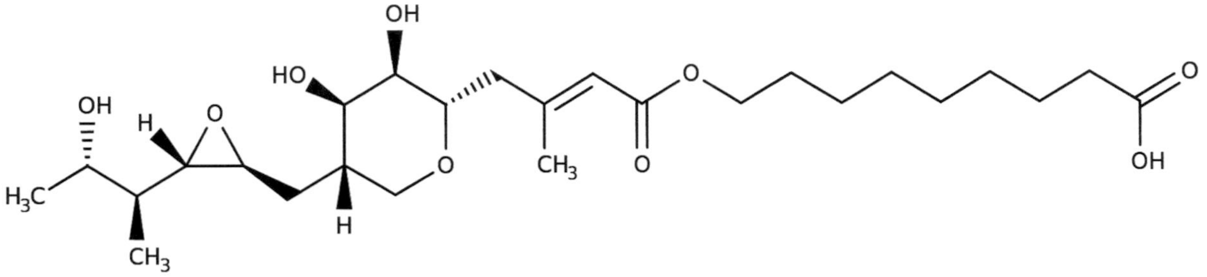

GENERAL

Mupirocin is a natural crotonic acid derivative and antibiotic extracted from a strain of *Pseudomonas fluorescens*. It inhibits bacterial protein synthesis by specific reversible binding to bacterial isoleucyl tRNA synthase. With excellent activity against gram-positive staphylococci and streptococci, mupirocin is primarily used for the topical treatment of primary and secondary bacterial skin infections, nasal infections, and for wound healing. In pharmaceutical products, ointments usually contain mupirocin and creams mupirocin calcium (dihydrate) (CAS number 115074-43-6, EC number not available, molecular formula $C_{52}H_{90}CaO_{20}$) (1).

CONTACT ALLERGY

Patch testing in groups of selected patients

In the period 1990-2014, in Belgium, 188 patients suspected of iatrogenic contact dermatitis were patch tested with mupirocin 10% pet. and there were 3 (1.6%) positive reactions. 96% of the positive patch test reactions to *all* topical drugs and antiseptics were considered to be relevant (3). In Croatia, before 1999, 100 patients with leg ulcers were tested with mupirocin 2% (vehicle?) and 2 patients had a positive patch test reaction. Relevance was not mentioned (2). Considering the unwillingness of the manufacturer to provide mupirocin for patch testing (see below), it cannot be certain that the patch test concentration and vehicle of 10% pet. as stated in ref. 3 are correct.

Case reports

A 37-year-old woman applied a mupirocin cream to a benign fibrous histiocytoma of the thigh. Because of exacerbation of the lesions (which lesions????), she was treated with oral antibiotics and multiple topical medications. None of the prescriptions prevented gradual spread of the rash to the whole thigh and the appearance of nummular lesions of the trunk. A patch test was strongly positive to the mupirocin cream; she also had positive reactions to multiple other drugs and ingredients of topical pharmaceuticals. The manufacturer declined to provide the different ingredients of the mupirocin cream. The patient was therefore tested with an ointment of another brand, also containing mupirocin but with different excipients. An open test on the arm elicited extensive contact dermatitis of the entire arm, which confirmed the mupirocin allergy (4).

A 31-year-old woman presented with a 2-year history of pruritic lip dermatitis. Physical examination showed a 2-mm circumferential rim of erythema and edema on the upper and lower lips. Previous therapies included pimecrolimus cream and mupirocin ointment. Patch tests were positive on day 5 to 2 different mupirocin 2% ointments and pimecrolimus 1% cream. The mupirocin ointment formulations consisted of only mupirocin and polyethylene glycol; the patient did not react to polyethylene glycol 100%. Several of the excipients of the pimecrolimus cream were tested negative, as was the cream base. A ROAT with pimecrolimus cream was positive after 3 days. It was concluded that the patient had allergic contact dermatitis from mupirocin and from pimecrolimus (5).

Previously, two cases of allergic contact dermatitis from mupirocin ointment had already been described. In both reports, the patient reacted to mupirocin ointment but not to the ointment base (6,7). In all case reports, therefore, the diagnosis of mupirocin allergy has been made *per exclusionem*, necessitated by the unwillingness of the manufacturer to provide the active ingredient mupirocin for patch testing.

LITERATURE

1 The data in the section 'General' may have been obtained from literature discussed in this chapter, but mostly also or exclusively from one or more of the following online sources: ChemIDPlus Advanced, PubChem, DrugBank, RxList, Drug Central, Drugs.com, and Wikipedia
2 Marasovic D, Vuksic I. Allergic contact dermatitis in patients with leg ulcers. Contact Dermatitis1999;41:107-109
3 Gilissen L, Goossens A. Frequency and trends of contact allergy to and iatrogenic contact dermatitis caused by topical drugs over a 25-year period. Contact Dermatitis 2016;75:290-302
4 Assier H, Hirsch G, Wolkenstein P, Chosidow O. Severe contact allergy to mupirocin in a polysensitized patient. Contact Dermatitis 2019;80:388-389
5 Zhang AJ, Warshaw EM. Allergic contact dermatitis caused by mupirocin and pimecrolimus. Contact Dermatitis 2019;80:132-133
6 Zappi EG, Brancaccio RR. Allergic contact dermatitis from mupirocin ointment. J Am Acad Dermatol 1997;36(2Pt.1):266
7 Eedy DJ. Mupirocin allergy in the setting of venous ulceration. Contact Dermatitis 1995;32:240-241

Chapter 3.233 NAFTIFINE

IDENTIFICATION

Description/definition : Naftifine is the synthetic allylamine derivate that conforms to the structural formula shown below

Pharmacological classes : Antifungal agents

IUPAC name : (E)-N-Methyl-N-(naphthalen-1-ylmethyl)-3-phenylprop-2-en-1-amine

Other names : (E)-N-Cinnamyl-N-methyl-1-naphthalenemethylamine

CAS registry number : 65472-88-0

EC number : Not available

Merck Index monograph : 7709

Patch testing : 1% and 5% alcohol

Molecular formula : $C_{21}H_{21}N$

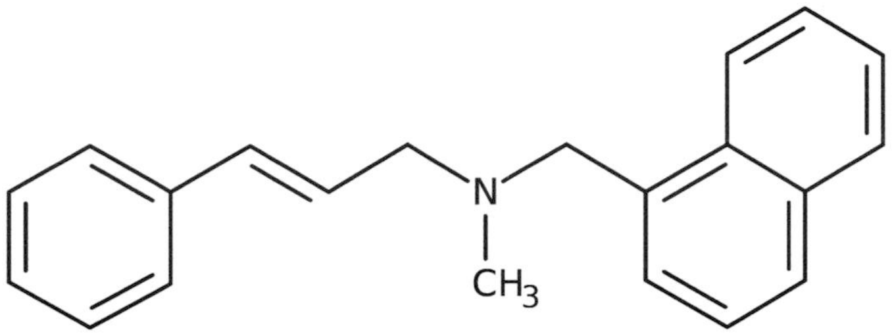

GENERAL

Naftifine is a synthetic allylamine derivate with broad-spectrum antifungal activity. It can be fungicidal or fungistatic depending on the concentration and the organisms involved. Naftifine is indicated for the topical treatment of tinea pedis, tinea cruris, and tinea corporis caused by the organisms *Trichophyton rubrum*, *Trichophyton mentagrophytes*, *Trichophyton tonsurans* and *Epidermophyton floccosum*. Naftifine is also effective against gram-negative and gram-positive bacteria and has anti-inflammatory activity by targeting the prostaglandin pathway. In pharmaceutical products, naftifine is employed as naftifine hydrochloride (CAS number 65473-14-5, EC number not available, molecular formula $C_{21}H_{22}ClN$) (1).

CONTACT ALLERGY

General

Nearly 20 cases of allergic contact dermatitis from naftifine HCl have been described, mostly from Germany in the period 1985-1990. In animal experiments, this antifungal drug was shown to be a moderate sensitizer (9), with a sensitizing capacity stronger than the imidazole antimycotics (13). There have been case series of 3 and 5 patients allergic to naftifine, attesting to the fact that sensitization was far from rare (8,12).

Case reports and case series

A 56-year-old woman had treated onychomycosis of the left great toe with naftifine HCl cream for 10 months. Two months before consultation an itchy and burning rash on the left foot appeared. Physical examination showed an eczematous, scaly dermatitis with some fissuring on her left foot and between the toes. Patch testing with the naftifine cream and *some* of its components yielded a positive (+) reaction to the cream. Ten controls were negative. A ROAT with the naftifine cream resulted in a pruritic, eczematous rash after 3 days. Although naftifine itself was not tested nor all excipients, the authors suggested naftifine to be the most likely allergenic culprit (2).

A 34-year-old atopic man was prescribed naftifine HCl solution for an erythematous dermatitis on the abdomen and genitals. After a few days, an acute pruritic eczematous eruption developed on the treated areas. Patch testing gave positive reactions to the solution and to its ingredient naftifine 1% and 5% alc. Previous use of allylamine antimycotics was denied, so the authors suggested that sensitization was due to 'inappropriate prolonged use of the antifungal', despite the fact that they wrote that the acute eczematous eruption already started after a few days (3).

A 12-year-old girl was prescribed naftifine 1% cream for two erythematosquamous patches on both lateral aspects of the neck. After a week an acute eczematous picture developed with very pruriginous vesicular and exudative lesions at 3 weeks. Patch tests with the cream and its components were positive to the cream and its ingredient naftifine at 1% and 5% in alcohol. Fifteen controls were negative. There was no reaction to the related allylamine terbinafine (4). A 49-year-old woman developed severe acute superinfected eczema of the feet and hands one week into self-treatment of suspected foot mycosis with naftifine cream. Patch tests were positive to the cream and to naftifine 5% alcohol (5).

In the years 1985-1990, at least 14 patients with allergic contact dermatitis from naftifine have been reported from Germany (6-12). Details are unknown.

Cross-reactions, pseudo-cross-reactions and co-reactions
Not to the related allylamine terbinafine (4).

LITERATURE

1   The data in the section 'General' may have been obtained from literature discussed in this chapter, but mostly also or exclusively from one or more of the following online sources: ChemIDPlus Advanced, PubChem, DrugBank, RxList, Drug Central, Drugs.com, and Wikipedia
2   Napolitano M, Fattore D, Fabbrocini G, Patruno C. Allergic contact dermatitis probably due to naftifine hydrochloride. Dermatitis 2019;30:231-233
3   Corazza M, Lauriola MM, Virgili A. Allergic contact dermatitis to naftifine. Contact Dermatitis 2005;53:302-303
4   Goday JJ, González-Güemes M, Yanguas I, Ilardia R, Soloeta R. Allergic contact dermatitis from naftifine in a child without cross-reaction to terbinafine. J Eur Acad Dermatol Venereol 1998;11:72-73
5   Willa-Craps C, Wyss M, Elsner P. Allergic contact dermatitis from naftifine. Contact Dermatitis 1995;32:369-370
6   Senff H, Tholen S, Stieler W, Reinel D, Hausen BM. Allergic contact dermatitis to naftifine. Report of two cases. Dermatologica 1989;178:107-108
7   Merkel M, Gall H. Kontaktallergie auf Naftifin. Akt Dermatol 1990;16:191-192 (Article in German)
8   Kleinhans D. Kontaktallergien auf Naftitin. 3 Fallbeobachtungen. Dtsch Dermatol 1986;34:1063-1068 (Article in German)
9   Hoting E, Küchmeister B, Hausen BM. Kontaktallergie auf das Antimykoticum Naftifin. Dermatosen 1987;35:124-127 (Article in German)
10   Tholen S. Allergisches Kontaktekzem auf Naftitin. Allergol 1987;10:428 (Article in German)
11   Raulin Ch, Frosch PH. Kontaktallergien auf Antimykotika. Z Hautkr 1987;62:1705-1709 (Article in German)
12   Mehringer A, Hartman AA, Pevny I, Burg G. Fünf Fälle einer Kontaktallergie durch Naftifin (Exoderil). Akt Dermatol 1990;16:193-195 (Article in German)
13   Hausen BM, Heesch B, Kiel U. Studies on the sensitizing capacity of imidazole derivatives. Am J Contact Dermat 1990;1:25-33 (Article in German)

Chapter 3.234 NAPHAZOLINE

IDENTIFICATION

Description/definition : Naphazoline is the imidazole derivative that conforms to the structural formula shown below

Pharmacological classes : Nasal decongestants; α-adrenergic agonists

IUPAC name : 2-(Naphthalen-1-ylmethyl)-4,5-dihydro-1*H*-imidazole

Other names : 2-(1-Naphthylmethyl)imidazoline

CAS registry number : 835-31-4

EC number : 212-641-5

Merck Index monograph : 7723

Patch testing : 1% pet.

Molecular formula : $C_{14}H_{14}N_2$

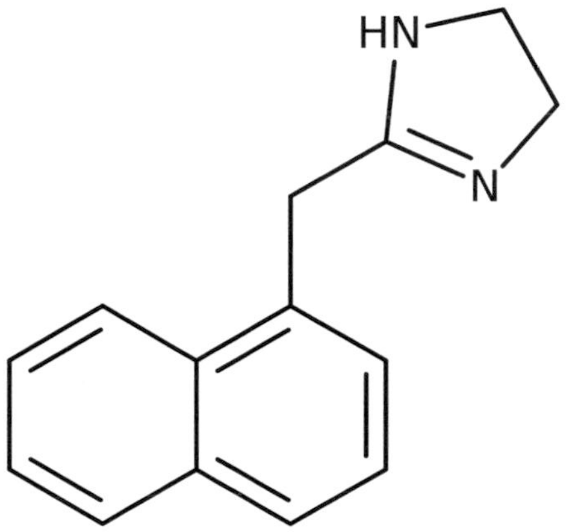

GENERAL

Naphazoline is an imidazole derivative and a direct-acting sympathomimetic amine with vasoconstrictive properties. Upon ocular administration, it acts on α-adrenergic receptors in the arterioles of the conjunctiva to produce vasoconstriction, resulting in decreased conjunctival congestion and diminished itching, irritation and redness. It has similar effect when applied to the nasal mucosa. Naphazoline is indicated for use as over-the-counter eye drops for ocular vasoconstriction and as a nasal preparation for nasal congestion. In pharmaceutical products, naphazoline is employed as naphazoline hydrochloride (CAS number 550-99-2, EC number 208-989-2, molecular formula $C_{14}H_{15}ClN_2$) (1).

CONTACT ALLERGY

Case report

An 89-year-old man presented with pruritic lesions over his left knee with linear extension down onto the lower leg, where a wound had been treated with povidone iodine for 18 days and an antiseptic for 2 days. Several years previously, he had developed eczema following the repeated use of an over-the-counter antiseptic. The current antiseptic contained benzalkonium chloride, dibucaine HCl, chlorpheniramine maleate, naphazoline HCl, and a mixture of fragrance ingredients. A patch test to the product was positive. In a second session, the patient reacted to dibucaine HCl 1% pet., naphazoline HCl 1% pet. and chlorpheniramine maleate 1% pet. The fragrance mix and benzalkonium chloride were negative. Twenty controls were negative to naphazoline 1% pet. (2).

LITERATURE

1   The data in the section 'General' may have been obtained from literature discussed in this chapter, but mostly also or exclusively from one or more of the following online sources: ChemIDPlus Advanced, PubChem, DrugBank, RxList, Drug Central, Drugs.com, and Wikipedia

2   Yamadori Y, Oiso N, Hirao A, Kawara S, Kawada A. Allergic contact dermatitis from dibucaine hydrochloride, chlorpheniramine maleate, and naphazoline hydrochloride in an over-the-counter topical antiseptic. Contact Dermatitis 2009;61:52-53

Chapter 3.235 NEOMYCIN

IDENTIFICATION

Description/definition   : Neomycin is an antibiotic complex produced by *Streptomyces fradiae*; it is composed of
                             neomycins A, B, and C; the structural formula of neomycin B is shown below
Pharmaceutical classes   : Anti-bacterial agents
IUPAC name   : (2R,3S,4R,5R,6R)-5-Amino-2-(aminomethyl)-6-[(1R,2R,3S,4R,6S)-4,6-diamino-2-
                             [(2S,3R,4S,5R)-4-[(2R,3R,4R,5S,6S)-3-amino-6-(aminomethyl)-4,5-dihydroxyoxan-2-yl]oxy-
                             3-hydroxy-5-(hydroxymethyl)oxolan-2-yl]oxy-3-hydroxycyclohexyl]oxyoxane-3,4-diol
Other names   : Framycetin; neomycin B (often used synonyms, but not entirely correct)
CAS registry number   : 1404-04-2
EC number   : 215-766-3
Merck Index monograph   : 7809
Patch testing   : Sulfate, 20.0% pet. (Chemotechnique, SmartPracticeCanada, SmartPracticeEurope);
                             late positive reactions (after D4) are frequent and readings at D7-D8 are imperative
Molecular formula   : $C_{23}H_{46}N_6O_{13}$

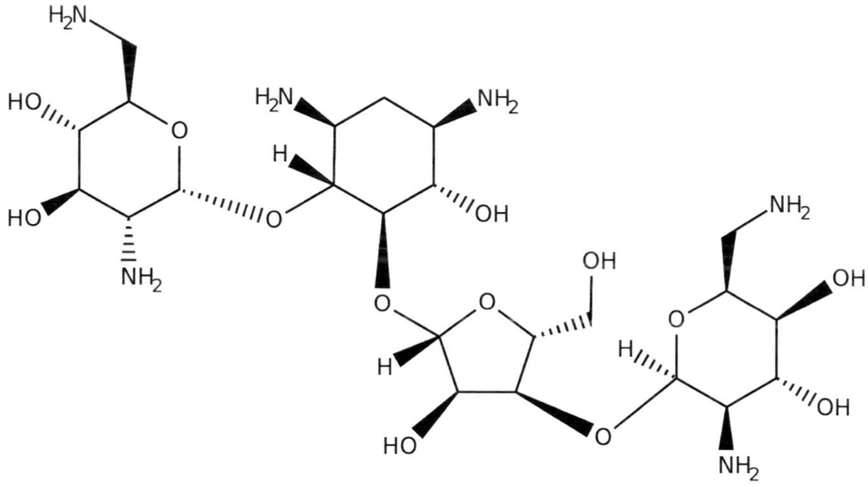

Chemical databases are not always consistent with nomenclature of neomycin, synonyms, CAS numbers (many numbers provided) and EC numbers. Neomycin is often used as synonymous with neomycin B, its main active ingredient. Framycetin consists mainly of neomycin B and consequently, neomycin, neomycin B and framycetin are often used as synonyms. See also Chapter 3.155 Framycetin.

GENERAL

Neomycin is the prototype of the aminoglycoside antibiotics and was first isolated in 1949 from the gram-positive bacillus *Streptomyces fradiae*. It consists of a variable mixture of two isomers, neomycin B (>88%) and neomycin C (<10%), along with small amounts of a degradation product, neamine or neomycin A (<2%). It exerts its antibacterial activity through irreversible binding of the nuclear 30S ribosomal subunit, thereby blocking bacterial protein synthesis. Neomycin is effective against most gram-negative organisms except *Pseudomonas aeruginosa* and anaerobic bacteria. Its activity against gram-positive microorganisms is more or less limited to staphylococci, but bacterial resistance supervenes after prolonged use (117,118).

Percutaneous absorption after topical application is minimal (140), and absorption through intact gastrointestinal mucosa is poor, ranging between 1% and 3%. Parenteral administration is associated with severe ototoxicity and nephrotoxicity, at times irreversible. Neomycin is therefore used: (1) in topical formulations for the prevention and treatment of superficial skin, ear, and eye infections; (2) in solutions for urinary instillations to prevent bacteriuria from indwelling catheters; and (3) in peritoneal irrigation solutions used to treat contaminations in abdominal surgery. Dentists may use a neomycin-containing paste in root canal treatment. Given orally in doses of 200 to 1,000 mg two to four times per day, neomycin is used to sterilize the gut before digestive tract surgery and is also used in the treatment of hepatic coma (to reduce the number of ammoniagenic intestinal bacteria). In addition, trace amounts of neomycin are present in numerous vaccines as a result of its use to prevent bacterial contamination during the vaccines' manufacture (117,118).

In some oral forms, neomycin base may be used, but topical pharmaceuticals contain neomycin as neomycin sulfate (CAS number 1405-10-3, EC number 215-773-1, molecular formula $C_{23}H_{48}N_6O_{17}S$) (1).

CONTACT ALLERGY

General
Sensitization of the skin to neomycin was first reported in the USA in 1952 (125,126). The first British cases were described in 1958 (127). Since then, contact allergy to neomycin has been the subject of a vast number of publications. Because of this, as neomycin sensitivity is generally well known and as data on contact allergy to/ allergic contact dermatitis from neomycin are often 'hidden' in scientific articles (i.e. that these data neither show from the title nor can they be found in targeted searches in online databases including PubMed), the author has not attempted to provide a full review of neomycin sensitization in this chapter; also, some data are described less detailed.

Neomycin is by far most frequently used in topical pharmaceuticals. In the USA (but also in other countries), neomycin is often combined with other antibiotics, notably bacitracin and polymyxin B, to improve the antibacterial spectrum. It is also often combined with topical corticosteroids to suppress inflammation and sometimes with lidocaine to suppress pain. In the USA, neomycin preparations are widely available as over-the-counter products. This explains the high frequencies of sensitization there of 7-11% in consecutive patients suspected of contact dermatitis (table 3.235.2) and the very frequent co-reactivity to bacitracin (129,130,132,143 [Finland],157,158 [Thailand], 188 [Finland]). Co-reactivities from combined presence or concomitant or successive sensitization can also regularly be observed to corticosteroids (143-149,151,172) and – to a lesser extent – lidocaine (132) or vice versa. Indeed, in one study, of 131 patients allergic to corticosteroids, 41 (31%) co-reacted to neomycin, versus 8% in a control group of consecutive dermatitis patients (147). In a very large IVDK study, multiple patch test reactions to standard allergens (polysensitization) was a very strong risk factor for neomycin allergy, and an association with corticosteroid allergy was also evident (172).

In European countries, neomycin preparations are available on prescription only, which results in far lower rates of sensitization, well below 2.5% (table 3.235.2). However, formerly, in Finland, neomycin over-the-counter preparations, often in combination with bacitracin, were widely available. As a result, in 1968, almost 19% of routine patients were allergic to neomycin, of whom 66% co-reacted to bacitracin (122). Previously, even 88% co-reactivity had been reported from this country (188).

Subgroups of patients at higher risk of sensitization include individuals with stasis dermatitis and leg ulcers, anogenital dermatitis, and otitis externa (117,118,172). Contact allergy to neomycin is slightly more prevalent in atopic individuals than in non-atopic patients, probably as a result of more frequent use of topical pharmaceutical products (92). Neomycin allergy is also seen more frequently in the older population. This can partly be explained by sensitization to neomycin from its use on stasis dermatitis and leg ulcers, but older age is also a risk factor for neomycin contact allergy independent of stasis dermatitis/leg ulcers (172).

The presenting clinical features of sensitivity usually occur in an existing inflammatory dermatosis and it is suspected when there is a deterioration of the condition or if it fails to respond as quickly as expected. A sudden severe deterioration is not so common and often the reaction is suppressed if the medicament also contains a corticosteroid. Consequently, it may go unrecognized (118).

Neomycin sulfate is typically used clinically at a concentration of 0.5% (0.35% neomycin) in antibiotic ointments, creams, and eye or eardrops, yet the standard patch test concentration is 20% in petrolatum. This high concentration is used due to the repeated observation that patch testing with a patient's own neomycin-containing cream or ointment, or with neomycin in petrolatum at in-use concentration, yields a high rate of false-negative results (127, 139). Even neomycin sulfate 20% in petrolatum can result in false-negative patch test reactions and intradermal tests may sometimes be necessary to correctly diagnose the sensitivity (128). These difficulties are probably due to neomycin's low penetration through intact skin (140). Use tests with the corticosteroid preparations may also be useful (154).

For the same reason of poor skin penetration, positive patch test reactions usually develop slowly (> D3-D4) and, consequently, a considerable number of positive patch test reactions to neomycin may be missed when readings are not performed at day 7 or D8 (74,98,99,100,164). Late readings are therefore imperative.

In the last decades, the frequency of sensitization to neomycin appears to decrease in several countries including Thailand (101), Belgium (105), Germany (172) and Canada (189), which may probably be explained by less frequent use and – over-the counter and prescription - availability of this aminoglycoside antibiotic in topical pharmaceuticals (101,189). A decline in the rates of sensitization, albeit somewhat modestly and only recently, can also be observed in North America in the NACDG studies (2,4), but this may well be caused by the decreasing rates in Canada and the increased representation of Canadian patients within the NACDG cohorts. Therefore, the actual frequency of sensitization in the USA is probably higher than reported in NACDG publications (189).

Contact allergy in the general population

In several investigations in European countries and in one from Australia, random samples of the population of certain age groups and subgroups have been patch tested with neomycin (table 3.235.1, data back to 1996). Most were performed In Denmark, where the prevalences of sensitization were invariably 0.3% or lower, sometimes even zero (82,83,85,86,88,95). Rates were also low (≤0.5%) in a study in 5 European countries (81), Norway (84,96; albeit 0.7% in women [84]), and Sweden (97). A German study scored 1.4% (women 1.2%, men 1.5%), but >50% of this population was atopic and may therefore have been exposed to neomycin more frequently from treatment of atopic dermatitis (87). The high prevalence of 1.8% in Australia (women 3.0%, men 0%) may conveniently be explained by the fact that the individuals in this group were self-selected volunteers (82).

In all these studies, T.R.U.E. test materials have been used. It should be realized that, in a number of these investigations, the results were read at D2 only. This may have resulted in an underestimation of the true prevalence of sensitization, as positive reactions at D3 or D4 (the usual time for the second reading) with negative result at D2 commonly occur with most haptens. Moreover, a considerable number (80%) of positive patch test reactions to the neomycin T.R.U.E. test may be missed when readings are not performed at day 7 (98).

Table 3.235.1 Contact allergy in the general population and in subgroups

Year and country	Selection and number tested	Prevalence of contact allergy			Comments	Ref.
		Total	Women	Men		
General population						
2008-11 five European countries	general population, random sample, 18-74 years, n=3119	0.4%	0.4%	0.4%	TRUE test	81
<2007 Norway	general population, random sample, 18-69 years, n=531	0%	0%	0%	TRUE test	96
2006 Denmark	general population, random sample, 18-69 years, n=3460	0.1%	0.1%	0%	patch tests were read on day 2 only; TRUE test	83
2005 Norway	general population, random sample, 18-69 years, n=1236	0.5%	0.7%	0.2%	TRUE test	84
1998 Denmark	general population, random sample, 15-41 years, n=469		0%	0%	patch tests were read on day 2 only; TRUE test	85
Subgroups						
2011-2013 Sweden	adolescents from a birth cohort, 15.8-18.9 years, n=2285	0.04%	0.0%	0.09%	TRUE test; patch tests were read at day 2 only	97
2010 Denmark	unselected population of 8th grade schoolchildren in Denmark, 15 years later; n=442	0%	0%	0%	follow-up study; TRUE test	82
<1999 Australia	self-selected adult healthy volunteers, 18-82 y., n=219	1.8%	3.0%	0%		94
1997-1998 Germany	adults 28-78 year, with a large percentage (>50%) of atopic individuals, n=1141	1.4%	1.2%	1.5%		87
1997-1998 Denmark	twins aged 20-44 years, n=1076	0.3%	0.3%	0.3%	TRUE test; 449 had self-reported hand eczema	95
1995-1996 Denmark	8th grade school children, 12-16 years, n=1146	0.2%	0.2%	0.2%		86,88

Estimates of the 10-year prevalence (1997-2006) of contact allergy to neomycin in the general population of Denmark based on the CE-DUR method ranged from 0.26% to 0.35% (89). In a similar study from Germany, the estimated prevalence in the general population in the period 1992-2000 ranged from 0.4% to 1.0% (90).

Also with the CE-DUR approach, the incidence of sensitization to neomycin in the German population was estimated to range from 17 to 69 cases/100,000/year in the period 1995-1999 and from 12 to 51 cases/100,000/year in the period 2000-2004 (102). In Germany, for the period 1995-2004, the population-based relative incidence (RI) of contact sensitization to neomycin (cases/100,000 defined daily doses (DDDs) per year) was estimated to be 59.4. In the group of antibiotics, the RI ranged from 1.6 (oxytetracycline) to 86.2 (framycetin) (103).

Patch testing in groups of patients

Results of patch testing neomycin in consecutive patients suspected of contact dermatitis (routine testing) back to 2000 are shown in table 3.235.2. Results of testing in groups of *selected* patients (e.g. patients with leg ulcers/stasis dermatitis, individuals with periorbital dermatitis/eyelid eczema, patients with anogenital dermatitis or otitis externa) are shown in table 3.235.3.

Patch testing in consecutive patients suspected of contact dermatitis: routine testing

As neomycin is present in most, if not all, baseline/routine/screening/standard series tested worldwide, data on testing this aminoglycoside antibiotic in consecutive patients (routine testing) is abundant. The results of over 40 such published investigations back to 2000 are shown in table 3.235.2. In North America, in the entire period, prevalences of sensitization have been high and fairly constant, generally in the 8.5% to 11% range. It would appear that, since 2009, the rates are gradually decreasing in the NACDG studies (USA + Canada). However, in Canada, a sharp drop in the frequency of 7.4% in 2001-2008 to 1.7% in the following period 2009-2013 has been observed (189).

Consequently, the decreasing rates of the NACDG may well be caused by the lower frequencies in Canada and the increased representation of Canadian patients within the NACDG cohorts and not indicate less frequent sensitization in the USA (189). Quite remarkably, relevance rates in the NACDG studies were always low with <20% definite or probable relevance.

The situation in Europe is entirely different. Prevalences of sensitization from 1999 to 2019 have ranged from 0.4% to a maximum of 3.0%, the result of limited prescribing of neomycin-containing topical drugs and the unavailability of over-the-counter preparations with neomycin. The higher rates were observed in the first decade of this century and very low frequencies were found in studies of the multinational ESSCA (European Surveillance System on Contact Allergies) (0.4% and 0.6% [13,59], albeit with considerable differences between countries), in The Netherlands (0.8% [98]) and in Belgium (0.7% in 2010-2014, decreasing trend [105]). Of the other countries, Thailand had the highest score with 4.8% in the period 2006-2018, but there also, an obvious decrease was visible from 10.8% in 2006-2008 to 4.1% in the period 2016-2018 (101). Data on relevance of the positive patch test reactions were hardly ever provided or specified for neomycin. Details on the incriminated products were never provided.

Table 3.235.2 Patch testing in groups of patients: Routine testing

Years and Country	Test conc. & vehicle	Number of patients tested | positive (%)		Selection of patients (S); Relevance (R); Comments (C)	Ref.
USA, Canada					
2015-2017 NACDG	20% pet.	5591	390 (7.0%)	R: definite + probable relevance: 17%	2
2007-2016 USA	20% pet.	2313	(9.4%)	R: not stated	179
2011-2015 USA	20% pet.	2574	201 (7.8%)	R: not stated	3
2013-2014 NACDG	20% pet.	4857	409 (8.4%)	R: definite + probable relevance: 16%	4
2009-2013 Canada	20% pet.	4485	(1.7%)	R: not stated; C: in the period 2001-2008, the rate of one of these 3 centers in Western Canada was still a high 7.4%, which had dropped to 1.7% in the current period 2009-2013	189
2011-2012 NACDG	20% pet.	4229	384 (9.1%)	R: definite + probable relevance: 16%	7
2009-2010 NACDG	20% pet.	4298	(8.7%)	R: definite + probable relevance: 19%	8
2006-2010 USA	20% pet.	3085	(10.3%)	R: 46%	9
2007-2008 NACDG	20% pet.	5072	(10.1%)	R: definite + probable relevance: 20%	14
2001-2008 Canada	20% pet.	1207	(7.4%)	R: not stated; C: in the period 2009-2013, the rate of this center and 2 other centers in Western Canada together would drop to 1.7%	189
2005-2006 NACDG	20% pet.	4439	(10.0%)	R: definite + probable relevance: 13%	15
2001-2005 USA	20% pet.	3834	(11.8%)	R: 50%	17
2003-2004 NACDG	20% pet.	5137	545 (10.6%)	R: not stated	20
2001-2002 NACDG	20% pet.	4900	(11.6%)	R: definite + probable relevance: 12%	26
1998-2000 USA	20% pet.	5822	(11.5%)	R: definite + probable relevance: 19%	30
1998-2000 USA	20% pet.	1322	(11.2%)	R: not stated	31
Europe, multinational					
2013-2014 12 European countries [b]	20% pet.	23,385	(1.3%)	R: not stated; C: range of positive reactions: 0% - 10.9%	5
2009-12, eight European countries [b]	20% pet.	28,569	(0.4%)	R: not stated; C: range per country: 0.4-2.8%	59
2007-2008 11 European countries [b]	20% pet.	25,181	160 (0.6%) [a]	R: not stated; C: prevalences ranged from 0% to 3.3%	13
2005-2006 10 European countries [b]	20% pet.	16,047	312 (1.9%)	R: not stated; C: prevalences were 1.1% in Central Europe, 1.9% in West, 3.8% in Northeast and 1.4% in South Europe	16
2004, 11 European countries [b]	20% pet.	9988	218 (2.1%) [a]	R: not stated; C: range positives per center: 0.0%-6.4%	19
2002-2003 Europe [b]	20% pet.	9672	(2.9%)	R: not stated	25

Table 3.235.2 Patch testing in groups of patients: Routine testing (continued)

Years and Country	Test conc. & vehicle	Number of patients tested | positive (%)		Selection of patients (S); Relevance (R); Comments (C)	Ref.
1996-2000 Europe	20% pet.	26,210	(3.0%)	R: not stated; C: ten centers, seven countries, EECDRG study	32
Europe, IVDK (Germany, Austria, Switzerland)					
2001-2004 IVDK	20% pet.	31,012	713 (2.3%)	R: not stated	21
1995-2004 IVDK	20% pet.	80,867	(2.5%)	R: not stated	102
1998-2003 IVDK	20% pet.	47,559	1208 (2.5%)	R: not specified, possibly around 40%; C: risk factors: leg dermatitis, older age, polysensitization; decrease in frequency of sensitization in the years 2002-2003	172
1996-1999 IVDK	20% pet.	34,822	(2.6%)	R: not stated; C: the frequency of sensitization rose from 1.8% in patients of 60 years or younger to 4.9% in the age group of 61-75 years and to 7.1% in patients over 75 years old; the increase was both dependent and independent of chronic leg ulcers	33
Europe, individual countries					
2008-18 Netherlands	TRUE test	3209	27 (0.8%)	R: not stated; C: >80% of the reactions may be missed when patch are not read at D7	98
2014-2016 Greece	20% pet.	1978	57 (2.9%)	R: not stated	93
1990-2014 Belgium	20% pet.	14,578	269 (1.8%)	R: 96% of the positive patch test reactions to all topical drugs and antiseptics were considered to be relevant; C: there was a significant decrease in the frequency of positive reactions from 3.2% in 1990-1994 to 0.7% in 2010-2014	105
2009 Sweden	20% pet.	3112	(0.4%)	R: not stated	11
1985-2005 Denmark	20% pet.	14,978	(2.8%)	R: not stated	18
2000-4 Switzerland	20% pet.	4094	81 (2.0%)	R: not stated	22
1999-2001 Sweden	20% pet.	3790	(2.4%)	R: not stated	27
1997-2001 Czech Rep.	20% pet.	12,058	234 (1.9%)	R: not stated	238
2000 Sweden	20% pet.	3112	(1.2%)	R: not stated	11
2000 United Kingdom	20% pet.	3063	(2.9%)	R: 50% (current and past relevance in one center)	29
Other countries					
2013-2019 Turkey	20% pet.	1309	16 (1.2%)	R: not stated	186
2006-2018 Thailand	20% pet.	2803	34 (4.8%)	R: not stated; C: decrease from 10.8% in 2006-2008 to 4.1% in the period 2016-2018	101
2009-2013 Singapore		2598	(2.5%)	R: present + past relevance: 20%; C: range of positive reactions per year 1.7 - 3.4%	6
2001-2010 Australia	20% pet.	5118	127 (2.5%)	R: 8%	10
2004-2009 China	20% pet.	2758	17 (0.6%)	R: 41% relevance for all positive patch test reactions together	12
1998-2004 Israel	20% pet.	2156	23 (1.1%)	R: not stated	23
1992-2004 Turkey	20% pet.	1038	25 (2.4%)	R: not stated	24

[a] age-standardized and sex-standardized proportions; [b] study of the ESSCA (European Surveillance System on Contact Allergies); EECDRG: European Environmental and Contact Dermatitis Research Group; IVDK: Information Network of Departments of Dermatology, Germany, Austria, Switzerland; NACDG: North American Contact Dermatitis group (USA, Canada)

Patch testing in groups of selected patients

Results of testing neomycin in groups of *selected* patients (e.g. patients with leg ulcers/stasis dermatitis, individuals with periorbital dermatitis/eyelid eczema, patients with anogenital dermatitis or otitis externa) are shown in table 3.235.3. In 16 studies in groups of patients tested because of leg ulcers or stasis dermatitis, prevalences of sensitization were generally high, ranging from 4% to 23%, with 13 studies scoring over 9% positive reactions. In studies with a control group, the rates in the target population were always (significantly) higher (34,46,47). The issue of relevance of the positive patch tests was hardly ever addressed.

In groups of patients tested for periorbital dermatitis/eyelid eczema, the sensitization frequencies were far lower, ranging from 0.7% to 10% with the great majority <5%. In some studies, the frequencies in control groups were lower (49,53,55), but in others equal (56) or even higher (50). Probably, neomycin is not very widely used in ophthalmic medications in the countries where the studies were performed.

In studies where patients with perianal, genital or anogenital were investigated, rates of sensitization varied from 1.5% to 12.1%, but most had scores between 3% and 6.7%. In 3 investigations with a control group, the frequencies in the target population were either not significantly different from the control group (64,72) or even lower (71). The relevance rates were high in most studies, >50% to 100% (60,61,62,65,66,72,73), but sometimes very low (64,70).

In groups of patients with chronic otitis externa, positive reactions to neomycin ranged from 5.8% to 21%.

Table 3.235.3 Patch testing in groups of patients: Selected patient groups

Years and Country	Test conc. & vehicle	Number of patients tested	positive (%)	Selection of patients (S); Relevance (R); Comments (C)	Ref.
Patients with leg ulcers and stasis dermatitis					
2012-2017 Lithuania	20% pet.	145	25 (17.2%)	S: patients with longstanding ulceration from venous insufficiency (65%) or other etiologies; R: not stated	44
2003-2014 IVDK	20% pet.	5202	(5.0%)	S: patients with stasis dermatitis/chronic leg ulcers; R: not stated; C: percentage of reactions significantly higher than in a control group of routine testing (2.4%)	34
2005-2008 France	20% pet.	423	39 (9.2%)	S: patients with leg ulcers; R: not stated	35
<2008 Poland	20% pet.	50	10 (20%)	S: patients with venous leg ulcers; R: not stated	39
2006-2007 Canada		100	4 (4%)	S: patients with leg ulcers or venous disease; R: not stated	38
<2007 Croatia	20% pet.	60	14 (23%)	S: patients with venous leg ulcers with or without (allergic contact) dermatitis: R: not stated	45
<2005 Poland	20% pet.	50	10 (20%)	S: patients with chronic venous leg ulcers; R: not stated	36
<2004 NACDG	20% pet.	52	7 (14%)	S: patients with past or present leg ulcers with or without dermatitis; R: definite + probable relevance 14%	37
2001-2002 France		106	10 (9.4%)	S: patients hospitalized for leg ulcers; R: not mentioned; C: the frequency of sensitization in a dermatitis control group was 1.5%	40
2000-2002 Serbia	20% pet.	75	8 (11%)	S: patients with chronic venous leg ulcers; R: not stated; C: 4% reactions in a (small) control group (difference not significant)	47
<1999 Croatia	20% pet.	100	10 (10%)	S: patients with leg ulcers; R: not stated	42
1996-1997 U.K.		109	13 (11.9%)	S: mostly patients with leg ulcers and stasis dermatitis; R: not stated	41
1995-7 Switzerland	20% pet.	153	28 (18%)	S: patients with chronic leg ulcers; R: not stated; C: in a control group of dermatitis patient the sensitization frequency was 4%	46
1992-1997 IVDK	20% pet.	3220	346 (10.7%)	S: patients with leg ulcers and/or leg eczema; R: not stated	48
1992-1997 Germany	20% pet.	131	9 (6.5%)	S: patients with leg ulcers and/or leg eczema; R: not stated	48
<1994 U.K.	20% pet.	85	(21%)	S: patients with longstanding venous ulceration or eczema complicating leg ulcers; R: not stated	43
Patients with periorbital dermatitis/eyelid eczema					
2001-2010 IVDK	20% pet.	1112	66 (5.9%)	S: patients with periorbital dermatitis tested with an ophthalmic tray; R: not stated	109
2000-2010 IVDK		3042	128 (3.8%)	S: patients with periorbital dermatitis; R: not stated; C: the frequency in patients suspected of eye medicament allergy was 4.6% and in a control group of dermatitis patients 1.9%	49
1990-2006 USA	20% pet.	266	9 (3.4%)	S: patients with periorbital dermatitis; R: 22%; C: the frequency was lower than in controls (5.1%)	50
1996-2005 Hungary	20% pet.	401	3 (0.7%)	S: patients with periorbital dermatitis; R: not stated	54
1999-2004 Germany		88	4 (5%)	S: patients with periorbital eczema; R: not stated	51
1994-2004 USA		46	5 (10%)	S: patients with allergic contact dermatitis of the eyelids; R: these were relevant reactions, but the causative products were not mentioned; C: this number included 'other aminoglycosides'	58
1997-2001 USA	20% pet.	203	20 (10%)	S: patients with eyelid dermatitis; R: not stated	57
1995-1999 IVDK	20% pet.	969	(3.8%)	S: patients with *allergic* periorbital contact dermatitis; C: the frequency was significantly higher than in a control group	55
1994-1998 U.K.	20% pet.	232	9 (3.9%)	S: patients with periorbital contact dermatitis; R:67%; C: the frequency in a control group was 3.8%	56
1990-1994 IVDK		585	33 (5.6%)	S: patients with periorbital eczema; R: not stated; C: frequency of sensitization in dermatitis controls was 3.3%	52
1990-1991 Italy	20% pet.	150	4 (2.7%)	S: patients with eyelid dermatitis; R: not stated; C: the rate of positive reactions in a control dermatitis group without eyelid involvement was 1.1%	53
Patients with anogenital dermatitis					
2005-2016 NACDG	20% pet.	449	17 (3.8%)	S: patients with only anogenital dermatitis; R: all positives represent relevant reactions; C: the frequency was not significantly different from a control group	72

Table 3.235.3 Patch testing in groups of patients: Selected patient groups (continued)

Years and Country	Test conc. & vehicle	Number of patients tested	positive (%)	Selection of patients (S); Relevance (R); Comments (C)	Ref.
2004-2016 Spain		124	15 (12.1%)	S: patients with perianal dermatitis lasting >4 weeks; R: 73%	65
2013-2015 Ireland	20% pet.	99	4 (4%)	S: patients patch tested for perianal and/or genital symptoms; R: all reactions to medicaments were relevant	73
2005-2015 U.K.	20% pet.	150	7 (4.7%)	S: patients with perianal dermatoses and/or pruritus ani; R: 14%	70
2003-2010 USA	20% pet.	90	3 (3%)	S: women with (predominantly) vulvar symptoms; R: 67%	66
2003-2010 USA	20% pet.	55	3 (6%)	S: women with (ichty) vulvar dermatoses; R: not stated	68
2004-2008 IVDK			(1.4%)	S: patients with anogenital dermatoses tested with a medicament series; R: not stated; C: number of patients tested unknown	69
1994-2004 NACDG	20% pet.	345	15 (4.4%)	S: patients with anogenital signs or symptoms; R: 10%; C: the frequency in a control group was 4.3%	64
1999-2003 Austria		58	1 (1.7%)	S: patients with perianal eczema suspected of allergic contact dermatitis; R: not stated; C: the frequency was lower than in a control group	71
1990-2003 Belgium	20% pet.	92	3 (3%)	S: women suffering from vulval complaints referred for patch testing; R: 33%	91
1991-1995 U.K.		121	8 (6.6%)	S: women with pruritus vulvae and primary vulval dermatoses suspected of secondary allergic contact dermatitis; R: 49% for all positive reactions together	61
1992-1994 U.K.	20% pet.	69	3 (4%)	S: women with 'vulval problems'; R: 58% for all allergens together	62
1986-1990 U.K.	20% pet.	135	9 (6.7%)	S: patients referred to a vulva clinic for patch testing; R: 7/9	60
<1990 U.K.		50	11 (22%)	S: women with pruritus vulvae; R: not stated	67

Patients with chronic otitis externa

<2004 Turkey	20% pet.	66	11 (17%)	S: patients with chronic eczematous otitis externa; R: not stated; C: 0% reactions in a small control group	77
1985-2002 U.K	20% pet.	179	37 (21%)	S: patients with chronic inflammatory ear disease; R: 92%; C: significant reduction in the frequency of contact allergy in the second half of the study period	74
<2002 U.K.		149	19 (12.8%)	S: patients with chronic otitis media or otosclerosis scheduled for middle ear surgery; R: not stated	104
1995-9 Netherlands	20% pet.	79	12 (15%)	S: patients with chronic otitis externa; R: 18% of all reactions to ingredients of topical medications were relevant	75
1992-1997 IVDK		139	8 (5.8%)	S: patients with chronic otitis externa; R: not stated; C: the frequency in a control group was 4.1%	76
<1982 U.K.	20% pet.	40	6 (15%)	S: patients with otitis externa, chronic suppurative otitis media or discharging mastoid cavities >1 year; R: not stated	185

Other patient groups

2000-2010 Canada		100	29 (29%)	S: charts reviewed and included in the study when there was at least one positive reaction to a topical drug; R: not stated; C: the high percentages to all drugs are obviously the result of the – rather unusual – selection procedure targeting at topical drug allergy	110
2000-6 Netherlands		90	7 (8%)	S: patients patch tested for suspected contact allergy to ophthalmological drugs; R: 69% for all reactions to drugs together	80
1978-2005 Taiwan		603	32 (5.3%)	S: patients suspected of contact allergy to medicaments; R: 65% of the reactions to all medicaments were considered to be relevant	106
1995-2004 IVDK	20% pet.	3711	(3.6%)	S: patients patch tested for suspected contact allergy to ophthalmological drugs; R: not stated; C: the estimated number of sensitizations per 100,000 prescriptions was 8.3	79
<1993 India	20% pet.	50	26 (52%)	S: patients suspected of footwear dermatitis; R: not specified; C: in a very small control group the frequency of neomycin sensitization was 47%	107

IVDK: Information Network of Departments of Dermatology (Germany, Austria, Switzerland); NACDG: North American Contact Dermatitis Group (USA, Canada)

Relevance, where mentioned, was 92% (74) and 18% (75). The rates were higher than in control groups (76,77). Of the other patient groups an older study from India stands out with 52% positive reactions to neomycin in 50 patients suspected of footwear dermatitis (107). However, in a control group, 47% was sensitized. It may be assumed that in that country, neomycin-containing preparations were freely available and very widely used, and that the selection procedure only included patients who were very strongly suspected of allergic contact dermatitis.

Case reports and case series
This section contains a few examples of case series and (literature references of) case reports; a full literature search was not attempted.

Case series
In Leuven, Belgium, between 1990 and 2017, 16,065 patients were investigated for contact allergy and 118 (0.7%) showed positive patch test reactions to topical ophthalmic medications and/or to their ingredients. Eighty-four individuals (71%) reacted to an active principle. Neomycin was tested in 15,712 patients and was the allergen in eye medications in 19. There were also 135 reactions to neomycin in other types of medications (78). Of 8 patients allergic to a combination preparation of neomycin – gramicidin – nystatin – triamcinolone acetonide. 2 had positive patch tests to the ingredient neomycin (190).

Case reports
A 71-year old woman had allergic contact dermatitis of the eyelids superimposed on discoid lupus erythematosus from neomycin in an unspecified topical pharmaceutical product (and from tosylamide/formaldehyde resin in nail varnish) (111). Despite the poor penetration of neomycin in normal skin (140), in one patient patch testing induced a flare-up of previous sites of allergic contact dermatitis from neomycin (systemic contact dermatitis) (131).

In a 39-year-old man who was previously shown to be allergic to neomycin and bacitracin, systemic contact dermatitis developed from a root canal paste filling containing both antibiotics. Symptoms included generalized itching, erythema and edema of the face, inflammation of the oral mucosa with difficulty in swallowing and breathing, and aggravation of previously existing eczema on the amputation stump of a leg (where he had become sensitized to neomycin and bacitracin). Oral treatment with bacitracin had considerably aggravated the dermatitis and oral provocation with neomycin resulted in itching of the skin in the popliteal folds, follicular eczema at those locations and edema of the face (138). This case of systemic contact dermatitis was caused by resorption of neomycin and bacitracin from the dental root canal paste.

A 63-year-old man developed allergic contact dermatitis from neomycin in an antibiotic ointment applied to the skin eruptions of herpes zoster. The contact dermatitis was confined to the area surrounding the sores, sparing the lesions and their periphery. It was postulated that a decrease in Langerhans cell activity in the herpes zoster lesions and their peripheral area was primarily responsible for this phenomenon (142).

An 8-year-old boy was treated for bilateral secretory otitis media with 3 different topical aural medicaments, but was not altogether symptomless. Patch tests were positive to tixocortol pivalate (marker for hydrocortisone sensitivity; hydrocortisone was present in an ear ointment), neomycin (present in ear drops) and Amerchol L-101, present in a corticosteroid ointment used by the patient. When the child stopped using the 3 medicaments, the symptoms ceased (151). Two patients, who were patch tested because of suspected allergy to bandages, had positive reactions to neomycin and bacitracin, both of which (probably in a combination product) had been applied under adhesive bandages following biopsies and had resulted in pruritic plaques extending beyond the bandage (157).

A 52-year-old male to female transgender patient presented with genital dermatitis. She complained of a red, papular, itchy rash around the newly constructed vulva since undergoing gender affirmation surgery in Thailand. She was prescribed a topical antibacterial ointment with neomycin and bacitracin immediately after surgery. Later, she started using a cream with the anesthetic pramoxine. Physical examination showed excellent anatomical reconstruction with diffuse erythema of the introitus, the labia majora and inguinal skin with scattered 2-3 mm erosions. These findings were markedly improved from patient-provided post-operative photos. Day 5 patch testing results showed positive reactions to the antibacterial ointment, the anesthetic cream, neomycin, pramoxine, and bacitracin (158).

A 70-year-old woman had been treated for some time with eardrops for seborrheic dermatitis in the right ear canal, when she developed worsening of itchy erythematous, edematous and vesicular dermatitis in the ear canal and on the right ear. She stopped the use of the eardrops, with clearance of the lesions within a week. Patch tests with the eardrops and all of its ingredients showed positive (++) patch test reactions at 2 and 4 days to the eardrops and its active ingredients neomycin sulfate 20% pet. and furaltadone HCl 1% pet.(160). Lichenoid allergic contact dermatitis (clinically and histopathologically) to neomycin has been observed. The positive patch test to neomycin was apparently of the lichenoid type, as were the patch tests to cross-reacting aminoglycosides (181).

Other case reports of allergic contact dermatitis to neomycin can be found in refs. 141,144,145,146,149,150, 152 (erythema multiforme-like), 155,163,165,167,168 (possibly systemic contact dermatitis from ophthalmic ointment), 170, 171, 182, 184, and 187 (examples, not a full review).

<u>Occupational contact dermatitis</u>
Occupational dermatitis involving the hands (and sometimes the face) can occur in (dental) nurses (166), physicians, pharmacists, animal feed mill workers (162), farmers (169), oculists (156), veterinarians (161) and dentists (132).

Cross-reactions, pseudo-cross-reactions and co-reactions
In patients sensitized to neomycin, the percentage of cross-reactions to other aminoglycoside antibiotics is high: to paromomycin and butirosin (183) about 90%, to framycetin 70%, to ribostamycin 70% (112), to tobramycin and kanamycin 60%, to gentamicin 50%, to amikacin 30% (180), and to sisomycin (115,122) 20%. However, streptomycin cross-reacts (or co-reacts more likely) in only about 4% of the patients (74,108,112-124,164,173). A high degree of co-reactivity between framycetin and neomycin can be expected, as framycetin consists mostly of neomycin B, which is also the main constituent of neomycin (112). However, neomycin B is not the only sensitizer in either framycetin or neomycin, as a considerable number of patients allergic to neomycin do not react to framycetin and vice versa (113,114,173).

The cross-sensitivity pattern in patients primarily sensitized to other aminoglycosides has not been well investigated. Primary sensitization to gentamicin may infrequently be accompanied by cross-sensitization to neomycin. In a group of 25 patients sensitized to gentamicin present in bone cement used for arthroplasty, only 2 (8%) co-reacted to neomycin (174). There are also indications that patients primarily sensitized to tobramycin in eye drops or eardrops often do *not* cross-react to neomycin (175-178).

Four of 5 patients who had developed allergic contact dermatitis from paromomycin also had positive reactions to 20% neomycin sulfate and the same four also reacted to kanamycin sulfate 10% pet. It was not mentioned whether these patients had used either aminoglycoside before (159).

Cutaneous adverse drug reactions from systemic administration caused by type IV (delayed-type) hypersensitivity

Systemic contact dermatitis
In an individual previously sensitized by topical exposure, the small amount of neomycin absorbed from the gastrointestinal tract (or from an overdose [134]) may be enough to trigger widespread dermatitis or a flare-up at the site of prior contact dermatitis, which is called systemic contact dermatitis (133-135). Such reactions may also occur when cross-reacting chemicals are administered, Thus, subconjunctival injection of framycetin after cataract surgery caused an acute, widespread, and weeping dermatitis in a neomycin-sensitive patient (136) and systemically administered gentamicin induced erythroderma in an individual previously shown to be allergic to neomycin (137).

Immediate contact reactions
Immediate contact reactions (contact urticaria) to neomycin are presented in Chapter 5.

LITERATURE
1   The data in the section 'General' may have been obtained from literature discussed in this chapter, but mostly also or exclusively from one or more of the following online sources: ChemIDPlus Advanced, PubChem, DrugBank, RxList, Drug Central, Drugs.com, and Wikipedia
2   DeKoven JG, Warshaw EM, Zug KA, Maibach HI, Belsito DV, Sasseville D, et al. North American Contact Dermatitis Group patch test results: 2015-2016. Dermatitis 2018;29:297-309
3   Veverka KK, Hall MR, Yiannias JA, Drage LA, El-Azhary RA, Killian JM, et al. Trends in patch testing with the Mayo Clinic standard series, 2011-2015. Dermatitis 2018;29:310-315
4   DeKoven JG, Warshaw EM, Belsito DV, Sasseville D, Maibach HI, Taylor JS, et al. North American Contact Dermatitis Group Patch Test Results: 2013-2014. Dermatitis 2017;28:33-46
5   Uter W, Amario-Hita JC, Balato A, Ballmer-Weber B, Bauer A, Belloni Fortina A, et al. European Surveillance System on Contact Allergies (ESSCA): results with the European baseline series, 2013/14. J Eur Acad Dermatol Venereol 2017;31:1516-1525
6   Ochi H, Cheng SWN, Leow YH, Goon ATJ. Contact allergy trends in Singapore – a retrospective study of patch test data from 2009 to 2013. Contact Dermatitis 2017;76:49-50
7   Warshaw EM, Maibach HI, Taylor JS, Sasseville D, DeKoven JG, Zirwas MJ, et al. North American Contact Dermatitis Group patch test results: 2011-2012. Dermatitis 2015;26:49-59

8   Warshaw EM, Belsito DV, Taylor JS, Sasseville D, DeKoven JG, Zirwas MJ, et al. North American Contact Dermatitis Group patch test results: 2009 to 2010. Dermatitis 2013;24:50-59

9   Wentworth AB, Yiannias JA, Keeling JH, Hall MR, Camilleri MJ, Drage LA, et al. Trends in patch-test results and allergen changes in the standard series: a Mayo Clinic 5-year retrospective review (January 1, 2006, to December 31, 2010). J Am Acad Dermatol 2014;70:269-275

10   Toholka R, Wang Y-S, Tate B, Tam M, Cahill J, Palmer A, Nixon R. The first Australian Baseline Series: Recommendations for patch testing in suspected contact dermatitis. Australas J Dermatol 2015;56:107-115

11   Fall S, Bruze M, Isaksson M, Lidén C, Matura M, Stenberg B, Lindberg M. Contact allergy trends in Sweden – a retrospective comparison of patch test data from 1992, 2000, and 2009. Contact Dermatitis 2015;72:297-304

12   Yin R, Huang XY, Zhou XF, Hao F. A retrospective study of patch tests in Chongqing, China from 2004 to 2009. Contact Dermatitis 2011;65:28-33

13   Uter W, Aberer W, Armario-Hita JC, , Fernandez-Vozmediano JM, Ayala F, Balato A, et al. Current patch test results with the European baseline series and extensions to it from the 'European Surveillance System on Contact Allergy' network, 2007-2008. Contact Dermatitis 2012;67:9-19

14   Fransway AF, Zug KA, Belsito DV, Deleo VA, Fowler JF Jr, Maibach HI, et al. North American Contact Dermatitis Group patch test results for 2007-2008. Dermatitis 2013;24:10-21

15   Zug KA, Warshaw EM, Fowler JF Jr, Maibach HI, Belsito DL, Pratt MD, et al. Patch-test results of the North American Contact Dermatitis Group 2005-2006. Dermatitis 2009;20:149-160

16   Uter W, Rämsch C, Aberer W, Ayala F, Balato A, Beliauskiene A, et al. The European baseline series in 10 European Countries, 2005/2006 – Results of the European Surveillance System on Contact Allergies (ESSCA). Contact Dermatitis 2009;61:31-38

17   Davis MD, Scalf LA, Yiannias JA, Cheng JF, El-Azhary RA, Rohlinger AL, et al. Changing trends and allergens in the patch test standard series. Arch Dermatol 2008;144:67-72

18   Carlsen BC, Menné T, Johansen JD. 20 Years of standard patch testing in an eczema population with focus on patients with multiple contact allergies. Contact Dermatitis 2007;57:76-83

19   ESSCA Writing Group. The European Surveillance System of Contact Allergies (ESSCA): results of patch testing the standard series, 2004. J Eur Acad Dermatol Venereol 2008;22:174-181

20   Warshaw EM, Belsito DV, DeLeo VA, Fowler JF Jr, Maibach HI, Marks JG, et al. North American Contact Dermatitis Group patch-test results, 2003-2004 study period. Dermatitis 2008;19:129-136

21   Worm M, Brasch J, Geier J, Uter W, Schnuch A. Epikutantestung mit der DKG-Standardreihe 2001-2004. Hautarzt 2005;56:1114-1124 (Article in German)

22   Janach M, Kühne A, Seifert B, French FE, Ballmer-Weber B, Hofbauer GFL. Changing delayed-type sensitizations to the baseline series allergens over a decade at the Zurich University Hospital. Contact Dermatitis 2010;63:42-48

23   Lazarov A. European Standard Series patch test results from a contact dermatitis clinic in Israel during the 7-year period from 1998 to 2004. Contact Dermatitis 2006;55:73-76

24   Akyol A, Boyvat A, Peksari Y, Gurgey E. Contact sensitivity to standard series allergens in 1038 patients with contact dermatitis in Turkey. Contact Dermatitis 2005;52:333-337

25   Uter W, Hegewald J, Aberer W et al. The European standard series in 9 European countries, 2002/2003 – First results of the European Surveillance System on Contact Allergies. Contact Dermatitis 2005;53:136-145

26   Pratt MD, Belsito DV, DeLeo VA, Fowler JF Jr, Fransway AF, Maibach HI, et al. North American Contact Dermatitis Group patch-test results, 2001-2002 study period. Dermatitis 2004;15:176-183

27   Lindberg M, Edman B, Fischer T, Stenberg B. Time trends in Swedish patch test data from 1992 to 2000. A multi-centre study based on age- and sex-adjusted results of the Swedish standard series. Contact Dermatitis 2007;56:205-210

28   Machovcova A, Dastychova E, Kostalova D, et al. Common contact sensitizers in the Czech Republic. Patch test results in 12,058 patients with suspected contact dermatitis. Contact Dermatitis 2005;53:162-166

29   Britton JE, Wilkinson SM, English JSC, Gawkrodger DJ, Ormerod AD, Sansom JE, et al. The British standard series of contact dermatitis allergens: validation in clinical practice and value for clinical governance. Br J Dermatol 2003;148:259-264

30   Marks JG Jr, Belsito DV, DeLeo VA, Fowler JF Jr, Fransway AF, Maibach HI, et al. North American Contact Dermatitis Group patch-test results, 1998–2000. Am J Contact Dermat 2003;14:59-62

31   Wetter DA, Davis MDP, Yiannias JA, Cheng JF, Connolly SM, el-Azhary RA, et al. Patch test results from the Mayo Contact Dermatitis Group, 1998–2000. J Am Acad Dermatol 2005;53:416-421

32 Bruynzeel DP, Diepgen TL, Andersen KE, Brandão FM, Bruze M, Frosch PJ, et al (EECDRG). Monitoring the European Standard Series in 10 centres 1996–2000. Contact Dermatitis 2005;53:146-152

33 Uter W, Geier J, Pfahlberg A, Effendy I. The spectrum of contact allergy in elderly patients with and without lower leg dermatitis. Dermatology 2002;204:266-272

34 Erfurt-Berge C, Geier J, Mahler V. The current spectrum of contact sensitization in patients with chronic leg ulcers or stasis dermatitis - new data from the Information Network of Departments of Dermatology (IVDK). Contact Dermatitis 2017;77:151-158

35 Barbaud A, Collet E, Le Coz CJ, Meaume S, Gillois P. Contact allergy in chronic leg ulcers: results of a multicentre study carried out in 423 patients and proposal for an updated series of patch tests. Contact Dermatitis 2009;60:279-287

36 Zmudzinska M, Czarnecka-Operacz M, Silny W, Kramer L. Contact allergy in patients with chronic venous leg ulcers – possible role of chronic venous insufficiency. Contact Dermatitis 2006;54:100-105

37 Saap L, Fahim S, Arsenault E, Pratt M, Pierscianowski T, Falanga V, Pedvis-Leftick A. Contact sensitivity in patients with leg ulcerations: a North American study. Arch Dermatol 2004;140:1241-1246

38 Smart V, Alavi A, Coutts P, Fierheller M, Coelho S, Holness LD, et al. Contact allergens in persons with leg ulcers: a Canadian study in contact sensitization. Int J Low Extrem Wounds 2008;7:120-125

39 Zmudzinska M, Czarnecka-Operacz M, SilnyW. Contact allergy to glucocorticosteroids in patients with chronic venous leg ulcers, atopic dermatitis and contact allergy. Acta Dermatovenerol Croat 2008;16:72-78

40 Machet L, Couhe C, Perrinaud A, Hoarau C, Lorette G, Vaillant L. A high prevalence of sensitization still persists in leg ulcer patients: a retrospective series of 106 patients tested between 2001 and 2002 and a meta-analysis of 1975-2003. Br J Dermatol 2004;150:929-935

41 Gooptu C, Powell SM. The problems of rubber hypersensitivity (types I and IV) in chronic leg ulcer and stasis eczema patients. Contact Dermatitis 1999;41:89-93

42 Marasovic D, Vuksic I. Allergic contact dermatitis in patients with leg ulcers. Contact Dermatitis1999;41:107-109

43 Zaki I, Shall L, Dalziel KL. Bacitracin: a significant sensitizer in leg ulcer patients? Contact Dermatitis1994;31:92-94

44 Raudonis T, Vankeviciute RA, Lideikaite A, Grigaityte AG, Grigaitiene J. Contact sensitization in patients with chronic leg ulcers: Results of a 5-year retrospective analysis. Adv Skin Wound Care 2019;32:558-562

45 Tomljanović-Veselski M, Lipozencić J, Lugović L. Contact allergy to special and standard allergens in patients with venous ulcers. Coll Antropol 2007;31:751-756

46 Perrenoud D, Ramelet AA. Chronic leg ulcers and eczema. Curr Probl Dermatol 1999;27:165-169

47 Jankićević J, Vesić S, Vukićević J, Gajić M, Adamic M, Pavlović MD. Contact sensitivity in patients with venous leg ulcers in Serbia: comparison with contact dermatitis patients and relationship to ulcer duration. Contact Dermatitis 2008;58:32-36.

48 Renner R, Wollina U. Contact sensitization in patients with leg ulcers and/or leg eczema: comparison between centers. Int J Low Extrem Wounds 2002;1:251-255

49 Landeck L, John SM, Geier J. Periorbital dermatitis in 4779 patients – patch test results during a 10-year period. Contact Dermatitis 2014;70:205-212

50 Landeck L, Schalock PC, Baden LA, Gonzalez E. Periorbital contact sensitization. Am J Ophthalmol 2010;150:366-370

51 Feser A, Plaza T, Vogelgsang L, Mahler V. Periorbital dermatitis – a recalcitrant disease: causes and differential diagnoses. Brit J Dermatol 2008;159:858-863

52 Ockenfels H, Seemann U, Goos M. Contact allergy in patients with periorbital eczema: an analysis of allergens. Dermatology 1997;195:119-124

53 Valsecchi R, Imberti G, Martino D, Cainelli T. Eyelid dermatitis: an evaluation of 150 patients. Contact Dermatitis 1992;27:143-147

54 Temesvári E, Pónyai G, Németh I, Hidvégi B, Sas A, Kárpáti S. Periocular dermatitis: a report of 401 patients. J Eur Acad Dermatol Venereol 2009;23:124-128

55 Herbst RA, Uter W, Pirker C, Geier J, Frosch PJ. Allergic and non-allergic periorbital dermatitis: patch test results of the Information Network of the Departments of Dermatology during a 5-year period. Contact Dermatitis 2004;51:13-19

56 Cooper SM, Shaw S. Eyelid dermatitis: an evaluation of 232 patchtest patients over 5 years. Contact Dermatitis 2000;42:291-293

57 Guin JD. Eyelid dermatitis: experience in 203 cases. J Am Acad Dermatol 2002;47:755-765

58 Amin KA, Belsito DV. The aetiology of eyelid dermatitis: a 10-year retrospective analysis. Contact Dermatitis 2006;55:280-285

59 Uter W, Spiewak R, Cooper SM, Wilkinson M, Sánchez Pérez J, Schnuch A, et al. Contact allergy to ingredients of topical medications: results of the European Surveillance System on Contact Allergies (ESSCA), 2009-2012. Pharmacoepidemiol Drug Saf 2016;25:1305-1312

60 Marren P, Wojnarowska F, Powell S. Allergic contact dermatitis and vulvar dermatoses. Br J Dermatol 1992;126:52-56

61 Lewis FM, Shah M, Gawkrodger DJ. Contact sensitivity in pruritus vulvae: patch test results and clinical outcome. Dermatitis 1997;8:137-140

62 Lewis FM, Harrington CI, Gawkrodger DJ. Contact sensitivity in pruritus vulvae: a common and manageable problem. Contact Dermatitis 1994;31:264-265

63 Morris SD, Rycroft RJ, White IR, Wakelin SH, McFadden JP. Comparative frequency of patch test reactions to topical antibiotics. Br J Dermatol 2002;146:1047-1051

64 Warshaw EM, Furda LM, Maibach HI, Rietschel RL, Fowler JF Jr, Belsito DV, et al. Anogenital dermatitis in patients referred for patch testing: retrospective analysis of cross-sectional data from the North American Contact Dermatitis Group, 1994-2004. Arch Dermatol 2008;144:749-755

65 Agulló-Pérez AD, Hervella-Garcés M, Oscoz-Jaime S, Azcona-Rodríguez M, Larrea-García M, Yanguas-Bayona JI. Perianal dermatitis. Dermatitis 2017;28:270-275

66 O'Gorman SM, Torgerson RR. Allergic contact dermatitis of the vulva. Dermatitis 2013;24:64-72

67 Doherty VR, Forsyth A, MacKie RM. Pruritus vulvae: a manifestation of contact hypersensitivity? Br J Dermatol 1990:123(Suppl.37):26-27

68 Lucke TW, Fleming CJ, McHenry P, Lever R. Patch testing in vulval dermatoses: how relevant is nickel? Contact Dermatitis 1998;38:111-112

69 Bauer A. Contact sensitization in the anal and genital area. Curr Probl Dermatol 2011;40:133-141

70 Abu-Asi MJ, White IR, McFadden JP, White JM. Patch testing is clinically important for patients with peri-anal dermatoses and pruritus ani. Contact Dermatitis 2016;74:298-300

71 Kränke B, Trummer M, Brabek E, Komericki P, Turek TD, Aberer W. Etiologic and causative factors in perianal dermatitis: results of a prospective study in 126 patients. Wien Klin Wochenschr 2006;118(3-4):90-94 (Article in German)

72 Warshaw EM, Kimyon RS, Silverberg JI, Belsito DV, DeKoven JG, Maibach HI, et al. Evaluation of patch test findings in patients with anogenital dermatitis. JAMA Dermatol 2019;156:85-91

73 Foley CC, White S, Merry S, Nolan U, Moriarty B, et al. Understanding the role of cutaneous allergy testing in anogenital dermatoses: a retrospective evaluation of contact sensitization in anogenital dermatoses. Int J Dermatol 2019;58:806-810

74 Millard TP, Orton DI. Changing patterns of contact allergy in chronic inflammatory ear disease. Contact Dermatitis 2004;50:83-86

75 Devos SA, Mulder JJ, van der Valk PG. The relevance of positive patch test reactions in chronic otitis externa. Contact Dermatitis 2000;42:354-355

76 Hillen U, Geier J, Goos M. Contact allergies in patients with eczema of the external ear canal. Results of the Information Network of Dermatological Clinics and the German Contact Allergy Group. Hautarzt 2000;51:239-243 (Article in German)

77 Yariktas M, Yildirim M, Doner F, Baysal V, Dogru H. Allergic contact dermatitis prevalence in patients with eczematous external otitis. Asian Pac J Allergy Immunol 2004;22:7-10

78 Gilissen L, De Decker L, Hulshagen T, Goossens A. Allergic contact dermatitis caused by topical ophthalmic medications: Keep an eye on it! Contact Dermatitis 2019;80:291-297

79 Uter W, Menezes de Pádua C, Pfahlberg A, Nink K, Schnuch A, Behrens-Baumann W. Contact allergy to topical ophthalmological drugs - epidemiological risk assessment. Klin Monbl Augenheilkd 2009;226:48-53 (Article in German)

80 Wijnmaalen AL, van Zuuren EJ, de Keizer RJ, Jager MJ. Cutaneous allergy testing in patients suspected of an allergic reaction to eye medication. Ophthalmic Res 2009;41:225-229

81 Diepgen TL, Ofenloch RF, Bruze M, Bertuccio P, Cazzaniga S, Coenraads P-J, et al. Prevalence of contact allergy in the general population in different European regions. Br J Dermatol 2016;174:319-329

82 Mortz CG, Bindslev-Jensen C, Andersen KE. Prevalence, incidence rates and persistence of contact allergy and allergic contact dermatitis in The Odense Adolescence Cohort Study: a 15-year follow-up. Brit J Dermatol 2013;168:318-325

83   Thyssen JP, Linneberg A, Menné T, Nielsen NH, Johansen JD. Contact allergy to allergens of the TRUE-test (panels 1 and 2) has decreased modestly in the general population. Br J Dermatol 2009;161:1124-1129

84   Dotterud LK, Smith-Sivertsen T. Allergic contact sensitization in the general adult population: a population-based study from Northern Norway. Contact Dermatitis 2007;56:10-15

85   Nielsen NH, Linneberg A, Menné T, Madsen F, Frølund L, Dirksen A, et al. Allergic contact sensitization in an adult Danish population: two cross-sectional surveys eight years apart (the Copenhagen Allergy Study). Acta Derm Venereol 2001;81:31-34

86   Mortz CG, Lauritsen JM, Bindslev-Jensen C, Andersen KE. Contact allergy and allergic contact dermatitis in adolescents: prevalence measures and associations. Acta Derm Venereol 2002;82:352-358

87   Schäfer T, Böhler E, Ruhdorfer S, Weigl L, Wessner D, Filipiak B, et al. Epidemiology of contact allergy in adults. Allergy 2001;56:1192-1196

88   Mortz CG, Lauritsen JM, Bindslev-Jensen C, Andersen KE. Prevalence of atopic dermatitis, asthma, allergic rhinitis, and hand and contact dermatitis in adolescents. The Odense Adolescence Cohort Study on Atopic Diseases and Dermatitis. Br J Dermatol 2001;144:523-532

89   Thyssen JP, Uter W, Schnuch A, Linneberg A, Johansen JD. 10-year prevalence of contact allergy in the general population in Denmark estimated through the CE-DUR method. Contact Dermatitis 2007;57:265-272

90   Schnuch A, Uter W, Geier J, Gefeller O (for the IVDK study group). Epidemiology of contact allergy: an estimation of morbidity employing the clinical epidemiology and drug-utilization research (CE-DUR) approach. Contact Dermatitis 2002;47:32-39

91   Nardelli A, Degreef H, Goossens A. Contact allergic reactions of the vulva: a 14-year review. Dermatitis 2004;15:131-136

92   Teo Y, McFadden JP, White IR, Lynch M, Banerjee P. Allergic contact dermatitis in atopic individuals: Results of a 30-year retrospective study. Contact Dermatitis 2019;81:409-416

93   Tagka A, Stratigos A, Lambrou GI, Nicolaidou E, Katsarou A, Chatziioannou A. Prevalence of contact dermatitis in the Greek population: A retrospective observational study. Contact Dermatitis 2019;81:460-462

94   Greig JE, Carson CF, Stuckey MS, Riley TV. Prevalence of delayed hypersensitivity to the European standard series in a self-selected population. Australas J Dermatol 2000;41:86-89

95   Bryld LE, Hindsberger C, Kyvik KO, Agner T, Menné T. Risk factors influencing the development of hand eczema in a population-based twin sample. Br J Dermatol 2003;149:1214-1220

96   Dotterud LK. The prevalence of allergic contact sensitization in a general population in Tromsø, Norway. Int J Circumpolar Health 2007;66:328-334

97   Lagrelius M, Wahlgren CF, Matura M, Kull I, Lidén C. High prevalence of contact allergy in adolescence: results from the population-based BAMSE birth cohort. Contact Dermatitis 2016;74:44-51

98   van Amerongen CCA, Ofenloch R, Dittmar D, Schuttelaar MLA. New positive patch test reactions on day 7—The additional value of the day 7 patch test reading. Contact Dermatitis 2019;81:280-287.

99   Madsen JT, Andersen KE. Outcome of a second patch test reading of TRUE Tests® on D6/7. Contact Dermatitis 2013;68:94-97

100 Chaudhry HM, Drage LA, El-Azhary RA, Hall MR, Killian JM, Prakash AV, et al. Delayed patch-test reading after 5 days: an update from the Mayo Clinic Contact Dermatitis Group. Dermatitis 2017;28:253-260

101 Sukakul T, Chaweekulrat P, Limphoka P, Boonchai W. Changing trends of contact allergens in Thailand: A 12-year retrospective study. Contact Dermatitis 2019;81:124-129

102 Menezes de Padua CA, Uter W, Schnuch A. Contact allergy to topical drugs: prevalence in a clinical setting and estimation of frequency at the population level. Pharmacoepidemiol Drug Saf 2007;16:377-384

103 Menezes de Padua CA, Schnuch A, Nink K, Pfahlberg A, Uter W. Allergic contact dermatitis to topical drugs – Epidemiological risk assessment. Pharmacoepidemiol Drug Saf 2008;17:813-821

104 Yung MW, RajendraT. Delayed hypersensitivity reaction to topical aminoglycosides in patients undergoing middle ear surgery. Clin Otolaryngol Allied Sci 2002;27:365-368

105 Gilissen L, Goossens A. Frequency and trends of contact allergy to and iatrogenic contact dermatitis caused by topical drugs over a 25-year period. Contact Dermatitis 2016;75:290-302

106 Shih Y-H, Sun C-C, Tseng Y-H, Chu C-Y. Contact dermatitis to topical medicaments: a retrospective study from a medical center in Taiwan. Dermatol Sinica 2015;33:181-186

107 Saha M, Srinivas CR, Shenoi SD, Balachandran C, Acharya S. Sensitivity to topical medicaments among suspected cases of footwear dermatitis. Contact Dermatitis 1993;28:44-45

108 Bajaj AK, Gupta SC, Chatterjee AK. Contact sensitivity to topical aminoglycosides in India. Contact Dermatitis 1992;27:204-205

109 Landeck L, John SM, Geier J. Topical ophthalmic agents as allergens in periorbital dermatitis. Br J Ophthalmol 2014;98:259-262

110 Spring S, Pratt M, Chaplin A. Contact dermatitis to topical medicaments: a retrospective chart review from the Ottawa Hospital Patch Test Clinic. Dermatitis 2012;23:210-213

111 Trindade MA, Alchorne AO, da Costa EB, Enokihara MM. Eyelid discoid lupus erythematosus and contact dermatitis: a case report. J Eur Acad Dermatol Venereol 2004;18:577-579

112 Samsoen M, Metz R, Melchior E, Foussereau J. Cross-sensitivity between aminoside antibiotics. Contact Dermatitis 1980;6:141

113 Carruthers JA, Cronin E. Incidence of neomycin and framycetin sensitivity. Contact Dermatitis 1976;2:269-270

114 Kirton, V, Munro-Ashman D. Contact dermatitis from neomycin and framycetin. Lancet 1965;i:138-139

115 Förström L, Pirilä V, Pirilä L. Cross-sensitivity within the neomycin group of antibiotics. Acta Derm Venereol (Stockh) 1979;59(Suppl.):67-69

116 Pirilä V, Pirilä L. Sensitization to the neomycin group of antibiotics. Patterns of cross-sensitivity as a function of polyvalent sensitization to different portions of the neomycin molecule. Acta Derm Venereol (Stockh) 1966;46:489-496

117 Sasseville D. Neomycin. Dermatitis 2010;21:3-7

118 MacDonald RH, Beck M. Neomycin: a review with particular reference to dermatological usage. Clin Exp Dermatol 1983;8:249-258

119 Gehrig KA, Warshaw EM. Allergic contact dermatitis to topical antibiotics: epidemiology, responsible allergens and management. J Am Acad Dermatol 2008;58:1-21

120 Kimura M, Kawada A. Contact sensitivity induced by neomycin with cross-sensitivity to other aminoglycoside antibiotics. Contact Dermatitis 1998;39:148-150

121 Schorr WF, Ridgway HB. Tobramycin-neomycin cross-sensitivity. Contact Dermatitis 1977;3:133-137

122 Förström L, Pirilä V. Cross-sensitivity within the neomycin group of antibiotics. Contact Dermatitis 1978;4:312

123 Rudzki E, Zakrzewski Z, Rebandel P, Grzywa Z, Hudymowicz W. Cross reactions between aminoglycoside antibiotics. Contact Dermatitis 1988;18:314-316

124 Rudzki E, Rebandel P. Cross-reactions with 4 aminoglycoside antibiotics at various concentrations. Contact Dermatitis 1996;35:62

125 Baer RL, Ludwig JS. Allergic eczematous sensitization to neomycin. Ann Allergy 1952;10:136-137

126 Kile RL, Rockwell EM, Schwartz J. Use of neomycin in dermatology. JAMA 1952;148:339-343

127 Calnan CD, Sarkany I. Contact dermatitis from neomycin. Br J Pharmacol 1958;70:435-445

128 Epstein E. Contact dermatitis to neomycin with false negative patch tests: allergy established by intradermal and usage tests. Contact Dermatitis 1980;6:219-220

129 Sood A, Taylor JS. Bacitracin: allergen of the year. Am J Contact Dermat 2003;14:3-4

130 Grandinetti PJ, Fowler JF. Simultaneous contact allergy to neomycin, bacitracin and polymyxin. J Am Acad Dermatol 1990;23:646-647

131 Jacob SE, Barland C, El Saie ML. Patch-test-induced "flare-up" reactions to neomycin at prior biopsy sites. Dermatitis 2008;19:E46-E48

132 Phillips DK. Neomycin sulfate. In: Guin JD, Ed. Practical contact dermatitis. New York: McGraw-Hill Inc., 1995:167-177

133 Ekelund AG, Möller H. Oral provocation in eczematous contact allergy to neomycin and hydroxyquinolines. Acta Derm Venereol (Stockh) 1969;49:422-426

134 Menné T, Weismann K. Hämatogenes Kontaktekzem nach oraler Gabe von Neomyzin. Hautarzt 1984;35:319-320 (Article in German)

135 Bouffioux B, Heid E. Eczéma endogène à la néomycine. Nouv Dermatol 1990;9:25 (Article in French, data cited in ref. 117)

136 Morton CA, Evans CD, Douglas WS. Allergic contact dermatitis following subconjunctival injection of framycetin. Contact Dermatitis 1993;29:42-43

137 Guin JD, Phillips D. Erythroderma from systemic contact dermatitis: a complication of systemic gentamicin in a patient with contact allergy to neomycin. Cutis 1989;43:564-567

138 Pirilä V, Rantanen AV. Root canal treatment with bacitracin-neomycin as a cause of flare-up of allergic eczema. Oral Surg 1960;13:589-593

139 Epstein S. Contact dermatitis from neomycin due to dermal delayed (tuberculin-type) sensitivity. Dermatologica 1956;113:191-201

140 Panzer JD, Epstein WL. Percutaneous absorption following topical application of neomycin. Arch Dermatol 1970;102:536-539

141 Cooper SM, Shaw S. Contact allergy to nystatin: an unusual allergen. Contact Dermatitis 1999;41:120

142 Katayama H, Karube S, Ueki Y, Yaota H. Contact dermatitis sparing the eruption of herpes zoster and its periphery. Dermatologica 1990;181:65-67

143 Reitamo S, Lauerma AI, Stubb S, Käyhkö K, Visa K, Förström L. Delayed hypersensitivity to topical corticosteroids. J Am Acad Dermatol 1986;14:582-589

144 Goh CL. Cross-sensitivity to multiple topical corticosteroids. Contact Dermatitis 1989;20:65-67

145 Kooij R. Hypersensitivity to hydrocortisone. Br J Dermatol 1959;71:392-394

146 Rivara G, Tomb RR, Foussereau J. Allergic contact dermatitis from topical corticosteroids. Contact Dermatitis 1989;21:83-91

147 Burden AD, Beck MH. Contact hypersensitivity to topical corticosteroids. Br J Dermatol 1992;127:497-501

148 Jagodzinski LJ, Taylor JS, Oriba H. Allergic contact dermatitis from topical corticosteroid preparations. Am J Contact Dermat 1995;6:67-74

149 Lyon CC, Beck MH. Allergic contact dermatitis reactions to corticosteroids in periorbital inflammation and conjunctivitis. Eye (Lond) 1998;12(Pt.1):148-149

150 Boujnah-Khouadja A, Brändle I, Reuter G, Foussereau J. Allergy to 2 new corticoid molecules. Contact Dermatitis 1984;11:83-87

151 Isaksson M. Triple sensitization in a child with chronic otitis externa. Contact Dermatitis 2002;47:172

152 Moreno Escobosa MC, Moya Quesada MC, Cruz Granados S, Amat López J. Contact dermatitis to antibiotic ointments. J Investig Allergol Clin Immunol 2009;19:510-511

153 Meneghini CL, Angelini G. Secondary polymorphic eruptions in allergic contact dermatitis. Dermatologica 1981;163:63-70

154 Prystowsky SD, Nonomura JH, Smith RW, Allen AM. Allergic hypersensitivity to neomycin. Relationship between patch test reactions and "use" tests. Arch Dermatol 1979;115:713-715

155 Calnan CD. Oxypolyethoxydodecane in an ointment. Contact Dermatitis 1978;4:168-169

156 Rebandel P. Rudzki. Occupational contact sensitivity in oculists. Contact Dermatitis 1986;15:92

157 Widman TJ, Oostman H, Storrs FJ. Allergic contact dermatitis from medical adhesive bandages in patients who report having a reaction to medical bandages. Dermatitis 2008;19:32-37

158 Schlarbaum JP, Kimyon RS, Liou YL, Becker O'Neill L, Warshaw EM. Genital dermatitis in a transgender patient returning from Thailand: A diagnostic challenge. Travel Med Infect Dis 2019;27:134-135

159 Veraldi S, Benzecry V, Faraci AG, Nazzaro G. Allergic contact dermatitis caused by paromomycin. Contact Dermatitis 2019;81:393-394

160 Sánchez-Pérez J, Córdoba S, del Río MJ, García-Díes A. Allergic contact dermatitis from furaltadone in eardrops. Contact Dermatitis 1999;40:222

161 Falk ES, Hektoen H, Thune PO. Skin and respiratory tract symptoms in veterinary surgeons. Contact Dermatitis 1985;12:274-278

162 Mancuso G, Staffa M, Errani A, Berdondini RM, Fabbri P. Occupational dermatitis in animal feed mill workers. Contact Dermatitis 1990;22:37-41

163 Jensen OC, Allen HJ, Mordecain LR. Neomycin contact dermatitis superimposed on otitis externa. JAMA 1966;195:131-133

164 Van Ginkel CJ, Bruintjes TD, Huizing EH. Allergy due to topical medications in chronic otitis externa and chronic otitis media. Clin Otolaryngol Allied Sci 1995;20:326-328

165 Sánchez-Pérez J, Abajo P, Córdoba S, García-Díez A. Allergic contact dermatitis from sodium metabisulfite in an antihemorrhoidal cream. Contact Dermatitis 2000;42:176-177

166 Kanerva L, Miettinen P, Alanko K, Estlander T, Jolanki R. Occupational allergic contact dermatitis from glyoxal, glutaraldehyde and neomycin sulfate in a dental nurse. Contact Dermatitis 2000;42:116-117

167 Mariani R, Tardio M, Bassi R, Alessandrini F. Allergic contact conjunctivitis without eyelid involvement. Contact Dermatitis 1991;24:227

168 Black H. Allergy to cycloheximide (Actidione). Contact Dermatitis Newsletter 1971;10:243

169 Simpson JR. Dermatitis from neomycin in a calf-drench. Contact Dermatitis Newsletter 1974;15:447

170 Freeman S, Stephens R. Cheilitis: analysis of 75 cases referred to a contact dermatitis clinic. Am J Contact Dermat 1999;10:198-200

171 Audicana M, Echechipia S, Fernández E, Bernaola G, Munoz D, Fernandez de Corres L. Contact dermatitis from phenylephrine. Am J Cont Derm 1993;4:225-228

172 Menezes de Padua CA, Schnuch A, Lessmann H, Geier J, Pfahlberg A, Uter W. Contact allergy to neomycin sulfate: results of a multifactorial analysis. Pharmacoepidemiol Drug Safe 2005;14:725-733

173 Kruyswijk MR, van Driel LM, Polak BC, Go-Sennema AA. Contact allergy following administration of eyedrops and eye ointments. Doc Ophthalmol 1980;48:251-253

174 Thomas B, Kulichova D, Wolf R, Summer B, Mahler V, Thomas P. High frequency of contact allergy to implant and bone cement components, in particular gentamicin, in cemented arthroplasty with complications: usefulness of late patch test reading. Contact Dermatitis 2015;73:343-349

175 Hagen S, Grey K, Warshaw E. Tobramycin sensitivity is not consistently detected by neomycin on patch testing. Dermatitis 2016;27:152-155

176 Alvarez MS, Brancaccio RR. Periorbital dermatitis. Dermatitis 2006;17:43-44

177 Caraffini S, Assalve D, Stingeni L, Lisi P. Allergic contact conjunctivitis and blepharitis from tobramycin. Contact Dermatitis 1995;32:186-187

178 Menéndez Ramos F, Llamas Martín R, Zarco Olivo C, Dorado Bris JM, Merino Luque MV. Allergic contact dermatitis from tobramycin. Contact Dermatitis 1990;22:305-306

179 Tam I, Schalock PC, González E, Yu J. Patch testing results from the Massachusetts General Hospital Contact Dermatitis Clinic, 2007-2016. Dermatitis 2020;31:202-208

180 Jerez J, Rodríguez F, Jiménez I, Martín-Gil D. Cross-reactions between aminoside antibiotics. Contact Dermatitis 1987;17:325

181 Lembo G, Balato N, Patruno C, Pini D, Ayala F. Lichenoid contact dermatitis due to aminoglycoside antibiotics. Contact Dermatitis 1987;17:122-123

182 Izumi AK, Shmunes E, Wood MG. Familial benign chronic pemphigus: the role of trauma including contact sensitivity. Arch Dermatol 1971;104:177-181

183 Schorr WF, Wenzel FJ, Hededus SI. Cross-sensitivity and aminoglycoside antibiotics. Arch Dermatol 1973;107:533-539

184 Seok Oh C, Young Lee J. Contact allergy to various ingredients of topical medicaments. Contact Dermatitis 2003;49:49-50

185 Holmes RC, Johns AN, Wilkinson JD, Black MM, Rycroft RJ. Medicament contact dermatitis in patients with chronic inflammatory ear disease. J R Soc Med 1982;75:27-30

186 Boyvat A, Kalay Yildizhan I. Patch test results of the European baseline series among 1309 patients in Turkey between 2013 and 2019. Contact Dermatitis 2020 Jul 3. doi: 10.1111/cod.13653. Online ahead of print.

187 Pevny I, Brennenstuhl, Razmskas G. Patch testing in children (II). Results and case reports. Contact Dermatitis 1984;11:302-310

188 Pirilä V, Rouhunkoski S. On cross-sensitization between neomycin, bacitracin, kanamycin and framycetin. Dermatologica 1960;121:335-342.

189 Elliott J, Abbas M, Hull P, de Gannes G, Toussi R, Milani A. Decreasing rates of neomycin sensitization in Western Canada. J Cutan Med Surg 2016;20:446-452

190 Larsen WG. Allergic contact dermatitis to the perfume in Mycolog cream. J Am Acad Dermatol 1979;1:131-133

Chapter 3.236 NETICONAZOLE

IDENTIFICATION

Description/definition	: Neticonazole is the vinyl imidazole that conforms to the structural formula shown below
Pharmacological classes	: Antifungal agents
IUPAC name	: 1-[(E)-2-Methylsulfanyl-1-(2-pentoxyphenyl)ethenyl]imidazole
Other names	: (E)-1-(2-(Methylthio)-1-(o-(pentyloxy)phenyl)vinyl)imidazole
CAS registry number	: 130726-68-0
EC number	: Not available
Merck Index monograph	: 7833
Patch testing	: 1% pet.; alcohol or MEK (methyl ethyl ketone) is usually preferred for patch testing imidazole antifungals
Molecular formula	: $C_{17}H_{22}N_2OS$

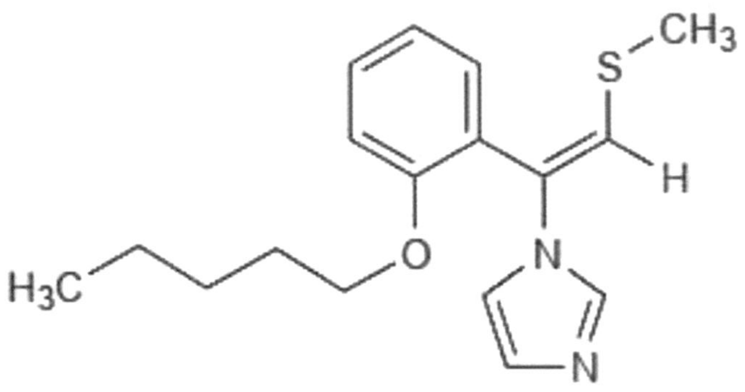

GENERAL

Neticonazole is an imidazole antifungal drug. It inhibits P450-dependent C-14α-demethylation of lanosterol, thereby preventing conversion of lanosterol to ergosterol and inhibiting fungal cell wall synthesis. Neticonazole is used in Japan as a topical antifungal drug for the treatment of superficial skin infections. In pharmaceutical products neticonazole is employed as neticonazole hydrochloride (CAS number 130773-02-3, EC number not available, molecular formula $C_{17}H_{23}ClN_2OS$) (1).

CONTACT ALLERGY

Case reports

A 65-year-old man presented with dermatitis on the feet. He had been using lanoconazole ointment to treat tinea pedis for 10 months. Patch testing with lanoconazole ointment and lanoconazole (10%, 1%, 0.1% pet.) was positive. Patch testing with other imidazoles including neticonazole was negative at that time. He then started using neticonazole ointment, but, 4 months later, he again developed dermatitis on the feet. Patch testing now was positive to neticonazole ointment and neticonazole (1%, 0.3%, 0.1% pet.; in the table concentrations were given as 10%, 1% and 0.1% pet.) and diethyl sebacate, an emulsifier in neticonazole ointment (2).

A 39-year-old woman developed dermatitis on the inner feet and soles, where she had applied neticonazole cream for 18 months. Patch tests were positive to the cream 'as is' and to neticonazole 1% and 0.5% pet., but negative to 0.1% pet. Ten controls were negative to neticonazole 1% pet. Eleven other imidazole antifungal creams were also tested and the patient had a positive reaction to sulconazole cream, but this was not considered to be a cross-reaction (3).

A 38-year-old man had started using 1% neticonazole cream for tinea pedis. Six months later, he had erythema and scaling on the toes of both feet. When patch tested, he reacted to the cream 'as is', to neticonazole 1% and 10% pet. and to commercial econazole and sulconazole creams, which were considered to be cross-reactions (4).

Allergic contact dermatitis from neticonazole has been reported in 8 other Japanese patients (5,6; details unknown, articles in Japanese, data cited in ref. 3).

Cross-reactions, pseudo-cross-reactions and co-reactions
Not to other imidazoles (2,3,6). Possibly cross-sensitivity to econazole and sulconazole in a patient sensitized to neticonazole (4)

LITERATURE

1   The data in the section 'General' may have been obtained from literature discussed in this chapter, but mostly also or exclusively from one or more of the following online sources: ChemIDPlus Advanced, PubChem, DrugBank, RxList, Drug Central, Drugs.com, and Wikipedia
2   Umebayashi Y, Ito S. Allergic contact dermatitis due to both lanoconazole and neticonazole ointments. Contact Dermatitis 2001;44:48-49
3   Shono M. Allergic contact dermatitis from neticonazole hydrochloride. Contact Dermatitis 1997;37:136-137
4   Kawada A, Hiruma M, Fujioka A, Tajima S, Ishibashi A, Kawada I. Contact dermatitis from neticonazole. Contact Dermatitis 1997;36:106-107
5   Kusunoki T. 7 Cases of contact dermatitis from Atolant ®. Jpn J Dermatol 1997;107:43 (Article in Japanese, cited in ref. 3).
6   Natsuaki M. A case of contact dermatitis from neticonazole hydrochloride associated with auto-sensitization dermatitis. Jpn J Dermatoallergol 1996;4:60 (Article in Japanese, cited in ref. 3).

Chapter 3.237　NICOBOXIL

IDENTIFICATION

Description/definition　　　: Nicoboxil is the aromatic carboxylic acid and member of pyridines that conforms to the structural formula shown below

Pharmacological classes　: Rubefacient

IUPAC name　　　　　　: 2-Butoxyethyl pyridine-3-carboxylate

Other names　　　　　　: Butoxyethyl nicotinate

CAS registry number　　　: 13912-80-6

EC number　　　　　　　: 237-684-7

Patch testing　　　　　　: 1% and 2.5% pet.

Molecular formula　　　　: $C_{12}H_{17}NO_3$

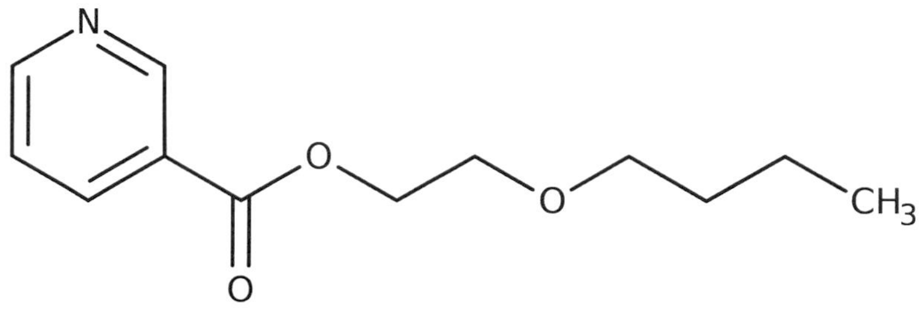

GENERAL

Nicoboxil is an aromatic carboxylic acid and a member of pyridines with topical vasodilatory (rubefacient) properties. It is indicated for use as an active ingredient in combination with nonivamide in topical analgesics for the temporary relief of the pain of rheumatism, arthritis, lumbago, muscular aches, sprains and strains, sporting injuries, and other conditions where local warmth is beneficial (1).

CONTACT ALLERGY

Case reports

A 38-year-old man presented with an eczematous reaction on both knees after topical application of a cream containing 2.5% butoxyethyl nicotinate (nicoboxil) cream for joint pain. Patch tests were positive to the cream, to its active ingredient nicoboxil 1% and 0.1% pet. (+++) and to the fragrances in the cream (2).

A 43-year-old man presented with numerous vesicles and blisters, some of 3 cm in diameter, around his left wrist. The whole hand and forearm up to his elbow was intensely erythematous and edematous, with a sensation of heat, pain, and itching. A few hours before the onset of dermatitis, the patient had applied a cream with 2.5% nicoboxil under a bandage on his sprained wrist, which he previously had used without problems. Open patch tests were strongly positive to nicoboxil 2.5% pet. and benzyl nicotinate 2.5% pet. Closed patch tests with a series of structurally-related pyridine derivatives were positive to 3-(aminomethyl)pyridine 1% water (3).

Cross-reactions, pseudo-cross-reactions and co-reactions

A patient sensitized to nicoboxil (butoxyethyl nicotinate) cross-reacted to benzyl nicotinate and 3-(aminomethyl)-pyridine (3).

Immediate contact reactions

Immediate contact reactions (contact urticaria) to nicoboxil are presented in Chapter 5.

LITERATURE

1　The data in the section 'General' may have been obtained from literature discussed in this chapter, but mostly also or exclusively from one or more of the following online sources: ChemIDPlus Advanced, PubChem, DrugBank, RxList, Drug Central, Drugs.com, and Wikipedia

2　Bilbao I, Aguirre A, Zabala R, González R, Ratón J, Diaz Pérez JL. Allergic contact dermatitis from butoxyethyl nicotinic acid and *Centella asiatica* extract. Contact Dermatitis 1995;33:435-436

3　Audicana M, Schmidt R, Fernández de Corres L. Allergic contact dermatitis from nicotinic acid esters. Contact Dermatitis 1990;22:60-61

Chapter 3.238 NICOTINE

IDENTIFICATION

Description/definition   : Nicotine is the plant alkaloid that conforms to the structural formula shown below
Pharmacological classes : Ganglionic stimulants; nicotinic agonists; smoking cessation agents
IUPAC name   : 3-[(2S)-1-Methylpyrrolidin-2-yl]pyridine
Other names   : 1-Methyl-2-(3-pyridyl)pyrrolidine
CAS registry number   : 54-11-5
EC number   : 200-193-3
Merck Index monograph   : 7879
Patch testing   : 10% water
Molecular formula   : $C_{10}H_{14}N_2$

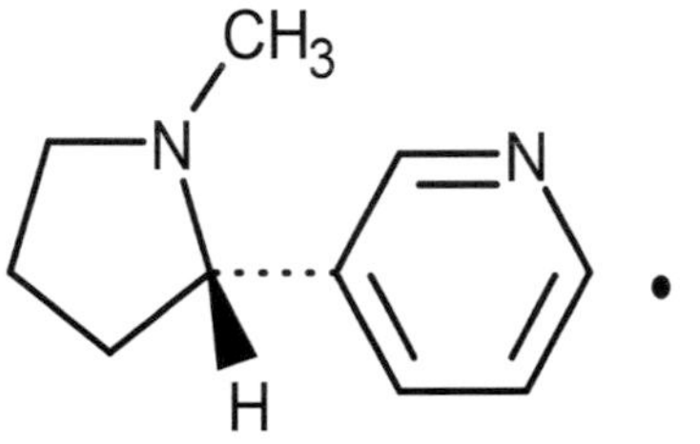

Nicotine 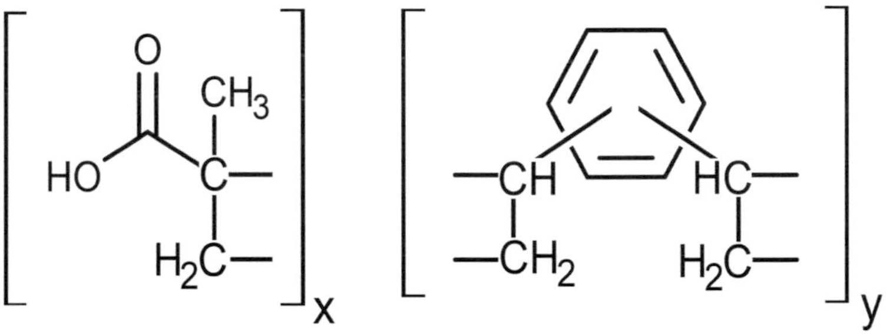   polacrilex

GENERAL

Nicotine is the primary and a highly toxic alkaloid in tobacco products. It binds stereo-selectively to nicotinic-cholinergic receptors on autonomic ganglia, the adrenal medulla, neuromuscular junctions and in the brain. Nicotine exerts two effects, a stimulant effect exerted at the locus coeruleus and a reward effect in the limbic system. Therefore, nicotine is a highly addictive substance. This agent also induces peripheral vasoconstriction, tachycardia and elevated blood pressure. Nicotine in inhalers and patches is indicated for the relief of nicotine withdrawal symptoms and as an aid to smoking cessation. In pharmaceutical products, both nicotine base and nicotine complexed with methacrylic acid polymer and divinylbenzene (nicotine polacrilex; CAS number 96055-45-7, EC number not available, molecular formula not available) may be employed (1).

CONTACT ALLERGY

Case series

In a double-blind study involving 183 smokers, of 7 patients who had a papulovesicular rash to nicotine TTS, 6 were patch tested with a nicotine TTS, a placebo TTS (without nicotine), matrix material, and gauze. Five of these patients had positive patch test reactions to the nicotine patch but not the placebo patch and they were diagnosed with contact allergy to nicotine (2).

Fourteen individuals, who had previously experienced cutaneous side effects of nicotine TTS (10 men, 4 women, mean age 39, range 23-65 year) were tested with the individual components of the TTS for immediate- and delayed-type reactions. Patch tests were performed with the adhesive and the matrix layer of the TTS, aqueous nicotine base 1%, 10% and 50%, and aqueous nicotine sulfate 5%. Five individuals had positive allergic patch test reactions to nicotine base: one to all 3 concentrations, 3 to 10% and 50%, and one only to 50% nicotine base. This concentration proved to be irritant in all subjects, but in this particular case, an allergic reaction was supported by conventional histology and immunohistochemistry. Nicotine sulfate 5% pet. was positive in one patient only (6).

From Italy, 5 cases of allergic contact dermatitis from nicotine patches were reported (7). The best patch test material was nicotine base 10% water. Quite curiously, 3 of the 5 patients developed symptoms after the first application, indicating that they were already allergic to nicotine at that moment. The authors did not comment on this finding. Short summaries of the case reports from this study are shown in table 3.238.1.

Case reports

A 54-year-old woman started treatment with a nicotine patch (14 mg/24 h) in order to stop smoking. On the 25th day, she noticed red, swollen itchy skin under the patch. She stopped applying them, but the skin reaction worsened and similar lesions began to appear at previous patch sites. Examination showed widespread eczematous lesions corresponding to the previous patch sites, but also a papulovesicular rash in other places. At this point, she tried nicotine-containing chewing gum. Her skin symptoms and lesions continued to progress despite topical treatment, and also spread to her legs and feet. Patch testing was performed with a nicotine patch, a placebo patch and a serial dilution of nicotine base in water, yielding positive reactions to the active patch and nicotine base 30%, 10% and 3% water, but negative to lower concentrations. She has continued smoking and not noticed any adverse effects (5). This was a case of allergic contact dermatitis and systemic allergic contact dermatitis, both from the nicotine chewing gum and from transcutaneous resorption of nicotine.

Short summaries of other case reports of allergic contact dermatitis from nicotine (7) are shown in table 3.238.1.

Table 3.238.1 Short summaries of case reports of allergic contact dermatitis from nicotine

Year and country	Sex	Age	Positive patch tests	Clinical data and comments	Ref.
1993 Italy	F	46	TTS; nicotine base 10% water	erythema and vesicles on all application sites; open test with nicotine 10% positive after 3 days	7
	F	40	nicotine base 1% and 10% water	pruritic erythematovesicular eruptions at application sites	
	F	46	nicotine base 10% water	itching and erythema on all application sites; open test produced erythema only	
	M	36	nicotine base 10% water	itchy erythema at the application sites	
	M	36	nicotine base 10% water	one month after daily application, >20 round eczematous patches developed simultaneously at both recent and former TTS applications	

Erythema is a common effect of nicotine on the skin from its pharmacological effect of vasodilatation (8). It should not be mistaken for a contact allergic reaction (3). In one patient, allergic reactions to the nicotine TTS were not caused by contact allergy to nicotine, but by an excipient, most likely methyl methacrylate (4).

Smoking does not appear to cause cutaneous adverse reactions in patients allergic to nicotine (6,7).

LITERATURE

1   The data in the section 'General' may have been obtained from literature discussed in this chapter, but mostly also or exclusively from one or more of the following online sources: ChemIDPlus Advanced, PubChem, DrugBank, RxList, Drug Central, Drugs.com, and Wikipedia
2   Eichelberg D, Stolze P, Block M, Buchkremer G. Contact allergies induced by TTS treatment. Methods Find Exp Clin Pharmacol 1989;11:223-225
3   Von Bahr B, Wahlberg JE. Reactivity to nicotine patches wrongly blamed on contact allergy. Contact Dermatitis 1997;37:44-45
4   Dwyer CM, Forsyth A. Allergic contact dermatitis from methacrylates in a nicotine transdermal patch. Contact Dermatitis 1994;30:309-310
5   Färm G. Contact allergy to nicotine from a nicotine patch. Contact Dermatitis 1993;29:214-215
6   Bircher AJ, Howald H, Rufli T. Adverse skin reactions to nicotine in a transdermal therapeutic system. Contact Dermatitis 1991;25:230-236
7   Vincenzi C, Tosti A, Cirone M, Guarrera M, Cusano F. Allergic contact dermatitis from transdermal nicotine systems. Contact Dermatitis 1993;29:104-105
8   Wolf R, Tüzün B, Tüzün Y. Adverse skin reactions to the nicotine transdermal system. Clin Dermatol 1998;16:617-623

Chapter 3.239 NIFURATEL

IDENTIFICATION

Description/definition : Nifuratel is the member of furans and C-nitro compound that conforms to the structural formula shown below

Pharmacological classes : Antitrichomonal agents; antifungal agents

IUPAC name : 5-(Methylsulfanylmethyl)-3-[(E)-(5-nitrofuran-2-yl)methylideneamino]-1,3-oxazolidin-2-one

Other names : 5-((Methylthio)methyl)-3-((5-nitrofurfurylidene)amino)-2-oxazolidinone; methylmercadone

CAS registry number : 4936-47-4

EC number : 225-576-2

Merck Index monograph : 7886

Patch testing : 1% acetone

Molecular formula : $C_{10}H_{11}N_3O_5S$

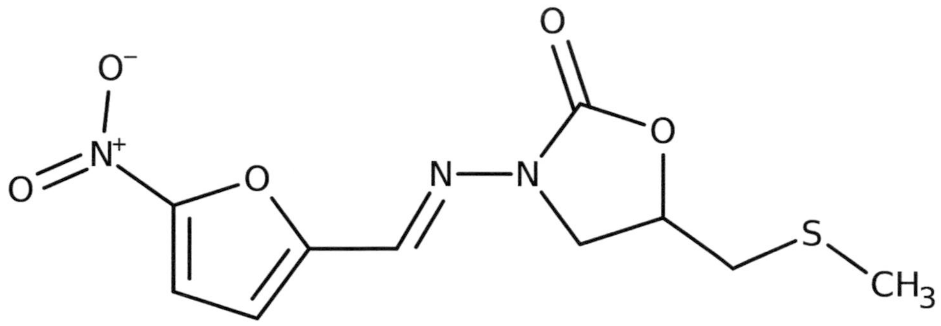

GENERAL

Nifuratel is a member of the furans that has antiprotozoal, antifungal and antibacterial activities. It is used as an ingredient in topical medicaments for the treatment of vulvovaginal infections (1). It is (or was) also used as antiseptic in anti-hemorrhoidal ointments and suppositories (3).

CONTACT ALLERGY

General

Only 7 cases of allergic contact dermatitis from nifuratel have been reported. Although it is mostly prescribed to women for vulvovaginal infections, only two of the patients with ACD from nifuratel were female (2,3). The 5 other cases (4-7) concerned men who had either used topical nifuratel in antihemorrhoidal preparations (4,6) or had acquired 'connubial' (or 'consort') allergic contact dermatitis from having sexual intercourse with partners using nifuratel (5,7). Nearly all reports came from Italy (3-7).

Case reports

A 40-year-old woman presented with a severe, partly erosive, vulvitis and vaginitis. Additionally, she showed a symmetrically distributed hemorrhagic erythema and induration on the inner thighs. She had used vaginal suppositories containing nifuratel and an antimycotic cream for 8 days for the treatment of a fungal vaginitis. A patch test to the suppository was positive, to the cream negative. The ingredients could not be tested separately, but nifuratel was indeed the most likely allergen (2).

Short summaries of other case reports of allergic contact dermatitis from nifuratel are shown in table 3.239.1.

Table 3.239.1 Short summaries of case reports of allergic contact dermatitis from nifuratel

Year and country	Sex	Age	Positive patch tests	Clinical data and comments	Ref.
1992 Italy	F	53	nifuratel 1% acetone	itchy red edematous dermatitis of the vulva and inner thighs; similar episode previously from nifuratel pessaries	3
1990 Italy	M	37	nifuratel ointment 'as is'; nifuratel 1% acetone	severe edema and erythema of the penis and genital region spreading to the buttocks, upper thighs and abdomen from nifuratel ointment against genital itch	4
	M	32	nifuratel ointment 'as is'; nifuratel 1% acetone	severe contact dermatitis in the perianal area from ointment and suppositories for burning and pain from hemorrhoids	
1990 Italy	M	38	nifuratel ointment 'as is'; nifuratel 1% acetone	severe eczema of the genital region spreading to the abdomen and upper thighs; 'connubial' ACD from his wife using nifuratel ointment for vulvovaginitis	5
1987 Italy	M	54	2 nifuratel ointments 'as is'	severe eczematous dermatitis in the perianal region spreading to buttocks and upper thighs from ointment; nifuratel itself was not tested, but also reaction to other nifuratel ointment containing other excipients	6
1983 Italy	M	43	nifuratel ointment 'as is'; nifuratel 1% acetone	marked edema of the penis and scrotum 5 hours after application of nifuratel ointment; also periorbital edema; previous sensitization from sexual intercourse with partner using vaginal suppositories ('connubial' ACD)	7

ACD: allergic contact dermatitis

LITERATURE

1   The data in the section 'General' may have been obtained from literature discussed in this chapter, but mostly also or exclusively from one or more of the following online sources: ChemIDPlus Advanced, PubChem, DrugBank, RxList, Drug Central, Drugs.com, and Wikipedia
2   Helbig D, Grabbe S, Hillen U. Vulvovaginal allergic contact dermatitis from nifuratel: report of a case and review of the literature. Contact Dermatitis 2008;58:251-252
3   Corazza M, Virgili A, Mantovani L. Vulvar contact dermatitis from nifuratel. Contact Dermatitis 1992;27:273-274
4   Valsecchi R, Imberti GL, Cainelli T. Nifuratel contact dermatitis. Contact Dermatitis 1990;23:187
5   Di Prima TM, De Pasquale R, Nigro MA. Connubial contact dermatitis from nifuratel. Contact Dermatitis 1990;22:117-118
6   Cusano F, Capozzi M, di Giulio P, Errico G. Contact dermatitis from nifuratel. Contact Dermatitis 1987;16:37
7   Bedello PG, Goitre M, Cane D, Fogliano MR. Contact dermatitis from nifuratel. Contact Dermatitis 1983;9:166

Chapter 3.240 NIFUROXIME

IDENTIFICATION

Description/definition : Nifuroxime is the nitrofuran-derivative that conforms to the structural formula shown
below
Pharmacological classes : Dermatological drugs; anti-infective agents
IUPAC name : (*NE*)-*N*-[(5-Nitrofuran-2-yl)methylidene]hydroxylamine
Other names : 5-Nitro-2-furaldoxime; 5-nitro-2-furaldehyde oxime
CAS registry number : 6236-05-1
EC number : 228-349-6
Merck Index monograph : 7890
Patch testing : No data available; suggested: 0.5%, 2% and 5% pet.
Molecular formula : $C_5H_4N_2O_4$

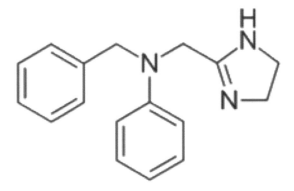

GENERAL

Nifuroxime is a topical anti-infective agent that is or was used in protozoal and fungal infections, e.g. for vaginal
infections (1).

CONTACT ALLERGY

Case reports

A woman probably had widespread dermatitis from contact allergy to nifuroxime and/or furazolidone in vaginal
suppositories. A patch test with shavings from a suppository was positive, but the ingredients were not tested
separately (2).

Another woman, aged 32, used the same brand of vaginal suppositories for discharge. The patient stated that
initially she had no complaints from their use, but about two weeks later, severe itching developed around the
vagina. Within another two weeks, generalized urticaria developed which resulted in hospitalization due to the
threat of laryngeal edema. After a long, stormy treatment course involving systemic corticosteroids and a variety of
antihistamines and tranquilizers, the patient has finally become able to work normally, although she occasionally has
urticaria. Skin tests (not further specified) with nifuroxime gave a positive reaction in the form of blistering and
dermatitis. A basophil degranulation test was reported positive for allergy to nifuroxime (3).

Both patients may have suffered from systemic contact dermatitis: sensitization from mucous membrane
contact with nifuroxime (and/or furazolidone in the first patient), followed by widespread dermatitis and urticaria
from systemic absorption.

LITERATURE

1 The data in the section 'General' may have been obtained from literature discussed in this chapter, but mostly
also or exclusively from one or more of the following online sources: ChemIDPlus Advanced, PubChem,
DrugBank, RxList, Drug Central, Drugs.com, and Wikipedia
2 Goette DK, Odom RB. Vaginal medications as a cause for varied widespread dermatitides. Cutis 1980;26:406-409
3 Aaronson CM. Generalized urticaria from sensitivity to nifuroxime. JAMA 1969;210:557-558

Chapter 3.241 NIFURPRAZINE

IDENTIFICATION

Description/definition : Nifurprazine is the nitrofuran derivative that conforms to the structural formula shown below
Pharmacological classes : Anti-infective agent
IUPAC name : 6-[(*E*)-2-(5-Nitrofuran-2-yl)ethenyl]pyridazin-3-amine
Other names : 3-Amino-6-(2-(5-nitro-2-furyl)vinyl)pyradizine
CAS registry number : 1614-20-6
EC number : 216-563-2
Patch testing : No data available; suggested: 0.5%, 2% and 5% pet.
Molecular formula : $C_{10}H_8N_4O_3$

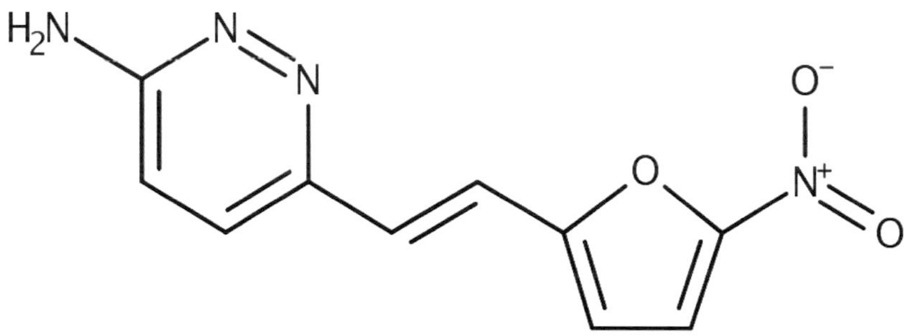

GENERAL

Nifurprazine is a nitrofuran derivative with antibacterial activity that was formerly used as topical anti-infective agent.

CONTACT ALLERGY

Case series

Nifurprazine allergy has been observed in 5 patients; details are not available to the author (2).

Cross-reactions, pseudo-cross-reactions and co-reactions

Of 41 patients sensitized to nitrofurazone, 2 cross-reacted to nifurprazine 0.2% water (2); in another study, none of 9 nitrofurazone-allergic individuals had positive reactions to nifurprazine 0.2% and 1% pet. (3). Conversely, of 3 patients primarily sensitized to nifurprazine, none cross-reacted to nitrofurazone 0.2% pet. (2).

LITERATURE

1   De Groot AC, Conemans JM. Contact allergy to furazolidone. Contact Dermatitis 1990;22:202-205
2   Braun W. Kontaktallergien durch Nifurprazin (Carofur). Deutsche Med Wochenschrift 1969;94:1685-1687 (Article in German, data cited in ref. 1)
3   Bleumink E, te Lintum JCA, Nater JP. Kontaktallergie durch Nitrofurazon (Furacin) und Nifurprazin (Carofur). Hautarzt 1974;25:403-406 (Article in German, data cited in ref. 1)

Chapter 3.242 NITROFURAZONE

IDENTIFICATION

Description/definition : Nitrofurazone is the nitrofuran that conforms to the structural formula shown below
Pharmacological classes : Anti-infective agents
IUPAC name : [(E)-(5-Nitrofuran-2-yl)methylideneamino]urea
Other names : Nitrofural; hydrazinecarboxamide, 2-[(5-nitro-2-furanyl)methylene]-; 5-nitro-2-furaldehyde semicarbazone; Furacin ®
CAS registry number : 59-87-0
EC number : 200-443-1
Merck Index monograph : 7957
Patch testing : 1.0% pet. (Chemotechnique, SmartPracticeCanada, SmartPracticeEurope)
Molecular formula : $C_6H_6N_4O_4$

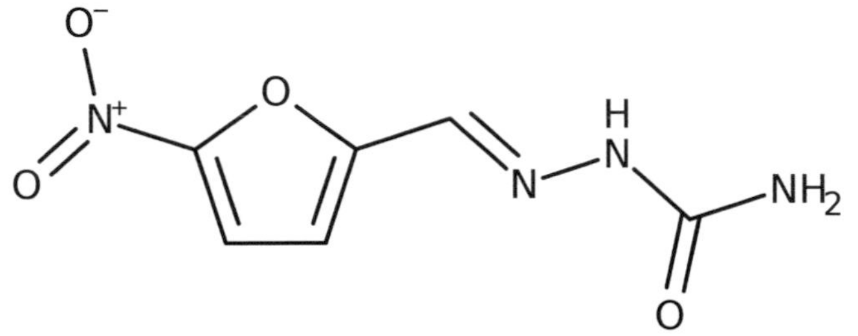

GENERAL

Nitrofurazone is a nitrofuran topical antibacterial agent with bactericidal activity against a number of pathogens, including *Staphylococcus aureus* and *Escherichia coli*; it does also have significant activity against *Pseudomonas aeruginosa*, *Proteus mirabilis*, and *Serratia marcescens*. Nitrofurazone is indicated for the topical treatment of bacterial skin infections including pyodermas, infected dermatoses and infections of cuts, wounds, burns and ulcers caused by susceptible organisms. Nitrofurazone was formerly used orally in humans. Veterinary use is mainly in the treatment and prophylaxis of coccidiosis in poultry and necrotic enteritis in pigs, both by administering the drug systemically and adding it to animal feed (1).

CONTACT ALLERGY

General

Nitrofurazone was introduced in 1945 and allergic reactions to it were already known some years later (19). Contact allergy used to be frequent in some European countries, where nitrofurazone soluble dressings were widely used for the treatment of leg ulcers (14,23). Large case series were reported from Germany and The Netherlands (10,14,23). Many cases of sensitization to nitrofurazone have also been reported from Belgium (4,6) and India (7,36,37,39). The antimicrobial appears to be a strong sensitizer: in human maximization tests, 14 of 24 subjects were sensitized (32).

Because the sensitizing capacity of nitrofurazone was well-known, its use in Europe and the USA has been largely abandoned. In some countries including Turkey and India, however, because of its efficacy and low price, it is still widely used for chronic leg ulcers, superficial skin infections, ulcers, burns and chronic dermatitis and is in fact, in Turkey, the most prescribed topical antibacterial agent (17,18). In India, in the early 1980s, sensitization to nitrofurazone was extremely frequent in patients suspected of allergic contact dermatitis to medicaments (7,39,42). Although more recently a low frequency of sensitization was observed in patients with venous leg ulcers (3), routine testing some 20 years ago still yielded high percentages (6-6.8%) of positive patch test reactions to nitrofurazone (36,37). This chemical was also an important cause of allergic contact dermatitis of the feet in Turkey (20).

The soluble dressing, solution and ointment contain 0.2% nitrofurazone and polyethylene glycols and many patients who become sensitized prove to be allergic to both nitrofurazone and polyethylene glycol (PEG) (18,21,22, 34, 35,40).

Patch testing in groups of patients

Results of studies in which consecutive patients suspected of contact dermatitis were patch tested (routine testing) and of patch testing in groups of selected patients are shown in table 3.242.1. In 3 studies from India, prevalences of sensitization were high with 2%, 6% and 6.8%, resp. (36-38). In 8 studies in which groups of *selected* patients were tested with nitrofurazone (patients with leg ulcers, individuals suspected of medicament allergy), frequencies of positive patch tests ranged from 3% to 56%. Extremely high figures were found in the late 1970s and beginning of the 1980s, but these patients were patch tested with commercial preparations, i.e. that sensitizations to polyethylene glycols but not nitrofurazone were also counted as positive (7,39,42). Most recently, the prevalence appears to have dropped in India, with 3.5% positive reactions in patients with venous leg ulcers (3).

Table 3.242.1 Patch testing in groups of patients

Years and Country	Test conc. & vehicle	Number of patients tested	positive (%)	Selection of patients (S); Relevance ©; Comments ©	Ref.
Routine testing					
1997-2006 India	2% pet.	1000	60　(6.0%)	R: 60% past relevance	37
<2004 India	2% pet.	220	15　(6.8%)	R: 82% of positive reactions to all haptens tested were relevant	36
<2002 India	1% pet.	100	2　(2%)	R: not stated; C: not enough patients for reliable data in the case of routine testing	38
Testing in groups of selected patients					
<2017 India	1% pet.	172	6　(3.5%)	S: patients with venous leg ulcers of over 6 weeks duration; R: 'the majority of reactions were relevant'	3
1990-2014 Belgium	20% pet. (?)	551	52　(9.4%)	S: patients suspected of iatrogenic contact dermatitis and tested with a pharmaceutical series and their own products; R: 96% of the positive patch test reactions to all topical drugs and antiseptics were considered to be relevant	4
1978-2005 Taiwan		603	18　(3.0%)	S: patients suspected of contact allergy to medicaments; R: 65% of the reactions to all medicaments were considered to be relevant	5
<2005 India	1% pet.	50	4　(8%)	S: patients with therapy-resistent nummular eczema; R: the reactions indicated 'past exposure'	24
<1999 Croatia	2% pet.	100	3　(3%)	S: patients with leg ulcers; R: not stated	2
1980-1983 India		390	141 (36.2%)	S: patients suspected of contact allergy to antimicrobial agents; R: not stated; C: probably patch tested with commercial preparations, not nitrofurazone itself	39
<1982 India		112	52　(46%)	S: patients suspected of allergy to antimicrobial agents; R: unknown; C: probably patch tested with commercial preparations, not nitrofurazone itself	7
<1981 India		101	56　(56%)	S: patients suspected of allergy to antimicrobial agents; R: unknown; C: probably patch tested with commercial preparations, not nitrofurazone itself	42

Case series

In the period 1985-1997, in Leuven, Belgium, 8521 patients were patch tested, and 29 positive reactions were observed to nitrofurazone. It was not stated, however, how many patients had been tested with this allergen and how many of the reactions were relevant (6). In three University hospitals in Turkey, between 2013 and 2017, 58 patients (46 men, 12 women; ages ranging from 14-85 years, mean 51) were hospitalized because of 'allergic contact dermatitis from nitrofurazone', apparently caused by 'topical nitrofurazone soluble dressing cream'. Quite curiously, patch tests were performed in only 15 patients and they were all negative! Yet, allergic contact dermatitis caused by nitrofurazone was diagnosed in all 58 patients (17).

Five patients from Spain developed allergic contact dermatitis from Furacin ®. When patch tested, 4 reacted to nitrofurazone 1% pet., of who 2 also had positive reactions to polyethylene glycol (22). Six cases of allergic contact dermatitis to nitrofurazone were reported in 1977 from South Africa; most patients had leg ulcers or stasis dermatitis (26).

In older literature, large case series of 58 (12), 37 (23), 23 (10), and 14 (14) patients sensitized to nitrofurazone have been reported from Germany and The Netherlands (examples, not a full literature review).

Case reports

A 67-year-old man developed a pruritic symmetrical non-follicular pustular eruption initially involving the dorsa of his hands and rapidly spreading to his forearms, arms, face, and trunk. Some days before, the patient had accidentally burned his hands. The wounds had been treated topically with two silver sulfadiazine creams, a nitrofurazone ointment and an impregnated gauze dressing. Patch tests were positive to all topicals and to their ingredients nitrofurazone, PEG 300, propylene glycol, polysorbate 80 and castor oil (Ricinus communis seed oil).

This was a case of acute generalized exanthematous pustulosis-like allergic contact dermatitis. Which ingredient or ingredients were responsible is unknown (40).

A 44-year-old man used nitrofurazone in polyethylene glycol to treat chronic leg ulcers and developed local eczematous lesions on his legs 24 hours after application (21). A 43-year-old woman developed eczematous lesions on her hand, face, and neck 24 hours after applying the same ointment to treat a burn on her hand. In both patients, patch tests were strongly positive to nitrofurazone 1% pet. and PEG 4% pet. (21).

A 46-year-old man treated hyperkeratotic eczema of his fingertips with a liquid adhesive containing nitrofurazone, acrylates copolymer and fragrances. Patch tests were positive to the adhesive 5% and 10% water and to nitrofurazone (no test details provided) (31). A man aged 70 with chronic conjunctivitis received 3 brands of eye drops. After several weeks treatment, he developed marked erythema, edema and pruritus affecting initially the eyelids, spreading to the forehead and temples. Patch tests with all active ingredients and polyethylene glycols were positive to nitrofurazone 0.2% pet. only. An open test with it was positive at day 3. Avoiding nitrofurazone eye drops cleared the eczema (33).

A 59-year-old man, who had previously used nitrofurazone ointment to burns on the feet, two years later started treating itching eczematous plaques on his left leg with nitrofurazone ointment (containing polyethylene glycols, PEGs). In 1-2 days after the first application, edema, erythema, pruritus and severe worsening of the previous skin lesions occurred. Patch tests were positive to the ointment, nitrofurazone 1% pet. and to various polyethylene glycols (PEG 200, 300 [present in the ointment], 400 and PEG-mix) (34).

One patient from Italy had conjunctivitis from contact allergy to nitrofurazone in eye drops; details were not provided (8). Other case reports of contact allergy to/allergic contact dermatitis from nitrofurazone, adequate data of which are not available to the author, can be found in refs. 27,28,29 and 41 (pustular allergic contact dermatitis).

Occupational allergic contact dermatitis
A 34-year-old cattle breeder gave a 3-year history of erythematous vesicular lesions on the sides of the fingers of both hands, on the forearms and the face. The lesions cleared when not working with cattle, when he also handled feeds and medicaments. Patch tests were positive to a uterine ovule for cattle and its ingredient nitrofurazone 1% pet. (25).

For 5 years, a 50-year-old male horse keeper at a race track had a symmetrical hand eczema affecting palms and dorsa, which occasionally spread to the feet in a pompholyx pattern. He used to bath the animals and apply several veterinary medicaments with his bare hands. Nitrofurazone ointment was often used to treat superficial infections and minor wounds. Patch tests were positive to nitrofurazone 0.2% (33). Airborne occupational allergic contact dermatitis has been caused by nitrofurazone from its presence in a powdered aquarium water additive (30).

Two cases of occupational sensitization from nitrofurazone in animal feed were reported from the USA (15,16). In one (16), no patch tests were performed, but nitrofurazone allergy was highly likely.

Cross-reactions, pseudo-cross-reactions and co-reactions
Of 41 patients sensitized to nitrofurazone, 2 cross-reacted to nifurprazine 0.2% water (12). Of 7 patients sensitized to nitrofurazone, 6 cross-reacted to nitrofurantoin powder 50%; it is unknown whether proper controls have been tested as well (13). In an individual occupationally sensitized to nitrofurazone, oral administration of nitrofurantoin caused a drug rash, suggesting cross-sensitivity; a patch test to nitrofurantoin was positive (11).

Primary sensitization to furazolidone may have caused cross-sensitivity to nitrofurazone in one patient (10). Patients sensitized to nitrofurazone often co-react (concomitant sensitization) to polyethylene glycols, which are excipients of the topical pharmaceuticals containing nitrofurazone (Furacin ®) (18,21,22,34,35).

LITERATURE

1 The data in the section 'General' may have been obtained from literature discussed in this chapter, but mostly also or exclusively from one or more of the following online sources: ChemIDPlus Advanced, PubChem, DrugBank, RxList, Drug Central, Drugs.com, and Wikipedia
2 Marasovic D, Vuksic I. Allergic contact dermatitis in patients with leg ulcers. Contact Dermatitis1999;41:107-109
3 Rai R, Shenoy MM, Viswanath V, Sarma N, Majid I, Dogra S. Contact sensitivity in patients with venous leg ulcer: A multi-centric Indian study. Int Wound J 2018;15:618-622
4 Gilissen L, Goossens A. Frequency and trends of contact allergy to and iatrogenic contact dermatitis caused by topical drugs over a 25-year period. Contact Dermatitis 2016;75:290-302
5 Shih Y-H, Sun C-C, Tseng Y-H, Chu C-Y. Contact dermatitis to topical medicaments: a retrospective study from a medical center in Taiwan. Dermatol Sinica 2015;33:181-186
6 Goossens A, Claes L, Drieghe J, Put E. Antimicrobials, preservatives, antiseptics and disinfectants. Contact Dermatitis 1998;39:133-134
7 Bajaj AK, Govil DC, Bajaj S, Govil M, Tewari AN. Contact hypersensitivity to topical antimicrobial and antifungal agents. Ind J Dermatol Venereol Leprol 1982;48:330-332
8 Tosti A, Tosti G. Allergic contact conjunctivitis due to ophthalmic solution. In: Frosch PJ, Dooms-Goossens A, Lachapelle JM, Rycroft RJG, Scheper RJ (eds). Current Topics in Contact Dermatitis. Berlin: Springer-Verlag, 1989: 269-272
9 De Groot AC, Conemans JM. Contact allergy to furazolidone. Contact Dermatitis 1990;22:202-205

10 Laubstein H, Niedergesäss G. Untersuchungen über Gruppensensibilisierungen bei Nitrofuranderivaten. Derm Monatsschrift 1970;156:1-8 (Article in German, data cited in ref. 9)

11 Jirasek L, Kalensky J. Kontakni alergicky ekzem z krmnych smesi v zivocisne vyrobe. Ceskoslovenska Dermatologie 1975;50:217 (Article in Czech, cited in ref. 9).

12 Braun W. Kontaktallergien durch Nifurprazin (Carofur). Deutsche Medizinische Wochenschrift 1969;94:1685-1687 (Article in German, data cited in ref. 9)

13 Behrbohm P, Zschunke E. Ekzem durch Nifucin (Nitrofural). Deutsche Gesundheitswesen 1967;22:273-275 (Article in German, data cited in ref. 9)

14 Bleumink E, te Lintum JCA, Nater JP. Kontaktallergie durch Nitrofurazon (Furacin) und Nifurprazin (Carofur). Hautarzt 1974;25:403-406 (Article in German, data cited in ref. 9)

15 Caplan RM. Cutaneous hazards posed by agricultural chemicals. J Iowa Med Soc 1969;59:295-299 (cited in ref. 9)

16 Neldner KH. Contact dermatitis from animal feed additives. Arch Dermatol 1972;106:722-723

17 Bilgili SG, Ozaydin-Yavuz G, Yavuz IH, Bilgili MA, Karadag AS. Cutaneous reactions caused by nitrofurazone. Postepy Dermatol Alergol 2019;36:398-402

18 Özkaya E, Kılıç S. Polyethylene glycol as marker for nitrofurazone allergy: 20 years of experience from Turkey. Contact Dermatitis 2018;78:211-215

19 Downing JG, Brecker FW. Further studies in the use of Furacin in dermatology. N Engl J Med 1948;239:862-864

20 Özkaya E, Polat Ekinci A. Foot contact dermatitis: nitrofurazone as the main cause in a retrospective, cross-sectional study over a 16-year period from Turkey. Int J Dermatol 2016;55:1345-1350

21 Moreno Escobosa MC, Moya Quesada MC, Cruz Granados S, Amat López J. Contact dermatitis to antibiotic ointments. J Investig Allergol Clin Immunol 2009;19:510-511

22 Prieto A, Baeza ML, Herrero T, Barranco R, De Castro FJ, Ruiz J, et al. Contact dermatitis to Furacin. Contact Dermatitis 2006;54:126

23 Braun W, Schütz R. Kontaktallergie gegen nitrofurazone (Furacin). Dtsch Med Wschr 1968;32:1524-1526 (Article in German)

24 Krupa Shankar DS, Shrestha S. Relevance of patch testing in patients with nummular dermatitis. Indian J Dermatol Venereol Leprol 2005;71:406-408

25 Condé-Salazar L, Guimaraens D, Gonzalez MA, Molina A. Occupational allergic contact dermatitis from nitrofurazone. Contact Dermatitis 1995;32:307-308

26 Hull PR, De Beer HA. Topical nitrofurazone: a potent sensitizer of the skin and mucosae. South African Medical Journal 1977;52:189-190

27 Miranda A, Quinones PA. Un caso de dermatitis alergica de contacto medicamentosa. Actas Dermo-Sif 1980;7-8:301-302 (Article in Spanish)

28 Perez Chacon V, Valeron P, Henriquez MC, Hernandez B. Eczema de contacto por nitrofurazona. Actas Dermo-Sif 1985;76:96-97 (Article in Spanish)

29 Lobos Bert P. Dermatitis de contacto alergica por nitrofurazona. Dermosur 1993;3:121-122 (Article in Spanish)

30 Lo JS, Taylor JS, Oriba H. Occupational allergic contact dermatitis to airborne nitrofurazone. Dermatol Clin 1990;8:165-168

31 Ballmer-Weber BK, Elsner P. Contact allergy to nitrofurazone. Contact Dermatitis 1994;31:274-275

32 Kligman AM. The identification of contact allergens by human assay. J Invest Dermatol 1966;47:393-409

33 Ancona A. Allergic contact dermatitis to nitrofurazone. Contact Dermatitis 1985;13:35

34 Córdoba Guijarro S, Sánchez-Pérez J, García-Díez A. Allergic contact dermatitis to polyethylene glycol and nitrofurazone. Am J Contact Dermatit 1999;10:226-227

35 Stenveld HJ, Langendijk PN, Bruynzeel DP. Contact sensitivity to polyethylene glycols. Contact Dermatitis 1994;30:184-185

36 Sharma VK, Sethuraman G, Garg T, Verma KK, Ramam M. Patch testing with the Indian standard series in New Delhi. Contact Dermatitis 2004;51:319-321

37 Bajaj AK, Saraswat A, Mukhija G, Rastogi S, Yadav S. Patch testing experience with 1000 patients. Indian J Dermatol Venereol Leprol 2007;73:313-318

38 Narendra G, Srinivas CR. Patch testing with Indian standard series. Indian J Dermatol Venereol Leprol 2002;68:281-282

39 Bajaj AK, Gupta SC. Contact hypersensitivity to topical antibiotics. Int J Dermatol 1986;25:103-105

40 González-Cantero A, Gatica-Ortega M, Pastor-Nieto M, Martínez-Lorenzo E-R, Gómez-Dorado BA, Mollejo-Villanueva M, et al. Acute generalized exanthematous pustulosis (AGEP)-like contact dermatitis resulting from topical therapy in a polysensitized patient. Contact Dermatitis 2019;80:329-333

41 Burkhart CG. Pustular allergic contact dermatitis: a distinct clinical and pathological entity. Cutis 1981;27:630-631, 638 (data cited in ref. 40)

42 Pasricha JS, Guru B. Contact hypersensitivity to local antibacterial agents. Indian J Dermatol Venereol Leprol 1981;47:27-30

Chapter 3.243 NITROGLYCERIN

IDENTIFICATION

Description/definition : Nitroglycerin is the organic nitrate that conforms to the structural formula shown below
Pharmacological classes : Vasodilator agents
IUPAC name : 1,3-Dinitrooxypropan-2-yl nitrate
Other names : 1,2,3-Propanetriol, trinitrate; glyceryl nitrate; glyceryl trinitrate
CAS registry number : 55-63-0
EC number : 200-240-8
Merck Index monograph : 7964
Patch testing : 2% pet.
Molecular formula : $C_3H_5N_3O_9$

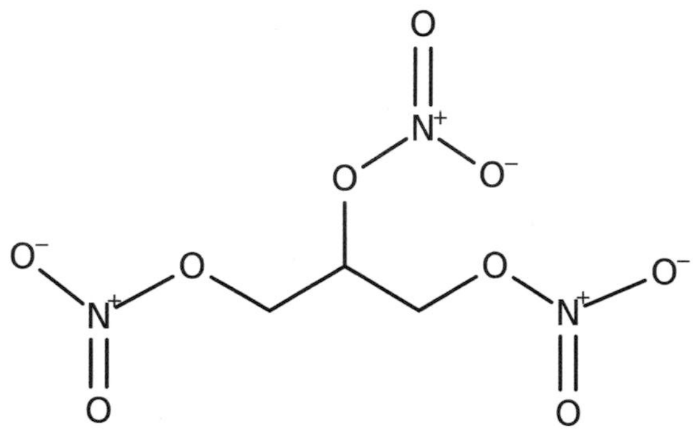

GENERAL

Nitroglycerin is an organic nitrate with vasodilator activity. Dilatation of the veins results in decreased venous return to the heart, thereby decreasing left ventricular volume (reduced preload) and decreasing myocardial oxygen requirements. Arteriolar relaxation reduces arteriolar resistance (reduced afterload), thereby also decreasing myocardial oxygen demands. In addition, nitroglycerin causes coronary artery dilatation, which improves myocardial blood distribution. Nitroglycerin is indicated for the prevention and treatment of angina pectoris. It is also used for the treatment of perioperative hypertension, to produce controlled hypotension during surgical procedures, to treat hypertensive emergencies, and to treat congestive heart failure associated with myocardial infarction (1).

Nitroglycerin was synthesized in 1846 and it became generally used as an explosive when Alfred Nobel (1833-1896) developed dynamite in 1867. Since 1879, sublingual nitroglycerin has been used for the treatment of angina pectoris (21).

CONTACT ALLERGY

General

Nitroglycerin has been used orally since 1879 and hardly caused sensitization. However, allergic contact dermatitis from nitroglycerin ointment (used since the mid-1950s) and nitroglycerin TTS (since early 1980s) has been reported repeatedly, yet appears to be very infrequent, with no larger case series having been reported (11,13-18 [n=2],19-22,25,26 [n=2],27-29,31,32,34 (n=2),35,36). Lesions of ACD are generally limited to the application site, but more extensive dermatitis on the trunk (17,29), severe dermatitis (31), generalized allergic contact dermatitis (13) and erythema multiforme (15) have all been described. The risk of spreading is increased if discontinuation of the use of nitroglycerin TTS in patients with cutaneous reactions to the device is not carried out promptly (13). Oral or sublingual use of nitroglycerin in sensitized individuals does not produce any allergic reactions (11,12,13,19,22,32, 34,35,36). ACD may be followed by post-inflammatory hyperpigmentation (11,19,28).

As nitroglycerin is a vasodilator, erythema under a nitroglycerin TTS is common and not indicative of contact allergy. Burns at the site of nitroglycerin discs have been described, caused by heating of the metallic component of the TTS provoked by radiation from a microwave oven (9) and from defibrillation (10).

Non-allergic and allergic cutaneous reactions to nitroglycerin have been reviewed in 2006 (12). Cutaneous reactions to transdermal therapeutic systems have been reviewed in several articles (3-8). Occupational allergic contact dermatitis to nitroglycerin in explosives has also been observed (21).

Case series

In Leuven, Belgium, in the period 1990-2014, iatrogenic contact dermatitis was diagnosed in 2600 individuals (17% of the total patch test population). 96% of all positive patch test reactions to topical drugs and antiseptics were considered to be relevant. Nitroglycerin (1% pet.) was tested in 9 patients and there were 4 positive reactions to it (2). Twenty-two patients (17 men, 5 women, 41-72 years) who had taken nitroglycerin tablets every day for the last 10 years were patch tested to nitroglycerin 2% pet. At the 96 h reading, 21 were negative but one was positive (++) with a persistent, itchy, allergic response. This 66-year-old woman had, however, never had any skin symptoms or signs (33).

Case reports

A 62-year-old woman with coronary insufficiency developed pruritus, edema, erythema and vesiculation, well demarcated at the application sites of nitroglycerin TTS on the trunk and upper arms, 6 days after initiation of this treatment. Lesions healed with post-inflammatory hyperpigmentation. The patches were substituted with another brand, but the pruritus and eczema did not decrease. Patch tests were positive to nitroglycerin 1%, 0.5% and 0.1% in petrolatum, but negative to the adhesive. The patient subsequently received sublingual nitroglycerin without adverse effects (11).

A 76-year-old man presented with a generalized eczematous eruption affecting the entire skin except the palms and soles, with erythema, edema, vesicles, lichenification, and excoriations. The eruption had started after 1 year of treatment with a nitroglycerin transdermal system with itchy red plaques under the patches. Although these were substituted with another brand, the pruritus and eczema gradually increased and spread during the next 10 months. Patch tests were positive to the solutions in the patches and to nitroglycerin 0.2 mg/ml in water and 0.5 mg/ml in alcohol but negative to the adhesives. An oral challenge with sublingual nitroglycerin was well tolerated (13).

A 61-year-old woman presented with a widespread itchy cutaneous eruption that had been present for 5 days. This had started on the application site of a nitroglycerin patch, and had remained most severe in this area. The patient had been on nitroglycerin patches for the last 20 days. Cutaneous examination showed a large erythematous and edematous patch on the chest and multiple smaller papulo-edematous lesions over the remainder of the trunk, resembling erythema multiforme. A skin biopsy was consistent with erythema multiforme. A patch test with the nitroglycerin patch was positive, with a nitroglycerin spray (0.4 mg/dose) negative and with the adhesive of the TSS also negative. The patient applied a nitroglycerin device once more, which resulted in a similar local eruption and widespread pruritus. Nitroglycerin itself was not tested (15).

Short summaries of other case reports of allergic contact dermatitis from nitroglycerin are shown in table 3.243.1.

Table 3.243.1 Short summaries of case reports of allergic contact dermatitis from nitroglycerin

Year and country	Sex	Age	Positive patch tests	Clinical data and comments	Ref.
2002 Malta	F	58	TTS; negative to placebo	localized dermatitis; possibly cross-reaction to isosorbide dinitrate, but this chemical was not patch tested	14
2000 United Kingdom	M	57	NITRO ointment; NITRO 2% pet.	perianal eczema and dermatitis of the hand from an ointment with NITRO, probably for fissure ani	16
1999 France	M	85	TTS (n=2) and NITRO spray	ACD from TTS, worsening by another TTS; the 2 TTS and the NITRO spray patch tested had only NITRO as common component	17
1992 Portugal	F	53	TTS; negative to placebo; NITRO 2% and 0.2% pet.	Localized ACD with post-inflammatory hyperpigmentation; 10 controls negative to NITRO 2% pet.; sublingual NITRO was well tolerated	19
1991 Finland	F	75	TTS; negative to placebo; NITRO 0.02% water	localized ACD; also 4 cases of occupational ACD in this report; 20 controls were negative to NITRO 2% pet.	21
1989 United Kingdom	M	72	TTS (n=2); two placebo patches negative	localized ACD; changing to another NITRO TTS did not help	22
1984 USA	M	59	NITRO ointment; NITRO solution 0.2 mg/ml	localized ACD after 1 year NITRO ointment and 4 months NITRO TTS; reapplying the ointment caused dermatitis after 3 weeks (low concentration NITRO?)	32
1979 USA	M	57	2% NITRO ointments (n= 2); NITRO 2% and 9.1% alc.; many other test materials with NITRO	localized ACD from 2% NITRO ointment; negative in 3 controls; pre-application of a strong dermatocorticosteroid prevented or minimized the development of allergic contact dermatitis	34
	M	71			
1979 USA	F	60	2% NITRO ointment; 1.6% crushed tablets in pet.; NITRO 2% pet.	localized ACD after 9 months NITRO ointment + sublingual	35
1978 USA	M	60	2 NITRO ointments; NITRO 0.2 mg/ml water	localized ACD after 3 months application of 2% NITRO ointment	36

ACD: allergic contact dermatitis; NITRO: nitroglycerin; placebo: nitroglycerin TTS without nitroglycerin; TTS: nitroglycerin transdermal therapeutic system;

Other case reports of contact allergy to/allergic contact dermatitis from nitroglycerin, adequate data of which are not available to the author, can be found in refs. 18 (2 patients),20,25,26 (2 patients),27,29,30,and 31 (2 patients). A report in which five cases of allergic reactions to nitroglycerin ointment were presented (four patients with cutaneous eruption and one patient with an anaphylactoid reaction) did not document that the allergen was nitroglycerin by patch testing (27, details unknown, text cited in ref. 32)

In some cases of allergic reactions to a nitroglycerin TTS, a chemical other than nitroglycerin may have been the hapten. In one such case, a patch test with the TTS was positive, with the placebo negative, but the reactions to nitroglycerin 0.1%, 1%, 2% and 3% pet. were also negative (23). In another case, both the TTS and the placebo TTS were positive, but nitroglycerin itself was not tested (24). In a third such case report, the patient did have positive patch tests to the TTS and to nitroglycerin, but also to the placebo patch, suggesting the presence of another hapten besides nitroglycerin (28).

Cross-reactions, pseudo-cross-reactions and co-reactions
A patient presumably sensitized to isosorbide dinitrate from intravenous administration may have cross-reacted to nitroglycerin, but patch tests with isosorbide dinitrate were not performed (14). In a patient with occupational allergic contact dermatitis from nitroglycerin, cross-reactivity to isosorbide dinitrate may also have occurred (21).

LITERATURE

1 The data in the section 'General' may have been obtained from literature discussed in this chapter, but mostly also or exclusively from one or more of the following online sources: ChemIDPlus Advanced, PubChem, DrugBank, RxList, Drug Central, Drugs.com, and Wikipedia
2 Gilissen L, Goossens A. Frequency and trends of contact allergy to and iatrogenic contact dermatitis caused by topical drugs over a 25-year period. Contact Dermatitis 2016;75:290-302
3 Bershow A, Warshaw E. Cutaneous reactions to transdermal therapeutic systems. Dermatitis 2011;22:193-203
4 Musel AL, Warshaw EM. Cutaneous reactions to transdermal therapeutic systems. Dermatitis 2006;17:109-122
5 Murphy M, Carmichael AJ. Transdermal drug delivery systems and skin sensitivity reactions. Incidence and management. Am J Clin Dermatol 2000;1:361-368
6 Carmichael AJ. Skin sensitivity and transdermal drug delivery. A review of the problem. Drug Saf 1994;10:151-159
7 Hogan DJ, Maibach HI. Adverse dermatologic reactions to transdermal drug delivery systems. J Am Acad Dermatol 1990;22:811-814
8 Holdiness MR. A review of contact dermatitis associated with transdermal therapeutic systems. Contact Dermatitis 1989;20:3-9
9 Murray KB. Hazard of microwave ovens to transdermal delivery system. N Engl J Med 1984;310:721
10 Wrenn K. The hazards of defibrillation through nitroglycerin patches. Ann Emerg Med 1990;19:1327-1328
11 Roche Gamón E, De la Cuadra Oyanguren J, Pérez Ferriols A, Fortea Baixauli JM. Contact dermatitis from nitroglycerin in a transdermal therapeutic system. Acta Derm Venereol 2006;86:544-545
12 Ramey JT, Lockey RF. Allergic and nonallergic reactions to nitroglycerin. Allergy Asthma Proc 2006;27:273-280
13 Pérez-Calderón R, Gonzalo-Garijo MA, Rodríguez-Nevado I. Generalized allergic contact dermatitis from nitroglycerin in a transdermal therapeutic system. Contact Dermatitis 2002;46:303
14 Aquilina S, Felice H, Boffa MJ. Allergic reactions to glyceryl trinitrate and isosorbide dinitrate demonstrating cross-sensitivity. Clin Exp Dermatol 2002;27:700-702
15 Silvestre JF, Betlloch I, Guijarro J, Albares MP, Vergara G. Erythema-multiforme-like eruption on the application site of a nitroglycerin patch, followed by widespread erythema multiforme. Contact Dermatitis 2001;45:299-300
16 McKenna KE. Allergic contact dermatitis from glyceryl trinitrate ointment. Contact Dermatitis 2000;42:246
17 Machet L, Martin L, Toledano C, Jan V, Lorette G, Vaillant L. Allergic contact dermatitis from nitroglycerin contained in 2 transdermal systems. Dermatology 1999;198:106-107
18 De la Fuente R, Armentis A, Dies JM. Contact dermatitis from nitroglycerin. Ann Allergy 1994;72:344-346
19 Torres V, Lopes JC, Leite L. Allergic contact dermatitis from nitroglycerin and estradiol transdermal therapeutic systems. Contact Dermatitis 1992;26:53-54
20 Laine R, Kanerva L, Tarvainen K, Jolanki R, Estlander T, Helander I. Nitroglycerin-induced allergic contact dermatitis. Duodecim 1991;107:41-46 (Article in Finnish)

21 Kanerva L, Laine R, Jolanki R, Tarvainen K, Estlander T, Helander I. Occupational allergic contact dermatitis caused by nitroglycerin. Contact Dermatitis 1991;24:356-362

22 Carmichael AJ, Foulds IS. Allergic contact dermatitis from transdermal nitroglycerin. Contact Dermatitis 1989;21:113-114

23 Di Landro A, Valsecchi R, Cainelli T. Contact dermatitis from Nitroderm. Contact Dermatitis 1989;21:115-116

24 Letendre PW, Barr C, Wilkens K. Adverse dermatologic reaction to transdermal nitroglycerin. Drug Intell Clin Pharm 1984;18:69-70

25 Apted J. Percutaneous nitroglycerin patches. Med J Aust 1988;148:482

26 Topaz O, Abraham D. Severe allergic contact dermatitis secondary to nitroglycerin in a transdermal therapeutic system. Ann Allergy 1987;59:365-366

27 Chandraratna PAN, O'Dell RE. Allergic reactions to nitroglycerin ointment: report of five cases. Curr Ther Res 1987;42:481-484

28 Harari Z, Sommer I, Knobel B. Multifocal contact dermatitis to nitroderm TTS 5 with extensive postinflammatory hypermelanosis. Dermatologica 1987;174:249-252

29 Weickel R, Frosch PJ. Contact allergy to glycerol trinitrate (Nitroderm TTS). Hautarzt 1986;37:511-512 (Article in German)

30 Fisher AA. Dermatitis due to transdermal therapeutic systems. Cutis 1984;34:526-531

31 Fischer RG, Tyler M. Severe contact dermatitis due to nitroglycerin patches. South Med J 1985;78:1523-1524

32 Rosenfeld AS, White WB. Allergic contact dermatitis secondary to transdermal nitroglycerin. Am Heart J 1984;108(4Pt.1):1061-1062

33 Camarasa JG, Perez M. Nitroglycerine sensitivity. Contact Dermatitis 1983;9:320-321

34 Hendricks AA, Dec GW Jr. Contact dermatitis due to nitroglycerin ointment. Arch Dermatol 1979;115:853-855

35 Zugerman C, Zheutlin T, Giacobetti R. Allergic contact dermatitis secondary to nitroglycerin in Nitro-Bid ointment. Contact Dermatitis 1979;5:270-271

36 Sausker WF, Frederick FD. Allergic contact dermatitis secondary to topical nitroglycerin. JAMA 1978;239:1743-1744

Chapter 3.244 NONOXYNOL-9

IDENTIFICATION

Description/definition   : Nonoxynol-9 is the polyethylene glycol and nonylphenyl ether that conforms to the structural formula shown below
Pharmacological classes   : Spermatocidal agents; surface-active agents
IUPAC name   : 2-[2-[2-[2-[2-[2-[2-[2-(4-Nonylphenoxy)ethoxy]ethoxy]ethoxy] ethoxy]ethoxy]-ethoxy]ethoxy]ethanol
Other names   : PEG-9 nonyl phenyl ether; polyoxyethylene (9) nonyl phenyl ether; nonoxynol, n=9
CAS registry number   : 26571-11-9
EC number   : 247-816-5
Merck Index monograph   : 8034 (Nonoxynol)
Patch testing   : 2%, 1%, 0.5% and 0.1% water; perform controls, irritant reactions are likely at concentrations of 1% and higher
Molecular formula   : $C_{33}H_{60}O_{10}$

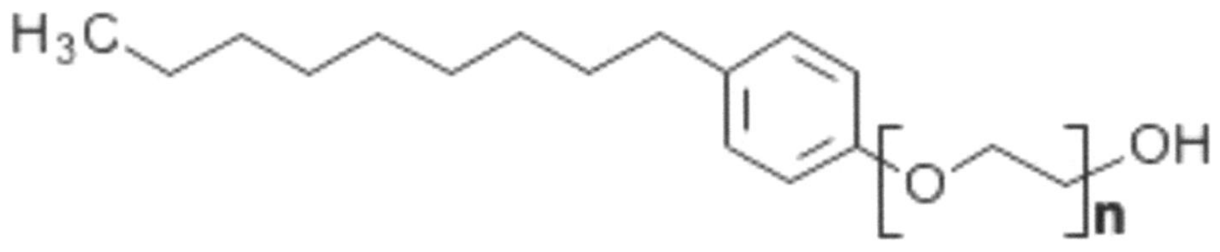

Nonoxynol 9 (n=9)

GENERAL

Nonoxynol-9 is a non-ionic surfactant used as a vaginal spermicide. Spermicides are locally acting non-hormonal contraceptives. When present in the vagina during intercourse, they immobilize/inactivate/damage and/or kill sperms without eliciting systemic effects (1). Nonoxynol-9 has been in use for more than 30 years as an over-the-counter drug in creams, gels, foams and condom lubricants. It is the most commonly used spermicidal contraceptive in the United Kingdom and the USA, although it has certain drawbacks and limitations (1).

CONTACT ALLERGY

Case reports

In a 26-year-old woman, sensitization to nonoxynol-9 present in a vaginal spermicide resulted in vaginal discomfort (itching and burning) and a malodorous white discharge, diagnosed as bacterial vaginosis. There were no changes in the vaginal mucosa or vulva suggestive of an allergic contact reaction (2). A 32-year-old man experienced allergic contact dermatitis of the penis from nonoxynol-9 present in the spermicide product used by his female partner (contact dermatitis *by proxy*) (3).

Contact allergy in non-pharmaceutical products

Eleven cases of allergic contact dermatitis from nonoxynol-9 present in two brands of antiseptic lotions were reported from Belgium in 1989 (4).

Cross-reactions, pseudo-cross-reactions and co-reactions

Cross-reactions between various nonoxynols (-6, -8.3, -9, -10, -14, -18) have been observed (4), undoubtedly because they contain the same allergenic ingredients (pseudo-cross-reactivity).

LITERATURE

1   Iyer V, Poddar SS. Update on nonoxynol-9 as vaginal spermicide. Eur J Contracept Reprod Health Care 2008;13:339-350
2   Haye KR, Mandal D. Allergic vaginitis mimicking bacterial vaginosis. Int J STD AIDS 1990;1:440-442
3   Fisher AA. Allergic contact dermatitis to nonoxynol-9 in a condom. Cutis 1994;53:110-111
4   Dooms-Goossens A, Deveylder H, de Alam HG, Lachapelle JM, Tennstedt D, Degreef H. Contact sensitivity to nonoxynols as a cause of intolerance to antiseptic preparations. J Am Acad Dermatol 1989;21(4Pt.1):723-727

Chapter 3.245 NORETHISTERONE

IDENTIFICATION

Description/definition   : Norethisterone is the synthetic progestin that conforms to the structural formula shown
                            below
Pharmacological classes : Contraceptive agents, female
IUPAC name   : (8R,9S,10R,13S,14S,17R)-17-Ethynyl-17-hydroxy-13-methyl-1,2,6,7,8,9,10,11,12,14,15,16-
                            dodecahydrocyclopenta[a]phenanthren-3-one
Other names   : Norethindrone
CAS registry number   : 68-22-4
EC number   : 200-681-6
Merck Index monograph   : 8056 (Norethindrone)
Patch testing   : 1% alcohol
Molecular formula   : $C_{20}H_{26}O_2$

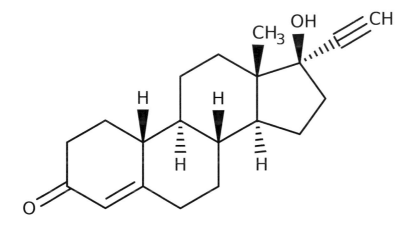

GENERAL

Norethisterone is a synthetic progestational hormone with actions similar to those of progesterone but functioning
as a more potent inhibitor of ovulation. It has weak estrogenic and androgenic properties. The hormone has been
used in treating amenorrhea, functional uterine bleeding, endometriosis, and for contraception. In pharmaceutical
products, norethisterone is employed as norethisterone acetate (CAS number 51-98-9, EC number 200-132-0,
molecular formula $C_{22}H_{28}O_3$) (1).

CONTACT ALLERGY

Case report

A 50-year-old postmenopausal woman, using 2x weekly transdermal patches containing 5 mg estradiol and 15 mg
norethisterone, developed itchy discoid indurated erythema under the 3rd patch applied, followed by a bullous
reaction. Because of this, treatment was switched to the estradiol-containing gel, but acute eczema also developed
on sites of application of this gel. Patch tests yielded strongly positive reactions to norethisterone acetate 0.1% and
1% alcohol and to 17β-estradiol 1% and 0.1% alc. Twenty controls were negative. (2).

LITERATURE

1   The data in the section 'General' may have been obtained from literature discussed in this chapter, but mostly
    also or exclusively from one or more of the following online sources: ChemIDPlus Advanced, PubChem,
    DrugBank, RxList, Drug Central, Drugs.com, and Wikipedia
2   Koch P. Allergic contact dermatitis from estradiol and norethisterone acetate in a transdermal hormonal patch.
    Contact Dermatitis 2001;44:112-113

Chapter 3.246 NYLIDRIN

IDENTIFICATION

Description/definition : Nylidrin is the phenylpropane that conforms to the structural formula shown below
Pharmacological classes : Sympathomimetics; β-adrenergic agonists; tocolytic agents; vasodilator agents
IUPAC name : 4-[1-Hydroxy-2-(4-phenylbutan-2-ylamino)propyl]phenol
Other names : Buphenine; 1-p-hydroxyphenyl-2-(1'-methyl-3'-phenylpropylamino)-1-propanol
CAS registry number : 447-41-6
EC number : 207-182-2
Merck Index monograph : 8091
Patch testing : 1% alcohol
Molecular formula : $C_{19}H_{25}NO_2$

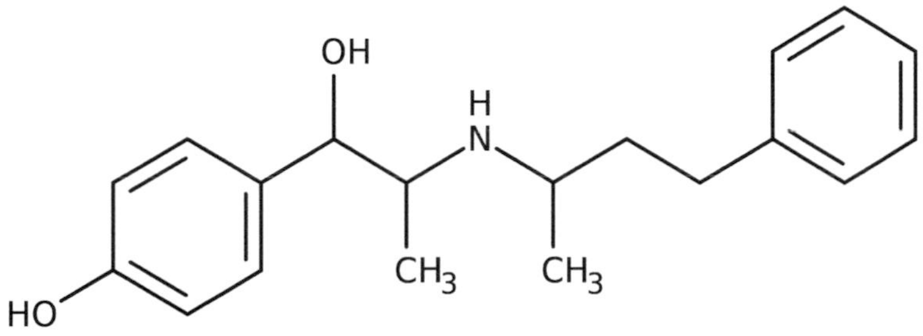

GENERAL

Nylidrin, also known as buphenine, is a β-adrenergic agonist with peripheral vasodilator properties. It is or has been utilized to treat disorders that may benefit from increased blood flow, for example certain mental disorders, blood vessel disease due to diabetes, frostbite, night leg cramps, and certain types of ulcers. Some studies have shown evidence of improving cognitive impairment in selected individuals, such as geriatric patients with mild to moderate symptoms of cognitive, emotional and physical impairment. However, FDA has considered nylidrin as 'lacking substantial evidence of effectiveness' in cerebral ischemia, cerebral arteriosclerosis, and other cerebral circulatory insufficiencies and has withdrawn nylidrin from the U.S. market. In pharmaceutical products, both nylidrin and nylidrin hydrochloride (CAS number 849-55-8, EC number 212-701-0, molecular formula $C_{19}H_{26}ClNO_2$) may be employed (1).

CONTACT ALLERGY

Case reports

A 50-year-old man with chronic venous insufficiency had eczema on the right leg after using a topical vasodilator gel containing nylidrin (also known as buphenine). The dermatitis became generalized in patches over the trunk and extremities. Patch tests showed a positive reaction to the gel 'as is' and its ingredient nylidrin HCl 1% alcohol. While the patch tests were in place, the generalized dermatitis reappeared. Two weeks days later, after stopping medication, the dermatitis again reappeared. The patient had been using an oral vasodilator agent for 2 years without any problem. This contained a closely related substance, isoxsuprine. A patch test with this vasodilator 1% alcohol gave a +++ reaction at D2 and D4 (2). It appears that isoxsuprine cross-reacted to nylidrin and caused a systemic contact dermatitis.

Two other patients also developed extensive dermatitis after using a gel containing 1% nylidrin on varicose veins/stasis dermatitis. They both had positive patch tests to the cream and its active ingredient nylidrin (buphenin) tested 'as is' (3). As the reaction to the 'undiluted' nylidrin was weaker than that of the gel, it seems likely that the manufacturer had supplied the nylidrin in the same concentration as in the gel.

Cross-reactions, pseudo-cross-reactions and co-reactions

Cross-reactivity from nylidrin to isoxsuprine or vice versa (2).

LITERATURE

1 The data in the section 'General' may have been obtained from literature discussed in this chapter, but mostly also or exclusively from one or more of the following online sources: ChemIDPlus Advanced, PubChem, DrugBank, RxList, Drug Central, Drugs.com, and Wikipedia
2 Alomar A. Buphenine sensitivity. Contact Dermatitis 1984;11:315
3 Pevny I. Buphenin allergy. Contact Dermatitis 1984;11:52-53

Chapter 3.247 NYSTATIN

IDENTIFICATION

Description/definition   : Nystatin is a macrolide antifungal antibiotic complex produced by *Streptomyces noursei*, other *Streptomyces* species, and *S. aureus*; it conforms to the structural formula shown below

Pharmacological classes   : Anti-bacterial agents; ionophores; antifungal agents

IUPAC name   : (4*E*,6*E*,8*E*,10*E*,14*E*,16*E*,18*S*,19*R*,20*R*,21*S*,35*S*)-3-[(2*S*,3*S*,4*S*,5*S*,6*R*)-4-Amino-3,5-dihydroxy-6-methyloxan-2-yl]oxy-19,25,27,29,32,33,35,37-octahydroxy-18,20,21-trimethyl-23-oxo-22,39-dioxabicyclo[33.3.1]nonatriaconta-4,6,8,10,14,16-hexaene-38-carboxylic acid

CAS registry number   : 1400-61-9

EC number   : 215-749-0

Merck Index monograph   : 8095

Patch testing   : 2% pet. (SmartPracticeCanada, SmartPracticeEurope); testing in polyethylene glycol 400 (100,000 IU nystatin/gr) may be preferable over petrolatum (11)

Molecular formula   : $C_{47}H_{75}NO_{17}$

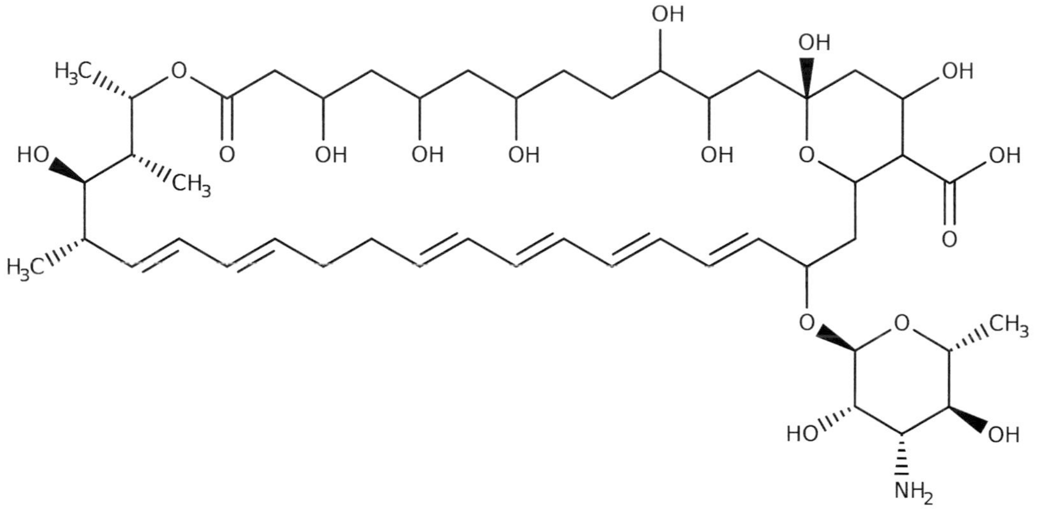

GENERAL

Nystatin is a macrolide antifungal antibiotic complex produced by *Streptomyces noursei, S. aureus*, and other *Streptomyces* species. The biologically active components are nystatin A1, A2, and A3. Nystatin is a topical and oral antifungal agent with activity against many species of yeast and *Candida albicans*, which is used largely to treat skin and oropharyngeal candidiasis. Nystatin is poorly absorbed from the gut. This agent acts by binding to sterols in the cell membrane of susceptible species resulting in a change in membrane permeability and the subsequent leakage of intracellular components. Nystatin is indicated for treatment of cutaneous or mucocutaneous mycotic infections caused by *Candida* species (1).

CONTACT ALLERGY

General

Sporadic case reports of allergic contact dermatitis from nystatin have been published (table 3.247.2). Consequently, it is often stated that contact allergy to nystatin is rare. Nevertheless, one author observed 4 cases (13-15) and at St. John's Hospital in London, five cases were discovered in the period 1971-1976 (20). It is conceivable therefore that nystatin allergy is not that rare, but goes unrecognized. There are three arguments for this hypothesis. In many published cases of nystatin allergy, Mycolog ® cream/ointment was implicated as the source of contact with nystatin (13-19). This widely used product contains or contained, depending on the period when used, geographic location and version of the product, a combination of nystatin, neomycin, triamcinolone acetonide, gramicidin and ethylenediamine. By far most allergic reactions to this cream are caused by neomycin and to a lesser extent by the excipient ethylenediamine. Neomycin is present in all routine series and ethylenediamine formerly also was. Allergy to Mycolog cream/ointment was therefore mostly detected by routine testing and the allergy would be ascribed to either constituent. However, co-sensitizations to nystatin (co-sensitization to various ingredients in topical pharmaceuticals is common) would not be detected, as nystatin is not tested separately. Indeed, in almost all cases of sensitization to nystatin caused by Mycolog cream, the patients co-reacted to ethylenediamine or neomycin (13,14,15,18,19).

The second argument is that petrolatum, which is commonly used for patch testing, is probably not the best vehicle for patch testing nystatin and may result in false-negative reactions (19); polyethylene glycol may be preferable for patch testing this antifungal drug (11). Thirdly, the presence of the corticosteroid triamcinolone acetonide in Mycolog cream may suppress the patch test and result in a false-negative reaction; a contact allergy to nystatin would then go unrecognized.

Contact allergy in the general population

In Germany, for the period 1995-2004, the population-based relative incidence (RI) of contact sensitization to nystatin (cases/100,000 defined daily doses (DDDs) per year) was estimated to be 0.9. In the group of all topical drugs, the RI ranged from 0.3 (dexamethasone sodium phosphate and pilocarpine) to 413.9 (benzocaine) (3).

Patch testing in groups of patients

Results of studies in which groups of selected patients were patch tested with nystatin are shown in table 3.247.1. Generally, the prevalences of sensitization were low, 0.3% or 0.4%. A higher rate of 2.9% was found in Belgium, in a University hospital specialized in allergy to cosmetics and drugs, probably from targeted testing with nystatin; the absolute numbers were low, 14 sensitized patients in 24 years (4).

Table 3.247.1 Patch testing in groups of patients

Years and Country	Test conc. & vehicle	Number of patients tested | positive (%)		Selection of patients (S); Relevance (R); Comments (C)	Ref.
1990-2014 Belgium	2% pet.	489	14 (2.9%)	S: patients suspected of iatrogenic contact dermatitis and tested with a pharmaceutical series and their own products; R: 96% of the positive patch test reactions to all topical drugs and antiseptics were considered to be relevant	4
2004-2008 IVDK			(0.3%)	S: patients with anogenital dermatoses tested with a medicament series; R: not stated; C: number of patients tested unknown	8
2002-2004 IVDK	2% pet.	2780	(0.4%)	S: not stated; R: not stated	2
2001-2002 IVDK	2% pet.	1177	3 (0.3%)	S: patients tested with an imidazole antifungal series; the selection procedure was not well described; R: not stated	5
1982-1998 U.K.	10% pet.	385	1 (0.3%)	S: not stated; R: the reaction was relevant (see Case reports)	9

IVDK: Information Network of Departments of Dermatology (Germany, Austria, Switzerland)

Case series

In a hospital in London, in the period 1971-1976, three men and two women were found to be allergic to nystatin. The men had used the topical medicament for eczema of the genitals, pruritus ani and intertrigo of the groins; one woman had applied it to her hand eczema and the other to a stasis ulcer. Patch tests were positive to 100,000-350,000 IU/g in Plastibase or petrolatum. The woman with hand eczema had noticed that it was exacerbated by a nystatin ointment. Subsequently she was given nystatin orally and after 4 days developed a generalized eruption (systemic contact dermatitis); her mouth was unaffected (20).

Case reports

A 44-year-old woman had a long history of seborrheic dermatitis affecting the scalp, flexures and natal cleft. She also had recurrent vaginal candidiasis. She had used a variety of antifungal pessaries and creams, including nystatin pessaries, nystatin cream with hydrocortisone, a corticosteroid and a corticosteroid with nystatin and neomycin. The patient reported an exacerbation of the dermatitis of the natal cleft after applying the latter ointment. Patch tests were positive to this ointment, nystatin 10% pet. and neomycin 20% pet. Subsequent retesting confirmed the contact allergy to nystatin (D2 - / D4 +) (9).

Short summaries of other case reports of allergic contact dermatitis from nystatin are shown in table 3.247.2.

Table 3.247.2 Short summaries of case reports of allergic contact dermatitis from nystatin

Year and country	Sex	Age	Positive patch tests	Clinical data and comments	Ref.
1993 United Kingdom	F	69	NYS 100,000 IU/gr pet.; NYS cream 'as is'	acute dermatitis of the antecubital fossae from nystatin - hydrocortisone cream	10
1990 The Netherlands	F	18	NYS ointment 'as is'; NYS 30,000 IU and 90,000 IU in PEG-400	hand eczema spreading to the face and neck from nystatin ointment; nystatin in pet. was only weakly positive, in water, methyl ethyl ketone and liquid paraffin negative; ROAT with nystatin in pet. strongly positive	11

Table 3.247.2 Short summaries of case reports of allergic contact dermatitis from nystatin (continued)

Year and country	Sex	Age	Positive patch tests	Clinical data and comments	Ref.
1985 Germany	M	54	NYS cream 'as is'; NYS 100,000 IU/gr pet.	papular eczema on the lower legs spreading to the neck and face from NYS-tolnaftate cream; co-sensitization to tolnaftate	12
1978 USA	F	16	NYS cream 'as is'; NYS 3% pet.	diffuse dermatitis after using nystatin vaginal suppositories; eczema of the fingers resistant to nystatin-triamcinolone; later multiple corticosteroid allergies	13
1971 USA	M	48	NYS cream 'as is'; NYS 100,000 IU/gr Velvachol [a]	erythematous, fissured, eroded eruption involving the perianal area from treatment of pruritus ani with NYS cream	14
1971 USA	M	26	NYS cream 'as is'; NYS 100,000 IU/gr Velvachol [a]	after 4 weeks' treatment of an eruption in the groin with NYS cream exacerbation and spreading to the abdomen, scrotum and penis	15
	F	19	NYS 100,000 IU/gr Velvachol [a]	widespread erythematous papular eruption from vaginal suppositories, probably systemic contact dermatitis; later worsening of hand eczema from NYS cream	
1971 France	F	3	NYS ointment 'as is'; diluted NYS tablet	eczema from nystatin ointment; patch test to 'excipient' of the ointment was negative	16, 17
1970, 1971 USA	F	35	NYS cream and ointment 'as is'; NYS 100,000 IU/gr alcohol 70%	worsening of vulvar dermatitis from 2 preparations containing nystatin; later, propylene glycol was found to be a sensitive vehicle for patch testing, whereas petrolatum gave false-negative reactions to nystatin at concentrations of 50,000 IU/gr or lower (19)	18, 19

NYS: nystatin; PEG: polyethylene glycol

[a] Velvachol is a commercial emollient; for (current) constituents see https://www.ndrugs.com/?s=velvachol

Cross-reactions, pseudo-cross-reactions and co-reactions

Patients sensitized to (oral) nystatin sometimes cross-react to the structurally related amphotericin B (6,7).

Cutaneous adverse drug reactions from systemic administration caused by type IV (delayed-type) hypersensitivity

Cutaneous adverse drug reactions from systemic administration of nystatin caused by type IV (delayed-type) hypersensitivity, including acute generalized exanthematous pustulosis (AGEP) (21-23), maculopapular exanthema (24,28), systemic contact dermatitis (25,29), micropapular eruption (26), and possibly drug eruption with eosinophilia and systemic symptoms (DRESS) (27), are planned to be discussed in Volume IV of the *Monographs in Contact Allergy* series on Systemic drugs.

LITERATURE

1 The data in the section 'General' may have been obtained from literature discussed in this chapter, but mostly also or exclusively from one or more of the following online sources: ChemIDPlus Advanced, PubChem, DrugBank, RxList, Drug Central, Drugs.com, and Wikipedia

2 Menezes de Padua CA, Uter W, Schnuch A. Contact allergy to topical drugs: prevalence in a clinical setting and estimation of frequency at the population level. Pharmacoepidemiol Drug Saf 2007;16:377-384

3 Menezes de Padua CA, Schnuch A, Nink K, Pfahlberg A, Uter W. Allergic contact dermatitis to topical drugs – Epidemiological risk assessment. Pharmacoepidemiol Drug Saf 2008;17:813-821

4 Gilissen L, Goossens A. Frequency and trends of contact allergy to and iatrogenic contact dermatitis caused by topical drugs over a 25-year period. Contact Dermatitis 2016;75:290-302

5 Menezes de Pádua CA, Uter W, Geier J, Schnuch A, Effendy I; German Contact Dermatitis Research Group (DKG); Information Network of Departments of Dermatology (IVDK). Contact allergy to topical antifungal agents. Allergy 2008;63:946-947

6 Barranco R, Tornero P, de Barrio M, de Frutos C, Rodríguez A, Rubio M. Type IV hypersensitivity to oral nystatin. Contact Dermatitis 2001;45:60

7 Martínez FV, Muñoz Pamplona MP, García EC, Urzaiz AG. Delayed hypersensitivity to oral nystatin. Contact Dermatitis 2007;57:200-201

8 Bauer A. Contact sensitization in the anal and genital area. Curr Probl Dermatol 2011;40:133-141

9 Cooper SM, Shaw S. Contact allergy to nystatin: an unusual allergen. Contact Dermatitis 1999;41:120

10 Hills RJ, Ive FA. Contact sensitivity to nystatin in Timodine. Contact Dermatitis 1993;28:48

11 De Groot AC, Conemans JM. Nystatin allergy. Petrolatum is not the optimal vehicle for patch testing. Dermatologic Clinics 1990;8:153-155

12 Lang E, Goos M. Combined allergy to tolnaftate and nystatin. Contact Dermatitis 1985;12:182

13 Coskey RJ. Contact dermatitis due to multiple corticosteroid creams. Arch Dermatol 1978;114:115-117

14 Coskey RJ. Contact dermatitis due to nystatin. Arch Dermatol 1971;103:228

15 Coskey RJ. Contact dermatitis due to nystatin (Reply). Arch Dermatol 1971;104:438

16 Foussereau J, Limam-Mestiri S, Khochnevis A. Contact allergy to nystatin. Contact Dermatitis Newsletter 1971;10:221

17 Foussereau J, Limam- Mestiri S, Khochnevis A, et al. L'Allergie à l'association thérapeutique locale (nystatine, néomycine et acetonide de triamcinolone). Bull Soc Franc Derm Syph 1971;78:457-459 (Article in French)

18 Wasilewski C. Allergic contact dermatitis from nystatin. Arch Dermatol 1971;104:437

19 Wasilewski C Jr. Allergic contact dermatitis from nystatin. Arch Dermatol 1970;102:216-217

20 Cronin E. Contact Dermatitis. Edinburgh: Churchill Livingstone, 1980:232-233

21 Küchler A, Hamm H, Weidenthaler-Barth B, Kämpgen E, Bröcker EB. Acute generalized exanthematous pustulosis following oral nystatin therapy: a report of three cases. Br J Dermatol 1997;137:808-811

22 Poszepczynska-Guigne E, Viguier M, Assier H, Pinquier L, Hochedez P, Dubertret L. Acute generalized exanthematous pustulosis induced by drugs with low-digestive absorption: acarbose and nystatin. Ann Dermatol Venereol 2003;130:439-442 (Article in French)

23 Rosenberger A, Tebbe B, Treudler R, Orfanos CE. Acute generalized exanthematous pustulosis, induced by nystatin. Hautarzt 1998;49:492-495 (Article in German)

24 Martínez FV, Muñoz Pamplona MP, García EC, Urzaiz AG. Delayed hypersensitivity to oral nystatin. Contact Dermatitis 2007;57:200-201

25 Vega F, Ramos T, Las Heras P, Blanco C. Concomitant sensitization to inhaled budesonide and oral nystatin presenting as allergic contact stomatitis and systemic allergic contact dermatitis. Cutis 2016;97:24-27

26 Barranco R, Tornero P, de Barrio M, de Frutos C, Rodríguez A, Rubio M. Type IV hypersensitivity to oral nystatin. Contact Dermatitis 2001;45:60

27 Cooper SM, Reed J, Shaw S. Systemic reaction to nystatin. Contact Dermatitis 1999;41:345-346

28 Quirce S, Parra F, Lázaro M, Gómez MI, Sánchez Cano M. Generalized dermatitis due to oral nystatin. Contact Dermatitis 1991;25:197-198

29 Lechner T, Grytzmann B, Baurle G. Hämatogenes allergisches Kontaktekzem nach oraler Gabe von Nystatin. Mykosen 1987;30:143-146 (Article in German)

Chapter 3.248 OLAFLUR

IDENTIFICATION

Description/definition   : Olaflur is the substituted amine salt that conforms to the structural formula shown below
Pharmacological classes : Caries prophylactic agents
IUPAC name   : 2-[3-[bis(2-Hydroxyethyl)amino]propyl-octadecylamino]ethanol
Other names   : *N'*-Octadecyltrimethylendiamine-*N,N,N'*-tris(2-ethanol)-dihydrofluoride; stearyl
                           trihydroxyethyl propylenediamine dihydrofluoride; ethanol, 2,2'-[[3-[(2-
                           hydroxyethyl)octadecylamino]propyl]imino] bis-, dihydrofluoride; C27-amine fluoride
CAS registry number   : 6818-37-7
EC number   : 229-891-6
Merck Index monograph   : 8182
Patch testing   : 0.9% water
Molecular formula   : $C_{27}H_{60}F_2N_2O_3$

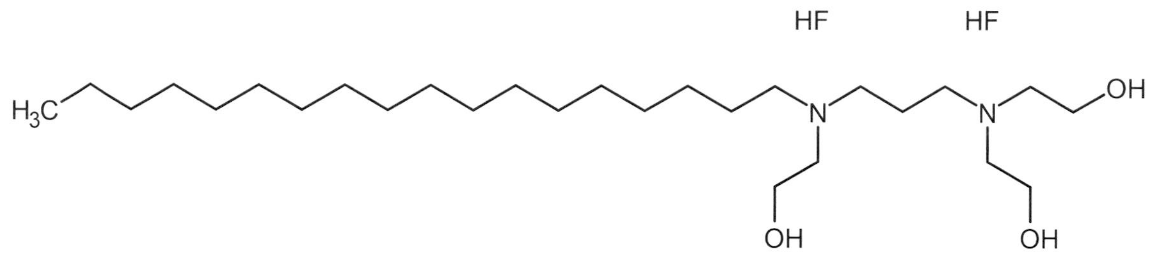

GENERAL

Amine fluorides, used in toothpastes to deliver fluoride to the teeth and combat caries, were developed at the end of the 1950's. The most widely used amine fluorides are olaflur (though only in one brand) and octadecenylammonium fluoride (synonym: dectaflur). In some articles, the exact nature of the 'amine fluoride' was not ascertained (1,3).

CONTACT ALLERGY

Case reports

A woman had suffered from itchy, red and dry lips with painful angular cheilitis for 9 months. She only applied Vaseline®, and did not use nail varnish. Patch testing revealed a positive reaction to the toothpaste, tested 1% water. Nineteen coded constituents were obtained from the manufacturer in the same concentrations as in the finished product. On patch testing, there was again a positive reaction to the toothpaste Itself (tested 'as is' this time) and also one of the constituents reacted, which was, according to the manufacturer, olaflur. Requests for more material for serial dilution and control testing were denied by the manufacturer. After the patient had changed to another brand of toothpaste, the symptoms disappeared (2).

Previously, in Italy, a patient was described with allergic contact cheilitis caused by 'amine fluoride' in a toothpaste (1). She had a positive patch test reaction to the toothpaste 3% pet., and a positive repeated open application test result with the toothpaste 'as is', and reacted later to the ingredient 'amine fluoride' 5% water. Ten controls were negative. Neither the brand of the toothpaste nor the exact nature of the amine fluoride was mentioned (1). However, the first author of this article was contacted, and she confirmed that the toothpaste in this report was the same as used by the patient described above, which means that the 'amine fluoride' very likely was olaflur. In both cases, the reported concentration of amine fluoride in the test substance was 5% water. However, the 5% concentration proved to be 5% of a stock solution used by the manufacturer, and the actual test concentration of olaflur was 0.9%, containing 700 ppm fluoride (2).

A young boy had cheilitis, papules and plaques around the mouth and vesicles in the oral mucosa, which were ascribed to the use of an anti-caries gel (same brand as the toothpastes in the cases described above), He had positive patch test reactions to the gel, diluted with 0.9% NaCl solution and tested at 50%, 10%, 1% and 0.1% and to 'amine fluoride' 3.7% water, which was the concentration in the gel. Five control patients were negative. The amine fluoride was a mixture of olaflur and dectaflur; these chemicals could not be tested separately (3).

Cross-reactions, pseudo-cross-reactions and co-reactions
Not to fluoride (2).

LITERATURE

1   Foti C, Romita Paolo, Ficco D, Bonamonte D, Angelini G. Allergic contact cheilitis to amine fluoride in a toothpaste. Dermatitis 2014;25:209
2   De Groot A, Tupker R, Hissink D, Woutersen M. Allergic contact cheilitis caused by olaflur in toothpaste. Contact Dermatitis 2017;76:61-62
3   Ganter G, Disch R, Borelli S, Simon D. Contact dermatitis and stomatitis due to amine fluoride. Contact Dermatitis 1997;37:248

Chapter 3.249 OXICONAZOLE

IDENTIFICATION

Description/definition   : Oxiconazole is the phenylethyl imidazole derivative that conforms to the structural
                             formula shown below
Pharmacological classes : Antifungal agents
IUPAC name   : (Z)-1-(2,4-Dichlorophenyl)-N-[(2,4-dichlorophenyl)methoxy]-2-imidazol-1-ylethanimine
Other names   : 2',4'-Dichloro-2-imidazol-1-ylacetophenone (Z)-(O-(2,4-dichlorobenzyl)oxime)
CAS registry number   : 64211-45-6
EC number   : Not available
Merck Index monograph : 8306
Patch testing   : 1% alcohol or MEK (methyl ethyl ketone)
Molecular formula   : $C_{18}H_{13}Cl_4N_3O$

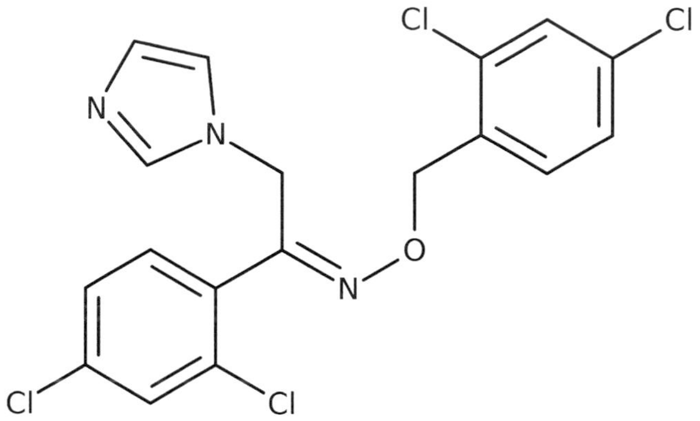

GENERAL

Oxiconazole is a broad-spectrum imidazole derivative that has fungicidal or fungistatic activity against a large number of pathogenic dermatophytes and yeasts. This agent inhibits the cytochrome P450-dependent demethylation of lanosterol, preventing the synthesis of ergosterol, which is a crucial component of fungal cell membrane. Subsequent altered cell membrane permeability promotes loss of essential intracellular components and eventually inhibits fungal cell growth. Oxiconazole is indicated for the treatment of fungal infection of the skin. In pharmaceutical products oxiconazole is employed as oxiconazole nitrate (CAS number 64211-46-7, EC number 264-730-3, molecular formula $C_{18}H_{14}Cl_4N_4O_4$) (1).

CONTACT ALLERGY

Case reports and case series

Between 1977 and 1986, 9 patients with contact allergy to imidazole antimycotics were investigated in the University clinic of Heidelberg, Germany. Six patients were allergic to miconazole, 3 to clotrimazole, 3 to econazole, 3 to isoconazole, and one to oxiconazole. The active ingredient at 1% in alcohol seems to be the most suitable choice for routine patch testing, petrolatum is less effective (3).

A man of 58 years presented with weeping erosions in the genital and inguinal area after using many different antifungal preparations. He had a positive patch test to oxiconazole cream 'as is'. There were also positive patch tests to creams with croconazole, clotrimazole, and tioconazole, which may have been used before, and to croconazole and tioconazole 1% and clotrimazole 5% pet. (5).

Another 58-year-old man had an acute vesicular contact dermatitis on the dorsa of both feet spreading to the lower legs. Two weeks earlier, he had started treatment of a fungal infection with oxiconazole cream. In the previous 6 months he had applied various medicaments for stasis dermatitis of the lower legs, including a clotrimazole containing cream. When patch tested, the patient reacted to the oxiconazole cream 'as is', oxiconazole 1% in alcohol and MEK (methyl ethyl ketone), but not to oxiconazole 1% in petrolatum. There were also positive patch tests to clotrimazole 1% in alcohol and MEK, but negative to the antifungal drug in petrolatum. Five controls were negative (2).

In a literature review up to 1994, the authors identified 105 reported patients who had a positive patch test reaction to at least one imidazole derivative, ranging from 51 reactions to miconazole to one reaction each to bifonazole and enilconazole. The number of reported reactions to oxiconazole was 4 (4).

LITERATURE

1   The data in the section 'General' may have been obtained from literature discussed in this chapter, but mostly also or exclusively from one or more of the following online sources: ChemIDPlus Advanced, PubChem, DrugBank, RxList, Drug Central, Drugs.com, and Wikipedia
2   Raulin C, Frosch PJ. Contact allergy to oxiconazole. Contact Dermatitis 1987;16:39-40
3   Raulin C, Frosch PJ. Contact allergy to imidazole antimycotics. Contact Dermatitis 1988;18:76-80
4   Dooms-Goossens A, Matura M, Drieghe J, Degreef H. Contact allergy to imidazoles used as antimycotic agents. Contact Dermatitis 1995;33:73-77
5   Brans R, Wosnitza M, Baron JM, Merk HF. Contact sensitization to azole antimycotics. Hautarzt 2009;60:372-375 (Article in German)

Chapter 3.250 OXYBUPROCAINE

IDENTIFICATION

Description/definition   : Oxybuprocaine is the benzoate ester that conforms to the structural formula shown
                               below
Pharmacological classes   : Anesthetics, local
IUPAC name   : 2-(Diethylamino)ethyl 4-amino-3-butoxybenzoate
Other names   : Benoxinate; 4-amino-3-butoxy-2-(diethylamino)ethyl ester benzoic acid
CAS registry number   : 99-43-4
EC number   : Not available
Merck Index monograph   : 2319
Patch testing   : 1% and 5% pet.
Molecular formula   : $C_{17}H_{28}N_2O_3$

GENERAL

Oxybuprocaine is a benzoate ester and local anesthetic, which is used especially in ophthalmology and otolaryngology. Oxybuprocaine binds to sodium channels and reversibly stabilizes the neuronal membrane which decreases
its permeability to sodium ions. Oxybuprocaine is used to temporarily numb the front surface of the eye so that the
eye pressure can be measured (tonometry) or a foreign body removed. In pharmaceutical products, oxybuprocaine is
employed as oxybuprocaine hydrochloride (CAS number 5987-82-6, EC number 227-808-8, molecular formula
$C_{17}H_{29}ClN_2O_3$) (1).

CONTACT ALLERGY

Case series

In Leuven, Belgium, in the period 1990-2014, iatrogenic contact dermatitis was diagnosed in 2600 individuals (17% of
the total patch test population). 96% of all positive patch test reactions to topical drugs and antiseptics were
considered to be relevant. Oxybuprocaine (5% pet.) was tested in 93 patients and there were 3 positive reactions to
it (2).

In the period 1995-1999, the Information Network of the Departments of Dermatology (Germany, Austria,
Switzerland) diagnosed 1053 patients with allergic periorbital contact dermatitis. In 42 cases (4%), allergens were
identified by testing of the patients' own substances, mostly beta-blockers, oxybuprocaine and dexpanthenol. Details
were not provided (6).

In Kuopio, Finland, in the period 1982 to 1984, 25 patients suffering from a prolonged or chronic conjunctivitis,
keratoconjunctivitis or blepharoconjunctivitis while receiving an adequate topical medication, were tested for
contact allergy with a battery of ophthalmic drugs. One patient (4%) reacted to oxybuprocaine hydrochloride 0.4%,
but relevance was not discussed (3).

Case reports

Two women aged 64 (patient 1) and 71 years (patient 2), underwent regular tonometry for longstanding glaucoma,
using several local anesthetics. Patient 1 was exposed to three eye medications containing oxybuprocaine, and

patient no. 2 used one eye solution. When patch tested, patient 1 reacted to all three medications and – later – to oxybuprocaine 0.5% pet. All excipients were also tested, but were negative. Patient 2 had a positive patch test to the medication and to oxybuprocaine 0.5% pet. Several other ester-type and amide-type local anesthetics were also tested, with negative result, but benzocaine (also a benzoic acid ester) was not tested (4,5).

A 70-year old woman with glaucoma was exposed to 5 ophthalmic medications for measurement of intraocular pressure, which resulted in edema and erythema of the lower eyelids. There was a weakly positive (D2 -, D3 ?+, D5 +) patch test reaction to eye drops containing oxybuprocaine and chlorhexidine only. When tested with oxybuprocaine 0.5% pet., identical reactions were observed (D2 -, D3 ?+, D5 +). Chlorhexidine was negative, as was the reaction to the related benzoic acid ester benzocaine (5).

LITERATURE

1   The data in the section 'General' may have been obtained from literature discussed in this chapter, but mostly also or exclusively from one or more of the following online sources: ChemIDPlus Advanced, PubChem, DrugBank, RxList, Drug Central, Drugs.com, and Wikipedia
2   Gilissen L, Goossens A. Frequency and trends of contact allergy to and iatrogenic contact dermatitis caused by topical drugs over a 25-year period. Contact Dermatitis 2016;75:290-302
3   Hätinen A, Teräsvirta M, Fräki JE. Contact allergy to components in topical ophthalmologic preparations. Acta Ophthalmol (Copenh) 1985;63:424-426
4   Blaschke V, Fuchs T. Periorbital allergic contact dermatitis from oxybuprocaine. Contact Dermatitis 2001;44:198
5   Blaschke V, Fuchs T. Relevant allergens by periorbital allergic contact dermatitis. Oxybuprocain, an underestimated allergen. Ophthalmologe 2003;100:628-632 (Article in German)
6   Herbst RA, Uter W, Pirker C, Geier J, Frosch PJ. Allergic and non-allergic periorbital dermatitis: patch test results of the Information Network of the Departments of Dermatology during a 5-year period. Contact Dermatitis 2004;51:13-19

Chapter 3.251 OXYPHENBUTAZONE

IDENTIFICATION

Description/definition   : Oxyphenbutazone is the hydroxylated metabolite of phenylbutazone that conforms to the
                             structural formula shown below
Pharmacological classes   : Anti-inflammatory agents, non-steroidal
IUPAC name   : 4-Butyl-1-(4-hydroxyphenyl)-2-phenylpyrazolidine-3,5-dione
Other names   : 1-(p-Hydroxyphenyl)-2-phenyl-4-butyl-3,5-pyrazolidinedione
CAS registry number   : 129-20-4
EC number   : 204-936-2
Merck Index monograph   : 8341
Patch testing   : 1% pet.
Molecular formula   : $C_{19}H_{20}N_2O_3$

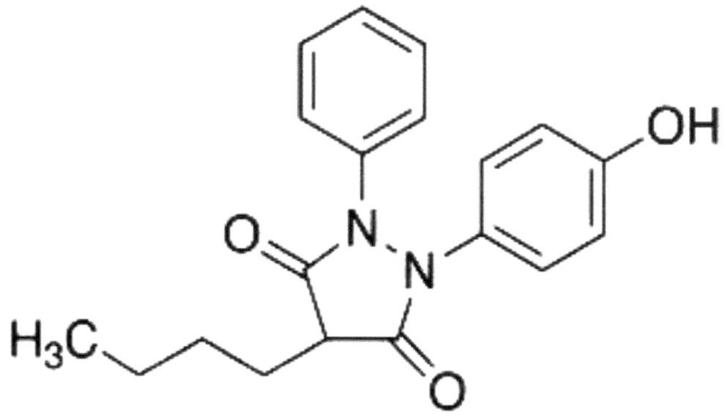

GENERAL

Oxyphenbutazone is a pyrazolidine nonsteroidal anti-inflammatory drug (NSAID) and a hydroxylated and active metabolite of phenylbutazone. Oxyphenbutazone eye drops have been used in the management of postoperative ocular inflammation, superficial eye injuries, and episcleritis. This NSAID was formerly used by mouth in rheumatic disorders such as ankylosing spondylitis, osteoarthritis, and rheumatoid arthritis but such use is no longer considered justified owing to the risk of severe hematological adverse effects and Stevens-Johnson syndrome. It was withdrawn in the mid-1980s in most countries but is apparently still available in China (www.drugs.com). The name oxyphenbutazone is also used for oxyphenbutazone monohydrate (CAS 7081-38-1, EC number not available, molecular formula $C_{19}H_{22}N_2O_4$), which is (or was) the usual form in drugs (1).

CONTACT ALLERGY

General

In Italy, topical preparations of oxyphenbutazone (and phenylbutazone) were apparently withdrawn from sale in the early 1990s due to its sensitizing properties (4).

Case series

In Italy, before 1993, the members of the GIRDCA Multicentre Study Group diagnosed 102 patients (49 men, 53 women), aged 16 to 66 years (mean 37 years), with (photo)dermatitis induced by systemic or topical NSAIDs. Oxyphenbutazone caused 7 contact allergic and zero photocontact allergic reactions (2).

  In a university clinic in Milan, Italy, in a 10-month period before 1985, 5 cases of sensitization to oxyphenbutazone cream were seen. One is described in the section 'Case reports' below and had systemic contact dermatitis, the other 4 subjects had typical allergic contact dermatitis at the site of the topical application. All patients were patch tested with oxyphenbutazone, phenylbutazone, and pyrazolone (all 1% pet.) and the excipients of the cream. The only positive reactions at D2 and D3 were to oxyphenbutazone in all 5 patients (6).

  In a university Hospital in Coimbra, Portugal, in the period 1977-1985, 11 cases of contact sensitivity to oxyphenbutazone cream were observed. Eight of these patients were patch tested with the cream, oxyphenbutazone, phenylbutazone, feprazone and suxibuzone, all tested 1% pet., as well as with the cream base. One was negative to all substances after being positive to the cream 7 years ago. Five were positive to the cream base and negative to oxyphenbutazone. In 2 patients, patch tests were strongly positive to both oxyphenbutazone and the other 3 NSAIDs tested (5).

Case reports

A 61-year-old woman had developed severe dermatitis of her face, arms and legs after using a cream containing oxyphenbutazone for the treatment of rheumatic pain in a leg. Patch tests were positive to the cream, to oxyphenbutazone 1% pet. and to phenylbutazone 1% pet., but not to other NSAIDs (4).

A 48-year-old man who had applied 5% oxyphenbutazone ointment several times to osteoarthritic painful joints, developed allergic contact dermatitis at the sites of application. About 3 months later, he used oxyphenbutazone suppositories which resulted in extensive dermatitis with erythema, papules and wheals, which then progressed into desquamative erythroderma (systemic contact dermatitis). Patch tests were positive to oxyphenbutazone, and in addition a wheal appeared within 30 minutes of application (6).

Another case of systemic contact dermatitis from oxyphenbutazone suppositories, with unknown primary sensitization source, was presented from Italy in 1983 (7). A 76-year-old man was treated with a suppository of oxyphenbutazone for rheumatic pains in the legs. After 6-7 hours, he developed pruritus, swelling of the face, and a generalized red, papular, vesicular and edematous rash which became erythrodermic in a few hours. When patch tested, there was, after 6-8 hours, an itchy, red, edematous, urticarial reaction at the oxyphenbutazone and suppository test sites, which became eczematous after 2 days (7).

A 67-year-old woman with a post-traumatic ulcer of the right leg was treated with a cream containing oxyphenbutazone. After 2 weeks she developed contact dermatitis of the leg, neck and face, with palpebral edema and itching. Patch tests were positive to oxyphenbutazone (test concentration not mentioned) and to phenylbutazone 1% and 10% pet. (8). In a clinical trial series, a 64-year-old man with stasis dermatitis on the lower legs was treated with oxyphenbutazone cream 5%. After two weeks the dermatitis became worse and spread to the legs and arms. Patch tests were positive to oxyphenbutazone 1% with a cross-reaction to phenylbutazone 1% pet. (9)

Cross-reactions, pseudo-cross-reactions and co-reactions

Patients sensitized to oxyphenbutazone may (4,5,7,8,9) or may not (6) cross-react to phenylbutazone. Conversely, patients sensitized to phenylbutazone may (11,12) or may not (9,10) cross-react to oxyphenbutazone. Two patients who had allergic contact dermatitis from topical oxyphenbutazone co-reacted to feprazone and suxibuzone (5). Of 10 patients with photoallergic contact dermatitis from ketoprofen, one co-reacted to oxyphenbutazone. This was considered to be a concomitant sensitization, not a cross-reaction (3).

Immediate contact reactions

Immediate contact reactions (contact urticaria) to oxyphenbutazone are presented in Chapter 5.

LITERATURE

1   The data in the section 'General' may have been obtained from literature discussed in this chapter, but mostly also or exclusively from one or more of the following online sources: ChemIDPlus Advanced, PubChem, DrugBank, RxList, Drug Central, Drugs.com, and Wikipedia
2   Pigatto PD, Mozzanica N, Bigardi AS, Legori A, Valsecchi R, Cusano F, et al. Topical NSAID allergic contact dermatitis. Italian experience. Contact Dermatitis 1993;29:39-41
3   Adamski H, Benkalfate L, Delaval Y, Ollivier I, le Jean S, Toubel G, et al. Photodermatitis from non-steroidal anti-inflammatory drugs. Contact Dermatitis 1998;38:171-174
4   Cameli N, Vincenzi C, Morelli R, Bardazzi F, Tardio M. Contact allergy to oxyphenbutazone. Contact Dermatitis 1991;24:75-76
5   Figueiredo A, Gonçalo S, Freitas JD. Contact sensitivity to pyrazolone compounds. Contact Dermatitis 1985;13:271
6   Pigatto PD, Riboldi A, Morelli M, Altomare GF, Polenghi MM. Allergic contact dermatitis from oxyphenbutazone. Contact Dermatitis 1985;12:236-237
7   Valsecchi R, Tornaghi A, Falgheri G, Rossi A, Cainelli T. Drug reaction from oxyphenbutazone. Contact Dermatitis. 1983;9:419
8   Valsecchi R, Serra M, Foiadelli L, Cainelli T. Contact sensitivity to oxyphenbutazone. Contact Dermatitis 1981;7:157
9   Krook G. Contact sensitivity to oxyphenbutazone (Tanderil®) and cross-sensitivity to phenylbutazone (Butazolidin®). Contact Dermatitis 1975;1:262 (in PubMed the same publication is cited to be present on pages 385-386, but these appear not to exist)
10   Meneghini CL, Angelini G. Contact allergy to antirheumatic drugs. Contact Dermatitis 1979;5:197-198
11   Fernández de Corres L, Bernaola G, Lobera T, Leanizbarrutia I, Muñoz D. Allergy from pyrazoline derivatives. Contact Dermatitis 1986;14:249-250
12   Vooys RC, van Ketel WG. Allergic drug eruption from pyrazolone compounds. Contact Dermatitis 1977;3:57-58

Chapter 3.252 OXYQUINOLINE

IDENTIFICATION

Description/definition	: Oxyquinoline is a heterocyclic phenol and derivative of quinoline that it conforms to the structural formula shown below
Pharmacological classes	: Dermatological agents; antiseptics and disinfectants
IUPAC name	: Quinolin-8-ol
Other names	: 8-Hydroxyquinoline; 8-quinolinol; phenopyridine; hydroxybenzopyridine
CAS registry number	: 148-24-3
EC number	: 205-711-1
Merck Index monograph	: 6151
Patch testing	: No data available; suggested: 5% pet.
Molecular formula	: C_9H_7NO

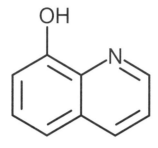

GENERAL

Oxyquinoline is a heterocyclic phenol and derivative of quinoline. It is an antiseptic and disinfectant with mild fungistatic, bacteriostatic, anthelmintic, and amebicidal properties. Oxyquinoline is used as a biocidal component of several over-the-counter products, mainly marketed for the purposes of inhibiting abnormal biological growth in the vagina and restoring natural pH. Oxyquinoline is also used as a stabilizer for hydrogen peroxide, as a reagent and metal chelator, and as a carrier for radio-indium for diagnostic purposes. In pharmaceutical products, both oxyquinoline, oxyquinoline sulfate (CAS number 134-31-6, EC number 205-137-1, molecular formula $C_{18}H_{16}N_2O_6S$) and other derivatives (hydrochloride, citrate, salicylate) may be employed (1).

CONTACT ALLERGY

Case series

In Leuven, Belgium, in the period 1990-2014, iatrogenic contact dermatitis was diagnosed in 2600 individuals (17% of the total population). 96% of all positive patch test reactions to topical drugs and antiseptics were considered to be relevant. Oxyquinoline (sulfate) (10% pet.) was tested in 154 patients and there were 3 positive reactions to it (3).

Oxyquinoline was also responsible for out 1 of 399 cases of *cosmetic* allergy where the causal allergen was identified in a study of the NACDG, USA, 1977-1983 (2).

Cross-reactions, pseudo-cross-reactions and co-reactions

Cross-reactions between oxyquinoline and halogenated hydroxyquinolines may occur (4).

LITERATURE

1 The data in the section 'General' may have been obtained from literature discussed in this chapter, but mostly also or exclusively from one or more of the following online sources: ChemIDPlus Advanced, PubChem, DrugBank, RxList, Drug Central, Drugs.com, and Wikipedia
2 Adams RM, Maibach HI, Clendenning WE, Fisher AA, Jordan WJ, Kanof N, et al. A five-year study of cosmetic reactions. J Am Acad Dermatol 1985;13:1062-1069
3 Gilissen L, Goossens A. Frequency and trends of contact allergy to and iatrogenic contact dermatitis caused by topical drugs over a 25-year period. Contact Dermatitis 2016;75:290-302
4 Leifer W, Steiner K. Studies in sensitization to halogenated hydroxyquinolines and related compounds. J Invest Dermatol 1951;17:233-240

Chapter 3.253 OXYTETRACYCLINE

IDENTIFICATION

Description/definition : Oxytetracycline is a tetracycline analog isolated from the actinomycete *Streptomyces rimosus* that conforms to the structural formula shown below
Pharmacological classes : Anti-bacterial agents
IUPAC name : (4S,4aR,5S,5aR,6S,12aR)-4-(Dimethylamino)-1,5,6,10,11,12a-hexahydroxy-6-methyl-3,12-dioxo-4,4a,5,5a-tetrahydrotetracene-2-carboxamide
Other names : 5-Hydroxytetracycline
CAS registry number : 79-57-2
EC number : 201-212-8
Merck Index monograph : 8345
Patch testing : 3% pet. (SmartPracticeCanada, SmartPracticeEurope)
Molecular formula : $C_{22}H_{24}N_2O_9$

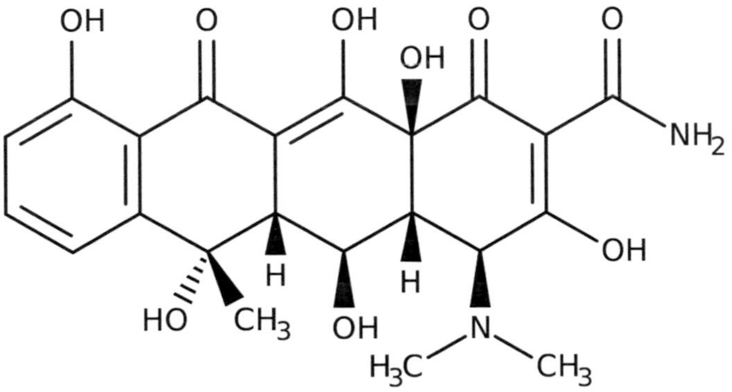

GENERAL

Oxytetracycline is a tetracycline analog isolated from the actinomycete *Streptomyces rimosus* with broad-spectrum antibacterial properties. This antibiotic is indicated for treatment of infections caused by a variety of gram-positive and gram-negative microorganisms including *Mycoplasma pneumoniae*, *Pasteurella pestis*, *Escherichia coli*, *Haemophilus influenzae* (respiratory infections), and *Diplococcus pneumoniae*. Oxytetracycline is used topically in the treatment of acne vulgaris, ophthalmic infections, and in the prevention or treatment of skin infections (11). In pharmaceutical products, both oxytetracycline and oxytetracycline hydrochloride (CAS number 2058-46-0, EC number 218-161-2, molecular formula $C_{22}H_{25}ClN_2O_9$) may be employed.

CONTACT ALLERGY

Contact allergy in the general population

In Germany, for the period 1995-2004, the population-based relative incidence (RI) of contact sensitization to oxytetracycline (cases/100,000 defined daily doses (DDDs) per year) was estimated to be 1.6. In the group of antibiotics, the RI ranged from 1.6 (oxytetracycline) to 86.2 (framycetin) (4).

Patch testing in groups of selected patients

In the period 1990-2014, in a University clinic in Leuven, Belgium, 495 patients with suspected iatrogenic contact dermatitis were tested with oxytetracycline 3% pet. and there were 7 (1.4%) positive reactions. 96% of the positive patch test reactions to *all* topical drugs and antiseptics were considered to be relevant (5). In the period 1995-2004, the IVDK (Germany, Austria, Switzerland) patch tested 10,661 patients with oxytetracycline 3% pet. and there were 11 (0.1%) positive reactions. The mode of selecting the patients and the relevance of the positive reactions were not mentioned (3).

Case series

In Leuven, Belgium, between 1990 and 2017, 16,065 patients were investigated for contact allergy and 118 (0.7%) showed positive patch test reactions to topical ophthalmic medications and/or to their ingredients. Eighty-four individuals (71%) reacted to an active principle. Oxytetracycline was tested in 518 patients and was the allergen in eye medications in 3. There were also 4 reactions to oxytetracycline in other types of medications (2).

Ten patients in Sweden, 8 women and 2 men, all suffering from hypostatic ulcer and/or stasis dermatitis, were sensitized to a combination ointment preparation containing oxytetracycline chloride 3 g, polymyxin B sulfate 106 IU, liquid paraffin and white petrolatum to 100 g. Nine reacted to oxytetracycline 3% pet., all 10 had positive patch tests to polymyxin B 3% pet. and all reacted (some weakly) to the ointment 'as is'. Adequate controls were negative (8). Three of these patients have been described before. These 3 had been tested with polymyxin B 'as is' (1/31 controls positive) (9).

In Poland, a combination preparation of oxytetracycline and a corticosteroid in the forms of ointment, aerosol and ophthalmic ointment is or was very popular, which led to many cases of oxytetracycline allergy. In Warsaw, in 1975, 1 of 66 patients (1.5%) with atopic dermatitis, 5 of 68 (7.3%) with stasis dermatitis and 3 of 476 (0.6%) had positive patch tests to oxytetracycline. In the period 1995-1997, 12 of 111 (10.8%) patients with stasis dermatitis were allergic to oxytetracycline, 5 of 276 patients with conjunctivitis (1.8%) and 6 of 832 patients with contact dermatitis (0.7%) were sensitized to this antibiotic (7).

In India, 101 patients suspected to have developed contact dermatitis due to antibacterial agents were patch tested with commercial antibiotic preparations. Positive patch tests were obtained with oxytetracycline in 13 cases. The active ingredients were (highly likely) not tested separately (12). In a similar study from India, performed in the period 1980-1983, 390 patients suspected of contact allergy to topical antibacterial ointments were patch tested with chlortetracycline ointment 3% and 86 (22%) had positive reactions. Contact allergy to base ingredients was not excluded and oxytetracycline itself was not tested (1).

Case reports
A 43-year-old woman had used many medications on varicose ulcers including a spray containing oxytetracycline. Patch tests were positive to oxytetracycline 5% pet. and to the spray 'as is'. Another female patient, aged 59, had used oxytetracycline ointment on hypostatic ulcers; she had a positive patch test to oxytetracycline 1% pet. (11).

A 34-year-old woman was treated with many topical pharmaceuticals for earlobe dermatitis from gold contact allergy, including oxytetracycline ointment. It was found that the patient had become sensitized to various corticosteroids, neomycin, oxytetracycline ointment and oxytetracycline (tested 3% and 5% pet.) (13).

Occupational allergic contact dermatitis
In a group of 107 workers in the pharmaceutical industry with dermatitis, investigated in Warsaw, Poland, before 1989, 5 reacted to oxytetracycline, tested 10% pet. (6). Also in Warsaw, Poland, in the period 1979-1983, 27 pharmaceutical workers, 24 nurses and 30 veterinary surgeons were diagnosed with occupational allergic contact dermatitis from antibiotics. The numbers that had positive patch tests to oxytetracycline (10% pet.) were 3, 0, and 1, respectively, total 4 (10).

Cross-reactions, pseudo-cross-reactions and co-reactions
Three patients sensitized to oxytetracycline had cross-reactions to 'other tetracyclines' (8), but their nature is unknown (9). Two patients who had allergic contact dermatitis from oxytetracycline did not react to chlortetracycline (11).

LITERATURE
1   Bajaj AK, Gupta SC. Contact hypersensitivity to topical antibacterial agents. Int J Dermatol 1986; 25: 103-105
2   Gilissen L, De Decker L, Hulshagen T, Goossens A. Allergic contact dermatitis caused by topical ophthalmic medications: Keep an eye on it! Contact Dermatitis 2019;80:291-297
3   Menezes de Padua CA, Uter W, Schnuch A. Contact allergy to topical drugs: prevalence in a clinical setting and estimation of frequency at the population level. Pharmacoepidemiol Drug Saf 2007;16:377-384
4   Menezes de Padua CA, Schnuch A, Nink K, Pfahlberg A, Uter W. Allergic contact dermatitis to topical drugs – Epidemiological risk assessment. Pharmacoepidemiol Drug Saf 2008;17:813-821
5   Gilissen L, Goossens A. Frequency and trends of contact allergy to and iatrogenic contact dermatitis caused by topical drugs over a 25-year period. Contact Dermatitis 2016;75:290-302
6   Rudzki E, Rebandel P, Grzywa Z. Contact allergy in the pharmaceutical industry. Contact Dermatitis 1989;21:121-122
7   Rudzki E, Rebandel P. Sensitivity to oxytetracycline. Contact Dermatitis 1997;37:136
8   Möller H. Eczematous contact allergy to oxytetracycline and polymyxin B. Contact Dermatitis 1976;2:289-290
9   Bojs G, Möller H. Eczematous contact allergy to oxytetracycline with cross-sensitivity to other tetracyclines. Berufsdermatosen 1974;22:202-208
10   Rudzki E, Rebendel P. Contact sensitivity to antibiotics. Contact Dermatitis 1984;11:41-42
11   Cronin E. Contact Dermatitis. Edinburgh: Churchill Livingstone, 1980:225
12   Pasricha JS, Guru B. Contact hypersensitivity to local antibacterial agents. Indian J Dermatol Venereol Leprol 1981;47:27-30
13   Hisa T, Katoh J, Yoshioka K, Taniguchi S, Mochida K, Nishimura T, et al. Contact allergies to topical corticosteroids. Contact Dermatitis 1993;28:174-179

Chapter 3.254 PAROMOMYCIN

IDENTIFICATION

Description/definition : Paromomycin is the 4,5-disubstituted 2-deoxystreptamine that conforms to the structural formula shown below

Pharmacological classes : Antiprotozoal agents; anti-bacterial agents

IUPAC name : (2S,3S,4R,5R,6R)-5-Amino-2-(aminomethyl)-6-[(2R,3S,4R,5S)-5-[(1R,2R,3S,5R,6S)-3,5-diamino-2-[(2S,3R,4R,5S,6R)-3-amino-4,5-dihydroxy-6-(hydroxymethyl)oxan-2-yl]oxy-6-hydroxycyclohexyl]oxy-4-hydroxy-2-(hydroxymethyl)oxolan-3-yl]oxyoxane-3,4-diol

Other names : Aminosidin; neomycin E

CAS registry number : 7542-37-2; 1263-89-4

EC number : 231-423-0

Merck Index monograph : 8416

Patch testing : 10-20% water and petrolatum; perform controls tests

Molecular formula : $C_{23}H_{45}N_5O_{14}$

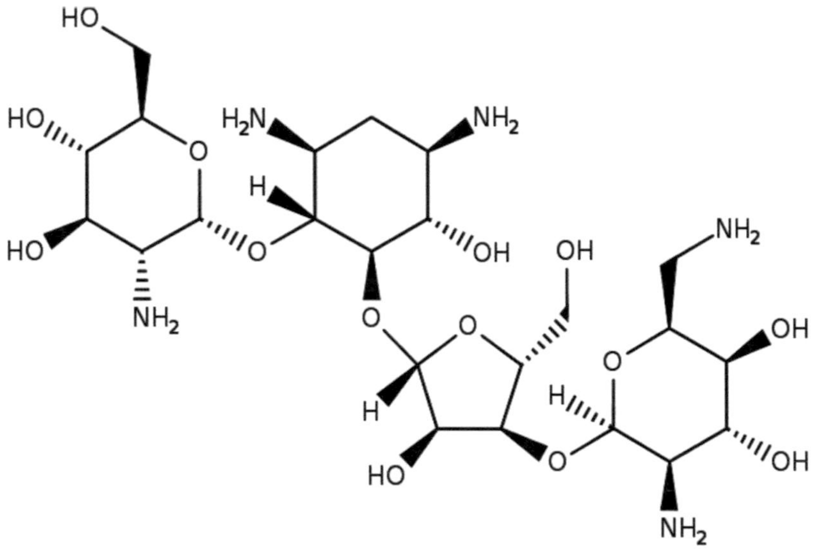

GENERAL

Paromomycin is an aminoglycoside antibiotic that was isolated in the 1950s from *Streptomyces krestomuceticus* and later from other *Streptomyces* species including *Streptomyces rimosus* var. *paromomycinus*. Its antibacterial spectrum of activity includes gram-negative and gram-positive bacteria, *Mycobacterium tuberculosis*, and non-tubercular mycobacteria; it also has strong amebicidal activity. Paromomycin is used worldwide for the topical and systemic treatment of visceral, mucocutaneous and cutaneous leishmaniasis. Topically, it is usually employed at a concentration of 15% along with 12% methylbenzethonium chloride in white petrolatum, and is applied once or twice daily for 3 to 4 weeks. It is also indicated for the treatment of acute and chronic intestinal amebiasis (2).

In pharmaceutical products, paromomycin is employed as paromomycin sulfate (CAS number 1263-89-4, EC number 215-031-7, molecular formula $C_{23}H_{47}N_5O_{18}S$) (1).

CONTACT ALLERGY

Case series

Five patients with cutaneous leishmaniasis (1 woman, 4 men, age range 40-72 years) had been treated with 15% paromomycin sulfate ointment by two applications daily for 3 weeks under occlusive dressings. They developed dermatitis characterized by erythema (one patient), erythema and vesicles (two patients), erythema and erosions (one patient), itching (four patients), and burning sensation (one patient). When patch tested, all reacted to paromomycin sulfate (++ or +++) 10% and 50% both in water and in petrolatum, and had negative results with the other compounds of the ointment. Five controls were negative. Four patients also had positive reactions to 20% neomycin sulfate and the same four also reacted to kanamycin sulfate 10% pet. (2). Curiously, the authors warn for the irritative effects of paromomycin, yet use 50% for patch testing!

In Leuven, Belgium, between 1990 and 2017, 16,065 patients were investigated for contact allergy and 118 (0.7%) showed positive patch test reactions to topical ophthalmic medications and/or to their ingredients. Eighty-four individuals (71%) reacted to an active principle. There were seven positive patch test reactions to paromomycin, but these were all the result of cross-sensitivity to other aminoglycosides (1).

Cross-reactions, pseudo-cross-reactions and co-reactions
In patients sensitized to neomycin, about 90% cross-reacts to paromomycin (Chapter 3.235 Neomycin). Seven patients had positive patch reactions to paromomycin as a result of cross-sensitivity to other (unspecified) aminoglycosides (1). Four of 5 patients who had developed allergic contact dermatitis from paromomycin also had positive reactions to 20% neomycin sulfate and the same four also reacted to kanamycin sulfate 10% pet. It was not mentioned whether these patients had used either aminoglycoside before (2).

LITERATURE
1   Gilissen L, De Decker L, Hulshagen T, Goossens A. Allergic contact dermatitis caused by topical ophthalmic medications: Keep an eye on it! Contact Dermatitis 2019;80:291-297
2   Veraldi S, Benzecry V, Faraci AG, Nazzaro G. Allergic contact dermatitis caused by paromomycin. Contact Dermatitis 2019;81:393-394

Chapter 3.255 PECILOCIN

IDENTIFICATION

Description/definition	: Pecilocin is the *N*-acylpyrrolidine that conforms to the structural formula shown below
Pharmacological classes	: Antifungal agents
IUPAC name	: 1-[(2*E*,4*E*,6*E*,8*R*)-8-Hydroxy-6-methyldodeca-2,4,6-trienoyl]pyrrolidin-2-one
Other names	: 1-(8-Hydroxy-6-methyl-1-oxo-2,4,6-dodecatrienyl)-2-pyrrolidinone; Variotin ®
CAS registry number	: 19504-77-9
EC number	: 243-116-9
Merck Index monograph	: 8438
Patch testing	: 1%, 5% and 10% pet.; perform controls
Molecular formula	: $C_{17}H_{25}NO_3$

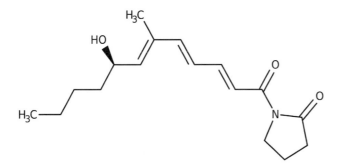

GENERAL

Pecilocin is a fungicidal antibiotic isolated from *Paecilomyces varioti* Bainier var. *antibioticus* with topical activity against common dermatophytes. It was introduced in 1959 in Japan and is probably not used anymore (1).

CONTACT ALLERGY

Case series

In 1975, in one practice in Denmark, there were 44 patients with a history of treatment with pecilocin (Variotin). These patients were patch tested with pecilocin, and seven were found to be allergic to this drug. In three of them the skin disease had been caused or exacerbated by pecilocin. In the other four patients the allergy was regarded as an 'immunological scar' (no details known) (3). Previously, the same author had reported another two patients with allergy to pecilocin (no data available, article in Danish) (4).

Case reports

A 25-year-old farmer treated a ringworm infection in the right axilla with an ointment containing pecilocin (10 mg = 3000 E per gram ointment), which resulted in acute dermatitis. Patch tests were strongly positive to the ointment and to pecilocin 10% and 50% pet., but 2% did not evoke a positive response. Eleven controls were tested with various ingredients of the ointment base, but it was not specifically stated that pecilocin 10% and 50% were negative in controls (2). A man who treated a fungal infection of the feet with a cream containing pecilocin developed dermatitis of the feet which spread to the hands. Patch tests were positive to the cream, pecilocin 1% and 0.5% pet. but negative to 0.1% pet. (7).

The first case of pecilocin sensitization may have been reported in 1969 from Iceland (5). Another case was published one year later (6). This patient had a positive patch test to pecilocin 3000 units/gram, which was negative in 12 controls (6).

LITERATURE

1 The data in the section 'General' may have been obtained from literature discussed in this chapter, but mostly also or exclusively from one or more of the following online sources: ChemIDPlus Advanced, PubChem, DrugBank, RxList, Drug Central, Drugs.com, and Wikipedia
2 Groen J, Bleumink E, Nater JP. Variotin® sensitivity. Contact Dermatitis Newsletter 1974;15:456
3 Nørgaard O. Pecilocin allergy. Hautarzt 1977;28:35-36 (Article in German)
4 Nørgaard O. 2 Cases of allergy to pecilocine (Variotin). Ugeskr Laeger 1970;132:2448-2449 (Article in Danish)
5 Gudjónsson H, Molin L. Variotin sensitization. Mykosen 1969;12:445-446
6 Sundararajan V. Variotin sensitivity. Contact Dermatitis Newsletter 1970;8:188
7 Cronin E. Contact Dermatitis. Edinburgh: Churchill Livingstone, 1980:230

Chapter 3.256 PENICILLAMINE

IDENTIFICATION

Description/definition	: Penicillamine is a β-dimethyl analog of the amino acid cysteine and a degradation product of penicillin antibiotics; it conforms to the structural formula shown below
Pharmacological classes	: Antirheumatic agents; chelating agents; antidotes
IUPAC name	: (2S)-2-Amino-3-methyl-3-sulfanylbutanoic acid
Other names	: D-Valine, 3-mercapto-; 3-mercapto-D-valine
CAS registry number	: 52-67-5
EC number	: 200-148-8
Merck Index monograph	: 8467
Patch testing	: 1% pet. (SmartPracticeCanada); 2.5% water
Molecular formula	: $C_5H_{11}NO_2S$

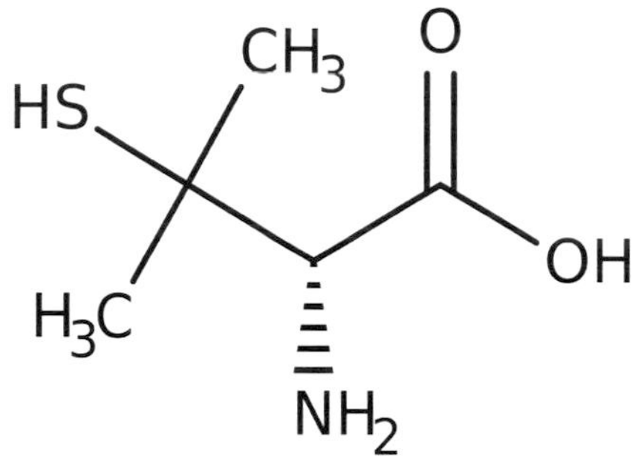

GENERAL

Penicillamine is the most characteristic degradation product of the penicillin antibiotics, but it has no antibacterial activity. Its pharmaceutical form is D-penicillamine, as L-penicillamine is toxic (it inhibits the action of pyridoxine). Penicillamine is a copper-chelating agent used in the treatment of Wilson's disease. It is also used to reduce cystine excretion in cystinuria. Another indication of penicillamine is in the treatment of rheumatoid arthritis. It works by immunosuppression: reducing numbers of T-lymphocytes, inhibiting macrophage function, decreasing IL-1, and decreasing rheumatoid factor. This agent also prevents collagen from cross-linking (1). In ophthalmology, penicillamine used to be administered as eye drops in a 3% solution in water to prevent corneal fibrosis after chemical trauma (2,3). Currently, such eye drops are probably are not in use anymore.

CONTACT ALLERGY

Case reports

A 65-year-old painter accidentally splashed lime in his face. Immediately, he rinsed the eyes with water. The ophthalmologist prescribed 4 different medications including penicillamine and atropine eye drops. Six weeks later, the patient was referred to the dermatology department because he had developed erythema, edema and vesicles of the eyelids and surrounding skin. Patch tests showed strongly positive reactions to penicillamine 0.15 M water (D3 +++, D4 ++++) and a weak reaction (D2 +, D4 +) to atropine 1% in saline. All medications had been stopped, hydrocortisone cream was prescribed and the skin healed within a few days (2).

A 43-year-old man accidentally splashed caustic soda in his right eye. The eye was immediately rinsed with water, and subsequently treated with atropine eye drops, corticosteroid eye drops and antibiotic ointment. Three months later, the patient received penicillamine eye drops, which caused severe edema and redness of the upper and lower eyelids. Because of corneal vascularization and ulceration, a cornea transplantation and lens extraction, with implantation of an artificial lens, was performed. Treatment with indomethacin, atropine and chloramphenlcol eye drops seemed to cause an allergic reaction. Patch testing showed positive reactions to atropine sulfate 1% water and d-penicillamine 1%, 2.5% and 3% water (3). Twenty-one controls were negative.

The same authors also describe a 44-year-old man who accidentally splashed a few drops of a very alkaline drainpipe unblocking agent in both eyes. After rinsing with water, his eyes were treated with dexamethasone, atropine, scopolamine, vitamin C and d-penicillamine eye drops. Two weeks later, both eyelids and conjunctivae became red and swollen. He was treated with prednisolone eye drops and betamethasone cream, while penicillamine was stopped. A week later, the redness and swelling had disappeared. Patch testing revealed very strongly positive reactions to penicillamine 1%, 2.5% and 3% water (21 controls negative) (3).

LITERATURE
1   The data in the section 'General' may have been obtained from literature discussed in this chapter, but mostly also or exclusively from one or more of the following online sources: ChemIDPlus Advanced, PubChem, DrugBank, RxList, Drug Central, Drugs.com, and Wikipedia
2   De Moor A, Van Hecke E, Kestelyn P. Contact allergy to penicillamine in eye drops. Contact Dermatitis 1993;29:155-156
3   Coenraads PJ, Woest TE, Blanksma LJ, Houtman WA. Contact allergy to d-penicillamine. Contact Dermatitis 1990;23:371-372

Chapter 3.257 PENICILLINS, UNSPECIFIED

IDENTIFICATION
Description/definition   : The penicillins are a group of antibiotics that contain 6-aminopenicillanic acid with a side
                            chain (R) attached to the 6-amino group; the penicillin nucleus is the chief structural
                            requirement for biological activity; the side-chain structure determines many of the
                            antibacterial and pharmacological characteristics
Pharmacological classes : Anti-bacterial agents
CAS registry number   : 1406-05-9
EC number   : 215-794-6

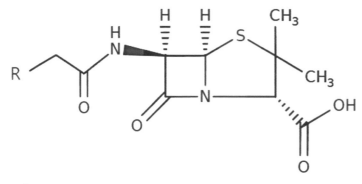

General
Penicillin is the generic name of the whole group of natural and semi-synthetic penicillins. Penicillin was originally obtained from the fungus *Penicillium chrysogenum* (old name: *Penicillium notatum*) and was discovered in 1928 by Sir Alexander Fleming, a Scottish researcher. All penicillins contain 6-aminopenicillanic acid with a side chain attached to the 6-amino group, which determines many of the antibacterial and pharmacological characteristics. As this is a historical topical allergen, the subject, of which there is abundant (often early) literature, will be discussed only very briefly.

CONTACT ALLERGY
When applied topically, penicillin is a potent sensitizer (5). The first case of (occupational, partly airborne) allergic contact dermatitis reported was that of an Army medical officer in Chicago who, shortly after beginning to prepare and give injections of penicillin, developed an acute dermatitis on his face and then his hands (6). The facility of topical penicillin to sensitize was realized during the 1939-1945 war and thereafter when it was used in the treatment of war wounds. Many cases of allergic contact dermatitis were reported (e.g. 7-11,13,14). The early literature was reviewed in 1948 (15).

Because of its strong sensitizing potential, the prescribing of topical penicillin was abandoned in many countries. However, in Malaysia, in 1976, penicillin preparations were still available over the counter and at that time penicillin was the most frequent cause of allergic contact dermatitis from antibiotics in that country (16). Also, in Bari, Italy, the prevalence of sensitization to penicillin was still 4.6% in the period 1968-1977, but dropped to 0.6% in the years 1978-1983 (17). In India, before 1981, 101 patients suspected to have developed contact dermatitis due to antibacterial agents were patch tested with commercial antibiotic preparations. Positive patch tests were obtained with penicillin in 12 cases. The active ingredients were (probably) not tested separately (18).

Accidental topical contact with penicillin has also caused a considerable number of occupational sensitizations in pharmaceutical employees and health care workers (e.g. 1,2,3,4,12).

Immediate contact reactions
Immediate contact reactions (contact urticaria) to penicillin are presented in Chapter 5.

LITERATURE
1   Rudzki E, Rebandel P, Grzywa Z. Contact allergy in the pharmaceutical industry. Contact Dermatitis 1989;21:121-
     122
2   Gielen K, Goossens A. Occupational allergic contact dermatitis from drugs in healthcare workers. Contact
     Dermatitis 2001;45:273-279

3   Rudzki E, Rebendel P. Contact sensitivity to antibiotics. Contact Dermatitis 1984;11:41-42

4   Goodman H. Dermatitis due to preparation and administration of penicillin solution. Arch Derm Syphilol 1946;54:206-208

5   Cronin E. Contact Dermatitis. Edinburgh: Churchill Livingstone, 1980:216

6   Pyle HD, Rattner H. Contact dermatitis from penicillin. JAMA 1944;125:903

7   Goldman L, Friend F, Mason LM. Dermatitis from penicillin. JAMA 1946;131:883-890

8   Friedlaender S, Watrous RM, Feinberg SM. Contact dermatitis from penicillin; the source of the antigen. Arch Derm Syphilol 1946;54:517-523

9   Loveman AB, Marks PL, Stritzler C. Contact dermatitis from penicillin with ocular manifestations. Med Bull U S Army Force Europe 1945;30:49-52

10 Schmidt OE. Dermatitis venenata due to a penicillin-containing ointment. Arch Derm Syphilol 1947;55:582

11 Markson LS. Dermatitis venenata following use of penicillin ointment. Arch Derm Syphilol 1945;52:384

12 Marsh WC, New WN. Dermatitis due to the preparation and administration of penicillin solution. U S Nav Med Bull 1948;48:391-394

13 Michie W, Bailie HW. Penicillin reaction. Br Med J 1945;1(4398):554

14 Vickers HR. Contact dermatitis caused by penicillin. Lancet 1946;1(6392):307

15 Brown EA. Reactions to penicillin; a review of the literature, 1943-1948. Ann Allergy 1948;6:723-746

16 Nagreh DS. Contact dermatitis from proprietary preparations in Malaysia. Int J Dermatol 1976;15:34-35

17 Angelini G, Vena GA, Meneghini CL. Allergic contact dermatitis to some medicaments. Contact Dermatitis 1985;12:263-269

18 Pasricha JS, Guru B. Contact hypersensitivity to local antibacterial agents. Indian J Dermatol Venereol Leprol 1981;47:27-30

Chapter 3.258 PHENIRAMINE

IDENTIFICATION

Description/definition   : Pheniramine is a tertiary amino compound and a member of pyridines that conforms to
                                        the structural formula shown below
Pharmacological classes : Histamine H1 antagonists; antipruritics; anti-allergic agents
IUPAC name   : *N,N*-Dimethyl-3-phenyl-3-pyridin-2-ylpropan-1-amine
Other names   : Prophenpyridamine
CAS registry number   : 86-21-5
EC number   : 201-656-2
Merck Index monograph : 8619
Patch testing   : 1% water
Molecular formula   : $C_{16}H_{20}N_2$

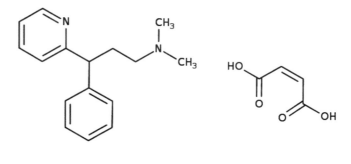

Pheniramine   Pheniramine maleate

GENERAL

Pheniramine is a first generation antihistamine in the alkylamine class. It is used in some over-the-counter allergy as well as cold & flu products in combination with other drugs. Pheniramine's use as an anti-allergy medication for hay fever, rhinitis, allergic dermatoses, and pruritus has largely been supplanted by second generation antihistamines. It is also used in eye drops for the treatment of allergic conjunctivitis. In pharmaceutical products, pheniramine is employed as pheniramine maleate (CAS number 132-20-7, EC number 205-051-4, molecular formula $C_{20}H_{24}N_2O_4$) (1).

CONTACT ALLERGY

Case series

In Ferrara, Italy, over a 65-month period before 2005, 50 patients affected by periorbital dermatitis while using topical ocular products were patch tested, including with their own ophthalmic medications. There was one reaction to eye drops containing pheniramine and tetrahydrozoline. The active ingredients were not tested separately, but contact allergy to the excipients and preservatives was excluded by patch testing (2).

Case report

A 30-year-old woman, with a history of allergic oculorhinitis from Gramineae pollen, presented with bilateral conjunctivitis and eyelid dermatitis for the past 4 months. When patch tested, she reacted to commercial eye drops containing pheniramine maleate and tetrazoline chloride. Further patch testing with the individual constituents of the eye drops revealed a positive reaction to pheniramine maleate 1% water (D2+/D3++), and the patient completely recovered after abandoning the eye drops. One month later, the patient was patch tested to other antihistamines. Positive reactions were found with dexchlorpheniramine, chlorpheniramine (both 1% water) and diphenhydramine 2% pet. A careful history revealed that, in the past, the patient had occasionally used a cream containing chlorpheniramine maleate without problems (3).

Cross-reactions, pseudo-cross-reactions and co-reactions

Possible cross-reactivity between pheniramine, chlorpheniramine and dexchlorpheniramine (3).

Cutaneous adverse drug reactions from systemic administration caused by type IV (delayed-type) hypersensitivity
Cutaneous adverse drug reactions from systemic administration of pheniramine caused by type IV (delayed-type) hypersensitivity, including an urticarial and maculopapular eruption (4), are planned to be discussed in Volume IV of the *Monographs in Contact Allergy* series on Systemic drugs.

LITERATURE

1   The data in the section 'General' may have been obtained from literature discussed in this chapter, but mostly also or exclusively from one or more of the following online sources: ChemIDPlus Advanced, PubChem, DrugBank, RxList, Drug Central, Drugs.com, and Wikipedia
2   Corazza M, Massieri LT, Virgili A. Doubtful value of patch testing for suspected contact allergy to ophthalmic products. Acta Derm Venereol 2005;85:70-71
3   Parente G, Pazzaglia M, Vincenzi C, Tosti A. Contact dermatitis from pheniramine maleate in eyedrops. Contact Dermatitis 1999;40:338
4   Epstein E. Dermatitis due to antihistaminic agents. J Invest Dermatol 1949;12:151-152

Chapter 3.259 PHENYLBUTAZONE

IDENTIFICATION

Description/definition   : Phenylbutazone is the synthetic pyrazolone derivative that conforms to the structural
                              formula shown below
Pharmacological classes : Anti-inflammatory agents, non-steroidal
IUPAC name   : 4-Butyl-1,2-diphenylpyrazolidine-3,5-dione
Other name(s)   : 1,2-Diphenyl-3,5-dioxo-4-butylpyrazolidine
CAS registry number(s)   : 50-33-9
EC number(s)   : 200-029-0
Merck Index monograph : 8660
Patch testing   : 10.0% pet. (Chemotechnique, SmartPracticeCanada, SmartPracticeEurope)
Molecular formula   : $C_{19}H_{20}N_2O_2$

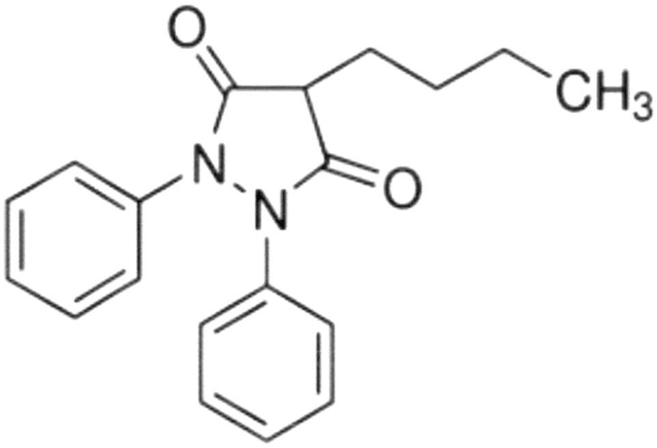

GENERAL

Phenylbutazone is a synthetic pyrazolone derivative and nonsteroidal anti-inflammatory drug (NSAID) with anti-inflammatory, antipyretic, and analgesic activities. Phenylbutazone was formerly used for the treatment of backache, ankylosing spondylitis, rheumatoid arthritis, and reactive arthritis. Because of serious systemic side effects and cutaneous adverse drug reactions such as Stevens-Johnson syndrome and toxic epidermal necrolysis, it is probably hardly used anymore, except for therapy-resistant ankylosing spondylitis. It does, however, still have applications in veterinary medicine (1). Topical use was formerly recommended for superficial phlebitis and some inflammatory diseases in the muscles and connective tissues (16).

See also Chapter 3.292 Pyrazinobutazone, Chapter 3.264 Piperazine and Chapter 3.251 Oxyphenbutazone.

CONTACT ALLERGY

General

In Italy, topical preparations containing phenylbutazone (and oxyphenbutazone) were apparently withdrawn from sale in Italy in the early 1990s due to its sensitizing properties (4). In the mid-1980s, phenylbutazone was one of the most frequent causes of contact allergy to topically applied medicaments in the urban population of Malmö, Sweden (12). In 1979, phenylbutazone was the second most frequent contact allergen in female hospitalized women in Budapest, Hungary, and the 7th most frequent in male hospitalized patients (18).

Case series

In Leuven, Belgium, in the period 1990-2014, iatrogenic contact dermatitis was diagnosed in 2600 individuals (17% of the total patch test population). 96% of all positive patch test reactions to topical drugs and antiseptics were considered to be relevant. Phenylbutazone (1% pet.) was tested in 7 patients and there were 2 positive reactions to it (2). In the period 1996-2001, in 2 hospitals in Spain, one patient was diagnosed with contact allergy and one with photocontact allergy to phenylbutazone. The accumulated incidence per million inhabitants (catchment population of the hospitals) of both side effects together was 2.4 (17).

Four cases of allergic contact dermatitis to phenylbutazone ointment were reported from Denmark in 1978. The patients were 3 women and one man aged 19-74 years. All patients had a pre-existing skin disease: two had leg ulcers, one had palmoplantar pustulosis and one irritant dermatitis. They all developed generalized dermatitis following the use of the ointment. In three individuals, patch tests were positive to phenylbutazone 0.5% and 5% pet. and negative to the ointment base. The 4th was tested with the ointment only and had a positive patch test to it. 118 controls were negative to phenylbutazone 5% pet. (16).

Case reports

A 65-year-old woman developed an extensive erythema multiforme-like allergic contact dermatitis after the use of phenylbutazone cream. Ten years earlier, she had used another brand also containing this NSAID and had suffered a cutaneous reaction to it. Patch tests were positive to phenylbutazone cream, the cream used 10 years ago and phenylbutazone 2% pet., but negative to oxyphenbutazone 5% pet. (3).

A 19-year-old girl applied phenylbutazone cream to a painful left ankle and, one day later, developed acute contact dermatitis at the site of application and on both legs. Patch tests were positive to the cream, but the other ingredients (except thimerosal and parabens, which were negative) were not tested, including phenylbutazone (13).

A 55-year-old man had previously suffered dermatitis of the face after using a cream containing phenylbutazone, while the area (arms or legs) where the ointment had been applied remained unaffected. After a nasal operation, he inserted a phenylbutazone-piperazine suppository. Ten hours later, he showed erythema and swelling on the face. He thought the handling of the suppository was the cause. Besides anal irritation, erythema with mild itching spread over his body with a later scaling over a month. Patch tests were positive to phenylbutazone 5% pet., oxyphenbutazone 5% pet., piperazine 1% water, 3-methylaminopyridine 1% olive oil and 3-(aminomethyl)pyridine 1% water (14). This was a case of systemic contact dermatitis.

A 35-year-old woman was treated with phenylbutazone ointment for rheumatic pains in the legs. After 10 days she presented with allergic contact dermatitis of the legs and trunk. Patch tests were positive to phenylbutazone 1% pet. and negative to isopyrine (synonym: ramifenazone, the other active substance in the ointment), the ointment base and oxyphenbutazone 1% pet. (10).

One patient suffered purpuric contact dermatitis associated with urticaria and an erythema multiforme-like eruption from a cream containing mephenesin and phenylbutazone. Patch tests were positive to both mephenesin (a muscle relaxant) and phenylbutazone (11). An author from Sweden in 1975 mentioned to have observed two cases of allergic contact dermatitis to phenylbutazone cream. Clinical details were not provided (9).

PHOTOSENSITIVITY

In the period 1996-2001, in 2 hospitals in Spain, one patient was diagnosed with photocontact allergy and one with contact allergy to phenylbutazone. The accumulated incidence per million inhabitants (catchment population of the hospitals) of both side effects together was 2.4 (17).

Cross-reactions, pseudo-cross-reactions and co-reactions

Patients sensitized to phenylbutazone may (14,15) or may not (9,10) cross-react to oxyphenbutazone. Conversely, individuals sensitized to oxyphenbutazone may (4,5,7,8,9) or may not (6) cross-react to phenylbutazone. A patient sensitized to phenylbutazone co-reacted to methylaminopyridine (1% olive oil) and 3-(aminomethyl)pyridine 1% water (14). Why these chemicals were tested and whether the sensitizations were considered to be cross-reactions was not discussed. Ik heb dit al gecontroleerd met oxyphenbutazone

Cutaneous adverse drug reactions from systemic administration caused by type IV (delayed-type) hypersensitivity

Cutaneous adverse drug reactions from systemic administration of phenylbutazone caused by type IV (delayed-type) hypersensitivity, including a maculopapular rash evolving into erythroderma (15), are planned to be discussed in Volume IV of the *Monographs in Contact Allergy* series on Systemic drugs.

LITERATURE

1   The data in the section 'General' may have been obtained from literature discussed in this chapter, but mostly also or exclusively from one or more of the following online sources: ChemIDPlus Advanced, PubChem, DrugBank, RxList, Drug Central, Drugs.com, and Wikipedia

2   Gilissen L, Goossens A. Frequency and trends of contact allergy to and iatrogenic contact dermatitis caused by topical drugs over a 25-year period. Contact Dermatitis 2016;75:290-302

3   Kerre S, Busschots A, Dooms-Goossens A. Erythema-multiforme-like contact dermatitis due to phenylbutazone. Contact Dermatitis 1995;33:213-214

4   Cameli N, Vincenzi C, Morelli R, Bardazzi F, Tardio M. Contact allergy to oxyphenbutazone. Contact Dermatitis 1991;24:75-76

5   Figueiredo A, Gonçalo S, Freitas JD. Contact sensitivity to pyrazolone compounds. Contact Dermatitis 1985;13:271

6   Pigatto PD, Riboldi A, Morelli M, Altomare GF, Polenghi MM. Allergic contact dermatitis from oxyphenbutazone. Contact Dermatitis 1985;12:236-237

7   Valsecchi R, Tornaghi A, Falgheri G, Rossi A, Cainelli T. Drug reaction from oxyphenbutazone. Contact Dermatitis. 1983;9:419

8   Valsecchi R, Serra M, Foiadelli L, Cainelli T. Contact sensitivity to oxyphenbutazone. Contact Dermatitis 1981;7:157

9   Krook G. Contact sensitivity to oxyphenbutazone (Tanderil®) and cross-sensitivity to phenylbutazone (Butazolidin®). Contact Dermatitis 1975;1:262 (in PubMed the same publication is cited to be present on pages 385-386, but these appear not to exist)

10 Meneghini CL, Angelini G. Contact allergy to antirheumatic drugs. Contact Dermatitis 1979;5:197-198

11 Bachmeyer C, Blum L, Picard 0, Cabane I, Imbert JC. Dermite de contact au Traumalgyl ®. Ann Derm Venereol 1994;121:93 (Article in French, data cited in ref. 3)

12 Edman B, Möller H. Medicament contact allergy. Derm Beruf Umwelt 1986;34:139-143 (Article in German)

13 Nayar M, Ng SK. Contact sensitivity to phenylbutazone (Butazolidine) cream. Contact Dermatitis 1991;25:263-264

14 Fernández de Corres L, Bernaola G, Lobera T, Leanizbarrutia I, Muñoz D. Allergy from pyrazoline derivatives. Contact Dermatitis 1986;14:249-250

15 Vooys RC, van Ketel WG. Allergic drug eruption from pyrazolone compounds. Contact Dermatitis 1977;3:57-58

16 Thormann J, Kaaber K. Contact sensitivity to phenylbutazone ointment (Butazolidine). Contact Dermatitis 1978;4:235-236

17 Diaz RL, Gardeazabal J, Manrique P, Ratón JA, Urrutia I, Rodríguez-Sasiain JM, Aguirre C. Greater allergenicity of topical ketoprofen in contact dermatitis confirmed by use. Contact Dermatitis 2006;54:239-243

18 Korossy S, Nebenführer L, Vincze E. Frequency, relevance and latency of chemical allergy in hospitalized patients in Budapest. Derm Beruf Umwelt 1983;31:39-44 (Article in German)

Chapter 3.260 PHENYLEPHRINE

IDENTIFICATION

Description/definition : Phenylephrine is the sympathomimetic amine chemically related to adrenaline and ephedrine that conforms to the structural formula shown below

Pharmacological classes : Mydriatics; nasal decongestants; α_1-adrenergic receptor agonists; sympathomimetics; vasoconstrictor agents; cardiotonic agents

IUPAC name : 3-[(1R)-1-Hydroxy-2-(methylamino)ethyl]phenol

Other names : (R)-3-Hydroxy-α-((methylamino)methyl)benzenemethanol; metaoxedrine; m-methylaminoethanolphenol

CAS registry number : 59-42-7

EC number : 200-424-8

Merck Index monograph : 8668

Patch testing : Hydrochloride, 10% water (SmartPracticeCanada, SmartPracticeEurope)

Molecular formula : $C_9H_{13}NO_2$

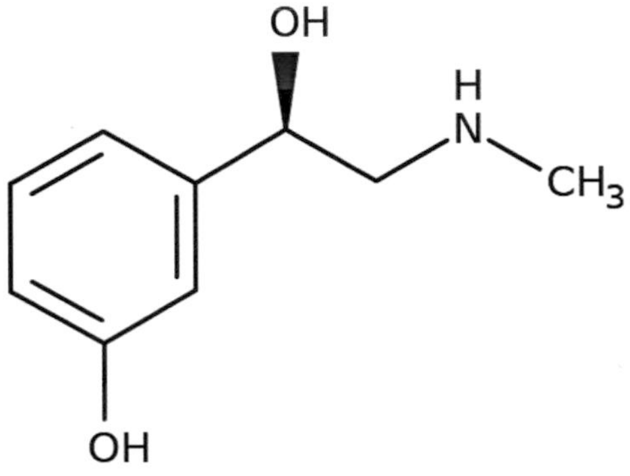

GENERAL

Phenylephrine (PE) is a direct-acting sympathomimetic amine chemically related to epinephrine and ephedrine with potent vasoconstrictor property. It is a post-synaptic α-adrenergic receptor agonist that causes vasoconstriction, increases systolic/diastolic pressures, reflex bradycardia, and stroke output. Phenylephrine is mainly used to treat nasal congestion due to the common cold or hay fever, sinusitis, or other upper respiratory problems. In eye drops, this agent is indicated to produce dilation of the pupil prior to intraocular surgery (mydriatic) and diagnostic examinations. It is also found in ear preparations and in rectal and vascular ointments (20). Oral phenylephrine, together with other drugs, may be used to treat certain diseases of the upper respiratory tract. In pharmaceutical products, phenylephrine (PE) is employed as phenylephrine hydrochloride (CAS number 61-76-7, EC number 200-517-3, molecular formula $C_9H_{14}ClNO_2$) (1).

CONTACT ALLERGY

General

Phenylephrine (PE) appears to be the most frequent cause of contact allergic reactions among the ophthalmological drugs (36,48). In recent years, the high frequency of usage of phenylephrine during repeated intravitreal injections of ranibizumab and follow-up consultations may have increased the numbers of sensitization (25). By far, most cases of allergic contact dermatitis are caused by eye drops containing PE, a few were the result of the presence of PE in a phlebological ointment (54), rectal ointment (56,69) and eardrops (61). Occupational allergic contact dermatitis in nurses instilling the drugs in patients' eyes have been observed occasionally (51,59) as have patients with systemic contact dermatitis from absorption of phenylephrine (29,69) and in one case from oral pseudoephedrine (57).

Allergic blepharoconjunctivitis caused by phenylephrine is usually an acute reaction. As symptoms arise often within a few hours and sometimes even within 30 minutes (43,65) after instillation of the eye drops, these type IV allergic reactions can be mistaken for immediate-type allergic reactions (30). In several cases, positive patch test reactions to PE have persisted for long time, ranging from 2-7 months (29, 39 [scratch-patch tests], 47,52).

Just as with beta-blockers, phenylephrine has been shown sometimes to cause a false-negative patch test reaction. Scratch-patch tests may then result in positive reactions (39); alternatives may be conjunctival challenge tests (23, 27, 43, 48) or intradermal tests read after 2 days (57).

Contact allergy in the general population

In Germany, for the period 1995-2004, the population-based relative incidence (RI) of contact sensitization to phenylephrine (cases/100,000 defined daily doses (DDDs) per year) was estimated to be 21.6. In the group of ophthalmic drugs, the RI ranged from 0.3 (pilocarpine) to 21.6 (phenylephrine) (12).

Patch testing in groups of selected patients

The results of studies in which groups of selected patients (patients with periorbital dermatitis, individuals suspected of eye medicament allergy) have been tested with phenylephrine are shown in table 3.260.1. Six of the nine investigations were performed by the IVDK (Information Network of Departments of Dermatology: Germany, Austria, Switzerland). Prevalences of positive patch tests to phenylephrine have ranged from 1.5% to 5.4%. The highest rates (5.4%, 4.1%) were observed in the 1990s (4,6). The relevance of the reactions was hardly ever specified. In studies with a control group of dermatitis patients, the frequency of sensitization in the investigated group was always (far) higher than in the control group (2,4,6).

Table 3.260.1 Patch testing in groups of patients: Selected patient groups

Years and Country	Test conc. & vehicle	Number of patients tested	positive (%)	Selection of patients (S); Relevance (R); Comments (C)	Ref.
2001-2010 IVDK	10% water	1144	34 (3.0%)	S: patients with periorbital dermatitis tested with an ophthalmic tray; R: not stated	14
2000-2010 IVDK		1822	41 (1.7%)	S: patients with periorbital dermatitis; R: not stated; C: the frequency in patients suspected of eye medicament allergy was 2.1% and in a control group of dermatitis patients 0.8%	2
2000-6 Netherlands		65	2 (3%)	S: patients patch tested for suspected contact allergy to ophthalmological drugs; R: 69% for all reactions to drugs together	10
1996-2005 Hungary	10% coca	133	2 (1.5%)	S: patients with periorbital dermatitis; R: the reactions to eye medicaments were relevant	5
1999-2004 Germany		88	3 (3%)	S: patients with periorbital eczema; R: not stated	3
1995-2004 IVDK	10% water	3879	(1.9%)	S: patients patch tested for suspected contact allergy to ophthalmological drugs; R: not stated; C: the estimated number of sensitizations per 100,000 prescriptions was 21.6	9
1995-2004 IVDK	10% water	4286	(1.9%)	S: not stated; R: not stated	11
1995-1999 IVDK	1% water	580	(4.1%)	S: patients with *allergic* periorbital contact dermatitis; the frequency was significantly higher than in a control group	6
1990-1994 IVDK		277	15 (5.4%)	S: patients with periorbital eczema; R: not stated; C: frequency of sensitization in dermatitis controls: 2.7%	4

IVDK: Information Network of Departments of Dermatology (Germany, Austria, Switzerland)

Case series

In Leuven, Belgium, between 1990 and 2017, phenylephrine was tested in 55 patients and was the allergen in eye medications in seven. There was also one reaction to phenylephrine in other types of medications (7, overlap with ref. 13). In Ghent, Belgium, from March 2012 to January 2014, 16 patients who developed a stinging sensation and skin reaction around the eye within 24 hours after intravitreal injection of ranibizumab, were tested with all topical drugs used (except ranibizumab), and nine had positive patch tests to phenylephrine HCl 10% water (25).

In Denmark, in the period 2000-2014, in a hospital in Odense and in 37 private practices, 764 patients suspected of having contact allergy to eye drops were tested with phenylephrine HCl 10% water. In the private practices, of 419 individuals patch tested, 130 (31%) had a positive reaction. In Odense university hospital, 345 subjects were tested and 104 (30%) had a positive reaction to PE. The majority were ++ or +++ reactions. It was the impression of the authors that the vast majority of patients had undergone regular ophthalmological examination prior to the development of acute periorbital dermatitis (26).

In Leuven, Belgium, in the period 1990-2014, iatrogenic contact dermatitis was diagnosed in 2600 individuals. Phenylephrine (1% water) was tested in 53 patients and there were 8 positive reactions to it; virtually all were relevant (13, overlap with ref. 7). In 2004, ten patients seen in a university hospital in Madrid, Spain, who had previously experienced local allergic contact dermatitis (blepharoconjunctivitis) after administration of eye drops containing phenylephrine, were patch tested with PE 10% pet.; nine of these subjects had a positive patch test. The one with a negative patch test had a positive ocular challenge test. Cross-reactions to other sympathomimetics, which were also patch tested, are shown below (23).

In Ferrara, Italy, over a 65-month period before 2005, 50 patients affected by periorbital dermatitis while using topical ocular products were patch tested, including with their own ophthalmic medications (n=210). There were two reactions to phenylephrine eye drops. The active ingredients were not tested separately, but contact allergy to the excipients and preservatives was excluded by patch testing. The authors concluded that patch testing with commercial eye drops has doubtful value (28). In Odense, Denmark, in the period 1995-2003, 19 patients with periorbital dermatitis and a clinical suspicion of allergic contact dermatitis were patch tested with PE hydrochloride 10% water and there were 6 positive and 2 doubtful positive reactions. All positives were relevant (36). In Bari, Italy, in the period 1998-2001, 306 patients referred for allergological evaluation of the eyes were patch tested. Seven reacted to two commercial preparations containing phenylephrine hydrochloride, but the active drug itself was not tested. Six of them had previously suffered from conjunctivitis after the use of phenylephrine eye drops (15).

In Mendaro, Spain, in the period 1993 to 1997, 35 patients (14 men, 21 women, mean age 68 year) referred from the ophthalmology department because of reactions (acute conjunctivitis) after administration of mydriatic eye drops, were evaluated by patch testing. The symptoms were very similar in all patients and included pruritus, lacrimation, edematous erythema, and sometimes blepharitis; the symptoms would last for at least a week (7-30 days). The symptoms started a few minutes to 24 hours after instillation, with an average of 7.5 hours. Patients were patch tested with – amongst others – PE 1% pet. and 10% water. At D4, there were 14 positive reactions to 1% pet. and 21 to 10% water. Twelve patients, who had a negative patch tests to PE, were challenged in the eye (conjunctival challenge) with PE eye drops and 7 of them were positive. In this study, PE was by far (>80%) the most frequent allergen in ophthalmic mydriatic drugs. It was concluded that eye challenge is necessary in cases of negative patch tests to PE when allergy is suspected (48).

In Tokyo, Japan, in the period january 1987 to December 1995, 141 patients were patch tested with eye drops and 49 individuals (35%) reacted positively and were diagnosed with allergic contact dermatitis. In 36 cases ingredient patch testing was performed and there were three reactions to phenylephrine HCl (8). In two hospitals in Spain, before 1993, five patients suspected of contact allergy to PE were investigated. All had developed ocular complaints 12-24 hours after eye examination by an ophthalmologist. When patch tested, 4 reacted to PE 10% water. A fifth was negative, but had a positive intradermal test to PE (1 mg/ml) after 2 days, which was also the case in the other 4 patients. Twenty controls were negative to the intradermal tests (57). One of the patients allergic to PE had taken a tablet containing pseudoephedrine sulfate for rhinitis and within 8 hours experiences itchy and red eyes, generalized itching and a maculopapular eruption. This was considered to result from cross-reactivity to pseudo-ephedrine, although a patch and intradermal test to it were negative (57).

In Israel, before 1993, eight patients (five women, 3 men, mean age 74.2 years) who had previously developed acute allergic blepharoconjunctivitis due to phenylephrine hydrochloride eye drops, were investigated. Patch tests were not performed. In the lymphocyte proliferation assay, in all patients peripheral blood lymphocytes showed enhanced specific proliferation on stimulation with phenylephrine hydrochloride (SI= 2.4-19), in contrast to those from the healthy controls (SI= 1.0-1.8). It was concluded that the results demonstrate that sensitivity to phenyl-ephrine may be mediated by a cellular immune response (58). Why these authors did not ask their colleague dermatologists to perform patch testing, is an absolute mystery to the author of this book.

In Pamplona, Spain, in one year's time (1992), 13 patients were diagnosed with contact allergy to ophthalmic medications, nearly 3% of all patients investigated for suspected contact dermatitis. There were 7 reactions to phenylephrine 5% pet., all in individuals who had previously used phenylephrine eye drops (16). In Bologna, Italy, before 1989, in 136 patients suspected of contact conjunctivitis from eye drops (including contact lens solutions), 2 had positive patch tests to phenylephrine eye drops and phenylephrine itself (72). In an early study from France, 2.6% of hospitalized patients had contact allergy to phenylephrine in eye drops. Details are not available to the author (33).

Case reports

A 53-year-old Japanese woman developed periocular edema and erythema with pruritus within 24 hours after an operation for retinal detachment. From the following day, pruritic erythema, papules and vesicles spread over the face, neck and chest. Allergic contact dermatitis caused by items used perioperatively, such as antiseptics or eye drops was suspected. Patch tests were strongly positive to the PE eye drops and – later – to phenylephrine HCl 5% pet. The patch test reactions both persisted for two months and left residual pigmentation. The patient had previously used eye drops containing phenylephrine, but she had not shown any symptoms at that time. The authors concluded that this was a case of both local and systemic (from absorption of PE) allergic contact dermatitis (29).

A 65-year-old man developed a severe symmetrical blepharoconjunctivitis only 2 hours after instillations of four eye drops containing phenylephrine, tetracaine, fluorescein and tropicamide for a protocol study of fovea degene-ration. These symptoms lasted 5 months in spite of using antiallergic ophthalmic topicals. The patient had received the same eye drops 3 years previously and had experienced ocular lacrimation with a conjunctival injection for 2 weeks. Skin prick-tests and intradermal tests with the eye drops containing 10% phenylephrine pure and diluted at 1:10 in water were negative at 20 minutes but positive at 8 hour, lasting for 7 days. Later, patch testing was positive for phenylephrine HCl 10% water (29). The same authors also reported the case of a 72-year-old-woman who presented an episode of blepharoconjunctivitis 10 hours after an intraocular examination using mydriatic eye drops containing phenylephrine. Patch testing was positive to the eye drops and negative to the excipients. There was no long-lasting positive patch test reaction, but the blepharoconjunctivitis persisted for 8 months and was complicated by keratitis and trichiasis (30).

A 59-year old woman suffered from ocular itching, tearing, conjunctival redness, and swelling of the eyelids 30 minutes after using several unknown mydriatic agents. She was treated with topical corticosteroids, and the symptoms disappeared in about 20 days. Two months later, skin prick tests and patch tests were performed. The skin prick test with phenylephrine 10% eye drops was negative. Patch tests with phenylephrine 10% pet. and 1% pet., however, were strongly positive. Patch tests with preservatives and other amines (naphazoline 0.08%, epinephrine 0.1%, and ephedrine 0.1%) were also negative. A single-blind, placebo-controlled conjunctival challenge, with saline irrigation as placebo, was performed. Four hours after the challenge test with one drop of phenylephrine, the patient developed ocular pruritus and a few minutes later redness and tearing (43).

A man aged 84 developed conjunctival itching and hyperemia, eyelid edema and desquamation after instillation of eye drops containing phenylephrine, tropicamide, and double anesthetic (oxybuprocaine, tetracaine) for diagnostic purposes. Patch tests with the eye drops and phenylephrine 10% water and 20% pet. were negative. A conjunctival challenge test using the eye drops involved were positive for commercial phenylephrine eye drops and caused substantial itching, periocular erythema, conjunctival hyperemia, and eyelid edema 8 hours after the instillation accompanied by eyelid desquamation in the following days. These findings were considered proof for the existence of allergic contact dermatitis to phenylephrine (27).

A 76-year-old man presented with 2 episodes of blepharoconjunctivitis and periocular dermatitis 2 weeks apart. In each case, symptoms had arisen within a few hours of ophthalmological examination. Prick tests were negative after 20 minutes). Patch tests showed positive reactions to PE 10% water and – on tape-stripped skin – PE 5% eye drops. At D2, an eczematous reaction was observed at the prick test site with the phenylephrine-containing eye drops. One day later, dermatitis had spread over the proximal half of the patient's forearm. At the same time, a flare-up dermatitis, involving the forehead and periocular regions, developed. In the lymphocyte proliferation assay, the patient's cells showed enhanced specific proliferation on stimulation with phenylephrine hydrochloride (SI=7.2), in contrast to those from the healthy controls (SI 1.2 to 2.8) (49).

Table 3.260.2 Short summaries of case reports of allergic contact dermatitis from phenylephrine

Year	Sex	Age	Clinical picture	Positive patch tests	Comments	Ref.
2019	F	1	papulovesicular, pustular, and oozy lesions on erythematous base over bilateral eyes, ears, face, and neck	phenylephrine and tropicamide containing eye drops; negative to tropicamide eye drops	diagnosis made *per exclusionem*; phenylephrine itself was not tested	17
2018	F	51	periorbital dermatitis, severe conjunctival involvement, pseudomembrane formation	phenylephrine 0.5% eye drops, phenylephrine 1% and 10% water; other constituents negative	also symblepharon	18
2016	F	64	conjunctivitis	PE 100 mg/ml eye drops; PE itself was not tested; other ingredients negative	prick and intradermal tests were positive at D2 with cross-reactivity to epinephrine	19
2011	F	53	bilateral nasal congestion and discharge, dysphagia and dyspnea 1 hour after nasal spray containing PE, maximum after 12-24 hours	PE nasal spray, PE 10% eye drops; other ingredients not tested; possible cross-reaction to ethylephrine		20
2008	F	40	edema and erythema of both eyes; conjunctival congestion and eczema of the cheeks	commercial 10% PE eye drops	allergy to the preservative chlorobutanol not excluded	31
2007	M	56	blepharoconjunctivitis	patch tests were negative, but prick and intradermal tests were positive at the 48 and 72 hours reading		21
2007	M	25	conjunctival congestion, vesicular rash in a linear fashion from the left eye to to the left ear lobule	commercial 10% PE eye drops	allergy to excipients 'excluded' by a negative patch test to 'similar' eye drops not containing phenylephrine	32
2007	F	70	blepharoconjunctivitis, eyelid eczema	commercial phenylephrine eye drops without preservatives		34
2004	M	63	edematous erythema, papules, conjunctival hyperemia	two commercial eye drops containing 5% and 0.5% phenylephrine	PE was the only ingredient common to the 2	37
2003	M	70	conjunctivitis, edematous erythema	PE eye drops 5%, PE 0.1%, 0.5%, 1%, 2% and 5% pet.	reactions became positive only in a scratch-patch test; they persisted for 3 months	39
2002	F	65	periorbital dermatitis, conjunctival injection	+++ reaction to PE 2.5% and 10%, vehicle not mentioned		41

Table 3.260.2 Short summaries of case reports of allergic contact dermatitis from phenylephrine (continued)

Year	Sex	Age	Clinical picture	Positive patch tests	Comments	Ref.
2002	M	62	symmetrical eyelid dermatitis	PE eye drops 5%, PE 10% water	tape stripping was performed for testing the eye drops	42
1998	M	?	pure eyelid edema	PE eye drops negative; PE 10% water strongly positive		45
1998	M	69	periorbital dermatitis and conjunctivitis	PE 5% eye drops ++, on tape-stripped skin ++++	PE itself was not tested, nor the other eye drop ingredients	46
1998	F	75	periocular edema, erythema and pruritus	PE eye drops; later PE 1% pet.; the other constituents were negative	the first positive patch test persisted for 2 months and healed with hyperpigmentation	47
1997	F	65	eyelid dermatitis	PE 1% pet.	the patch test persisted for 7 months (2nd time 5 months) and healed with residual pigmentation	52
1997	F	54	symmetrical eyelid dermatitis	PE 1% pet.	the patch test persisted for 5 months (2nd time: 4 month)	52
1996	M	49	relapsing erythema, edema itching and burning of the eyes, eyelids and upper cheeks	PE 1% water	the patient was also allergic to the cyclopentolate eye drops that he used	73
1995	F	54	dermatitis of the lower legs	phlebological ointment with 0.5% PE (10 controls negative); PE 0.5% pet.		54
1995	F	61	acute eczema of the eyelids and periorbital region after one instillation of PE eye drops	reactions to PE 10% water en PE eye drops were regarded as irritant; negative to PE 1% water	probably wrong conclusion of the author; the patients most likely was allergic to PE	55
1993	M	50	perianal dermatitis	rectal ointment with 0.1% PE; PE 1% pet.		56
1991	F	?	dermatitis on the fingers	PE eye drops; PE (test concentration & vehicle unknown (article not read)	occupational contact dermatitis in a nurse instilling eye drops into patients' eyes for fundoscopy	59
1991	F	61	conjunctival injection, erythema, edema and vesicles on the eyelids	PE eye drops; PE 10% pet.	the ocular challenge test was strongly positive; severe flare-up of eczema one day later	60
1991	M	67	conjunctival injection, eczema of the eyelids spreading to the face	two PE eye drops; PE itself was not tested	phenylephrine was the only ingredient present in both eye drop preparations	60
1991	M	50	worsening of otitis externa	PE 5% pet.	PE was present in eardrops	61
1988	F	58	acute eczema of the eyelids and periorbital areas	PE eye drops; PE 1% water	there were lines of dermatitis from the eyes to the neck (tracks of the eye drops)	63
1988	F	0	edematous erythema, periorbital papules and vesicles	2 commercial eye drops with PE	contact allergy to another hapten in the eye drops not excluded	35
1987	M	66	burning eye, lacrimation, acute conjunctivitis	PE 0.5% eye drops; PE 5% pet.		64
1986	M	76	acute dermatitis of the periorbital area	PE eye drops; PE 1%, 5% and 10% in both petrolatum and water	symptoms began already after 30 minutes	65
1984	F	64	red, itchy edema of the eyelids and upper face	PE 10% eye drops; PE 1% and 10% pet.		66
1983	F	66	conjunctival injection and edema, periorbital eczema	eye drops 10%, PE 1%, 2.5%, 5% and 10% water and pet.	recurrence of dermatitis and conjunctivitis during patch testing	67
1979	F	75	chronic conjunctivitis and periorbital erythema	PE eye drops 10%; PE 10% water, neg. at 5% and 1%		68
1976	M	23	rapid increase in size of leg ulcer and intense pain	PE 5% and 10% water	the product used was a rectal ointment	69, 70
1976	M	54	widespread dermatitis, not at the site of application (perianally for hemorrhoids)	PE 0.5% and 1% water	the product was a rectal ointment; the absence of local eczema may be explained by its ingredient betamethasone valerate; this may have been a case of systemic contact dermatitis	69

PE: Phenylephrine

A 43-year-old nurse, working in both the ophthalmology and otolaryngology departments of a hospital, over the last 3 years had developed an itchy erythematous dermatitis on her face, especially the eyelids, related to her occupational duty of applying topical ophthalmic drugs, mainly phenylephrine and homatropine. Lesions improved when she worked in otolaryngology, and almost disappeared on holidays. Patch tests revealed positive reactions to phenylephrine 10% water and homatropine 1% water. Patch tests in 10 controls were negative. She left her work in ophthalmology and has since remained symptomless (51).

Other case reports of contact allergic reactions to phenylephrine are summarized in table 3.260.2. Additional case reports of contact allergy to phenylephrine, adequate data of which are not available to the author, can be found in refs. 24,38 (10 cases from Japan), 40,44,50,53,62, and 71 .

Cross-reactions, pseudo-cross-reactions and co-reactions

Nine patients who had experienced allergic contact dermatitis (blepharoconjunctivitis) after administration of eye drops containing phenylephrine, and who had positive patch tests to PE 10% pet., were tested with a battery of related sympathomimetic drugs. Three (33%) co-reacted to pseudoephedrine 10% pet., 5 (55%) to ephedrine 20%, 4 (44%) to phenylpropanolamine 10%, 2 (22%) to methoxamine 1% and 4 (44%) to oxymetazoline 10%, all in dimethyl sulfoxide (DMSO). An unspecified number of controls were negative. The reactions were considered to be the result of cross-reactivity to PE (23). In patients with allergic contact dermatitis from phenylephrine, cross-sensitization may have occurred to ethylephrine (20), fepradinol (23), epinephrine (69; reaction may have been irritant), and pseudoephedrine (22,57).

Patients sensitized to pseudoephedrine (23) or ephedrine (74) may have cross-reacted to phenylephrine.

LITERATURE

1   The data in the section 'General' may have been obtained from literature discussed in this chapter, but mostly also or exclusively from one or more of the following online sources: ChemIDPlus Advanced, PubChem, DrugBank, RxList, Drug Central, Drugs.com, and Wikipedia

2   Landeck L, John SM, Geier J. Periorbital dermatitis in 4779 patients – patch test results during a 10-year period. Contact Dermatitis 2014;70:205-212

3   Feser A, Plaza T, Vogelgsang L, Mahler V. Periorbital dermatitis – a recalcitrant disease: causes and differential diagnoses. Brit J Dermatol 2008;159:858-863

4   Ockenfels H, Seemann U, Goos M. Contact allergy in patients with periorbital eczema: an analysis of allergens. Dermatology 1997;195:119-124

5   Temesvári E, Pónyai G, Németh I, Hidvégi B, Sas A, Kárpáti S. Periocular dermatitis: a report of 401 patients. J Eur Acad Dermatol Venereol 2009;23:124-128

6   Herbst RA, Uter W, Pirker C, Geier J, Frosch PJ. Allergic and non-allergic periorbital dermatitis: patch test results of the Information Network of the Departments of Dermatology during a 5-year period. Contact Dermatitis 2004;51:13-19

7   Gilissen L, De Decker L, Hulshagen T, Goossens A. Allergic contact dermatitis caused by topical ophthalmic medications: Keep an eye on it! Contact Dermatitis 2019;80:291-297

8   Aoki J. [Allergic contact dermatitis due to eye drops. Their clinical features and the patch test results]. Nihon Ika Daigaku Zasshi 1997;64:232-237 (Article in Japanese)

9   Uter W, Menezes de Pádua C, Pfahlberg A, Nink K, Schnuch A, Behrens-Baumann W. Contact allergy to topical ophthalmological drugs - epidemiological risk assessment. Klin Monbl Augenheilkd 2009;226:48-53 (Article in German)

10 Wijnmaalen AL, van Zuuren EJ, de Keizer RJ, Jager MJ. Cutaneous allergy testing in patients suspected of an allergic reaction to eye medication. Ophthalmic Res 2009;41:225-229

11 Menezes de Padua CA, Uter W, Schnuch A. Contact allergy to topical drugs: prevalence in a clinical setting and estimation of frequency at the population level. Pharmacoepidemiol Drug Saf 2007;16:377-384

12 Menezes de Padua CA, Schnuch A, Nink K, Pfahlberg A, Uter W. Allergic contact dermatitis to topical drugs – Epidemiological risk assessment. Pharmacoepidemiol Drug Saf 2008;17:813-821

13 Gilissen L, Goossens A. Frequency and trends of contact allergy to and iatrogenic contact dermatitis caused by topical drugs over a 25-year period. Contact Dermatitis 2016;75:290-302

14 Landeck L, John SM, Geier J. Topical ophthalmic agents as allergens in periorbital dermatitis. Br J Ophthalmol 2014;98:259-262

15 Ventura MT, Di Corato R, Di Leo E, Foti C, Buquicchio R, Sborgia C, et al. Eyedrop induced allergy: clinical e valuation and diagnostic protocol. Immunopharmacol Immunotoxicol 2003;25:529-538

16 Tabar AI, García BE, Rodríguez A, Quirce S, Olaguibel JM. Etiologic agents in allergic contact dermatitis caused by eyedrops. Contact Dermatitis 1993;29:50-51

17 Lokhande AJ, Soni R, D'souza P, Yadav Y, Goel R. Allergic contact dermatitis to phenylephrine eye drops in an infant. Pediatr Dermatol 2019;36:975-977

18 Kato M, Nitta K, Kano Y, Yamada M, Ishii N, Hashimoto T, et al. Case of phenylephrine hydrochloride-induced periorbital contact dermatitis with fulminant keratoconjunctivitis causing pseudomembrane formation. J Dermatol 2018;45:e27-e28

19 Gutiérrez Fernández D, de la Varga Martínez R, Lasa Luaces EM, Foncubierta Fernández A, Andrés García JA, Medina Varo F. Allergic contact conjunctivitis and cross-reaction between phenylephrine and epinephrine due to phenylephrine eye drops. Ann Allergy Asthma Immunol 2016;117:564-565

20 Bobadilla-González P, Pérez-Rangel I, García-Menaya JM, Sánchez-Vega S, Cordobés-Durán C, Zambonino-Carreiras MA. Type IV reaction due to phenylephrine administered nasally with cross-reactivity with ethylephrine. J Investig Allergol Clin Immunol 2011;21:e69-e72

21 Botelho C, Rodríguez J, Castel-Branco MG. Allergic contact blepharoconjunctivitis with phenylephrine eyedrops – the relevance of late readings of intradermal tests. Allergol et Immunopathol 2007;35:157-158 (Article in Spanish)

22 Gonzalo-Garijo MA, Pérez-Calderón R, de Argila D, Rodríguez Nevado MI. Erythrodermia to pseudoephedrine in a patient with contact allergy to phenylephrine. Allergol et Immunopathol 2002;30:239-242 (Article in Spanish)

23 Barranco R, Rodríguez A, de Barrio M, Trujillo MJ, de Frutos C, Matheu V, et al. Sympathomimetic drug allergy: cross-reactivity study by patch test. Am J Clin Dermatol. 2004;5:351-355

24 Quirce Gancedo S, Compaired Villa JA, Fernández Rivas M, Losada Cosmes E. Hypersensitivity to phenylephrine in topical ocular administration. Med Clin (Barc) 1991;96:317-318

25 Veramme J, de Zaeytijd J, Lambert J, Lapeere H. Contact dermatitis in patients undergoing serial intravitreal injections. Contact Dermatitis 2016;74:18-21

26 Madsen JT, Andersen KE. Phenylephrine is a frequent cause of periorbital allergic contact dermatitis. Contact Dermatitis 2015;73:64-65

27 Haroun-Díaz E, Ruíz-García M, De Luxán de la Lastra S, Pastor-Vargas C, De las Heras M, Sastre Domínguez J, et al. Contact dermatitis to both tropicamide and phenylephrine eye drops. Dermatitis 2014;25:149-150

28 Corazza M, Massieri LT, Virgili A. Doubtful value of patch testing for suspected contact allergy to ophthalmic products. Acta Derm Venereol 2005;85:70-71

29 Tamagawa-Mineoka R, Katoh N, Yoneda K, Cho Y, Kishimoto S. Systemic allergic contact dermatitis due to phenylephrine in eyedrops, with a long-lasting allergic patch test reaction. Eur J Dermatol 2010;20:125-126

30 Raison-Peyron N, Du Thanh A, Demoly P, Guillot B. Long-lasting allergic contact blepharoconjunctivitis to phenylephrine eyedrops. Allergy 2009;64:657-658

31 Singal A, Rohatgi J, Pandhi D. Allergic contact dermatitis to phenylephrine. Indian J Dermatol Venereol Leprol 2008;74:298

32 Garg P, Rajiv G, Singh L, Malhotra R, Prasad R. Allergy to phenylephrine hydrochloride eyedrops: A case report. Indian J Allergy Asthma Immunol 2007;21:73-76

33 Vadot E, Piasentin D. Incidence of allergy to eyedrops. Results of a prospective survey in a hospital milieu. J Fr Ophtalmol 1986;9:41-43 (Article in French)

34 Dewachter P, Mouton-Faivre C. Anaesthesists should be aware of delayed hypersensitivity to phenylephrine. Acta Anaesthesiol Scand 2007;51:637-639

35 Aihara M, Ikezawa Z. Neonatal allergic contact dermatitis. Contact Dermatitis 1988;36:105

36 Borch JE, Elmquist JS, Bindslev-Jensen C, Andersen KE. Phenylephrine and acute periorbital dermatitis. Contact Dermatitis 2005;53:298-299

37 Yamamoto A, Harada S, Nakada T, Iijima M. Contact dermatitis to phenylephrine hydrochloride eyedrops. Clin Exp Dermatol 2004;29;200-201

38 Suzuki R, Matsunaga K, Suzuki K, et al. Ten cases of contact dermatitis due to phenylephrine hydrochloride. Rinsho Hifuka 1999;53:504-506 (Article in Japanese).

39 Akita H, Akamatsu H, Matsunaga K. Allergic contact dermatitis due to phenylephrine hydrochloride, with an unusual patch test reaction. Contact Dermatitis 2003;49:232-235

40 Zucchi A, Antonaccio F, de Panfilis G, Allegra F. Contact dermatitis caused by eye drops containing phenylephrine. G Ital Dermatol Venereol 1990;125:155-156 (Article in Italian)

41 Narayan S, Prais L, Foulds IS. Allergic contact dermatitis caused by phenylephrine eyedrops. Am J Contact Dermat 2002;13:208-209

42 Erdmann SM, Sachs B, Merk HF. Allergic contact dermatitis from phenylephrine in eyedrops. Am J Contact Dermat 2002;13:37-38

43 Almeida L, Ortega N, Dumpierrez AG, Castillo R, Blanco C, Navarro L, et al. Conjunctival allergic contact hypersensitivity. Allergy 2001;56:785

44 Resano A, Esteve C, Fernández Benítez M. Allergic contact blepharoconjunctivitis due to phenylephrine eye drops. J Investig Allergol Clin Immunol 1999;9:55-57

45 Blum A, Brummer C, Lischka G. Edematous swelling of the eyelids caused by contact allergy. Hautarzt 1998;49:651-653 (Article in German)

46 Wigger-Alberti W, Elsner P, Wüthrich B. Allergic contact dermatitis to phenylephrine. Allergy 1998;53:217-218
47 Rafael M, Pereira F, Faria MA. Allergic contact blepharoconjunctivitis caused by phenylephrine, associated with persistent patch test reaction. Contact Dermatitis 1998;39:143-144
48 Villarreal O. Reliability of diagnostic tests for contact allergy to mydriatic eyedrops. Contact Dermatitis 1998;38:150-154
49 Thomas P, Rueff F, Przybilla B. Severe allergic contact blepharoconjunctivitis from phenylephrine in eyedrops, with corresponding T-cell hyper-responsiveness *in vitro*. Contact Dermatitis 1998;38:41-43
50 Moreno-Ancillo A, Munoz-Robles ML, Cabañas R, Barranco P, Lopez-Serrano MC. Allergic contact reactions due to phenylephrine hydrochloride in eyedrops. Ann Allergy Asthma Immunol 1997;78:569-572
51 Marcos ML, Garcés MM, Alonso L, Juste S, Carretero P, Blanco J, et al. Occupational allergic contact dermatitis from homatropine and phenylephrine eyedrops. Contact Dermatitis 1997;37:189
52 Mancuso G, Reggiani M, Staffa M. Long-lasting allergic patch test reaction to phenylephrine. Contact Dermatitis 1997;36:110-111
53 Buzo-Sánchez G, Martín-Muñoz MR, Navarro-Pulido AM, Orta-Cuevas JC. Stereoisomeric cutaneous hypersensitivity. Ann Pharmacother 1997;31:1091
54 Thomas P, Rueff F, Przybilla B. Allergic contact dermatitis from phenylephrine in a phlebological ointment. Contact Dermatitis 1995;32:249-250
55 Urbani CE. Probable irritant contact dermatitis from phenylephrine in eyedrops with irritant (false-positive) phenylephrine patch test reactions. Am J Cont Derm 1995;6:49-51
56 Wilkinson SM, Kingston TP, Beck MH. Allergic contact dermatitis from phenylephrine in a rectal ointment. Contact Dermatitis 1993;29:100-101
57 Audicana M, Echechipia S, Fernández E, Bernaola G, Munoz D, Fernandez de Corres L. Contact dermatitis from phenylephrine. Am J Cont Derm 1993;4:225-228
58 Geyer O, Neudorfer M, Lazar M, Dayan M, Mozes E. Cellular sensitivity in allergic blepharoconjunctivitis due to phenylephrine eye drops. Graefes Arch Clin Exp Ophthalmol 1993;231:748-750
59 Okamoto H, Kawai S. Allergic contact sensitivity to mydriatic agents on a nurse's fingers. Cutis 1991;47:357-358
60 Añíbarro B, Barranco P, Ojeda JA. Allergic contact blepharoconjunctivitis caused by phenylephrine eyedrops. Contact Dermatitis 1991;25:323-324
61 Bardazzi F, Tardio M, Mariani R, Rapacchiale S, Valenti R. Phenylephrine in eardrops causing contact dermatitis. Contact Dermatitis 1991;24:56
62 Geyer O, Yust I, Lazer M. Allergic blepharoconjunctivitis due to phenylephrine. J Ocul Pharmacol 1988;40:123-126
63 Milpied B, Fleischmann M, Berre F, Sourisse M, Litoux P. Another case of allergic contact dermatitis from phenylephrine in eyedrops. Contact Dermatitis 1988;19:146-147
64 Tosti A, Bardazzi F, Tosti G, Colombati S. Contact dermatitis to phenylephrine. Contact Dermatitis 1987;17:110-111
65 Ducombs G, de Casamayor J, Verin P, Maleville J. Allergic contact dermatitis to phenylephrine. Contact Dermatitis 1986;15:107-108
66 Camarasa JG. Contact dermatitis to phenylephrine. Contact Dermatitis 1984;10:182
67 Barber KA. Allergic contact eczema to phenylephrine. Contact Dermatitis 1983;9:274-277
68 Mathias CG, Maibach HI, Irvine A, Adler W. Allergic contact dermatitis to echothiopate iodide and phenylephrine. Arch Ophthalmol 1979;97:286-287
69 Roed-Petersen J. Contact sensitivity to metaoxedrine. Contact Dermatitis 1976;2:235-236
70 Roed-Petersen J. Metaoxedrine sensitivity. Contact Dermatitis Newsletter 1973;14:392
71 Hanna C, Brainard J, Augspurger KD, Roy FH, Fox MJ. Allergic dermatoconjunctivitis caused by phenylephrine. Am J Ophthalmol 1983;95:703-704
72 Tosti A, Tosti G. Allergic contact conjunctivitis due to ophthalmic solution. In: Frosch PJ, Dooms-Goossens A, Lachapelle JM, Rycroft RJG, Scheper RJ (eds). Current Topics in Contact Dermatitis. Berlin: Springer-Verlag, 1989: 269 272
73 Camarasa JG, Pla C. Allergic contact dermatitis from cyclopentolate. Contact Dermatitis 1996;35:368-369
74 Maul LV, Streit M, Grabbe J. Ephedrine-induced maculopapular rash. Contact Dermatitis 2018;79:193-194

Chapter 3.261 PIKETOPROFEN

IDENTIFICATION

Description/definition : Piketoprofen is the non-steroidal anti-inflammatory drug that conforms to the structural formula shown below

Pharmacological classes : Anti-inflammatory drugs, non-steroidal

IUPAC name : 2-(3-Benzoylphenyl)-N-(4-methylpyridin-2-yl)propanamide

Other names : Benzeneacetamide, 3-benzoyl-α-metyl-N-(4-metyl-2-pyridinyl)-

CAS registry number : 60576-13-8

EC number : Not available

Merck Index monograph : 8804

Patch testing : 1% pet. (SmartPracticeCanada)

Molecular formula : $C_{22}H_{20}N_2O_2$

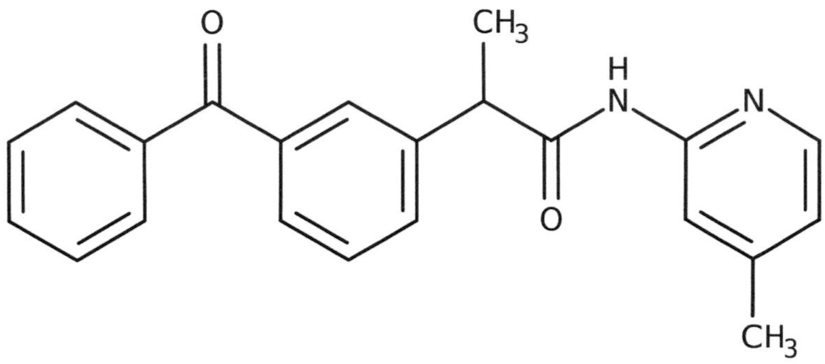

GENERAL

Piketoprofen is a nonsteroidal anti-inflammatory drug (NSAID) from the arylpropionic acid group, which prevents the synthesis of prostaglandins and other prostanoids through competitive and reversible inhibition of cyclooxygenase. It is used in topical preparations in a few countries including Spain and Portugal for the treatment of pain from musculoskeletal, rheumatic or traumatic origin. In pharmaceuticals, piketoprofen may be employed as piketoprofen hydrochloride (CAS number 59512-37-7, EC number not available, molecular formula $C_{22}H_{21}ClN_2O_2$) (1).

CONTACT ALLERGY

Case reports and case series

In the period 1996-2001, in 2 hospitals in Spain, one patient was diagnosed with contact allergy and 4 with photocontact allergy to piketoprofen. The accumulated incidence per million inhabitants (catchment population of the hospitals) of both side effects together was 6.0 (10).

A 73-year-old woman, after having applied a cream containing piketoprofen 2x per day on her left leg for venous insufficiency for 5 days, developed an acute eczematous reaction. She denied sun exposure at that time. Patch testing with the standard series, a NSAID battery and the active principle of the cream showed positive reactions only to piketoprofen 1% and 2% pet. at D2 and D4. Fifteen controls were negative (3).

A 27-year-old woman presented with itchy erythematous vesicular and edematous lesions on the hands, forearms and back. She had been applying a cream containing piketoprofen for back pain the preceding 3 days. Patch tests were positive to the cream and piketoprofen 1% and 5% pet., but negative to its excipients and a series of other NSAIDs. Five controls were negative (7).

PHOTOSENSITIVITY

Case series

In the period 2004-2005, in a Spanish multicenter study performing photopatch testing, 2 relevant positive photopatch tests were observed to piketoprofen 2.5% pet. It was not mentioned how many patients had been tested with this chemical (9). In the period 1996-2001, in 2 hospitals in Spain, 4 patients were diagnosed with photocontact allergy and 1 with contact allergy to piketoprofen. The accumulated incidence per million inhabitants (catchment population of the hospitals) of both side effects together was 6.0 (10). In 2 Spanish hospitals, in the period 1985-1996, 8 patients with photoallergy due to piketoprofen have been observed among 153 patients; no details are available (6).

Case reports

A 30-year-old man had developed vesiculobullous and intensely itchy lesions localized on the palms of his hands and on his fingers some days after applying a gel containing piketoprofen to his girlfriend's neck for 3 days. His hands had been exposed to sunlight. The patient had not used the gel himself before, but he knew that he was allergic to ketoprofen. Patch testing with the standard series and a NSAID series was negative, but photopatch tests using UVA irradiation 5 J/cm^2 showed positive reactions to piketoprofen 2% pet., ketoprofen 1% pet. and dexketoprofen 1% and 2% pet. (2). The reactions to piketoprofen and dexketoprofen may have been cross-reactions to primary photosensitization to ketoprofen.

A 46-year-old man in the summer developed an acute eczema on the back of his neck after having applied a gel containing piketoprofen for 3 days. Patch tests were negative. Photopatch tests using 5 J/cm^2 UVA irradiation were positive to the gel, piketoprofen (1% and 2% pet., 1%, 2% and 5% alc.) but negative to the other ingredients of the gel and a series of NSAIDs. Ten controls were negative (5).

A 14-year-old girl had applied ketoprofen gel on her knee for a trauma. After exposure to sunlight, erythema, edema, local hyperthermia and small macular lesions with pruritus and burning developed in the area of application. Some months later, and hours after having applied the same preparation to another person, she developed pruriginous vesicles on the fingers of her hands, which progressed to intense edema and erythema. The third episode occurred after the application of a gel containing piketoprofen on her neck, and again consisted in an intense local skin reaction of similar characteristics. Patch tests were positive to the fragrance mix and 'lavender' (oil?), which was present in both gels, but negative to keto- and piketoprofen. In a second session, photopatch tests, however, were positive to ketoprofen and piketoprofen 2% and 5% pet. The non-irradiated sites of both NSAIDs now also became positive, but the authors did not comment on this. They did, however, state, that the allergy to the fragrance mix and lavender had no relationship to the skin eruptions, whereas both gels contained essence of lavender!! (11).

Cross-reactions, pseudo-cross-reactions and co-reactions

Photocross-reactions may occur between ketoprofen, piketoprofen and dexketoprofen (2,4,8,11).

LITERATURE

1   The data in the section 'General' may have been obtained from literature discussed in this chapter, but mostly also or exclusively from one or more of the following online sources: ChemIDPlus Advanced, PubChem, DrugBank, RxList, Drug Central, Drugs.com, and Wikipedia
2   Fernández-Jorge B, Goday Buján JJ, Paradela S, Mazaira M, Fonseca E. Consort photocontact dermatitis from piketoprofen. Contact Dermatitis 2008;58:113-115
3   Rodríguez-Lozano J, Goday Buján JJ, Del Pozo J, Fonseca E. Allergic contact dermatitis from topical piketoprofen. Contact Dermatitis 2005;52:110-111
4   Valenzuela N, Puig L, Barnadas MA, Alomar A. Photocontact dermatitis due to dexketoprofen. Contact Dermatitis 2002;47:237
5   Goday Buján JJ, oleaga Morante JM, González Güemes MG, Del Pozo Losada J, Fonseca Capdevila E. Photoallergic contact dermatitis from piketoprofen. Contact Dermatitis 2000;43:315
6   Ortiz de Frutos, Alomar. Third Latin Meeting of Allergic Contact Dermatitis, unpublished communication. Data cited in ref. 5
7   Navarro LA, Jorro G, Morales C, Peláez A. Allergic contact dermatitis due to piketoprofen. Contact Dermatitis 1995;32:181
8   Subiabre-Ferrer D, Esteve-Martínez A, Blasco-Encinas R, Sierra-Talamantes C, Pérez-Ferriols A, Zaragoza-Ninet V. European photopatch test baseline series: A 3-year experience. Contact Dermatitis 2019;80:5-8
9   De La Cuadra-Oyanguren J, Perez-Ferriols A, Lecha-Carrelero M, et al. Results and assessment of photopatch testing in Spain: towards a new standard set of photoallergens. Actas DermoSifiliograficas 2007;98:96-101
10 Diaz RL, Gardeazabal J, Manrique P, Ratón JA, Urrutia I, Rodríguez-Sasiain JM, Aguirre C. Greater allergenicity of topical ketoprofen in contact dermatitis confirmed by use. Contact Dermatitis 2006;54:239-243
11 García Bara MT, Matheu V, Pérez A, Díaz MP, Martínez MI, Zapatero L. Contact photodermatitis due to ketoprofen and piketoprofen. Allergol Inmunol Clin 1999;14:148-150

Chapter 3.262 PILOCARPINE

IDENTIFICATION

Description/definition : Pilocarpine is a natural alkaloid extracted from plants of the genus *Pilocarpus* that conforms to the structural formula shown below

Pharmacological classes : Muscarinic agonists; miotics

IUPAC name : (3S,4R)-3-Ethyl-4-[(3-methylimidazol-4-yl)methyl]oxolan-2-one

Other names : 2(3H)-Furanone, 3-ethyldihydro-4-[(1-methyl-1H-imidazol-5-yl)methyl]-, (3S,4R)-

CAS registry number : 92-13-7

EC number : 202-128-4

Merck Index monograph : 8806

Patch testing : Hydrochloride 1% water (SmartPracticeCanada); HCl 1% pet. (SmartPracticeEurope)

Molecular formula : $C_{11}H_{16}N_2O_2$

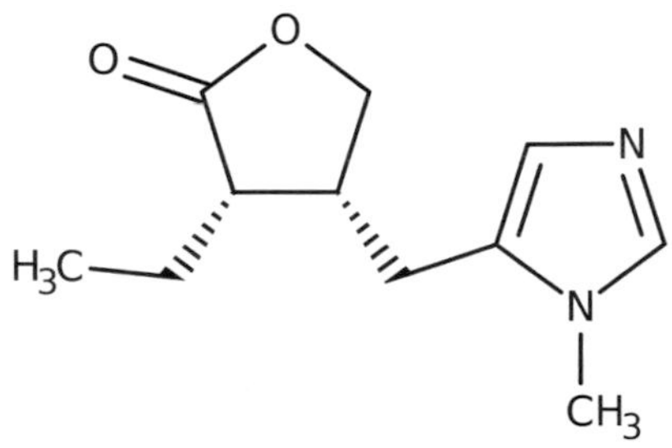

GENERAL

Pilocarpine is a natural alkaloid extracted from plants of the genus *Pilocarpus* with cholinergic agonist activity. As a cholinergic parasympathomimetic agent, it predominantly binds to muscarinic receptors, thereby inducing exocrine gland secretion and stimulating smooth muscle in the bronchi, urinary tract, biliary tract, and intestinal tract. When applied topically to the eye, this agent stimulates the sphincter pupillae to contract, resulting in miosis. It also stimulates the ciliary muscle to contract, resulting in spasm of accommodation and may cause a transitory rise in intraocular pressure followed by a more persistent fall due to opening of the trabecular meshwork and an increase in the outflow of aqueous humor. Pilocarpine is indicated for the treatment of radiation-induced dry mouth (xerostomia) and symptoms of dry mouth in patients with Sjögren's syndrome. It is also used as a miotic in the treatment of glaucoma. In pharmaceutical products, pilocarpine is employed as pilocarpine hydrochloride (CAS number 54-71-7, EC number 200-212-5, molecular formula $C_{11}H_{17}ClN_2O_2$) (1).

CONTACT ALLERGY

Contact allergy in the general population

In Germany, for the period 1995-2004, the population-based relative incidence (RI) of contact sensitization to pilocarpine (cases/100,000 defined daily doses (DDDs) per year) was estimated to be 0.3. In the group of ophthalmic drugs, the RI ranged from 0.3 (pilocarpine) to 21.6 (phenylephrine) (5).

Patch testing in groups of selected patients

In 4 studies testing selected patient groups with pilocarpine, low prevalences of sensitization of 0.4% to 0.8% were found (table 3.262.1).

Table 3.262.1 Patch testing in groups of patients: Selected patient groups

Years and Country	Test conc. & vehicle	Number of patients tested | positive (%)		Selection of patients (S); Relevance (R); Comments (C)	Ref.
2001-2010 IVDK	1% water	1144	5 (0.4%)	S: patients with periorbital dermatitis tested with an ophthalmic tray; R: not stated	6
1996-2005 Hungary	1% water	133	1 (0.8%)	S: patients with periorbital dermatitis; R: the reactions to eye medicaments were relevant	2
1995-2004 IVDK	1% water	3877	(0.4%)	S: patients patch tested for suspected contact allergy to ophthalmological drugs; R: not stated; C: the estimated number of sensitizations per 100,000 prescriptions was 0.3	3
1995-2004 IVDK	1% water	4284	(0.4%)	S: not stated; R: not stated	4

IVDK: Information Network of Departments of Dermatology (Germany, Austria, Switzerland)

Case reports

A woman aged 72 had dermatitis of both eyelids and periorbital areas, while using various medications for her glaucoma. Patch tests were positive to two commercial pilocarpine-containing eyedrop preparations, but the active principle itself was not tested. Nevertheless, the patient was diagnosed with pilocarpine allergy. She also reacted to levobunolol eye drops and levobunolol 1% in water, but to the latter only when applied to tape-stripped skin (7).

A 56-year-old woman was seen with a 1-year history of eyelid eczema and conjunctivitis of the left eye. She had suffered from severe glaucoma since about the age of 40 years. In the last 3 years, she had used only 2 brands of eye drops: timolol maleate and pilocarpine hydrochloride 2%. Patch testing with these 2 products (as is), benzalko-nium chloride 0.1% water and sodium EDTA 1% pet. (both present in the pilocarpine eye drops) was negative at 2 and 3 days. A ROAT was then performed with both eye drops, resulting in erythema and vesicles on the application site of pilocarpine after about 10 days. A ROAT with eye drops containing pilocarpine 1% in a quite different vehicle was also positive (9).

A 62-year-old man had suffered for 3 months from an outbreak of erythematous edematous plaques involving his eyelids. For 1 year, he had been applying 4% pilocarpine eye drops. Previously, he was shown to be allergic to timolol eye drops and eye drops containing thimerosal, but ingredient patch testing had not been performed. Patch testing revealed positive reactions to the 4% pilocarpine eye drops, pilocarpine hydrochloride 2% and 4% pet. and thimerosal 0.1% pet. (past relevance). Twelve controls were negative to pilocarpine HCl 2%, 4%, 10% and 20% pet. (10).

A (possible) early case report of allergic contact dermatitis from pilocarpine was published in 1955 (11, no data available). Another case of allergic contact dermatitis to pilocarpine, but aggravated by sunlight (photo-aggravation) is discussed below in the section Photosensitivity (8).

PHOTOSENSITIVITY

A 68-year-old man presented with a 5-month history of pruritic erythematous scaly dermatitis of both upper and lower eyelids. He also had irritation and redness of his conjunctivae and sclerae. The patient had used 4% pilocarpine eye drops for one year. He had noticed that the dermatitis worsened when he was exposed to the sun. Patch tests to 2 brands of 4% pilocarpine eye drops were negative. A ROAT became positive after 3 weeks. Next, patch and photopatch tests were performed with pilocarpine nitrate 2% eye drops, pilocarpine HCl 4% eye drops and pilocarpine 4% pet. Seven days after the patches were applied and 5 days after irradiation with UVA, there were positive reactions to 2% pilocarpine nitrate and 4% pilocarpine HCl both on the non-irradiated side and the irradiated side, but stronger on the UVA-exposed skin. Pilocarpine 4% pet. was positive on the irradiated side, but negative on the non-irradiated side. It was concluded that the patient had photoaggravated allergic contact dermatitis from pilocarpine, although the negative patch test to pilocarpine 4% could not be explained (8).

LITERATURE

1   The data in the section 'General' may have been obtained from literature discussed in this chapter, but mostly also or exclusively from one or more of the following online sources: ChemIDPlus Advanced, PubChem, DrugBank, RxList, Drug Central, Drugs.com, and Wikipedia

2   Temesvári E, Pónyai G, Németh I, Hidvégi B, Sas A, Kárpáti S. Periocular dermatitis: a report of 401 patients. J Eur Acad Dermatol Venereol 2009;23:124-128

3   Uter W, Menezes de Pádua C, Pfahlberg A, Nink K, Schnuch A, Behrens-Baumann W. Contact allergy to topical ophthalmological drugs - epidemiological risk assessment. Klin Monbl Augenheilkd 2009;226:48-53 (Article in German)

4   Menezes de Padua CA, Uter W, Schnuch A. Contact allergy to topical drugs: prevalence in a clinical setting and estimation of frequency at the population level. Pharmacoepidemiol Drug Saf 2007;16:377-384

5   Menezes de Padua CA, Schnuch A, Nink K, Pfahlberg A, Uter W. Allergic contact dermatitis to topical drugs – Epidemiological risk assessment. Pharmacoepidemiol Drug Saf 2008;17:813-821

6   Landeck L, John SM, Geier J. Topical ophthalmic agents as allergens in periorbital dermatitis. Br J Ophthalmol 2014;98:259-262

7   Koch P. Allergic contact dermatitis due to timolol and levobunolol in eyedrops, with no cross-sensitivity to other ophthalmic beta-blockers. Contact Dermatitis 1995;33:140-141

8   Helton J, Storrs FJ. Pilocarpine allergic contact and photocontact dermatitis. Contact Dermatitis 1991;25:133-134

9   Cusano F, Luciano S, Capozzi M, Verrilli DA. Contact dermatitis from pilocarpine. Contact Dermatitis 1993;29:99

10   Ortiz FJ, Postigo C, Ivars J, Ortiz PL, Merino V. Allergic contact dermatitis from pilocarpine and thimerosal. Contact Dermatitis 1991;25:203-204

11   Holmberg A. Eczema of the eyelids caused by pilocarpine. Acta Ophthalmol (Copenh) 1955;33:371-375

Chapter 3.263 PIMECROLIMUS

IDENTIFICATION

Description/definition	: Pimecrolimus is the 33-epi-chloro-derivative of the ascomycin macrolactam that conforms to the structural formula shown below
Pharmacological classes	: Calcineurin inhibitors; dermatological agents; anti-inflammatory agents, non-steroidal; immunosuppressive agents
IUPAC name	: (1R,9S,12S,13R,14S,17R,18E,21S,23S,24R,25S,27R)-12-[(E)-1-[(1R,3R,4S)-4-Chloro-3-methoxycyclohexyl]prop-1-en-2-yl]-17-ethyl-1,14-dihydroxy-23,25-dimethoxy-13,19,21,27-tetramethyl-11,28-dioxa-4-azatricyclo[22.3.1.0^{4,9}]octacos-18-ene-2,3,10,16-tetrone
Other names	: 33-epi-Chloro-33-desoxyascomycin
CAS registry number	: 137071-32-0
EC number	: 603-999-7
Merck Index monograph	: 8811
Patch testing	: 1% pet.; if not available, test 1% pimecrolimus cream and its base
Molecular formula	: $C_{43}H_{68}ClNO_{11}$

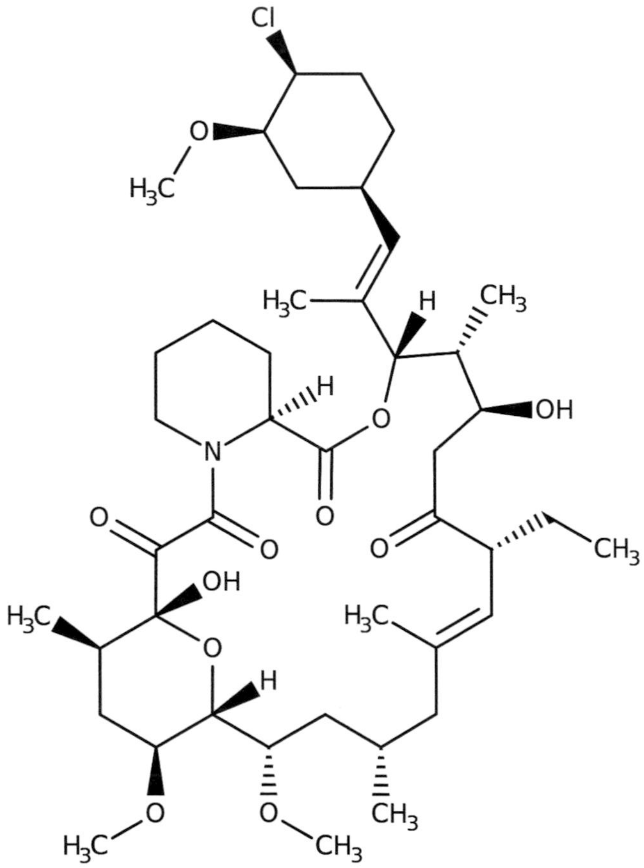

GENERAL

Pimecrolimus is a derivative of the macrolactam ascomycin with immunosuppressant and immunomodulating properties. Its mechanism of action involves calcineurin inhibition, blockage of T-cell activation, blocking signal transduction pathways in T-cells, reducing the ability of mast cells to release chemicals that promote inflammation, and inhibition of the synthesis of inflammatory cytokines, specifically Th1- and Th2-type cytokines. Pimecrolimus is indicated for topical treatment of mild to moderate atopic dermatitis (1).

CONTACT ALLERGY

General

Only two cases of allergic contact dermatitis to pimecrolimus have been reported. In both cases, the diagnosis was made *per exclusionem*: positive patch test to the cream, negative to the placebo-cream (same cream, but without pimecrolimus) (2,5). The reason for this suboptimal situation is the refusal of the manufacturer to provide pimecro-limus for patch testing.

Contact allergy to the excipient oleyl alcohol in the commercial pimecrolimus cream has been described (6). This cream also contains the well-known sensitizer propylene glycol (PG), but in patients allergic to PG, pimecrolimus cream showed a very low potential to elicit allergic skin reactions (8). In one case, a patient had a positive patch test to the commercial cream but not to any of its ingredients (7).

Case reports

A 31-year-old woman presented with a 2-year history of pruritic lip dermatitis. Physical examination showed a 2-mm circumferential rim of erythema and edema on the upper and lower lips. Previous therapies included pimecrolimus cream and mupirocin ointment. Patch tests were positive on day 5 to pimecrolimus cream and to 2 different mupirocin 2% ointments. Several of the excipients of the pimecrolimus cream were tested negative, as was the cream base. A ROAT with pimecrolimus cream was positive after 3 days. The mupirocin ointment formulations consisted of only mupirocin and polyethylene glycol; the patient did not react to polyethylene glycol 100%. It was concluded that the patient had allergic contact dermatitis from pimecrolimus and from mupirocin (2).

A 70-year-old woman had treated chronic stasis dermatitis with pimecrolimus 1% cream, when she developed a severe exacerbation of the dermatitis on her legs, feet and hands. A patch test to the cream was positive (++) on 3 occasions. Later, several of the inactive ingredients were patch test negative, as was the entire cream base (tested 3 times). The active ingredient was not tested, as the manufacturer 'could not' provide the active ingredient (5).

Cross-reactions, pseudo-cross-reactions and co-reactions

A patient sensitized to tacrolimus cross-reacted to pimecrolimus (3). Such cross-sensitization has also been observed by others (4).

LITERATURE

1   The data in the section 'General' may have been obtained from literature discussed in this chapter, but mostly also or exclusively from one or more of the following online sources: ChemIDPlus Advanced, PubChem, DrugBank, RxList, Drug Central, Drugs.com, and Wikipedia
2   Zhang AJ, Warshaw EM. Allergic contact dermatitis caused by mupirocin and pimecrolimus. Contact Dermatitis 2019;80:132-133
3   Shaw DW, Maibach HI, Eichenfield LF. Allergic contact dermatitis from pimecrolimus in a patient with tacrolimus allergy. J Am Acad Dermatol 2007;56:342-345
4   Schmutz JL, Barbaud A, Tréchot P. Contact allergy with tacrolimus then pimecrolimus. Ann Dermatol Venerol 2008;135:89 (Article in French)
5   Saitta P, Brancaccio R. Allergic contact dermatitis to pimecrolimus. Contact Dermatitis 2007;56:43-44
6   Andersen KE, Broesby-Olsen S. Allergic contact dermatitis from oleyl alcohol in Elidel cream. Contact Dermatitis 2006;55:354-356
7   Neczyporenko F, Blondeel A. Allergic contact dermatitis to Elidel cream itself? Contact Dermatitis 2010;63:171-172
8   Fowler JF Jr, Fowler L, Douglas JL, Thorn D, Parneix-Spake A. Skin reactions to pimecrolimus cream 1% in patients allergic to propylene glycol: a double-blind randomized study. Dermatitis 2007;18:134-139

Chapter 3.264 PIPERAZINE

IDENTIFICATION

Description/definition : Piperazine is the cyclic organic compound that conforms to the structural formula shown below

Pharmacological classes : Antinematodal agents
IUPAC name : Piperazine
Other names : 1,4-Diazacyclohexane; 1,4-diethylenediamine; hexahydropyrazine
CAS registry number : 110-85-0
EC number : 203-808-3
Merck Index monograph : 8846
Patch testing : 1% pet. (SmartPracticeCanada, SmartPracticeEurope)
Molecular formula : $C_4H_{10}N_2$

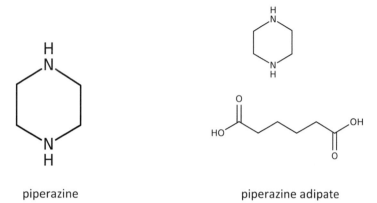

piperazine   piperazine adipate

GENERAL

Piperazine is an organic compound that was introduced to medicine as a solvent for uric acid and later as an anthelmintic (antinematodal agent). It produces a neuromuscular block leading to flaccid muscle paralysis in susceptible worms, which are then dislodged from the gut and expelled in feces. Piperazine is used as alternative treatment for ascariasis caused by *Ascaris lumbricoides* (roundworm) and enterobiasis (oxyuriasis) caused by *Enterobius vermicularis* (pinworm). In pharmaceutical products, piperazine may be employed as piperazine adipate (CAS number 142-88-1, EC number 205-569-0, molecular formula $C_{10}H_{20}N_2O_4$) (1). Piperazine also has numerous non-pharmaceutical applications. In this chapter, only allergic reactions to piperazine from pharmaceutical use are presented.

See also Chapter 3.292 Pyrazinobutazone and Chapter 3.259 Phenylbutazone.

CONTACT ALLERGY

General

Piperazine (diethylenediamine) is metabolized in the human body into ethylenediamine, which means that patients who have previously been sensitized to ethylenediamine (mostly from one brand of topical corticosteroids containing triamcinolone acetonide, ethylenediamine and neomycin, sometimes from aminophylline [= theophylline + ethylenediamine]) may develop systemic contact dermatitis from oral administration of piperazine (5,6,9,10).

Case reports and case series

A 71-year-old man presented with a bilateral acute knee eczema after applying phenylbutazone-piperazine gel for rheumatic pain. Patch testing showed positive reactions to the gel and piperazine 5% water. Phenylbutazone was not tested separately (3).

A 55-year-old man had previously suffered dermatitis of the face after using a cream containing phenylbutazone. After a nasal operation, he inserted a phenylbutazone-piperazine suppository. Ten hours later, he showed erythema and swelling on the face. He thought the handling of the suppository was the cause. Besides anal irritation, erythema with mild itching spread over his body with a later scaling over a month. Patch tests were positive to phenylbutazone 5% and piperazine 1% water (2). This was a case of systemic contact dermatitis. It is unknown whether and how much the allergy to piperazine has contributed to the clinical manifestations in this case.

Cross-reactions, pseudo-cross-reactions and co-reactions
Patients allergic to ethylenediamine may cross-react to systemic piperazine (5,6,10), as piperazine (diethylenediamine) is metabolized in the human body into ethylenediamine.

Cutaneous adverse drug reactions from systemic administration caused by type IV (delayed-type) hypersensitivity
Cutaneous adverse drug reactions from systemic administration of piperazine caused by type IV (delayed-type) hypersensitivity, including erythroderma (6), angioedema (5), and morbilliform drug rash (10,11), as well as occupational allergic contact dermatitis (3,4,7,8,9,12), are planned to be discussed in Volume IV of the *Monographs in Contact Allergy* series on Systemic drugs.

LITERATURE

1   The data in the section 'General' may have been obtained from literature discussed in this chapter, but mostly also or exclusively from one or more of the following online sources: ChemIDPlus Advanced, PubChem, DrugBank, RxList, Drug Central, Drugs.com, and Wikipedia
2   Fernández de Corres L, Bernaola G, Lobera T, Leanizbarrutia I, Muñoz D. Allergy from pyrazoline derivatives. Contact Dermatitis 1986;14:249-250
3   Brandão FM, Foussereau J. Contact dermatitis to phenylbutazone-piperazine suppositories (Carudol) and piperazine gel (Carudol). Contact Dermatitis 1982;8:264-265
4   Rudzki E, Rebandel P, Grzywa Z. Contact allergy in the pharmaceutical industry. Contact Dermatitis 1989;21:121-122
5   Eedy DJ. Angioneurotic oedema following piperazine ingestion in an ethylenediamine-sensitive subject. Contact Dermatitis 1993;28:48-49
6   Price ML, Hall-Smith SP. Allergy to piperazine in a patient sensitive to ethylenediamine. Contact Dermatitis 1984;10:120
7   Rudzki E, Grzywa Z. Occupational piperazine dermatitis. Contact Dermatitis 1977;3:216
8   Calnan CD. Occupational piperazine dermatitis. Contact Dermatitis 1975;1:126
9   Fregert S. Respiratory symptoms with piperazine patch testing. Contact Dermatitis 1976;2:61-62
10 Burry JN. Ethylenediamine sensitivity with a systemic reaction to piperazine citrate. Contact Dermatitis 1978;4:380
11 Wright S, Harman RR. Ethylenediamine and piperazine sensitivity. Br Med J (Clin Res Ed) 1983;287(6390):463-464
12 Foussereau J. La piperazine, allergène de contact chez le personnel soignant. Rev Franc Allerg 1963;3:236-240

Chapter 3.265 PIRENOXINE

IDENTIFICATION

Description/definition   : Pirenoxine is the phenoxazine derivative that conforms to the structural formula shown
                               below
Pharmacological classes   : Antioxidant
IUPAC name   : 1,5-Dioxo-4H-pyrido[3,2-a]phenoxazine-3-carboxylic acid
Other names   : Pirfenoxone (unknown term used in the article); 1-hydroxy-5-oxo-5H-pyrido(3,2-
                               a)phenoxazine-3-carboxylic acid
CAS registry number   : 1043-21-6
EC number   : 213-872-4
Merck Index monograph : 8873
Patch testing   : 1% water
Molecular formula   : $C_{16}H_8N_2O_5$

GENERAL

Pirenoxine is a phenoxazine derivative used to inhibit the development of cataracts in patients considered at risk. Its mechanism of action is interacting with calcium and selenite ions, which are known to play a central role in the formation of lens cataracts (1).

CONTACT ALLERGY

Case report

A 77-year-old woman had applied eye drops containing 0.005% pirenoxine when itchy erythema appeared around both eyes. Patch tests were positive to the eye drops as is and to pirenoxine 0.005% and 1% water (D2 ?+, D3 +). After withdrawal of the drug, the eruption gradually disappeared (2). The authors mention one previous case of contact dermatitis from pirenoxine in the same brand of eye drops (3, details unknown).

LITERATURE

1 The data in the section 'General' may have been obtained from literature discussed in this chapter, but mostly also or exclusively from one or more of the following online sources: ChemIDPlus Advanced, PubChem, DrugBank, RxList, Drug Central, Drugs.com, and Wikipedia
2 Inui S, Ozawa K, Song M, Itami S, Katayama I. Contact dermatitis due to pirfenoxone. Contact Dermatitis 2004;50:375-376
3 Washizaki K, Koseki M, Kantoh H. A case of contact dermatitis due to Catalin K eye drop. Skin Res 1993;35:337-343 (Article in Japanese).

Chapter 3.266 PIROXICAM

IDENTIFICATION

Description/definition   : Piroxicam is the nonsteroidal oxicam that conforms to the structural formula shown
                             below
Pharmacological classes   : Anti-inflammatory agents, non-steroidal; cyclooxygenase inhibitors
IUPAC name   : 4-Hydroxy-2-methyl-1,1-dioxo-N-pyridin-2-yl-2H-1,2-benzothiazine-3-carboxamide
CAS registry number   : 36322-90-4
EC number   : 252-974-3
Merck Index monograph   : 8889
Patch testing   : 1.0% pet. (Chemotechnique, SmartPracticeCanada)
Molecular formula   : $C_{15}H_{13}N_3O_4S$

GENERAL

Piroxicam is an oxicam derivative with anti-inflammatory, antipyretic and analgesic properties. As a non-selective, nonsteroidal anti-inflammatory drug (NSAID), piroxicam binds and chelates both isoforms of cyclooxygenases (COX1 and COX2), thereby stalling phospholipase A2 activity and conversion of arachidonic acid into prostaglandin precursors. This results in inhibition of prostaglandin biosynthesis. As a second, independent effect, piroxicam inhibits the activation of neutrophils. Piroxicam is indicated for treatment of osteoarthritis and rheumatoid arthritis, musculoskeletal disorders, dysmenorrhea and postoperative pain. It is also used in topical formulations for treating pain and swelling due to strains, sprains, backache or arthritis (1).

CONTACT ALLERGY

General

Allergic contact dermatitis and allergic photocontact dermatitis from topical piroxicam has infrequently been reported. Far more often, oral administration of piroxicam has led to fixed drug eruptions and photoallergic dermatitis. Many patients who are sensitized to thiosalicylic acid (one of the components of the preservative thimerosal) develop a *photo*allergic eruption after taking oral piroxicam and most patients showing a photosensitivity reaction to oral piroxicam have positive patch tests to thimerosal (9,10,13,14,15,16,17,22). Piroxicam also has phototoxic properties and differentiation between photoallergic and phototoxic photopatch tests may be difficult (20).

Testing in groups of patients

In the period 2008-2011, in a multinational and multicenter European study, 1031 patients suspected of photoallergic contact dermatitis were patch and photopatch tested with piroxicam 1% pet. There was one (0.1%) positive patch test, the relevance of which was not mentioned. Five patients had photocontact allergy (21). In an Italian multicenter investigation performed in the period 2004 to 2006, 1082 patients were patch and photopatch tested with piroxicam 10% pet. and there were 3 (0.3%) positive patch tests, which were all photoaugmented. Their relevance was not mentioned. Ten patients in this study had photocontact allergy to piroxicam (20).

Case series

In the period 1996-2001, in 2 hospitals in Spain, 12 patients were diagnosed with contact allergy and 9 with photocontact allergy to topical piroxicam. The accumulated incidence per million inhabitants (catchment population of the hospitals) of both side effects together was 25.3. Seven out of 9 patients with photocontact allergy had contact allergy to thimerosal (22).

In Italy, before 1993, the members of the GIRDCA Multicentre Study Group diagnosed 102 patients (49 men, 53 women), aged 16 to 66 years (mean 37 years), with (photo)dermatitis induced by systemic or topical NSAIDs. Piroxicam caused 2 contact allergic and one photocontact allergic (from systemic administration) reactions (2).

Case reports

A 42-year-old man was prescribed a cream containing 1% piroxicam to treat a painful left foot. After 10 days' application, the patient developed severe itchy dermatitis around the left medial malleolus; within 6 days, this acute dermatitis spread to the legs with erythema, edema, vesicles and exudation. Patch tests were positive (+++) to the cream 'as is' and – later – to piroxicam 1% pet. There were no reactions to the other ingredients of the cream or to an NSAID series. Six controls were negative to piroxicam 1% pet. (4).

A 58-year-old man developed a rash on the back of his neck 2 days after applying piroxicam 1% gel to this area for pain relief. Within 5 days, the rash had extended to involve the trunk and limbs as a severe excoriated eczema. It was decided not to proceed to patch testing 'in view of the severity of the eruption following only limited application of the gel' (5).

A 30-year-old physical therapist presented with dermatitis localized to both hands lasting for 2 years and characterized by erythema and vesicles, with a long-term course that never completely cleared. In her work, the patient performed iontophoresis and sonophoresis and was exposed to piroxicam, naproxen and diclofenac gels. Patch tests were positive to piroxicam 1% and 5% pet. (negative in 5 controls). ROATs with all 3 gels was positive to piroxicam gel only (11). This was a case of occupational allergic contact dermatitis.

PHOTOSENSITIVITY

Photopatch testing in groups of patients

In 4 studies in which patients suspected of photoallergic contact dermatitis, photoaggravated facial dermatitis or systemic photosensitivity were photopatch tested with piroxicam, 3 had low prevalences of photosensitization ranging from 0.5% to 1.8% (19-21) (table 3.266.1). In the fourth, performed in Portugal in the period 2003-2007, 9 of 30 patients (30%) had positive photopatch tests to piroxicam 1% pet. (17). All reactions were attributed to systemic photosensitization and all co-reacted to thimerosal and thiosalicylic acid (which is the usual co-reactivity). It was not mentioned what the criteria for testing with NSAIDs were. The high frequency of photosensitization was thought to result from the widespread use of piroxicam in the population investigated and a high prevalence of thimerosal sensitization (17).

Table 3.266.1 Photopatch testing in groups of patients

Years and Country	Test conc. & vehicle	Number of patients tested	positive (%)	Selection of patients (S); Relevance (R); Comments (C)	Ref.
2008-2011 Europe	1% pet.	1031	5 (0.5%)	S: patients suspected or photoallergic contact dermatitis; R: not stated	21
2003-2007 Portugal	1% pet.	30	9 (30%)	S: subgroup of 83 patients with suspected photoaggravated facial dermatitis or systemic photosensitivity; R: all reactions were relevant after previous photosensitivity from oral piroxicam; co-reactivity to thimerosal and thiosalicylic acid in all 9; how the subgroup was selected was not mentioned	17
2004-2006 Italy	10% pet.	1082	10 (0.9%)	S: patients with histories and clinical features suggestive of photoallergic contact dermatitis; R: not specified; C: there were also 3 photoaugmented contact allergic reactions and 6 irritative/phototoxic reactions	20
2004-2005 Spain	1% pet.	224	4 (1.8%)	S: not stated; R: 100%	19

Case reports and case series

In a multicenter study in Italy, performed in the period 1985-1994, 3 photopatch test reactions were seen to piroxicam; the patch test concentration used was not mentioned and relevance was not specified (78% for all photoallergens together) (18). In the period 1996-2001, in 2 hospitals in Spain, 9 patients were diagnosed with photocontact allergy and 12 with contact allergy to topical piroxicam. The accumulated incidence per million inhabitants (catchment population of the hospitals) of both side effects together was 25.3. Seven out of 9 patients with photocontact allergy had contact allergy to thimerosal (22).

Cross-reactions, pseudo-cross-reactions and co-reactions

Patients who have a fixed drug eruption to oral piroxicam and a positive patch test on post-lesional skin may (or may not) show cross-reactions to tenoxicam (3,7,8,12), meloxicam (3,6), or droxicam (7). A man sensitized to thiosalicylic acid who developed photosensitivity from oral piroxicam, had positive photopatch tests to piroxicam and co-reactions to tenoxicam, droxicam and meloxicam (26).

Many patients who are sensitized to thiosalicylic acid (one of the components of thimerosal) develop a *photo*allergic eruption after taking oral piroxicam and most patients showing a photosensitivity reaction to oral piroxicam have positive patch tests to thimerosal (9,10,13,14,15,16,17,22).

Cutaneous adverse drug reactions from systemic administration caused by type IV (delayed-type) hypersensitivity

Cutaneous adverse drug reactions from systemic administration of piroxicam caused by type IV (delayed-type) hypersensitivity, including fixed drug eruption (3,6,7,8), photoallergic dermatitis (9,23,25,26) and acrovesicular dermatitis (dyshidrosiform dermatitis) (24), are planned to be discussed in Volume IV of the *Monographs in Contact Allergy* series on Systemic drugs.

LITERATURE

1   The data in the section 'General' may have been obtained from literature discussed in this chapter, but mostly also or exclusively from one or more of the following online sources: ChemIDPlus Advanced, PubChem, DrugBank, RxList, Drug Central, Drugs.com, and Wikipedia

2   Pigatto PD, Mozzanica N, Bigardi AS, Legori A, Valsecchi R, Cusano F, et al. Topical NSAID allergic contact dermatitis. Italian experience. Contact Dermatitis 1993;29:39-41

3   Andrade P, Brinca A, Gonçalo M. Patch testing in fixed drug eruptions – a 20-year review. Contact Dermatitis 2011;65:195-201

4   Valsecchi R, Pansera B, di Landro A, Cainelli T. Contact sensitivity to piroxicam. Contact Dermatitis 1993;29:167

5   Green C, Lowe JG. Contact allergy to piroxicam gel. Contact Dermatitis 1992;27:261

6   Ben Romdhane H, Ammar H, BenFadhel N, Chadli Z, Ben Fredj N, Boughattas NA. Piroxicam-induced fixed drug eruption: Cross-reactivity with meloxicam. Contact Dermatitis 2019;81:24-26

7   Ordoqui E, De Barrio M, Rodríguez VM, Herrero T, Gil PJ, Baeza ML. Cross-sensitivity among oxicams in piroxicam-caused fixed drug eruption: two case reports. Allergy 1995;50:741-744

8   Oliveira HS, Gonçalo M, Reis JP, Figueiredo A. Fixed drug eruption to piroxicam. Positive patch tests with cross-sensitivity to tenoxicam. J Dermatol Treat 1999;10:209-212

9   McKerrow KJ, Greig DE. Piroxicam-induced photosensitive dermatitis. J Am Acad Dermatol 1986;15:1237-1241

10 Vasconcelos C, Magina S, Quirino P, Barros MA, Mesquita-Guimarães J. Cutaneous drug reactions to piroxicam. Contact Dermatitis 1998;39:145

11 Arévalo A, Blancas R, Ancona A. Occupational contact dermatitis from piroxicam. Am J Contact Dermat 1995;6:113-114

12 Gastaminza G, Echechipía S, Navarro JA, Fernández de Corrés L. Fixed drug eruption from piroxicam. Contact Dermatitis 1993;28:43-44

13 Gonçalo M, Figueiredo A, Tavares P, Ribeiro CA, Teixeira F, Baptista AP. Photosensitivity to piroxicam: absence of cross-reaction with tenoxicam. Contact Dermatitis 1992;27:287-290

14 De Castro JL, Freitas JP, Brandão FM, Themido R. Sensitivity to thimerosal and photosensitivity to piroxicam. Contact Dermatitis 1991;24:187-192

15 Serrano G, Bonillo J, Aliaga A, Cuadra J, Pujol C, Pelufo C, et al. Piroxicam-induced photosensitivity and contact sensitivity to thiosalicylic acid. J Am Acad Dermatol 1990;23(3Pt.1):479-483

16 De la Cuadra J, Pujol C, Aliaga A. Clinical evidence of cross-sensitivity between thiosalicylic acid, a contact allergen, and piroxicam, a photoallergen. Contact Dermatitis 1989;21:349-351

17 Cardoso J, Canelas MM, Gonçalo M, Figueiredo A. Photopatch testing with an extended series of photoallergens: a 5-year study. Contact Dermatitis 2009;60:325-329

18 Pigatto PD, Legori A, Bigardi AS, Guarrera M, Tosti A, Santucci B, et al. Gruppo Italiano recerca dermatiti da contatto ed ambientali Italian multicenter study of allergic contact photodermatitis: epidemiological aspects. Am J Contact Dermatitis 1996;17:158-163

19 De La Cuadra-Oyanguren J, Perez-Ferriols A, Lecha-Carrelero M, et al. Results and assessment of photopatch testing in Spain: towards a new standard set of photoallergens. Actas DermoSifiliograficas 2007;98:96-101

20 Pigatto PD, Guzzi G, Schena D, Guarrera M, Foti C, Francalanci, S, Cristaudo A, et al. Photopatch tests: an Italian multicentre study from 2004 to 2006. Contact Dermatitis 2008;59:103-108

21 The European Multicentre Photopatch Test Study (EMCPPTS) Taskforce. A European multicentre photopatch test study. Br J Dermatol 2012;166:1002-1009

22 Diaz RL, Gardeazabal J, Manrique P, Ratón JA, Urrutia I, Rodríguez-Sasiain JM, Aguirre C. Greater allergenicity of topical ketoprofen in contact dermatitis confirmed by use. Contact Dermatitis 2006;54:239-243

23 Youn JI, Lee HG, Yeo UC, Lee YS. Piroxicam photosensitivity associated with vesicular hand dermatitis. Clin Exp Dermatol 1993;18:52-54

24 Piqué E, Pérez JA, Benjumeda A. Oral piroxicam-induced dyshidrosiform dermatitis. Contact Dermatitis 2004;50:382-383

25 Erdmann S, Sachs B, Merk HF. Photosensibilisierung durch Piroxicam. Z Hautkr 2001;76:180-182 (Article in German)

26 Trujillo MJ, de Barrio M, Rodríguez A, Moreno-Zazo M, Sánchez I, Pelta R, et al. Piroxicam-induced photodermatitis. Cross-reactivity among oxicams. A case report. Allergol Immunopathol (Madr) 2001;29:133-136

Chapter 3.267 PIROXICAM CINNAMATE

IDENTIFICATION

Description/definition   : Piroxicam cinnamate is the cinnamate ester of piroxicam that conforms to the structural
                             formula shown below
Pharmacological classes : Anti-inflammatory agents, non-steroidal
IUPAC name   : [2-Methyl-1,1-dioxo-3-(pyridin-2-ylcarbamoyl)-1λ^6,2-benzothiazin-4-yl] (E)-3-phenylprop-
                             2-enoate
Other names   : Cinnoxicam; 4-hydroxy-2-methyl-N-2-pyridyl-2H-1,2-benzothiazine-3-carboxamide 1,1-
                             dioxide, cinnamate (ester)
CAS registry number   : 87234-24-0
EC number   : Not available
Merck Index monograph : 8889 (Piroxicam)
Patch testing   : No data available; suggested: 1% and 2.5% pet.; perform controls
Molecular formula   : $C_{24}H_{19}N_3O_5S$

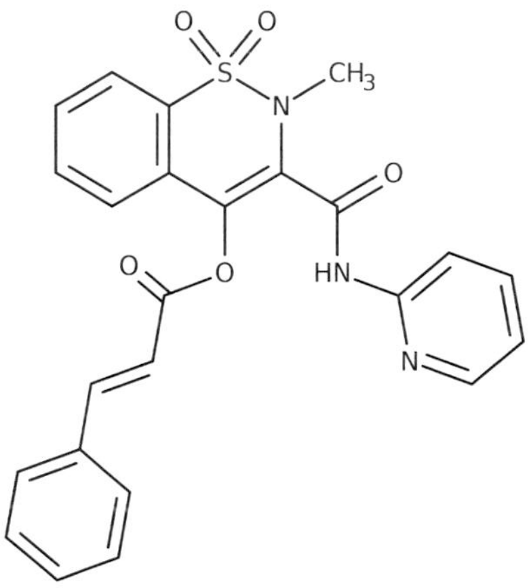

GENERAL

Piroxicam cinnamate is a nonsteroidal oxicam derivative with anti-inflammatory, antipyretic and analgesic properties and a prodrug of piroxicam. As a non-selective, nonsteroidal anti-inflammatory drug (NSAID), piroxicam binds and chelates both isoforms of cyclooxygenases (COX1 and COX2), thereby inhibiting prostaglandin biosynthesis. Piroxicam cinnamate is indicated for treatment of osteoarthritis and rheumatoid arthritis, musculoskeletal disorders, dysmenorrhea and postoperative pain (1). Little information can be found on this drug. It was used in Italy systemically for many years before 1995 and topically since a few years before 1995 (2). It is currently available in oral form in Argentina and Italy and possibly other countries (www.drugs.com).

CONTACT ALLERGY

Case reports and case series
In Italy, before 1993, the members of the GIRDCA Multicentre Study Group diagnosed 102 patients (49 men, 53 women), aged 16 to 66 years (mean 37 years), with (photo)dermatitis induced by systemic or topical NSAIDs. Cinnoxicam (piroxicam cinnamate) caused one contact allergic and zero photocontact allergic reactions (3).

A 26-year-old woman presented with very itchy erythema, edema, vesicles and exudation on the neck, left shoulder and arm after having used 1.5% cinnoxicam (piroxicam cinnamate) cream to treat a painful left shoulder joint for 10 days. Patch tests were positive to the cream, tested 'as is'. Later, patch tests were negative to its ingredients, but cinnoxicam itself was not tested as it could not be obtained. There were also no reactions to an NSAID series consisting of ibuproxam 5% pet., ketoprofen 2.5% pet., ibuprofen 5% pet., naproxen 5% pet., piroxicam 1% pet., flurbiprofen 5% pet., fenoprofen 5% pet. and tiaprofenic acid 5% pet. (2).

LITERATURE

1   The data in the section 'General' may have been obtained from literature discussed in this chapter, but mostly also or exclusively from one or more of the following online sources: ChemIDPlus Advanced, PubChem, DrugBank, RxList, Drug Central, Drugs.com, and Wikipedia
2   Valsecchi R, Pansera B, Di Landro A, Cainelli T. Contact allergy to cinnoxicam. Contact Dermatitis 1995;32:63
3   Pigatto PD, Mozzanica N, Bigardi AS, Legori A, Valsecchi R, Cusano F, et al. Topical NSAID allergic contact dermatitis: Italian experience. Contact Dermatitis 1993;29:39-41

Chapter 3.268 POLIDOCANOL

IDENTIFICATION

Description/definition	: Polidocanol is the polyethylene glycol ether of lauryl alcohol that conforms to the structural formula shown below
Pharmaceutical classes	: Sclerosing agents for local injection; topical anesthetic; dermatological agent
IUPAC name	: 2-[2-[2-[2-[2-[2-[2-[2-(2-Dodecoxyethoxy)ethoxy]ethoxy]ethoxy]ethoxy] ethoxy]ethoxy]ethanol
Other names	: Dodecan-1-ol, ethoxylated; hydroxypolyethoxydodecane; laureth-9; dodecylnonaoxy-ethylene glycol monoether; dodecyl alcohol, ethoxylated; lauromacrogol 400; polyoxyethylene 9-lauryl ether
CAS registry number	: 3055-99-0
EC number	: 221-284-4
Merck Index monograph	: 8945
Patch testing	: 3% pet. (SmartPracticeCanada, SmartPracticeEurope)
Molecular formula	: $C_{30}H_{62}O_{10}$

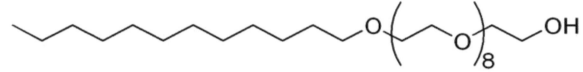

GENERAL

Polidocanol is an addition polymer of lauryl alcohol and an average of 9 units of ethylene oxide. In medicine, it is used as a sclerosing agent to treat uncomplicated spider veins (varicose veins ≤1 mm in diameter) and reticular veins (varicose veins 1 to 3 mm in diameter) of the legs, as a topical anaesthetic and antipruritic (1). Polidocanol is also widely used as an emulsifier and surfactant in cosmetics and cleaning products. In this chapter, only side effects of polidocanol from topical application are discussed, thereby excluding complications of sclerotherapy (mostly type-I reactions [6,12]).

CONTACT ALLERGY

Contact allergy in the general population

In Germany, for the period 1995-2004, the population-based relative incidence (RI) of contact sensitization to polidocanol (cases/100,000 defined daily doses (DDDs) per year) was estimated to be 3.1. In the group of local anesthetics, the RI ranged from 1.5 (lidocaine) to 413.9 (benzocaine) (11).

Patch testing in groups of patients

Results of studies testing polidocanol in consecutive patients suspected of contact dermatitis (routine testing) and in groups of selected patients are shown in table 3.268.1. In an older German study, routine testing with polidocanol yielded 1.5% positive reactions. However, the test concentration is unknown and the relevance of the positive patch tests was not stated (9). In three studies in selected patients, all performed by the IVDK (Information Network of Departments of Dermatology: Germany, Austria, Switzerland), rates of sensitization ranged from 1.2% to 2.1% (3,10,13). In patients tested with a topical drugs series, over half of the reactions were of current or past relevance. Many positive individuals had lower leg dermatitis and co-reactions to other topical drugs (3). In the 2 other studies, details were not available (10,13).

Case reports and case series

Sixteen patients with positive patch test reactions to polidocanol were observed in Germany. The majority suffered from chronic dermatitis and had further contact sensitizations to antibiotics and to vehicle constituents. The clinical relevance of positive reactions to polidocanol remained unclear in many cases. A positive reaction was reproduced in only two of six patients at a follow-up examination (2).

A 52-year-old woman developed ACD from a polidocanol containing ointment for treatment of a leg ulcer. She had previously been treated with the chemical for varicose veins; after injection, she had felt pain and nearly collapsed (probably unrelated to contact allergy). Patch tests were positive to polidocanol 3% in water and negative to the other ingredients (7).

A woman aged 33 had anogenital pruritus 5 years before consultation and it resolved after 18 months. Six weeks ago the trouble recurred. She had used a wide variety of local applications. Examination showed an extensive lichenified and excoriated dermatitis of the perianal and vulval skin. *Candida albicans* was grown from both sites. Patch tests were positive to a rectal ointment containing polidocanol and – later – to its ingredient polidocanol 5% pet. Fifteen controls were negative. The patient was also allergic to neomycin and a cream containing this antibiotic and to clioquinol and clioquinol-hydrocortisone cream (8).

Table 3.268.1 Patch testing in groups of patients

Years and Country	Test conc. & vehicle	Number of patients tested	positive (%)	Selection of patients (S); Relevance (R); Comments (C)	Ref.
Routine testing					
<1970 Germany	?	2551	38 (1.5%)	R: not stated (data cited in ref. 8)	9
Testing in groups of selected patients					
2004-2008 IVDK			(1.2%)	S: patients with anogenital dermatoses tested with a medicament series; R: not stated; C: number of patients tested unknown	13
1995-2004 IVDK	1% pet.	8191	(2.0%)	S: not stated; R: not stated	10
1992-1999 IVDK	0.5% water	3186	(1.2%)	S: patients tested with a topical drug patch test series;	3
	3% pet.	6202	(2.1%)	R: 53% had current or past relevance; C: high frequency in elderly people with lower leg dermatitis and other topical drugs as co-sensitizers	

IVDK: Information Network of Departments of Dermatology (Germany, Austria, Switzerland)

Allergic contact dermatitis from *cosmetic* products has been caused by polidocanol present in a shampoo (1), a cosmetic itch relief cream (4) and a moisturizing lotion (5). In the latter case, a repeated open application test (ROAT) with the lotion was positive at D2 already. When tested with its ingredients, the patient had a ?+ reaction to polidocanol 3% in water. Patch testing with a dilution series of polidocanol at 0.3%, 3%, and 10% resulted in positive reactions to all three concentrations, which were negative in 20 controls (5).

Cross-reactions, pseudo-cross-reactions and co-reactions
Co-reaction to laureth-2, which was present in the same product (1).

LITERATURE

1	Grills CE, Cooper SM. Polidocanol: a potential contact allergen in shampoo. Contact Dermatitis 2007;56:178
2	Frosch PJ, Schulze-Dirks A. Contact allergy caused by polidocanol. Hautarzt 1989;40:146-149
3	Uter W, Geier J, Fuchs T. Contact allergy to polidocanol, 1992 to 1999. J Allergy Clin Immunol 2000;106:1203-1204
4	Fairhurst D,Wilkinson M. Independent sensitization to polidocanol and trometamol or glycerol within same product. Contact Dermatitis 2007;56:179
5	Gallo R, Basso M, Voltolini S, Guarrera M. Allergic contact dermatitis from laureth-9 and polyquaternium-7 in a skin-care product. Contact Dermatitis 2001;45:356-357
6	Henriquez-Santana A, Fernandez-Guarino M, González de Olano D, Gonzalez-Cervera J, Huertas-Barbudo B, Aldanondo I. Urticaria induced by Etoxisclerol (polidocanol). J Eur Acad Dermatol Venereol 2008;22:261-262
7	Huber-Riffeser G. Allergic contact dermatitis to polidocanol (Thesit). Contact Dermatitis 1978;4:245
8	Calnan CD. Oxypolyethoxydodecane in an ointment. Contact Dermatitis 1978;4:168-169
9	Hartung J, Rudolph PO. Z Haut- und Geschlechtskrankheiten 1970;45:457 (data cited in ref. 8) (Article in German)
10	Menezes de Padua CA, Uter W, Schnuch A. Contact allergy to topical drugs: prevalence in a clinical setting and estimation of frequency at the population level. Pharmacoepidemiol Drug Saf 2007;16:377-384
11	Menezes de Padua CA, Schnuch A, Nink K, Pfahlberg A, Uter W. Allergic contact dermatitis to topical drugs – Epidemiological risk assessment. Pharmacoepidemiol Drug Saf 2008;17:813-821
12	Stricker BH, van Oijen JA, Kroon C, Ovink AH. Anaphylaxis following use of polidocanol. Ned Tijdschr Geneeskd 1990;134:240-242 (article in Dutch)
13	Bauer A. Contact sensitization in the anal and genital area. Curr Probl Dermatol 2011;40:133-141

Chapter 3.269 POLYMYXIN B

IDENTIFICATION

Description/definition : Polymyxin B is a mixture of the polypeptides, polymyxins B1 and B2, which are obtained from *Bacillus polymyxa* strains; their structural formulas are shown below

Pharmacological classes : Anti-bacterial agents

IUPAC name : *N*-[4-Amino-1-[[1-[[4-amino-1-oxo-1-[[6,9,18-tris(2-aminoethyl)-15-benzyl-3-(1-hydroxyethyl)-12-(2-methylpropyl)-2,5,8,11,14,17,20-heptaoxo-1,4,7,10,13,16,19-heptazacyclotricos-21-yl]amino]butan-2-yl]amino]-3-hydroxy-1-oxobutan-2-yl]amino]-1-oxobutan-2-yl]-6-methyloctanamide;sulfuric acid

CAS registry number : 1404-26-8

EC number : 215-768-4

Merck Index monograph : 8963

Patch testing : Sulfate 5.0% pet. (Chemotechnique); sulfate 3% pet. (SmartPracticeCanada, SmartPracticeEurope)

Molecular formula : $C_{56}H_{98}N_{16}O_{13}$ (empirical)

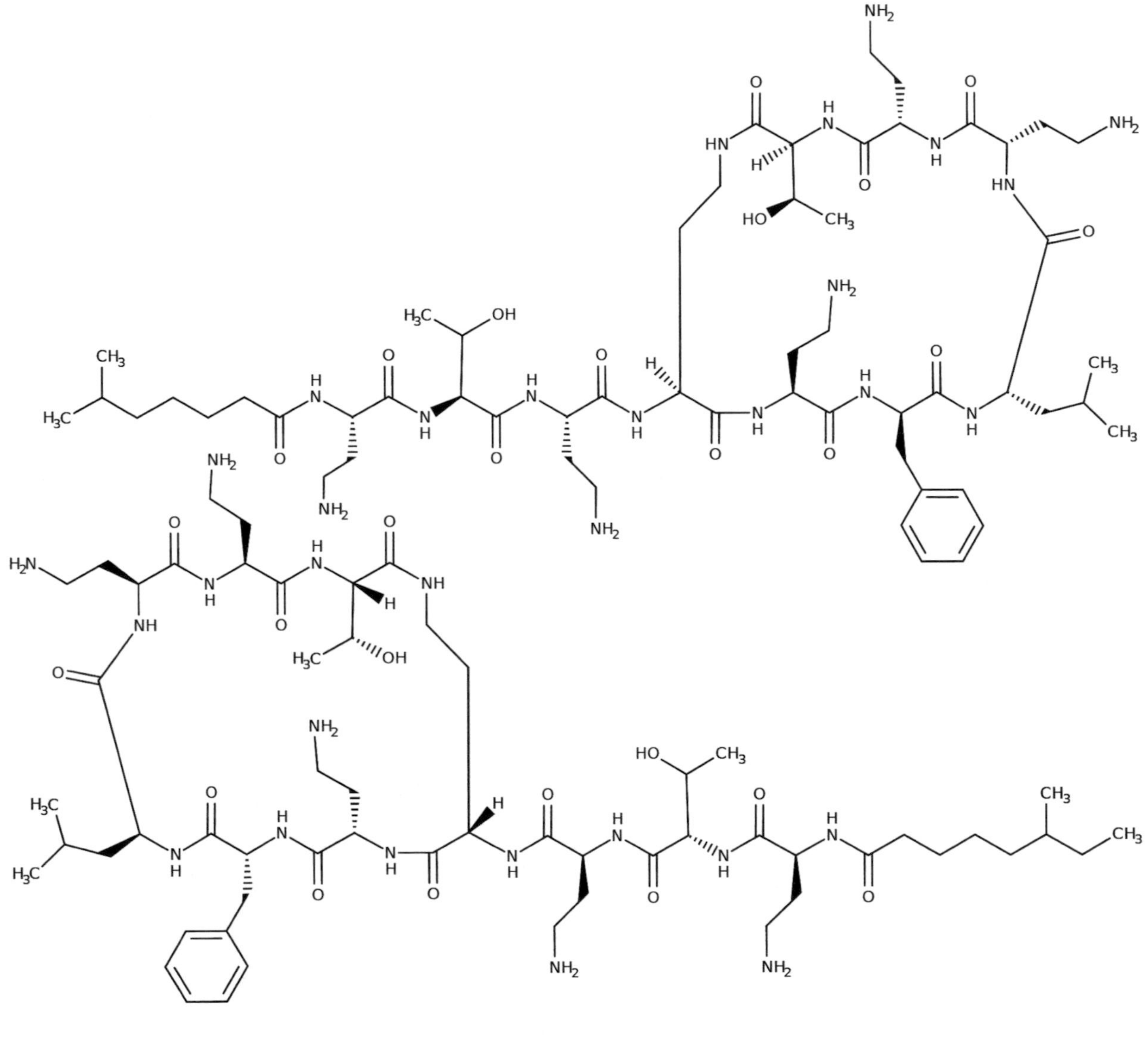

GENERAL

Polymyxin B is a mixture of polymyxins B1 and B2, obtained from *Bacillus polymyxa* strains, with antibacterial activity. They are basic polypeptides of about eight amino acids and have cationic detergent action on cell membranes. Polymyxin B is used for infections with gram-negative organisms, notably for treatment of infections of the urinary tract, meninges, and blood stream, caused by susceptible strains of *Pseudomonas aeruginosa* (1). It is often used in combination (compounded) with other antibiotics such as bacitracin or neomycin, especially in ophthalmic and ear preparations.

In pharmaceutical products, polymyxin B is employed as polymyxin B sulfate (CAS number 1405-20-5, EC number 215-774-7, molecular formula $C_{56}H_{100}N_{16}O_{17}S$ [empirical]).

CONTACT ALLERGY

General

Contact allergy to and allergic contact dermatitis from polymyxin B is not infrequent. A plausible explanation is that this antibiotic in topical pharmaceuticals is often combined with bacitracin and neomycin, which are both well-known sensitizers. Their combined presence probably facilitates sensitization, which also explains the frequent co-reactivity to bacitracin and neomycin in patients who have positive patch tests to polymyxin B. Polymyxin B sulfate 3% pet. was added to the American Contact Dermatitis Society core allergen series in 2017 (25).

Contact allergy in the general population

With the CE-DUR approach, the incidence of sensitization to polymyxin B sulfate in the German population was estimated to range from 1 to 3 cases/100,000/year in the period 1995-1999 and from 1 to 3 cases/100,000/year in the period 2000-2004 (10). Also in Germany, for the period 1995-2004, the population-based relative incidence (RI) of contact sensitization to polymyxin B (cases/100,000 defined daily doses (DDDs) per year) was estimated to be 8.2. In the group of antibiotics, the RI ranged from 1.6 (oxytetracycline) to 86.2 (framycetin). For polymyxin B used in ophthalmic drugs, the RI was 1.4 (15).

Patch testing in consecutive patients suspected of contact dermatitis: routine testing

In Canada, between March 2014 and November 2015, 795 adult consecutive patients referred to the contact dermatitis clinic were patch tested with polymyxin B sulfate 3% pet. and there were 18 (2.3%) positive reactions. The group consisted of 12 women and 6 men with an average age of 51 years. The eruptions, as reported by the referred patients, were localized to almost all parts of the body with the face being the most commonly affected site. An isolated reaction to polymyxin B was seen in 9 (50%) patients whereas concomitant reactions to bacitracin and polymyxin B were seen in the other 9 (50%) patients. One patient had reactions to bacitracin, polymyxin B, and neomycin. Most (12/18) of the reactions represented previous exposure to polymyxin B (past relevance) (19,20).

Patch testing in groups of selected patients

The results of studies in which groups of selected patients were patch tested with polymyxin B are shown in table 3.269.1. In 11 such studies, prevalences of sensitization have ranged from 0.7% to 17%. In patients with otitis externa or media, the rates were 17% (5), 5% (24) and 4.2% (6). Higher frequencies were also observed in patients with leg ulcers/stasis dermatitis: 10.1% (23), 7% (3) and 4% (2). Relevance figures – where mentioned – were generally high.

Table 3.269.1 Patch testing in groups of patients: Selected patient groups

Years and Country	Test conc. & vehicle	Number of patients tested	positive (%)	Selection of patients (S); Relevance (R); Comments (C)	Ref.
2013-2015 Ireland	3% pet.	99	1 (1%)	S: patients patch tested for perianal and/or genital symptoms; R: all reactions to medicaments were relevant	4
1990-2014 Belgium	3% pet.	537	17 (3.2%)	S: patients suspected of iatrogenic contact dermatitis and tested with a pharmaceutical series and their own products; R: 96% of the positive patch test reactions to all topical drugs and antiseptics were considered to be relevant	11
2001-2010 IVDK	3% pet.	1142	19 (1.7%)	S: patients with periorbital dermatitis tested with an ophthalmic tray, R: not stated	12
2006-2007 Canada		100	7 (7%)	S: patients with leg ulcers or venous disease; R: not stated	3
2000-6 Netherlands		90	4 (4%)	S: patients patch tested for suspected contact allergy to ophthalmological drugs; R: 69% for all reactions to drugs together	9
1995-2004 IVDK	10% pet.	3876	(0.7%)	S: patients patch tested for suspected contact allergy to ophthalmological drugs; R: not stated; C: the estimated number of sensitizations per 100,000 prescriptions was 1.4	8
2000-2001 IVDK	3% pet.	7600	(0.9%)	S: not stated; R: not stated	10
1993-4 Netherlands	5% pet.	34	5 (17%)	S: patients with chronic otitis externa or media; R: not stated	5

Table 3.269.1 Patch testing in groups of patients: Selected patient groups (continued)

Years and Country	Test conc. & vehicle	Number of patients tested | positive (%)		Selection of patients (S); Relevance (R); Comments (C)	Ref.
1984-1987 Germany	5% pet.	317	32 (10.1%)	S: patients with leg ulcers; in most cases, the patients had used polymyxin B ointment	23
<1985 Finland		142	6 (4.2%)	S: patients with chronic otitis externa; C: details unknown	6
<1982 U.K.	10^5 IU/gr	40	2 (5%)	S: patients with otitis externa, chronic suppurative otitis media or discharging mastoid cavities >1 year; R: not stated	24
1976-1978 Finland	50% pet.	74	3 (4%)	S: patients with leg ulcers or stasis dermatitis; R: not stated	2

IVDK: Information Network of Departments of Dermatology (Germany, Austria, Switzerland)

Case series

In Leuven, Belgium, between 1990 and 2017, 16,065 patients were investigated for contact allergy and 118 (0.7%) showed positive patch test reactions to topical ophthalmic medications and/or to their ingredients. Eighty-four individuals (71%) reacted to an active principle. Polymyxin B sulfate was tested in 561 patients and was the allergen in eye medications in 12. There were also 5 reactions to polymyxin B sulfate in other types of medications (7).

Ten patients in Sweden, 8 women and 2 men, all suffering from hypostatic ulcer and/or stasis dermatitis, were sensitized to a combination ointment preparation containing polymyxin B sulfate 10^6 IU, oxytetracycline chloride 3 g, liquid paraffin and white petrolatum to 100 g. All reacted to the ointment and to polymyxin B 3% pet. and nine reacted to oxytetracycline 3% pet. Adequate controls were negative (13). Three of these patients had been described before. These 3 had been tested with polymyxin B 'as is' (1/31 controls positive) (14).

Case reports

A 13-year-old boy presented with a pruritic, erythematous, and weepy rash along his right arm and trunk occurring 3 days after using a polymyxin B-containing cream on an abrasion caused by a fall. He had used this product before without any problem. Positive patch test reactions were seen to the cream (+) and the polymyxin B sulfate (++) on D7 of testing (18).

A 39-year-old woman had developed an acute, erythematous, vesicular dermatitis on the right arm, perioral region, and right side of the forehead, while being treated with an ointment containing polymyxin B, neomycin and bacitracin for impetigo. Patch tests were positive to polymyxin B sulfate 5% pet., neomycin 20% pet. and bacitracin 1% pet. (21). A 53-year-old woman had a subacute, scaly, edematous dermatitis of the upper lip and nares of several months' duration. It began after she had been given an antibiotic ointment for skin irritation that resulted from nasal oxygen administration. Patch tests were positive to neomycin, bacitracin and polymyxin B sulfate. It was not mentioned whether the antibiotic ointment contained all 3 antibiotics (21).

A female patient aged 28 years was treated for tinea pedis with an ointment containing polymyxin B sulfate (10.000 U/gr) and bacitracin (500 U/gr), which resulted in contact dermatitis. Patch tests were positive to the ointment, to polymyxin B sulfate and to bacitracin, both tested at 1%, 5% and 30% pet. The patient also reacted to colistin (polymyxin E), which was considered to be a cross-reaction to polymyxin B (16).

A woman aged 20 had periocular dermatitis. When patch tested, she reacted to chloramphenicol, to an ointment containing polymyxin B and to polymyxin B sulfate 30% pet. (17). A 65-year-old man was hospitalized for otitis externa. Multiple treatment attempts with topical agents (polymyxin B sulfate-neomycin sulfate-hydrocortisone) and oral agents (cephalexin and ciprofloxacin) were unsuccessful. On examination, a severe red, itchy, weepy reaction on both ears was noted. Positive patch test reactions were found to polymyxin, bacitracin, and dexamethasone (22).

Cross-reactions, pseudo-cross-reactions and co-reactions

A patient sensitized to polymyxin B sulfate also had a positive patch test to colistin (polymyxin E), which was considered to be a cross-reaction (16). A 56-year-old woman was patch tested for periauricular dermatitis. She reacted to chloramphenicol and polymyxin B sulfate 30% pet. It was uncertain whether she had previously used polymyxin B, but she had a positive patch test to a pharmaceutical containing colistin, which is polymyxin E. Possibly, the positive polymyxin B patch test was a cross-reaction to polymyxin E (17).

Patients with allergy to polymyxin B often co-react to bacitracin and to a lesser degree with neomycin. These are not cross-reactions, but result from concomitant sensitivity, as these 3 antibiotics are often present together in topical pharmaceuticals, their combined presence probably facilitating sensitization.

LITERATURE

1   The data in the section 'General' may have been obtained from literature discussed in this chapter, but mostly also or exclusively from one or more of the following online sources: ChemIDPlus Advanced, PubChem, DrugBank, RxList, Drug Central, Drugs.com, and Wikipedia

2   Fräki JE, Peltonen L, Hopsu-Havu VK. Allergy to various components of topical preparations in stasis dermatitis and leg ulcer. Contact Dermatitis 1979;5:97-100

3   Smart V, Alavi A, Coutts P, Fierheller M, Coelho S, Holness LD, et al. Contact allergens in persons with leg ulcers: a Canadian study in contact sensitization. Int J Low Extrem Wounds 2008;7:120-125

4   Foley CC, White S, Merry S, Nolan U, Moriarty B, et al. Understanding the role of cutaneous allergy testing in anogenital dermatoses: a retrospective evaluation of contact sensitization in anogenital dermatoses. Int J Dermatol 2019;58:806-810

5   Van Ginkel CJ, Bruintjes TD, Huizing EH. Allergy due to topical medications in chronic otitis externa and chronic otitis media. Clin Otolaryngol Allied Sci 1995;20:326-328

6   Fräki JE, Kalimo K, Tuohimaa P, Aantaa E. Contact allergy to various components of topical preparations for treatment of external otitis. Acta Otolaryngol 1985;100:414-418

7   Gilissen L, De Decker L, Hulshagen T, Goossens A. Allergic contact dermatitis caused by topical ophthalmic medications: Keep an eye on it! Contact Dermatitis 2019;80:291-297

8   Uter W, Menezes de Pádua C, Pfahlberg A, Nink K, Schnuch A, Behrens-Baumann W. Contact allergy to topical ophthalmological drugs - epidemiological risk assessment. Klin Monbl Augenheilkd 2009;226:48-53 (Article in German)

9   Wijnmaalen AL, van Zuuren EJ, de Keizer RJ, Jager MJ. Cutaneous allergy testing in patients suspected of an allergic reaction to eye medication. Ophthalmic Res 2009;41:225-229

10 Menezes de Padua CA, Uter W, Schnuch A. Contact allergy to topical drugs: prevalence in a clinical setting and estimation of frequency at the population level. Pharmacoepidemiol Drug Saf 2007;16:377-384

11 Gilissen L, Goossens A. Frequency and trends of contact allergy to and iatrogenic contact dermatitis caused by topical drugs over a 25-year period. Contact Dermatitis 2016;75:290-302

12 Landeck L, John SM, Geier J. Topical ophthalmic agents as allergens in periorbital dermatitis. Br J Ophthalmol 2014;98:259-262

13 Möller H. Eczematous contact allergy to oxytetracycline and polymyxin B. Contact Dermatitis 1976;2:289-290

14 Bojs G, Möller H. Eczematous contact allergy to oxytetracycline with cross-sensitivity to other tetracyclines. Berufsdermatosen 1974;22:202 (Article in German)

15 Menezes de Padua CA, Schnuch A, Nink K, Pfahlberg A, Uter W. Allergic contact dermatitis to topical drugs – Epidemiological risk assessment. Pharmacoepidemiol Drug Saf 2008;17:813-821

16 Van Ketel WG. Polymixine B-sulfate and bacitracin. Contact Dermatitis Newsletter 1974;15:445

17 Van Joost T, Dikland W, Stolz E, Prens E. Sensitization to chloramphenicol; a persistent problem. Contact Dermatitis 1986;14:176-178

18 Jiaravuthisan MM, DeKoven JG. Contact dermatitis to polymyxin B. Contact Dermatitis 2008;59:314-316

19 Alfalah M, Zargham H, Moreau L, Stanciu M, Sasseville D. Contact allergy to polymyxin B among patients referred for patch testing. Dermatitis 2016;27:119-122

20 Alfalah M, Zargham H, Moreau L, Stanciu L, Sasseville D. Prevalence of allergy to polymyxin B among patients referred for patch testing. Dermatitis 2016;27(6):e1

21 Grandinetti PJ, Fowler JF Jr. Simultaneous contact allergy to neomycin, bacitracin, and polymyxin. J Am Acad Dermatol 1990;23:646-647

22 Vilaça S, Lobo I, Selores M. Simultaneous contact allergy to neomycin, bacitracin, polymyxin, and dexamethasone: an often unrecognized cause of allergic contact dermatitis in patients using topical antibiotics. J Am Acad Dermatol 2012;66(4 Suppl.1):AB74 (Poster presentation)

23 Kleinhans D. Bacitracin and polymyxin B: Important contact allergens in patients with leg ulcers. In: Frosch PJ, Dooms-Goossens A, Lachapelle JM, Rycroft RJG, Scheper RJ (eds). Current Topics in Contact Dermatitis. Berlin: Springer-Verlag, 1989: 258-260

24 Holmes RC, Johns AN, Wilkinson JD, Black MM, Rycroft RJ. Medicament contact dermatitis in patients with chronic inflammatory ear disease. J R Soc Med 1982;75:27-30

25 Schalock PC, Dunnick CA, Nedorost S, Brod B, Warshaw E, Mowad C. American Contact Dermatitis Society core allergen series: 2017 update. Dermatitis 2017;28:141-143

Chapter 3.270 POVIDONE-IODINE

IDENTIFICATION

Description/definition	: Povidone-iodine is a stable chemical complex of polyvinylpyrrolidone (povidone, PVP) and elemental iodine; it contains from 9.0% to 12.0% available iodine and conforms to the structural formula shown below
Pharmacological classes	: Anti-infective agents, local
IUPAC name	: 1-Ethenylpyrrolidin-2-one;molecular iodine
Other names	: Polyvinylpyrrolidone iodine; PVP-iodine; 2-pyrrolidinone, 1-ethenyl-, homopolymer, compd. with iodine; Betadine; PVP-I
CAS registry number	: 25655-41-8
EC number	: 607-771-8
Merck Index monograph	: 9085
Patch testing	: 10% water (SmartPracticeCanada); 10% pet. (SmartPracticeEurope); considerable risk of false-positive, irritant reactions; alternative: 2% and 5% water (12)
Molecular formula	: $C_6H_9I_2NO$

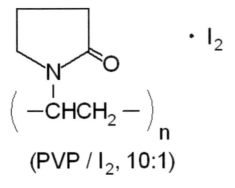

(PVP / I_2, 10:1)

GENERAL

Povidone-iodine (PVP-I) is a stable chemical complex of polyvinylpyrrolidone (povidone, PVP) and elemental iodine. Povidone-iodine directly causes *in vivo* protein denaturation and precipitation of bacteria resulting in the death of pathogenic microorganisms. It can kill viruses, bacteria, spores, fungi, and protozoa. Povidone-iodine aqueous solution has strong pharmacological activity against *Staphylococcus aureus*, *Neisseria gonorrhoeae*, *Pseudomonas aeruginosa*, *Treponema pallidum*, hepatitis B virus, HIV, and *Trichomonas vaginalis*. PVP-I is an effective and safe disinfectant agent. It has many applications in medicine: as a surgical scrub, in surgical drapes, in preoperative and postoperative skin cleansing, and for the treatment and prevention of infections in wounds and ulcers. For these purposes, povidone-iodine has been formulated at concentrations of 7.5-10.0% in various pharmaceutical forms including solution, ointment, dressing and scrub. Povidone-iodine is the most commonly used topical anti-infective agent worldwide because of its potent germicidal activity with relatively low irritancy and toxicity. A 10% PVP-I solution contains 1% releasable iodine, but only 0.001% free iodine, which is why PVP-I is less toxic and irritant than iodine, without being less antiseptic (1,35,42).

CONTACT ALLERGY

General

Diagnosing contact allergy to povidone-iodine is problematic and a reliable test concentration and vehicle have yet to be established (12,13,14,15,28,32). For patch testing, povidone-iodine 10% in water is generally used and is commercially available (12). The results of patch tests with lower concentrations, dilution series and ROATs suggest that a considerable proportion of 'positive' patch test reactions to PVP-I 10% water are in fact false-positive, irritant (12,13,14). PVP-I 5% and 10% and commercial solutions tested 'as is' in one study caused up to 50% ?+ or + patch test reactions in unexposed controls, clearly indicating irritancy (13). In another investigation, 29 of 80 (36%) patch tests with PVP-I solutions were scored as irritant (37).

In a recent study, only the strong reactions (++, +++) had probable or proven clinical relevance in terms of allergic contact dermatitis caused by povidone-iodine solution. Also, strong reactions had a remarkable concordance with positive patch test reactions to iodine 0.5% pet.: of the 5 strong reactors to povidone-iodine, 4 also had positive reactions to iodine (12). A concentration of 1% is too low to detect contact sensitization; using PVP-I 2% (13,15) or 2% and 5% water (12) has been suggested, but false-negative reactions cannot be excluded and 5% water may also result in irritant reactions (22). Other suggestions for patch testing have included dried 10% PVP-I solution (27,32), 10% PVP-I gel (32), and PVP-I 10% pet. (22).

The exact allergenic part of povidone-iodine is unknown. In some studies, positive patch tests to iodine have been observed (which may also induce irritant responses, depending on the patch test concentration and vehicle) (12,13,35,42,43,46), but not in others (13,40,44,45). Povidone (polyvinylpyrrolidone) does not appear to be the allergenic moiety (13,35). Most often, there is only a positive reaction to PVP-I (as its components have infrequently been tested). Some commercial PVP-I preparations contain nonoxynol-9 (polyoxyethylene (9) nonyl phenyl ether), which may also cause allergic reactions (13) (see Chapter 3.244 Nonoxynol-9).

In the light of the potential irritant properties of both the commercial povidone-iodine preparations and the commonly used 10% aquatic solution of PVP-I for patch testing, it may safely be assumed that a (considerable ?) number of cases diagnosed as contact allergy to/allergic contact dermatitis from PVP-I in published case reports and case series have in fact been irritant in the clinical and patch test setting.

Related to the extremely widespread use of povidone-iodine, contact allergy to PVP-I is very rare. Several cases of occupational sensitization have been observed (10,30,36,39,40,45) and 2 patients with erythema multiforme-like allergic contact dermatitis (38,41).

Patch testing in groups of patients

The results of routine testing with povidone-iodine and patch testing in groups of selected patients (patients with stasis dermatitis/leg ulcers, individuals suspected of iatrogenic contact dermatitis, patients suspected of occupational contact dermatitis) are shown in table 3.270.1. In Belgium, 14 of 500 consecutive patients (2.8%) reacted to PVP-I 1%. All performed a ROAT with PVP-I 10% water and only 2 had a positive patch reaction; these 2 were the only ones diagnosed as allergic to PVP-I (14).

In studies in which groups of selected patients were patch tested, the prevalences of positive patch tests ranged from 1% to 15.8% and were highest in individuals with leg ulcers/stasis dermatitis (2,3,4) and in nurses with occupational contact dermatitis (48). Relevance data were not provided.

Table 3.270.1 Patch testing in groups of patients

Years and Country	Test conc. & vehicle	Number of patients tested	positive (%)	Selection of patients (S); Relevance (R); Comments (C)	Ref.
Routine testing					
<2005 Belgium	1% water	500	14 (2.8%)	R: 2 were relevant; C: all performed a ROAT with povidone-iodine 10% solution and only 2 were positive and accepted as allergic	14
Testing in groups of selected patients					
2004-2016 Spain		124	2 (1.6%)	S: patients with perianal dermatitis lasting >4 weeks; R: 50%	6
2003-2014 IVDK	10% water	1944	(10.3%)	S: patients with stasis dermatitis/chronic leg ulcers; R: not stated	3
1990-2014 Belgium	2% water	775	60 (7.7%)	S: patients suspected of iatrogenic contact dermatitis and tested with a pharmaceutical series and their own products; R: 96% of the positive patch test reactions to all topical drugs and antiseptics were considered to be relevant	8
2003-2012 IVDK		1723	150 (8.7%)	S: nurses with occupational contact dermatitis; R: not stated	48
2006-2008 Germany		95	(15.8%)	S: patients with leg ulcers; R: not stated; C: it was mentioned that irritant reactions to this material may occur	4
2005-2008 France	10% water	423	54 (12.8%)	S: patients with leg ulcers; R: not stated	2
1978-2005 Taiwan		603	15 (2.5%)	S: patients suspected of contact allergy to medicaments; R: 65% of the reactions to all medicaments were considered to be relevant	9
<1999 Croatia	10%	100	2 (2%)	S: patients with leg ulcers; R: not stated	5
1974-1988 Finland	PVP-I 7.5% sol. 'as is'	200	2 (1%)	S: patients suspected to have occupational skin disease; R: both had occupational allergic contact dermatitis from PVP-I; twenty controls were negative	36

IVDK: Information Network of Departments of Dermatology (Germany, Austria, Switzerland)

Case series

In Leuven, Belgium, between 1990 and 2017, 16,065 patients were investigated for contact allergy. PVP-I was tested in 120 patients and there were 72 positive reactions to it (7). Also in Belgium, 7 patients were sensitized to iodine (tested 0.25% alc.) and/or povidone-iodine (tested 5% water or with the commercial Betadine ® solution). The reactions were considered to be responsible for acute dermatitis during surgical interventions (n=2), for perilesional dermatitis resulting from the treatment of a wound (n=4), and for occupational dermatitis in one of the patients (a midwife with hand dermatitis sensitized to PVP-I in the hospital) (10).

Seven patients with postsurgical allergic contact dermatitis from PVP-I were reported from Spain. The diagnosis was based on the clinical manifestations, a history of exposure, the site of the lesions, and the results of the skin tests. All had positive patch tests to PVP-I 10% and 5% water, one also reacted to 1% water. All 7 also had positive patch tests to PVP-I 10% pet., 4 reacted to 5% pet. and zero to 1% pet. Control testing showed PVP-I 5% and 10% water to be irritant, but 30 controls were negative to PVP-I 10%, 5% and 1% pet. ROATs with commercial PVI-I solution 10% were negative in all patients (only executed for 1 week) (22). Based on these data, the pictures shown (sharp edges, clinical picture, localization) and the fact that occlusive dressing had been used in most patients, it is likely that a number of these patients had suffered irritant rather than allergic contact dermatitis.

In Ghent, Belgium, from March 2012 to January 2014, 16 patients (7 women, nine men, mean age 70 years, range 49-87 years) who underwent intravitreal injections with ranibizumab, underwent patch testing. They all complained of a burning and stinging sensation upon instillation of povidone-iodine ophthalmic solution, and redness and swelling of both eyelids (without extension to the face), starting within 24 hours after the injection. The mean number of intravitreal injections before a reaction occurred was 12 (range 1-35). The skin reaction cleared spontaneously after 4-5 days, without scaling. Ranibizumab itself was not tested because of its very high cost. Five patients (31%) had positive patch tests to 5% PVP-I ophthalmic solution. Thirty-two controls were negative, which seems to exclude irritancy. However, none of the 5 patients had positive reactions to another 10% PVP-I containing product used on the skin (11).

In a two-year-period (2001-2003) the members of a French Dermato-Allergology Network known as Revidal collected 14 cases of contact allergy to PVP-I (commercial solution tested 10% water, data for PVP-I not stated), of which one was occupational allergic contact dermatitis in a cattle farmer. Clinical details were not provided (30). In Japan, 10 patients had onset of contact dermatitis during the application of povidone-iodine preparations and had a positive patch test to them. By serial dilution testing with PVP-I (1%, 2%, 5% and 10% water), five were accepted as being allergic to it and the other five as irritant. Only one of the 5 allergic individuals reacted to iodine, none to povidone (polyvinylpyrrolidone) (13). Of 80 patients undergoing hemodialysis in Spain and patch tested with commercial 10% PVP-I solution, 3 had allergic reactions to it, but not clinical dermatitis. At the same time, there were 29 irritant patch test reactions to the solution (37).

In The Netherlands, in a 5-year-period before 1990, 8 patients were investigated who had developed contact dermatitis from povidone-iodine-containing preparations (solution, ointment, scrub). Three patients had been treated with gauzes soaked in PVP-I solution, 2 were ophthalmic surgeons with (occupational) hand dermatitis, two had stomas treated with the solution and one was a laboratory worker who 'probably' washed his hands with PVP-I solution. Patch tests were positive to one or more of the following: 10% PVP-I solution, 10% ointment, 7.5% scrub 2% water, or PVP-I 10% pet. There were no positive reactions to potassium iodide (5%,10%,15%,20% pet.) or iodine tincture (open test) (40). It is very likely that some patients had irritant instead of allergic contact dermatitis and that patch tests had been false-positive. Of four patients with positive reactions to the 10% PVP-I solution, for example, only 1 also reacted to the ointment, whereas both contain 10% PVP-I!

Case reports

A 53-year old woman presented with an acute vesicular dermatitis on her left hand, palm and dorsal surface, and interdigital spaces. Four days previously, she had had a carpal tunnel release after which PVP-I 10% solution had been applied under a cotton bandage. Patch tests were positive to the solution diluted to 1% water and negative to its excipients. A ROAT with the solution was positive (26).

A 68-year-old man developed an erythematous, blistering rash in a linear distribution over the hip, 4 hours after having had a hip replacement. The site of surgery had been covered by an antimicrobial incise drape impregnated with a iodophor. Patch tests were positive to PVP-I 10% water (+++) and to the drape (++) (24). It was not stated (although indirectly suggested) that the iodophor in the drape was povidone-iodine.

A 58-year-old man was treated with a 10% PVP-I solution for secondary infection of an ulcer on the right lower leg. After about 1 month, the size of the ulcer increased, and a pruritic eruption developed around it, but the therapy was continued, resulting in spreading of the dermatitis. Physical examination revealed scattered erythematous papules and erythema with scales on the neck, trunk and both arms, and erythema with serous papules around the ulcer on the right lower leg. Patch tests were positive to dried 10% PVP-I solution and PVP-I 2% water (27).

Short summaries of other case reports of allergic contact dermatitis from povidone-iodine are shown in table 3.270.2. Additional case reports, adequate data of which are not available to the author, can be found in refs. 31 and 43. Some cases of alleged allergy to PVP-I have been published, in which confirmatory patch tests were not performed (23,25,29). In one of these reports (23), irritant dermatitis was far more likely.

Occupational allergic contact dermatitis
In the period 2003-2012, the IVDK tested 1723 nurses with occupational dermatitis with povidone-iodine 10% water and there were 150 (8.7%) positive reactions. Their relevance was not mentioned, but as these were all nurses with occupational allergic contact dermatitis, as many will have had regular contact with this anti-infective material, and with a very high score of 8.7% positive reactions, it may, despite the lack of a control group and the fact that povidone-iodine 10% water may cause (many) irritant reactions, be concluded that a certain proportion of this group suffered from occupational allergic contact dermatitis to povidone-iodine (48). Several other cases of occupational sensitization have been observed (10,30,36,39,40,45).

Table 3.270.2 Short summaries of case reports of allergic contact dermatitis from povidone-iodine

Year and country	Sex	Age	Positive patch tests	Clinical data and comments	Ref.
2001 France	F	58	PVP-I solution 'as is'; PVP-I 10% water	acute spreading dermatitis around surgical incision; also (stronger) contact allergy to another antiseptic used	34
1999 Germany	M	16	PVP-I cream; PVP-I 5% and 10% water and pet.; iodine 0.5% pet.	postoperative dermatitis on the distal right foot; PVP-I cream 10% had been used after surgery on the big toe nail	35
	M	62	PVP-I solution 'as is'; PVP-I 10% water; iodine 0.5% pet.	allergic contact dermatitis on the right flank where PVP-I solution had been used prior to surgery on the left kidney	
1994 Italy	F	46	PVP-I 10% pet.	erythema multiforme-like dermatitis starting at the vulvar region and spreading to the inguinal folds and upper thighs from using PVP-I vaginal solution for vulvovaginitis	38
1990 Italy	F	44	PVP-I solution 'as is'; PVP-I 10% pet.	masseuse with dermatitis of the hands and arms, spreading to the face, neck and trunk; occupational allergic contact dermatitis	39
	F	67	PVP-I 10% ointment 'as is'; PVP-I 10% pet.	acute dermatitis at skin biopsy site after applying 10% PVP-I ointment	
1990 Japan	M	30	PVP-I 7.5% aqueous solution 'as is'	surgical wound treated with PVP-I solution daily; after one week acute dermatitis developed with erythema multiforme ike spread to the limbs and trunk with target-like lesions with dusky centers; patch test read only at D2 (unreliable)	41
1988 Japan	F	57	PVP-I 10% water 'as is'; sugar/PVP-I (3%); potassium iodide 5% water	traumatic venous ulcer treated with 3% PVP-I in sugar; worsening of ulcer	42
	F	89	PVP-I 10% water 'as is'; sugar/PVP-I (3%); potassium iodide 5% water	traumatic venous ulcer treated with 3% PVP-I in sugar; itching and erythema developed around the ulcer	
1985 Mexico	F	40	PVP-I 5% and 10% water	erythematous vesicular rash where PVP-I solution had concentrated; no reaction where PVP-I solution had been applied; classic irritant, not allergic, contact dermatitis!	44
1984 Belgium	M	41	PVP-I 10% solution 'as is'	butcher washing his hands frequently with PVP-I 10%; vesicular dermatitis of the back of both hands and forearms; negative reaction to iodine; ROAT positive with PVP-I 10% water after 6 applications; occupational sensitization	45
1982 USA	M	40	PVP-10% solution; iodine 0.5% isopropyl alcohol	markedly pruritic eruption around a wound treated with PVP-I dressing; also reaction to iodine	46
	F	65	PVP-I solution; iodine 0.5% isopropyl alcohol	dermatitis at the site of PVP-I application on 3 occasions; history or iodine allergy; positive patch test to iodine	

PVP-I: povidone-iodine (polyvinylpyrrolidone-iodine)

Cross-reactions, pseudo-cross-reactions and co-reactions
Patients sensitized to iodine may cross-react (or more accurately: pseudo-cross-react) to PVP-I and vice versa (43,47). In a group of 7 patients sensitized to iodine (tested 0.25% alc.) and/or PVP-iodine (tested 5% water or with the commercial solution), a co-reaction was found to iodopropynyl butylcarbamate (IPBC), for which no relevance could be found. The authors suspected that free iodine, released from IPBC, had caused the positive patch test reaction to this preservative (10).

IRRITANT CONTACT DERMATITIS
Although povidone-iodine (PVP-I) has a relatively low irritancy potential, irritant contact dermatitis has been reported quite frequently (16,17,18,19) and even chemical burns and skin ulceration have occurred (20,21) from the use of PVP-I solution. Liquid preparations are known to produce skin irritation by releasing $(I_3)^-$, contrary to gel formulations. The relevant literature has been reviewed *in extenso* in 2016 (16). Risk factors for irritant contact

dermatitis are long surgical procedures (irritation due to PVP-I is time-dependent [33]), the presence of irritating excipients in the solution, occlusion with plastic drapes, waterproof dressings, or medical devices, prolonged skin contact with PVP-I-saturated cotton pads or drapes during surgery, and improper use of PVP-I (inadequate drying, pool formation beneath the body).

Four different clinical patterns have been described as a result of 'PVP-I misuse' (16). The first and most distinctive one consists of a double lumbar parallel pattern. This pattern is the consequence of the folding and maintenance of moist PVP-I during a surgical procedure in which the patient is maintained in the supine position. A second pattern is due to embedded cotton or gauze pads used to protect some of the medical devices employed during the procedure, as has been described with the use of tourniquets. The third clinical pattern is observed when the lesion delineates a terminal or a device glued to the skin of the patient, as reported in dermatitis occurring after spinal anesthesia. The fourth is a random pattern that follows the folds or grooves of clinical drapes. PVP-I irritant dermatitis can easily be prevented by allowing the solution in contact with the skin to adequately dry (16).

LITERATURE

1   The data in the section 'General' may have been obtained from literature discussed in this chapter, but mostly also or exclusively from one or more of the following online sources: ChemIDPlus Advanced, PubChem, DrugBank, RxList, Drug Central, Drugs.com, and Wikipedia

2   Barbaud A, Collet E, Le Coz CJ, Meaume S, Gillois P. Contact allergy in chronic leg ulcers: results of a multicentre study carried out in 423 patients and proposal for an updated series of patch tests. Contact Dermatitis 2009;60:279-287

3   Erfurt-Berge C, Geier J, Mahler V. The current spectrum of contact sensitization in patients with chronic leg ulcers or stasis dermatitis - new data from the Information Network of Departments of Dermatology (IVDK). Contact Dermatitis 2017;77:151-158

4   Reich-Schupke S, Kurscheidt J, Appelhans C Kreuter A, Altmeyer P, Stücker M. Patch testing in patients with leg ulcers with special regard to modern wound products. Hautarzt 2010;61:593-597 (Article in German)

5   Marasovic D, Vuksic I. Allergic contact dermatitis in patients with leg ulcers. Contact Dermatitis1999;41:107-109

6   Agulló-Pérez AD, Hervella-Garcés M, Oscoz-Jaime S, Azcona-Rodríguez M, Larrea-García M, Yanguas-Bayona JI. Perianal dermatitis. Dermatitis 2017;28:270-275

7   Gilissen L, De Decker L, Hulshagen T, Goossens A. Allergic contact dermatitis caused by topical ophthalmic medications: Keep an eye on it! Contact Dermatitis 2019;80:291-297

8   Gilissen L, Goossens A. Frequency and trends of contact allergy to and iatrogenic contact dermatitis caused by topical drugs over a 25-year period. Contact Dermatitis 2016;75:290-302

9   Shih Y-H, Sun C-C, Tseng Y-H, Chu C-Y. Contact dermatitis to topical medicaments: a retrospective study from a medical center in Taiwan. Dermatol Sinica 2015;33:181-186

10 Vanhoutte C, Goossens A, Gilissen L, Huygens S, Vital-Durand D, Dendooven E, et al. Concomitant contact-allergic reactions to iodopropynyl butylcarbamate and iodine. Contact Dermatitis 2019;81:17-23

11 Veramme J, de Zaeytijd J, Lambert J, Lapeere H. Contact dermatitis in patients undergoing serial intravitreal injections. Contact Dermatitis 2016;74:18-21

12 Amschler K, Fuchs T, Geier J, Buhl T. In search of a better patch test concentration for povidone-iodine. Contact Dermatitis 2017;77:346-347

13 Nishioka K, Seguchi T, Yasuno H, Yamamoto T, Tominaga K. The results of ingredient patch testing in contact dermatitis elicited by povidone-iodine preparations. Contact Dermatitis 2000;42:90-94

14 Lachapelle JM. Allergic contact dermatitis from povidone-iodine: a re-evaluation study. Contact Dermatitis 2005;52:9-10

15 Tsunoda T, Kawamura M, Aoki E, Deguchi M, Manome H, Iguchi M. 22 cases of contact allergy due to povidone iodine. Rinsho Derma (Tokyo) 1998;52:201-205 (Article in Japanese, data cited in ref. 13)

16 Borrego L, Hernández N, Hernández Z, Peñate Y. Povidone-iodine induced post-surgical irritant contact dermatitis localized outside of the surgical incision area. Report of 27 cases and a literature review. Int J Dermatol 2016:55:540-545

17 Vandergriff TW, Wasko CA, Schwartz MR, Hsu S. Irritant contact dermatitis from exposure to povidone-iodine may resemble toxic epidermal necrolysis. Dermatol Online J 2006;12:12

18 Iijima S, Kuramochi M. Investigation of irritant skin reaction by 10% povidone-iodine solution after surgery. Dermatology 2002;204(Suppl.1):103-108

19 Okano M. Irritant contact dermatitis caused by povidone-iodine. J Am Acad Dermatol 1989;20:860

20 Corazza M, Bulciolu G, Spisani L, Virgili A. Chemical burns following irritant contact with povidone-iodine. Contact Dermatitis 1997;36:115-116

21 Mochida K, Hisa T, Yasunaga C, Nishimura T, Nakagawa K, Hamada T. Skin ulceration due to povidone-iodine. Contact Dermatitis 1995;33:61-62

22   De la Cuadra-Oyanguren J, Zaragozá-Ninet V, Sierra-Talamantes C, Alegre de Miquel V. Postsurgical contact dermatitis due to povidone iodine: a diagnostic dilemma. Actas Dermosifiliogr2014;105:300-304

23   Reyazulla MA, Gopinath AL, Vaibhav N, Raut RP. An unusual complication of late onset allergic contact dermatitis to povidone iodine in oral & maxillofacial surgery - a report of 2 cases. Eur Ann Allergy Clin Immunol 2014;46:157-159

24   Zokaie S, White IR, McFadden JD. Allergic contact dermatitis caused by iodophor-impregnated surgical incise drape. Contact Dermatitis 2011;65:309

25   Rahimi S, Lazarou G. Late-onset allergic reaction to povidone-iodine resulting in vulvar edema and urinary retention. Obstet Gynecol 2010;116(Suppl.2):562-564

26   Velázquez D, Zamberk P, Suárez R, Lázaro P. Allergic contact dermatitis to povidone-iodine. Contact Dermatitis 2009;60:348-349

27   Sowa J, Tsuruta D, Nakanishi T, Kobayashi H, Ishii M. Generalized dermatitis with eosinophilia resulting from allergic contact dermatitis due to povidone iodine. Contact Dermatitis 2006;54:174-176

28   Lee SK, Zhai H, Maibach HI. Allergic contact dermatitis from iodine preparations: a conundrum. Contact Dermatitis 2005;52:184-187

29   Yavascan O, Kara OD, Sozen G, Aksu N. Allergic dermatitis caused by povidone iodine: an uncommon complication of chronic peritoneal dialysis treatment. Adv Perit Dial 2005;21:131-133

30   Barbaud A, Vigan M, Delrous JL, Assier H, Avenel-Audran M, Collet E, et al; Membres du Groupe du REVIDAL. Contact allergy to antiseptics: 75 cases analyzed by the dermato-allergovigilance network (Revidal). Ann Dermatol Venereol 2005;132(12Pt.1):962-965 (Article in French)

31   Borja JM, Galindo PA, Gomez E, Feo F. Contact dermatitis due to povidone-iodine: allergic or irritant? J Investig Allergol Clin Immunol 2003;13:131-132

32   Kozuka T. Patch testing to exclude allergic contact dermatitis caused by povidone-iodine. Dermatology 2002;204(Suppl.1):96-98

33   Dukes PJ, Marks R. An evaluation of the irritancy potential of povidone-iodine solution: Comparison of subjective and objective assessment techniques. Clin Exp Dermatol 1992;17:246-249

34   Reichert-Pénétrat S, Barbaud A, Pénétrat E, Granel F, Schmutz JL. Allergic contact dermatitis from surgical paints. Contact Dermatitis 2001;45:116-117

35   Erdmann S, Hertl M, Merk HF. Allergic contact dermatitis from povidone-iodine. Contact Dermatitis 1999;40:331-332

36   Kanerva L, Estlander T. Occupational allergic contact dermatitis caused by povidone-iodine (Betadine). Environ Dermatol 1999;6:101-104 (data cited in ref. 32)

37   Gonzalo MA, Revenga F, Caravaca F, Pizarro JL. Epidemiologic study of contact dermatitis in hemodialysis patients. J Investig Allergol Clin Immunol 1997;7:20-23

38   Vincenzi C, Stinchi C, Guerra L, Piraccini BM, Bardazzi F, Tosti A. Erythema multiformlike contact dermatitis: Report of four cases. Am J Contact Dermat 1994;5:90-93

39   Tosti A, Vincenzi C, Bardazzi F, Mariani R. Allergic contact dermatitis due to povidone-iodine. Contact Dermatitis 1990;23:197-198

40   Van Ketel WG, Berg WHHW. Sensitization to povidone-iodine. Dermatologic Clinics 1990;8:107-109

41   Torinuki W. Generalized erythema-multiforme-like eruption following allergic contact dermatitis. Contact Dermatitis 1990;23:202-203

42   Kudo H, Takahashi K, Suzuki Y, Tanaka T, Miyachi Y, Imamura S. Contact dermatitis from a compound mixture of sugar and povidone-iodine. Contact Dermatitis 1988;18:155-157

43   Böckers M, Bork K. Contact dermatitis caused by PVP-iodine. Dtsch Med Wochenschr 1986;111(28-29):1110-1112 (Article in German)

44   Ancona A, Suárez de la Torre R, Macotela E. Allergic contact dermatitis from povidone-iodine. Contact Dermatitis 1985;13:66-68

45   Lachapelle JM. Occupational allergic contact dermatitis to povidone-iodine. Contact Dermatitis 1984;11:189-190

46   Marks JG Jr. Allergic contact dermatitis to povidone-iodine. J Am Acad Dermatol 1982;6(4Pt.1):473-475

47   Kunze J, Kaiser HJ, Petres J. Relevanz einer Iodallergie bei handelsüblichen Polyvidon-Jod-Zubereitungen. Z Hautkr 1983;58:255 (Article in German)

48   Molin S, Bauer A, Schnuch A, Geier J. Occupational contact allergy in nurses: results from the Information Network of Departments of Dermatology 2003-2012. Contact Dermatitis 2015;72:164-171

Chapter 3.271 PRAMOXINE

IDENTIFICATION

Description/definition   : Pramoxine is the morpholine derivative that conforms to the structural formula shown
                            below
Pharmacological classes : Anesthetics, local
IUPAC name   : 4-[3-(4-Butoxyphenoxy)propyl]morpholine
Other names   : Pramocaine
CAS registry number   : 140-65-8
EC number   : 205-425-7
Merck Index monograph   : 9099
Patch testing   : Hydrochloride 2.0% pet. (Chemotechnique)
Molecular formula   : $C_{17}H_{27}NO_3$

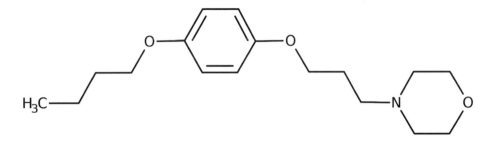

GENERAL

Pramoxine is a morpholine derivative with local anesthetic and antipruritic activities. It is indicated for temporary relief of pain and pruritus from minor lip and skin irritations as well as for temporary relief of pain, burning, itching and discomfort associated with hemorrhoids and other anorectal/anogenital disorders. In pharmaceutical products, pramoxine is employed as pramoxine hydrochloride (CAS number 637-58-1, EC number 211-293-1, molecular formula $C_{17}H_{28}ClNO_3$) (1).

CONTACT ALLERGY

General

The literature on contact allergy to pramoxine was recently (2020) reviewed (9).

Patch testing in groups of patients

In Alberta, Canada, 495 consecutive patients seen in a tertiary patch test clinic between May 2014 and August 2015 were patch tested with pramocaine 2% in pet. and 15 (3%) subjects had a positive reaction. There was no significant gender difference. The mean age of positive patients was 39 years (range 11-69). Four patients showed concomitant responses to another anesthetic agent: two to dibucaine and two to lidocaine. Relevance was not addressed (3).

Case series

From a University clinic in the USA, 8 cases of allergic contact dermatitis to pramoxine, investigated in the period 2016-2019, were reported (9). There were 5 women and 3 men, ages ranging from 29 to 73 years, mean 48 years. Four had diffuse dermatitis, 4 localized. Six had positive patch tests to one or more preparations containing pramoxine, including antibiotic, anti-itch and antihemorrhoidal pharmaceuticals. Four of these were tested with pramoxine 2% pet. and all had positive reactions. Two patients reacted to pramoxine in the medicament series, but they did not use pramoxine-containing topicals and the reactions were 'likely' of past relevance (9). This series shows that contact allergy to pramoxine may be more frequent than the limited number of case reports previously published suggests.

Case reports

A 52-year-old male to female transgender patient complained of a red, papular, itchy rash around the newly constructed vulva since undergoing gender affirmation surgery in Thailand. She had been prescribed a topical antibacterial ointment with neomycin and bacitracin immediately after surgery. Later, she started using a cream with the anesthetic pramoxine. Physical examination showed diffuse erythema of the introitus, the labia majora and inguinal skin with scattered 2-3mm erosions. These findings were markedly improved from patient-provided post-operative photos. Day 5 patch testing results showed positive reactions to the anesthetic cream and its ingredient pramoxine. There were also positive reactions to the antibacterial ointment and its active components neomycin and bacitracin (2).

A 56-year-old man treated a surgical wound with an antibacterial ointment containing neomycin, bacitracin, polymyxin B, and pramoxine (inactive ingredient: petrolatum) and with lidocaine spray. The wound itself healed slowly, but the patient developed a pruritic erythematous rash of the left knee. Just before consultation the patient ran out of the previously described ointment and unknowingly purchased a preparation by the same manufacturer that lacked pramoxine but was otherwise identical. Examination revealed eczematous eruptions of the left knee and leg, hands, as well as scattered areas of the trunk. Patch tests were positive only to the patient's triple antibiotic ointment with pramoxine but not to the very same formulation without the topical anesthetic. Pertinent negatives included bacitracin, neomycin, polymyxin, and lidocaine. It was concluded *per exclusionem* that the patient had contact allergy to and allergic contact dermatitis from pramoxine (4).

A 23-year-old woman had treated hemorrhoids for a long time with a local anesthetic product containing pramoxine HCl. This resulted in severe eczema, spreading from the perianal region to her buttocks and upper thighs. A patch test with the ointment was positive. Pramoxine itself was not tested, but, as the excipients were all patch test negative, the authors suggested that these results strongly indicated that the drug was responsible for the eruption (6).

In a period of 2 years, an investigator in The Netherlands described three patients with allergic contact dermatitis from pramoxine HCl, all in the same ointment (7,8). A 36-year-old woman had a relapse of perianal dermatitis from pramoxine (test concentration and vehicle not mentioned) in an ointment and lidocaine in another topical pharmaceutical (7). Two patients with contact eczema were seen after using the same ointment containing pramoxine. Patch tests were positive to the ointment and its active ingredient pramoxine HCl 1% water in both (8).

Cross-reactions, pseudo-cross-reactions and co-reactions

A patient sensitized to prilocaine co-reacted to pramoxine. As he had apparently never had contact with pramoxine before, the authors considered this to be a cross-reaction, although the structural formulas of the two chemicals are quite different (5). No cross-reactions to lidocaine or benzocaine in 8 patients allergic to pramoxine (9).

Immediate contact reactions

Immediate contact reactions (contact urticaria) to pramoxine are presented in Chapter 5.

LITERATURE

1   The data in the section 'General' may have been obtained from literature discussed in this chapter, but mostly also or exclusively from one or more of the following online sources: ChemIDPlus Advanced, PubChem, DrugBank, RxList, Drug Central, Drugs.com, and Wikipedia
2   Schlarbaum JP, Kimyon RS, Liou YL, Becker O'Neill L, Warshaw EM. Genital dermatitis in a transgender patient returning from Thailand: A diagnostic challenge. Travel Med Infect Dis 2019;27:134-135
3   Abbas M, Suzuki K, Elliott JF. Contact sensitivity to topical anesthetic pramocaine occurs in three percent of patch test patients. Dermatitis 2016;27(6):e1
4   Hylwa SA, Warshaw E. Contact allergy to pramoxine (pramocaine): the importance of testing to personal products. Dermatitis 2014;25:147-148
5   García F, Iparraguirre A, Blanco J, Alloza P, Vicente J, Báscones O, et al. Contact dermatitis from prilocaine with cross-sensitivity to pramocaine and bupivacaine. Contact Dermatitis 2007;56:120-121
6   Cusano F, Luciano S. Contact dermatitis from pramoxine. Contact Dermatitis 1993;28:39
7   Van Ketel WG. Contact allergy to different antihaemorrhidal anaesthetics. Contact Dermatitis 1983;9:512-513
8   Van Ketel WG. Allergy to pramoxine (pramocaine). Contact Dermatitis 1981;7:49
9   Kimyon RS, Schlarbaum JP, Liou YL, Hylwa SA, Warshaw EM. Allergic contact dermatitis to pramoxine (pramocaine). Dermatitis 2020 May 12. doi: 10.1097/DER.0000000000000606. Online ahead of print.

Chapter 3.272 PREDNICARBATE

IDENTIFICATION

Description/definition	: Prednicarbate is a synthetic non-halogenated double-ester derivative of the corticosteroid prednisolone that conforms to the structural formula shown below
Pharmacological classes	: Anti-inflammatory agents; glucocorticoids
IUPAC name	: [2-[(8S,9S,10R,11S,13S,14S,17R)-17-Ethoxycarbonyloxy-11-hydroxy-10,13-dimethyl-3-oxo-7,8,9,11,12,14,15,16-octahydro-6H-cyclopenta[a]phenanthren-17-yl]-2-oxoethyl]
Other names	: 11β,17,21-Trihydroxypregna-1,4-diene-3,20-dione 17-(ethyl carbonate) 21-propionate
CAS registry number	: 73771-04-7
EC number	: 277-590-3
Merck Index monograph	: 9110
Patch testing	: 1% alc. (SmartPracticeCanada); late readings (6-10 days) are strongly recommended
Molecular formula	: $C_{27}H_{36}O_8$

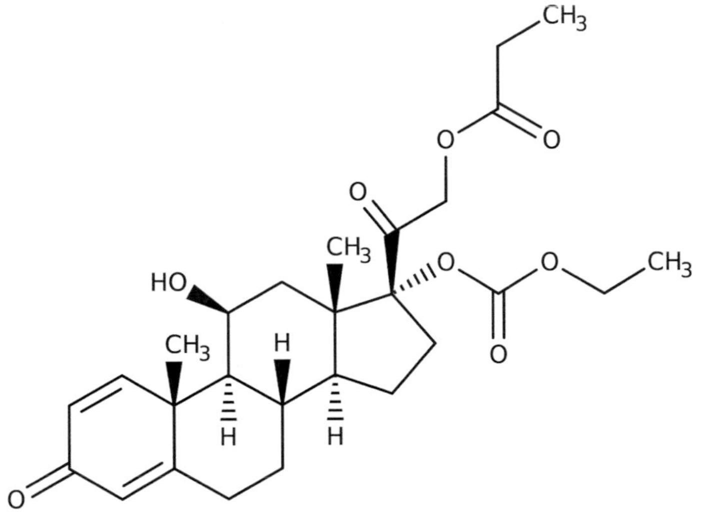

GENERAL

General aspects of corticosteroids used on the skin and mucous membranes are discussed in Chapter 2.4. A practical guideline for diagnosing allergic reactions to corticosteroids is presented in ref. 1.

CONTACT ALLERGY

Patch testing in consecutive patients suspected of contact dermatitis: Routine testing

The results of routine testing with prednicarbate are shown in table 3.272.1. In 3 studies (of which one was performed in two countries with separate results [12]), prevalences of positive reactions have ranged from 0.4% to 0.7%. In two investigations, no relevance data were provided, in the third, the reactions were scored as not relevant and interpreted as cross-reactions (5).

Table 3.272.1 Patch testing in groups of patients

Years and Country	Test conc. & vehicle	Number of patients tested	positive (%)	Selection of patients (S); Relevance (R); Comments (C)	Ref.
2015-2016 Spain	1% alc.	3699	17 (0.5%)	R: not stated	11
1998 Italy	1% alc. and pet.	325	2 (0.6%)	R: not relevant, considered to be cross-reactions	5
1991 Belgium	1% alc.	610	4 (0.7%)	R: not stated	12
1991 The Netherlands	1% alc.	533	2 (0.4%)	R: not stated	12

Patch testing in groups of selected patients

In a group of 203 patients with eyelid dermatitis, two patients reacted to prednicarbate (test concentration and vehicle not mentioned); the relevance was not specified and it is unknown how many individuals were tested with this material (2).

Case series

From January 1990 to June 2008, in Leuven, Belgium, 315 patients were diagnosed with contact allergy to/allergic contact dermatitis from corticosteroids (CSs) from routine patch testing with a baseline series including tixocortol pivalate, budesonide, hydrocortisone butyrate and prednisone caproate, patch testing with patients' own CS preparations, and testing those with proven contact allergy to a corticosteroid or strongly suspected of CS allergy later with a series of 66 CSs, including two sex hormones (progesterone and testosterone). 71% of the patients had relevant reactions, but these were not specified. In this group of 315 CS allergic patients, 111 had positive patch tests to prednicarbate 1% alc. (10). As this corticosteroid has never been used in pharmaceuticals in Belgium, these positive reactions must all be considered cross-reactions to other corticosteroids.

Case reports

A 37-year-old woman working as waitress presented with desquamative lesions on the backs of both hands that had been resistant to treatment with various topical (including prednicarbate) and systemic corticosteroids for a year. Skin biopsy showed spongiotic dermatitis. Patch tests were positive to budesonide and hydrocortisone 17-butyrate in the baseline series (not previously used by the patient), and to prednicarbate cream 0.25%. When patch tested with the active and inactive ingredients of prednicarbate cream, positive ring-shaped reactions to prednicarbate 0.1% and 1% alcohol (++) and 1% water (+) were seen at D2, D4, and D7 (3).

A 53-year-old atopic woman had a history of dyshidrotic hand eczema and self-limited outbreaks of itchy dermatitis in her face and neck. She presented with itchy, confluent red macules and papules that had evolved into sharply demarcated erythematous plaques localized on her forehead and neck after having applied prednicarbate cream 3 days previously. The clinical diagnosis was lupus erythematosus. A biopsy taken from a lesion on the forehead showed hydropic degeneration of the basal layer, edema, and a lymphocytic infiltrate around the vessels and hair follicles. Immunofluorescence showed IgM and C3 at the dermo-epidermal junction. ANA, anti-ds-DNA and complement were normal or negative. Acute signs disappeared in a few weeks after avoiding the cream. Patch tests were positive to budesonide, prednicarbate cream (ROAT also positive), and prednicarbate 1% pet. and 1% alcohol. The patient was diagnosed with allergic contact dermatitis to prednicarbate, presenting with the clinical and histopathological features of lupus erythematosus (4).

Short summaries of other case reports of allergic contact dermatitis from prednicarbate are shown in table 3.272.2.

Table 3.272.2 Short summaries of case reports of allergic contact dermatitis from prednicarbate

Year and country	Sex	Age	Positive patch tests	Clinical data and comments	Ref.
1998 Spain	M	68	PC cream 'as is'; PC 1% pet.	edema of the penis and scrotum with erythema and vesicles after 3 days' application of prednicarbate cream to dermatitis caused by hydrocortisone	6
1997 Spain	M	58	PC cream 'as is'; PC 0.2% (vehicle unknown)	worsening of eczema on the forearms and spreading to the legs and trunk	7
1991 Germany	FF	??	prednicarbate	two women had allergic contact dermatitis from prednicarbate; no cross-reactions; details unavailable to the author	8
1991 Germany	F	77	prednicarbate 'as is'	erythema and burning after several days' use of prednicarbate ointment; patch test positive twice; also contact allergy to benzyl alcohol in the corticosteroid ointment; same patient as next one	9
1991 Germany	F	77	PC cream 'as is'; prednicarbate 100% (?)	contact dermatitis around leg ulcers; same patient as previous one	13

PC: prednicarbate

Cross-reactions, pseudo-cross-reactions and co-reactions

Cross-reactions between corticosteroids are discussed in Chapter 2.8.

LITERATURE

1   Baeck M, Goossens A. Immediate and delayed allergic hypersensitivity to corticosteroids: practical guidelines. Contact Dermatitis 2012;66:38-45

2   Guin JD. Eyelid dermatitis: experience in 203 cases. J Am Acad Dermatol 2002;47:755-765

3   Otero-Rivas MM, Ruiz-González I, Pérez-Bustillo A, Rodríguez-Prieto MÁ. Allergic contact dermatitis caused by prednicarbate presenting as chronic hand eczema. Contact Dermatitis 2015;73:51-52

4   Sánchez-Pérez J, Gala SP, Jiménez YD, Fraga J, Diez AG. Allergic contact dermatitis to prednicarbate presenting as lupus erythematosus. Contact Dermatitis 2006;55:247-249

5   Stingeni L, Lisi P. Contact allergy to prednicarbate: frequency of positive reactions in consecutively-patch-tested patients. Contact Dermatitis 1999;40:286-287

6   Miranda-Romero A, Sánchez-Sambucety P, Bajo C, Martinez M, Garcia-Munõz M. Genital oedema from contact allergy to prednicarbate. Contact Dermatitis 1998;38:228-229

7   Villas Martínez F, Navarro Echeverría JA, Joral Badás A, Garmendia Goitia FJ. Prednicarbate contact allergy. Contact Dermatitis 1997;37:299-300

8   Senff H, Kunz R, Köllner A, Kunze J. Allergic contact dermatitis due to prednicarbate. Hautarzt 1991;42:53-55 (Article in German)

9   Dunkel FG, Elsner P, Burg G. Allergic contact dermatitis from prednicarbate. Contact Dermatitis 1991;24:59-60

10 Baeck M, Chemelle JA, Terreux R, Drieghe J, Goossens A. Delayed hypersensitivity to corticosteroids in a series of 315 patients: clinical data and patch test results. Contact Dermatitis 2009;61:163-175

11 Mercader-García P, Pastor-Nieto MA, García-Doval I, Giménez-Arnau A, González-Pérez R, Fernández-Redondo V, et al. GEIDAC. Are the Spanish baseline series markers sufficient to detect contact allergy to corticosteroids in Spain? A GEIDAC prospective study. Contact Dermatitis 2018;78:76-82

12 Dooms-Goossens A, Meinardi MM, Bos JD, Degreef H. Contact allergy to corticosteroids: the results of a two-centre study. Br J Dermatol 1994;130:42-47

13 Dunkel FG, Elsner P, Burg G. Contact allergies to topical corticosteroids: 10 cases of contact dermatitis. Contact Dermatitis 1991;25:97-103

Chapter 3.273 PREDNISOLONE

IDENTIFICATION

Description/definition   : Prednisolone is the synthetic glucocorticoid that conforms to the structural formula
                             shown below
Pharmacological classes : Antineoplastic agents, hormonal; glucocorticoids; anti-inflammatory agents
IUPAC name   : (8S,9S,10R,11S,13S,14S,17R)-11,17-Dihydroxy-17-(2-hydroxyacetyl)-10,13-dimethyl-
                             7,8,9,11,12,14,15,16-octahydro-6H-cyclopenta[a]phenanthren-3-one
Other names   : 11β,17,21-Trihydroxypregna-1,4-diene-3,20-dione
CAS registry number   : 50-24-8
EC number   : 200-021-7
Merck Index monograph   : 9111
Patch testing   : 1% pet. (SmartPracticeCanada, SmartPracticeEurope)
Molecular formula   : $C_{21}H_{28}O_5$

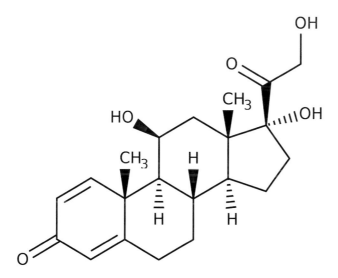

GENERAL

General aspects of corticosteroids used on the skin and mucous membranes are discussed in Chapter 2.4. A practical
guideline for diagnosing allergic reactions to corticosteroids is presented in ref. 9. Prednisolone base has long been
used in tablets only, but since 2018 also in an ointment combined with neomycin (www.drugbank.ca). Esters used in
other applications include prednisolone acetate (Chapter 3.274), prednisolone caproate (Chapter 3.275) ,
prednisolone hemisuccinate (Chapter 3.276), prednisolone pivalate (Chapter 3.277), prednisolone sodium metazoate
(Chapter 3.278), and prednisolone valerate acetate (Chapter 3.279).

Prednisolone *base* is virtually always used as tablet, which implies that by far most allergic reactions to
'prednisolone' have in fact been the result of sensitization to an ester of prednisolone or of cross-reactivity to
another corticosteroid. It is also likely that there has been confusion in some publications on the correct forms of the
drugs used, e.g. that prednisolone was mentioned where in fact an ester form should have been mentioned.

CONTACT ALLERGY

Contact allergy in the general population

In Germany, for the period 1995-2004, the population-based relative incidence (RI) of contact sensitization to
prednisolone (cases/100,000 defined daily doses (DDDs) per year) was estimated to be 0.5. In the group of
corticosteroids, the RI ranged from 0.3 (dexamethasone sodium phosphate) to 43.2 (budesonide) (2).

Patch testing in groups of patients

The results of patch testing prednisolone in consecutive patients suspected of contact dermatitis (routine testing)
and of testing in groups of *selected* patients are shown in table 3.273.1. In both types of studies, frequencies of posi-

tive patch tests were low, 0.4% or 0.5%. Relevance data were either not provided (1,5), 100% (7) or 0% (6), but in the latter two studies, there was only one patient allergic to prednisolone.

Table 3.273.1 Patch testing in groups of patients

Years and Country	Test conc. & vehicle	Number of patients tested	positive (%)	Selection of patients (S); Relevance (R); Comments (C)	Ref.
Routine testing					
<2005 Poland	1% pet.	257	1 (0.4%)	R: 0%	6
Testing in groups of selected patients					
2002-2004 IVDK	1% pet.	2349	9 (0.4%)	S: unknown; R: not stated	1
1995-2004 IVDK	1% pet.	2259	9 (0.4%)	S: patients tested with a corticosteroid series; R: not stated	5
1995-9 Netherlands	1% alc. 97%	216	1 (0.5%)	S: not stated; R: definite 100%	7

IVDK: Information Network of Departments of Dermatology (Germany, Austria, Switzerland)

Case series

From January 1990 to June 2008, in Leuven, Belgium, 315 patients were diagnosed with contact allergy to/allergic contact dermatitis from corticosteroids (CSs) from routine patch testing with a baseline series including tixocortol pivalate, budesonide, hydrocortisone butyrate and prednisone caproate, patch testing with patients' own CSs preparations, and testing those with proven contact allergy to a corticosteroid or strongly suspected of CS allergy later with a series of 66 CSs, including two sex hormones (progesterone and testosterone). 71% of the patients had relevant reactions, but these were not specified. In this group of 315 CS allergic patients, 55 had positive patch tests to prednisolone 0.1% alc. (3, overlap with ref. 4). It is unknown how many of these reactions were caused by the use of a pharmaceutical product containing prednisolone and how many were cross-reactions to other corticosteroids, but it may be assumed that (virtually) all have been cross-sensitizations.

In the period 1990-2008, in the same clinic in Leuven, Belgium, 315 patients were diagnosed with contact allergy to/allergic contact dermatitis from corticosteroids. Eighteen subjects (5.7%) presented with allergic manifestations (conjunctivitis, eczema of the face, periocular skin or eyelids) caused by the use of CS-containing ocular preparations. Three patients had used ophthalmic preparations containing prednisolone and one prednisolone pivalate (4, overlap with ref. 3). The former is probably incorrect, as prednisolone in eye drops is used as an ester (acetate, pivalate, disodium phosphate).

In reviewing the Japanese literature up to 1994, 43 patients with allergic contact dermatitis from topical corticosteroids were identified, including one caused by prednisolone (10).

In Tokyo, Japan, between 1967 and 1988, 69 patients had positive patch tests to topical corticosteroid preparations used by them. 44 were sensitive to vehicle components or non-steroid active ingredients, 18 were allergic to the corticosteroid itself. In 7 cases, the allergen could not be detected. The most frequently implicated CSs were prednisolone (esters) (n=7) and betamethasone valerate (n=5) (8).

Cross-reactions, pseudo-cross-reactions and co-reactions

Cross-reactions between corticosteroids are discussed in Chapter 2.8.

Cutaneous adverse drug reactions from systemic administration caused by type IV (delayed-type) hypersensitivity

Cutaneous adverse drug reactions from systemic administration of prednisolone caused by type IV (delayed type) hypersensitivity, including exanthemas/systemic contact dermatitis (11-15,17,18,19 [baboon syndrome]), maculopapular rash (20), 'rash' (21,22) and acute generalized exanthematous pustulosis (AGEP) (16), are planned to be discussed in Volume IV of the *Monographs in Contact Allergy* series on Systemic drugs.

LITERATURE

1   Menezes de Padua CA, Uter W, Schnuch A. Contact allergy to topical drugs: prevalence in a clinical setting and estimation of frequency at the population level. Pharmacoepidemiol Drug Saf 2007;16:377-384
2   Menezes de Padua CA, Schnuch A, Nink K, Pfahlberg A, Uter W. Allergic contact dermatitis to topical drugs – Epidemiological risk assessment. Pharmacoepidemiol Drug Saf 2008;17:813-821
3   Baeck M, Chemelle JA, Terreux R, Drieghe J, Goossens A. Delayed hypersensitivity to corticosteroids in a series of 315 patients: clinical data and patch test results. Contact Dermatitis 2009;61:163-175
4   Baeck M, De Potter P, Goossens A. Allergic contact dermatitis following ocular use of corticosteroids. J Ocul Pharmacol Ther 2011;27:83-92

5   Uter W, de Pádua CM, Pfahlberg A, Nink K, Schnuch A, Lessmann H. Contact allergy to topical corticosteroids –
     results from the IVDK and epidemiological risk assessment. J Dtsch Dermatol Ges 2009;7:34-41
6   Reduta T, Laudanska H. Contact hypersensitivity to topical corticosteroids – frequency of positive reactions in
     patch-tested patients with allergic contact dermatitis. Contact Dermatitis 2005;52:109-110
7   Devos SA, Van der Valk PG. Relevance and reproducibility of patch-test reactions to corticosteroids. Contact
     Dermatitis 2001;44:362-365
8   Sasaki E. Corticosteroid sensitivity and cross-sensitivity. A review of 18 cases 1967-1988. Contact Dermatitis
     1990;23:306-315
9   Baeck M, Goossens A. Immediate and delayed allergic hypersensitivity to corticosteroids: practical guidelines.
     Contact Dermatitis 2012;66:38-45
10   Oh-i T. Contact dermatitis due to topical steroids with conceivable cross reactions between topical steroid
     preparations. J Dermatol 1996;23:200-208
11   Rytter M, Walther T, Süss E, Haustein UF. Allergic reactions of the immediate and delayed type following
     prednisolone medication. Dermatol Monatsschr 1989;175:44-48 (Article in German)
12   English JS, Ford G, Beck MH, Rycroft RJ. Allergic contact dermatitis from topical and systemic steroids. Contact
     Dermatitis 1990;23:196-197
13   Bircher AJ, Levy F, Langauer S, Lepoittevin JP. Contact allergy to topical corticosteroids and systemic contact
     dermatitis from prednisolone with tolerance of triamcinolone. Acta Derm Venereol 1995;75:490-493
14   Isaksson M, Persson LM. Contact allergy to hydrocortisone and systemic contact dermatitis from prednisolone
     with tolerance of betamethasone. Am J Contact Dermat 1998;9:136-138
15   McKenna DB, Murphy GM. Contact allergy to topical corticosteroids and systemic allergy to prednisolone.
     Contact Dermatitis 1998;38:121-122
16   Buettiker U, Keller M, Picheler WJ, Braathen LR, Yamalkar N. Oral prednisolone induced acute generalized
     exanthematous pustulosis due to corticosteroids of group A confirmed by epicutaneous testing and lymphocyte
     transformation tests. Dermatology 2006;213:40-43
17   Quirce S, Alvarez MJ, Olaguibel JM, Tabar AI. Systemic contact dermatitis from oral prednisolone. Contact
     Dermatitis 1994;30:53-54
18   De Benito V, Ratón JA, Palacios A, Garmendia M, Gardeazábal J. Systemic contact dermatitis to prednisone: a
     clinical model approach to the management of systemic allergy to corticosteroids. Clin Exp Dermatol
     2012;37:680-681
19   Treudler R, Simon J. Symmetric, drug-related, intertriginous, and flexural exanthema in a patient with polyvalent
     intolerance to corticosteroids. J Allergy Clin Immunol 2006;118:965-967
20   Bursztejn AC, Tréchot P, Cuny JF, Schmutz JL, Barbaud A. Cutaneous adverse drug reactions during
     chemotherapy: consider non-antineoplastic drugs. Contact Dermatitis 2008;58:365-368
21   Mathelier-Fusade P, Marinho E, Aissaoui M, Mounedji N, Chabane MH, Leynadier F. Toxicoderma caused by
     prednisone and prednisolone. Value of skin tests in the screening of cross sensitivity. Ann Dermatol Venereol
     1996;123:453-455 (Article in French)
22   Harris A, McFadden JP. Dermatitis following systemic prednisolone: patch testing with prednisolone eye drops.
     Australas J Dermatol 2000;41:124-125

Chapter 3.274 PREDNISOLONE ACETATE

IDENTIFICATION

Description/definition : Prednisolone acetate is the acetate ester of prednisolone that conforms to the structural formula shown below

Pharmacological classes : Anti-inflammatory agents; glucocorticoids

IUPAC name : [2-[(8S,9S,10R,11S,13S,14S,17R)-11,17-Dihydroxy-10,13-dimethyl-3-oxo-7,8,9,11,12,14, 15,16-octahydro-6H-cyclopenta[a]phenanthren-17-yl]-2-oxoethyl] acetate

Other names : Prednisolone 21-acetate; 11β,17,21-trihydroxypregna-1,4-diene-3,20-dione 21-acetate

CAS registry number : 52-21-1

EC number : 200-134-1

Merck Index monograph : 911 (Prednisolone)

Patch testing : In general, corticosteroids may be tested at 0.1% and 1% in alcohol; late readings (6-10 days) are strongly recommended

Molecular formula : $C_{23}H_{30}O_6$

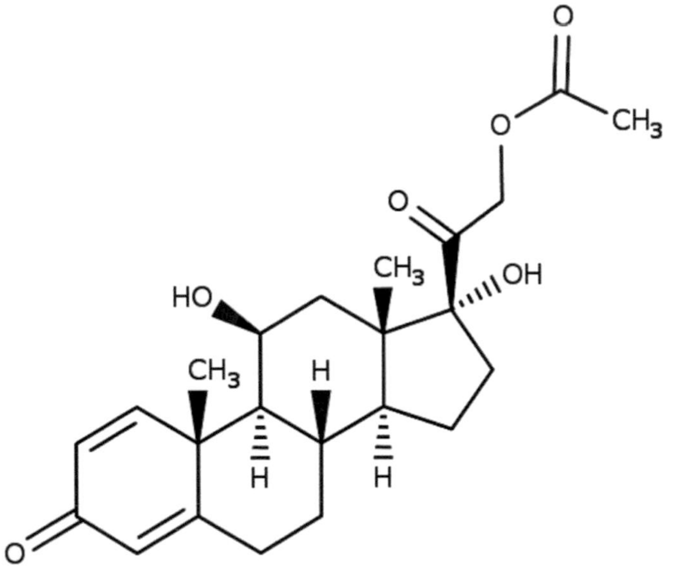

GENERAL

General aspects of corticosteroids used on the skin and mucous membranes are discussed in Chapter 2.4. A practical guideline for diagnosing allergic reactions to corticosteroids is presented in ref. 1. See also prednisolone (Chapter 3.273), prednisolone caproate (Chapter 3.275), prednisolone hemisuccinate (Chapter 3.276), prednisolone pivalate (Chapter 3.277), prednisolone sodium metazoate (Chapter 3.278), and prednisolone valerate acetate (Chapter 3.279).

CONTACT ALLERGY

Case series

In Leuven, Belgium, between 1990 and 2017, 16,065 patients were investigated for contact allergy and 118 (0.7%) showed positive patch test reactions to topical ophthalmic medications and/or to their ingredients. Eighty-four individuals (71%) reacted to an active principle. Prednisolone acetate was tested in 57 patients and was the allergen in eye medications in 4. There were no reactions to prednisolone acetate in other types of medications (2).

In the period 2008-2011, 10 patients (ages 40-75 years, 8 males) were investigated in Bern, Switzerland, because of severe cutaneous drug eruptions following application of topical antimycotics with or without corticosteroids. Seven to 21 days after initiation of antifungal therapy, the patients developed widespread eczema as well as erythematous, maculopapular, erythema multiforme-like and blistering eruptions, that occurred in addition to intense eczematous reactions at application sites associated with peripheral blood eosinophilia, suggesting drug reactions. After the eruptions had resolved, patch testing was performed. Two patients showed sensitization to both clotrimazole and corticosteroids, five were sensitized to tixocortol pivalate or prednisolone. The most severe, erythema multiforme-like and blistering eruptions were associated with corticosteroid allergy, whereas eczematous and maculopapular exanthemas were associated with clotrimazole allergy (3). It was not specified which corticosteroids were present in the antifungal-corticosteroid preparations, but commercially available are preparations with prednisolone acetate, hydrocortisone acetate and betamethasone dipropionate.

Case reports
A 13-year-old boy with mild atopic eczema treated 2 solitary, erythematous lesions on his face with a cream containing prednisolone acetate and clotrimazole. During this, the eczema worsened and spread to the neck and upper thorax. Patch tests were positive to the cream (+++). Additional patch testing to all 10 ingredients revealed positive reactions to prednisolone acetate 1% pet. and to 4 other ingredients (hexamidine diisethionate, PEG-1500-stearate, glycerol-poly-(oxyethylene)-6-alkanoate and clotrimazole) (4).

A 40-year-old woman reported the recent occurrence of an itchy maculopapular rash, originating in the main body folds, with subsequent involvement of the trunk, arms, and legs, without any systemic complaints. It had started 2-3 days following treatment with a nasal treatment containing prednisolone acetate. No skin lesions on the face or around the orifices (nostrils and nose), nasal congestion or rhinitis had been present. The patient had experienced a very similar generalized eruption several years ago following the use of ear drops, compounded by her local pharmacist containing either prednisolone or methylprednisolone. Furthermore, the patient recalled having had a localized skin reaction during childhood, with worsening of a pre-existent dermatitis, after applying a corticosteroid cream. Patch tests were negative to the steroid markers in the baseline series, but positive to prednisolone 1% alc., methylprednisolone 1% and 0.1% alc., and hydrocortisone 0.5% alc./DMSO. The patient was diagnosed with systemic allergic contact dermatitis from resorption of prednisolone acetate in nasal drops, after having been sensitized to another prednisolone ester or a cross-reacting corticosteroid in her youth (5).

A 38-year-old man was treated for an axillary dermatitis with a preparation containing prednisolone acetate, which resulted in an exacerbation. Two years later, he again used this cream and suffered from a disseminated eczema within a few days. After treatment with oral prednisone he developed a severe generalized eczematous eruption which slowly healed. Patch tests were positive to prednisolone and prednisolone acetate 1% pet. (6).

Cross-reactions, pseudo-cross-reactions and co-reactions
Cross-reactions between corticosteroids are discussed in Chapter 2.8.

Cutaneous adverse drug reactions from systemic administration caused by type IV (delayed-type) hypersensitivity
Cutaneous adverse drug reactions from systemic administration of prednisolone acetate caused by type IV (delayed-type) hypersensitivity, including a local reaction to intra-articular injection (7), are planned to be discussed in Volume IV of the *Monographs in Contact Allergy* series on Systemic drugs.

LITERATURE
1   Baeck M, Goossens A. Immediate and delayed allergic hypersensitivity to corticosteroids: practical guidelines. Contact Dermatitis 2012;66:38-45
2   Gilissen L, De Decker L, Hulshagen T, Goossens A. Allergic contact dermatitis caused by topical ophthalmic medications: Keep an eye on it! Contact Dermatitis 2019;80:291-297
3   Tang MM, Corti MA, Stirnimann R, Pelivani N, Yawalkar N, Borradori L, et al. Severe cutaneous allergic reactions following topical antifungal therapy. Contact Dermatitis 2013;68:56-67
4   Brand CU, Ballmer-Weber BK. Contact sensitivity to 5 different ingredients of a topical medicament (Imacort cream). Contact Dermatitis 1995;33:137
5   Faber MA, Sabato V, Ebo DG, Verheyden M, Lambert J, Aerts O. Systemic allergic dermatitis caused by prednisone derivatives in nose and ear drops. Contact Dermatitis 2015;73:317-320
6   Bircher AJ, Bigliardi P, Zaugg T, Mäkinen-Kiljunen S. Delayed generalized allergic reactions to corticosteroids. Dermatology 2000;200:349-351
7   Gall HM, Paul E. A case of corticosteroid allergy. Hautarzt 2001;52:891-894 (Article in German)

Chapter 3.275 PREDNISOLONE CAPROATE

IDENTIFICATION

Description/definition : Prednisolone caproate is caproate ester of the synthetic glucocorticoid prednisolone that conforms to the structural formula shown below

Pharmacological classes : Glucocorticoids

IUPAC name : [2-[(8S,9S,10R,11S,13S,14S,17R)-11,17-Dihydroxy-10,13-dimethyl-3-oxo-7,8,9,11,12, 14,15,16-octahydro-6H-cyclopenta[a]phenanthren-17-yl]-2-oxoethyl] hexanoate

Other names : 11β,17,21-Trihydroxypregna-1,4-diene-3,20-dione 21-hexanoate

CAS registry number : 69164-69-8

EC number : 273-898-7

Patch testing : In general, corticosteroids may be tested at 0.1% and 1% in alcohol; late readings (6-10 days) are strongly recommended

Molecular formula : $C_{27}H_{38}O_6$

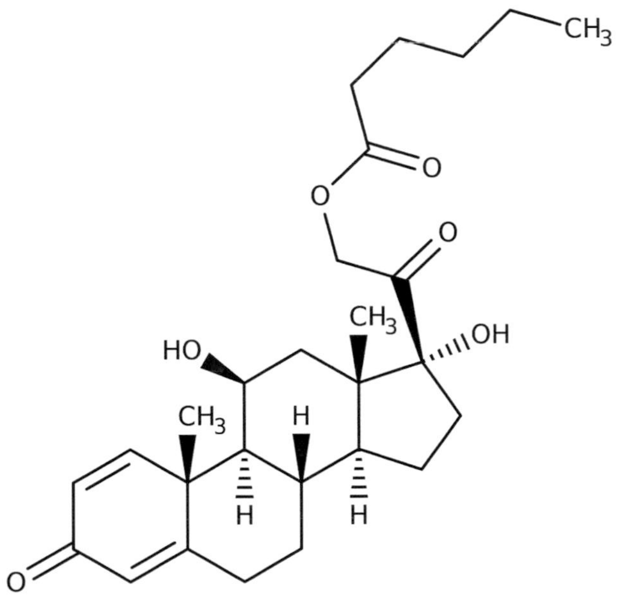

GENERAL

General aspects of corticosteroids used on the skin and mucous membranes are discussed in Chapter 2.4. A practical guideline for diagnosing allergic reactions to corticosteroids is presented in ref. 3. Prednisolone caproate seems to be rarely used in pharmaceuticals, but it is or was available in a suppository with dibucaine hydrochloride for the short term symptomatic relief of perianal discomfort, inflammation and itching caused by thrombosed haemorrhoids, anal fissure and pruritus ani (Scheriproct suppositories).

See also prednisolone (Chapter 3.273), prednisolone acetate (Chapter 3.274), prednisolone hemisuccinate (Chapter 3.276), prednisolone pivalate (Chapter 3.277), prednisolone sodium metazoate (Chapter 3.278), and prednisolone valerate acetate (Chapter 3.279).

CONTACT ALLERGY

Case series

From January 1990 to June 2008, in Leuven, Belgium, 315 patients were diagnosed with contact allergy to/allergic contact dermatitis from corticosteroids (CSs) from routine patch testing with a baseline series including tixocortol pivalate, budesonide, hydrocortisone butyrate and prednisone caproate, patch testing with patients' own CS preparations, and testing those with proven contact allergy to a corticosteroid or strongly suspected of CS allergy later with a series of 66 CSs, including two sex hormones (progesterone and testosterone). 71% of the patients had relevant reactions, but these were not specified. In this group of 315 CS allergic patients, 70 had positive patch tests to prednisolone caproate 0.1% alc. (2, overlap with ref. 1). It is unknown how many of these reactions were caused by the use of a pharmaceutical product containing prednisolone caproate and how many were cross-reactions to other corticosteroids.

Between 1990 and 2003, in Leuven, Belgium, 92 patients were patch tested for chronic vulval complaints. Fifteen of these women were tested with a series of topical drugs; there was one reaction to prednisolone caproate 1% alc. which was considered to be relevant (1, overlap with ref. 2).

Cross-reactions, pseudo-cross-reactions and co-reactions
Cross-reactions between corticosteroids are discussed in Chapter 2.8.

LITERATURE
1   Nardelli A, Degreef H, Goossens A. Contact allergic reactions of the vulva: a 14-year review. Dermatitis 2004;15:131-136
2   Baeck M, Chemelle JA, Terreux R, Drieghe J, Goossens A. Delayed hypersensitivity to corticosteroids in a series of 315 patients: clinical data and patch test results. Contact Dermatitis 2009;61:163-175
3   Baeck M, Goossens A. Immediate and delayed allergic hypersensitivity to corticosteroids: practical guidelines. Contact Dermatitis 2012;66:38-45

Chapter 3.276 PREDNISOLONE HEMISUCCINATE

IDENTIFICATION

Description/definition : Prednisolone hemisuccinate is the hemisuccinate ester of the synthetic glucocorticoid prednisolone that conforms to the structural formula shown below

Pharmacological classes : Glucocorticoids

IUPAC name : 4-[2-[(8S,9S,10R,11S,13S,14S,17R)-11,17-Dihydroxy-10,13-dimethyl-3-oxo-7,8,9,11,12, 14,15,16-octahydro-6H-cyclopenta[a]phenanthren-17-yl]-2-oxoethoxy]-4-oxobutanoic acid

Other names : Prednisolone succinate; 11β,17,21-Trihydroxypregna-1,4-diene-3,20-dione 21-(hydrogen succinate); prednisolone 21-(hydrogen succinate)

CAS registry number : 2920-86-7

EC number : 220-861-8

Patch testing : In general, corticosteroids may be tested at 0.1% and 1% in alcohol; late readings (6-10 days) are strongly recommended

Molecular formula : $C_{25}H_{32}O_8$

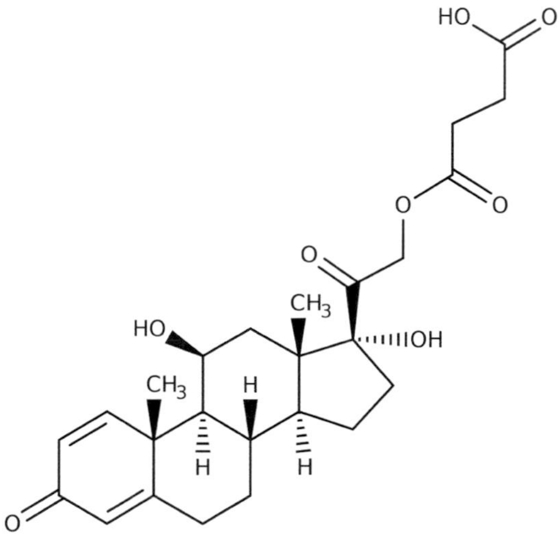

GENERAL

General aspects of corticosteroids used on the skin and mucous membranes are discussed in Chapter 2.4. A practical guideline for diagnosing allergic reactions to corticosteroids is presented in ref. 1. See also prednisolone (Chapter 3.273), prednisolone acetate (Chapter 3.274), prednisolone caproate (Chapter 3.275), prednisolone pivalate (Chapter 3.277), prednisolone sodium metazoate (Chapter 3.278), and prednisolone valerate acetate (Chapter 3.279).

CONTACT ALLERGY

Case series

From January 1990 to June 2008, in Leuven, Belgium, 315 patients were diagnosed with contact allergy to/allergic contact dermatitis from corticosteroids (CSs) from routine patch testing with a baseline series including tixocortol pivalate, budesonide, hydrocortisone butyrate and prednisone caproate, patch testing with patients' own CS preparations, and testing those with proven contact allergy to a corticosteroid or strongly suspected of CS allergy later with a series of 66 CSs, including two sex hormones (progesterone and testosterone). 71% of the patients had relevant reactions, but these were not specified. In this group of 315 CS allergic patients, 13 had positive patch tests to prednisolone hemisuccinate 1% alc. (2). It is unknown how many of these reactions were caused by the use of a pharmaceutical product containing this drug and how many were cross-reactions to other corticosteroids.

Cross-reactions, pseudo-cross-reactions and co-reactions

Cross-reactions between corticosteroids are discussed in Chapter 2.8.

LITERATURE

1   Baeck M, Goossens A. Immediate and delayed allergic hypersensitivity to corticosteroids: practical guidelines. Contact Dermatitis 2012;66:38-45

2   Baeck M, Chemelle JA, Terreux R, Drieghe J, Goossens A. Delayed hypersensitivity to corticosteroids in a series of 315 patients: clinical data and patch test results. Contact Dermatitis 2009;61:163-175

Chapter 3.277 PREDNISOLONE PIVALATE

IDENTIFICATION

Description/definition : Prednisolone pivalate is the pivalate ester of the glucocorticoid prednisolone that conforms to the structural formula shown below

Pharmacological classes : Anti-inflammatory agents; glucocorticoids

IUPAC name : [2-[(8S,9S,10R,11S,13S,14S,17R)-11,17-Dihydroxy-10,13-dimethyl-3-oxo-7,8,9,11,12, 14,15,16-octahydro-6H-cyclopenta[a]phenanthren-17-yl]-2-oxoethyl] 2,2-dimethyl-propanoate

Other names : Prednisolone 21-trimethylacetate; prednisolone 21-pivalate

CAS registry number : 1107-99-9

EC number : 214-172-1

Merck Index monograph : 9111 (Prednisolone)

Patch testing : In general, corticosteroids may be tested at 0.1% and 1% in alcohol; late readings (6-10 days) are strongly recommended

Molecular formula : $C_{26}H_{36}O_6$

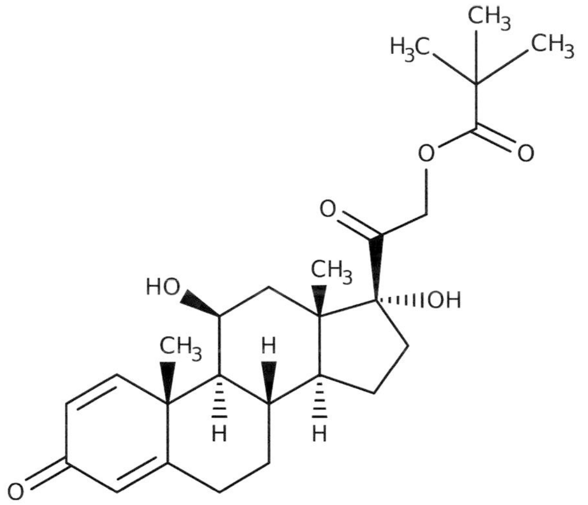

GENERAL

General aspects of corticosteroids used on the skin and mucous membranes are discussed in Chapter 2.4. A practical guideline for diagnosing allergic reactions to corticosteroids is presented in ref. 1. Prednisolone pivalate is used mostly in eye ointments for the treatment of allergic conjunctivitis, of keratitis and uveitis. See also prednisolone (Chapter 3.273), prednisolone acetate (Chapter 3.274), prednisolone caproate (Chapter 3.275), prednisolone hemisuccinate (Chapter 3.276), prednisolone sodium metazoate (Chapter 3.278), and prednisolone valerate acetate (Chapter 3.279).

CONTACT ALLERGY

Case reports and case series

In Leuven, Belgium, between 1990 and 2017, 16,065 patients were investigated for contact allergy and 118 (0.7%) showed positive patch test reactions to topical ophthalmic medications and/or to their ingredients. Eighty-four individuals (71%) reacted to an active principle. Prednisolone pivalate was tested in 471 patients and was the allergen in eye medications in 2. There was also one reaction to prednisolone pivalate in other types of medications (2, overlap with refs. 4 and 5).

From January 1990 to June 2008, in the same clinic in Leuven, Belgium, 315 patients were diagnosed with contact allergy to/allergic contact dermatitis from corticosteroids (CSs) from routine patch testing with a baseline series including tixocortol pivalate, budesonide, hydrocortisone butyrate and prednisone caproate, patch testing with patients' own CS preparations, and testing those with proven contact allergy to a corticosteroid or strongly suspected of CS allergy later with a series of 66 CSs, including two sex hormones (progesterone and testosterone). 71% of the patients had relevant reactions, but these were not specified. In this group of 315 CS allergic patients, 46 had positive patch tests to prednisolone pivalate 0.1% alc. It is unknown how many of these reactions were caused by the use of a pharmaceutical product containing prednisolone pivalate and how many were cross-reactions to other corticosteroids (4, overlap with refs. 2 and 5).

In the period 1990-2008, in Leuven, Belgium, 315 patients were diagnosed with contact allergy to/allergic contact dermatitis from corticosteroids. Eighteen subjects (5.7%) presented with allergic manifestations (conjunctivitis, eczema of the face, periocular skin or eyelids) caused by the use of CS-containing ocular preparations. Three patients had used ophthalmic preparations containing prednisolone and one prednisolone pivalate (5, overlap with refs. 2 and 4).

A 64-year-old woman developed periorbital eczema after eye treatment with a corticosteroid ointment containing prednisolone pivalate. Patch tests were positive to prednisolone pivalate (= prednisolone trimethyl acetate); cross-reactions were not observed, not even to the closely related prednisolone 21-acetate. Details are not available to the author (3).

Cross-reactions, pseudo-cross-reactions and co-reactions
Cross-reactions between corticosteroids are discussed in Chapter 2.8.

LITERATURE

1   Baeck M, Goossens A. Immediate and delayed allergic hypersensitivity to corticosteroids: practical guidelines. Contact Dermatitis 2012;66:38-45
2   Gilissen L, De Decker L, Hulshagen T, Goossens A. Allergic contact dermatitis caused by topical ophthalmic medications: Keep an eye on it! Contact Dermatitis 2019;80:291-297
3   Schmoll M, Hausen BM. Allergic contact dermatitis to prednisolone-21-trimethyl acetate. Z Hautkr 1988;63:311-313 (Article in German)
4   Baeck M, Chemelle JA, Terreux R, Drieghe J, Goossens A. Delayed hypersensitivity to corticosteroids in a series of 315 patients: clinical data and patch test results. Contact Dermatitis 2009;61:163-175
5   Baeck M, De Potter P, Goossens A. Allergic contact dermatitis following ocular use of corticosteroids. J Ocul Pharmacol Ther 2011;27:83-92

Chapter 3.278 PREDNISOLONE SODIUM METAZOATE

IDENTIFICATION

Description/definition   : Prednisolone sodium metazoate is the analog of the synthetic glucocorticoid
                               prednisolone that conforms to the structural formula shown below
Pharmacological classes   : Glucocorticoids
IUPAC name   : Sodium;3-[2-[(8S,9S,10R,11S,13S,14S,17R)-11,17-dihydroxy-10,13-dimethyl-3-oxo-
                               7,8,9,11,12,14,15,16-octahydro-6H-cyclopenta[a]phenanthren-17-yl]-2-oxoethoxy]
                               carbonylbenzenesulfonate
Other names   : Prednisolone methylsulfobenzoate; prednisolone sodium metasulfobenzoate;
                               prednisolone 21-(3-sodium-sulphobenzoate);
CAS registry number   : 630-67-1
EC number   : 211-141-4
Merck Index monograph : 9111 (Prednisolone)
Patch testing   : In general, corticosteroids may be tested at 0.1% and 1% in alcohol; late readings (6-10
                               days) are strongly recommended
Molecular formula   : $C_{28}H_{31}NaO_9S$

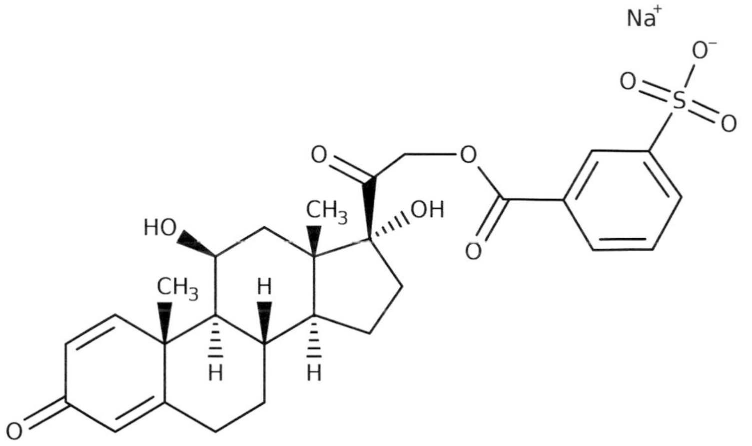

GENERAL

General aspects of corticosteroids used on the skin and mucous membranes are discussed in Chapter 2.4. A practical guideline for diagnosing allergic reactions to corticosteroids is presented in ref. 1. Prednisolone sodium metazoate is used in enema's for application into the rectum for the treatment of colitis ulcerosa in some countries (www.drugs.com). See also prednisolone (Chapter 3.273), prednisolone acetate (Chapter 3.274), prednisolone caproate (Chapter 3.275), prednisolone hemisuccinate (Chapter 3.276), prednisolone pivalate (Chapter 3.277), and prednisolone valerate acetate (Chapter 3.279).

CONTACT ALLERGY

Case reports and case series

From January 1990 to June 2008, in Leuven, Belgium, 315 patients were diagnosed with contact allergy to/allergic contact dermatitis from corticosteroids (CSs) from routine patch testing with a baseline series including tixocortol pivalate, budesonide, hydrocortisone butyrate and prednisone caproate, patch testing with patients' own CS preparations, and testing those with proven contact allergy to a corticosteroid or strongly suspected of CS allergy later with a series of 66 CSs, including two sex hormones (progesterone and testosterone). 71% of the patients had relevant reactions, but these were not specified. In this group of 315 CS allergic patients, 48 had positive patch tests to prednisolone sodium metasulfobenzoate (prednisolone sodium metazoate) 0.1% alc. (2). As this corticosteroid has never been used in pharmaceuticals in Belgium, these positive reactions must all be considered cross-reactions to other corticosteroids.

A 53-year-old woman who had had ulcerative colitis for 11 years experienced worsening of her diarrhea with prednisolone metasulfobenzoate sodium enema. She had positive patch tests to prednisolone 0.1% and 1.0% in isopropyl alcohol. The patient was unable to attend for further tests (3). The authors also described a 40-year-old man who had used prednisolone metasulfobenzoate sodium enema's in the past which had worsened his diarrhea. He had also developed a rash from intravenous hydrocortisone on two occasions. He had positive patch tests to tixocortol pivalate, a hydrocortisone acetate 1% enema, and a positive reaction at 48 hours to intradermal prednisolone acetate enema (3). A third patient in this series was a 39-year-old man with ulcerative colitis for 13 years who had experienced local irritation from prednisolone metasulfobenzoate sodium enema and therefore avoided further use of it. He had a positive patch test to budesonide and a 7-mm reaction to intradermal prednisolone acetate enema at 48 hours (3). Although patch tests with the incriminated enema nor its active ingredient were performed, it is highly likely that these patients had suffered an allergic reaction from prednisolone metasulfobenzoate sodium enema's.

Cross-reactions, pseudo-cross-reactions and co-reactions
Cross-reactions between corticosteroids are discussed in Chapter 2.8.

LITERATURE
1   Baeck M, Goossens A. Immediate and delayed allergic hypersensitivity to corticosteroids: practical guidelines. Contact Dermatitis 2012;66:38-45
2   Baeck M, Chemelle JA, Terreux R, Drieghe J, Goossens A. Delayed hypersensitivity to corticosteroids in a series of 315 patients: clinical data and patch test results. Contact Dermatitis 2009;61:163-175
3   Malik M, Tobin AM, Shanahan F, O'Morain C, Kirby B, Bourke J. Steroid allergy in patients with inflammatory bowel disease. Br J Dermatol 2007;157:967-969

Chapter 3.279 PREDNISOLONE VALERATE ACETATE

IDENTIFICATION

Description/definition   : Prednisolone valerate acetate is the valerate acetate ester of prednisolone that conforms to the structural formula shown below

Pharmacological classes : Anti-inflammatory agents; glucocorticoids

IUPAC name   : [(8S,9S,10R,11S,13S,14S,17R)-17-(2-Acetyloxyacetyl)-11-hydroxy-10,13-dimethyl-3-oxo-7,8,9,11,12,14,15,16-octahydro-6H-cyclopenta[a]phenanthren-17-yl] pentanoate

Other names   : 11β,17α,21-Trihydroxy-1,4-pregnadiene-3,20-dione 21-acetate 17-valerate

CAS registry number   : 72064-79-0

EC number   : 276-312-8

Merck Index monograph : 9111 (Prednisolone)

Patch testing   : In general, corticosteroids may be tested at 0.1% and 1% in alcohol; late readings (6-10 days) are strongly recommended

Molecular formula   : $C_{28}H_{38}O_7$

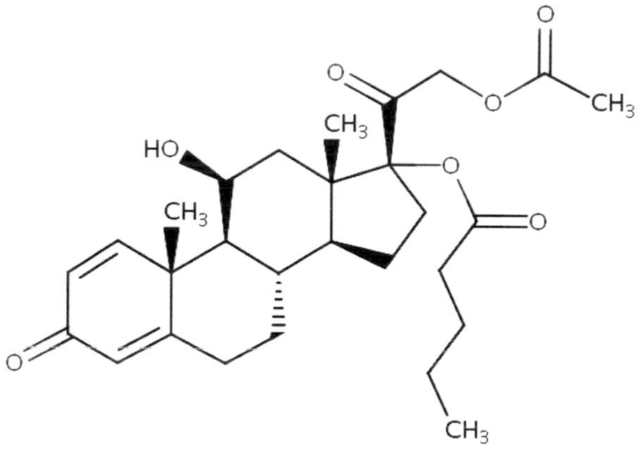

GENERAL

General aspects of corticosteroids used on the skin and mucous membranes are discussed in Chapter 2.4. A practical guideline for diagnosing allergic reactions to corticosteroids is presented in ref. 1. See also prednisolone (Chapter 3.273), prednisolone acetate (Chapter 3.274), prednisolone caproate (Chapter 3.275), prednisolone hemisuccinate (Chapter 3.276), prednisolone pivalate (Chapter 3.277), and prednisolone sodium metazoate (Chapter 3.278).

CONTACT ALLERGY

Case reports

A 67-year-old female patient developed itchy erythema on her head. When she was treated with a lotion containing prednisolone acetate valerate, the eruption exacerbated. Patch testing showed positive reactions to the lotion and to prednisolone valerate acetate. There was a cross-reaction to hydrocortisone butyrate propionate (2). Several other case reports of contact allergy to/allergic contact dermatitis from prednisolone valerate acetate appear to have been published in Japanese literature, but their details are not available (3-5).

Cross-reactions, pseudo-cross-reactions and co-reactions

A patient sensitized to prednisolone valerate acetate may have cross-reacted to hydrocortisone butyrate propionate (hydrocortisone probutate) (2). Cross-reactions between corticosteroids are discussed in Chapter 2.8.

LITERATURE

1   Baeck M, Goossens A. Immediate and delayed allergic hypersensitivity to corticosteroids: practical guidelines. Contact Dermatitis 2012;66:38-45

2   Ogura, Natsuaki M, Hirano A, Yasugi Y, Miyata A, Yamanishi K. A case of contact dermatitis due to prednisolone valerate acetate. Hifu no kagaku 2005;4:111-115 (Article in Japanese)

3   Tamiya Y, Matsumura E, Sasaki E, Ishinaga M. Allergic contact dermatitis due to sulconazole nitrate, streptomycin sulfate and prednisolone valerate acetate. Skin Res 1989;31(Suppl.):206-211 (Article in Japanese)

4   Mizuno E. Contact dermatitis due to prednisolone valerate acetate. Published 1994, cited in semanticscholar.org; bibliographic data unknown, most likely a Japanese journal

5   Matsumoto C. A case of contact allergy due to prednisolone valerate acetate. Published 1995, cited in semanticscholar.org; bibliographic data unknown, most likely a Japanese journal

Chapter 3.280 PREDNISONE

IDENTIFICATION

Description/definition	: Prednisone is the synthetic glucocorticoid that conforms to the structural formula shown below
Pharmacological classes	: Anti-inflammatory agents; glucocorticoids; antineoplastic agents, hormonal
IUPAC name	: (8S,9S,10R,13S,14S,17R)-17-Hydroxy-17-(2-hydroxyacetyl)-10,13-dimethyl-6,7,8,9,12, 14,15,16-octahydrocyclopenta[a]phenanthrene-3,11-dione
Other names	: 17,21-Dihydroxypregna-1,4-diene-3,11,20-trione; dehydrocortisone
CAS registry number	: 53-03-2
EC number	: 200-160-3
Merck Index monograph	: 9112
Patch testing	: In general, corticosteroids may be tested at 0.1% and 1% in alcohol; late readings (6-10 days) are strongly recommended
Molecular formula	: $C_{21}H_{26}O_5$

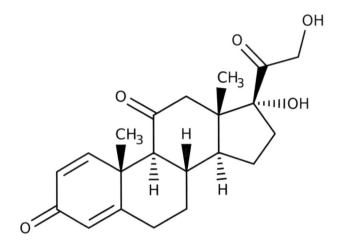

GENERAL

General aspects of corticosteroids used on the skin and mucous membranes are discussed in Chapter 2.4. A practical guideline for diagnosing allergic reactions to corticosteroids is presented in ref. 1. Prednisone is used only orally. This means that by far most allergic reactions to 'prednisone' have in fact been cross-reactions to other corticosteroids.

CONTACT ALLERGY

Patch testing in groups of patients

In the period 200-2005, in the Mayo Clinic in the USA, 1187 patients suspected of corticosteroid allergy were patch tested with prednisone 5% pet. and there were 6 (0.5%) positive reactions; in 4 cases (67%), these were judged to be relevant (2).

Case series

From January 1990 to June 2008, in Leuven, Belgium, 315 patients were diagnosed with contact allergy to/allergic contact dermatitis from corticosteroids (CSs) from routine patch testing with a baseline series including tixocortol pivalate, budesonide, hydrocortisone butyrate and prednisone caproate, patch testing with patients' own CS preparations, and testing those with proven contact allergy to a corticosteroid or strongly suspected of CS allergy later with a series of 66 CSs, including two sex hormones (progesterone and testosterone). 71% of the patients had relevant reactions, but these were not specified. In this group of 315 CS allergic patients, 43 had positive patch tests to prednisone 0.1% alc. (3). As prednisone is used only orally, it may be assumed that virtually all reactions were the result of cross-reactivity to other corticosteroids.

Cross-reactions, pseudo-cross-reactions and co-reactions
Cross-reactions between corticosteroids are discussed in Chapter 2.8.

Cutaneous adverse drug reactions from systemic administration caused by type IV (delayed-type) hypersensitivity
Cutaneous adverse drug reactions from systemic administration of prednisone caused by type IV (delayed-type) hypersensitivity, including systemic contact dermatitis (4,5,6) and a bullous eruption (7), are planned to be discussed in Volume IV of the *Monographs in Contact Allergy* series on Systemic drugs.

LITERATURE

1   Baeck M, Goossens A. Immediate and delayed allergic hypersensitivity to corticosteroids: practical guidelines. Contact Dermatitis 2012;66:38-45

2   Davis MD, El-Azhary RA, Farmer SA. Results of patch testing to a corticosteroid series: a retrospective review of 1188 patients during 6 years at Mayo Clinic. J Am Acad Dermatol 2007;56:921-927

3   Baeck M, Chemelle JA, Terreux R, Drieghe J, Goossens A. Delayed hypersensitivity to corticosteroids in a series of 315 patients: clinical data and patch test results. Contact Dermatitis 2009;61:163-175

4   Quirce S, Alvarez MJ, Olaguibel JM, Tabar AI. Systemic contact dermatitis from oral prednisone. Contact Dermatitis 1994;30:53-54

5   De Benito V, Ratón JA, Palacios A, Garmendia M, Gardeazábal J. Systemic contact dermatitis to prednisone: a clinical model approach to the management of systemic allergy to corticosteroids. Clin Exp Dermatol 2012;37:680-681

6   Bircher AJ, Bigliardi P, Zaugg T, Mäkinen-Kiljunen S. Delayed generalized allergic reactions to corticosteroids. Dermatology 2000;200:349-351

7   Lew DB, Higgins GC, Skinner RB, Snider MD, Myers LK. Adverse reaction to prednisone in a patient with systemic lupus erythematosus. Pediatr Dermatol 1999;16:146-150

Chapter 3.281 PRILOCAINE

IDENTIFICATION

Description/definition	: Prilocaine is the α-amino acid amide that conforms to the structural formula shown below
Pharmacological classes	: Anesthetics, local
IUPAC name	: N-(2-Methylphenyl)-2-(propylamino)propanamide
Other names	: Propitocaine
CAS registry number	: 721-50-6
EC number	: 211-957-0
Merck Index monograph	: 9132
Patch testing	: Hydrochloride 5.0% pet.(Chemotechnique)
Molecular formula	: $C_{13}H_{20}N_2O$

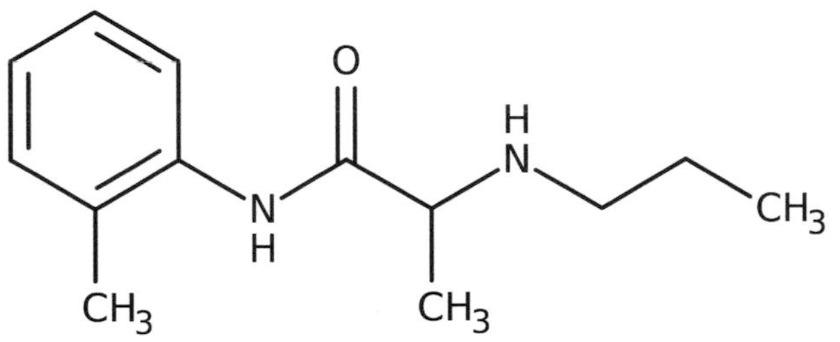

GENERAL

Prilocaine is an intermediate-acting local anesthetic of the amide type chemically related to lidocaine. It is used for local anesthesia by infiltration and is the most often used local anesthetic in dentistry. In pharmaceutical products, both prilocaine and prilocaine hydrochloride (CAS number 1786-81-8, EC number 217-244-0, molecular formula $C_{13}H_{21}ClN_2O$) may be employed (1). It is present, together with lidocaine, in a frequently used anesthetic cream for surface anesthesia called EMLA ® (4,5,7,8,11,12,13,22).

CONTACT ALLERGY

General

Although few cases of allergic contact dermatitis from topical prilocaine have been reported, repeated use of lidocaine-prilocaine cream not infrequently leads to sensitization, (nearly) always from the prilocaine component. Indeed, of 75 patients non-selected dialysis patients, 8 (11%) had positive patch tests reactions to the cream, in all cases caused by prilocaine sensitization (22).

Non-allergic cutaneous reactions to this prilocaine-lidocaine cream include irritant contact dermatitis (23,24), potentially leading to diagnostic errors (29) and purpuric/petechial reactions (25-28).

Testing in groups of patients

In the hemodialysis unit of a hospital in Marseille, France, 75 non-selected hemodialysis patients (41 men, 34 women, mean age 65 years), with a mean 3.8 years under dialysis, were patch tested. There were 8 positive reactions (11%) to lidocaine-prilocaine cream, tested 5% in water, of which 7 were relevant. These patients had symptoms including localized pruritus at the fistula, or eczema at the vascular access previously anesthetized by the cream. Lidocaine-prilocaine cream was the primary allergen in this study. Eleven per cent of the hemodialysis population developed delayed sensitization after an average period of use of 4 years, affecting 37% of patients with positive patch tests (to any allergen) and 13% of patients which had previously used lidocaine–prilocaine cream (22).

In the USA and Canada, in 2001-2002, 4890 patients were routinely tested with prilocaine 2.5% pet. and there were 4 (0.1%) positive reactions, two of which were considered to be relevant (2). In Belgium, during 1990-2014, 100 patients suspected of iatrogenic contact dermatitis were patch tested with prilocaine 5% pet. and 3 patients (3%) had a positive reaction. 96% of the positive reactions to all allergen in this study were considered to be relevant (3).

Case reports

An 80-year-old woman with persistent local pain on her left thorax after clinical clearance of herpes zoster started to apply an anesthetic cream containing prilocaine and lidocaine once a day. After ten days, she developed erythema and inflammation at the application site. Patch tests were positive to the cream, to lidocaine 1% pet. (on 2 occasions) and to prilocaine 5% pet. and negative to the other constituents of the cream (4).

Short summaries of other case reports of allergic contact dermatitis from prilocaine are shown in table 3.281.1.

Table 3.281.1 Short summaries of case reports of allergic contact dermatitis from prilocaine

Year and country	Sex	Age	Positive patch tests	Clinical data and comments	Ref.
2007 Spain	M	76	cream 'as is'; PRILO 5% pet.	itchy vesicular eczematous eruption from cream on leg ulcer; co-reaction to pramoxine and bupivacaine	5
2006 Spain	?	67	cream 'as is'; PRILO 0.5%, 1% and 5% pet.	itchy erythematous and desquamative lesion on the fore-arm from cream before cannulation for hemodialysis	7
2005 United Kingdom	M	9	cream 'as is'; PRILO 5% pet.	eczematous rash at site of cream application before needle insertion	8
1998 United Kingdom	F	35	PRILO 1% pet.	worsening of sunburn from PRILO-lidocaine cream resulting in blistering; also contact allergy to prilocaine and type-I anaphylactic reaction to lidocaine buccal injection	20
1997 Finland	F	56	4 prilocaine products	red, edematous weeping skin 3 days after infiltration anesthesia with prilocaine; also positive intracutaneous test (delayed reading); cross-reaction to articaine	10
1996 France	F	39	cream 'as is'; PRILO 1% and 5% pet.	itchy, erythematous edematous vesicular lesions on the arm from anesthetic cream before hemodialysis cannulation	11
1995 USA	F	73	cream 'as is'; PRILO 1% and 2% pet.	erythematous, warm, indurated vesiculated and excoriated lesions from anesthetic cream for post-herpetic neuralgia	12
1994 Belgium	M	78	cream 'as is'; PRILO 5% pet.	dermatitis over the back of the left foot from anesthetic cream applied to painful arterial ulcer; positive intradermal test to prilocaine solution at D2	13

PRILO: prilocaine

Another case report or information on contact allergy to prilocaine, adequate data of which are not available to the author, can be found in ref. 15.

Cutaneous adverse drug reactions from systemic administration caused by type IV (delayed-type) hypersensitivity
Cutaneous adverse drug reactions from systemic administration of prilocaine caused by type IV (delayed-type) hypersensitivity, including cutaneous and mucosal injections for infiltration anesthesia (17,21,30), are planned to be discussed in Volume IV of the *Monographs in Contact Allergy* series on Systemic drugs.

Cross-reactions, pseudo-cross-reactions and co-reactions
Several patients sensitized to lidocaine have cross-reacted to prilocaine (6,14,16,17,18,19; Chapter 3.201 Lidocaine). Conversely, patients sensitized to prilocaine do not seem to cross-react – and even hardly ever co-react from their combined presence in a cream (4) - to lidocaine (5,7,8,10,11,12,13). A patient sensitized to prilocaine co-reacted to pramoxine and mepivacaine (5). These were considered to be cross-reactions, but pramoxine has a quite different structure. A patient sensitized to mepivacaine probably cross-reacted to prilocaine (9).

LITERATURE

1   The data in the section 'General' may have been obtained from literature discussed in this chapter, but mostly also or exclusively from one or more of the following online sources: ChemIDPlus Advanced, PubChem, DrugBank, RxList, Drug Central, Drugs.com, and Wikipedia
2   Pratt MD, Belsito DV, DeLeo VA, Fowler JF Jr, Fransway AF, Maibach HI, et al. North American Contact Dermatitis Group patch-test results, 2001-2002 study period. Dermatitis 2004;15:176-183
3   Gilissen L, Goossens A. Frequency and trends of contact allergy to and iatrogenic contact dermatitis caused by topical drugs over a 25-year period. Contact Dermatitis 2016;75:290-302
4   Timmermans MW, Bruynzeel DP, Rustemeyer T. Allergic contact dermatitis from EMLA cream: concomitant sensitization to both local anesthetics lidocaine and prilocaine. J Dtsch Dermatol Ges 2009;7:237-238
5   Garcia F, Iparraguirre A, Blanco J, Alloza P, Vicente J, Bascones O, et al. Contact dermatitis from prilocaine with cross-sensitivity to pramocaine and bupivacaine. Contact Dermatitis 2007;56:120-122
6   Langan SM, Collins P. Photocontact allergy to oxybenzone and contact allergy to lignocaine and prilocaine. Contact Dermatitis 2006;54:173-174
7   Pérez-Pérez LC, Fernández-Redondo V, Ginarte-Val M, Paredes-Suárez C, Toribio J. Allergic contact dermatitis from EMLA cream in a hemodialyzed patient. Dermatitis 2006;17:85-87

8   Ismail F, Goldsmith PC. Emla cream-induced allergic contact dermatitis in a child with thalassaemia major. Contact Dermatitis 2005;52:111

9   Kanerva L, Alanko K, Estlander T, Jolanki R. Inconsistent intracutaneous and patch test results in a patient allergic to mepivacaine and prilocaine. Contact Dermatitis 1998;39:197-199

10 Suhonen R, Kanerva L. Contact allergy and cross-reactions caused by prilocaine. Am J Cont Dermat 1997;8:231-235

11 Le Coz CJ, Cribier BJ, Heid E. Patch testing in suspected allergic contact dermatitis due to Emla cream in haemodialyzed patients. Contact Dermatitis 1996;35:316-317

12 Thakur BK, Murali MR. EMLA cream-induced allergic contact dermatitis: a role for prilocaine as an immunogen. J Allergy Clin Immunol 1995;95:776-778

13 Van den Hove J, Decroix J, Tennstedt D, Lachapelle JM. Allergic contact dermatitis from prilocaine, one of the local anaesthetics in EMLA cream. Contact Dermatitis 1994;30:239

14 Black RJ, Dawson TA, Strang WC. Contact sensitivity to lignocaine and prilocaine. Contact Dermatitis 1990;23:117-118

15 Aldrete TA, O'Higgins TW. Evaluation of patients with a history of allergy to local anaesthetics. South Med J 1971;64:1115 (data cited in ref. 14)

16 Fregert S, Tegner E, Thelin I. Contact allergy to lidocaine. Contact Dermatitis 1979;5:185-188

17 Curley RK, Macfarlane AW, King CM. Contact sensitivity to the amide anesthetics lidocaine, prilocaine, and mepivacaine. Arch Dermatol 1986;122:924-926

18 Bircher AJ, Messmer SL, Surber C, Rufli T. Delayed-type hypersensitivity to subcutaneous lidocaine with tolerance to articaine: confirmation by in vivo and in vitro tests. Contact Dermatitis 1996;34:387-389

19 Bassett I, Delaney T, Freeman S. Can injected lignocaine cause allergic contact dermatitis? Australas J Dermatol 1996;37:155-156

20 Downs AM, Lear JT, Wallington TB, Sansom JE. Contact sensitivity and systemic reaction to pseudoephedrine and lignocaine. Contact Dermatitis 1998;39:33

21 Trautmann A, Stoevesandt J. Differential diagnosis of late-type reactions to injected local anaesthetics: inflammation at the injection site is the only indicator of allergic hypersensitivity. Contact Dermatitis 2019;80:118-124

22 Gaudy-Marqueste C, Jouhet C, Castelain M, Brunet P, Berland Y, Grob JJ, Richard MA. Contact allergies in haemodialysis patients: a prospective study of 75 patients. Allergy 2009;64:222-228

23 Dong H, Kerl H, Cerroni L. EMLA cream-induced irritant contact dermatitis. J Cutan Pathol 2002;29:190-192

24 Kluger N, Raison-Peyron N, Michot C, Guillot B, Bessis D. Acute bullous irritant contact dermatitis caused by EMLA® cream. Contact Dermatitis 2011;65:181-183

25 Roldán-Marín R, de-la-Barreda Becerril F. Petechial and purpuric eruption induced by lidocaine/prilocaine cream: a rare side effect. J Drugs Dermatol 2009;8:287-288

26 Neri I, Savoia F, Guareschi E, Medri M, Patrizi A. Purpura after application of EMLA cream in two children. Pediatr Dermatol 2005;22:566-568

27 Calobrisi SD, Drolet BA, Esterly NB. Petechial eruption after the application of EMLA cream. Pediatrics 1998;101(3Pt.1):471-473

28 De Waard-van der Spek FB, Oranje AP. Purpura caused by Emla is of toxic origin. Contact Dermatitis 1997;36:11-13

29 Schmutz JL, Trechot P. Note the irritant effect of Emla® cream potentially leading to diagnostic errors. Ann Dermatol Venereol 2012;139:82-83 (Article in French)

30 Spornraft-Ragaller P, Stein A. Contact dermatitis to prilocaine after tumescent anesthesia. Dermatol Surg 2009;35:1303-1306

Chapter 3.282 PROCAINE

IDENTIFICATION

Description/definition : Procaine is the *p*-aminobenzoic acid derivative that conforms to the structural formula shown below
Pharmacological classes : Anesthetics, local
IUPAC name : 2-(Diethylamino)ethyl 4-aminobenzoate
Other names : *p*-Aminobenzoyldiethylaminoethanol; Novocain ®
CAS registry number : 59-46-1
EC number : 200-426-9
Merck Index monograph : 9145
Patch testing : Hydrochloride, 1.0% pet.(Chemotechnique, SmartPracticeCanada, SmartPracticeEurope); hydrochloride, 2% pet. (SmartPracticeCanada, SmartPracticeEurope)
Molecular formula : $C_{13}H_{20}N_2O_2$

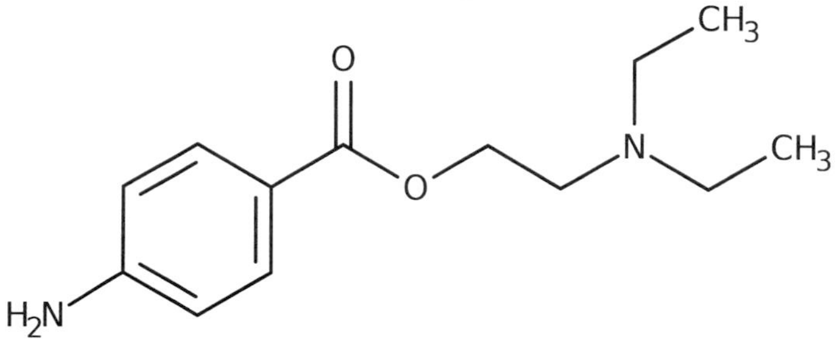

GENERAL

Procaine is a local anesthetic of the ester type that has a slow onset and a short duration of action. It is mainly used for production of local or regional anesthesia, particularly for oral surgery. Procaine (like cocaine) has the advantage of constricting blood vessels which reduces bleeding, unlike other local anesthetics such as lidocaine. In pharmaceutical products, both procaine and procaine hydrochloride (CAS number 51-05-8, EC number 200-077-2, molecular formula $C_{13}H_{21}ClN_2O_2$) may be employed (1).

CONTACT ALLERGY

General

Procaine, an ester-type local anesthetic, was formerly widely used, especially in dentistry. The first description of contact sensitivity to procaine was in 1921 in three dentists (16). While further reports followed (17-19), procaine remained popular in dentistry and was also a common constituent of preparations such as eye drops. Occupational allergic contact dermatitis in dentists and physicians became well-known (21).

In Czechoslovakia, in the mid-1950s, over 10% of routinely tested dermatitis patients reacted to procaine. The prevalence fell dramatically to 0.5% in the period 1963-1967, when procaine had largely been replaced (20). Later, some cases of allergic contact dermatitis appeared from the use of Gerovital ® hair lotions and creams containing procaine, imported from or purchased in Rumania (6,7,8). Some patients developed occupational allergic contact dermatitis from veterinary medications (23,24).

The early literature on procaine allergy has been reviewed in refs. 14 and 15. Because procaine can largely be considered a historical allergen, the subject of contact allergy to it is presented in brief format

Contact allergy in the general population

In Germany, for the period 1995-2004, the population-based relative incidence (RI) of contact sensitization to procaine (cases/100,000 defined daily doses (DDDs) per year) was estimated to be 3.2. In the group of local anesthetics, the RI ranged from 1.5 (lidocaine) to 413.9 (benzocaine) (3).

Patch testing in groups of selected patients

Results of patch testing procaine in groups of selected patients (individuals with anogenital or vulval dermatoses, patients suspected to allergy to topical medicaments) are shown in table 3.282.1. From 1990 on, low prevalences of sensitization to procaine ranging from 0.5% to 1.5% were observed. The relevance of the positive reactions was either not mentioned (2,12,23) or not specified for procaine (4.11), leaving to possibility open that a number of reactions have been the result of cross-sensitivity, e.g. to benzocaine or *p*-phenylenediamine.

Table 3.282.1 Patch testing in groups of patients: Selected patient groups

Years and Country	Test conc. & vehicle	Number of patients tested	positive (%)	Selection of patients (S); Relevance (R); Comments (C)	Ref.
2004-2008 IVDK			(0.8%)	S: patients with anogenital dermatoses tested with a medicament series; R: not stated; C: number of patients tested unknown	12
1997-2007 Finland	1% pet.	620	4 (0.6%)	S: patients suspected of allergy to topical therapies; R: not stated	23
1978-2005 Taiwan		603	3 (0.5%)	S: patients suspected of contact allergy to medicaments; R: 65% of the reactions to all medicaments were considered to be relevant	4
1995-2004 IVDK	2% pet.	720	(1.5%)	S: not stated; R: not stated	2
1992-1994 U.K.	1% pet.	69	1 (1%)	S: women with 'vulval problems'; R: 58% for all allergens together	11
1967-1970 Poland	2% water	600	29 (4.8%)	S: unknown; R: not stated	22

IVDK: Information Network of Departments of Dermatology (Germany, Austria, Switzerland)

Case report

A 43-year-old man developed a perianal eruption 2 hours after applying an ointment containing procaine HCl 1%, benzocaine 0.25%, and enoxolone 0.7% as active ingredients, which he had previously used for hemorrhoids for 2 years without problems. Patch tests were positive to the ointment 'as is' and 10%, to procaine HCl 1% pet., benzocaine 1% pet., and enoxolone 10% pet. (9).

Occupational contact dermatitis

A 44-year-old man had worked in a pigsty for 12 years. He had used injectable antibiotics in his 1400 animal pigsty to treat suspected infections in animals. He presented with swollen eyelids, redness and flaking in the face, in the upper part of the body, on the arms and behind the knees. Patch tests were positive to penicillin and procaine, with both of which he came in contact during his work. There were also positive reactions to benzocaine, tetracaine, p-phenylenediamine (all para-amino compounds, cross-reactions), to dibucaine (not related) and to the caine mix III (containing benzocaine, dibucaine and tetracaine). Penicillin-containing and procaine-containing medicines were replaced by other preparations, and since then the patient has been working in his pigsty without skin symptoms (23). Occupational allergic contact dermatitis to procaine was also found in a 47-year-old veterinary obstetrician from regularly using procaine penicillin G in his work (24).

In a group of 107 workers in the pharmaceutical industry with dermatitis, investigated in Warsaw, Poland, before 1989, one reacted to procaine, tested 2% pet. (5). In Norway, before 1985, of ten veterinary surgeons with occupational allergic contact dermatitis with chronic or relapsing eczema of the hands as the main complaint, 3 had positive patch tests to procaine 2% pet. (10). In 1921, occupational allergic contact dermatitis to procaine was described in three dentists (16). Such reactions in dentists and physicians (27) later became well-known (21,26) and procaine remained an important occupational allergen until the mid-1970s (25).

Allergic contact dermatitis from procaine in cosmetics

There have been several case reports of allergic contact dermatitis from a brand of cosmetics/cosmeceuticals from Rumania containing procaine: a hair growth stimulating lotion and a cream (6,7,8). In one case the contact dermatitis was caused by the hair lotion used by the patient's partner, a so-called 'connubial allergic contact dermatitis' (6). A woman who had previously suffered allergic gingivostomatitis following the application of a procaine-based local anesthetic for treatment of a decayed tooth, developed an intense, red, swollen and vesicular eruption on the face after using a cosmetic cream containing 1% procaine (same brand as the lotion) (8). In those days, procaine was already prohibited in cosmetics in the EEC (European Economic Community).

Cross-reactions, pseudo-cross-reactions and co-reactions

Procaine, the 2-(diethylamino)ethyl ester of p-aminobenzoic acid (PABA) is a 'para-amino' compound and therefore may cross-react with other ester-type local anesthetics including benzocaine (ethyl p-aminobenzoate) (23,27), tetracaine (23,27,28), butacaine (27), and other para-compounds including p-phenylenediamine (23,27), p-aminoazobenzene, p-aminobenzoic acid (PABA) (27), p-aminophenol (6), and sulfonamide (27).

Cutaneous adverse drug reactions from systemic administration caused by type IV (delayed-type) hypersensitivity

Cutaneous adverse drug reactions from systemic administration of procaine caused by type IV (delayed-type) hypersensitivity, including reactions from subcutaneous injections (13) and exanthemas from injections of penicillin

G procaine (14,15), are planned to be discussed in Volume IV of the *Monographs in Contact Allergy* series on Systemic drugs.

LITERATURE

1   The data in the section 'General' may have been obtained from literature discussed in this chapter, but mostly also or exclusively from one or more of the following online sources: ChemIDPlus Advanced, PubChem, DrugBank, RxList, Drug Central, Drugs.com, and Wikipedia

2   Menezes de Padua CA, Uter W, Schnuch A. Contact allergy to topical drugs: prevalence in a clinical setting and estimation of frequency at the population level. Pharmacoepidemiol Drug Saf 2007;16:377-384

3   Menezes de Padua CA, Schnuch A, Nink K, Pfahlberg A, Uter W. Allergic contact dermatitis to topical drugs – Epidemiological risk assessment. Pharmacoepidemiol Drug Saf 2008;17:813-821

4   Shih Y-H, Sun C-C, Tseng Y-H, Chu C-Y. Contact dermatitis to topical medicaments: a retrospective study from a medical center in Taiwan. Dermatol Sinica 2015;33:181-186

5   Rudzki E, Rebandel P, Grzywa Z. Contact allergy in the pharmaceutical industry. Contact Dermatitis 1989;21:121-122

6   Dooms-Goossens A, Swinnen E, VanderMaesen J, Marien K, Dooms M. Connubial dermatitis from a hair lotion. Contact Dermatitis 1987;16:41-42

7   Förström L, Hannuksela M, Idänpään-Heikkilä J, Salo OP. Hypersensitivity reactions to Gerovital. Dermatologica 1977;154:367-369

8   Goitre M, Bedello PG, Cane D, Roncarolo G. Contact dermatitis from novocaine in Gerovital® cream. Contact Dermatitis 1985;12:234-235

9   Tanaka S, Otsuki T, Matsumoto Y, Hayakawa R, Sugiura M. Allergic contact dermatitis from enoxolone. Contact Dermatitis 2001;44:192

10 Falk ES, Hektoen H, Thune PO. Skin and respiratory tract symptoms in veterinary surgeons. Contact Dermatitis 1985;12:274-278

11 Lewis FM, Harrington CI, Gawkrodger DJ. Contact sensitivity in pruritus vulvae: a common and manageable problem. Contact Dermatitis 1994;31:264-265

12 Bauer A. Contact sensitization in the anal and genital area. Curr Probl Dermatol 2011;40:133-141

13 Trautmann A, Stoevesandt J. Differential diagnosis of late-type reactions to injected local anaesthetics: inflammation at the injection site is the only indicator of allergic hypersensitivity. Contact Dermatitis 2019;80:118-124

14 Fernström AI. Studies on procaine allergy with reference to urticaria due to procaine penicillin treatment. B. The medical uses of procaine and reactions due to procaine penicillin. Review of the literature. Acta Derm Venereol 1960;40:19-34.

15 Fernström AI. Studies on procaine allergy with reference to urticaria due to procaine penicillin treatment. C. Reactions following procaine due to direct skin contact or injection. Hypersensitivity to procaine as an expression of cross-sensitization. Effects of procaine penicillin in persons with clinically verified sensitivity to procaine or chemically related substances. Review of the literature. Acta Derm Venereol 1960;40:175-205

16 Lane CG. Occupational dermatitis in dentists: Susceptibility to procaine. Arch Dermatol 1921;3:235-244

17 James BM. Procaine dermatitis. JAMA 1931;97:440-443

18 Waldron GW. Hypersensitivity to procaine. Mayo Clin Proc 1934;9:254-256

19 Goodman MH. Cutaneous hypersensitivity to the procaine anesthetics. J Invest Dermatol 1939;2:53-66

20 Heygi E. Procain sensitivity 1948-1967. Contact Dermatitis Newsletter 1969;5:95

21 Lane CG, Luikart R. Dermatitis from local anesthetics, with a review of one hundred and seven cases from the literature. J Am Med Assoc 1951;146:717-720

22 Rudzki E, Kleniewska D. The epidemiology of contact dermatitis in Poland. Br J Dermatol 1970;83:543-545

23 Jussi L, Lammintausta K. Sources of sensitization, cross-reactions, and occupational sensitization to topical anaesthetics among general dermatology patients. Contact Dermatitis 2009;60:150-154

24 Bruijn MS, Lavrijsen APM, van Zuuren EJ. An unusual case of contact dermatitis to procaine. Contact Dermatitis 2009;60:182-183

25 Hensten-Pettersen A, Jacobsen N. The role of biomaterials as occupational hazards in dentistry. Int Dent J 1990;40:159-166

26 Laden EL, Wallace DA. Contact dermatitis due to procaine; a common occupational disease of dentists. J Invest Dermatol 1949;12:299-306

27 Peck SM, Feldman FF. Contact allergic dermatitis due to the procaine fraction of procaine penicillin. J Invest Dermatol 1949;13:109

28 Kalveram K, Günnewig W, Wehling K, Forck G. Tetracaine allergy: cross-reactions with para-compounds? Contact Dermatitis 1978;4:376

Chapter 3.283 PROCINONIDE

IDENTIFICATION

Description/definition : Procinonide is the synthetic glucocorticoid that conforms to the structural formula shown below

Pharmacological classes : Glucocorticoids

IUPAC name : [2-[(1S,2S,4R,8S,9S,11S,12R,13S,19S)-12,19-Difluoro-11-hydroxy-6,6,9,13-tetramethyl-16-oxo-5,7-dioxapentacyclo[10.8.0.0^{2,9}.0^{4,8}.0^{13,18}]icosa-14,17-dien-8-yl]-2-oxoethyl] propanoate

Other names : Fluocinolone acetonide 21-propionate; 6α,9-difluoro-11β,16α,17,21-tetrahydroxypregna-1,4-diene-3,20-dione cyclic 16,17-acetal with acetone, 21-propionate

CAS registry number : 58497-00-0

EC number : 261-289-9

Patch testing : Generally, corticosteroids may be tested at 0.1% and 1% in alcohol; late readings (6-10 days) are strongly recommended

Molecular formula : $C_{27}H_{34}F_2O_7$

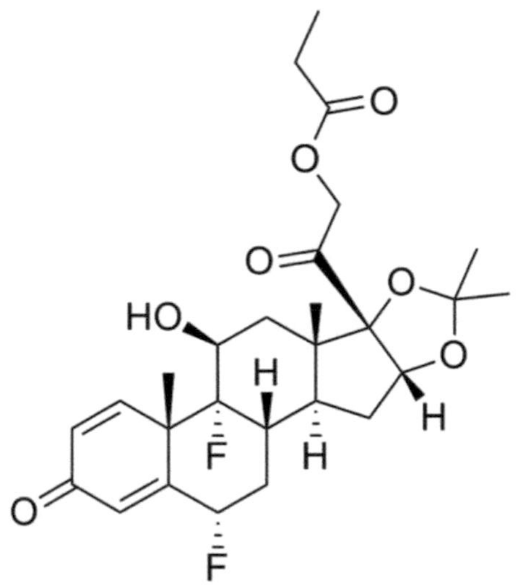

GENERAL

General aspects of corticosteroids used on the skin and mucous membranes are discussed in Chapter 2.4. A practical guideline for diagnosing allergic reactions to corticosteroids is presented in ref. 1.

CONTACT ALLERGY

Case series

From January 1990 to June 2008, in Leuven, Belgium, 315 patients were diagnosed with contact allergy to/allergic contact dermatitis from corticosteroids (CSs) from routine patch testing with a baseline series including tixocortol pivalate, budesonide, hydrocortisone butyrate and prednisone caproate, patch testing with patients' own CS preparations, and testing those with proven contact allergy to a corticosteroid or strongly suspected of CS allergy later with a series of 66 CSs, including two sex hormones (progesterone and testosterone). 71% of the patients had relevant reactions, but these were not specified. In this group of 315 CS allergic patients, 29 had positive patch tests to procinonide 0.1% alc. (2). As this corticosteroid has never been used in pharmaceuticals in Belgium, these positive reactions must all be considered cross-reactions to other corticosteroids.

Cross-reactions, pseudo-cross-reactions and co-reactions

Cross-reactions between corticosteroids are discussed in Chapter 2.8.

LITERATURE

1   Baeck M, Goossens A. Immediate and delayed allergic hypersensitivity to corticosteroids: practical guidelines. Contact Dermatitis 2012;66:38-45

2   Baeck M, Chemelle JA, Terreux R, Drieghe J, Goossens A. Delayed hypersensitivity to corticosteroids in a series of 315 patients: clinical data and patch test results. Contact Dermatitis 2009;61:163-175

Chapter 3.284 PROFLAVINE

IDENTIFICATION

Description/definition	: Proflavine is the acridine derivative that conforms to the structural formula shown below
Pharmacological classes	: Anti-infective agents, local
IUPAC name	: Acridine-3,6-diamine
CAS registry number	: 92-62-6
EC number	: 202-172-4
Merck Index monograph	: 9157
Patch testing	: Hydrochloride or hemisulfate 0.5% pet.
Molecular formula	: $C_{13}H_{11}N_3$

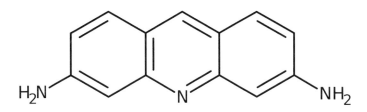

GENERAL

Proflavine is an acridine derivative which has bacteriostatic properties against many gram-positive bacteria. It is (or was) used as topical anti-infective agent, mainly in solutions and wound dressings. In pharmaceutical products, proflavine is employed as proflavine hemisulfate (CAS number 1811-28-5, EC number 217-320-3, molecular formula $C_{26}H_{24}N_6O_4S$) (1).

CONTACT ALLERGY

General

Proflavine was commonly used in an aqueous solution as an antiseptic in the tropics and developing countries, where more effective and expensive antiseptics and antibiotics were not readily available. Proflavine was a very common sensitizer in Singapore in the 1980s, with frequencies of sensitization of 6-7% in consecutive patients patch tested for suspected contact dermatitis (table 3.284.1). This relatively cheap antiseptic, which was shown to be a strong sensitizer in animal experiments (9), was available in many OTC drugs and also commonly found in first aid boxes (3,7). It often caused acute and severe dermatitis, in about 20% of the patients with secondary spread. Erythema multiforme-like eruptions with urticarial papules, plaques, target lesions and purpura have been observed (5,11). In recent years, the frequency of sensitization in Singapore (there are no data from other countries) has dropped considerably (2), probably because proflavine has largely been replaced with more modern and effective antiseptics.

Patch testing in groups of patients

The results of studies performing routine testing with proflavine HCl, all from Singapore (with considerable overlap) are shown in table 3.284.1. In the 1980s, frequencies of sensitization were 6-7%, but the rate had dropped to 1.2% in the period 2009-2013 (2).

Table 3.284.1 Patch testing in groups of patients: Routine testing

Years and Country	Test conc. & vehicle	Number of patients tested	positive (%)	Relevance (R); Comments (C)	Ref.
2009-2013 Singapore		2598	(1.2%)	R: present + past relevance: 8%; C: range of positive reactions per year 0.5 - 2.5%; the test material was proflavine hemisulfate	2
1986-1990 Singapore	0.5% pet.	5557	361 (6.5%)	R: not stated	6
1985-1987 Singapore	0.5% pet.	3145	190 (6.0%)	R: not stated; C: the frequency of sensitization was significantly higher in older (>50 y) patients	3
1985-1986 Singapore	0.5% pet.	1685	119 (7.1%)	R: not stated; C: frequency in women 4.2%, in men 9.9%	7
1984-1985 Singapore	1.0% pet.	1892	119 (6.3%)	R: not stated; C: frequency in women 5.0%, in men 7.3%	8

Case series

In Singapore, between January 1983 and July 1984, 45 patients were seen in a hospital in Singapore with sensitivity to proflavine, as shown by a positive patch test to proflavine hemisulfate 1% pet. (12). Proflavine sensitivity was more common in men (73%) than in women (27%). All age groups were equally affected. Fifteen (33%) and 19 (42%) of the 45 patients presented with acute and subacute dermatitis, respectively. Most of the lesions had onset distributed over the arms and legs (94%), areas where trauma was most frequently encountered. Ten (22%) had secondary spread to other parts of the body. Proflavine sensitivity was of current relevance in 64% and of past relevance in 11% of the cases. The majority of the patients had used proflavine for cuts and abrasions (61%); 17% used it for itch and insect bites. Two patients developed dermatitis after proflavine was used postoperatively for wound dressing. There were frequent co-reactivities, e.g. to neomycin, lanolin alcohol and clioquinol (12).

Also in Singapore, a non-eczematous eruption associated with allergic contact dermatitis to proflavine was described in four men, aged 22 to 41 years, who had used proflavine lotion for 1 week (n=3) to 2 months to a skin abrasion, when subacute dermatitis developed at the application site. One to 14 days later, erythematous urticarial papular and/or plaque eruptions would appear around the primary contact site and often at distant sites. Some lesions were target-like, and mimicked lesions of erythema multiforme, but the histology was different. The eruption persisted longer than the primary eczematous lesions and tended to persist after clearance of the initial dermatitis. All patients had a positive patch test to proflavine 0.5% pet. The author suggested to use the term 'urticarial papular and plaque eruption of contact allergy' to describe the eruption, which was usually termed 'erythema multiforme-like eruption' (5). One of these patients had previously been presented as having an 'erythema multiforme-like and purpuric eruption' (11).

During the years 1955 to 1958, proflavine dihydrochloride was used at a Veterans Hospital in Vancouver, British Columbia, Canada, as a preoperative surgical application to the skin. Many dozens of cases of allergic eczematous pigmented contact dermatitis were observed from the its use. The number of patients exposed and the number of patients affected were not documented. Some individuals developed dermatitis on two and three occasions after successive operations (13).

Case reports

A 53-year-old man complained of swelling and numbness of both lips 12 hours after a dental extraction. Local anesthetics had been used and the mouth had been rinsed with 0.1% acriflavine solution (which is an antiseptic mixture containing 50% proflavine). The patient had used proflavine ointment 2 years before for an open wound and had developed a rash around the wound. Patch tests showed a positive reaction to proflavine 0.5% pet. Acriflavine (0.1% aqueous solution) was then painted on the lips. The patient had subjective numbness of the lips after 16 hours but no discernible edema. Provocation with oral acriflavine rinse (0.1% aqueous solution) evoked small blisters at the vermilion border of the lips, together with gross edema of the lips after 12 hours. The patient had, according to the authors, allergic contact dermatitis from proflavine. This may well be, but the acriflavine itself and the non-proflavine component of acriflavine were not patch tested (4).

A 58-year-old man presented with recurrent pruritus of his hands and forearms of 3 months duration. He had been an aquarium keeper in a fishery research station for the past 6 months. His work involved cleaning and changing the water of fish tanks, which included adding acriflavine powder (containing proflavine) as an antiparasitic, into the tanks regularly. After that, he would dip his arms into the tank to disperse the powder. Because of the itching and scratching, the patient self-medicated with topical acriflavine lotion, resulting in acute dermatitis on his arms after 3 days. Patch tests showed positive reactions to proflavine HCl 0.5% pet. and to neomycin (past relevance). He stopped working with acriflavine and when reviewed 8 weeks later, his pruritus had cleared (10). This was a case of occupational allergic contact dermatitis.

A 24-year-old Chinese man had applied proflavine lotion to an abrasion on his left leg and developed a rash the next day. Purpura and blisters appeared 5 days later and similar lesions developed centrifugally on areas where the proflavine had not been applied. Clinical examination showed a purpuric, vesiculobullous eruption and eczematous lesions on the left leg and knee. Several target lesions were seen. The rest of the body was normal. A biopsy showed changes of bullous erythema multiforme with capillaritis. When patch tested, a ++ reaction was observed to proflavine 0.5% pet. (11).

One or more cases of allergic contact dermatitis from proflavine were also reported in ref. 14, but details are not available to the author.

LITERATURE

1   The data in the section 'General' may have been obtained from literature discussed in this chapter, but mostly also or exclusively from one or more of the following online sources: ChemIDPlus Advanced, PubChem, DrugBank, RxList, Drug Central, Drugs.com, and Wikipedia

2   Ochi H, Cheng SWN, Leow YH, Goon ATJ. Contact allergy trends in Singapore – a retrospective study of patch test data from 2009 to 2013. Contact Dermatitis 2017;76:49-50

3   Goh CL. Contact sensitivity to topical antimicrobials (I). Epidemiology in Singapore. Contact Dermatitis 1989;21:46-48

4   Lim J, Goh CL, Lee CT. Perioral and mucosal oedema due to contact allergy to proflavine. Contact Dermatitis 1991;25:195-196

5   Goh CL. Urticarial papular and plaque eruptions. A noneczematous manifestation of allergic contact dermatitis. Int J Dermatol 1989;28:172-176

6   Lim JT, Goh CL, Ng SK, Wong WK. Changing trends in the epidemiology of contact dermatitis in Singapore. Contact Dermatitis 1992;26:321-326

7   Goh CL. Contact sensitivity to topical medicaments. Int J Dermatol 1989;28:25-28

8   Goh CL. Epidemiology of contact allergy in Singapore. Int J Dermatol 1988;27:308-311

9   Goh CL. Contact sensitivity to topical antimicrobials. (II). Sensitizing potentials of some topical antimicrobials. Contact Dermatitis 1989;21:166-171

10   Goh CL. Occupational dermatitis from proflavine. Contact Dermatitis 1987;17:256

11   Goh CL. Erythema multiforme-like and purpuric eruption due to contact allergy to proflavine. Contact Dermatitis 1987;17:53-54

12   Goh CL. Contact sensitivity to proflavine. Int J Dermatol 1986;25:449-451

13   Mitchell JC. Contact dermatitis from proflavine dihydrochloride. Arch Dermatol 1972;106:924

14   Morgan JK. Iatrogenic epidermal sensitivity. Br J Clin Practice 1968;22:261-268 (cited in ref. 12)

Chapter 3.285 PROMESTRIENE

IDENTIFICATION

Description/definition : Promestriene is the 3-propyl and 17β-methyl ether of estradiol; it conforms to the structural formula shown below

Pharmacological classes : Estrogens

IUPAC name : (8R,9S,13S,14S,17S)-17-Methoxy-13-methyl-3-propoxy-6,7,8,9,11,12,14,15,16,17-decahydrocyclopenta[a]phenanthrene

Other names : 3-Propoxy-17β-methoxy-1,3,5(10)-estratriene

CAS registry number : 39219-28-8

EC number : 254-361-6

Patch testing : 0.1% alcohol

Molecular formula : $C_{22}H_{32}O_2$

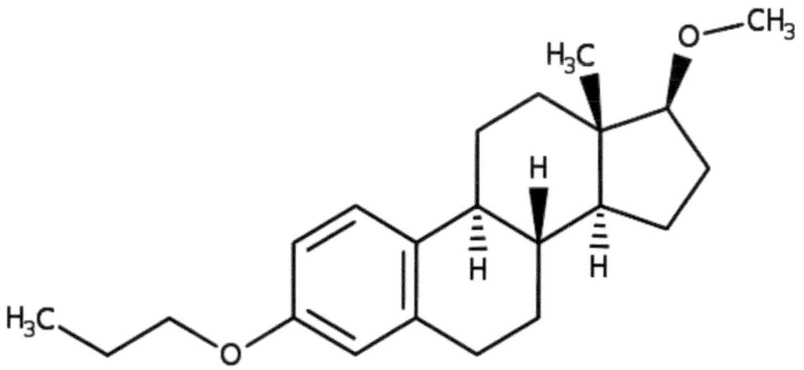

GENERAL

Promestriene is a 3-propyl and 17β-methyl ether of estradiol that may be used intravaginally to relieve vaginal atrophy and its associated symptoms such as dryness and itching. Promestriene has also been used in trials studying the prevention of hypospadias (1).

CONTACT ALLERGY

Case report

A 31-year-old woman who had previously become sensitized to clioquinol from an antiseptic ointment to treat otitis externa, suffered vaginal erythema, itching, and edema, and eczematous lesions in the abdominal and thoracic areas one day after application of a vaginal ovule containing chlorquinaldol and promestriene. Patch tests were positive to promestriene 0.1% alcohol and to chlorquinaldol and clioquinol (both 5% pet.) (2). This was a case of local and systemic contact dermatitis.

LITERATURE

1 The data in the section 'General' may have been obtained from literature discussed in this chapter, but mostly also or exclusively from one or more of the following online sources: ChemIDPlus Advanced, PubChem, DrugBank, RxList, Drug Central, Drugs.com, and Wikipedia

2 Rodríguez A, Cabrerizo S, Barranco R, de Frutos C, de Barrio M. Contact cross-sensitization among quinolines. Allergy 2001;56:795

Chapter 3.286 PROMETHAZINE

IDENTIFICATION

Description/definition   : Promethazine is the phenothiazine-derivative that conforms to the structural formula
                             shown below
Pharmacological classes   : Histamine H1 antagonists; anti-allergic agents; antipruritics
IUPAC name   : *N,N*-Dimethyl-1-phenothiazin-10-ylpropan-2-amine
CAS registry number   : 60-87-7
EC number   : 200-489-2
Merck Index monograph   : 9171
Patch testing   : Hydrochloride, 0.1% pet. (Chemotechnique); hydrochloride, 2% pet.
                             (SmartPracticeCanada)
Molecular formula   : $C_{17}H_{20}N_2S$

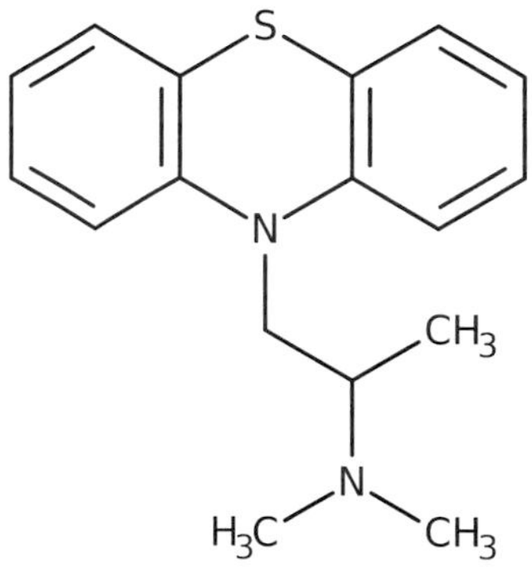

GENERAL

Promethazine is a phenothiazine derivative with antihistaminic, sedative and antiemetic properties. It selectively blocks peripheral H1 receptors, thereby diminishing the effects of histamine on effector cells. Promethazine also blocks the central histaminergic receptors, thereby depressing the reticular system causing sedative and hypnotic effects. In addition, this agent has centrally acting anticholinergic properties and probably mediates nausea and vomiting by acting on the medullary chemoreceptive trigger zone. Promethazine is used as an antiallergic, in the treatment of pruritus, for sedation and to prevent and treat nausea and vomiting, e.g. from motion sickness. In pharmaceutical products, promethazine is employed as promethazine hydrochloride (CAS number 58-33-3, EC number 200-375-2, molecular formula $C_{17}H_{21}ClN_2S$). It is available as tablets, syrup, injection fluid, suppository and in some countries as cream for the treatment of itch and insect bites (1).

CONTACT ALLERGY

General

In early literature, many cases of allergic, photoallergic and photoaugmented contact dermatitis to topical and – to a lesser degree - systemic promethazine have been reported. See the section 'General' under 'Photosensitivity' below. It has not been attempted to fully review the literature on the subject.

Patch testing in groups of patients

The results of patch testing with promethazine in consecutive patients (routine testing) and in groups of selected patients (mostly patch and photopatch tested because of suspected photosensitivity) are shown in table 3.286.1. In routine testing in Italy, in the period 1968-1977, 4.9% of the patients had positive patch tests to promethazine. Unfortunately, it was not mentioned how many of these reactions were relevant. In the following period 1978-1983, the prevalence of sensitization had dropped considerably to 1.6%, possibly because the cream causing most reactions had been withdrawn or indications for its use had been limited (3).

In studies in which groups of selected patients were patch tested with promethazine, the frequencies of positive patch tests have ranged from 0.3% to 5.2%, but most scored 0.3-1.6% reactions. The high 5.2% positive patch tests figure was observed in Italy in 1970-1973 in patients with leg ulcers and stasis dermatitis (2), but this percentage was nearly identical to the 4.9% that the same authors found in that period with routine testing (3). A 4.4% rate of positive patch tests in Canada in the period 2001-2010 is remarkable (just as the 8.1% positive photopatch tests in that study). The authors did not discuss this issue nor did they provide data on relevance (7).

Table 3.286.1 Patch testing in groups of patients

Years and Country	Test conc. & vehicle	Number of patients tested	positive (%)		Selection of patients (S); Relevance (R); Comments (C)	Ref.
Routine testing						
1978-1983 Italy	2% pet.	4472		(1.6%)	R: not stated; C; significant decrease in sensitization compared to the 1968-1977 period	3
1978-1981 Belgium		1823	7	(0.4%)	R: not stated	28
1968-1977 Italy	2% pet.	3758		(4.9%)	R: not stated	3
Testing in groups of selected patients						
2001-2010 Canada		160	7	(4.4%)	S: patients with suspected photosensitivity and patients who developed pruritus or a rash after sunscreen application; R: not stated; C: there were also 13 photoallergic reactions	7
2004-2006 Italy	10% pet.	1082	3	(0.3%)	S: patients with histories and clinical features suggestive of photoallergic contact dermatitis; R: 100%; C: all reactions were photoaugmented; there were also 3 (pure) photoallergic reactions	24
2000-2005 USA	1% pet.	177	2	(1.1%)	S: patients photopatch tested for suspected photodermatitis; R: 50%	9
1983-1998 U.K.	1% pet.	2715	9	(0.3%)	S: patients suspected of photosensitivity or with (a history of) dermatitis at exposed sites; R: not established; C: there were also 10 photocontact allergic reactions	22
1990-1994 France	1% pet.	370	4	(1.1%)	S: patients with suspected photodermatitis; R: not stated; C: there were also 32 photocontact allergic reactions	6
1986-1989 Italy	1% pet.	128	2	(1.6%)	S: not stated; R: not stated; C: there were also 7 photocontact allergic reactions	23
1980-1981 4 Scandinavian countries	1% pet.	745	3	(0.4%)	S: patients suspected of sun-related skin disease; R: not specified; C: there were also 24 photopatch tests reactions	13
1970-1973 Italy	2% pet.	231	12	(5.2%)	S: patients with stasis dermatitis with (n=223) or without leg ulcers; R: not stated	2

Case reports and case series

In the period 1978-1997, in a University clinic in Leuven, Belgium, 12 patients had positive patch test reactions and 2 photopatch tests to promethazine. The relevance of these reactions was not mentioned. Phototoxic reactions were not included, it was not mentioned how many of such reactions were observed ('many weak phototoxic and irrelevant reactions...') (28).

A 55 year-old man had been applying an antihemorrhoidal cream for variable periods of time. He presented with a 1-month history of pruritic, erythematous scaly plaques, starting in the perianal area and spreading in a few days to the arms and occipital area. Patch tests were positive to the cream and the following ingredients: promethazine 1% pet., neomycin, sodium metabisulfite and propylparaben (27).

A 72-year-old man had photoaggravated allergic contact dermatitis from promethazine cream (43). A 24-year-old man developed eczema in the antecubital fossa of the left arm, later spreading to the other arm, face, abdomen and legs from photoaugmented contact allergy to promethazine and to contact allergy to lanolin alcohol, which was present in the promethazine cream and a cosmetic moisturizer used (44).

Promethazine has caused erythema multiforme-like allergic contact dermatitis (25). Early case reports of allergic contact dermatitis (<1970) are not discussed here.

PHOTOSENSITIVITY

General

Early literature has established promethazine as an important cause of allergic contact dermatitis and − far more frequently - photoallergic contact dermatitis. Many patients had a combination: photoaugmented (photoaggravated) contact allergy. Photosensitization not infrequently led to persistent light reactions (26,30-42,46,47). One cream in particular (Phenergan ® cream) caused many reactions. In one hospital in Paris, France, in a period of 3 years, this cream was suspected as being a causal or contributory factor in 262 cases of eczema (30). However, patch testing in 83 of these patients showed that promethazine probably was the allergen in only 7. Many patients reacted to the cream and the cream vehicle, partly probably from irritant reactions, partly from contact allergy to an unknow excipient (30,45). Triethanolamine was one of the allergenic ingredients: 8 of 22 patients contact allergic to the cream (or at least with positive patch test reactions, some of which may have been irritant) reacted to promethazine 2% in lanettewax cream and 4 had positive patch test reactions to triethanolamine (38). In various countries, the use of this very popular promethazine-containing cream was abandoned.

In patients sensitized to topical promethazine, oral administration has resulted in systemic contact dermatitis with reactivation of the (previous) eruption, spreading of the eczema with development of erythroderma as well as systemic manifestations such as fever, chills, intestinal upset and sometimes even syncope (30,31). Such systemic contact dermatitis could also be caused by oral administration of the related phenothiazine chlorpromazine in promethazine-allergic individuals (30).

Photosensitive eruptions were also caused by oral administration of promethazine without previous sensitization from topical application (33). Occupational sensitization to promethazine has been described various times, e.g. in a nurse handling promethazine (and chlorpromazine) injectable solutions (30).

Photopatch testing in consecutive patients suspected of contact dermatitis: Routine testing

In Belgium, in the period 1978-1981, 1823 consecutive patients suspected of contact dermatitis were patch and photopatch tested with promethazine (test concentration and vehicle not mentioned) and there was one positive photopatch test (0.05%) and there were 7 positive patch tests (0.4%). Their relevance was not mentioned (28).

Photopatch testing in groups of selected patients

The results of photopatch testing with promethazine in groups of selected patients (mostly patch and photopatch tested because of suspected photosensitivity) are shown in table 3.286.2. In quite a few studies, high percentages (>5%, maximum 25%) of positive photopatch tests have been observed. Usually, their relevance was either not mentioned or unknown. It is well known that promethazine can induce phototoxic reactions, depending on the concentration used and the irradiation parameters. When photopatch tested at 1% pet. and with an UVA irradiation of 11 J/cm^2, for example, it caused 20% phototoxic reactions in 1129 patients (11). It may be assumed that many of the positive photopatch tests in the studies in table 3.286.2. with high prevalences have been phototoxic rather than photoallergic, which was also suggested by some of the investigators themselves (6,12,13,14,16,20,22).

Table 3.286.2 Photopatch testing in groups of patients: Selected patient groups

Years and Country	Test conc. & vehicle	Number of patients tested	positive (%)	Selection of patients (S); Relevance (R); Comments (C)	Ref.
2001-2010 Canada		160	13 (8.1%)	S: patients with suspected photosensitivity and patients who developed pruritus or a rash after sunscreen application; R: not stated	7
1993-2009 USA	1% pet.	30	2 (7%)	S: patients with chronic actinic dermatitis; R: not stated	18
2003-2007 Portugal	0.1% pet.	83	7 (8.4%)	S: patients with suspected photoaggravated facial dermatitis or systemic photosensitivity; R: all reactions were relevant	8
2004-2006 Italy	10% pet.	1082	3 (0.3%)	S: patients with histories and clinical features suggestive of photoallergic contact dermatitis; R: 100%; C: there were also 3 photoaugmented contact allergic reactions and 6 irritative/phototoxic reactions	24
1993-2006 USA	1% pet.	76	9 (12%)	S: not stated; R: 21% of all reactions to medications were considered 'of possible relevance'	17
1992-2006 Greece	0.1% pet.	207	(25%)	S: patients suspected of photosensitivity; R: not stated; C: 'some reactions may have been phototoxic'	14
2004-2005 Spain	0.5% pet.	224	10 (4.5%)	S: not stated; R: 0%	19
2000-2005 USA	1% pet.	177	13 (7.3%)	S: patients photopatch tested for suspected photodermatitis; R: 23%	9
1994-9 Netherlands	1% pet., later 0.1%	99	2 (2%)	S: patients suspected of photosensitivity disorders; R: not stated; C: only reactions with the 1% concentration; it was suggested that these may have been phototoxic	16
1983-1998 U.K.	1% pet.	2715	10 (0.4%)	S: patients suspected of photosensitivity or with (a history of) dermatitis at exposed sites; R: not established; C: the authors suggested that these reactions were phototoxic and have removed it from the photopatch test series	22
1991-97 Germany, Austria, Switzerland	0.1% pet.	1261	(0.4%)	S: patients suspected of photosensitivity; R: not stated; C: many phototoxic reactions not counted as positive	15
1990-1994 France	1% pet.	370	32 (8.6%)	S: patients with suspected photodermatitis; R: not stated; C: some reactions may have been phototoxic, according to the authors, from using high UVA-doses	6
1985-1994 Italy		1050	87 (8.3%)	S: patients with histories or clinical pictures suggestive of allergic contact photodermatitis; R: 78% for all photoallergens together	10

Table 3.286.2 Photopatch testing in groups of patients: Selected patient groups (continued)

Years and Country	Test conc. & vehicle	Number of patients tested | positive (%)			Selection of patients (S); Relevance (R); Comments (C)	Ref.
1991-1993 Singapore	1% pet.	62	3	(5%)	S: patients with clinical features suggestive of photosensitivity; R: 0%; C: the authors suggested that (a number of) the reactions may have been phototoxic	20
1987-1989 Thailand	1% pet.	274	10	(3.6%)	S: patients suspected of photosensitivity; R: 0%	21
1986-1989 Italy	1% pet.	128	7	(5.5%)	S: not stated; R: not stated	23
1980-85 Germany, Austria, Switzerland	1% pet.	1129		(0.7%)	S: patients suspected of photoallergy, polymorphic light eruption, phototoxicity and skin problems with photo-distribution; R: not stated; C: 20% phototoxic reactions	11
1980-1985 USA	2% pet.	70	8	(11%)	S: not stated; R: none; C: it was suggested that these were phototoxic reactions	12
1980-1981 4 Scandinavian countries	1% pet.	745	24	(3.2%)	S: patients suspected of sun-related skin disease; R: not specified; C: 'the reactions to phenothiazines were frequently phototoxic'	13

Case reports and case series

In the period 1978-1997, in a University Clinic in Leuven, Belgium, 2 patients had positive photopatch test reactions and 2 positive patch tests to promethazine. Their relevance was not mentioned. Phototoxic reactions were not included, it was not mentioned how many of such reactions were observed ('many weak phototoxic and irrelevant reactions...') (28).

Older case reports of photoallergic contact dermatitis or photoaugmented allergic contact dermatitis from promethazine are not discussed here.

Cross-reactions, pseudo-cross-reactions and co-reactions

Promethazine has commonly (photo)cross-reacted with the related phenothiazine chlorpromazine (30,33). Allergic sensitization to promethazine was often associated with allergy to the related phenothiazine thiazinamium metilsulfate (Multergan ®) (30). The same authors stated that of 128 promethazine-intolerant patients 39 (31%) also reacted to p-phenylenediamine. Some of them also had a sensitivity to other p-amino compounds: procaine and other ester-type anesthetics and sulfonamides. However, the reverse phenomenon was not seen: subjects sensitized to p-phenylenediamine as a rule did not react to promethazine (30).

Four patients sensitized to chlorproethazine in an ointment were patch and photopatch tested with promethazine 0.1% pet. The patch tests were negative, but three of the 4 had positive *photo*patch tests (photocross-reactivity) (5). A patient photosensitized to topical dioxopromethazine photocross-reacted to promethazine (29). Photocross-reactions between the phenothiazines isothipendyl, promethazine and chlorpromazine may occur (43).

Immediate contact reactions

Immediate contact reactions (contact urticaria) to promethazine are presented in Chapter 5.

Cutaneous adverse drug reactions from systemic administration caused by type IV (delayed-type) hypersensitivity

Cutaneous adverse drug reactions from systemic administration of promethazine caused by type IV (delayed-type) hypersensitivity including fixed drug eruption (4), photosensitivity (33) and systemic contact dermatitis (30,31), as well as occupational allergic contact dermatitis (30) are planned to be discussed in Volume IV of the *Monographs in Contact Allergy* series on Systemic drugs.

LITERATURE

1 The data in the section 'General' may have been obtained from literature discussed in this chapter, but mostly also or exclusively from one or more of the following online sources: ChemIDPlus Advanced, PubChem, DrugBank, RxList, Drug Central, Drugs.com, and Wikipedia

2 Angelini G, Rantuccio F, Meneghini CL. Contact dermatitis in patients with leg ulcers. Contact Dermatitis 1975;1:81-87

3 Angelini G, Vena GA, Meneghini CL. Allergic contact dermatitis to some medicaments. Contact Dermatitis 1985;12:263-269

4 Lee AY. Topical provocation in 31 cases of fixed drug eruption: change of causative drugs in 10 years. Contact Dermatitis 1998;38:258-260

5   Barbaud A, Collet E, Martin S, Granel F, Trechot P, Lambert D, et al. Contact sensitization to chlorproethazine can induce persistent light reaction and cross-photoreactions to other phenothiazines. Contact Dermatitis 2001;44:373-374

6   Journe F, Marguery M-C, Rakotondrazafy J, El Sayed F, Bazex J. Sunscreen sensitization: a 5-year study. Acta Derm Venereol (Stockh) 1999;79:211-213

7   Greenspoon J, Ahluwalia R, Juma N, Rosen CF. Allergic and photoallergic contact dermatitis: A 10-year experience. Dermatitis 2013;24:29-32

8   Cardoso J, Canelas MM, Gonçalo M, Figueiredo A. Photopatch testing with an extended series of photoallergens: a 5-year study. Contact Dermatitis 2009;60:325-329

9   Scalf LA, Davis MDP, Rohlinger AL, Connolly SM. Photopatch testing of 182 patients: A 6-year experience at the Mayo Clinic. Dermatitis 2009;20:44-52

10   Pigatto PD, Legori A, Bigardi AS, Guarrera M, Tosti A, Santucci B, et al. Gruppo Italiano recerca dermatiti da contatto ed ambientali Italian multicenter study of allergic contact photodermatitis: epidemiological aspects. Am J Contact Dermatitis 1996;17:158-163

11   Hölzle E, Neumann N, Hausen B, Przybilla B, Schauder S, Hönigsmann H, et al. Photopatch testing: the 5-year experience of the German, Austrian and Swiss Photopatch Test Group. J Am Acad Dermatol 1991;25:59-68

12   Menz J, Muller SA, Connolly SM. Photopatch testing: a 6-year experience. J Am Acad Dermatol 1988;18:1044-1047

13   Wennersten G, Thune P, Brodthagen H, Jansen C, Rystedt I, Crames M, et al. The Scandinavian multicenter photopatch study. Contact Dermatitis 1984;10:305-309

14   Katsarou A, Makris M, Zarafonitis G, Lagogianni E, Gregoriou S, Kalogeromitros D. Photoallergic contact dermatitis: the 15-year experience of a tertiary referral center in a sunny Mediterranean city. Int J Immunopathol Pharmacol 2008;21:725-727

15   Neumann NJ, Hölzle E, Plewig G, Schwarz T, Panizzon RG, Breit R, et al. Photopatch testing: The 12-year experience of the German, Austrian and Swiss Photopatch Test Group. J Am Acad Dermatol 2000;42(2Pt/1):183-192

16   Bakkum RS, Heule F. Results of photopatch testing in Rotterdam during a 10-year period. Br J Dermatol 2002;146:275-279

17   Victor FC, Cohen DE, Soter NA. A 20-year analysis of previous and emerging allergens that elicit photoallergic contact dermatitis. J Am Acad Dermatol 2010;62:605-610

18   Que SK, Brauer JA, Soter NA, Cohen DE. Chronic actinic dermatitis: an analysis at a single institution over 25 years. Dermatitis 2011;22:147-154

19   De La Cuadra-Oyanguren J, Perez-Ferriols A, Lecha-Carrelero M, et al. Results and assessment of photopatch testing in Spain: towards a new standard set of photoallergens. Actas DermoSifiliograficas 2007;98:96-101

20   Leow YH, Wong WK, Ng SK, Goh CL. 2 years' experience of photopatch testing in Singapore. Contact Dermatitis 1994;31:181-182

21   Gritiyarangsan P. A three-year photopatch study in Thailand. J Dermatol Sci 1991;2:371-375

22   Darvay A, White I R, Rycroft R J G, Jones A B, Hawk J L M, McFadden J P. Photoallergic contact dermatitis is uncommon. Br J Dermatol 2001;145:597-601

23   Guarrera M. Photopatch testing: a three-year experience. J Am Acad Dermatol 1989;21:589-591

24   Pigatto PD, Guzzi G, Schena D, Guarrera M, Foti C, Francalanci, S, Cristaudo A, et al. Photopatch tests: an Italian multicentre study from 2004 to 2006. Contact Dermatitis 2008;59:103-108

25   Meneghini CL, Angelini G. Secondary polymorphic eruptions in allergic contact dermatitis. Dermatologica 1981;163:63-70

26   Soto JM. Promethazine photosensitivity. Contact Dermatitis Newsletter 1968;3:53

27   Sánchez-Pérez J, Abajo P, Córdoba S, García-Díez A. Allergic contact dermatitis from sodium metabisulfite in an antihemorrhoidal cream. Contact Dermatitis 2000;42:176-177

28   Goossens A, Linsen G. Contact allergy to antihistamines is not common. Contact Dermatitis 1998;39:38

29   Schauder S. Dioxopromethazine-induced photoallergic contact dermatitis followed by persistent light reaction. Am J Contact Dermat 1998;9:182-187

30   Sidi E, Hincky M, Gervais A. Allergic sensitization and photosensitization to Phenergan cream. J Invest Dermatol 1955;24:345-352

31   Sidi E, Melki GR. Rapport entre dermites de cause externe et sensibilisation par voie interne. Semaine des Hopitaux 1954; No. 25, April 14 (Article in French)

32   Tzanck A, Sidi E, Mazalton G, et al. Sur 2 cas de dermite au Phenergan avec photosensibilisation. Bull Soc Franç Derm Syph 1951;58:433 (Article in French)

33 Epstein S, Rowe R. Photoallergy and photocross-sensitivity to Phenergan. J Invest Dermatol 1957;29:319-326

34 Epstein ST. Allergic photocontact dermatitis from promethazine (Phenergan). Arch Dermatol 1960;81:53-58

35 Lopes G, Doussy J. Sur l'allergie médicamenteuse au groupe phénothiazine. Rev Franc d'Allergie 1961;3:185 (Article in French)

36 LeCoulant P, Texier L, Malville J, et al. Sensibilisation à une crème antihistaminique. Soc Derm et Syph 1964;71:234-237 (Article in French)

37 Suurmond D. Skin reactions to Phenergan cream. Dermatologica 1964;128:87-89

38 Suurmond D. Patch test reactions to Phenergan cream: Promethazine and triethanolamine. Dermatologica 1966;133:503-506

39 Tzanck A, Sidi E, Melki GR. Dermites artificielles aux crèmes antihistaminiques. Bull Soc Franc Derm Syph 1951;58:282 (Article in French)

40 Tzanck A, Sidi E, Longueville R. Conséquences des sensibilisations aux crèmes antihistaminiques. Semaine des Hôpitaux 1951; No. 81, November 2, 1951 (Article in French)

41 Sidi E, Melki GR, Longueville R. Dermites aux pommades antihistaminiques. Acta Allergologica 1952;5:292-303 (Article in French)

42 Schulz KH, Wiskemann A, Wulf K. Klinische und experimentelle Untersuchungen über die photodynamische Wirksamkeit von Phenothiazinderivaten, insbesondere von Megaphen. Arch klin u exper Dermat 1956;202:285-298 (Article in German)

43 Cariou C, Droitcourt C, Osmont MN, Marguery MC, Dutartre H, Delaunay J, et al. Photodermatitis from topical phenothiazines: A case series. Contact Dermatitis 2020;83:19-24

44 Arrue I, Rosales B, Ortiz de Frutos FJ, Vanaclocha F. Photoaggravated eczema due to promethazine cream. Actas Dermosifiliogr 2007;98:717-718

45 Zina G, Bonu G. Phenergan cream (role of base constituents). Contact Dermatitis Newsletter 1969;6:117

46 Di PriscoJ, Soto JM, Herrera E. Phenergan sensitivity. Contact Dermatitis Newsletter 1968;4:63

47 Epstein S. Allergic photocontact dermatitis from promethazine (Phenergan). Arch Dermatol Syphilol 1960;81;175-177

Chapter 3.287 PROPANOCAINE

IDENTIFICATION

Description/definition : Propanocaine is the benzoic acid derivative that conforms to the structural formula shown below
Pharmacological classes : Anesthetics, local
IUPAC name : [3-(Diethylamino)-1-phenylpropyl] benzoate
Other names : α-(2-Diethylaminoethyl)benzyl benzoate
CAS registry number : 493-76-5
EC number : 207-778-2
Merck Index monograph : 437
Patch testing : 1% pet.
Molecular formula : $C_{20}H_{25}NO_2$

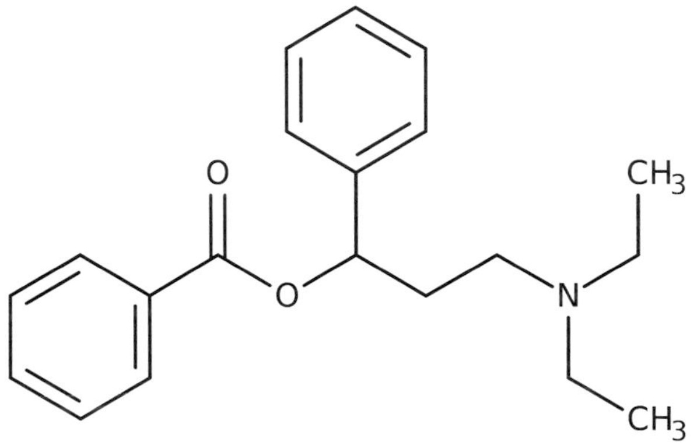

GENERAL

Propanocaine is a benzoic acid derivative that was formerly (from 1960 on) used as a local anesthetic, but is hardly, if at all, utilized anymore (1).

CONTACT ALLERGY

Case series

From Strasbourg, France, 4 cases of allergic contact dermatitis from propanocaine were reported, 3 women and one man, in age ranging from 19 to 57 years. They had used creams containing propanocaine for genitocrural itching, itching on the back of the hands or itch from insect bites. Contact dermatitis developed after 2 and 3 days in two of them (duration of use in the others not mentioned). In all 4, patch tests were positive to the cream 'as is' and to propanocaine 1% pet. (2). The author states to have seen 9 other patients with contact dermatitis from propanocaine creams, but in these individuals, patch tests with propanocaine itself were not performed (2).

LITERATURE

1 The data in the section 'General' may have been obtained from literature discussed in this chapter, but mostly also or exclusively from one or more of the following online sources: ChemIDPlus Advanced, PubChem, DrugBank, RxList, Drug Central, Drugs.com, and Wikipedia
2 Foussereau J. Contact dermatitis from propanocaine. Contact Dermatitis 1986;15:40

Chapter 3.288 PROPANTHELINE

IDENTIFICATION

Description/definition   : Propantheline is the quaternary ammonium compound that conforms to the structural
                             formula shown below
Pharmacological classes : Muscarinic antagonists; anti-ulcer agents; anticholinergic agents
IUPAC name   : Methyl-di(propan-2-yl)-[2-(9H-xanthene-9-carbonyloxy)ethyl]azanium
CAS registry number   : 298-50-0
EC number   : 206-063-2
Merck Index monograph : 9189 (Propantheline bromide)
Patch testing   : Bromide, 1% water (1); 5% water (2); 10% water (5); 5% pet. (6) (all as bromide)
Molecular formula   : $C_{23}H_{30}NO_3$

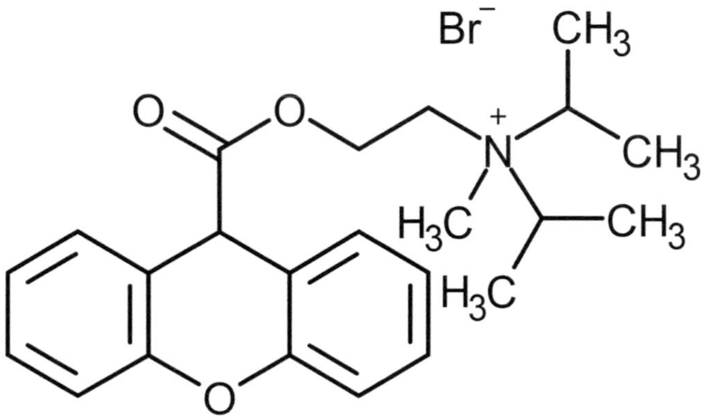

GENERAL

Propantheline is a quaternary ammonium compound structurally related to belladonna alkaloids. It competitively antagonizes acetylcholine activity mediated by muscarinic receptors at neuroeffector sites on smooth muscle and exocrine gland cells. The blocking effect leads to a reduction of exocrine glands secretions, to relax bronchial muscle and reduce tone and motility of intestinal smooth muscle. Propantheline was formerly used to treat gastrointestinal conditions associated with intestinal spasm, to decrease secretions during anesthesia, and in treating rhinitis, urinary incontinence and ulcers. In topical medications it was utilized in antiperspirants to decrease axillary sweating (10). In pharmaceutical drugs, propantheline is employed as propantheline bromide (CAS number 50-34-0, EC number 200-030-6, molecular formula $C_{23}H_{30}BrNO_3$). It is probably hardly used anymore (10).

CONTACT ALLERGY

Case reports and case series
Three patients (1) and another 3 (7) had allergic contact dermatitis from propantheline bromide in antiperspirants. Out of 14 patients with axillary dermatitis from an antiperspirant, 11 reacted to propantheline bromide (2). Of seven patients with eczema in the axillae after the use of antiperspirants, 6 had positive patch tests to propantheline bromide (5). Five single case reports of contact allergy to propantheline bromide in an antiperspirant have been reported (3,4,6,8,9). All these cases were caused by a commercial antiperspirant containing 5% propantheline bromide, 0.25% triclocarban, 90% propylene glycol and 4.75% water. One or more cases were also reported from Germany in 1996; details are unknown (11).

Some authors considered xanthene-9-carboxylic acid, which is a part of the propantheline bromide molecule, to be the primary sensitizer (3,6).

LITERATURE
1   Skog E. Incidence of cosmetic dermatitis. Contact Dermatitis 1980;6:449-451
2   Ågren-Jonsson S, Magnusson, B. Sensitization to propantheline bromide, trichlorocarbanilide and propylene glycol in an antiperspirant. Contact Dermatitis 1976;2:79-80
3   Fregert S, Möller H. Allergic contact dermatitis from propantheline bromide. Contact Dermatitis Newsletter 1967;1:12

4   Wereide K. Contact allergy to propantheline bromide. Contact Dermatitis Newsletter 1968;4:61
5   Hannuksela, M. Allergy to propantheline in an antiperspirant (Ercoril® lotion). Contact Dermatitis 1975;1:244
6   Osmundsen PE. Concomitant contact allergy to propantheline bromide and TCC. Contact Dermatitis 1975;1:251-252
7   Przybilla B, Schwab U, Hölzle E, Ring J. Kontaktsensibilisierung durch ein Antiperspirant mit dem Wirkstoff Propanthelinbromid. Hautarzt 1983;34:459-462 (Article in German)
8   Fregert S, Möller H. Allergic contact dermatitis from propantheline bromide. Contact Dermatitis Newsletter 1967;1:12
9   Gall H, Kempf E. Kontaktallergie auf das lokale Antiperspirant Propanthelinbromid. Dermatosen 1982;30:55-57 (Article in German)
10   The data in the section 'General' may have been obtained from literature discussed in this chapter, but mostly also or exclusively from one or more of the following online sources: ChemIDPlus Advanced, PubChem, DrugBank, RxList, Drug Central, Drugs.com, and Wikipedia
11   Jansen T, Plewig G, Hölzle E. Allergic contact dermatitis due to propantheline bromide. Dtsch Med Wochenschr 1996;121:41-42

Chapter 3.289 PROPARACAINE

IDENTIFICATION

Description/definition : Proparacaine is the benzoate ester that that conforms to the structural formula shown below
Pharmacological classes : Anesthetics, local
IUPAC name : 2-(Diethylamino)ethyl 3-amino-4-propoxybenzoate
Other names : Proxymetacaine
CAS registry number : 499-67-2
EC number : 207-884-9
Merck Index monograph : 9191
Patch testing : 2% water
Molecular formula : $C_{16}H_{26}N_2O_3$

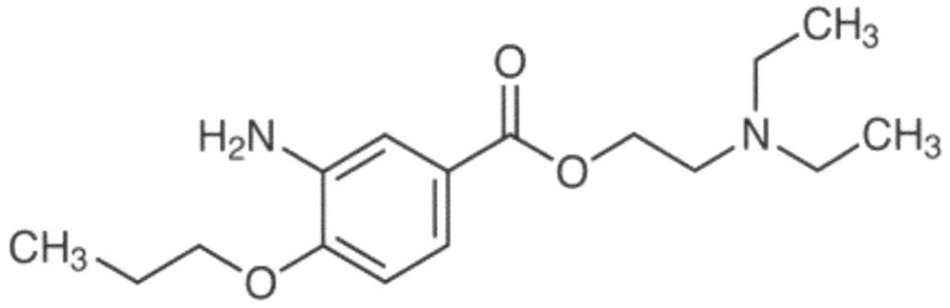

GENERAL

Proparacaine (often termed proxymetacaine) is a benzoic acid ester and topical anesthetic drug. It is used in ophthalmic solutions for local anesthesia as the hydrochloride salt (CAS number 5875-06-9, EC number 227-541-7, molecular formula $C_{16}H_{27}ClN_2O_3$) (1).

CONTACT ALLERGY

General

Nine patients suffering from allergic contact dermatitis to proparacaine eye drops have been reported. Five of them had occupational contact dermatitis of the fingertips, 4 ophthalmologists (4,5,8,9) and one worker in an animal laboratory (6).

Case reports

A 60-year-old woman complained of an allergic reaction to ophthalmic anesthetic drops instilled before laser eye surgery. On examination, there was an eczematous dermatitis of the upper and lower eyelids. She had positive patch tests to proparacaine and tetracaine eye drops, to which she had been exposed, and to tetracaine 1% pet. Both ophthalmic solutions were preserved with benzalkonium chloride, which was negative in patch testing at 0.01% water (4).

A man had recurrent bouts of periocular dermatitis and conjunctivitis, following measurements of his intraocular pressure by an ophthalmologist. The patient had become sensitized to the topical ophthalmic anesthetic proparacaine, which was used prior to each measurement (7).

A female doctor aged 52 presented with a 9-months history of blepharitis. She had been prescribed many ointments and solutions. Patch testing showed a positive reaction only to eye drops containing proparacaine and benzalkonium chloride. Later, she had a strongly positive patch test to proparacaine 2% water, but she did not react to the preservative (3).

One case of allergic contact dermatitis from proparacaine (proxymetacaine) in an ophthalmic medication was reported from Germany in 1974. The patient reacted to proparacaine 0.5% water. Further details are not known (2).

Occupational allergic contact dermatitis

A 58-year-old ophthalmologist presented with bilateral fissuring and scaling of the finger pads for 3 years. His thumbs were most affected. There was a strongly positive patch test reaction to eye drops containing 0.5% proparacaine. The preservative was not tested separately. A diagnosis of occupational ACD was made and the patient was instructed to change to tetracaine, to which he had a negative patch test. He continued to work, but later began to again suffer from sporadic bouts of painful fissuring of his fingertips. Repeat testing confirmed the allergy to proparacaine, but he now also reacted to tetracaine 1% eye drops. The ophthalmologist continued his work using tetracaine drops without glove protection and had persistent mild finger pad dermatitis (4).

A right-handed 45-year-old female ophthalmologist began to suffer itching of the middle fingertip on her left hand. The problem progressed and caused thickening of the periungual skin and onycholysis of the distal nail plate. The fingertip became increasingly erythematous with painful fissuring, crusting and occasional bleeding. The affected finger was used to hold down the lower eyelid whilst applying drops to patients' eyes. When the ophthalmologist began using her index finger for holding the eyelid down whilst applying eye drops, similar symptoms started to develop on this finger within 7 days. Patch testing showed a positive reaction to proparacaine eye drops 0.5% at days 2 and 4. She changed to using oxybuprocaine eye drops. After that, the skin significantly improved within 2 weeks, although over half a year later the finger remains sensitive and prone to breaking down and the skin at the nail bed remains swollen. There is persistent deformity of the nail (8). It must be mentioned that the topical anesthetic itself was not tested in the patient and no mention was made of any excipients present in the eye drops and whether these were patch tested.

A similar case of occupational allergic contact dermatitis of the fingers from proparacaine in an ophthalmologist was already reported in the late 1960s (5). More recently, another ophthalmologist developed fingertip dermatitis from proparacaine eye drops. It took almost 3 years before the offending agent proparacaine was identified and removed. In the meantime, the patient had been treated with oral prednisone, methotrexate, UV-B phototherapy and Grenz ray (9).

An investigator involved in a program measuring intraocular pressure in rabbits, dogs and cats, using proparacaine as the topical anesthetic, developed local contact dermatitis with dryness and fissuring of the fingertips. Patch tests with proparacaine 0.316%, 1% and 3.16% gave very strong reactions to the 2 highest concentrations (6).

Cross-reactions, pseudo-cross-reactions and co-reactions
It has been suggested that tetracaine may be able to cross-react to proparacaine (4).

LITERATURE
1   The data in the section 'General' may have been obtained from literature discussed in this chapter, but mostly also or exclusively from one or more of the following online sources: ChemIDPlus Advanced, PubChem, DrugBank, RxList, Drug Central, Drugs.com, and Wikipedia
2   Maucher OM. Periorbitalekzem als iatrogene Erkrankung. Klin Monatsbl Augenheilkd 1974;164:350-356 (Article in German)
3   Bandmann H-J, Breit R, Mutzeck E. Allergic contact dermatitis front proxymetacaine. Contact Dermatitis Newsletter 1974;15:451
4   Dannaker CJ, Maibach HI, Austin E. Allergic contact dermatitis to proparacaine with subsequent cross-sensitization to tetracaine from ophthalmic preparations. Dermatitis 2001;12:177-179
5   March C, Greenwood MA. Allergic contact dermatitis to proparacaine. Arch Ophthalmol 1968;79:159-160
6   Lorenzetti OJ. Proparacaine contact dermatitis. Arch Dermatol 1969;100(4):489
7   Brancaccio RR, Milburn PB, Silvi E. Iatrogenic contact dermatitis to proparacaine: an ophthalmic topical anesthetic. Cutis 1993;52:296-298
8   Riddell CE, Reed J, Shaw S, Duvall-Young J. Allergic contact fingertip dermatitis secondary to proxymetacaine in an ophthalmologist. Eye 2000;14:907-908
9   Liesegang TJ, Perniciaro C. Fingertip dermatitis in an ophthalmologist caused by proparacaine. Am J Ophthalmol 1999;127:240-241

Chapter 3.290 PROPIPOCAINE

IDENTIFICATION

Description/definition : Propipocaine is the piperidine derivative that conforms to the structural formula shown below

Pharmacological classes : Local anesthetics

IUPAC name : 3-Piperidin-1-yl-1-(4-propoxyphenyl)propan-1-one

Other names : Propoxypiperocaine; 3-piperidino-4'-propoxypropiophenone; Falicaine (propipocaine hydrochloride)

CAS registry number : 3670-68-6

EC number : Not available

Merck Index monograph : 907

Patch testing : 1% pet.

Molecular formula : $C_{17}H_{25}NO_2$

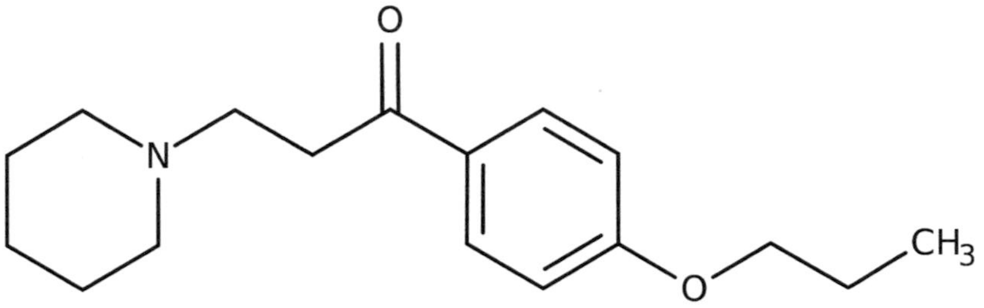

GENERAL

Propipocaine is a piperidine derivative that was formerly used as a local anesthetic, probably only or mostly in Germany, for infiltration anesthesia, epidural anesthesia and as surface anesthetic in otolaryngology. It was also present in hemorrhoidal ointments, suppositories, nose drops, antimicrobials and lotions for oral application. In pharmaceutical products, propipocaine was employed as propipocaine hydrochloride (CAS number 1155-49-3, EC number not available, molecular formula $C_{17}H_{26}ClNO_2$) (1).

CONTACT ALLERGY

Case series

In an out-patient department in Berlin, in 'the last few years' before 1975, 35 cases of allergic contact dermatitis from topical pharmaceuticals containing propipocaine HCl were observed. Most cases appeared after the treatment of perianal dermatitis; sometimes, the dermatitis was generalized. Contact allergy to propipocaine in nose drops resulted in dermatitis of the face that resembled erysipelas. Patch tests were positive to propipocaine 1% pet. The local anesthetic was then replaced with lidocaine (2).

Also in another German publication, one or more cases of contact allergy to propipocaine have been reported (no details available) (3).

LITERATURE

1   The data in the section 'General' may have been obtained from literature discussed in this chapter, but mostly also or exclusively from one or more of the following online sources: ChemIDPlus Advanced, PubChem, DrugBank, RxList, Drug Central, Drugs.com, and Wikipedia

2   Behrbohm P, Lenzner M. Sensitivity to falicain (propoxypiperocainhydrochloride). Contact Dermatitis 1975;1:187-188

3   Scholz A, von Richter G. Zur Allergie gegen Falikain (Propipokainhydrochlorid). Derm Monatsschr 1977;163:966-969 (Article in German)

Chapter 3.291 PROPRANOLOL

IDENTIFICATION

Description/definition   : Propranolol is the secondary alcohol that conforms to the structural formula shown below
Pharmacological classes : β-adrenergic antagonists; anti-arrhythmia agents; antihypertensive agents;
                           vasodilator agents
IUPAC name   : 1-Naphthalen-1-yloxy-3-(propan-2-ylamino)propan-2-ol
Other names   : 2-Propanol, 1-[(1-methylethyl)amino]-3-(1-naphthalenyloxy)-
CAS registry number   : 525-66-6
EC number   : 208-378-0
Merck Index monograph   : 8223
Patch testing   : Hydrochloride, 2% pet. (SmartPracticeCanada)
Molecular formula   : $C_{16}H_{21}NO_2$

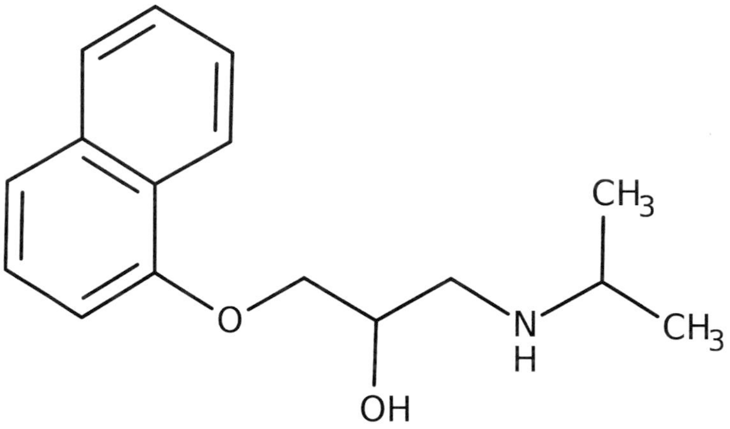

GENERAL

Propranolol is a synthetic non-cardioselective β-adrenergic receptor blocker with antianginal, antiarrhythmic, and antihypertensive properties. It competitively antagonizes β-adrenergic receptors, thereby inhibiting β-adrenergic reactions, such as vasodilation, and negative chronotropic and inotropic effects. Propranolol is used in the treatment or prevention of many disorders including acute myocardial infarction, arrhythmias, angina pectoris, hypertension and hypertensive emergencies, hyperthyroidism, migraine, pheochromocytoma, menopause, and anxiety. Since 2008, oral – and later topical - propranolol has been shown to be effective in the treatment of infantile hemangiomas. In pharmaceutical products, propranolol is employed as propranolol hydrochloride (CAS number 318-98-9, EC number 206-268-7, molecular formula $C_{16}H_{22}ClNO_2$) (1).

CONTACT ALLERGY

Case report

The superficial hemangioma on the right hand of a 5-month-old girl was treated with propranolol 1% in lipid cream base twice daily. One month later, the hemangioma was less red, but two months thereafter, the parents reported worsening associated with pruritus. On examination, scaling was evident on the tumor and vesicles and scratch marks were visible at its periphery. The application of propranolol was stopped and a corticosteroid was applied daily. One month after the complete resolution of dermatitis, patch tests showed positive reactions to propranolol 1% in petrolatum and in water. There was no co-reaction to timolol. Ten healthy controls were negative (3).

Cross-reactions, pseudo-cross-reactions and co-reactions

A patient with allergic contact dermatitis from timolol had a positive patch test to propranolol 2% water; the authors did not comment on this finding (2). One patient sensitized to metoprolol also had a positive skin test to propranolol 1% water (14).

Cutaneous adverse drug reactions from systemic administration caused by type IV (delayed-type) hypersensitivity

Cutaneous adverse drug reactions from systemic administration of propranolol caused by type IV (delayed-type) hypersensitivity, including toxic epidermal necrolysis (13), and in addition occupational allergic contact dermatitis (6-12,15), are planned to be discussed in Volume IV of the *Monographs in Contact Allergy* series on Systemic drugs.

LITERATURE

1   The data in the section 'General' may have been obtained from literature discussed in this chapter, but mostly also or exclusively from one or more of the following online sources: ChemIDPlus Advanced, PubChem, DrugBank, RxList, Drug Central, Drugs.com, and Wikipedia

2   Horcajada-Reales C, Rodríguez-Soria VJ, Suárez-Fernández R1. Allergic contact dermatitis caused by timolol with cross-sensitivity to levobunolol. Contact Dermatitis 2015;73:368-369

3   Bonifazi E, Milano A, Foti C. Allergic contact dermatitis caused by topical propranolol in a 5-month-old baby. Contact Dermatitis 2014;71:250-251

4   Léauté-Labrèze C, Dumas de la Roque E, Hubiche T, Boralevi F, Thambo JB, Taïeb A. Propranolol for severe hemangiomas of infancy. N Engl J Med 2008;358:2649-2651

5   Schneider M, Reimer A, Cremer H, Ruef P. Topical treatment with propranolol gel as a supplement to the existing treatment of hemangiomas. World J Pediatr 2014;10:313-317

6   Ali FR, Shackleton DB, Kingston TP, Williams JD. Occupational exposure to propranolol: A rarely recognised cause of allergic contact dermatitis. Int J Occup Med Environ Health 2015;28:639-640

7   Swinnen I, Ghys K, Kerre S, Constandt L, Goossens A. Occupational airborne contact dermatitis from benzodiazepines and other drugs. Contact Dermatitis 2014;70:227-232

8   Pereira F, Dias M, Pacheco FA. Occupational contact dermatitis from propranolol, hydralazine and bendroflumethiazide. Contact Dermatitis 1996;35:303-304

9   Mitchell JC, Maibach HI. Allergic contact dermatitis from phenoxybenzamine hydrochloride: cross-sensitivity to some related haloalkylamine compounds. Contact Dermatitis 1975;1:363-366

10 Mitchell JC. Allergic contact dermatitis from alpha- and beta-adrenergic receptor blocking agents (dibenzyline and propranolol). Contact Dermatitis Newsletter 1974;16:488

11 Rebandel P, Rudzki E. Dermatitis caused by epichlorohydrin, oxprenolol hydrochloride and propranolol hydrochloride. Contact Dermatitis 1990;23:199

12 Valsecchi R, Leighissa P, Piazzolla S, Naldi L, Cainelli T. Occupational contact dermatitis from propranolol. Contact Dermatitis 1994;30:177

13 Van Ketel WG, Soesman A. Een op de zlekte van Lyell gelijkende eruptie door propranolol. Ned T Geneeskd 1977;121:1475-1476 (Article in Dutch)

14 Van Joost T, Middelkamp Hup J, Ros FE. Dermatitis as a side effect of long-term topical treatment with certain beta-blocking agents. Br J Dermatol 1979;101:171-176

15 Rudzki E. Occupational dermatitis among health service workers. Derm Beruf Umwelt 1979;27:112-115

Chapter 3.292 PYRAZINOBUTAZONE

IDENTIFICATION

Description/definition   : Pyrazinobutanone is the combination of phenylbutazone and piperazine that conforms to
                            the structural formula shown below
Pharmacological classes : Anti-inflammatory drugs, non-steroidal
IUPAC name   : 4-Butyl-1,2-diphenylpyrazolidine-3,5-dione;piperazine
Other names   : Pyrasanone; 4-butyl-1,2-diphenyl-3,5-pyrazolidinedione compd. with piperazine (1:1);
                            phenylbutazone piperazium
CAS registry number   : 4985-25-5
EC number   : 225-639-4
Patch testing   : 1% and 5% pet.
Molecular formula   : $C_{23}H_{30}N_4O_2$

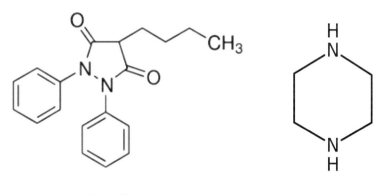

Phenylbutazone   Piperazine

GENERAL

Pyrazinobutanone is a nonsteroidal anti-inflammatory drug (NSAID) which has anti-inflammatory properties and is an equimolar salt of phenylbutazone and piperazine. Its clinical use is similar to other related NSAIDs, but its digestive tolerance is said to be superior. It is uncertain whether this compound is used as drug anymore (1).
See also Chapter 3.259 Phenylbutazone and Chapter 3.264 Piperazine.

CONTACT ALLERGY

Case reports

A 35-year-old food handler presented with a 10-day history of eczematous papules and vesicles over the right hip and on both hands, beginning 3 days after he had started applying a gel containing pyrazinobutazone to his right hip and taking 3 capsules of pyrazinobutazone daily by mouth. From 18 to 22 years of age, he had worked in a pharmaceutical laboratory handling, among other drugs, tablets of phenylbutazone – prednisone – meprobamate. The development of erythema and itching on the backs of his hands and fingers and an isolated episode of asthma, at that time, had forced him to change his occupation. Patch tests the gel 'as is', its active substance pyrazinobuta-zone, piperazine hexahydrate, phenylbutazone and the remaining components of the gel and capsules gave positive reactions to the gel 'as is', pyrazinobutazone 1% (+) and 5% (++) pet. (negative to 0.5%), piperazine hexahydrate 5% water (negative to 1%) and to phenylbutazone 1% (++) and 5% (+++) pet. The patient then took 1 capsule of 300 mg pyrazinobutazone every 12 hours. After 3 days, he developed erythemato-edematous vesicular lesions, with hemorrhagic features, symmetrically on the palms and on the dorsolateral aspects of the first 3 digits, as well as a flare-up at the site of the positive patch test to the gel a month previously. The patient had probably become sensitized to phenylbutazone during his work in a pharmaceutical laboratory (2). This was a case of systemic contact dermatitis.

   A 27-year-old woman, working in a pharmaceutical laboratory, was seen with a hand eczema which she suspected to be due to phenylbutazone-piperazine suppositories, which she had contact with when packing them.

During week-ends and holidays, the lesions almost healed, but immediately reappeared every time she returned to work. There were strongly positive patch tests to the suppositories 'as is', pyrazinobutazone 1% pet. and piperazine 5% water, but phenylbutazone was not tested (3). The same authors describe a patient who developed allergic contact dermatitis from the gel with pyrazinobutazone. A patch test with the gel was positive, as was piperazine 5% water, but, again, phenylbutazone was not tested (3).

A 55-year-old man had previously suffered dermatitis of the face after using a cream containing phenylbutazone. After a nasal operation, he inserted a phenylbutazone-piperazine (pyrazinobutazone) suppository. Ten hours later, he showed erythema and swelling on the face. He thought the handling of the suppository was the cause. Besides anal irritation, erythema with mild itching spread over his body with a later scaling over a month. Patch tests were positive to phenylbutazone 5% and piperazine 1% water (4). This was a case of systemic contact dermatitis. It is unknown whether and how much the allergy to piperazine has contributed to the clinical manifestations in this case.

Cross-reactions, pseudo-cross-reactions and co-reactions
Most patients allergic to pyrazinobutazone appear to be sensitized to both components; phenylbutazone and piperazine (2,4). A patient sensitized to phenylbutazone co-reacted to methylaminopyridine (1% olive oil) and 3-(aminomethyl)pyridine 1% water (4). Why these chemicals were tested and whether the sensitizations were considered to be cross-reactions was not discussed. See also Chapter 3.259 Phenylbutazone and Chapter 3.264 Piperazine.

LITERATURE
1   The data in the section 'General' may have been obtained from literature discussed in this chapter, but mostly also or exclusively from one or more of the following online sources: ChemIDPlus Advanced, PubChem, DrugBank, RxList, Drug Central, Drugs.com, and Wikipedia

2   Dorado Bris JM, Aragues Montañes M, Sols Candela M, Garcia Diez A. Contact sensitivity to pyrazinobutazone (Carudol) with positive oral provocation test. Contact Dermatitis 1992;26:355-356

3   Menezes-Brandao F, Foussereau J. Contact dermatitis to phenylbutazone-piperazine suppositories (Carudol®) and piperazine gel (Carudol®). Contact Dermatitis 1982;8:264-265

4   Fernandez de Corres L, Bernaola G, Lobera T, Leanizbarrutia I, Munoz D. Allergy from pyrazoline derivatives. Contact Dermatitis 1986;14:249-250

Chapter 3.293 PYRIDOXINE

IDENTIFICATION

Description/definition   : Pyridoxine is the 4-methanol form of vitamin B_6 that conforms to the structural formula shown below
Pharmacological classes   : Vitamin B complex
IUPAC name   : 4,5-bis(Hydroxymethyl)-2-methylpyridin-3-ol
Other names   : Vitamin B_6 (erroneous according to ChemIDPlus); 5-hydroxy-6-methyl-3,4-pyridine-dimethanol
CAS registry number   : 65-23-6
EC number   : 200-603-0
Merck Index monograph : 9365
Patch testing   : 1% and 10% pet.; perform controls
Molecular formula   : $C_8H_{11}NO_3$

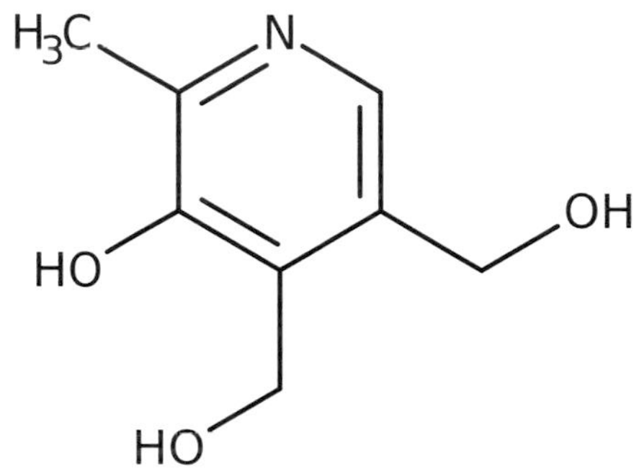

GENERAL

Pyridoxine is the 4-methanol form of vitamin B_6, an important water-soluble vitamin that is naturally present in many foods. As its classification as a vitamin implies, vitamin B_6 (and pyridoxine) are essential nutrients required for normal functioning of many biological systems within the body. Pyridoxine is converted to pyridoxal phosphate, which is a coenzyme for synthesis of amino acids, neurotransmitters (serotonin, norepinephrine), sphingolipids, and aminolevulinic acid. Although pyridoxine and vitamin B_6 are frequently used as synonyms, this practice is, according to some sources, erroneous (ChemIDPlus). In this database, it is stated that vitamin B_6 refers to several picolines, especially pyridoxine, pyridoxal and pyridoxamine. Pyridoxine is indicated for the treatment of vitamin B_6 deficiency and for the prophylaxis of isoniazid-induced peripheral neuropathy. In pharmaceutical products, pyridoxine is employed as pyridoxine hydrochloride (CAS number 58-56-0, EC number 200-386-2, molecular formula $C_8H_{12}ClNO_3$) (1).

CONTACT ALLERGY

Case reports

A 65-year-old woman, with a 30-year history of venous insufficiency, had had a leg ulcer on the inner aspect of the right ankle for 19 years. Over the last few weeks she had applied a new cream, which produced an acute severely itchy vesicular dermatitis around an enlarging ulcer. Patch tests were positive to the cream 'as is', pyridoxine hydrochloride (vitamin B_6) 10% pet., fragrances and parabens. The latter 2 were present in the cream, but the manufacturer would not disclose details of the composition, so it is uncertain whether pyridoxine was actually present in the cream (2).

    A 52-year-old man used 2 steroid creams, one with hydrocortisone 17-butyrate, the other with dexamethasone and pyridoxine HCl for the treatment of a mild burn. Erythema and vesicles developed around the initial lesion 3 days later followed by edema and erosions. By the 7th day the patient started to develop similar lesions on other parts of the body. Patch tests showed positive reactions to the cream with hydrocortisone 17-butyrate at D4 and the other cream at D10. A second test with the components of the creams showed positive reactions at D4 to hydrocortisone 17-butyrate 0.1% and pyridoxine HCl 1%. Finally, several steroids and vitamin B_6 derivatives were tested. At D4, there were positive reactions to 1% pyridoxine HCl, 1% pyridoxal HCl and 1% pyridoxal 5-phosphate, but pyridoxamine was negative. It was suggested that the allergy to pyridoxine was caused by active sensitization, as the patch test had become positive at D10 only (3). This is certainly possible, but nowadays it is known that reactions at D10 need not necessarily indicate patch test sensitization.

Cross-reactions, pseudo-cross-reactions and co-reactions
A patient sensitized to pyridoxine HCl (possibly actively sensitized by a patch test) cross-reacted to pyridoxal HCl and pyridoxal 5-phosphate, but not to pyridoxamine (3).

Patch test sensitization
A patch test with a steroid cream may possibly have caused active sensitization to its ingredient pyridoxine HCl; its concentration in the cream was unknown (3).

Cutaneous adverse drug reactions from systemic administration caused by type IV (delayed-type) hypersensitivity
Cutaneous adverse drug reactions from systemic administration of pyridoxine caused by type IV (delayed type) hypersensitivity, including photoallergic reactions (5-8), and in addition occupational allergic contact dermatitis (4,5), are planned to be discussed in Volume IV of the *Monographs in Contact Allergy* series on Systemic drugs.

LITERATURE

1 The data in the section 'General' may have been obtained from literature discussed in this chapter, but mostly also or exclusively from one or more of the following online sources: ChemIDPlus Advanced, PubChem, DrugBank, RxList, Drug Central, Drugs.com, and Wikipedia
2 Camarasa JG, Serra-Baldrich E, Lluch M. Contact allergy to vitamin B6. Contact Dermatitis 1990;23:115
3 Yoshikawa K, Watanabe K, Mizuno N. Contact allergy to hydrocortisone 17-butyrate and pyridoxine hydrochloride. Contact Dermatitis 1985;12:55-56
4 Córdoba S, Martínez-Morán C, García-Donoso C, Borbujo J, Gandolfo-Cano M. Non-occupational allergic contact dermatitis from pyridoxine hydrochloride and ranitidine hydrochloride. Dermatitis 2011;22:236-237
5 Bajaj AK, Rastogi S, Misra A, Misra K, Bajaj S. Occupational and systemic contact dermatitis with photosensitivity due to vitamin B6. Contact Dermatitis 2001;44:184
6 Murata Y, Kumano K, Ueda T, Araki N, Nakamura T, Tani M. Photosensitive dermatitis caused by pyridoxine hydrochloride. J Am Acad Dermatol 1998;39(2Pt.2):314-317
7 Tanaka M, Niizeki H, Shimizu S, Miyakawa S. Photoallergic drug eruption due to pyridoxine hydrochloride. J Dermatol 1996;23:708-709
8 Morimoto K, Kawada A, Hiruma M, Ishibashi A. Photosensitivity from pyridoxine hydrochloride (vitamin B6). J Am Acad Dermatol 1996: 35: 304-305.

Chapter 3.294 PYRILAMINE

IDENTIFICATION

Description/definition   : Pyrilamine is the histamine H1 receptor inverse agonist that conforms to the structural
                             formula shown below
Pharmacological classes : Anti-allergic agents; histamine H1 antagonists; sleep aids, pharmaceutical
IUPAC name   : N'-[(4-Methoxyphenyl)methyl]-N,N-dimethyl-N'-pyridin-2-ylethane-1,2-diamine
Other names   : Mepyramine; pyranisamine
CAS registry number   : 91-84-9
EC number   : 202-102-2
Merck Index monograph   : 9367 (Pyrilamine)
Patch testing   : 2% pet.
Molecular formula   : $C_{17}H_{23}N_3O$

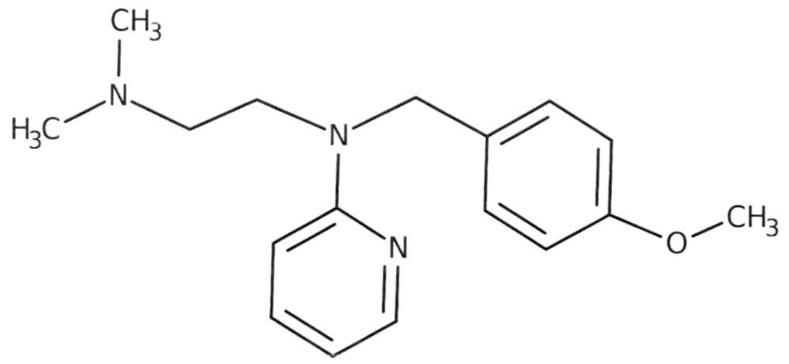

Pyrilamine 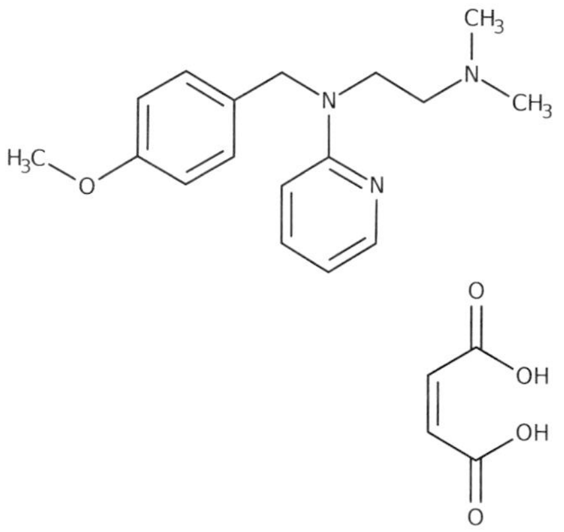   Pyrilamine maleate

GENERAL

Pyrilamine is a first-generation histamine H1 antagonist and ethylenediamine derivative with mild sedative and
hypnotic properties and some local anesthetic action. It is used in treating allergies and pruritic skin disorders and is
often present in over-the-counter combination products for colds and menstrual symptoms. In topical products,
pyrilamine is used to treat a variety of itching skin disorders. In pharmaceutical products, pyrilamine is employed as
pyrilamine maleate (CAS number 59-33-6, EC number 200-422-7, molecular formula $C_{21}H_{27}N_3O_5$).

CONTACT ALLERGY

Case reports and case series

In the period 1971-1976 inclusive, at St. John's Hospital, London, two men and 3 women were seen who had been
sensitized to one proprietary cream containing 2% pyrilamine (mepyramine) maleate. The patients had used the
cream for various itching dermatoses. Each patient had a positive patch test to the cream, tested 'as is'. Two were
tested with pyrilamine maleate 2% pet. and both reacted (3).

A 59-year-old woman was stung by several European hornets, and developed redness, edema and oozing around
the stings, measuring up to 4-5 cm. She treated these with 5 non-prescription topical itch-relieving products. During
the next 2 days, the skin reaction progressed and expanded to the face, neck, and upper extremities, as widespread
dermatitis with vesicles. Patch tests showed extreme positive reactions (+++) to three of the patient's itch-relieving
products, which all contained mepyramine (pyrilamine). Supplementary tests were negative to a number of their
ingredients, but positive to pyrilamine maleate 1% in pet. on D4 (2).

Cross-reactions, pseudo-cross-reactions and co-reactions

A patient who had allergic contact dermatitis from tripelennamine (pyribenzamine) may have cross-reacted to
pyrilamine maleate (1).

LITERATURE

1   Sherman WB, Cooke RA. Dermatitis following the use of pyribenzamine and antistine. J Allergy 1950;21:63-67
2   Winther AH, Andersen KE, Mortz CG. Allergic contact dermatitis caused by mepyramine in topical products.
    Contact Dermatitis 2015;73:255-256
3   Cronin E. Contact Dermatitis. Edinburgh: Churchill Livingstone, 1980:236

Chapter 3.295 PYRITHIONE ZINC

IDENTIFICATION

Description/definition	: Pyrithione zinc is a coordination complex of the zinc ion and pyrithione, a derivative of the naturally occurring antibiotic aspergillic acid; it conforms to the structural formula shown below
Pharmacological classes	: Keratolytic agents
IUPAC name	: Zinc;1-oxidopyridine-2-thione
Other names	: Zinc pyrithione; pyridine-2-thiol-1-oxide, zinc complex (2:1)
CAS registry number	: 13463-41-7
EC number	: 236-671-3
Merck Index monograph	: 9377 (Pyrithione)
Patch testing	: 1% pet. (Chemotechnique); 0.1% pet. (SmartPracticeCanada)
Molecular formula	: $C_{10}H_8N_2O_2S_2Zn$

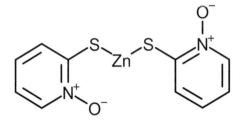

GENERAL

Pyrithione zinc is a coordination complex of the zinc ion and pyrithione, a derivative of the naturally occurring antibiotic aspergillic acid with antimicrobial, antifungal and anti-seborrheic actions. This agent is indicated for the treatment of dandruff and seborrheic dermatitis and is present in many over-the-counter products including shampoos.

CONTACT ALLERGY

Patch testing in groups of patients

Results of studies in which consecutive patients suspected of contact dermatitis have been patch tested with zinc pyrithione (routine testing) and of studies testing groups of selected patients are shown in table 3.295.1. In routine testing, in Denmark before 1985, there were 3 positive reactions in a group of 1652 patients tested (0.2%); one of these three had allergic contact dermatitis from zinc pyrithione in a shampoo (5). In groups of selected patients, frequencies of sensitization have ranged from 0.1% to 1.2%. The few reactions found were either not relevant or relevance was not mentioned (11,16,19,20).

Table 3.295.1 Patch testing in groups of patients

Years and Country	Test conc. & vehicle	Number of patients tested	positive (%)	Selection of patients (S); Relevance (R); Comments (C)	Ref.
Routine testing					
<1985 Denmark	1% pet.	1652	3 (0.2%)	R: 1/3 had allergic contact dermatitis from a shampoo containing zinc pyrithione	5
Testing in groups of selected patients					
1997-2001 USA		203	1 (0.5%)	S: patients with eyelid dermatitis; R: not stated; C: unknown in how many patients this material was tested	21
<1995 Spain	1% pet.	171	2 (1.2%)	S: hairdressers and clients of hairdressers; R: not relevant	11
1990-1994 IVDK	0.1% pet.	7782	6 (0.1%)	S: patients tested with a preservative series; R: not stated	16
1981 France	1% pet.	465	2 (0.4%)	S: patients suspected of allergy to cosmetics, drugs, industrial products, or clothes; R: not stated	19

IVDK: Information Network of Departments of Dermatology (Germany, Austria, Switzerland)

Case reports and case series

In a group of 119 patients with allergic contact dermatitis from cosmetics, investigated in The Netherlands in 1986-1987, one case was caused by zinc pyrithione in a shampoo (17,18). Two cases (11) and one case each (5,10,12,13) of allergic reactions from zinc pyrithione in shampoos have been reported. One patient had worsening of scalp psoriasis and eyelid dermatitis from contact allergy to zinc pyrithione and cocamide DEA in an anti-dandruff shampoo (2). Another patient developed allergic contact dermatitis from zinc pyrithione in shampoo, which led to worsening of existing psoriasis (Köbner-phenomenon) and development of pustular psoriasis (3). A similar case had been reported previously (9). Two patients had allergic contact dermatitis from zinc pyrithione, one by its presence in a hair cream, the other from the same hair cream and from an antidandruff lotion (6). In another patient, contact allergic sensitivity to zinc pyrithione in a shampoo was followed by the photosensitivity dermatitis and actinic reticuloid syndrome (7). One patient reacted to zinc pyrithione in a hair cream and a shampoo (8). One patient was allergic to zinc pyrithione, of whom details are unknown (1).

Cross-reactions, pseudo-cross-reactions and co-reactions
Not to sodium pyrithione (14). Cross-reaction to sodium pyrithione (assumed on theoretical grounds, zinc pyrithione itself not tested) (15).

PHOTOSENSITIVITY
One patient had a positive photopatch test to zinc pyrithione; details are unknown (4). In another individual, contact allergy to zinc pyrithione in a shampoo was followed by the photosensitivity dermatitis and actinic reticuloid syndrome (7).

LITERATURE
1   Calnan CD, cited by Fisher AA. Highlights of the First International Symposium on Contact Dermatitis. Cutis 1976;18:645-662 (pages probably incorrect)
2   De Groot AC, de Wit FS, Bos JD, Weyland JW. Contact allergy to cocamide DEA and lauramide DEA in shampoos. Contact Dermatitis 1987;16:117-118
3   Jo J-H, Jang H-S, Ko H-C, Kim M-B, Oh C-K, Kwon Y-W, Kwon K-S. Pustular psoriasis and the Köbner phenomenon caused by allergic contact dermatitis from zinc pyrithione-containing shampoo. Contact Dermatitis 2005;52:142-144
4   Osmundsen PE. Contact photodermatitis due to tribromsalicylanilide (cross-reaction pattern). Dermatologica 1970;140:65-74
5   Brandrup F, Menné T. Zinc pyrithione allergy. Contact Dermatitis 1985;12:50
6   Muston HL, Messenger AG, Byrne JPH. Contact dermatitis from zinc pyrithione, an antidandruff agent. Contact Dermatitis 1979;5:276-277
7   Yates VM, Finn OA. Contact allergic sensitivity to zinc pyrithione followed by the photosensitivity dermatitis and actinic reticuloid syndrome. Contact Dermatitis 1980;6:349-350
8   Goh CL, Lim KB. Allergic contact dermatitis to zinc pyrithione. Contact Dermatitis 1984;11:120
9   Nielsen N, Menné T. Allergic contact dermatitis caused by zinc pyrithione associated with pustular psoriasis. Am J Contact Dermat 1997;8:170–171
10   Hsieh C-W, Tu M-E, Wu Y-H. Allergic contact dermatitis induced by zinc pyrithione in shampoo: a case report. Dermatologica Sinica 2010;28:163-166
11   Gonzalez Perez R, Aguirre A, Raton JA, Eizaguirre X, Fiaz-Pérez JL. Positive patch tests to zinc pyrithione. Contact Dermatitis 1995;32:118-119
12   Pereira F, Fernandes C, Dias M, Lacerda MH. Allergic contact dermatitis from zinc pyrithione. Contact Dermatitis 1995;33:131
13   Nigam PK, Tyagi S, Saxena AK, Misra RS. Dermatitis from zinc pyrithione. Contact Dermatitis 1988;19:219
14   Isaksson, M. Delayed diagnosis of occupational contact dermatitis from sodium pyrithione in a metalworking fluid. Contact Dermatitis 2002;47:248-249
15   Le Coz C-J. Allergic contact dermatitis from sodium pyrithione in metalworking fluid. Contact Dermatitis 2001;45:58-59
16   Schnuch A, Geier J, Uter W, Frosch PJ. Patch testing with preservatives, antimicrobials and industrial biocides. Results from a multicentre study. Br J Dermatol 1998;138:467 476
17   De Groot AC, Bruynzeel DP, Bos JD, van der Meeren HL, van Joost T, Jagtman BA, Weyland JW. The allergens in cosmetics. Arch Dermatol 1988;124:1525-1529
18   De Groot AC. Adverse reactions to cosmetics. PhD Thesis, University of Groningen, The Netherlands: 1988, chapter 3.4, pp.105-113
19   Meynadier JM, Meynadier J, Colmas A, Castelain PY, Ducombs G, Chabeau G, et al. Allergy to preservatives. Ann Dermatol Venereol 1982;109:1017-1023
20   Guin JD. Eyelid dermatitis: experience in 203 cases. J Am Acad Dermatol 2002;47:755-765

Chapter 3.296 PYRROLNITRIN

IDENTIFICATION

Description/definition   : Pyrrolnitrin is the pyrrole that conforms to the structural formula shown below
Pharmacological classes   : Antifungal agents
IUPAC name   : 3-Chloro-4-(3-chloro-2-nitrophenyl)-1*H*-pyrrole
CAS registry number   : 1018-71-9
EC number   : 213-812-7
Merck Index monograph   : 9402
Patch testing   : 1% pet.
Molecular formula   : $C_{10}H_6Cl_2N_2O_2$

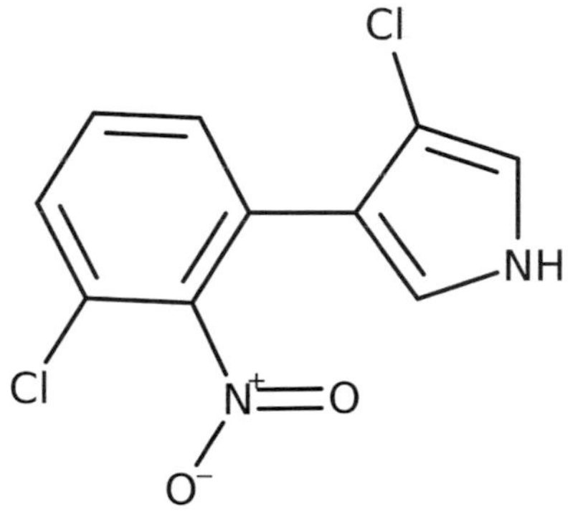

GENERAL

Pyrrolnitrin is a pyrrole antifungal agent isolated from several *Pseudomonas* species including *Pseudomonas pyrrocinia*. It is effective mainly against *Trichophyton*, *Microsporum*, *Epidermophyton*, and *Penicillium*. It was formerly marketed in Italy, Spain and Japan (10), but is nowadays probably hardly, if at all, used anymore in topical antifungal preparations (1).

CONTACT ALLERGY

General

In the 1970s and 1980s, many cases of allergic contact dermatitis from pyrrolnitrin have been observed, virtually all in Italy, with one report from Spain of 5 patients (12) and one from Japan with one sensitized individual (9). There have been case series of 62 (2, including the 41 patients in ref. 10), 41 (10, also included in ref. 2), 9 (8), 8 (11), 7 (4), 5 (12) and 3 (3) patients sensitized to pyrrolnitrin. The clinical picture most often had an atypical manifestation with erythematous urticarial papules and plaques (with or without vesiculation), spreading from the primary site to the arms and/or legs and/or trunk, which was characteristic enough to suggest pyrrolnitrin as the cause of the eruption on the base of clinical examination alone (10,11). Some authors described it as an erythema multiforme-like contact dermatitis (2,3,10). The large number of sensitized patients seen in individual clinics and the short period from first application of the pyrrolnitrin creams to the appearance of contact dermatitis suggest that pyrrolnitrin is a strong sensitizer. There have been no more reports of sensitization to pyrrolnitrin since 1990.

Case series

In Bari, Italy, in the period 1968-1983, 62 patients were seen with allergic contact dermatitis from pyrrolnitrin cream, often with features resembling erythema multiforme; they all had positive patch tests to the cream and to pyrrolnitrin 1% pet. (2, overlap with ref. 10). Previously, the authors had presented 41 of these 62 patients. They had a widespread red edematous vesicular eruption, appearing in patients with various forms of tinea (38 tinea cruris, 2 tinea manuum, 1 tinea versicolor) after 8-20 days of treatment with either a cream containing pyrrolnitrin or pyrrolnitrin and betamethasone 17-valerate. All patients who had applied the cream for tinea showed quite distinct morphological features. The eruption would start in the inguinocrural areas and spread bilaterally on to the thighs and trunk. The individual lesions were characterized by erythemato-edematous bullous patches either isolated or in many cases confluent, which imitated an erythema multiforme eruption. Some patches showed evident vesiculation. The authors stated that the appearance of the lesions was sufficiently characteristic and different from the usual erythemato-papulo-vesicular eruption of allergic contact dermatitis, to suggest pyrrolnitrin as the cause of the rash on clinical examination. All had positive patch tests to the their antifungal preparation(s) and pyrrolnitrin 1% pet. (10).

In 1982, 9 patients (aged 23 to 65 years) were seen in a University clinic in Naples, Italy, with a widespread rash after treatment of tinea cruris or erythrasma with a cream containing pyrrolnitrin or pyrrolnitrin and betamethasone

17-valerate. After 10-15 days of treatment, an intensely itchy, red, edematous, vesicular eruption appeared in the groins and spread rapidly to the arms and trunk. All patients were patch test positive to pyrrolnitrin cream and pyrrolnitrin 1% pet. (8).

In 1981, Italian researchers from Bergamo presented 8 new cases of ACD from pyrrolnitrin (11) after first having published 7 patients in Italian literature (4). Patch tests to the antifungal cream used and pyrrolnitrin 1% were positive in all. The special features observed in this study were: (a) the morphology is typical; at first the severely itching erythemato-edematous, urticarial-like lesions appear, followed by microvesicles 5-7 days later; (b) the lesions are initially localized to the sites of application, and then rapidly spread to adjacent areas and most of the body. The clinical appearances were so typical that etiology can be diagnosed on clinical examination alone (11).

In 1980, 5 patients who had developed ACD from pyrrolnitrin creams were reported from Spain. In only one individual did the dermatitis spread from the groin to thighs, abdomen, buttocks and lower part of the back with erythematous and edematous lesions. Patch tests were positive to the creams and pyrrolnitrin 1% alcohol; one also reacted to the ingredient betamethasone 17-valerate 0.1% alc. (12). Three patients developed secondary erythema multiforme-like spreading of allergic contact dermatitis to pyrrolnitrin (3).

Case reports

A woman treated dermatitis on her left foot with pyrrolnitrin lotion and honeybee royal jelly, which aggravated the condition. Patch tests were positive to the jelly, the lotion and pyrrolnitrin 0.5%, 0.1% and 0.01% pet. (9). A 54-year-old woman was treated for an inguinal dermatitis, presumably fungal in nature, with a cream containing pyrrolnitrin for 10 days, when she developed an itchy erythematous rash on both thighs, spreading to the trunk, arms and legs. It consisted of large erythematous, edematous vesicular elements with a tendency to coalesce. Patch tests were positive to the cream and pyrrolnitrin 1% pet. A 24-year-old man had similar lesions from treatment of tinea cruris with pyrrolnitrin, but limited to the inguinal region and the medial aspects of the thighs. Patch tests were positive to the cream and pyrrolnitrin 1% pet. (13).

Other case reports/series or information on contact allergy to pyrrolnitrin, adequate data of which are not available to the author, can be found in refs. 4 (7 cases), 5,6, and 7.

Cross-reactions, pseudo-cross-reactions and co-reactions

No cross-reactivity to substances chemically related to pyrrolnitrin: pyrrole, chlorobenzene, 2-nitrochlorobenzene, nitrobenzene, p-chloronitrobenzene, 1-chloro-2,4-dinitrobenzene (DNCB) and chloramphenicol (10). Not to chloramphenicol and DNCB (11).

LITERATURE

1 The data in the section 'General' may have been obtained from literature discussed in this chapter, but mostly also or exclusively from one or more of the following online sources: ChemIDPlus Advanced, PubChem, DrugBank, RxList, Drug Central, Drugs.com, and Wikipedia
2 Angelini G, Vena GA, Meneghini CL. Allergic contact dermatitis to some medicaments. Contact Dermatitis 1985;12:263-269
3 Meneghini CL, Angelini G. Secondary polymorphic eruptions in allergic contact dermatitis. Dermatologica 1981;163:63-70
4 Valsecchi R, Rozzoni M, Serra M, Cainelli T. Dermatite allergica da contatto alla pirrolnitrina. Chronica Dermatologica 1979;1:352 (Article in Italian); the bibliography was cited elsewhere as Chronica Dermatologica 1979: 10:21-25
5 Seidenari S, Di Nardo A, Motolese A, Pincelli C. Erythema multiforme associated with contact sensitization. Description of 6 clinical cases. G Ital Dermatol Venereol 1990;125:35-40 (Article in Italian)
6 Valsecchi R, Falgheri G, Serra M, Cainelli T. Sui problema della sensibilizzazione crociata tra dinitroclorobenzene, cloramfenicolo e pirrolnitrina. Giornale Italiano di Dermatologia -Minerva Dermatologica 1981;116 :119-121 (Article in Italian, cited in ref. 8)
7 Fabbri P, Sertoli A, Di Fonzo E, Donati E. Contact dermatitis from pyrrolnitrin. Italian General Review of Dermatology 1976;13:97-100 (Article in Italian, cited in ref. 8)
8 Balato N, Lembo G, Cusano F, Ayala F. Contact dermatitis from pyrrolenitrin. Contact Dermatitis 1983;9:238
9 Takahashi M, Matsuo I, Ohkido M. Contact dermatitis due to honeybee royal jelly. Contact Dermatitis 1983;9:452-455
10 Meneghini CL, Angelini G. Contact dermatitis from pyrrolnitrin. Contact Dermatitis 1982;8:55-58
11 Valsecchi R, Foiadelli L, Cainelli T. Contact dermatitis from pyrrolnitrin. Contact Dermatitis 1981;6:340
12 Romaguera C, Grimalt F. Five cases of contact dermatitis from pyrrolnitrine. Contact Dermatitis 1980;6:352-353
13 Meneghini CL, Angelini G. Contact dermatitis from pyrrolnitrin (an antimycotic agent). Contact Dermatitis 1975;1:288-292

Chapter 3.297 QUINOLINE MIX

The quinoline mix is a mixture used for diagnostic patch testing and currently consists of 3% clioquinol and 3% chlorquinaldol in petrolatum. The quinoline mix is also available as T.R.U.E. TEST ® (www.smartpracticecanada.com). It contains 190 microgram/cm^2 active ingredients, corresponding to 154 microgram/patch, equal parts clioquinol and chlorquinaldol. In this chapter, only the results of testing with the mix are presented. The individual ingredients are monographed in Chapter 3.72 (Clioquinol) and Chapter 3.66 (Chlorquinaldol).

CONTACT ALLERGY

General
A considerable number of positive patch test reactions to the quinoline mix (T.R.U.E. test) may be missed when readings are not performed at day 7 (29). The quinoline mix has never been part of the European baseline series or of the NACDG screening series. The former contained clioquinol 5% pet., but this yielded positive results so infrequently, that it was removed from the European baseline series in 2019 (33). Because of yellow staining, limited efficacy and the availability of safer and more effective pharmaceuticals, clioquinol and chlorquinaldol are hardly used anymore in western countries.

Contact allergy in the general population
Estimates of the 10-year prevalence (1997-2006) of contact allergy to the quinoline mix in the general population of Denmark based on the CE-DUR method ranged from 0.18% to 0.24% (22). In a similar study from Germany, the estimated prevalence in the general population in the period 1992-2000 ranged from 1.8 to 4.2% (23).

Patch testing in the general population and in subgroups
In several investigations, random samples of the population of certain age groups have been patch tested with the quinoline mix (table 3.297.1, data back to 1996). Generally, the prevalences of sensitization to the mix have ranged from 0% to 0.2%. The highest rates were found in Norway just before 2007: 0.7% in women and 0.6% in men. However, the population investigated was small (n=531) and data on relevance were not provided (27).

Table 3.297.1 Contact allergy in the general population and in subgroups

Year and country	Selection and number tested	Prevalence of contact allergy			Comments	Ref.
		Total	Women	Men		
General population						
2008-11 five European countries	general population, random sample, 18-74 years, n=3119	0.2%	0.2%	0.1%	TRUE test	15
<2007 Norway	general population, random sample, 18-69 years, n=531	0.6%	0.7%	0.6%	TRUE test	27
2006 Denmark	general population, random sample, 18-69 years, n=3460	0.1%	0%	0.1%	patch tests were read on day 2 only; TRUE test	17
2005 Norway	general population, random sample, 18-69 years, n=1236	0.1%	0%	0.2%	TRUE test	18
<1999 Australia	self-selected adult healthy volunteers, 18-82 y., n=219	0%	0%	0%		25
1998 Denmark	general population, random sample, 15-41 years, n=469		0%	0%	patch tests were read on day 2 only; TRUE test	19
Subgroups						
2011-2013 Sweden	adolescents from a birth cohort, 15.8-18.9 years, n=2285	0.04%	0.08%	0.0%	TRUE test; patch tests were read at day 2 only	28
2010 Denmark	unselected population of 8th grade schoolchildren in Denmark, 15 years later; n=442	0%	0%	0%	follow-up study; TRUE test	16
1997-1998 Denmark	twins aged 20-44 years, n=1076	0%	0%	0%	TRUE test; 449 had self-reported hand eczema	26
1995-1996 Denmark	8th grade school children, 12-16 years, n=1146	0%	0%	0%		20,21

In all these studies, T.R.U.E. test materials have been used. It should be realized that, in a number of these investigations, the results were read at D2 only. This may have resulted in an underestimation of the true prevalence of sensitization, as positive reactions at D3 or D4 (the usual time for the second reading) with negative result at D2 commonly occur with most haptens. Moreover, a considerable number (>25%) of positive patch test reactions to the T.R.U.E. test quinoline mix may be missed when readings are not performed at day 7 (29).

Patch testing in groups of patients

Results of patch testing the quinoline mix in consecutive patients suspected of contact dermatitis (routine testing) back to 2000 are shown in table 3.297.2. Results of testing in groups of *selected* patients (e.g. patients patch tested for perianal and/or genital symptoms, individuals suspected of contact allergy to medicaments, patients with chronic inflammatory ear disease, individuals with leg ulcers and stasis dermatitis) are shown in table 3.297.3.

In routine testing in 8 studies, frequencies of sensitization were invariably low and have ranged from 0.2% to 0.8%. Relevance data were provided in one investigation only: the positive patch tests to the quinoline mix were considered to be relevant in 11 of 22 patients in a 2000 study from the United Kingdom (5).

Table 3.297.2 Patch testing in groups of patients: Routine testing

Years and Country	Test conc. & vehicle	Number of patients tested	positive (%)	Selection of patients (S); Relevance (R); Comments (C)	Ref.
2008-18 Netherlands [a]		3218	7 (0.2%)	R: not stated; C: >25% of the reactions may be missed when patch are not read at D7	29
2014-2016 Greece	5% pet.	1978	10 (0.5%)	R: not stated	24
2009-2013 Singapore		2598	(0.4%)	R: present + past relevance: 0%; C: range of positive reactions per year 0% - 1.3%	1
2009 Sweden	6% pet.	3112	(0.2%)	R: not stated	2
1992-2004 Turkey	5% pet.	1038	9 (0.8%)	R: not stated	3
1999-2001 Sweden	5% pet.	3790	(0.5%)	R: not stated	4
2009 Sweden	6% pet.	3112	(0.5%)	R: not stated	2
2000 United Kingdom	6% pet.	3063	22 (0.7%)	R: 50% (current and past relevance in one center); C: range of positive reactions per center 0.3% - 1.2%	5

[a] T.R.U.E. test

In studies testing the quinoline mix in groups of *selected* patients, prevalences of positive reactions have ranged from 1% to 18.3% (table 3.297.3). The very high percentage of 18.3 was in a study from the United Kingdom performed in the period 1988-1991 (32). The population tested consisted of patients with contact allergy to one or more corticosteroids. Co-sensitization in this group to other ingredients of topical pharmaceuticals, both active drugs and excipients, is well known and, presumably, many of these patients had used a corticosteroid-clioquinol combination product (32). High frequencies of sensitization were also found, as can be expected, in patients with leg ulcers/stasis dermatitis (6,7) and otitis externa (13,14). In most studies, no relevance data were provided, but in those that did, relevance figures were high with 60-100% (9-12,30).

Table 3.297.3 Patch testing in groups of patients: Selected patient groups

Years and Country	Test conc. & vehicle	Number of patients tested	positive (%)	Selection of patients (S); Relevance (R); Comments (C)	Ref.
2013-2015 Ireland	6% pet.	99	1 (1%)	S: patients patch tested for perianal and/or genital symptoms; R: all reactions to medicaments were relevant	12
1978-2005 Taiwan		603	10 (1.7%)	S: patients suspected of contact allergy to medicaments; R: 65% of the reactions to all medicaments were considered to be relevant	30
1985-2002 U.K		179	8 (4.5%)	S: patients with chronic inflammatory ear disease; R: not stated	13
1996-1997 U.K.		109	8 (7.3%)	S: mostly patients with leg ulcers and stasis dermatitis; R: not stated	6
<1994 U.K.		85	12 (14%)	S: patients with longstanding venous ulceration or eczema complicating leg ulcers; R: not stated	7
1993-4 Netherlands	6% pet.	34	2 (6%)	S: patients with chronic otitis externa or media; R: not stated; C: the frequency was 0.2% in a control group	14
1992-1994 IVDK	6% pet.	807	27 (3.3%)	S: patients with leg ulcers and/or leg eczema; R: not stated	8
1992-1994 U.K.	6% pet.	69	2 (3%)	S: women with 'vulval problems'; R: 58% for all allergens together	9

Table 3.297.3 Patch testing in groups of patients: Selected patient groups (continued)

Years and Country	Test conc. & vehicle	Number of patients tested | positive (%)			Selection of patients (S); Relevance (R); Comments (C)	Ref.
1988-1991 U.K.		131	24	(18.3%)	S: patients with contact allergy to one or more corticosteroids; R: not stated; C: the frequency in all patients seen for routine testing was 1.9%	32
1986-1990 U.K.	6% pet.	135	3	(2.2%)	S: patients referred to a vulva clinic for patch testing; R: 2/3	11
1988-1989 U.K.	6% pet.	815	3	(4%)	S: patients with leg ulcers; R: all reactions were considered to be relevant	10

IVDK: Information Network of Departments of Dermatology (Germany, Austria, Switzerland)

Case series

In the period 1985-1997, in Leuven, Belgium, 8521 patients were patch tested, and 28 positive reactions were observed to the quinoline mix. It was not stated, however, how many patients had been tested with this allergen and how many of the reactions were relevant (31).

LITERATURE

1   Ochi H, Cheng SWN, Leow YH, Goon ATJ. Contact allergy trends in Singapore – a retrospective study of patch test data from 2009 to 2013. Contact Dermatitis 2017;76:49-50

2   Fall S, Bruze M, Isaksson M, Lidén C, Matura M, Stenberg B, Lindberg M. Contact allergy trends in Sweden – a retrospective comparison of patch test data from 1992, 2000, and 2009. Contact Dermatitis 2015;72:297-304

3   Akyol A, Boyvat A, Peksari Y, Gurgey E. Contact sensitivity to standard series allergens in 1038 patients with contact dermatitis in Turkey. Contact Dermatitis 2005;52:333-337

4   Lindberg M, Edman B, Fischer T, Stenberg B. Time trends in Swedish patch test data from 1992 to 2000. A multi-4 centre study based on age- and sex-adjusted results of the Swedish standard series. Contact Dermatitis 2007;56:205-210

5   Britton JE, Wilkinson SM, English JSC, Gawkrodger DJ, Ormerod AD, Sansom JE, et al. The British standard series of contact dermatitis allergens: validation in clinical practice and value for clinical governance. Br J Dermatol 2003;148:259-264

6   Gooptu C, Powell SM. The problems of rubber hypersensitivity (types I and IV) in chronic leg ulcer and stasis eczema patients. Contact Dermatitis 1999;41:89-93

7   Zaki I, Shall L, Dalziel KL. Bacitracin: a significant sensitizer in leg ulcer patients? Contact Dermatitis1994;31:92-94

8   Renner R, Wollina U. Contact sensitization in patients with leg ulcers and/or leg eczema: comparison between centers. Int J Low Extrem Wounds 2002;1:251-255

9   Lewis FM, Harrington CI, Gawkrodger DJ. Contact sensitivity in pruritus vulvae: a common and manageable problem. Contact Dermatitis 1994;31:264-265

10 Wilson CL, Cameron J, Powell SM, Cherry G, Ryan TJ. High incidence of contact dermatitis in leg-ulcer patients – implications for management. Clin Exp Dermatol 1991;16:250-253

11 Marren P, Wojnarowska F, Powell S. Allergic contact dermatitis and vulvar dermatoses. Br J Dermatol 1992;126:52-56

12 Foley CC, White S, Merry S, Nolan U, Moriarty B, et al. Understanding the role of cutaneous allergy testing in anogenital dermatoses: a retrospective evaluation of contact sensitization in anogenital dermatoses. Int J Dermatol 2019;58:806-810

13 Millard TP, Orton DI. Changing patterns of contact allergy in chronic inflammatory ear disease. Contact Dermatitis 2004;50:83-86

14 Van Ginkel CJ, Bruintjes TD, Huizing EH. Allergy due to topical medications in chronic otitis externa and chronic otitis media. Clin Otolaryngol Allied Sci 1995;20:326-328

15 Diepgen TL, Ofenloch RF, Bruze M, Bertuccio P, Cazzaniga S, Coenraads P-J, et al. Prevalence of contact allergy in the general population in different European regions. Br J Dermatol 2016;174:319-329

16 Mortz CG, Bindslev-Jensen C, Andersen KE. Prevalence, incidence rates and persistence of contact allergy and allergic contact dermatitis in The Odense Adolescence Cohort Study: a 15-year follow-up. Brit J Dermatol 2013;168:318-325

17 Thyssen JP, Linneberg A, Menné T, Nielsen NH, Johansen JD. Contact allergy to allergens of the TRUE-test (panels 1 and 2) has decreased modestly in the general population. Br J Dermatol 2009;161:1124-1129

18 Dotterud LK, Smith-Sivertsen T. Allergic contact sensitization in the general adult population: a population-based study from Northern Norway. Contact Dermatitis 2007;56:10-15

19 Nielsen NH, Linneberg A, Menné T, Madsen F, Frølund L, Dirksen A, et al. Allergic contact sensitization in an adult Danish population: two cross-sectional surveys eight years apart (the Copenhagen Allergy Study). Acta Derm Venereol 2001;81:31-34

20 Mortz CG, Lauritsen JM, Bindslev-Jensen C, Andersen KE. Contact allergy and allergic contact dermatitis in adolescents: prevalence measures and associations. Acta Derm Venereol 2002;82:352-358

21 Mortz CG, Lauritsen JM, Bindslev-Jensen C, Andersen KE. Prevalence of atopic dermatitis, asthma, allergic rhinitis, and hand and contact dermatitis in adolescents. The Odense Adolescence Cohort Study on Atopic Diseases and Dermatitis. Br J Dermatol 2001;144:523-532

22 Thyssen JP, Uter W, Schnuch A, Linneberg A, Johansen JD. 10-year prevalence of contact allergy in the general population in Denmark estimated through the CE-DUR method. Contact Dermatitis 2007;57:265-272

23 Schnuch A, Uter W, Geier J, Gefeller O (for the IVDK study group). Epidemiology of contact allergy: an estimation of morbidity employing the clinical epidemiology and drug-utilization research (CE-DUR) approach. Contact Dermatitis 2002;47:32-39

24 Tagka A, Stratigos A, Lambrou GI, Nicolaidou E, Katsarou A, Chatziioannou A. Prevalence of contact dermatitis in the Greek population: A retrospective observational study. Contact Dermatitis 2019;81:460-462

25 Greig JE, Carson CF, Stuckey MS, Riley TV. Prevalence of delayed hypersensitivity to the European standard series in a self-selected population. Australas J Dermatol 2000;41:86-89

26 Bryld LE, Hindsberger C, Kyvik KO, Agner T, Menné T. Risk factors influencing the development of hand eczema in a population-based twin sample. Br J Dermatol 2003;149:1214-1220

27 Dotterud LK. The prevalence of allergic contact sensitization in a general population in Tromsø, Norway. Int J Circumpolar Health 2007;66:328-334

28 Lagrelius M, Wahlgren CF, Matura M, Kull I, Lidén C. High prevalence of contact allergy in adolescence: results from the population-based BAMSE birth cohort. Contact Dermatitis 2016;74:44-51

29 van Amerongen CCA, Ofenloch R, Dittmar D, Schuttelaar MLA. New positive patch test reactions on day 7—The additional value of the day 7 patch test reading. Contact Dermatitis 2019;81:280-287.

30 Shih Y-H, Sun C-C, Tseng Y-H, Chu C-Y. Contact dermatitis to topical medicaments: a retrospective study from a medical center in Taiwan. Dermatol Sinica 2015;33:181-186

31 Goossens A, Claes L, Drieghe J, Put E. Antimicrobials, preservatives, antiseptics and disinfectants. Contact Dermatitis 1998;39:133-134

32 Burden AD, Beck MH. Contact hypersensitivity to topical corticosteroids. Br J Dermatol 1992;127:497-501

33 Wilkinson M, Gonçalo M, Aerts O, Badulici S, Bennike NH, Bruynzeel D, et al. The European baseline series and recommended additions: 2019. Contact Dermatitis 2019;80:1-4

Chapter 3.298 RESORCINOL

IDENTIFICATION

Description/definition   : Resorcinol is the phenol that conforms to the structural formula shown below
Pharmacological classes   : Dermatological drugs
IUPAC name   : Benzene-1,3-diol
Other names   : CI 76505; 1,3-benzenediol
CAS registry number   : 108-46-3
EC number   : 203-585-2
Merck Index monograph   : 9546
Patch testing   : 1% pet. (Chemotechnique, SmartPracticeCanada); 2% pet. (SmartPracticeEurope, SmartPracticeCanada)
Molecular formula   : $C_6H_6O_2$

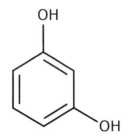

GENERAL

Resorcinol is a 1,3-isomer (or meta-isomer) of benzenediol that formerly had widespread pharmaceutical use in anti-acne preparations, anti-wart ointments, antipsoriatic creams, exfoliative preparations, suppositories, eye drops and Castellani's paint, because of its (alleged) antiseptic, itch relieving, keratolytic and antifungal properties. Currently, it is used as a coupler in hair dyes and in many industrial applications. Consumer products that may contain resorcinol include hair dyes, adhesives and sealants, building/construction materials, wood and engineered wood products, electrical and electronic products, fabric, textile, and leather products, paints and coatings, personal care products, plastic and rubber products, and rubber tires (36,37). Resorcinol may also be present in some patch test marking inks (23).

In this chapter, only contact allergy to resorcinol is – briefly - discussed. A full account of the literature on contact allergy to resorcinol and its other side effects (including irritant contact dermatitis, other non-eczematous contact reactions, systemic side effects) have recently been reviewed (38).

CONTACT ALLERGY

General

Resorcinol has been classified as a moderate (24) and strong (25) sensitizer. Despite this, contact allergy to this chemical was (31) and is very infrequent. Most cases have been caused in the past by Castellani's paint, anti-wart treatments in France and Belgium, topical anti-acne preparations and a few cases by its presence as a coupler in hair dye.

Patch testing in groups of patients

In three studies in which routine testing with resorcinol was performed in consecutive patients suspected of contact dermatitis, low rates of sensitization of 0.1% to 0.5% have been found (9,17,26). Most allergies were caused by topical medicaments, including anti-wart treatments (9,26). In the majority of studies in which resorcinol was patch tested in groups of selected patients, rates of sensitization were low (<1%) (38). High frequencies of sensitization (4-6.3%) were found in 4 studies in groups of patients with allergic contact cheilitis (19), individuals suspected of hair dye allergy (1), patients with longstanding venous ulcerations or eczema complicating leg ulcers (7) and hairdressers with contact dermatitis (18). However, two studies were very small (7,18) and in 3 of 4 the relevance of the observed positive patch tests was either not mentioned (1,7) or not specified (18).

Case reports and case series

Of eight early cases of contact allergy to resorcinol, 5 were caused by topical medicaments and one by resorcinol in eye drops; in two patients, the allergy was probably of past relevance from previous use of resorcinol-containing topical drugs (4). In France, in the period 1992-1999, contact sensitization to resorcinol was found in 24 patients, who all but one had previously used an anti-wart ointment containing resorcinol. All developed contact dermatitis at the site of application of the ointment, with generalized urticaria (4 cases), pompholyx (1 case), and generalized papulovesicular rash with pompholyx (6 cases) (26,36). Four patients from the Netherlands had allergic contact dermatitis from resorcinol. Two were sensitized by an anti-acne cream and one by a suppository containing resorcinol and causing perianal dermatitis; the fourth patient probably had been primarily sensitized by oral hexylresorcinol and later developed facial dermatitis from a cream containing resorcinol (34).

A man developed a bullous allergic contact dermatitis of the legs from contact allergy to resorcinol in an anti-wart ointment used by his wife on the soles of her feet (connubial contact dermatitis, contact dermatitis by proxy). Because of cold feet, the woman would habitually rub her feet against the legs of her husband in bed. The patient had suffered from allergic contact dermatitis as a child from a topical anti-wart treatment (35). One patient had periorbital and eyelid ACD from resorcinol in a pharmaceutical eye ointment (6). Resorcinol in an anti-acne product caused ACD in one patient (7). Resorcinol ointment caused ACD in another individual superimposed on psoriasis (13).

A woman developed allergic contact dermatitis from resorcinol in an anti-wart treatment (23). Two patients had anal dermatitis from contact allergy to resorcinol in anal suppositories (32). Three young women became sensitized to resorcinol from their anti-acne medication containing 2% resorcinol (33). During 1975 and 1976, three cases of contact allergy to resorcinol in Castellani paint have been observed in one clinic in London, U.K. (30). Several single case reports of contact allergy to resorcinol in Castellani's paint have been published (14,27,28,29).

Cosmetics
Resorcinol was responsible for 3 out of 399 cases of cosmetic allergy where the causal allergen was identified in a study of the NACDG, USA, 1977-1983 (2). Two individuals developed allergic contact dermatitis from hair dyes caused by resorcinol (23). A hairdresser developed occupational hand dermatitis after being sensitized to resorcinol from dyeing her own hair (10). An unknown number of patients was cited to have developed allergic contact dermatitis from resorcinol used as a dye coupler (20,21, data cited in ref. 22).

Other products
Two patients reacted to resorcinol in Castellani's paint used as radiotherapy marker dye (8,15) and 4 other patients had allergic contact dermatitis from resorcinol in Castellani's paint used as marker for patch tests (12,23).

Cross-reactions, pseudo-cross-reactions and co-reactions
(Possible) cross-reactions are summarized in table 3.298.1.

Table 3.298.1 Possible cross-reactions to/from resorcinol

Chemical	Frequency of cross-reactions and references [a]
Catechol	3, cited in ref. 5
Hexylresorcinol	2/7 cases (4); 34; *not* to hexylresorcinol (26)
Hydroquinone	6/17 cases (26); 4/8 cases (4); 1/3 cases (36); 11; 32; 3, cited in ref. 5
Hydroxyhydroquinone	1/4 cases (4)
Orcinol	4/8 cases (4)
Phenol	3/8 cases (4); 8
p-Phenylenediamine	16
Phloroglucinol	1/5 cases (4)
Pyrocatechol	7/14 cases (26); 3/8 cases (4); 4/4 cases (36); 11; 32
Pyrogallol	5/8 cases (4); 9/19 cases (26)
Resorcinol acetate	7/8 cases (4); 34
Resorcinol benzoate	6/6 cases (36); 15/19 cases (26); 7; 23
Salicylaldehyde	1/2 cases (37); 2/17 cases (26)

[a] if only references are given, the number of cross-reactions is either one or unknown

LITERATURE
1   Basketter DA, English J. Cross-reactions among hair dye allergens. Cut Ocular Toxicol 2009;28:104-106
2   Adams RM, Maibach HI, Clendenning WE, Fisher AA, Jordan WJ, Kanof N, et al. A five-year study of cosmetic reactions. J Am Acad Dermatol 1985;13:1062-1069
3   Bloch B. Ekzem Pathogenese. Arch Derm Syph (Berl.) 1924;145:34-82 (Article in German)
4   Keil H. Group reactions in contact dermatitis due to resorcinol. Arch Dermatol 1962;86:212-216
5   Hemmer W, Focke M, Wolf-Abdolvahab S, Bracun R, Wantke F, et al. Group allergy to tri- and ortho-diphenols (catechols) in a patient sensitized by propyl gallate. Contact Dermatitis 1996;35:110-112

6 Massone L, Anonide A, Borghi S, Usiglio D. Contact dermatitis of the eyelids from resorcinol in an ophthalmic ointment. Contact Dermatitis 1993;29:49

7 Nakagawa M, Kawai K, Kawai K. Cross-sensitivity between resorcinol, resorcinol monobenzoate and phenyl salicylate. Contact Dermatitis 1992;27:199

8 Pecegueiro M. Contact dermatitis due to resorcinol in a radiotherapy dye. Contact Dermatitis 1992;26:273

9 Estatica do GPEDC. Boletim Informativo 1991, no. 5 (cited in ref. 9)

10 Vilaplana J, Romaguera C, Grimalt F. Contact dermatitis from resorcinol in a hair dye. Contact Dermatitis 1991;24:151-152

11 Caron GA, Calnan CD. Studies in contact dermatitis. XIV. Resorcin. Trans St John's Hosp Dermatol Soc 1962;48:149-156 (data cited in ref. 12)

12 Langeland T, Braathen LR. Allergic contact dermatitis from resorcinol. Contact Dermatitis 1987;17:126

13 Waddell MM, Finn DA. Sensitivity to resorcin. Contact Dermatitis 1981;7:216

14 Cronin E. Resorcin in Castellani's paint. Contact Dermatitis Newsletter 1973;14:401

15 Marks JG, West GW. Allergic contact dermatitis to radiotherapy dye. Contact Dermatitis 1978;4:1-2

16 Basketter DA, English J. Cross-reactions among hair dye allergens. Cut Ocular Toxicol 2009;28:104-106

17 Søsted H, Rustemeyer T, Gonçalo M, Bruze M, Goossens A, Giménez-Arnau AM, et al. Contact allergy to common ingredients in hair dyes. Contact Dermatitis 2013;69:32-39

18 Katsarou A, Koufou B, Takou K, Kalogeromitros D, Papanayiotou G, Vareltzidis A. Patch test results in hairdressers with contact dermatitis in Greece (1985-1994). Contact Dermatitis 1995;33:347-348

19 O'Gorman SM, Torgerson RR. Contact allergy in cheilitis. Int J Dermatol 2016;55:e386-e391

20 Borelli S. Die Verträglichkeit gebräuchlicher Haarfärbungspraparate, Farbstoffsgrundsubstanzen und verwandter chemischer Verbindungen. Hautarzt 1958;9:19-25 (data cited in ref. 41) (Article in German)

21 Connor DS, Ritz HL, Ampulski RS, Kowollik HG, Lim P, Thomas DW, Parkhurst R. Identification of certain sultones as the sensitizers in alkyl ethoxy sulfate. Fette, Seifen, Anstrichmittel 1975;77:25-29 (data cited in ref. 41, possibly incorrect) (Article in German)

22 Eskelinen A, Molitor C, Kanerva L. Allergic contact dermatitis from 2,7-dihydroxynaphthalene in hair dye. Contact Dermatitis 1997;36:312-313

23 Darcis J, Goossens A. Resorcinol: a strong sensitizer but a rare contact allergen in the clinic. Contact Dermatitis 2016;74:310-312

24 Kern PS, Gerberick F, Ryan CA, Kimber I, Aptula A, Basketter DA. Local lymph node data for the evaluation of skin sensitization alternatives: a second compilation. Dermatitis 2010;21:8-32

25 SCCS (Scientific Committee on Consumer Safety). Opinion on: Resorcinol (A11), 23 March 2010. Available at: https://ec.europa.eu/health/scientific_committees/consumer_safety/docs/sccs_o_015.pdf

26 Barbaud A, Reichert-Penetrat S, Trechot P et al. Sensitization to resorcinol in a prescription verrucide preparation: unusual systemic clinical features and prevalence. Ann Dermatol Venereol 2001;128:615-618 (Article in French)

27 Komericki P, Kränke B, Aberer W. Allergische Kontaktdermatitis auf die Epikutantest-Markierungslösung Solutio Castellani. Dermatologie in Beruf und Umwelt (Dermatosen) 1997;45:176-178 (Article in German)

28 Foti C, Romita P, Ettorre G, Angelini, G, Bonamonte D. Allergic contact dermatitis caused by resorcinol and sodium dehydroacetate in a patient with leg ulcers. Contact Dermatitis 2016;74:383-384

29 Dave VK. Contact dermatitis due to resorcin in Castellani's paint. Contact Dermatitis Newsletter 1973;13:384

30 Cronin E. Contact Dermatitis. Edinburgh: Churchill Livingstone, 1980:265

31 Fisher AA. Resorcinol, a rare sensitizer. Cutis 1982;29:331-332

32 Mitchell JH. Resorcin anal dermatitis due to resorcin in anusol suppositories. JAMA 1933;101:1067

33 Serrano G, Fortea JM, Millan F, Botella R, Latasa JM. Contact allergy to resorcinol in acne medications: report of three cases. J Am Acad Dermatol 1992;26(3 Pt.2):502-504

34 Van Ketel WG. Allergic contact eczema caused by resorcinol. Ned Tijdschr Geneeskd 1970;114:905-907 (Article in Dutch)

35 Kanny G, Blanchard N, Morisset M, Nominé V, Moneret-Vautrin DA. Bullous skin eruption to resorcin by proxy. Rev Med Interne 2004;25:324-327 (article in French)

36 Barbaud A, Modiano P, Cocciale M, Reichert S, Schmutz JL. The topical application of resorcinol can provoke a systemic allergic reaction. Br J Dermatol 1996;135:1014-1015

37 These data are obtained from one or more of the following online sources: ChemIDPlus Advanced, PubChem, DrugBank, RxList, Drug Central, Drugs.com, and Wikipedia

38 De Groot AC. Monographs in contact allergy Volume I. Non-fragrance allergens in cosmetics (Part I and Part 2). Boca Raton, Fl, USA: CRC Press Taylor and Francis Group, 2018: chapter 2.397, pages 1100--1106

Chapter 3.299 RETAPAMULIN

IDENTIFICATION

Description/definition   : Retapamulin is the semisynthetic pleuromutilin antibiotic that conforms to the structural formula shown below

Pharmacological classes   : Anti-bacterial agents

IUPAC name   : [(1S,2R,3S,4S,6R,7R,8R,14R)-4-Ethenyl-3-hydroxy-2,4,7,14-tetramethyl-9-oxo-6-tricyclo[5.4.3.0^{1,8}]tetradecanyl] 2-[[(1S,5R)-8-methyl-8-azabicyclo[3.2.1]octan-3-yl]sulfanyl]acetate

CAS registry number   : 224452-66-8

EC number   : 9551

Merck Index monograph   : Not available

Patch testing   : Retapamulin 1% pet.; when not available: 1% retapamulin ointment 'as is'; exclude contact allergy to BHT, when indicated

Molecular formula   : $C_{30}H_{47}NO_4S$

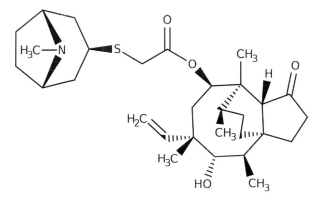

GENERAL

Retapamulin is a semisynthetic pleuromutilin antibiotic. This drug is usually bacteriostatic in action, but may become bactericidal at higher concentrations. It inhibits protein synthesis by (I) binding a component of the bacterial ribosome that affects normal 50S subunit formation, (II) blocking ribosomal P-site interactions, and (III) inhibiting peptidyl transferase. The interaction with bacterial ribosomes is unique from other topical antibiotics and prevents cross-resistance (3). Retapamulin is indicated for the topical treatment of impetigo due to *Staphylococcus aureus* (methicillin-susceptible isolates only) or *Streptococcus pyogenes* and secondarily infected traumatic lesions (1).

CONTACT ALLERGY

Case reports

Allergic contact dermatitis due to retapamulin ointment has been described in 4 patients (2,4). Dermatitis improved after cessation of the use of the antibiotic ointment. Patch tests were positive to the ointment (no controls performed). In one patient, a ROAT with the ointment was positive. Retapamulin itself was not tested, but the ointment base in the USA consists of white petrolatum only, which means that the diagnosis of retapamulin contact allergy can be made *per exclusionem*. Three patients co-reacted to neomycin, two to bacitracin (2,4). These are not cross-reactions, but the result of previous sensitization.

In the EU, the base of retapamulin ointment also contains BHT (2), which means that contact allergy to the antioxidant must be excluded in patients with a positive patch test reaction to the ointment. Thus far, no reports from Europe have emerged. In clinical trials, 'contact dermatitis' was reported in the US at <1% with no further explanation (2).

LITERATURE

1   The data in the section 'General' may have been obtained from literature discussed in this chapter, but mostly also or exclusively from one or more of the following online sources: ChemIDPlus Advanced, PubChem, DrugBank, RxList, Drug Central, Drugs.com, and Wikipedia

2   Schalock PC. Allergic contact dermatitis to retapamulin ointment. Contact Dermatitis 2009;61:126

3   Schram SE, Warshaw EM. Retapamulin: an emerging antimicrobial allergen. Dermatitis 2010;21:160-161

4   Warshaw EM, Toby Mathias CG, Baker DR. Allergic contact dermatitis from retapamulin ointment. Dermatitis 2009;20:220-221

Chapter 3.300 RETINOL

IDENTIFICATION

Description/definition : Retinol is the fat-soluble vitamin A, that conforms to the structural formula shown below
Pharmaceutical classes : Vitamins
IUPAC name : (2*E*,4*E*,6*E*,8*E*)-3,7-Dimethyl-9-(2,6,6-trimethylcyclohexen-1-yl)nona-2,4,6,8-tetraen-1-ol
Other names : Vitamin A
CAS registry number : 68-26-8; 11103-57-4
EC number : 200-683-7; 234-328-2
Merck Index monograph : 11481
Patch testing : 1% and 10% acet.
Molecular formula : $C_{20}H_{30}O$

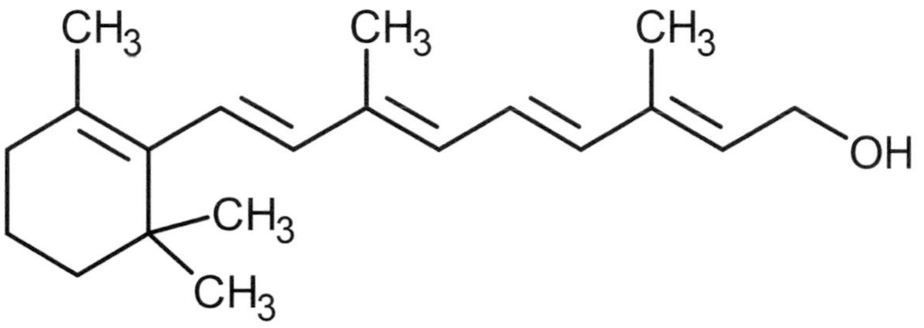

GENERAL

Retinol or vitamin A is a fat-soluble vitamin that plays a vital role in vision, epithelial differentiation, growth, reproduction, pattern formation during embryogenesis, bone development, hematopoiesis, brain development and modulation of immune function. Dietary retinol is derived from a variety of carotenoids found in plants. It is enriched in the liver, egg yolks, and the fat component of dairy products. Pharmaceutical retinol is indicated for the treatment of vitamin A deficiency. In cosmetics, retinol may be present as skin conditioning agent (1). In pharmaceutical products, retinol is mostly employed as retinyl palmitate (Chapter 3.301) or retinyl acetate (CAS number127-47-9, EC number 204-844-2, molecular formula $C_{22}H_{32}O_2$).

CONTACT ALLERGY

Case reports

A woman developed allergic contact dermatitis from a pharmaceutical cream containing vitamin A (retinol) 100,000 IU, vitamin E (tocopherol) 5% and urea 10%. She reacted to the cream, and to retinol 200,000 IU/100 g in corn oil and tocopherol 2.5% pet. on two occasions, while 4 controls were negative (4).

Cosmetics
A young boy had atopic dermatitis on the face and was intermittently treated with a corticosteroid and an emollient cream. After 3 weeks, erythematovesicular, intensely pruritic papules and patches developed on his cheeks. When patch tested with its ingredients, the patient reacted to retinol 1% and 10% in acetone, but not to retinol 0.1% pet. (3). In another report from the same authors on the same product, it is suggested that the allergenic ingredient is not retinol but retinyl palmitate (2).

LITERATURE

1 The data in the section 'General' may have been obtained from literature discussed in this chapter, but mostly also or exclusively from one or more of the following online sources: ChemIDPlus Advanced, PubChem, DrugBank, RxList, Drug Central, Drugs.com, and Wikipedia
2 Manzano D, Aguirre A, Gardeazabal J, Eizaguirre X, Pérez J LD. Allergic contact dermatitis form tocopheryl acetate (vitamin E) and retinol palmitate (vitamin A) in a moisturizing cream. Contact Dermatitis 1994;31:324
3 Sanz De Galdeano C, Aguirre A, Ratón JA, Zabala R, Landa N, Díaz-Perez JL. Contact dermatitis from a moisturizing cream. Contact Dermatitis 1994;30:50-51
4 Bazzano C, de Angeles S, Kleist G, Macedo N. Allergic contact dermatitis from topical vitamins A and E. Contact Dermatitis 1996;35:261-262

Chapter 3.301 RETINYL PALMITATE

IDENTIFICATION

Description/definition   : Retinyl palmitate is the ester of retinol (vitamin A) and palmitic acid that conforms to the structural formula shown below
Pharmaceutical classes   : Anticarcinogenic agents; antioxidants
IUPAC name   : [(2E,4E,6E,8E)-3,7-Dimethyl-9-(2,6,6-trimethylcyclohexen-1-yl)nona-2,4,6,8-tetraenyl] hexadecanoate
Other names   : Retinol palmitate; vitamin A palmitate; retinol, hexadecanoate
CAS registry number   : 79-81-2
EC number   : 201-228-5
Merck Index monograph   : 11481 (vitamin A, retinol)
Patch testing   : 5% pet.; this concentration may cause occasional irritant reactions (5); 1% and 10% MEK (methyl ethyl ketone)
Molecular formula   : $C_{36}H_{60}O_2$

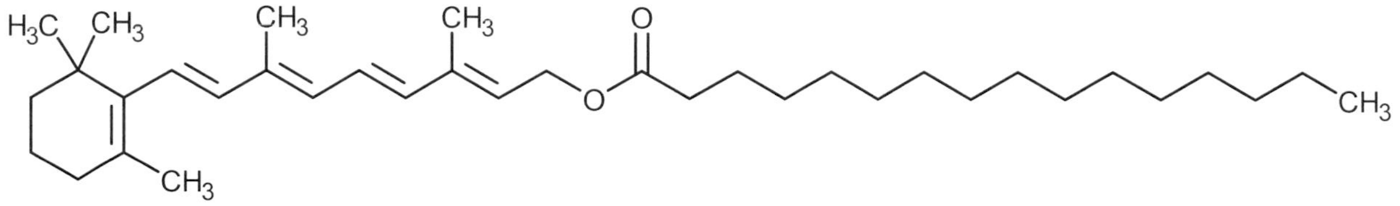

GENERAL

Retinyl palmitate is a naturally occurring phenyl analog of retinol (vitamin A) with potential antineoplastic and chemopreventive activities. As the most common form of vitamin A taken for dietary supplementation, retinyl palmitate binds to and activates retinoid receptors, thereby inducing cell differentiation and decreasing cell proliferation. This agent also inhibits carcinogen-induced neoplastic transformation, induces apoptosis in some cancer cell types, and exhibits immunomodulatory properties. Retinyl palmitate is a common vitamin supplement, available in both oral and injectable forms for treatment of vitamin A deficiency. In cosmetics it may be used as skin conditioning agent (1).

CONTACT ALLERGY

Case reports

A 55-year-old woman had a melanoma on her back removed and replaced by a skin graft. At the donor site on the right side, she developed eczematous dermatitis after the use of several topical applications. Patch tests were positive to a magistral pharmaceutical preparation containing retinyl palmitate. Further patch testing revealed positive reactions to retinyl palmitate oily solution 10^6 U/gr (negative to its antioxidants) and to pure retinyl palmitate (or 10^6 U/gr ?) (3).

A woman was given intramuscular water-soluble retinyl palmitate injections for lichen sclerosus et atrophicus. Two weeks after receiving a second injection (50,000 IU) she noted a red itchy area on her right buttock over the injection site. This grew into an oval erythematous scaly plaque with a size of 7.5 x 9.0 cm and a hemorrhagic component; it slowly disappeared over three weeks. An intradermal skin test to retinyl palmitate (0.1%) was positive at 48 hours, showing a red nodule of 1.2 cm. Similar intradermal tests with all other constituents of the water soluble retinol injection gave negative results. Five control patients tested with the same material showed no reaction. A retinol patch test gave negative results and oral challenge with vitamin A was without incident (6).

Cosmetics

A woman developed dermatitis of the face after a few days use of a new anti-wrinkle cream. A ROAT was positive after 5 applications. Patch tests were positive to the cream and – later - to a retinyl palmitate - polycaprolactone (PCL) complex and to retinyl palmitate 5% pet., but negative to PCL 5% pet. A ROAT with retinyl palmitate 5% pet. was positive (5). One patient had contact dermatitis from allergy to retinyl palmitate in a moisturizing cream (2). In another publication from the same authors on the same product, it was suggested that it contains retinol rather than retinyl palmitate (4).

LITERATURE

1 The data in the section 'General' may have been obtained from literature discussed in this chapter, but mostly also or exclusively from one or more of the following online sources: ChemIDPlus Advanced, PubChem, DrugBank, RxList, Drug Central, Drugs.com, and Wikipedia

2 Manzano D, Aguirre A, Gardeazabal J, Eizaguirre X, Pérez J LD. Allergic contact dermatitis form tocopheryl acetate (vitamin E) and retinol palmitate (vitamin A) in a moisturizing cream. Contact Dermatitis 1994;31:324

3 Blondeel A. Contact allergy to vitamin A. Contact Dermatitis 1984;11:191-192

4 Sanz De Galdeano C, Aguirre A, Ratón JA, Zabala R, Landa N, Díaz-Perez JL. Contact dermatitis from a moisturizing cream. Contact Dermatitis 1994;30:50-51

5 Clemmensen A, Thormann J, Andersen, KE. Allergic contact dermatitis from retinyl palmitate in polycaprolactone. Contact Dermatitis 2007;56:288-289

6 Shelley WB, Shelley ED, Talanin NY. Hypersensitivity to retinol palmitate injection. BMJ 1995;311(6999):232

Chapter 3.302 RIFAMYCIN

IDENTIFICATION

Description/definition   : Rifamycin is the antibiotic produced by certain strains of *Streptomyces mediterranei*
                            or synthetically that conforms to the structural formula shown below
Pharmacological classes   : Anti-infective agents
IUPAC name   : (7S,9E,11S,12R,13S,14R,15R,16R,17S,18S,19E,21Z)-2,15,17,27,29-Pentahydroxy-11-
                            methoxy-3,7,12,14,16,18,22-heptamethyl-6,23-dioxo-8,30-dioxa-24-azatetra-
                            cyclo[23.3.1.1^{4,7}.0^{5,28}]triaconta-1(28),2,4,9,19,21,25(29),26-octaen-13-yl acetate
Other names   : Rifamycin SV
CAS registry number   : 6998-60-3
EC number   : 230-273-3
Merck Index monograph   : 9613
Patch testing   : 2.5% pet.
Molecular formula   : $C_{37}H_{47}NO_{12}$

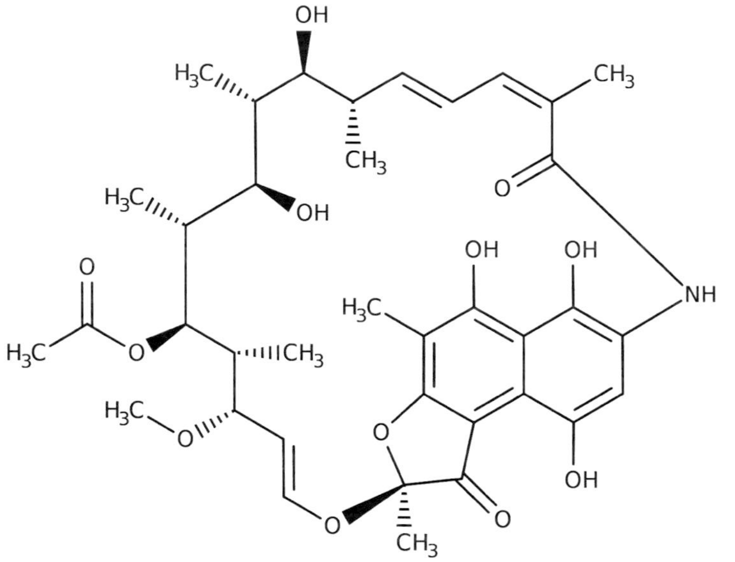

GENERAL

Rifamycin is the prime member of the rifamycin family which are represented by drugs that are a product of fermentation from the gram-positive bacterium *Amycolatopsis mediterranei*, also known as *Streptomyces mediterranei*. Rifamycin has an activity spectrum against gram-positive and gram-negative bacteria, including *Mycobacterium* species (especially *M. tuberculosis*). Rifamycin is indicated for the treatment of adult patients with travelers' diarrhea caused by non-invasive strains of *E. coli*. It is also used in tuberculosis in association with other agents to overcome resistance (1). In pharmaceutical products, rifamycin is employed as rifamycin sodium (CAS number 14897-39-3, EC number 238-965-7, molecular formula $C_{37}H_{46}NNaO_{12}$) (1). It is (or was) also used in topical pharmaceuticals in Italy and France as a 0.5% solution, often used on leg ulcers (4,5). Topical rifamycin is currently available in Brazil for infected wounds (8), (with lidocaine) in Taiwan (www.drugs.com) and possibly in other countries.

CONTACT ALLERGY

Case series

In Leuven, Belgium, between 1990 and 2017, 16,065 patients were investigated for contact allergy and 118 (0.7%) showed positive patch test reactions to topical ophthalmic medications and/or to their ingredients. Eighty-four individuals (71%) reacted to an active principle. Rifamycin was tested in 50 patients and was the allergen in eye medications in one (2, overlap with ref. 3). Also in Leuven, Belgium, in the period 1990-2014, iatrogenic contact dermatitis was diagnosed in 2600 individuals (17% of the total patch test population). 96% of all positive patch test reactions to topical drugs and antiseptics were considered to be relevant. Rifamycin (5% pet.) was tested in 47 patients and there was one positive reaction to it (3, overlap with ref. 2).

Three cases of allergic contact dermatitis were reported from Italy in 1991 (4). A 68-year-old man had been applying a topical solution of rifamycin 0.5% in water on a venous ulcer of his left leg for 2 months of treatment, when itching, erythema and edema developed around the wound. A 58-year-old man had been treated with topical rifamycin for a leg ulcer. After 4 weeks of therapy, eczematous lesions appeared around the treated site. A 34-year-old man had had a complex fracture of the left external malleolus surgically reset. Topical rifamycin was prescribed for the post-surgical wound. After one week, he developed eczema with vesicles and edema at the site of rifamycin application. Patch tests were positive to the rifamycin solution and to rifamycin 2.5% pet. in all three patients (4).

Case reports

A 30-year-old woman had applied topical rifamycin 0.5% in water on a leg ulcer that had started 10 months before. About 3 weeks later, she developed eczema located at the site of application. Patch tests were positive to the solution and to rifamycin 0.5% pet. Ten controls were negative (5). A 27-year-old woman applied topical rifamycin 0.5% solution twice a day on the excision area of a benign lesion on the back; after 2 weeks, eczema appeared at the site. Patch tests were positive to the solution and to rifamycin 0.5% pet. (6).

An 11-year-old boy injured his left elbow with lacerations and contusions, which were treated with topical mercurochrome and rifamycin. Five months later, the boy injured himself for a second time, now on the right leg, and his parents again applied mercurochrome and rifamycin. This resulted in dermatitis with erythema and blisters at the site of application, later extending to the antecubital folds. Patch tests were positive to rifamycin, mercurochrome and 2 other mercurial compounds (test concentrations and vehicles not mentioned) (7).

A 70-year-old man was admitted to hospital for open thoracostomy drainage of pleural effusion in the left hemithorax. After surgery, he was treated with topical rifamycin solution twice daily for 2 months. At an outpatient follow-up visit, the patient complained of itching erythema at the site of application. Patch tests were positive to rifamycin 1%, 10%, and 30% in petrolatum at D3 (+) and D7 (++). Ten controls were negative (8).

A 25-year-old woman known to be allergic to virginiamycin presented with bilateral otitis associated with eczema of the external auditory meatus. She was treated with an auricular solution containing rifamycin; secondarily, oral pristinamycin was added. Eczema on the ears rapidly increased, and facial edema and generalized erythema appeared. She also developed low blood pressure, fever (39°C), arthralgia and myalgia. Patch tests were positive to virginiamycin, pristinamycin (cross-reacting to virginiamycin and causing systemic contact dermatitis) and rifamycin (concentration and vehicle not mentioned) (9).

Immediate contact reactions

Immediate contact reactions (contact urticaria) to rifamycin are presented in Chapter 5.

LITERATURE

1   The data in the section 'General' may have been obtained from literature discussed in this chapter, but mostly also or exclusively from one or more of the following online sources: ChemIDPlus Advanced, PubChem, DrugBank, RxList, Drug Central, Drugs.com, and Wikipedia
2   Gilissen L, De Decker L, Hulshagen T, Goossens A. Allergic contact dermatitis caused by topical ophthalmic medications: Keep an eye on it! Contact Dermatitis 2019;80:291-297
3   Gilissen L, Goossens A. Frequency and trends of contact allergy to and iatrogenic contact dermatitis caused by topical drugs over a 25-year period. Contact Dermatitis 2016;75:290-302
4   Guerra L, Adamo F, Venturo N, Tardio M. Contact dermatitis due to rifamycin. Contact Dermatitis 1991;25:328
5   Balato N, Lembo G, Patruno G, Ayala F. Allergic contact dermatitis from rifamycin. Contact Dermatitis 1988;19:310
6   Milpied B, van Wassenhove L, Larousse C, Barriere H. Contact dermatitis from rifamycin. Contact Dermatitis 1986;14:252-253
7   Riboldi A, Pigatto PD, Morelli M, Altomare GF, Polenghi MM. Allergy to mercurochrome and rifamycin. Contact Dermatitis 1985;12:180
8   Teixeira FM, Vasconcelos LM, Araújo Tda S, Vasconcelos AM, de Almeida TL, Nagao-Dias AT. Rifamycin-associated postoperative allergic contact dermatitis in a 70-year-old patient. J Investig Allergol Clin Immunol 2013;23:282-283
9   Michel M, Dompmartin A, Szczurko C, Castel B, Moreau A, Leroy D. Eczematous-like drug eruption induced by synergistins. Contact Dermatitis 1996;34:86-87

Chapter 3.303 RIPASUDIL

IDENTIFICATION

Description/definition	: Ripasudil is the isoquinoline derivative that conforms to the structural formula shown below
Pharmacological classes	: Rho kinase inhibitor
IUPAC name	: 4-Fluoro-5-[[(2S)-2-Methyl-1,4-diazepan-1-yl]sulfonyl]isoquinoline
CAS registry number	: 223645-67-8
EC number	: Not available
Merck Index monograph	: 11878
Patch testing	: 1% and 10% pet.
Molecular formula	: $C_{15}H_{18}FN_3O_2S$

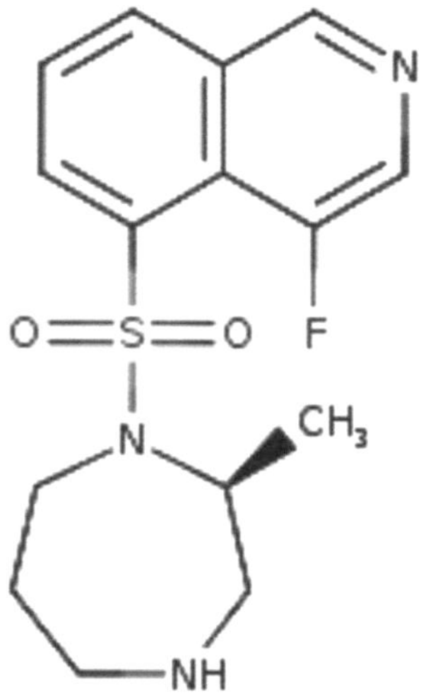

GENERAL

Ripasudil is a rho kinase (rho-associated coiled-coil-containing protein kinase/ROCK) inhibitor, which decreases intraocular pressure (IOP) by increasing outflow facility. An ophthalmic solution containing 0.4% ripasudil hydrochloride hydrate was approved in Japan in 2014 for treatment of glaucoma and ocular hypertension (2). In pharmaceutical products, ripasudil is employed as ripasudil hydrochloride dihydrate (CAS number 887375-67-9, EC number not available, molecular formula $C_{15}H_{23}ClFN_3O_4S$) (1).

CONTACT ALLERGY

Case report

A 74-year-old woman with glaucoma had been treated with eye drops for 13 years. She had applied 4 types of eye drops, of which one was recently replaced with ripasudil solution. Three months later, painful and itchy erythema appeared on both of her periocular regions. A patch test with the ophthalmic solution 'as is' was positive (D2 ++, D3 ++), and patch testing with each of its ingredient was positive (D2 ++ /D3 ++) for ripasudil at 10% and 1% pet., but negative to the preservative benzalkonium chloride. No reactions to ripasudil were seen in 3 healthy controls. After replacing ripasudil solution with another antiglaucoma medication and applying topical hydrocortisone butyrate ointment, the woman's periocular erythema resolved (2).

LITERATURE

1   The data in the section 'General' may have been obtained from literature discussed in this chapter, but mostly also or exclusively from one or more of the following online sources: ChemIDPlus Advanced, PubChem, DrugBank, RxList, Drug Central, Drugs.com, and Wikipedia
2   Kusakabe M, Imai Y, Natsuaki M, Yamanishi K. Allergic contact dermatitis due to ripasudil hydrochloride hydrate in eye-drops: A case report. Acta Derm Venereol 2018;98:278-279

Chapter 3.304 ROTIGOTINE

IDENTIFICATION

Description/definition   : Rotigotine is the tetralin that conforms to the structural formula shown below
Pharmacological classes   : Dopamine agonists
IUPAC name   : (6S)-6-[Propyl(2-thiophen-2-ylethyl)amino]-5,6,7,8-tetrahydronaphthalen-1-ol
CAS registry number   : 99755-59-6
EC number   : Not available
Merck Index monograph   : 9674
Patch testing   : 10% pet. and water
Molecular formula   : $C_{19}H_{25}NOS$

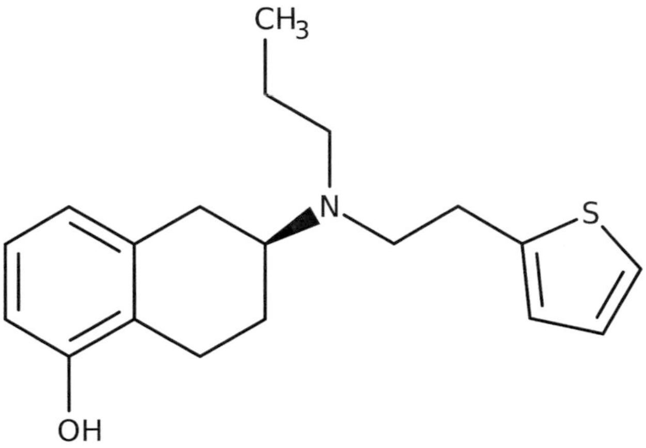

GENERAL

Rotigotine is a non-ergot dopamine receptor agonist at all five dopamine receptor subtypes (D1-D5) but binds to the D_3 receptor with the highest affinity. In addition, it is an antagonist at α_2-adrenergic receptors and an agonist at the 5HT1A receptors. Rotigotine, formulated as a once-daily transdermal patch, is indicated for treatment of Parkinson's disease and moderate-to-severe primary restless legs syndrome. Like other dopamine agonists, rotigotine has been shown to possess antidepressant effects and may be useful in the treatment of depression as well (1).

CONTACT ALLERGY

Case report

A 69-year-old patient presented with well-demarcated itchy red plaques with vesicles and secondary desquamation at the application sites of a rotigotine TTS for 3 months, with extension on the neckline and the upper back. These reactions began 2 weeks after the patient had started using this TTS for early signs of Parkinson's disease. Patch tests were negative to sodium metabisulfite (an excipient in the TTS) and the placebo patch, but positive to pure rotigotine at 10% pet. and water. Five controls were negative. The patient applied the placebo TTS once daily at different application sites for 2 weeks, without any reaction (2).

LITERATURE

1   The data in the section 'General' may have been obtained from literature discussed in this chapter, but mostly also or exclusively from one or more of the following online sources: ChemIDPlus Advanced, PubChem, DrugBank, RxList, Drug Central, Drugs.com, and Wikipedia
2   Raison-Peyron N, Guillot B. Allergic contact dermatitis caused by rotigotine in a transdermal therapeutic system. Contact Dermatitis 2016;75:121-122

Chapter 3.305 RUBIDIUM IODIDE

IDENTIFICATION

Description/definition	: Rubidium iodide is the salt with the chemical formula RbI, that can be formed by the reaction of rubidium with iodine (2 Rb + 2 I → 2 RbI)
Pharmacological classes	: Iodine source
IUPAC name	: Rubidium(1+);iodide
Other names	: Rubidium jodatum
CAS registry number	: 7790-29-6
EC number	: 232-198-1
Merck Index monograph	: 9689
Patch testing	: 1% pet.
Molecular formula	: IRb

$$Rb^+ \qquad\qquad I^-$$

GENERAL

Rubidium iodide is (or was) used in ophthalmic preparations to retard or prevent the formation of cataracts, by interacting with biochemical processes, such as oxidation processes and cellular permeability, involved in crystalline lens opacification (2). Rubidium iodide probably is not used in pharmaceuticals anymore.

CONTACT ALLERGY

Case report

A 70-year-old man presented with severe dermatitis of the face, eyelids and periorbital areas. His symptoms had appeared 5 months after starting on an ophthalmic preparation for the treatment of cataracts. Patch tests with the patient's eye drops and their individual constituents revealed a strongly positive reaction to rubidium iodide 1% pet., but no reaction to the eye drops themselves. Patch testing with rubidium iodide 10% pet. in 20 healthy volunteers was negative (1).

LITERATURE

1   Cameli N, Bardazzi F, Morelli R, Tosti A. Contact dermatitis from rubidium iodide in eyedrops. Contact Dermatitis 1990;23:377-378

Chapter 3.306 SALICYLAMIDE

IDENTIFICATION

Description/definition	: Salicylamide is the 1-hydroxy-4-unsubstituted benzenoid that conforms to the structural formula shown below
Pharmacological classes	: Anti-inflammatory agents, non-steroidal
IUPAC name	: 2-Hydroxybenzamide
Other names	: Benzamide, 2-hydroxy-; 2-carbamoylphenol
CAS registry number	: 65-45-2
EC number	: 200-609-3
Merck Index monograph	: 9735
Patch testing	: 2% pet. (SmartPracticeCanada, SmartPracticeEurope)
Molecular formula	: $C_7H_7NO_2$

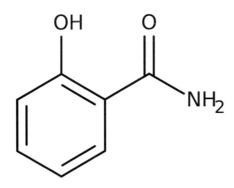

GENERAL

Salicylamide is a non-prescription drug with analgesic and antipyretic properties. Its medicinal uses are similar to those of acetylsalicylic acid (Aspirin). It is often used in combination with acetylsalicylic acid and caffeine in the over-the-counter pain remedies (1). In France, it is used in a topical pharmaceutical product containing hydroxyethyl (glycol) salicylate, dexamethasone acetate, and salicylamide marketed for treatment of benign joint conditions such as mild tendinitis, small joint arthritis and sprains (2).

CONTACT ALLERGY

Case series

In a period of 11 years (2000-2010), the French Pharmacovigilance received 53 reports of adverse skin reactions to a pharmaceutical product containing hydroxyethyl (glycol) salicylate, dexamethasone acetate, and salicylamide marketed for treatment of benign joint conditions such as mild tendinitis, small joint arthritis and sprains. The main cutaneous side effect (n=41) was contact dermatitis with secondary extension in 15 cases. Onset was immediate in 12 cases, delayed in 32 cases and unspecified in eight cases. Twelve patients were hospitalized. Allergological tests were performed in 14 cases and were positive for the drug itself (eight cases, ingredients probably not tested), salicylamide (n=6), glycol salicylate (n=7), dexamethasone (n=3), and propylene glycol (n=2) (2).

LITERATURE

1   The data in the section 'General' may have been obtained from literature discussed in this chapter, but mostly also or exclusively from one or more of the following online sources: ChemIDPlus Advanced, PubChem, DrugBank, RxList, Drug Central, Drugs.com, and Wikipedia
2   Remy C, Barbaud A, Lebrun-Vignes B, Perrot JL, Beyens MN, Mounier G, et al. Skin toxicity related to Percutalgine (®): analysis of the French pharmacovigilance database. Ann Dermatol Venereol 2012;139:350-354 (Article in French)

Chapter 3.307 SALICYLIC ACID

IDENTIFICATION

Description/definition : Salicylic acid is a β-hydroxy acid that conforms to the structural formula shown below
Pharmacological classes : Anti-infective agents; antifungal agents; keratolytic agents
IUPAC name : 2-Hydroxybenzoic acid
Other names : Benzoic acid, 2-hydroxy-; 2-hydroxybenzenecarboxylic acid
CAS registry number : 69-72-7
EC number : 200-712-3
Merck Index monograph : 9739
Patch testing : 5% pet. (SmartPracticeCanada)
Molecular formula : $C_7H_6O_3$

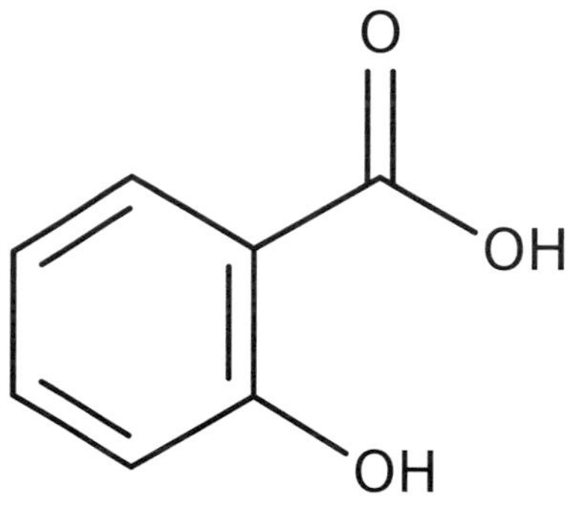

GENERAL

Salicylic acid is a β-hydroxy acid obtained from the bark of the white willow and wintergreen leaves, and also prepared synthetically. It has bacteriostatic, fungicidal, and keratolytic actions. Because of its keratolytic and exfoliating effects, salicylic acid is a key additive in topical treatments for acne, psoriasis, calluses, corns, keratosis pilaris and warts (1).

CONTACT ALLERGY

Patch testing in groups of selected patients

In Rotterdam, The Netherlands, 47 patients hospitalized for psoriasis were patch tested with an extended baseline series and a topical treatment series. There was one reaction to salicylic acid 1% pet. Relevance was not addressed (2). In another Dutch study, from the period 1995-1999, 79 patients with chronic otitis externa were patch tested with salicylic acid 5% pet. and there were 3 (3.8%) positive reactions. Relevance was not specified, but in this study, 18% of all reactions to ingredients of topical medications were considered to be relevant (3).

In view of the widespread use of salicylic acid-containing preparations in Singapore, salicylic acid 5% w/w pet. was included in the standard patch test battery from 1979 to 1983 to determine its prevalence of allergy. Of 9701 patch tests performed during this period, 11 positives to salicylic acid 5% pet. were recorded, all of doubtful relevance. The 11 patients were recalled in 1984 for repeat patch tests to serial dilutions of salicylic acid 5%, 2%, 1% and 0.5% w/w pet. Eight patients responded to the recall; of these, only 1 showed a positive reaction to the 5%, 2% and 1% dilutions. This patient had a history of immediate type hypersensitivity to oral salicylates. The authors concluded that salicylic acid is a very weak skin sensitizer (4).

Case series

In Warsaw, Poland, 5% salicylic acid in petrolatum was routinely tested in the early 1970s. Five patients had a positive patch test reaction to it. None had allergic contact dermatitis upon presentation, but all previously had used 2% salicylic acid in alcohol (n=4) or 5% salicylic ointment (n=1). The latter patient recalled that in the last period of treatment for psoriasis redness and itching occurred after every application of 5% unguentum salicylicum. Of the 4 patients having used salicylic acid in alcohol, 3 noticed that the use of this preparation caused 'reactions'. All patients would occasionally take Aspirin (acetyl salicylic acid), but none of them found it to cause exacerbations or relapses of dermatitis (5).

Immediate contact reactions
Immediate contact reactions (contact urticaria) to salicylic acid are presented in Chapter 5.

LITERATURE

1   The data in the section 'General' may have been obtained from literature discussed in this chapter, but mostly also or exclusively from one or more of the following online sources: ChemIDPlus Advanced, PubChem, DrugBank, RxList, Drug Central, Drugs.com, and Wikipedia
2   Heule F, Tahapary GJ, Bello CR, van Joost T. Delayed-type hypersensitivity to contact allergens in psoriasis. A clinical evaluation. Contact Dermatitis 1998;38:78-82
3   Devos SA, Mulder JJ, van der Valk PG. The relevance of positive patch test reactions in chronic otitis externa. Contact Dermatitis 2000;42:354-355
4   Goh CL, Ng SK. Contact allergy to salicylic acid. Contact Dermatitis 1986;14:114
5   Rudzki E, Koslowska A. Sensitivity to salicylic acid. Contact Dermatitis 1976;2:178

Chapter 3.308 SCOPOLAMINE

IDENTIFICATION

Description/definition   : Scopolamine is the tropane alkaloid that conforms to the structural formula shown below
Pharmacological classes   : Adjuvants, anesthesia; mydriatics; antiemetics; cholinergic antagonists; muscarinic
     antagonists
IUPAC name   : [(1R,2R,4S,5S)-9-Methyl-3-oxa-9-azatricyclo[3.3.1.0^{2,4}]nonan-7-yl] (2S)-3-hydroxy-2-
     phenylpropanoate
Other names   : Hyoscine; 6,7-epoxytropine tropate
CAS registry number   : 51-34-3
EC number   : 200-090-3
Merck Index monograph : 9813
Patch testing   : 1% pet.
Molecular formula   : $C_{17}H_{21}NO_4$

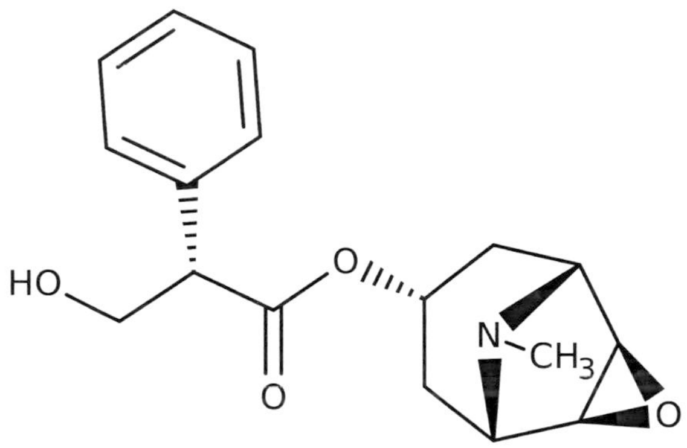

GENERAL

Scopolamine (hyoscine) is a tropane alkaloid derived from various plants of the nightshade family (Solanaceae), especially *Datura metel* L. and *Scopola carniolica*, with anticholinergic, antiemetic and antivertigo properties. Structurally similar to acetylcholine, scopolamine antagonizes acetylcholine activity mediated by muscarinic receptors located on structures innervated by postganglionic cholinergic nerves as well as on smooth muscles that respond to acetylcholine but lack cholinergic innervation. This agent induces mydriasis and cycloplegia, controls the secretion of saliva and gastric acid, slows gut motility (antispasmodic), and prevents vomiting. Scopolamine is indicated for the treatment of excessive salivation (sialorrhea), colicky abdominal pain, bradycardia, diverticulitis, irritable bowel syndrome and motion sickness. In ophthalmology it is utilized to induce mydriasis and cycloplegia before diagnostic procedures. Scopolamine given in transdermal therapeutic systems is widely used to prevent motion sickness (1).

In pharmaceutical products, both scopolamine (in transdermal delivery systems) and scopolamine hydrobromide (CAS number 114-49-8, EC number 204-050-6, molecular formula $C_{17}H_{22}BrNO_4$) may be employed (1).

CONTACT ALLERGY

Case series

A total of 164 male naval crew members were treated for seasickness with transdermal scopolamine for several months (range 1.5 to 15 months). Allergic contact dermatitis caused by the drug was diagnosed in 16 men (10%). They all had pruritus and erythema at the site of the patch. Placing the patch behind the other ear produced an identical local reaction. Removal of the patch was followed by regression of the lesion. Total resolution took up to 14 days, depending on the severity of the lesion. In all cases the allergic reaction reappeared when a new patch was applied. All lesions were confined to the site of application. Clinical examination of the lesions showed circular areas of erythema, edema, and vesiculobullous or eczematous response in various stages of resolution. Placebo patches containing all the components of the patches apart from scopolamine were applied in all of the men; no local skin reactions were observed (6), which makes contact allergy to scopolamine – although not tested separately – in these individuals highly likely.

In Pamplona, Spain, in one year's time (1992), 13 patients were diagnosed with contact allergy to ophthalmic medications, nearly 3% of all patients investigated for suspected contact dermatitis. There was one reaction to scopolamine 2.5% water in an individual who had previously used scopolamine eye drops (3).

In Kuopio, Finland, in the period 1982 to 1984, 25 patients suffering from a prolonged or chronic conjunctivitis, keratoconjunctivitis or blepharoconjunctivitis while receiving an adequate topical medication, were tested for contact allergy with a battery of ophthalmic drugs. One patient (4%) reacted to scopolamine hydrobromide 0.25%, but relevance was not discussed (2).

Case reports

A 53-year-old woman, who had suffered persistent vertigo and dizziness for many years, was given scopolamine transdermal device, which provided prompt relief of her symptoms, and she began using it on a daily, uninterrupted basis. One year later, erythema and pruritus were noted beneath the transdermal device at the site of application. Examination revealed a transdermal device in the proper location on the glabrous skin behind the left ear. This device had been in place about 18 hours and was surrounded by a distinct zone of erythema. Areas of eczema with a circular configuration and in various stages of resolution were noted behind both ears. Patch tests revealed positive reactions to the transdermal device and scopolamine 1.8% pet. She did not react to a placebo transdermal device or its ingredients. Fifteen controls were negative (5).

A 10-year-old mentally retarded and spastic boy received transdermal scopolamine for severe chronic ptyalism. After 2 weeks, a reaction was observed at the retroauricular application site. Patch tests showed strongly positive patch tests to the scopolamine device and to scopolamine 1%, 0.5% and 0.25% water (7). A woman had used a scopolamine TTS for 3 weeks to prevent motion sickness when she developed a pruritic eruption under the TTS. The patient proved to be patch test positive to scopolamine 2% pet. (9).

A 63-year-old man was treated with 0.5% idoxuridine ointment and 0.25% scopolamine hydrobromide solution for herpes simplex keratitis of the right eye, which produced definite improvement. Later, he developed new eye symptoms and was noted to have signs of herpes simplex infection of the left eye, the nares, and the upper lip. The topical combination therapy was then used in both eyes. No improvement occurred and the patient was suspected of idoxuridine allergy, because cultures for herpes simplex virus had been consistently negative. A patch test to the 0.5% idoxuridine ointment, however, was negative at 48 hours, and the patient was instructed to continue using the topically applied medications. Sometime later he presented with acute allergic contact dermatitis of the face. All medications were stopped, and patch tests to 5% idoxuridine in dimethylacetamide and preservative-free scopolamine hydrobromide both produced a strongly positive reaction at 48 hours. Thus, the patient's dermatitis may have been caused by scopolamine hydrobromide, idoxuridine, or both medications (8).

Cross-reactions, pseudo-cross-reactions and co-reactions

Possible cross-reactivity to scopolamine, homatropine and belladonna in a patient with allergic contact dermatitis from atropine (4).

LITERATURE

1   The data in the section 'General' may have been obtained from literature discussed in this chapter, but mostly also or exclusively from one or more of the following online sources: ChemIDPlus Advanced, PubChem, DrugBank, RxList, Drug Central, Drugs.com, and Wikipedia
2   Hätinen A, Teräsvirta M, Fräki JE. Contact allergy to components in topical ophthalmologic preparations. Acta Ophthalmol (Copenh) 1985;63:424-426
3   Tabar AI, García BE, Rodríguez A, Quirce S, Olaguibel JM. Etiologic agents in allergic contact dermatitis caused by eyedrops. Contact Dermatitis 1993;29:50-51
4   Decraene T, Goossens A. Contact allergy to atropine and other mydriatic agents in eye drops. Contact Dermatitis 2001;45:309-310
5   Trozak OJ. Delayed hypersensitivity to scopolamine delivered by a transdermal device. J Am Acad Dermatol 1985;13:247-251
6   Gordon CR, Shupak A, Doweck I, Spitzer O. Allergic contact dermatitis caused by transdermal hyoscine. BMJ 1989;298(6682):1220-1221
7   Van der Willigen AH, Oranje AP, Stolz E, Van Joost T. Delayed hypersensitivity to scopolamine in transdermal therapeutic systems. J Am Acad Dermatol 1988;18:146-147
8   Amon RB, Lis AW, Hanifin JM. Allergic contact dermatitis caused by idoxuridine. Patterns of cross reactivity with other pyrimidine analogues. Arch Dermatol 1975;111:1581-1584
9   Fisher AA. Dermatitis due to transdermal therapeutic systems. Cutis 1984;34:526-531

Chapter 3.309 SELENIUM SULFIDE

IDENTIFICATION

Description/definition   : Selenium sulfide is the inorganic salt that conforms to the structural formula shown below
Pharmacological classes   : Dermatological agents
Other names   : Selenium disulfide
CAS registry number   : 7488-56-4
EC number   : 231-303-8
Merck Index monograph   : 9843
Patch testing   : 2% pet.; perform controls
Molecular formula   : S_2Se

$$S=Se=S$$

GENERAL

Selenium sulfide is an antifungal agent highly active in inhibiting the growth of *Malassezia* species (formerly known as *Pityrosporum*), micro-organisms which play an important role in the development of dandruff. It is also a proven cytostatic agent, slowing the growth of both hyperproliferative and normal cells. Topical selenium sulfide has been used for decades in the treatment of dandruff, seborrheic dermatitis and tinea versicolor. Selenium sulfide is available in various topical formulations including shampoo, lotion, cream, foam, and suspension.

In this chapter, only contact allergy to selenium sulfide is discussed. A full review of its other side effects can be found in ref. 3.

CONTACT ALLERGY

Case reports and case series

In a group of 119 patients with allergic contact dermatitis from cosmetics, investigated in The Netherlands in 1986-1987, one case was caused by selenium sulfide in a shampoo (1,2).

Three patients had contact dermatitis of the earlobes, external auditory canal and one of them also the neck, from the use of selenium sulfide shampoo. The symptoms were reproduced by application of undiluted or 'properly' diluted selenium sulfide shampoo. It was not specified whether patch tests have been performed with the shampoo, but certainly not with the active ingredient selenium sulfide; thus, contact allergy to selenium sulfide was not proven. With proper attention to plugging the ears and thorough rinsing of the scalp and ears after use of the preparation, the patients were able to continue using selenium sulfide once every three weeks (4).

LITERATURE

1   De Groot AC, Bruynzeel DP, Bos JD, van der Meeren HL, van Joost T, Jagtman BA, Weyland JW. The allergens in cosmetics. Arch Dermatol 1988;124:1525-1529
2   De Groot AC. Adverse reactions to cosmetics. PhD Thesis, University of Groningen, The Netherlands: 1988, chapter 3.4, pp.105-113
3   De Groot AC. Monographs in Contact Allergy Volume I. Non-Fragrance Allergens in Cosmetics (Part I and Part 2). Boca Raton, Fl, USA: CRC Press Taylor and Francis Group, 2018:119-1120
4   Eisenberg BC. Contact dermatitis from selenium sulfide shampoo. Arch Derm Syph 1955;72:71-72

Chapter 3.310 SERTACONAZOLE

IDENTIFICATION

Description/definition : Sertaconazole is the synthetic imidazole derivative that conforms to the structural formula shown below

Pharmacological classes : Antifungal agents

IUPAC name : 1-[2-[(7-Chloro-1-benzothiophen-3-yl)methoxy]-2-(2,4-dichlorophenyl)ethyl]imidazole

Other names : 1-(2,4-Dichloro-β-((7-chlorobenzo(b)thien-3-yl)methoxy)phenethyl)imidazole

CAS registry number : 99592-32-2

EC number : Not available

Merck Index monograph : 9874

Patch testing : 1% and 5% pet.; imidazoles are usually tested 1% alcohol or MEK (methyl ethyl ketone

Molecular formula : $C_{20}H_{15}Cl_3N_2OS$

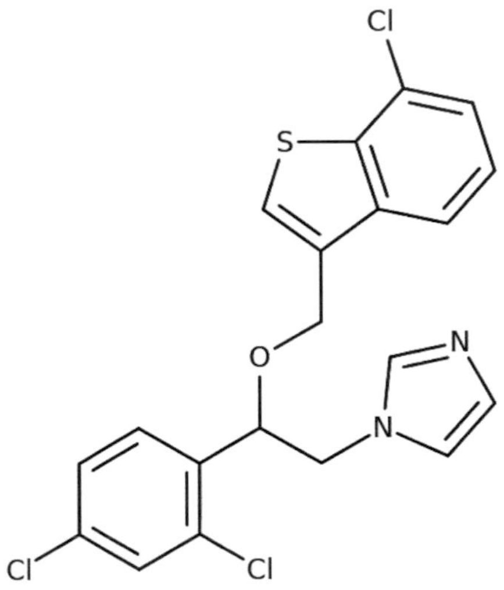

GENERAL

Sertaconazole is a synthetic imidazole derivative with antifungal, antibacterial, anti-inflammatory and anti-pruritic activity. Besides its ability to inhibit the synthesis of ergosterol, the benzothiophene ring of sertaconazole is able to insert into the fungal cell wall instead of tryptophan. This increases the permeability of the cell wall. In addition, sertaconazole suppresses the release of cytokines. This drug is indicated for the topical treatment of interdigital tinea pedis caused by *Trichophyton rubrum*, *Trichophyton mentagrophytes*, and *Epidermophyton floccosum*. In pharmaceutical products, sertaconazole is employed as sertaconazole nitrate (CAS number 99592-39-9, EC number not available, molecular formula $C_{20}H_{16}Cl_3N_3O_4S$) (1).

CONTACT ALLERGY

Case report

A 26-year-old man presented with a 1-month history of impetiginized eczema on the dorsum of the right hand, following the use of sertaconazole cream for an injury. He denied previous use of antifungals. The patient was patch tested with the cream and its constituents and reacted to the cream 'as is' (++), sertaconazole 1% (++) and 5% (+++) pet., but not to any other ingredient. Twenty controls were negative. Later testing with an imidazole series yielded positive reactions to miconazole and econazole 2% pet. (2).

Cross-reactions, pseudo-cross-reactions and co-reactions

A patient sensitized to sertaconazole, who had never used antifungal preparations before, cross-reacted to miconazole and econazole (2). The question has been raised (but not answered) whether contact allergy to sertaconazole and other (nitro)imidazoles may be overrepresented in patients allergic to methylchloroisothiazolinone/methylisothiazolinone (MCI/MI) (4).

LITERATURE

1   The data in the section 'General' may have been obtained from literature discussed in this chapter, but mostly also or exclusively from one or more of the following online sources: ChemIDPlus Advanced, PubChem, DrugBank, RxList, Drug Central, Drugs.com, and Wikipedia

2   Goday JJ, Yanguas I, Aguirre A, Ilardia R, Soloeta R. Allergic contact dermatitis from sertaconazole with cross-sensitivity to miconazole and econazole. Contact Dermatitis 1995;32:370-371

3   Dooms-Goossens A, Matura M, Drieghe J, Degreef H. Contact allergy to imidazoles used as antimycotic agents. Contact Dermatitis 1995;33:73-77

4   Stingeni L, Rigano L, Lionetti N, Bianchi L, Tramontana M, Foti C, et al. Sensitivity to imidazoles/nitroimidazoles in subjects sensitized to methylchloroisothiazolinone/methylisothiazolinone: A simple coincidence? Contact Dermatitis 2019;80:181-183

Chapter 3.311 SILVER NITRATE

IDENTIFICATION

Description/definition	: Silver nitrate is the nitrate salt of silver that conforms to the structural formula shown below
Pharmacological classes	: Anti-infective agents, local
IUPAC name	: Silver;nitrate
CAS registry number	: 7761-88-8
EC number	: 231-853-9
Merck Index monograph	: 9927
Patch testing	: 1% water (Chemotechnique, SmartPracticeCanada); 1% pet. (SmartPracticeEurope); in older solutions, irritant effects may increase; with an 'edge effect', it may be difficult to differentiate an allergic from an irritant patch test (5)
Molecular formula	: AgNO₃

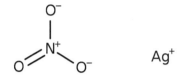

GENERAL

The inorganic compound silver nitrate has been used topically as an antiseptic agent, e.g. to prevent ophthalmia neonatorum (conjunctivitis of the newborn caused by *Neisseria gonorrhoeae* or *Chlamydia trachomatis*). It can also be used as a caustic stick to stop bleeding of the skin or mucous membranes, and (formerly) as a patch test marker (4). Silver is also present in sulfadiazine silver, very commonly used to treat burns (Chapter 3.322 Sulfadiazine silver) and is used in silver sodium carboxymethylcellulose dressing, silver charcoal and nanocrystalline silver dressing for its antimicrobial properties (3). Non-medical applications for silver are in making photographic films (1).

CONTACT ALLERGY

Patch testing in groups of patients

In Serbia, in the period 2000-2002, 75 patients with chronic venous leg ulcers were patch tested with 1% silver nitrate and there were 9 (12%) positive reactions. Their relevance was not mentioned, but in a (small) control group, the prevalence of sensitization to silver nitrate was only 3.6% (difference statistically significant) (2).

Case report

A 26-year-old man was patch tested for the evaluation of dermatitis involving his feet. The test sites were circled with 10% silver nitrate. Within one day, the patient developed erythema, edema, and vesiculation at the site of silver nitrate application. Confirmatory patch testing with silver compounds resulted in positive reactions to 2-week-old 10% silver nitrate, silver foil, and argyrols (silver protein antimicrobial). However, there was no reaction to freshly prepared 10% silver nitrate. It was thought that this patient reacted to ionizable silver, considering that he only reacted to the aged solution, which would have had a chance to decompose into small amounts of colloidal silver, silver nitrate, and nitric acid following exposure to light and air. The patient may previously have become sensitized to silver by the application of silver nitrate solution to foot dermatitis, which had exacerbated the dermatitis (4).

LITERATURE

1   The data in the section 'General' may have been obtained from literature discussed in this chapter, but mostly also or exclusively from one or more of the following online sources: ChemIDPlus Advanced, PubChem, DrugBank, RxList, Drug Central, Drugs.com, and Wikipedia
2   Jankićević J, Vesić S, Vukićević J, Gajić M, Adamic M, Pavlović MD. Contact sensitivity in patients with venous leg ulcers in Serbia: comparison with contact dermatitis patients and relationship to ulcer duration. Contact Dermatitis 2008;58:32-36
3   Group A, Lea A. Contact dermatitis with a highlight on silver: a review. Wounds 2010;22:311-315
4   Gaul LE, Underwood GB. The effect of aging a solution of silver nitrate on its cutaneous reaction. J Invest Derm 1948;11:7
5   Iliev D, Elsner P. Unusual edge effect in patch testing with silver nitrate. Am J Contact Dermat 1998;9:57-59

Chapter 3.312 SIROLIMUS

IDENTIFICATION

Description/definition : Sirolimus is a macrolide compound obtained from *Streptomyces hygroscopicus* that conforms to the structural formula shown below

Pharmacological classes : Immunosuppressive agents; anti-bacterial agents; antifungal agents; antibiotics, antineoplastic

IUPAC name : (1R,9S,12S,15R,16E,18R,19R,21R,23S,24E,26E,28E,30S,32S,35R)-1,18-Dihydroxy-12-[(2R)-1-[(1S,3R,4R)-4-hydroxy-3-methoxycyclohexyl]propan-2-yl]-19,30-dimethoxy-15,17,21,23,29,35-hexamethyl-11,36-dioxa-4-azatricyclo[30.3.1.0^{4,9}]hexatriaconta-16,24,26,28-tetraene-2,3,10,14,20-pentone

Other names : Rapamycin

CAS registry number : 53123-88-9

EC number : 610-965-5

Merck Index monograph : 9502

Patch testing : Sirolimus 0.1% oral solution; also patch test with excipients, if sirolimus itself is unavailable; test controls

Molecular formula : $C_{51}H_{79}NO_{13}$

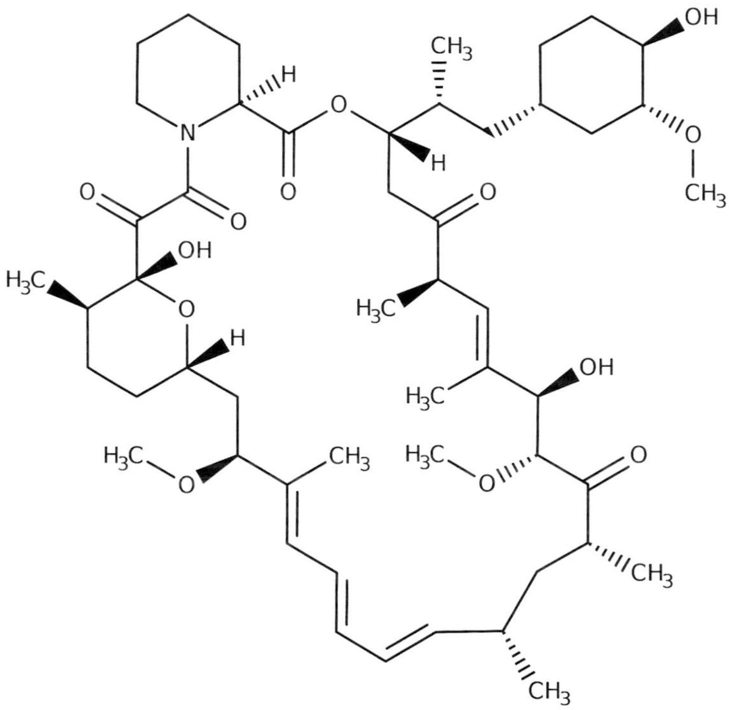

GENERAL

Sirolimus is a natural macrocyclic lactone produced by the bacterium *Streptomyces hygroscopicus* that has immunosuppressant properties. In cells, sirolimus binds to the immunophilin FK Binding Protein-12 (FKBP-12) to generate an immunosuppressive complex that binds to and inhibits the activation of the mammalian Target Of Rapamycin (mTOR), a key regulatory kinase. This results in inhibition of T-lymphocyte activation and proliferation that occurs in response to antigenic and cytokine (IL-2, IL-4, and IL-15) stimulation and inhibition of antibody production. Sirolimus is indicated for the prophylaxis of organ rejection in patients receiving renal transplants (1). Because of its antiangiogenic properties, topical use of sirolimus as an adjuvant for the laser treatment of port wine stains has been suggested (2).

CONTACT ALLERGY

Case report

A 32-year-old healthy woman had a port wine stain (PWS) on her left flank treated with laser therapy, directly followed by topical application of sirolimus under occlusion and left *in situ* for 7 days. This treatment was repeated five times, with 2-week intervals. After the second and third treatments, the patient developed mild dermatitis (erythema) at the sirolimus application sites. The treatment was repeated two more times, but, because of increased itching and dermatitis, the occlusive bandages had to be removed after 2 days on both occasions. After the last treatment, the patient developed an itching rash on her entire upper body, in addition to the erythematous skin reaction at the application site. Patch tests were positive to the commercial topical sirolimus 1 mg/ml) 'as is' (+++), 1:3 in olive oil (++), and 1:3 in water (+). Some excipients were tested negative, others were not patch tested but considered to be 'unlikely allergens'. Controls were not performed (2).

Cross-reactions, pseudo-cross-reactions and co-reactions
Not to other macrolide antibiotics: azithromycin, clarithromycin, and erythromycin (2).

LITERATURE
1   The data in the section 'General' may have been obtained from literature discussed in this chapter, but mostly also or exclusively from one or more of the following online sources: ChemIDPlus Advanced, PubChem, DrugBank, RxList, Drug Central, Drugs.com, and Wikipedia
2   Greveling K, Kunkeler AC, Prens EP, van Doorn MB. Allergic contact dermatitis caused by topical sirolimus used as an adjuvant for laser treatment of port wine stains. Contact Dermatitis 2016;75:184-185

Chapter 3.313 SISOMICIN

IDENTIFICATION

Description/definition	: Sisomicin is an aminoglycoside antibiotic produced from the fermentation broth of the bacillus *Micromonospora inyoensis*; it conforms to the structural formula shown below
Pharmacological classes	: Anti-bacterial agents; protein synthesis inhibitors
IUPAC name	: (2R,3R,4R,5R)-2-[(1S,2S,3R,4S,6R)-4,6-Diamino-3-[[(2S,3R)-3-amino-6-(aminomethyl)-3,4-dihydro-2H-pyran-2-yl]oxy]-2-hydroxycyclohexyl]oxy-5-methyl-4-(methylamino)oxane-3,5-diol
Other names	: Sisomycin; Extramycin (sisomicin sulfate)
CAS registry number	: 32385-11-8
EC number	: 251-018-2
Patch testing	: 20% pet.
Molecular formula	: $C_{19}H_{37}N_5O_7$

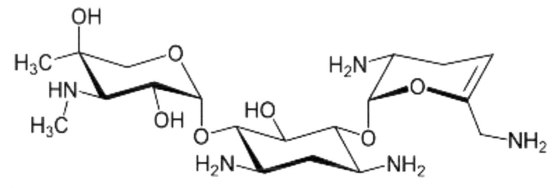

GENERAL

Sisomicin is a broad-spectrum aminoglycoside antibiotic produced by the bacillus *Micromonospora Inyoensis*. It is closely related to gentamicin C1A, one of the components of the gentamicin complex, but has a unique unsaturated diamino sugar. Of the aminoglycoside antibiotics, sisomicin has the greatest activity against gram-positive bacteria, and is also effective against most strains of *Pseudomonas aeruginosa*. It is available in eye drops and in cream to treat bacterial infections of the eye and the skin. In pharmaceutical products, sisomicin is employed as sisomicin sulfate (Extramycin; CAS number 53179-09-2, EC number 258-414-4, molecular formula $C_{38}H_{84}N_{10}O_{34}S_5$) (1).

CONTACT ALLERGY

Case report

A 32-year-old woman presented with an itchy eruption on the trunk. Recently, she had used a newly prescribed eye lotion of sisomicin for a recurring eye infection, for which she had previously been treated with gentamicin eye drops on various occasions. Physical examination revealed miliary-sized reddish papules spread over the chest and right popliteal fossa. Ill-defined erythema was also noted on the right periorbital area. A provocation test with two drops of sisomicin twice a day resulted in similar reddish papules appearing again on the chest and right popliteal fossa within 24 hours. Patch tests were performed with sisomicin eye lotion, benzalkonium chloride, and other antibiotics. An immediate wheal reaction to variously diluted preparations of sisomicin was observed at 60 minutes after application, and a papuloerythematous reaction was subsequently observed 48 hours later. Similar results were obtained with gentamicin sulfate. The patient had probably previously become sensitized to gentamicin and now cross-reacted to the related sisomicin leading to systemic contact dermatitis from transmucosal absorption (1).

Cross-reactions, pseudo-cross-reactions and co-reactions

A patient sensitized to gentamicin may have cross-reacted to sisomicin (2). Patients sensitized to neomycin may show cross-reactivity to sisomicin (Chapter 3.235 Neomycin).

LITERATURE

1   The data in the section 'General' may have been obtained from literature discussed in this chapter, but mostly also or exclusively from one or more of the following online sources: ChemIDPlus Advanced, PubChem, DrugBank, RxList, Drug Central, Drugs.com, and Wikipedia
2   Katayama I, Nishioka K. Systemic contact dermatitis medicamentosa induced by topical eye lotion (sisomicin) in a patient with corneal allograft. Arch Dermatol 1987;123:436-437

Chapter 3.314 SODIUM FLUORIDE

IDENTIFICATION

Description/definition   : Sodium fluoride is the inorganic salt of fluoride that conforms to the structural formula shown below

Pharmacological classes : Cariostatic agents

IUPAC name   : Sodium;fluoride

CAS registry number   : 7681-49-4

EC number   : 231-667-8

Merck Index monograph : 10024

Patch testing   : 0.5% water; perform controls

Molecular formula   : FNa

$$Na^+ \qquad F^-$$

GENERAL

Sodium fluoride is an inorganic salt of fluoride used topically or in municipal water fluoridation systems to prevent dental caries. Fluoride appears to bind to calcium ions in the hydroxyapatite of surface tooth enamel, preventing corrosion of tooth enamel by acids. This agent may also inhibit acid production by commensal oral bacteria. When topical fluoride is applied to hypersensitive exposed dentin, the formation of insoluble materials within the dentinal tubules blocks transmission of painful stimuli (1).

CONTACT ALLERGY

Case report

The teeth of an 8-year-old atopic girl had been treated topically by her stomatologist with sodium fluoride 1 year, 4 months and 1 month previously. Seven hours after the 3rd application, a very severe disseminated acute urticaria and facial angioedema appeared, requiring hospital admission. The commercial product applied was composed of 50 mg sodium fluoride (22.6 mg fluorine) in an alcoholic solution of natural resins. An extensive history clarified that the patient had been intolerant of certain fluoride toothpastes in the past. Patch testing showed a weakly positive reaction (+) with erythema and edema at 2 days to the commercial sodium fluoride product. At 4 days, only a slight erythema persisted. Fourteen controls were negative. The diagnosis was contact urticaria from sodium fluoride (2).

This author disagrees with the diagnosis contact urticaria. There were apparently no signs of immediate contact reactions in the mouth and the appearance of an extensive urticarial rash after 7 hours does not fit the concept of contact urticaria. One possibility is that the patient was contact allergic to sodium fluoride and the urticarial rash was caused by resorption of fluoride through the oral mucosa causing systemic contact dermatitis. However, patch test results were inadequate to ascertain a diagnosis of fluoride allergy.

LITERATURE

1   The data in the section 'General' may have been obtained from literature discussed in this chapter, but mostly also or exclusively from one or more of the following online sources: ChemIDPlus Advanced, PubChem, DrugBank, RxList, Drug Central, Drugs.com, and Wikipedia

2   Camarasa JG, Serra-Baldrich E, Lluch M, Malet A. Contact urticaria from sodium fluoride. Contact Dermatitis 1993;28:294

Chapter 3.315 SPAGLUMIC ACID

IDENTIFICATION

Description/definition : Spaglumic acid is the hybride peptide that conforms to the structural formula shown below

Pharmacological classes : Neuroprotective agents; anti-allergic agents; neurotoxins; bronchodilator agents; Histamine H1 antagonists

IUPAC name : (2S)-2-[[(2S)-2-Acetamido-3-carboxypropanoyl]amino]pentanedioic acid

Other names : N-Acetyl-L-aspartyl-L-glutamic acid

CAS registry number : 4910-46-7

EC number : Not available

Patch testing : 1% pet.

Molecular formula : $C_{11}H_{16}N_2O_8$

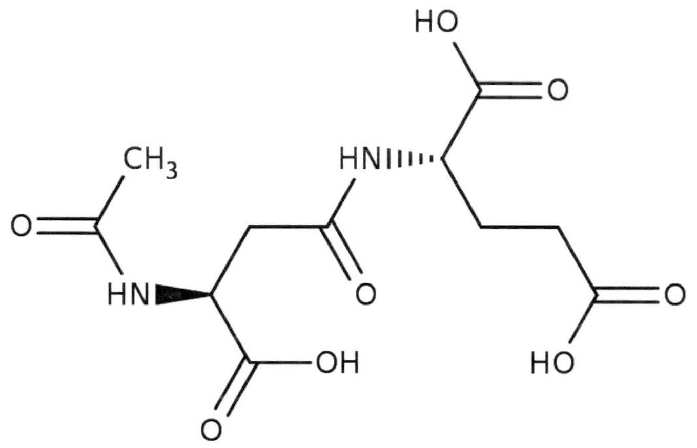

GENERAL

Spaglumic acid is a naturally occurring dipeptide, found in the brain of many mammals. This substance inhibits mast cell degranulation, the synthesis of certain leukotrienes, and the activation of the complement cascade. It has been reported to be effective in the treatment of seasonal and perennial allergic rhinitis and conjunctivitis (2). In pharmaceutical products, spaglumic acid is employed as magnesium spaglumate (CAS number 135981-31-6, EC number not available, molecular formula $C_{22}H_{26}Mg_3N_4O_{16}$) or as sodium spaglumate (CAS number 57096-29-4, EC number not available, molecular formula $C_{11}H_{13}N_2Na_3O_8$) (1).

Case report

A 19-year-old man was treated for perennial conjunctivitis with N-acetyl aspartyl glutamic acid (spaglumic acid) in eye drops. Several days after beginning treatment and 3 hours after the last dose was given, his symptoms worsened with ocular pruritus, hyperemia, chemosis of the conjunctiva, ocular redness, lacrimation, and eyelid edema. These symptoms subsided within a few days after withdrawal of the eye drops. Other eye drops were well tolerated. The patient was patch tested with the eye drops 'as is' and its ingredients, obtained from the manufacturer. There were positive reactions to the commercial eye drops and to spaglumic acid 1% pet. (D2 +, D4 ++), but not to any excipient. Five controls were negative (2).

LITERATURE

1 The data in the section 'General' may have been obtained from literature discussed in this chapter, but mostly also or exclusively from one or more of the following online sources: ChemIDPlus Advanced, PubChem, DrugBank, RxList, Drug Central, Drugs.com, and Wikipedia

2 Blanco R. Allergic conjunctivitis from eyedrops. Allergy 1997;52:686-687

Chapter 3.316 SPIRONOLACTONE

IDENTIFICATION

Description/definition   : Spironolactone is the synthetic steroid lactone that conforms to the structural formula
                            shown below
Pharmacological classes : Mineralocorticoid receptor antagonists; diuretics
IUPAC name   : S-[(7R,8R,9S,10R,13S,14S,17R)-10,13-Dimethyl-3,5'-dioxospiro[2,6,7,8,9,11,12,14,15,16-
                            decahydro-1H-cyclopenta[a]phenanthrene-17,2'-oxolane]-7-yl] ethanethioate
Other names   : Pregn-4-ene-21-carboxylic acid, 7-(acetylthio)-17-hydroxy-3-oxo-, γ-lactone, (7α,17α)-
CAS registry number   : 52-01-7
EC number   : 200-133-6
Merck Index monograph   : 10157
Patch testing   : 1% pet. and alcohol; if not available: crushed tablets 10-30% pet. and saline
Molecular formula   : $C_{24}H_{32}O_4S$

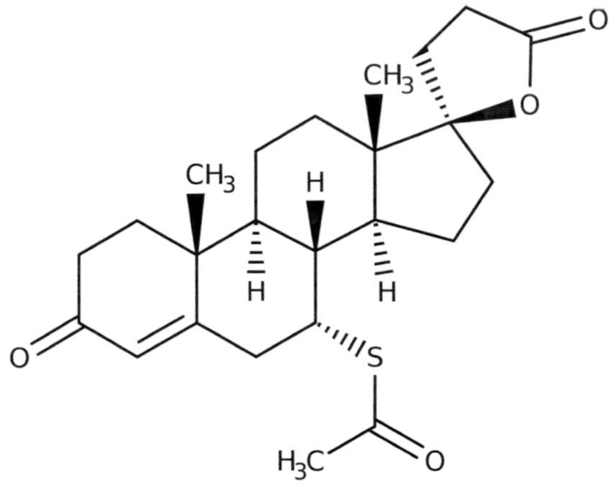

GENERAL

Spironolactone is a synthetic corticosteroid with potassium-sparing diuretic, antihypertensive, and antiandrogen activities. Spironolactone competitively inhibits adrenocortical hormone aldosterone activity in the distal renal tubules, myocardium, and vasculature. Spironolactone is used mainly in the treatment of refractory edema in patients with congestive heart failure, nephrotic syndrome, or hepatic cirrhosis, of primary hyperaldosteronism and hypertension. Off-label uses of spironolactone involving its antiandrogenic activity include hirsutism, female pattern hair loss, and adult acne vulgaris. It is also frequently used in medical gender transition (1). Topical spironolactone shows antiandrogenic effects by competitive inhibition of dihydrotestosterone receptors, and has been used to treat acne vulgaris, idiopathic hirsutism and androgenic alopecia (3).

CONTACT ALLERGY

Patch testing in groups of patients

In Italy, before 1986, 204 patients (43 male and 161 female; average age 19.4 years), under treatment for at least 2 months for acne, were patch tested with a battery of topical antiacne drugs. There were 7 positive reactions: 4 to spironolactone 5% pet., 2 to benzoyl peroxide 5% and 2% pet. (both concentrations cause irritant reactions) and one to tretinoin 0.05% pet. and alc. (both irritant materials). In a number of cases (not well described), dilution series were performed with positive results. ROATs with the commercial products were positive in all. All reactions were considered to be relevant (2).

Case reports

A 30-year-old woman presented with acute facial dermatitis after the use of spironolactone cream for acne vulgaris. Patch tests were positive to the cream and to spironolactone 1% pet. and alc., 5% alc., saturated aquatic solution and dry powder pure (3). Another female patient developed allergic contact dermatitis from spironolactone after having used 5% spironolactone cream for hirsutism for one month (8); details are not available to the author.

   Short summaries of other case reports of allergic contact dermatitis from spironolactone are shown in table 3.316.1.

Table 3.316.1 Short summaries of case reports of allergic contact dermatitis from spironolactone

Year and country	Sex	Age	Positive patch tests	Clinical data and comments	Ref.
1994 Italy	F	26	SPIRO 5% cream 'as is'; SPIRO 1% and 5% pet.	erythematous, edematous, vesicular pruritic eruption of the face after 5 months' SPIRO cream application; ROAT pos. after 4 days; 10 controls negative to SPIRO 1% pet.	4
1994 Spain	F	44	SPIRO 1% and 0.1% alc.	allergic contact dermatitis on the head after applying 2% SPIRO hydroalcoholic solution for 14 days for androgenic alopecia	5
1994 Spain	F	41	SPIRO 2% hydro-alcoholic solution; SPIRO 1% alc. and petrolatum	erythema, edema, vesiculation and intense pruritus of the scalp after applying 2% SPIRO hydro-alcoholic solution for 1 month, with occasional edema of the eyelids	6
1993 Italy	F	29	SPIRO 5% cream 'as is'; SPIRO saturated aq. sol., 1% alc. and powder pure	erythematovesicular dermatitis of the face after 11 month' application of SPIRO 5% cream for acne; 20 controls were negative	7

SPIRO: spironolactone

Cutaneous adverse drug reactions from systemic administration caused by type IV (delayed-type) hypersensitivity
Cutaneous adverse drug reactions from systemic administration of spironolactone caused by type IV (delayed-type) hypersensitivity, including drug reaction with eosinophilia and systemic symptoms (DRESS) (10,11), photosensitivity (12), and in addition occupational allergic contact dermatitis (9) are planned to be discussed in Volume IV of the *Monographs in Contact Allergy* series on Systemic drugs.

LITERATURE

1   The data in the section 'General' may have been obtained from literature discussed in this chapter, but mostly also or exclusively from one or more of the following online sources: ChemIDPlus Advanced, PubChem, DrugBank, RxList, Drug Central, Drugs.com, and Wikipedia
2   Balato N, Lembo G, Cuccurullo FM, Patruno C, Nappa P, Ayala F. Acne and allergic contact dermatitis. Contact Dermatitis 1996;34:68-69
3   Corazza M, Strumìa R, Lombardi AR, Virgili A. Allergic contact dermatitis from spironolactone. Contact Dermatitis 1996;35:365-366
4   Balato N, Patruno C, Lembo G, Cuccurullo FM, Ayala F. Allergic contact dermatitis from spironolactone. Contact Dermatitis 1994;31:203
5   Aguirre A, Manzano D, Zabala R, Eizaguirre X, Díaz-Pérez JL. Allergic contact dermatitis from spironolactone. Contact Dermatitis 1994;30:312
6   Fernandez-Vozmediano JM, Gil-Tocados G, Manrique-Plaza A, Romero-Cabrera MA, Picazo-Sanchez J, Nieto-Montesinos I. Contact dermatitis due to topical spironolactone. Contact Dermatitis 1994;30:118-119
7   Vincenzi C, Trevisi P, Farina P, Stinchi C, Tosti A. Facial contact dermatitis due to spironolactone in an anti-acne cream. Contact Dermatitis 1993;29:277-278
8   Gomez F, Ramelet AA, Rüedi B, Mühlemann M. Lack of effect of a spironolactone-containing cream on hair growth in hirsute women [letter]. Dermatologica 1987;174:102-103
9   Klijn J. Contact dermatitis from spironolactone. Contact Dermatitis 1984;10:105
10   Fernandes R-A, Regateiro FS, Faria E, Martinho A, Gonçalo M, Todo-Bom A. Drug reaction with eosinophilia and systemic symptoms caused by spironolactone: Case report. Contact Dermatitis 2018;79:255-256
11   Ghislain P, Bodarwe A, Vanderdonckt O, Tennstedt D, Marot L, Lachapelle J. Drug-induced eosinophilia and multisystemic failure with positive patch-test reaction to spironolactone: DRESS syndrome. Acta Derm Venereol 2004;84:65-68
12   Schwarze HP, Albes B, Marguery MC, Loche F, Bazex J. Evaluation of drug-induced photosensitivity by UVB photopatch testing. Contact Dermatitis 1998;39:200

Chapter 3.317 STANNOUS FLUORIDE

IDENTIFICATION

Description/definition   : Stannous fluoride is the inorganic salt with the molecular formula SnF2/F2Sn
Pharmaceutical classes   : Cariostatic agents
IUPAC name   : Difluorotin
Other names   : Tin difluoride; Tn(II) fluoride
CAS registry number   : 7783-47-3
EC number   : 231-999-3
Merck Index monograph   : 10181
Patch testing   : Stannous fluoride 0.5% pet.; to confirm sensitization to tin: stannous oxalate 1% pet. and tin 50% pet. (both commercially available); to confirm allergy to fluoride: sodium fluoride 0.5% water
Molecular formula   : SnF_2; F_2Sn

GENERAL

Stannous fluoride is a compound commonly used in toothpastes for the prevention of gingivitis, dental infections, cavities, and to relieve dental hypersensitivity. Although similar in function and activity to sodium fluoride (NaF), which is the conventionally added ingredient in toothpastes, stannous fluoride has been shown to be more effective at stopping and reversing dental lesions. It manages and prevents dental caries and gingivitis by promoting enamel mineralization, reducing gingival inflammation and bleeding through its potential broad-spectrum antibiotic effect and modulation of the microbial composition of the dental biofilm. Also, a stable acid-resistant layer is deposited on the tooth surfaces which is composed of calcium fluoride produced when stannous fluoride converts the calcium mineral apatite into fluorapatite (1).

CONTACT ALLERGY

Case reports

A 69-year-old man presented with cheilitis. He reported recurrent swelling with small blisters and red spots intra-orally and on his tongue, and in addition crustae on the lips since six months. He was using a toothpaste containing 0.454% w/w stannous (tin) fluoride. Patch testing in three sessions showed positive reactions to tin (50% pet.), the toothpaste 100%, 50%, 30%, 10% and 3% water, stannous oxalate 1%, stannous chloride 1% pet. (may cause irritant reactions), and stannous fluoride 0.5%, 0.15% and 0.05% pet. (weaker reactions in water). Sodium fluoride was negative. The patient was diagnosed with allergic contact cheilitis from tin, present as stannous fluoride, in toothpaste (2). The authors also present a second patient allergic to the same toothpaste, who had very similar patch test results to tin, the toothpaste, and the tin salts. She later had a recurrence of her complaints, which was probably caused by linalool present in the product, to which she was allergic (2).

Two years earlier, a 50-year-old woman had been reported using the same toothpaste and suffering from angioedema-like lip dermatitis and aphthous stomatitis. When patch tested, she reacted to the toothpaste, tin, tin oxalate and tin chloride. Possibly, she had signs of systemic contact dermatitis with flexural erythema and perianal itching from resorption of tin from the toothpaste (3).

In another patient, a woman aged 55 years, recurrent idiopathic urticaria and cheilitis were attributed to contact allergy to tin present in a stannous fluoride-containing toothpaste (other brand). Patch tests showed only a very weak, hardly convincing, reaction to tin 50% pet. on D4. The toothpaste itself was not patch tested, and no control tests were performed. However, avoidance of the toothpaste led to resolution of the patient's symptoms, whereas accidental re-use caused a flare-up (4).

LITERATURE

1   The data in the section 'General' may have been obtained from literature discussed in this chapter, but mostly also or exclusively from one or more of the following online sources: ChemIDPlus Advanced, PubChem, DrugBank, RxList, Drug Central, Drugs.com, and Wikipedia

2   Van Amerongen CCA, de Groot A, Volkering R, Schuttelaar MLA. Cheilitis caused by contact allergy to toothpaste containing stannous (tin) - two cases. Contact Dermatitis, 2020;83:126-129

3   Toma N, Horst N, Dandelooy J, Romaen E, Leysen J, Aerts O. Contact allergy caused by stannous fluoride in toothpaste. Contact Dermatitis 2018;78:304-306

4   Enamandram M, Das S, Chaney KS. Cheilitis and urticaria associated with stannous fluoride in toothpaste. J Am Acad Dermatol 2014;71:e75-76

Chapter 3.318 STEPRONIN

IDENTIFICATION

Description/definition   : Stepronin is the *N*-acyl-amino acid and sulfhydryl compound that conforms to the
                             structural formula shown below
Pharmacological classes   : Antiviral agents; expectorants; immunosuppressive agents
IUPAC name   : 2-[2-(Thiophene-2-carbonylsulfanyl)propanoylamino]acetic acid
Other names   : 2-(α-Thenoylthio)propionylglycine
CAS registry number   : 72324-18-6
EC number   : 276-587-4
Merck Index monograph   : 10206
Patch testing   : 180 mg/ml saline; intradermal tests (1.8 mg/ml saline) read at D1
Molecular formula   : $C_{10}H_{11}NO_4S_2$

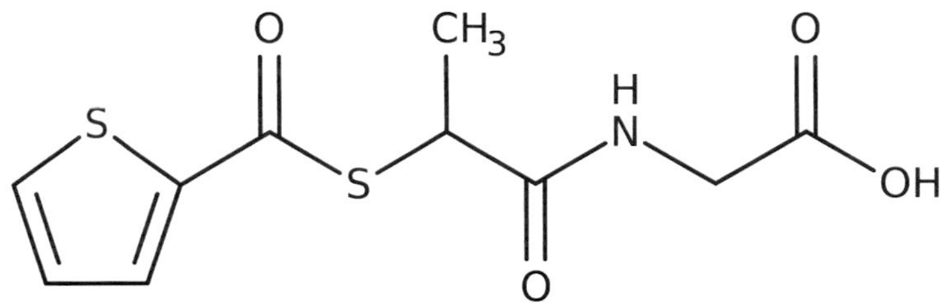

GENERAL

Stepronin is a mucolytic (expectorant) drug, an agent which dissolves thick mucus, and is usually used to help relieve respiratory difficulties. Stepronin promotes drainage of mucus from the lungs by thinning it and lubricating the irritated respiratory tract (1). It has also been used in the treatment of acute and chronic hepatitis (2).

CONTACT ALLERGY

Case reports

A 20-year-old man was hospitalized for bronchospasm that had occurred approximately 8 hours after the third inhalation of an aerosol product containing stepronin for acute bronchitis. At the time of admission, the patient presented a hacking cough, dyspnea, and wheezing, as well as a maculopapular eruption in the perioral region (the area that had been covered by the aerosol mask) and conjunctival injection. Prick and intradermal skin tests were then performed with stepronin (1.8 mg/ml in normal saline). A positive reaction was observed 8 hours after intradermal injection of 0.02 ml. At 24 h, an indurated, erythematous zone 22 mm in diameter was present at the test site and persisted for 3 days. A patch test using 0.1 ml of a normal saline solution containing stepronin 180 mg/ml was also positive at the 48 and 72 hour readings. Control tests in five volunteers were negative. A bronchial aerosol challenge resulted in a progressive decrease in the FEV (forced expiratory volume) of 33% after 4 hours, at which point the patient began to complain of dyspnea and moderate weakness, the blood pressure dropped to 90/50 mm Hg and facial erythema, conjunctival injection, and fever developed. Wheezing, previously absent, could be heard over all lung fields. It was concluded that the obstructive bronchial reaction with a maculopapular eruption and conjunctival injection were caused by delayed hypersensitivity to stepronin (2).

The authors had previously reported a similar patient with an obstructive bronchial reaction and perioral maculopapular rash that developed on the second day of stepronin aerosol therapy (3). This first case was characterized by symptom onset more than 24 hours after initiation of treatment, patch-test positivity, weak intradermal reactivity first noted at 36 hours, and marked reduction in FEV, beginning 6 hours after stepronin aerosol challenge (3).

LITERATURE

1   The data in the section 'General' may have been obtained from literature discussed in this chapter, but mostly also or exclusively from one or more of the following online sources: ChemIDPlus Advanced, PubChem, DrugBank, RxList, Drug Central, Drugs.com, and Wikipedia
2   Romano A, Di Fonso M, Mormile F, Quaratino D, Giuffreda F, Venuti A. Accelerated cell-mediated broncho-obstructive reaction to inhaled stepronin: a case report. Allergy 1996;51:269-271
3   Romano A, Pietrantonio F, di Fonso M, Pocobelli D, Venuti A. Delayed hypersensitivity to stepronin: a case report. Contact Dermatitis 1993;29:166

Chapter 3.319 SULBENTINE

IDENTIFICATION

Description/definition   : Sulbentine is the thiadiazinane that conforms to the structural formula shown below
Pharmacological classes   : Antifungal drugs, topical
IUPAC name   : 3,5-Dibenzyl-1,3,5-thiadiazinane-2-thione
Other names   : Dibenzthione; 3,5-dibenzyl-2-thion-tetrahydro-1,3,5-thiadiazin; carbothialdine
CAS registry number   : 350-12-9
EC number   : 206-497-2
Merck Index monograph   : 10294
Patch testing   : 3% pet.
Molecular formula   : $C_{17}H_{18}N_2S_2$

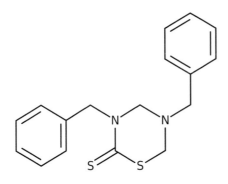

GENERAL

Sulbentine (dibenzthione) is a broad-spectrum antifungal agent that was formerly used in the topical treatment of dermatomycoses (1).

CONTACT ALLERGY

Case reports and series

In a 'fairly short time' before 1975, 5 patients with side effects of an antimycotic ointment containing sulbentine were observed in the University hospital of Umea, Sweden. Four were patch tested and reacted to the ointment 50% pet. Three were tested with the ingredients and two had positive reactions to sulbentine 3% pet., and the third to the base material lauromacrogol 'as is'. Seventeen controls were negative to sulbentine 3% pet. (2).

Apparently, sulbentine-containing antimycotics in the mid-1970s were one of the most frequent causes of iatrogenic contact dermatitis in former East Germany (5, no details known, data cited in ref. 6). The first report of contact allergy to sulbentine seems to have been published in 1965 (4).

Experimental investigations in patients with sulbentine allergy (number and clinical data unknown) and in guinea pigs suggested that benzyl isothiocyanate is the hapten in cases of sulbentine allergy and is a potent sensitizer (3).

Cross-reactions, pseudo-cross-reactions and co-reactions

Cross-reactions may have occurred with 'thiadiazine and triazine compounds' (4).

LITERATURE

1 The data in the section 'General' may have been obtained from literature discussed in this chapter, but mostly also or exclusively from one or more of the following online sources: ChemIDPlus Advanced, PubChem, DrugBank, RxList, Drug Central, Drugs.com, and Wikipedia
2 Lindén S, Göransson K. Contact allergy to dibenzthion. Contact Dermatitis 1975;1:258
3 Richter G, Heidelbach U, Heidenbluth I, Kadner H. Group-allergic reaction spectrum and sensitization potency of benzylisothiocyanate. Berufsdermatosen 1976;24:147-151 (Article in German)
4 Behrbohm P, Zschunke E. Allergisches Ekzem durch das Antimykotikum 'Afungin (Dibenzthion). Derm Wschrift 1965;151:1447 (Article in German, cited in refs. 2 and 7)
5 Richter G. Zur Kritik der externen Dermatika unter dem Aspekt ihrer allergischen Nebenwirkungen. Derm Mschr 1975;161:384-387 (Article in German)
6 Richter G. Allergic contact dermatitis from methylisothiocyanate in soil disinfectants. Contact Dermatitis 1980;6:183-186
7 Cronin E. Contact Dermatitis. Edinburgh: Churchill Livingstone, 1980:228

Chapter 3.320 SULCONAZOLE

IDENTIFICATION

Description/definition   : Sulconazole is the phenylethyl imidazole derivative that conforms to the structural
                                                formula shown below
Pharmacological classes : Antifungal agents
IUPAC name   : 1-[2-[(4-Chlorophenyl)methylsulfanyl]-2-(2,4-dichlorophenyl)ethyl]imidazole
Other names   : 1-(2,4-Dichlor-β-((4-chlorbenzyl)thio)phenethyl)imidazole
CAS registry number   : 61318-90-9
EC number   : Not available
Merck Index monograph   : 10295
Patch testing   : 1% alcohol or MEK (methyl ethyl ketone)
Molecular formula   : $C_{18}H_{15}Cl_3N_2S$

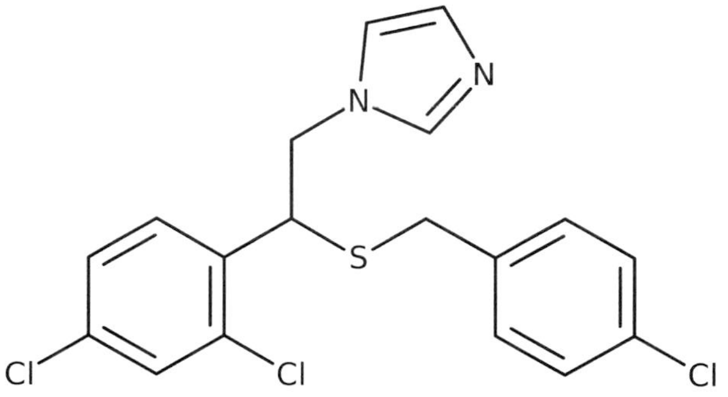

GENERAL

Sulconazole is a broad-spectrum imidazole antifungal agent with some antibacterial activity. It is used for the topical treatment of tinea cruris and tinea corporis caused by *Trichophyton rubrum*, *Trichophyton mentagrophytes*, *Epidermophyton floccosum*, and *Microsporum canis* and for the treatment of tinea (pityriasis) versicolor. Sulconazole has not been proven effective in treating tinea pedis (athlete's foot).

    In pharmaceutical products, sulconazole is employed as sulconazole nitrate (CAS number 61318-91-0, EC number not available, molecular formula $C_{18}H_{16}Cl_3N_3O_3S$) (1).

CONTACT ALLERGY

Case series

In Japan, in the period 1984 to 1994, 3049 outpatients were patch tested for suspected contact dermatitis and 218 of these with topical antifungal preparations. Thirty-five were allergic to imidazoles, including sulconazole in 16 individuals. The reason that sulconazole induced most positive reactions was probably that this imidazole was prescribed most frequently in the hospital in which the study was performed during that period. In 60% of the cases, there were cross-reactions between imidazoles (2).

    In a literature review up to 1994, the authors identified 105 reported patients who had a positive patch test reaction to at least one imidazole derivative, ranging from 51 reactions to miconazole to one reaction each to bifonazole and enilconazole. The number of reported reactions to sulconazole was 13 (4).

Case reports

In a 55-year-old man, three days after starting treatment with miconazole cream for tinea pedis, an itchy vesicular eruption appeared on the dorsa of his feet. He had been treated with sulconazole cream 2 months before, without adverse effects clinically, but never before with any other imidazoles. Patch tests were positive to sulconazole cream (considered to be the primary sensitizer) and miconazole cream (considered to be a cross-reaction). However, sulconazole and miconazole themselves were not tested 'because patch tests with the commercial topical imidazoles are more frequently positive than those with the ingredients tested individually' (5).

    A 53-year-old man presented with itchy exudative dermatitis of the groins, glans and scrotum, following the twice daily application for many days of sulconazole cream for suspected von Hebra's erythema marginatum. Patch testing was positive to the cream only. Sulconazole itself was not tested, but patch tests with the excipients and preservatives in the cream were negative. No cross-reactions to other antifungal preparations were observed (6).

Other case reports or information on contact allergy to sulconazole, adequate data of which are not available to the author, can be found in refs. 8,9 and 10.

Cross-reactions, pseudo-cross-reactions and co-reactions

Three of 6 patients who had developed sensitization to croconazole also reacted to sulconazole nitrate 1% or 5% pet., which the authors considered to be cross-reactions (3). Statistically significant associations have been found in patient data in co-sensitization between sulconazole, miconazole, and econazole (4,7).

LITERATURE

1   The data in the section 'General' may have been obtained from literature discussed in this chapter, but mostly also or exclusively from one or more of the following online sources: ChemIDPlus Advanced, PubChem, DrugBank, RxList, Drug Central, Drugs.com, and Wikipedia

2   Yoneyama E. Allergic contact dermatitis due to topical imidazole antimycotics. The sensitizing ability of active ingredients and cross-sensitivity. Nippon Ika Daigaku Zasshi 1996;63:356-364 (Article in Japanese)

3   Shono M, Hayashi K, Sugimoto R. Allergic contact dermatitis from croconazole hydrochloride. Contact Dermatitis 1989;21:225-227

4   Dooms-Goossens A, Matura M, Drieghe J, Degreef H. Contact allergy to imidazoles used as antimycotic agents. Contact Dermatitis 1995;33:73-77

5   Machet L, Vaillant L, Muller C, Cochelin N, Lorette G. Contact dermatitis and cross-sensitivity from sulconazole nitrate. Contact Dermatitis 1992;26:352-353

6   Bigardi AS, Pigatto PD, Altomare G. Allergic contact dermatitis due to sulconazole. Contact Dermatitis 1992;26:281-282

7   Carmichael AJ, Foulds IS. Imidazole cross-sensitivity to sulconazole. Contact Dermatitis 1988;19:237-238

8   Tani A, Hamada T, Kanzaki T. A case of allergic contact dermatitis due to lanoconazole and sulconazole. Environ Dermatol 1997;4:148 (Article in Japanese)

9   Tamiya Y, Matsumura E, Sasaki E, Ishinaga M. Allergic contact dermatitis due to sulconazole nitrate, streptomycin sulfate and prednisolone valerate acetate. Skin Res 1989;31(Suppl.):206-211 (Article in Japanese)

10 Matsumura E, Iizumi Y, Hata M, Yajima J, Hattori S, Honda M. Allergic contact dermatitis to imidazole antimycotics. Rinsho Derma 1987;29:673-677 (Article in Japanese)

Chapter 3.321 SULFACETAMIDE

IDENTIFICATION

Description/definition   : Sulfacetamide it the synthetic sulfanylacetamide derivative that conforms to the structural formula shown below
Pharmacological classes : Anti-bacterial agents; anti-infective agents, local; anti-infective agents, urinary
IUPAC name   : N-(4-Aminophenyl)sulfonylacetamide
Other names   : Acetosulfamine
CAS registry number   : 144-80-9
EC number   : 205-640-6
Merck Index monograph : 10301
Patch testing   : 5% pet.
Molecular formula   : $C_8H_{10}N_2O_3S$

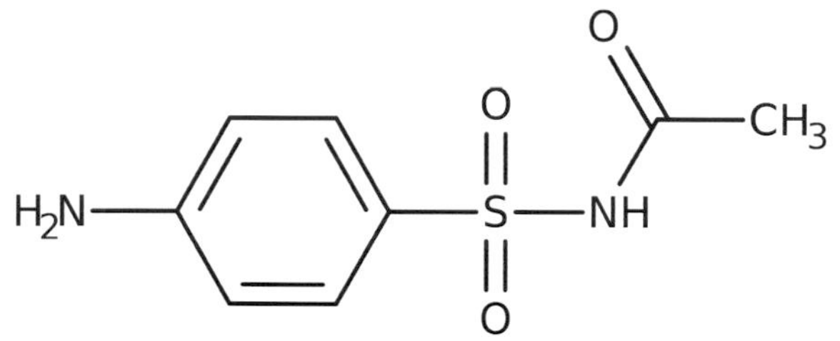

GENERAL

Sulfacetamide is a synthetic sulfonamide antibiotic with bacteriostatic activity. It inhibits bacterial folic acid synthesis by competing with p-aminobenzoic acid. With broad-spectrum activity against most gram-positive and many gram-negative organisms, it is used as an anti-infective topical agent to treat skin infections, bacterial vaginitis, keratitis, acute conjunctivitis, and blepharitis and as an oral agent for treating urinary tract infections. In pharmaceutical preparations, sulfacetamide is employed as sulfacetamide sodium (CAS number 127-56-0, EC number 204-848-4, molecular formula $C_8H_9N_2NaO_3S$) (1).

CONTACT ALLERGY

Patch testing in groups of selected patients

In a group of 203 patients with eyelid dermatitis, two patients reacted to sodium sulfacetamide (test concentration and vehicle not mentioned); the relevance was not specified and it is unknown how many individuals were tested with this material (2).

Case reports

Two women were sensitized by sulfacetamide in eye drops containing 10% of this sulfonamide; the had positive patch tests to sulfacetamide 5% pet. (4).

Cross-reactions, pseudo-cross-reactions and co-reactions

Nine patients allergic to sulfanilamide were patch tested with a battery of 25 sulfonamides (all tested 5% pet.) to detect cross-sensitization and there were 2 positive reactions to sulfacetamide sodium (3). Sulfonamides may cross-react to each other or other p-amino compounds, including p-phenylenediamine.

LITERATURE

1   The data in the section 'General' may have been obtained from literature discussed in this chapter, but mostly also or exclusively from one or more of the following online sources: ChemIDPlus Advanced, PubChem, DrugBank, RxList, Drug Central, Drugs.com, and Wikipedia
2   Guin JD. Eyelid dermatitis: experience in 203 cases. J Am Acad Dermatol 2002;47:755-765
3   Degreef H, Dooms-Goossens A. Patch testing with silver sulfadiazine cream. Contact Dermatitis 1985;12:33-37
4   Cronin E. Contact Dermatitis. Edinburgh: Churchill Livingstone, 1980:223

Chapter 3.322 SULFADIAZINE SILVER

IDENTIFICATION

Description/definition : Sulfadiazine silver is the silver salt sulfonamide derivative that conforms to the structural formula shown below

Pharmacological classes : Anti-infective agents

IUPAC name : Silver;(4-aminophenyl)sulfonyl-pyrimidin-2-ylazanide

Other names : Silver sulfadiazine; 4-amino-N-(2-pyrimidinyl)benzenesulfonamide silver salt;

CAS registry number : 22199-08-2

EC number : 244-834-5

Merck Index monograph : 10305 (Sulfadiazine)

Patch testing : 5% pet.; test also silver nitrate 1% water and sulfadiazine 5% pet.; test propylene glycol

Molecular formula : $C_{10}H_9AgN_4O_2S$

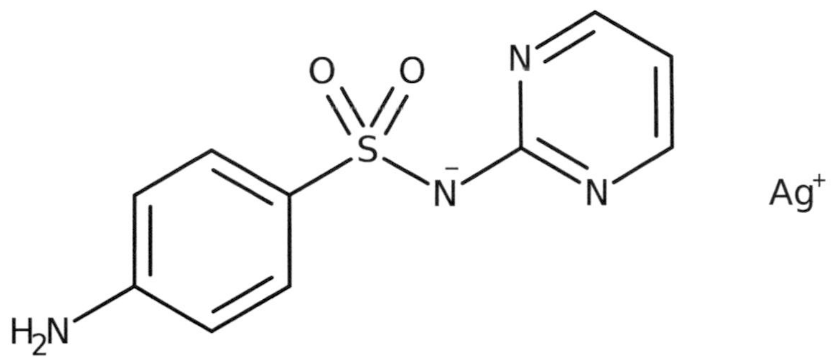

GENERAL

Sulfadiazine silver is a sulfonamide-based topical agent which is bactericidal for many gram-negative and gram-positive bacteria as well as being effective against yeast. Silver sulfadiazine may act through a combination of the activity of silver and sulfadiazine. Sulfadiazine silver is indicated as an adjunct for the prevention and treatment of wound infection and sepsis in patients with second- and third-degree burns (1). Side effects of silver sulfadiazine have been reviewed in 2009 (4).

CONTACT ALLERGY

General

Contact allergy to silver sulfadiazine appears to be extremely rare. In fact, there are no well documented cases where patients were tested with all ingredients of sulfadiazine cream and the allergenic culprit was clearly shown to be the active ingredient, be it silver sulfadiazine, silver or sulfadiazine. Most reactions to the cream are probably caused by the excipient propylene glycol (9,10).

Patch testing in groups of selected patients

In Leuven, Belgium, in the period 1990-2014, iatrogenic contact dermatitis was diagnosed in 2600 individuals (17% of the total patch test population). Silver sulfadiazine 10% pet. was tested in 335 patients and there were 3 (0.9%) positive reactions to it. Relevance was not specified for individual allergens, but 96% of all positive patch test reactions to topical drugs and antiseptics were considered to be relevant (2).

Case reports

A 26-year-old woman had suffered a second degree burn of 10×3 cm diameter caused by hot oil and located on her right forearm and wrist. The patient started treatment with topical silver sulfadiazine 1% cream, but after 15 days she presented progressive worsening of the burnt area, with appearance of peri-lesional erythema and blisters. Patch tests were positive to the cream and to silver nitrate 1% water. Patch tests to most excipients were negative, but sulfadiazine was not tested (3).

A 42-year-old man developed, while using silver sulfadiazine cream 1% for a thermal burn on his left foot, redness and weeping on the foot spreading to the legs, trunk, arms, and face. A ROAT with the cream was positive. Patch tests could not be performed with the active ingredients and most of the excipients. Five years later, the patient was patch tested for hand eczema and showed a strong allergic vesicular reaction to the material used to mark the test sites after removing the test materials. The marker consisted of methylrosanilin 1%, silver nitrate 10%, and alcohol and aqua in equal parts to 100%. After 3 months, a second patch test with the ingredients of the skin marker showed ++ reactions to silver nitrate 1% pet. and negative to gentian violet (1%, 0.1%, and 0.01% aqueous). It was concluded that the patient had previously become sensitized to silver from silver sulfadiazine cream (5).

A 27-year-old man who had been treated with silver sulfadiazine for burn injuries developed a severe drug eruption while using sulfamethoxazole-trimethoprim (ST), administered orally for the treatment of a urinary infection. Patch tests were positive to 10% ST solution and silver sulfadiazine cream diluted with water. An undefined number of healthy controls was negative. Histopathology of the patch test reactions was consistent with allergic contact dermatitis. Additionally, a lymphocyte-stimulation test with ST and a leukocyte migration-inhibition test with silver sulfadiazine used for the treatment of this patient both showed positive reactions. It was assumed that the patient had become sensitized to silver sulfadiazine (presumably the sulfadiazine moiety) and later cross-reacted to sulfamethoxazole (8).

Another patient was allergic to silver and the excipient cetyl alcohol in silver sulfadiazine cream (6). Quite remarkably, nothing was said about allergy to sulfadiazine, neither in the title nor in the discussion. However, after a positive open test to the cream, the patient was tested with 1 per cent sulfadiazine in an emulsifying ointment, which produced itching within 8 hours and a strongly positive reaction, that lasted for over 2 weeks. The emulsifying ointment alone produced no reaction. This would indicate that this patient was also allergic to sulfadiazine (6).

Silver sulfadiazine cream of the brand Flamazine contains propylene glycol, which is probably in most cases the cause of positive patch tests or contact allergic reaction to the cream (9,10).

Cross-reactions, pseudo-cross-reactions and co-reactions
Nine patients allergic to sulfanilamide and one to sulfathioureum were patch tested with silver sulfadiazine 5% pet. and there were no positive reactions (7). Possible cross-sensitivity to sulfamethoxazole in a patient sensitized to sulfadiazine silver (8).

LITERATURE
1   The data in the section 'General' may have been obtained from literature discussed in this chapter, but mostly also or exclusively from one or more of the following online sources: ChemIDPlus Advanced, PubChem, DrugBank, RxList, Drug Central, Drugs.com, and Wikipedia
2   Gilissen L, Goossens A. Frequency and trends of contact allergy to and iatrogenic contact dermatitis caused by topical drugs over a 25-year period. Contact Dermatitis 2016;75:290-302
3   García AA, Rodríguez Martín AM, Serra Baldrich E, Manubens Mercade E, Puig Sanz L. Allergic contact dermatitis to silver in a patient treated with silver sulphadiazine after a burn. J Eur Acad Dermatol Venereol 2016;30:365-366
4   Fuller FW. The side effects of silver sulfadiazine. J Burn Care Res 2009;30:464-470
5   Ozkaya E. A rare case of allergic contact dermatitis from silver nitrate in a widely used special patch test marker. Contact Dermatitis 2009;61:120-122
6   Fraser-Moodie A. Sensitivity to silver in a patient treated with silver sulphadiazine (Flamazine). Burns 1992;18:74-75
7   Degreef H, Dooms-Goossens A. Patch testing with silver sulfadiazine cream. Contact Dermatitis 1985;12:33-37
8   Sawada Y. Adverse reaction to sulphonamides in a burned patient – a case report. Burns Incl Therm Inj 1985;12:127-131
9   Van der Horst JC, Van Ketel WG. Problems in allergological studies of a patient with a probable Flammazine cream allergy. Ned Tijdschr Geneeskd 1982;126:841-843 (Article in Dutch)
10 Rasmussen I. Patch test reactions to Flamazine. Contact Dermatitis 1984;11:133-134

Chapter 3.323 SULFANILAMIDE

IDENTIFICATION

Description/definition   : Sulfanilamide is the sulfonamide that conforms to the structural formula shown below
Pharmacological classes   : Anti-bacterial agents
IUPAC name   : 4-Aminobenzenesulfonamide
Other names   : Benzenesulfonamide, 4-amino-; sulphanilamide; *p*-aminobenzene sulfonamide
CAS registry number   : 63-74-1
EC number   : 200-563-4
Merck Index monograph   : 10327
Patch testing   : 5.0% pet. (Chemotechnique, SmartPracticeCanada, SmartPracticeEurope)
Molecular formula   : $C_6H_8N_2O_2S$

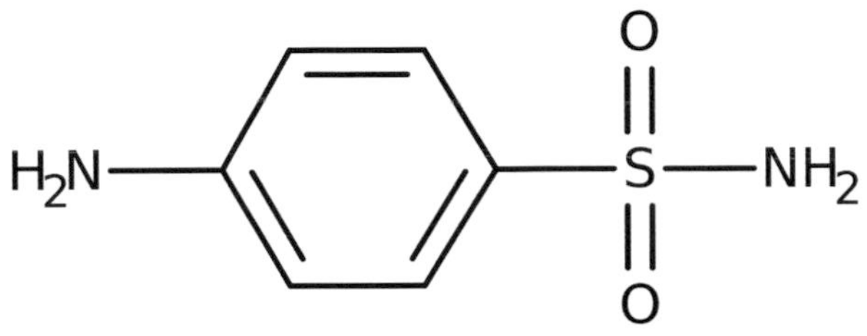

GENERAL

Sulfanilamide is a short-acting sulfonamide antibiotic. It is bacteriostatic against most gram-positive and many gram-negative organisms, but many strains of an individual species may be resistant. Sulfanilamide competes with *p*-aminobenzoic acid (PABA) for the bacterial enzyme dihydropteroate synthase, thereby preventing the incorporation of PABA into dihydrofolic acid, the immediate precursor of folic acid. This leads to an inhibition of bacterial folic acid synthesis and *de novo* synthesis of purines and pyrimidines, ultimately resulting in cell growth arrest and cell death. Sulfanilamide is used in vaginal cream for the treatment of vulvovaginitis caused by *Candida albicans.* The active agent sulfanilamide is present in a specially compounded base buffered to the pH (about 4.3) of the normal vagina to encourage the presence of the normally occurring Döderlein's bacilli in the vagina (1). In Belgium (and probably other countries), it is also available in an ointment for wound treatment.

CONTACT ALLERGY

General

Sulfanilamide and other sulfonamides were extensively used in the 1940s and 1950s in the treatment of all types of infected skin conditions and some patients showed allergic contact dermatitis, but many more would develop photoallergic contact dermatitis (11,13-17,19). The early literature is not reviewed here.

Patch testing in groups of selected patients

Results of studies in which patch testing with sulfanilamide in groups of selected patients (patients suspected of photodermatoses, individuals with possible iatrogenic contact dermatitis, patients with chronic otitis externa) was performed are shown in table 3.323.1. Prevalences of sensitization ranged from 0.1% to a staggering 41% (selection?), with a large number (n=161) of sensitized patients (4). The authors of this study have confirmed that patch testing was done very selectively, testing mostly patients who had used pharmaceuticals containing sulfanilamide. In the initial period of this study, a powder containing pure sulfanilamide was freely available in Belgium, but it was withdrawn from the market some years ago. Nevertheless, an ointment with sulfanilamide for wound treatment can still be obtained in Belgium (An Goossens, Email communication, September 2020).

In the studies among photosensitive individuals, far more positive photopatch tests were observed than conventional patch tests (see the section 'Photosensitivity' below). The issue of relevance of the positive patch tests was hardly ever addressed.

Table 3.323.1 Patch testing in groups of patients: Selected patient groups

Years and Country	Test conc. & vehicle	Number of patients tested	positive (%)	Selection of patients (S); Relevance (R); Comments (C)	Ref.
2005-2014 China	5% water	6012	165 (2.7%)	S: patients suspected of photodermatoses; R: not stated; C: there were also 350 photoallergic reactions and 2 combined contact allergic and photoallergic reactions; overlap with ref. 8	9
1990-2014 Belgium	5% pet.	393	161 (41.0%)	S: patients suspected of iatrogenic contact dermatitis and tested with a pharmaceutical series and their own products; R: 96% of the positive patch test reactions to all topical drugs and antiseptics were considered to be relevant	4
2006-2012 China	5% water	3717	72 (1.9%)	S: patients with suspected photodermatoses; R: not stated; C: there were also 333 photoallergic reactions; overlap with ref. 9	8
1995-2004 IVDK	5% pet.	10,659	(0.1%)	S: not stated; R: not stated	3
1993-4 Netherlands	5% pet.	34	1 (3%)	S: patients with chronic otitis externa or media; R: not stated	2
1987-1989 Thailand	5% pet.	55	4 (7%)	S: patients suspected of photosensitivity; R: not stated; C: there were also 10 positive photopatch test reactions; this antibiotic is widely used in topical preparations in Thailand	10
1967-1970 Poland	5% pet.	621	5 (0.8%)	S: not stated; R: not stated	12

IVDK: Information Network of Departments of Dermatology (Germany, Austria, Switzerland)

Case reports and case series
In the 1980s, a powder containing 25% p-aminobenzene sulfonamide (sulfanilamide) was a popular drug in Poland for the treatment of wounds and infections. In a dermatology department in Warsaw, about 5 cases of allergic contact dermatitis to sulfanilamide were seen each year (5).

A 41-year-old man had used sulfanilamide powder on an erosion from uncomfortable underwear in the groins. The next day, redness and itching appeared and increased, but he kept using the sulfanilamide for 3 more days. On the 5th day after the first application, the patient had a disseminated eruption involving the whole body, mainly the trunk and hands with some exudation. The eyelids, penis and scrotum showed severe edema. Patch tests were positive to sulfathiazole 2% and 5% pet. and to 1% p-phenylenediamine. After 10 days, patch tests were performed with 32 aromatic amines and he reacted to 13 of them, including p-toluidine, p-nitroaniline, aniline, p-aminophenol, p-toluenediamine sulfate, p-nitro-o-aminophenol, and p-aminobenzoic acid diethylaminoethyl ester (Novocaine). The authors admitted that some reactions may have been false-positive from the angry back syndrome, but stated that the others were probably the result of cross-sensitization (5).

Cross-reactions, pseudo-cross-reactions and co-reactions
Nine patients allergic to sulfanilamide were patch tested with a battery of 25 sulfonamides (all tested 5% pet.) to detect cross-sensitization and there were 3 reactions to sodium sulfadiazine, 2 to sodium sulfacetamide, one to sulfaguanidine, one to sodium sulfadimidine and one to sodium sulfamerazine (6).

Sulfonamides may cross-react to each other or other p-amino compounds, including p-phenylenediamine (see the data in ref. 5 in the section 'Case reports and case series' above.

PHOTOSENSITIVITY
Sulfanilamide and other sulfonamides were extensively used in the 1940s and 1950s in the treatment of all types of infected skin conditions and some patients showed allergic contact dermatitis, but many more would develop photoallergic contact dermatitis (11,13-17,19). The early literature is not reviewed here.

Photopatch testing in groups of selected patients
Results of studies in which photopatch testing with sulfanilamide in groups of selected patients (patients suspected of photodermatoses) was performed, are shown in table 3.323.2. Prevalences of sensitization ranged from 0.6% in the USA (7) to a high 18% (10). The latter was observed in a study performed in Thailand in the second half of the 1980s, when sulfanilamide was apparently widely used in topical preparations there; 80% of the reactions were relevant (10). Two overlapping studies from China also had high rates of photocontact allergy with large numbers of patients, but the relevance of the positive photopatch tests was not mentioned (8,9).

Table 3.323.2 Photopatch testing in groups of patients

Years and Country	Test conc. & vehicle	Number of patients tested | positive (%)		Selection of patients (S); Relevance (R); Comments (C)	Ref.
2005-2014 China	5% water	6012	350 (5.8%)	S: patients suspected of photodermatoses; R: not stated; C: there were also 165 contact allergic reactions and 2 combined contact allergic and photoallergic reactions; overlap with ref. 8	9
2006-2012 China	5% water	4836	333 (6.9%)	S: patients with suspected photodermatoses; R: not stated; C: there were also 3 'plain' contact allergic reactions; overlap with ref. 9	8
1993-2006 USA	1% pet.	76	2 (3%)	S: patients photopatch tested for suspected photosensitivity; R: 94% of all reactions to antimicrobials had unknown relevance	18
2000-2005 USA	10% pet.	178	1 (0.6%)	S: patients photopatch tested for suspected photodermatitis; R: 0%	7
1987-1989 Thailand	5% pet.	55	10 (18%)	S: patients suspected of photosensitivity; R: 80%; C: this antibiotic is widely used in topical preparations in Thailand; there were also 4 'plain' contact allergic reactions	10

LITERATURE

1   The data in the section 'General' may have been obtained from literature discussed in this chapter, but mostly also or exclusively from one or more of the following online sources: ChemIDPlus Advanced, PubChem, DrugBank, RxList, Drug Central, Drugs.com, and Wikipedia

2   Van Ginkel CJ, Bruintjes TD, Huizing EH. Allergy due to topical medications in chronic otitis externa and chronic otitis media. Clin Otolaryngol Allied Sci 1995;20:326-328

3   Menezes de Padua CA, Uter W, Schnuch A. Contact allergy to topical drugs: prevalence in a clinical setting and estimation of frequency at the population level. Pharmacoepidemiol Drug Saf 2007;16:377-384

4   Gilissen L, Goossens A. Frequency and trends of contact allergy to and iatrogenic contact dermatitis caused by topical drugs over a 25-year period. Contact Dermatitis 2016;75:290-302

5   Rudzki E, Rebandel P. Primary sensitivity to sulphonamide and secondary sensitization to aromatic amines. Contact Dermatitis 1987;17:49

6   Degreef H, Dooms-Goossens A. Patch testing with silver sulfadiazine cream. Contact Dermatitis 1985;12:33-37

7   Scalf LA, Davis MDP, Rohlinger AL, Connolly SM. Photopatch testing of 182 patients: A 6-year experience at the Mayo Clinic. Dermatitis 2009;20:44-52

8   Gao L, Hu Y, Ni C, Xu Y, Ma L, Yan S, Dou X. Retrospective study of photopatch testing in a Chinese population during a 7-year period. Dermatitis 2014;25:22-26 (overlap with ref. 9)

9   Hu Y, Wang D, Shen Y, Tang H. Photopatch testing in Chinese patients over 10 years. Dermatitis 2016;27:137-142 (overlap with ref. 8)

10 Gritiyarangsan P. A three-year photopatch study in Thailand. J Dermatol Sci 1991;2:371-375

11 Peterkin GAG. Skin eruptions due to the local application of sulphonamides. Br J Dermatol 1945;57:1-9

12 Rudzki E, Kleniewska D. The epidemiology of contact dermatitis in Poland. Br J Dermatol 1970;83:543-545

13 Greenwood AM. Skin manifestations due to sulfanilamide and its derivatives. New Engl J Med 1941;224:237-238

14 Burckhardt W. Untersuchungen über die Photoaktivität einiger Sulfanilamide. Dermatologica 1941;83:63 (Article in German)

15 Burckhardt W. Photoallergische Ekzeme durch Sulfanilamidsalben. Dermatologica 1948;96:280-285 (Article in German)

16 Epstein S. Photoallergy and primary phototoxicity to sulfanilamide. J Invest Dermatol 1939;2:43-51

17 Burckhardt W. Sulfanilamide photoallergy, a new form of physicochemical hypersensitivity. Int Arch Allergy Appl Immunol 1950;1(Suppl.):9-10 (Article in German)

18 Victor FC, Cohen DE, Soter NA. A 20-year analysis of previous and emerging allergens that elicit photoallergic contact dermatitis. J Am Acad Dermatol 2010;62:605-610

19 Kooij R, Van Vloten TJ. Epidermal sensitization due to sulphonamide drugs; groups- and cross-sensitivity; photosensitivity; passive transfer of antibodies; leftwich reaction. Dermatologica 1952;104:151-167

Chapter 3.324 SULFIRAM

IDENTIFICATION

Description/definition   : Sulfiram is the organosulfur compound that conforms to the structural formula shown
                            below
Pharmacological classes : Antiparasitic products, insecticides and repellents
IUPAC name   : Diethylcarbamothioyl *N,N*-diethylcarbamodithioate
Other names   : Monosulfiram; tetraethylthiuram monosulfide
CAS registry number   : 95-05-6
EC number   : 202-387-3
Merck Index monograph   : 10350
Patch testing   : 1% pet.
Molecular formula   : $C_{10}H_{20}N_2S_3$

GENERAL

Sulfiram is a fungicide, bactericide, and wood preservative that has been used for the treatment of scabies. Soap containing sulfiram (Tetmosol ® medicated soap) is still available 'for the treatment and prevention of scabies and other skin-related problems' (Google seach, many websites). In formal medicine, it is probably not used for this indication anymore, but it may still be used as an antiparasitic product in veterinary medicine (1).

CONTACT ALLERGY

Case reports

A 41-year-old woman developed an acute exudative dermatitis on her right forearm, left upper arm and left side of neck. The cause of the eruption was not discovered, but it settled quickly with topical corticosteroids. Six months later, she had a further episode with a very similar distribution. Patch testing was positive to the thiuram-mix (tetraethylthiuram disulfide, tetramethylthiuram disulfide, tetramethylthiuram monosulfide, dipentamethylene-thiuram disulfide). Further enquiry revealed that she had treated her 2 dogs with a proprietary ear mite treatment before both episodes of dermatitis. When applying the drops, she had held the dogs' head between her arms and neck, causing some smearing of the medication onto her skin. This product contained 5% sulfiram (monosulfiram, tetraethylthiuram monosulfide). An open test of the ear drops on her forearm produced a marked spreading dermatitis. It was concluded that the patient had suffered an allergic contact dermatitis from exposure to sulfiram in the ear mite preparation (3). Sulfiram itself was not tested, but the conclusion is almost certainly valid.

A 47-year-old woman developed an exanthema with diffuse and considerable dermal edema affecting all of her body, though initially it was less severe on the face and neck. On her admission the red areas were rapidly becoming covered by large numbers of small pustules which were coalescing to form lakes of subcorneal pus. The horny layer now started to disintegrate and to be shed in vast sheets- the one below her right knee like a stocking in one piece. The pustules were distributed diffusely, and not in groups or in rings. Mucous membranes were unaffected. She had a fever and was severely ill. A cytological smear from the pustules showed a great quantity of polymorph neutrophils but no bacteria. No organisms were cultured. This exanthema was diagnosed as toxic epidermal necrolysis. It had evolved from a rash starting 3 hours after the application of a 25% alcoholic solution of sulfiram diluted with 3 parts of water for the treatment of scabies. Previously, she had become sensitized to the related rubber compound tetramethylthiuram disulfide from her work in a munition factory. Patch tests were positive to tetramethylthiuram disulfide and sulfiram 14% water (or the 25% alcoholic solution diluted with six parts of water?), the latter inducing a very severe eczematous reaction within three hours of application (2). It should be mentioned that the diagnosis was based solely on the clinical picture, histopathology was not performed. There were also features of acute generalized exanthematous pustulosis, which at that time was unknown.

Cross-reactions, pseudo-cross-reactions and co-reactions
Cross-reactivity between tetramethylthiuram disulfide and tetraethylthiuram monosulfide (sulfiram) (2).

LITERATURE
1   The data in the section 'General' may have been obtained from literature discussed in this chapter, but mostly also or exclusively from one or more of the following online sources: ChemIDPlus Advanced, PubChem, DrugBank, RxList, Drug Central, Drugs.com, and Wikipedia
2   Monckton Copeman PW. Toxic epidermal necrolysis caused by skin hypersensitivity to monosulfiram. Br Med J 1968;1(5592):623-624
3   Dwyer CM, Ormerod AD. Allergic contact dermatitis from thiuram in a veterinary medication. Contact Dermatitis 1997;37:132

Chapter 3.325 SUPROFEN

IDENTIFICATION

Description/definition   : Suprofen is the arylpropionic acid derivative that conforms to the structural formula
                              shown below
Pharmacological classes : Cyclooxygenase inhibitors; anti-inflammatory agents, non-steroidal
IUPAC name   : 2-[4-(Thiophene-2-carbonyl)phenyl]propanoic acid
Other names   : (+-)-2-(p-(2-Thenoyl)phenyl)propionic acid
CAS registry number   : 40828-46-4
EC number   : 255-096-9
Merck Index monograph : 10404
Patch testing   : 1% pet. (perform photopatch tests)
Molecular formula   : $C_{14}H_{12}O_3S$

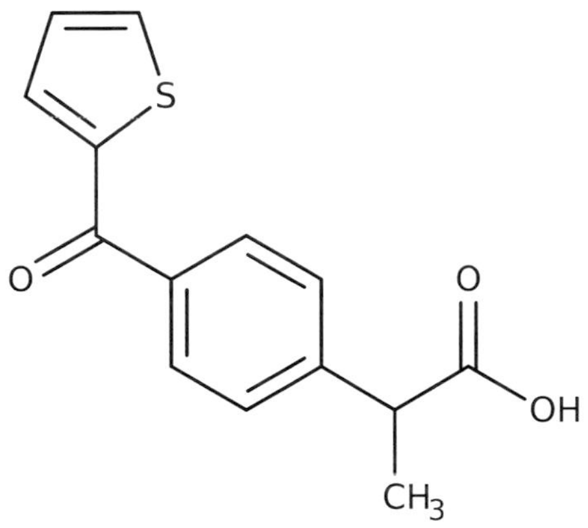

GENERAL

Suprofen is a nonsteroidal anti-inflammatory drug (NSAID) with anti-inflammatory, analgesic and antipyretic activities. It binds to the cyclooxygenase-1 (COX-1) and cyclooxygenase-2 (COX-2) isoenzymes, preventing the synthesis of prostaglandins and reducing the inflammatory response. Currently, it appears not to be used systemically, but suprofen is utilized in eye drops to inhibit the miosis (pupil constriction) that may occur during ocular surgery (1). It is also available since 1989 in Japan in an ointment containing 1% suprofen (1,3). All reports of (photo)contact allergy thus far have come from Japan, with the exception of one patient who was sensitized during a clinical trial (12).

CONTACT ALLERGY

Case reports and case series

In Japan, before 1994, a group of 74 patients with severe refractory atopic dermatitis involving the face were patch tested. There was one relevant reaction to suprofen (1%). It is uncertain whether all patients were tested with this hapten and the test concentration and vehicle were not mentioned (2).

A 32-year-old female patient had received suprofen 5% in a cream base as part of a clinical trial in order to treat an irritant dermatitis on her right hand. She developed acute eczematous lesions after repeated applications. The patient had not been previously treated with arylalcanoic acids, either systemically or topically. Patch tests were performed with the cream, its ingredients and several chemical analogs of suprofen, tested at 2% and 5% pet. The patient reacted to suprofen cream and suprofen 2% and 5% pet., but not to other ingredients of the cream. Testing with the suprofen-analogs revealed cross-sensitivity with ibuprofen, ketoprofen, naproxen, fenoprofen, alclofenac, metiazinic acid, and chlorambucil. Three controls were negative to all compounds (12).

PHOTOSENSITIVITY

Up to 1994, nine cases of photocontact allergy due to suprofen have been reported in Japanese literature (5-10) and 6 in English literature (3,4).

Case series

In a university hospital in Tokyo, in the period March to May 1990, 5 patients (4 women, one man, ages 13-67) were seen with pruritic erythematous edematous and vesicular lesions on their faces after sun exposure. All patients had been using 1% suprofen ointment from 2 weeks to 3 months for lesions on their faces due to either atopic dermatitis or contact dermatitis. Patch tests were negative to the ointment and suprofen. Photopatch test results showed that all patients reacted positively to the ointment and to suprofen at concentrations down to 0.1% to 0.01% pet. with UVA, and 3 patients tested showed positive reactions to suprofen down to 1% to 0.1% pet. with UVB. Thirteen controls were negative to the UVA-tests. The patients were also photopatch tested with other NSAIDs. Five were positive to tiaprofenic acid 10% pet. (positive in 1/9 controls), of whom 4 also to 1% (negative in controls). None reacted to ketoprofen or other NSAIDs (tested as commercial ointments) (4).

Case reports

A 79-year-old woman began applying suprofen 1% ointment for seborrheic eczema on her face and for asteatosis on her extremities. Diffuse edematous erythema and an itchy sensation occurred on her face, the V-shaped area of her upper chest, and her forearms, followed by vesicles, 2 months later in the summer. Patch tests were negative, but photopatch tests were positive (++) to the ointment, tested 'as is' and to suprofen 1%, 0.1%, 0.01% and 0.001% pet. (+) The patient also had weakly positive (+) photopatch test reactions to ketoprofen 3% pet. and tiaprofenic acid 5% pet. (the latter possibly phototoxic). Six controls were negative to all suprofen materials used. The biological action spectrum for photodermatitis extended from UVA into UVB and into the UVC range in this case (3).

Cross-reactions, pseudo-cross-reactions and co-reactions

Virtually all patients photosensitized to ketoprofen photocross-react to suprofen (11). A patient photosensitized to suprofen possibly photocross-reacted to ketoprofen and tiaprofenic acid (3). Of 5 patients photosensitized to suprofen, 4 and possibly 5 photocross-reacted to tiaprofenic acid, which is structurally closely related to suprofen (4). Four other ketoprofen-photosensitized patients all photocross-reacted to suprofen (13).

A patient who was sensitized to suprofen during a clinical trial and who had never been previously treated with arylalcanoic acids, either systemically or topically, was tested with a large number of suprofen-analogs. There were positive reactions to ibuprofen, ketoprofen, fenoprofen, naproxen, alclofenac, metiazinic acid and chlorambucil, which were considered to be cross-reactions. The antigenic determinant seemed to correspond to a very well-defined structure: the carboxyl group (-COOH) has to be separated from the aryl structure by at least one C atom (substituted or not). Furthermore, the aryl group must be substituted with different chemical groups but not in the ortho-positions (12).

LITERATURE

1 The data in the section 'General' may have been obtained from literature discussed in this chapter, but mostly also or exclusively from one or more of the following online sources: ChemIDPlus Advanced, PubChem, DrugBank, RxList, Drug Central, Drugs.com, and Wikipedia
2 Tada J, Toi Y, Arata J. Atopic dermatitis with severe facial lesions exacerbated by contact dermatitis from topical medicaments. Contact Dermatitis 1994;31:261-263
3 Kuno Y, Numata T. Photocontact allergy due to suprofen. J Dermatol 1994;21:352-357
4 Kurumaji Y, Oshiro Y, Miyamoto C, Keong CH, Katoh T, Nishioka K. Allergic photocontact dermatitis due to suprofen; Photopatch testing and cross-reaction study. Contact Dermatitis 1991;25:218-223
5 Nakazawa A, Matsuo I. Photocontact dermatitis due to suprofen photoallergy. Rinsho Hifuka 1990;44:1249-1251 (Article in Japanese, data cited in ref. 3)
6 Tanaka K, Watanabe M, Ohgami T, Nonaka S, Yoshida H. A case of photocontact dermatitis due to suprofen. Nishinihon J Dermatol 1991;53:695-698 (Article in Japanese, data cited in ref. 3)
7 Asai T, Fujioka A, Yonemoto K. Photocontact dermatitis due to suprofen. Hifubyoh-Shinryoh 1991;13:907-910 (Article in Japanese, data cited in ref. 3)
8 Tamura T, Iizuka H, Kishiyama K. A case of photocontact dermatitis caused by suprofen. Rinsho Derma (Tokyo) 1991;33:1565-1568 (Article in Japanese, data cited in ref. 3)
9 Tanaka T, Ohtubo H, Narisawa Y, Kohda H. Two cases of allergic contact dermatitis due to suprofen. Nishinihon J Dermatol 1993;55:653-657 (Article in Japanese, data cited in ref. 3)
10 Morishima Y, Suehiro M, Yamada K, Hirano S, Yasuno H. Three cases of photocontact dermatitis from suprofen ointment. Skin Research 1993;35(Suppl.16):101-107 (Article in Japanese, data cited in ref. 3)
11 Devleeschouwer V, Roelandts R, Garmyn M, Goossens A. Allergic and photoallergic contact dermatitis from ketoprofen: results of (photo) patch testing and follow-up of 42 patients. Contact Dermatitis 2008;58:159-166
12 Dooms-Goossens A, Dooms M, Van Lint L, Degreef H. Skin sensitizing properties of arylalcanoic acids and their analogues. Contact Dermatitis 1979;5:324-328
13 Matsushita T, Kamide R. Five cases of photocontact dermatitis due to topical ketoprofen: photopatch testing and cross-reaction study. Photodermatol Photoimmunol Photomed 2001;17:26-31

Chapter 3.326 TACALCITOL

IDENTIFICATION

Description/definition : Tacalcitol is the vitamin D3 derivative that conforms to the structural formula shown
                          below
Pharmacological classes : Anti-inflammatory agents; dermatological agents
IUPAC name : (1R,3S,5Z)-5-[(2E)-2-[(1R,3aS,7aR)-1-[(2R,5R)-5-Hydroxy-6-methylheptan-2-yl]-7a-methyl-
              2,3,3a,5,6,7-hexahydro-1H-inden-4-ylidene]ethylidene]-4-methylidenecyclohexane-1,3-
              diol
Other names : 1α,24(R)-Dihydroxyvitamin D3; (1α,3β,5Z,7E,24R)-9,10-secocholesta-5,7,10(19)-triene-
              1,3,24-triol
CAS registry number : 57333-96-7
EC number : Not available
Merck Index monograph : 10422
Patch testing : 0.0002% alcohol (2 µg/ml)
Molecular formula : $C_{27}H_{44}O_3$

GENERAL

Tacalcitol is a vitamin D_3 derivative with anti-inflammatory activity that is used in topical pharmaceuticals for the
treatment of psoriasis. In such products, both tacalcitol and tacalcitol monohydrate (CAS number 93129-94-3, EC
number not available, molecular formula $C_{27}H_{46}O_4$) may be employed (1).

CONTACT ALLERGY

Case report
A 78-year-old woman developed itchy redness of the skin while treating her psoriasis with tacalcitol ointment. On
examination, edematous scaly erythema extended over initially psoriatic lesions on her face, arms and legs. Patch
testing with the ointment and its ingredients was carried out and showed positive reactions to tacalcitol ointment 'as
is' (D3 +, D7 +) and tacalcitol 0.0002% alcohol (2 µg/ml) (D3 -, D7 +) and a cross-reaction to calcitriol 0.0002% alcohol
(D3 -, D7 +); there were no reactions to the excipients liquid paraffin and di-isopropyl adipate. Patch tests with
tacalcitol in 3 healthy controls were negative. Lymphocyte stimulation tests were positive to both tacalcitol and
calcitriol (2).

Cross-reactions, pseudo-cross-reactions and co-reactions
A patient sensitized to tacalcitol cross-reacted to calcitriol (2). Four patients sensitized to calcipotriol did not cross-
react to tacalcitol (3,4).

LITERATURE
1   The data in the section 'General' may have been obtained from literature discussed in this chapter, but mostly
     also or exclusively from one or more of the following online sources: ChemIDPlus Advanced, PubChem,
     DrugBank, RxList, Drug Central, Drugs.com, and Wikipedia
2   Kimura K, Katayama I, Nishioka K. Allergic contact dermatitis from tacalcitol. Contact Dermatitis 1995;33:441-442
3   Foti C, Carnimeo L, Bonamonte D, Conserva A, Casulli C, Angelini G. Tolerance to calcitriol and tacalcitol in three
     patients with allergic contact dermatitis to calcipotriol. J Drugs Dermatol 2005;4:756-759
4   Zollner TM, Ochsendorf FR, Hensel O, Thaci D, Diehl S, Kalveram CM, et al. Delayed-type reactivity to calcipotriol
     without cross-sensitization to tacalcitol. Contact Dermatitis 1997;37:251

Chapter 3.327 TACROLIMUS

IDENTIFICATION

Description/definition	: Tacrolimus is the macrolide lactam that conforms to the structural formula shown below
Pharmacological classes	: Immunosuppressive agents; calcineurin inhibitors
IUPAC name	: (1R,9S,12S,13R,14S,17R,18E,21S,23S,24R,25S,27R)-1,14-Dihydroxy-12-[(E)-1-[(1R,3R,4R)-4-hydroxy-3-methoxycyclohexyl]prop-1-en-2-yl]-23,25-dimethoxy-13,19,21,27-tetramethyl-17-prop-2-enyl-11,28-dioxa-4-azatricyclo[22.3.1.0^{4,9}]octacos-18-ene-2,3,10,16-tetrone;hydrate
Other names	: Tacrolimus hydrate; tacrolimus monohydrate; Tsukubaenolide hydrate
CAS registry number	: 109581-93-3
EC number	: Not available
Merck Index monograph	: 10425
Patch testing	: 5% alcohol
Molecular formula	: $C_{44}H_{71}NO_{13}$

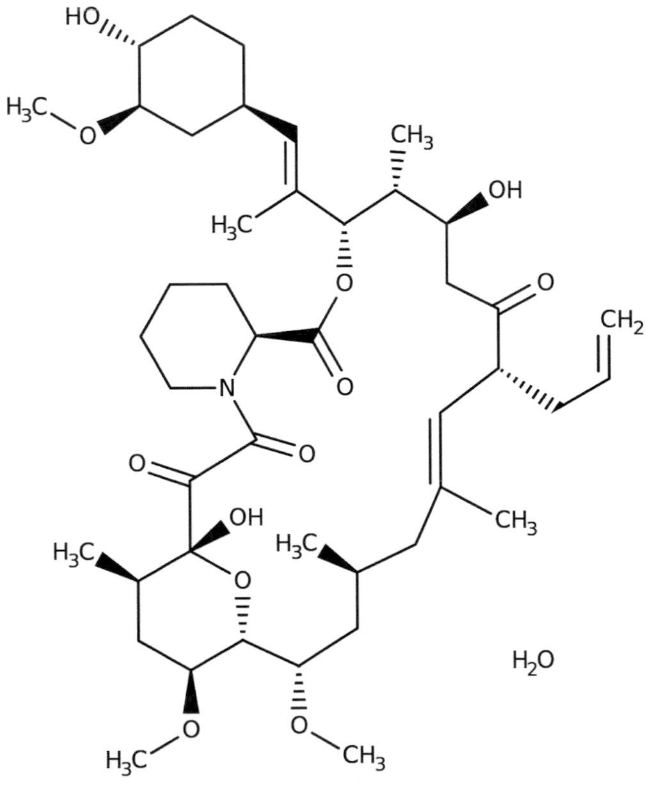

GENERAL

Tacrolimus is a macrolide immunosuppressive drug obtained from the fermentation broth of a Japanese soil sample that contained the bacterium *Streptomyces tsukubaensis*. Tacrolimus binds to the FKBP-12 protein and forms a complex with calcium-dependent proteins, thereby inhibiting calcineurin phosphatase activity and resulting in decreased cytokine production. This agent exhibits potent immunosuppressive activity *in vivo* and prevents the activation of T-lymphocytes in response to antigenic or mitogenic stimulation. Tacrolimus is used orally after allogenic organ transplantation for immunosuppression to reduce the risk of organ rejection. It is also widely utilized topically for the treatment of atopic dermatitis, severe refractory uveitis after bone marrow transplantation, and vitiligo (1).

CONTACT ALLERGY

Case reports

A 9-year-old boy with a history of hay fever and asthma was treated with tacrolimus 0.1% ointment for atopic dermatitis of the antecubital and popliteal fossae, lower legs, and scattered areas of the trunk, which gradually improved his dermatitis. After 8 months, the patient developed new severe erythematous, edematous, oozing plaques at sites of tacrolimus application on the arms, hands, and legs. Initial patch testing was negative with tacrolimus ointment 0.1% 'as is,' tacrolimus 1%, 0.1%, 0.01%, and 0.001% in alcohol, and the excipients of the ointment. Subsequent patch testing with tacrolimus 5% and 2.5% alcohol was strongly positive. Forty controls tested with tacrolimus 5% alc. and 10 with tacrolimus 5% pet. were negative at D2 and D4 or D5.

Later a double-blind ROAT with tacrolimus 0.1% ointment on the right arm and the inactive ointment vehicle left was negative after 6 weeks. At this point, the patient was instructed to begin twice-daily applications of the test materials to right and left preauricular facial skin and to continue the antecubital applications. After one week of preauricular and 7 weeks of antecubital application, mild dermatitis appeared on the tacrolimus-treated side at both locations. The skin was completely normal on the side treated with the inactive vehicle (4). The authors hypothesized that low percutaneous absorption through normal extra-facial skin, due in part to the high molecular weight of tacrolimus, explains the long delay in the positive antecubital provocative use test and the need for a relatively high patch test concentration on the back (4). In addition, the immunosuppressive effects of tacrolimus may also suppress the T cell-mediated inflammatory response of allergic contact dermatitis and positive patch test reactions (5).

Another case of allergic contact dermatitis to tacrolimus was reported from France in 2008, but details are unavailable to the author (3).

Cross-reactions, pseudo-cross-reactions and co-reactions
A patient sensitized to tacrolimus cross-reacted to pimecrolimus (2). Such cross-sensitization has also been observed by others (3).

Cutaneous adverse drug reactions from systemic administration caused by type IV (delayed-type) hypersensitivity
Cutaneous adverse drug reactions from systemic administration of tacrolimus caused by type IV (delayed-type) hypersensitivity, including symmetric drug-related intertriginous and flexural exanthema (SDRIFE) (6), and eczematous drug eruption (7), are planned to be discussed in Volume IV of the *Monographs in Contact Allergy* series on Systemic drugs.

Immediate contact reactions
Immediate contact reactions (contact urticaria) to tacrolimus are presented in Chapter 5.

LITERATURE
1   The data in the section 'General' may have been obtained from literature discussed in this chapter, but mostly also or exclusively from one or more of the following online sources: ChemIDPlus Advanced, PubChem, DrugBank, RxList, Drug Central, Drugs.com, and Wikipedia
2   Shaw DW, Maibach HI, Eichenfield LF. Allergic contact dermatitis from pimecrolimus in a patient with tacrolimus allergy. J Am Acad Dermatol 2007;56:342-345
3   Schmutz JL, Barbaud A, Tréchot P. Contact allergy with tacrolimus then pimecrolimus. Ann Dermatol Venerol 2008; 135:89 (Article in French)
4   Shaw DW, Eichenfield LF, Shainhouse T, Maibach HI. Allergic contact dermatitis from tacrolimus. J Am Acad Dermatol 2004;50:962-965
5   Belsito D, Wilson DC, Warshaw E, Fowler J, Ehrlich A, Anderson B, et al. A prospective randomized clinical trial of 0.1% tacrolimus ointment in a model of chronic allergic contact dermatitis. J Am Acad Dermatol 2006;55:40-46
6   Scherrer M, Araujo MG, Farah K. Tacrolimus-induced symmetric drug-related intertriginous and flexural exanthema (SDRIFE). Contact Dermatitis 2018;78:414-416
7   Saito R, Sawada Y, Nakamura M. Two cases of eczematous drug eruption caused by oral tacrolimus administration. Contact Dermatitis 2017;77:128-130

Chapter 3.328 TERCONAZOLE

IDENTIFICATION

Description/definition : Terconazole is the synthetic triazole derivative that conforms to the structural formula shown below

Pharmacological classes : Antifungal agents

IUPAC name : 1-[4-[[(2R,4S)-2-(2,4-Dichlorophenyl)-2-(1,2,4-triazol-1-ylmethyl)-1,3-dioxolan-4-yl]methoxy]phenyl]-4-propan-2-ylpiperazine

CAS registry number : 67915-31-5

EC number : 267-751-6

Merck Index monograph : 10573

Patch testing : No data available; suggestion: 1% pet., alcohol and MEK (methyl ethyl ketone)

Molecular formula : $C_{26}H_{31}Cl_2N_5O_3$

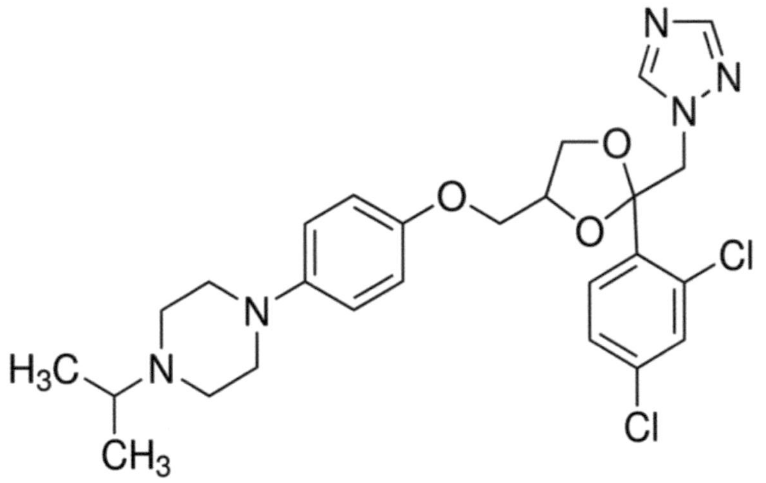

GENERAL

Terconazole is a synthetic triazole derivative structurally related to fluconazole. This antifungal seems to disrupt cell wall synthesis by inhibiting biosynthesis of ergosterol or other sterols, damaging the fungal cell membrane, altering its permeability, and promoting loss of essential intracellular elements. Terconazole is active against *Candida* species and is available in cream and suppository forms (1).

CONTACT ALLERGY

Patch testing in groups of selected patients

In the Mayo Clinic, Rochester, USA, in the period 2003-2010, 56 women with predominantly vulvar symptoms were tested with terconazole 'as is' and there were 4 positive patch test reactions; all were considered to be relevant. Thirty-four patients were patch tested with 'terconazole 3 cream 0.8%' and 3 (9%) reactions were seen, of which 2 were considered to be relevant. It was not stated whether the same patients reacted to the cream and terconazole 'as is' and it is unclear what exactly 'as is' means (pure powder?). In addition, the reactions to terconazole 3 cream may well have been caused by an excipient. It must be concluded that this publication does not seem to prove that one or more patients actually had allergic contact dermatitis from terconazole (2).

LITERATURE

1 The data in the section 'General' may have been obtained from literature discussed in this chapter, but mostly also or exclusively from one or more of the following online sources: ChemIDPlus Advanced, PubChem, DrugBank, RxList, Drug Central, Drugs.com, and Wikipedia

2 O'Gorman SM, Torgerson RR. Allergic contact dermatitis of the vulva. Dermatitis 2013;24:64-72

Chapter 3.329 TESTOSTERONE

IDENTIFICATION

Description/definition   : Testosterone is the androgenic steroid and major product secreted by the Leydig cells of
                             the testis that conforms to the structural formula shown below
Pharmacological classes : Androgens
IUPAC name   : (8R,9S,10R,13S,14S,17S)-17-Hydroxy-10,13-dimethyl-1,2,6,7,8,9,11,12,14,15,16,17-
                             dodecahydrocyclopenta[a]phenanthren-3-one
Other names   : 17β-Hydroxyandrost-4-en-3-one
CAS registry number   : 58-22-0
EC number   : 200-370-5
Merck Index monograph : 10594
Patch testing   : Testosterone esters (enanthate, undecanoate, propionate) 1% alcohol and pet.
Molecular formula   : $C_{19}H_{28}O_2$

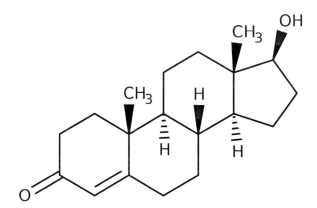

GENERAL

Testosterone is a potent androgenic steroid sex hormone and major product secreted by the Leydig cells of the testis. *In vivo*, testosterone is irreversibly converted to dihydrotestosterone (DHT) in target tissues by the enzyme 5α-reductase. In addition, testosterone is metabolized to estradiol by the enzyme complex aromatase, particularly in the liver and adipose tissue. Testosterone and DHT promote the development and maintenance of male sex characteristics related to the internal and external genitalia, skeletal muscle, and hair follicles; estradiol promotes epiphyseal maturation and bone mineralization (1).

Therapeutic testosterone is a synthetic form of the endogenous hormone; due to rapid metabolism by the liver, therapeutic testosterone is generally administered as an ester derivative. It is used as hormone replacement or substitution of diminished or absent endogenous testosterone. In males testosterone is indicated for management of congenital or acquired hypogonadism, hypogonadism associated with HIV infection, and male climacteric symptoms (andropauze). In females testosterone is utilized for palliative treatment of androgen-responsive, advanced, inoperable, metastatic (skeletal) carcinoma of the breast. Testosterone esters may also be used in combination with estrogens in the management of vasomotor symptoms associated with menopause. It is available in a transdermal therapeutic system (1).

CONTACT ALLERGY

Case series

From January 1990 to June 2008, in Leuven, Belgium, 315 patients were diagnosed with contact allergy to/allergic contact dermatitis from corticosteroids (CSs) from routine patch testing with a baseline series including tixocortol pivalate, budesonide, hydrocortisone butyrate and prednisone caproate, patch testing with patients' own CSs preparations, and testing those with proven contact allergy to a corticosteroid or strongly suspected of CS allergy later with a series of 66 CSs, including two sex hormones (progesterone and testosterone). 71% of the patients had relevant reactions, but these were not specified. In this group of 315 CS allergic patients, 38 had positive patch tests to testosterone propionate 2% alc. (7).

Between 1990 and 2003, in Leuven, Belgium, 92 patients were patch tested for chronic vulval complaints. Fifteen of these women were tested with a series of topical drugs; there was one reaction to testosterone 1% alc. which was relevant (2).

Case reports

A 74-year-old man with the diagnosis of hypogonadism presented with a rash at the site of a 5 mg/24 hour testosterone patch application. Patch tests to testosterone in an ethanol vehicle at various concentrations (1%, 2.5%, 5%, and 10%) were negative, but positive to the same concentrations of testosterone in petrolatum. The patient was switched to testosterone intramuscular injections without any systemic or local dermatitis (3).

A man aged 70 was treated for low testosterone levels and consequent lethargy with testosterone patches 5 mg/24 hour. Several days after his treatment began, he developed erythema at the patch sites, often with induration and blistering, which would take up to 4 weeks to settle. After 3 months' use of the patches, the patient developed generalized eczema, possibly representing systemic contact dermatitis. Patch tests were positive to the testosterone patch, testosterone enanthate and testosterone undecanoate, both 1% in alcohol, but not to testosterone propionate. There were also positive reactions to 2 estradiol patches, which was considered to be the result of cross-sensitivity. Pre-treatment of the skin with a corticosteroid cream 30 minutes before patch application has enabled the patient to continue to use the testosterone patches with minimal skin reaction (4).

A 37-year-old man with Klinefelter's syndrome had been taking testosterone undecanoate 40 mg p.o. daily for 7 years, when his therapy was switched to testosterone TTS 2.5 mg daily. From the 10th day of treatment, he noticed erythema and vesiculation under the patches. Each previous application site flared when a patch was applied elsewhere. Patch tests were positive to the patch (negative to the dummy patch without testosterone and to its ingredients), testosterone i.m injection fluid, testosterone implant, testosterone propionate 1% alcohol and 2.5% in oil, testosterone undecanoate 4% oil and progesterone 1% alcohol. In spite of this, intramuscular injections with testosterone were well-tolerated (5). Whether progesterone cross-reacted was not discussed.

A patient with hypogonadism from Klinefelter's syndrome had been treated with testosterone injections, but was placed on testosterone patches. The first application resulted in irritation followed by a blister and a thick eschar which had the size and configuration of the hormone patch. The lesion had the clinical appearance of an old, untreated full thickness burn. The eschar was excised and the full thickness wound was debrided and covered with a meshed split skin graft. The author considered it likely that this was the result of allergic contact dermatitis, but patch tests were not performed (6). However, irritant dermatitis from the patches is not uncommon (9) and has even caused skin ulceration (8).

Cross-reactions, pseudo-cross-reactions and co-reactions

A patient sensitized to testosterone had positive reactions to 2 patches containing estradiol. *In vivo*, testosterone is metabolized to estradiol by the enzyme complex aromatase, particularly in the liver and adipose tissue. Therefore, cross-reactivity could be anticipated (4). Patients sensitized to corticosteroids may cross-react to testosterone (7).

LITERATURE

1   The data in the section 'General' may have been obtained from literature discussed in this chapter, but mostly also or exclusively from one or more of the following online sources: ChemIDPlus Advanced, PubChem, DrugBank, RxList, Drug Central, Drugs.com, and Wikipedia

2   Nardelli A, Degreef H, Goossens A. Contact allergic reactions of the vulva: a 14-year review. Dermatitis 2004;15:131-136

3   Ta V, Chin WK, White AA. Allergic contact dermatitis to testosterone and estrogen in transdermal therapeutic systems. Dermatitis 2014;25:279

4   Shouls J, Shum KW, Gadour M, Gawkrodger DJ. Contact allergy to testosterone in an androgen patch: control of symptoms by pre-application of topical corticosteroid. Contact Dermatitis 2001;45:124-125

5   Buckley DA, Wilkinson SM, Higgins EM. Contact allergy to a testosterone patch. Contact Dermatitis 1998;39:91-92

6   Bennett NJ. A burn-like lesion caused by a testosterone transdermal system. Burns 1998;24:478-480

7   Baeck M, Chemelle JA, Terreux R, Drieghe J, Goossens A. Delayed hypersensitivity to corticosteroids in a series of 315 patients: clinical data and patch test results. Contact Dermatitis 2009;61:163-175

8   Lawrentschuk N, Fleshner N. Severe irritant contact dermatitis causing skin ulceration secondary to a testosterone patch. ScientificWorldJournal 2009;9:333-338

9   Jordan WP Jr. Allergy and topical irritation associated with transdermal testosterone administration: A comparison of scrotal and nonscrotal transdermal systems. Am J Contact Dermat 1997;8:108-113

Chapter 3.330 TETRACAINE

IDENTIFICATION

Description/definition   : Tetracaine is the 2-(dimethylamino)ethyl ester of *p*-butylaminobenzoic acid; it conforms
                                       to the structural formula shown below
Pharmacological classes : Anesthetics, local
IUPAC name   : 2-(Dimethylamino)ethyl 4-(butylamino)benzoate
Other names   : Benzoic acid, 4-(butylamino)-, 2-(dimethylamino)ethyl ester; amethocaine
CAS registry number   : 94-24-6
EC number   : 202-316-6
Merck Index monograph : 10603
Patch testing   : Hydrochloride 5.0% pet. (Chemotechnique); hydrochloride 1% pet. (SmartPracticeCanada,
                                       SmartPracticeEurope)
Molecular formula   : $C_{15}H_{24}N_2O_2$

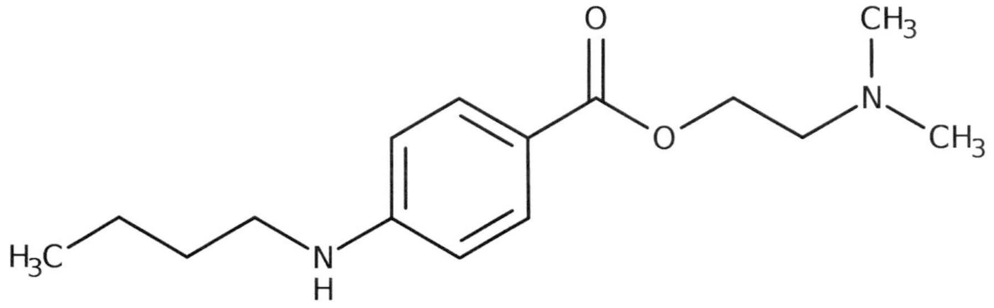

GENERAL

Tetracaine is a benzoic acid ester-type local anesthetic. It is present in many antihemorrhoidal creams and ointments, eardrops, eye drops and preparations for analgesia of the skin and mucosae (e.g. lubricant gels). In pharmaceutical products, both tetracaine and tetracaine hydrochloride (CAS number 136-47-0, EC number 205-248-5, molecular formula $C_{15}H_{25}ClN_2O_2$) may be employed (1).

CONTACT ALLERGY

General

In animal experiments, tetracaine was found to be a strong sensitizer (38). Indeed, tetracaine (amethocaine) is a well-known cause of allergic contact dermatitis, particularly in patients using ointments for pruritus ani and pruritus vulvae (14,15,24,25,30). Other products containing tetracaine which have caused allergic contact dermatitis include lubricating gels and ointments for probes, catheters etc. (15,33,34,35), ear drops (29,31) and eye drops (5,8,13,30).

Systemic contact dermatitis has been observed in some individuals from resorption of tetracaine through the perianal skin and anal mucosa (14), and through the mucosa of the urethra (15). Even a digital rectal exam with a urological lubricant containing tetracaine has caused a systemic reaction (35).

Occupational allergic contact dermatitis from tetracaine in health personnel has been observed several times, including in an otorhinolaryngologist (16), ophthalmologist (13), two nurses (17,18), a chiropodist (21), two oculists (26) and especially dentists (19,20,22,24). It is likely that tetracaine has been gradually replaced with less sensitizing anesthetics such as lidocaine and prilocaine, which should reflect on the number of sensitizations.

Contact allergy in the general population

In Germany, for the period 1995-2004, the population-based relative incidence (RI) of contact sensitization to tetracaine (cases/100,000 defined daily doses (DDDs) per year) was estimated to be 107.8. In the group of local anesthetics, the RI ranged from 1.5 (lidocaine) to 413.9 (benzocaine) (7).

Patch testing in groups of patients

The results of studies testing tetracaine HCl in consecutive patients suspected of contact dermatitis (routine testing) and of patch testing in groups of *selected* patients (individuals with chronic otitis externa, women with 'vulval problems') are shown in table 3.330.1.Tetracaine HCl 1% pet. was part of the NACDG screening series in the USA and Canada in the period 2001-2004 and yielded 0.2% and 0.6% positive reactions in routine testing (2,3). Probably because of these low prevalences and also because the reactions were hardly ever relevant, tetracaine was removed from the screening series again.

In 3 studies with selected patient groups, the frequencies of sensitization ranged from 0.8% (6) to 9% (4) (table 3.330.1). This high rate was seen in a small series of 33 patients suffering from chronic otitis externa. It was not mentioned whether the 3 positive patch tests to tetracaine were relevant to the patients' complaints (4).

Table 3.330.1 Patch testing in groups of patients

Years and Country	Test conc. & vehicle	Number of patients tested | positive (%)		Selection of patients (S); Relevance (R); Comments (C)	Ref.
Routine testing					
2003-4 USA, Canada	1% pet.	5137	10 (0.2%)	R: not stated	2
2001-2004 USA		10,061	38 (0.4%)	R: definite + probable relevance : 5%	36
2001-02 USA, Canada	1% pet.	4891	(0.6%)	R: definite + probable relevance: 4%	3
Testing in groups of selected patients					
1995-2004 IVDK	5% pet.	7576	(0.8%)	S: not stated; R: not stated	6
1992-1997 IVDK		33	3 (9%)	S: patients with chronic otitis externa; R: not stated	4
1992-1994 UK	1% pet.	69	1 (1%)	S: women with 'vulval problems'; R: 58% for all allergens together	9

Case series

In Leuven, Belgium, between 1990 and 2017, 16,065 patients were investigated for contact allergy and 118 (0.7%) showed positive patch test reactions to topical ophthalmic medications and/or to their ingredients. Tetracaine was tested in 107 patients and was the allergen in eye medications in one. There were no reactions to tetracaine in other types of medications (5).

In Pamplona, Spain, in one year's time (1992), 13 patients were diagnosed with contact allergy to ophthalmic medications, nearly 3% of all patients investigated for suspected contact dermatitis. There were 2 reactions to tetracaine 1% pet., both in individuals who had previously used tetracaine eye drops (8).

From Milan, Italy, a small case series of 2 patients with allergic contact dermatitis from tetracaine was reported in 1999 (30). A 53-year-old woman had a fissure and an erythematous and edematous lesion of the perianal skin after using an ointment containing tetracaine. Patch tests were positive to the ointment and tetracaine 5% pet. A man aged 59 also developed erythematous and edematous lesions of the (peri)anal region from application of an ointment containing tetracaine. Previously, he had suffered edema of the eyelids after having used tetracaine eye drops. Patch tests were positive to the ointment and tetracaine 5% pet. (30).

In a study in the United Kingdom in the period 1988-1998, 63 patients reacting to a mix of benzocaine, tetracaine and dibucaine were tested with the three active ingredients and there were 23 reactions to tetracaine, 22 to benzocaine, and 28 to dibucaine. Of the entire group, 55% were interpreted as either of current or of past relevance (10).

Of 13 patients reacting to an investigational 'caine mix' in a study from the EECDRG and   tested   with   its anesthetic components, 7 reacted to dibucaine 2.5% pet., 5 to benzocaine 5% pet., and 2 to amethocaine (tetracaine) 2.5% pet. Relevance was not specified (11). In a similar study from the United Kingdom from before 1988, 40 patients reacting to a 'caine mix' containing 5% benzocaine, 1% dibucaine (cinchocaine) and 1% tetracaine (amethocaine) were tested with its ingredients and there were 16 reactions to tetracaine, 19 to benzocaine and 12 to dibucaine. The most involved primary sites were the legs (29%) and the anogenital region (27%). Relevance for the entire group was a little over 50%, past relevance 19% (12).

Sixteen patients allergic to tetracaine were reported from Spain in 1981. Six had occupational allergic contact dermatitis, mostly dentists (20). Between 1971 and 1976, in a hospital in London, 23 patients (13 women) were found to be sensitive to tetracaine. Nineteen had used a cream containing 0.8% tetracaine and 2 other anesthetics. Six of thirteen women had used the cream for pruritus vulvae and 9 of the 10 men for pruritus ani (24).

In the first half of the 1960s, one investigator from the United Kingdom patch tested 20 patients with dermatitis due to local anesthetic ointments with tetracaine 1% pet. Thirteen of the 16 with anogenital pruritus and three out of 4 with other patterns of eczema were positive (25).

Case reports

A 56-year-old man presented with an indurated, brightly erythematous and well-demarcated plaque with small vesicles and pustules on the buttocks and perianal skin, a diffuse symmetrical erythematous papular eruption in the large body folds, and erythematous exudative plaques with vesicles on the trunk, two days after the application of tetracaine 1% ointment on the perianal area for the treatment of haemorrhoids. Two weeks before, the patient had applied dibucaine 1% ointment for the same purpose. He had used several anti-hemorrhoidal ointments for many years. Patch tests were positive to both ointments, tetracaine HCl 1% pet., dibucaine HCl 5% pet. and caine mixes II and III, both of which contain tetracaine and dibucaine. It was concluded that this was a case of systemic contact dermatitis manifesting as the baboon syndrome, most likely caused by tetracaine HCl absorbed through the perianal skin and anal mucosa (14).

Another case of systemic contact dermatitis (with limited manifestations) was seen in a 71-year-old man, who had previously experienced severe pruritus and fissures in the perianal area after using antihemorrhoidal ointments. He developed dermatitis in the groin area 2 days after insertion of a probe lubricated with tetracaine ointment in the urethra. Patch tests were positive to the ointment and a caine mix containing tetracaine (15). The authors diagnosed 'systemic contact-type dermatitis due to caine mix', which is obviously incorrect.

A third case of systemic allergic contact dermatitis was in a 72-year-old man who presented with erythematous and edematous papules and plaques with superficial desquamation involving the neck, superior thoracic region, both axillae, elbow flexures, and perianal area. The day before the eruption began, a digital rectal exam with a urological lubricant had been performed. Patch tests were positive to the caine mix (cinchocaine [dibucaine] 1%, tetracaine 1%, benzocaine 5%) and tetracaine. Tetracaine was contained in the urological lubricant (35).

A 60-year-old woman complained of an allergic reaction to ophthalmic anesthetic drops instilled before laser eye surgery. On examination, there was an eczematous dermatitis of the upper and lower eyelids. She had positive patch tests to proparacaine and tetracaine eye drops, to which she had been exposed, and to tetracaine 1% pet. Both ophthalmic solutions were preserved with benzalkonium chloride, but this had been negative in patch testing at 0.01% water (13).

A boy and a girl, both 12 years old, had dermatitis of the pinnae and skin around the ears. They had used ear drops containing tetracaine. Patch tests were positive to the ear drops 'as is' and to tetracaine, apparently tested pure. The female patient co-reacted to cornecaine (hydroxycaine; [3-(dimethylamino)-2-hydroxypropyl] 4-(propylamino)ben-zoate) and stadacaine (butoxycaine hydrochloride; 2-(diethylamino)ethyl 4-butoxybenzoate;hydrochloride). These are both benzoic acid esters and the reactions were considered to be cross-reactions (29).

One patient in a group of 23 with otitis externa had allergic contact dermatitis from tetracaine in a combination preparation with a corticosteroid, neomycin and polymyxin B (28). An 89-year-old woman presented with eczematous lesions affecting the ear canal with secondary spread to the face and neck that had begun after she had started using ear drops containing tetracaine HCl. Patch tests were positive to these ear drops, to a caine mix (benzocaine, tetracaine HCl, dibucaine HCl) and tetracaine HCl 5% pet., but negative to the excipients of the product (31).

A 52-year-old woman with a 3-month' history of pruritus vulvae had been repeatedly operated on because of urinary incontinence and frequently catheterized. She presented with erythematous scaly plaques on the vulva, perineal and perianal regions, and inner thighs. Patch tests were positive to an urological lubricant containing tetracaine, to the caine mix and to tetracaine HCl and negative to all topical medicaments previously used by the patient. She greatly improved when the lubricant was substituted. When, accidentally, a new nurse used the original lubricant again, the symptoms returned (32). Allergy to tetracaine in the same ointment caused allergic contact balanitis in a 41-year-old man, who had a urethral stenosis which was treated by periodic dilatations with a PVC dilator lubricated with the ointment (34).

A 58-year-old man who had a urostoma in the supraumbilical area developed allergic contact dermatitis with pruritus, erythema and scaly, exudative plaques in the peristomal skin. This proved to be caused by tetracaine present in a urological ointment used to lubricate the urine drain, which was inserted in the stoma 3-5 times/day (33).

An additional case of allergic contact dermatitis to tetracaine (with unusual presentation) has been reported in Italian literature (39) and a case of systemic contact dermatitis from the local anesthetic in a lubricant gel for digital rectal examination in French literature (40), but details are unknown. Case reports and case series of allergy to tetracaine published in early literature can be found in ref. 37.

Occupational allergic contact dermatitis

A 46-year-old otorhinolaryngologist presented with severe eczema on the second and third fingers of his right hand which was very troubling for his work as a surgeon. He reported that, prior to surgery, he would put – without protective gloves – nose drops containing tetracaine HCl 2% and phenylephrine HCl 0.01% on a tampon, which was then inserted into the nose of the patient. Patch tests were positive to the nose drops and to tetracaine HCl 5% pet., but negative to phenylephrine HCl 10% water, benzocaine and lidocaine. The patient was diagnosed with occupational allergic contact dermatitis caused by tetracaine (16).

A 58-year-old ophthalmologist, known with occupational allergic contact dermatitis of the hands from proparacaine, was advised to change to tetracaine, to which he had a negative patch test. However, later, he began to suffer from sporadic bouts of painful fissuring of his fingertips again. Repeat testing confirmed the allergy to proparacaine, but he now also reacted to tetracaine 1% eye drops. He continued his work using tetracaine drops without glove protection and had persistent mild finger pad dermatitis (13).

Other cases of occupational allergic contact dermatitis from tetracaine, usually of the fingers, have included a nurse in the beauty industry who applied tetracaine gel before collagen injections (17), a laser-clinic nurse (18), a chiropodist (21), 2 oculists (26) and – most frequently – dentists (19,20,22,24)

Cross-reactions, pseudo-cross-reactions and co-reactions
Cross-reactions to other 'para-compounds' (e.g. *p*-phenylenediamine, *p*-aminobenzoic acid, toluene-2,5-diamine) are infrequent (20,21,22,23,27,32,34). However, cross-reactions to procaine (27) and other PABA ester local anesthetics including benzocaine, hydroxycaine and butoxycaine hydrochloride (29) may occur.

LITERATURE

1 The data in the section 'General' may have been obtained from literature discussed in this chapter, but mostly also or exclusively from one or more of the following online sources: ChemIDPlus Advanced, PubChem, DrugBank, RxList, Drug Central, Drugs.com, and Wikipedia

2 Warshaw EM, Belsito DV, DeLeo VA, Fowler JF Jr, Maibach HI, Marks JG, et al. North American Contact Dermatitis Group patch-test results, 2003-2004 study period. Dermatitis 2008;19:129-136

3 Pratt MD, Belsito DV, DeLeo VA, Fowler JF Jr, Fransway AF, Maibach HI, et al. North American Contact Dermatitis Group patch-test results, 2001-2002 study period. Dermatitis 2004;15:176-183

4 Hillen U, Geier J, Goos M. Contact allergies in patients with eczema of the external ear canal. Results of the Information Network of Dermatological Clinics and the German Contact Allergy Group. Hautarzt 2000;51:239-243 (article in German)

5 Gilissen L, De Decker L, Hulshagen T, Goossens A. Allergic contact dermatitis caused by topical ophthalmic medications: Keep an eye on it! Contact Dermatitis 2019;80:291-297

6 Menezes de Padua CA, Uter W, Schnuch A. Contact allergy to topical drugs: prevalence in a clinical setting and estimation of frequency at the population level. Pharmacoepidemiol Drug Saf 2007;16:377-384

7 Menezes de Padua CA, Schnuch A, Nink K, Pfahlberg A, Uter W. Allergic contact dermatitis to topical drugs – Epidemiological risk assessment. Pharmacoepidemiol Drug Saf 2008;17:813-821

8 Tabar AI, García BE, Rodríguez A, Quirce S, Olaguibel JM. Etiologic agents in allergic contact dermatitis caused by eyedrops. Contact Dermatitis 1993;29:50-51

9 Lewis FM, Harrington CI, Gawkrodger DJ. Contact sensitivity in pruritus vulvae: a common and manageable problem. Contact Dermatitis 1994;31:264-265

10 Sidhu SK, Shaw S, Wilkinson JD. A 10-year retrospective study on benzocaine allergy in the United Kingdom. Am J Cont Dermat 1999;10:57-61

11 Wilkinson JD, Andersen KE, Lahti A, Rycroft RJ, Shaw S, White IR. Preliminary patch testing with 25% and 15% 'caine'-mixes. The EECDRG. Contact Dermatitis 1990;22:244-245

12 Beck MH, Holden A. Benzocaine—an unsatisfactory indicator of topical local anaesthetic sensitization for the U.K. Br J Dermatol 1988;118:91-94

13 Dannaker CJ, Maibach HI, Austin E. Allergic contact dermatitis to proparacaine with subsequent cross-sensitization to tetracaine from ophthalmic preparations. Dermatitis 2001;12:177-179

14 Matos-Pires E, Pina-Trincão D, Brás S, Lobo L. Baboon syndrome caused by anti-haemorrhidal ointment. Contact Dermatitis 2018;78:170-171

15 Huerta Brogeras M, Avilés JA, González-Carrascosa M, de la Cueva P, Suárez R, Lázaro P. Dermatitis sistémica de contacto por tetracaínas [Tetracaine-induced systemic contact-type dermatitis]. Allergol Immunopathol (Madr) 2005;33:112-114 (Article in Spanish)

16 Opstrup MS, Sørensen HB, Zachariae C. Occupational allergic contact dermatitis caused by tetracaine in an otorhinolaryngologist. Contact Dermatitis 2017;76:55-57

17 Connolly M, Mehta A, Sansom JE, Dunnill MG. Allergic contact dermatitis from tetracaine in the beauty industry. Contact Dermatitis 2004;51:95-96

18 Dawe RS, Watt D, O'Neill S, Forsyth A. A laser-clinic nurse with allergic contact dermatitis from tetracaine. Contact Dermatitis 2002;46:306

19 Gall H. Allergien auf zahnärztliche Werkstoffe und Dentalpharmaka. Hautarzt 1983;34:326-331 (Article in German, data cited in ref. 18)

20 García-Pérez A, Conde Salazar L, Guimaraens D, García-Bravo B, López Correcher B. La sensibilidad de conacto a ametocaina. [Contact sensitivity to amethocaine]. Actas Dermosifiliogr 1981;72:441-448 (Article in Spanish, data cited in ref. 18)

21 Condé-Salazar L, Llinás MG, Guimaraens D, Romero L. Occupational allergic contact dermatitis from amethocaine. Contact Dermatitis 1988;19:69-70

22 Condé-Salazar L, Guimaraens D, Romero L, García-Pérez A. Dermatitis alérgica de contacto profesional por anestésicos en dentistas. Medicina y Seguridad del Trabajo 1982:XXX:39-42 (Article in Spanish, data cited in ref. 21)

23 Calnan CD, Stevenson CJ. Studies in contact dermatitis: dental materials. Transactions of the St. John's Hospital Dermatological Society 1963;49:9-26 (data cited in ref. 21).

24 Cronin E. Contact Dermatitis. Edinburgh: Churchill Livingstone, 1980:194-195

25 Wilson HTH. Dermatitis from anaesthetic ointments. Practitioner 1966;197:673-677 (Data cited in ref. 24)

26 Rebandel P, Rudzki E. Occupational contact sensitivity in oculists. Contact Dermatitis 1986;15:92

27 Kalveram K, Günnewig W, Wehling K, Forck G. Tetracaine allergy: cross-reactions with para-compounds? Contact Dermatitis 1978;4:376

28 Lembo G, Nappa P, Balato N, Pucci V, Ayala F. Contact sensitivity in otitis externa. Contact Dermatitis 1988;19:64-65

29 Pevny I, Brennenstuhl, Razmskas G. Patch testing in children (II). Results and case reports. Contact Dermatitis 1984;11:302-310

30 Lodi A, Ambonati M, Coassini A, Kouhdari Z, Palvarini M, Crosti C. Contact allergy to 'caines' caused by anti-hemorrhoidal ointments. Contact Dermatitis 1999;41:221-222

31 García-Gavín J, Alonso-González J, Gutiérrez-González E, Álvarez-Pérez A, Fernández-Redondo V, Toribio J. Allergic contact dermatitis caused by tetracaine contained in otic drops. Contact Dermatitis 2011;65:175-176

32 Villarreal LC, de Frutos FJ, Del Prado Sánchez Caminero M, Sebastián FV. Perineal allergic contact dermatitis due to tetracaine in a urological lubricant. Contact Dermatitis 2004;51:321-322

33 Fernández-Redondo V, León A, Santiago T, Toribio J. Allergic contact dermatitis from local anaesthetic on peristomal skin. Contact Dermatitis 2001;45:358

34 Sánchez-Pérez J, Córdoba S, Cortizas CF, García-Díez A. Allergic contact balanitis due to tetracaine (amethocaine) hydrochloride. Contact Dermatitis 1998;39:268

35 Caro-Gutiérrez D, Gómez-de la Fuente E, Pampín-Franco A, Ascanio-Armada L, López-Estebaranz JL. Systemic contact dermatitis due to amethocaine following digital rectal examination. Dermatol Online J 2015 Mar 1;21(5)

36 Warshaw EM, Schram SE, Belsito DV, DeLeo VA, Fowler JF Jr, Maibach HI, et al. Patch-test reactions to topical anesthetics: retrospective analysis of cross-sectional data, 2001 to 2004. Dermatitis 2008;19:81-85

37 Lane CG, Luikart R. Dermatitis from local anesthetics, with a review of one hundred and seven cases from the literature. J Am Med Assoc 1951;146:717-720

38 Kalveram K, Semmelmann J, Forck G. Experimental animal study of the allergenicity of tetracaine. Contact Dermatitis 1978;4:374

39 Bruscino N, Corradini D, Francalanci S, Palleschi GM. A case of allergic contact dermatitis due to tetracaine with unusual presentation. G Ital Dermatol Venereol 2015;150:266-267 (Article in Italian)

40 Schmutz JL. Generalised contact eczema: Care is needed in selecting a lubricant for digital rectal examination. Ann Dermatol Venereol 2015;142:624-625.

Chapter 3.331 TETRACHLORODECAOXIDE

IDENTIFICATION

Description/definition : Tetrachlorodecaoxide is the non-metal chlorite that conforms to the structural formula shown below
Pharmacological classes : Radiation-protective agents; wound-healing agent
IUPAC name : Molecular oxygen;tetrachlorite;hydrate
Other names : Oxoferin ®; chlorine dioxide generated from tetrachlorodecaoxide complex (tcdo) by acidification; tetrachlorodecaoxygen
CAS registry number : 92047-76-2
EC number : 420-970-2
Patch testing : Commercial product 'as is' (2)
Molecular formula : $Cl_4H_2O_{11}^{-4}$

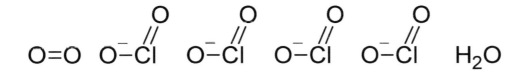

GENERAL

Tetrachlorodecaoxide is an immunomodulating drug that consists of chlorine-oxygen complexes and that may be effective in treating certain cancers. It is being investigated for use/treatment in acquired immune deficiency syndrome (AIDS) and aids-related infections, cancer/tumors (unspecified), HIV infection, and inflammatory disorders (unspecified). It is also used as a stimulant of wound healing (2). Tetrachlorodecaoxide-containing pharmaceuticals may be available in a few countries (1).

CONTACT ALLERGY

Case report

A 77-year-old woman with post-thrombotic syndrome had a large leg ulcer which was treated with topical tetra-chlorodecaoxide twice daily. On the 45th day of this therapy, the ulcer was reduced to approximately half of the original size, but then there was no further improvement. The ulcer gradually increased in size, and the surrounding skin was inflamed. Within a few days, itchy maculopapular lesions developed. A swab from the base of the ulcer grew *Staphylococcus aureus*. A patch test with 0.05 ml of tetrachlorodecaoxide yielded a positive reaction (D2 +, D3 and D4 ++) (2).

LITERATURE

1 The data in the section 'General' may have been obtained from literature discussed in this chapter, but mostly also or exclusively from one or more of the following online sources: ChemIDPlus Advanced, PubChem, DrugBank, RxList, Drug Central, Drugs.com, and Wikipedia
2 Schallreuter KU, Gupta MA. Allergic contact dermatitis from tetrachlordecaoxide (Oxofèrin). Contact Dermatitis 1987;17:253-254

Chapter 3.332 TETRACYCLINE

IDENTIFICATION

Description/definition : Tetracycline is the naphthacene antibiotic that conforms to the structural formula shown below

Pharmacological classes : Anti-bacterial agents; protein synthesis inhibitors

IUPAC name : (4S,4aS,5aS,6S,12aR)-4-(Dimethylamino)-1,6,10,11,12a-pentahydroxy-6-methyl-3,12-dioxo-4,4a,5,5a-tetrahydrotetracene-2-carboxamide

CAS registry number : 60-54-8

EC number : 200-481-9

Merck Index monograph : 10611

Patch testing : Hydrochloride, 2% pet. (SmartPracticeCanada, SmartPracticeEurope)

Molecular formula : $C_{22}H_{24}N_2O_8$

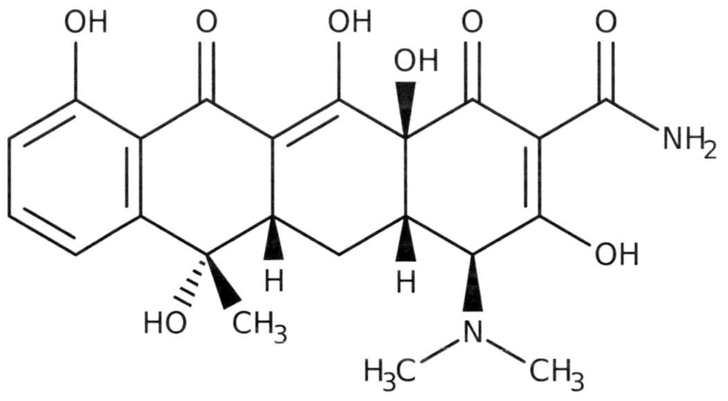

GENERAL

Tetracycline is a broad-spectrum naphthacene antibiotic produced semisynthetically from chlortetracycline, an antibiotic isolated from the bacterium *Streptomyces aureofaciens*. It exerts a bacteriostatic effect on bacteria by binding reversibly to the bacterial 30S ribosomal subunit and blocking incoming aminoacyl tRNA from binding to the ribosome acceptor site. It also binds to some extent to the bacterial 50S ribosomal subunit and may alter the cytoplasmic membrane causing intracellular components to leak from bacterial cells. Tetracycline is used to treat a wide variety of infections caused by susceptible bacteria and is also widely utilized in the treatment of acne vulgaris and acne conglobata. In pharmaceutical products, tetracycline is employed as tetracycline hydrochloride (CAS number 64-75-5, EC number 200-593-8, molecular formula $C_{22}H_{25}ClN_2O_8$) (1).

CONTACT ALLERGY

Patch testing in groups of selected patients

In the period 2004-2008, the IVDK (Information Network of Departments of Dermatology: Germany, Austria, Switzerland) tested patients with anogenital dermatoses (number unspecified) with a medicament series and 3.7% had a positive reaction to tetracycline. The relevance of the positive patch tests was not mentioned (4). In Taiwan, between 1978 and 2005, 603 patients suspected of contact allergy to topical medicaments were patch tested with tetracycline (test concentration not stated) and there were 5 (0.8%) positive reactions. The relevance was not specified for individual drugs, but 65% of the reactions to all medicaments together were considered to be relevant (2).

Occupational allergic contact dermatitis

In a group of 107 workers in the pharmaceutical industry with dermatitis, investigated in Warsaw, Poland, before 1989, 6 reacted to tetracycline, tested 10% pet. (3). Eralier in Warsaw, Poland, in the period 1979-1983, of 27 pharmaceutical workers diagnosed with occupational allergic contact dermatitis from antibiotics, seven had positive patch tests to tetracycline (ampoule content) (5).

Possibly, occupational allergy to tetracycline was also described in German literature (6).

LITERATURE

1   The data in the section 'General' may have been obtained from literature discussed in this chapter, but mostly also or exclusively from one or more of the following online sources: ChemIDPlus Advanced, PubChem, DrugBank, RxList, Drug Central, Drugs.com, and Wikipedia

2   Shih Y-H, Sun C-C, Tseng Y-H, Chu C-Y. Contact dermatitis to topical medicaments: a retrospective study from a medical center in Taiwan. Dermatol Sinica 2015;33:181-186

3   Rudzki E, Rebandel P, Grzywa Z. Contact allergy in the pharmaceutical industry. Contact Dermatitis 1989;21:121-122

4   Bauer A. Contact sensitization in the anal and genital area. Curr Probl Dermatol 2011;40:133-141

5   Rudzki E, Rebendel P. Contact sensitivity to antibiotics. Contact Dermatitis 1984;11:41-42

6   Schwarting HH. Occupational tetracycline allergy. Derm Beruf Umwelt 1983;31:130 (Article in German)

Chapter 3.333 TETRAHYDROZOLINE

IDENTIFICATION

Description/definition	: Tetrahydrozoline is the member of the imidazolines and a carboxamidine that conforms to the structural formula shown below
Pharmacological classes	: Nasal decongestants; sympathomimetics; ophthalmic solutions
IUPAC name	: 2-(1,2,3,4-Tetrahydronaphthalen-1-yl)-4,5-dihydro-1H-imidazole
Other names	: Tetryzoline
CAS registry number	: 84-22-0
EC number	: 201-522-3
Merck Index monograph	: 10634
Patch testing	: No data available; suggested 1% in pet., water and alcohol
Molecular formula	: $C_{13}H_{16}N_2$

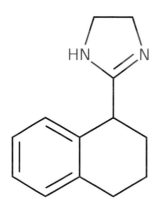

GENERAL

Tetrahydrozoline is an imidazole derivative with sympathomimetic activity. Applied locally to the eye or nose, tetrahydrozoline binds to and activates alpha-adrenergic receptors, resulting in vasoconstriction and decreased nasal and ophthalmic congestion. It is indicated for temporary relief of discomfort and redness of the eye or nose due to minor irritations. In pharmaceutical products, tetrahydrozoline is employed as tetrahydrozoline hydrochloride (CAS number 522-48-5, EC number 208-329-3, molecular formula $C_{13}H_{17}ClN_2$) (1).

CONTACT ALLERGY

Case reports and case series

In Ferrara, Italy, over a 65-month period before 2005, 50 patients affected by periorbital dermatitis while using topical ocular products were patch tested, including with their own ophthalmic medications (n=210). There was one reaction to tetrahydrozoline (tetryzoline) and pheniramine. The active ingredients were not tested separately, but contact allergy to the excipients and preservatives was excluded by patch testing (2).

LITERATURE

1   The data in the section 'General' may have been obtained from literature discussed in this chapter, but mostly also or exclusively from one or more of the following online sources: ChemIDPlus Advanced, PubChem, DrugBank, RxList, Drug Central, Drugs.com, and Wikipedia
2   Corazza M, Massieri LT, Virgili A. Doubtful value of patch testing for suspected contact allergy to ophthalmic products. Acta Derm Venereol 2005;85:70-71

Chapter 3.334 THIABENDAZOLE

IDENTIFICATION

Description/definition : Thiabendazole is the 2-substituted benzimidazole that conforms to the structural formula shown below
Pharmacological classes : Anthelmintics
IUPAC name : 4-(1*H*-Benzimidazol-2-yl)-1,3-thiazole
Other names : 2-(4-Thiazolyl)-1*H*-benzimidazole
CAS registry number : 148-79-8
EC number : 205-725-8
Merck Index monograph : 10710
Patch testing : 2% pet.
Molecular formula : $C_{10}H_7N_3S$

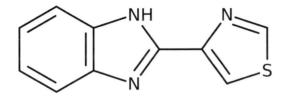

GENERAL

Thiabendazole is a 2-substituted benzimidazole with fungicidal and anthelminthic (parasiticidal) properties. It is also a chelating agent, which means that this agent is used medicinally to bind metals in cases of metal poisoning, such as lead, mercury or antimony. Thiabendazole is indicated for the treatment of strongyloidiasis (threadworm), cutaneous larva migrans (creeping eruption), visceral larva migrans, and trichinosis.

CONTACT ALLERGY

Case report

A 40-year-old man presented with severe acute eczema on sun-exposed areas of the face and neck. He had been treated for rosacea with topical thiabendazole 10% in o/w emulsion for 3 months. The localization of the dermatitis coincided more or less with the areas where the patient had applied the thiabendazole cream. The MEDs for UVA and UVB were normal. Subsequently, patch and photopatch testing was performed, yielding positive reactions to the 10% thiabendazole cream on both patch and photopatch testing. Subsequently, the patient was tested with thiabendazole 2% pet. and an imidazole series to detect any cross-reactivity. Only the thiabendazole 2% pet. patch and photopatch tests were positive, but clearly stronger on the irradiated areas. The patient was therefore diagnosed with photoaugmented allergic contact dermatitis to thiabendazole (2).

Occupational allergic contact dermatitis
In Italy, during 1986-1988, 204 animal feed mill workers (191 men, 13 women) were patch tested with a large number of animal feed additives. There were two reactions to thiabendazole 4% pet. in a group of 36 subjects with clinical complaints (dermatitis or pruritus sine materia) and one reaction in the group of 168 individuals without skin complaints. All reactions were considered to be relevant. In one patient with evident contact dermatitis, there was a positive 'stop-start test' of his working activity (2,3).

Cross-reactions, pseudo-cross-reactions and co-reactions

Not to econazole, bifonazole, metronidazole, mebendazole, miconazole and ketoconazole (all 2% pet.) (1).

PHOTOSENSITIVITY

See the section 'Case report' above for a case of photoaugmented allergic contact dermatitis to thiabendazole.

LITERATURE

1 Izu R, Aguirre A, Goicoechea A, Gardeazabal J, Díaz Pérez JL. Photoaggravated allergic contact dermatitis due to topical thiabendazole. Contact Dermatitis 1993;28:243-244
2 Mancuso G, Staffa M, Errani A, Berdondini RM, Fabbri P. Occupational dermatitis in animal feed mill workers. Contact Dermatitis 1990;22:37-41
3 Mancuso G. Topical thiabendazole allergy. Contact Dermatitis 1994;31:207

Chapter 3.335 THIAMINE

IDENTIFICATION

Description/definition	: Thiamine is the essential vitamin, belonging to the vitamin B family, that conforms to the structural formula shown below
Pharmacological classes	: Vitamin B complex
IUPAC name	: 2-[3-[(4-Amino-2-methylpyrimidin-5-yl)methyl]-4-methyl-1,3-thiazol-3-ium-5-yl]ethanol
Other names	: Vitamin B_1
CAS registry number	: 70-16-6; 59-43-8 (chloride)
EC number	: 200-425-3 (chloride)
Merck Index monograph	: 10717
Patch testing	: 10% water
Molecular formula	: $C_{12}H_{17}N_4OS+$

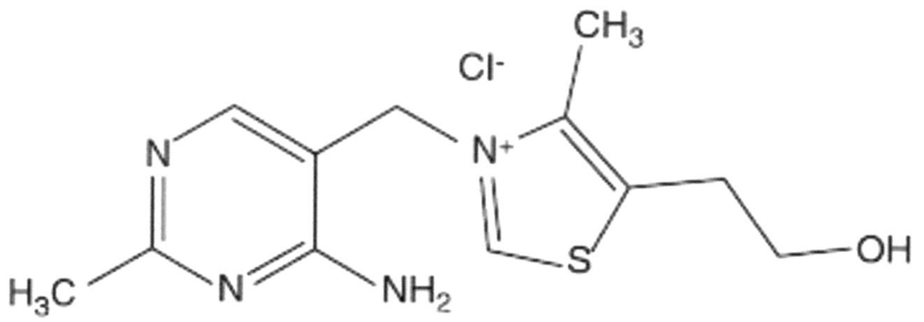

According to ChemIDPlus, thiamine is the chloride salt ($C_{12}H_{17}ClN_4OS$)

GENERAL

Thiamine is an essential vitamin, belonging to the vitamin B family, with antioxidant, erythropoietic, mood modulating, and glucose-regulating activities. Thiamine plays an important role in intracellular glucose metabolism, the conversion of carbohydrates and fat into energy, it is essential for normal growth and development and helps to maintain proper functioning of the heart and the nervous and digestive systems. Pharmaceutical thiamine is indicated for the treatment of thiamine and niacin deficiency states, Korsakov's alcoholic psychosis, Wernicke-Korsakov syndrome, delirium, and peripheral neuritis. In pharmaceuticals, thiamine is most often present as thiamine hydrochloride (CAS number 67-03-8, EC number 200-641-8, molecular formula $C_{12}H_{18}Cl_2N_4OS$); other derivatives used are thiamine mononitrate and thiamine (di)chloride (1).

CONTACT ALLERGY

Case report

A 46-year-old woman developed a pruritic micropapular erythematous rash on the right shoulder after topical application of diclofenac cream and a commercial solution containing lidocaine, dexamethasone, cyanocobalamin (vitamin B_{12}) and thiamine (vitamin B_1) by iontophoresis. Patch tests and prick and intradermal tests with these medicaments were negative. One hour after the intramuscular injection of the solution (the usual method of administration), the patient developed skin itching, and 24 hr later, erythematous plaques were noticed in the forearms and right shoulder (the application area of the iontophoresis treatment). Eight hours after oral administration of a multivitamin containing vitamins B_1, B_{12} and B_6, the patient developed a pruritic micropapular erythematous rash on the buttocks and back. Finally, patch tests were performed with this multivitamin and its components, which gave positive results to the multivitamin 'as is' (+++), and vitamin B_1 (thiamine hydrochloride 10% water, ++) at D4. An intradermal test with thiamine gave a positive result at 24 hours. Ten controls were negative to thiamine 10% water. The patient had allergic contact dermatitis from thiamine and systemic contact dermatitis from an oral challenge and intramuscular injection (2).

Occupational allergic contact dermatitis

A 54-year-old man employed in a pharmaceutical plant filled and packed thiamine hydrochloride in a dusty process. After one month, he developed an itchy eczema on the forearms and dorsa of the hands, with some spread to the face. Patch tests were positive to thiamine 10% and 5% water: thiamine 1% water showed erythema only. Ten controls were negative. Positive reactions were also recorded to thiothiamine 5% and 1% water. This patient had occupational (possibly partly airborne) allergic contact dermatitis from thiamine (3).

A 32-year-old man, after 3 months of employment as a process worker in the same pharmaceutical plant as the previous patient, developed an itchy eczema on the hands and legs, which spread to the rest of the body. The patient was dismissed from work. The skin gradually cleared after treatment with betamethasone ointment. After discontinuation of topical treatment, however, an intermittent itchy dermatitis persisted on the dorsa of the feet, on the face, and occasionally at other sites. Further questioning revealed that he took oral vitamins containing thiamine daily. Patch testing was positive to thiamine 10% water. This was a case of systemic contact dermatitis (3).

Cases of allergic contact dermatitis due to thiamine had previously been reported among pharmaceutical workers. Two patients filling ampoules developed allergic eczema of the hands and arms, and in one also of the eyelids (4). A pharmaceutical worker developed eczema of exposed sites while working with thiamine hydrochloride. The patch test was positive to thiamine HCl in this patient and in 9 other workers in the factory. None of these reacted to the vitamin orally (5). A 17-year-old girl was sensitized to thiamine from filling ampoules with vitamins. A flare was observed after she had resumed her work. Oral provocation with 200 mg thiamine and intracutaneous tests with 10 mg also induced relapses (systemic contact dermatitis) (6).

Cross-reactions, pseudo-cross-reactions and co-reactions

A patient sensitized to thiamine co-reacted to thiothiamine 5% and 1% water (3). Another co-reacted to co-carboxylase (10% and 1% pet.), which is a co-enzyme in the cellular metabolism and is formed by esterification of thiamine with pyrophosphoric acid after intestinal resorption (6).

LITERATURE

1   The data in the section 'General' may have been obtained from literature discussed in this chapter, but mostly also or exclusively from one or more of the following online sources: ChemIDPlus Advanced, PubChem, DrugBank, RxList, Drug Central, Drugs.com, and Wikipedia
2   Arruti N, Bernedo N, Audicana MT, Villarreal O, Uriel O, Muñoz D. Systemic allergic dermatitis caused by thiamine after iontophoresis. Contact Dermatitis 2013;69:375-376
3   Ingemann Larsen A, Riis Jepsen J, Thulin H. Allergic contact dermatitis from thiamine. Contact Dermatitis 1989;20:387-388
4   Combes FC, Groopman J. Contact dermatitis due to thiamine. Arch Dermatol Syph 1950;61:558-559
5   Dalton JE, Pierce JD. Dermatological problems among pharmaceutical workers. Arch Dermatol 1951;64:667-675
6   Hjorth N. Contact dermatitis from vitamin B1 (thiamine). J Invest Dermatol 1958;30:261-264

Chapter 3.336 THIOCOLCHICOSIDE

IDENTIFICATION

Description/definition   : Thiocolchicoside is the semisynthetic derivative of colchicine that conforms to the structural formula shown below

Pharmacological classes : Central nervous system depressants; muscle relaxants

IUPAC name   : N-[(7S)-1,2-Dimethoxy-10-methylsulfanyl-9-oxo-3-[(2S,3R,4S,5S,6R)-3,4,5-trihydroxy-6-(hydroxymethyl)oxan-2-yl]oxy-6,7-dihydro-5H-benzo[a]heptalen-7-yl]acetamide

Other names   : 2,10-Di(demethoxy)-2-glucosyloxy-10-methylthiocolchicine

CAS registry number   : 602-41-5

EC number   : 210-017-7

Merck Index monograph : 10747 (Thiocolchicine)

Patch testing   : 0.5%, 1.5% and 5% pet. and water

Molecular formula   : $C_{27}H_{33}NO_{10}S$

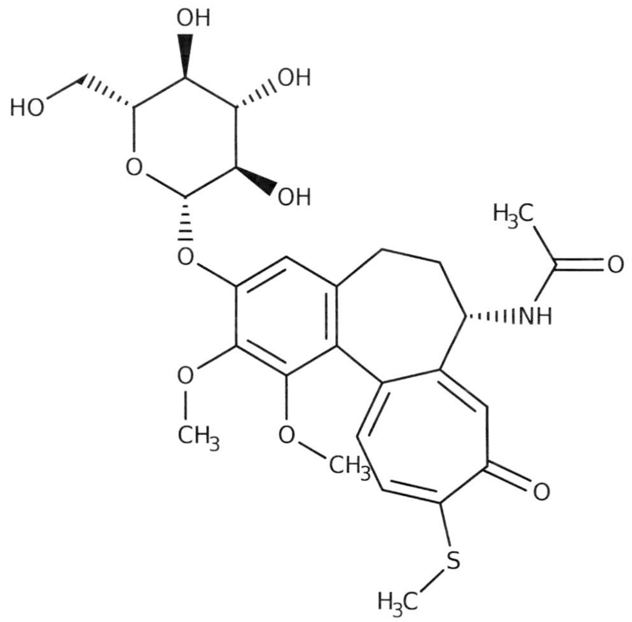

GENERAL

Thiocolchicoside is a semisynthetic derivative of colchicine, a natural anti-inflammatory glycoside which originates from the flower seeds of *Superba gloriosa*. It is a muscle relaxant with anti-inflammatory and analgesic effects. Thiocolchicoside is used as an adjuvant drug in the treatment of painful muscle spasms and in acute spinal pathology. Some other conditions that may benefit from this medication are acute and chronic lumbar and sciatic pain, cervicobrachial neuralgia, persistent torticollis, and post-traumatic and post-operative pain. It may be administered orally, parenterally or topically (1).

CONTACT ALLERGY

Case reports

A 27-year-old woman had developed an itching eruption immediately after the application of a cream containing 0.25% thiocolchicoside (TCD) for sciatica pain relief on the right thigh. She had previously used the same cream without any adverse reaction. Open tests with the cream on normal back skin and on previously affected but clinically normal thigh skin were negative after 20, 40, and 60 minutes. Prick tests, however, were strongly positive to the cream and to TCD 0.25% water. The wheals persisted for approximately 30 to 60 minutes but disappeared without apparent sequelae in less than 2 hours. However, about 2 days later, the patient noticed an eczematous reaction at the sites of the two positive prick tests that persisted for a few days. Patch tests with the cream and its components showed positive reactions at D2 and D4 to the cream (as is) and to TCD 1% water (negative to 0.25%). Eleven controls were negative. Epicutaneous application of the cream and TCD 0.25% water on slightly eczematous skin of the chest resulted in urticarial reactions after 20 minutes, but were negative on normal skin of the chest (2). This was a case of combined delayed-type and immediate-type (contact urticaria) hypersensitivity.

A 47-year old physiotherapist had a one-month history of erythematous desquamative dermatitis on the hands and forearms, accompanied by pruritus and burning. She suspected topical thiocolchicoside cream, handled in her work, as the cause. Patch tests were positive to the cream 'as is' and to thiocolchicoside in a dilution series of 0.25% - 0.5% - 1% - 2% and 5% pet. Concurrently with her positive patch tests to thiocolchicoside, the patient had a flare-up of dermatitis on the hands and forearms. Eleven controls were negative to thiocolchicoside 1% pet. (3). This was a case of occupational allergic contact dermatitis to thiocolchicoside.

A 27-year-old woman presented with an acute erythematovesicular dermatitis on the right shoulder. For 15 days previously she had applied an ointment containing thiocolchicoside for backache twice daily. Patch tests were positive to the ointment 'as is', thiocolchicoside 1%, 3% and 5% pet. (negative to 0.5% pet.) and the excipient hydrogenated lanolin, but negative to the other excipients (4).

PHOTOSENSITIVITY

Case reports
A 46-year-old man presented with an acute erythematovesicular dermatitis of the legs, face, neck and, to a lesser extent, the trunk. For 10 days previously, he had applied a cream containing thiocolchicoside to a sprained right ankle, where the rash had first started. Patch tests with the cream, as is, and with its individual components (including thiocolchicoside 0.2% pet. and 0.2% and 0.1% water) were positive only to the cream and to the lanolin alcohol it contained (which explains the positive reaction to the cream). In view of the dermatitis being most prominent at photo-exposed areas, the same substances were photopatch tested giving positive reactions not only to the cream 'as is', but also to all 3 thiocolchicoside test materials (++). 25 controls were negative to thiocolchicoside 0.2% water (5).

A 66-year-old woman presented with acute eczema at the contact area of an elbow strap on the right distal elbow she had been using for relief of her epicondylitis for 6 weeks. The elbow strap had a neoprene pressure pad in the center. Topical nonsteroidal anti-inflammatory drugs were used concomitantly at the beginning, but not any longer during the last 2-3 weeks. Patch testing showed a +++ positive reaction to the inner part of the used neoprene pad and +++ positive reactions to the previously used anti-inflammatory creams 'as is', containing thiocolchicoside or etofenamate. Patch testing with a new piece of pressure band of an unused elbow strap of the same brand showed no reaction, suggesting the role of the retained NSAIDs in the used pad. Patch testing could not be performed with thiocolchicoside or etofenamate separately, but was negative with the tested inactive ingredients: parabens, lanolin alcohol and lavender oil (6).

Immediate contact reactions
Immediate contact reactions (contact urticaria) to thiocolchicoside are presented in Chapter 5.

LITERATURE
1   The data in the section 'General' may have been obtained from literature discussed in this chapter, but mostly also or exclusively from one or more of the following online sources: ChemIDPlus Advanced, PubChem, DrugBank, RxList, Drug Central, Drugs.com, and Wikipedia
2   Mancuso G. Immediate and delayed contact hypersensitivity to thiocolchicoside. Acta Dermatovenerol Alp Pannonica Adriat 2018;27:203-205
3   Mancuso G, Berdondini RM. Occupational allergic contact dermatitis from thiocolchicoside. Contact Dermatitis 2000;43:180-181
4   Foti C, Cassano N, Mazzarella F, Bonamonte D, Veña GA. Contact allergy to thiocolchicoside. Contact Dermatitis 1997;37:134
5   Foti C, Vena GA, Angelini G. Photocontact allergy due to thiocolchicoside. Contact Dermatitis 1992;27:201-202
6   Özkaya E. Patch testing with used and unused personal products : a practical way to show contamination with contact allergens. Contact Dermatitis 2016;75:328-330

Chapter 3.337 THIOCTIC ACID

IDENTIFICATION

Description/definition   : Thioctic acid is the heterocyclic thia fatty acid that conforms to the structural formula
                              shown below
Pharmacological classes   : Antioxidants; vitamin B complex
IUPAC name   : 1,2-Dithiolane-3-pentanoic acid, (+/-)-
Other names   : α-Lipoic acid; 1,2-dithiolane-3-valeric acid; 5-(dithiolan-3-yl)valeric acid
CAS registry number   : 1077-28-7
EC number   : 214-071-2
Merck Index monograph   : 10749
Patch test allergens   : 1% and 5% pet.
Molecular formula   : $C_8H_{14}O_2S_2$

GENERAL

Thioctic acid (α-lipoic acid) is an essential cofactor in metabolic reactions through mitochondrial-specific pathways, and it is synthesized in small amounts in humans. Thioctic acid shows antioxidant and metal-chelating activity. Hence, it is widely used in a variety of conditions, including diabetes, insulin resistance, atherosclerosis, neuropathy, neurodegenerative diseases, and ischemia-reperfusion (2). Other suggested indications include diseases of the eye such as cataracts, diabetic retinopathy, and age-related macular degeneration (4) and a variety of other conditions (1). Finally, thioctic acid is used in cosmetic anti-ageing products, such as 'anti-wrinkle' creams (2).

CONTACT ALLERGY

Case reports

A woman had a 7-month history of periorbital swelling and dermatitis. Patch testing yielded a positive reaction to her eye drops, tested 'as is' on day D2 and D4. When tested with its ingredients, the patient now reacted to thioctic acid on D2 and D4. Thirty controls were negative (4).

Thioctic acid has also caused several cases of allergic contact dermatitis from its presence in cosmetics (2,3).

Systemic administration

A woman suffering from shoulder pain resulting from cervical disc herniation developed a pruritic maculopapular rash on the face and scalp after 10 days of treatment with a dietary supplement containing thioctic acid and other components and with two oral and one intramuscular NSAIDs. Prick tests with the supplement and the three drugs (powdered drug dissolved in saline) were negative. Patch tests gave a very strong reaction to the commercial dietary supplement. Later, patch tests were performed with its ingredient, and the patient now reacted to the supplement 10% pet. and to thioctic acid 5% pet., 2.5% pet. and 0.025% pet. Twelve controls were negative (1).

LITERATURE

1   Rizzi A, Nucera E, Buonomo A, Schiavino D. Delayed hypersensitivity to α-lipoic acid: look at dietary supplements. Contact Dermatitis 2015;73:62-63
2   Bergqvist-Karlsson A, Thelin I, Bergendorff O. Contact dermatitis to α-lipoic acid in an anti-wrinkle cream. Contact Dermatitis 2006;55:56-57
3   Leysen J, Aerts O. Further evidence of thioctic acid (α-lipoic acid) being a strong cosmetic sensitizer. Contact Dermatitis 2016;74:182-184
4   Craig S, Urwin R, Wilkinson M. Contact allergy to thioctic acid present in Hypromellose® eye drops. Contact Dermatitis 2017;76:361-362

Chapter 3.338 TIMOLOL

IDENTIFICATION

Description/definition : Timolol is the propanolamine derivative that conforms to the structural formula shown below

Pharmacological classes : Antihypertensive agents; β-adrenergic antagonists; anti-arrhythmia agents

IUPAC name : (2S)-1-(tert-Butylamino)-3-[(4-morpholin-4-yl-1,2,5-thiadiazol-3-yl)oxy]propan-2-ol

CAS registry number : 26839-75-8

EC number : 248-032-6

Merck Index monograph : 10871

Patch testing : 10% pet.

Molecular formula : $C_{13}H_{24}N_4O_3S$

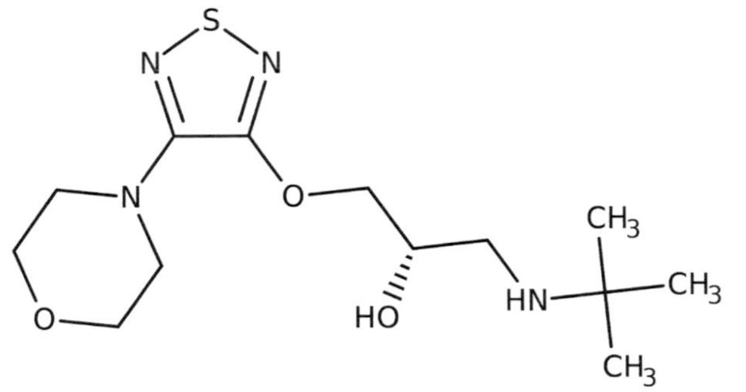

Timolol

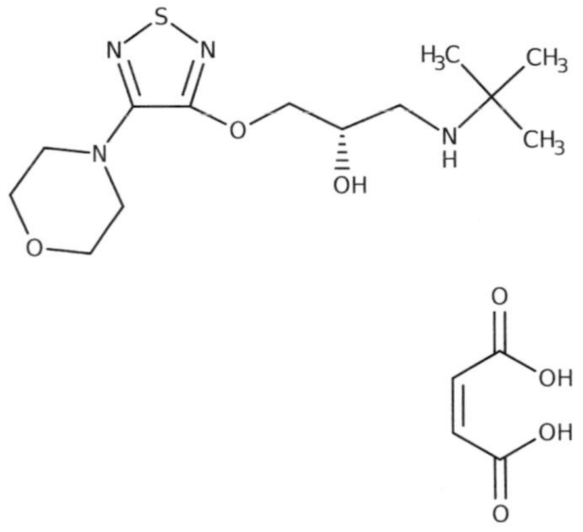

Timolol maleate

GENERAL

Timolol is a propanolamine derivative and a non-selective β-adrenergic antagonist with antihypertensive property. It competitively binds to β1-adrenergic receptors in the heart and vascular smooth muscle and β2-receptors in the bronchial and vascular smooth muscle, resulting in a decrease in β-adrenergic stimulation. This leads to a decrease in resting and exercise heart rate and cardiac output, and a decrease in both systolic and diastolic blood pressure. β2-blockade results in an increase in peripheral vascular resistance. The ultimate results include vasodilation and negative chronotropic and inotropic cardiac effects. In addition, timolol reduces intra-ocular pressure, possibly by decreasing aqueous humor production from reduction of blood flow to the ciliary processes and reduced cAMP synthesis. The oral form of timolol is used to treat high blood pressure and prevent heart attacks, and occasionally to prevent migraine headaches. Ophthalmic timolol is indicated for the treatment of open-angle and occasionally secondary glaucoma and is the most widely used anti-glaucoma drug (1).

In pharmaceutical products, timolol is mostly employed as timolol maleate (CAS number 26921-17-5, EC number 248-111-5, molecular formula $C_{17}H_{28}N_4O_7S$); sometimes, the hemihydrate is used (1).

CONTACT ALLERGY

General

Patch tests with commercial eye drop preparations containing beta-blockers may not infrequently be false-negative (12,14,29,30,31,32,33,35,38). The likely explanation is that, whereas the anatomic and physiologic properties of eyelid skin cause a lower threshold to development of sensitization and allergic contact dermatitis, the low concentration of the allergens (usually 0.25% to 0.5%) in commercial products is insufficient to elicit an allergic reaction in the far thicker skin of the back. When contact allergy is strongly suspected, but patch tests are negative, the following alternative diagnostic methods have been suggested: patch test on adhesive tape-stripped (17,32), pricked (10x with a prick-test lancet) (34) or scarified skin (32); patch test after pretreatment with 0.5% aqueous sodium lauryl sulfate solution for 24 hours (37); enlarge the patch test area (31); or perform intradermal testing (19).

By far the best method, however, is to patch test the active principle itself at a concentration higher than the one present in the commercial preparation (12,14,29,35), up to 10% (14,39). ROATs are usually (and therefore not always [40]) negative (12,35), but provocative use testing may be positive and aid in diagnosis (30,33,36).

Of all beta-blockers, timolol has by far caused most contact allergic reactions, but is also the most frequently used.

Case series

In Leuven, Belgium, between 1990 and 2017, 16,065 patients were investigated for contact allergy and 118 (0.7%) showed positive patch test reactions to topical ophthalmic medications and/or to their ingredients. Eighty-four individuals (71%) reacted to an active principle. Timolol was tested in 50 patients and was the allergen in eye medications in one. There were no reactions to timolol in other types of medications (2).

In a hospital in Slovenia, in an undefined period before 2015, 55 patients with suspected contact allergy to topical drugs used for treatment of glaucoma were retrospectively analyzed. Eight of 55 patients (5 women, 3 men) had positive patch tests to one or more products. Six patients were positive to beta-blockers, of which 5 to timolol (26). It is highly likely that these patients have not been tested with the active drugs but only with the commercial eye drops and that the diagnosis of allergic contact dermatitis to the active ingredients was, at best, made *per exclusionem*.

In Leuven, Belgium, in the period 1990-2014, iatrogenic contact dermatitis was diagnosed in 2600 individuals (17% of the total patch test population). 96% of all positive patch test reactions to topical drugs and antiseptics were considered to be relevant. Timolol (5% water) was tested in 48 patients and there were 4 positive reactions to it (4).

In Ferrara, Italy, over a 65-month period before 2005, 50 patients affected by periorbital dermatitis while using topical ocular products were patch tested, including with their own ophthalmic medications (n=210). Only 15 positive reactions were detected in 12 subjects, including 14 reactions to commercial eye drops. There were two reactions to timolol eye drops. The active ingredients were not tested separately, but contact allergy to the excipients and preservatives was excluded by patch testing. The authors concluded that patch testing with commercial eye drops has doubtful value (25).

Out of 112,430 patients patch tested by the IVDK between 1993 and 2004, 332 had been tested with their own topical anti-glaucoma eye drops containing different β-blockers because of suspected allergic contact dermatitis. 189 subjects were tested with timolol eye drops and there were 21 (11%) positive reactions. The patients were not tested with the active substance, but reactions to the (possible) adjuvants benzalkonium chloride, sodium EDTA and sodium bisulfite were excluded (3).

Case reports

A 10-month old infant developed dermatitis from timolol eye drops 0.5% applied to an infantile haemangioma. A patch test with the commercial eye drops was positive, with its preservative benzalkonium chloride negative. Timolol itself, however, was not tested (5).

A 17-year-old adolescent boy with Sturge-Weber syndrome and a choroidal hemangioma in the left eye developed persistent left-sided periorbital dermatitis. He had been prescribed timolol maleate 0.25% eye drops, containing benzalkonium chloride 0.01% as a preservative, for the last 3 months for glaucoma in the left eye after fractional radiotherapy for a diffuse choroidal hemangioma. Patch tests were positive to the commercial eye drops but negative to benzalkonium chloride. A diagnosis of timolol contact allergy was made and the dermatitis disappeared after stopping the use of the eye drops. Timolol itself, however, was not patch tested (6).

A 61-year-old man, who had long been treated for open-angle glaucoma after corneal transplantation with eye drops containing the β-blocker timolol, developed eyelid dermatitis associated with conjunctival hyperemia. Patch tests gave a positive reaction to the commercial preparation. Further tests showed reactions to two other preparations with timolol, timolol 1% water, levobunolol 1% water and a commercial preparation containing levobunolol. As it was ascertained that the patient had never used levobunolol before, the allergy to this beta-blocker was considered to be a cross-reaction to timolol. There was also a positive patch test to propranolol 2% water, but this was not commented upon (7).

A 70-year-old man with glaucoma had been treated with eye drops containing timolol maleate and brimonidine tartrate for more than 2 years, when he developed itchy erythematous, edematous and scaling dermatitis of the eyelids and the cheeks. There were weak-positive patch tests to the commercial preparation and other eye drops containing only timolol. Later, the patient was patch tested with timolol maleate 1% water and 2%, 5% and 10% pet., brimonidine tartrate 0.2%, 0.4%, 1% water, 1% pet., and 10% pet., and benzalkonium chloride 0.1% water. Results were positive only for timolol maleate 5% pet. (+) and 10% pet. (+) on D2 and D4. Five controls were negative (8).

An 80-year-old lady presented with conjunctival inflammation and severe eczematous rash affecting the eyelids and cheeks. Her eye drops for glaucoma had been changed to a commercial preparation containing dorzolamide hydrochloride 2%, timolol maleate 0.5% and benzalkonium chloride 3 weeks before presentation. Patch tests and a ROAT with these eye drops, however, were negative. A supply of pure dorzolamide and timolol was obtained and there were positive patch tests to timolol 10% pet. (negative to 5% and 2.5% pet.) and dorzolamide. Ten controls were negative (12).

Short summaries of other case reports of allergic contact dermatitis from timolol are shown in table 3.338.1.

Table 3.338.1 Short summaries of other case reports of allergic contact dermatitis from timolol

Year	Sex	Age	Clinical picture	Positive patch tests	Comments	Ref.
2020	M	72	'pruritic eruption''	eye drops 'as is'; timolol emulsion 0.25% 'as is'; timolol itself not tested	topical treatment of Kaposi sarcoma; positive ROAT; positive provocation test	22
2014	F	27	pruritic dermatitis around the left eye	combination eye drops; timolol dilution series 0.0078% - 5% pet.		9
2014	M	62	bilateral redness and itching of periorbital skin	timolol eye drops and combination eye drops with timolol	timolol itself was not tested	9
2011	?	?	blepharoconjunctivitis	timolol eye drops; negative to all excipients	diagnosis *per exclusionem*, timolol itself not tested	20
2009	M	40	eyelid erythema, edema, burning, lacrimation	timolol eye drops; timolol 1% water	no cross-reaction to levobunolol	11
2000	M	60	eyelid dermatitis, itching, edema, conjunctival chemosis	timolol eye drops; timolol (conc. and vehicle not mentioned)	possible cross-reactivity to levobunolol	13
1998	M	58	dermatitis around the eyes	timolol eye drops	timolol itself not tested, the preservative was negative	14
1995	F	69	dermatitis of both eyelids and periorbital areas	2 commercial eye drop products; timolol base 1% water	the patch test to timolol was positive only on stripped skin	17
1995	F	75	dermatitis of both eyelids and periorbital areas	commercial eye drop product; timolol 1% water	the patch test to timolol was positive only on stripped skin	17
1994	M	61	eczema on the face, palms and forearms	timolol eye drops; timolol 1% water	the patient was also allergic to befunolol	18
1993	M	70	bilateral upper and lower eyelid dermatitis	commercial eye drops; negative to timolol maleate 0.5%, but positive to an *intracutaneous* test at D2 and D4	the patient later became sensitized to betaxolol and to metipranolol	19
1991	M	68	follicular conjunctivitis, erythema and swelling of the eyelids, redness of conjunctivae	commercial eye drops; timolol 0.5% water		21
1986	F	62	conjunctivitis, bilateral red scaly eyelids and edema	two commercial timolol preparations; timolol 5%, 2.5%, 1%, 0.5%, and 0.25% water	patch test with the excipient and control tests were negative	22
1986	M	58	pruritus, rash and edema of the (operated) left eye	commercial eye drops	timolol itself was not tested; negative test to the preservative benzalkonium chloride	23

Other case reports of contact allergy to timolol, adequate data of which are not available to the author, can be found in refs. 15, 24, and possibly 42.

Cross-reactions, pseudo-cross-reactions and co-reactions

Cross-reactions between beta-blockers appear to be infrequent (3). Two patients allergic to timolol cross-reacted to levobunolol (7,13). Possible cross-reaction to carteolol in a timolol-sensitized man (15). An individual sensitized to betaxolol may have cross-reacted to timolol (10). Possibly cross-reaction to timolol in a patient previously sensitized to levobunolol (16). One patient sensitized to metoprolol also had a positive patch test to timolol 0.5% water (28).

LITERATURE

1   The data in the section 'General' may have been obtained from literature discussed in this chapter, but mostly also or exclusively from one or more of the following online sources: ChemIDPlus Advanced, PubChem, DrugBank, RxList, Drug Central, Drugs.com, and Wikipedia

2   Gilissen L, De Decker L, Hulshagen T, Goossens A. Allergic contact dermatitis caused by topical ophthalmic medications: Keep an eye on it! Contact Dermatitis 2019;80:291-297

3   Jappe U, Uter W, Menezes de Pádua CA, Herbst RA, Schnuch A. Allergic contact dermatitis due to beta-blockers in eye drops: a retrospective analysis of multicentre surveillance data 1993-2004. Acta Derm Venereol 2006;86:509-514

4   Gilissen L, Goossens A. Frequency and trends of contact allergy to and iatrogenic contact dermatitis caused by topical drugs over a 25-year period. Contact Dermatitis 2016;75:290-302

5   Sacchelli L, Vincenzi C, La Placa M, Piraccini BM, Neri I. Allergic contact dermatitis caused by timolol eyedrop application for infantile haemangioma. Contact Dermatitis 2019;80:255-256

6   Koumaki D, Orton D. Unilateral allergic dermatitis to timolol in eye drops for treating glaucoma in a patient with Sturge-Weber syndrome and a choroidal hemangioma. Dermatitis 2019;30:373-374

7   Horcajada-Reales C, Rodríguez-Soria VJ, Suárez-Fernández R1. Allergic contact dermatitis caused by timolol with cross-sensitivity to levobunolol. Contact Dermatitis 2015;73:368-369

8   Otero-Rivas MM, Ruiz-González I, Valladares-Narganes LM, Delgado-Vicente S, Rodríguez-Prieto MÁ. A case of contact dermatitis caused by timolol in anti-glaucoma eyedrops. Contact Dermatitis 2015;73:256-257

9   Chernoff KA, Zippin JH. Allergic contact dermatitis to timolol: a report of 2 cases and review of the literature. Dermatitis 2014;25:41-42

10 Nino M, Napolitano M, Scalvenzi M. Allergic contact dermatitis due to the beta-blocker betaxolol in eyedrops, with cross-sensitivity to timolol. Contact Dermatitis 2010;62:319-320

11 Buquicchio R, Foti C, Cassano N, Ventura M, Vena GA. Allergic contact dermatitis from timolol complicating choroidal melanoma-related glaucoma. Eur J Dermatol 2009;19:74-75

12 Kalavala M, Statham BN. Allergic contact dermatitis from timolol and dorzolamide eye drops. Contact Dermatitis 2006;54:345

13 Quiralte J, Florido F, de San Pedro BS. Allergic contact dermatitis from carteolol and timolol in eyedrops. Contact Dermatitis 2000;42:245

14 De Groot AC, van Ginkel CJ, Bruynzeel DP, Smeenk G, Conemans JM. [Contact allergy to eyedrops containing beta-blockers]. Ned Tijdschr Geneeskd. 1998;142:1034-1036 (Article in Dutch)

15 Giordano-Labadie F, Lepoittevin JP, Calix I, Bazex J. [Contact allergy to beta blockaders in eye drops: cross allergy?]. Ann Dermatol Venereol 1997;124:322-324 (Article in French)

16 Förster W. Allergisches Kontaktekzem auf Levobunolol und Timolol in der Glaukombehandlung. Derm (Praktische Dermatologie) 1997;3:130-131 (Article in German)

17 Koch P. Allergic contact dermatitis due to timolol and levobunolol in eyedrops, with no cross-sensitivity to other ophthalmic beta-blockers. Contact Dermatitis 1995;33:140-141

18 Vincenzi C, Ricci C, Peluso AM, Tosti A. Allergic contact dermatitis caused by β-blockers in eyedrops. Am J Contact Derm 1994;5:102-103

19 O'Donnell BF, Foulds IS. Contact allergy to beta-blocking agents in ophthalmic preparations. Contact Dermatitis 1993;28:121-122

20 Urbancek S, Kuklova M. Allergic blepharoconjunctivitis caused by antiglaucomatics—Central Slovakia experiences. American Contact Dermatitis Society 22nd Annual Meeting, New Orleans, USA, February 3, 2011. Meeting program, poster presentations, pages 32-33. Available at: https://www.contactderm.org/files/2011_ACDS_Abstracts.pdf

21 Cameli N, Vicenzi C, Tosti A. Allergic contact conjunctivitis due to timolol in eyedrops. Contact Dermatitis 1991;25:129-130

22 Romaguera C, Grimalt F, Vilaplana J. Contact dermatitis by timolol. Contact Dermatitis 1986;14:248

23 Fernandez-Vozmediano JM, Blasi NA, Romero-Cabrera MA, Carrascosa-Cerquero A. Allergic contact dermatitis to timolol. Contact Dermatitis 1986;14:252

24 Baldone JAM, Hankin JF, Zimmerman TJ. Allergic conjunctivitis associated with timolol therapy in an adult. Ann Ophthalmol 1978;10:847-850

25 Corazza M, Massieri LT, Virgili A. Doubtful value of patch testing for suspected contact allergy to ophthalmic products. Acta Derm Venereol 2005;85:70-71

26 Slavomir U, Kuklova Bielikova M. Allergic reactions due to antiglaucomatics. Dermatitis 2015;26(2):e12-13

27 Feiler-Ofry V, Godel V, Lazar M. Nail pigmentation following timolol maleate therapy. Ophthalmologica 1981;182:153-156

28 Van Joost T, Middelkamp Hup J, Ros FE. Dermatitis as a side effect of long-term topical treatment with certain beta-blocking agents. Br J Dermatol 1979;101:171-176

29 De Groot AC, Conemans J. Contact allergy to metipranolol. Contact Dermatitis 1988;18:107-108

30 Corazza M, Virgili A, Mantovani L, Masieri LT. Allergic contact dermatitis from cross-reacting beta-blocking agents. Contact Dermatitis 1993;28:188-189

31 Gailhofer G, Ludvan M. 'Beta-blockers': sensitizers in periorbital allergic contact dermatitis. Contact Dermatitis 1990;23:262

32 Frosch PJ, Weickel R, Schmitt T, Krastel H. [Side effects of external ophthalmologic drugs]. Z Hautkr 1988;63:126, 129-132, 135-136 (Article in German)

33 Gonzalo-Garijo MA, Zambonino MA, Pérez-Calderón R, Pérez-Rangel I, Sánchez-Vega S. Allergic contact dermatitis due to carteolol, with good tolerance to betaxolol. Dermatitis 2011;22:232-233

34 Wilkinson SM. False-negative patch test with levobunolol. Contact Dermatitis 2001;44:264

35 Statham BN. Failure of patch testing with levobunolol eyedrops to detect contact allergy. Contact Dermatitis 2000;43:365-366

36 Sánchez-Pérez J, Jesús Del Río M, Fernández-Villalta MJ, García-Díez A. Positive use test in contact dermatitis from betaxolol hydrochloride. Contact Dermatitis 2002;46:313-314

37 Corazza M, Virgili A. Allergic contact dermatitis from ophthalmic products: can pre-treatment with sodium lauryl sulfate increase patch test sensitivity? Contact Dermatitis 2005;52:239-241

38 Corazza M, Levratti A, Zampino MR, Virgili A. Conventional patch tests are poor detectors of contact allergy from ophthalmic products. Contact Dermatitis 2002;46:298-299

39 Hashimoto Y, Aragane Y, Kawada A. Allergic contact dermatitis due to levobunolol in an ophthalmic preparation. J Dermatol 2006;33:507-509

40 Vandebuerie L, Kerre S. Allergic contact dermatitis due to betablocker agents in eye drops: relevance of patch testing versus open use testing (ROAT). Ned Tijdschr Derm Venereol 2011;21:328-330 (article in Dutch)

41 Pérez-Tato B, Polimón I, Marinero S, Lozano-Masdemont B, de la Cruz E, Galván C. Allergic contact dermatitis caused by timolol eyedrop application for classic Kaposi sarcoma. Australas J Dermatol 2020 Jun 16. doi: 10.1111/ajd.13353. Online ahead of print.

42 Lazarov A, Amichai B. Skin reactions due to eye drops: report of two cases. Cutis 1996;58:363-364

Chapter 3.339 TIOCONAZOLE

IDENTIFICATION

Description/definition   : Tioconazole is the phenethyl imidazole derivative that conforms to the structural
formula shown below
Pharmacological classes   : Antifungal agents; 14α-demethylase inhibitors
IUPAC name   : 1-[2-[(2-Chlorothiophen-3-yl)methoxy]-2-(2,4-dichlorophenyl)ethyl]imidazole
CAS registry number   : 65899-73-2
EC number   : 265-973-8
Merck Index monograph   : 10878
Patch testing   : 1.0% pet. (Chemotechnique)
Molecular formula   : $C_{16}H_{13}Cl_3N_2OS$

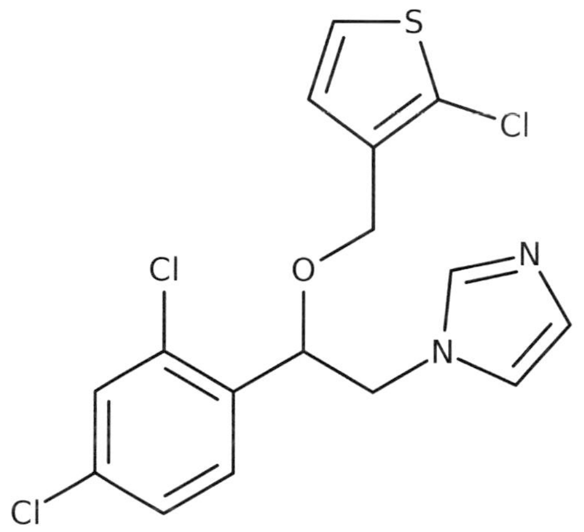

GENERAL

Tioconazole is a broad spectrum imidazole antifungal used to treat fungal and yeast infections. Tioconazole interacts with 14-alpha demethylase, a cytochrome P-450 enzyme that converts lanosterol to ergosterol, an essential component of the fungal cell membrane. In this way, tioconazole inhibits ergosterol synthesis, resulting in increased cellular permeability. Tioconazole is indicated for the local treatment of vulvovaginal candidiasis, dermatophyte skin and nail infections and tinea (pityriasis) versicolor (1).

CONTACT ALLERGY

General

Publications on allergic contact dermatitis caused by tioconazole became common following its introduction (mid-1980s) in the years 1989 to 1996, and many were related to exposure to a medicated nail solution containing 28% tioconazole. Apart from single case reports (11,13,14,16,18,19,20), case series of 72 (10), 16 (9), 14 (15), and 3 (12) patients with allergic contact dermatitis have been reported. These large numbers, especially reported from Finland, may be related to a high sensitizing potential of tioconazole (although it had only a weak sensitizing capacity in animal experiments [17]), its widespread use (50% of all antifungal prescriptions in 1990 in Finland; 36,000 prescription treatments in the region of Helsinki University Central Hospital, which has 1.2 million inhabitants) and high concentrations of tioconazole in nail solution (28%) and ointment (6.5%), or a combination of these factors (10,15).

Since 1996, up to 2019 (8,21), no new cases of tioconazole contact allergy have been reported in the English-language literature. One explanation is that the nail solution is rarely used nowadays. Indeed, in Spain, in 2017, only 1.7% of all antifungal prescriptions were for tioconazole (8). Another explanation, of course, is that contact allergy to tioconazole is so well known that it needs not be published anymore.

In retrospect, the manufacturer could have anticipated problems of an allergological nature from tioconazole (2). In a monograph produced by the company, it was mentioned that 5 out of 113 patients treated with topical tioconazole had cutaneous side-effects of itching and rash. Of these, 3 were patch tested, 2 proving to be allergic to tioconazole but not to the vehicle (2).

Patch testing in groups of patients

In Finland, over a 3.5-year period from 1991 to 1994, tioconazole 2% in 70% alcohol was included in the standard series for patch testing. 4816 patients were routinely tested and 54 (1.1%) had positive reactions to the antifungal. In addition to tioconazole-allergic patients thus detected, a further 18 individuals with suspected imidazole allergy were shown to be tioconazole-allergic when tested with an imidazole series, containing tioconazole in both alcohol and petrolatum (which vehicles were equally suitable). The group consisted of 37 men and 35 women with an average age of 53 and 56 years, respectively. The most common clinical diagnoses were eczema, tinea pedis and onychomycosis. Tioconazole allergy was suspected in 57% of cases. In 34 of 52 (65%) patients tested with an imidazole series, there were co-reactions to other imidazoles (not specified). The authors suggested that the high incidence could be the result of a high sensitizing potential of tioconazole, the high concentrations used (28% in a nail solution, 6.5% in an ointment) and/or widespread use of tioconazole preparations (in 1990, tioconazole had a market share of alsmost 50% of antifungal preparations in Finland) (10).

Case series

Two women aged 24 and 31 years, respectively, and a 25-year-old man presented with perionyxis, itchy vesicles, and a purulent exudate from their big toes. All 3 patients had been applying, for a period ranging from a few weeks to 3 months, a nail lacquer containing tioconazole and undecylenic acid to treat what they suspected to be a fungal nail infection. Patch tests were positive to the nail lacquer 'as is' and tioconazole 1% alc. in all 3 individuals. At follow-up within 6 weeks of resolution of the acute dermatitis, all patients showed onychomadesis of the previously involved big toes (21).

Eight patients sensitized to tioconazole nail solution were diagnosed between February 2010 and March 2016 in 2 hospitals in Spain. Their mean age was 58 years and 7 were female. The patients developed erythematosquamous or edematous and blistering eruptions on their fingernails (4 patients) or toenails (4 patients), particularly involving the periungual areas and occasionally spreading to the dorsal hands or feet, lower legs, or back. All 8 patients had positive patch tests to tioconazole 1% pet. and 7 reacted to the 28% tioconazole nail solution 'as is' (8).

Three patients from Portugal (F48, F55, M24) had treated onychomycosis of the feet with tioconazole 28% solution for 1-3 months, when they developed pruritic erythematovesicular lesions with edema in the areas of application. Patch tests were positive to the solution and to tioconazole 1% and 10% pet. Twenty controls were negative (12). In 2 hospitals in Portugal, 'in the last few years' before 1996, 16 patients were seen (12 women, 4 men, age range 16-75 years) with allergic contact dermatitis from 28% tioconazole nail solution manifesting as pruritic erythema, edema and vesicles around the fingernails (9 patients) or toenails (7 patients). Strongly positive reactions were observed in all patients to the nail solution and tioconazole 1% and 10% pet. (9).

In Bologna, Italy, in 'the few years' before 1995, 20 cases of allergic contact dermatitis due to imidazole preparations were found. However, patch tests with the relevant imidazoles were positive in only 12, including 5 to tioconazole. This high number of reactions was ascribed to the use of a nail solution containing 28% tioconazole (4). In Japan, in the period 1984 to 1994, 3049 outpatients were patch tested for suspected contact dermatitis and 218 of these with topical antifungal preparations. Thirty-five were allergic to imidazoles, including tioconazole in 3 individuals. In 60% of the cases, there were cross-reactions between imidazoles (3).

In a period of 15 months in 1990 and 1991, in Helsinki, Finland, 14 patients were seen with contact allergy to tioconazole: 9 women, 5 men, age range 33-76 years, mean age 51 years. Nine had used tioconazole 28% solution, and 5 a cream containing 1% of the imidazole, of whom one also had applied a 6.5% ointment. Most had used the pharmaceuticals for several months to 1-2 years to onychomycosis (n=10) or fungal skin infections, mostly tinea pedis. Quite curiously, the authors did not provide clinical data on the symptoms of sensitization, other than 'clinical symptoms were exacerbated during treatment with various tioconazole preparations'. The patients all had positive patch tests to tioconazole 2% in 70% alcohol. There were 8 co-reactions to miconazole (including 4 ?+ reactions), but most of these patients had used miconazole before. Of 7 patients co-reactive to econazole, only one had used this imidazole before; the other 6 were considered to be cross-reactions. The authors suggested to limit the use of the 28% nail solution (15).

In the period 1976-1988, 13 cases of allergic contact dermatitis to imidazole antifungal preparations were seen in a private dermatological practice in Saverne, France, and 2 in the Université Catholique de Louvain, Belgium. Most had been prescribed for tinea cruris, pedis or corporis, some for anal pruritus, dermatitis and even a leg ulcer. The responsible antifungals (tested 2% pet.) were tioconazole in one case, miconazole in 6 cases, isoconazole in 5 cases, and econazole in 2 cases; one patient reacted to a base ingredient (7).

In a literature review up to 1994, the authors identified 105 reported patients who had a positive patch test reaction to at least one imidazole derivative, ranging from 51 reactions to miconazole to one reaction each to bifonazole and enilconazole. The number of reported reactions to tioconazole was 33 (6).

Case reports
Short summaries of case reports of allergic contact dermatitis from tioconazole are shown in table 3.339.1.

Table 3.339.1 Short summaries of case reports of allergic contact dermatitis from tioconazole

Year and country	Sex	Age	Positive patch tests	Clinical data and comments	Ref.
2009 Germany	M	58	TIO cream 'as is'; TIO 1% pet.	weeping erosions in the groins and on scrotum; treated with various antifungal products; co-reactions to clotrimazole, oxiconazole and croconazole	22
1994 Ireland	M	44	28% nail solution 'as is'; TIO 10% and 20% pet.	dermatitis of periungual skin toenail	11
1993 Switzerland	M	56	1% cream 'as is'; tioconazole 2% pet.	subacute dermatitis of left hand and both feet after using tioconazole cream for one month	13
1992 U.K.	F	63	28% nail solution 'as is'; TIO 1% and 10% pet.	solution to right great toe; exacerbation of foot dermatitis spreading to trunk and arms	14
	F	42	28% nail solution 'as is'; tioconazole 10% pet.	after 7 month solution to toenails localized rash on the toes followed by secondary spread	
1992 Italy	F	37	25% nail solution 'as is'; TIO 2% and 5% pet.	acute eczema of the right thumb and ring finger after some weeks' application of nail solution for paronychia	16
1992 U.K.	F	22	25% nail solution 'as is'; TIO 1% and 10% pet.	after 6 weeks' use of nail solution to fingers erythema, blistering, extreme pain and debilitation	18
1992 Spain	F	30	1% cream 'as is'; TIO 1%, 10%, 20% and 50% pet.	severe dermatitis of the feet 2 days after starting TIO cream; co-reactions to (unrelated) metronidazole and bifonazole	19
1990 U.K.	M	49	TIO 1%, 10%, 20% and 50% pet.	erythema, edema and blistering of the skin area of application of nail solution for leukonychia	20

TIO: tioconazole

Cross-reactions, pseudo-cross-reactions and co-reactions
Statistically significant associations have been found in patient data between isoconazole and tioconazole (6). In 52 patients sensitized to tioconazole and tested with an imidazole series, 34 (65%) had co-reactivities to other imidazoles (not specified). It was not specifically stated that these were cross-reactions, but it was suggested that tioconazole may be a useful marker for imidazole allergy (10). Of 15 patients sensitized to tioconazole, 7 co-reacted to econazole, of whom 6 had never used econazole before (15). The question has been raised (but not answered) whether contact allergy to tioconazole and other (nitro)imidazoles may be overrepresented in patients allergic to methylchloroisothiazolinone/methylisothiazolinone (MCI/MI) (5).

LITERATURE
1   The data in the section 'General' may have been obtained from literature discussed in this chapter, but mostly also or exclusively from one or more of the following online sources: ChemIDPlus Advanced, PubChem, DrugBank, RxList, Drug Central, Drugs.com, and Wikipedia
2   Jevons S, Lees L. In: Trosyd (tioconazole), clinical experience: A monograph. Pfizer Ltd. New York: Academy Professional Information Services Inc., 1985
3   Yoneyama E. [Allergic contact dermatitis due to topical imidazole antimycotics. The sensitizing ability of active ingredients and cross-sensitivity]. Nippon Ika Daigaku Zasshi 1996;63:356-364 (article in Japanese)
4   Guidetti MS, Vincenzi C, Guerra L, Tosti A. Contact dermatitis due to imidazole antimycotics. Contact Dermatitis 1995;33:282
5   Stingeni L, Rigano L, Lionetti N, Bianchi L, Tramontana M, Foti C, et al. Sensitivity to imidazoles/nitroimidazoles in subjects sensitized to methylchloroisothiazolinone/methylisothiazolinone: A simple coincidence? Contact Dermatitis 2019;80:181-183
6   Dooms-Goossens A, Matura M, Drieghe J, Degreef H. Contact allergy to imidazoles used as antimycotic agents. Contact Dermatitis 1995;33:73-77
7   Jelen G, Tennstedt D. Contact dermatitis from topical imidazole antifungals: 15 new cases. Contact Dermatitis 1989;21:6-11
8   Pérez-Mesonero R, SchnellerPavelescu L, Ochando-Ibernón G, Vergara-Sánchez A, Sánchez-Herreros C, Martín-Alcalde E, et al. Is tioconazole contact dermatitis still a concern? Bringing allergic contact dermatitis caused by topical tioconazole back into the spotlight. Contact Dermatitis 2019;80:168-169

9   Faria A, Gonçalo S, Gonçalo M, Freitas C, Baptista PP. Allergic contact dermatitis from tioconazole. Contact Dermatitis 1996;35:250-252

10   Heikkilä H, Stubb S, Reitamo S. A study of 72 patients with contact allergy to tioconazole. Br J Dermatol 1996;134:678-680

11   Gibson G, Buckley A, Murphy GM. Allergic contact dermatitis from tioconazole without cross-sensitivity to other imidazoles. Contact Dermatitis 1994;30:308

12   Quirino AP, Barros MA. Contact dermatitis from tioconazole. Contact Dermatitis 1994;30:240-241

13   Piletta P, Pasche-Koo F, Saurat JH. Contact dermatitis from tioconazole mimicking "one hand two feet syndrome". Contact Dermatitis 1993;28:308

14   Onayemi O, Aldridge RD, Shaw S. Allergic contact dermatitis from tioconazole. A report of 2 cases. Contact Dermatitis 1992;26:193-194

15   Stubb S, Heikkilä H, Reitamo S, Förström L. Contact allergy to tioconazole. Contact Dermatitis 1992;26:155-158

16   Brunelli D, Vincenzi C, Morelli R, Tosti A. Contact dermatitis from tioconazole. Contact Dermatitis 1992;27:120

17   Hausen BM, Angel M. Studies on the sensitizing capacity of imidazole and triazole derivatives. Part II. Amer J Contact Dermat 1992;3: 95-101

18   Marren P, Powell S. Contact sensitivity to tioconazole and other imidazoles. Contact Dermatitis 1992;27:129-130

19   Izu R, Aguirre A, González M, Díaz-Pérez JL. Contact dermatitis from tioconazole with cross-sensitivity to other imidazoles. Contact Dermatitis 1992;26:130-131

20   Jones SK, Kennedy CT. Contact dermatitis from tioconazole. Contact Dermatitis 1990;22:122-123

21   Romita P, Guarneri F, De Prezzo S, Dimauro D, Ambrogio F, Bonamonte1 D, Foti C. Onychomadesis secondary to allergic contact dermatitis to tioconazole contained in a nail lacquer: Description of three cases. Contact Dermatitis 2020;82:242-243

22   Brans R, Wosnitza M, Baron JM, Merk HF. Contact sensitization to azole antimycotics. Hautarzt 2009;60:372-375 (Article in German)

Chapter 3.340 TIOPRONIN

IDENTIFICATION

Description/definition : Tiopronin is the acylated sulfhydryl-containing derivative of glycine that conforms to the structural formula shown below
Pharmacological classes : Urologicals; mucolytic agents
IUPAC name : 2-(2-Sulfanylpropanoylamino)acetic acid
Other names : (2-Mercaptopropionyl)glycine; thiolpropionamidoacetic acid
CAS registry number : 1953-02-2
EC number : 217-778-4
Merck Index monograph : 10879
Patch testing : 10% pet.
Molecular formula : $C_5H_9NO_3S$

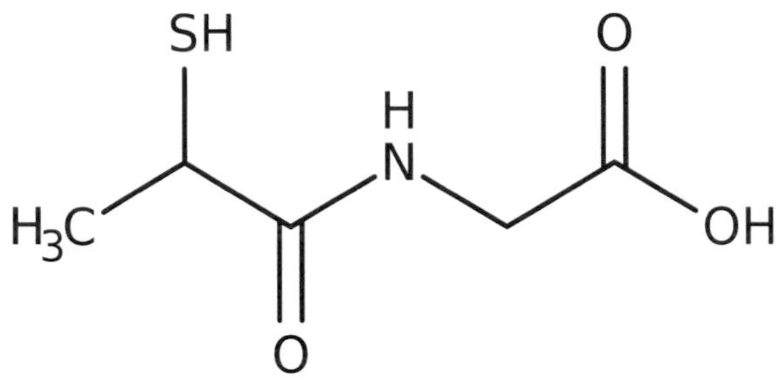

GENERAL

Tiopronin is an acylated sulfhydryl-containing derivative of glycine with reducing and complexing properties. It breaks the disulfide bond of cystine (an oxidized dimeric form of cysteine) and binds the sulfhydryl group of the resultant cysteine monomers to form a soluble tiopronin-cysteine-mixed disulfide, which is more water-soluble than cystine and is readily excreted. This leads to a reduction in urinary cystine concentration and subsequently reduces cystine stone formation. Tiopronin is indicated as a second-line for the prevention of kidney stone formation in patients with severe homozygous cystinuria. This drug may also be used as a mucolytic drug and to bind metal nanoparticles in Wilson's disease, which is an overload of copper in the body (1,2).

CONTACT ALLERGY

Case report

A 46-year-old woman developed a maculopapular facial rash and angioedema of the lips approximately 8 hours after aerosol treatment with tiopronin, initiated the day before the reaction for acute bronchitis. A prick test with 0.66 mg/ml of tiopronin in normal saline was negative, but intradermal injection of 0.02 ml of the same solution was positive (erythema and infiltrate 12 mm in diameter) after 12 hours. A patch test with 0.1 ml of a solution containing 6.66 mg/ml tiopronin was positive (++) at D2 and D3. Six controls were negative (2).

Cross-reactions, pseudo-cross-reactions and co-reactions

A patient with a lichenoid drug eruption to tiopronin and a positive reaction to this drug, co-reacted to captopril and D-penicillamine, with which the patient had never been treated before. All three compounds have a sulfhydryl group, and the authors suggested this to play a role in the positive patch tests (4). Co-reactivity to captopril has also been observed by other investigators (5).

Cutaneous adverse drug reactions from systemic administration caused by type IV (delayed-type) hypersensitivity

Cutaneous adverse drug reactions from systemic administration of tiopronin caused by type IV (delayed-type) hypersensitivity, including erythema multiforme-like and lichenoid drug reactions (3,4,5), are planned to be discussed in Volume IV of the *Monographs in Contact Allergy* series on Systemic drugs.

LITERATURE

1   The data in the section 'General' may have been obtained from literature discussed in this chapter, but mostly also or exclusively from one or more of the following online sources: ChemIDPlus Advanced, PubChem, DrugBank, RxList, Drug Central, Drugs.com, and Wikipedia

2   Romano A, Pietrantonio F, di Fonso M, Venuti A, Fabrizi G. Contact allergy to tiopronin: a case report. Contact Dermatitis 1995;33:269

3   Kitamura K, Aihara M, Osawa J, Naito S, Ikezawa Z. Sulfhydryl drug-induced eruption: a clinical and histological study. J Dermatol 1990;17:44-51

4   Kurumaji Y, Miyazaki K. Tiopronin-induced lichenoid eruption in a patient with liver disease and positive patch test reaction to drugs with sulfhydryl group. J Dermatol 1990;17:176-181

5   Piérard E, Delaporte E, Flipo RM, Duneton-Bitbol V, Dejobert Y, Piette F, Bergoend H. Tiopronin-induced lichenoid eruption. J Am Acad Dermatol 1994;31:665-667

Chapter 3.341 TIOXOLONE

IDENTIFICATION

Description/definition : Tioxolone is the 1,3-benzoxathiole that conforms to the structural formula shown below
Pharmacological classes : Dermatological agents
IUPAC name   : 6-Hydroxy-1,3-benzoxathiol-2-one
Other names   : Thioxolone; 6-hydro-2-oxo-1,3-benzoxathiole
CAS registry number   : 4991-65-5
EC number   : 225-653-0
Merck Index monograph : 10882
Patch testing   : 0.1% and 0.5% alcohol; 1% pet.
Molecular formula   : $C_7H_4O_3S$

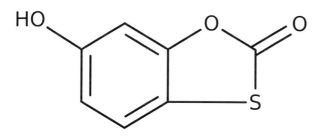

GENERAL

Tioxolone is a 1,3-benzoxathiole derivative having a hydroxy substituent at the 6-position. It has a role as an antiseborrheic agent in anti-dandruff shampoo and is or probably was also used to treat acne vulgaris and psoriasis, often in combination products (1,4).

CONTACT ALLERGY

Case reports

A 37-year-old woman with seborrhea capitis developed intense erythema, edema and vesicles on the scalp and forehead, 8 hours after applying a product containing tioxolone to treat the seborrhea. Patch tests were positive to the product 'as is' and to tioxolone 1% pet., but negative to the other ingredients. Ten controls were negative (2).

A 58-year-old woman with psoriasis of the scalp presented with an acute vesicular dermatitis on the frontal area extending to the forehead, parts of the cheek, neck, and hands. Six days prior to the onset of dermatitis she had started treatment with a tincture containing clocortolone 21-pivalate, salicylic acid, panthenol, and tioxolone. Patch testing with the product and its ingredients showed positive reactions to the tincture 'as is', and to tioxolone 0.1% and 0.01% alcohol (both ++/++). Five controls were negative to tioxolone 0.01%. In view of the widespread use, the authors considered tioxolone to be a rare sensitizer (4).

A 26-year-old female patient with seborrhea capitis had suffered vesiculation on the scalp with severe pruritus after daily use of a cosmeceutical product containing tioxolone. She improved spontaneously when she avoided the use of the product, but after reuse an acute relapse appeared. Patch tests with the product 'as is' and its ingredients were positive to the product and to tioxolone (+++) 1% and 2% pet. Twenty controls were negative (5).

A girl treated her facial acne with a powder containing tioxolone for 10 days and then developed dermatitis of her face, hands and arms. Patch testing revealed positive reactions to the powder and to lotion and solution of the same brand, and to tioxolone in alcohol 0.5% and dilutions to 0.005%, but not to 0.001%. Twenty-five controls were negative to tioxolone 0.5% (6). A patient used a 2% solution containing tioxolone for facial acne and developed eczema of the face with widespread nummular patches. The solution was chromatographed and the paper strip used for patch testing. The patient reacted to tioxolone 0.005% pet., but not to phenols (7).

One more case of allergic contact dermatitis to tioxolone, details of which are not available to the author, has been reported in Portuguese literature (3).

LITERATURE

1   The data in the section 'General' may have been obtained from literature discussed in this chapter, but mostly also or exclusively from one or more of the following online sources: ChemIDPlus Advanced, PubChem, DrugBank, RxList, Drug Central, Drugs.com, and Wikipedia

2   Villas Martinez F, Joral Badas A, Garmendia Goitia JF. Contact dermatitis from thioxolone. Contact Dermatitis 1993;29:96

3   Menezes Brandao F. Dermite de contacto por tioxolona. Grupo Portuges de Estudo das Dermites de Contacto. Boletim Informativo 1992;6:43-44 (Article in Portuguese; cited in ref. 2)

4   Näher H, Frosch PJ. Contact dermatitis to thioxolone. Contact Dermatitis 1987;17:250-251

5   Camarasa JG. Contact dermatitis to thioxolone. Contact Dermatitis 1981;7:213-214

6   Wahlberg JE. Sensitization of thioxolone - used for topical treatment of acne. Contact Dermatitis Newsletter 1971;10:222

7   Blohm G, Rajka G. A simple method for combined chemical and dermatological analysis of chemical mixtures by paper chromatography. Acta Derm Venereol 1966;46:432-435

Chapter 3.342 TIXOCORTOL PIVALATE

IDENTIFICATION

Description/definition   : Tixocortol pivalate is the pivalate thioester of tixocortol that conforms to the structural
                             formula shown below
Pharmacological class   : Glucocorticoids; anti-allergic agents
IUPAC name   : S-[2-[(8S,9S,10R,11S,13S,14S,17R)-11,17-Dihydroxy-10,13-dimethyl-3-oxo-2,6,7,8,9,11,
                             12,14,15,16-decahydro-1H-cyclopenta[a]phenanthren-17-yl]-2-oxoethyl] 2,2-
                             dimethylpropanethioate
Other names   : (11β)-21-[(2,2-Dimethyl-1-oxopropyl)thio]-11,17-dihydroxypregn-4-ene-3,20-dione;
                             tixocortol 21-pivalate
CAS registry number(s)   : 55560-96-8
EC number(s)   : 259-706-4
Merck Index monograph   : 10911 (Tixocortol)
Patch testing   : 1.0% pet. (Chemotechnique, SmartPracticeCanada, SmartPracticeEurope); 0.1% pet.
                             (Chemotechnique, SmartPracticeCanada); late reading at D6-D10 is strongly
                             recommended to avoid false-negative reactions (45,62,65)
Molecular formula   : $C_{26}H_{38}O_5S$

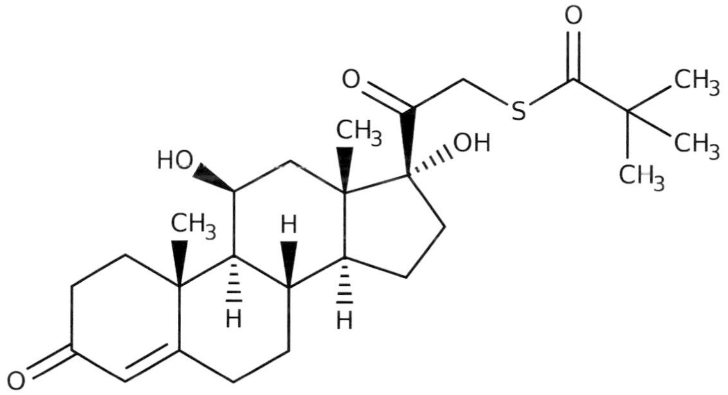

GENERAL

General aspects of corticosteroids used on the skin and mucous membranes are discussed in Chapter 2.4. A practical guideline for diagnosing allergic reactions to corticosteroids is presented in ref. 49. Tixocortol pivalate (TP) 0.1% pet. is included in the European baseline series (48) and at a concentration of 1% pet. in the American core allergen series (www.smartpracticecanada.com) as screening agent for corticosteroid – notably hydrocortisone (50,51,70,71,72) - hypersensitivity. In early investigations, a nasal spray/emulsion containing 1% tixocortol pivalate was used to screen for allergy to corticosteroids (50,51). A considerable number of positive patch test reactions to tixocortol pivalate will be missed when readings are not performed at day 6-10 (45,62,65).

Which concentration is preferable (1% or 0.1%) has been much debated. In all but one study, in which both concentrations were tested simultaneously, TP 1% pet. identified (considerably) more sensitized patients than TP 0.1% (4,11,14,52,60,61). In only one study, a European multicenter study performed by the EECDRG, patch testing with TP 1% and 0.1% in pet. detected the same number of positive reactions. A strong feature of this study was that most centers performed late readings (D6-D7), minimizing the risk of missing - frequently occurring – late positive reactions (62). In the same year, the EECDRG, that decides on the composition of the European standard/baseline series, advised to include TP in the standard series. To minimize the risk of patch test sensitization (which is extremely rare), and because tixocortol pivalate is also rather expensive, the lower concentration of 0.1% was recommended. However, when tixocortol pivalate allergy is strongly suspected and testing with the routine concentration (0.1%) is negative, additional testing with the higher concentration of 1.0% should be done (62).

CONTACT ALLERGY

General

Tixocortol pivalate (TP) is both a sensitive and a specific marker for hypersensitivity to hydrocortisone (50,51,70,71, 72). Hydrocortisone and TP differ only in the side chain attached at the C_{21} position. TP can also induce contact allergy itself, mostly from its presence in nasal drops/spray (55,56,68,73,74,75). Contact allergy to tixocortol pivalate is more prevalent in atopic individuals than in non-atopic patients, probably as a result of more frequent use of topical pharmaceutical products (43). Sensitization in 2 sisters has been observed (53).

In animal experiments, TP was found to be a strong sensitizer (57,58,59) and – as expected – to cross-react to hydrocortisone but not to but not amcinonide, budesonide, or hydrocortisone butyrate (58).

Patch testing in the general population and in subgroups

The results of patch testing tixocortol pivalate in (subgroups of) the general population are shown in table 3.342.1.In adolescents from Sweden, the prevalence of positive patch tests was <0.3%, 0.34% in women and 0.18% in men (44) and in 8th grade school children from Denmark, not a single positive reaction was observed (41). Of adults from 5 European countries, 0.3% had positive reactions to tixocortol pivalate, 0.3% of the women and 0.4% of the men (40).

It should be appreciated that the reactions were read at D3 or even once at D2 only. It is well known that many patch test reactions to corticosteroids first develop after D3-4, which means that a (possibly considerable) number of positive patch test reactions may have gone unnoticed.

Table 3.342.1 Contact allergy in the general population and in subgroups

Year and country	Selection and number tested	Prevalence of contact allergy			Comments	Ref.
		Total	Women	Men		
General population						
2008-11 five European countries	general population, random sample, 18-74 years, n=3119	0.3%	0.3%	0.4%	TRUE test	40
Subgroups						
2011-2013 Sweden	adolescents from a birth cohort, 15.8-18.9 years, n=2285	0.27%	0.34%	0.18%	TRUE test; patch tests were read at day 2 only	44
2010 Denmark	unselected population of 8th grade schoolchildren in Denmark, 15 years later; n=442	0%	0%	0%	follow-up study; TRUE test	41

Patch testing in groups of patients

Results of patch testing tixocortol pivalate in consecutive patients suspected of contact dermatitis (routine testing) back to 2000 are shown in table 3.342.2. Results of testing in groups of *selected* patients (individuals with perianal dermatitis, leg ulcers, inflammatory ear diseases, patient suspected of corticosteroid allergy) are shown in table 3.342.3.

Patch testing in consecutive patients suspected of contact dermatitis: routine testing

As tixocortol pivalate is part of both the European standard/baseline series (at 0.1% pet.) and of the screening series of the North American Contact Dermatitis Group (NACDG, USA + Canada) (at 1% pet.), there is an abundance of data on routine testing with this corticosteroid and indicator of corticosteroid allergy. The results of 35 such investigations are shown in table 3.342.2. In European countries, prevalences of sensitization have ranged from 0.2% to 1.6%. The higher rates were all observed before 2010. In Belgium, the frequency of sensitization decreased significantly from 1.7% in the period 1990-1994 to 1.1% in 2010-2014 (46). Relevance data were provided in 3 studies only and the rate of relevant reactions was >90% in all three (22,46,60). Very low rates were observed in Turkey (0.2% [90]) and Singapore (0.3% [5]).

The prevalences of sensitization in the USA and Canada, however, were considerably and consistently higher, ranging from 1.8% to 3.0%. Relevance rates (in the NACDG studies definite + probable relevance) generally were in the 40%-60% range. Possible explanations for the higher rates in the USA/Canada include higher exposure to tixocortol and/or related corticosteroids than in Europe, stricter selection of patients (which is probably a fact, most [[or all?] members of the NACDG work in tertiary referral centers) and patch testing with 1% tixocortol pivalate (versus 0.1% in Europe), which tends to detect more cases of sensitization.

An incredible 9.2% positive patch tests was observed in China (15). Other haptens also scored very high in this study. This either means that the patients had been very strictly selected (for routine testing) or suggests suboptimal patch test procedures and interpretation of the results.

Table 3.342.2 Patch testing in groups of patients: Routine testing

Years and Country	Test conc. & vehicle	Number of patients tested | positive (%)		Selection of patients (S); Relevance (R); Comments (C)	Ref.
European countries					
2008-18 Netherlands	0.1% pet.	3277	27 (0.8%)	R: not stated; C: 15% of the reactions may be missed when	45
2015-2016 Spain		3699	11 (0.3%)	R: not stated; TRUE test or 0.1% pet. tested	82
2013-2014 12 European countries [b]	0.1% pet. 1% pet.	14,130 8517	(0.5%) (0.9%)	R: not stated; C: range of positive reactions: 0% - 1% R: not stated; C: range of positive reactions: 0% - 1.4%	4
1990-2014 Belgium	0.1% pet.	14,582	238 (1.6%)	R: 96% of the positive patch test reactions to all topical drugs and antiseptics were considered to be relevant; C: there was a significant decrease in the frequency of positive reactions from 1.7% in 1990-1994 to 1.1% in 2010-2014	46
2009-12, five European countries [a]	1% pet. 0.1% pet.	23,542 11,962	(1.0%) (0.7%)	R: not stated; C: range per country: 0.6%-1.2% R: not stated; C: range per country: 0.2%-2.6%	52
2009 Sweden	0.1% pet.	3112	(1.2%)	R: not stated	10
2007-2008 11 European countries [b]	0.1% pet. 1% pet.	20,458 22,541	43 (0.2%) [a] 114 (0.5%) [a]	R: not stated; C: prevalences ranged from 0% to 1.0% R: not stated; C: prevalences ranged from 0% to 1.2%	11
2005-2008 Denmark	0.1% pet.	3594	30 (0.8%)	R: not stated	84
2005-2006 10 European countries [b]	1% pet. 0.1% pet.	6211 4758	85 (1.3%) 43 (0.9%)	R: not stated; C: prevalences were 0.9% in Central Europe, 1.3% in West (tested 1% pet.), 1.7% in Northeast and 0.1% in South Europe	14
2004-2005 U.K.	1% pet. 0.1% pet.	3747 3747	33 (0.9%) 28 (0.7%)	R: all were of current or past relevance; no significant differences between testing with 0.1% and 1%	60
2004, 11 European countries [b]	0.1% pet.	2162	11 (0.5%) [a]	R: not stated; C: range positives per center: 0.0% - 2.4%	17
2002-2003 Europe [b]	0.1% pet.	2320	(1.3%)	R: not stated	19
1999-2001 Sweden	0.1% pet.	3790	(1.6%)		21
2000 Sweden	0.1% pet.	3112	(1.6%)	R: not stated	10
2000 United Kingdom	0.1% pet.	3063	(1.5%)	R: 90% (current and past relevance in one center); C: range of positive reactions per center 0% - 4.2%	22
1996-2000 Europe	0.1% pet.	26,210	(1.3%)	R: not stated; C: ten centers, seven countries, EECDRG study	25
USA and Canada					
2015-2017 NACDG	1% pet.	5591	117 (2.1%)	R: definite + probable relevance: 56%	1
2007-2016 USA	0.1% pet.	2314	(2.4%)	R: not stated	89
2011-2015 USA	1% pet.	2574	47 (1.8%)	R: not stated	2
2013-2014 NACDG	1% pet.	4859	102 (2.1%)	R: definite + probable relevance: 47%	3
2007-2014 NACDG	1% pet.	17,978	407 (2.3%)	R: definite + probable relevance: 59%	83
2011-2012 NACDG	1% pet.	4230	99 (2.3%)	R: definite + probable relevance: 55%	6
2009-2010 NACDG	1% pet.	4304	(2.0%)	R: definite + probable relevance: 66%	7
2006-2010 USA	1% pet.	3091	(2.3%)	R: 59%	8
2007-2008 NACDG	1% pet.	5083	(2.5%)	R: definite + probable relevance: 64%	12
2005-2006 NACDG	1% pet.	4437	(2.7%)	R: definite + probable relevance: 52%	13
2001-2005 USA	1% pet.	3844	(3.0%)	R: 73%	16
2003-2004 NACDG	1% pet.	5142	138 (2.7%)	R: not stated	18
2001-2002 NACDG	1% pet.	4901	(3.0%)	R: definite + probable relevance: 39%	20
1998-2000 USA	1% pet.	5807	(2.7%)	R: definite + probable relevance: 42%	23
1998-2000 USA	1% pet.	1319	(2.9%)	R: not stated	24
Other countries					
2013-2019 Turkey	0.1% pet.	1309	2 (0.2%)	R: not stated	90
2009-2013 Singapore		2598	(0.3%)	R: present + past relevance: 0%; C: range of positive reactions per year 0% - 0.7%	5
2001-2010 Australia	1% pet.	5128	51 (1.0%)	R: 24%	9
2001-2006 China	1% pet.	1354	(9.2%)	R: not stated; C: all other tested haptens also had very high prevalence scores, suggesting that the patients were highly selected for (routine) patch testing	15

[a] age-standardized and sex-standardized proportions; [b] study of the ESSCA (European Surveillance System on Contact Allergies); EECDRG: European Environmental and Contact Dermatitis Research Group; NACDG: North American Contact Dermatitis Group (USA, Canada)

Patch testing with tixocortol pivalate 1% pet. *before* 2000 yielded 0.8% positive reactions in Germany and Austria in 1996-1997 (76), 2.9% in the USA in 1992-1996 (77), 1.9% in a multicenter study in Europe before 1989 (79; range per center: 0.2%-3.4%) and 2.3% positive reactions to tixocortol pivalate in 1992-1993 in Canada (78).

Patch testing in groups of selected patients

In groups of patients with leg ulcers/stasis dermatitis (table 3.342.3), prevalences of positive patch tests to tixocortol pivalate have ranged from 1.7% to 16%, the latter in two Polish studies (28,30). However, the number of patients in these investi-gations was very small (n=50), relevance was not discussed and the groups were probably the same. In groups of patients with perianal, genital or anogenital dermatitis, the frequencies of sensitization were modest with 1% to 2.4% positive reactions, mostly not higher than can be expected in routine testing in those countries.

High rates of sensitization have been observed in patients suspected of corticosteroid allergy (5% [80]) and in a - very small – study from The Netherlands with 5 positive patch tests in 34 patients (15%) with chronic otitis externa or media (39), which is explainable. The high percentage of 19 in a Canadian study, however, is an artifact, as patients were selected on the basis of previously established contact allergy to medicaments (47).

Table 3.342.3 Patch testing in groups of patients: Selected patient groups

Years and Country	Test conc. & vehicle	Number of patients tested	positive (%)	Selection of patients (S); Relevance (R); Comments (C)	Ref.
Patients with leg ulcers/stasis dermatitis					
2003-2014 IVDK	1% pet.	<1133	(1.9%)	S: patients with stasis dermatitis/chronic leg ulcers; R: not stated	26
2005-2008 France	1% pet.	423	7 (1.7%)	S: patients with leg ulcers; R: not stated	27
<2008 Poland	1% pet.	50	8 (16%)	S: patients with venous leg ulcers; R: not stated	30
<2005 Poland	1% pet.	50	8 (16%)	S: patients with chronic venous leg ulcers; R: not stated	28
1997-2001 U.K.		200	8 (4.0%)	S: patients with venous or mixed venous/arterial leg ulcers; R: all reactions to topical drugs were considered to be of probable, past or current relevance	29
Patients with perianal, genital or anogenital dermatitis					
2005-2016 NACDG	1% pet.	449	6 (1.3%)	S: patients with only anogenital dermatitis; R: all positives represent relevant reactions; C: the frequency was not significantly different from a control group	36
2004-2016 Spain		124	3 (2.4%)	S: patients with perianal dermatitis lasting >4 weeks; R: 67%	33
2013-2015 Ireland	1% pet.	99	1 (1%)	S: patients patch tested for perianal and/or genital symptoms; R: all reactions to medicaments were relevant	37
2003-2010 USA	1% pet.	90	1 (1%)	S: women with (predominantly) vulvar symptoms; R: 100%	34
2003-2010 USA	1% pet.	55	1 (2%)	S: women with (ichty) vulvar dermatoses; R: not stated	35
1994-2004 NACDG	1% pet.	345	8 (2.3%)	S: patients with anogenital signs or symptoms; R: only relevant reactions were included; C: the frequency in a control group was also 2.3%	32
1990-2003 Belgium	0.1% pet.	92	2 (2%)	S: women suffering from vulval complaints referred for patch testing; R: relevant as corticosteroid allergy marker	42
1992-1994 UK	1% pet.	69	1 (1%)	S: women with 'vulval problems'; R: 58% for all allergens together	64
Other patient groups					
2000-2010 Canada		100	19 (19%)	S: charts reviewed and included in the study when there was at least one positive reaction to a topical drug; R: not stated; C: the high percentages to all drugs are obviously the result of the – rather unusual – selection procedure targeting at previously established topical drug allergy	47
2000-2005 USA	1% pet.	1172	59 (5.0%)	S: patients suspected of corticosteroid allergy; R: 93%	80
1985-2002 U.K		179	2 (1.1%)	S: patients with chronic inflammatory ear disease; R: not stated	38
1994-1998 U.K.	1% pet.	232	3 (1.3%)	S: patients with periorbital contact dermatitis; R: 100%; C: the frequency in a control group was 1.6%	31
1993-4 Netherlands	1% pet.	34	5 (15%)	S: patients with chronic otitis externa or media; R: not stated	39

IVDK: Information Network of Departments of Dermatology (Germany, Austria, Switzerland); NACDG: North American Contact Dermatitis Group (USA, Canada)

Case series

From January 1990 to June 2008, in Leuven, Belgium, 315 patients were diagnosed with contact allergy to/allergic contact dermatitis from corticosteroids (CSs) from routine patch testing with a baseline series including tixocortol pivalate, budesonide, hydrocortisone butyrate and prednisone caproate, patch testing with patients' own CS preparations, and testing those with proven contact allergy to a corticosteroid or strongly suspected of CS allergy later with a series of 66 CSs, including two sex hormones (progesterone and testosterone). 71% of the patients had relevant reactions, but these were not specified. In this group of 315 CS allergic patients, 135 had positive patch tests to tixocortol pivalate 0.1% pet. (81). It is unknown how many of these reactions were caused by the use of a pharmaceutical product containing tixocortol pivalate and how many were cross-reactions to other corticosteroids.

In the period 1980-1983, in a hospital in Strasbourg, France, 6 women were investigated who proved to be allergic to tixocortol pivalate. Five had used tixocortol-containing nose drops for rhinorrhea, one after removal of a nasal polyp and one had used it to treat conjunctivitis. They all developed acute dermatitis of the face, which had started after 1 to 8 days exposure. Patch tests were positive to the nasal drops and to tixocortol pivalate 0.1% pet. Testing with a large battery of other corticosteroids yielded only negative results (56).

Two years later, 9 more patients (7 women, 2 men) were reported from the same hospital, seen in 1984 and 1985. The responsible agent was TP nose spray in 5 and ear drops in 4 individuals. In each case, contact dermatitis was localized on the face, especially around the nose or ears. Patch tests with 15 other corticosteroids were always negative safe one reaction to dexamethasone acetate. Patch testing with hydrocortisone 25% pet. was negative in 11/11 patients tested and positive to 20% in alcohol in only one individual (75).

Case reports

A 47-year-old woman used a nasal spray containing 1% tixocortol pivalate (TP) for a non-specific pharyngitis. After 11 days she developed acute rhinitis, with itching, sneezing and rhinorrhea, and therefore increased the dose to 3x daily, which worsened her symptoms. Five days later, she had a burning nasal sensation and developed a pruritic papulovesicular dermatitis around the nose, on the upper lip and on both ear lobes. Patch tests were positive to the spray, tixocortol pivalate 0.1% and 1% pet., hydrocortisone 1% and 2% alc. and to aqueous cetylpyridinium chloride, the preservative in the spray (55).

A 71-year-old woman was treated orally for bronchitis with cefuroxime and tixocortol pivalate plus bacitracin chewing tablets. Eight days after the onset of therapy, she developed pruritus and edema of the left side of the face, with normal overlying skin, pain in the right side of the face, excessive salivation and laryngeal edema. Examination revealed left jugal mucosal edema and erythema. Patch tests were positive to tixocortol pivalate 10% pet. and TP 1% nasal suspension and negative to bacitracin and cefuroxime. The patient was diagnosed with allergic contact stomatitis from tixocortol pivalate (68).

A 38-year-old woman was treated with 1% tixocortol pivalate spray for chronic rhinitis on 2 occasions and both times developed severe local mucosal intolerance without manifestations of the surrounding skin. Patch tests were positive to TP 1% pet. and nickel sulfate. She had previously treated nickel dermatitis many times with various corticosteroids (74). A 60-year-old woman developed facial dermatitis from contact allergy to tixocortol pivalate used in a nasal inhalation (85). A 66-year-old woman developed periorbital edema and eyelid eczema 2 days after she started using tixocortol pivalate-containing nasal spray. Patch tests were positive to the spray and to TP 1% pet. (86). A 31-year-old woman started using tixocortol pivalate nasal spray for allergic rhinitis . Three weeks later she noticed pruritic vesicles at the nostrils with a worsening of her nasal symptoms. Subsequently she developed an itchy patch of eczema on the eyelids. Patch tests were positive to the spray and to TP 1% pet. (86).

A 41-year-old woman with a history of aphthous stomatitis was treated with tixocortol pivalate lozenges. After 2 days she developed deterioration of her oral symptoms with swelling of the upper lip. One day later angioedema of one side of the face developed. Ten days later she again took a TP lozenge and within 1 day experienced pruritic swelling of the lips with respiratory difficulties. Patch tests were positive to tixocortol pivalate 1% pet. (86). One patient developed acute eczema on the nose and cheeks from using TP nasal spray for a week (88).

Two other patients were sensitized to tixocortol pivalate in an auricular preparation (73) and in a nasal spray (91); details are not available to the author. In a patient who developed an allergic reaction to tixocortol pivalate in nasal spray, the symptoms were misinterpreted as infectious complications resulting in hospitalization and considerable medical expenses (87).

Cross-reactions, pseudo-cross-reactions and co-reactions

Cross-reactions between corticosteroids are discussed in Chapter 2.8. See also the sections 'General' above.

Patch test sensitization

A woman who had apparently never used topical corticosteroids had a positive patch test reaction to tixocortol pivalate 1% pet., which had appeared at D10. The authors diagnosed probable active sensitization. Whereas this may well have been the case, late reactions to corticosteroids are well known, and it cannot be excluded that the reaction had begun earlier dan D10 but was not noticed by the patient at that time (54).

PHOTOSENSITIVITY
No evidence has been found for tixocortol pivalate being a photosensitizer by photopatch testing patients with dermatitis localized on light-exposed skin (66).

Miscellaneous side effects
A 16-year-old girl with type II hereditary angioedema developed a flare of her angioedema, causing severe swelling of her face, but no associated laryngeal edema, within 24 hour of being patch tested for facial dermatitis. The only positive reaction was to tixocortol pivalate. The patient's sensitivity to hydrocortisone was thought to be clinically relevant (an unexpected and unsubstantiated claim by the authors). However, it was unclear whether this specific contact allergy or a non-specific effect of patch testing was relevant in triggering the attack of angioedema (67). May be it was just a co-incidence.

LITERATURE
1   DeKoven JG, Warshaw EM, Zug KA, Maibach HI, Belsito DV, Sasseville D, et al. North American Contact Dermatitis Group patch test results: 2015-2016. Dermatitis 2018;29:297-309
2   Veverka KK, Hall MR, Yiannias JA, Drage LA, El-Azhary RA, Killian JM, et al. Trends in patch testing with the Mayo Clinic standard series, 2011-2015. Dermatitis 2018;29:310-315
3   DeKoven JG, Warshaw EM, Belsito DV, Sasseville D, Maibach HI, Taylor JS, et al. North American Contact Dermatitis Group Patch Test Results: 2013-2014. Dermatitis 2017;28:33-46
4   Uter W, Amario-Hita JC, Balato A, Ballmer-Weber B, Bauer A, Belloni Fortina A, et al. European Surveillance System on Contact Allergies (ESSCA): results with the European baseline series, 2013/14. J Eur Acad Dermatol Venereol 2017;31:1516-1525
5   Ochi H, Cheng SWN, Leow YH, Goon ATJ. Contact allergy trends in Singapore – a retrospective study of patch test data from 2009 to 2013. Contact Dermatitis 2017;76:49-50
6   Warshaw EM, Maibach HI, Taylor JS, Sasseville D, DeKoven JG, Zirwas MJ, et al. North American Contact Dermatitis Group patch test results: 2011-2012. Dermatitis 2015;26:49-59
7   Warshaw EM, Belsito DV, Taylor JS, Sasseville D, DeKoven JG, Zirwas MJ, et al. North American Contact Dermatitis Group patch test results: 2009 to 2010. Dermatitis 2013;24:50-59
8   Wentworth AB, Yiannias JA, Keeling JH, Hall MR, Camilleri MJ, Drage LA, et al. Trends in patch-test results and allergen changes in the standard series: a Mayo Clinic 5-year retrospective review (January 1, 2006, to December 31, 2010). J Am Acad Dermatol 2014;70:269-275
9   Toholka R, Wang Y-S, Tate B, Tam M, Cahill J, Palmer A, Nixon R. The first Australian Baseline Series: Recommendations for patch testing in suspected contact dermatitis. Australas J Dermatol 2015;56:107-115
10 Fall S, Bruze M, Isaksson M, Lidén C, Matura M, Stenberg B, Lindberg M. Contact allergy trends in Sweden – a retrospective comparison of patch test data from 1992, 2000, and 2009. Contact Dermatitis 2015;72:297-304
11 Uter W, Aberer W, Armario-Hita JC, , Fernandez-Vozmediano JM, Ayala F, Balato A, et al. Current patch test results with the European baseline series and extensions to it from the 'European Surveillance System on Contact Allergy' network, 2007-2008. Contact Dermatitis 2012;67:9-19
12 Fransway AF, Zug KA, Belsito DV, Deleo VA, Fowler JF Jr, Maibach HI, et al. North American Contact Dermatitis Group patch test results for 2007-2008. Dermatitis 2013;24:10-21
13 Zug KA, Warshaw EM, Fowler JF Jr, Maibach HI, Belsito DL, Pratt MD, et al. Patch-test results of the North American Contact Dermatitis Group 2005-2006. Dermatitis 2009;20:149-160
14 Uter W, Rämsch C, Aberer W, Ayala F, Balato A, Beliauskiene A, et al. The European baseline series in 10 European Countries, 2005/2006 – Results of the European Surveillance System on Contact Allergies (ESSCA). Contact Dermatitis 2009;61:31-38
15 Cheng S, Cao M, Zhang Y, Peng S, Dong J, Zhang D, et al. Time trends of contact allergy to a modified European baseline series in Beijing between 2001 and 2006. Contact Dermatitis 2011;65:22-27
16 Davis MD, Scalf LA, Yiannias JA, Cheng JF, El-Azhary RA, Rohlinger AL, et al. Changing trends and allergens in the patch test standard series. Arch Dermatol 2008;144:67-72
17 ESSCA Writing Group. The European Surveillance System of Contact Allergies (ESSCA): results of patch testing the standard series, 2004. J Eur Acad Dermatol Venereol 2008;22:174-181
18 Warshaw EM, Belsito DV, DeLeo VA, Fowler JF Jr, Maibach HI, Marks JG, et al. North American Contact Dermatitis Group patch-test results, 2003-2004 study period. Dermatitis 2008;19:129-136
19 Uter W, Hegewald J, Aberer W et al. The European standard series in 9 European countries, 2002/2003 – First results of the European Surveillance System on Contact Allergies. Contact Dermatitis 2005;53:136-145

20 Pratt MD, Belsito DV, DeLeo VA, Fowler JF Jr, Fransway AF, Maibach HI, et al. North American Contact Dermatitis Group patch-test results, 2001-2002 study period. Dermatitis 2004;15:176-183

21 Lindberg M, Edman B, Fischer T, Stenberg B. Time trends in Swedish patch test data from 1992 to 2000. A multi-centre study based on age- and sex-adjusted results of the Swedish standard series. Contact Dermatitis 2007;56:205-210

22 Britton JE, Wilkinson SM, English JSC, Gawkrodger DJ, Ormerod AD, Sansom JE, et al. The British standard series of contact dermatitis allergens: validation in clinical practice and value for clinical governance. Br J Dermatol 2003;148:259-264

23 Marks JG Jr, Belsito DV, DeLeo VA, Fowler JF Jr, Fransway AF, Maibach HI, et al. North American Contact Dermatitis Group patch-test results, 1998–2000. Am J Contact Dermat 2003;14:59-62

24 Wetter DA, Davis MDP, Yiannias JA, Cheng JF, Connolly SM, el-Azhary RA, et al. Patch test results from the Mayo Contact Dermatitis Group, 1998–2000. J Am Acad Dermatol 2005;53:416-421

25 Bruynzeel DP, Diepgen TL, Andersen KE, Brandão FM, Bruze M, Frosch PJ, et al (EECDRG). Monitoring the European Standard Series in 10 centres 1996–2000. Contact Dermatitis 2005;53:146-152

26 Erfurt-Berge C, Geier J, Mahler V. The current spectrum of contact sensitization in patients with chronic leg ulcers or stasis dermatitis - new data from the Information Network of Departments of Dermatology (IVDK). Contact Dermatitis 2017;77:151-158

27 Barbaud A, Collet E, Le Coz CJ, Meaume S, Gillois P. Contact allergy in chronic leg ulcers: results of a multicentre study carried out in 423 patients and proposal for an updated series of patch tests. Contact Dermatitis 2009;60:279-287

28 Zmudzinska M, Czarnecka-Operacz M, Silny W, Kramer L. Contact allergy in patients with chronic venous leg ulcers – possible role of chronic venous insufficiency. Contact Dermatitis 2006;54:100-105

29 Tavadia S, Bianchi J, Dawe RS, McEvoy M, Wiggins E, Hamill E, et al. Allergic contact dermatitis in venous leg ulcer patients. Contact Dermatitis 2003;48:261-265

30 Zmudzinska M, Czarnecka-Operacz M, SilnyW. Contact allergy to glucocorticosteroids in patients with chronic venous leg ulcers, atopic dermatitis and contact allergy. Acta Dermatovenerol Croat 2008;16:72-78

31 Cooper SM, Shaw S. Eyelid dermatitis: an evaluation of 232 patchtest patients over 5 years. Contact Dermatitis 2000;42:291-293

32 Warshaw EM, Furda LM, Maibach HI, Rietschel RL, Fowler JF Jr, Belsito DV, et al. Anogenital dermatitis in patients referred for patch testing: retrospective analysis of cross-sectional data from the North American Contact Dermatitis Group, 1994-2004. Arch Dermatol 2008;144:749-755

33 Agulló-Pérez AD, Hervella-Garcés M, Oscoz-Jaime S, Azcona-Rodríguez M, Larrea-García M, Yanguas-Bayona JI. Perianal dermatitis. Dermatitis 2017;28:270-275

34 O'Gorman SM, Torgerson RR. Allergic contact dermatitis of the vulva. Dermatitis 2013;24:64-72

35 Lucke TW, Fleming CJ, McHenry P, Lever R. Patch testing in vulval dermatoses: how relevant is nickel? Contact Dermatitis 1998;38:111-112

36 Warshaw EM, Kimyon RS, Silverberg JI, Belsito DV, DeKoven JG, Maibach HI, et al. Evaluation of patch test findings in patients with anogenital dermatitis. JAMA Dermatol 2019;156:85-91

37 Foley CC, White S, Merry S, Nolan U, Moriarty B, et al. Understanding the role of cutaneous allergy testing in anogenital dermatoses: a retrospective evaluation of contact sensitization in anogenital dermatoses. Int J Dermatol 2019;58:806-810

38 Millard TP, Orton DI. Changing patterns of contact allergy in chronic inflammatory ear disease. Contact Dermatitis 2004;50:83-86

39 Van Ginkel CJ, Bruintjes TD, Huizing EH. Allergy due to topical medications in chronic otitis externa and chronic otitis media. Clin Otolaryngol Allied Sci 1995;20:326-328

40 Diepgen TL, Ofenloch RF, Bruze M, Bertuccio P, Cazzaniga S, Coenraads P-J, et al. Prevalence of contact allergy in the general population in different European regions. Br J Dermatol 2016;174:319-329

41 Mortz CG, Bindslev-Jensen C, Andersen KE. Prevalence, incidence rates and persistence of contact allergy and allergic contact dermatitis in The Odense Adolescence Cohort Study: a 15-year follow-up. Brit J Dermatol 2013;168:318-325

42 Nardelli A, Degreef H, Goossens A. Contact allergic reactions of the vulva: a 14-year review. Dermatitis 2004;15:131-136

43 Teo Y, McFadden JP, White IR, Lynch M, Banerjee P. Allergic contact dermatitis in atopic individuals: Results of a 30-year retrospective study. Contact Dermatitis 2019;81:409-416

44 Lagrelius M, Wahlgren CF, Matura M, Kull I, Lidén C. High prevalence of contact allergy in adolescence: results from the population-based BAMSE birth cohort. Contact Dermatitis 2016;74:44-51

45 van Amerongen CCA, Ofenloch R, Dittmar D, Schuttelaar MLA. New positive patch test reactions on day 7—The additional value of the day 7 patch test reading. Contact Dermatitis 2019;81:280-287.

46 Gilissen L, Goossens A. Frequency and trends of contact allergy to and iatrogenic contact dermatitis caused by topical drugs over a 25-year period. Contact Dermatitis 2016;75:290-302

47 Spring S, Pratt M, Chaplin A. Contact dermatitis to topical medicaments: a retrospective chart review from the Ottawa Hospital Patch Test Clinic. Dermatitis 2012;23:210-213

48 Wilkinson M, Gonçalo M, Aerts O, Badulici S, Bennike NH, Bruynzeel D, et al. The European baseline series and recommended additions: 2019. Contact Dermatitis 2019;80:1-4

49 Baeck M, Goossens A. Immediate and delayed allergic hypersensitivity to corticosteroids: practical guidelines. Contact Dermatitis 2012;66:38-45

50 Wilkinson SM, English JSC. Hydrocortisone sensitivity: clinical features of 59 cases. J Am Acad Dermatol 1992;27:683-687

51 Wilkinson SM, Cartwright PH, English JSC. Hydrocortisone: an important cutaneous allergen. Lancet 1991;337:761-762

52 Uter W, Spiewak R, Cooper SM, Wilkinson M, Sánchez Pérez J, Schnuch A, et al. Contact allergy to ingredients of topical medications: results of the European Surveillance System on Contact Allergies (ESSCA), 2009-2012. Pharmacoepidemiol Drug Saf 2016;25:1305-1312

53 Gibson-Smith B, Fleming CJ, Forsyth A. Contact sensitivity to tixocortol pivalate in sisters. Contact Dermatitis 1998;38:351-352

54 Goldsmith PC, White IR, Rycroft RJ, McFadden JP. Probable active sensitization to tixocortol pivalate. Contact Dermatitis 1995;33:429

55 Bircher AJ. Short induction phase of contact allergy to tixocortol pivalate in a nasal spray. Contact Dermatitis 1990;22:237-238

56 Boujnah-Khouadja A, Brändle I, Reuter G, Foussereau J. Allergy to 2 new corticoid molecules. Contact Dermatitis 1984;11:83-87

57 Hausen BM, Foussereau J. The sensitizing capacity of tixocortol pivalate. Contact Dermatitis 1988;18:63-64

58 Frankild S, Lepoittevin JP, Kreilgaard B, Andersen KE. Tixocortol pivalate contact allergy in the GPMT: frequency and cross-reactivity. Contact Dermatitis 2001;44:18-22

59 Bruze M, Björkner B, Dooms-Goossens A. Sensitization studies with mometasone furoate, tixocortol pivalate, and budesonide in the guinea pig. Contact Dermatitis 1996;34:161-164

60 Kalavala M, Statham BN, Green CM, King C, Ormerod AD, Sansom J, et al. Tixocortol pivalate: what is the right concentration? Contact Dermatitis 2007;57:44-46

61 Chowdhury MM, Statham BN, Sansom JE, Foulds IS, English JS, Podmore P, et al. Patch testing for corticosteroid allergy with low and high concentrations of tixocortol pivalate and budesonide. Contact Dermatitis 2002;46:311-312

62 Isaksson M, Andersen K E, Brandão FM, Goossens A. Patch testing with corticosteroid mixes in Europe. Contact Dermatitis 2000;42:27-35

63 Isaksson M, Brandão FM, FM, Bruze M, Goossens A. Recommendation to include budesonide and tixocortol pivalate in the European standard series. Contact Dermatitis 2000;43:41-42.

64 Lewis FM, Harrington CI, Gawkrodger DJ. Contact sensitivity in pruritus vulvae: a common and manageable problem. Contact Dermatitis 1994;31:264-265

65 Isaksson M, Bruze M, Björkner B, M Hindsén, L Svensson. The benefit of patch testing with a corticosteroid at a low concentration. Am J Contact Dermat 1999;10:31-33

66 Wilkinson SM, Beck MH. Is tixocortol pivalate a photoallergen? Contact Dermatitis 1995;33:55-56

67 Parry EJ, Lever RS. An unusual complication of patch testing. Contact Dermatitis 1995;32:179-180

68 Callens A, Vaillant L, Machet L, Pelucio-Lopez C, de Calan S, Lorette G. Contact stomatitis from tixocortol pivalate. Contact Dermatitis 1993;29:161

69 Lauerma AI, Tarvainen K, Forström L, Reitamo S. Contact hypersensitivity to hydrocortisone-free-alcohol in patients with allergic patch test reactions to tixocortol pivalate. Contact Dermatitis 1993;28:10-14

70 Wilkinson SM, English JSC. Hydrocortisone sensitivity: an investigation into the nature of the allergen. Contact Dermatitis 1991;25:178-181

71 Wilkinson SM, English JSC. Hydrocortisone sensitivity: a prospective study into the value of tixocortol pivalate and hydrocortisone acetate as patch test markers. Contact Dermatitis 1991;25:132-133

72 Lauerma AI, Tarvainen K, Forström L, Reitamo S. Contact hypersensitivity to hydrocortisone-free-alcohol in patients with allergic patch test reactions to tixocortol pivalate. Contact Dermatitis 1993;28:10-14

73 Jelen G, Grosshans E, Foussereau J. Une autre forme d'allergie au pivalate de tixocortol. La lettre du GERDA 1985;5:16 (Article in French)

74 Camarasa JG, Malet A, Serra-Baldrich E, Lluch M. Contact allergy to tixocortol pivalate. Contact Dermatitis 1988;19:147-148

75 Foussereau J, Jelen G. Tixocortol pivalate – an allergen closely related to hydrocortisone. Contact Dermatitis 1986;15:37-38

76 Uter W, Geier J, Richter G, Schnuch A; IVDK Study Group, German Contact Dermatitis Research Group. Patch test results with tixocortol pivalate and budesonide in Germany and Austria. Contact Dermatitis 2001;44:313-314

77 Lutz ME, el-Azhary RA, Gibson LE, Fransway AF. Contact hypersensitivity to tixocortol pivalate. J Am Acad Dermatol 1998;38:691-695

78 Morton CE, Dohil MA. A survey of patch test results with tixocortol pivalate in Vancouver. Am J Contact Dermat 1995;6:17-18

79 Dooms-Goossens A, Andersen KE, Burrows D, Camarasa JG, Ducombs G, Frosch PJ, et al. A survey of the results of patch tests with tixocortol pivalate. Contact Dermatitis 1989;20:158

80 Davis MD, El-Azhary RA, Farmer SA. Results of patch testing to a corticosteroid series: a retrospective review of 1188 patients during 6 years at Mayo Clinic. J Am Acad Dermatol 2007;56:921-927

81 Baeck M, Chemelle JA, Terreux R, Drieghe J, Goossens A. Delayed hypersensitivity to corticosteroids in a series of 315 patients: clinical data and patch test results. Contact Dermatitis 2009;61:163-175

82 Mercader-García P, Pastor-Nieto MA, García-Doval I, Giménez-Arnau A, González-Pérez R, Fernández-Redondo V, et al. GEIDAC. Are the Spanish baseline series markers sufficient to detect contact allergy to corticosteroids in Spain? A GEIDAC prospective study. Contact Dermatitis 2018;78:76-82

83 Pratt MD, Mufti A, Lipson J, Warshaw EM, Maibach HI, Taylor JS, et al. Patch test reactions to corticosteroids: Retrospective analysis from the North American Contact Dermatitis Group 2007-2014. Dermatitis 2017;28:58-63

84 Vind-Kezunovic D, Johansen JD, Carlsen BC. Prevalence of and factors influencing sensitization to corticosteroids in a Danish patch test population. Contact Dermatitis 2011;64:325-329

85 Baeck M, Pilette C, Drieghe J, Goossens A. Allergic contact dermatitis to inhalation corticosteroids. Eur J Dermatol 2010;20:102-108

86 Bircher AJ, Pelloni F, Langauer Messmer S, Müller D. Delayed hypersensitivity reactions to corticosteroids applied to mucous membranes. Br J Dermatol 1996;135:310-313

87 Bircher AJ, Hirsbrunner P, Tschopp K, Wildermuth V. Allergic contact dermatitis from tixocortol pivalate in a nasal spray masquerading as infectious complication of sinusitis. J Otorhinolaryng Relat Spec 1995:57:54-56

88 Jeune R. Eczéma de contact à la Pivalone (nouvelle observation). La Lettre du GERDA 1991;8:60 (Article in French)

89 Tam I, Schalock PC, González E, Yu J. Patch testing results from the Massachusetts General Hospital Contact Dermatitis Clinic, 2007-2016. Dermatitis 2020;31:202-208

90 Boyvat A, Kalay Yildizhan I. Patch test results of the European baseline series among 1309 patients in Turkey between 2013 and 2019. Contact Dermatitis 2020 Jul 3. doi: 10.1111/cod.13653. Online ahead of print.

91 Foussereau J, Jelen G. Une cause méconnue d'eczéma allergique de la face: un corticoide, le pivalate de tixocortol. La Presse Medicale 1987;16:832 (Article in French)

Chapter 3.343 TOBRAMYCIN

IDENTIFICATION

Description/definition : Tobramycin is an aminoglycoside antibiotic produced by *Streptomyces tenebrarius* that conforms to the structural formula shown below
Pharmacological classes : Anti-bacterial agents
IUPAC name : (2S,3R,4S,5S,6R)-4-Amino-2-[(1S,2S,3R,4S,6R)-4,6-diamino-3-[(2R,3R,5S,6R)-3-amino-6-(aminomethyl)-5-hydroxyoxan-2-yl]oxy-2-hydroxycyclohexyl]oxy-6-(hydroxymethyl)-oxane-3,5-diol
Other names : D-Streptamine, *O*-3-amino-3-deoxy-α-D-glucopyranosyl-(1→6)-*O*-[2,6-diamino-2,3,6-trideoxy-α-D-ribohexopyranosyl-(1→4)]-2-deoxy-
CAS registry number : 32986-56-4
EC number : 251-322-5
Merck Index monograph : 10917
Patch testing : 20.0% pet. (Chemotechnique, SmartPracticeCanada); late readings are advisable (7)
Molecular formula : $C_{18}H_{37}N_5O_9$

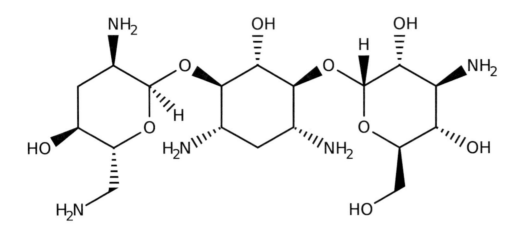

GENERAL

Tobramycin is an aminoglycoside broad-spectrum antibiotic produced by *Streptomyces tenebrarius* with bacterio-static activity. It is effective against gram-negative bacteria, especially the *Pseudomonas* species. Tobramycin is a 10% component of the antibiotic complex, nebramycin, produced by the same species. This agent is indicated for the treatment of *Pseudomonas aeruginosa* lung infections. Topically, it is mostly used in eye drops to treat infectious conjunctivitis and keratitis caused by *Pseudomonas* and in otic medications (eardrops). In pharmaceutical products, both tobramycin and tobramycin sulfate (CAS number 49842-07-1, EC number 256-499-2, molecular formula $C_{18}H_{39}N_5O_{13}S$) may be employed (1).

CONTACT ALLERGY

General

Although tobramycin is widely used in ophthalmological drugs, contact allergy appears to be very infrequent with only a few case reports and one small case series (7) published. Twenty-seven patients with allergic contact dermatitis from tobramycin in eye medications were reported from Leuven, Belgium, but this was in a 27-year-period (3). High frequencies of positive patch tests to tobramycin were found in some studies in selected patient groups (2,4,5), but many were probably cross-reactions to primary neomycin sensitization.

Patch testing in groups of patients

Patch testing with tobramycin has been performed in selected patient groups in 3 studies (table 3.343.1). High percentages (29-36%) positive reactions were found in patients suspected of iatrogenic contact dermatitis (5), suspected of contact allergy to ophthalmological drugs (4) and individuals with chronic otitis externa or media (2). However, in one of these investigations from The Netherlands, all positive reactions represented cross-allergies, as the drug was at that time not available in that country (2). It is highly probable that, also in the other 2 studies, cross-reactivity must have played an significant role.

Table 3.343.1 Patch testing in groups of patients: Selected patient groups

Years and Country	Test conc. & vehicle	Number of patients tested | positive (%)		Selection of patients (S); Relevance (R); Comments (C)	Ref.
1990-2014 Belgium	20% pet.	186	63 (34%)	S: patients suspected of iatrogenic contact dermatitis and tested with a pharmaceutical series and their own products; R: 96% of the positive patch test reactions to all topical drugs and antiseptics were considered to be relevant (overlap with reference 3, from which shows that a number of the positive reactions must have been cross-reactions)	5
2000-6 Netherlands		22	8 (36%)	S: patients patch tested for suspected contact allergy to ophthalmological drugs; R: 69% for all reactions to drugs together	4
1993-4 Netherlands	20% pet.	34	19 (29%)	S: patients with chronic otitis externa or media; R: as tobramycin at that time was not used in The Netherlands, these reactions must represent cross-reactions to neomycin	2

Case series

In Leuven, Belgium, between 1990 and 2017, 16,065 patients were investigated for contact allergy and 118 (0.7%) showed positive patch test reactions to topical ophthalmic medications and/or to their ingredients. Eighty-four individuals (71%) reacted to an active principle. Tobramycin was tested in 198 patients and was the allergen in eye medications in 27. There were also 11 reactions to tobramycin in other types of medications (3, overlap with ref. 5).

A series of 3 patients with allergic contact dermatitis from tobramycin was reported from the USA in 2016 (7). A 66-year-old man presented with a 1-year history of upper eyelid dermatitis. He had been treated with various topical pharmaceuticals including tobramycin 0.3% - dexamethasone ophthalmic ointment (TDOO). A 35-year-old woman presented with a 2-year history of eyelid dermatitis initially associated with makeup. Her symptoms worsened after a stenting procedure with silicone tubing for nasolacrimal duct obstruction. Her postoperative course was complicated by local infection. After treatment with TDOO, she developed dermatitis of the cheeks and chin. A 59-year-old woman presented with a 1-year history of erythematous, scaly eyelid dermatitis. Previous treatments included tacrolimus ointment and TDOO. Patch tests were positive in all 3 patients to tobramycin 20% pet. and in 2 to TDOO tested 'as is'. Neomycin yielded ?+ reactions in 2 individuals and was negative in the third (7).

Case reports

A 56-year-old man had a left total knee arthroplasty in which tobramycin-containing bone cement was used. Three years later, after arthroscopic lysis of adhesions was performed, 'left knee allergic contact dermatitis' developed. Patch testing showed a positive reaction to gentamicin, 'which cross reacts with tobramycin. 'Tobramycin patch testing is pending' (6). This author has not been able to find a formal article following this Abstract.

A 90-year-old woman presented with severe bilateral erythematous and edematous periorbital dermatitis of 1 month' duration. She had been treated with tobramycin 0.3% ophthalmic solution and ointment for bilateral bacterial conjunctivitis and after approximately 1 month of use the rash started. Dermatitis had persisted despite additional treatment with a topical corticosteroid. Patch tests were positive to the ophthalmic solution and ointment at D2 and D4. There was no reaction to neomycin or the vehicle ingredients of the tobramycin preparations (8).

A 59-year-old woman used various ophthalmics for bilateral open angle glaucoma and had ocular surgery (left eye cataract removal) a month ago. She was then prescribed tobramycin 0.3% - dexametasone eye drops and atropine. Several days later, the patient developed severe erythema, edema and exudation of the left eye and eyelid, ocular pruritus and conjunctival hyperemia. Patch tests were positive to tobramycin 0.3% pet. and negative to the (diluted) eye drops, dexamethasone and the excipients (9).

A 70-year-old woman, while using various ophthalmics for bilateral dacryocystitis, developed severe erythema, edema and exudation of the eyelids and zygomatic regions, with conjunctival hyperemia. Patch tests were positive to 0.3% tobramycin ophthalmic ointment 'as is' and tobramycin sulfate 5% pet. (D2 +, D3 ++) and 25% pet. (+++). Twenty controls were negative. Neomycin and other aminoglycosides were also negative (10).

A 32-year-old woman developed severe dermatitis of the external auditory meatus after the instillation of eardrops containing betamethasone and sulfamethazine, mixed with an ampoule of tobramycin. Patch tests were positive to tobramycin, kanamycin, ribostamycin and sisomycin (all 20% water). Neomycine 20% pet., however, was negative (11).

Cross-reactions, pseudo-cross-reactions and co-reactions

In patients sensitized to neomycin, about 60% cross-react to tobramycin (Chapter 3.235 Neomycin). The cross-sensitivity pattern between aminoglycoside antibiotics in patients primarily sensitized to tobramycin has not been well investigated. However, there are indications that such patients often do *not* cross-react to neomycin (7,8,10,11). One patient allergic to tobramycin but not to neomycin co-reacted to kanamycin, ribostamycin and sisomycin (all 20% water) (11).

LITERATURE

1   The data in the section 'General' may have been obtained from literature discussed in this chapter, but mostly also or exclusively from one or more of the following online sources: ChemIDPlus Advanced, PubChem, DrugBank, RxList, Drug Central, Drugs.com, and Wikipedia

2   Van Ginkel CJ, Bruintjes TD, Huizing EH. Allergy due to topical medications in chronic otitis externa and chronic otitis media. Clin Otolaryngol Allied Sci 1995;20:326-328

3   Gilissen L, De Decker L, Hulshagen T, Goossens A. Allergic contact dermatitis caused by topical ophthalmic medications: Keep an eye on it! Contact Dermatitis 2019;80:291-297

4   Wijnmaalen AL, van Zuuren EJ, de Keizer RJ, Jager MJ. Cutaneous allergy testing in patients suspected of an allergic reaction to eye medication. Ophthalmic Res 2009;41:225-229

5   Gilissen L, Goossens A. Frequency and trends of contact allergy to and iatrogenic contact dermatitis caused by topical drugs over a 25-year period. Contact Dermatitis 2016;75:290-302

6   Sawchuk MA, Pratt M, Lipson J. Allergic contact dermatitis caused by tobramycin contained in bone cement. Dermatitis 2017;27(5):e1

7   Hagen S, Grey K, Warshaw E. Tobramycin sensitivity is not consistently detected by neomycin on patch testing. Dermatitis 2016;27:152-155

8   Alvarez MS, Brancaccio RR. Periorbital dermatitis. Dermatitis 2006;17:43-44

9   González-Mendiola MR, Balda AG, Delgado MC, Montaño PP, De Olano DG, Sánchez-Cano M. Contact allergy from tobramycin eyedrops. Allergy 2005;60:527-528

10   Caraffini S, Assalve D, Stingeni L, Lisi P. Allergic contact conjunctivitis and blepharitis from tobramycin. Contact Dermatitis 1995;32:186-187

11   Menéndez Ramos F, Llamas Martín R, Zarco Olivo C, Dorado Bris JM, Merino Luque MV. Allergic contact dermatitis from tobramycin. Contact Dermatitis 1990;22:305-306

Chapter 3.344 TOCOPHEROL

IDENTIFICATION

Description/definition	: Tocopherol is a racemic mixture of naturally occurring tocopherols, including α-tocopherol, β-tocopherol, γ-tocopherol and δ-tocopherol; their structural formulas are shown below
Pharmaceutical classes	: Vitamins; antioxidants
IUPAC name	: 3,4-Dihydro-2,5,7,8-tetramethyl-2-(4,8,12-trimethyltridecyl)-2H-benzopyran-6-ol
Other names	: Vitamin E; natural vitamin E; mixed tocopherols; DL-α-tocopherol
CAS registry number	: 1406-66-2 (tocopherols); 59-02-9 (α-tocopherol); 10191-41-0 (DL-α-tocopherol); 148-03-8 (β-tocopherol); 54-28-4 (γ-tocopherol); 119-13-1 (δ-tocopherol)
EC number	: 233-466-0; 218-197-9; 200-412-2; 205-708-5; 204-299-0; 200-201-5
Merck Index monograph	: 10923 (α-tocopherol); 10924 (β-tocopherol); 10925 (γ-tocopherol); 10926 (δ-tocopherol
Patch testing	: DL-α-Tocopherol 100% (Chemotechnique, SmartPracticeCanada)
Molecular formula	: $C_{29}H_{50}O_2$ (α-tocopherol; DL-α-tocopherol); $C_{28}H_{48}O_2$ (β-tocopherol; γ-tocopherol); $C_{27}H_{46}O_2$ (δ-tocopherol)

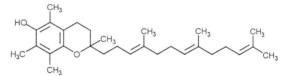

α-tocopherol

α-tocotrienol

β-tocopherol

β-tocotrienol

γ-tocopherol

γ-tocotrienol

δ-tocopherol

δ-tocotrienol

GENERAL

Vitamin E (tocopherol) is a term that encompasses a group of potent lipid-soluble antioxidants. Structural analyses have revealed that molecules having vitamin E antioxidant activity include four tocopherols (α, β, γ, δ) and four tocotrienols (α, β, γ, δ). α-Tocopherol is the most abundant form in nature and has the highest biological activity (20,23). This vitamin is considered essential for the stabilization of biological membranes, especially those with high amounts of polyunsaturated fatty acids. It is a potent peroxyl radical scavenger and inhibits noncompetitively cyclooxygenase activity in many tissues, resulting in a decrease in prostaglandin production. Vitamin E also inhibits angiogenesis and tumor dormancy through suppressing vascular endothelial growth factor (VEGF) gene transcription.

Vitamin E is found in many dietary products. Pharmaceutical tocopherol can be used as a dietary supplement for patients with a deficit of vitamin E. Vitamin E deficiency is rare, and it is primarily found in premature babies of very low birth weight, patients with fat malabsorption and patients with a-β-lipoproteinemia, which is a rare, inherited disorder that causes poor absorption of dietary fat.

As tocopherol purportedly minimizes (photo)aging, increases stratum corneum hydration, and helps wound healing, manufacturers commonly include it in cosmetics and skin care products. Frequently, patients apply vitamin E to scars; oral intake is also popular among the health conscious public. Indeed, the practice of prescribing vitamin E after surgery for scar prevention and treatment is widespread and increasingly popular among both the public and clinicians, although there is not yet sufficient evidence that monotherapy with topical vitamin E has a significant beneficial effect on scar appearance to justify its widespread use (34).

Being a strong antioxidant, tocopherol is unstable and is usually esterified to facilitate stability; tocopheryl acetate is the most widely used form (21). It should be realized that, in several articles on contact allergy, the term 'vitamin E' was used in the title, whereas the allergenic component was in fact tocopheryl acetate (12,22,35). These articles are discussed in Chapter 3.345 Tocopheryl acetate.

A review of contact allergy to tocopherol and vitamin E derivatives was published in 2010 (1). Most allergic reactions to tocopherol are from cosmetic applications rather than from its use in pharmaceuticals. Delayed-type allergy in a young boy to tocopherol in foods may have caused widespread allergic dermatitis on the face, neck, scalp and back, that had for years been diagnosed as atopic dermatitis (42).

CONTACT ALLERGY

Patch testing in groups of patients

Results of patch testing tocopherol in consecutive patients suspected of contact dermatitis (routine testing) and of testing groups of selected patients are shown in table 3.344.1. Tocopherol has been part of the screening series of the NACDG since 2001; its members publish their patch test results biannually. There is only one (early) study with tocopherol routine testing from outside the USA (+ Canada), a very small investigation from Denmark (13). In North America, in 13 studies performing routine testing, rates of sensitization to tocopherol have always been low and ranged from 0.5% to 1.1%. 'Definite' or 'probable' relevance has ranged from 25% to 59% in the NACDG studies (table 3.344.1). In three groups of *selected* patients (patients with anogenital dermatitis, individuals tested with a cosmetic series, patients with chronic cheilitis), rates were 0.9% (41), 1.3% (4) and 1.6% (40).

Table 3.344.1 Patch testing in groups of patients

Years and Country	Test conc. & vehicle	Number of patients tested	positive (%)		Selection of patients (S); Relevance (R); Comments (C)	Ref.
Routine testing						
2015-2017 NACDG	pure	5590	38	(0.7%)	R: definite + probable relevance: 42%	36
2011-2015 USA	pure	2574	13	(0.5%)	R: not stated	37
2013-2014 NACDG	pure	4859	29	(0.6%)	R: definite + probable relevance: 52%	38
2011-2012 NACDG	pure	4230	30	(0.7%)	R: definite + probable relevance: 63%	25
2009-2010 NACDG	pure	4304	43	(1.0%)	R: definite + probable: relevance 44%	26
2006-2010 USA	pure	2974		(0.8%)	R: 36%	33
2007-2008 NACDG	pure	5082		(1.0%)	R: definite + probable relevance: 59%	15
1987-2007 USA	various	2950	18	(0.6%)	R: all reactions were considered relevant; C: test concentrations: 1987-1999: various; 2000-2005 tocopheryl acetate 10% pet; .2005 pure vitamin E	21
2005-2006 NACDG	pure	4435		(0.7%)	R: definite + probable relevance: 33%	14
2000-2005 USA	10% pet.	3389		(0.5%)	R: 44%	5
2003-2004 NACDG	pure	5139	57	(1.1%)	R: not stated	16
2001-2002 NACDG	pure	4881		(1.1%)	R: definite + probable relevance: 25%	31
1998-2000 USA	10% pet.	711		(0.8%)	R: not stated	32
<1976 Denmark	20% pet.	116	1	(0.9%)	R: not found	13
Testing in groups of selected patients						
2005-2016 NACDG	100% pet.	449	4	(0.9%)	S: patients with only anogenital dermatitis; R: all positives represent relevant reactions; C: the frequency was not significantly different from a control group	41
2000-2007 USA	pure	318	5	(1.3%)	S: patients tested with a supplemental cosmetic screening series; R: 75%; C: weak study: a. high rate of macular erythema and weak reactions; b. relevance figures included 'questionable' and 'past' relevance	4
2001-2006 Italy	20% pet.	129	2	(1.6%)	S: patients with chronic cheilitis; R: 0 reactions were relevant	40

Case reports (pharmaceutical products)

Occupational allergic contact dermatitis from vitamin E in a medicine developed in a veterinary surgeon (29). A woman suffered from allergic contact dermatitis to 'pure vitamin E oil' applied to abort vitiliginous lesions (8). Another young female patient had allergic contact dermatitis from vitamin E in a pharmaceutical vitamin cream containing 5% tocopherol (30). A woman developed erythema multiforme-like allergic contact dermatitis from DL-α-tocopherol in pure vitamin E oil and two cosmetic products (17). Another female individual developed redness, swelling, stinging, and infection of the face requiring antibiotics after the application of a mixture of the content of pure 'vitamin E' capsule with vegetable fat for facial care after a full-face chemical peel had been performed. Patch tests were not performed (7). A young woman developed allergic contact dermatitis from synthetic α-tocopherol in a vitamin cream (11). A similar case was reported from the USA (27).

Allergic contact dermatitis from tocopherol in cosmetic products

Tocopherol was responsible for 6 out of 959 cases of non-fragrance cosmetic allergy where the causal allergen was identified, Belgium, 2000-2010 (3). Tocopherol was responsible for 2 out of 399 cases of cosmetic allergy where the causal allergen was identified in a study of the NACDG, USA, 1977-1983 (2). Among 2193 patients with (presumed) cosmetic allergy seen by the NACDG between 2001-2004, 32 had a positive patch test reaction to dl-α-tocopherol 100% (6). Cosmetic allergy has been caused by tocopherol present in (aerosol) deodorants (9,10,27), a night cream (24), skin care products (28,39), creams (17), sunscreens (18) and a moisturizing cream (19).

Cross-reactions, pseudo-cross-reactions and co-reactions

In a NACDG study, all 12 patients reacting to tocopheryl acetate also reacted to tocopherol (31).

Immediate contact reactions

Immediate contact reactions (contact urticaria) to tocopherol are presented in Chapter 5.

LITERATURE

1   Kosari P, Alikhan A, Sockolov M, Feldman SR. Vitamin E and allergic contact dermatitis. Dermatitis 2010;21:148-153

2   Adams RM, Maibach HI, Clendenning WE, Fisher AA, Jordan WJ, Kanof N, et al. A five-year study of cosmetic reactions. J Am Acad Dermatol 1985;13:1062-1069

3   Travassos AR, Claes L, Boey L, Drieghe J, Goossens A. Non-fragrance allergens in specific cosmetic products. Contact Dermatitis 2011;65:276-285

4   Wetter DA, Yiannias JA, Prakash AV, Davis MD, Farmer SA, el-Azhary RA, et al. Results of patch testing to personal care product allergens in a standard series and a supplemental cosmetic series: an analysis of 945 patients from the Mayo Clinic Contact Dermatitis Group, 2000-2007. J Am Acad Dermatol 2010;63:789-798

5   Davis MD, Scalf LA, Yiannias JA, Cheng JF, El-Azhary RA, Rohlinger AL, et al. Changing trends and allergens in the patch test standard series. Arch Dermatol 2008;144:67-72

6   Warshaw EM, Buchholz HJ, Belsito DV et al. Allergic patch test reactions associated with cosmetics: Retrospective analysis of cross-sectional data from the North American Contact Dermatitis Group, 2001-2004. J Am Acad Dermatol 2009;60:23-38

7   Hunter D, Frumkin A. Adverse reactions to vitamin E and Aloe vera preparations after dermabrasion and chemical peel. Cutis 1991;47:193-196

8   Goldman MP, Rapaport M. Contact dermatitis to vitamin E oil. J Am Acad Dermatol 1986;14:133-134

9   Minkin W, Cohen HI, Frank SB. Contact dermatitis from deodorants. Arch Dermatol 1973;107:774-775

10   Aeling JL, Panagotacos PI, Andreozzi RJ. Allergic contact dermatitis to vitamin E aerosol deodorant. Arch Dermatol 1973;108:579-580

11   Brodkin RH, Bleiberg J. Sensitivity to topically applied vitamin E. Arch Dermatol 1965;92:76-77

12   Corazza M, Minghetti S, Borghi A, Bianchi, A, Virgili A. Vitamin E contact allergy: a controversial subject. Dermatitis 2012;23:167-169

13   Roed-Petersen J, Hjorth N. Contact dermatitis from antioxidants. Br J Dermatol 1976;94:233-241

14   Zug KA, Warshaw EM, Fowler JF Jr, Maibach HI, Belsito DL, Pratt MD, et al. Patch-test results of the North American Contact Dermatitis Group 2005-2006. Dermatitis 2009;20:149-160

15   Fransway AF, Zug KA, Belsito DV, Deleo VA, Fowler JF Jr, Maibach HI, et al. North American Contact Dermatitis Group patch test results for 2007-2008. Dermatitis 2013;24:10-21

16   Warshaw EM, Belsito DV, DeLeo VA, Fowler JF Jr, Maibach HI, Marks JG, et al. North American Contact Dermatitis Group patch-test results, 2003-2004 study period. Dermatitis 2008;19:129-136

17 Saperstein H, Rapaport M, Rietschel RL. Topical vitamin E as a cause of erythema-multiforme-like eruption. Arch Dermatol 1984;120:906-908

18 Simonsen AB, Koppelhus U, Sommerlund M, Deleuran M. Photosensitivity in atopic dermatitis complicated by contact allergy to common sunscreen ingredients. Contact Dermatitis 2016;74:56-58

19 Ramírez Santos A, Fernández-Redondo V, Pérez Pérez L, Concheiro Cao J, Toribio J. Contact allergy from vitamins in cosmetic products. Dermatitis 2008;19:154-156

20 Caraffa AL, Varvara G, Spinas E, Kritas SK, Lessiani G, Ronconi G, et al. Is vitamin E an anti-allergic compound? J Biol Regul Homeost Agents 2016;30:11-15

21 Adams AK, Connolly SM. Allergic contact dermatitis from vitamin E: the experience at Mayo Clinic Arizona, 1987 to 2007. Dermatitis 2010;21:199-202

22 Harris, BD, Taylor JS. Contact allergy to vitamin E capsules: false-negative patch tests to vitamin E? Contact Dermatitis 1997;36:273

23 Brigelius-Flohe R, Traber MG. Vitamin E: function and metabolism. FASEB J 1999;13:1145-1155

24 De Groot AC, Rustemeyer T, Hissink D, Bakker M. Contact allergy to capryloyl salicylic acid. Contact Dermatitis 2014;71:185-187

25 Warshaw EM, Maibach HI, Taylor JS, Sasseville D, DeKoven JG, Zirwas MJ, et al. North American Contact Dermatitis Group patch test results: 2011-2012. Dermatitis 2015;26:49-59

26 Warshaw EM, Belsito DV, Taylor JS, Sasseville D, DeKoven JG, Zirwas MJ, et al. North American Contact Dermatitis Group patch test results: 2009 to 2010. Dermatitis 2013;24:50-59

27 Maibach HI, Marzulli F. Personnal communication to the authors of ref. 13, 1974

28 Fisher AA. Cosmetic warning: this product may be detrimental to your purse. Cutis 1987;39:23-24

29 Hjorth N. Contact dermatitis from vitamin E and from combelen (Bayer) in a veterinary surgeon. Contact Dermatitis Newsletter 1974;15:434

30 Bazzano C, de Angeles S, Kleist G, Macedo N. Allergic contact dermatitis from topical vitamins A and E. Contact Dermatitis 1996;35:261-262

31 Pratt MD, Belsito DV, DeLeo VA, Fowler JF Jr, Fransway AF, Maibach HI, et al. North American Contact Dermatitis Group patch-test results, 2001–2002 study period. Dermatitis 2004;15:176-183

32 Wetter DA, Davis MDP, Yiannias JA, Cheng JF, Connolly SM, el-Azhary RA, et al. Patch test results from the Mayo Contact Dermatitis Group, 1998–2000. J Am Acad Dermatol 2005;53:416-421

33 Wentworth AB, Yiannias JA, Keeling JH, Hall MR, Camilleri MJ, Drage LA, et al. Trends in patch-test results and allergen changes in the standard series: a Mayo Clinic 5-year retrospective review (January 1, 2006, to December 31, 2010). J Am Acad Dermatol 2014;70:269-275

34 Tanaydin V, Conings J, Malyar M, van der Hulst R, van der Lei B. The role of topical vitamin E in scar management: A systematic review. Aesthet Surg J 2016;36:959-965

35 Garcia-Bravo B, Mozo P. Generalized contact dermatitis from vitamin E. Contact Dermatitis 1992;26:280

36 DeKoven JG, Warshaw EM, Zug KA, Maibach HI, Belsito DV, Sasseville D, et al. North American Contact Dermatitis Group patch test results: 2015-2016. Dermatitis 2018;29:297-309

37 Veverka KK, Hall MR, Yiannias JA, Drage LA, El-Azhary RA, Killian JM, et al. Trends in patch testing with the Mayo Clinic standard series, 2011-2015. Dermatitis 2018;29:310-315

38 DeKoven JG, Warshaw EM, Belsito DV, Sasseville D, Maibach HI, Taylor JS, et al. North American Contact Dermatitis Group Patch Test Results: 2013-2014. Dermatitis 2017;28:33-46

39 Guin JD. Eyelid dermatitis: A report of 215 patients. Contact Dermatitis 2004;50:87-90

40 Schena D, Fantuzzi F, Girolomoni G. Contact allergy in chronic eczematous lip dermatitis. Eur J Dermatol 2008;18:688-692

41 Warshaw EM, Kimyon RS, Silverberg JI, Belsito DV, DeKoven JG, Maibach HI, et al. Evaluation of patch test findings in patients with anogenital dermatitis. JAMA Dermatol 2019;156:85-91

42 Chen R, Raffi J, Murase JE. Tocopherol allergic dermatitis masquerading as lifelong atopic dermatitis. Dermatitis 2020;31:e3-e4

Chapter 3.345 TOCOPHERYL ACETATE

IDENTIFICATION

Description/definition   : Tocopheryl acetate is the ester of tocopherol and acetic acid that conforms to the
                             structural formula shown below
Pharmacological classes   : Vitamins; antioxidants
IUPAC name   : [2,5,7,8-Tetramethyl-2-(4,8,12-trimethyltridecyl)-3,4-dihydrochromen-6-yl] acetate
Other names   : Vitamin E acetate; tocopherol acetate
CAS registry number   : 7695-91-2; 58-95-7
EC number   : 231-710-0; 200-405-4
Merck Index monograph   : 10923 (α-tocopherol)
Patch testing   : 10% pet. (Chemotechnique)
Molecular formula   : $C_{31}H_{52}O_3$

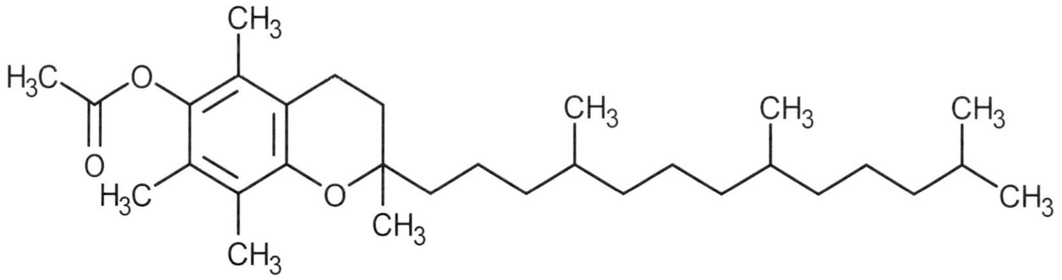

GENERAL

Tocopheryl acetate is the acetate ester of tocopherol (vitamin E). Being a strong antioxidant, tocopherol is unstable and is usually esterified to facilitate stability; tocopheryl acetate is the most widely used form. The primary health-related use for which tocopheryl acetate is formally indicated is as a dietary supplement for patients who demonstrate a genuine vitamin E deficiency. At the same time, vitamin E deficiency is generally quite rare but may occur in premature babies of very low birth weight (< 1500 grams), individuals with fat-malabsorption disorders (as fat is required for the digestive tract to absorb vitamin E), or individuals with a-β-lipoproteinemia, which is a rare, inherited disorder that causes poor absorption of dietary fat (1).

As vitamin E is perceived to minimize photo-aging, increase stratum corneum hydration, and help wound healing, manufacturers commonly include it in cosmetics and skin care products, often as tocopheryl acetate (3).

CONTACT ALLERGY

General

Patch test reactions to tocopheryl acetate are sometimes unexpectedly negative (false-negative), which has been ascribed to the anti-inflammatory action of tocopherols (3,6). A review of contact allergy to vitamin E and vitamin E derivatives was published in 2010 (16).

Patch testing in groups of patients

In 2001-2002, in the USA, the members of the North American Contact Dermatitis Group (NACDG) patch tested 4874 consecutive patients suspected of contact dermatitis (routine testing) with tocopheryl acetate 100% and there were 24 (0.5%) reactions. Forty per cent were considered to be of 'definite' or 'probable' relevance. All allergic patients also reacted to tocopherol and, as tocopherol was already part of the NACDG screening series, tocopheryl acetate was removed from it (9).

Case reports (pharmaceutical products)

One patient had allergic contact dermatitis from tocopheryl acetate in a topical pharmaceutical product (2). Another individual had erythema multiforme-like contact dermatitis from vitamin E oil used to prevent scars, which consisted exclusively of tocopheryl acetate (5). A patient had a reaction to the contents of a vitamin E capsule (in the form of dl-α-tocopheryl acetate) applied to the skin. A patch test with it was negative, but a ROAT positive. Contact allergy to soybean oil, gelatin, and glycerin, the other constituents of the vitamin E capsule could not be excluded, but the patient had repeatedly developed skin reactions to cosmetics containing tocopherol (6).

Allergic contact dermatitis from tocopherol in cosmetic products

Tocopheryl acetate was stated to be the (or an) allergen in one patient in a group of 603 individuals suffering from cosmetic dermatitis, seen in the period 2010-2015 in Leuven, Belgium (10). Among 2193 patients with (presumed) cosmetic allergy seen by the NACDG between 2001 and 2004, 9 had a positive patch test reaction to dl-α-tocopheryl acetate 100% associated with a moisturizer. The total number of positive reactions was not mentioned and tocopheryl acetate had been tested in 2 years of the 4- year-period only (11).

Four patients had dermatitis from contact allergy to tocopheryl acetate in cosmetic creams (4). One patient had contact dermatitis from allergy to tocopheryl acetate in a moisturizing cream (7). Another individual had generalized ACD from tocopheryl acetate in a 'soapy oats lotion' (8). An unspecified number of patients with eyelid dermatitis had allergic reactions to tocopheryl acetate in personal care products they used (15). Two women may have had allergic reactions to tocopheryl acetate in cosmetic products, one of which was an oil consisting entirely of tocopheryl acetate (3). Eight patients reacted to a cosmetic cream ('lipogel') and had positive patch tests to the product. Testing with the ingredients, however, was negative (14). The possibility of compound allergy was suggested, as this had been observed previously (positive reactions to the product, negative to the constituents, formation of a new allergen in the product) (12,13).

Cross-reactions, pseudo-cross-reactions and co-reactions

Cross-reaction to tocopheryl nicotinate (4). In one NACDG study, all 12 patients reacting to tocopheryl acetate also reacted to tocopherol (9).

LITERATURE

1   The data in the section 'General' may have been obtained from literature discussed in this chapter, but mostly also or exclusively from one or more of the following online sources: ChemIDPlus Advanced, PubChem, DrugBank, RxList, Drug Central, Drugs.com, and Wikipedia

2   Matsumura T, Nakada T, Iijima M. Widespread contact dermatitis from tocopherol acetate. Contact Dermatitis 2004;51:211-212

3   Corazza M, Minghetti S, Borghi A, Bianchi, A, Virgili A. Vitamin E contact allergy: a controversial subject. Dermatitis 2012;23:167-169

4   De Groot AC, Berretty PJM, Van Ginkel CJW, Den Hengst CW, Van Ulsen J, Weyland JW. Allergic contact dermatitis from tocopheryl acetate in cosmetic cream. Contact Dermatitis 1991;25:302-304

5   Saperstein H, Rapaport M, Rietschel RL. Topical vitamin E as a cause of erythema multiforme-like eruption. Arch Dermatol 1984;120:906-908

6   Harris, BD, Taylor JS. Contact allergy to vitamin E capsules: false-negative patch tests to vitamin E? Contact Dermatitis 1997;36:273

7   Manzano D, Aguirre A, Gardeazabal J, Eizaguirre X, Pérez JLD. Allergic contact dermatitis from tocopheryl acetate (vitamin E) and retinol palmitate (vitamin A) in a moisturizing cream. Contact Dermatitis 1994;31:324

8   Garcia-Bravo B, Mozo P. Generalized contact dermatitis from vitamin E. Contact Dermatitis 1992;26:280

9   Pratt MD, Belsito DV, DeLeo VA, Fowler JF Jr, Fransway AF, Maibach HI, et al. North American Contact Dermatitis Group patch-test results, 2001–2002 study period. Dermatitis 2004;15:176-183

10 Goossens A. Cosmetic contact allergens. Cosmetics 2016, 3, 5; doi:10.3390/cosmetics3010005

11 Warshaw EM, Buchholz HJ, Belsito DV et al. Allergic patch test reactions associated with cosmetics: Retrospective analysis of cross-sectional data from the North American Contact Dermatitis Group, 2001-2004. J Am Acad Dermatol 2009;60:23-38

12 Schianchi S, Arcangeli F, Calista D. Compound allergy to vea oil. Contact Dermatitis 2003;49:222

13 Corazza M, Ricci M, Minghetti, S et al. Compound allergy to a lipophilic gel containing vitamin E acetate and cyclopentasiloxane. Dermatitis 2013;24:198-199

14 Milanesi N, Gola M, Francalanci S. Allergic contact dermatitis caused by VEA® lipogel: compound allergy?. Contact Dermatitis 2016;75:243-244

15 Guin JD. Eyelid dermatitis: A report of 215 patients. Contact Dermatitis 2004;50:87-90

16 Kosari P, Alikhan A, Sockolov M, Feldman SR. Vitamin E and allergic contact dermatitis. Dermatitis 2010;21:148-153

Chapter 3.346 TOLAZOLINE

IDENTIFICATION

Description/definition   : Tolazoline is the benzene derivative that conforms to the structural formula shown below
Pharmacological classes : α-Adrenergic antagonists; vasodilator agents; antihypertensive agents
IUPAC name   : 2-Benzyl-4,5-dihydro-1*H*-imidazole
Other names   : 2-Benzyl-2-imidazoline; 4,5-dihydro-2-(phenylmethyl)-1*H*-imidazole
CAS registry number   : 59-98-3
EC number   : 200-448-9
Merck Index monograph   : 10936
Patch testing   : Hydrochloride 1%, 2%, 5% and 10% water; perform controls
Molecular formula   : $C_{10}H_{12}N_2$

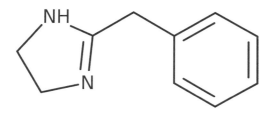

GENERAL

Tolazoline is a non-selective competitive α-adrenergic receptor antagonist with vasodilator activity. It is or most likely was used in treatment of persistent pulmonary hypertension of the newborn and was also used to treat spasms of peripheral blood vessels, e.g. in acrocyanosis. Topically, it was used for ophthalmologic indications such as circulatory changes of the retina, chorioid and optic tract, caustic trauma and degenerative changes of the cornea and as an adjunct in the treatment of various inflammatory corneal affections. Products containing tolazoline were withdrawn from the U.S. market by the (or a) producer in 2002. The drug is however used in veterinary medicine, to reverse xylazine-induced sedation in horses. In pharmaceutical products (mostly veterinarian), tolazoline is employed as tolazoline hydrochloride (CAS number 59-97-2, EC number 200-447-3, molecular formula $C_{10}H_{13}ClN_2$) (1).

CONTACT ALLERGY

Case report

A 43-year-old man had developed a severe itchy dermatitis of the right periorbital region after applying 2 eye ointments, containing resp. chloramphenicol and 10% tolazoline, for a chalazion. Previously, he had suffered from a similar dermatitis after applying idoxuridine in DMSO to treat herpes labialis. Patch tests were positive to the eye ointment containing tolazoline and to tolazoline hydrochloride 10% water. The patient was also allergic to chloramphenicol and to idoxuridine. Three patients served as controls and all showed some erythema at D1, one of them also at D2, but negative at D3 to tolazoline 10% water (2). This indicates that tolazoline 10% water is (slightly) irritant in patch testing.

LITERATURE

1 The data in the section 'General' may have been obtained from literature discussed in this chapter, but mostly also or exclusively from one or more of the following online sources: ChemIDPlus Advanced, PubChem, DrugBank, RxList, Drug Central, Drugs.com, and Wikipedia
2 Frosch PJ, Olbert D, Weickel R. Contact allergy to tolazoline. Contact Dermatitis 1985;13:272

Chapter 3.347 TOLCICLATE

IDENTIFICATION

Description/definition : Tolciclate is the monothiocarbamic ester that conforms to the structural formula shown below

Pharmacological classes : Antifungal agents

IUPAC name : O-(4-Tricyclo[6.2.1.0^{2,7}]undeca-2(7),3,5-trienyl) N-methyl-N-(3-methylphenyl)carba-mothioate

Other names : 1,2,3,4-Tetrahydro-1,4-methanonaphthalen-6-yl N-methyl-N-(m-tolyl)carbamothioate

CAS registry number : 50838-36-3

EC number : 256-792-5

Merck Index monograph : 10939

Patch testing : 1% pet.

Molecular formula : C$_{20}$H$_{21}$NOS

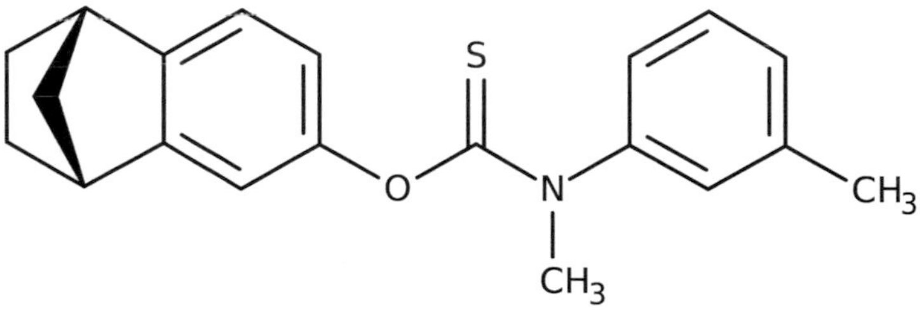

GENERAL

Tolciclate is a monothiocarbamic ester and a broad-spectrum antifungal agent. It is used in the treatment of cutaneous infections caused by dermatophytes, *Candida albicans* and *Malassezia* species (tinea versicolor) (1).

CONTACT ALLERGY

Case reports

A 30-year-old man was prescribed 1% tolciclate cream for mycosis of the groins. Ten days after starting therapy, the patient was referred with erythematous, edematous and vesicular lesions of the lower abdomen, scrotum and inguinal and intergluteal folds, accompanied by pruritus and burning. Patch tests to the cream and its ingredients were positive to the antifungal cream and tolciclate 1% pet. Eight controls were negative (2). The first case of tolciclate allergy was reported from Germany in 1980. In a 60-year-old man a mycosis of hands and feet spreading to other regions of the skin was treated with various antimycotic ointments. During treatment the status of the skin did not improve. Patch testing showed delayed-type hypersensitivity to tolciclate, miconazole, and econazole; details are unknown (3).

LITERATURE

1   The data in the section 'General' may have been obtained from literature discussed in this chapter, but mostly also or exclusively from one or more of the following online sources: ChemIDPlus Advanced, PubChem, DrugBank, RxList, Drug Central, Drugs.com, and Wikipedia

2   Veraldi S, Schianchi-Veraldi R. Allergic contact dermatitis from tolciclate. Contact Dermatitis 1991;24:315

3   Mücke C. Ein Fall von Überempfindlichkeit gegen Econazol und Tolciclat. Dermatosen 1980;28:118 (Article in German)

Chapter 3.348 TOLNAFTATE

IDENTIFICATION

Description/definition   : Tolnaftate is the thiocarbamate derivative that conforms to the structural formula shown below
Pharmacological classes   : Antifungal agents
IUPAC name   : O-Naphthalen-2-yl N-methyl-N-(3-methylphenyl)carbamothioate
Other names   : 2-Naphthyl N-methyl-N-(3-tolyl)thionocarbamate
CAS registry number   : 2398-96-1
EC number   : 219-266-6
Merck Index monograph : 10948
Patch testing   : 1% pet.
Molecular formula   : $C_{19}H_{17}NOS$

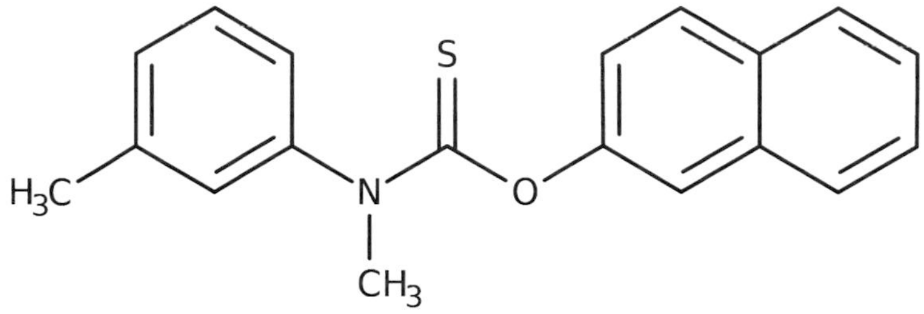

GENERAL

Tolnaftate is a thiocarbamate derivative with either fungicidal or fungistatic properties. It is a selective, reversible and non-competitive inhibitor of membrane-bound squalene 2,3-epoxidase, an enzyme involved in the biosynthesis of ergosterol. Inhibition leads to the accumulation of squalene and a deficiency in ergosterol, an essential component of fungal cell walls, thereby increasing membrane permeability, disrupting cellular organization and causing cell death. In addition, this agent may also distort the hyphae and stunts mycelial growth in susceptible fungi. Tolnaftate is (or was) used to treat fungal skin infections caused by dermatophytes; it is not effective against *Candida albicans* (1). In the 1970s, over 75% of the prescriptions written in the U.S.A. for an antifungal agent specified tolnaftate (6).

CONTACT ALLERGY

Case reports

After 2 weeks treatment for axillary intertrigo with a cream containing tolnaftate, clioquinol, gentamicin and betamethasone valerate, a 47-year-old man developed acute eczema in both axillae. Patch tests were positive to the cream, tolnaftate 0.1% and 1% pet., and to clioquinol 5% pet.; all other active ingredients and excipients tested negative (2). A 54-year-old man had treated an itchy rash on his right leg with an antifungal cream containing tolnaftate and nystatin, which he had used a year earlier for treatment of tinea pedum. In a period of 4 weeks, the lesions became worse resulting in papular dermatitis of the lower legs spreading to the neck and face. Patch tests were positive to the cream 'as is', to tolnaftate 1% and 10% pet., and to nystatin 100,000 IU/gr pet. (3).

A 48-year-old man developed a weeping dermatitis on the dorsal aspect of the left foot, where he had treated a suspected fungal infection with tolnaftate cream, as he had done previously a few times. Patch tests were positive to the cream and to tolnaftate 1% and 0.1% pet., but negative to 0.01% pet. There were no reactions to the excipients of the cream (4). Another patient, a woman of 52 years, developed a very extensive dermatitis from using tolnaftate cream on a fungal infection and as a cosmetic cream on the face. A patch test with the cream was positive, but tolnaftate was not tested separately (5).

A 43-year-old overweight man had groin intertrigo for two years. Many pharmaceuticals had been tried with inconstant success. Tolnaftate solution 1% was used for 3 weeks, but the patient discontinued it when his intertrigo worsened. Two successive trials with tolnaftate resulted in worsening of dermatitis, pruritus and inflammation within a few days. Patch tests were positive to the solution and to tolnaftate in a dilution series (0.01%, 0.05%, 0.1%,

1%). The patch tests to tolnaftate remained pruritic and dermatitic for two months. Patch tests to the other ingredients of tolnaftate solution were negative (6,7). One or more cases of allergic contact dermatitis from tolnaftate had already been reported in 1967 from The Netherlands (8).

LITERATURE

1   The data in the section 'General' may have been obtained from literature discussed in this chapter, but mostly also or exclusively from one or more of the following online sources: ChemIDPlus Advanced, PubChem, DrugBank, RxList, Drug Central, Drugs.com, and Wikipedia

2   González Pérez R, Aguirre A, Oleaga JM, Eizaguirre X, Díaz Pérez JL. Allergic contact dermatitis from tolnaftate. Contact Dermatitis 1995;32:173

3   Lang E, Goos M. Combined allergy to tolnaftate and nystatin. Contact Dermatitis 1985;12:182

4   Emmett EA, Marrs JM. Allergic contact dermatitis from tolnaftate. Arch Dermatol 1973;108:98-99

5   Bowyer A. Tolnaftate. Contact Dermatitis Newsletter 1972;12:339

6   Gellin GA, Maibach HI, Wachs GN. Contact allergy to tolnaftate. Arch Dermatol 1972;106:715-716

7   Gellin GA, Maibach HI. Contact allergy to tolnaftate. Contact Dermatitis Newsletter 1972;11:293

8   Van Ketel WH. Allergy to Tineafax-ointment. Contact Dermatitis Newsletter 1967;2:20

Chapter 3.349 TRAVOPROST

IDENTIFICATION

Description/definition   : Travoprost is the isopropyl ester prodrug and synthetic prostaglandin F2α-analog
                             that conforms to the structural formula shown below
Pharmacological classes : Antihypertensive agents
IUPAC name   : Propan-2-yl (Z)-7-[(1R,2R,3R,5S)-3,5-dihydroxy-2-[(E,3R)-3-hydroxy-4-[3-(trifluoromethyl)-
                             phenoxy]but-1-enyl]cyclopentyl]hept-5-enoate
CAS registry number   : 157283-68-6
EC number   : Not available
Merck Index monograph : 11008
Patch testing   : No data available; in general, active ingredients in ophthalmic medications should be
                             tested in (far) higher concentrations than in the medication itself to avoid false-negative
                             reactions
Molecular formula   : $C_{26}H_{35}F_3O_6$

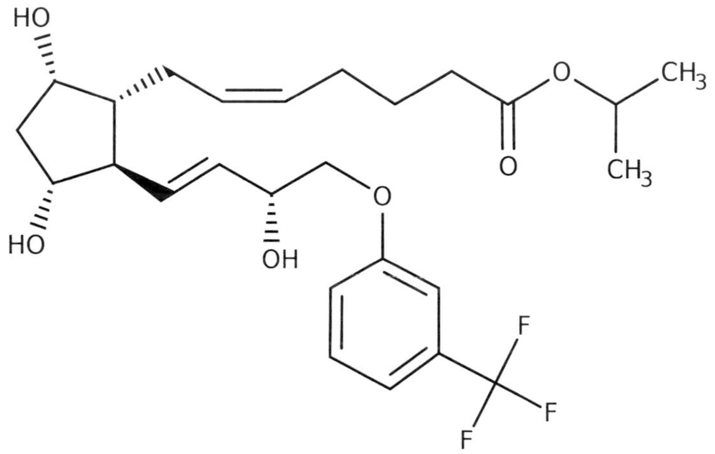

GENERAL

Travoprost is a synthetic lipophilic cloprostenol derivative. It is a prodrug of the active compound travoprost free
acid, a prostaglandin F2α-analog with anti-glaucoma property. Upon administration, travoprost is hydrolyzed to the
free acid by corneal esterases. It then selectively stimulates the prostaglandin F receptor, thereby increasing the
uveoscleral outflow which leads to a reduction in intra-ocular pressure. Travoprost ophthalmic solution is indicated
for the reduction of elevated intraocular pressure in patients with open-angle glaucoma or ocular hypertension (1).

CONTACT ALLERGY

Case report

A patient from Slovakia (age and sex not mentioned) presented with blepharoconjunctivitis. This individual reacted
upon patch testing to eye drops containing travoprost. The active principle itself was not tested, but there were no
reactions to the vehicle constituents and its preservative. After stopping the use of the eye drops and changing to a
chemically different antiglaucoma drug, all complaints disappeared (2).

LITERATURE

1   The data in the section 'General' may have been obtained from literature discussed in this chapter, but mostly
     also or exclusively from one or more of the following online sources: ChemIDPlus Advanced, PubChem,
     DrugBank, RxList, Drug Central, Drugs.com, and Wikipedia
2   Urbancek S, Kuklova M. Allergic blepharoconjunctivitis caused by antiglaucomatics—Central Slovakia
     experiences. American Contact Dermatitis Society 22nd Annual Meeting, New Orleans, USA, February 3, 2011.
     Meeting program, poster presentations, pages 32-33. Available at: https://www.contactderm.org/files/2011_
     ACDS_Abstracts.pdf

Chapter 3.350 TRETINOIN

IDENTIFICATION

Description/definition : Tretinoin is the vitamin A derivative (retinoid) that conforms to the structural formula shown below
Pharmacological classes : Keratolytic agents; antineoplastic agents
IUPAC name : (2E,4E,6E,8E)-3,7-Dimethyl-9-(2,6,6-trimethylcyclohexen-1-yl)nona-2,4,6,8-tetraenoic acid
Other names : *trans*-Retinoic acid; 3,7-dimethyl-9-(2,6,6-trimethyl-1-cyclohexen-1-yl)-2,4,6,8-nonatetra-enoic acid; vitamin A acid
CAS registry number : 302-79-4
EC number : 206-129-0
Merck Index monograph : 9558
Patch testing : 0.005% alcohol; if higher (0.01% or 0.02%), perform adequate controls; petrolatum is not suitable for patch testing tretinoin (12,13)
Molecular formula : $C_{20}H_{28}O_2$

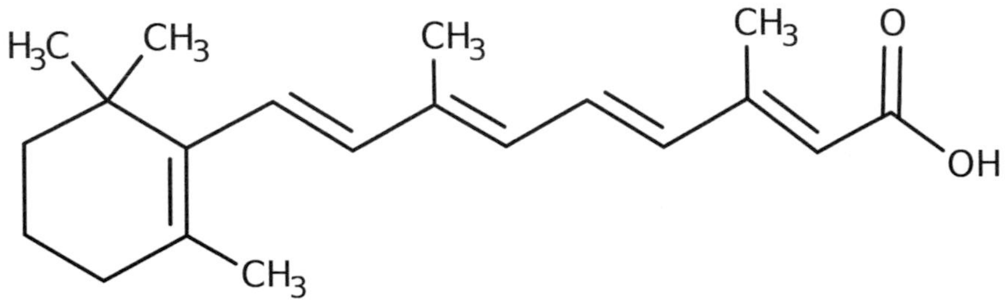

GENERAL

Tretinoin, also known as all-*trans*-retinoic acid, is a naturally occurring derivative of vitamin A (retinol). Tretinoin binds to and activates retinoic acid receptors, thereby inducing changes in gene expression that lead to cell differentiation, decreased cell proliferation, and inhibition of tumorigenesis. Tretinoin is indicated for the induction of remission of acute promyelocytic leukemia. In topical preparations it is used to treat acne vulgaris, flat warts, psoriasis, ichthyosis, keratoderma palmare et plantare, senile comedones, solar keratosis, keratosis follicularis Darier and basal cell carcinomas. In addition, its application to the skin may improve fine wrinkling, mottled hyper-pigmentation, and roughness of the skin associated with photodamage (1).

CONTACT ALLERGY

Patch testing in groups of selected patients

In Italy, before 1986, 204 patients (43 male and 161 female; average age 19.4 years), under treatment for at least 2 months for acne, were patch tested with a battery of topical antiacne drugs. There were 7 positive reactions: one to tretinoin 0.05% pet. and alc. (both irritant concentrations), 4 to spironolactone 5% pet., and 2 to benzoyl peroxide 5% and 2% pet. (5% causes many irritant reactions). In a number of cases (not well described), dilution series were performed with positive results. ROAT with the commercial products were positive in all. All reactions were considered to be relevant (2).

Case series

A small case series of 3 patients (no details on sex and age) with allergic contact dermatitis from tretinoin was reported in 1976 (13). The patients were treated with tretinoin 0.05% cream, two for hereditary keratoderma and one for Fox-Fordyce disease. Patient 1 developed fairly acute itching and erythema on the treated areas after 14 weeks. Patient 2 showed acute vesicular eczema on the hands, feet, inside of the thighs, and face after seven weeks' treatment. The patient with Fox-Fordyce disease developed itching and erythema of the treated axilla after only one week. Patch tests were positive to the 0.05% cream, 2/2 to another brand 0.05% cream, and to tretinoin 0.05% and 0.005% in alc. 96%. One also reacted to the cream base from allergy to stearyl alcohol. The 0.05% preparations caused irritant reactions in controls. The weaker alcoholic solution (0.005%) caused no primary reactions. In three out of 10 controls, however, erythema developed after 5-6 days at the site of the test; in two of these, papules appeared, and in one there was also itching (13). Patient 3 may have had irritant contact dermatitis and false-positive patch tests.

Case reports

A 20-year-old woman presented with bilateral periocular dermatitis. She had used various topical preparations for ocular sequelae of toxic epidermal necrolysis, including a compounded preparation of retinoic acid 0.05% in simple eye ointment (wool fat 10%, liquid paraffin 10%, made up to 100% with yellow soft paraffin) at night to the inner lower eyelids. The patient had remained on a similar treatment regime for many years. Patch testing was performed to the wool fat and to a commercial preparation of retinoic acid cream, revealing a significant reaction to the retinoic acid cream with no reaction to the wool fat. It was concluded that the patient was allergic to retinoic acid. It was not mentioned why the patient was not tested with her original tretinoin preparations and why the reactions were read at D2 only. Also, no controls test with the commercial tretinoin preparation were performed (4).

A 19-year-old woman with facial acne vulgaris had been applying 0.05% tretinoin cream daily. During the second month of treatment, she developed eczema of the face. Patch tests were positive to the cream 'as is' and tretinoin 0.05% in both petrolatum and ethanol. Ten controls were negative (5). This report was challenged and it was suggested that the patch test had been irritant (6). In a reply, the authors stated that the patient had been retested with tretinoin at 0.025% and 0.01% petrolatum, alcohol and acetone with positive (++) reactions which persisted for 7 days. A ROAT with tretinoin 0.05% was strongly positive after 4 days of application. They did admit that testing with tretinoin 0.05% alcohol in patients with facial acne resulted in 63/204 (31%) irritant reactions (7,8).

Two male prison volunteers, participating in a three-week predictive patch test study for evaluation of irritancy, were found to have edematous, spreading, vesiculo-bullous responses (+++) to three formulations of tretinoin 0.05% at the first 48 hour change. Three control products (product minus tretinoin) were negative. They had never been treated with tretinoin before. Patch tests (some 2x) were positive to tretinoin 0.01% in petrolatum, alcohol and gel, 0.05% in commercial cream and petrolatum and 0.00625% in commercial lotion vehicle, diluted with alcohol. Tretinoin inhibited the migration of the leukocytes of these 2 men and the authors considered that this supported the concept that their responses to patch testing were due to delayed hypersensitivity. As these individuals had never been in contact with tretinoin, the authors further suggested 'that these responses were cross-reactions to a prior sensitizer, perhaps unique to this (prison ?) environment' (14).

Short summaries of other case reports of allergic contact dermatitis from tretinoin are shown in table 3.350.1. It should be realized that tretinoin in a concentration of 0.05% causes many irritant reactions.

Table 3.350.1 Short summaries of case reports of allergic contact dermatitis from

Year and country	Sex	Age	Positive patch tests	Clinical data and comments	Ref.
1992 Italy	F	26	0.05% crème; TRE 0.1% pet. and 0.005% alc.	itching, erythema and swelling of the face; clinical symptoms of ACD could be confused with irritant dermatitis	8
1992 France	M	34	'tretinoin in low concentrations'	contact dermatitis; co-sensitization to isotretinoin; details not available to the author	9
1980 Spain	F	?	TRE 0.1% pet.	microvesicles, erythema, edema, itching face, ears, neck and hairline; co-reaction to benzoyl peroxide	10
1978 Poland	F	21	0.05% cream 'as is'; TRE 0.05% in alcohol and propylene glycol	treatment of verrucae vulgares et planae on the legs; after 3 weeks, dermatitis developed on both legs; provocation test with cream positive after 2 days	11
1977 USA	F	26	commercial liquid 0.05% diluted 2x and 8x; TRE 0.01% and 0.02% alcohol and acetone	treatment of folliculitis on both lower legs; intense itching and development of erythematous and vesicular eruption; patch test with tretinoin in petrolatum negative; ten controls were negative	12

TRE : tretinoin

Cross-reactions, pseudo-cross-reactions and co-reactions

Two patients most likely sensitized to retinoic acid (tretinoin) from topical application cross-reacted to isotretinoin (3,9). No cross-reactions to retinol, retinal, and retinyl palmitate (14).

LITERATURE

1   The data in the section 'General' may have been obtained from literature discussed in this chapter, but mostly also or exclusively from one or more of the following online sources: ChemIDPlus Advanced, PubChem, DrugBank, RxList, Drug Central, Drugs.com, and Wikipedia

2   Balato N, Lembo G, Cuccurullo FM, Patruno C, Nappa P, Ayala F. Acne and allergic contact dermatitis. Contact Dermatitis 1996;34:68-69
3   Auffret N, Bruley C, Brunetiere RA, Decot MC, Binet O. Photoaggravated allergic reaction to isotretinoin. J Am Acad Dermatol 1990;23:321-322
4   Anderson A, Gebauer K. Periorbital allergic contact dermatitis resulting from topical retinoic acid use. Australas J Dermatol 2014;55:152-153
5   Balato N, Patruno C, Lembo G, Cuccurullo FM, Ayala F. Allergic contact dermatitis from retinoic acid. Contact Dermatitis 1995;32:51
6   Serup J. Allergic contact dermatitis from retinoic acid. Contact Dermatitis 1995;33:142 (Letter)
7   Balato N. Allergic contact dermatitis from retinoic acid. Reply. Contact Dermatitis 1995;33:142 (Letter)
8   Tosti A, Guerra L, Morelli R, Piraccini BM. Contact dermatitis due to topical retinoic acid. Contact Dermatitis 1992;26:276-277
9   Tomb R, Dolfus A, Couppie P. Eczema caused by contact allergy to tretinoin. Ann Dermatol Venereol 1992;119:761-764 (Article in French)
10 Romaguera C, Grimalt F. Sensitization to benzoyl peroxide, retinoic acid and carbon tetrachloride. Contact Dermatitis 1980;6:442
11 Rudzki E, Grzywa Z. Dermatitis from retinoic acid. Contact Dermatitis 1978;4:305-306
12 Nordqvist BC, Mehr K. Allergic contact dermatitis to retinoic acid. Contact Dermatitis 1977;3:55-56
13 Lindgren S, Groth O, Molin L. Allergic contact response to vitamin A acid. Contact Dermatitis 1976;2:212-217
14 Jordan WP Jr, Higgins M, Dvorak J. Allergic contact dermatitis to all-*trans*-retinoic acid; epicutaneous and leukocyte migration inhibition testing. Contact Dermatitis 1975;1:306-310

Chapter 3.351 TRIAFUR

IDENTIFICATION

Description/definition   : Triafur is the thiadiazole derivative, C-nitro compound and a member of the furans that
                             conforms to the structural formula shown below
Pharmacological classes : Unknown
IUPAC name   : 5-(5-Nitrofuran-2-yl)-1,3,4-thiadiazol-2-amine
Other names   : 2-(5-Nitro-2-furyl)-5-amino-1,3,4-thiadiazole
CAS registry number   : 712-68-5
EC number   : 211-925-6
Merck Index monograph   : 973
Patch testing   : 1% pet.
Molecular formula   : $C_6H_4N_4O_3S$

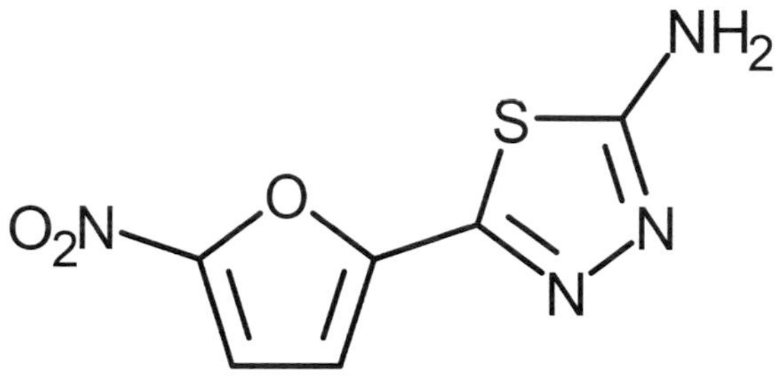

GENERAL

In the 1960s, an ointment and suppositories containing triafur were used in Sweden to treat hemorrhoids and
pruritus ani. It was probably withdrawn when the chemical was found to have carcinogenic properties (1).

CONTACT ALLERGY

Case series

Five patients used either an ointment or suppositories containing 0.7% triafur for hemorrhoids or pruritus ani and
developed contact dermatitis. Patch tests were positive in all to the medicament and to triafur 1% pet. By studying
the cross-reaction patterns, it was established that the nitro group on the furan ring and the close bonding of the
furan and thiadiazole rings were necessary to elicit a reaction. Tests with nitrofurantoin and nitrofurazone, which
have no thiadiazole ring, were negative (2).

Cross-reactions, pseudo-cross-reactions and co-reactions
No cross-reactivity to nitrofurantoin and nitrofurazone (2).

LITERATURE
1   The data in the section 'General' may have been obtained from literature discussed in this chapter, but mostly
    also or exclusively from one or more of the following online sources: ChemIDPlus Advanced, PubChem,
    DrugBank, RxList, Drug Central, Drugs.com, and Wikipedia
2   Fregert S. Cross-sensitization among nitrofurylaminothiadiazoles. Acta Derm Venereol 1968;48:106-109 (data
    cited in ref. 3)
3   Cronin E. Contact Dermatitis. Edinburgh: Churchill Livingstone, 1980:268

Chapter 3.352 TRIAMCINOLONE

IDENTIFICATION

Description/definition	: Triamcinolone is the synthetic glucocorticoid that conforms to the structural formula shown below
Pharmacological classes	: Glucocorticoids; anti-inflammatory agents
IUPAC name	: (8S,9R,10S,11S,13S,14S,16R,17S)-9-Fluoro-11,16,17-trihydroxy-17-(2-hydroxyacetyl)-10,13-dimethyl-6,7,8,11,12,14,15,16-octahydrocyclopenta[a]phenanthren-3-one
Other names	: 11β,16α,17α,21-Tetrahydroxy-9α-fluoro-1,4-pregnadiene-3,20-dione; fluoxyprednisolone
CAS registry number	: 124-94-7
EC number	: 204-718-7
Merck Index monograph	: 11027
Patch testing	: In general, corticosteroids may be tested at 0.1% and 1% in alcohol; late readings (6-10 days) are strongly recommended
Molecular formula	: $C_{21}H_{27}FO_6$

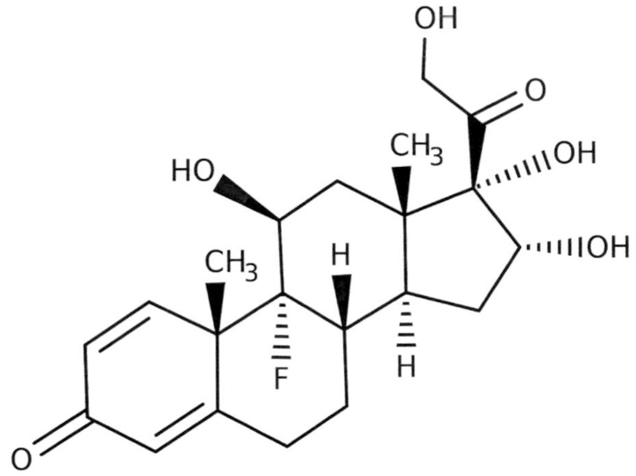

GENERAL

General aspects of corticosteroids used on the skin and mucous membranes are discussed in Chapter 2.4. A practical guideline for diagnosing allergic reactions to corticosteroids is presented in ref. 1. Triamcinolone base (alcohol) is used in tablets only. In other applications, esters are used: triamcinolone acetonide (Chapter 3.353), triamcinolone diacetate (Chapter 3.354) or triamcinolone hexacetonide (Chapter 3.355). In topical preparations, triamcinolone acetonide is (virtually) always used. As triamcinolone base is used in oral preparations only, most positive patch test reactions to this corticosteroid must have been the result of sensitization to one of its esters or of cross-sensitization to another corticosteroid.

CONTACT ALLERGY

Case series

From January 1990 to June 2008, in Leuven, Belgium, 315 patients were diagnosed with contact allergy to/allergic contact dermatitis from corticosteroids (CSs) from routine patch testing with a baseline series including tixocortol pivalate, budesonide, hydrocortisone butyrate and prednisone caproate, patch testing with patients' own CS preparations, and testing those with proven contact allergy to a corticosteroid or strongly suspected of CS allergy later with a series of 66 CSs, including two sex hormones (progesterone and testosterone). 71% of the patients had relevant reactions, but these were not specified. In this group of 315 CS allergic patients, 65 had positive patch tests to triamcinolone 0.1% alc. (3). As triamcinolone is used in tablets only, these reactions must have been the result of sensitization to a triamcinolone ester or of cross-reactivity to an other corticosteroid.

Case report

A 26-year-old woman presented with a pruritic eruption on her arms that had begun 3 to 4 days after initiation of 0.1% topical triamcinolone (this was a compounded formulation, which possibly may have contained triamcinolone base) in petrolatum to treat a pruritic dermatitis on her right forearm. Physical examination showed well-demarcated erythematous, annular, edematous plaques on both upper arms and forearms resembling erythema multiforme (EM). These plaques then coalesced to form larger serpiginous plaques with central clearing over the bilateral antecubital fossae. The patient was subsequently patch tested to 0.05% desonide and 0.1% triamcinolone and was found to have a positive reaction to desonide at D2 and D3 but was negative to triamcinolone itself. A ROAT, after 4 to 5 days, evoked the identical EM-like reaction that she had on initial presentation. The patient was diagnosed with erythema multiforme-like allergic contact dermatitis from triamcinolone (2).

Cross-reactions, pseudo-cross-reactions and co-reactions
Cross-reactions between corticosteroids are discussed in Chapter 2.8.

LITERATURE

1   Baeck M, Goossens A. Immediate and delayed allergic hypersensitivity to corticosteroids: practical guidelines. Contact Dermatitis 2012;66:38-45
2   Smart DR, Powell DL. Erythema multiforme-like allergic contact reaction to topical triamcinolone. Dermatitis 2014;25:89-90
3   Baeck M, Chemelle JA, Terreux R, Drieghe J, Goossens A. Delayed hypersensitivity to corticosteroids in a series of 315 patients: clinical data and patch test results. Contact Dermatitis 2009;61:163-175

Chapter 3.353 TRIAMCINOLONE ACETONIDE

IDENTIFICATION

Description/definition : Triamcinolone acetonide is the 16,17 acetonide ester of the synthetic glucocorticoid triamcinolone that conforms to the structural formula shown below

Pharmacological classes : Glucocorticoids; immunosuppressive agents; anti-inflammatory agents

IUPAC name : (1S,2S,4R,8S,9S,11S,12R,13S)-12-Fluoro-11-hydroxy-8-(2-hydroxyacetyl)-6,6,9,13-tetramethyl-5,7-dioxapentacyclo[10.8.0.0^{2,9}.0^{4,8}.0^{13,18}]icosa-14,17-dien-16-one

Other names : 9α-Fluoro-11β,21-dihydroxy-16α,17α-isopropylidenedioxypregna-1,4-diene-3,20-dione

CAS registry number : 76-25-5

EC number : 200-948-7

Merck Index monograph : 11028

Patch testing : 1.0% pet. (Chemotechnique, SmartPracticeCanada, SmartPracticeEurope); 0.1% pet. (SmartPracticeCanada); the lower concentration may be preferable (20); late readings (6-10 days) are strongly recommended

Molecular formula : $C_{24}H_{31}FO_6$

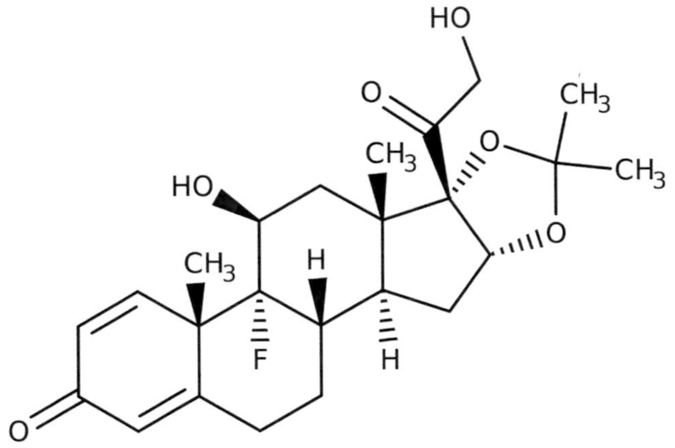

GENERAL

General aspects of corticosteroids used on the skin and mucous membranes are discussed in Chapter 2.4. A practical guideline for diagnosing allergic reactions to corticosteroids is presented in ref. 15. Triamcinolone acetonide 1% pet. is included in the American core allergen series (www.smartpracticecanada.com). Intradermal tests with triamcinolone acetonide may detect more cases of sensitization to triamcinolone acetonide than patch tests (21,34).

See also triamcinolone (Chapter 3.352), triamcinolone diacetate (Chapter 2.254) and triamcinolone hexacetonide (Chapter 3.355).

CONTACT ALLERGY

Contact allergy in the general population

In Germany, for the period 1995-2004, the population-based relative incidence (RI) of contact sensitization to triamcinolone acetonide (cases/100,000 defined daily doses (DDDs) per year) was estimated to be 1.4. In the group of corticosteroids, the RI ranged from 0.3 (dexamethasone sodium phosphate) to 43.2 (budesonide) (14).

Patch testing in groups of patients

Results of patch testing triamcinolone acetonide in consecutive patients suspected of contact dermatitis (routine testing) back to 1990 are shown in table 3.353.1. Results of testing in groups of *selected* patients (patients tested with a corticosteroid series, individuals with anogenital dermatitis, patients with chronic otitis externa) are shown in table 3.353.2.

Patch testing in consecutive patients suspected of contact dermatitis: routine testing

Triamcinolone acetonide has been included in the screening series of the North American Contact Dermatitis group (NACDG) from 2003 on, but was deleted again 2008, presumably because of disappointing results. Indeed, in the USA

+ Canada, prevalences of sensitization to triamcinolone acetonide have invariably been low, ranging from 0.1% to 0.5% (table 3.353.1). In the 1990s, low rates had also been observed in Belgium (31,35), The Netherlands (31) and Israel (30). Only one study from Australia had a score higher than 1% (3); about one-third of the reactions were scored as relevant. In the USA, relevance or 'definite + probable relevance' ranged from 27% to 69% (table 3.353.1).

Table 3.353.1 Patch testing in groups of patients: Routine testing

Years and Country	Test conc. & vehicle	Number of patients tested	positive (%)	Selection of patients (S); Relevance (R); Comments (C)	Ref.
2011-2015 USA	1% pet.	2561	2 (0.1%)	R: not stated	1
2006-2010 USA	1% pet.	3085	(0.5%)	R: 64%	2
2001-2010 Australia	1% pet.	1280	14 (1.1%)	R: 36%	3
2007-2008 NACDG	1% pet.	5085	14 (0.3%)	R: definite + probable relevance 43%	24
2007-2008 NACDG	1% pet.	5081	(0.3%)	R: definite + probable relevance: 43%	4
2005-2006 NACDG	1% pet.	4438	(0.3%)	R: definite + probable relevance: 27%	5
2001-2005 USA	1% pet.	3848	(0.4%)	R: 69%	6
2003-2004 NACDG	1% pet.	5141	17 (0.1%)	R: not stated	7
1998-2000 USA	1% pet.	713	(0.4%)	R: not stated	8
1995-1998 Israel		660	1 (0.2%)	R: not stated	30
1991 Belgium	1% alc.	610	3 (0.5%)	R: not stated	31
1991 The Netherlands	1% alc.	533	3 (0.6%)	R: not stated	31
1988-1990 Belgium	1% alc. 94%	1947	6 (0.3%)	R: not stated	35

NACDG: North American Contact Dermatitis Group (USA, Canada)

Patch testing in groups of selected patients

In groups of selected patients, prevalences of sensitization have ranged from 0.2% to 5% (table 3.353.2). The highest rates were found in a small group of patients *strongly* suspected of corticosteroid allergy (33; 5%), individuals with chronic otitis externa (11; 2.5%) and patients with exclusively anogenital dermatitis (10; 1.7%). Somewhat surprising, the frequencies of sensitization in groups of patients tested with a corticosteroid series/suspected of steroid allergy, were not or only marginally higher than seen in routine testing (23,26,28,32). Relevance data were hardly ever provided (11,23).

Table 3.353.2 Patch testing in groups of patients: Selected patient groups

Years and Country	Test conc. & vehicle	Number of patients tested	positive (%)	Selection of patients (S); Relevance (R); Comments (C)	Ref.
2005-2016 NACDG	1% pet.	117	2 (1.7%)	S: patients with only anogenital dermatitis; R: all positives represent relevant reactions; C: the frequency was significantly higher than in a control group	10
2000-2005 USA	1% pet.	1176	9 (0.8%)	S: patients suspected of corticosteroid allergy; R: 89%	23
1995-2004 IVDK	0.1% pet.	6141	(0.5%)	S: not stated; R: not stated	13
1995-2004 IVDK	0.1% pet.	5936	31 (0.5%)	S: patients tested with a corticosteroid series; R: not stated	26
1990-2003 IVDK		193	2 (1.0%)	S: patients with perianal dermatoses; R: not stated; C: the frequency was not higher than in a control group	9
1995-9 Netherlands	0.1% pet.	79	2 (2.5%)	S: patients with chronic otitis externa; R: 18% of all reactions to ingredients of topical medications were relevant	11
1995-9 Netherlands	0.1% pet.	343	1 (0.3%)	S: not stated; R: possible	29
1996-1997 IVDK	0.1% pet.	608	3 (0.5%)	S: patients tested with a corticosteroid series; R: not stated	28
1988-1991 U.K.	20% pet.	528	1 (0.2%)	S: patients with a positive patch test to tixocortol pivalate or suspected of corticosteroid allergy; R: not stated	32
1985-1990 Finland	0.1% and 1% alc. and pet.	66	3 (5%)	S: patients very likely to be corticosteroid-allergic; R: not stated	33

IVDK: Information Network of Departments of Dermatology, Germany, Austria, Switzerland; NACDG: North American Contact Dermatitis Group (USA, Canada)

Case series

From January 1990 to June 2008, in Leuven, Belgium, 315 patients were diagnosed with contact allergy to/allergic contact dermatitis from corticosteroids (CSs) from routine patch testing with a baseline series including tixocortol pivalate, budesonide, hydrocortisone butyrate and prednisone caproate, patch testing with patients' own CS preparations, and testing those with proven contact allergy to a corticosteroid or strongly suspected of CS allergy later with a series of 66 CSs, including two sex hormones (progesterone and testosterone). 71% of the patients had

relevant reactions, but these were not specified. In this group of 315 CS allergic patients, 51 had positive patch tests to triamcinolone acetonide 0.1% alc. (22, overlap with ref. 12 and 25). It is unknown how many of these reactions were caused by the use of a pharmaceutical product containing triamcinolone acetonide and how many were cross-reactions to other corticosteroids.

In the period 1990-2008, in the same university clinic in Leuven, Belgium, of the 315 patients diagnosed with contact allergy to/allergic contact dermatitis from corticosteroids, 18 (5.7%) presented with allergic manifestations (conjunctivitis, eczema of the face, periocular skin or eyelids) caused by the use of CS-containing ocular preparations. One of these patients had used ophthalmic preparations containing triamcinolone acetonide (25, overlap with refs. 12 and 22).

Between 1990 and 2003, in Leuven, Belgium, 92 patients were patch tested for chronic vulval complaints. Fifteen of these women were tested with a series of topical drugs; there were 4 reactions to triamcinolone acetonide 1% alc. of which 3 were relevant (12, overlap with refs. 22 and 25).

Case reports

A 49-year-old woman with otitis externa that did not respond well to triamcinolone acetonide solution, later was patch test positive to the corticosteroid 1% pet. (27). A 30-year-old woman sensitized to budesonide from nasal application developed pain in the nose and an acneiform facial rash two days after initiating triamcinolone acetonide nose spray. She used it intermittently thereafter and each time developed the same symptoms (identical to those caused by budesonide allergy). Later, its use resulted in systemic contact dermatitis with flexural eczema. Though highly likely caused by triamcinolone acetonide cross-reacting to budesonide, patch tests with triamcinolone acetonide 0.01% alc. and the nasal spray 'as is' were negative (27).

An 81-year-old woman presented with eyelid dermatitis following several applications of triamcinolone acetonide ointment. Patch testing was positive to the ointment and triamcinolone acetonide 0.1% pet. (38). A 16-year-old girl was treated in the course of several years with various corticosteroids for seborrheic dermatitis of the axillae, face or scalp and became sensitized to fluocinonide, desonide, and triamcinolone acetonide in these topical pharmaceuticals (46). A woman had been treated with various corticosteroid preparations for dermatitis of the axillae and noticed worsening from triamcinolone acetonide cream. Patch tests were positive to the cream 'as is' and to triamcinolone acetonide 1%, 0.1% and 0.01% pet. and 1% in propylene glycol (47).

Other (early or somewhat older) case reports, some in German and French literature, of allergic contact dermatitis from triamcinolone acetonide can be found in refs. 17,37,39-44 and 48.

Cross-reactions, pseudo-cross-reactions and co-reactions

Cross-reactions between corticosteroids are discussed in Chapter 2.8.

Cutaneous adverse drug reactions from systemic administration caused by type IV (delayed-type) hypersensitivity

Cutaneous adverse drug reactions from systemic administration of triamcinolone acetonide caused by type IV (delayed-type) hypersensitivity including local allergic reaction from intralesional injection (45,51,53), erythema multiforme-like allergic dermatitis from intra-articular injection (18), generalization of dermatitis (19,36), systemic contact dermatitis (16,50,54), morbilliform and partially persistent urticarial dermatitis (52), and maculopapular eruption (49), are planned to be discussed in Volume IV of the *Monographs in Contact Allergy* series on Systemic drugs.

LITERATURE

1 Veverka KK, Hall MR, Yiannias JA, Drage LA, El-Azhary RA, Killian JM, et al. Trends in patch testing with the Mayo Clinic standard series, 2011-2015. Dermatitis 2018;29:310-315
2 Wentworth AB, Yiannias JA, Keeling JH, Hall MR, Camilleri MJ, Drage LA, et al. Trends in patch-test results and allergen changes in the standard series: a Mayo Clinic 5-year retrospective review (January 1, 2006, to December 31, 2010). J Am Acad Dermatol 2014;70:269-275
3 Toholka R, Wang Y-S, Tate B, Tam M, Cahill J, Palmer A, Nixon R. The first Australian Baseline Series: Recommendations for patch testing in suspected contact dermatitis. Australas J Dermatol 2015;56:107-115
4 Fransway AF, Zug KA, Belsito DV, Deleo VA, Fowler JF Jr, Maibach HI, et al. North American Contact Dermatitis Group patch test results for 2007-2008. Dermatitis 2013;24:10-21
5 Zug KA, Warshaw EM, Fowler JF Jr, Maibach HI, Belsito DL, Pratt MD, et al. Patch-test results of the North American Contact Dermatitis Group 2005-2006. Dermatitis 2009;20:149-160
6 Davis MD, Scalf LA, Yiannias JA, Cheng JF, El-Azhary RA, Rohlinger AL, et al. Changing trends and allergens in the patch test standard series. Arch Dermatol 2008;144:67-72

7   Warshaw EM, Belsito DV, DeLeo VA, Fowler JF Jr, Maibach HI, Marks JG, et al. North American Contact
     Dermatitis Group patch-test results, 2003-2004 study period. Dermatitis 2008;19:129-136

8   Wetter DA, Davis MDP, Yiannias JA, Cheng JF, Connolly SM, el-Azhary RA, et al. Patch test results from the Mayo
     Contact Dermatitis Group, 1998–2000. J Am Acad Dermatol 2005;53:416-421

9   Kügler K, Brinkmeier T, Frosch PJ, Uter W. Anogenital dermatoses—allergic and irritative causative factors.
     Analysis of IVDK data and review of the literature. J Dtsch Dermatol Ges 2005;3:979-986

10   Warshaw EM, Kimyon RS, Silverberg JI, Belsito DV, DeKoven JG, Maibach HI, et al. Evaluation of patch test
     findings in patients with anogenital dermatitis. JAMA Dermatol 2019;156:85-91

11   Devos SA, Mulder JJ, van der Valk PG. The relevance of positive patch test reactions in chronic otitis externa.
     Contact Dermatitis 2000;42:354-355

12   Nardelli A, Degreef H, Goossens A. Contact allergic reactions of the vulva: a 14-year review. Dermatitis
     2004;15:131-136

13   Menezes de Padua CA, Uter W, Schnuch A. Contact allergy to topical drugs: prevalence in a clinical setting and
     estimation of frequency at the population level. Pharmacoepidemiol Drug Saf 2007;16:377-384

14   Menezes de Padua CA, Schnuch A, Nink K, Pfahlberg A, Uter W. Allergic contact dermatitis to topical drugs –
     Epidemiological risk assessment. Pharmacoepidemiol Drug Saf 2008;17:813-821

15   Baeck M, Goossens A. Immediate and delayed allergic hypersensitivity to corticosteroids: practical guidelines.
     Contact Dermatitis 2012;66:38-45

16   Brambilla L, Boneschi V, Chiappino G, Fossati S, Pigatto PD. Allergic reactions to topical desoxymethasone and
     oral triamcinolone. Contact Dermatitis 1989;21:272-274.

17   Bandmann H-J, Huber-Riffeser G, Woyton A. Kontaktallergie gegen triamcinolonacetonid. Hautarzt 1966;17:183-
     185 (Article in German)

18   Valsecchi R, Reseghetti A, Leghissa P, Cologni L, Cortinovis R. Erythema-multiforme-like lesions from
     triamcinolone acetonide. Contact Dermatitis 1998;38:362-363

19   Stingeni L, Caraffini S, Assalve D, Lapomarda V, Lisi P. Erythema-multiforme-like contact dermatitis from
     budesonide. Contact Dermatitis 1996;34:154-155

20   Isaksson M, Bruze M, Lepoittevin J-P, Goossens A. Patch testing with serial dilutions of budesonide, its R and S
     diastereomers, and potentially cross-reacting substances. Am J Contact Dermat 2001;12:170-176

21   Ferguson AD, Emerson RM, English JS. Cross-reactivity patterns to budesonide. Contact Dermatitis 2002;47:337-
     340

22   Baeck M, Chemelle JA, Terreux R, Drieghe J, Goossens A. Delayed hypersensitivity to corticosteroids in a series of
     315 patients: clinical data and patch test results. Contact Dermatitis 2009;61:163-175

23   Davis MD, El-Azhary RA, Farmer SA. Results of patch testing to a corticosteroid series: a retrospective review of
     1188 patients during 6 years at Mayo Clinic. J Am Acad Dermatol 2007;56:921-927

24   Pratt MD, Mufti A, Lipson J, Warshaw EM, Maibach HI, Taylor JS, et al. Patch test reactions to corticosteroids:
     Retrospective analysis from the North American Contact Dermatitis Group 2007-2014. Dermatitis 2017;28:58-63

25   Baeck M, De Potter P, Goossens A. Allergic contact dermatitis following ocular use of corticosteroids. J Ocul
     Pharmacol Ther 2011;27:83-92

26   Uter W, de Pádua CM, Pfahlberg A, Nink K, Schnuch A, Lessmann H. Contact allergy to topical corticosteroids –
     results from the IVDK and epidemiological risk assessment. J Dtsch Dermatol Ges 2009;7:34-41

27   Isaksson M. Systemic contact allergy to corticosteroids revisited. Contact Dermatitis 2007;57:386-388

28   Uter W, Geier J, Richter G, Schnuch A; IVDK Study Group, German Contact Dermatitis Research Group. Patch test
     results with tixocortol pivalate and budesonide in Germany and Austria. Contact Dermatitis 2001;44:313-314

29   Devos SA, Van der Valk PG. Relevance and reproducibility of patch-test reactions to corticosteroids. Contact
     Dermatitis 2001;44:362-365

30   Weltfriend S, Marcus-Farber B, Friedman-Birnbaum R. Contact allergy to corticosteroids in Israeli patients.
     Contact Dermatitis 2000;42:47

31   Dooms-Goossens A, Meinardi MM, Bos JD, Degreef H. Contact allergy to corticosteroids: the results of a two-
     centre study. Br J Dermatol 1994;130:42-47

32   Burden AD, Beck MH. Contact hypersensitivity to topical corticosteroids. Br J Dermatol 1992;127:497-501

33   Lauerma AI. Contact hypersensitivity to glucocorticosteroids. Am J Contact Dermat 1992;3:112-132

34   Wilkinson SM, English JS. Patch tests are poor detectors of corticosteroid allergy. Contact Dermatitis 1992;26:67-
     68

35   Dooms-Goossens A, Morren M. Results of routine patch testing with corticosteroid series in 2073 patients.
     Contact Dermatitis 1992;26:182-191

36 English JS, Ford G, Beck MH, Rycroft RJ. Allergic contact dermatitis from topical and systemic steroids. Contact Dermatitis 1990;23:196-197

37 Wulf K. Beitrag zur Triamcinolon-Kontaktallergie. Z Hautkr 1967;42:765-768 (Article in German)

38 Rivara G, Tomb RR, Foussereau J. Allergic contact dermatitis from topical corticosteroids. Contact Dermatitis 1989;21:83-91

39 Wiegel O. Kontaktallergie durch Kortikosteroidhaltige externa (Triamcinolon-Acetonid und Dexamethason). Med Welt 1968;13:828-829 (Article in German)

40 Foussereau J, Liman-Mestiri S, Khochnevis A, Basset A. l'Allergie á l'association thérapeutique locale "nystatine, néomycine et acétonide de triamcinolone". Bull Soc Franç Derm 1971;78:457-459 (same as ref. 43, Article in French)

41 Fisher AA. Allergic reactions to intralesional and multiple topical corticosteroids. Cutis 1979;23:564,708-709

42 Esser B, Beitrag zur Kortison-Allergie. Zeitschrift für Hautkrankheiten 1983;58:29-32 (Article in German)

43 Foussereau J, Limam-Mesfiri S, Khochnevis A. Contact allergy to nystatin. Contact Dermatitis Newsletter 1971;10:221 (same as ref. 40)

44 Guin JD. Contact sensitivity to topical corticosteroids. J Am Acad Dermatol 1984;10(5Pt.1):773-782

45 Kark EC. Sensitivity to fluorinated steroids presenting as a delayed hypersensitivity. Contact Dermatitis 1980;6:214-216

46 Coskey RJ. Contact dermatitis due to multiple corticosteroid creams. Arch Dermatol 1978;114:115-117

47 Tegner E. Contact allergy to corticosteroids. Int J Dermatol 1976;15:520-523

48 Maucher OM, Knipper H, Faber M. Drug-induced dermatitis with acetonides of corticoids. In: Frosch PJ, Dooms-Goossens A, Lachapelle JM, Rycroft RJG, Scheper RJ (eds): Current topics in contact dermatitis. Berlin Heidelberg: Springer-Verlag, 1989:238-243

49 Santos-Alarcón S, Benavente-Villegas FC, Farzanegan-Miñano R, Pérez-Francés C, Sánchez-Motilla JM, Mateu-Puchades A. Delayed hypersensitivity to topical and systemic corticosteroids. Contact Dermatitis 2018;78:86-88

50 Bianchi L, Marietti R, Tramontana M, Hansel K, Stingeni L. Systemic allergic dermatitis from intra-articular triamcinolone acetonide: Report of two cases with unusual clinical manifestations. Contact Dermatitis 2020 Jul 16. doi: 10.1111/cod.13667. Online ahead of print.

51 Brancaccio RR, Zappi EG. Delayed type hypersensitivity to intralesional triamcinolone acetonide. Cutis 2000;65:31-33

52 Ijsselmuiden OE, Knegt-Junk KJ, van Wijk RG, van Joost T. Cutaneous adverse reactions after intra-articular injection of triamcinolone acetonide. Acta Derm Venereol 1995;75:57-58

53 Kreeshan FC PHP. Delayed hypersensitivity reaction to intralesional triamcinolone acetonide following treatment for alopecia areata. Intradermal testing. Dermatol Case Rep 2015;9:107-109

54 Gumaste P, Cohen D, Stein J. Bullous systemic contact dermatitis caused by an intra-articular steroid injection. Br J Dermatol 2015;172:300-302

Chapter 3.354 TRIAMCINOLONE DIACETATE

IDENTIFICATION

Description/definition : Triamcinolone diacetate is the 16,21 diacetate ester of the synthetic glucocorticoid triamcinolone that conforms to the structural formula shown below
Pharmacological classes : Glucocorticoids
IUPAC name : [2-[(8S,9R,10S,11S,13S,14S,16R,17S)-16-Acetyloxy-9-fluoro-11,17-dihydroxy-10,13-dimethyl-3-oxo-6,7,8,11,12,14,15,16-octahydrocyclopenta[a]phenanthren-17-yl]-2-oxoethyl] acetate
Other names : 16α,21-Diacetoxy-9α-fluoro-11β,17α-dihydroxy-1,4-pregnadiene-3,20-dione; triamcinolone 16α,21-di(acetate)
CAS registry number : 67-78-7
EC number : 200-669-0
Merck Index monograph : 11027 (Triamcinolone)
Patch testing : In general, corticosteroids may be tested at 0.1% and 1% in alcohol; late readings (6-10 days) are strongly recommended
Molecular formula : $C_{25}H_{31}FO_8$

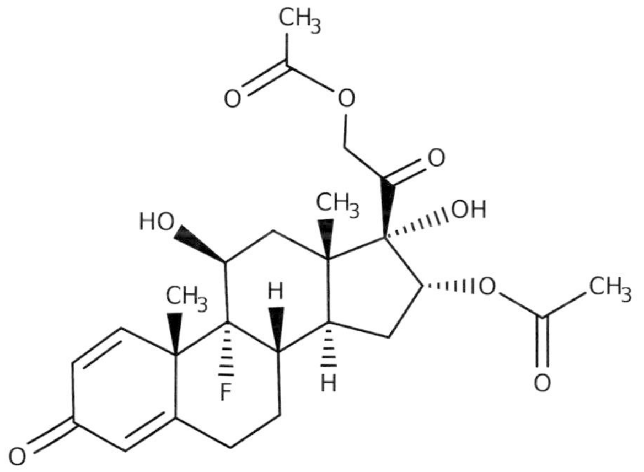

GENERAL

General aspects of corticosteroids used on the skin and mucous membranes are discussed in Chapter 2.4. A practical guideline for diagnosing allergic reactions to corticosteroids is presented in ref. 1. See also triamcinolone (Chapter 3.352), triamcinolone acetonide (Chapter 3.353), and triamcinolone hexacetonide (Chapter 3.355).

CONTACT ALLERGY

Case series

From January 1990 to June 2008, in Leuven, Belgium, 315 patients were diagnosed with contact allergy to/allergic contact dermatitis from corticosteroids (CSs) from routine patch testing with a baseline series including tixocortol pivalate, budesonide, hydrocortisone butyrate and prednisone caproate, patch testing with patients' own CS preparations, and testing those with proven contact allergy to a corticosteroid or strongly suspected of CS allergy later with a series of 66 CSs, including two sex hormones (progesterone and testosterone). 71% of the patients had relevant reactions, but these were not specified. In this group of 315 CS allergic patients, 27 had positive patch tests to triamcinolone diacetate 0.1% alc. (2). As this corticosteroid has never been used in pharmaceuticals in Belgium, these positive reactions must all be considered cross-reactions to other corticosteroids.

Cross-reactions, pseudo-cross-reactions and co-reactions

Cross-reactions between corticosteroids are discussed in Chapter 2.8.

LITERATURE

1 Baeck M, Goossens A. Immediate and delayed allergic hypersensitivity to corticosteroids: practical guidelines. Contact Dermatitis 2012;66:38-45
2 Baeck M, Chemelle JA, Terreux R, Drieghe J, Goossens A. Delayed hypersensitivity to corticosteroids in a series of 315 patients: clinical data and patch test results. Contact Dermatitis 2009;61:163-175

Chapter 3.355 TRIAMCINOLONE HEXACETONIDE

IDENTIFICATION

Description/definition : Triamcinolone hexacetonide is the hexacetonide ester of the synthetic glucocorticoid triamcinolone that conforms to the structural formula shown below

Pharmacological classes : Anti-inflammatory agents

IUPAC name : [2-[(1S,2S,4R,8S,9S,11S,12R,13S)-12-Fluoro-11-hydroxy-6,6,9,13-tetramethyl-16-oxo-5,7-dioxapentacyclo[10.8.0.0^{2,9}.0^{4,8}.01^{3,18}]icosa-14,17-dien-8-yl]-2-oxoethyl] 3,3-dimethyl-butanoate

Other names : (11β,16α)-21-(3,3-Dimethyl-1-oxobutoxy)-9-fluoro-11-hydroxy-16,17-((1-methylethyl-idene)bis(oxy))pregna-1,4-diene-3,20-dione

CAS registry number : 5611-51-8

EC number : 227-031-4

Merck Index monograph : 11029

Patch testing : In general, corticosteroids may be tested at 0.1% and 1% in alcohol; late readings (6-10 days) are strongly recommended

Molecular formula : $C_{30}H_{41}FO_7$

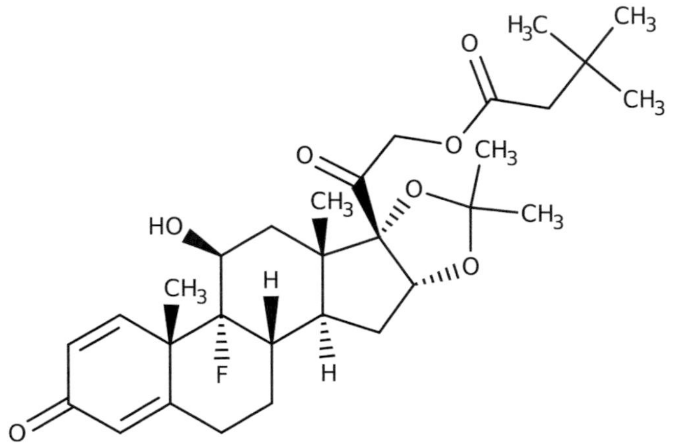

GENERAL

General aspects of corticosteroids used on the skin and mucous membranes are discussed in Chapter 2.4. A practical guideline for diagnosing allergic reactions to corticosteroids is presented in ref. 1. See also triamcinolone (Chapter 3.352), triamcinolone acetonide (Chapter 3.353), and triamcinolone diacetate (Chapter 3.354).

CONTACT ALLERGY

Case series

From January 1990 to June 2008, in Leuven, Belgium, 315 patients were diagnosed with contact allergy to/allergic contact dermatitis from corticosteroids (CSs) from routine patch testing with a baseline series including tixocortol pivalate, budesonide, hydrocortisone butyrate and prednisone caproate, patch testing with patients' own CS preparations, and testing those with proven contact allergy to a corticosteroid or strongly suspected of CS allergy later with a series of 66 CSs, including two sex hormones (progesterone and testosterone). 71% of the patients had relevant reactions, but these were not specified. In this group of 315 CS allergic patients, 39 had positive patch tests to triamcinolone hexacetonide 1% alc. (2). As this corticosteroid has never been used in pharmaceuticals in Belgium, these positive reactions must all be considered cross-reactions to other corticosteroids.

Cross-reactions, pseudo-cross-reactions and co-reactions

Cross-reactions between corticosteroids are discussed in Chapter 2.8.

LITERATURE

1 Baeck M, Goossens A. Immediate and delayed allergic hypersensitivity to corticosteroids: practical guidelines. Contact Dermatitis 2012;66:38-45

2 Baeck M, Chemelle JA, Terreux R, Drieghe J, Goossens A. Delayed hypersensitivity to corticosteroids in a series of 315 patients: clinical data and patch test results. Contact Dermatitis 2009;61:163-175

Chapter 3.356 TRIBENOSIDE

IDENTIFICATION

Description/definition : Tribenoside is the glycoside that conforms to the structural formula shown below
Pharmacological classes : Anti-inflammatory agents, non-steroidal
IUPAC name : (3R,4R,5R)-5-[(1R)-1,2-bis(Phenylmethoxy)ethyl]-2-ethoxy-4-phenylmethoxyoxolan-3-ol
Other names : D-Glucofuranoside, ethyl 3,5,6-tris-O-(phenylmethyl)-; ethyl 3,5,6-tri-O-benzyl-D-glucofuranoside; tribenzoside
CAS registry number : 10310-32-4
EC number : 233-687-2
Merck Index monograph : 11039
Patch testing : 1% and 5% pet.
Molecular formula : $C_{29}H_{34}O_6$

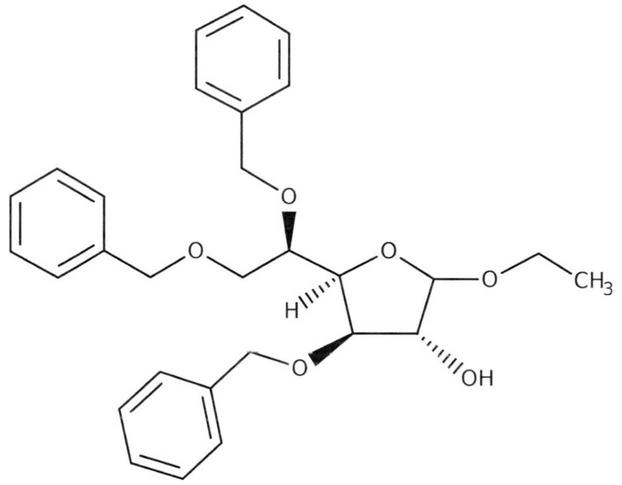

GENERAL

Tribenoside is a glycoside with anti-inflammatory and mild analgesic properties. Pharmaceuticals with tribenoside are available in many countries, e.g. as cream/ointment, tablets and as suppositories, often combined with lidocaine, to treat irritation from hemorrhoids (1).

CONTACT ALLERGY

Case report

A 25-year-old woman complained of erythema and itching around her anus. She had applied tribenoside ointment for 2 weeks for treatment of hemorrhoids. Patch tests were strongly positive to the ointment, tested 'as is' (+++) and to its main active ingredient tribenoside 1% and 5% pet. (+++ at D2 and D3). Three controls were negative (2).

Cutaneous adverse drug reactions from systemic administration caused by type IV (delayed-type) hypersensitivity

Cutaneous adverse drug reactions from systemic administration of tribenoside caused by type IV (delayed-type) hypersensitivity, including erythema multiforme (3) and drug hypersensitivity syndrome (4), are planned to be discussed in Volume IV of the *Monographs in Contact Allergy* series on Systemic drugs.

LITERATURE

1   The data in the section 'General' may have been obtained from literature discussed in this chapter, but mostly also or exclusively from one or more of the following online sources: ChemIDPlus Advanced, PubChem, DrugBank, RxList, Drug Central, Drugs.com, and Wikipedia
2   Inoue A, Tamagawa-Mineoka R, Katoh N, Kishimoto S. Allergic contact dermatitis caused by tribenoside. Contact Dermatitis 2009;60:349-350
3   Endo H, Kawada A, Yudate T, Aragane Y, Yamada H, Tezuka T. Drug eruption due to tribenoside. Contact Dermatitis 1999;41:223
4   Hashizume H, Takigawa M. Drug-induced hypersensitivity syndrome associated with cytomegalovirus reactivation: immunological characterization of pathogenic T cell. Acta Derm Venereol 2005;85:47-50

Chapter 3.357 TRIETHANOLAMINE POLYPEPTIDE OLEATE CONDENSATE

IDENTIFICATION

Description/definition : Triethanolamine polypeptide oleate condensate is a condensation product of triethanolamine, peptides and oleic fatty acids

Pharmacological classes : Cerumenolytics

Other names : Trolamine polypeptide oleate condensate; Cerumenex ®; Xerumenex ®

CAS registry number : Not available

EC number : Not available

Patch testing : Oleyl polypeptide 25% and 50% alc. (perform controls); triethanolamine polypeptide oleate condensate 1% pet.; the eardrops undiluted may cause irritant reactions (12,14) and can be tested at 25% in petrolatum (14)

Molecular formula : Not available

GENERAL

Triethanolamine polypeptide oleate condensate (TPOC) is a condensation product of triethanolamine, peptides and oleic fatty acids. It is used as a cerumenolytic drug, a compound to dissolve earwax (1). The name of this chapter is the name used by the manufacturer of eardrops containing the material (Cerumenex ®, Xerumenex ®). Of this compound, no details can be found. It is probably the same as TEA-oleoyl hydrolyzed collagen (synonyms: TEA-oleoyl hydrolyzed animal protein, triethanolamine oleoyl hydrolyzed animal protein). In the INCI (International Nomenclature Cosmetic Ingredients) database of the Personal Care Products Council this material is described as 'the triethanolamine salt of the condensation product of oleic acid chloride and hydrolyzed collagen' (https://www.per-sonalcarecouncil. org).

CONTACT ALLERGY

General

Several cases of contact allergy from a brand of ear drops (Cerumenex, Xerumenex) containing 10% triethanolamine polypeptide oleate condensate (TPOC) have been reported, often causing severe and spreading allergic contact dermatitis (3,4,6-14). The allergen in most cases was the ingredient oleyl polypeptide, as supplied by the manufacturer. In two earlier cases (11,12), ingredient testing (also supplied by the manufacturer) was positive to triethanolamine oleyl polypeptide (with a negative reaction to triethanolamine in one [12]) and in the first report of sensitization to the eardrops, in 1960, 'triethanolamine condensate' tested positive (13).

Allergic contact dermatitis to TPOC may be less rare than the limited number of reported patients suggest. In a hospital in Bristol, U.K, 3 cases of sensitization were observed in a 6-month period (9). According to the instructions, the eardrops should remain in the ear canals for 15-30 minutes only. Failure to comply (eardrops in the canal too long, frequent applications) may have increased the risk of sensitization (9). In addition, in 1988 the Belgian Contact Dermatitis Group had collected 15 cases of contact allergy to TPOC-containing eardrops (5).

In several cases, patients have experienced allergic contact dermatitis after the first application of the eardrops, without prior use. In these cases, sensitization may have occurred from contact with TPOC, oleyl polypeptide or related, cross-reactive chemicals, present as surfactant in cosmetics or topical pharmaceuticals (3,4,7,8,10). Sensitization from one single application has occurred (11).

Case series

Fifteen cases of contact allergy to triethanolamine polypeptide oleate condensate-containing eardrops were registered by the Belgian Contact Dermatitis group in 1988. No clinical or patch testing details were provided (5). Three patients with allergic contact dermatitis from the eardrops were seen in a hospital in Bristol, United Kingdom, in a 6-month period. Failure to comply with the manufacturer's recommendations may have promoted sensitization (9). Short summaries are shown in table 3.357.1.

Case report

One patient had allergic contact dermatitis from triethanolamine present in both ear drops and shampoo (2). Details are not available to the author, but it is likely, that the implicated eardrops contained triethanolamine polypeptide oleate condensate, from which free triethanolamine had been liberated.

A 62-year-old woman developed a severe, edematous and oozing dermatitis of the right ear canal, with extension to the external ear and to the right side of her face and neck. This occurred within 3 days of a first exposure to eardrops containing 10% TPOC. Eight years previously, the patient had experienced a similarly severe dermatitis of the scalp, face and neck after using a shampoo with 12% TPOC, which was manufactured by the same company as the ear drops. Patch tests showed positive reactions to the ear drops 'as is', TPOC 10% and to its breakdown product oleyl polypeptide 8.7% (3).

A 45-year-old woman had redness, swelling and edema with itching on the pinnae, sides of the neck and eyelids while using TPOC-containing eardrops. A 19-year-old girl developed severe eczematous lesions on the pinna, cheek, neck and eyelid after using TPOC-drops in her right ear. Both had positive patch tests to the eardrops 'as is' and were further tested with its ingredients (chlorobutanol 25% alc., propylene glycol 2% pet., triethanolamine (TEA) 2.5% pet., oleyl polypeptide 'as is', oleyl polypeptide 25% and 50% alc.). Both reacted to oleyl polypeptide (all 3 concentrations; 10 controls were negative) and to 3 components of oleyl polypeptide condensate: oleyl polypeptide condensate of TEA containing coconut fatty acids, oleyl polypeptide condensate of potassium containing coconut fatty acids and oleyl polypeptide condensate of potassium, containing coconut fatty acids (doubtful whether this is correct; test concentrations and vehicles unclearly described) (6).

Short summaries of other case reports of allergic contact dermatitis from triethanolamine polypeptide oleate condensate are shown in table 3.357.1.

Table 3.357.1 Short summaries of case reports of allergic contact dermatitis from triethanolamine polypeptide oleate condensate

Year and country	Sex	Age	Positive patch tests	Clinical data and comments	Ref.
1989 Italy	F	8	TPOC drops 'as is'; oleyl polypeptide 100%, 50% and 25% (petrolatum?)	severe, itching eczematous dermatitis of the left ear, cheek, pinna and neck	4
1986 Italy	F	67	TPOC drops 'as is'; oleyl polypeptide 25% and 50% alcohol	redness, swelling and edema of the left ear, cheek, neck and eyelid one day after using eardrops; primary sensitizer possibly cosmetic or topical drug with oleyl polypeptide	7
1985 Germany	M	6	TPOC-drops 'as is'; oleyl polypeptide 'as is'	itchy papular lesions on the pinnae and sides of the neck one day after irrigation of both ears	8
	F	30	TPOC-drops 'as is' oleyl polypeptide 25% and 50% alcohol	massive swelling of both pinnae and retro-auricular eczema	
1984 Germany	M	7	TPOC-drops 'as is'; polypeptide oleate 1% and 5% pet.	acute vesicular dermatitis of the ears, and neck	15
1984 United Kingdom	M	5	TPOC-drops 25% pet.	acute severe eczema affecting both ears, face and neck	9
	F	28	TPOC-drops 25% pet.	acute facial dermatitis; mother of previous patient; incorrect use of eye drops	
	F	43	TPOC-drops 25% pet.	acute severe dermatitis of both ears, face, neck, shoulders and upper back	
1981 Denmark	F	11	TPOC-drops 'as is'; oleyl polypeptide 1% water	bilateral acute otitis externa with spreading to the neck, face, chest and back; marked edema periorbital and cheeks	10
1976 United Kingdom	F	38	TPOC-drops, 'as is' and 25% prop. glycol; TEA-oleyl polypeptide 10% and 1% in prop. glycol	itch and yellow discharge from the ears, redness and swelling around the external auditory meati, spreading to the pinnae and cheeks; followed by bilateral periorbital edema, weeping and crusting pinnae, maculopapular erythema down both sides of the neck; sensitization from one single application	11
1972 United Kingdom	M	43	TPOC-drops 'as is'; TPOC 1% pet; negative to triethanolamine	bilateral acute otitis externa with spread of dermatitis to the face, neck and upper chest with marked swelling of the eyelids	12
1960 USA	F	17	TPOC-drops 'as is'; triethanolamine condensate (test conc./veh. ?)	marked erythema, edema and exudation of the tissues of left ear, maculopapular eruption of the left cheek and erythema and edema left eye and anterior chest	13
	F	33	not patch tested	dermatitis; the 2 cases were discovered during 'a limited clinical trial' of the drug at the Mayo Clinic, Rochester	

conc.: concentration; TPOC: triethanolamine polypeptide oleate condensate; veh.: vehicle

Two patients were seen at St. John's Hospital in London in 1975 and 1976 with allergic contact dermatitis from the eardrops. Both reacted strongly to the drops and to TPOC, probably at 1% pet. It was mentioned that the undiluted drops are mildly irritant and that several such reactions have been observed; these patients did not react when tested with the separate ingredients (14).

Cross-reactions, pseudo-cross-reactions and co-reactions

A woman who probably had become sensitized to TPOC-containing eardrops later had positive reactions to TEA-oleyl polypeptide (oleyl polypeptide is the allergen in the eardrops and in TPOC), potassium coco-hydrolyzed animal protein (causing allergic contact dermatitis from its presence in a shampoo), TEA-coco-hydrolyzed animal protein, and potassium undecylenoyl hydrolyzed animal protein (5).

LITERATURE

1   The data in the section 'General' may have been obtained from literature discussed in this chapter, but mostly also or exclusively from one or more of the following online sources: ChemIDPlus Advanced, PubChem, DrugBank, RxList, Drug Central, Drugs.com, and Wikipedia

2   Schmutz JL, Barbaud A, Tréchot P. Allergie de contact à la triéthanolamine contenue dans des gouttes auriculaires et dans un shampooing (Contact allergy to triethanolamine in ear drops and shampoo). Ann Dermatol Venereol 2007;134:105 (Article in French)

3   Sasseville D, Moreau L. Allergic contact dermatitis from triethanolamine polypeptide oleate condensate in eardrops and shampoo. Contact Dermatitis 2005;52:233

4   Balato N, Lembo G, Patruno C, Ayala F. Allergic contact dermatitis from Cerumenex in a child. Contact Dermatitis 1989;21:348-349

5   Dooms-Goossens A, Debusschère K, Dupré K, Degreef H. Can eardrops induce a shampoo dermatitis? A case study. Contact Dermatitis 1988;19:143-145

6   Valsecchi R, Cainelli T. Contact allergy to Cerumenex. Contact Dermatitis 1988;18:312

7   De Padova MP, Bardazzi F, Vassilopoulou A, Lama L. Contact dermatitis to Cerumenex. Contact Dermatitis 1986;15:43

8   Pevny I. Cerumenex allergy. Contact Dermatitis 1985;12:51-52

9   Grattan CE, Burton JL. Facial contact dermatitis from Xerumenex ear drops. J R Coll Gen Pract 1984;34(263):336

10 Kroon S. Contact dermatitis to oleylpolypeptide in Xerumenex ™ eardrops. Contact Dermatitis 1981;7:271-272

11 Boxley JD, Dawber RP. Contact dermatitis to one ingredient of Xerumenex ear drops. Contact Dermatitis 1976;2:233-234

12 Grice K, Johnstone CI. Contact dermatitis from Xerumenex. Br Med J 1972;1(5798):508

13 Perry HO, McBean JB. Dermatitis venenata from cerumenex, a new otic solution. AMA Arch Otolaryngol 1960;71:653-655

14 Cronin E. Contact Dermatitis. Edinburgh, Churchill Livingstone 1980:269

15 Pevny I, Brennenstuhl, Razmskas G. Patch testing in children (II). Results and case reports. Contact Dermatitis 1984;11:302-310

Chapter 3.358 TRIFLURIDINE

IDENTIFICATION

Description/definition   : Trifluridine is the fluorinated thymidine analog that conforms to the structural formula shown below

Pharmacological classes   : Antimetabolites; antiviral agents

IUPAC name   : 1-[(2R,4S,5R)-4-Hydroxy-5-(hydroxymethyl)oxolan-2-yl]-5-(trifluoromethyl)pyrimidine-2,4-dione

Other names   : Trifluorothymidine; 2'-deoxy-5-(trifluoromethyl)uridine; 5-(trifluoromethyl)deoxyuridine

CAS registry number   : 70-00-8

EC number   : 200-722-8

Merck Index monograph   : 11125

Patch testing   : 1%, 5% and 10% pet.

Molecular formula   : $C_{10}H_{11}F_3N_2O_5$

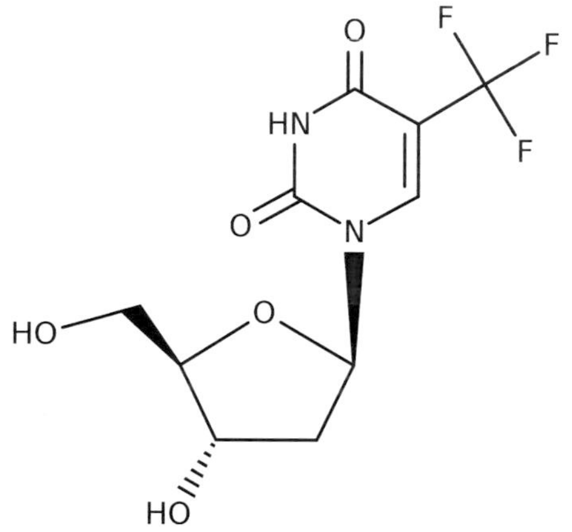

GENERAL

Trifluridine is a fluorinated pyrimidine nucleoside with antiviral activity against *Herpes simplex* virus type 1 and 2 and vacciniavirus and with potential antineoplastic activity. In anticancer therapy, trifluridine is incorporated into DNA and inhibits thymidylate synthase, resulting in inhibition of DNA synthesis, inhibition of protein synthesis, and apoptosis. Trifluridine in ophthalmic solutions is indicated for the treatment of primary keratoconjunctivitis and recurrent epithelial keratitis due to *Herpes simplex* virus types 1 and 2. Oral trifluridine, in combination with tipiracil, is indicated for the treatment of metastatic colorectal cancer (1).

CONTACT ALLERGY

Case series

In Japan, before 1987, 69 patients were treated with trifluridine (trifluorothymidine) for herpes keratitis. Patch tests were routinely done when patients exhibited contact dermatitis. Of the patients treated with trifluridine, 7 (10%) showed contact dermatitis. These side effects were resolved by switching to another anti-herpetic drug without the occurrence of cross-allergy. Clinical details and results of patch testing are unknown (2).

Case reports

A 70-year-old man had a 3-rnonth history of eczema of both left eyelids. He had been using various eye drops and ophthalmologic products for the past year for a corneal ulcer of the left eye. Patch tests showed positivity only to 1% trifluridine eye drops, indicated for herpetic eye infections. Later, all ingredients of this product were tested separately, and there were positive reactions to trifluridine 10% and 5% pet., but not to 1% or 0.5%. Fifty controls were negative to the 5% and 10% preparations. On stopping the eye drops and providing symptomatic treatment, the lesions receded (3).

     A man aged 63 who had recurrent *Herpes simplex* keratitis of his right eye developed dermatitis around the right orbit, which spread within a few days to the entire face, the trunk and proximal part of the extremities. For a recent recurrence, the patient had been treated for a week with 1% trifluridine eye drops and scopolamine 0.5% eye drops.

There was a positive patch test to the trifluridine but not to the scopolamine eye drops. Trifluridine itself was not tested, but an additional positive reaction to trifluridine ointment and negative reaction to the eyedrop base without trifluridine make contact allergy to the active drug highly likely (4,5).

A 15-year-old atopic patient developed contact allergy to trifluridine ointment. Details are not available to the author (7).

Cross-reactions, pseudo-cross-reactions and co-reactions

A patient primarily sensitized to idoxuridine later developed allergic contact dermatitis from trifluridine eye drops. The authors considered this to be cross-reactivity, as the structural formulas are virtually identical, except for the fact that idoxuridine has iodine at position 5 and trifluorthymidine CF3. The patient, however, had been treated – albeit only for a week - with trifluridine eye drops, so sensitization from exposure (successive sensitization) cannot be ruled out (4,5).

Three patients primarily sensitized to the antiviral agent ibacitabine were patch tested with several chemicals to detect cross-allergy, including trifluridine; there were no positive reactions to trifluridine (6).

LITERATURE

1   The data in the section 'General' may have been obtained from literature discussed in this chapter, but mostly also or exclusively from one or more of the following online sources: ChemIDPlus Advanced, PubChem, DrugBank, RxList, Drug Central, Drugs.com, and Wikipedia
2   Naito T, Shiota H, Mimura Y. Side effects in the treatment of herpetic keratitis. Curr Eye Res 1987;6:237-239
3   Millán-Parrilla F, de la Cuadra J. Allergic contact dermatitis from trifluoridine in eyedrops. Contact Dermatitis 1990;22:289
4   Cirkel PK, van Ketel WG. Allergic contact dermatitis to trifluorothymidine eyedrops. Contact Dermatitis 1981;7:49-50
5   Cirkel PK, van Ketel WG. A patient with contact allergy for various virostatic agents used in herpetic keratitis. Ned Tijdschr Geneeskd 1982;126:1453-1454 (Article in Dutch)
6   Foussereau J, Tomb R. Cross-allergy between 5-iodo-2'-deoxycytidine and idoxuridine. J Am Acad Dermatol 1987;1:145-147
7   Gailhofer G, Ludvan M, Posawetz-Kresbach M. Allergisches Kontaktekzem durch Trifluorthymidin. Wien Klin Wochenschr 1987;99:192-194 (Article in German)

Chapter 3.359 TRIMEBUTINE

IDENTIFICATION

Description/definition   : Trimebutine is the trihydroxybenzoic acid derivative that conforms to the structural
                             formula shown below (trimebutine maleate)
Pharmacological classes   : Gastrointestinal agents; parasympatholytics
IUPAC name   : [2-(Dimethylamino)-2-phenylbutyl] 3,4,5-trimethoxybenzoate
Other names   : Benzoic acid, 3,4,5-trimethoxy-, 2-(dimethylamino)-2-phenylbutyl ester
CAS registry number   : 39133-31-8
EC number   : 254-309-2
Merck Index monograph   : 11138
Patch testing   : 0.5% and 1% water and pet.
Molecular formula   : $C_{22}H_{29}NO_5$

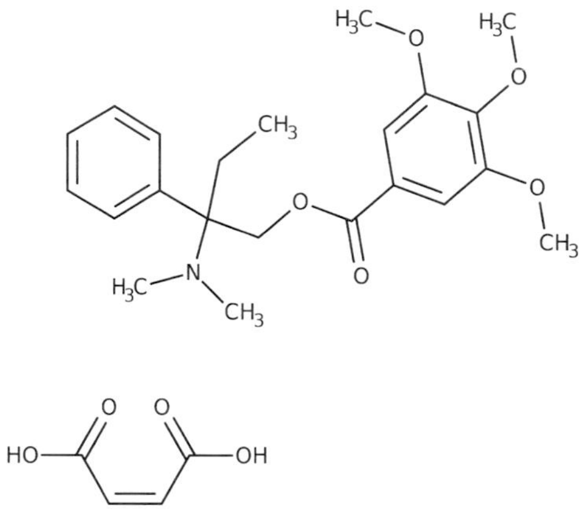

GENERAL

Trimebutine is a trimethoxybenzoic acid and spasmolytic agent that regulates intestinal and colonic motility and relieves abdominal pain with antimuscarinic and weak mu opioid agonist effects. This drug is indicated for symptomatic treatment of irritable bowel syndrome and treatment of postoperative paralytic ileus following abdominal surgery (1). It is also used rectally and topically for anal fissures and hemorrhoids (2). In pharmaceutical products, both trimebutine and trimebutine maleate (CAS number 34140-59-5, EC number 251-845-9, molecular formula $C_{26}H_{33}NO_9$) may be employed (1).

CONTACT ALLERGY

Case report

A 40-year-old man presented with a pruritic perianal eruption characterized by erythema, papules, vesicles and exudation. He had been applying a cream containing trimebutine and ruscogenin to hemorrhoids for 5 days. Patch tests were positive to the cream 'as is' (+++) at D2 and D4. Patch tests were then performed with trimebutine powder contained in a commercial suspension in water at 4.8, 2.4, 1.2, 0.6, 0.3 and 0.15 mg/ml. All concentrations of trimebutine were +++ at D2 and D4. Later, patch testing with trimebutine in cream-gel at concentrations of 2%, 1%, 0.5% and 0.1%, and the excipients of the cream-gel, was performed. All trimebutine tests were positive at D2 and D4, but the excipients tested negative. Twenty controls were negative (2).

A 40-year-old woman presented with generalized pruritic and erythematous hives without angioedema, 7 days after initiating therapy with a cream for hemorrhoids containing trimebutine. Patch tests were positive to the cream 'as is' and to trimebutine (test concentration and vehicle not mentioned). An oral provocation test with trimebutine (dosage and schedule not mentioned) resulted in generalized urticaria after 3 days (3). This patient had systemic contact dermatitis presenting as generalized urticaria from absorption of trimebutine from the cream for hemorrhoids, followed by systemic contact dermatitis from oral provocation.

LITERATURE

1   Martin-Garcia C, Martinez-Borque N, Martinez-Bohigas N, Torrecillas-Toro M, Palomeque-Rodrìguez MT.
    Delayed reaction urticaria due to trimebutine. Allergy 2004;59:789-790
2   Reyes JJ, Fariña MC. Allergic contact dermatitis due to trimebutine. Contact Dermatitis 2001;45:164

Chapter 3.360 TRIPELENNAMINE

IDENTIFICATION

Description/definition : Tripelennamine is the 2-benzylaminopyridine that conforms to the structural formula shown below

Pharmacological classes : Histamine H1 antagonists; anti-allergic agents

IUPAC name : *N'*-Benzyl-*N,N*-dimethyl-*N'*-pyridin-2-ylethane-1,2-diamine

Other names : Pyribenzamine

CAS registry number : 91-81-6

EC number : 202-100-1

Merck Index monograph : 11180

Patch testing : 2% pet.

Molecular formula : $C_{16}H_{21}N_3$

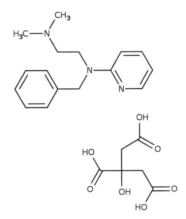

Tripelennamine   Tripelennamine citrate

GENERAL

Tripelennamine is an ethylenediamine-type histamine H1 antagonist. It is used to treat asthma, hay fever, urticaria, and rhinitis, and also in veterinary applications. Tripelennamine is administered by various routes, including topically. In pharmaceutical products, it may be employed as tripelennamine citrate (CAS number 6138-56-3, EC number 228-121-6, molecular formula $C_{22}H_{29}N_3O_7$) or as tripelennamine hydrochloride (CAS number 154-69-8, EC number 205-833-5, molecular formula $C_{16}H_{22}ClN_3$) (1).

Case series

In Leuven, Belgium, in the period 1990-2014, iatrogenic contact dermatitis was diagnosed in 2600 individuals (17.4% of the total population). 96% of all positive patch test reactions to topical drugs and antiseptics were considered to be relevant. Tripelennamine (1% pet.) was tested in one patient and there was a positive reaction to it (5).

In Bari, Italy, in the period 1968-1983, 403 selected patients (selection procedure unknown) were patch tested with tripelennamine (pyribenzamine) 2% pet. and there were 2 (0.5%) positive reactions; relevance was not discussed (2).

Case reports

Of ninety patients treated with tripelennamine 2% and 5% cream for a variety of dermatoses, 2 were found to have developed an 'allergy of the eczematous contact-type'. They both reacted to patch tests with the 2% and 5% tripelennamine cream but were negative to the cream bases without the active ingredient (4).

A man of 50 years old had been under the care of an otolaryngologist for an eruption involving the ear and had been given an ointment consisting of 2% pyribenzamine (tripelennamine) in a hydrophilic base. Instead of improving, the eruption became worse, finally presenting an acute, edematous, oozing dermatitis of the entire face. Patch tests were strongly positive to the cream. Subsequent tests showed strongly positive patch tests to tripelennamine 2% in pet. and in two water-miscible bases with different formulas, whereas the bases used were negative (3).

A 46-year-old woman complained of dermatitis of the eyelids of 6 weeks' duration after the use of tripelennamine (pyribenzamine) ointment for itching of the eyelids. Patch tests were positive to pyribenzamine cream and ointment and to pure pyribenzamine powder. There were no reactions to 5 other antihistamines. Oral pyribenzamine was well-tolerated (6). The authors also describe a 56-year-old woman who treated dermatitis of the neck and eyelids caused by cosmetics including nail polish with pyribenzamine ointment. During the following 2 weeks the eruption disappeared, then suddenly recurred. Patch testing yielded positive reactions to pyribenzamine cream, ointment and pyribenzamine powder pure. Three controls were negative to the powder. The patient also reacted to pyrilamine maleate and methapyrilene, possibly from cross-reactivity (6).

A 54-year-old woman was prescribed oral tripelennamine (pyribenzamine) for possible allergic contact dermatitis of the eyelids caused by antazoline eye drops. Within 24 hours she presented with marked edema of the eyelids with an area of red, oozing, vesicular dermatitis covering the lids and upper parts of the cheeks. The skin of the back and chest showed a diffuse maculopapular rash most marked below the axillae. Patch tests were positive to antazoline, pyribenzamine and various other antihistamines (test concentrations/vehicles not mentioned). The maculopapular rash was probably systemic contact dermatitis caused by oral tripelennamine (6).

Cross-reactions, pseudo-cross-reactions and co-reactions

A patient who had allergic contact dermatitis from tripelennamine may have cross-reacted to pyrilamine maleate and methapyrilene; all three are antihistamines of the ethylenediamine-type (6).

LITERATURE

1   The data in the section 'General' may have been obtained from literature discussed in this chapter, but mostly also or exclusively from one or more of the following online sources: ChemIDPlus Advanced, PubChem, DrugBank, RxList, Drug Central, Drugs.com, and Wikipedia

2   Angelini G, Vena GA, Meneghini CL. Allergic contact dermatitis to some medicaments. Contact Dermatitis 1985;12:263-269

3   Strauss MJ. Eczematous contact-type allergy to pyribenzamine. J Invest Dermat 1948;11:155

4   Sulzberger MB, Baer RL, Levin HB. Local therapy with pyribenzamine hydrochloride. J Invest Dermat 1948;10:41-42

5   Gilissen L, Goossens A. Frequency and trends of contact allergy to and iatrogenic contact dermatitis caused by topical drugs over a 25-year period. Contact Dermatitis 2016;75:290-302

6   Sherman WB, Cooke RA. Dermatitis following the use of pyribenzamine and antistine. J Allergy 1950;21:63-67

Chapter 3.361 TROMANTADINE

IDENTIFICATION

Description/definition   : Tromantadine is the cyclic amine and secondary carboxamide that conforms to the
                               structural formula shown below
Pharmacological classes   : Antiviral agents
IUPAC name   : N-(1-Adamantyl)-2-[2-(dimethylamino)ethoxy]acetamide
Other names   : 2-(2-(Dimethylamino)ethoxy)-N-tricyclo(3.3.1.1^{3,7})dec-1-ylacetamide
CAS registry number   : 53783-83-8
EC number   : 258-770-0
Merck Index monograph : 11220
Patch testing   : 1% pet.
Molecular formula   : $C_{16}H_{28}N_2O_2$

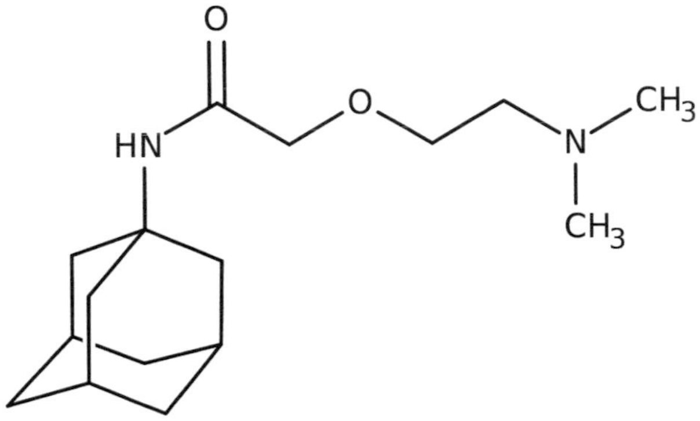

GENERAL

Tromantadine is a cyclic amine derived from amantadine with activity against *Herpes simplex* virus. It inhibits absorption of virions to cell surfaces, as well as penetration and uncoating of the virus. It is used in the topical treatment of herpes simplex. In pharmaceutical products, tromantadine is employed as tromantadine hydrochloride (CAS number 41544-24-5, EC number 255-434-5, molecular formula $C_{16}H_{29}ClN_2O_2$) (1).

CONTACT ALLERGY

General

Contact allergy to and allergic contact dermatitis from tromantadine HCl was formerly far from rare. There are strong indications that, of patients using commercial preparations with tromantadine frequently and repeatedly, >5% will become sensitized to this antiviral drug (14,23). Up to 1990, >110 cases were described (3-14,16-24). Next to single case histories (3,7,8,9,12,16,17), case series of 37 (13), 26 (23; these include the 12 patients presented in ref. 14), 14 (20), 8 (24), 6 (2), 4 (11,22), and 3 (5,6) patients with allergic contact dermatitis from topical tromantadine HCl have been published. Cross-sensitivity with amantadine has been observed repeatedly (9,10,18,19,21). In animal experiments, tromantadine was shown to be a moderate sensitizer, with a sensitizing capacity far stronger than other antivirals (15).

   Because of questionable efficacy and the many cases of allergic contact dermatitis, tromantadine has been banned in many countries. In the last 30 years, only a few publications on the issue have appeared (16,17). This may reflect lesser use of the antiviral, but may also indicate that the sensitizing potential is so well known that it needs not be reported anymore.

Patch testing in groups of patients

In Germany, 119 patients who were patch tested because of various dermatoses, tromantadine gel and the gel base were patch tested additionally. Two positive reactions (1.7%) to the active substance and two to the gel base only were found. One of the latter patients also had a positive reaction to parabens, which was a constituent of the gel (20). It is likely that these were consecutive patients, not a selected patient group.

Case series

In Leuven, Belgium, in the period 1990-2014, iatrogenic contact dermatitis was diagnosed in 2600 individuals (17% of the total patch test population). 96% of all positive patch test reactions to topical drugs and antiseptics were considered to be relevant. Tromantadine (2% pet.) was tested in 37 patients and there were 6 positive reactions to it (2).

In the university clinic of Rotterdam, The Netherlands, in a period of one year (1986-1987), 8 patients (7 women) with allergic contact dermatitis from tromantadine ointment were observed (24). In Bari, Italy, in 1986, 2 women with herpes simplex labialis developed oozing dermatitis of the face after 7 days' use of tromantadine ointment. A 42-year-old man treated herpes zoster with the ointment and presented with widespread dermatitis after 10 days. Patch tests were positive in all 3 to the ointment and to tromantadine HCl 1% pet. and negative to the base (5).

In Bologna, Italy, in a period of a few months before 1985, 3 patients with ACD from tromantadine were investigated. Two had recurrent herpes labialis and one developed cheilitis and the other severe dermatitis of the lips, the cheeks and the chin from tromantadine ointment. The third patient had swelling and a bullous eruption of the genital area after using the ointment for recurrent vulvar herpes simplex. All had positive patch tests to the ointment and to tromantadine HCl 1% water (6).

In Germany, before 1984, 19 patients with dermatitis after application of tromantadine ointment were patch tested and in 14 of them, contact allergy to tromantadine was observed (20). In Nijmegen, The Netherlands, 4 patients (3 male) developed dermatitis of the face, of who 2 with secondary bacterial infection, from the use of tromantadine ointment. They all had positive patch tests to the ointment and to tromantadine HCl 1% pet. (22).

From a hospital in Lisbon, the capital city of Portugal, 4 patients (3 women and 1 man) with contact dermatitis to tromantadine were reported in 1982. They had a similar clinical picture, having had recurrent labial herpes simplex for which they regularly used tromantadine. Suddenly one of the herpes episodes became much more violent, the lesions spreading to the other lip, being exudative and itchy. After stopping the ointment and using topical anti-eczematous treatment the patients were promptly cured. All 4 were patch test positive to tromantadine HCl 1% pet. (11). In 1980, 37 cases of allergic contact dermatitis to tromantadine were reported from Germany; details are not available to the author (13).

In Austria, before 1976, 240 patients with herpes simplex were treated with tromantadine 1% ointment. Twenty individuals (12 women, 8 men), who had suffered from frequent, recurring herpetic eruptions for several years, showed local side effects after long-term use of this agent. Whereas all female patients had herpes simplex on the lips or face, in four male patients the genital region, predominantly the glans penis, was affected. Those patients seen in the acute phase of the dermatitis showed considerable inflammation also in the skin area adjacent to the herpetic eruption. In all but one patient, in whom the eczematous response had occurred after only 10 days, the allergic reaction developed after many months of repeated well-tolerated applications. Patch tests were positive in twelve patients (5% of all treated; 9 women) to the ointment and to tromantadine hydrochloride, tested pure and negative to the ointment base. In one patient, generalized dermatitis followed patch testing (14). Two years later, the number of treated patients had risen to 387 and in 26 (6.7%), contact allergy to the antiviral was established (23).

Case reports

An 18-year-old woman was referred with an eczematous reaction on her left cheek which developed after 5 days' application of a tromantadine hydrochloride-containing ointment for recurrent herpes simplex. She had previously used the ointment for other relapses without any cutaneous reactions. Patch tests were positive to the antiviral ointment, tromantadine HCl 10% pet. but negative to the ointment base (3).

Table 3.361.1 Short summaries of single case reports of allergic contact dermatitis from tromantadine

Year and country	Sex	Age	Positive patch tests	Clinical data and comments	Ref.
1985 Italy	M	17	TRO 0.5% and 1% pet.	acute allergic contact dermatitis after TRO ointment for herpes labialis	7
1984 Italy	F	26	TRO ointment 'as is'	contact dermatitis on left cheek; TRO itself not tested, not even the base or all its ingredients	8
1984 Italy	F	40	TRO 0.5% and 0.8% pet.	contact dermatitis; cross-reaction to amantadine HCl 1% pet.	9
1982 The Netherlands	F	30	TRO ointment; TRO 1% pet; ointment base	allergic contact dermatitis after applying TRO ointment to herpes simplex on the left cheek; allergen in ointment base unknown, the composition was not available	12

TRO: tromantadine hydrochloride

A 40-year-old man presented with erythematous pruritic lesions of the glans. One week before presentation, he had had unprotected oral sex with a new female partner. Physical examination showed small vesicles on an erythematous background on the glans penis. The patient stated that, following intercourse, he had been applying a topical cream containing tromantadine twice daily to prevent any HSV infection. An open application test with a small amount of the antiviral cream was applied to an area 1 cm^2 in size on the upper arm with a positive reaction at D2 and D3. The patient was diagnosed with allergic contact dermatitis caused by tromantadine (which is, of course, an unsubstantiated diagnosis, as tromantadine itself was not tested) (16).

Short summaries of other single case reports of allergic contact dermatitis from tromantadine HCl are shown in table 3.361.1.

Cross-reactions, pseudo-cross-reactions and co-reactions
Of 15 patients sensitized by tromantadine, 13 (87%) cross-reacted to amantadine (10,19). Cross-reactivity has also been observed by other authors (9,18,21). Oral administration of amantadine in these patients may or might result in systemic contact dermatitis (18).

LITERATURE

1   The data in the section 'General' may have been obtained from literature discussed in this chapter, but mostly also or exclusively from one or more of the following online sources: ChemIDPlus Advanced, PubChem, DrugBank, RxList, Drug Central, Drugs.com, and Wikipedia

2   Gilissen L, Goossens A. Frequency and trends of contact allergy to and iatrogenic contact dermatitis caused by topical drugs over a 25-year period. Contact Dermatitis 2016;75:290-302

3   Patruno C, Auricchio L, Mozzillo R, Brunetti B. Allergic contact dermatitis due to tromantadine hydrochloride. Contact Dermatitis 1990;22:187

4   Miranda A, Gomez S, del Pozo LJ, Quiñones PA. Allergic contact dermatitis to tromantadine. Contact Dermatitis 1987;17:55-56

5   Angelini G, Vena GA, Meneghini CL. Contact allergy to antiviral agents. Contact Dermatitis 1986;15:114-115

6   Tosti A, Melino M, Veronesi S, Labanca M. Contact dermatitis to tromantadine. Contact Dermatitis 1985;13:339

7   Valsecchi R, Foiadelli L, Cainelli T. Contact allergy from tromantadine hydrochloride. Contact Dermatitis 1985;13:341

8   Lembo G, Balato N, Cusano F, Ayala F. Allergic dermatitis from Viruserol ointment probably due to tromantadine hydrochloride. Contact Dermatitis 1984;10:317

9   Santucci B, Picardo M, Cristaudo A. Contact dermatitis to tromantadine. Contact Dermatitis 1984;10:317-318

10 Przybilla B. Allergic contact dermatitis to tromantadine. J Am Acad Dermatol 1983;9:165

11 Brandao FM, Pecegueiro M. Contact dermatitis to tromantadine hydrochloride. Contact Dermatitis 1982;8:140-141

12 Van Ketel WG. Allergic dermatitis from Viru-Merz ointment, tromantadine hydrochloride and serol base. Contact Dermatitis 1982;8:71

13 Przybilla B, Balda R. Allergische Kontaktdermatitis durch Tromantadin. Münch Med Wschr 1980;122:1195-1198 (Article in German)

14 Fanta D, Mischer P. Contact dermatitis from tromantadine hydrochloride. Contact Dermatitis 1976;2:282-284

15 Hausen BM, Schulze R. Comparative studies of the sensitizing capacity of drugs used in herpes simplex. Derm Beruf Umwelt 1986;34:163-170 (Article in German)

16 Maatouk I. Contact balanitis to tromantadine. Clin Exp Dermatol 2016;41:926

17 Jáuregui I, Urrutia I, Gamboa PM, Antépara I. Allergic contact dermatitis from tromantadine. J Investig Allergol Clin Immunol 1997;7:260-261

18 Van Ketel WG. Systemic contact-type dermatitis by derivatives of adamantane? Derm Beruf Umwelt 1988;36:23-24 (Article in German)

19 Przybilla B, Wagner-Grösser G, Balda BR. Kontaktallergische Kreuzreaktion von Tromantadin und Amantadin. Dtsch Med Wochenschr 1983;108:172-175 (Article in German)

20 Agathos M, Remien C, Mutzeck E. Contact allergy to tromantadine. Derm Beruf Umwelt 1984;32:157-160 (Article in German)

21 Klaschka F. Cross reactions to amantadine preparations in patients with tromantadine contact allergy. Dtsch Med Wochenschr 1983;108:1735-1736 (Article in German)

22 Van der Walle HB, Malten KE, Waegemakers TH. Contact allergy to tromantadine (Viru-Merz) in the treatment of herpes simplex infections. Ned Tijdschr Geneeskd 1982;126:1033-1035 (Article in Dutch)

23 Mischer P, Fanta D. The tromantadine contact eczema. Hautarzt 1978;29:337-339 (Article in German)

24 Van Joost T, Stolz E, Piket-van Ulsen J. Sensitization to tromantadine (Viru-Merz) in herpes simplex infections. Ned Tijdschr Geneeskd 1987;131:21-22 (Article in Dutch)

Chapter 3.362 TROMETHAMINE

IDENTIFICATION

Description/definition   : Tromethamine is the organic amine that conforms to the structural formula shown below
Pharmacological classes : Buffers; excipients
IUPAC name   : 2-Amino-2-(hydroxymethyl)propane-1,3-diol
Other names   : Aminotris(hydroxymethyl)methane; tris(hydroxymethyl)aminomethane; trometamol; trisamine; trisaminol
CAS registry number   : 77-86-1
EC number   : 201-064-4
Merck Index monograph   : 11221
Patch testing   : 1% water
Molecular formula   : $C_4H_{11}NO_3$

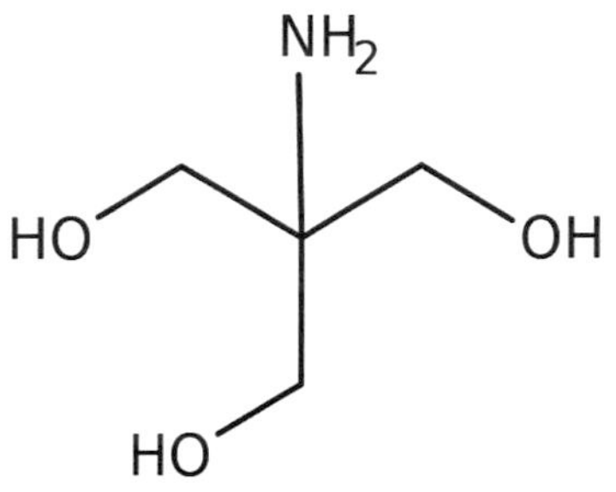

GENERAL

Tromethamine is an organic amine proton acceptor. It is used in the synthesis of surface-active agents and pharmaceuticals, as an emulsifying agent for mineral oil and paraffin wax emulsions, and as a biological buffer. In medicine, it is employed as an alkalizer indicated for the prevention and correction of metabolic acidosis. Tromethamine has been used in cardioplegic solutions, liver transplantation and chemolysis of renal calculi. In cosmetics, tromethamine is used for buffering/pH adjusting and for masking unpleasant odors (1,2).

CONTACT ALLERGY

Case reports

A woman presented with a 2-year history of perianal irritation, which had recently worsened and spread to the vulval area. She had been treated with topical corticosteroids, emollients, and a variety of over-the-counter preparations including an 'itch relief cream'. When examined, she had marked erythema and edema of the perianal area, perineum, and natal cleft. She was patch tested and reacted positively to the itch relief cream. Subsequent testing with its ingredients showed positive reactions to tromethamine 1% water and to the original preparation (1).

A 71-year-old woman developed an itchy edematous eczema of the periorbital region and cheeks. She also had ocular itching, tearing and conjunctival injection. The oculist prescribed prednisolone ophthalmologic ointment and advised her to continue with her previous therapy consisting of retinol ophthalmologic gel and polyvidone eyedrops. However, the symptoms persisted. Patch tests yielded a +++ reaction to the retinol eye gel only. Later, she reacted to its ingredient tromethamine 0.5% water and – in a dilution series – to tromethamine 1%, 0.5% and 0.1% water. Withdrawal of and re-exposure to the retinol ophthalmologic gel resulted in clearing and rapid relapse of contact dermatitis (2).

Another possible case of contact allergy to tromethamine was presented in 2007 (3), but tromethamine itself was not tested and allergy not proven, not even *per exclusionem* (3).

LITERATURE

1   Singh M, Winhoven SM, Beck MH. Contact sensitivity to octyldodecanol and trometamol in an anti-itch cream. Contact Dermatitis 2007;56:89-90
2   Bohn S, Hurni M, Bircher AJ. Contact allergy to trometamol. Contact Dermatitis 2001;44:319
3   Fairhurst D, Wilkinson M. Independent sensitization to polidocanol and trometamol or glycerol within same product. Contact Dermatitis 2007;56:179

Chapter 3.363 TROPICAMIDE

IDENTIFICATION

Description/definition	: Tropicamide is the phenylacetamide that conforms to the structural formula shown below
Pharmacological classes	: Muscarinic antagonists; mydriatics
IUPAC name	: N-Ethyl-3-hydroxy-2-phenyl-N-(pyridin-4-ylmethyl)propanamide
Other names	: N-Ethyl-2-phenyl-N-(4-pyridylmethyl)hydracrylamide
CAS registry number	: 1508-75-4
EC number	: 216-140-2
Merck Index monograph	: 11229
Patch testing	: 1% pet.
Molecular formula	: $C_{17}H_{20}N_2O_2$

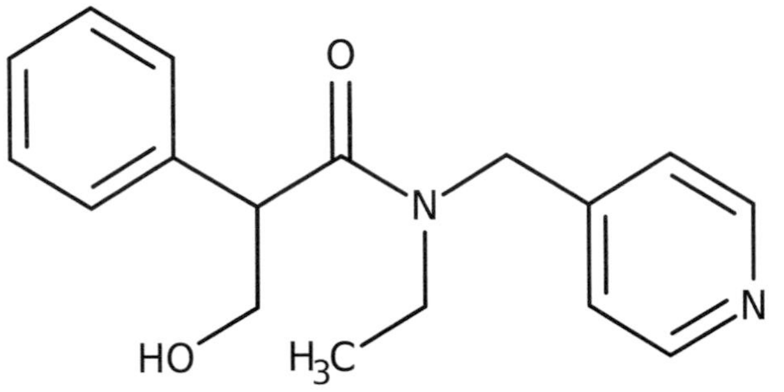

GENERAL

Tropicamide is a synthetic muscarinic antagonist with actions similar to atropine and with an anticholinergic property. Upon ocular administration, tropicamide binds to and blocks the muscarinic receptors in the sphincter and ciliary muscle in the eye. This inhibits the responses from cholinergic stimulation, producing dilation of the pupil and paralysis of the ciliary muscle. Tropicamide in eye drops is indicated to induce mydriasis and cycloplegia in diagnostic procedures, such as measurement of refractive errors and examination of the fundus of the eye (1).

CONTACT ALLERGY

Case series

In Leuven, Belgium, in the period 1990-2014, iatrogenic contact dermatitis was diagnosed in 2600 individuals (17% of the total patch test population). 96% of all positive patch test reactions to topical drugs and antiseptics were considered to be relevant. Tropicamide (1% pet.) was tested in 8 patients and there was 1 positive reaction to it (2).

In Ferrara, Italy, over a 65-month period before 2005, 50 patients affected by periorbital dermatitis while using topical ocular products were patch tested, including with their own ophthalmic medications (n=210). Only 15 positive reactions were detected in 12 subjects, including 14 reactions to commercial eye drops. There were two reactions to tropicamide and phenylephrine. The active ingredients were not tested separately, but contact allergy to the excipients and preservatives was excluded by patch testing. The authors concluded that patch testing with commercial eye drops has doubtful value (4).

In Pamplona, Spain, in one year's time (1992), 13 patients were diagnosed with contact allergy to ophthalmic medications, nearly 3% of all patients investigated for suspected contact dermatitis. There was one reaction to tropicamide 5% water in an individual who had previously used tropicamide eye drops (3).

In Bologna, Italy, before 1989, 136 patients (35 men, 101 women, range of age 14-84 years, average 36 years) suspected of contact conjunctivitis from eye drops (including contact lens solutions), were patch tested. In sixty, there was also eyelid contact dermatitis. In 75 individuals, the causative agents were found. The great majority was caused by thimerosal (n=52), followed by benzalkonium chloride (n=7), both present in contact lens solutions. There was one reaction to tropicamide eye drops and tropicamide itself (6).

Case reports

A man aged 84 developed conjunctival itching and hyperemia, eyelid edema and desquamation after instillation of eye drops containing tropicamide, phenylephrine, and double anesthetic (oxybuprocaine, tetracaine) for diagnostic purposes. Patch tests with the eye drops and phenylephrine 10% water and 20% pet. were negative. A conjunctival challenge test using the eye drops involved were positive for commercial tropicamide eye drops and phenylephrine eye drops, which caused substantial itching, periocular erythema, conjunctival hyperemia, and eyelid edema 8 hours after the instillation accompanied by eyelid desquamation in the following days. These findings were considered proof for the existence of allergic contact dermatitis to tropicamide and phenylephrine (8).

A 63-year-old woman presented with severe erythematous scaly dermatitis of the face, eyelids and periorbital areas. Her symptoms had appeared hours after using a mydriatic ophthalmic preparation for diagnostic purposes. Patch tests were positive to tropicamide ophthalmic solution. Further testing with the patient's ophthalmics, their constituents and other environmental agents revealed a strongly positive reaction to tropicamide 1% pet. only. Twenty controls were negative (9).

A nurse whose work in the ophthalmology department of a hospital included the instillation of eye drops containing tropicamide and phenylephrine into the eyes of patients undergoing routine funduscopic examination, developed an itchy rash on her fingers. Examination showed well-demarcated brownish erythema with scaling on the second and third fingers of her left hand, which are used for opening the eyes of patients and are very often subjected to contact with leaking mydriatic eye drops. Patch testing revealed contact allergy to the mydriatic product and its ingredients tropicamide and phenylephrine. A diagnosis of occupational allergic contact dermatitis was made (5).

A 48-year-old man developed severe dermatitis 2 days after photocoagulation therapy of his retina. Topical drugs used at that time were atropine and 2 antibacterial agents. He had previously used eye drops containing tropicamide and phenylephrine for an eye examination. Discontinuation of all eye drops and application of a steroid ointment resulted in resolution within a few days. Patch tests showed positive reactions to atropine, homatropine 1%, tropicamide 1% and phenylephrine 1%. Thus, the dermatitis had been caused by contact allergy to adrenaline (epinephrine) and the reactions to tropicamide and phenylephrine were of past relevance (7).

LITERATURE

1   The data in the section 'General' may have been obtained from literature discussed in this chapter, but mostly also or exclusively from one or more of the following online sources: ChemIDPlus Advanced, PubChem, DrugBank, RxList, Drug Central, Drugs.com, and Wikipedia
2   Gilissen L, Goossens A. Frequency and trends of contact allergy to and iatrogenic contact dermatitis caused by topical drugs over a 25-year period. Contact Dermatitis 2016;75:290-302
3   Tabar AI, García BE, Rodríguez A, Quirce S, Olaguibel JM. Etiologic agents in allergic contact dermatitis caused by eyedrops. Contact Dermatitis 1993;29:50-51
4   Corazza M, Massieri LT, Virgili A. Doubtful value of patch testing for suspected contact allergy to ophthalmic products. Acta Derm Venereol 2005;85:70-71
5   Okamoto H, Kawai S. Allergic contact sensitivity to mydriatic agents on a nurse's fingers. Cutis 1991;47:357-358
6   Tosti A, Tosti G. Allergic contact conjunctivitis due to ophthalmic solution. In: Frosch PJ, Dooms-Goossens A, Lachapelle JM, Rycroft RJG, Scheper RJ (eds). Current Topics in Contact Dermatitis. Berlin: Springer-Verlag, 1989: 269-272
7   Yoshikawa K, Kawahara S. Contact allergy to atropine and other mydriatic agents. Contact Dermatitis 1985;12:56-57
8   Haroun-Díaz E, Ruíz-García M, De Luxán de la Lastra S, Pastor-Vargas C, De las Heras M, Sastre Domínguez J, et al. Contact dermatitis to both tropicamide and phenylephrine eye drops. Dermatitis 2014;25:149-150
9   Boukhman MP, Maibach HI. Allergic contact dermatitis from tropicamide ophthalmic solution. Contact Dermatitis 1999;41:47-48

Chapter 3.364 TROXERUTIN

IDENTIFICATION

Description/definition : Troxerutin is the flavonoid that conforms to the structural formula shown below
Pharmacological classes : Anticoagulants; vasoprotective agents
IUPAC name : 2-[3,4-bis(2-Hydroxyethoxy)phenyl]-5-hydroxy-7-(2-hydroxyethoxy)-3-[(2S,3R,4S,5S,6R)-3,4,5-trihydroxy-6-[[(2R,3R,4R,5R,6S)-3,4,5-trihydroxy-6-methyloxan-2-yl]oxymethyl]oxan-2-yl]oxychromen-4-one
Other names : Trioxyethylrutin; vitamin P4
CAS registry number : 7085-55-4; 55965-63-4
EC number : 230-389-4
Merck Index monograph : 11239
Patch testing : 2% pet.
Molecular formula : $C_{33}H_{42}O_{19}$

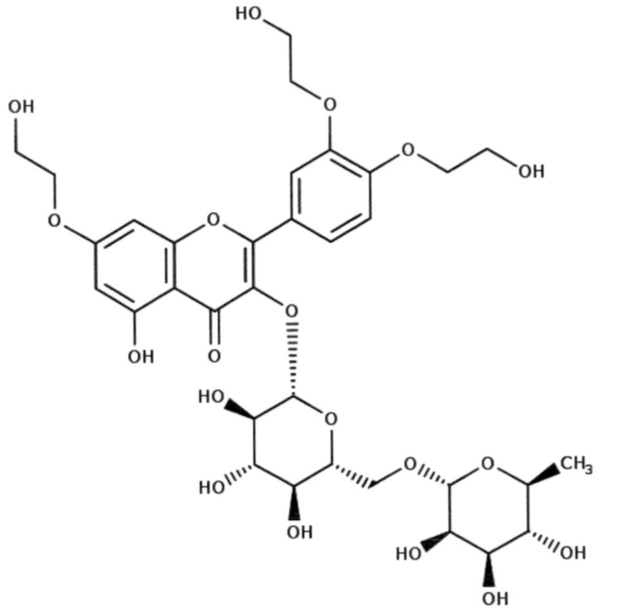

GENERAL

Troxerutin is a rutoside, a naturally occurring flavonoid. Flavonoids are polyphenolic compounds that are present in most fruits and vegetables. Although flavonoids are devoid of classical nutritional value, they are increasingly viewed as beneficial dietary components that act as potential protectors against human diseases such as coronary heart disease, cancers, and inflammatory bowel disease. Troxerutin has been used in trials studying the treatment of chronic venous insufficiency. It is also used in topical products, together with Ginkgo biloba leaf extract, as analgesic for temporary relief of pain, swelling and bruising.

CONTACT ALLERGY

Patch testing in groups of selected patients

In Italy, in the period 1970-1973, 100 patients with stasis dermatitis with or without leg ulcers were patch tested with troxerutin 2% pet. and there was one positive reaction; its relevance was not discussed (1).

Case series

In Leuven, Belgium, in the period 1990-2014, iatrogenic contact dermatitis was diagnosed in 2600 individuals (17% of the total patch test population). 96% of all positive patch test reactions to topical drugs and antiseptics were considered to be relevant. Troxerutin (2% pet.) was tested in 6 patients and there were two positive reactions to it (2).

LITERATURE

1   Angelini G, Rantuccio F, Meneghini CL. Contact dermatitis in patients with leg ulcers. Contact Dermatitis 1975;1:81-87
2   Gilissen L, Goossens A. Frequency and trends of contact allergy to and iatrogenic contact dermatitis caused by topical drugs over a 25-year period. Contact Dermatitis 2016;75:290-302

Chapter 3.365 TYROTHRICIN

IDENTIFICATION

Description/definition	: Tyrothricin is an antibiotic peptide complex produced and extracted from the aerobic gram-positive bacterium *Brevibacillus parabrevis*
Pharmacological classes	: Antibacterial agents; anti-infective agents, local
IUPAC name	: 3-[(3*R*,6*S*,9*S*,12*S*,15*S*,17*S*,20*S*,22*R*,25*S*,28*S*)-20-(2-Amino-2-oxoethyl)-9-(3-aminopropyl)-3,22,25-tribenzyl-15-[(4-hydroxyphenyl)methyl]-6-(2-methylpropyl)-2,5,8,11,14,18,21,24,27-nonaoxo-12-propan-2-yl-1,4,7,10,13,16,19,23,26-azabicyclo[26.3.0]hentriacontan-17-yl]propanoic acid (PubChem) *
CAS registry number	: 1404-88-2
EC number	: 215-771-0
Merck Index monograph	: 11293
Patch testing	: 1%, 2.5%, 5% and 10% pet.; perform controls
Molecular formula	: $C_{65}H_{85}N_{11}O_{13}$ (PubChem) *

* the composition of the mixture is variable

GENERAL

Tyrothricin is an antibiotic peptide complex produced and extracted from the aerobic gram-positive bacterium *Brevibacillus parabrevis*, which was previously categorized as *Bacillus brevis* and *Bacillus aneurinolyticus* . This complex is a mixture comprised of 60% tyrocidine cationic cyclic decapeptides (consisting largely of the six predominant tyrocidines, TrcA/A1, TrcB/B1, TrcC/C1, and other more minor contributors) and 40% neutral linear gramicidins (where valine-gramicidin A is often the major gramicidin present). Tyrothricin possesses broad spectrum gram-positive antibacterial and antifungal activity. The antibiotic mixture is very toxic to blood, liver, kidneys, meninges, and the olfactory apparatus, but is sometimes used topically in sore throat medications and in agents for the healing of infected superficial and small-area wounds (1).

CONTACT ALLERGY

Case reports and case series

In India, in the early 1980s, 101 patients suspected to have developed contact dermatitis due to antibacterial agents were patch tested with commercial antibiotic preparations. Positive patch tests were obtained with tyrothricin in 12 cases. The active ingredients were (probably) not tested separately (3).

A 43-year-old woman developed an ulcer on the medial aspect of the left ankle two years previously. Various types of treatment for a period of 6 months were of no avail until application of tyrothricin solution compresses healed the lesion. When the ulcer recurred, tyrothricin solution was again used and the ulcer began to heal. After about 6 weeks of this treatment an acute eczematous dermatitis of the lower leg developed. Penicillin ointment was then used without relief and with perhaps further aggravation of the dermatitis. Physical examination showed acute dermatitis involving the lower one third of the left leg and the middle portion of the plantar surface and entire dorsum of the left foot. The skin of these areas was edematous, erythematous and eczematous in character. There was also a moderately severe erythematous papulovesicular dermatitis of the arms. Patch tests were positive to tyrothricin solution 25 mg/ml, tyrothricin ointment 0.5 mg/gr and penicillin ointment and solution (2).

LITERATURE

1   The data in the section 'General' may have been obtained from literature discussed in this chapter, but mostly also or exclusively from one or more of the following online sources: ChemIDPlus Advanced, PubChem, DrugBank, RxList, Drug Central, Drugs.com, and Wikipedia

2   Goldman L, Feldman MD, Altemeier WA. Contact dermatitis from topical tyrothricin and associated with polyvalent hypersensitivity to various antibiotics; report of a case. J Invest Dermatol 1948;11:243-244

3   Pasricha JS, Guru B. Contact hypersensitivity to local antibacterial agents. Indian J Dermatol Venereol Leprol 1981;47:27-30

Chapter 3.366 UNDECYLENIC ACID

IDENTIFICATION

Description/definition : Undecylenic acid is the undecanoic fatty acid that conforms to the structural formula
shown below
Pharmacological classes : Dermatological agents; antifungal agents
IUPAC name : Undec-10-enoic acid
Other names : 10-Hendecenoic acid
CAS registry number : 112-38-9
EC number : 203-965-8
Merck Index monograph : 11303
Patch testing : Undecylenic acid and zinc undecylenate 5% pet.
Molecular formula : $C_{11}H_{20}O_2$

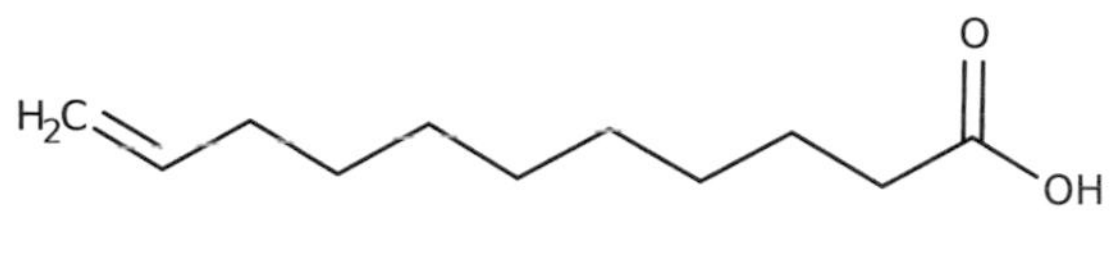

Undecylenic acid 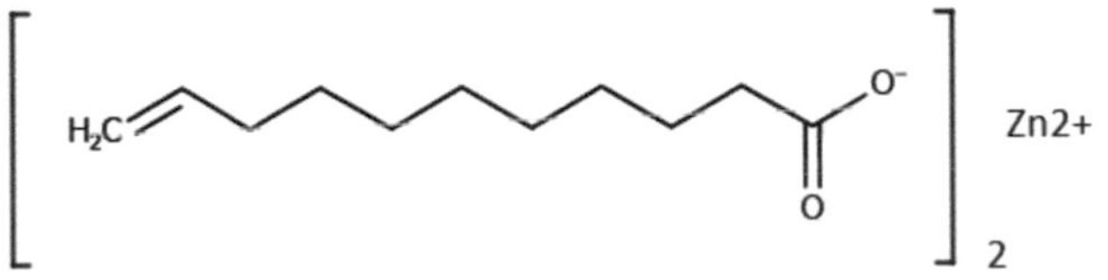   Zinc undecylenate

GENERAL

Undecylenic acid is a natural (derived from castor oil) or synthetic fatty acid with antifungal activity against
dermatophytes and *Candida albicans*. It is used topically as a zinc salt (zinc undecylenate) in preparations for the
treatment of superficial fungal infections. Zinc undecylenate acts as a fungistatic agent but fungicidal activity may be
observed with chronic exposure in high concentrations. The zinc molecules provide an astringent action, reducing
rawness and irritation. Currently, undecylenic acid is added to tioconazole nail solution to promote its antifungal
effects (1,2).

Undecylenic acid is found in black elderberry and also in human sweat. It is not only used as a medicament, but
also as flavoring ingredient, as a precursor in the manufacture of aromatic chemicals, polymers and modified
silicones. Zinc undecylenate is used as a linking molecule to conjugate other biomolecules such as proteins and
serves as an acid moiety for the anabolic steroid boldenone. In pharmaceutical preparations, undecylenic acid is
often used as or combined with zinc undecylenate (CAS number 557-08-4, EC number 209-155-0, molecular formula
$C_{22}H_{38}O_4Zn$) (1).

CONTACT ALLERGY

Case reports

A 61-year-old woman had used a solution for treating onychomycosis for 6 months, when she developed acute
dermatitis at the periungual aspect of the fingers of both hands. A patch test to the solution was positive, that to a
cream of the same brand negative. When she was tested with its 3 ingredients (tioconazole, ethyl acetate,
undecylenic acid), provided by the manufacturer, there was a positive reaction only to undecylenic acid 4% pet. (2%
?+, 1% neg.). Twenty controls were negative (2).

A 64-year-old woman was referred because of chronic eczema on the dorsa of the toes, which had not improved
in spite of the application of topical corticosteroids. Previously, she had applied multiple antifungal creams and
solutions for suspected onychomycosis to the nail plates and periungual skin of all toes. A ROAT with all topical
pharmaceuticals used was positive only to a nail solution at D4, the same product as in the previous patient. Patch
testing with tioconazole, related imidazoles and undecylenic acid showed positive (++/+++) reactions to undecylenic
acid 1%, 2% and 4% pet. Twenty controls were negative. After the application of the nail solution had been stopped,
the lesions improved rapidly. Multiple cultures of nail plates were negative. The patient was diagnosed with
onychodystrophy and allergic contact dermatitis caused by undecylenic acid (3).

A 28-year-old nurse had developed dermatitis of the feet from using an antifungal ointment, powder and liquid
containing undecylenic acid and zinc undecylenate. Patch tests with the products, tested 'as is', were positive. Later,
ingredient patch testing showed positive reactions to undecylenic acid pure, 5% undecylenic acid in Aquaphor
(petrolatum with modified cholesterol), zinc undecylenate pure and 20% in talcum (4).

LITERATURE

1   The data in the section 'General' may have been obtained from literature discussed in this chapter, but mostly also or exclusively from one or more of the following online sources: ChemIDPlus Advanced, PubChem, DrugBank, RxList, Drug Central, Drugs.com, and Wikipedia

2   Anguita JL, Escutia B, Marí JI, Rodríguez M, De La Cuadra J, Aliaga A. Allergic contact dermatitis from undecylenic acid in a commercial antifungal nail solution. Contact Dermatitis 2002;46:109

3   Ochando-Ibernón G, Schneller-Pavelescu L, Vergara de Caso EF, Silvestre-Salvador JF. Allergic contact dermatitis caused by undecylenic acid in an antifungal nail solution. Contact Dermatitis 2019;80:313-314

4   Gelfarb M, Leider M. Allergic eczematous contact dermatitis. Report of a case caused by sensitization to undecylenic acid and its zinc salt. Arch Dermatol 1960;82:642-643

Chapter 3.367 VANCOMYCIN

IDENTIFICATION

Description/definition
: Vancomycin is the branched tricyclic glycosylated peptide antibiotic obtained from *Streptomyces orientalis* that conforms to the structural formula shown below

Pharmacological classes
: Anti-bacterial agents

IUPAC name
: (1S,2R,18R,19R,22S,25R,28R,40S)-48-[(2S,3R,4S,5S,6R)-3-[(2S,4S,5S,6S)-4-Amino-5-hydroxy-4,6-dimethyloxan-2-yl]oxy-4,5-dihydroxy-6-(hydroxymethyl)oxan-2-yl]oxy-22-(2-amino-2-oxoethyl)-5,15-dichloro-2,18,32,35,37-pentahydroxy-19-[[(2R)-4-methyl-2-(methylamino)pentanoyl]amino]-20,23,26,42,44-pentaoxo-7,13-dioxa-21,24,27,41,43-pentazaoctacyclo[26.14.2.2^{3,6}.2^{14,17}.1^{8,12}.1^{29,33}.0^{10,25}.0^{34,39}]pentaconta-3,5,8,10,12(48),14,16,29(45),30,32,34(39),35,37,46,49-pentadecaene-40-carboxylic acid

CAS registry number
: 1404-90-6

EC number
: Not available

Merck Index monograph
: 11386

Patch testing
: Hydrochloride, 10.0% water (Chemotechnique)

Molecular formula
: $C_{66}H_{75}Cl_2N_9O_{24}$

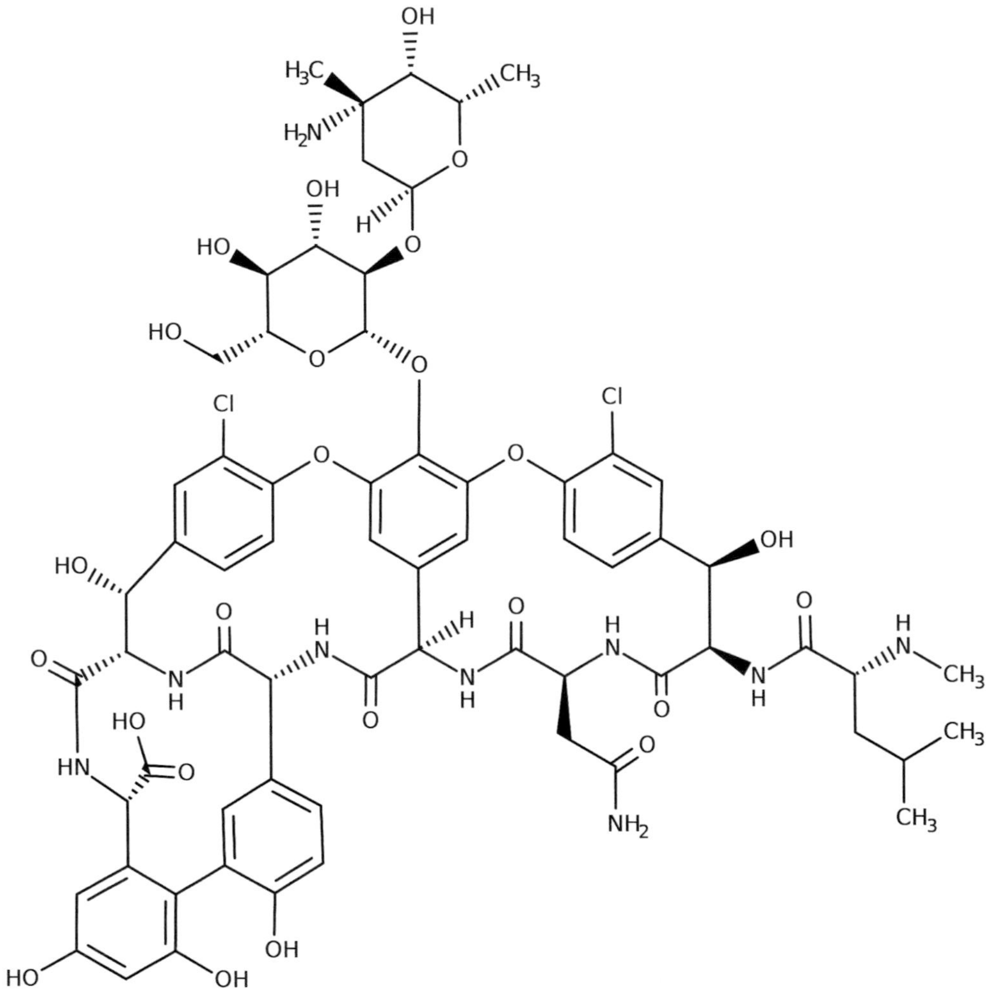

GENERAL

Vancomycin is a branched tricyclic glycosylated peptide obtained from *Streptomyces orientalis* with antibacterial properties. This antibiotic has bactericidal activity against most organisms and bacteriostatic effect on enterococci. It activates autolysins that destroy the bacterial cell wall, alters the permeability of bacterial cytoplasmic membranes and may selectively inhibit RNA synthesis. Vancomycin is indicated for the treatment of serious or severe infections caused by susceptible strains of methicillin-resistant and β-lactam-resistant staphylococci. In addition, an oral liquid preparation is indicated for the treatment of *Clostridium difficile*-associated diarrhea and enterocolitis caused by *Staphylococcus aureus*, including methicillin-resistant strains (1). In ophthalmology, topical or intravenous vancomycin is currently used to treat sight-threatening bacterial infections of the eyes, including infectious keratitis and endophthalmitis (3). In pharmaceutical products, vancomycin is employed as vancomycin hydrochloride (CAS number 1404-93-9, EC number 604-193-8, molecular formula $C_{66}H_{76}Cl_3N_9O_{24}$) (1).

CONTACT ALLERGY

Case report
A 76-year-old man presented with progressive pruritus, soreness, burning, photophobia, and tearing in his left eye 3 days after completing 2 weeks of treatment with 5% vancomycin eye drops for treatment of recurrent endoph-thalmitis. On examination, a severe periorbital erosive skin rash and swelling in the left eyelid were noted. Slit-lamp microscope examination revealed severe congestion of the left conjunctiva, central epithelial defects, stromal edema, and Descemet membrane striae of the left cornea. Intradermal injections with 0.005% and 5% vancomycin reconstituted with sterile water were positive at immediate (15 minutes; caused by non-allergic histamine release) and delayed (48 hours) readings. Patch tests with vancomycin 0.005% and 5% were positive after 48 hours (3).

Cross-reactions, pseudo-cross-reactions and co-reactions
Patients sensitized to vancomycin may cross-react to teicoplanin (2,7).

Cutaneous adverse drug reactions from systemic administration caused by type IV (delayed-type) hypersensitivity
Cutaneous adverse drug reactions from systemic administration of vancomycin caused by type IV (delayed-type) hypersensitivity, including macular erythematous exanthema (5), maculopapular rashes (7) and drug reaction with eosinophilia and systemic symptoms (DRESS) (4,6), are planned to be discussed in Volume IV of the *Monographs in Contact Allergy* series on Systemic drugs.

LITERATURE
1   The data in the section 'General' may have been obtained from literature discussed in this chapter, but mostly also or exclusively from one or more of the following online sources: ChemIDPlus Advanced, PubChem, DrugBank, RxList, Drug Central, Drugs.com, and Wikipedia
2   Bernedo N, Gonzalez I, Gastaminza G, Audicana M, Fernández E, Muñoz D. Positive patch test in vancomycin allergy. Contact Dermatitis 2001;45:43
3   Hwu JJ, Chen KH, Hsu WM, Lai JY, Li YS. Ocular hypersensitivity to topical vancomycin in a case of chronic endophthalmitis. Cornea 2005;24:754-756
4   Wendland T, Daubner B, Pichler WJ. Ceftobiprole associated agranulocytosis after drug rash with eosinophilia and systemic symptoms induced by vancomycin and rifampicin. Br J Clin Pharmacol 2011;71:297-300
5   Bernedo N, Gonzalez I, Gastaminza G, Audicana M, Fernández E, Muñoz D. Positive patch test in vancomycin allergy. Contact Dermatitis 2001;45:43
6   Liippo J, Pummi K, Hohenthal U, Lammintausta K. Patch testing and sensitization to multiple drugs. Contact Dermatitis 2013;69:296-302
7   Perrin-Lamarre A, Petitpain N, Trechot P, Cuny J-F, Schmutz J-L, Barbaud A. Glycopeptide-induced cutaneous adverse reaction: results of an immunoallergic investigation in eight patients. Ann Dermatol Venereol 2010;137:101-105

Chapter 3.368 VIRGINIAMYCIN

IDENTIFICATION

Description/definition : Virginiamycin is a cyclic polypeptide antibiotic complex from *Streptomyces virginiae*, *S. loidensis*, *S. mitakaensis*, *S. pristinaspiralis*, *S. ostreogriseus*, and others; it consists of 2 major components, virginiamycin factor M1 and virginiamycin factor S1

Pharmacological classes : Anti-bacterial agents

IUPAC name : *N*-[(3*S*,6*S*,12*R*,15*S*,16*R*,19*S*,22*S*)-3-benzyl-12-ethyl-4,16-dimethyl-2,5,11,14,18,21,24-heptaoxo-19-phenyl-17-oxa-1,4,10,13,20-pentazatricyclo[20.4.0.0^{6,10}]hexacosan-15-yl]-3-hydroxypyridine-2-carboxamide;(10*R*,11*R*,12*E*,17*E*,19*E*,21*S*)-21-hydroxy-11,19-dimethyl-10-propan-2-yl-9,26-dioxa-3,15,28-triazatricyclo[23.2.1.0^{3,7}]octacosa-1(27),6,12,17,19, 25(28)-hexaene-2,8,14,23-tetrone

Other names : Staphylomycin ®

CAS registry number(s) : 11006-76-1

EC number : 234-244-6

Merck Index monograph : 11470

Patch testing : 5% and 10% pet.

Molecular formula : C$_{71}$H$_{84}$N$_{10}$O$_{17}$

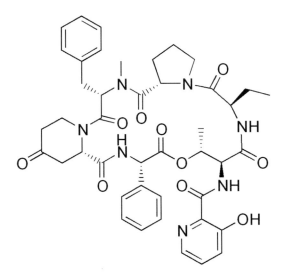

Virginiamycin factor M1
(= pristinamycin IIA)

Virginiamycin factor S1

GENERAL

Virginiamycin is a streptogramin antibiotic similar to pristinamycin and quinupristin/dalfopristin. It is a combination of pristinamycin IIA (virginiamycin M1) and virginiamycin S1. Virginiamycin binds to and inhibits ribosome assembly in susceptible bacteria, thereby preventing protein synthesis. It is active against gram-positive bacteria. Virginiamycin is currently only used in veterinary practice, both to combat infections and as as a growth promoter in cattle, swine, and poultry. It is also employed in the fuel ethanol industry to prevent microbial contamination (1). It was still used in human medicine in France in 1996 (4).

The data provided in various online databases on virginiamycin and pristinamycin are very confusing, overlapping and sometimes probably inaccurate.

CONTACT ALLERGY

General

Virginiamycin ointment (Staphylomycine ®) was formerly widely used in the treatment of skin infections and caused a number of sensitizations (4-11). It appears that the ointment is not used anymore, presumably because it was found that virginiamycin cross-reacts to pristinamycin, which is an 'essential' oral antibiotic for treatment of methicillin-resistant staphylococci.

Patch testing in groups of patients

In Leuven, Belgium, in the period 1990-2014, virginiamycin 5% pet. was patch tested in 357 patients suspected of iatrogenic contact dermatitis and there were 19 (5.3%) positive reactions to it. Relevance was not specified for individual allergens, but 96% of all positive patch test reactions to topical drugs and antiseptics were considered to be relevant (2).

Case series

In 1974, eight cases of contact allergy to virginiamycin were reported. The patients were patch tested with factor M and factor S of virginiamycin as well as fraction IA and fraction IIA of pristinamycin. All eight patients were positive to factor M of virginiamycin 1% pet and fraction IIA of pristinamycin 1% pet. (5).

In 1973, five cases of virginiamycin sensitivity to factor M of virginiamycin (5% pet.) and to pristinamycin (5% pet., individual fractions not tested) were reported from Belgium. Two of the five patients were also sensitized to neomycin, presumably because virginiamycin was often combined with neomycin sulfate in topical antibiotic preparations in Belgium at that time (13).

A case series of four patients sensitized to virginiamycin ointment was described in 1996 in France. A 46-year-old atopic man presented with a left eyelid stye which was treated with an ointment containing virginiamycin. After 1 week's treatment, an impetiginized eczema appeared on the left cheek. A 25-year-old atopic woman had presented with contact dermatitis from virginiamycin. A 23-year-old woman had developed facial contact dermatitis from virginiamycin ointment. When they were given oral pristinamycin, all 3 patients developed systemic contact dermatitis manifesting as generalised maculovesicular erythema with fever, generalized erythema and facial edema, resp. erythema of the trunk, fever and headache. They all had positive patch tests to Staphylomycine ® ointment containing 0.5% virginiamycin. Two of the patients were patch tested with pristinamycin (Pyostacine® 1 tablet crushed in 1 ml water) and reacted positively (4). The fourth patient is described below (Case reports, systemic contact dermatitis).

Case reports

A 40-year-old woman was treated for a second degree burn with virginiamycin 0.5% ointment, which resulted in complete healing after 10 days. One year later, she developed impetigo of the face, and the same ointment was prescribed. Within a few days her face became hot, edematous and severely itchy. Patch tests were positive to the ointment, virginiamycin powder 2% in lactose basis and negative to the excipients of the ointment and many other antibiotics (9). Later, the patient was patch tested again and was now positive to virginiamycin factor M 2% and 5% pet. (negative to 0.5%) and negative to factor S. There were cross-reactions to pristinamycin 0.5%, 2% and 5% pet. (10).

A girl aged 18 was prescribed virginiamycin ointment for an 'infection' of her axillae with good result. One year later, she had a relapse and was treated with the same ointment, which caused an extensive erythematous, itching vesiculobullous dermatitis, spreading from the axillae to the arms. Patch tests were positive to virginiamycin powder 5% and 10% pet. (negative to 2% pet.), to virginiamycin factor M 5% and 10% pet. (negative to 2% and negative to factor S) and to pristinamycin 5% and 10% pet. (11).

Dutch investigators in 1972 reported one case of contact allergy to virginiamycin factor M in a burn patient. The patient reacted to 2% and 5% concentrations but not 0.5%. This individual was also sensitive to pristinamycin at all tested concentrations (individual fractions not tested) (12).

One or more cases of sensitization to virginiamycin were probably reported from France in 1971; details are not available to the author (14).

Systemic contact dermatitis

A 52-year-old man was applying topical virginiamycin ointment on eczema of the hands and legs. As the lesions spread, his doctor prescribed oral virginiamycin. Bullous eczema rapidly appeared on the hands and elbows, together with facial edema, pruritus and generalized erythema. With topical corticosteroids, the lesions cleared within 15 days. Patch tests were positive to the virginiamycin 0.5% ointment 'as is' (4).

A man sensitized to virginiamycin in a topical preparation took one tablet of 250 mg pristinamycin and 4 hours later had a reaction with stupor, urticaria and vomiting. Patch tests with virginiamycin, pristinamycin, factors M and IIA, each 1% pet., were all positive. He developed transient edema of his eyes and lips, and wheals adjacent to a positive patch test (5).

<u>Occupational allergic contact dermatitis</u>

A man aged 31 worked for 6 months in a pharmaceutical factory as a warehouseman. Two or three times daily he entered a dusty room where a food additive, virginiamycin, for pigs and poultry was prepared and stored in barrels. After 3 months the patient developed a pruritic erythematosquamous eruption of the face around the eyes with some edema of the eyelids. He had no previous history of using virginiamycin as an antibiotic either locally or systemically. Moreover, he had never used other antibiotics of the same chemical group, such as pristinamycin. Patch tests were positive to the food additive 20% pet., virginiamycin 5% pet., virginiamycin factor M 5% pet. (which is also present in pristinamycin) and pristinamycin 5% pet., but negative to virginiamycin factor S (7). This was a case of airborne occupational allergic contact dermatitis.

Cross-reactions, pseudo-cross-reactions and co-reactions

Patients sensitized to pristinamycin (from oral administration) may cross-react to virginiamycin (3). Conversely, patients sensitized to virginiamycin may cross-react to pristinamycin (4,5,7,10,11) and may develop systemic contact dermatitis when given pristinamycin orally (4,6,8).

Cross-sensitivity is easily explained by structural similarities between these 2 antibiotics: virginiamycin factor M1 is the same as pristinamycin IIA and virginiamycin factor S1 is virtually identical to pristinamycin IA. The most important sensitizer appears to be virginiamycin factor M1 (= pristinamycin IIA) (7,10,11)

LITERATURE

1   The data in the section 'General' may have been obtained from literature discussed in this chapter, but mostly also or exclusively from one or more of the following online sources: ChemIDPlus Advanced, PubChem, DrugBank, RxList, Drug Central, Drugs.com, and Wikipedia

2   Gilissen L, Goossens A. Frequency and trends of contact allergy to and iatrogenic contact dermatitis caused by topical drugs over a 25-year period. Contact Dermatitis 2016;75:290-302

3   Barbaud A, Trechot P, Weber-Muller F, Ulrich G, Commun N, Schmutz JL. Drug skin tests in cutaneous adverse drug reactions to pristinamycin: 29 cases with a study of cross-reactions between synergistins. Contact Dermatitis 2004;50:22-26

4   Michel M, Dompmartin A, Szczurko C, Castel B, Moreau A, Leroy D. Eczematous-like drug eruption induced by synergistins. Contact Dermatitis 1996;34:86-87

5   Baes H. Allergic contact dermatitis to virginiamycin. False cross-sensitivity with pristinamycin. Dermatologica 1974;149:231-235

6   Pillette M, Claudel JP, Muller C, Lorette G. Contact dermatitis caused by pristinamycin after sensitization to topical virginiamycin. Allerg Immunol (Paris) 1990;22:197 (Article in French)

7   Tennstedt D, Dumont-Fruytier M, Lachapelle JM. Occupational allergic contact dermatitis to virginiamycin, an antibiotic used as a food additive for pigs and poultry. Contact Dermatitis 1978;4:133-134

8   Mathivon F, Petit A, Mourier C, Sigal Nahum M. Toxidermie à la pristinamycine après réaction de contact à la virginiamycine. Rev Eur Dermatol MST 1991;3:527-529 (Article in French)

9   Nater JP. Sensitization to staphylomycine ®. Contact Dermatitis Newsletter 1971;10:238

10   Bleumink E, Nater JP. Sensitization to Staphylomycin ® (virginiamycin). Contact Dermatitis Newsletter 1972;11:306

11   Bleumink E, Nater JP. Allergic contact dermatitis to virginiamycin (Staphylomycin ®) and pristinamycin (Stapyocin®). Contact Dermatitis Newsletter 1972;12:337

12   Bleumink E, Nater JP. Allergic contact dermatitis to virginiamycin. Dermatologica 1972;144:253-256

13   Lachapelle JM, Lamy F. On allergic contact dermatitis to virginiamycin. Dermatologica 1973;146:320-322

14   Castelain PY. A rare sensitization: that due to staphylomycin. Bull Soc Fr Dermatol Syphiligr 1971;78:526-527 (Article in French)

Chapter 3.369 XANTOCILLIN

IDENTIFICATION

Description/definition   : Xantocillin is the antibiotic belonging to the group of (neo)lignans and a cyanide
                              compound that conforms to the structural formula shown below
Pharmacological classes   : Anti-bacterial agents
IUPAC name   : 4-[(1Z,3Z)-4-(4-Hydroxyphenyl)-2,3-diisocyanobuta-1,3-dienyl]phenol
Other names   : Xanthocillin; 1,4-di-p-oxyphenyl-2,3-di-isonitrilo-1,3-butadiene
CAS registry number   : 580-74-5
EC number   : 234-271-3
Merck Index monograph   : 11529
Patch testing   : 1% and 10% pet.
Molecular formula   : $C_{18}H_{12}N_2O_2$

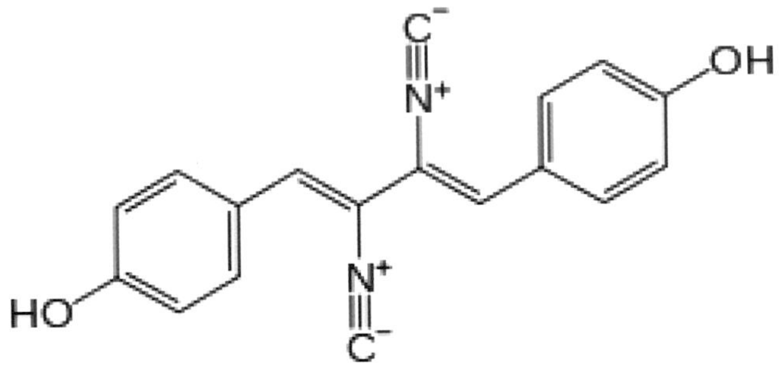

GENERAL

Xantocillin is an antibiotic complex which was first isolated from *Penicillium notatum* (current name: *Penicillium chrysogenum*) in 1950 and subsequently from several other sources. It consists of xantocillin X (CAS number 580-74-5), xantocillin Y_1 (CAS number 38965-69-4) and xantocillin Y_2 (CAS number 38965-70-7). It is active against most of the common gram-positive and gram-negative pathogens, including *Proteus* and *Pseudomonas aeruginosa*, and also has some suppressive action against *Mycobacterium tuberculosis* and certain pathogenic fungi and yeasts. Xantocillin has not been used systemically because of toxicity and poor absorption. However, it has been employed in topical pharmaceuticals, often combined with tyrothricin, but is probably not used anymore (1,2).

CONTACT ALLERGY

Case reports and case series
In the 1960s, several articles have documented allergic contact dermatitis to xantocillin (3-7).

LITERATURE
1   The data in the section 'General' may have been obtained from literature discussed in this chapter, but mostly
     also or exclusively from one or more of the following online sources: ChemIDPlus Advanced, PubChem,
     DrugBank, RxList, Drug Central, Drugs.com, and Wikipedia
2   Bettley FR. Xanthocillin cream for local treatment. Br Med J 1959;1(5131):1226-1227
3   Cramer HJ. Xanthocillin allergy. Allerg Asthma (Leipz) 1961;7:336-339 (Article in German)
4   Bandmann HJ, Dohn W. Xanthocillin as a contact allergen. Hautarzt 1962;13:84-87 (Article in German)
5   Langer H. Sensitization of skin disease patients by the local use of the antibiotics xanthocillin and
     chloramphenicol. Z Haut Geschlechtskr 1962;33:210-215 (Article in German)
6   Schubert H. Xanthocillin as a local antibiotic and allergen. Allerg Asthma (Leipz) 1969;15:61-66 (Article in
     German)
7   Heijer A. Sensitization to xantocillin in salve. Acta Derm Venereol 1961;41:201-204

Chapter 4 ALLERGIC CONTACT DERMATITIS TO INGREDIENTS OTHER THAN THE ACTIVE DRUG

4.1 INTRODUCTION

In cases of allergic contact dermatitis from topical pharmaceuticals, the allergenic culprit may be either the active drug or one or more of the other components, for example vehicle ingredients, preservatives, antioxidants, fragrances or other chemicals present in the topical drug product. Such reactions to non-drug components are far from rare. In eye drops, for example, many allergic reactions have been reported to the preservatives thimerosal and benzalkonium chloride. In creams and ointments, especially those used for herpes simplex (acyclovir), fungal infections (ketoconazole), and in topical corticosteroids, propylene glycol was often found to be the allergenic culprit. It should be realized, though, that these chemicals tend to cause irritant patch test reactions, and there can be no doubt that a number of the 'allergic' reactions reported have, in fact, been false-positive.

In this chapter, an overview is given of non-drug components of topical pharmaceuticals that have caused allergic contact dermatitis. However, a complete literature review of the topical was not attempted. The data below were found during the preparation of this book and in the two previous volumes in the *Monographs in Contact Allergy* series (1,2). It can safely be assumed that most relevant data published in *Contact Dermatitis*, the *American Journal of Contact Dermatitis* and *Dermatitis* are included in the information provided here. However, the references in relevant articles were not always screened for further data on this topic and a systematic literature search has not been performed, also because much of the information is 'hidden' in articles and cannot be found with the aid of key words or phrases in digital databases.

The data from a relevant investigation performed in Leuven, Belgium, on this topic, are presented in full, the other information on non-drug components of topical pharmaceuticals that have caused allergic contact dermatitis by their presence in these products is summarized in tabular format in table 4.3 (vehicle components), table 4.4 (preservatives and antioxidants), table 4.5 (fragrances and essential oils) and table 4.6 (other chemicals).

4.2 CASE SERIES

In a university center in Leuven, Belgium, 17,367 patients were patch tested between 1990 and 2013, of whom 2513 (14.5%) presented with an iatrogenic allergic contact dermatitis (3). The non-drug ingredients responsible for allergic contact dermatitis to topical pharmaceuticals in this study are shown in table 4.1 (vehicle components) and table 4.2 (preservatives, antioxidants and fragrances). Obviously, the data do not reliably represent the actual numbers of sensitization. In the case of the vehicle components, for example, the high number of reactions to lanolin alcohol are also the result of its presence in the routine series.

Table 4.1 Vehicle components of topical pharmaceuticals responsible for allergic contact dermatitis in Leuven, Belgium, 1990-2013 (3)

Ingredient	Number positive	Ingredient	Number positive
Wool (lanolin) alcohol	354	PEG-20 glyceryl oleate	3
Propylene glycol	146	Sorbitan monolaurate	2
Sorbitan sesquioleate	132	Polysorbate 80	2
Cetyl alcohol	77	Petrolatum	2
Nonoxynol-9	66	Eucerin	2
Lauramine oxide	21	Almond oil (sweet)	2
Benzyl alcohol	14	Lactic acid	2
Polyethylene glycol	12	Laureth-9	2
Olive oil	12	Propylene glycol monostearate	2
Stearyl alcohol	10	Beeswax	2
Monosodium sulforicinoleate	9	Polyethylene glycol distearate	1
Modified colophonium	8	Hydroxyethyl methacrylate	1
Sesame oil	7	Butylene glycol	1
Isopropyl myristate	7	Glyceryl rosinate	1
Glyceryl monostearate	5	Cetearyl glucoside	1
Triethanolamine	4	Myristyl alcohol	1
Polyethylene glycol stearate	4	Castor oil	1
Hexylene glycol	4	Diethanolamine	1
Polysorbate 60	3	Ethylene glycol monostearate	1

Propylene glycol, sorbitan sesquioleate and cetyl alcohol were probably tested in an extended baseline series or an additional series that was very frequently tested. The same most likely applies to the preservatives thimerosal, ethylenediamine, the antioxidants sodium metabisulfite and BHA, and certainly to the parabens and the fragrance Myroxylon pereirae resin (balsam of Peru), which are included in the European baseline series. It may be assumed that only a small minority of all topical drugs that had caused allergic contact dermatitis have been tested with all ingredients, which means that the rarer allergenic ingredients (not identified by a positive patch test in the routine series or an additional series), go unnoticed.

Table 4.2 Preservatives, antioxidants and fragrances in topical pharmaceuticals responsible for allergic contact dermatitis in Leuven, Belgium, 1990-2013 (3)

Ingredient	Number positive	Ingredient	Number positive
Preservative agents		Disodium edetate (EDTA)	4
Thimerosal	81	Butylated hydroxytoluene (BHT)	2
Parabens	65		
Benzoic acid	33	**Fragrances** [b]	
Phenylmercuric borate	27	Fragrance mix I [c]	88
Benzalkonium chloride	26	Myroxylon pereirae resin (Balsam of Peru)	68
Chlorocresol	15		
Sorbic acid	12	Lavender oil	31
Imidazolidinyl urea	10	Perfume (unspecified)	15
Formaldehyde	10	Geranium oil	10
Potassium sorbate	9	Orange flower oil	8
Phenylmercuric nitrate	5	Eucalyptus oil	6
Phenoxyethanol	3	Terpineol	4
Diazolidinyl urea	3	Fragrance-mix II [c]	4
Quaternium-15	1	Pine tree oil	3
Propyl gallate	1	Thyme oil	2
Phenylmercuric acetate	1	Niaouli oil	2
		Rose oil	2
Antioxidants		Jasmin oil	1
Ethylenediamine [a]	62	Clove oil	1
Sodium metabisulfite	38	Nutmeg oil	1
Butylated hydroxyanisole (BHA)	21	Rosemary oil	1

[a] a stabilizer and preservative rather than antioxidant; [b] some fragrances/essential oils are also used for purposes other than perfuming, e.g. for their antiseptic properties; [c] not an ingredient, but an indicator of fragrance allergy present in the routine series

4.3 NON-DRUG INGREDIENTS THAT HAVE CAUSED CONTACT ALLERGY/ALLERGIC CONTACT DERMATITIS BY THEIR PRESENCE IN TOPICAL PHARMACEUTICALS

During the preparation of the monographs in Chapter 3 and by searching in the first 2 volumes of the *Monographs in Contact Allergy* series (1,2), the author found 149 non-drugs chemicals, that have caused contact allergy/allergic contact dermatitis by their presence in topical drugs: 46 vehicle ingredients, 35 preservatives and antioxidants, 44 fragrances and essential oils and 24 other chemicals (tables 4.3-4.6). Some additional haptens, not present there, may be found in tables 4.1 and 4.2. In the category vehicle ingredients, the most frequently implicated chemicals were propylene glycol, followed by lanolin (alcohol) and cetyl and stearyl alcohol. Among the preservatives and antioxidants, thimerosal and benzalkonium chloride caused many cases of sensitization, especially from their presence in eye drops. Frequent reactions have also been noted to sodium metabisulfite and benzyl alcohol, followed by propyl gallate, chlorocresol and the parabens, the latter often in older publications from their presence in corticosteroids and antibiotic preparations. Fragrances and essential oils are not frequently implicated, although lavender essence/oil has caused several cases of (photo)allergic contact dermatitis from its presence in ketoprofen gel (4,41,42). Of the other chemicals, only alcohol in estradiol transdermal therapeutic system and colophonium (notably in wart treatments) have caused several cases, and ethylenediamine in Mycolog ® cream many cases of allergic contact dermatitis. The other ingredients, some of which cannot properly be identified on the base of their names as provided in the original publications, mostly caused single reactions only. However, the actual nature of the non-drug constituents in topical pharmaceuticals, and their (absolute and relative) frequencies remain largely unknown, as only a very small minority of all patients reacting to a topical drug will be patch tested with all its constituents. Testing bias (haptens present or not in the routine series or additional test series) and publication bias (well-known allergens are not published) further complicate this issue.

Table 4.3 Ingredients other than the active drug that have caused contact allergy/allergic contact dermatitis in topical pharmaceuticals: Vehicle ingredients

Ingredient	Topical pharmaceutical product	References
1,3-Butylene glycol	antimycotic	181
Castor oil	antihemorrhoidal	252
	cerumenolytic	307,308
	OTC zinc and castor oil cream	309
	wart treatment	162,232
Cera alba (white wax)	corticosteroid	76
Cetearyl alcohol	antibiotic, antiseptic	70
	antihistamine	79
	antimycotic	91,129,131,213
	antiviral	74
	corticosteroid	64,89,157,214
	improvement of blood flow	212
	treatment of thrombophlebitis	56
	unspecified topical pharmaceutical	211
	wound treatment	209
Cetyl alcohol	antibiotic	94,95,215
	antimycotic	103,118,220
	antipruritic	90
	corticosteroid	9,77,79,217,219
	corticosteroid – antibiotic	216
	unspecified topical pharmaceutical	218
	wound treatment	209
Chloroacetamide	treatment of thrombophlebitis	54
Cocamide DEA	antifungal	231
	antiseptic	81,231
Diethyl sebacate (ethyl sebacate)	antifungal	19,20,106-113
	corticosteroid	114
Diisopropyl sebacate	antibiotic	160
Ethylene glycol monomethyl ether	treatment of actinic keratosis	44,45
Glycerin	transdermal therapeutic system	243
1,2,6-Hexanetriol	corticosteroid	165
Hydrogenated lanolin	corticosteroid	78
	muscle relaxant and analgesic	59
Isopropyl myristate	antibiotic – corticosteroid	245
	antimycotic	244
	corticosteroid	277
Isopropyl palmitate	corticosteroid	61
Lanette SX ®	wound treatment	210
Lanolin	antihemorrhoid	247
	antimycotic	246,252 (by proxy)
	corticosteroid	246
	NSAID	246
	treatment of actinic keratosis	98,249
	treatment of thrombophlebitis	56,248
	unspecified topical pharmaceutical	250-251
Lanolin alcohol	antibiotic cream	8,83
	antihistamine	27,79
	corticosteroid	63,64,66,69,86
	muscle relaxant and analgesic cream	58
	ophthalmologic ointment	253
	treatment of actinic keratosis	98
	treatment of thrombophlebitis	55
	vasodilator	88
Laureth-4	anti-acne	254

Table 4.3 Ingredients other than the active drug that have caused contact allergy/allergic contact dermatitis in topical pharmaceuticals: Vehicle ingredients (continued)

Ingredient	Topical pharmaceutical product	References
Lauromacrogol (unspecified)	antimycotic	133
Methyl glucose dioleate	antibiotic	257,258
	antimycotic	259
Myristyl alcohol	treatment of thrombophlebitis	56
	unknown type of topical pharmaceutical	267
Octyldodecanol	antimycotic	21,22,23
	antipruritic	185
Oleyl alcohol	calcineurin-inhibitor	136
PEG-6 (polyethylene glycol 300)	antibiotic	272-275
PEG-300	antibiotic	25
PEG-155 stearate	corticosteroid – antifungal	116
PEG-1500 stearate	corticosteroid – antifungal	75
Pentylene glycol	anti-inflammatory	276
Petrolatum	antibiotic	220
	corticosteroid	154
	keratolytic	277
	treatment of thrombophlebitis	248
Polawax NF ®	corticosteroid	80
Polyethylene glycol	antibiotic	82,135
Polyethylene glycol stearate	antimycotic	118
Polysorbate 40	unspecified topical pharmaceutical	279
Polysorbate 60	NSAID	119 (photoallergy)
Polysorbate 80	antibiotic	25
	antimycotic	103
	corticosteroid (inhalation)	176
	antiviral	280
Propylene glycol	antibiotic	11,12,15,25,74,94,294
	anti-inflammatory	97
	antimycotic	104,105,117,244,286
	antipsoriatic	14
	antiviral	82,120-129,285
	cerumenolytic	291
	corticosteroid	61,62,66,72,77,82,132,239,281-284
	lubricant	115
	NSAID	10,292
	treatment of actinic keratosis	98,99,177,249
	treatment of hair loss (minoxidil)	82,96,148.287-290
	unspecified topical pharmaceutical	293
	wound treatment	295
Sesame oil	unknown, probably wound treatment	310
Sodium cetearyl sulfate	wound treatment	209 (Lanette E ®)
Sorbitan monolaurate	corticosteroid	152,153,154
Sorbitan oleate	treatment of actinic keratosis	99
Sorbitan sesquioleate	corticosteroid	65,333,334
Sorbitan stearate	NSAID	119 (photoallergy)
Stearyl alcohol	anti-acne	93
	antimycotic	118,220
	corticosteroid	79,215,282,335,337,338
	treatment of actinic keratosis	99,172,336
	unspecified topical pharmaceutical	218
	wound treatment	209
Triethanolamine	antihistamine	344-346
	antimycotic	101
	unknown/unspecified	347,348
Triethanolamine palmitate	antiseptic	81

Table 4.4 Ingredients other than the active drug that have caused contact allergy/allergic contact dermatitis in topical pharmaceuticals: Preservatives and antioxidants

Ingredient	Topical pharmaceutical product	References
Benzalkonium chloride	anti-asthma inhalation	138
	antibiotic – corticosteroid	43
	antifungal	168
	antihistamine	137
	corticosteroid	68
	corticosteroid – antifungal	174
	corticosteroid nasal spray/inhalation	7,67
	eye drops/other eye medicaments	8,17,24,26,140 (Review),142,145, 150,189-196
	inhalation	197 (by proxy)
	wound treatment	198
Benzoic acid	antihistamine	137
	antimycotic	118
Benzoxonium chloride	corticosteroid	200
Benzyl alcohol (also used as fragrance ingredient)	antibiotic	180
	anti-inflammatory	358
	antimycotic	4,91,167,180,354-356
	antipruritic	351
	calcineurin-inhibitor	352
	corticosteroid	4,61,66,74,180,353
	corticosteroid – antimycotic – antibiotic	349
	degreaser	4
	disinfectant	4
	treatment of actinic keratosis	45
	wound healing	4
BHA (butylated hydroxy-anisole)	antimycotic	100,201
	corticosteroid	201
	corticosteroid – antimycotic	171
2-Bromo-2-nitropropane-1,3-diol	antibiotic	173,204
	treatment of thrombophlebitis	205
Butylparaben	corticosteroid	206,207
Cetalkonium chloride	eye drops/other eye medicaments	140,208
Cetylpyridinium chloride	corticosteroid nasal inhalation	84
Chloroacetamide	antipruritic	221
	antiviral	221
	treatment of thrombophlebitis	54,221
Chlorhexidine digluconate	eye medicament	24
Chlorocresol	corticosteroid	64,70,74,132,224-228
	unknown type of topical pharmaceutical	229
Chloroxylenol	unspecified topical pharmaceutical	23
Disodium EDTA	corticosteroid	61,241
	ophthalmic medication	238,240
Dodecyl gallate	keratolytic	239
Ethylparaben	corticosteroid	206,242
Formaldehyde	anti-scabies	178
Hexamidine diisethionate	corticosteroid – antifungal	75,116
Methyl(chloro)isothiazolinone	treatment of hair loss (minoxidil)	255
Methyldibromo glutaronitrile	NSAID	256
Methylisothiazolinone	NSAID	260
Methylparaben	analgesic	262
	antibiotic	13,263,264,265
	corticosteroid	72,206,207,242

Table 4.4 Ingredients other than the active drug that have caused contact allergy/allergic contact dermatitis in topical pharmaceuticals: Preservatives and antioxidants (continued)

Ingredient	Topical pharmaceutical product	References
Methylparaben (continued)	treatment of thrombophlebitis	56
	unknown type of topical pharmaceutical	266
	wound treatment	210
Parabens (unspecified)	antibiotic	70,261
	antiviral	92
	condom retarding gel	5
	corticosteroid	60,62,64,66,71,269,270
	eye drops	145,170
	NSAID	30
	unspecified topical pharmaceutical	261,271
	wound healing	53,268,269
Phenoxyethanol	antiglaucoma	18
	wound treatment	182
Phenylmercuric acetate	eye medicaments	24
Polyaminopropyl biguanide	wound treatment	278
Propyl gallate	antibiotic	296,298,299,301,302,304
	anti-inflammatory	276
	corticosteroid	73,295
	wound treatment	303,306
Propylparaben	antibiotic	13,261,263
	antihemorrhoidal	29
	corticosteroid	206,242
	unspecified topical pharmaceutical	261
	wound treatment	261
Sodium bisulfite	antifungal	311
	eye drops	139
Sodium dehydroacetate	wound treatment	210
Sodium metabisulfite	antibiotic	57,318,320,321,323
	antihemorrhoidal	29
	antimycotic	313,315
	corticosteroid	313
	corticosteroid – antibiotic – antimycotic	164,175
	corticosteroid – antimycotic	314,317,318,319
	eye drops	149,313,316,322
	topical anesthetic	313
	unspecified topical pharmaceutical	324
Sodium sulfite	antimycotic	102, 103,325,326,327
Sorbic acid	antibiotic	332
	corticosteroid	329,330
	unspecified topical pharmaceutical	328
	wound treatment	331
Thimerosal	antibiotic – corticosteroid	343
	eye drops and other eye medicaments	24,140 (Review),141,143,144,145, 146,147,339-341
	vasoconstrictor	342
1,3,5-Trihydroxyethylhexa-hydrotriazine	treatment of thrombophlebitis	56

Table 4.5 Ingredients other than the active drug that have caused contact allergy/allergic contact dermatitis in topical pharmaceuticals: Fragrances and essential oils

Ingredient	Topical pharmaceutical product	References
Amyl cinnamal	corticosteroid – antimycotic – antibiotic	349
Benzoin	degreaser	4
	wound healing	4
Benzyl benzoate	antiseptic/disinfectant	4
Bergamot fruit oil	antibiotic	4
	corticosteroid	4
	wound healing	4
Cajeput oil	inhalation	4
Camphor	anti-itch/analgesic	4
	inhalation	4
Cinnamal	ophthalmological pharmaceutical	195
Cinnamyl alcohol	corticosteroid	4
	corticosteroid – antimycotic – antibiotic	349
Citronella oil	corticosteroid	4
Coumarin	wound treatment	357
Cypress oil	analgesic/anti-inflammatory	4
	inhalation	4
Dwarf pine needle oil	NSAID	360
Elemi oil	wound healing	4
Essence of lilies	anti-inflammatory	358
Eucalyptol	inhalation	4
Eucalyptus oil	anti-inflammatory	169
	inhalation	4
	NSAID	4,42
Gardenia perfume oil	antibiotic	6
Geranium oil	antibiotic	4
	wound healing	4
Hydroxycitronellal	corticosteroid	4
	corticosteroid – antimycotic – antibiotic	349
Ionone	corticosteroid	4
Laurel oil	wound healing	4
Lavender essence	NSAID	38,362
Lavender oil	antibiotic	4
	antihistamine	4,361
	antipruritic	4
	antiseptic/disinfectant	4
	corticosteroid	4
	inhalation	4
	NSAID	4,16,39,40,41,42,184 (photoallergy)
	vascular disorders	4
	wound healing	4
Lemongrass oil	wound healing	4
Linalool	corticosteroid	4
Menthol	anti-inflammatory	358,359
	anti-itch/analgesic	4
	inhalation	4
	NSAID	4,186
	transcutaneous therapeutic system	179
	wound healing	4
Musk ketone	corticosteroid	4
	inhalation	4
Myroxylon pereirae	anti-hemorrhoids	4
	disinfectant	4
	wound healing	4
Neroli oil	antibiotic	4

Table 4.5 Ingredients other than the active drug that have caused contact allergy/allergic contact dermatitis in topical pharmaceuticals: Fragrances and essential oils (continued)

Ingredient	Topical pharmaceutical product	References
Neroli oil (continued)	corticosteroid	4
	NSAID	4,39,41
	vascular disorders	4
Niaouli oil	wound healing	4
Nutmeg oil	inhalation	4
Oakmoss synthetic	corticosteroid – antimycotic – antibiotic	349
Orange flower oil	NSAID	42
Origanum oil	wound healing	4
Peppermint oil	inhalation	4
	transdermal therapeutic system	179
Perfume (unspecified)	antibiotic – corticosteroid	4
	anti-itch/analgesic	4
	local anesthetic	159
	NSAID	4
Pine needle oil	disinfectant	4,81
	inhalation	4
	NSAID	4,42
Piperonal	corticosteroid	4
Rose oil	antimycotic	4
	degreaser	4
	wound healing	4
Terpineol	antiseptic/disinfectant	4
Thyme oil	wound healing	4
Thymol	inhalation	4
	wound healing	4
Turpentine oil	inhalation	4
	wound healing	4
Vanillin	wound treatment	357

Table 4.6 Ingredients other than the active drug that have caused contact allergy/allergic contact dermatitis in topical pharmaceuticals: Miscellaneous chemicals

Ingredient	Topical pharmaceutical product	References
Alcohol	estradiol transdermal therapeutic system	46-48,151,187,188
Alkanamido propylbetaine [a]	antiseptic	81
Alkyl dimethylcarboxy methylamine [a]	antiseptic	81
Cocamidopropyl betaine	antiseptic	81
Colophonium	keratolytic	4
	transdermal therapeutic system	134
	wart treatment	161,232-237
Dibutyl phthalate	corticosteroid	73
	corticosteroid – antifungal	174
Diisopropanolamine	NSAID	31-37
Dimethyl sulfoxide	anti-inflammatory	358
Dioctyl sodium sulfosuccinate (docusate sodium)	corticosteroid	163
Emulgade F ®	wound treatment	155,156,159,209
Emulgin RO/40®	wound treatment	155
Ethylenediamine	corticosteroid – antimycotic – antibiotic	many reported cases, notably from Mycolog ® cream (e.g. 60,349)
Ethylhexyl acrylate	transdermal therapeutic system	28
Glycerol-poly-(oxyethylene-6-alkanoate [a]	corticosteroid – antifungal	75,116

Table 4.6 Ingredients other than the active drug that have caused contact allergy/allergic contact dermatitis in topical pharmaceuticals: Miscellaneous chemicals (continued)

Ingredient	Topical pharmaceutical product	References
Glyceryl hydrogenated rosinate	NSAID (tape)	186
	transdermal therapeutic system	183
Hydroxypropyl cellulose	transdermal therapeutic system	49
Isopropanolamine	NSAID	166
Lactic acid	corticosteroid	61
	wart treatment	162
Methyl methacrylate	transdermal therapeutic system	51
Montapal [a]	antiseptic	81
Oleoyl polyoxyl-6 glycerides (Labrafil ® M 1944)	antimycotic	117
Oramide [a]	antiseptic	81
Palmitoyl collagen amino acids	wound treatment (keloid)	158
Polyisobutylene	transdermal therapeutic system	50
Sodium lauryl sulfate	antimycotic	130
	corticosteroid	312

[a] (unknown chemical)

LITERATURE

1   De Groot AC. Monographs in Contact Allergy Volume I. Non-Fragrance Allergens in Cosmetics (Part I and Part 2). Boca Raton, Fl, USA: CRC Press Taylor and Francis Group, 2018 (ISBN 978-1-138-57325-3 and 9781138573383)

2   De Groot AC. Monographs in Contact Allergy, Volume II. Fragrances and Essential Oils. Boca Raton, Fl, USA: CRC Press Taylor and Francis Group, 2019 (ISBN 9780367149802)

3   Goossens A. Allergic contact dermatitis from the vehicle components of topical pharmaceutical products. Immunol Allergy Clin North Am 2014:34:663-670

4   Nardelli A, D'Hooghe E, Drieghe J, Dooms M, Goossens A. Allergic contact dermatitis from fragrance components in specific topical pharmaceutical products in Belgium. Contact Dermatitis 2009;60:303-313

5   Foti C, Bonamonte D, Antelmi A, Conserva A, Angelini G. Allergic contact dermatitis to condoms: description of a clinical case and analytical review of current literature. Immunopharm Immunotoxol 2004;26:479-483

6   Martins C, Freitas JD, Gonçalo M, Gonçalo S. Allergic contact dermatitis from erythromycin. Contact Dermatitis 1995;33:360

7   Lechien JR, Costa de Araujo P, De Marrez LG, Halloy JL, Khalife M, Saussez S. Contact allergy to benzalkonium chloride in patients using a steroid nasal spray: A report of 3 cases. Ear Nose Throat J 2018;97:E20-E22

8   Lewis FM, Gawkrodger DJ, Bleehen SS, Nelson ME. Multiple contact sensitivity to eyedrops. Contact Dermatitis 1993;28:246-247

9   Hausen BM, Kulenkamp D. Contact allergy to fludroxycortid and cetyl alcohol. Derm Beruf Umwelt 1985;33:27-28 (Article in German)

10   Lynde CB, Pierscianowski TA, Pratt MD. Allergic contact dermatitis caused by diclofenac cream. CMAJ 2009;181:925-926

11   Rasmussen I. Patch test reactions to Flamazine. Contact Dermatitis 1984;11:133-134

12   Van der Horst JC, Van Ketel WG. Problems in allergological studies of a patient with a probable Flammazine cream allergy. Ned Tijdschr Geneeskd 1982;126:841-843 (Article in Dutch)

13   Schamberg IL. Allergic contact dermatitis to methyl and propyl paraben. Arch Dermatol 1967;95:626-628

14   Fisher DA. Allergic contact dermatitis to propylene glycol in calcipotriene ointment. Cutis 1997;60:43-44

15   El Sayed F, Bayle-Lebey P, Marguery MC, Bazex J. Contact dermatitis from propylene glycol in Rifocine. Contact Dermatitis 1995;33:127-128

16   García Bara MT, Matheu V, Pérez A, Díaz MP, Martínez MI, Zapatero L. Contact photodermatitis due to ketoprofen and piketoprofen. Allergol Inmunol Clin 1999;14:148-150

17   Aoki J. [Allergic contact dermatitis due to eye drops. Their clinical features and the patch test results]. Nihon Ika Daigaku Zasshi 1997;64:232-237 (article in Japanese)

18   Moore A, Kempers S, Murakawa G, Weiss J, Tauscher A, Swinyer L, et al. Long-term safety and efficacy of once-daily topical brimonidine tartrate gel 0.5% for the treatment of moderate to severe facial erythema of rosacea: results of a 1-year open-label study. J Drugs Dermatol 2014;13:56-61

19   Berlin AR, Miller OF. Allergic contact dermatitis from ethyl sebacate in haloprogin cream. Arch Dermatol 1976;112:1563-564

20   Moss HV. Allergic contact dermatitis due to Halotex solution. Arch Dermatol 1974;109:572

21 Dawn G, Forsyth A. Genital swelling caused by octyldodecanol contact dermatitis. Clin Exp Dermatol 2003;28:228-229

22 Dharmagunawardena B, Chrales-Holmes R. Contact dermatitis due to octyldodecanol in clotrimazole cream. Contact Dermatitis 1997;36:231

23 Tucker WFG. Contact dermatitis to Eutanol G. Contact Dermatitis 1983;9:88-89

24 Lyon CC, Beck MH. Allergic contact dermatitis reactions to corticosteroids in periorbital inflammation and conjunctivitis. Eye (Lond) 1998;12(Pt.1):148-149

25 González-Cantero A, Gatica-Ortega M, Pastor-Nieto M, Martínez-Lorenzo E-R, Gómez-Dorado BA, Mollejo-Villanueva M, et al. Acute generalized exanthematous pustulosis (AGEP)-like contact dermatitis resulting from topical therapy in a polysensitized patient. Contact Dermatitis 2019;80:329-333

26 Romita P, Stingeni L, Barlusconi C, Hansel K, Foti C. Allergic contact dermatitis in response to ketotifen fumarate contained in eye drops. Contact Dermatitis 2020;83:35-37

27 Arrue I, Rosales B, Ortiz de Frutos FJ, Vanaclocha F. Photoaggravated eczema due to promethazine cream. Actas Dermosifiliogr 2007;98:717-718

28 Navarro-Triviño FJ, Ruiz-Villaverde R. Allergic contact dermatitis caused by 2-ethylhexyl acrylate in a rivastigmine transdermal therapeutic system. Contact Dermatitis 2020;83:143-145

29 Sánchez-Pérez J, Abajo P, Córdoba S, García-Díez A. Allergic contact dermatitis from sodium metabisulfite in an antihemorrhoidal cream. Contact Dermatitis 2000;42:176-177

30 Cusano F, Rafenelli A, Bacchilega R, Errico G. Photo-contact dermatitis from ketoprofen. Contact Dermatitis 1987;17:108-109

31 Nabeya RT, Kojima T, Fujita M. Photocontact dermatitis from ketoprofen with an unusual clinical feature. Contact Dermatitis 1995;32:52-53

32 Oiso N, Fukai K, Ishii M. Triple allergic contact sensitivities due to ferbinac, crotamiton and diisopropanolamine. Contact Dermatitis 2003;49:261-263

33 Fujimoto K, Hashimoto S, Kozuka T, Yoshikawa K. Contact dermatitis due to diisopropanolamine. Contact Dermatitis 1989;21:56

34 Rind T, Oiso N, Hirao A, Kawada A. Allergic contact dermatitis with diffuse erythematous reaction from diisopropanolamine in a compress. Case Rep Dermatol 2010;2:50-54

35 Umebayashi Y. Two cases of contact dermatitis due to diisopropanolamine. J Dermatol 2005;32:145-146 (article in Japanese) (data cited in ref. 34)

36 Umebayashi Y. Contact dermatitis due to diisopropanolamine. Rinsho Derma (Tokyo) 2000;42:526-527 (article in Japanese) (data cited in ref. 34)

37 Hosokawa K, Mitsuya K, Nishijima S, Horio T, Asada Y. Photocontact dermatitis from a non-steroidal anti-inflammatory drug (Sector Lotion). Skin Research 1993;35:26-32 (article in Japanese) (data cited in ref. 34)

38 Fernández de Corrès L, Díez JM, Audicana M, García M, Muñoz D, Fernández E, Etxenagusía M. Photodermatitis from plant derivatives in topical and oral medicaments. Contact Dermatitis 1996;35:184-185

39 Baudot S, Milpied B, Larousse C. Cutaneous side effects of ketoprofen gels: results of a study based on 337 cases. Therapie 1993;53:137-144

40 Hindsén M, Zimerson E, Bruze M. Photoallergic contact dermatitis from ketoprofen in southern Sweden. Contact Dermatitis 2006;54:150-157

41 Matthieu L, Meuleman L, Van Hecke E, Blondeel A, Dezfoulian B, Constandt L, Goossens A. Contact and photocontact allergy to ketoprofen. The Belgian experience. Contact Dermatitis 2004;50:238-241

42 Devleeschouwer V, Roelandts R, Garmyn M, Goossens A. Allergic and photoallergic contact dermatitis from ketoprofen: results of (photo) patch testing and follow-up of 42 patients. Contact Dermatitis 2008;58:159-166

43 Robinson M. Contact sensitivity to gentamicin-hydrocortisone ear drops. J Laryngol Otol 1988;102:577-578

44 Taibjee SM, Prais L, Foulds IS. Allergic contact dermatitis from polyethylene glycol monomethyl ether 350 in Solaraze gel. Contact Dermatitis 2003;49:170-171

45 Kleyn CE, Bharati A, King CM. Contact dermatitis from 3 different allergens in Solaraze gel. Contact Dermatitis 2004;51:215-216

46 Ducros B, Bonnin JP, Navaranne A, et al. Eczema due to contact with ethanol in oestradiol transdermal patch (Estraderm TTS 50). Nouvelles Dermatologiques 1989;8:21-22 (Article in French)

47 Grebe SKG, Adams JD, Feek CM. Systemic sensitization to ethanol by transdermal estrogen patches. Arch Dermatol 1993;129:379-380

48 Pecquet C, Pradalier A, Dry J. Allergic contact dermatitis from ethanol in a transdermal estradiol patch. Contact Dermatitis 1992;27:275-276

49 Schwarz BK, Clendenning WE. Allergic contact dermatitis from hydroxypropyl cellulose in a transdermal estradiol patch. Contact Dermatitis 1988;18:106-107

50 Growth H, Vetter H, Knuesel J, Vetter W. Allergic skin reactions to transdermal clonidine. Lancet 1983;2(8354):850-851

51 Dwyer CM, Forsyth A. Allergic contact dermatitis from methacrylates in a nicotine transdermal patch. Contact Dermatitis 1994;30:309-310

52 Green C, Lowe JG. Contact allergy to piroxicam gel. Contact Dermatitis 1992;27:261

53 Camarasa JG, Serra-Baldrich E, Lluch M. Contact allergy to vitamin B6. Contact Dermatitis 1990;23:115

54 Smeenk G, Prins FJ. Allergic contact eczema due to chloracetamide. Dermatologica 1972;144:108-114

55 Smeenk G, Kerckhoffs HP, Schreurs PH. Contact allergy to a reaction product in Hirudoid cream: an example of compound allergy. Br J Dermatol 1987;116:223-231

56 Pecegueiro M, Brandão M, Pinto J, Conçalo S. Contact dermatitis to Hirudoid cream. Contact Dermatitis 1987;17:290-293

57 Milpied B, van Wassenhove L, Larousse C, Barriere H. Contact dermatitis from rifamycin. Contact Dermatitis 1986;14:252-253

58 Foti C, Vena GA, Angelini G. Photocontact allergy due to thiocolchicoside. Contact Dermatitis 1992;27:201-202

59 Foti C, Cassano N, Mazzarella F, Bonamonte D, Veña GA. Contact allergy to thiocolchicoside. Contact Dermatitis 1997;37:134

60 Coskey RJ. Contact dermatitis due to multiple corticosteroid creams. Arch Dermatol 1978;114:115-117

61 Feldman SB, Sexton FM, Buzas J, Marks JG Jr. Allergic contact dermatitis from topical steroids. Contact Dermatitis 1988;19:226-228

62 Rytter M, Walther T, Süss E, Haustein UF. Allergic reactions of the immediate and delayed type following prednisolone medication. Dermatol Monatsschr 1989;175:44-48 (Article in German)

63 Goh CL. Cross-sensitivity to multiple topical corticosteroids. Contact Dermatitis 1989;20:65-67

64 Burden AD, Beck MH. Contact hypersensitivity to topical corticosteroids. Br J Dermatol 1992;127:497-501

65 Green C, Kenicer KJ. A case of "contact allergy to corticosteroid". Contact Dermatitis 1993;28:39-40

66 Jagodzinski LJ, Taylor JS, Oriba H. Allergic contact dermatitis from topical corticosteroid preparations. Am J Contact Dermat 1995;6:67-74

67 Bennett ML, Fountain JM, McCarty MA, Sherertz EF. Contact allergy to corticosteroids in patients using inhaled or intranasal corticosteroids for allergic rhinitis or asthma. Am J Contact Dermat 2001;12:193-196

68 Isaksson M. Systemic contact allergy to corticosteroids revisited. Contact Dermatitis 2007;57:386-388

69 Brandao FM, Camarasa FM. Contact allergy to hydrocortisone 17-butyrate. Contact Dermatitis 1979;5:354-356

70 Brown R. Allergy to hydrocortisone-17-butyrate. Contact Dermatitis 1980;6:504-505

71 Yoshikawa K, Watanabe K, Mizuno N. Contact allergy to hydrocortisone 17-butyrate and pyridoxine hydrochloride. Contact Dermatitis 1985;12:55-56

72 Yoshikawa K, Watanabe K, Mizuno N. Contact allergy to hydrocortisone 17-butyrate and pyridoxine hydrochloride. Contact Dermatitis 1985;12:55-56

73 Wilkinson SM, Beck MH. Allergic contact dermatitis from dibutyl phthalate, propyl gallate and hydrocortisone in Timodine. Contact Dermatitis 1992;27:197

74 Dunkel FG, Elsner P, Burg G. Allergic contact dermatitis from prednicarbate. Contact Dermatitis 1991;24:59-60

75 Tang MM, Corti MA, Stirnimann R, Pelivani N, Yawalkar N, Borradori L, et al. Severe cutaneous allergic reactions following topical antifungal therapy. Contact Dermatitis 2013;68:56-67

76 Aranzana A, Gines E, Garcia-Bravo B, Camacho F. Allergic contact dermatitis from fluocortin butylester. Contact Dermatitis 1994;31:271-272

77 Goossens A, Huygens S, Matura M, Degreef H. Fluticasone propionate: a rare contact sensitizer. Eur J Dermatol 2001;11:29-34

78 Bunney MH. Contact dermatitis due to Betamethasone 17-valerate (Betnovate). Contact Dermatitis Newsletter 1972;12:318

79 Dooms-Goossens A, Vanhee J, Vanderheyden D, Gevers D, Willems L, Degreef H. Allergic contact dermatitis to topical corticosteroids: clobetasol propionate and clobetasone butyrate. Contact Dermatitis 1983;9:470-478

80 Hayakawa R, Matsunaga K, Ukei C, Hosokawa K. Allergic contact dermatitis from amcinonide. Contact Dermatitis 1985;12:213-214

81 Barbaud A, Vigan M, Delrous JL, Assier H, Avenel-Audran M, Collet E, et al; Membres du Groupe du REVIDAL. Contact allergy to antiseptics: 75 cases analyzed by the dermato-allergovigilance network (Revidal). Ann Dermatol Venereol 2005;132(12Pt.1):962-965 (Article in French)

82 Özkaya E, Kılıç S. Polyethylene glycol as marker for nitrofurazone allergy: 20 years of experience from Turkey. Contact Dermatitis 2018;78:211-215

83 Hogan DJ. Widespread dermatitis after topical treatment of chronic leg ulcers and stasis dermatitis. CMAJ 1988;138:336-338

84 Mirando-Romero A, Drake M, Asumendi L. Dermatitis alergica de contacto por fepradinol. XXXI Meeting of GEIDC. Santander, Spain. May 1993 (oral communication). Data cited in ref. 85

85 Ortiz-Frutos FJ, Hergueta JP, Quintana I, Zarco C, Iglesias L. Allergic contact dermatitis from fepradinol: report of 4 cases and review of the literature. Contact Dermatitis 1994;31:193-195

86   Isaksson M. Triple sensitization in a child with chronic otitis externa. Contact Dermatitis 2002;47:172

87   Bircher AJ. Short induction phase of contact allergy to tixocortol pivalate in a nasal spray. Contact Dermatitis 1990;22:237-238

88   McKenna KE. Allergic contact dermatitis from glyceryl trinitrate ointment. Contact Dermatitis 2000;42:246

89   Rademaker M, Wood B, Greig D. Contact dermatitis from cetostearyl alcohol. Australas J Dermatol 1997;38: 220-221

90   Oiso N, Fukai K, Ishii M. Concomitant allergic reaction to cetyl alcohol and crotamiton. Contact Dermatitis 2003;49:261

91   Jager SU, Pönninghaus JM, Koch P. Allergic contact dermatitis from cyclopiroxolamine? Contact Dermatitis 1995;33:349-350

92   Agathos M, Remien C, Mutzeck E. Contact allergy to tromantadine. Derm Beruf Umwelt 1984;32:157-160 (Article in German)

93   Lindgren S, Groth O, Molin L. Allergic contact response to vitamin A acid. Contact Dermatitis 1976;2:212-217

94   Degreef H, Dooms-Goossens A. Patch testing with silver sulfadiazine cream. Contact Dermatitis 1985;12:33-37

95   Fraser-Moodie A. Sensitivity to silver in a patient treated with silver sulphadiazine (Flamazine). Burns 1992;18:74-75

96   La Placa M, Balestri R, Bardazzi F, Vincenzi C. Scalp psoriasiform contact dermatitis with acute telogen effluvium due to topical minoxidil treatment. Skin Appendage Disord 2016;1:141-143

97   Remy C, Barbaud A, Lebrun-Vignes B, Perrot JL, Beyens MN, Mounier G, et al. Skin toxicity related to Percutalgine (®): analysis of the French pharmacovigilance database. Ann Dermatol Venereol 2012;139:350-354 (Article in French)

98   Sams WM. Untoward response with topical fluorouracil. Arch Dermatol 1968;97:14-23

99   Meijer BU, de Waard-van der Spek FB. Allergic contact dermatitis because of topical use of 5-fluorouracil (Efudix cream). Contact Dermatitis 2007;57:58-60

100 Degreef H, Verhoeve L. Contact dermatitis to miconazole nitrate. Contact Dermatitis 1975;1:269-270

101 Samsoen M, Jelen G. Allergy to Daktarin gel. Contact Dermatitis 1977;3:351-352

102 Vissers-Croughs KJM, Van der Kley AMJ, Vulto AG, Hulsmans RFHJ. Allergic contact dermatitis from sodium sulfite. Contact Dermatitis 1988;18:252-253

103 Garcia-Bravo B, Mazuecos J, Rodriguez-Pichardo A, Navas J, Camacho F. Hypersensitivity to ketoconazole preparations: study of 4 cases. Contact Dermatitis 1989;21:346-348

104 Romaguera C, Ferrando J, Lecha M, et al. Dermatitis de contacto al propilenglicol del excipiente de un preparado de ketoconazol. Piele 1989;4:6-8 (Article in Spanish)

105 Santucci B, Cannistraci C, Cristaudo A, Picardo M. Contact dermatitis from ketoconazole cream. Contact Dermatitis 1992;27:274-275

106 Schneider KW. Contact dermatitis due to diethyl sebacate. Contact Dermatitis 1980;6:506-507

107 Sasaki E, Hata M, Aramaki J, Honda M. Allergic contact dermatitis due to diethyl sebacate. Contact Dermatitis 1997;36:172

108 Kimura M, Kawada A. Contact dermatitis due to diethyl sebacate. Contact Dermatitis 1999;40:48-49

109 Tanaka M, Kobayashi S, Murata T, Tanikawa A, Nishikawa T. Allergic contact dermatitis from diethyl sebacate in lanoconazole cream. Contact Dermatitis 2000;43:233-234

110 Soga F, Katoh N, Kishimoto S. Contact dermatitis due to lanoconazole, cetyl alcohol and diethyl sebacate in lanoconazole cream. Contact Dermatitis 2004;50:49-50

111 Moss HV. Allergic contact dermatitis due to Halotex solution. Arch Dermatol 1974;109:572

112 Berlin AR, Miller OF. Allergic contact dermatitis from ethyl sebacate in haloprogin cream. Arch Dermatol 1976;112:1563-1564

113 Umebayashi Y, Ito S. Allergic contact dermatitis due to both lanoconazole and neticonazole ointments. Contact Dermatitis 2001;44:48-49

114 Kabasawa Y, Kanzaki T. Allergic contact dermatitis from ethyl sebacate. Contact Dermatitis 1990;22:226

115 Fisher AA. Consort contact dermatitis. Cutis 1979;24:595-596

116 Brand CU, Ballmer-Weber BK. Contact sensitivity to 5 different ingredients of a topical medicament (Imacort cream). Contact Dermatitis 1995;33:137

117 Jelen G, Tennstedt D. Contact dermatitis from topical imidazole antifungals: 15 new cases. Contact Dermatitis 1989;21:6-11

118 Raulin C, Frosch PJ. Contact allergy to imidazole antimycotics. Contact Dermatitis 1988;18:76-80

119 Giménez-Arnau A, Gilaberte M, Conde D, Espona M, Pujol RM. Combined photocontact dermatitis to benzydamine hydrochloride and the emulsifiers, Span 60 and Tween 60 contained in Tantum cream. Contact Dermatitis 2007;57:61-62

120 Ozkaya E, Topkarci Z, Ozarmağan G. Allergic contact cheilitis from a lipstick misdiagnosed as herpes labialis: Subsequent worsening due to Zovirax contact allergy. Australas J Dermatol 2007;48:190-192

121 Kim YJ, Kim JH. Allergic contact dermatitis from propylene glycol in Zovirax cream. Contact Dermatitis 1994;30:119-120

122 Bourezane Y, Girardin P, Aubin F, Vigan M, Adessi B, Humbert P, et al. Allergic contact dermatitis to Zovirax cream. Allergy 1996;51:755-756

123 Claverie F, Giordano-Labadie F, Bazex J. Contact eczema induced by propylene glycol. Concentration and vehicle adapted for patch tests. Ann Dermatol Venereol 1997;124:315-317 (Article in French)

124 Corazza M, Virgili A, Mantovani L, La Malfa W. Propylene glycol allergy from acyclovir cream with cross-reactivity to hydroxypropyl cellulose in a transdermal estradiol system? Contact Dermatitis 1993;29:283-284

125 Hernández N, Hernández Z, Liuti F, Borrego L. Intolerance to cosmetics as key to the diagnosis in a patient with allergic contact dermatitis caused by propylene glycol contained in a topical medication. Contact Dermatitis 2017;76:246-247

126 Bayrou O, Gaouar H, Leynadier F. Famciclovir as a possible alternative treatment in some cases of allergy to acyclovir. Contact Dermatitis 2000;42:42

127 Vernassiere C, Barbaud A, Trechot PH, Weber-Muller F, Schmutz JL. Systemic acyclovir reaction subsequent to acyclovir contact allergy: which systemic antiviral drug should then be used? Contact Dermatitis 2003;49:155-157

128 Gola M, Francalanci S, Brusi C, Lombardi P, Sertoli A. Contact sensitization to acyclovir. Contact Dermatitis 1989;20:394-395

129 Lammintausta K, Mäkelä L, Kalimo K. Rapid systemic valaciclovir reaction subsequent to aciclovir contact allergy. Contact Dermatitis 2001;45:181

130 Aguirre A, Manzano D, Izu R, Gardeazabal J, Díaz Pérez JL. Allergic contact cheilitis from mandelic acid. Contact Dermatitis 1994;31:133-134

131 Goday J, Aguirre A, Ibarra NG, Eizaguirre X. Allergic contact dermatitis from acyclovir. Contact Dermatitis 1991;24:380-381

132 Oleffe J A, Blondeel A, Coninck A D. Allergy to chlorocresol and propylene glycol in a steroid cream. Contact Dermatitis 1979;5:53-54

133 Lindén S, Göransson K. Contact allergy to dibenzthion. Contact Dermatitis 1975;1:258

134 Tennstedt D, Lachapelle JM. Allergic contact dermatitis from colophony in a nitroglycerin transdermal therapeutic system. Contact Dermatitis1990:23:254-255

135 Daly BM. Bactroban allergy due to polyethylene glycol. Contact Dermatitis 1987;17:48-49

136 Andersen K E, Broesby-Olsen S. Allergic contact dermatitis from oleyl alcohol in Elidel cream®. Contact Dermatitis 2006;55:354-356

137 Leroy A, Baeck M, Tennstedt D. Contact dermatitis and secondary systemic allergy to dimethindene maleate. Contact Dermatitis 2011;64:170-171

138 Smeenk G, Burgers GJ, Teunissen PC. Contact dermatitis from salbutamol. Contact Dermatitis 1994;31:123

139 Nagayama H, Hatamochi A, Shinkai H. A case of contact dermatitis due to sodium bisulfite in an ophthalmic solution. J Dermatol 1997;24:675-677

140 Herbst RA, Maibach HI. Contact dermatitis caused by allergy to ophthalmic drugs and contact lens solutions. Contact Dermatitis 1991;25:305-312

141 Ortiz FJ, Postigo C, Ivars J, Ortiz PL, Merino V. Allergic contact dermatitis from pilocarpine and thimerosal. Contact Dermatitis 1991;25:203-204

142 Orsini D, D'Arino A, Pigliacelli F, Assorgi C, Latini A, Cristaudo A. Allergic contact dermatitis to dorzolamide and benzalkonium chloride. Postepy Dermatol Alergol 2018;35:538-539

143 Camarasa JG, Serra-Baldrich E, Monreal P, Soller J. Contact dermatitis from sodium-cromoglycate-containing eyedrops. Contact Dermatitis 1997;36:160-161

144 Tabar AI, García BE, Rodríguez A, Quirce S, Olaguibel JM. Etiologic agents in allergic contact dermatitis caused by eyedrops. Contact Dermatitis 1993;29:50-51

145 Tosti A, Tosti G. Allergic contact conjunctivitis due to ophthalmic solution. In: Frosch PJ, Dooms-Goossens A, Lachapelle JM, Rycroft RJG, Scheper RJ (eds). Current Topics in Contact Dermatitis. Berlin: Springer-Verlag, 1989: 269-272

146 Corazza M, Massieri LT, Virgili A. Doubtful value of patch testing for suspected contact allergy to ophthalmic products. Acta Derm Venereol 2005;85:70-71

147 Ventura MT, Di Corato R, Di Leo E, Foti C, Buquicchio R, Sborgia C, et al. Eyedrop induced allergy: clinical evaluation and diagnostic protocol. Immunopharmacol Immunotoxicol 2003;25:529-538

148 Camarasa JG, Serra-Baldrich E, Garcia-Bravo B, Vozmediano JF. Contact dermatitis to minoxidil. In: Frosch PJ, Dooms-Goossens A, Lachapelle JM, Rycroft RJG, Scheper RJ (eds). Current topics in contact dermatitis. Berlin: Springer-Verlag, 1989:261-263

149 Veramme J, de Zaeytijd J, Lambert J, Lapeere H. Contact dermatitis in patients undergoing serial intravitreal injections. Contact Dermatitis 2016;74:18-21

150 Fisher AA, Stillman MA. Allergic contact sensitivity to benzalkonium chloride: cutaneous, ophthalmic, and general medical implications. Arch Dermatol 1972;106:169-171

151 Gata I, Garcia Bravo B, Rodriguez Pichardo A, Ortega Resina M, Camacho F. Allergic contact dermatitis to ethanol in a transdermal estradiol patch. Am J Cont Dermat 1994;5:221-222

152 Finn OA, Forsyth A. Contact dermatitis due to sorbitan monolaurate. Contact Dermatitis 1975;1:318

153 Boyle J, Kennedy CT. Contact urticaria and dermatitis to Alphaderm. Contact Dermatitis 1984;10:178

154 Lawrence CM, Smith AG. Ampliative medicament allergy: concomitant sensitivity to multiple medicaments including yellow soft paraffin, white soft paraffin, gentian violet and Span 20. Contact Dermatitis 1982;8:240-245

155 Jones SK, Kennedy CT. Contact dermatitis from Emulgin RO/40, an emulsifier in Hioxyl cream. Contact Dermatitis 1988;18:108-110.

156 Houghton M, Mann R J. Contact dermatitis from Hioxyl cream. Contact Dermatitis 1986;2:243-244

157 Marston S. Contact dermatitis from cetostearyl alcohol in hydrocortisone butyrate lipocream, and from lanolin. Contact Dermatitis 1991;24:372

158 Bordalo O, Brandão FM. Contact allergy to palmitoyl collagenic acid in an anti-keloid cream. Contact Dermatitis 1991;24:316-317

159 Garioch JJ, Forsyth A, Chapman RS. Allergic contact dermatitis from the perfume in Locan cream. Contact Dermatitis 1989;20:61-62

160 De Groot AC, Conemans JM, Schutte T. Contact allergy to di-isopropyl sebacate in Zineryt lotion. Contact Dermatitis 1991;25:260-261

161 Di Landro A, Pansera B, Valsecchi R, Caineli T. Allergic contact dermatitis from a wart remover solution. Contact Dermatitis 1995;32:178-179

162 Tabar A, Muro MD, Quirce S, Olaguibel JM. Contact dermatitis due to sensitization to lactic acid and castor oil in a wart remover solution. Contact Dermatitis 1993l;29:49-50

163 Lee AY, Lee KH. Allergic contact dermatitis from dioctyl sodium sulfosuccinate in a topical corticosteroid. Contact Dermatitis 1998;38:355-356

164 Tucker SC, Yell JA, Beck MH. Allergic contact dermatitis from sodium metabisulfite in Trimovate cream. Contact Dermatitis 1999;40:164

165 Miura Y, Hata M, Yuge M, Numano K, Iwakiri K. Allergic contact dermatitis from 1,2,6-hexanetriol in fluocinonide cream. Contact Dermatitis 1999;41:118-119

166 Cooper SM, Shaw S. Contact allergy to isopropanolamine in Traxam gel. Contact Dermatitis 1999;41:233-234

167 Podda M, Zollner T, Grundmann-Kollmann M, Kaufmann R, Boehncke WH. Allergic contact dermatitis from benzyl alcohol during topical antimycotic treatment. Contact Dermatitis 1999;41:302-303

168 Park HJ, Kang HA, Lee JY, Kim HO. Allergic contact dermatitis from benzalkonium chloride in an antifungal solution. Contact Dermatitis 2000;42:306-307

169 Vilaplana J, Romaguera C. Allergic contact dermatitis due to eucalyptol in an anti-inflammatory cream. Contact Dermatitis 2000;43:118

170 Vilaplana J, Romaguera C. Contact dermatitis from parabens used as preservatives in eyedrops. Contact Dermatitis 2000;43:248

171 Orton DI, Shaw S. Allergic contact dermatitis from pharmaceutical grade BHA in Timodine, with no patch test reaction to analytical grade BHA. Contact Dermatitis 2001;44:191-192

172 Yesudian PD, King CM. Allergic contact dermatitis from stearyl alcohol in Efudix cream. Contact Dermatitis 2001;45:313-314

173 Choudry K, Beck MH, Muston HL. Allergic contact dermatitis from 2-bromo-2-nitropropane-1,3-diol in Metrogel. Contact Dermatitis 2002;46:60-61

174 Chowdhury MM, Statham BN. Allergic contact dermatitis from dibutyl phthalate and benzalkonium chloride in Timodine cream. Contact Dermatitis 2002;46:57

175 Harrison DA, Smith AG. Concomitant sensitivity to sodium metabisulfite and clobetasone butyrate in Trimovate cream. Contact Dermatitis 2002;46:310

176 Isaksson M, Jansson L. Contact allergy to Tween 80 in an inhalation suspension. Contact Dermatitis 2002;47:312-313

177 Farrar CW, Bell HK, King CM. Allergic contact dermatitis from propylene glycol in Efudix cream. Contact Dermatitis 2003;48:345

178 Kaminska R, Mörtenhumer M. Nummular allergic contact dermatitis after scabies treatment. Contact Dermatitis 2003;48:337

179 Foti C, Conserva A, Antelmi A, Lospalluti L, Angelini G. Contact dermatitis from peppermint and menthol in a local action transcutaneous patch. Contact Dermatitis 2003;49:312-313

180 Sestini S, Mori M, Francalanci S. Allergic contact dermatitis from benzyl alcohol in multiple medicaments. Contact Dermatitis 2004;50:316-317

181 Oiso N, Fukai K, Ishii M. Allergic contact dermatitis due to 1,3-butylene glycol in medicaments. Contact

Dermatitis 2004;51:40-41

182 Gallo R, Marro I, Sorbara S. Contact allergy from phenoxyethanol in Fitostimoline gauzes. Contact Dermatitis 2005;53:241

183 Foti C, Bonamonte D, Conserva A, Casulli C, Angelini G. Allergic contact dermatitis to glyceryl-hydrogenated rosinate in a topical plaster. Contact Dermatitis 2006;55:120-121

184 Goiriz R, Delgado-Jiménez Y, Sánchez-Pérez J, García-Diez A. Photoallergic contact dermatitis from lavender oil in topical ketoprofen. Contact Dermatitis 2007;57:381-382

185 Singh M, Winhoven SM, Beck MH. Contact sensitivity to octyldodecanol and trometamol in an anti-itch cream. Contact Dermatitis 2007;56:289-290

186 Ota T, Oiso N, Iba Y, Narita T, Kawara S, Kawada A. Concomitant development of photoallergic contact dermatitis from ketoprofen and allergic contact dermatitis from menthol and rosin (colophony) in a compress. Contact Dermatitis. 2007;56:47-48

187 Pitarch G, de la Cuadra Précis J. Patch testing with ethyl alcohol. Dermatitis 2010;21:120-121

188 Barbaud A, Trechot P, Reichert-Penetrat S, Schmutz J. Contact dermatitis due to ethyl alcohol: how to perform patch tests? Ann Dermatol Venereol 2000;127:484-487 (Article in French)

189 Afzelius H, Thulin H. Allergic reactions to benzalkonium chloride. Contact Dermatitis 1979;5:60

190 Tosti A, Tosti G. Thimerosal: a hidden allergen in ophthalmology. Contact Dermatitis 1988;18:268-273

191 Cox NH. Allergy to benzalkonium chloride simulating dermatomyositis. Contact Dermatitis 1994;31:50

192 Klein GF, Sepp N, Fritsch P. Allergic reactions to benzalkonium chloride? Do the use test! Contact Dermatitis 1991;25:269-270

193 Haetinen A, Teraesvirta M, Fraeki IE. Contact allergy to components in topical ophthalmologic preparations. Acta Dermato-venerologica 1985:63:424-426

194 Svensson Å, Möller H. Eyelid dermatitis: the rôle of atopy and contact allergy. Contact Dermatitis 1986;15:178-182

195 Frosch PI, Weickel R, Schmitt T, Krastel H. Nebenwirkungen von ophthalmologischen Externa. Z Hautkr 1988;63:126-136 (Article in German)

196 Chiambaretta F, Pouliquen P, Rigal D. Allergy and preservatives. A propos of 3 cases of allergy to benzalkonium chloride. J Fr Ophtalmol 1997;20:8-16 (Article in French)

197 Benjamin B, Chris F, Salvador G, et al. Visual and confocal microscopic interpretation of patch tests to benzethonium chloride and benzalkonium chloride. Skin Res Technol 2012;18:272-277

198 Wahlberg JE. Two cases of hypersensitivity to quaternary ammonium compounds. Acta Derm Venereol (Stockh.) 1962;42:230-234

199 Fräki JE, Kalimo K, Tuohimaa P, Aantaa E. Contact allergy to various components of topical preparations for treatment of external otitis. Acta Otolaryngol 1985;100:414-418

200 Díaz-Ramón L, Aguirre A, Ratón-Nieto JA, de Miguel M. Contact dermatitis from benzoxonium chloride. Contact Dermatitis 1999;41:53-54

201 Tosti A, Bardazzi F, Valeri F, Russo R. Contact dermatitis from butylated hydroxyanisole. Contact Dermatitis 1987;17:257-258

202 Dever TT, Herro EM, Jacob SE. Butylhydroxytoluene – from jet fuels to cosmetics? Dermatitis 2012;23:90-91

203 Bardazzi F, Misciali C, Borrello P, Capoblanco, C. Contact dermatitis due to antioxidants. Contact Dermatitis 1988;19:385-386

204 Choudry K, Beck MH, Muston HL. Allergic contact dermatitis from 2-bromo-2-nitropropane-1,3-diol in Metrogel®. Contact Dermatitis 2002;46:60-61

205 Frosch PJ, Weickel R. Kontaktallergie auf das Konservierungsmittel Bronopol. Hautarzt 1987;38:267-270 (Article in German)

206 Schorr WF. Paraben allergy. A cause of intractable dermatitis. JAMA 1968;204:859-862

207 Fisher AA. Cortaid cream dermatitis and the 'paraben paradox'. J Am Acad Dermatol 1982;6:116-117

208 Maucher OM. Periorbitalekzem als iatrogene Erkrankung. Klin Monatsbl Augenheilkd 1974;164:350-356 (Article in German)

209 Dissanayke M, Powell SM. Hioxyl® sensitivity. Contact Dermatitis 1990;22:242-243

210 Milpied B, Collet E, Genillier N, Vigan M. Allergic contact dermatitis caused by sodium dehydroacetate, not hyaluronic acid, in Ialuset® cream. Contact Dermatitis 2011;65:359-361

211 Pasche-Koo F, Piletta P-A, Hunziker N, Hauser C. High sensitization rate to emulsifiers in patients with chronic leg ulcers. Contact Dermatitis 1994;31:226-228

212 Armengot-Carbo MA, Rodríguez-Serna M, Taberner-Bonastre P, Miquel-Miquel J. Allergic contact dermatitis from cetearyl alcohol in Thrombocid® ointment. Dermatol Online J 2016 Jul 15;22(7). pii: 13030/qt8ht9300r

213 Raulin C, Frosch PJ. Contact allergies to antifungal agents. Z Hautkr 1987;62:1705-1709 (Article in German)

214 Aerts O, Naessens T, Dandelooy J, Leysen J, Lambert J, Apers S. Allergic contact dermatitis caused by wet wipes containing steareth-10: Is stearyl alcohol to blame? Contact Dermatitis 2017;77:117-119

215 Ishiguro N, Kawishima M. Contact dermatitis from impurities in alcohol. Contact Dermatitis 1991;25:257

216 Komamura H, Dor T, Inui S, Yoshikawa K. A case of contact dermatitis due to impurities of cetyl alcohol. Contact Dermatitis 1997;36:44-46

217 Van Ketel, WG. Allergy to cetylalcohol. Contact Dermatitis 1984;11:125-126

218 Gaul LE. Dermatitis from cetyl and stearyl alcohols. Arch Dermatol 1969;99:593

219 Schmoll M, Hausen BM. Allergic contact dermatitis to prednisolone-21-trimethyl acetate. Z Hautkr 1988;63:311-313 (Article in German).

220 Kang H, Choi J, Lee A Y. Allergic contact dermatitis to white petrolatum. J Dermatol 2004;31:428-430

221 Prins FI, Smeenk G. Contacteczeem door Hirudoidzalf. Ned Tijdschr Geneeskd 1971;115:1934-1938 (Article in Dutch)

222 Wantke F, Demmer CM, Götz M, Jarisch R. Sensitization to chloroacetamide. Contact Dermatitis 1993;29:213-214

223 Detmar U, Agathos, M. Contact allergy to chloroacetamide. Contact Dermatitis 1988;19:66-67

224 Archer CB, MacDonald DM. Chlorocresol sensitivity induced by treatment of allergic contact dermatitis with steroid creams. Contact Dermatitis 1984;11:144-145

225 Lewis PG, Emmet EA. Irritant dermatitis from tributyl tin oxide and contact allergy from chlorocresol. Contact Dermatitis 1987;17:129-132

226 Gómez de la Fuente E, Andreu-Barasoain M, Nuño-González A, López-Estebaranz JL. Allergic contact dermatitis due to chlorocresol in topical corticosteroids. Actas Dermosifiliogr 2013;104:90-92

227 Camarasa JG, Serra-Baldrich E. Dermatitis de contacto por cremas conteniendo propionato de clobetasol. Med Cutan Ibero Lat Am 1988;16:328-330 (Article in Spanish)

228 Salim A, Powell S, Wojnarowska F. Allergic contact dermatitis of the vulva - An overlooked diagnosis. J Obstet Gynaecol 2002;22:447

229 Calnan CD. Contact dermatitis from drugs. Proceedings of the Royal Society of Medicine 1962;55:39-42

230 Storrs FJ. Para-chloro-meta-xylenol allergic contact dermatitis in seven individuals. Contact Dermatitis 1975;1:211-213

231 Badaoui A, Amsler E, Raison-Peyron N, Vigan M, Pecquet C, Frances C, et al. An outbreak of contact allergy to cocamide diethanolamide? Contact Dermatitis 2015;72:407-409

232 Lodi A, Leuchi S, Mancini L, Chiarelli G, Crosti C. Allergy to castor oil and colophony in a wart remover. Contact Dermatitis 1992;26: 266-267

233 Cameli N, Vassilopoulou A, Vincenzi C. Contact allergy to colophony in a wart remover. Contact Dermatitis 1991;24:315

234 Monk B. Allergic contact dermatitis to colophony in a wart remover. Contact Dermatitis 1987;17:242

235 Lachapelle JM, Leroy B. Allergic contact dermatitis to colophony included in the formulation of flexible collodion BP, the vehicle of a salicylic and lactic acid wart paint. Dermatol Clin 1990;8:143-146

236 Moss C, Berry K. Allergy to colophony. BMJ 1995;310(6979):603

237 O'Brien TJ. Colophony in collodion. Australas J Dermatol 1986;27:142-143

238 Raymond J Z, Gross P R. EDTA preservative dermatitis. Arch Dermatol 1969;100:436-440

239 Kuznetsov AV, Erlenkeuser-Uebelhoer I, Thomas P. Contact allergy to propylene glycol and dodecyl gallate mimicking seborrheic dermatitis. Contact Dermatitis 2006;55:307-308

240 Raymond JZ, Gross PR. EDTA: preservative dermatitis. Arch Dermatol 1969;100:436-440

241 De Groot AC. Contact allergy to EDTA in a topical corticosteroid preparation. Contact Dermatitis 1986;15:250-252

242 Schorr WP, Mohajerin AH. Paraben sensitivity. Arch Dermatol 1966;93:721-723

243 Kounis NG, Zavras GM, Papadaki PJ, Soufras GD, Poulos EA, Goudevenos J, et al. Allergic reactions to local glyceryl trinitrate administration. Br J Clin Pract 1996;50:437-439

244 Guidetti MS, Vincenzi C, Guerra L, Tosti A. Contact dermatitis due to imidazole antimycotics. Contact Dermatitis 1995;33:282

245 Calnan CD. Isopropyl myristate sensitivity. Contact Dermatitis Newsletter 1968;3:41

246 Giorgini S, Melli MC, Sertoli A. Comments on the allergenic activity of lanolin. Contact Dermatitis 1983;9:425-426

247 Calnan CD. Oxypolyethoxydodecane in an ointment. Contact Dermatitis 1978;4:168-169

248 Prins FJ, Smeenk G. Contacteczeem door Hirudoid zalf. Ned T Geneesk 1971;115:1935-1938 (Article in Dutch)

249 Sams WM. Untoward response with topical fluorouracil. Arch Dermatol 1968;97:14-22

250 Fraser K, Pratt M. Polysensitization in recurrent lip dermatitis. J Cutan Med Surg 2015;19:77-80

251 Sulzberger MB, Lazar MP, Furman D. A study of the allergenic constituents of lanolin (wool fat). J Invest Dermatol 1950;15:453-458

251 Pauluzzi P, Rizzi GM. Contact dermatitis to wool alcohols: an unusual manifestation. Am J Cont Dermat 1994;5:113-114

252 Leysen J, Goossens A, Lambert J, Aerts O. Polyhexamethylene biguanide is a relevant sensitizer in wet wipes. Contact Dermatitis 2014: 70: 323-325

253 Higgins CL, Nixon RL. Periorbital allergic contact dermatitis caused by lanolin in a lubricating eye ointment. Australas J Dermatol 2016;57:68-69

254 Svensson Å. Allergic contact dermatitis to laureth-4. Contact Dermatitis 1988;18:113-114

255 Martin-Falero AA, Calderòn Gutierrez MJ, Díaz-Pérez JL. Contact allergy to the preservative Kathon CG. Contact Dermatitis 1990;22:185

256 Amaro C, Cravo M, Fernandes C, Santos R, Gonçalo M. Undisclosed methyldibromo glutaronitrile causing allergic contact dermatitis in a NSAID cream. Contact Dermatitis 2012;67:173-174

257 Foti C, Vena GA, Mazzarella F, Angelini G. Contact allergy due to methyl glucose dioleate. Contact Dermatitis 1995;32:303-304

258 Schianchi S, Calista D, Landi G. Widespread contact dermatitis due to methyl glucose dioleate. Contact Dermatitis 1996;35:257-258

259 Corazza M, Levratti A, Virgili A. Allergic contact dermatitis due to methyl glucose dioleate. Contact Dermatitis 2001;45:308

260 Macias VC, Fernandes S, Amaro C, Santos R, Cardoso J. Sensitization to methylisothiazolinone in a group of methylchloroisothiazolinone/methylisothiazolinone allergic patients. Cutan Ocul Toxicol 2013;32:99-101

261 Sagara R, Nakada T, Iijima M. Paraben allergic contact dermatitis in a patient with livedo reticularis. Contact Dermatitis 2008;58:53-54

262 Sánchez-Pérez J, Ballesteros Diez M, Alonso Pérez A, Delgado Jiménez Y, Diez G. Allergic and systemic contact dermatitis to methylparaben. Contact Dermatitis 2006;54:117-118

263 Sarkany I. Contact dermatitis from paraben. Br J Dermatol 1960;72:345-347

264 Hjorth N, Trolle-Lassen C. Skin reactions to ointment bases. Trans St John Hosp Derm Soc 1963;49:127-140

265 Wuepper KD. Paraben contact dermatitis. JAMA 1967;202:579-581

266 Batty KT. Hypersensitivity to methylhydroxybenzoate: a case for additive labeling of pharmaceuticals. Med J Aust 1986;144:107-108

267 Edman B, Möller H. Medicament contact allergy. Derm Beruf Umwelt 1986;34:139-142 (Article in German)

268 Dejobert Y, Delaporte E, Piette F, Thomas P. Vesicular eczema and systemic contact dermatitis from sorbic acid. Contact Dermatitis 2001;45:291

269 Malten KE. Sensitization to solcoseryl and methylanisate (fragrance ingredient). Contact Dermatitis 1977;3:219

270 Fisher AA. Allergic paraben and benzyl alcohol hypersensitivity relationship of the "delayed" and "immediate" varieties. Contact Dermatitis 1975;1:281-284

271 Verhaeghe I, Dooms-Goossens A. Multiple sources of allergic contact dermatitis from parabens. Contact Dermatitis 1997;36:269

272 Braun W. Contact allergies to polyethylene glycols. Z Haut und Geschl Krankh 1969;44:385-388 (Article in German)

273 Fisher AA. Contact urticaria due to polyethylene glycol. Cutis 1977;19:409-412

274 Fisher AA. Immediate and delayed allergic contact reactions to polyethylene glycol. Contact Dermatitis 1978;4:135-138

275 Guijarro SC, Sánchez-Pérez J, García-Díez A. Allergic contact dermatitis to polyethylene glycol and nitrofurazone. Am J Contact Dermat 1999;10:226-227

276 Foti C, Bonamonte D, Cassano N, Conserva A, Vena GA. Allergic contact dermatitis to propyl gallate and pentylene glycol in an emollient cream. Australas J Dermatol 2010;51:147-148

277 Grimalt F, Romaguera C. Sensitivity to petrolatum. Contact Dermatitis 1978;4:377

278 Bervoets A, Aerts O. Polyhexamethylene biguanide in wound care products: a non-negligible cause of peri-ulcer dermatitis. Contact Dermatitis 2016;74:53-55

279 Tosti A, Guerra L, Morelli R, Bardazzi F. Prevalence and sources of sensitization to emulsifiers: a clinical study. Contact Dermatitis 1990;23:68-72

280 Lucente P, Iorizzo M, Pazzaglia M. Contact sensitivity to Tween 80 in a child. Contact Dermatitis 2000;43:172

281 Fisher AA, Brancaccio RR. Allergic contact sensitivity to propylene glycol in a lubricant jelly. Arch Dermatol 1979;115:1451

282 Shore BN, Shelley WB. Contact dermatitis from stearyl alcohol and propylene glycol in flucinonide cream. Arch Dermatol 1974;109:397-399

283 Aguirre A, Gardeazábal J, Izu R, Antonio Ratón J, Diaz-Pérez JL. Allergic contact dermatitis due to plant extracts in a multisensitized patient. Contact Dermatitis 1993;28:186-187

284 Fowler JF Jr. Contact allergy to propylene glycol in topical corticosteroids. Am J Cont Dermat 1993;4:37-38

285 Piletta P, Pasche-Koo F, Saurat J-H, Hauser C. Contact dermatitis to propylene glycol in topical Zovirax cream. Am J Contact Dermat 1994;5:168-169

286 Eun HC, Kim YC. Propylene glycol allergy from ketoconazole cream. Contact Dermatitis 1989;21:274-275

287 Friedman ES, Friedman PM, Cohen DE, Washenik K. Allergic contact dermatitis to topical minoxidil solution: etiology and treatment. J Am Acad Dermatol 2002;46:309-312

288 Van Der Willigen AH, Dutree-Meulenberg RO, Stolz E, Geursen-Reitsma AM, van Joost T. Topical minoxidil sensitization in androgenic alopecia. Contact Dermatitis 1987;17:44-45

289 Fisher AA. Use of glycerin in topical minoxidil solutions for patient allergic to propylene glycol. Cutis 1990;45:81-82

290 Scheman AJ, West DP, Hordinksy MK, Osburn AH, West LE. Alternative formulation for patients with contact reactions to topical 2% and 5% minoxidil vehicle ingredients. Contact Dermatitis 2000;42:241

291 Frosch PJ, Pekar U, Enzmann H. Contact allergy to propylene glycol – do we use the appropriate test concentration. Dermatol Clin 1990;8:111-113

292 Lamb SR, Ardley HC, Wilkinson SM. Contact allergy to propylene glycol in brassiere padding inserts. Contact Dermatitis 2003;48:224-225

293 Angelini G, Meneghini CL. Contact allergy from propylene glycol. Contact Dermatitis 1981;7:197-198

294 Ortega MEG, Nieto MAP, Camacho MM, Moya AIS, Palma OA, Monné CB, et al. Allergic contact dermatitis from propylene glycol: a case with past and present relevance in relation to several owned products. Contact Dermatitis 2016;75(Suppl.1):74

295 Hernández N, Assier-Bonnet H, Terki N, Revuz J. Allergic contact dermatitis from propyl gallate in desonide cream (Locapred®). Contact Dermatitis 1997;36:111

296 Corazza M, Mantovani L, Roveggio C, Virgili A. Allergic contact dermatitis from propyl gallate. Contact Dermatitis 1994;31:203-204

297 Hausen BM, Beyer W. The sensitizing capacity of the antioxidants propyl, octyl, and dodecyl gallate and some related gallic acid esters. Contact Dermatitis 1992;26:253-258

298 Pansera B, Valsecchi R, Tornaghi A. Dermatitis allergica da contatto a! propile gallato. Proceedings of the 3rd GIRDCA Meeting, Bergamo, 30 April 1983 (in Italian, cited in ref. 300)

299 Soro A. Dermatitis allergica da contatto al propile gallato. Proceedings of the 3rd GIRDCA Meeting, Bergamo, 30 April 1983 (in Italian, cited in ref. 300)

300 De Groot A C, Gerkens F. Occupational airborne contact dermatitis from octyl gallate. Contact Dermatitis 1990;23:184-186

301 Valsecchi R, Cainelli T. Contact allergy to propyl gallate. Contact Dermatitis 1988;19:380-381

302 Pigatto PD, Boneschi V, Riva F, Altomare GF. Allergy to propylgallate, with unusual clinical and histological features. Contact Dermatitis 1984;11:43

303 Cusano F, Capozzi M, Errico G. Safety of propyl gallate in topical products. J Am Acad Dermatol 1987;17(2 Pt. 1):308-310

304 Pansera B, Valsecchi R, Tornaghi A, et at. Dermatite allergica da contatto al propile gallato. Proceedings of the 3rd GIRDCA Meeting, Bergamo, April 30, 1983 (in Italian, data cited in ref. 303)

305 Soro A. Dermatite allergica da contatto con propile gallato. Proceedings of the 3rd GIRDCA Meeting, Bergamo, April 30, 1983 (in Italian, data cited in ref. 303)

306 Rudzki E, Baranowska E. Reactions to gallic acid esters. Contact Dermatitis 1975;1:393 (reference cannot be found, data cited in ref. 297)

307 Caralli ME, Rodríguez MS, Rojas Pérez-Ezquerra P, Pelta Fernández R, De Barrio Fernández M. Palpebral angioedema and allergic contact dermatitis caused by a cerumenolytic. Contact Dermatitis 2015;73:376-377

308 Sánchez-Guerrero IM, Huertas AJ, López MP, Carreño A, Ramírez M, Pajarón, M. Angioedema-like allergic contact dermatitis to castor oil. Contact Dermatitis 2010;62:318-319

309 Wakelin SH, Harris AJ, Shaw S. Contact dermatitis from castor oil in zinc and castor oil cream. Contact Dermatitis 1996;35:259

310 Neering H, Vitanyi BEJ, Malten KE, Ketel WG van, Dijk E van. Allergens in sesame oil contact dermatitis. Acta Dermato-venereol 1975;55:31-34

311 Köhler A, Gall H. Kontaktallergisches Ekzem auf Natriumdisulfit in einer antimykotischen Ketoconazol-creme. Dermatosen Beruf Umwelt 2000;48:11-12 (Article in German)

312 Sams WM, Smith G. Contact dermatitis due to hydrocortisone ointment. Report of a case of sensitivity to emulsifying agents in a hydrophilic ointment base. JAMA 1957;164:1212-1213

313 Garcia-Gavin J, Parente J, Goossens A. Allergic contact dermatitis caused by sodium metabisulfite: a challenging allergen. A case series and literature review. Contact Dermatitis 2012;67:260-269

314 Madan V, Walker SL, Beck MH. Sodium metabisulfite allergy is common but is it relevant? Contact Dermatitis 2007;57:173-176

315 Dooms-Goossens A, de Alam AG, Degreef H, Kochuyt A. Local anesthetic intolerance due to metabisulfite. Contact Dermatitis 1989;20:124-126

316 Seitz CS, Bröcker EB, Trautmann A. Eyelid dermatitis due to sodium metabisulfite. Contact Dermatitis 2006;55:249-250

317 Harrison DA, Smith AG. Concomitant sensitivity to sodium metabisulfite and clobetasone butyrate in trimovate cream. Contact Dermatitis 2002;46:310

318 Sánchez-Pérez J, Abajo P, Córdoba S, García-Díez A. Allergic contact dermatitis from sodium metabisulfite in an antihemorrhoidal cream. Contact Dermatitis 2000;42:176-177

319 Tucker SC, Yell JA, Beck MH. Allergic contact dermatitis from sodium metabisulfite in trimovate cream. Contact Dermatitis 1999;40:164

320 Giorgini S, Brusi C, Melle MC, Sertoli A. Contact dermatitis by sodium metabisulfite. Med Staff Dermatol 1998;44:16-17

321 Vestergaard L, Andersen KE. Allergic contact dermatitis from sodium metabisulfite in a topical preparation. Am J Contact Dermat 1995;6:174-175

322 Veramme J, de Zaeytijd J, LamberTJ. and Lapeere H. Contact dermatitis in patients undergoing serial intravitreal injections. Contact Dermatitis 2016;74:18-21

323 Milpied B, Wassenhove LV, Larousse C, Barriere H. Contact dermatitis from rifamycin. Contact Dermatitis 1986;14:252-253

324 Ralph N, Verma S, Merry S. What is the relevance of contact allergy to sodium metabisulfite and which concentration of the allergen should we use? Dermatitis 2015;26:162-165

325 Petersen CS, Menné T. Consecutive patch testing with sodium sulphite in eczema patients. Contact Dermatitis 1992;27:344-345

326 Lodi A, Chiarelli G, Mancini LL, Crosti C. Contact allergy to sodium sulfite contained in an antifungal preparation. Contact Dermatitis 1993;29:97

327 Ikehata K, KatoJ, Kuwano A, Mita T, Sugai T. Two cases of allergic contact dermatitis caused by sodium sulfite in the vehicle of antifungal cream. Skin Research 1996;38:198-202 (Article in Japanese)

328 Giordano Labadie F, Pech-Ormieres C, Bazex J. Systemic contact dermatitis from sorbic acid. Contact Dermatitis 1996;34:61-62

329 Raison-Peyron N, Meynadier JM, Meynadier J. Sorbic acid: an unusual cause of systemic contact dermatitis in an infant. Contact Dermatitis 2000;43:247-248

330 Ramsing DW, Menné T. Contact sensitivity to sorbic acid. Contact Dermatitis 1993;28:124-125

331 Patrizi A, Orlandi C, Vincenzi C, Bardazzi F. Allergic contact dermatitis caused by sorbic acid: rare occurrence. Am J Contact Dermat 1999;10:52

332 Grange-Prunier A, Bezier M, Perceau G, Bernard P. Tobacco contact dermatitis caused by sensitivity to sorbic acid. Ann Dermatol Venereol 2008;135:135-138 (Article in French)

333 Hald M, Menné T, Johansen JD, Zachariae C. Allergic contact dermatitis caused by sorbitan sesquioleate imitating severe glove dermatitis in a patient with filaggrin mutation. Contact Dermatitis 2013;69:313-315

334 Mallon E, Powell SM. Sorbitan sesquioleate – a potential allergen in leg ulcer patients. Contact Dermatitis1994;30:180

335 Thormann H, Kollander M, Andersen KE. Allergic contact dermatitis from dichlorobenzyl alcohol in a patient with multiple contact allergies. Contact Dermatitis 2009;60:295-296

336 DeBerker D, Marren P, Powell SM, Ryan TJ. Contact sensitivity to the stearyl alcohol in Efudix cream (5-fluorouracil). Contact Dermatitis 1992;26:138

337 Black H. Contact dermatitis from stearyl alcohol in Metosyn (Fluocinonide) Cream. Contact Dermatitis 1975;1:125

338 Shore RN, Shelley WB. Contact dermatitis from stearyl alcohol and propylene glycol. Arch Derm 1974;110:636

339 Bardazzi F, Manuzzi P, Riguzzi G, Veronesi S. Contact dermatitis with rosacea. Contact Dermatitis 1987;16:298

340 Zemtsov A, Bolton GG. Thimerosal-induced bullous contact dermatitis. Contact Dermatitis 1994;30:57

341 Iliev D, Wüthrich B. Conjunctivitis to thimerosal mistaken as hay fever. Allergy 1998;53:333-334

342 Landa N, Aguirre A, Goday J, Ratón JA, Díaz-Pérez JL. Allergic contact dermatitis from a vasoconstrictor cream. Contact Dermatitis 1990;22:290-291

343 Aschenbeck KA, Warshaw EM. Clinically relevant reactions to thimerosal (the "nonallergen") exist! Dermatitis 2018;29:44-45

344 Suurmond D. Patch test reactions to Phenergan cream, promethazine and triethanolamine. Dermatologica 1966;133:503-506

345 Zina G, Bonu G. Phenergan cream (role of base constituents). Contact Dermatitis Newsletter 1969;6:117

346 Foussereau J, Sengel D. Un allergène habituellement méconnu: la triéthanolamine. Strasbourg Med 1965;10:873-880 (Article in French)

347 Schmutz J-L, Barbaud A, Tréchot P. Allergie de contact à la triéthanolamine contenue dans des gouttes auriculaires et dans un shampooing. Ann Dermatol Venereol 2007;134:105 (Article in French)

348 Curtis G, Netherton EW. Cutaneous hypersensitivity to triethanolamine. Arch Derm Syph 1940;41:729-731

349 Larsen WG. Allergic contact dermatitis to the perfume in Mycolog cream. J Am Acad Dermatol 1979;1:131-133

350 Larsen WG. Perfume dermatitis. A study of 20 patients. Arch Dermatol 1977;113:623-626

351 Corazza M, Mantovani L, Maranini C, Virgli A. Allergic contact dermatitis from benzyl alcohol. Contact Dermatitis 1996;34:74-75

352 Jacob SE, Stechschulte S. Eyelid dermatitis associated with balsam of Peru constituents: benzoic acid and benzyl alcohol. Contact Dermatitis 2008;58:111-112

353 Lazzarini S. Contact allergy to benzyl alcohol and isopropyl palmitate, ingredients of topical corticosteroid. Contact Dermatitis 1982;8:349-350

354 Shoji A. Allergic reaction to benzyl alcohol in an antimycotic preparation. Contact Dermatitis 1983;9:510

355 Würbach G, Schubert H, Phillipp I. Contact allergy to benzyl alcohol and benzyl paraben. Contact Dermatitis 1993;28:187-188

356 Li M, Gow E. Benzyl alcohol allergy. Australas J Dermatol 1995;36:219-220

357 Van Ketel WG. Allergy to cumarin and cumarin-derivatives. Contact Dermatitis Newsletter 1973;13:355

358 Aguirre A, Oleaga JM, Zabala R, Izu R, Díaz-Pérez JL. Allergic contact dermatitis from Reflex® spray. Contact Dermatitis 1994;30:52-53

359 Nakagawa S, Tagami H, Aiba S. Erythema multiforme-like generalized contact dermatitis to *l*-menthol contained in anti-inflammatory medical compresses as an ingredient. Contact Dermatitis 2009;61:178-179

360 Knöll R, Ulrich R, Spallek W. Allergic contact eczema to etofenamate and dwarf pine oil. Sportverletz Sportschaden 1990;4(2):96-98 (Article in German)

361 Le Coulant P, Texier L, Malleville J, Doussy NN. Sensibilization à une crème antihistaminique. Bull Soc Franç Derm Syph 1964;71:234-237 (Article in French)

362 Rademaker M. Allergic contact dermatitis from lavender fragrance in Difflam gel. Contact Dermatitis 1994;31:58-59

Chapter 5   IMMEDIATE CONTACT REACTIONS (CONTACT URTICARIA)

Topical drugs that have been reported to cause immediate contact reactions (contact urticaria) are shown in table 5.1. The entire spectrum of symptoms and signs of the contact urticaria syndrome can result from reactions to drugs including erythema and wealing, localized urticaria, generalized urticaria, gastrointestinal symptoms, pulmonary symptoms, drop in blood pressure, and anaphylactic shock (12,83). Not included are cases where drugs applied to non-intact skin (e.g. leg ulcers) caused signs of immediate-type hypersensitivity (e.g. anaphylactic shock) without local contact urticaria, and where the diagnosis was made by scratch, prick or intradermal tests (e.g. refs. 59,60,61). In other words, the occurrence of contact urticaria from local application to intact skin, either in the clinical or the diagnostic setting, is a prerequisite for topical drugs to be included in table 5.1 The references provided are or may be *examples* of cases of immediate contact reactions, that the author came across while writing this book; a full literature review was not attempted.

Table 5.1 Topical drugs that have caused immediate contact reactions (contact urticaria)

Drug	References	Drug	References
Aminolevulinic acid [k]	33, 38	Hydrocortisone	51 [f]
Atropine	41	Ketoprofen	63
Bacitracin [c]	15-18	Lidocaine	62
Benzocaine	36 [l], 66 [h]	Loxoprofen	63
Benzoyl peroxide	52	Mechlorethamine [d]	27-32, 77
Benzyl nicotinate	11, 59	Methyl aminolevulinate [k]	39, 40
Buserelin acetate	6	Methyl nicotinate	1-4, 8
Chloramphenicol	80 [i]	Methyl salicylate	10 [b]
Chlorproethazine	7 [a]	Neomycin	66 [h], 84 [l]
Chlorpromazine	7 [a]	Nicoboxil	11
Clioquinol	18, 66 [h], 81	Nicotine	34, 35
Clobetasol propionate	49	Oxyphenbutazone	53
Clobetasone butyrate	51 [f]	Penicillin	83-85 [i]
Cyclopentolate	5	Polymyxin B	16
Diclofenac	63	Povidone-iodine	45-48 [e]
Diethylstilbestrol	10 [b]	Pramoxine	43
Diethyl toluamide	88-92	Promethazine	64
Disodium cromoglycate (cromolyn sodium)	82	Rifamycin	57, 58
		Salicylic acid	42, 65, 87 [i]
Etofenamate	54, 55 [g]	Sulfur	85 [i]
Fluticasone propionate	50	Tacrolimus	9 [m]
Fusidic acid	44	Thiocolchicoside	56 [j]
Gentamicin	83 [i]	Tocopherol	13, 14, 37
Hexamidine diisethionate	78	Virginiamycin	86 [i]

[a] *photo*contact urticaria

[b] no details provided, only tabulated

[c] there have been numerous reports of anaphylactic reactions to bacitracin, mostly from irrigation during operations and from application to leg ulcers or other wounds; in these cases, no contact urticaria was observed, neither during the event nor during testing (usually prick testing, sometimes no diagnostic tests at all, never open skin application or patch tests read at 20 minutes) (19-26,67-76)

[d] immediate contact reactions in 6-10% of patients with mycosis fungoides treated with mechlorethamine HCl solutions

[e] in refs. 45-47, the allergenic culprit proved to be povidone (polyvinylpyrrolidone, PVP)

[f] uncertain whether the commercial preparation or the pure corticosteroid was tested positive

[g] not classic immediate contact reaction

[h] routine testing in investigational study of immediate contact reactions

[i] details unknown, cited in ref. 79

[j] only positive reaction on slightly eczematous skin (not on normal skin), ergo strictly speaking not contact urticaria

[k] after irradiation during photodynamic therapy: photocontact urticaria; as the drug was not applied to intact skin, this reaction is *sensu stricto* not contact urticaria

[l] type-I reaction, but uncertain whether contact urticaria was present

[m] contact urticaria from tacrolimus ointment; the ingredients were not tested, but contact urticaria from one of the excipients is very unlikely

LITERATURE

1   Fergusson DA. Systemic symptoms associated with a rubefacient. BMJ 1988;297:1339

2   Ylipieti S, Lahti A. Effect of the vehicle on non-immunologic immediate contact reactions. Contact Dermatitis 1989;21:105-106

3   Larmi E, Lahti A, Hannuksela M. Effects of infra-red and neodymium yttrium aluminium garnet laser irradiation on non-immunologic immediate contact reactions to benzoic acid and methyl nicotinate. Derm Beruf Umwelt 1989;37:210-214 (Article in German)

4   Roussaki-Schulze AV, Zafiriou E, Nikoulis D, Klimi E, Rallis E, Zintzaras E. Objective biophysical findings in patients with sensitive skin. Drugs Exp Clin Res 2005;31(Suppl):17-24

5   Muñoz-Bellido FJ, Beltrán A, Bellido J. Contact urticaria due to cyclopentolate hydrochloride. Allergy 2000;55:198-199

6   Crijns MB, Jansen FW, van Praag MC, van der Schroeff JG. Immediate-type reaction to buserelin acetate in a nasal spray. Contact Dermatitis 1991;25:189

7   Loesche C, Dejobert Y, Thomas P. Immediate wheal after topical administration of chlorproethazine. Contact Dermatitis 1992;26:278

8   Coverly J, Peters L, Whittle E, Basketter DA. Susceptibility to skin stinging, non-immunologic contact urticaria and acute skin irritation; is there a relationship? Contact Dermatitis 1998;38:90-95

9   Darlenski R. Probable contact urticaria caused by tacrolimus-containing ointment in the treatment of atopic dermatitis. J Allergy Clin Immunol Pract 2019;7:1665-1667

10   De Groot AC. Patch Testing, 3rd Edition. Wapserveen, The Netherlands: acdegroot publishing, 2018

11   Audicana M, Schmidt R, Fernández de Corres L. Allergic contact dermatitis from nicotinic acid esters. Contact Dermatitis 1990;22:60-61

12   Gimenez-Arnau A, Maurer M, De La Cuadra J, Maibach H. Immediate contact skin reactions, an update of contact urticaria, contact urticaria syndrome and protein contact dermatitis -- "A never ending story". Eur J Dermatol 2010;20:552-562

13   Kassen B, Mitchell JC. Contact urticaria from vitamin E preparation in two siblings. Contact Dermatitis Newsletter 1974;16:482

14   Mitchell J, Kassen L. Contact urticaria from vitamin E preparation (vitamin E-vegetable oil in two siblings). Int J Dermatol 1975;14:246-247

15   Elsner P, Pevny I, Burg G. Anaphylaxis induced by topically applied bacitracin. Am J Contact Dermat 1990;1:162-164

16   Bommarito L, Mietta S, Cadario G. Anaphylaxis after application of topical bacitracin-neomycin powder. Ann Allergy Asthma Immunol 2015;115:74-75

17   Sharif S, Goldberg B. Detection of IgE antibodies to bacitracin using a commercially available streptavidin-linked solid phase in a patient with anaphylaxis to triple antibiotic ointment. Ann Allergy Asthma Immunol 2007;98:563-566

18   Palungwachira P. Contact urticaria syndrome and anaphylactoid reaction from topical clioquinol and bacitracin (Banocin): a case report. J Med Assoc Thai 1991;74:43-46

19   Desai M, Castells M. Anaphylactic shock to bacitracin irrigation during breast implant surgery. Ann Allergy Asthma Immunol 2019;122:217-218

20   Comaish JS, Cunliffe WJ. Absorption of drugs from varicose ulcers: a cause of anaphylaxis. Br J Clin Pract 1967;21:97-98

21   Roupe G, Strannegard O. Anaphylactic shock elicited by topical administration of bacitracin. Arch Dermatol 1969;100:450-452

22   Vale MA, Connolly A, Epstein AM, Vale MR. Bacitracin induced anaphylaxis. Arch Dermatol 1978;114:800.

23   Schecter JF, Wilkinson RD, Del Carpio J. Anaphylaxis following the use of bacitracin ointment. Arch Dermatol 1984;120:909-911

24   Katz BE, Fisher AA. Bacitracin: a unique topical antibiotic sensitizer. J Am Acad Dermatol 1987;17:1016-1024

25   Freiler JF, Steel KE, Hagan LL, Rathkopf MM, Roman-Gonzalez J. Intraoperative anaphylaxis to bacitracin during pacemaker change and laser lead extraction. Ann Allergy Asthma Immunol 2005;95:389-393

26   Schroer BC, Fox CC, Hauswirth DW. Skin testing after anaphylaxis to a topical Neosporin preparation. Ann Allergy Asthma Immunol 2008;101:444

27   Ramsay DL, Halperin PS, Zeleniuch-Jacquotte A. Topical mechlorethamine therapy for early stage mycosis fungoides. J Am Acad Dermatol 1988;19:684-691

28   Ramsay DL, Parnes RE, Dubin N. Response of mycosis fungoides to topical chemotherapy with mechlorethamine. Arch Dermatol 1984;120:1585-1590

29   Zachariae H, Thestrup-Pedersen K, Sogaard H. Topical nitrogen mustard in early mycosis fungoides. A 12-year experience. Acta Derm Venereol 1985;65:53-58

30 Grunnet E. Contact urticaria and anaphylactoid reaction induced by topical application of nitrogen mustard. Br J Dermatol 1976;94:101-103

31 Daughters D, Maibach HI. Urticarial and anaphylactoid reactions to the topical applications of nitrogen mustard. Contact Dermatitis Newsletter 1973;13:359

32 Daughters D, Zackheim H, Maibach H. Urticaria and anaphylactoid reactions. Arch Dermatol 1973;107:429-430

33 Yokoyama S, Nakano H, Nishizawa A, Kaneko T, Harada K, Hanada K. A case of photocontact urticaria induced by photodynamic therapy with topical 5-aminolaevulinic acid. J Dermatol 2005;32:843-847

34 Bircher AJ, Howald H, Rufli T. Adverse skin reactions to nicotine in a transdermal therapeutic system. Contact Dermatitis 1991;25:230-236

35 Abelin T, Buehler A, Muller P, Vesanen K, Imhof PR. Controlled trial of transdermal nicotine patch in tobacco withdrawal. Lancet 1989;1:7-10

36 Kleinhans D, Zwissler H. Anaphylaktischer Schock nach Anwendung einer Benzocainhaltigen salbe. Z Hautkr 1980;55:945-947 (Article in German)

37 Sanz-Sánchez T, Núñez Acevedo B, Rubio Flores C, Díaz-Díaz RM. Contact urticaria caused by tocopherol. Contact Dermatitis 2018;79:395

38 Kerr AC, Ferguson J, Ibbotson SH. Acute phototoxicity with urticarial features during topical 5-aminolaevulinic acid photodynamic therapy. Clin Exp Dermatol 2007;32:201-202

39 Kaae J, Philipsen PA, Haedersdal M, Wulf HC. Immediate whealing urticaria in red light exposed areas during photodynamic therapy. Acta Derm Venereol 2008;88:480-483

40 Wolfe CM, Green WH, Hatfield HK, Cognetta AB Jr. Urticaria after methyl aminolevulinate photodynamic therapy in a patient with nevoid basal cell carcinoma syndrome. J Drugs Dermatol 2012;11:1364-1365

41 Ventura MT, Di Corato R, Di Leo E, Foti C, Buquicchio R, Sborgia C, et al. Eyedrop induced allergy: clinical evaluation and diagnostic protocol. Immunopharmacol Immunotoxicol 2003;25:529-538

42 Norsworthy J, Bhatti Z, George T. Topical salicylic acid hypersensitivity. Am J Ther 2018;25:e568-e569

43 Kim SJ, Goldberg BJ. Anaphylaxis due to topical pramoxine. Ann Allergy Asthma Immunol 2015;114:72-73

44 Park MR, Kim DS, Kim J, Ahn K. Anaphylaxis to topically applied sodium fusidate. Allergy Asthma Immunol Res 2013;5:110-112

45 Adachi A, Fukunaga A, Hayashi K, Kunisada M, Horikawa T. Anaphylaxis to polyvinylpyrrolidone after vaginal application of povidone-iodine. Contact Dermatitis 2003;48:133-136

46 López Sáez MP, de Barrio M, Zubeldia JM, Prieto A, Olalde S, Baeza ML. Acute IgE-mediated generalized urticaria-angioedema after topical application of povidone-iodine. Allergol Immunopathol (Madr) 1998;26:23-26

47 Jeep S, Sterry W, Zuberbier T, Charite H. Sofottyp reaction auf polyvidon. Allergo J 1997;6:41 (Article in German)

48 Waran KD, Munsick RA. Anaphylaxis from povidone-iodine. Lancet 1995;345:1506

49 Gottman-Lückerath I. Kontakturtikaria nach Dermoxin. Derm Beruf Umwelt 1982;30:124 (Article in German)

50 Ventura MT. Cutaneous adverse drug reactions to fluticasone propionate and deflazacort in an asthmatic patient. Contact Dermatitis 2005;53:118

51 Peng YS, Shyur SK, Lin HY, Wang CY. Steroid allergy: a report of two cases. J Microbiol Immunol Infect 2001;34:150-154

52 Tkach J R. Allergic contact urticaria to benzoyl peroxide. Cutis 1982;29:187-188

53 Pigatto PD, Riboldi A, Morelli M, Altomare GF, Polenghi MM. Allergic contact dermatitis from oxyphenbutazone. Contact Dermatitis 1985;12:236-237

54 Piñol J, Carapeto FJ. Contact urticaria to etofenamate. Contact Dermatitis 1984;11:132-133

55 Piñol J, Navarro M, Carapeto FJ. Allergic contact dermatitis to etophenamate. Contact Dermatitis 1985;13:193

56 Mancuso G. Immediate and delayed contact hypersensitivity to thiocolchicoside. Acta Dermatovenerol Alp Pannonica Adriat 2018;27:203-205

57 Mancuso G, Masarà N. Contact urticaria and severe anaphylaxis from rifamycin SV. Contact Dermatitis 1992;27:124-125

58 Grob JJ, Pommier G, Robaglia A, Collet-Villette AM, Boneradi JJ. Contact urticaria from rifamycin. Contact Dermatitis 1987;16:284-285

59 Aguirre A, Manzano D, Zabala R, Eizaguirre X, Díaz-Pérez JL. Allergic contact dermatitis from spironolactone. Contact Dermatitis 1994;30:312

60 Jovanović M, Karadaglić D, Brkić S. Contact urticaria and allergic contact dermatitis to lidocaine in a patient sensitive to benzocaine and propolis. Contact Dermatitis 2006;54:124-126

61 Sanan N, Lee J, Baxter C, Jeskey J, Hostoffer R. Delayed and protracted allergic reaction to oral lidocaine. Ann Allergy Asthma Immunol 2019;123:413-414

62 Waton J, Boulanger A, Trechot PH, Schmutz JL, Barbaud A. Contact urticaria from Emla cream. Contact Dermatitis 2004;51:284-287

63 Suzuki T, Kawada A, Hashimoto Y, Isogai R, Aragane Y, Tezuka T. Contact urticaria due to ketoprofen. Contact Dermatitis 2003;48:284-285

64 Haustein UF. Anaphylactic shock and contact urticaria after the patch test with professional allergens. Allerg Immunol (Leipz) 1976;22:349-352 (Article in German)

65 Sukakul T, Kumpangsin T, Boonchai W. Contact urticaria caused by salicylic acid in a chemical peel solution. Contact Dermatitis 2020;82:121122

66 Katsarou A, Armenaka M, Ale I, Koufou V, Kalogeromitros D. Frequency of immediate reactions to the European standard series. Contact Dermatitis 1999;41:276-279

67 Saryan JA, Dammin TC, Bouras AE. Anaphylaxis to topical bacitracin zinc ointment. Am J Emerg Med 1998;16:512-513

68 Sprung J, Schedewie HK, Kampine JP. Intraoperative anaphylactic shock after bacitracin irrigation. Anesth Analg 1990;71:430-433

69 Blas M, Briesacher KS, Lobato EB. Bacitracin irrigation: a cause of anaphylaxis in the operating room. Anesth Analg 2000;91:1027-1028

70 Gall R, Blakley B, Warrington R, Bell DD. Intraoperative anaphylactic shock from bacitracin nasal packing after septorhinoplasty. Anesthesiology 1999;91:1545

71 Burnett GW, Meisner J, Hyman JB, Levin EJ. Bacitracin irrigation leading to anaphylaxis and cardiovascular collapse in the ambulatory surgery center setting. J Clin Anesth 2018;46:35-36

72 Caraballo J, Binkley E, Han I, Dowden A. Intraoperative anaphylaxis to bacitracin during scleral buckle surgery. Ann Allergy Asthma Immunol 2017;119:559-560

73 Damm S. Intraoperative anaphylaxis associated with bacitracin irrigation. Am J Health Syst Pharm 2011;68:323-327

74 Cronin H, Mowad C. Anaphylactic reaction to bacitracin ointment. Cutis 2009;83:127-129

75 Greenberg K, Espinosa J, Scali V. Anaphylaxis to topical bacitracin ointment. Am J Emerg Med 2007;25:95-96

76 Goh CL. Anaphylaxis from topical neomycin and bacitracin. Australas J Dermatol 1986;27:125-126

77 Vega FA, Halprin KM, Taylor JR, Woodyard C, Comerford M. Failure of periodic ultraviolet radiation treatments to prevent sensitization to nitrogen mustard: a case report. Br J Dermatol 1982;106:361-366

78 Mullins RJ. Systemic allergy to topical hexamidine. Med J Aust 2006;185:177

79 De Groot AC, Weyland JW, Nater JP. Unwanted effects of cosmetics and drugs used in dermatology, 3rd edition. Amsterdam: Elsevier Science BV, 1994

80 Kozáková M. Sub-shock brought on by epidermic skin test for chloramphenicol. Cs Derm 1976;51:62 (Article in Czech, cited in ref. 79)

81 Von Liebe V, Karge HJ, Burg G. Kontakturtikaria. Hautarzt 1979;30:544-546 (Article in German, cited in ref. 79)

82 Hutt N, Firdion O, Abbas F, Pauli G. A propos d'un cas d'allergie immédiate vraisemblable au cromoglycate disodique. Rev Fr Allergol 1986;26:147-148 (Article in French, cited in ref. 79)

83 Lahti A, Maibach HI. Immediate contact reactions. In: Menné T, Maibach HI (Eds). Exogenous dermatoses, environmental contact dermatitis. Boca Raton: CRC Press, 1990:21-35

84 Maucher OD. Anaphylaktische Reaktionen beim Epikutantest. Hautarzt 1972;23:139 (Article in German, cited in ref. 79)

85 Böttger EM, Mücke Chr, Tronnier H. Kontaktdermatitis auf neuere Antimykotika und Kontakturtikaria. Acta Dermatol 1981;7:70 (Article in German, cited in ref. 79)

86 Baes H. Allergic contact dermatitis to virginiamycin. False cross-sensitivity with pristinamycin. Dermatologica 1974;149:231-235

87 Odom RB, Maibach HI. Contact urticaria: a different contact dermatitis. Cutis 1976;18:672

88 Maibach HI, Johnson HL. Contact urticaria syndrome. Contact urticaria to diethyltoluamide (immediate-type hypersensitivity). Arch Dermatol 1975;111:726-730

89 Von Mayenburg J, Rakoski J. Contact urticaria to diethyltoluamide. Contact Dermatitis 1983;9:171

90 Vozmediano JM, Armario J, Gonzalez-Cabrerizo A. Immunologic contact urticaria from diethyltoluamide. Int J Dermatol 2000;39:876-877

91 Shutty B, Swender D, Chernin L, Tcheurekdjian H, Hostoffer R. Insect repellents and contact urticaria: differential response to DEET and picaridin. Cutis 2013;91:280-282

92 Wantke F, Focke M, Hemmer W, Götz M, Jarisch R. Generalized urticaria induced by a diethyltoluamide-containing insect repellent in a child. Contact Dermatitis 1996;35:186-187

Chapter 6   SYSTEMIC DRUGS AND DELAYED-TYPE HYPERSENSITIVITY: A PREVIEW

6.1 SYSTEMIC DRUGS AND DELAYED-TYPE HYPERSENSITIVITY REACTIONS

Already more than half a century ago it was discovered that some drugs that had caused an exanthema could induce a positive patch test reaction, pointing at the existence of delayed-type (type IV) hypersensitivity to it. In the last 25 years, the relationship between cutaneous adverse drug reactions and systemic drugs and this form of hypersensitivity (which causes allergic contact dermatitis when the allergens [haptens] are applied to the skin and is termed 'contact allergy' then) has been extensively investigated. It was found that certain drug eruptions are frequently associated with delayed-type hypersensitivity, of which the macular and maculopapular (morbilliform) exanthematous eruptions are by far the most frequent. Other mild, harmless, cutaneous adverse drug reactions that are often associated with delayed-type hypersensitivity are the fixed drug eruptions and 'symmetrical drug-related intertriginous and flexural exanthema' (SDRIFE). However, sensitization to systemic drugs may also – albeit infrequently – lead to serious and sometimes life-threatening drug reactions with systemic symptoms such as erythema multiforme-like eruption, Stevens-Johnson syndrome/toxic epidermal necrolysis (SJS/TEN), acute generalized exanthematous pustulosis (AGEP) or drug reaction with eosinophilia and systemic symptoms (DRESS) (1). The drugs that are responsible for these reactions is dependent on the type of cutaneous adverse drug reaction; common causes are shown in table 6.1.

Table 6.1 Common causes of cutaneous drug reactions (adapted from ref. 1)

Drug reaction	Common causes
Macular/maculopapular rash	aminopenicillins, cephalosporins, sulfamethoxazole, anticonvulsants
Fixed drug eruption	sulfonamides, NSAIDs, acetaminophen, tetracyclines, barbiturates, pseudoephedrine
SDRIFE	β-lactam antibiotics (mostly aminopenicillins)
EM-like drug eruption	sulfonamides, anticonvulsants, allopurinol
SJS/TEN	carbamazepine, allopurinol, sulfonamides, lamotrigine, nevirapine, NSAIDs (oxicams), phenobarbital, phenytoin
DRESS	carbamazepine, phenobarbital, lamotrigine, allopurinol, sulfasalazine, antibiotics (sulfonamides, fluoroquinolones, β-lactams), abacavir, nevirapine
AGEP	aminopenicillins, fluoroquinolones, macrolides, sulfonamide antibiotics, (hydroxy)chloroquine, terbinafine, diltiazem

AGEP: acute generalized exanthematous pustulosis; DRESS: drug reaction with eosinophilia and systemic symptoms; EM: erythema multiforme; SDRIFE: symmetrical drug-related intertriginous and flexural exanthema; SJS/TEN: Stevens-Johnson syndrome / toxic epidermal necrolysis

In these drug reactions, sensitization occurs from the systemic administration of the drug. However, sometimes patients have become sensitized to a particular drug from *topical* application. When this drug or a cross-reacting agent is administered systemically in the presensitized individual, a cutaneous adverse drug reaction may ensue which is called 'systemic contact dermatitis' or 'systemic allergic dermatitis' (paragraph 2.2.3). This reaction may manifest as a macular or maculopapular exanthema, extensive dermatitis/eczema, exacerbation of previous allergic contact dermatitis or positive patch tests, urticaria, the 'baboon syndrome' (which is called 'symmetrical drug-related intertriginous and flexural exanthema' [SDRIFE] when no previous sensitization has taken place) or otherwise. However, it does not induce serious reactions such as SJS/TEN, DRESS or AGEP (see above). The corticosteroids are frequent causes of systemic contact dermatitis.

Another manifestation of delayed-type allergy to systemic drugs is sensitization from inadvertent contact of the drug with the skin of individuals working with these drugs. These may be workers in the pharmaceutical industry where the drugs are being produced, or health care workers including pharmacists and nurses preparing the drugs for use by patients. Sensitization and continued or renewed contact then leads to allergic contact dermatitis of the hands and forearms. The face (eyelids) is also frequently affected by such occupational allergic contact dermatitis, when the drug powders, e.g. during synthesis or by breaking tablets, spread through the air (airborne allergic contact dermatitis) (2).

6.2 SYSTEMIC DRUGS TO BE DISCUSSED IN THE NEXT VOLUME OF 'MONOGRAPHS IN CONTACT ALLERGY'

Systemic drugs that the author has identified thus far as having caused delayed-type hypersensitivity reactions, either manifesting as cutaneous adverse drug reactions (systemic contact dermatitis, fixed drug eruption, maculopapular eruption, urticaria, DRESS, AGEP, SDRIFE, SJS/TEN, other exanthemas), as occupational allergic contact dermatitis or as both, are shown in table 6.2. The drugs that have caused occupational contact dermatitis are marked by an [a]. As the literature on systemic drugs has not yet fully been reviewed, additional drugs will undoubtedly be found and these will also be included for presentation in the next volume of this book series: *Monographs in Contact Allergy, Volume 4: Systemic drugs*. Alternatively, some drugs may be deleted when they prove to be unsuitable for inclusion. About 50 of the drugs shown below are used both in topical and systemic applications. Contact allergy to and allergic contact dermatitis from topical pharmaceuticals containing these drugs are presented in monographs in the current book; these are marked by the symbol [b].

Table 6.2 Systemic drugs that have caused type IV (delayed-type) hypersensitivity reactions

Abacavir [a]	Betamethasone [b]	Cimetidine
Aceclofenac [b]	Betamethasone sodium	Ciprofloxacin
Acetaminophen [a]	phosphate [b]	Cisplatin
Acetazolamide [b]	Bisoprolol [a]	Citiolone
Acetylsalicylic acid	Brivudine	Clarithromycin
Acexamic acid [b]	Bromazepam [a]	Clavulanic acid
Actarit	Bucillamine	Clindamycin [b]
Acyclovir [b]	Bupivacaine	Clobazam
Adalimumab	Captopril [a]	Clomipramine
Albendazole [a]	Carbamazepine	Clopidol [a]
Albuterol	Carbenicillin [a]	Clorazepate
Alendronic acid	Carbimazole [a]	Clotiazepam
Allopurinol	Carbocromen [a]	Cloxacillin
Alprazolam [a]	Carprofen [a]	Codeine [a]
Alprenolol [a]	Carvedilol [a]	Cotrimoxazole
Amantadine	Casanthranol [a]	Cyanocobalamin [a]
Ambroxol [b]	Cefaclor	Deflazacort
Amikacin [a]	Cefadroxil	Dexamethasone [b]
Aminocaproic acid [b]	Cefalexin	Dexamethasone phosphate [b]
Aminophenazone	Cefalotin	Dexamethasone sodium
Aminophylline [a]	Cefamandole [a]	phosphate [b]
Amitriptyline	Cefazolin [a]	Dexlansoprazole [a]
Amlexanox [b]	Cefcapene pivoxil	Diatrizoic acid [a]
Amobarbital	Cefoxime	Diazepam [a]
Amoxicillin [a]	Cefmetazole [a]	Diclofenac [b]
Amoxicillin - clavulanic acid	Cefodizime [a]	Dicloxacillin
Ampicillin [a]	Cefoperazone	Dihydrocodeine
Ampiroxicam	Ceforanide [a]	Dihydrostreptomycin [a]
Amprolium [a]	Cefotaxime [a]	Diltiazem [b]
Anileridine	Cefotetan [a]	Dimenhydrinate
Antipyrine salicylate	Cefotiam	Dimetridazole
Apomorphine [a]	Cefoxitin	Dipyridamole [a]
Apronalide	Cefpodoxime	Disulfiram [a]
Arecoline [a]	Cefradine [a]	Donepezil
Aripiprazole [a]	Ceftazidime [a]	Doxycycline [a]
Articaine	Ceftiofur [a]	Emtricitabine
Ascorbic acid	Ceftizoxime [a]	Enalapril [a]
Atorvastatin	Ceftriaxone [a]	Enoxacin
Avoparcin [a]	Cefuroxime [a]	Ephedrine [b]
Azaperone [a]	Cefuroxime axetil	Epirubicin
Azathioprine [a]	Celecoxib	Erythromycin [b]
Azithromycin [a,b]	Cetirizine	Esomeprazole [a]
Aztreonam	Chlorambucil [a]	Ethambutol [a]
Bacampicillin	Chloramphenicol [a,b]	Ethenzamide
Baclofen [a]	Chlormezanone	Ethylmorphine
Bendamustine	Chloroquine [a]	Etonogestrel
Bendroflumethiazide	Chlorpheniramine [b]	Etoricoxib
Benznidazole	Chlorpromazine [a]	Famotidine [a]
Benzylpenicillin [a]	Chlorprothixene	Fenofibrate

Table 6.2 Systemic drugs that have caused type IV (delayed-type) hypersensitivity reactions (continued)

Flavoxate	Lormetazepam [a]	Pantoprazole [a]
Fleroxacin	Meclofenoxate [a]	Paramethasone acetate
Flucloxacillin [a]	Mefenamic acid	Paroxetine
Fluconazole	Meloxicam	Penethamate [a]
Fluoxetine [a]	Melphalan [a]	Penicillin [a]
Flurbiprofen [b]	Menadione sodium bisulfite [a]	Penicillin G benzathine [a]
Flutamide	Mepivacaine	Pentamidine isethionate
Fluvastatin	Meprednisone	Perazine
Fluvoxamine	Meprobamate	Periciazine [a]
Foscarnet	Methoxsalen [b]	Perindopril [a]
Fulvestrant	Metronidazole [b]	Perphenazine [a]
Furazolidone [a]	Mequitazine	Phenacetin
Gadobutrol	Meropenem [a]	Phenazone
Ganciclovir	Mesalazine	Pheniramine [b]
Gentamicin [a,b]	Mesna [a]	Phenobarbital
Glatiramer	Metamizole/dipyrone	Phenoxybenzamine [a]
Granisetron	Metaproterenol [a]	Phenoxymethylpenicillin
Griseofulvin	Methacycline [a]	Phenylbutazone [b]
Heparins	Methotrexate	Phenytoin
Heroin [a]	Methylprednisolone [a,b]	Pindolol
Hyaluronidase	Methylprednisolone acetate [b]	Piperacillin
Hydralazine [a]	Methylprednisolone	Piperazine [a,b]
Hydrochlorothiazide	hemisuccinate [b]	Pirmenol
Hydrocortisone [b]	Metoprolol [a,b]	Piroxicam [b]
Hydromorphone	Mexiletine	Piroxicam betadex
Hydroxychloroquine [a]	Mezlocillin	Potassium aminobenzoate
Hydroxyprogesterone	Midecamycin [a]	Practolol
Hydroxyzine	Minocycline	Pravastatin
Ibuprofen	Misoprostol	Prednisolone [b]
Imatinib	Mitomycin C [a,b]	Prednisolone acetate [b]
Imipenem	Mofebutazone	Prednisone [b]
Insulin	Monensin [a]	Pregabalin
Iobitridol	Morantel	Prilocaine [b]
Iodixanol	Morphine [a]	Pristinamycin [a]
Iohexol	Naloxone	Procaine [a,b]
Iomeprol [a]	Naproxen	Procaine benzylpenicillin [a]
Iopamidol	Nebivolol	Progesterone
Iopromide	Nevirapine	Promazine
Ioversol	Nicergoline [a]	Promethazine [a,b]
Ioxaglic acid	Niflumic acid	Propacetamol [a]
Isoflurane [a]	Nifuroxazide [a]	Propanidid [a]
Isoniazid	Nimesulide	Propicillin
Isotretinoin	Nimodipine	Propionylpromazine [a]
Isoxsuprine	Norfloxacin	Propranolol [a,b]
Itraconazole	Nosiheptide [a]	Pseudoephedrine
Ketoprofen [a,b]	Nystatin [b]	Pyrazinamide
Kitasamycin [a]	Ofloxacin	Pyridoxine [a,b]
Lamivudine	Olanzapine [a]	Pyritinol [a]
Lamotrigine	Olaquindox [a]	Quinidine [a]
Lansoprazole [a]	Omeprazole [a]	Rabeprazole
Lepirudin	Ondansetron	Ramipril
Levocetirizine	Ornidazole	Ranitidine [a]
Levofloxacin	Oxacillin [a]	Retinyl acetate [a]
Levomepromazine	Oxaliplatin	Ribostamycin
Lidocaine [b]	Oxolamine [a]	Rifampicin [a]
Lisinopril [a]	Oxprenolol [a]	Risedronic acid
Lomefloxacin	Oxybutynin [a]	Risperidone [a]
Lorazepam [a]	Oxycodone [a]	Roxithromycin

Table 6.2 Systemic drugs that have caused type IV (delayed-type) hypersensitivity reactions (continued)

Secnidazole	Sotalol [a]	Succinylcholine
Secukinumab	Spectinomycin [a]	Sulbactam
Sertraline	Spiramycin	Sulfadiazine
Sevoflurane [a]	Spironolactone [b]	Sulfaguanidine
Simvastatin [a]	Streptomycin	Sulfamethoxazole
Sulfasalazine	Tiopronin [b]	Tylosin [a]
Tacrolimus [b]	Topiramate	Valaciclovir
Tamsulosin	Tosufloxacin tosilate	Valdecoxib
Tenoxicam	Tramadol	Valproic acid
Terbinafine	Tranexamic acid	Varenicline
Tetrazepam [a]	Trazodone [a]	Vancomycin [b]
Thioridazine	Triamcinolone	Verapamil
Tiaprofenic acid	Triamcinolone acetonide [b]	Vinburnine
Ticarcillin	Tribenoside [b]	Vincamine [a]
Ticlopidine	Triflusal	Ziprasidone
Tilactase	Trimeprazine	Zolpidem
Tilisolol	Trimethoprim	
Tinidazole	Tubocurarine	

[a] has caused occupational allergic contact dermatitis

[b] both used in topical (this book, Volume 3) and systemic applications (Volume 4)

LITERATURE

1   Brandt O, Bircher AJ. Delayed-type hypersensitivity to oral and parenteral drugs. J Dtsch Dermatol Ges 2017;15:1111-1132

2   Gilissen I, Boeckxstaens E, Geebelen J, Goossens A. Occupational allergic contact dermatitis from systemic drugs. Contact Dermatitis 2020;82:24-30

Index